Flexible Automation

and

Intelligent Manufacturing

1997

Proceedings of the Seventh International FAIM Conference
European Process Industries Competitiveness Centre
University of Teesside
Middlesbrough
England

June 25 - 27, 1997

Editors

M. Munir Ahmad, PhD., CEng. EurIng., P.E., Professor
EPICC Director of Research and Development
School of Science and Technology
University of Teesside
Middlesbrough, United Kingdom

William G. Sullivan, Ph.D., P.E., Professor
Department of Industrial and Systems Engineering
Virginia Polytechnic Institute and State University
Blacksburg, Virginia, U.S.A.

Begell House, Inc.
New York • Wallingford (U.K.)

Library of Congress Cataloging-in-Publication Data

International FAIM Conference (7th : 1997 : University of Teesside)
Flexible automation and intelligent manufacturing 1997 :
proceedings of the Seventh International FAIM Conference, European Process Industries Competitiveness Centre, University of Teesside, Middlesborough, England : June 25-27, 1997 / editors, M. Munir Ahmad, William G. Sullivan.
p. cm.
Includes bibliographical reference (p.).
ISBN 1-56700-089-4 (hardcover)
1. Automation—Congresses. 2. Flexible manufacturing systems—Congresses. 3. computer integrated manufacturing systems—Congresses. 4. Concurrent engineering—Congresses. I. Ahmad, Mohammed Munir. II. Sullivan, William G., 1942- . III. Title.
T59.5.I5714 1997
670.42'7—dc21 97-15432
CIP

Direct all inquiries to Begell House, Inc., 79 Madison Avenue, New York, NY 10016.

International Standard Book Number 1-56700-089-4

Printed in the United States of America 1 2 3 4 5 6 7 8 9 0

PREFACE

The International Conference on Flexible Automation and Intelligent Manufacturing (FAIM) has a rich tradition: The first conference, "Factory Automation and Information Management - FAIM," was hosted in March 1991 by the University of Limerick. Authors from more than 18 countries contributed over 100 papers, and each paper was thoroughly refereed. The second conference, "Flexible Automation and Information Management," was hosted by Virginia Tech in Washington, DC on June 30 - July 2, 1992. The third conference, "Flexible Automation and Integrated Manufacturing," was hosted in June 1993 by the University of Limerick, Ireland. The fourth conference was held at Virginia Tech in Blacksburg, Virginia in May 1994. The fifth conference was held at the Fraunhofer Institute for Manufacturing and Management, University of Stuttgart, Germany in June, 1995. Last year the sixth international conference was conducted at the Georgia Institute of Technology in Atlanta, Georgia. The conferences have been endorsed by numerous societies, associations, university affiliates, and international organisations. Based on the huge success of previous FAIM conferences, FAIM '97 is being hosted this year by the European Process Industries Competitiveness Centre (EPICC Ltd) at the University of Teesside in Middlesbrough, United Kingdom. The 1998 conference will take place at Portland State University in Portland, Oregon on July 1 - 3 and is expected to draw a large audience.

The preceding conferences have all been dedicated to the main topics of Flexible Automation and Intelligent Manufacturing. As a consequence of international activity in Production Research, which has led to many new developments, FAIM '97 promotes and advances the state-of-the-art in this domain by focusing on intelligent manufacturing. The FAIM conference series differs from many other conferences in that it presents an integrated picture of these related technologies, rather than focusing on one single technology.

This year, authors from approximately 20 countries have contributed more than 90 papers, and as in the case of previous FAIM conferences, each paper has been thoroughly refereed by at least two reviewers. The papers cover the following themes:

Manufacturing in a World Market
- Global product strategies
- International concurrent engineering
- International technology transfer
- Global manufacturing logistics
- World issues related to CIM technologies
- Lean/agile/virtual manufacturing
- Lean/agile/virtual manufacturing

Concurrent and Simultaneous Engineering
- Tools for product and process design
- Virtual reality
- Rapid prototyping technology and methods (RPT)
- CAD/CAE/CAM
- Design for manufacture, assembly, test, service (Dfx)
- Quality function deployment (QFD)
- Integrated product and process design (IPPD)

Factory Automation
- Sensors, sensor fusion, intelligent control, actuators
- Process automation and optimisation
- Cell control and shop floor logistics
- Monitoring, control and maintenance
- Economics of automation/integration

Information Technology and Manufacturing
- Electronic Data Interchange (EDI)
- Data capture automation
- IGES/PDES/STEP
- Database technology
- Factory communication
- AI/ES in design and manufacture

Competitiveness Issues Relevant to Process Industries
- Good Manufacturing Practice (GMP)
- Benchmarking
- Legislation and regulatory issues
- Safety, Health and Environment
- Cost effective maintenance of process plants
- odelling, design and engineering of process plants
- rocess control systems

The objectives of the Seventh International Conference on Flexible Automation and Intelligent Manufacturing (FAIM '97) are to support the international exchange of experience about outstanding industrial developments and the latest results of production research.

We hope you enjoy reading the Proceedings of FAIM '97. We believe the results of this conference will continue to be useful in your professional activities and development.

M. Munir Ahmad and William G. Sullivan

June 1997

CONTRIBUTORS

S.A. Abbas
Merryweather Management Systems Limited
Darlington
UNITED KINGDOM

Martin Adickes
University of Pittsburgh

M. Munir Ahmad
EPICC
University of Teesside
Middlesborough TS1 3BA
UNITED KINGDOM

Jane C. Ammons
School of Industrial and Systems Engineering
The Georgia Institute of Technology
Atlanta, GA 30332-0205

Richard C. Anderson
The George Washington University

C.J. Anumba
Construction Research Unit (SST)
University of Teesside
Middlesborough TS1 3BA
UNITED KINGDOM

E. Appleton
School of Engineering
University of Durham
Durham DH1 3LE
UNITED KINGDOM

J. Ashayeri
Department of Econometrics
Tilburg University
Tilburg
THE NETHERLANDS

Joze Balic
Faculty of Mechanical Engineering
Maribor
SLOVENIA

Massoud Bazargan-Lari
Faculty of Science and Technology
University of Western Sydney - Nepean
P.O. Box 10
Kingswood, NSW 2747
AUSTRALIA

Antonio Batochhio
DEF / FEM UNICAMP
Campinas SP
BRAZIL

Matthew D. Bauer
Systems Realization Laboratory
The George W. Woodruff School of
Mechanical Engineering
Georgia Institute of Technology
Atlanta, GA 30332-0405

Michael Bedwell
School of Engineering
Coventry University, CV1 5FB
UNITED KINGDOM

David Bedworth
Arizona State University

G. Bennett
Buckinghamshire College
UNITED KINGDOM

Bopaya Bidanda
University of Pittsburgh

William E. Biles
Department of Industrial Engineering
University of Louisville
Louisville, KY 40292

Richard E. Billo
University of Pittsburgh

P. Blackburn
British Steel PLC
UNITED KINGDOM

Grace M. Bochenek
U.S. Army Tank-Automotive & Armaments Command
Vetronics Technology Center
Warren, MI

C. Bocking
GEC

Vincent Boly
Ecole Nationale Supérieure en Génie des Systèmes Industriels
4, allée Pelletier Doisy
54603 Villers les Nancy
FRANCE

T.O. Boucher
Department of Industrial Engineering
Rutgers University
Piscataway, NJ 08855-0909

J. Brass
Department of Computing and Mathematics
Hartlepool College
Hartlepool TS24 7NT
UNITED KINGDOM

Carlos Bremer
Escola de Engenharia de São Carlos
Universidade de São Paulo
BRAZIL

R. Bröker
AWB Business Development B.V.
Eindhoven
THE NETHERLANDS

Iain G. Brownlie
Eutech Engineering Solutions Ltd.
Belasis Hall Technology Park
Billingham
Cleveland
UNITED KINGDOM

Agostino G. Bruzzone
Institute of Technology and Mechanical Plants
University of Genoa
via Opera Pia 15
16145 Genoa
ITALY

M. Burwood
Melbourne Business School
The University of Melbourne
AUSTRALIA

Andrew W. Campbell
Laboratory for Intelligent Systems Technology
School of Science and Technology
University of Teesside
UNITED KINGDOM

Tom Caporello
Arizona State University

John G. Casali
Grado Professor and Department Head
Human Factors Engineering Center
Industrial and Systems Engineering Department
Virginia Polytechnic Institute and State University
Blacksburg, VA 24061-0118

Alistair R. Clark
School of Business and Management
University of Teesside
Middlesborough TS1 3BA
UNITED KINGDOM

C.J. Connor
School of Science & Technology
University of Teesside
Middlesborough TS1 3BA
UNITED KINGDOM

C.S. Cox
Control Systems Centre
University of Sunderland
UNITED KINGDOM

Nashan Dawood
Division of Civil Engineering & Building
School of Science & Technology
The University of Teesside
Middlesborough TS1 3BA
UNITED KINGDOM

Soumya K. De
Purdue University

Alain Delchambre
Department of Applied Mechanics
University of Brussels (ULB)
BELGIUM

Abhijit V. Deshmukh
Department of Industrial Engineering
FAMU-FSU College of Engineering
Tallahassee, FL 23210-6046

J. Detand
Katholeke Universiteit Leuven
Department of Mechanical Engineering
Division P.M.A.
Celestijnenlaan 300B
B-3001 Leuven
BELGIUM

F. Dhailami
University of the West of England
Bristol
UNITED KINGDOM

Paulo C. de Carvalho Dias
Escola de Engenharia de São Carlos
Universidade de São Paulo
BRAZIL

S.C. Diplaris
National Tech. University of Athens
GREECE

S. Dover
Buckinghamshire College
UNITED KINGDOM

R.D. Dryden
Dean of Engineering
Portland State University
Portland, OR

Hua X. Du
School of Industrial and Systems Engineering
Georgia Institute of Technology
Atlanta, GA 30332-0205

Michael R. Duffey
The George Washington University

Duane D. Dunlap
Purdue University

Orlando Durán
FEAR Universidade de Passo Fundo
Passo Fundo (RS)
BRAZIL

Willi Dürig
ETH Zurich
Institute for Operations Research
CH - 8092 Zurich
SWITZERLAND

Holger Dürr
Department of Manufacturing Technology
Technische Universität Chemnitz-Zwickau
09107 Chemnitz
GERMANY

Anthony Eddison
Unit Leader - Virtual Environment Research Unit (VEDRU)
Course Leader - MA Advanced Digital Design (MA ADD)
University of Teesside
Middlesborough TS1 3BA
UNITED KINGDOM

Kimberly P. Ellis
Department of Industrial and Systems Engineering
Virginia Polytechnic Institute and State University
Blacksburg, VA 24061-0118

Bahram Emamizadeh
Industrial Engineering Department
Amirkabir University of Technology
Tehran
IRAN

N.F.O. Evbuomwan
Department of Civil Engineering
University of Newcastle
Newcastle-upon-Tyne NE1 7RU
UNITED KINGDOM

Reza Zanjirani Farahani
Industrial Engineering Department
Amirkabir University of Technology
Tehran
IRAN

John D. Foster
Department of National Defence
CANADA

I.G. French
EPICC
University of Teesside
Middlesborough TS1 3BA
UNITED KINGDOM

W.F. Gaughran
Department of Manufacturing and Operations Engineering
University of Limerick
IRELAND

A.M. Gerrard
Division of Chemical Engineering
University of Teesside
Middlesborough TS1 3BA
UNITED KINGDOM

Richard Giglio
Department of Mechanical and Industrial Engineering
University of Massachusetts
Amherst, MA 01002

Nabil N. Gindy
Department of Manufacturing Engineering and Operations Management
University of Nottingham
Nottingham NG7 2RD
UNITED KINGDOM

Hossein Golnabi
Institute of Water and Energy
Sharif University of Technology
P.O. Box 11365 - 8639
Tehran
IRAN

Timothy J. Greene
Oklahoma State University

K. Grefen
Fraunhofer Institut für Produktionstechnik und Automatisierung (IPA)
GERMANY

J. Gui
EPICC
University of Teesside
Middlesborough TS1 3BA
UNITED KINGDOM

Claudine Guidat
Ecole Nationale Supérieure en Génie des Systèmes Industriels
4, allée Pelletier Doisy
54603 Villers les Nancy
FRANCE

Per Gullander
Department of Production Engineering
Chalmers University of Technology
SE-412 96 Göteborg
SWEDEN

P.M. Hackney
Centre for Rapid Product Development
School of Engineering
University of Northumbria at Newcastle
UNITED KINGDOM

D. Hampel
Dresden University of Technology
Electronics Technology Laboratory
GERMANY

A. Hart
School of Science & Technology
University of Teesside
Middlesborough TS1 3BA
UNITED KINGDOM

Paul Higginbottom
School of Engineering
Coventry University, CV1 5FB
UNITED KINGDOM

C.K.S. Ho
Control Systems Centre
University of Sunderland
UNITED KINGDOM

B. Hobbs
Division of Civil Engineering & Building
School of Science & Technology
The University of Teesside
Middlesborough TS1 3BA
UNITED KINGDOM

S. Hogarth
Centre for Rapid Product Development
School of Engineering
University of Northumbria at Newcastle
UNITED KINGDOM

Dezhong Hong
Laboratory for Intelligent Systems Technology
School of Science and Technology
University of Teesside
UNITED KINGDOM

Azim Houshyar
Department of Industrial & Manufacturing Eng.
Western Michigan University
Kalamazoo, MI 49008-5061

Jueng-Shing Hwang
Industrial and Management System Engineering
University of South Florida
Tampa, FL 33620

Ilkka Ikonen
Department of Industrial Engineering
University of Louisville
Louisville, KY 40292

George Ioannou
Department of Industrial and Systems Engineering
Virginia Polytechnic Institute and State University
Blacksburg, VA 24061-0118

V.S. Issopoulos
National Tech. University of Athens
GREECE

Yoshiaki Iwata
Faculty of Science and Technology
Kinki University
Kowakae 3-4-1
Higashi-Osaka
Osaka 577
JAPAN

M.A. Jafari
Department of Industrial Engineering
Rutgers University
Piscataway, NJ 08855-0909

R. Jamieson
Regional Centre for Innovation in Engineering Design
University of Newcastle-upon-Tyne
UNITED KINGDOM

George W. Johnson
Bath Iron Works Quality Engineer (Photogrammetry)
Student, School of Industrial Technology
University of Southern Maine
Gorham, ME

A. Jones
ICI Films Ltd.
UNITED KINGDOM

Albert Jones
National Institute of Standards and Technology
Gaithersburg, MD 20899-0001

Gregory B. Jones
Navale Undersea Warfare Center
Newport, RI

Sanjay Joshi
The Pennsylvania State University
University Park, PA 16812

A. Kalian
Division of Civil Engineering & Building
School of Science & Technology
The University of Teesside
Middlesborough TS1 3BA
UNITED KINGDOM

J.M. Kamara
Construction Research Unit (SST)
University of Teesside
Middlesborough TS1 3BA
UNITED KINGDOM

Eman Kamel
Department of Industrial Engineering
University of Louisville
Louisville, KY 40292

Uwe Kaschka
Department of Manufacturing Technology
Technische Universität Chemnitz-Zwickau
09107 Chemnitz
GERMANY

R. Keij
Department of Econometrics
Tilburg University
Tilburg
THE NETHERLANDS

Mohammed Kazem Farhang Kermani
Industrial Engineering Department
Amirkabir University of Technology
Tehran
IRAN

Brian M. Kleiner
Department of Industrial and Systems Engineering
Virginia Polytechnic Institute and State University
Blacksburg, VA 24061-0118

Pär Klingstam
Department of Production Engineering
Chalmers University of Technology
SE-412 96 Göteborg
SWEDEN

Karl H.E. Kroemer
Professor and Director
Industrial Ergonomics Laboratory
Human Factors Engineering Center
Industrial and Systems Engineering Department
Virginia Polytechnic Institute and State University
Blacksburg, VA 24061-0118

Naga K.C. Krothapalli
Department of Industrial Engineering
FAMU-FSU College of Engineering
Tallahassee, FL 23210-6046

J.P. Kruth
Katholeke Universiteit Leuven
Department of Mechanical Engineering
Division P.M.A.
Celestijnenlaan 300B
B-3001 Leuven
BELGIUM

E.J. Kyriannakis
Intelligent Robotics and Automation Laboratory
Department of Electrical and Computer Engineering
National Technical University of Athens
15773, Zografou Campus
Athens
GREECE

M. Leck
Division of Chemical Engineering
School of Science and Technology
University of Teesside
Middlesborough TS1 3BA
UNITED KINGDOM

Pui-Mun Lee
Nanyang Technological University

Paulo Leitão
Assistente da ESTG do Instituto Politécnico de Bragança
Quinta Sta Apolónia
Apartado 134
5300 Bragança
PORTUGAL

Huw J. Lewis
University of Limerick
Limerick
IRELAND

C.S. Liu
Department of Industrial & Manufacturing Systems Engineering
Lehigh University
Bethlehem, PA

Brendan Lynch
Department of Computer Science and Information Systems and National Centre for Quality Management
University of Limerick
National Technological Park
Castleroy
Limerick
IRELAND

Kevin W. Lyons
National Institute of Standards and Technology

Y.H. Ma
Department of Industrial & Manufacturing Systems Engineering
Lehigh University
Bethlehem, PA

Bertrand Mareschal
Institute of Statistics
University of Brussels (ULB)
BELGIUM

Pär Mårtensson
Department of Manufacturing Systems
Royal Institute of Technology
SE-100 44 Stockholm
SWEDEN

A.J. Matchett
Department of Chemical Engineering
University of Teesside
Middlesborough TS1 3BA
UNITED KINGDOM

Fergus Maughan
University of Limerick
Limerick
IRELAND

L.P. McAlinden
Engineering Design Centre
University of Newcastle-upon-Tyne
UNITED KINGDOM

Leon F. McGinnis
School of Industrial and Systems Engineering
The Georgia Institute of Technology
Atlanta, GA 30332-0205

William A. Miller
Industrial and Management System Engineering
University of South Florida
Tampa, FL 33620

D.A. Mills
JADE Networks Ltd.

Edson dos S. Moreira
Instituto de Ciências Matemáticas de São Carlos
Universidade de São Paulo
BRAZIL

Laure Morel
Ecole Nationale Supérieure en Génie des Systèmes Industriels
4, allée Pelletier Doisy
54603 Villers les Nancy
FRANCE

Roberto Mosca
Institute of Technology and Mechanical Plants
University of Genoa
via Opera Pia 15
16145 Genoa
ITALY

Anne Mungwattana
Department of Industrial and Systems Engineering
Virginia Polytechnic Institute and State University
Blacksburg, VA 24061-0118

F. Nabhani
School of Science & Technology
University of Teesside
Middlesborough TS1 3BA
UNITED KINGDOM

J. Neugebauer
Fraunhofer Institut für Produktionstechnik und Automatisierung (IPA)
GERMANY

K.C.S. Ng
Eurotherm Controls Ltd.
UNITED KINGDOM

P.W. Norman
Engineering Design Centre
University of Newcastle-upon-Tyne
UNITED KINGDOM

N.G. Odrey
Department of Industrial & Manufacturing Systems Engineering
Lehigh University
Bethlehem, PA

S.J.T. Owen*
Dean of Engineering
Oregon State University
Corvallis, OR

S.D. Parvin
Division of Chemical Engineering
School of Science and Technology
University of Teesside
Middlesborough TS1 3BA
UNITED KINGDOM

D. Peel
Division of Chemical Engineering
School of Science and Technology
University of Teesside
Middlesborough TS1 3BA
UNITED KINGDOM

* Deceased.

Fabrice Pellichero
Department of Applied Mechanics
University of Brussels (ULB)
BELGIUM

F. Petit
Department of Mechanical Engineering
Université catholique de Louvain
BELGIUM

Husam Petros
Division of Civil Engineering & Building
School of Science & Technology
The University of Teesside
Middlesborough TS1 3BA
UNITED KINGDOM

Scott Pierce
Systems Realization Laboratory
The George W. Woodruff School of Mechanical Engineering
Georgia Institute of Technology
Atlanta, GA 30332-0405

António Quintas
Professor Associado do DEEC da Faculdade de Engenharia da Universidade do Porto
Rua dos Bragas
4099 Porto Codex
PORTUGAL

James M. Ragusa
University of Central Florida
College of Engineering
Orlando, FL

Bala Ram
North Carolina AT&T State University
Greensboro, NC 27411

Sabah Randhawa
Department of Industrial and Manufacturing Engineering
Oregon State University

B. Raucent
Department of Mechanical Engineering
Université catholique de Louvain
BELGIUM

A.H. Redford
University of Salford
Salford
UNITED KINGDOM

A. Redlein
Department of Automation
Vienna University of Technology
Treitlgasse 1
A-1040 Vienna
AUSTRIA

Jean Renaud
Ecole Nationale Supérieure en Génie des Systèmes Industriels
4, allée Pelletier Doisy
54603 Villers les Nancy
FRANCE

Amerdeep Riat
CAMM
The Industry Centre
University of Sunderland
Hylton Riverside West
Wessington Way
Sunderland SR5 3XB
UNITED KINGDOM

Ita Richardson
Department of Computer Science and Information Systems and National Centre for Quality Management
University of Limerick
National Technological Park
Castleroy
Limerick
IRELAND

Mark Richardson
Department of Mechanical Engineering and Institute for Systems Research
University of Maryland
College Park, MD 20742

G.G. Rigatos
Intelligent Robotics and Automation Laboratory
Department of Electrical and Computer Engineering
National Technical University of Athens
15773, Zografou Campus
Athens
GREECE

R. Rohrhofer
Department of Automation
Vienna University of Technology
Treitlgasse 1
A-1040 Vienna
AUSTRIA

David W. Rosen
Systems Realization Laboratory
The George W. Woodruff School of Mechanical Engineering
Georgia Institute of Technology
Atlanta, GA 30332-0405

Sameh. M. Saad
Department of Manufacturing Engineering and Operations Management
University of Nottingham
Nottingham NG7 2RD
UNITED KINGDOM

Budi Saleh
Department of Industrial and Manufacturing Engineering
Oregon State University

Sharon Sandhu
School of Engineering
Coventry University, CV1 5FB
UNITED KINGDOM

Thompson Sarkodie-Gyan
Laboratory for Intelligent Systems Technology
School of Science and Technology
University of Teesside
UNITED KINGDOM

M. Sarwar
Centre for Rapid Product Development
School of Engineering
University of Northumbria at Newcastle
UNITED KINGDOM

W. Sauer
Dresden University of Technology
Electronics Technology Laboratory
GERMANY

G. Schildt
Department of Automation
Vienna University of Technology
Treitlgasse 1
A-1040 Vienna
AUSTRIA

R.D. Schraft
Fraunhofer Institut für Produktionstechnik und Automatisierung (IPA)
GERMANY

Brane Semolic
Faculty of Mechanical Engineering
Maribor
SLOVENIA

M.M. Sfantsikopoulos
National Tech. University of Athens
GREECE

John Shewchuk
Department of Industrial and Systems Engineering
Virginia Polytechnic Institute and State University
Blacksburg, VA 24061-0118

Petia Sice
European Process Industries Competitiveness Centre
University of Teesside
UNITED KINGDOM

Marc Sielemann
Institute for Production Engineering and Machine Tools
Schloßwender Straße 5
30159 Hannover
GERMANY

P.J. Sitoh
Engineering Design Centre
University of Newcastle-upon-Tyne
UNITED KINGDOM

A.J.R. Smith
Department of Mechanical and Manufacturing Engineering
The University of Melbourne
AUSTRALIA

Manbir S. Sodhi
Industrial and Manufacturing Eng.
University of Rhode Island

G.A. Spolek
Chair of Mechanical Engineering
Portland State University
Portland, OR

Jacqueline M. Sprinkle
Department of Industrial Engineering
University of Louisville
Louisville, KY 40292

William G. Sullivan
Department of Industrial and Systems Engineering
Virginia Polytechnic Institute and State University
Blacksburg, VA 24061-0118

E. Summad
School of Engineering
University of Durham
Durham DH1 3LE
UNITED KINGDOM

Rena Surana
Department of Mechanical Engineering and Institute for Systems Research
University of Maryland
College Park, MD 20742

Janis P. Terpenny
Department of Industrial and Systems Engineering
Virginia Polytechnic Institute and State University
Blacksburg, VA 24061-0118

John T. Tester
Department of Industrial and Systems Engineering
Virginia Polytechnic Institute and State University
Blacksburg, VA 24061-0118

Hans Kurt Tönshoff
Institute for Production Engineering and Machine Tools
Schloßwender Straße 5
30159 Hannover
GERMANY

S.G. Tzafestas
Intelligent Robotics and Automation Laboratory
Department of Electrical and Computer Engineering
National Technical University of Athens
15773, Zografou Campus
Athens
GREECE

Heinz Ulrich
ETH Zurich
Institute for Operations Research
CH - 8092 Zurich
SWITZERLAND

John S. Usher
Department of Industrial Engineering
University of Louisville
Louisville, KY 40292

Fredy J. Valente
Instituto de Ciências Matemáticas de São Carlos
Universidade de São Paulo

G. Van Zeir
Katholeke Universiteit Leuven
Department of Mechanical Engineering
Division P.M.A.
Celestijnenlaan 300B
B-3001 Leuven
BELGIUM

Th. Vast
Department of Mechanical Engineering
Université catholique de Louvain
BELGIUM

Anthony Venus
CAMM
The Industry Centre
University of Sunderland
Hylton Riverside West
Wessington Way
Sunderland SR5 3XB
UNITED KINGDOM

M. Wake
School of Science & Technology
University of Teesside
Middlesborough TS1 3BA
UNITED KINGDOM

H. Fred Walker
Faculty Member
Department of Industrial Technology
University of Southern Maine
Gorham, ME

Jun Wei
School of Industrial and Systems Engineering
Georgia Institute of Technology
Atlanta, GA 30332-0205

G. Weigert
Dresden University of Technology
Electronics Technology Laboratory
GERMANY

Thomas West
Department of Industrial and Manufacturing Engineering
Oregon State University

S. Widdows
RapidCast (UK) Ltd.
Consett
Co Durham
UNITED KINGDOM

Philip M. Wolfe
Arizona State University

Victor Fay Wolfe
Computer Science and Statistics Department
University of Rhode Island

Andrew Wooster
Panasonic Technologies
Kyushu Matsushita Electric Research Lab
Research Triangle Park, NC

R.C. Wurl
Department of Industrial Engineering
Rutgers University
Piscataway, NJ 08855-0909

A. Yalcin
Department of Industrial Engineering
Rutgers University
Piscataway, NJ 08855-0909

Min-Chun Yu
Oklahoma State University

Guangming Zhang
Department of Mechanical Engineering and Institute for Systems Research
University of Maryland
College Park, MD 20742

Chen Zhou
School of Industrial and Systems Engineering
Georgia Institute of Technology
Atlanta, GA 30332-0205

TABLE OF CONTENTS

ACKNOWLEDGEMENTS

The organising committee of the FAIM '97 conference is indebted to many individuals and organisations for their contribution, support, and endorsement. We wish to thank all of the keynote speakers who shared their views and visions of flexible automation and intelligent manufacturing. We also wish to acknowledge, with many thanks, the original papers of all of the authors who submitted their manuscripts for publication and who presented their work at the conference. Our special thanks are extended to all FAIM '97 reviewers and session moderators who helped to ensure the high quality of the conference.

We wish to acknowledge, with great gratitude, the help and support received from Professor Derek Fraser, Vice Chancellor, University of Teesside; Ms. Helen Pickering, Pro-Vice-Chancellor, Research and External, University of Teesside; Dr. Ray Sheahan, Chief Executive, EPICC. We would also like to acknowledge the continuous support given by Dr. Paul Torgersen, President of Virginia Tech, USA and Dr. Edward Walsh, President of University of Limerick, Ireland.

Although there are many who deserve credit for their various contributions to the success of FAIM '97, there are some who deserve special recognition for their truly outstanding efforts. Dr. Ian French, Principal Lecturer, EPICC / University of Teesside; Ms. Jean Tennant, Corporate Services Manager, EPICC; Dr. Roy Grimwood, EPICC; Ms. Dot Cupp, Virginia Tech, USA were invaluable for the planning of FAIM '97. Without their hard work, this conference would have been impossible.

Finally, we wish to acknowledge the excellent editorial support of Mr. Jim Kelly of Begell House, Inc. His patience and enthusiasm for this edited work are sincerely appreciated.

COPING WITH DISTURBANCES A PERFORMANCE EVALUATION APPROACH IN VIRTUAL MANUFACTURING ENVIRONMENTS

Nabil N. Gindy and Sameh. M. Saad
Department of Manufacturing Engineering and Operations Management,
University of Nottingham, Nottingham NG7 2RD,
England

ABTSRACT: The aim of this paper is to report on an extensive investigation of one of the important problems that is frequently encountered during the operation of manufacturing system, the system's ability to make an appropriately balanced and rapid response to predictable and unpredictable changes. In this work this is termed as manufacturing "Responsiveness" and is considered as one of the major attributes success and profitability of manufacturing systems. The focus of the investigation is the "Responsiveness" of machining environments. Simulation is used to represent a machining facility belonging to a large industrial company and generic capability units termed *Resource Elements (REs) are developed* and used for representing the machining facilities as a set of virtual resources. The reported results show that significant improvements in manufacturing system performances and responsiveness can be achieved when a system is represented and scheduled using the proposed methodology.

INTRODUCTION

An important characteristic of any manufacturing system is its ability to maintain an immutable performance under changing operating conditions, the disturbances or unpredictable events that can influence the system's ability to achieve its performance objectives. Many research publications list machine breakdowns as a most important factor that can influence shop floor performance. Additional types of disturbances identified are: variations in production demand patterns and volume, unavailability of cutting tools and catastrophic tool breakdowns, transport system failures, etc. Farhoodi (1990) suggested the use of a causation tree for typing disturbances that can be traced back to inefficient scheduling, and other disturbance types that originate from activities such as process planning or resource planning. Chatima (1995) in his discussion on disturbances, proposed extensions to types of disturbances proposed by Farhoodi (1990).

In this work, disturbances are classified based upon their sources of origination (similar to Farohoodi (1990). Disturbances are first classified as internal and external disturbances depending on whether the origin of the disturbance is from within or originates from outside the boundary of the manufacturing system. Figure 1 outlines the types of disturbances considered and a brief summary is outlined as follows:

Internal disturbances

Machines.
1- Unavailability (breakdown, corrective maintenance, preventive maintenance).
2- Availability but with limitations on machine capacity.

Tools/Fixtures.
1- Unavailability (not requested, not earmarked for job, shared and used by resources).
2- Available but with constraints .

Transport.
1-Unavailability (e.g. breakdown etc.)
2- Available with limited capacity.

Operators
1- Unavailability (e.g. sickness, holiday, disputes, etc).
2- Available with constraints (e.g. lack of appropriate skills).

External disturbances

Demand Related.
1- Unexpected orders.
2- Expected orders with time delay.
3- Expected orders arriving early.
4- Change in order priority.
5- Quantity variation Lower/higher than planned.

Supplier Related.
1- Delivery at the wrong time.
2- Non-delivery of the required parts.

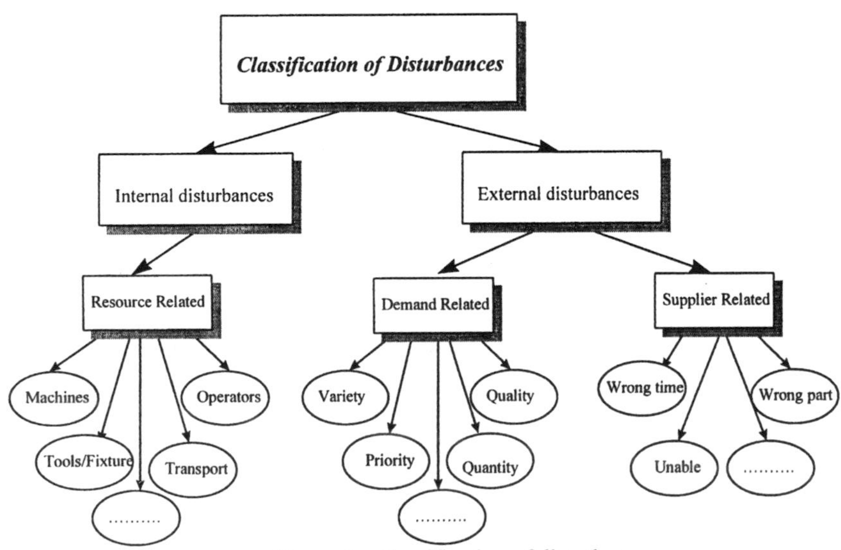

Figure 1. Classification of disturbances

The objective of this investigation is to examine the performance and responsiveness of a machining environment, based on the data obtained from a large industrial company. Three different models are used in this investigation: The first is the traditional machine-based representation and scheduling of manufacturing facilities, the second relates to the same machining facility represented and scheduled using the resource element approach operated as virtual resources and in the third the machining facility is represented using resource elements approach but divided into physical cells. The system performance is compared based upon it ability to cope with a variety of disturbances in its operating conditions.

RESOURCE ELEMENTS (RE) AND VIRTUAL MANUFACTURING

In this work the formation of a virtual machining cell is based upon assessing the machining requirements of a product set in relation to generic factory-wide capability units termed *Resource Elements* that collectively make up the virtual cell. *Resource Elements (RE)*, as defined in Gindy et al. (1996), are machining facility specific capability units, which capture information relating to the distribution (commonality and uniqueness) of form generating schemas among the machine tools available in the machining facility. A Form Generating Schemas (FGS) is a technologically meaningful combination of a cutting tool of specific geometry, a set of relative motions between a part and the cutting tool, and the typical levels of technological output associated with using that combination of tool and relative motions. The machine tools contained in manufacturing system are described using a set of *RE*. Each *RE* represents a collection of form generating schemas such that the exclusive and the over-lapping capabilities between all the available machine tools are uniquely identified. Full explanation of *RE* concept is beyond the scope of this paper, more details can be found in Gindy et al.(1996).

SYSTEM MODELS

Machine-Based System (M/C-based)

The model used for this study is based upon a machining facility containing 22 machine tools that perform a wide variety of operations such as turning, milling and drilling. Each machine tool has a limited buffer at which parts are allowed to wait. Parts and materials are transported in the system by a combination of forklift trucks and cranes. Each part entering the system is routed through the machine shop according to its processing requirements. Parts are dispatched to machine tools according to their machine-based routes.

Resource Element-Based System (RE-based)

The model is identical to the machine based model (MC), however, the machining facility is represented as a collection of resource elements operated as virtual resources (entities that can be scheduled individually). Parts are dispatched to machine tools for processing on individual RE's according to their RE-based requirements. No physical cells (fixed groupings of RE's) are identified in this model. A virtual cell based on the generic RE-based route attached to each part is identified. The virtual processing cell contains all the possible RE-based processing routes for the component (RE'S)

Resource Element-Based Cell System (REC-based)

The model used in this investigation is based on dividing the machining facility into four physical cells. Each cell has a target component set based on part similarity in terms of processing route. Each cell is operated independently from the other cells and part cross overs between cells is not allowed. The physical cells are represented by the set of RE's and operated as an independent unit in which parts are dispatched according to their RE processing requirements.

MODELLING THE MANUFACTURING ENVIRONMENT AND OPERATING ASSUMPTIONS

Three discrete simulation models were developed to represent the above three systems. The following operating assumptions are made:

- Deterministic processing time at each machine tool.
- Exponentially distribution of part arrival.
- Machine breakdown and corrective maintenance work occur during the production run.
- The mean time between failure and repair period is exponentially distributed.
- Set up time is not included in the operation time and represented separately.
- Single part transfer by handling by equipment.
- Transit time for the transfer of part between machines is considered.
- Part variety is twenty different jobs.
- One of a kind/ small batch production environment .
- Due-date calculated using processing plus waiting approach (PPW) see Gindy and Saad (1995).
- Dispatcher uses earliest due-date rule (EDD) see Gindy and Saad (1995).

METHODS OF ANALYSIS

A manufacturing system is generally modelled as a non-terminating system, which are commonly analysed on the basis of data collected when the system has achieved steady state [Saad (1994)]. In many situations, where production disturbances are pronounced, steady state conditions may never be reached due to practical operating conditions (machine breakdowns, changes in set-up times, different production requirements, etc.). It was essential therefore that during this investigation to distinguish between transient and steady state behaviour during the discrete simulations conducted. This provided the basis for comparative judgements since results have been collected under same conditions, free from the initial bias and where the mean value of a dependent observation parameter does not change significantly.

A batching method was applied to carry out the steady state analysis. Considering initial pilot runs, the statistics for the first 50,000 unit times of the resource element-based model, first 50,000 unit times of the resource element-based cell and first 100,000 unit times of the machine-based were discarded and samples were collected for another 100,000 unit time. The batch size was determined as 15 times as large as the largest lag in order to ensure the independence of each batch and the correlation between each observation remains significant.

EXPERIMENTAL DESIGN

Selecting the Factors to be Investigated

Planned machine utilisation (PU). Two levels of machine utilisation were selected at 60% and 80%. The First level represents the case when the system is not busy. While the second levels depict a busy system. The objective from selecting those two levels is to observe the performance of the systems under investigation in coping with disturbances, that might lead to useful conclusions to help the practitioners in deciding their system load in order to cope with disturbances.

Load contribution (L). A discrete probability distribution was chosen to assign the contribution of each part family. Two different contributions were selected, first, uniform load contribution where each part family contributes equally to the total production load (e.g. with the twenty part families, each contributes 1/20 of the total load), see Figure 2. Figure 3 shows the second case, skewed load contribution, in this case some of the families contributes different than the others with a condition that the cumulative value of the discrete distribution must be one

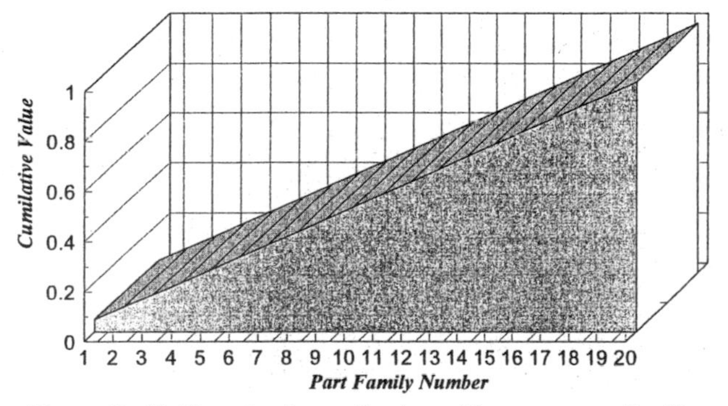

Figure 2. Uniform load contribution with twenty part families

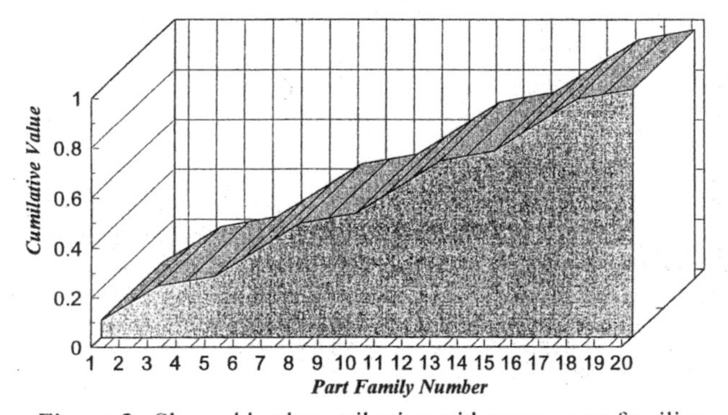

Figure 3. Skewed load contribution with twenty part families

Failure rate (FR). The experiment was limited to three levels of failure rate [reliable system (e.g. no failure), 5%, 10%] the percentage expresses the ratio MTTR/MTBF, where the time between failure and the repair period were modelled by assuming an exponential distribution for both, as this type of distribution is often used to model such events see Table 1.

Table 1
Failure Rate Ratio

Failure Rate %	MTTR	MTBF
5	exponential (30)	exponential (600)
10	exponential (60)	exponential (600)

System responsiveness (k). In general to calculate the due-date of a part, usually it is consists of two terms. First term is the effective processing time required to produce the part plus allowance. This allowance may represent handling time, waiting time in queues, or waiting time in the system for other reasons. Responsiveness of the system (*k*) depicts how the due-date is far from the processing time. For more detail refer to Gindy and Saad (1995) and (1996). In our experiments, system responsiveness tested corresponded to *k*=2, 5, 8, and 11. The experiments started with *k*=2, because at *k*=1 the calculated due-date value using the proposed due-date approach is equal to the processing time. This is an extreme case and quite difficult to achieve in practice. Selection of more than these levels will enlarge the number of experiments considerably

A complete four factorial experimental design involves 48 treatment combinations. The total number of experiments required for the three proposed systems were 144 experiments. Table 2 displays the detailed design of the 48 combinations.

Table 2.
The Factorial design

		Failure Rate FR											
		Reliable				*5%*				*10%*			
Planned Utilisation	***Load Contribution***	***Due-Date Tightness K***											
		2	5	8	11	2	5	8	11	2	5	8	11
60%	*Uniform*												
	Skewed												
80%	*Uniform*												
	Skewed												

Performance Measures of the Simulated Manufacturing Environment

With most real-word simulations, several measures of performance are of interest. Four performance measures were considered

1) Average Flow time in system (*AFT*)
2) Mean tardiness (*MT*)

3) Proportional tardy job (*PT*)
4) Machine utilisation (*U*) in case of skewed load contribution

RESULTS AND DISCUSSION

The purpose of the simulation experiments conducted was to analyse and compare the performance of the three proposed systems in order to investigate their responsiveness. Some representative results are shown graphically in Figures 4, 5, 6, 7 and 8. Figure 4 shows the relationship between the system's mean tardiness performance and responsiveness measure *k* under the three chosen failure rate (*FR*) levels, no breakdown, 5% *FR* and 10% *FR* in Figures 4-a, 4-b and 4-c respectively. As can be seen from these set of Figures RE-based system performs much better than M/C-based and even better than REC-based for all the three levels of *FR*.

It was expected that as the *FR* increases the performance of the three systems will get worse. Figure 5 displays the deterioration rate θ in the three systems caused by the machine breakdown. We can conclude that the ability of the RE system to cope with such disturbance (machine breakdown) is better than REC and M/C systems and the rate of deterioration even better in the RE system with higher failure rate than REC and M/C-based where $\theta_{RE} < \theta_{REC} < \theta_{MC}$ see Figure 5.

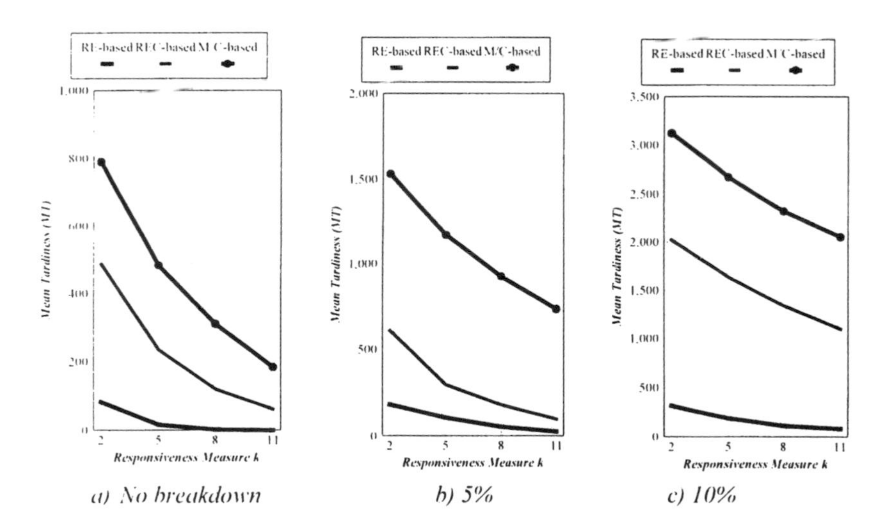

Figure 4. Mean tardiness performance of the three proposed systems (at PU=60%, Uniform load)

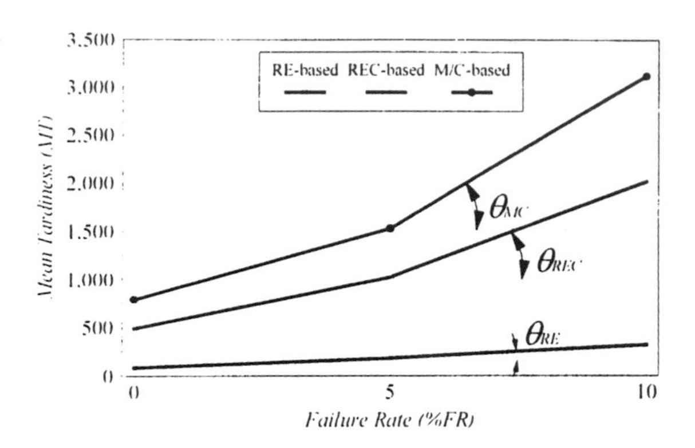

Figure 5. System's ability to cope with machines failure (at PU=60%, Uniform load, k=2)

As we stated above one of the objectives of these experiments was also to examine the ability of the proposed systems in coping with the changes in the demand pattern. Figure 6 shows the relationship between average flow time taken as a performance measure and the effect of changing the demand on the manufacturing facility from uniform to skewed load contribution. It can be observed from the figure that the average flow time for components in the system is lower in the RE-based system, compared with the REC-based and M/C-based systems and that the RE-based system is much more capable of coping with changes in the demand pattern.

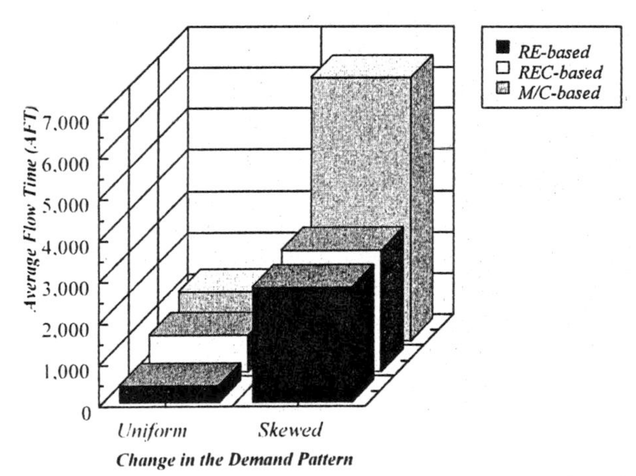

Figure 6. The effect of the demand pattern on the average flow time (at PU=60%, k=2, FR=0)

Figure 7 displays the percentage of job tardy *PT* in RE, REC, and MC systems under two different load contributions (uniform and skewed). As can be seen from the Figure, a larger percentage of tardy jobs was found in MC system than in other two systems under investigation.

Another interesting result is presented in Figure 8. One of the objectives of the experiments was to observe the systems performance in terms of machine shop utilisation under the skewed load contribution. Figure 8-a displays the behaviour of the machine shop utilisation for the RE, REC, and M/C-based systems under different failure rate in the case of 60% planned utilisation. At this level of planned utilisation the system is not busy, that is why the results do not show much difference between RE-based and REC-based systems. They are achieving quite similar utilisation levels in case of the skewed load contribution under different failure rate and there was not much difference between achieved levels and the planned utilisation (60%). While in M/C-based system there was a significant drop in the achieved utilisation either in the case of reliable system, 5% and 10% failure rates. Figure 8-b shows the case of 80% planned utilisation that represent a quite active system.

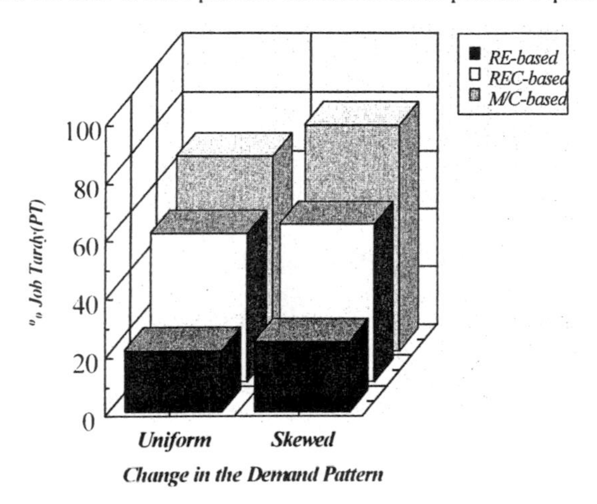

Figure 7. The effect of the demand pattern on the job tardy (at PU=60%, k=2, FR=0)

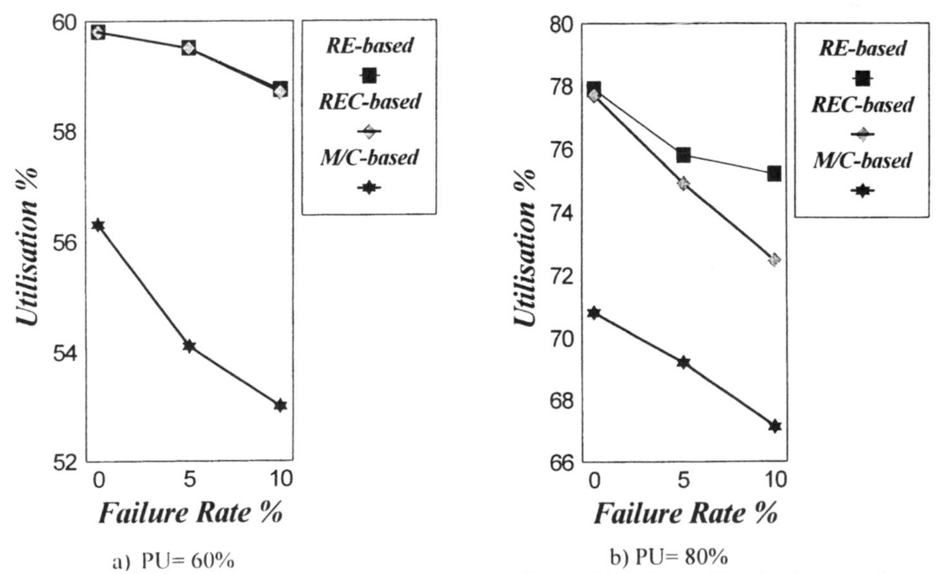

a) PU= 60%

b) PU= 80%

Figure 8. The effect of machine breakdown on the machine shop utilisation in case of skewed load contribution

The results here clarify the ability of RE-based system to cope with such disturbances. As can be seen at reliable system RE and REC systems very near to each other but as the failure rate increases the difference becomes bigger and also the drop in the achieved utilisation levels becomes remarkable especially in M/C-based system. The important thing that has been discovered during the experiments, the M/C-based system finds difficulty in reaching to the steady state particularly at 10% FR. This means that the system could not recover from the failure at the shop floor and also from the skewed load effect. On the other hand, the use of *Resource Elements* concept to describe the capabilities of machine tools and machining facilities allows the selection, loading and scheduling of machining resources to be carried out based on pre-knowledge regarding the total number of RE's in the system as well as the number of repeated resource elements contained in the machining facility. This is a much higher level of detail than what is available when a machining facility is described based on "whole" machine tools.

MEASURING THE SYSTEM RESPONSIVENESS

The system responsiveness was an important feature to be studied in this work. Figure 9 displays the responsiveness in term of responsiveness measure (k). Figure 9-a shows the values of k in case of 5%FR and 60% planned utilisation and Figure 9-b presents the case of 5% FR and 80% planned utilisation for the three systems under investigation. The values of k at which give approximately zero mean tardiness in the three different systems have been calculated. These values are shown in the figure; as can be seen, the smallest values were found in RE-based system followed by REC-based and then M/C-based. This means that RE-based system allows much tighter assignment due-dates compared with the M/C-based. Repercussion of this, RE-based system will operate at low cost, high efficiency and will be high competent system in the market having the ability to achieve what is needed in a shorter time.

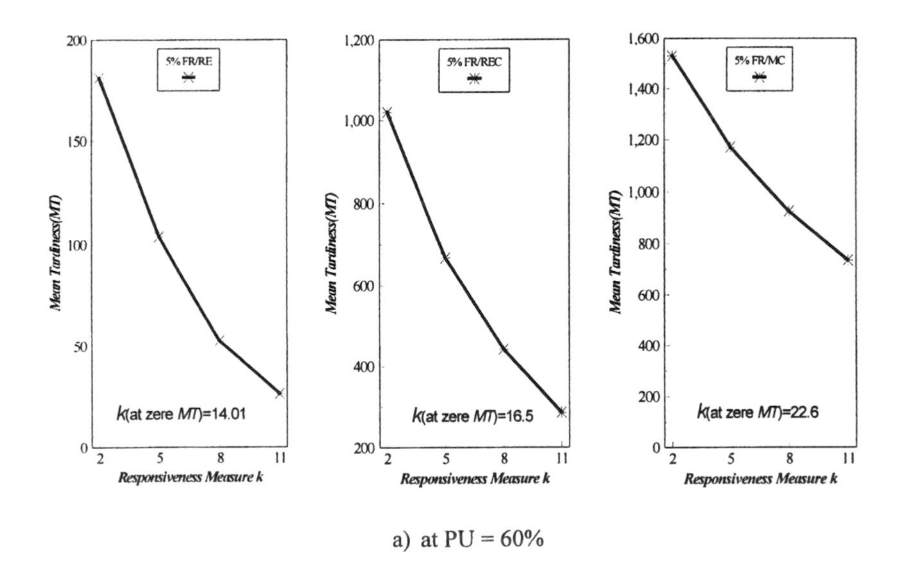

a) at PU = 60%

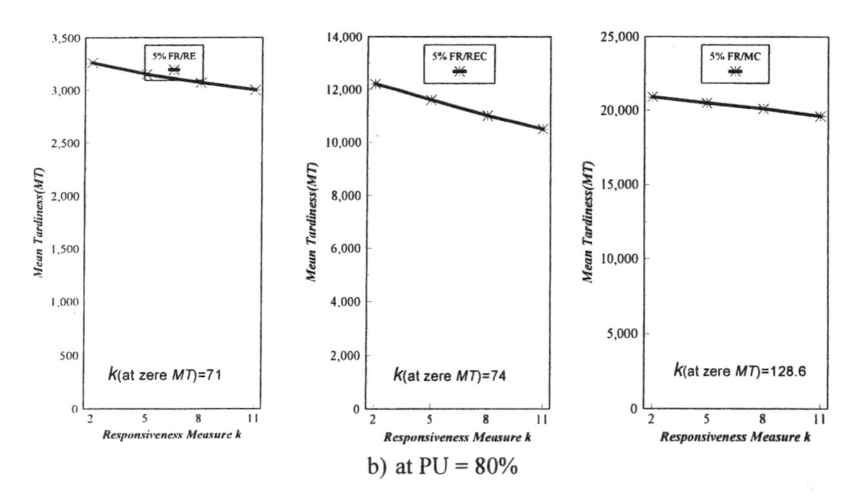

b) at PU = 80%

Figure 9. Responsiveness of the systems investigated under different planned utilisation (at FR=5%)

CONCLUSIONS

Comprehensive experimental work has carried out to analyse and compare the performance of three different systems. First, machine-based system (M/C-based) second, resource-element based system (RE-based) and the third was resource-element based cell (REC-based). The results clearly indicate that significant improvements in system performance measures can be achieved when a machine shop is represented as a set of generic capability units. Moreover, the results prove the ability of the RE system to cope with different type of disturbances such as machine breakdown and the changes in the demand pattern. These proposed three models could be used to select best design and control strategies for given environmental conditions. Another contribution of this work is in the proposed measure of the Responsiveness.

ACKNOWLEDGEMENT

The reported research is supported by the UK EPSRC research grant ref.: GR/J 90022. Their financial contribution is gratefully acknowledged.

REFERENCES

Chatima P., (1995) "Real-time operational control of flexible manufacturing systems". PhD thesis, Department of Manufacturing Engineering and Operations Management, University of Nottingham, England.

Farhoodi F., (1990) "A knowledge-based approach to dynamic job-shop scheduling", International Journal of Computer Integrated Manufacturing, Vol. 3, No. 2, pp. 84-95.

Gindy N. N. Z., Ratchev T. M., and Case K.,(1995) "Component Grouping for Cell Formation Using Resource Elements. *International Journal of Production Research,* 34(3) 729-759.

Gindy N. N. Z. and Saad S. M., (1995) "Factory Model Implementation, Experimental Investigation of Resource-Element-Based and Machine-based Scheduling", *Technical Report N0. 2-2 EPSRC/GR-J9002.*

Gindy N. N. Z. and Saad S. M., (1996) "Resource-Based Scheduling in Virtual Manufacturing Environments" *12th National Conference on CAD/CAM Robotics and Factory of the Future*, Middlesex University, London, 14-16 August 1996.
Saad, S. M., 1994, *Design and analysis of a flexible hybrid assembly system.* PhD thesis, Department of Manufacturing Engineering and Operations Management, University of Nottingham, England.

DEVELOPING A WEBSITE FOR THE FORMATION OF VIRTUAL ENTERPRISES

Carlos Bremer[1,a], Edson dos S. Moreira[2], Fredy J. Valente[2], Paulo C. de Carvalho Dias[1]
[1]Escola de Engenharia de São Carlos - Universidade de São Paulo;
[2]Instituto de Ciências Matemáticas de São Carlos - Universidade de São Paulo

ABSTRACT: Virtual enterprises are getting a higher acceptance within enterprises circles as an alternative to act in highly dynamic markets. The aim of this paper is to describe a proposal of an World Wide Web site to support the formation of virtual enterprises. The main characteristics of this site are the use of middleware technologies to allow a full integration between a web servers and the database of competence. The system also supports a customised and dynamic generation of web pages according to the interests of the user. An especial attention has been given to the process of partner searching. The site has been implemented as a prototype system to support the search of technology-oriented competencies. Other potentials of this approach are also described.

1 - INTRODUCTION

New information and communication technologies allow a unprecedent number of persons and organisations to manipulate (acquisition, storing and processing) an increasingly high amount of information [1]. Instead of doing everything in house, co-operation through the sharing of information (resulting in cuts of transaction costs) can be very profitable to the enterprises [2]. The co-ordination of economic activities are moving from hierarchical to market-oriented forms [3].

Enterprises are turning into boundaryless organisations, configured by smaller and more autonomous organisation units, with a high degree of interaction between each one of them and others, within or outside their corporation [2]. The aim is to achieve a high degree of agility, which can be defined as the "ability of an organisation to respond well, in fact thrive, in response to unexpected events" [4]. One major consequence is the enterprise strategy of concentrating in their core competencies [5,6]. For small and medium sized enterprises this is also a positive environment for developing and leveraging their business.

The continuous development of information and communication systems and services is also an important driver for the formation of more co-operation among enterprises, since they are built over a computer-supported collaborative environment, where the physical distances are no longer a hindrance. This scenario allows a very dynamic and global actuation of different actors within a business sector (customers, competitors, suppliers and so on) simultaneously. Therefore, the development of an efficient information infrastructure that allows an enterprise to act and response to this business dynamics can be considered as an important competitive advantage.

[a] The author would like to thank the Conselho Nacional de Desenvolvimento Científico e Tecnológico - CNPq for the financial support during his post doctor research activities

The aim of this paper is to describe the proposition of a Website that supports the formation of virtual enterprises. Envisaged enterprises are production companies, specifically from the mechanical sectors, although other users from related sectors, as for instance engineering offices are considered.

2 - VIRTUAL ENTERPRISES

Virtual enterprises can be defined as a way of organising business activities, where different and independent partners exploit a specific business opportunity by establishing an enterprise co-operation [7]. The virtuality is then the ability to offer customers a complete product or service, where the enterprise itself just owns some of its competencies. Other required competencies are achieved through co-operation. In fact the idea of virtual enterprises is nothing new and has been used in some business sectors for many years, but the changes in the society as a whole as well as the new information and communication technologies are making them strategically and economically feasible for a wider range of enterprises [8,9].

Characteristics and Life Cycle

Three main aspects can characterise virtual enterprises [10]:

- uniqueness - a virtual enterprise exists to explore one single business opportunity at a time,
- competence orientation - a virtual enterprise aims to join the best competencies in order to fulfil the opportunity, independently of their location, and
- modern information infrastructure - the use of modern and efficient information and communication technologies (information infrastructure) allows the formation and management of dynamic co-operations among different and also global partners.

Virtual enterprises have a life cycle similar to enterprise co-operations. The Agility Forum has proposed the Agile Virtual Enterprise Framework (AVEF), where five processes and four infrastructures are defined, as shown in **Figure 1** [11].

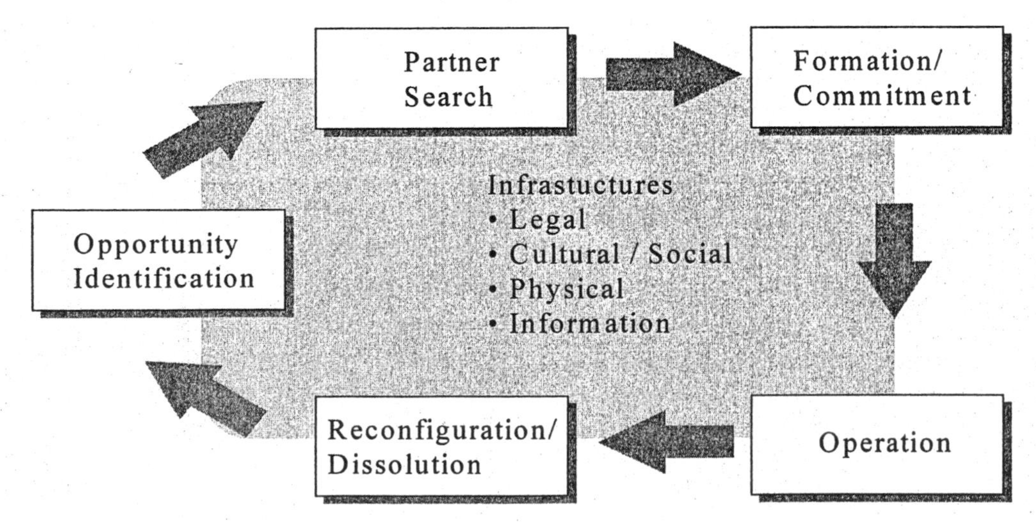

Figure 1. Processes and Infrastructures of the Agile Virtual Enterprise Framework

Although the AVEF defines formation as been the third process, in this paper the formation spreads over the initial 3 processes (*opportunity, partnership searching* and the *formation* itself). The information infrastructure for manufacturing enterprises is of great importance, mainly for the ones acting in global business environments.

Internet and the World Wide Web as the Information Infrastructure

The development of the Internet was supported by the Defence Advanced Research Projects Agency (DARPA, United States) in the seventies. In the beginning, the Internet was comprised by some universities, government and military networks. Nowadays the Internet has a world-wide span and represents one of the main information medium for enterprises to make their business.

The Internet architecture is mainly based on a connection-oriented transport service which is supported by the TCP/IP (Transport Control Protocol/Internet Protocol) protocol family. Services such as FTP (File Transfer Protocol) and the HTTP (HyperText Transfer Protocol) are implemented over TCP/IP. HTTP is used in World Wide Web (WWW) as a mean to transport HTML (Hypertext Markup Language) files which, after being interpreted by a Web browser (e.g. Netscape, MS Explorer) are called web pages [12]. The multimedia capabilities present in the Web browsers has made the Internet very popular.

3 - WEBSITE REQUIREMENTS

The requirements have been defined through an evaluation of similar sites as well as a literature review, concerning virtual enterprises, enterprise co-operation and enterprise integration.

The evaluation was done through the analysis of some Web sites designed to use the Internet as an information infrastructure to explore the concept of virtual enterprises. One of the sites analysed was the PartNet (*http://part.net*). The PartNet was funded by research contracts between the U.S. Department of Defense and the departments of Mechanical Engineering and Computer Science at the University of Utah. The main goal of the PartNet was the development of a Web site as a mean to connect designers and engineers with part suppliers. In this way the time spent by engineers in locating and acquiring design components can be reduced. The PartNet was first tested in the Sacramento Air Logistics Center using data from a variety of part suppliers. The PartNet provides a way to integrate in a single virtual information network, data from many dissimilar databases across many different companies via Intranet technology (*see definition below*). In this way, the user of the system can easily find the desired part somewhere in the PartNet using its preferred Web browser.

Virtual enterprises are based on the exploitation of the partners best competencies. Each partner contributes with its core competence. Therefore a method to define and represent competencies is required. This work adopts the approach defined by the Motion Project which is represented in **Figure 2** [13].

According to this definition, competencies are dynamic. An important requirement for the infrastructure is the possibility to cope with dynamic competencies, which means flexibility to store skills and adaptability to represent the competencies according to the tasks that are envisaged or requested.

Groupware used to be defined as "information technology used to help people work together more effectively". The generic benefits of groupware are: work flow improvement, time shifting, location shifting, improved communication through computer mediation, organising, indexing, storing and finding information and sharing hardware resources [14]. To implement this gains via a computer-supported co-operative work (CSCW), the need of a group collaboration software is primordial.

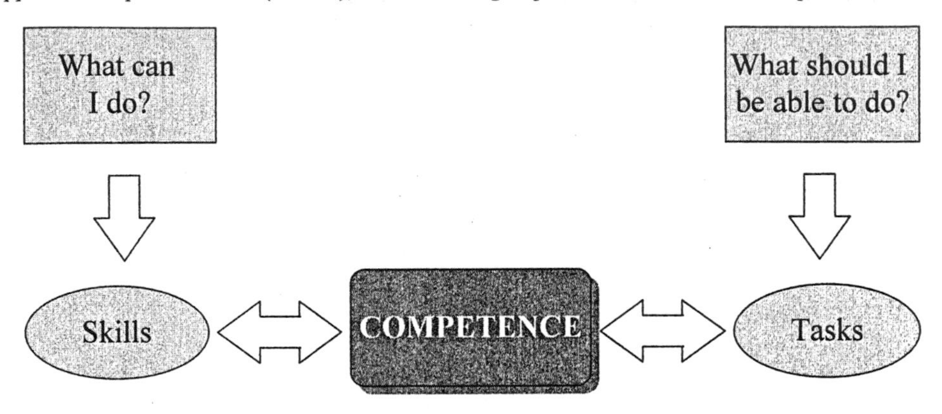

Figure 2. Competence as a relationship between tasks and skills.

According to Ellis there are six types of groupware applications [15]. These types can be used as an initial set of functions for the information infrastructure of virtual enterprises:

- Message Exchange Systems - targeting the asynchronous communication among users;
- Multi-user Editing Systems - aiming to support the collaborative and simultaneous work among different users in the same document;
- Decision Support Systems and Electronic Rooms - improving the productivity and quality of the collaborative work;
- Conferencing Systems - converging computer and telecommunication systems to support the collaborative work;
- Intelligent Agents Based Systems - taking over some of the decision making processes from the users and;
- Co-ordination Systems - increasing the management capability of group leaders.

In the other hand, the new developments in the Internet technology with the spread of WWW and the added security provided by firewalls (entrance doors), which restricts the access to Web sites and pages via passwords and smart cards, allowed the development of intra-organisational networks [16]. Such closed network is called Intranet [17]. Additionally, the integration of Web Servers with database systems via CGI (Common Gateway Interface) and Java allows the use of Web browsers as a front-end to existing distributed database systems, thus allowing the migration of customised corporate information systems onto an Intranet environment [18].

Roberts, comparing the adaptability of Groupware and Intranet technologies also adds some clues on the profile of the information infrastructure requirements [19]:

- use of standards - concerning here the idea of plug-and-play systems;
- having reliable security system- providing security for the information and documents flowing through the systems, externally as well as internally to the group which is co-operating;
- capacity of replication - allowing the sharing of information and documents independent of time and location;
- ease of application development - making possible that end users themselves generate their own applications as well as for systems managers and developers for generating critical applications;

- object oriented storage - storing and making accessible any type of information; providing a connection to other objects storage as well as managing them and;
- versatile messages processing models - offering the possibility to send messages linked to specific information on a database.

Regarding the aspect of integration, Vernadat defines six principles to be pursued by an integration infrastructure (where the information is one of its cores) [20]:

- application confinement - concerning the portability of the applications running on the infrastructure;
- modularity - providing an open system, which can be easily extended and maintained;
- interconnectivity - making the data transfer among different hardware possible;
- interoperability - allowing two or more systems to interact in performing one task;
- information neutrality - exchanging information among different systems need to be provided and
- scalability - relating with the possibility to extend the system in a distributed environment.

The specification of the server and the choice of technologies to implement it was guided by the requirements derived from the three groups mentioned before (virtual enterprises, enterprise co-operation and enterprise integration).

4 - THE DESIGN OF THE WEBSITE

This description is divided in two parts. One describes the site architecture and the other regards its adequacy to the defined requirements.

Site Architecture

The following architecture has been proposed to attend the mentioned requirements (**Figure 3**). The system consists of an Intranet environment in which clients (partners) can browse information.

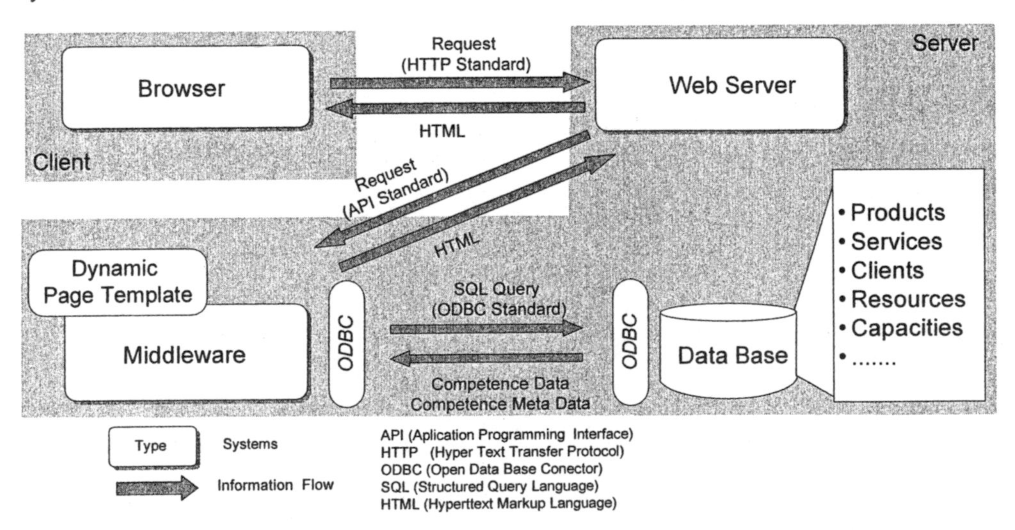

Figure 3. The Architecture of the Virtual Shop Floor System

The transactions flows following a sequence. A request is sent by the browser using the HTTP format. The server then passes the control to an engine (implemented in the middleware layer), eg. a CGI-bin process, which then deals with the request possibly querying the database using SQL, via ODBC standard [21]. Middleware, in this context, consists of a class of software which binds data objects between the web server and the database management system. The query is then processed by the database system and the result is sent back the to the middleware layer. Web pages, in the HTML format, are then generated dynamically, containing the requested data. These pages are published by the Web server which will then be read and displayed on the browser of the client.

Every connection made by the client's Web browser to a Web server is unique in the eyes of the server. However, the middleware is capable to keep track of a user as he moves through the pages of an application. This is called client state management, which is implemented in the middleware by creating a client record for each Web browser that requests a template. The middleware then creates a client variable database (stored in the system registry), and identifies each client record by an unique token which is stored in an HTTP cookie in the user's browser; i.e., the client state management implementation is made in the same way as the logical security application framework. Using client state management it is possible to store any client-related variable, making the system potentially fully customisable according to the user's preferences or often usage. It's possible, for example, to recognise that a certain user searches always using the same search algorithm. In further searches, the system could offer the frequently searched algorithm as a first option to the search.

The site was designated to work on the Internet at the same time as being part of an Intranet. To reach that purpose, security features has to be made to assure that only authorised people can use the system. The first security level is made by the Web server itself, which is built on the operation system level. The access control can be made by restricting the permitted IP addresses, requiring assigned user accounts and passwords, and limiting the user rights of these accounts in the operation system, assigning permissions to folders and files on the server. In addition to the operation system security features, the Web server can also be set to accept or not anonymous requests, and is able to support the Secure Sockets Layer (SSL) protocol, which securely encrypts data transmissions between clients and servers. SSL is an industry-standard protocol that uses the public-key technology to cryptograph data transmissions between clients and servers, assuring a secure way of data transfer.

Apart from the server built-in security, the middleware can be used as well to enhance security in a logical way (via software). A system can be built under an Web application framework, which are a collection of template files that work together providing advanced security enhancement. Every time a template file is requested, the middleware runs first an application template file, which is set to verify the presence of a specific cookie on the client's Web browser. Cookie is a mechanism in which server-side applications can use to store information in individual Web browsers. They are domain-specific (set and retrieved for a specific server reference) variables stored on the client's browser, supported by all major commercial Web browsers. This cookie is created only after the login procedure, expiring when the connection is terminated; and no template can be evaluated without the presence of this cookie. As the cookie is domain-specific, the middleware server can identify the user's connection analysing this (browser-specific) variable, and then decide if the user is allowed or not to run the requested template.

Site adequacy

The use of the middleware layer and the consequent dynamic generation of pages templates contribute to fulfil the requirement of representing the dynamic of competencies.

Concerning enterprise co-operation requirements following can be pointed out. The site architecture offers a reasonable environment for asynchronous communication. The middleware makes even the generation of dynamic electronic mails possible, in so far to automates short processes. For synchronous communication it is expected that the next generation of browsers will extend their capabilities in this field [22]. A requirement that is not satisfied is the possibility to work on the same document and the replication. The middleware, because of its customisation features, can assume part of the decision making process from users. Since the concept adopted to use this architecture is very decentralised (see next chapter), the management capabilities of group leaders have been not considered.

The use of standards is satisfied by the inherent characteristics of the Internet. The security features of the architecture need to be divided into the external and internal ones. For external a encryption schema should be provided, which is not the case here. Just the not allowed access has been taken into consideration. For internal security no special precaution has been made. The easy application development is provided just for critical applications by the middleware. Since no restrictions exists on the data base level, the object oriented storage requirement can be considered as satisfied.

For most of the requirements concerning the enterprise integration an adequacy consideration can not be made, since the integration of different systems running on this architecture are not yet made. The portability is partial, since the middleware is specific for a operation system, but it can access any ODBC Standard data base and works with any web server. Concerning the modularity it is satisfied because of the Internet architecture. The same consideration can be assumed for the requirements of interconnectivitiy.

Additionally other features like standards as Electronic Document Interface are being integrated to the Internet. It can be then expected that standards already accepted by industries will be adapted to the Internet.

5 - THE VIRTUAL SHOP FLOOR (VISHOF) SYSTEM: A CASE STUDY

The globalisation of manufacturing business enterprises is a reality and a need. Additionally, the increased dynamic of this business environment makes the virtual enterprise a reasonable way to conduct business [23]. The application of the described site architecture is done for the competence-oriented partner search, specifically for the technical-oriented competencies (**Figure 4**). The most important information objects of this type of competence are resources and processes. In the current system, called Virtual Shop Floor (VISHOF), only the resources are taken under consideration.

An information broker assumes the function of acquiring and distributing information for the existing or for the potential partners. Partners can be manufacturing enterprises as well as engineering offices and so on. Because of the global aspect of the Internet, partners can be located in different regions or countries. These partners can then offer and search technical competencies (or shop floor resources) using the VISHOF system.

The Information Broker, which can be an independent organisation as well as another type of partner (for instance an Internet provider) offers services to support the life cycle of virtual enterprises. It is also responsible for managing the Website, which involves the maintenance of the system parts including the communication networks, databases and security, and for the subscription and managing of partners.

One extension of this concept is the creation of regional Information Brokers to cover a specific region formed by one or more cities. Through the interconnection of such regional Information

Brokers it is possible to extend the virtual information and thus construct a virtual information network containing resource information of a greater region.

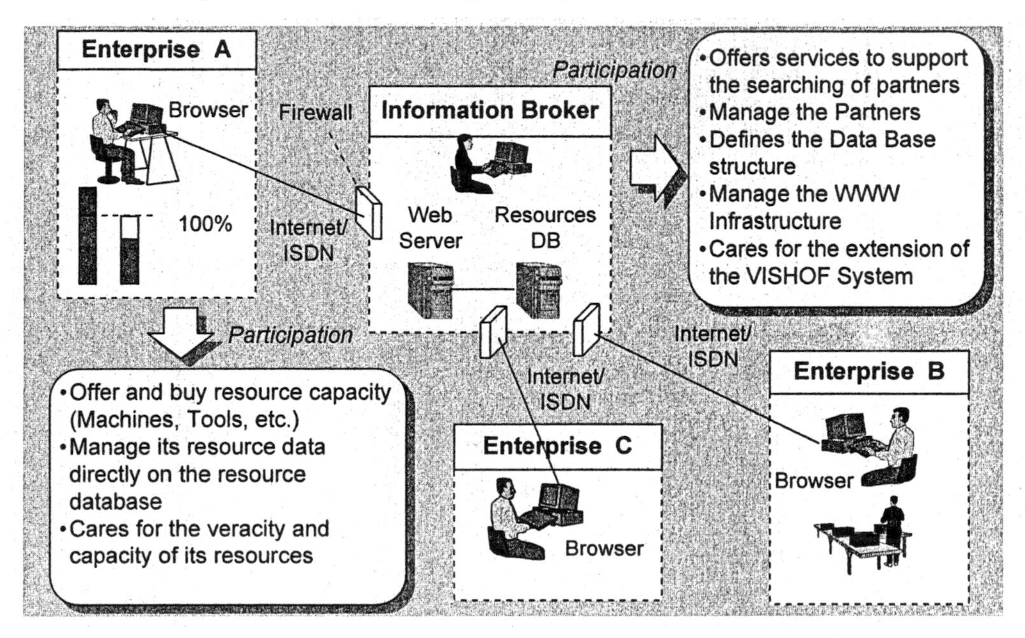

Figure 4. The Virtual Shop Floor Concept

To access the VISHOF infrastructure, the partner only needs a Web browser connected to the Internet. Through an access control provided by a firewall, the partner can access the VISHOF's database.

The system offers mainly three services: partner search, data management and statistics information. The search for technical competencies can be made in two ways: one where the user searches for resources themselves and other one where a product profile is given for the system, which then finds in its database which resource could produce the part. This second type of search is not yet implemented.

Within the search for resources two specialisation are offered. In one, the user knows the resource model he/her wants. Therefore when the group, class and type of the resources are chosen, the system shows all the machines that fits to this type. The second choice of searching for resources is based on dimensional data of the part to be produced. According to the group, class and type of resource chosen, the first input is the data that are absolutely critical to define the correct machine.

If after this first search, too many machines have been found, a detailed search is offered. In this refined search all the attributes concerning the chosen resource type are shown. The user can input the data he/her wants and filter it continuously until the number of resources found is satisfactory (normally between 1 and 5). Thereafter the resources attributes can be seen. Even visual information (eg. a picture) can be supplied. The resource capacity is only shown at this last level of representation. Finally the most appropriate resource is located and the information of the resource owner are given. The contact between the enterprises are still made by traditional forms.

The resource's data of an enterprise can be managed directly by the owner, considering that the system provides the necessary authorisation mechanisms for remote database administration. This means that the user can constantly change its information such as the inclusion of new resources, accessories, current free capacity and so on. This gives a high level of flexibility and autonomy to the users.

Additionally a statistic service is offered. It is possible to get information about the transactions at the Web server as well as about the resources accesses. Information such as the most searched resources can be provided by the service. Government agencies and other organisations can obtain an overview of the technologies used in a specific region as well as the identification of types of companies (and in which locations) could be looking for co-operation.

6 - CONCLUSION AND FINAL REMARKS

The Website and systems described above are useful tools to support the formation of virtual enterprises because of its capacity to find out partners based on their technical competencies. **Figure 5** summarises the potentials of the proposed Website.

The systems architecture, considering the users present some important benefits. It is easy to use because of the multimedia interface and portable since just a browser is needed. Therefore a low cost system is also provided. Additionally, the use of Internet guarantees an open standard architecture. The fact that the companies can manage their own data, without the intervention of intermediates, makes this system very flexible.

Considering the functionality to the enterprise a higher use of its resources is an important advantage of the system. Furthermore, the search of new technologies are possible, as well as finding partner for long term co-operations.

System Architecture:
•Easy to use (Multimedia and Hyper Text Interface).
•Portability (It is independent of Operational System and Hardware).
•Low cost for users (It is only necessary a Browser).
•Open Standard (It uses Internet Protocols)

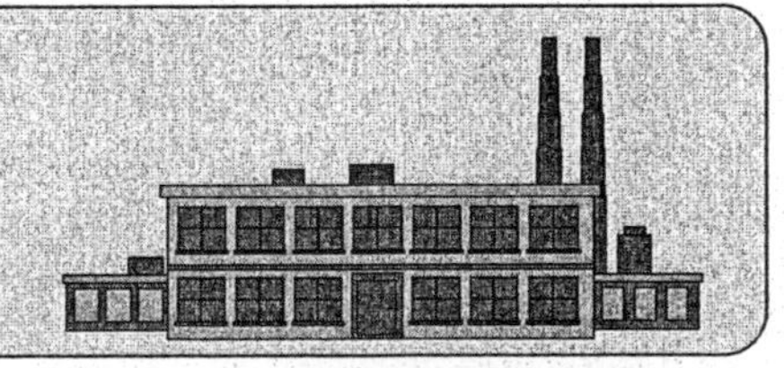

Functionality for the Enterprise:
•Higher the use of Enterprise's resources.
•Easy access to specific technologies or processes.
•Basis for future cooperation.

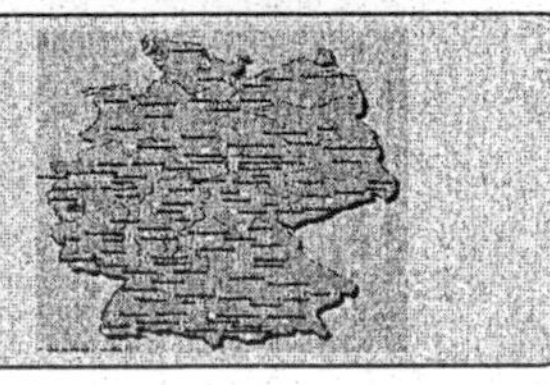

Functionality for Region:
•Statistics about region's resources available or not.
•Synergy effect providing a higher cooperation between enterprises.

Figure 5. Potentials of the Website and systems for the partner search

For a region that has this system and site, the possibility to get statistic information about the transactions flowing by the server can be seen as very useful. This statistic information can be used for defining policies for the region as well as providing services to increase the synergy among its enterprises.

The VISHOF system is now being extended to support:

- the partner search for whole competencies;
- the partner contact by asynchronous (e-mail) and synchronous (chat, video conferencing) services;
- the customisation of searching schemas;
- the opportunity identification within virtual enterprises and
- the implementation of databases in different locations, even in different countries, forming so a Global Virtual Enterprise.

This activities are planned to be developed within the Global Virtual Enterprise Project between the University of São Paulo, Brazil and the Aachen University of Technology, Germany.

7 - REFERENCES

1. **Baran, N.**, *Inside the Information Superhighway Revolution,* Coriolis Group Book, 1995, 253 pgs.
2. **Picot, A., Reichwald, R., and Wigand, R.**, *Die grenzenlose Unternehmung,* Gabler Verlag, 1996, 579 pgs.
3. **Rotering, J.**, *Zwischenbetriebliche Kooperation als alternative Organisationsform - Ein transatktionstheorethischer Erklärungsansatz.* Schäffer-Poeschel Verlag, Stuttgart 1993.
4. **Preiss, K., Goldman, S., and Nagel, R.**. *Cooperate to compete: Building Agile Business Relationships.* Van Nostrand Reinhold, 1996, 313 pgs.
5. **Eversheim, W. and Heyn, M.** Optimale Leistungstiefe. *VDI-Z,* 137 Nr. 11/12, 32-35, Nov-Dez, 1995.
6. **Hamell, G. and Prahalad, C.K.**, *Competindo pelo Futuro,* Editora Campus, Rio de Janeiro, 1995, 377p.
7. **Davidow, W.H. and Malone, M.S.**, *Das virtuelle Unternehmen: Der Kunde als Co-Produzent.* Frankfurt / New York: Campus Verlag, 1993.
8. **Groos, P.,** Wandel der Arbeitswelt: Vom Angestellten zum Protfolio-Worker. *Seminar Virtuelle Fabrik,* Universität St. Gallen, Vortrag 9, Februar 1997.
9. **Eversheim, W.**, Informationstechnologie als Wegbereiter für den Wertschöpfungsverbund. *Seminar Virtuelle Fabrik,* Universität St. Gallen, Vortrag 12, Februar 1997.
10. **Goldman, S., Nagel, R. and Preiss, K.**, *Agile competitors and virtual organizations: strategies for enriching the customer.* Van Nostrand Reinhold, 1995, 414 pgs.
11. **Goranson, T.**, Agile Virtual Enterprise: Best Agile Practice Reference Base, *Working Draft,* January, 1995, 11p.
12. **Raggett, D.,** HTML 3 - Electronic Publishing on the World Wide Web, Addison-Wesley, USA, 1996.
13. **Teerhag, O.; Dresse, S.; Kölscheid, W. and Nieder, A.**, Model for Transforming, Identifying and Optimizing Core Processes (MOTION), *Work Package 1.2 Final Report,* 8-10, 1996.
14. **Wood, C.**, Principles of secure information systems design with groupware examples. Computers&Security, 12, 1993, 663-678.

15. **Ellis, C.A.; Gibbs, S.J. and Rein, G.L.**, Groupware - some issues and experiences, *Communications of the ACM*, 34, 1991.
16. **Suma, H.A.**, Intranet: Ernst-zunehmende alternative in Sicht. *Focus 4*, , 1996, 20-24.
17. **Bernard, R.**, Corporate INTRANET, John Wiley & Sons, Inc., USA, 1996.
18. **Deep, J. and Holfelder, P.**, Developing CGI Applications with Perl, John Wiley & Sons, Inc., USA, 1996.
19. **Roberts, B.**., Notes x Web, *Byte Brasil*, Julho 1996, 40-53.
20. **Vernadat, F.B.**, *Enterprise Modelling and Integration: principles and applications,* Chapmann & Hall, London 1996.
21. **Lewis, T. G.**, Where is Client/Server Software Headed?, IEEE Computer, April 1995.
22. **Kumar, V.**, Mbone: Interactive Multimedia on the Internet. New Riders Publishing, 1996.
23. **Bremer, C., Erb, M., Kampmeyer, J., and Correa, G.**, Global Virtual Enterprise - A Worldwide Network Of Small And Medium Sized Production Companies. *XV Encontro Nacional de Engenharia de Produção - Brazil First Congrees of Industrial Engineering,* Universidade Federal de São Carlos, Setembro 1995.

FACILITY LOCATION AND RELOCATION IN GLOBAL MANUFACTURING STRATEGY

Bahram Emamizadeh, Reza Zanjirani Farahani
Industrial Engineering Department, Amirkabir University of Technology, Tehran, Iran
Tel.(9821)646-6497 Fax (9821)641-3025

ABSTRACT: In light of rapid developments in many geographical areas in the world, one issue of strategic importance is the determination of new facilities' locations. This can no longer be viewed in a static manner where the behavior of the weights, effecting this decision, remains constant through out the time. The facility location problem(FLP) discussed in this paper not only assumes the weights are time dependent, but also, if economically justified, the new facility can be relocated one or more times at predetermined time junctures within a finite planning horizon. Relocation times are the solution to a binary integer programming (BIP) model while optimal locations are found through employing procedures relevant to constant weights FLP. A computer program in PASCAL helps the analyst to solve the location problems and convert the input data to the appropriate coefficients of the BIP model. Using LINGO, subsequently, the optimal relocation times are found.

INTRODUCTION

Facility location is a strategic management decision. This decision is usually made, however, with respect to the current parameters, called weights, such as population, infrastructure, service requirements and etc. A new situation arises when weights are time dependent, and vary as time proceeds. Under this circumstance the decision maker must take the time dependent weights into consideration when making the location decision. Locating manufacturing centers in a dynamic economical region, such as Asia and Latin America are examples of this phenomenon. Sometimes the cost of maintaining the current location far exceeds the relocation of the facility to a more appropriate and cost efficient location. In this paper an optimal algorithm for determination of the new facility's location when weights are time dependent, and relocations can take place on pre-determined points in time is presented. It is also assumed that relocation cost is effected by the inflation rate. A computer program, henceforth, was written to help the analyst and decision maker to overcome the numerous weight calculations. The optimal solution is found via exploiting methods relevant to constant weights and also solving a Binary Integer Programming (BIP) problem. The latter is done using LINGO optimization software.

SINGLE FACILITY LOCATION PROBLEM

To locate a new facility among the known demand points, when the weights are deterministic and constant, the following objective cost function is minimized:

$$F(X) = \sum_{i=1}^{m} W_i \, d(X, p_i) \qquad (2.1)$$

W_i is the weight associated with the ith demand point, and $d(X,p_i)$ is the distance between the ith demand point, locating at $p_i=(a_i,b_i)$ and the new facility, locating at $X=(x,y)$. There are several ways for representing $d(X,p_i)$, among which rectilinear, squared euclidean euclidean distance are wellknown.

Calculation of the optimum location, (x,y), are easily obtained [4]. Now assume that the weights change in time. Then W_i will change to $w_i(t)$, where t is defined in the range of [0,T]. In related studies, Wesolowsky [9] extended the single facility location problem so that the facility location could change in a given time interval. In his paper, Wesolowsky presented an algorithm for calculating the optimal location when the costs change at specific times and locations were pre defined. Wesolowsky and Truscott [8] considered the problem when the weights vary but their effect is not continuous. Using dynamic programming their procedure decides whether or not to locate the facility in predetermined location at the end of each period. Chand [2] provides several decision/forecast horizon results for a single facility dynamic location/relocation problem; these results are helpful in finding optimal initial decisions for the infinite horizon problem by using information only for a finite horizon. Murthy [7] has modeled the allocation of n facilities to n sites as an assignment problem. In his model the assignment of the facilities to sites carry some costs which can change from period to period. The fixed relocation cost at the end of each period is included as well. Using 'Dual Ascent" procedure the mixed integer programming model has been solved for the location and relocation of the facility. Andreatta and Mason [1] proposed a perfect forward algorithm for the solution of a single facility dynamic location/relocation problem. In their paper, the problem is restated in terms of a shortest path problem in an acyclic network and give an obvious condition (which is both necessary and sufficient) for the existence of a finite forecast horizon for obtaining an optimal initial decision. Huang, Batta and Babu[6] assumed that the weights are function of location and not time. For the exponentially defined weights, they find the optimal location of the facility using euclidean distance. Hormozi and Khumawala [5] offered an algorithm for optimizing the problems with weights having different pre defined values in time and pre defined locations. Using mixed integer programming model and the dynamic programming approach the problem is sub divided into smaller simple problems.

The common factor among the previous studies were the assumption that the weights can change in time, but the change is not continuous. Drezner and Wesolowsky [3] paid attention to the continuous weight functions. They presented an optimal procedure for single facility location problem with a single relocation, when the weights are linear, $w_i(t) = u_i + v_i t$, and the distances are rectilinear. In their procedure the relocation time can be anywhere in the range of [0,T].

In this paper, however, the single facility location problem with multiple relocations and continuous weight functions is investigated. It is also assumed that there is a cost associated with the relocation and this cost is effected by the inflation rate. The procedure offered here is not constrained by the type of distance definition or the weight function. The location times, however, is assumed to be pre defined and thus the decision is to choose the best relocation times and new facility's relevant locations, in order to minimize the total location and relocation costs.

Time Dependent Weights

It is important to note that the weights, $w_i(t)$, within the period of [0,T] cannot be negative. For example for the linear weights the following holds [3]:

$$w_i(t) = u_i + v_i t \ , \quad w_i(t) \geq 0 \ , \quad t \in [0,T] \ \Rightarrow \ u_i \geq 0 \ , \ v_i \geq -\frac{u_i}{T} \tag{2.2}$$

The objective function for the time dependent weights can be represented as:

$$F = \int_0^T \left\{ \sum_{i=1}^{m} w_i(t) \ d(X, p_i) \right\} dt \tag{2.3}$$

which is equivalent to the following:

$$F = \sum_{i=1}^{m} d(X, p_i) \int_0^T w_i(t) \ dt \tag{2.4}$$

Integrating the weights will result in obtaining constant weights:

$$W_i = \int_0^T w_i(t) \ dt \qquad i = 1, \ldots, n \tag{2.5}$$

Now the location problem can be solved with constant weights.

FACILITY RELOCATION

Given the specific weight functions, it could arguably be more economical to relocate the new facility at some particular time in the future, such that the total location cost is minimized. It is assumed here that the relocation can take place only on predetermined points in time. This is due to the fact that relocation for an industrial center is not feasible at all times. The total location cost, therefore, is the sum of the location costs before and after the relocation plus the relocation cost at the given point in time. Minimization of this cost is subject to identification of the optimal relocation time and the facility's optimal locations prior and after the relocation. Before presenting the optimal algorithm, let's discuss the following lemma which play important role in the algorithm.

Lemma : Objective cost function is additive. Let's consider Figure 1. t_{j-1}, t_j and t_{j+1} are predetermined points in time. If the relocation of the new facility takes place at t_j where should its locations prior and after the relocation time be in order to minimize the total cost?

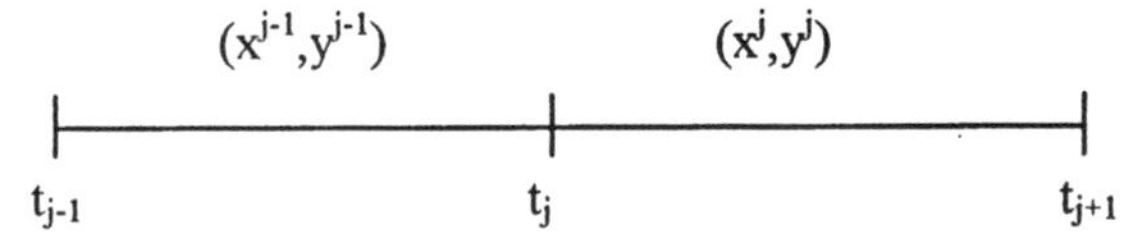

Figure 1. New facility locations and relocation time.

The objective function for this case is:

$$F = \int_{t_{j-1}}^{t_j} \sum_{i=1}^{m} d(X^{j-1,j}, p_i) . w_i(t) \, dt + \int_{t_j}^{t_{j+1}} \sum_{i=1}^{m} d(X^{j,j+1}, p_i) . w_i(t) \, dt \tag{3.1}$$

$$= g_1(x^{j-1}, y^{j-1}) + g_2(x^j, y^j)$$

where X^{jk} is the optimal facility location during the $[t_j,t_k)$. The result shows that the objective cost function is equal to the sum of two positive functions with independent variables. This verifies the additivity of the objective function. Therefore given t_j, optimal location of the new facility before and after the relocation time can be determined independently. The total location cost is thus the summation of the location costs before and after the relocation. The optimal location during each interval can be calculated after integrating the weights, using the constant weight procedures.

Determination of Relocation Points in Time

Given several points in time as candidates for executing relocation of a facility, the intention is to find those relocation points, and subsequently, facility's location such that the overall location cost is minimized. Figure 2. shows examples of the cost rate functions, f_{ij}, associated with the location of the facility in each time interval, $[t_i,t_j)$. The total cost will be the sum of location costs for each interval. The location cost is represented as the area under the cost functions.

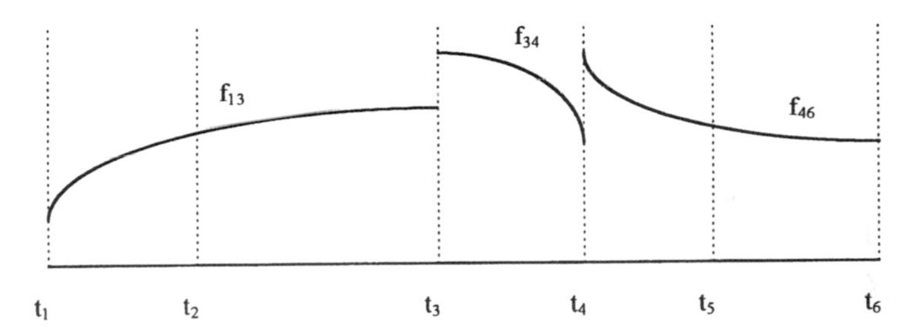

Figure 2. Examples of cost rate functions

In this case, the facility will have three locations, and the relocation points are t_3 and t_4. The cost rate function in each interval is defined as:

$$f_{13} = \sum_{i=1}^{m} d(X^{13}, p_i).w_i(t) \quad , \quad t \in [t_1, t_3)$$

$$f_{34} = \sum_{i=1}^{m} d(X^{34}, p_i).w_i(t) \quad , \quad t \in [t_3, t_4)$$

$$f_{46} = \sum_{i=1}^{m} d(X^{46}, p_i).w_i(t) \quad , \quad t \in [t_4, t_6)$$

The total cost function, thus will be:

$$F(X^{13}, X^{34}, X^{46}) = \int_{t_1}^{t_3} f_{13}\, dt + \int_{t_3}^{t_4} f_{34}\, dt + \int_{t_4}^{t_6} f_{46}\, dt$$

Based on this strategy, the facility will have an initial location at time t_1, the first relocation will take place at t_3 and the next one at t_4. Next, the algorithm for finding the optimal relocation times and the optimal location of the facility will be presented.

Optimal relocation times Algorithm

Step 1. There are n-2 discrete points in time, candidates for relocation. Considering t=0 and t=T, then there will be n points. The interval [0,T] is divided to n-1 subintervals. There are m existing facilities. Conduct the following calculations.

$$W_i^j = \int_{t_j}^{t_{j+1}} w_i(t)\, dt \qquad , i = 1,\ldots,m \qquad j = 1,\ldots,n-1 \tag{3.2}$$

Step 2. Let's define W_i^{jk} as follow:

$$W_i^{jk} = W_i^j + W_i^{j+1} + W_i^{j+2} + \ldots + W_i^{k-2} + W_i^{k-1} = \int_{t_j}^{t_k} w_i(t)\, dt \tag{3.3}$$

$$\text{where} \qquad i = 1,\ldots,m \ , \ j = 1,\ldots,n-1 \ , \ k = j+1,\ldots,n \ \text{ and } \ j < k$$

This is the sum of the weights from t_j to t_k. Calculate W_i^{jk} for all values of j and k and for every demand point. This will generate the integrated weight associated with the ith point if it is decided to have the same location for the new facility during the time interval of $[t_j,t_k)$.

Step 3. For each interval $[t_j,t_k)$ find the optimal location of the facility, (x^{jk},y^{jk}) as follow:

a) Rectilinear distance: Using the values of W_i^{jk} and the coordinates of the demand point, the optimal location is found by Weber procedure [4], the median locations.

b) Squared euclidean distance: Using the following relationships the optimal location is found:

$$x^{jk} = \frac{\sum_{i=1}^{m} W_i^{jk} a_i}{\sum_{i=1}^{m} W_i^{jk}} \quad , \quad y^{jk} = \frac{\sum_{i=1}^{m} W_i^{jk} b_i}{\sum_{i=1}^{m} W_i^{jk}} \tag{3.4}$$

c) Euclidean distance: Using Kuhn optimality conditions [4] and the following iterative procedure the optimal location is found:

$$g_i^{jk}(x^{jk},y^{jk}) = \frac{W_i^{jk}}{\sqrt{(x^{jk}-a_i)^2 + (y^{jk}-b_i)^2}}$$

$$x_h^{jk} = \frac{\sum_{i=1}^{m} g_i^{jk}(x_{h-1}^{jk},y_{h-1}^{jk}).a_i}{\sum_{i=1}^{m} g_i^{jk}(x_{h-1}^{jk},y_{h-1}^{jk})} \quad , \quad y_h^{jk} = \frac{\sum_{i=1}^{m} g_i^{jk}(x_{h-1}^{jk},y_{h-1}^{jk}).b_i}{\sum_{i=1}^{m} g_i^{jk}(x_{h-1}^{jk},y_{h-1}^{jk})} \tag{3.5}$$

$$\text{where} \quad j = 1,\ldots,n-1 \qquad \text{and} \qquad k = j+1,\ldots,n$$

In this procedure h shows the iteration number. The starting point for this step is the location found using the squared euclidean distance, and the iterative procedure is continued until the values of x^{jk} and y^{jk} converge. The convergence of the above relationships was shown by Weiszfeld [10].

Step 4. Calculate C_{jk}, the location cost, if the new facility has the same location during the time interval $[t_j,t_k)$, using the following relationship:

$$C_{jk} = \sum_{i=1}^{m} W_i^{jk} . d(X^{jk}, p_i) \tag{3.6}$$

where $d(X^{jk},p_i)$ is the distance between the optimal location of the new facility and the ith demand point when $t \in [t_j, t_k)$, and determination of X^{jk} was defined in step 3.

Step 5. Using the cost coefficients in step 4, the following model is used to find the optimal relocation times:

$$\text{Minimize} \quad F = \sum_{j=1}^{n-1} \sum_{k=j+1}^{n} C_{jk} . Z_{jk} + \sum_{j=2}^{n-1} \sum_{k=j+1}^{n} \frac{1}{(1+f)^{t_j}} S_j . Z_{jk}$$

Subject to:

$$\sum_{k=2}^{n} Z_{1k} = 1$$

$$\sum_{j=1}^{k-1} Z_{jk} = \sum_{l=k+1}^{n} Z_{kl} \quad , \quad k = 2, \ldots, n-1$$

$$\sum_{j=1}^{n-1} Z_{jn} = 1$$

$$Z_{jk} = 0 \text{ or } 1$$

where

$$Z_{jk} = \begin{cases} 1 & \text{if current relocation takes place at } t_j \text{ and the next one at } t_k \\ 0 & \text{otherwise} \end{cases}$$

S_j is the fixed relocation cost
f is the inflation rate

This is a Binary Integer Programming (BIP) model with *n* constraints and *n(n-1)/2* variables. The first constraint ensures that the relocation decision starts at time t_1=0. The second set enforces the consideration of the next relocation time exactly after the last relocation time. The last constraint guarantees that the decision making will continue to the end of planning horizon, T. The objective function defines the total location and relocation costs. The relocation costs are subject to the inflation consideration. The present value of the relocation costs is presented in the second term of the objective function. The BIP model can be solved using optimization softwares such as LINDO and LINGO. The total cost of this policy will be represented by F.

Example

The management of a sporting goods industrial cooperation is considering the installation of a new unit in one of the South Asian countries. The planning horizon is set to be for the next 10 years. The management has agreed to relocate the new unit at the end of t=[0,1,2,4,5.5,7,8,10] year, if the production rates of the raw material suppliers, and their relative costs justify the move. The suppliers are stationed in five regions. The distance between the new unit and the suppliers is measured based on euclidean norm. The suppliers are located at: $p_1=(1,1)$, $p_2=(0,3)$, $p_3=(3,9)$, $p_4=(4,4)$ and $p_5=(7,1)$. The production rate in these regions, w_i, i=1,...,5, are changing with time, due to population change and subsequent work force shifts. The functions by which the production rates are changing are as follow:

$$w_1(t) = t^2 - 10t + 26 \qquad w_2(t) = 2.6t + 2 \qquad w_3(t) = -2.8t + 30$$
$$w_4(t) = 28e^{-1.5t} \qquad w_5(t) = -0.1t^3 + t^2 + 10$$

The initial installation cost is estimated to be 2200 (local currency) and the relocation cost for each period is estimated to be, Sj=(34 ,18 , 55 , 45 , 25 , 65 , 0). The inflation rate is set at 2%. Management would like to find out if it would be economical to relocate the facility in the future, and if wanted to do so at what period the relocation should take place and where would the optimal location be. Figure 3. shows the demand rate functions, and how they change according to time.

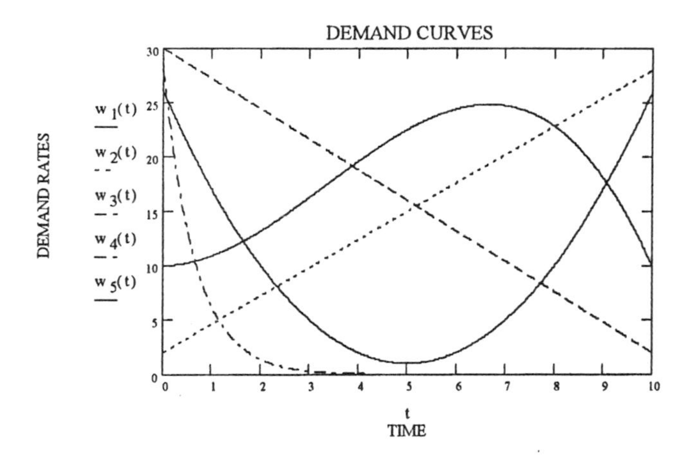

Figure 3. Demand rate function for the regions

Step 1. Calculate the weight of each facility for each time interval $[t_j,t_{j+1})$, W_i^j. Table 1. shows the results of this step.

Step 2. Calculate the cumulative weight of the existing facilities (suppliers), W_i^{jk}. The results are shown in Table 2.

Step 3. For each interval find the optimal location facility using the formulas presented for the euclidean distances. The results are presented in Table 3.

Table 1
Weights of the existing facility in each time interval

	t_1-t_2	t_2-t_3	t_3-t_4	t_4-t_5	t_5-t_6	t_6-t_7	t_7-t_8
W_1^j	21.3	13.3	10.66	1.875	4.125	7.3	34.66
W_2^j	3.3	5.9	19.6	21.525	27.375	21.5	50.8
W_3^j	28.6	25.8	43.2	25.05	18.75	9	9.6
W_4^j	14.5	3.327	0.883	0.042	0.002	0	0
W_5^j	10.31	11.96	32.67	32.65	36.72	23.96	35.07

Step 4. Calculate the location cost associated with placing the facility in its optimal location in the relevant time interval. Table 3. presents the results.

Step 5. Using the cost coefficients, the BIP model is formed and solved using LINGO. The BIP model and its output is presented in Figure 4 and Figure 5, respectively.

Table 2
Cumulative weights of existing facilities

t_j _ t_k	W_1^{jk}	W_2^{jk}	W_3^{jk}	W_4^{jk}	W_5^{jk}
1_2	21.33	3.3	28.6	14.5	10.31
1-3	34.67	9.2	54.4	17.74	22.27
1-4	45.3	28.8	97.6	18.62	54.93
1-5	47.21	50.33	122.65	18.66	87.59
1-6	51.33	77.7	141.4	18.66	124.31
1-7	58.67	99.2	150.4	18.66	148.26
1-8	93.3	150	160	18.66	183.33
2-3	13.33	5.9	25.8	3.24	11.96
2-4	24	25.5	69	4.12	44.63
2-5	25.87	47.03	94.05	4.16	77.28
2-6	30	74.4	112.8	4.16	114
2-7	37.33	95.9	121.8	4.16	137.96
2-8	72	146.7	131.4	4.16	173.03
3-4	10.66	19.6	43.2	0.88	32.67
3-5	12.54	41.13	68.25	0.92	65.32
3-6	16.67	68.5	87	0.92	102.04
3-7	24	90	96	0.92	126
3-8	58.67	140.8	105.6	0.92	161.07
4-5	1.87	21.53	25.05	0.045	32.65
4-6	6	48.9	43.8	0.049	69.37
4-7	13.33	70.4	52.8	0.05	93.33
4-8	48	121.2	62.4	0.05	128.4
5-6	4.13	27.37	18.75	0.002	36.72
5-7	11.46	48.87	27.75	0.002	60.68
5-8	46.12	99.67	37.35	0.002	95.75
6-7	7.33	21.5	9	0	23.96
6-8	42	72.3	18.6	0	59.03
7-8	34.66	50.8	9.6	0	35.07

Table 3
Calculation of the facility location and its cost in each time interval

t_i-t_k	(x^{jk},y^{jk})	C_{ik}
1-2	(3.6730,4.1023)	293.2603
1-3	(3.3285,4.3206)	552.1851
1-4	(3.2463,4.4727)	1030.0497
1-5	(3.3148,4.2868)	1394.4947
1-6	(3.2229,3.9056)	1802.7860
1-7	(2.8208,3.6524)	2014.5586
1-8	(2.0720,3.0607)	2468.7318
2-3	(2.9908,4.9438)	256.3221
2-4	(3.0891,4.8847)	733.4052
2-5	(3.2031,4.3918)	1099.8467
2-6	(3.2090,3.8996)	1480.5302
2-7	(3.0446,3.6020)	1732.8257
2-8	(2.1261,3.0325)	2194.2234
3-4	(3.1540,4.8568)	476.9398
3-5	(3.2946,4.2351)	842.2666
3-6	(3.3085,3.7058)	1220.1237
3-7	(3.1246,3.4164)	1437.4945
3-8	(1.9902,2.9242)	1924.2919
4-5	(3.6664,3.6081)	361.3012
4-6	(3.6220,3.1730)	733.0883
4-7	(3.2098,3.0095)	982.7101
4-8	(1.6538,2.6874)	1413.5560
5-6	(3.5161,2.8644)	369.7872
5-7	2.7575,2.7809)	615.7689
5-8	(1.3070,2.5230)	1024.6501
6-7	(1.8581,2.5860)	241.7077
6-8	(1.0559,2.3526)	635.0770
7-8	(0.9570,2.1910)	389.2160

```
MODEL:
  SETS : INDEX /1..8/ :   F_cost  , T  ; MATR (INDEX , INDEX) | &1 # LT
          # &2:   INFL_T_S , C , Z;   ENDSETS
  DATA : @file (test.lng); ENDDATA
       @FOR ( MATR(J,K):
     INFL_T_S (J,K)  =  (    (1+ INFLATION) ^ (-T(J))   ) * F_cost(J) ;
      MIN = @SUM   (   MATR :  (C + INFL_T_S)  *  Z  );
           Z(1,2) + Z(1,3) + Z(1,4) + Z(1,5) + Z(1,6) + Z(1,7) + Z(1,8) = 1  ;
           Z(1,2) = Z(2,3) + Z(2,4) + Z(2,5) + Z(2,6) + Z(2,7) + Z(2,8)   ;
           Z(1,3) + Z(2,3) = Z(3,4) + Z(3,5) + Z(3,6) + Z(3,7) + Z(3,8)   ;
           Z(1,4) + Z(2,4) + Z(3,4) = Z(4,5) + Z(4,6) + Z(4,7) + Z(4,8)   ;
           Z(1,5) + Z(2,5) + Z(3,5) + Z(4,5) = Z(5,6) + Z(5,7) + Z(5,8)   ;
           Z(1,6) + Z(2,6) + Z(3,6) + Z(4,6) + Z(5,6) = Z(6,7) + Z(6,8)   ;
           Z(1,7) + Z(2,7) + Z(3,7) + Z(4,7) + Z(5,7) + Z(6,7) = Z(7,8)   ;
           Z(1,8) + Z(2,8) + Z(3,8) + Z(4,8) + Z(5,8) + Z(6,8) + Z(7,8) = 1 ;
         @FOR (  MATR : @BIN(X);  ) ;
      END
```

Figure 4. Input file for LINGO

```
Optimal solution found at step:        10
Objective value:                 4646.455
Branch count:                           0

         VARIABLE        VALUE        REDUCED COST

         Z( 1, 2)   0.0000000E+00      17.51233
         Z( 1, 3)    1.000000        0.0000000E+00
         Z( 1, 4)   0.0000000E+00      24.34058
         Z( 1, 5)   0.0000000E+00     0.0000000E+00
         Z( 1, 6)   0.0000000E+00      13.17908
         Z( 1, 7)   0.0000000E+00      12.80194
         Z( 1, 8)   0.0000000E+00      22.27502
         Z( 2, 3)   0.0000000E+00      13.21106
         Z( 2, 4)   0.0000000E+00      36.78156
         Z( 2, 5)   0.0000000E+00      14.44104
         Z( 2, 6)   0.0000000E+00     0.0000000E+00
         Z( 2, 7)   0.0000000E+00      40.15289
         Z( 2, 8)   0.0000000E+00      56.84595
         Z( 3, 4)   0.0000000E+00      40.72150
         Z( 3, 5)   0.0000000E+00      17.26099
         Z( 3, 6)    1.000000        0.0000000E+00
         Z( 3, 7)   0.0000000E+00      5.222862
         Z( 3, 8)   0.0000000E+00      47.32605
         Z( 4, 5)   0.0000000E+00      23.33099
         Z( 4, 6)   0.0000000E+00     0.0000000E+00
         Z( 4, 7)   0.0000000E+00      37.47284
         Z( 4, 8)   0.0000000E+00      23.62598
         Z( 5, 6)   0.0000000E+00      15.02536
         Z( 5, 7)   0.0000000E+00      48.85822
         Z( 5, 8)   0.0000000E+00      13.04138
         Z( 6, 7)   0.0000000E+00      51.32685
         Z( 6, 8)    1.000000        0.0000000E+00
         Z( 7, 8)   0.0000000E+00     0.0000000E+00
```

Figure 5. Lingo output for the optimal solution

Based on the results, the optimal solution includes three locations for the police station and two relocation times. Table 4 shows the final results.

Table 4
Optimal solution

Time Interval	Location	Cost
t_1=0 to t_2=2	(3.3285,4.3206)	2752.1861
t_2=2 to t_3=7	(2.9908,4.9438)	1237.4247
t_3=7 to t_7=10	(3.1246,3.4164)	656.8470
Total		4646.455

CONCLUSIONS

Single facility location problems can be extended to include the time dependent weights. In such a case, the weights change as time proceeds. An optimal algorithm has been presented to identify the locations of the new facility during a finite time interval. It is assumed that the new facility can be relocated several times, given that a set of relocation times have been chosen in advance. With respect to the installation and relocation costs, inflation rate, the algorithm generates the optimal relocation times, and the optimal location of the facility after each relocation, such that the total present value of location and relocation costs is minimized. The important characteristic of the approach presented here is that the weight functions are continuos and the algorithm is not restricted to any specific distance definition or weight function. For the future research, one can consider multi-facility location and relocation problem.

REFERENCES

[1] **Andretta G. and Mason F. M.** "*A Perfect Forward Procedure for a Single Facility Dynamic Location/Relocation Problem,*" Operation Research Letters, Vol. 15, 1994, pp. 81-83

[2] **Chand S.** "*Decision/Forecast Horizon Results For a Single Facility Dynamic Location/Relocation Problem,*" Operations Research Letters, Vol. 7, No. 5, October 1988, pp. 247-251

[3] **Drezner Z. and Wesolowsky G. O.** "*Facility Location When Demand is Time Dependent,*" Naval Research Logistics, Vol. 38, pp. 763-771, 1991

[4] **Francis R.L. & White** ,'Facility Layout and Location", Prentice Hall, Inc., Englewood Cliffs, New Jersy

[5] **Hormozi A.M. and Khumawala** B.M. "*An Improved Algorithm For Solving A Multi-Period Facility Location Problem,*" IIE Transactions, Vol. 28, 1996, pp. 105-114

[6] **Huang W.V. and Batta R. and Babu A.J.G**. "*Relocation Promotion Problem with Euclidean Distance,*" European Journal of Operations Research, Vol. 46, 1990, pp. 61-72

[7] **Ishwar Murthy**, "*Solving The Multiperiod Assignment Problem with Start-up Costs Using Dual Ascent*", Naval Research Logistics, Vol. 40, pp. 325-344, 1993

[8] **Wesolowsky, G. O. & Truscott, W. G.** "*The Multiperiod Location and Allocation Problem with Relocation of Facility*", Management Science, pp. 57-65, 1975

[9] **Wesolowsky, G.O.**, "*Dynamic Facility Location*", Management Science, Pg. 1241-1248, 1973

[10] **Weiszfeld E**. "*Sur Le Point Pour Lequel La Somme Des Distances de n Points Donnes est Minimum,*" Tohoku Math, J., 43, 1936, pp. 355-386

AGILITY: MORE THAN A BUZZWORD?

Leon F. McGinnis
School of Industrial and Systems Engineering
The Georgia Institute of Technology

ABSTRACT: For the past twenty-five years, manufacturing has endured a seemingly endless assault by gurus promising salvation from the despair of competitive failure. The path to competitive success often has a catchy name, like "total quality," "zero inventories," "flexible manufacturing," or "lean production systems." "Agility" is among the newer "must have" attributes for globally competitive firms. What is agility, really? How would you know if you were agile? How would you set out to become agile? And why?

INTRODUCTION

I've heard it referred to as "buzzword engineering" and "management by magazine cover." It's the manufacturing firm's *haute couture*—the latest fashion that everyone "must have." It's what the leading firms are doing that makes them leaders. It's the absolutely essential attribute that all firms must have if they want to be competitive. In the sixties, it was EDP and MRP. In the seventies, it was JIT and ZD. In the eighties, it was FMS and MRPII. In the nineties it has been TQM, lean manufacturing, and now, agility. At least we've progressed from acronyms to real words.

The reason it's *haute couture* is that your firm probably won't recognize what "looks good" on you without the assistance of a "fashion expert," the "guru" who can translate the latest fashion idiom into a program specifically tailored to your firm. You can't do this for yourself, because you won't be able to understand exactly what the idiom *is*, even with the gurus trying to explain it to you. Take agility for example. Here's what the experts say about it:

> "Agility is dynamic, context-specific, aggressively change-embracing, and growth-oriented. It is not about improving efficiency, cutting costs or battening down the business hatches to ride out fearsome competitive "storms." It is about succeeding and about winning: about succeeding in emerging competitive arenas, and about winning profits, market share, and customers in the very center of the competitive storms many companies now fear." [1]

> "Agility is a response to a new competitive environment, one that requires an integration of people, technology, and organization that is appropriate to a particular market opportunity." [2]

> "a very seductive word. One that finds immediate and personal definition for almost everyone. ... can embrace almost any of our current competitiveness interests with considerable intuitive appeal. ...describes the missing characteristic in ... organizations

[that] could not adapt at the same pace as their changing environment... . ...is a core fundamental requirement for all organizations." [3]

"Confusion in this early stage of understanding is introduced principally from two sources: indiscriminate use of the word to promote narrowly defined technologies, and a seductive focus on the business strategy advantages and manifestations. The nature and reality of an Agile organization are determined by how it is organized--it is a systems and structural issue."[3]

"Being agile means being proficient at change—and allows an organization to do anything it wants to do whenever it wants to." [4]

"[Agility requires m]aintaining an ability to be nimble because of the rapidly changing manufacturing environment." [5]

"Agile manufacturing is marked by fast product turnaround and quickly moving market windows." [6]

"Agile manufacturing implies mass customization instead of mass production. It means producing highly customized products, where and when the customer wants them." [7]

"Agile manufacturing refers to the principle of producing relatively small lots to meet existing (rather than imagined) demand without compromising cost inventory, leadtime, quality, service or other strategic goals." [8]

"The agile manufacturing organization produces high quality products that are defect free in a short lead time, just-in-time manner." [9]

It's little wonder that most technical or management professionals have a difficult time translating agility into the terms of their specific firm's business.

So, is "agility" more than a buzzword? The answer, I think, is both "no" and "yes." The term "agility" is an icon for a complex set of capabilities and behaviors, that collectively allow a firm to remain successful in a particular type of competitive environment. At that level, agility is nothing more than a buzzword—it's a term that seems to convey considerable information, but in fact does not. We have a vague sense of understanding what agility is, yet we cannot define it, nor do we know how to measure it.

On the other hand, there is little question that many firms are achieving improved competitiveness through changes in their capabilities or behavior that they describe as "becoming more agile." While there might be quibbling over whether their success is due to "agility" or "flexibility" or "supplier partnerships" or some other change, one must admit that there is something happening that needs to be better understood. If its name is agility, then clearly agility is more than just a buzzword

If we want to understand agility as more than a buzzword, then we first must go back to the basics and ask, "What must a firm accomplish in order to be successful?" In exploring this question, we will find that there are two distinct drivers of change, in the context of agility—***customers*** and ***owners***. In order to understand the strategies that the firm may employ in responding to these change drivers, we will need a high level abstraction for the firm. It's from this high level abstraction that we can finally develop a basis for understanding agility as more than a buzzword.

BACK TO BASICS

A manufacturing firm exists in a competitive environment. Porter, for example, provides a richly detailed analysis of that competitive environment, and the strategies that firms may pursue. [10,11] A key element in the firm's environment is its customers, and a key to competitive success is the delivery of value to customers, or the creation of ***customer value***. At the same time, firms exist to create ***economic profit*** for their owners, in the long term. The market does not tolerate long term competitive strategies that involve long term economic losses for the owners.

Customer Value. What customers want can be succinctly described as "cost, quality, speed, and innovation." Whatever the good or service a customer purchases from a firm, it must be competitively priced. Even a firm like Intel, with a technology that no other firm can match, is compelled to consider cost in the customer value equation. And the fact that Intel enjoys such enormous margins on its product explains why it attracts "technology followers" who compete on the basis of duplicating Intel's functionality, albeit with delayed market entry.

"Quality" may be viewed as an umbrella concept, under which we group all the non-cost attributes and features of the product that directly impact the customer. We may speak of the quality of function, or the quality of form. When comparing the offerings from two different firms, the customer always considers quality in this way.

Today, speed is a very important contributor to customer value. The delay between desire and fulfillment must be as short as possible. Availability is a key element of speed, but also important are the speed with which a new product can be developed once a customer need is identified, and the speed with which production can ramp up to match demand.

The fourth key element of customer value is innovation—the delivery of some product attribute or feature before it is available from competitors. Innovation may take the form of intrinsic product attributes, or may be created through advertising. In either case, it represents something that competitors will attempt to match, although they probably can't duplicate it exactly. Even a cursory examination of the marketplace today leads to the conclusion that innovation is key to market leadership.

Customer value considerations drive the conceptualization and technical design of products, and are primary in the management of quality and logistics elements of the product realization process. However, customers don't really care **how** a firm delivers customer value; they don't care if the firm is efficient, flexible, lean, agile, or even profitable, as long as it delivers customer value. Especially in mature markets, brand loyalty is but a memory; customers only want to know "What can you do for me now?"

Customer value is not terribly difficult to assess, especially relative to the competition. Almost every manufacturing firm today employs some process for comparative analysis of competing products, which reveals technical differences in features and functions. One has only to look at sales data to determine how customers value a product. And clearly, customer value can be enhanced by reducing price. Perhaps the single most important fact about the customer value of a product is that it is ***relative*** to the alternative products ***currently available*** to the potential purchaser. When relative prices change, or new products come onto the market, customer value must be reassessed.

Economic Profit. By far the largest portion of manufacturing involves firms whose ownership is itself a commodity traded in a market. Why do people own shares in a firm? This is a question whose complete answer is beyond a brief exposition, but there is little doubt that share ownership is based upon some anticipation or expectation of future economic profit. People own shares in a firm, because they believe that owning those shares will yield a better economic profit than alternative investments. The two key observations about share ownership are that it is an ***investment***, and that it has a ***payoff in the future***. The return will come in one or both of two forms, dividends and gains from selling shares.

Competitive and successful firms behave in such a way that share owners observing current behavior are lead to conclude that future economic profits will be attractive, relative to alternative investments. Why else would people buy shares in a firm that clearly is "on hard times?" In the US, it is not at all uncommon to see a firm's share price rise in response to "downsizing," or "restructuring," which represent admissions of prior management failures, but signal a "change of heart."

The only time at which you may know for sure the economic profit from owning shares in a firm is when you sell the shares and are therefore no longer an owner. This creates a dilemma—how can one assess the future economic profit from share ownership while still owning the shares? The answer, of course, is that you can only guess about the future share price and dividend streams. You can observe the conditions and forces at work in the marketplace, and you can observe the strategies and current actions of the firm's management to better inform your guesswork, but it remains guesswork.

So, while it is relatively straightforward to determine a firm's success in creating customer value, it is much more difficult to assess the extent to which it is creating share owner economic profit, for those share owners who do not sell their shares. At best, the assessment involves educated guesswork about the future of the marketplace in which the firm competes, and the wisdom of current management strategy.

To be sure, share owners (and stockbrokers) have searched since the birth of the stockmarket for indices that will be reliable predictors of future economic profit, largely without lasting success. Today, the popular wisdom is that measures such as economic value added (EVA) or market value added (MVA) are good predictors of future success. While there is substantial evidence that EVA growth mirrors share price growth, its predictive power has not been proven over a long period of time.

A CONCEPTUAL FRAMEWORK

A firm is a participant in a competitive market. To be successful, the firm must behave in ways that are consistent with the market's requirements, namely, it must produce a good or service that meets or exceeds customers' expectations, and must do so in a manner that creates economic profit for its owners.

Over the past twenty-five years, the market for most firms has changed dramatically as a result of increasing globalization and the resulting intensification of competition. As Thurow has observed, this new market reality is characterized by:

- global transportation systems that allow natural resources to be sourced to countries that have none
- global information and financial systems that allow capital to move rapidly to any emerging market
- technologies that rapidly diffuse throughout global markets
- wages that fall toward the minimum demanded anywhere in the world. [12]

Along with these changes, many markets worldwide have seen product supply draw into balance with consumer demand, or even exceed it.

In this rapidly changing market environment, firms are required to change their behavior in order to remain competitive. Price and quality must match market levels. Products must be enhanced with the features and technologies expected in the market, leading to a broader product line and shorter product lifecycles. These behavioral changes impact the entire product realization process.

A firm's market behavior is the product of its external (market) environment and its internal organization. Its internal organization can be described as ***a portfolio of resources, deployed through a network of activities in order to produce goods and services for customers and create economic profit for owners***. In this context, *resources* has a broader definition than simply assets on a balance sheet; resources include people, organization structure, reputation, corporate culture, and other similar contributors to competitiveness. An important resource for any firm is its repertoire of practices, which include protocols for situation-specific actions, recipes for combining resources to produce a certain result, and other organizational "algorithms".

A firm's resources are deployed through a network of activities. An activity may directly contribute to the product realization process, or may support the process indirectly. Porter's "value chain" formulation is one useful way for viewing these activities. [10,11] Activities generate transactions, either between activities (internal) or with the market (external)

Thus, when a firm seeks to change the way it behaves in the market, it must change its external transactions. Rarely can such a change be implemented without a corresponding enabling change in the firm's resources, i.e., in its capital structure, its workforce structure, its organization, its practices, etc. These enabling changes are what allows a firm to become more competitive or to create more economic value.

In order to successfully manage change, two fundamental questions must be addressed:
1. What change in the firm's market behavior is required for continued success?
2. What is the most effective strategy for modifying the firm's portfolio of resources in order to enable these required behavioral changes?

These two questions rarely have simple or obvious answers. What should be clear, however, is that there is no single answer that is best for all firms, or even for all firms in a specific market. Rather, the answer for a specific firm depends upon the firm's current portfolio of resources.

BEHAVIORAL CHANGE

If a firm can remain successful without change, it should do so, since change consumes resources. In today's market, of course, firms cannot remain successful without changing. The two fundamental forces driving change are customer value expectations and owner economic profit expectations.

Customer Value Driven Change. Customers' expectations are centered on the product--its concept and design, its manufacturing execution, its price, and its availability and accessibility--which is to say that they are centered on an objectively observable entity. The product itself is the medium through which the firm interacts with customers and through which it can determine how well it is meeting their expectations. Over the past fifty years, a wide range of tools have been developed for use by firms in their efforts to design and manufacture products that meet customers' expectations.

In the design phase of the product realization process, quality function deployment, solid models, prototyping, focus groups, and test marketing are all examples of tools that firms may use to create and refine product concepts and designs while maintaining a clear focus on customer value. Thus, while design inherently may not be quantifiable, it is amenable to formal processes for improvement.

Cost and quality are aspects of the product which are amenable to quantitative metrics. In fact, cost and quality may be measured down to the unit process level, if so desired. Activity based costing emerged in the last decade precisely because firms wanted more accurate cost measurements. Likewise, statistical process control allows processes to be characterized in terms of defective parts per million (ppm) operations.

Quantification begets improvement; over the past twenty-five years, we have seen a number of improvements in manufacturing driven primarily by customer value considerations. Examples include all the quality-related programs, just-in-time and inventory reduction, automation and flexible manufacturing, and self-managed work groups and employee empowerment. An important observation about all these programs is that they have a directly measurable impact on product attributes about which the customer cares.

Economic Profit Driven Change. When a firm changes in response to the expectations of owners, it must modify its behavior in ways that its owners will value, i.e., it must change their perceptions of its future financial performance. The results of changing its behavior are not known except in the economists' "long run", and ultimately may be assessed only subjectively. However, managers must have feedback in the present, and there are two distinct sources of feedback. Managers may use share price as a current indicator of owners' perceptions, or they may compute some proxy that they believe reflects future financial performance, such as EVA or MVA.

Thus, the process of managing resource portfolio changes in order to meet owners' expectations does not focus directly on owners' expectations. Rather, a subjective judgment is required--"What kind of change to the resource portfolio will best meet our owners' expectations for future economic value creation?" Over the past twenty years, many answers to this question have been proposed-- focus on core competency, become a learning organization, become agile, etc. Each of these answers reflects a belief about the future and about the most effective resource portfolio for that future. And each of these answers permits the focus of assessment to shift from owners' expectations to the more limited scope of portfolio changes that enable a particular capability. In contrast to customer driven change, owner driven change may not have a directly measurable impact on the attributes of the firm that are valued by the owners. Owner driven change is much more subjective, and therefore more risky.

AGILITY

The fundamental issue here is that often the "capability" being proposed is not precisely defined, or it is not defined in operational terms, or in a way that leads directly to objective measurements. Take, for example, "agility." As a concept, agility is not terribly difficult to understand, and most informed professionals would probably agree that "An agile firm is one that is able to respond successfully to changes in its market." Each one of them would then quite likely embellish the statement in a unique way. So, we can discuss agility as a concept, or an ideal, but we can't measure it.

Instead of defining agility as a measurable attribute of the firm, the approach taken in the literature has been to enumerate the manifestations of agility in the firm's resource portfolio. For example, the agile firm might be characterized by the ability for electronic commerce, empowered employees, self-managed work groups, flexible manufacturing processes, lean production, short product development times, virtual manufacturing, etc. In fact, if one constructs the union of all the "agile" attributes that have been described in the literature, one has a rather lengthy list.

Unfortunately, such a list is not terribly useful for any specific firm. A firm may behave in ways that are consistent with the concept of agility without having "all" or even "many" of the attributes of agility that have been identified in the literature. The specific attributes that a firm needs to acquire in order to be able to behave in an agile fashion are determined by the firm's resource portfolio, its products, and its market. Furthermore, many of the "agile" attributes are themselves difficult to quantify objectively. Finally, as the old saying goes, "There's more than one way to skin a cat," i.e., there may be many different ways that a firm can achieve "agility".

So, where does this leave us in our search for an operational definition for "agility?" I believe that the firm's resource portfolio is a key to understanding agility. However, what is important is not the specific content of the portfolio, but the firm's ability to ***redeploy*** its resources as needed to respond to change or in anticipation of change in its environment. Having self-managed work groups is not enough to achieve agility; employees must be able to move quickly and effectively from one task to another, from one team to another, or across organizational boundaries. Having flexible machining centers is not enough to achieve agility; machines must be rapidly reconfigured to enable quickly changing from one product to another. Having EDI is not enough; the firm must be able to rapidly add new customers and new suppliers to its network.

Agility, in the end, is merely the byproduct of successful competition, not a goal in itself. Agile firms have identified the important changes in their markets, and have positioned their resource portfolios so that they can rapidly redeploy when change occurs or just before change occurs. Because every firm is unique, an attempt to understand agility by enumerating its manifestations in the resource portfolio of "agile" firms is certain to fail. We must focus, not on the "what" of the resource portfolio, but on its trajectory.

CONCLUSION

"Beauty is as beauty does" seems an apt description for the agile firm; you're agile if you behave as if you were agile, i.e., you remain successful in a market that is changing and to some degree unpredictable. If you're not agile now, and you believe you must become agile, then, like Dorothy in the Wizard of Oz, you face a challenging journey to a destination that may turn out to be less important than the journey itself.

Ultimately, agility is like personality—the result of both nature and nurture, and not something that can be copied from one individual to another. Achieving agility, like achieving personality change, may involve a deep and sometimes painful process of self-study and internal change. And any promise of a "one size fits all" solution is not likely to be of much value.

ACKNOWLEDGEMENTS

My examination of agility has come as the result of collaboration with the Electronic Agile Manufacturing Research Institute at the Rensselaer Polytechnic Institute. Especially helpful have been many discussions with Prof. Robert J. Graves, and Prof. Ron J. Guttman.

REFERENCES

1. Goldman, S., Nagel, R., and K. Preiss, Agile Competitors and Virtual Organizations: Strategies for Enriching the Customer, Van Nostrand Reinhold, New York, 1995.

2. Goldman, S., "Cooperating to compete: From alliances to virtual companies," CMA Magazine, March, 1994, pp. 13-17.

3. Dove, R., "Introduction to the Agile Practice Reference Base: Challenge Models and Benchmarks," in Agile Practice Reference Base, The Agility Forum, May, 1995

4. Dove, R., "The meaning of life & the meaning of agile," Production, November 1994, pp. 14-15.

5. Koepfer, G., "Agile: It's about machines too," Modern Machine Shop," January, 1995, p. 10.

6. Damian, J., "Agile manufacturing can revive US competitiveness, industry study says: A modest proposal," Electronics, February, 1992, p. 26.

7. Hormozi, A., "Agile Manufacturing," American Production and Inventory Control Society 1994 Conference Proceedings, pp. 216-218.

8. Porter, A., "Tight supplies challenge agile manufacturing," Purchasing, January 12, 1995, pp. 24-25.

9. James-Moore, S. M. R., "Agility is easy, but effective agile manufacturing is not," IEE Manufacturing Division Colloquium on Agile Manufacturing Proceedings, October 26, 1995.

10. Porter Michael E., Competitive Strategy: Techniques for Analyzing Industries and Competitors, The Free Press, New York, 1980

11. Porter, Michael E., Competitive Advantage: Creating and Sustaining Superior Performance, The Free Press, New York, 1985.

12. Thurow, Lester, Head to Head: The Coming Economic Battle Among Japan, Europe, and America, William Morrow and Company, New York, 1992.

ECONOMIC VALUE ADDED BY MANUFACTURING FIRMS

William G. Sullivan
Department of Industrial and Systems Engineering
Virginia Polytechnic Institute and State University
Blacksburg, Virginia 24061-0118 USA

ABSTRACT: Approximately two-thirds of the wealth created by a modern industrialized nation is from its manufacturing sector. This paper discusses a measure for estimating the wealth creation potential of capital investments in manufacturing. The measure, called "Economic Value Added," has recently become popular in the United States and is based on an after-tax analysis of cash flows generated by a capital investment.

INTRODUCTION

One of the most highly touted financial metrics of the 1990s is Economic Value Added, or EVA [2, 3, 4]. EVA is purportedly a measure of economic wealth creation that was devised in 1989 by Stern Stewart Management Services. For the past seven years, EVA has been adopted by many large manufacturing companies such as AT&T, Eastman Chemical, Coca-Cola and Eli Lilly, as a means of aligning executive compensation plans with corporate shareholder returns. This matching of management salaries and bonuses is possible because of the high statistical correlation that EVA has with wealth creation (i.e., common stock value) in <u>some</u> companies. Other popular equity valuation metrics used in corporate America include the price-earnings ratio, price to sales ratio, and accounting return on fixed assets [5]. Usually some combination of all the above metrics is used to tie corporate "performance" to management rewards.

So what's special about EVA? Given that some companies have discovered a very strong relationship between the EVA metric and the value of their common stock (a continuous variable), does this metric <u>really</u> reflect the potential for wealth creation? Can EVA also be used to evaluate the profit-earning potential of discrete capital investments that are of interest to manufacturing engineers?

Simply stated, EVA is the difference between a company's adjusted net operating profit (after taxes) in a particular year and its total cost of capital [2]. Another way to characterized EVA is "the spread between the cost of capital and the return on that capital" [3]. Statistical analysis of the

Standard and Poor's Compustat database in the U.S. confirms that EVA is a strong predictor of stockholder value for some companies but not for a broad randomly-selected population of companies [2]. But on a project-by-project basis (discrete investments rather than continuous ones), can the EVA metric also be used to predict profitability and hence wealth creation opportunity for manufacturing firms?

AN EXAMPLE FOR A MANUFACTURING INVESTMENT

Here we demonstrate that EVA can be applied to capital investments in manufacturing, such as building plant capacity or purchasing a new injection molding machine. Equation 1 defines EVA for this purpose:

$$EVA_k = (\text{Net Operating Profit After Taxes})_k - (\text{Cost of Capital Used To Produce Profit})_k \quad (1)$$

where k = an index for the year in question ($1 \leq k \leq N$)

N = the study period

Figure 1 allows us to relate EVA to after-tax cash flow (ATCF) for a proposed capital investment [1]. EVA can be obtained from Figure 1 by simply adding the entry in Column C for year k ($1 \leq k \leq N$) to the corresponding entry in Column D, and then subtracting the product of the project's tax-adjusted MARR and its beginning-of-year book value. Realistic estimates (forecasts) of the components of Figure 1 underlie believable predictions of ATCF (and EVA). Clearly, cost/revenue estimating accuracy is critical to any evaluation of wealth creation opportunity.

Let's demonstrate the use of Equation 1 and Figure 1 with a simple example. We'll consider the manufacturing investment below and determine (a) the year-by-year ATCF and (b) the net operating profit after taxes (NOPAT) and EVA for each year.

Example Manufacturing Investment

Capital Investment	=	$84,000
Salvage Value (end of year 4)	=	$0
0 & M/year	=	$30,000
Gross Revenues/year	=	$70,000
Depreciation (Tax and Book)	=	Straight Line
Life (Tax and Useful)	=	4 years
Effective Income Tax Rate	=	50%
After-Tax MARR	=	12% (25% of total monies borrowed at 8%, 75% remainder from equity sources that expect 14.67%/year return)

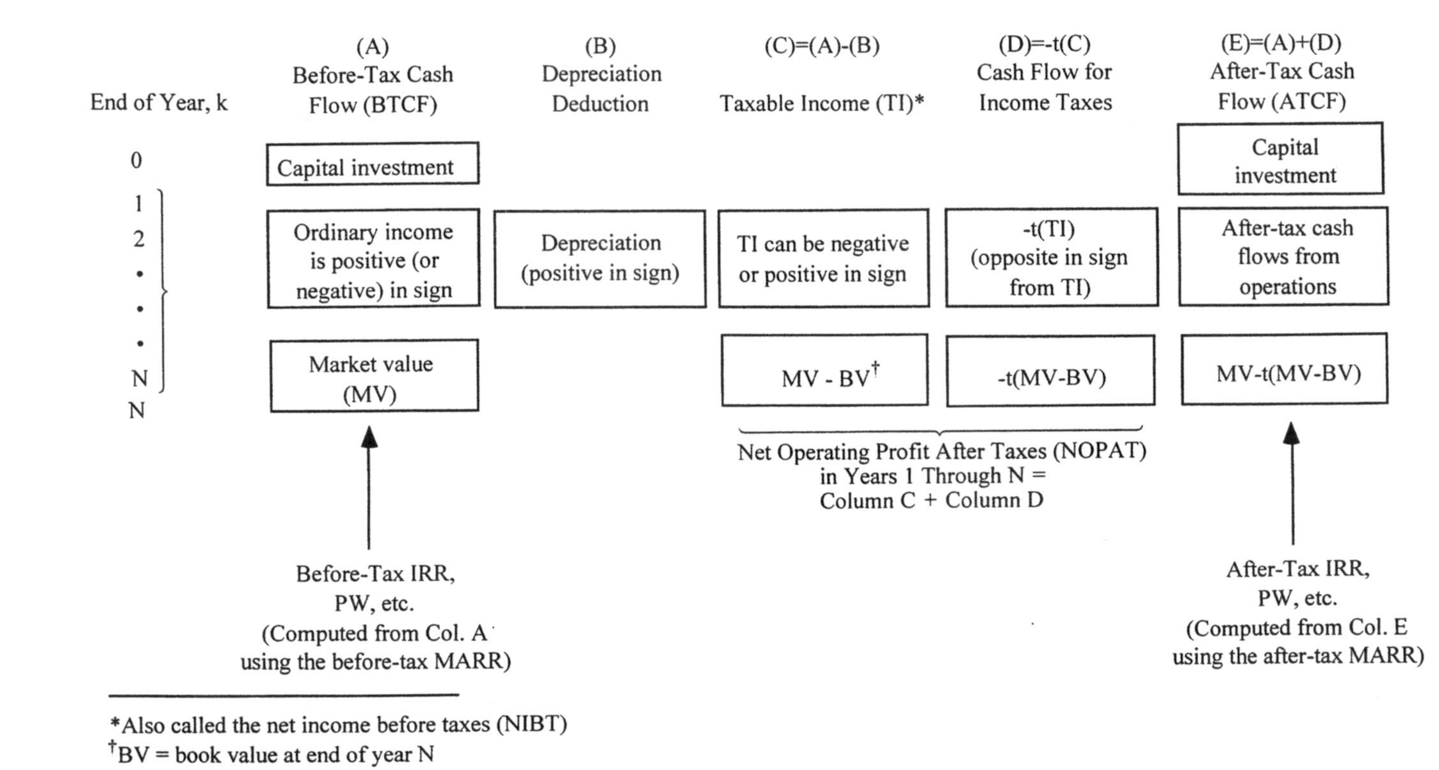

Figure 1. General Format (Worksheet) for Determining the ATCF and NOPAT

(a)

EOY	BTCF	Deprec.	Taxable Income	Income Taxes	ATCF
0	$-84,000	---	---	---	$-84,000
1	70,000-30,000	21,000	19,000	-9,500	30,500
2	70,000-30,000	21,000	19,000	-9,500	30,500
3	70,000-30,000	21,000	19,000	-9,500	30,500
4	70,000-30,000	21,000	19,000	-9,500	30,500

Net PW(12%) = $8,639; Net AW (12%) = $2,844; IRR = 16.8%

(b)

EOY	NOPAT	EVA
1	$19,000-$9,500 = $9,500	$9,500-0.12($84,000)=$ -580
2	= 9,500	9,500-0.12($63,000)= 1,940
3	= 9,500	9,500-0.12($42,000)= 4,460
4	= 9,500	9,500-0.12($21,000)= 6,980

PW(12%) of EVA = $8,639; Annual Equivalent EVA(12%) = $2,844

It can be seen that the AW of this project is identical to the annual equivalent EVA when MARR = 12% per year. Therefore, the annual equivalent EVA is simply the annual worth of a project.

EQUIVALENCE OF EVA AND OTHER PROFITABILITY MEASURES

It can be shown that EVA is equivalent to revenue requirements (RR) analysis used by investor-owned utilities and benefit-cost ratio analysis employed by most government agencies to evaluate the profitability of capital investments [6]. For instance, the previous example has been re-formatted in Table 1 in order to calculate the annual revenue requirements (RR) of the proposed project. Based on Column 7 of Table 1, the present worth of the RR at the tax-adjusted cost of capital (i′) is $195,339 and the annual equivalent RR is $64,311. Notice that gross revenues do not appear in Table 1 because the objective of revenue requirements analysis is to determine the minimum annual revenue that is needed to cover all capital costs and annual operating expenses. When evaluating mutually exclusive alternatives, the investment selected is the one having minimum equivalent RR in order to satisfy all functional requirements.

Because our estimated gross revenues for the investment being evaluated are $70,000 per year, we surpass the minimum equivalent annual RR of $64,311, and the project is profitable. Notice that the difference, $70,000 - $64,311, is the after-tax annual equivalent EVA determined from Figure 1

TABLE 1

Annual Revenue Requirement for the Proposed Investment

Year	(1) Unrecovered Investment	(2) Book and Tax Depreciation	(3) Annual Expenses	(4) Debt Return 0.25(0.08) x (Col. 1)	(5) Equity Return 0.75(0.1467) x (Col. 1)	(6) Income Taxes T=Col. 5[a]	(7)[b] RR = Cols. 2 + 3 + 4 + 5 + 6
1	$84,000	$21,000	$30,000	$1,680	$9,240	$9,240	$71,160
2	63,000	21,000	30,000	1,260	6,930	6,930	66,120
3	42,000	21,000	30,000	840	4,620	4,620	61.080
4	21,000	21,000	30,000	420	2,310	2,310	56,040

[a] T = Col. 5 only when t_e (effective income tax rate) = 0.50. When $t_e \neq 0.50$, Equation a below is applicable.

[b] Present worth of Col. 7 at i′ of 12% = $195,339, the annual equivalent is $64,311 (i′ = tax-adjusted weighted average cost of capital -- see Equation b).

Equation a: Income Tax = (t/1-t)[Equity Return - Book Depreciation - Tax Depreciation].

Equation b: After-Tax Cost of Capital $(i') = i_b(1-t_e)\lambda + e_a(1-\lambda)$, where λ = fraction of borrowed money in the firm's total capitalization, i_b = cost of borrowed capital (as a decimal) and e_a = return to equity capital (as a decimal).

multiplied by (1/1-t). Thus, we see that the difference of \$5,689 equals \$2,844 (1/1-0.5). EVA is therefore equivalent to the before-tax profit that results from a revenue requirements analysis.

A benefit-cost analysis, which is typically utilized by government agencies to evaluate whether a proposed project is worthwhile, can also be constructed from the results of Table 1. To demonstrate this, let's use the conventional B/C ratio computed from annual equivalent benefits and costs [1]:

$$B/C = \frac{\$70,000 \text{ (gross revenues = "benefits")}}{\$64,311 \text{ (annual equivalent RR = "costs")}} = 1.09$$

A ratio greater than or equal to one indicates a favorable project. Consequently, we see that the B/C ratio confirms the same recommendation made with RR analysis (and PW/AW analysis).

CONCLUSION

In conclusion, an after-tax cash flow analysis based on Figure 1 can be utilized to obtain EVA for discrete capital investment opportunities. Even though EVA has been applied predominantly to equity valuation of manufacturing companies, EVA can also be utilized to evaluate individual capital investments because it is equivalent to the annual worth (AW) of projected ATCF for a project.

It is also possible to conclude that after-tax engineering economy study results are used by many large corporations as a proxy for shareholder (equity) valuation purposes. Even though most engineers in industry perform before-tax engineering economy studies, there is widespread use of results from after-tax engineering economy studies as evidenced by their equivalence to the popular Economic Value Added measure of worth.

Profitability measures developed from Figure 1 and Table 1 are equivalent, so all metrics discussed in this paper can be obtained from Figure 1. Therefore, we conclude that EVA is "old wine in a new bottle" and is subject to the same limitations and interpretation as classical profitability metrics used for decades in after-tax engineering economy studies of manufacturing investments.

REFERENCES

1. **DeGarmo, E. P., W. G. Sullivan, J. A. Bontadelli and E. M. Wicks,** *Engineering Economy*, (Tenth Edition), Prentice Hall, Upper Saddle River, NJ, 1997. See Chapters 6, 7 and 13.
2. **Dodd, J. L. and S. Chen,** "EVA: A New Panacea?" *B&E Review*, July-September 1996, pp. 26-28.
3. **Freedman, W.,** "How Do You Add Up?" *Chemical Week*, October 9, 1996, pp. 31-34.
4. **Gressle, M.,** "How To Implement EVA and Make Share Prices Rise," *Corporate Cashflow*, March 1996, pp. 28-30.
5. **O'Shaughnessy, J. P.,** *What Works On Wall Street*, McGraw-Hill, New York, 1997.
6. **Ward, T. L. and W. G. Sullivan,** "Equivalence of the Present Worth and Revenue Requirements Methods of Capital Investment Analysis," *AIIE Transactions*, Vol. 13, No. 1, March 1981, pp. 29-40.

Product design methodological contribution based on product decisions : Proposition of reference decisionnal process

Jean RENAUD
Vincent BOLY, Valérie RAULT-JACQUOT, Laure MOREL, Claudine GUIDAT
Ecole Nationale Supérieure en Génie des Systèmes Industriels
4, allée Pelletier Doisy
54603 Villers les Nancy
FRANCE
Tél : 03-83-44-38-38
Fax : 03-83-44-27-29

ABSTRACT

Nowadays everyone agrees to say that it is important to improve time of design or manufacturing of products. The concurrent engineering concept is playing a great part offering a better control of associated decisions and know how during development or production. The article we are presenting should consider the improvement of a decisional process during manufacturing of product within Food Industry.
The approach proposed by L.R.G.S.I[1] is divided in four stages whose aim is to present an optimal decisional referential system as far as studied development is concerned; also a decisional know how cartography to allow contribution to research approach of new products.
We propose to introduce approach in a global manner as well as four steps with necessary implements. Then, to enable this approach, an industrial application has been made to autothenticate it. It concerns a food industry specialist in cheese production. These works are completing a study we have already started on decisions taken between process actors where research of a best formalization of as associated decisions and know how is opened.

KEYWORDS

Concurrent Engineering, Modelling Decisionnal, Know How, Methodological.

1. INTRODUCTION

Works we are presenting depend on discontinued process, that is to say associate the product preparation phase (continued process by batch) with the discontinous phase while manufacturing of product (discreet process). We try to solve the problem in proposing an approach in view of improving decisional processing starting from an evolutive referential. Then we shall connect associated know how to permit us to locate some potentially realizable products.

Our entire approach is to get centered on decision taken at each product transformation stage. This to eliminate uncertainty of decisions and to obtain the best decisional processing by consensus from different actors. Here we propose an approach based on audit starting from description implements; it is issued from R(Resource).A(Activity).R(Result) method developed at L.R.G.S.I. Starting from decisions descriptive formalism developed within our research, we intend then to equip actors with communication tools, with exchanges and

[1]- Research Laboratory in Industrial Engineering Systems

validation of this decisions process. The last step of our approach will present necessary tools allowing to formalize associated know how decisions.

2. DECISION ENGINEERING

Our fundamental research consideration is essentially based upon concept of "the decision". Throughout product preparation and manufacturing process in a discontinued context, a certain number of decisions[2] are taken allowing to associate an action, therefore an added value, with product. For example : choice of technological solution means for action to implement it. If we are interested in the decision, it is for the following reason :

- from one part, it formed the object of a study (programmed decisions or not, structured or not or hybrid [LE MOIGNE 74], dependent or independent decision [BERIOT 92], but the objective was to better understand production systems or systems in general.

- From the other part, the decisional aspect is associated with expert, therefore with its associated knowledge. The latter is the sense method we are interested in which is a real enterprise memory. The objective is to better formalize expert knowledge in order to memorize the principles.

As J.W FORRESTER points it out, the decision follows a process starting from knowledge usually based upon experience, rules, if they exist, actual position (state of product), present data or information. To make this decision clear, several samples have been proposed such as H. A. SIMON's restarted by LE MOIGNE as a tool of concept and diagnostic of complex states. B. ROY is proposing an example that refers to a multicriteria analysis; this model being restarted by POMEROL has the inconvenient not to integrate all actors concerned by the decision; also to believe that the decision belongs to one person only. BERIOT's model is the one corresponding best to our research situation : it identifies notions of definition of study context, collection of information research for best solution and validation of the decision. BERIOT's example has been imagined to help the decision head maker in his behaviour facing the decision and integrating all other actor's advice. The decision study is placed in up-stream phase; that means in searching phase of information.
Some other actors propose a characterization of the decision system [COQUET 89] as being consumer of asset and human resources (Can Do) integrating customer's request (Must Do) to give consumed resources in results (Can Do) and the product (Must Do). This system necessitates somme decisions orders (Want to Do). It is after we considered this idea mainly from BERIOT and COQUET that our research consists of proposing an identification model of decision.

So, the product can be defined as follows [RENAUD 94] :

$\textbf{Product} \longrightarrow \textbf{Process} = \sum_{i=1}^{n} \textbf{di(know-how)}$, where ***di*** is the ***decision i*** and ***n*** : the whole of process decisions. The product t+1 may be defined as :

$$\textbf{Product}(t+1) = \textbf{Final product} - \sum_{i=t+1}^{n} \textbf{decisions (i)}$$

[2]- B. ROY definies the decision as follows : a decision taken on a development of product pproduces a consumer action of know how , experience and knowledge. It generaly furnishe quantitative and qualitative information. This decision is usually taken from a not clear, multiple, evolutive, implicit, even unconscious reasoning.

In our precedent works [RENAUD 94] we have proposed a model to show actions and decisions. Formalism utilized consists of representing actions following horizontal axis and the decisions following vertical axis. The objective of our representation is to visualize the whole of actions engaged for process. From this diagram, a participative approach has been focused with panels and graphics materializing the decisional process in order to improve the proposed process. These works being done, we have tried to improve this representation patterns to make it more pertinent as far as description of decisions is concerned, and more operational for experts of decisional process, too. The experience made on several industrial cases allowed us to make a better adaptation of method to studied process.

3. PRINCIPE OF CONCURRENT ENGINEERING

We do not intend to explain concept of Concurrent Engineering consisting of conceiving and manufacturing product with making of product to reduce "Time to Market" searching for notion of parallelism.

So, our research is consisting to improve the parallelism of actions from the decisions that taking up-stream. This anticipation is integrating the experience of expert.

Basic principle of our approach is to get centered on decision in view to make decisional process more operational, but also to better formalize known knowledge associated to decisions. Basic audit formalism of decision is shown at figure 1. Which is supported by R.A.R model. Once this thinking over is realized, the aim is to propose an approach, thus allowing to get potentially realizable products (this phase in question is not shwn here). To make those principles operational, an approach made of four stages set up with tools is used (figure 2).

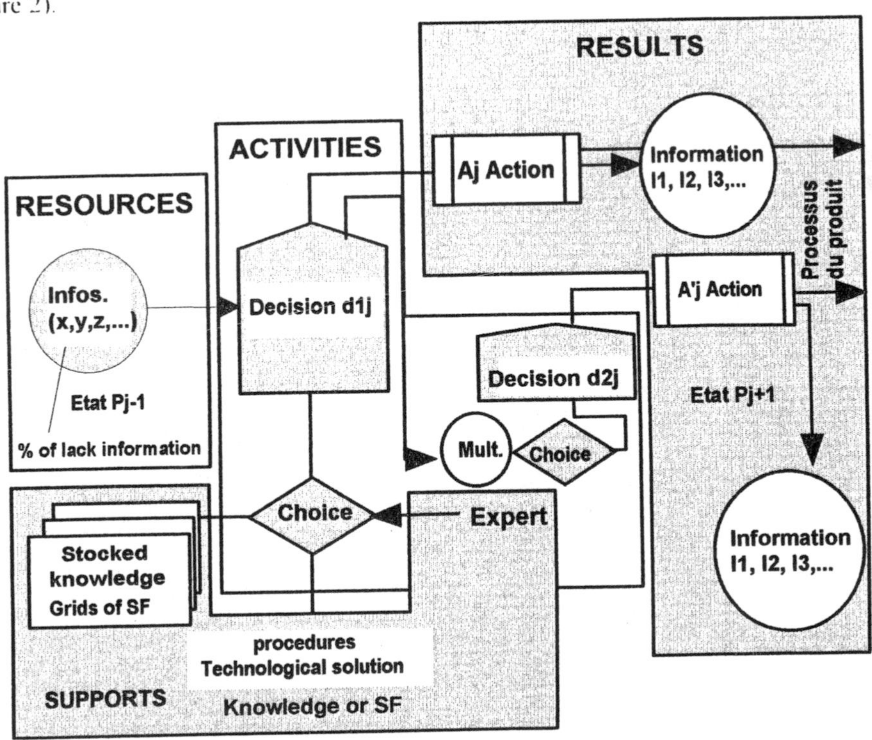

Figure 1. Decision represented from R.A.R principle

4. APPROACH IN FOUR STAGES

4.1 STUDY CONTEXT

If we do wish to better seize our approach within a Concurrent Engineering, we necessarily must bring on industrial validation. Therefore, to test our approach, we placed ourselves within an action research context. That means that through successive operations on site, of inductive type, we tried to utilized the action research which contribut the team-work taking into account grounds facts.

4.2 DEVELOPMENT OF PROSED APPROACH

The proposed approach is divided into four stages (figure 2). The first stage will describe the decisional process during preparation and manufacturing of product while audit stage with experts and through interviews (stage 1). Then from that audit, the decision should be formalized taking a proposed model issued from the principle of R.A.R model (Resources, Activities, Results) (stage 2).

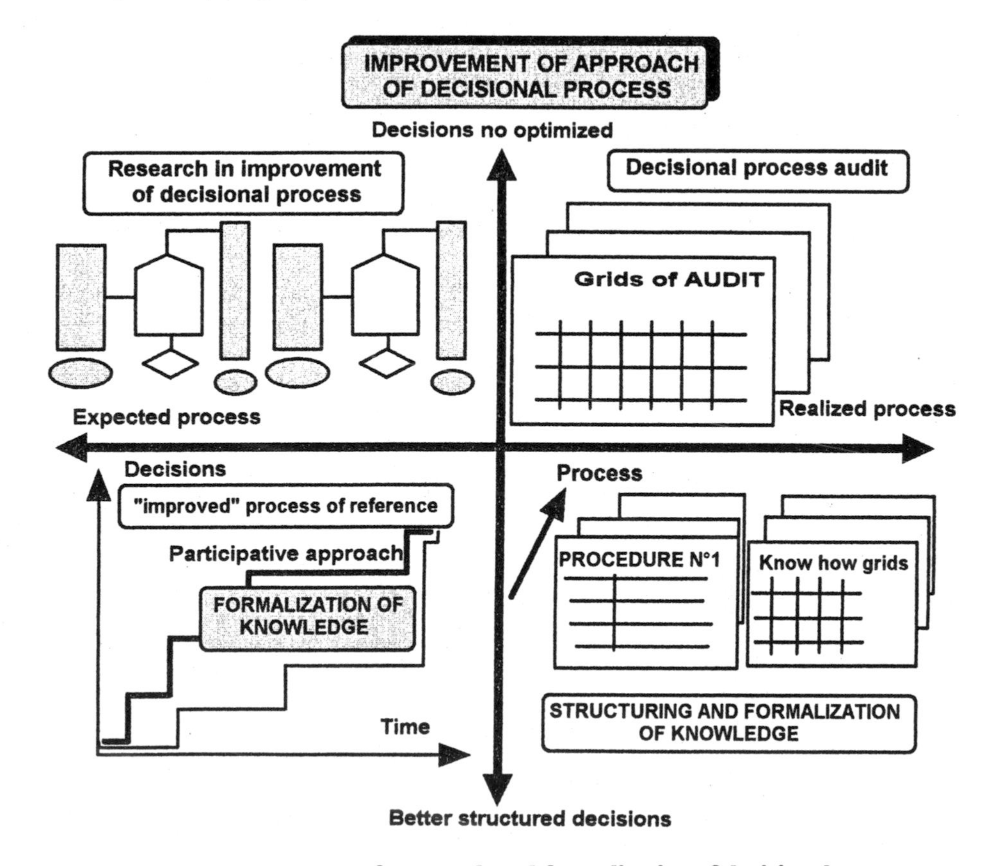

Figure 2. Improvement of approach and formalization of decisional proces

At the end of stage 2, searching for optimized decisional referential (stage 3) is proposed within a Concurrent Engineering context. Finally, after this decisional reference, an approach upon formalization of know how associated knowledge is proposed (stage 4). The approach in question should allow some propositions of innovating products. We positioned our approach following two axes, such as :

- ***axis of formalized decisions or not ,***
- ***axis of decisional process as it is materialized and wanted.***

4.21 STAGE 1 : AUDIT OF DECISIONAL PROCESS

To extract decisional knowledge from each actor directly stepping over process with interviews, four grids allow audit of decisional process as follows :

Decision activity : the question is to know the type of decision, encountered risk, the possible, the possible anticipation of the decision, repercussion on other decisions or on organization.

Decision resources : type, way and improvement type of necessary information to take the decision should be known : lack of information to take the decision, the risks are to be taken into account, too, while audit.

Decision base : know the procedure type, know how, experience for decision base.

Obtained results of decision : type of action over product should be known repercussion over organization , missing information or completing those previously taken, influence over other decisions, obtained information.

The objective of this step is to prepare formalization of decisions of process.

4.22 STAGE 2 : REPRESENTATION OF DECISIONAL PROCESS

The stage 2 consists of representing decisions from a model prepared to that effect and depending from R.A.R principle. But its characteristic is to be more complete, from one part, and to be used as means of decisions improvement through a participative approach with all actors from another part. Each part of the audit decision is reproduced on document (figure 3). So, the whole actors will be able to modify the decision environment to get it improved.

4.23 STAGE 3 : OPTIMALIZATION OF DECISIONAL PROCESS

The objective of this stage will consist of improving decisional process in utilizing documents prepared to that effect (figure 3). Different actors concerned will be able to modify process in trying to move decisions at the earlierst. Improvement goes along with a formalization of associated knowledge (writing of procedures, additional information). The charateristics of their stage 3 is to propose an improved and accepted process by actors concerned by decisions to be taken. The rules, here, are to make better know how to the detriment of acquired knowledge by the expert. During the stage, it is important to integrate a maximum of knowledge from experts. This knowledge which is a real "experience return" since it has acquired a certain validation must be collected at the soonest in view to allow the anticipation of actions to be engaged with less risk, as much as possible.

Furthermore, the case is not rare to notice that when decision is not clear and is based on knowledge to use, without any security, it is taken too late, generating some arrangements on process actions. When a validation of all taken decisions on process are accepted by all actors, then a capitalization of knowledge may be made up. It can become a real memory for searching of new products which means putting in place Concurrent Engineering. In words storage of knowledge will both contain :

- ***Technical information*** (professional knowledge) in the form of technical data controlled by enterprise. They are knowledge of enterprise (choice of a technical solution, choice of a know how in a definite position). This knowledge will have to be validated by all actors.

- **Proceedings**, which are the way to implement this technical information by the use of algorithms, heuristics,. ..). It concerns to implement know-how in the form of rules of experts corresponding to types of situations given. This kind of information is going to set up the " trade " knowledge memory of the firm. To structure and to formalize this knowledge will help to improve existing or future products processes.

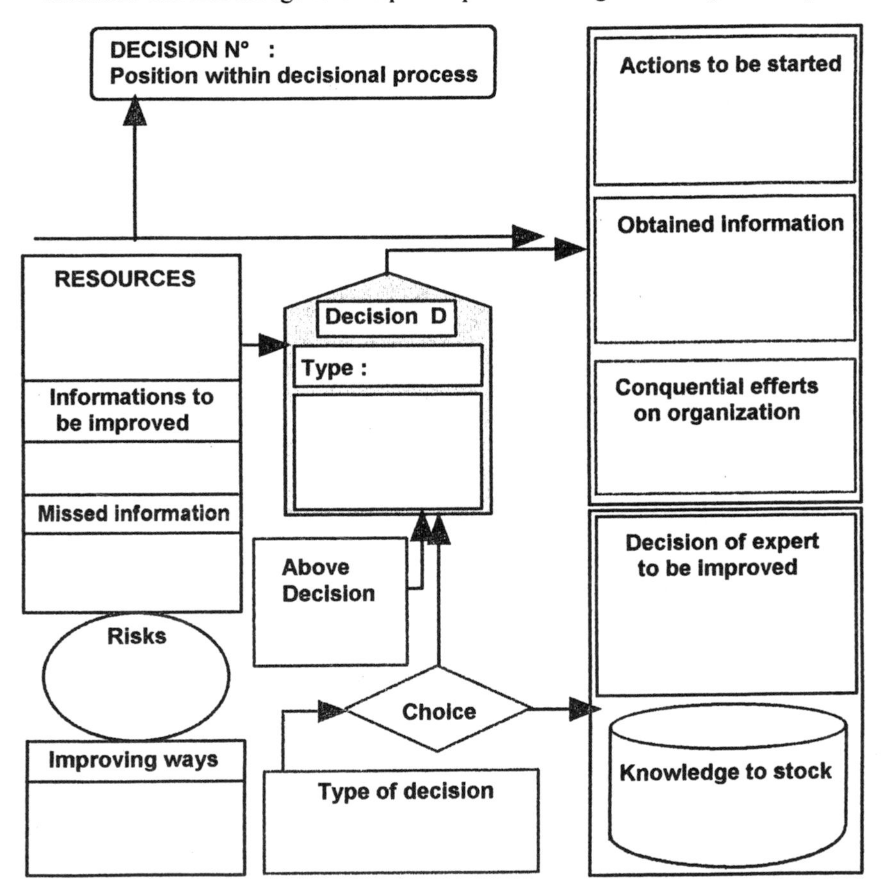

Figure 3. Work document for improvement of decisional process

4.24 STOP 4: STRUCTURING AND FORMALIZATION OF DECISIONS ASSOCIATED KNOWLEDGE

This step remains the most delicate to lead since it consists in drawing a know-how mapping based on each decision taken. Furthermore, this step is more complete than the previous one since it allows to validate the reference decisional process by constituting a knowledge memory. By the way, product design decision is easier.

5. INDUSTRIAL APPLICATION

To validate our proposal, we have tested it in a small enterprise and more precisely in a cheese dairy. The manufacturing process is a discontinuous one : preparation of the mature milk by batch and then manufacture of cheeses in continuous chain. This firm tries to better master its manufacture decisional process in order to reduce design and manufacture risks. It represents an important stake for the enterprise. Indeed this enterprise tries to innovate in potentially achievable products from mastered expert knowledge.

6. RESULTS OBTAINED

At the end of the step 1, the initial audit grids have been completely modified, because their utilization was too maladjusted. A reflection with all the actors and the seekers has allowed to propose a new type of grids. Results obtained on a part of the manufacture process have shown that many decisions have been taken from an expert know-how (acquired in the course of years) without a real validation and without a confrontation with the other decisions. The second important point of this audit is that a large number of decisions are taken too late in the process and therefore that they generate adjustments during manufacture. A will to better surround and to insure a best support of information and knowledge of the decision proves necessary. On the figure 4 one can note that a certain number of decisions corresponding to a part of the process of manufacture can be improved.

Observation under study decisions	d1	d2	d3	d4	d5	d6	d7	d8	d9
The decision is explit	Y	Y	Y	N	N	N	N	N	N
risk engage during the action-making	m	M	m	M	M	M	m	M	M
lack of information	Y	Y	Y	N	N	Y	Y	Y	Y
necessary information for the decision-making	Y	Y	Y	Y	Y	Y	Y	Y	Y
the decision can be taken upstream ?	Y	Y	N	Y	Y	N	N	N	N
an other decision can be taken to the place	N	Y	N	N	N	N	N	N	N
repercussion on the organization	Y	Y	N	Y	Y	N	Y	Y	N
influence on the future action	Y	Y	Y	Y	Y	Y	Y	Y	Y
the obtained result corresponds to what was anticipated ?	Y	Y	Y	B	B	B	B	B	B

m : minor; **M** : Major; **Y** : YES; **N** : NO; **B** : Better

Figure 4. Example of results concerning decisions

The step 2 has allowed to better apprehend each decision in order to help in the research of ways of improvement. A work of group was necessary to propose an improved process, what corresponds to the step 3. The step 4 is not exploited in this research work.

5 - CONCLUSION

Our experimental results allowed us to refine the first steps of our proposal. During the application, we have been able to observe that the audit step from interviews has to be very relevant and adapted to the type of process (manufacturer, continuous, discontinuous). Results depend largely on the share of experts:

- actors of the decisional process have to evolve from a situation of wait to a situation of cooperation in the audit level but also in the level of the improvement and the training of know-how and their application,

- They have to understand the importance of the decision-making, its content, its risks, the results to obtain,. .. (central element of the process)

REFERENCES

1. CARTER, D. and BAKER, B. Concurrent Engineering: The Product Development Environment for the 199O's. Addison-Wesley Publishing Company, Massachssets.

2. CHAMPETIER, M. et LOPPINET, T. Un systeme d'aide à la décision pour la conception de produits complexes dans un contexte d'Ingénierie Concourante. Congrès International de Génie Industriel de Montréal - La productivité dans un monde sans frontières, Montréal, Octobre 1995.

3. CHANAN S. SYAN, UNNY MENON. Concurrent Engineering - Chapman and Hall 1994.

4. MOREL, L. MEYER, F.; RENAUD, J. Decision-aided Modelling for Industrial System Engineering. Second International Artic Workshop on Industrial Managament - Lappeenranta University of Technology, June 17-18, 1996.

5. PARSAEI, H.R. and SULLIVAN(eds). Concurrent Engineering : Contemporary Issues and Modern Design Tools. Boundary Row, London; Chapman and Hall.

6. RENAUD, J.; BOLY, V.; GUIDAT, C.; RAULT-JACQUOT, V; MOREL, L. Contribution méthodologique pour l'amélioration de la réponse au marché à partir d'une démarche participative issue de l'Ingénierie Concourante. Congrés International de Génie Industriel de Montréal - La productivité dans un monde sans frontières, Montréal, Octobre 1995.

SIMULTANEOUS PRODUCT/PROCESS DESIGN FOR DISASSEMBLY

Matthew D. Bauer, David W. Rosen
Systems Realization Laboratory
The George W. Woodruff School of Mechanical Engineering
Georgia Institute of Technology
Atlanta, Georgia 30332-0405
(404) 894-9668 *Fax: (404) 894-9342* *david.rosen@me.gatech.edu*

ABSTRACT The research reported in this paper illustrates the integration of product and disassembly process design via virtual prototyping. In this context, a virtual prototype is an information model consisting of a product model and one or more process models. Integrated product/process design is accomplished by defining a parametric design problem in terms of constraints/goals and coupling virtual prototypes and a multiobjective optimization code for solution. Application of optimization techniques to solution has proven difficult. Problematic characteristics are highlighted, and an effective optimization algorithm is described. An automotive center console is used as an example. The focus of this paper is the formulation and solution of integrated product/process design problems, rather than the results of the center console design problem, *per se*.

1.0 INTRODUCTION

Life-cycle design (or Design for the Life Cycle, DFLC) is an evolution of concurrent engineering and consists of all life cycle phases (needs, recognition, development, production, distribution, usage, and retirement) being considered simultaneously throughout product development (Atling 1993). Our approach to DFLC centers around two main points: ① an intelligent partition of responsibilities between designer and computer based upon their abilities and capabilities, respectively, and ② an integrated decision support environment in which all design requirements can be considered simultaneously. In this work, this approach is illustrated through its application to simultaneous product/process design for disassembly.

1.1 Frame of Reference

Industrial countries are beginning to face one of the consequences of the rapid development of the last decade. Wide diffusion of consumer goods and shortening of product life cycles have given rise to an increasing quantity of used products being discarded. One of the most evident metrics of this trend is our decreasing supply of available landfill space. To address these issues, regulations inspired by the "polluter pays principle" are emerging, and Germany is leading the way because of its high scrap rate relative to its surface available for dumping (Jovane, *et al.* 1993). Whether motivated by a moral sense of obligation or recent legislative efforts, manufacturers (and specifically designers) will soon be responsible for their products' retirement. Obviously, product design for demanufacture[1], of which disassembly is a principal component, will be an integral part of this new responsibility.

---

[1] Demanufacturing characterizes the entire process involved in recycling, reuse, incineration, and/or disposal of a product after it has been taken back by a company.

The current state of **D**esign **f**or **D**isassembly (DFD) is that of *redesigning* a product for ease of disassembly. The typical approach to DFD is to define general design rules and databases of real-time standards to utilize in spreadsheets for analyses of detailed product models. For example, Beitz (1993) developed tables to keep track of the number of parts, component materials, access to parts and fasteners, tools needed for disassembly operations (among other measures) in order to evaluate product disassemblability in a manner similar to Boothroyd & Dewhurst's DFA method (Boothroyd and Dewhurst 1989). Similarly, Kroll *et al.* (1994, 1995) have developed an evaluation chart consisting of quantitative measures and procedures for evaluating a disassemblability for recycling. Dowie *et al.* (1995) have taken a more direct approach to DFD by adapting Boothroyd & Dewhurst's principle DFA questions (1989) for product disassembly costing. Lastly, Lowe and Niku (1995) have developed "a framework of concepts used to analyze disassembly [effort] during the design phase" with the goal being to improve disassemblability through improvement of overall layout and part attachment. While this is not a comprehensive literature review, the general approach is apparent.

While improvements in product disassembly can be realized through these methods, they are not ideal, and limitations do exist. First, the assessment fidelity is unnecessarily low because: ① disassembly criteria are generally evaluated on interval scales which are inherently subjective, ② extrapolation of empirical observations are commonly made to procedures which may not have been directly measured, and ③ current DFD criteria (such as time, force, number of parts, and accessibility, etc.) are incompatible and cannot be adequately represented by a single disassemblability metric. Second, the application of current DFD methods is generally limited to the latter stages of product development because they rely on detailed product models (preferably physical prototypes) upon which to perform analyses. Third, current DFD methods are limited in scope because they emphasis a narrow product-only approach to reducing disassembly effort, rather than an integrated product/process approach.

The goal in this research is to address the limitations of current DFD methods through an integrated product/process approach to DFD via virtual prototyping. Towards this goal, simultaneous product/process design for disassembly is presented in this work.

1.2 Simultaneous Product/Process Design

With the introduction of virtual prototyping into product development **S**imultaneous **P**roduct/**P**rocess **D**esign (SPPD) is enabled. SPPD refers to a design method for integrating product/process design via virtual prototyping in a decision support environment during parametric design[2]. SPPD is significant in that it enables:

- integrated product/process development, thereby ensuring favorable compromise between product and process requirements;
- process simulation to be performed directly on CAD-based product models, thereby decreasing reliance on physical prototypes and facilitating consideration of process issues earlier in design;
- process analysis to be performed directly on CAD-based product models, thereby increasing objectivity of evaluation, eliminating the use of empirical real-time data, and reducing product model re-entry;

The principal steps of SPPD are outlined in Figure 1 and described in the remainder of section. The 1st step in SPPD is the *definition of the problem.* From a set of design requirements, a specific problem is developed and expressed in the form of a problem statement. This problem statement

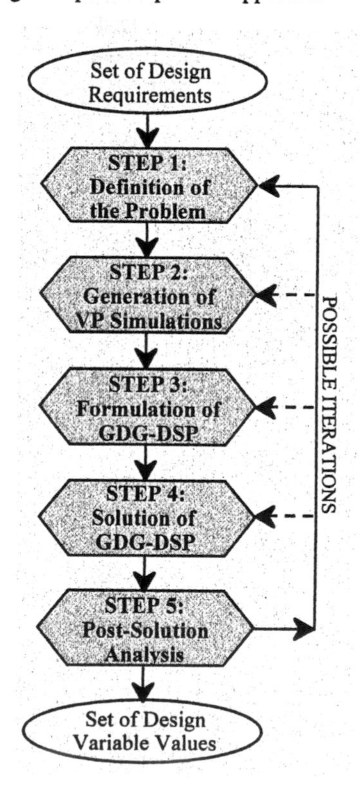

Figure 1. SPPD Steps

[2] Parametric design refers to the stage in product development where the system configuration and parameters are known but the specific values of those parameters have yet to be determined.

should articulate the specific design issues to be addressed in addition to the product and process requirements in the form of demands, wishes, and their preferences. *Generation of the virtual prototype simulations* is the 2nd step in SPPD. From a product model in CAD, a process model is developed in a virtual prototyping environment. By capturing disassembly motions in the virtual prototyping environment, a feasible (but not optimal) disassembly process is developed. This process is then parameterized and executed on the product model to form the virtual prototype simulations. The 3rd step is the *formulation of a goal-directed geometry Decision Support Problem* (gdg-DSP) which is a mathematical representation of the designer's demands, wishes, and preferences which are articulated in the problem statement. The 4th step is the *solution of the gdg-DSP*. Once the gdg-DSP is implemented on a computer, the virtual prototype simulations are utilized in the evaluation of product/process constraints and goals. Optimization techniques are applied to facilitate rapid exploration of the design space. Lastly, *post-solution analysis* is performed. As illustrated in Figure 1, the procedure may be iterative.

The supporting system architecture for SPPD is illustrated conceptually in Figure 2. As shown, the principal constituents of the architecture are: an integrated virtual prototype/CAD system, the goal-directed geometry Decision Support Problem, and a solution mechanism. Also illustrated in Figure 2 is our desire to port results from SPPD to traditional DFD methods. This architecture is implemented in ANSI C on a Silicon Graphics workstation.

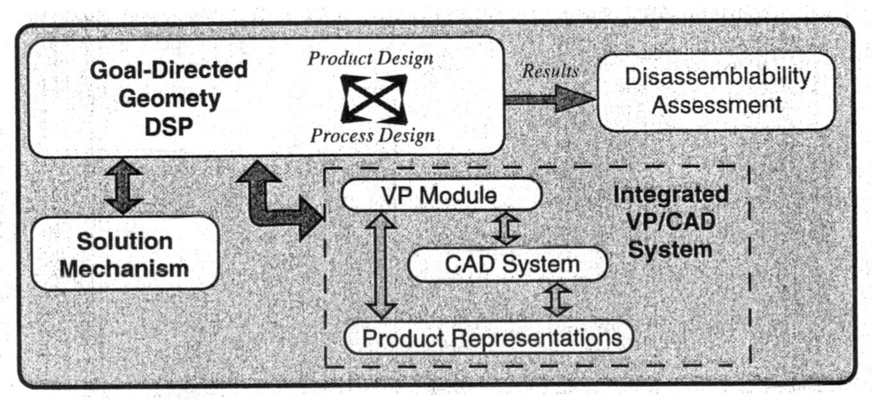

Figure 2. Conceptual System Architecture for SPPD

2.0 VIRTUAL PROTOTYPING IN SIMULTANEOUS PRODUCT/PROCESS DESIGN

A virtual prototype is a model of a product and a process that the product undergoes. The key features of a Virtual Prototype (VP) for disassembly are outlined in Figure 3, along with a VP illustrating virtual disassembly. A VP's product model contains geometric component models and assembly information, including a hierarchy, mating relationships among components, fasteners, and other spatial relationships. For disassembly, the process model contains information about operation sequences, tool changes, component disassembly paths, etc. Some elements of the process model can be generated automatically from an existing CAD model while other elements require human interaction and assistance. Generation of a product's disassembly process through virtual prototyping can be implemented throughout product development, from preliminary design to detail design, and the resulting VPs can be built upon as design progresses.

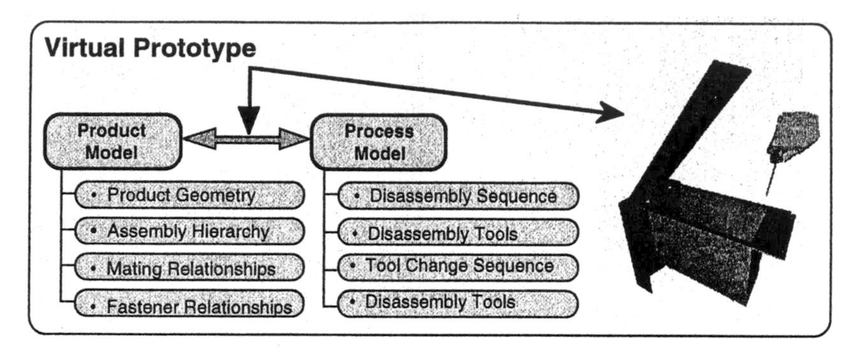

Figure 3. Definition of a Virtual Prototype

Our overall understanding of virtual prototyping is articulated in Figure 4. *Prototype generation* and *process simulation* are the main activities involved in virtual prototyping. The goal in prototype generation is the development of a VP from a CAD product model which is appropriate for analyses. *Process simulation* is the execution of the process model on the VP's product model. A brief overview of these activities is presented in the remainder of this section. For a detailed description of our work in this area, the reader is referred to Siddique (1996).

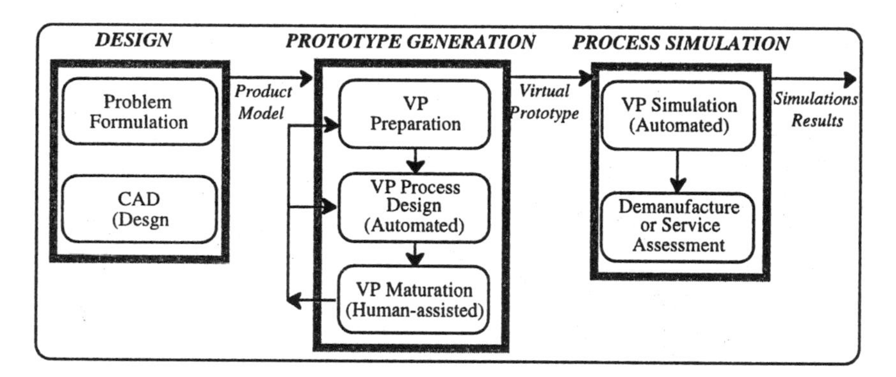

Figure 4. Principal Activities in Virtual Prototyping

As illustrated in Figure 4, VP generation consists of three steps: *VP preparation* in which product CAD models are converted into a representation suitable for prototyping, *VP process design* in which automated reasoning methods determine certain attributes of the process model, and *VP maturation* in which designers interact with the virtual prototype to fine-tune the process model.

Emerging from *VP generation* is a virtual prototype suitable for simulation and assessment. Two modes of virtual prototype simulation have been developed in this work: an *interactive graphical simulator* and a *CNG simulator* (pronounced "See No Graphics" simulator). While significant software overlaps exist between these two simulators, their functionality is very different, as outlined in the remainder of this section.

In the interactive graphical simulator, disassembly processes are graphically executed for inspection and manipulation. In addition, disassembly paths, along with their control vertices, can be displayed in the graphics window. More importantly, the system user can interact with the disassembly process through the manipulation of the path's control vertices, as illustrated in Figure 5. Overall, the interactive graphical simulator has proven to be an useful design tool for rapid development and visually assessment of processes.

The 2nd simulation mode is the CNG simulator whose purpose is to support objective evaluation of processes and solution of SPPD problems. Whereas the interactive graphical simulator utilizes graphical product models during simulation, the CNG simulator utilizes a product's solid models. While the use of solid models in the simulator is computationally intensive, it is required to support the evaluation of geometric constraints and goals in SPPD. For example, collision detection in this work is accomplished through the use of bounding boxes and Boolean intersections between product models which requires the use of the solid models.

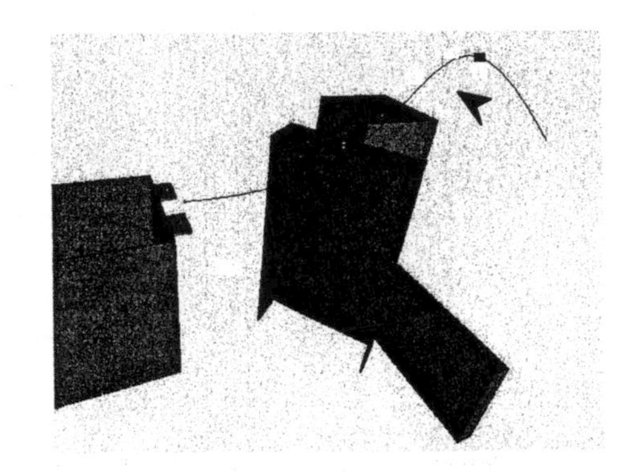

Figure 5. User Manipulation of Process Model

To execute a virtual prototype in either simulator, two descriptors of the disassembly process are needed: ① an ordered list of disassembly operations and ② a description of each disassembly operation. In our system, disassembly paths are represented mathematically as parametric curves and manipulated through control vertices. To date, three parametric curve have been implemented: Bezier, spline, and NURBS. For brevity, the reader is referred to (Mortenson 1985), (Press *et al.* 1992), and (Piegl and Tiller 1995) for the mathematics.

Results to date indicate that utilizing parametric curves to represent disassembly paths is effective in SPPD. However, any formal statement weighing the pros and cons of any particular curve formulation would be premature due to our limited knowledge base with this class of optimization problems. The only serious drawback appears to be that a constant step size in parametric space seldom corresponds to a constant step size in Cartesian space. As a result, a constant step in the parametric variable u may result in an unacceptably large or undesirably small step in Cartesian space. To minimize this characteristic, the average Cartesian translation and rotation step sizes are constrained to a maximum value based on each object's minimum geometric dimension.

3.0 DECISION SUPPORT IN SIMULTANEOUS PRODUCT/PROCESS DESIGN

Integral to SPPD for disassembly is the integration of product and disassembly process design in a decision support environment where trade-off assessments between product and process requirements can be performed efficiently and effectively. The **G**oal-**D**irected **G**eometry (GDG) framework provides such an environment in that it is intended to aid a designer confronted with *what if* questions, examining the changes to the solution if (for instance) new constraints are added, if old constraints are removed, or if parameter values are fixed. More specifically, GDG involves the development and solution of optimization problems involving geometry (Rosen *et al.* 1994). This framework has been applied in the parametric design stage, where the system configuration and parameters are known but the specific values of those parameters have yet been determined. GDG's mathematical foundation is the **g**oal-**d**irected **g**eometry **D**ecision **S**upport **P**roblem (gdg-DSP). The gdg-DSP refers to a class of constrained, multiobjective optimization problems involving geometry, where values of design variables are determined which satisfy a set of constraints and achieve as closely as possible a set of conflicting goals. The gdg-DSP as particularized for SPPD is described in Section 3.1, and the definition of rudimentary process constraints/goals is presented in Section 3.2.

3.1 Goal-Directed Geometry Decision Support Problem in SPPD

A pictorial representation of the gdg-DSP as particularized for SPPD is presented in Figure 6. Interpretation of the figure is as follows: given an initial virtual prototype model that is to be improved through modification, find values for independent product/process variables, such that they satisfy both product/process constraints and achieve the product/process goals to the greatest extent possible. The solution to a gdg-DSP is referred to as a satisficing solution since it is a feasible point that achieves the system goals to the extent that is possible.

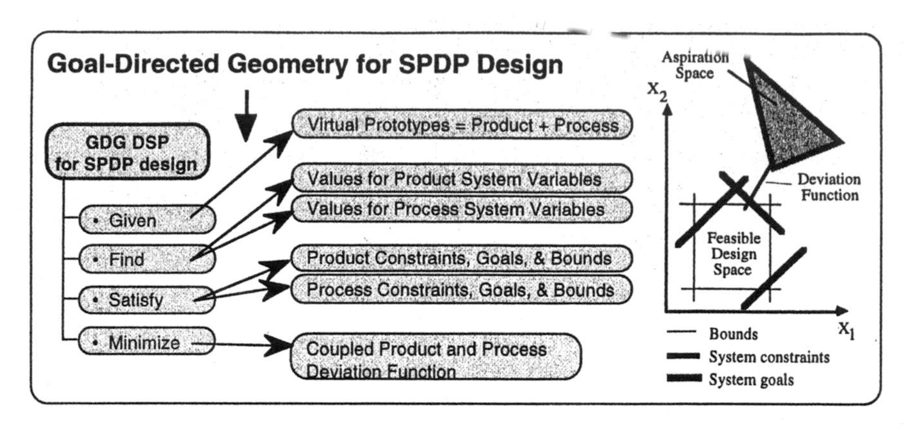

Figure 6. Pictorial Representation of gdg-DSP for SPDP

Virtual prototypes are included in the gdg-DSPs for SPPD to enable the direct evaluation of process constraints/goals, thus eliminating the need for physical prototypes. Process system variables are included to enable process modification/improvement and the performance of trade-off assessments between product/process requirements. Furthermore, the feasibility and goodness of a process is assessed through the process constraints/goals. The deviation from the target values for the process goals are combined with product deviation variables to form a coupled product/process deviation function which drives product/process improvement. With the aid of optimization tools, a solution is located which satisfies the product/process constraints and achieves to the extent possible the product/process goals as modeled in the deviation function. The deviation function can be implemented in two ways: preemptive and Archimedean forms - refer to (Mistree *et al.* 1993).

3.2 Definition of Process Constraints and Goals

In their most rudimentary form, disassembly process constraints and goals are formulated to force the disassembly process to a state of non-interference and minimum energy. Currently, non-interference is the standard disassembly process constraint. Therefore, a disassembly process is deemed feasible if it results in product disassembly and avoids physical interference between objects in the simulation environment. In addition to the non-interference constraint, three process goals are utilized to drive the process's energy to a minimum state: ① minimize path lengths, ② minimize component/tool rotations, and ③ minimize work.

Overall, process constraints and goals are limited only by one's imagination and evaluation capability (i.e., simulation capability) and computational resources. For instance if robotic disassembly is utilized, additional constraints would be added to govern the kinematics of the robot. The ability to develop, utilize, and evaluate detailed simulations is constrained by one's available computational resources, as well as the current level of design knowledge. In the early stages of design, simple process constraints/goals and simulations are advisable because high product uncertainty and multiple conceptual designs can strain resources. As the design progresses, detailed simulations can be developed which enable the evaluation of more informative process constraints and goals.

4 SOLUTION OF SIMULTANEOUS PRODUCT/PROCESS DESIGN PROBLEMS

In this section, solution of SPPD problems is discussed. An overview of information flow in the evaluation of product/process constraints and goals is presented in Section 4.1. Characteristics of gdg-DSPs which have been problematic to the solution of SPPD problems is presented in Section 4.2. Lastly, an optimization algorithm is described which has proven effective (but inefficient) at solving this class of gdg-DSPs.

4.1 Information Flow in Evaluation of Product/Process Constraints and Goals

Critical to our success in SPPD is the capability to evaluate product/process requirements via virtual prototypes and the application of optimization techniques to solution of integrated product/process design problems. To accomplish this, information must be exchanged between several stand-alone system modules (refer to Figures 2). The flow of information between these system modules during evaluation of product/process constraints and goals is modeled in Figure 7. Given an initial system variable set ($\underline{X}$), the values of those variables are assigned to VP/CAD parameters ($\underline{P}$) in a gdg-VP interface module. Propagation of this information in the VP/CAD system results in modification to a product's VPs. During VP execution, information ($\underline{V}$) pertaining to the product and its process are captured and associated with the appropriate gdg-DSP constraints/goals (g, $\underline{A}$) through the gdg-VP interface. Using this information, the feasibility and goodness (i.e., deviation from target: d^-, d^+) are determined, and the deviation function values ($\underline{Z}$) are calculated in the gdg-DSP template. Based on the values of the deviation function, new values for the system variable set are calculated by the gdg-DSP solver. The procedure is repeated until termination criteria are satisfied in the gdg-DSP solver.

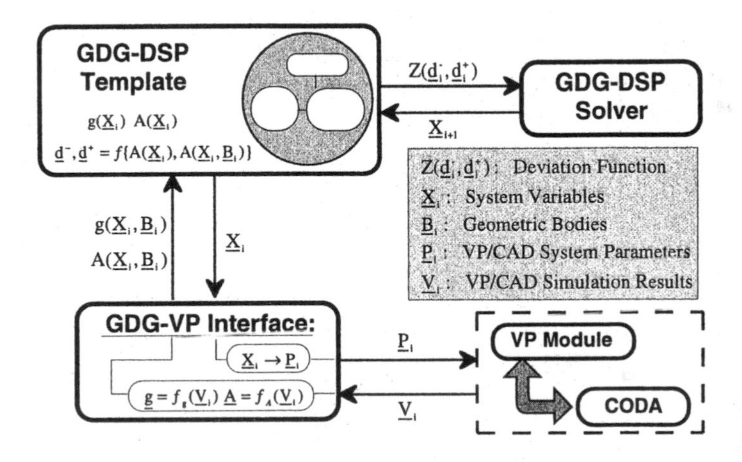

Figure 7. Information Flow during gdg-DSP Solution

4.2 Problematic Characteristics of SPPD Problems in Optimization

Multiobjective optimization techniques are applied to the solution of gdg-DSPs in SPPD to increase the efficiency and effectiveness of design space exploration. Unfortunately, solution of this class of gdg-DSPs using traditional optimization techniques has proven difficult because most methods employ a *one-variable-at-a-time* exploration at some level in their search routines. The characteristics of SPPD problems which have proven problematic in this regard are as follows:

1. Evaluation of nonlinear constraints/goals,
2. Product/process interactions, and
3. Size of design space.

Due to these characteristics, traditional optimization techniques have proven ineffective in solution. In particular, three algorithms have been explored: the **A**daptive **L**inear **P**rogramming (ALP) algorithm resident in DSIDES (Mistree *et al.* 1993), **H**ooke-**J**eeves (HJ) Pattern Search Method (Reklaitis, Ravindran *et al.* 1983), and the Method of Simulated Annealing with Down-hill Simplex (Press, Teukolsky *et al.* 1992). Of the three algorithms, the most successful has been a modified pattern search method. This algorithm is described in the next section.

4.3 Definition of an Effective Solution Method

The HJ Pattern Search Method is basically a combination of "exploratory" moves of the *one-variable-at-a-time* kind with "pattern" or acceleration moves regulated by some heuristic rules (Reklaitis *et al.* 1983). The exploratory moves examine the local behavior of the objective function and seek to locate the direction of any sloping valleys that might be present. The pattern moves utilize the information generated in the exploration to step “rapidly” along the valleys.

Like all direct search algorithms, the HJ Pattern Search Method requires at least N independent search directions, where N is the number of problem variables. The one-variable-at-a-time (or sectioning technique) is an elementary method that recursively uses a fixed set of search directions to examine the local behavior of the objective function. In the traditional HJ Pattern Search Method, this set of directions is chosen to be the coordinate directions in the space of the problem variables. Each of the coordinate directions is sequentially searched using a single-variable optimization scheme to determine a set of direction vectors to guide the pattern moves. A modification was proposed by Emery and O'Hagan (Emery and O'Hagan 1966) in which the exploratory phase is altered by using a set of orthogonal search directions whose orientation is redirected randomly after each iteration.

In our implementation, several modifications were made to the Hooke-Jeeves Pattern Search Method to support solution of gdg-DSPs. First, the method was extended to support solutions of DSPs involving multiple objectives, constraints, and bounds. This was accomplished by implementing the lexicographic minimum defined by Ignizio (Ignizio 1985):

> **LEXICOGRAPHIC MINIMUM:** Given an ordered array $\mathbf{f} = (f_1, f_2, \ldots, f_n)$ of nonnegative elements f_k, the solution given by $f^{(1)}$ is preferred to $f^{(2)}$ if
> $$f_k^{(1)} < f_k^{(2)}$$
> and all higher-order elements (i.e., $f_1, \ldots, f_{k-1}$) are equal. If no other solution is preferred to **f**, then **f** is the lexicographic minimum.

The second significant modification is the replacement of the one-variable-at-a-time optimization scheme with that of the method proposed by Emery and O'Hagan. In using orthogonal off-diagonal randomly oriented search directions, the goal is to define exploratory moves (and corresponding pattern moves) which couple the necessary system variables. As we illustrate in Section 5.0, these modification has proven worthwhile.

5.0 AN ILLUSTRATIVE EXAMPLE: AUTOMOTIVE CENTER CONSOLE

In this section, SPPD is illustrated through an application to a design of an automotive center console from a Chrysler LHS (Figure 9). This problem was selected because it is a practical example of a system which is candidate for demanufacture. In the remainder of this section, the steps of SPPD as outlined in Figure 1 are applied to the design of the center console's rear module and disassembly process.

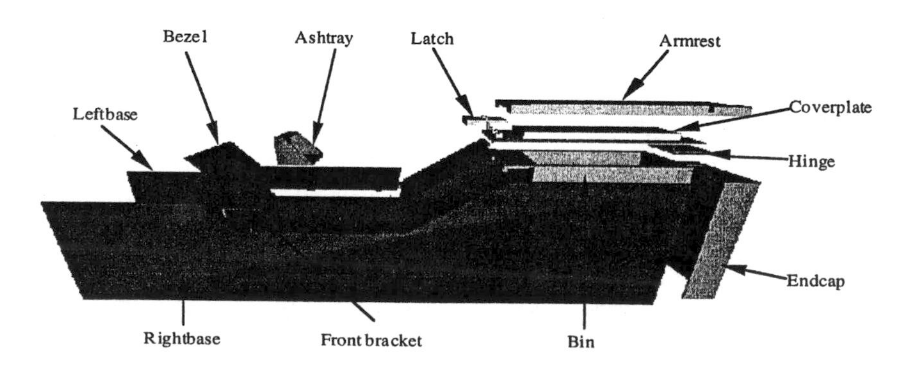

Figure 9. Conceptual Product Model of Center Console

5.1 Step 1: Definition of the Problem

Given the conceptual design of the center console as shown in Figure 9, the goal is to integrate the design of its rear module (i.e., bin, endcap, and armrest components) with the design of its disassembly process. The rear module is illustrated in Figure 3. In this scenario, the product design consists of determining optimal dimensions for the bin to maximize storage capacity and minimize the subassembly's material volume. Ensuring adequate wall thickness to support anticipated loading conditions and facilitate injection molding is the only product constraint explicitly defined. All other product constraints are implemented as bounds on the system variables. Process design consists of minimizing disassembly effort: minimizing disassembly path lengths, rotations, and work resulting from changes in potential energy while avoiding interference.

5.2 Step 2: Generation of Virtual Prototype Simulation

In the 2nd step of SPPD, the virtual prototype simulations of the center console's disassembly process are generated. In a virtual prototyping environment, a disassembly process is defined (refer to Table 2). The process is then parameterized for simulation and utilized to evaluate relevant product/process requirements, thus eliminating the need for physical prototypes during process evaluation.

Table 2

Center Console's Rear Module's Disassembly Process

Task #	Part Name	Description of Task	Removal Direction	Tool	Comments
1	Armrest	Open Armrest	[0, 0, 0]	None	Rotate Armrest 120 degree about y axis
2	Screw #1	Unfasten, Remove	[0, 0, 1] [0, 0,1]	Screwdriver	Rotate Screwdriver about +z axis and remove fastener in +z direction
3	Screw #2	Unfasten Remove	[0, 0, 1] [0, 0,1]	Screwdriver	Rotate Screwdriver about +z axis and remove fastener in +z direction
4	Armrest	Remove	[0, 0, 1]	None	Remove Armrest in the +z direction
5	Endcap	Remove	[-1, 0, 0]	None	Remove Endcap in the -x direction

5.3 Step 3: Formulation of Goal-Directed Geometry DSP

The 3rd step in SPPD consists of the formulation of the gdg-DSP which mathematically represents the product (center console's rear module), the process (disassembly of rear module), design requirements (product/process constraints and goals), as well as the designer's preferences (merit function). If we proceeded and formulated the gdg-DSP for the problem as presented thus far, the gdg-DSP would include (28) system variables, (3) constraints, and (15) goals. Due to limitations in space, the problem is simplified to just (2) components (the bin and endcap components) and a single disassembly operation (the removal of the endcap from the bin in 2D). The product/process constraints and goals remain as originally defined but at a more manageable level.

The gdg-DSP for this simplified problem is presented in Figure 10. Briefly, the goals are as follows: (1) maximize storage volume in the bin, (2) minimize material content in the assembly, (3) minimize the disassembly path for the endcap component, (4) minimize rotation of the endcap during disassembly, and (5) minimize work against gravity (implemented as minimize the endcap's positive vertical travel during disassembly), and (6) maintain parametric distribution. The mathematical form of the constraints/goals are omitted to save space. Finally, a two tier deviation function is used to model the designer's preferences with product and process goals at the 1st and 2nd levels, respectively. Now that the necessary virtual prototype simulations and mathematical form of the gdg-DSP have been developed, the SPPD problem is implemented on a computer and solved. Discussion of the solution results are given in next section.

Given:

Product model of center console's rear module:
- parameterized geometric models of rear module.
- mating relationships between components.

Flow path ratio: ratio of length of injection molding flow per unit wall thickness

Disassembly process model for center console's rear module:
- parameterized disassembly process for rear module

Find:

Values for dimensions defining bin's overall geometric shape:
- bin's length (L), bin's width (W), bin's height (H), and bin's wall thickness (WT)

Values for disassembly process variables: endcap's control vertices (ρ_1, x_1, z_1) (ρ_2, x_2, z_2)

Satisfy:

Bounds:

$24.0 \le L \le 30.0$ cm $\quad 10.5 \le W \le 16.5$ cm $\quad 13.5 \le H \le 19.5$ cm $\quad -2\pi \le \Delta\alpha \le 2\pi$ rad.

$-2\pi \le \rho_{1,2} \le 2\pi \quad -200 \le x_{1,2}, z_{1,2} \le 200$

Constraints:

- Non-interference constraint

$$\sum_{i=0}^{N_s} \left[\frac{volume(endcap_i \cap environment)}{volume(endcap)} \right] = \varnothing$$

where N_s = the number of disassembly steps
$endcap_i$ = the endcap at its ith step
$environment$ = all other objects

Goals:

- Bin's storage volume goal (maximize)

$$\frac{storage\_volume}{target_{storage\_volume}} - 1 = d_1^+ - d_1^-$$

- Assembly's material volume goal (minimize)

$$1 - \sum_{i=0}^{num\ comps} \left(\frac{target_{volume(comp_i)}}{volume(comp_i)} \right) = d_2^+ - d_2^-$$

- Disassembly path length goal (minimize)

$$1 - \left[\frac{target_{path\_length}}{path\_length} \right] = d_3^+ - d_3^-$$

- Disassembly reorientation goal (minimize)

$$1 - \left[\frac{target_{reorient}}{reorient} \right] = d_5^+ - d_5^-$$

- Disassembly work goal (minimize)

$$1 - \left[\frac{target_{work}}{work} \right] = d_5^+ - d_5^-$$

- Disassembly Path Parametric Distribution (maintain)

$$1 - \left[\frac{target_{param\_dist}}{param\_dist} \right] = d_6^+ - d_6^-$$

Minimize:

$$Z = \begin{bmatrix} 0.4 \cdot d_1^+ + 0.6 \cdot d_2^- \\ (d_3^+ + d_4^+ + d_5^+ + d_6^+ + d_6^-)/4 \end{bmatrix}$$

NOTE: The variable, d, represents the distance (deviation) between the aspiration level and the actual attainment of a goal. The deviation variable, d, can be positive or negative depending on whether over or under achievement has occurred Therefore, the deviation variable is replaced with two variables: $d = d_i^- - d_i^+$ where $d_i^- \cdot d_i^+ = 0$ and $d_i^-, d_i^+ \ge 0$. Accordingly, the objective in the C-DSP is to minimize the deviation from target.

Figure 10. Illustrative Example's Goal-Directed Geometry DSP

5.4 Step 4: Solution of Design Problem

Using the solution mechanism described in Section 4.3 and an injection molding flow path ratio of 115 (Cracknell and Dyson 1993), an "optimized" solution is located for the SPPD problem presented in Figure 10. Initial and final values for the design variables are given in Table 3.

The impact of our problem formulation and solution scheme on the design of the center console's rear module is described in Table 4. As shown, significant improvements were realized with respect to all but two of the key design drivers. The bin's interior volume goal dominated the assembly's material volume goals because of the weighting scenario in the first priority level of the deviation function (refer to Figure 10). Also, the endcap's rotational value decreased because its path length and work goals were more heavily weighted. The convergence of the deviation function is illustrated in Figure 11. Notice that the 2nd level fluctuates until the 1st level converges. Once the first level converges, the 2nd

Table 3
Design Variable Values

Design Variable	Initial Value	Final Value
L (cm.)	27.0	29.97
W (cm.)	13.5	16.485
H (cm.)	16.5	19.442
WT (cm.)	0.27	0.30
ρ_1 (rad.)	0.0	-0.64
x_1 (cm.)	-80.0	-99.5
z_1 (cm.)	30.0	-1.41
ρ_2 (rad.)	0.0	-52.7
x_2 (cm.)	-45.0	-52.7
z_2 (cm.)	30.0	14.2

Table 4
Product/Process Goals: Initial and Final Values

Design Goals	Initial Value	Final Value	Percent Change
Bin's Storage Vol. (cm^3)	5557.4	8892.3	60.0 %
Assy Material Volume (cm^3)	639.5	957.5	-49.7 %
Endcap's Path Length (cm)	119.7	103.3	13.7 %
Endcap's Rotation (rad.)	0.0	2.9	n/a
Endcap's Work: ΔZ (cm)	21.7	4.9	77.4 %

level is improved. This is the concept of lexicographic minimum as discussed in the previous section.

The proper selection of constraints/goals and the development of an accurate and representative deviation function is key to our success in applying SPPD usefully. If the goals, constraints, and deviation functions do not represent the designers' requirements and preferences, any modifications to the design are meaningless. Therefore, we are of the opinion that multiple problem formulations are warranted in order to determine a true solution, or better yet, a region of solutions. Consequently, the most difficult aspect of SPPD via virtual prototyping is not the development of the virtual prototypes or simulations but the development of a gdg-DSP problem which is representative of our design requirements and preferences.

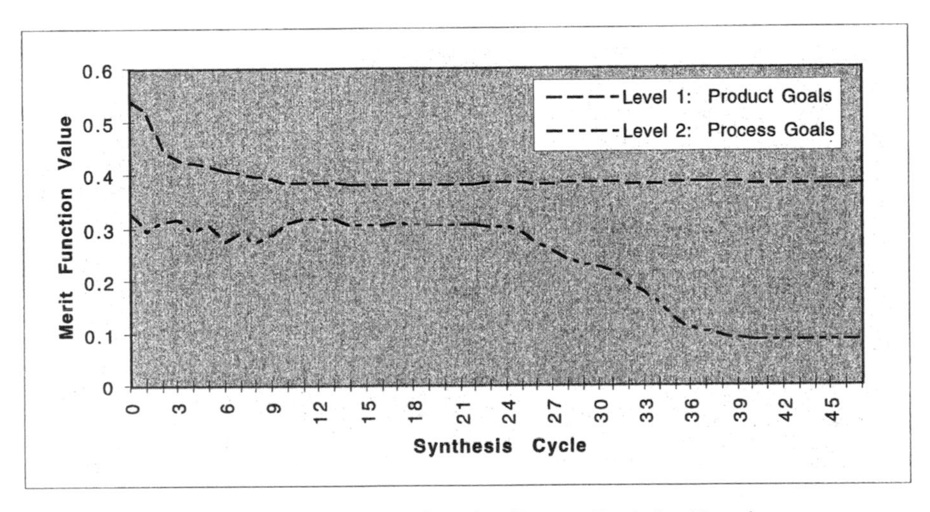

Figure 11. Convergence of Product/Process Deviation Function

6.0 CONCLUDING REMARKS

Within the context of product DFLC, disassembly plays an important role in product service and demanufacture. In this work, an approach to integrated product/process design for disassembly is proposed to facilitate rapid product development in light of these life cycle considerations. This integration is accomplished through the development of design problems which include both product and process design considerations. By first ensuring the satisfaction of product functionality, disassembly processes can be improved by utilizing remaining design freedoms to identify product parameter values that are favorable to disassembly. Central to the success of SPPD is the capability to evaluate product/process requirements via virtual prototyping and the application of optimization techniques to solution of integrated product/process design problems. Currently, SPPD is being applied to the design of an automotive center console. Results to date indicate that SPPD greatly facilitates concurrent improvement of both product and process. Unfortunately, solution of SPPD problems through traditional optimization techniques has proven difficult due to non-linearity of product/process constraints/goals, product/process interactions, and exploration of large, underconstrained design space. Of the optimization algorithms investigated, the most effective solver of SPPD problems has been an algorithm based on HJ Pattern Search Method. This algorithm has been modified to include a lexicographic minimum and orthogonal off-diagonal randomly oriented search direction. Future work will include fastener selection as part of the product/process problem formulation since fasteners strongly influence the ease of disassembly. Fortunately, the development of a coupled selection and compromise Decision Support Problem will enable fastener considerations as well as parametric synthesis of products and their disassembly processes.

ACKNOWLEDGMENTS

The support of the National Science Foundation through grants DMI-9420405 and DMI-9414715 is gratefully acknowledged. Matt Bauer is supported by the Department of Energy through the Integrated Manufacturing Fellowship.

REFERENCES

Atling, L. (1993) Designing for a Lifetime, *Manufacturing Breakthrough,* May/June, 29-33.

Beitz, W. (1993) Designing for Ease of Recycling - General Approach and Industrial Application, *9th International Conference on Engineering Design,* The Hauge, Netherlands, HEURISTS, 731-738.

Boothroyd, G. and Dewhurst, P. (1989) *Product Design for Assembly*, Boothroyd Dewhurst, Inc., Wakefield, RI.

Cracknell, P. S. and Dyson, R. W. (1993) *Handbook of Thermoplastics Injection Mould Design*, Blackie Academic & Professionals, New York.

Dowie, T., Simon, M. and Fogg, B. (1995) Product Disassembly Costing in a Life-Cycle Context, *CONCEPT - Clean Electronics Products and Technology, IEE.*

Emery, F. E. and O'Hagan, M. O. (1966) Optimal Design of Matching Networks for Microwave Transistor Amplifiers, *IEEE Trans. Microwave Theory Tech.*, 696-698.

Ignizio, J. P. (1985) *Introduction to Linear Goal Programming*, Sage University Papers, Beverly Hills, California.

Jovane, F., Alting, L., Armillotta, A., Eversheim, W., Feldmann, K., et al. (1993) A Key Issue in Product Life Cycle: Disassembly, *Annals of the CIRP,* 42(2), 651-658.

Kroll, E. (1995) *Ease-of-Disassembly Evaluation,* Technical Report, Texas A&M University.

Kroll, E., Beardsley, B., Parullan, A. and Berners, D. (1994) Evaluating Ease-of-Disassembly for Product Recycling, *ASME Materials and Design Technology*, 165-172.

Lowe, A. S. and Niku, S. B. (1995) A Methodology for Design for Disassembly, *ASME Design for Manufacturability*, 47-53.

Mistree, F., Hughes, O.F. and Bras, B.A. (1993) The Compromise Decision Support Problem and Adaptive Linear Programming Algorithm, *Structural Optimization: Status and Promise,* 249-289.

Mortenson, M. E. (1985) *Geometric Modeling*, John Wiley & Sons, Inc., New York.

Piegl, L. and Tiller, W. (1995) *The NURBS Book*, Spr-Verlag, New York.

Press, W. H., Teukolsky, S. A., Vetterling, W. T. and Flannery, B. P. (1992) *Numerical Recipes in C*, Cambridge University Press, Cambridge, Massachusetts.

Reklaitis, G. V., Ravindran, A. and Ragsdell, K. M. (1983) *Engineering Optimization: Methods and Applications*, John Wiley and Sons, New York.

Rosen, D. W., Chen, W., Coulter, S. and Vadde, S. (1994) Goal-Directed Geometry: Beyond Parametric and Variational Geometry CAD Technologies, *ASME Design Automation Conference,* DE-Vol. 69.1, 417-426.

Siddique, Z. and Rosen, D. W. (1996) An Approach to Virtual Prototyping for Product Disassembly, *ASME Computers in Engineering Conference*, J. M. McCarthy, Irvine, CA,

AN INTEGRATED SYSTEM FOR FUNCTIONAL MODELING AND CONFIGURATION IN CONCEPTUAL DESIGN[1]

Janis P. Terpenny
Department of Industrial and Systems Engineering
Virginia Polytechnic Institute and State University
Blacksburg, Virginia 24061-0118

ABSTRACT: Concurrent engineering has brought much attention in recent years to engineering design and its impact on issues such as costs, cycle-time, and quality. These issues, coupled with global markets and rapid technology advancements, have created significant need for improved methods and supporting tools for engineering design. To date, advances have focused predominantly on tasks that are well into the latter stages of product development. Advances for early design have been limited. In general, methodologies have taken either a top-down or a bottom-up approach to design, and as such, have virtually guaranteed the continued separation of abstraction and detail during conceptualization. This paper presents an integrated conceptual modeling system that supports a blended methodology from the earliest stages of engineering design. Modeling, synthesis, integration, knowledge capture, and reuse are considered. An example design problem from the domain of power conversion systems is included to demonstrate the utility of the system for technologically sophisticated products.

INTRODUCTION

Background and Motivation

Researchers have spent many years studying design theory and methodology seeking to understand and capture the *how* and *why* of experienced designers. With some variation in defining the exact stages of design, most agree that design processes can be characterized as hierarchical, with synthesis and transformations from abstraction to finalized specification known to be very iterative and information intense. Typical process definitions have included stages such as conceptualization, configuration, embodiment, and detailed design [1].

For many companies, the separation of design into distinct stages has been implemented in organizational structures where engineering disciplines and associated specialties are both physically and functionally grouped. This has led to design practices that are largely sequential, involving several hand-offs as well as transferals of responsibility. As a result, the iterative processes of design

[1] This work was supported by research grants from General Electric Motors and Industrial Systems (GEMIS) and Virginia's Center for Innovative Technology (Virginia's CIT).

have become difficult, costly, fraught with significant delays, and often suffer from poorly coordinated objectives. These practices are a frustration for designers and counter to remaining competitive in today's marketplace.

In truth, the definitive steps of sequential stages, though iterative, are contrary to natural human reasoning processes which require *frequent* movement between abstraction and detail. Conceptual designers naturally need to alternate between steps of reasoning in which detail is initially ignored to focus on high level functional abstractions, and steps in which the addition of detail is necessary for evaluation and design representation [2,3].

As evident in the volume of research literature, there is much interest in developing enabling technologies for improved product design processes. While showing great promise, tools have demonstrated the feasibility of automating particular tasks, but have not necessarily led to an increased understanding or improvement to *overall* product realization processes [4, 5]. To date, advances have most often been for specific tasks which are well into the latter stages of product development. Developments for early design processes still remain largely investigational, specialized, and rarely consider the interconnectedness of abstraction and requirements for detail necessary in a concurrently engineered development process. Much work remains to develop enabling technologies capable of supporting the concurrency and integration requirements so crucial to improved design processes for the early stages of conceptual design.

The Problem

The inability of previous research to adequately address the early stages of conceptual design becomes evident when one considers the apparent dichotomy in current methodologies. Largely, methodologies have approached design either from a top-down or a bottom-up approach, and as such, have virtually guaranteed the continued separation of abstraction and detail during conceptualization.

The top-down approach, as it implies, begins with a top-level view of the design problem. Ideally, based upon a requirements specification and functional characterization, design proceeds iteratively through successive levels of refinement toward the ultimate goal of identifying feasible alternatives. Inherently, this approach satisfies the functional requirements of the design problem. Unfortunately however, there is no assurance that the process will lead to realizable solutions (i.e., real-world embodiments). Frustrated with the time consuming and perhaps fruitless efforts of this approach, designers frequently abandon the effort seeking solutions in the combinatoric domain of detail.

The bottom-up approach to design is based on the selection and configuration of solutions from known components. This approach relies heavily on individual expertise and ingenuity. While realizable solutions are implicit, embodiments which meet functional requirements are not immediately assured. The process is known to be highly iterative. In the case of large or complex designs, the problem quickly leads to combinatorial explosion, inefficiency in the design process, and difficulties assessing the effectiveness and merits of design alternatives.

Overview

With an emphasis on improving the early stages of product development, this paper reports on an integrated system developed for a blended approach to engineering design. This blended approach, it is suggested, 1) improves the quality of design, 2) reduces human resistance to the rigor of top-down design, and 3) reduces the combinatorial explosion often present in bottom-up design. The sections that follow describe the system in terms of approach, uses, and architecture. Examples from the domain of power conversion systems are also included. Evaluating the system against the above mentioned assertions is considered along with conclusions that can be drawn from this research.

APPROACH

Several mechanisms are utilized in defining the framework for blending top-down and bottom-up approaches to engineering design. In particular, three mechanisms that are key to this definition include a:

- functional object modeling environment,
- component knowledge-base, and an
- integrated design domain.

Figure 1 provides a pictorial view of the integrated design domain. As shown, the conceptual modeling system (CMS) is central to the environment. It supports design processes in four important ways. Listed here, these include capabilities to:

(1) create and visualize a conceptual model representation,
(2) configure solutions that meet functional requirements (i.e., synthesis),
(3) collect and re-use knowledge, and
(4) interface with tools and data necessary for analysis and evaluation.

Based on an object-oriented paradigm and semantic reasoning, the framework has been designed in detail. A windows-based user interface has also been developed. The discussion that follows describes the framework and resultant system from a functional viewpoint (i.e., the uses of the system). The architectural view of the system is then considered.

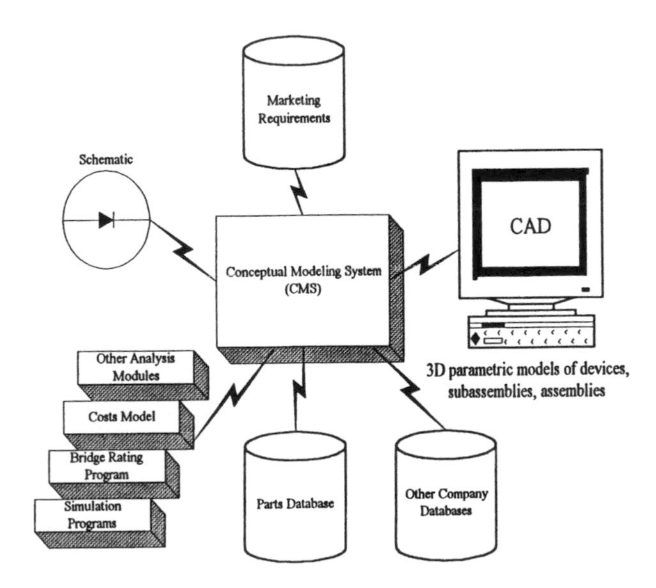

Figure 1. Conceptual design environment

Functional View

As shown in Figure 2, the conceptual modeling system can be described in terms of six primary use cases. Each use case describes a role, or function, that the system provides to the user. (Refer to [6] for more on use case modeling.) A textual overview of each of these use cases follows. Selected screens of the user interface are provided later in the context of example problems.

Use Case 1: Build Conceptual Model. The focus of this use case is on the interactions required to build (compose) a conceptual model. A screen work-area provides the graphical palette for a building-block approach to conceptual modeling. Within the work-area, design objects (building blocks representing design function/intent) may be placed, connected, related, and constrained. The building blocks used in modeling are selected from previously accumulated knowledge that has been classified and stored in terms of function category templates. Modeling actions required to delete functions or relations from the conceptual model as well as file save actions (permitting the exit and return to an in-process model during a later design session) are also supported.

Use Case 2: Maintain Functions (Knowledge Capture). This use case is concerned with the interactions related to the maintenance of function categories (i.e., the templates defining the building blocks available for modeling). This includes adding a new function solution to an existing function category as well as the creation of new categories. Functions are permitted to be basic/atomic (i.e., non-decomposable) or of a higher level. Conceptual models built within the modeling work area serve as the source for the definition of higher level functions.

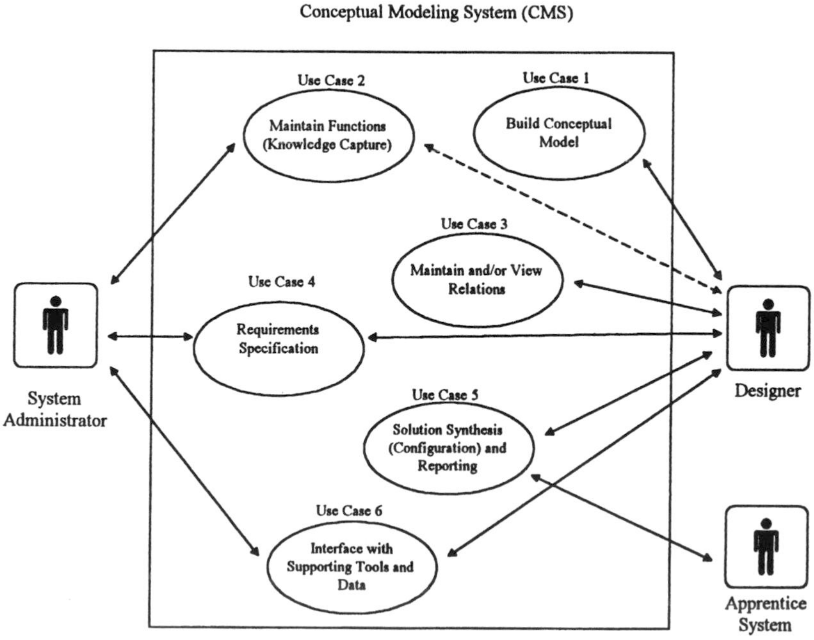

Figure 2. Use cases describing the conceptual modeling system

Use Case 3: Maintain and/or View Relations. This use case is concerned with maintaining and viewing relations available for conceptual modeling. In their simplest form, relations are used to name the connection placed between functional objects in the conceptual model. A relation may also possess additional responsibilities to perform some action when placed in the conceptual model.

Use Case 4: Requirements Specification. Requirements specification data provides the context for the configuration process (i.e., solution synthesis). Modeling tasks can precede the stipulation of this data, but it is imperative that the particular specification is provided prior to any requests for configuration solution. This use case is concerned with the interactions related to defining, maintaining, and viewing requirements specification data.

Use Case 5: Solution Synthesis (Configuration) and Reporting. This use case is concerned with user actions and system responses associated with the solution of the conceptual model. Users of the system have the option to request a configuration solution over a subset of functions present in the conceptual model, or alternatively, choose to configure the complete model. Solution synthesis is constrained by requirements specification data, optional additional constraints specified by the user, and/or implicit constraints imposed by relationships with neighboring functions.

Use Case 6: Interface with Supporting Tools and Data. The focus of this use case is on the temporary exit from the modeling environment to access other tools and/or data that come to bear on the conceptual design process. In the case of designing power conversion systems, resources outside of the conceptual modeling system (refer back to Figure 1) might include programs for simulation, bridge rating, costing, and data sources such as a parts database, marketing requirements, service records, graphical models, and electrical schematics.

Architectural View

Discussion of the architecture for the integrated conceptual design system will be limited in this presentation to the top-level class category diagram presented in Figure 3. Seemingly uncomplex,

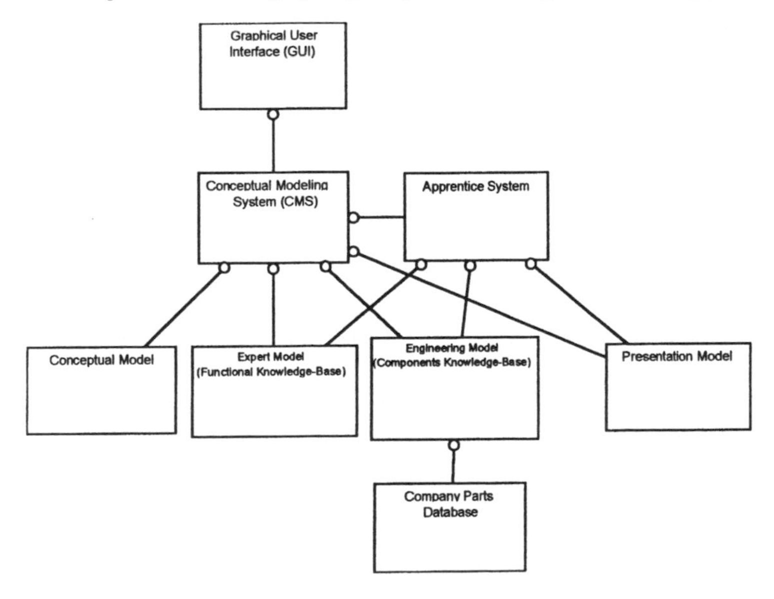

Figure 3. Top-level class category diagram

each class category in the diagram is in reality a subset, or collection, of classes collaborating to provide a set of services. As shown, the categories are arranged in layers with several connections indicating "uses" relationships between categories. The sections that follow provide an overview of these class categories describing the role, or service, that the category provides to the overall system.

Graphical User Interface. The graphical user interface (GUI) is at the highest level of abstraction in the top-level diagram. The role of the GUI is to provide the interface between the user and the conceptual modeling system. Microsoft Visual Basic [7] provided the application programming environment used to build over 20 interfaces for this application.

Conceptual Modeling System. The conceptual modeling system, as shown in the class category diagram, is central to the architecture for integrated conceptual design. The purpose of this category in the diagram is to represent the functional partitioning of the system as well as its relationship to other systems. As implemented, the graphical user interface manages the interactions with class categories shown in "uses" relationships with the conceptual modeling system. Detail of the participating components in the system are described in the class categories that follow.

Conceptual Model. The conceptual model class category is concerned with the model that is current in the work area. The primary role of this category is to manage the conceptual model as it is constructed. The conceptual model also provides input for configuration processes and input for updates to the Expert Model (Functional Knowledge-Base).

Expert Model (Functional Knowledge-Base). The expert model provides two primary services to the conceptual modeling system. First, it provides the foundation for knowledge capture and reuse. Second, the expert model is integral in functional decomposition and solution synthesis procedures.

Engineering Model (Components Knowledge-Base) and Company Parts Database. The engineering model category is concerned with the low-level functions (components) of design. The primary role of this class category is to support configuration processes of the system. The engineering model is used to specify components and interface with company databases in the part selection process.

Apprentice System and Presentation Model. The Apprentice System is a commercially available knowledge-based engineering tool. As shown in the class category diagram, Apprentice System works directly with the conceptual modeling system. Its purpose is to provide the reasoning mechanism (inferencing) for solution synthesis (configuration). At this point, Apprentice handles the presentation of solution synthesis results to the user.

EXAMPLES

Top-Down Approach

The design of power conversion systems, a complex and technically rich process, is included here to illustrate the utility of the integrated system. A top-down approach is considered first. The designer uses the function categories defined in system to locate a high level modeling object. Figure 4 shows an alphabetized list of function categories for the domain of power conversion systems. In this case, the designer selects "Power-bridge". The system opens and displays the template of previously defined solutions (functional knowledge contained in the Expert Model) associated with the power bridge function category (Figure 5). Suppose the designer chooses to let the system advise which type of bridge is appropriate based on the design problem requirements rather than choosing a specific solution type. Thus, the designer selects and places the general building block associated with the power bridge function category in the modeling work area (again, Figure 5).

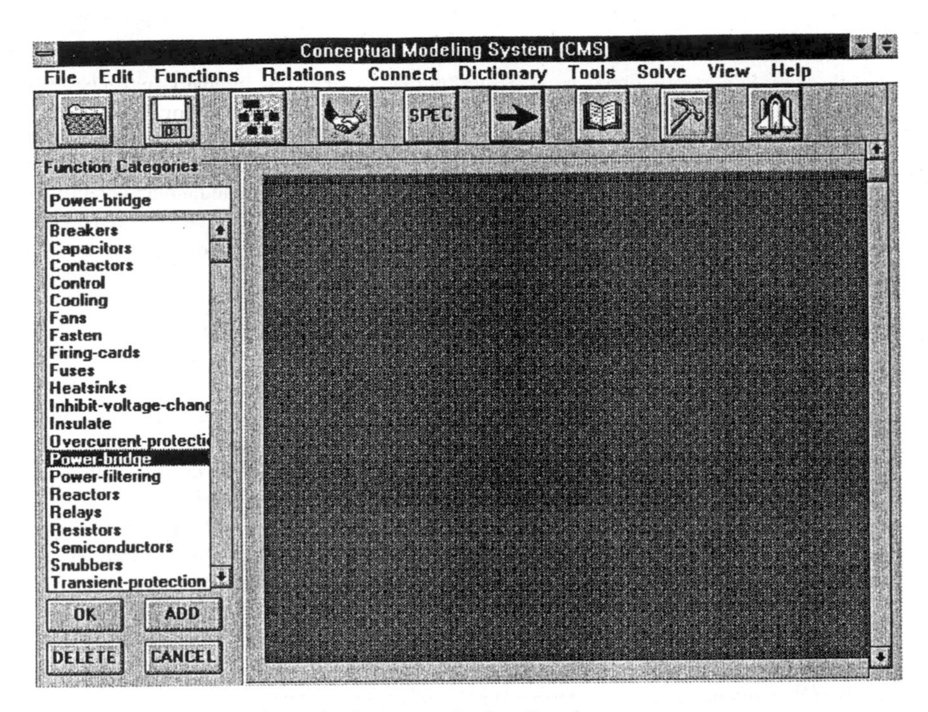

Figure 4. Locate and select function category

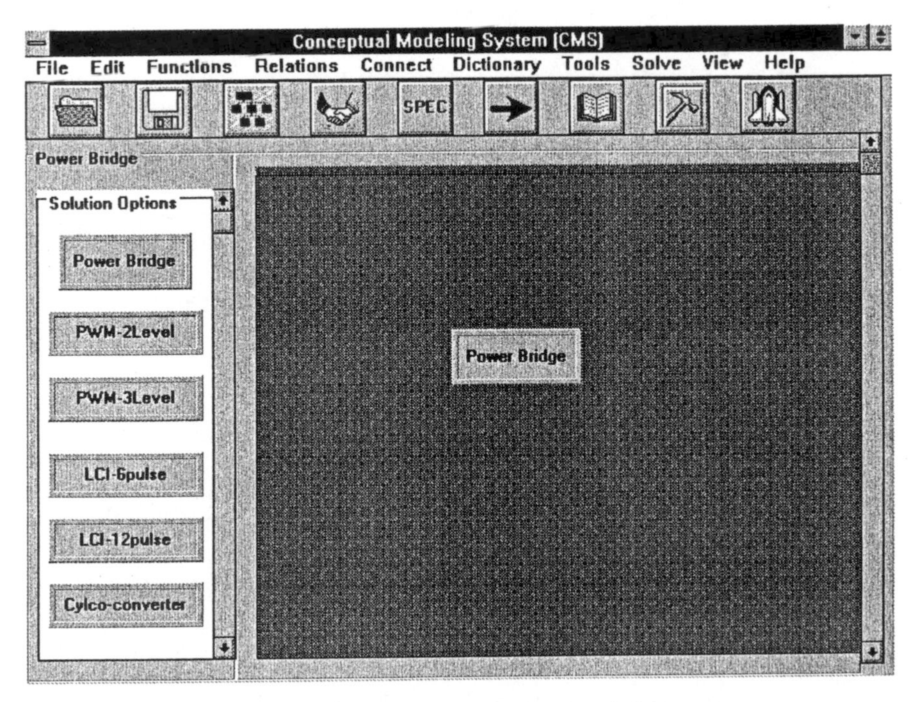

Figure 5. Top-down functional modeling

Next, the designer fills in requirement specification information, any additional selection criteria, and launches the configuration request. The system then invokes inferencing (Apprentice System) based on this Conceptual Model description.

In particular, suppose requirements specification indicate a high system performance is needed (i.e., greater than 3000 horse power (hp) and greater than 4160 volts) and an application requiring the control of a synchronous motor. Inferencing determines (using exists_when conditions of functional solutions in the Expert Model) that the bridge type should be an LCI (Load-Commutated Inverter).

Continuing, Figure 6 illustrates how each step in the functional decomposition process is an independent mapping between the function space and the solution space. With each mapping, the specific solution is a result of exists_when conditions and function-solution relationships present in the Expert Model. As depicted in the first mapping of Figure 6, from all known solutions for Power Bridge, the LCI is selected based on exists_when conditions (e.g., performance and motor requirements). Continuing with solution synthesis, the LCI is itself assembled from six functional building-blocks (transient protection, overcurrent protection, feedback, power conversion, power filtering, and control). Each of these building-blocks is then considered in the functional decomposition process. Figure 6 illustrates the continued decomposition of the power conversion building-block. As shown, it is assembled from power converter, control, cooling, and transient protection. Each of these functions is then considered. In the case of power converter, it is not decomposable (i.e., at the atomic/component level). Based on exists_when conditions (e.g., costs, application, rating) the thyristor solution is recommended as the solution for power converter. Still with a focus on the left side of the figure and a top-down approach, the decomposition process would continue through other sections, assembled functions, and components. In each case, decomposition continues until functional building-blocks are fully decomposed to the component level (atomic functions).

Turning attention to the right side of Figure 6, the Engineering Model refines the specification of component level functions based on sizing rules, selection rules, and/or rules defined for business practice standards. Specific results are dependent on the constraints prescribed in the requirements specification and any additional constraints provided before the solution request. Relationships with other building-blocks can also affect the results of the specification process.

This section has illustrated how the framework accommodates a top-down approach using a power bridge design example. This type of an approach provides the greatest degree of assistance to the design process, assures functional feasibility of solution instances, and poses the highest potential for automated processes. However, this approach does not accommodate solution concepts that deviate significantly from knowledge accumulated in the system. The next section examines how the framework accommodates a bottom-up approach to conceptual design.

Bottom-Up Approach

The conceptual modeling system cannot "invent" knowledge (functional capabilities that are not known) for complex systems such as power converters. It can, however, provide the building-block and integrated environment to aid the design process for such cases.

For instance, consider the situation when requirements come from marketing for a power converter to fill a range of powers where there is no current product capable of covering this range. Unfortunately, it is clear there is a market demand for this range of powers as evident in the increased market share of a competitor following the introduction of a new product offering for this range of powers. In response, a new topology (bridge type) must be developed quickly.

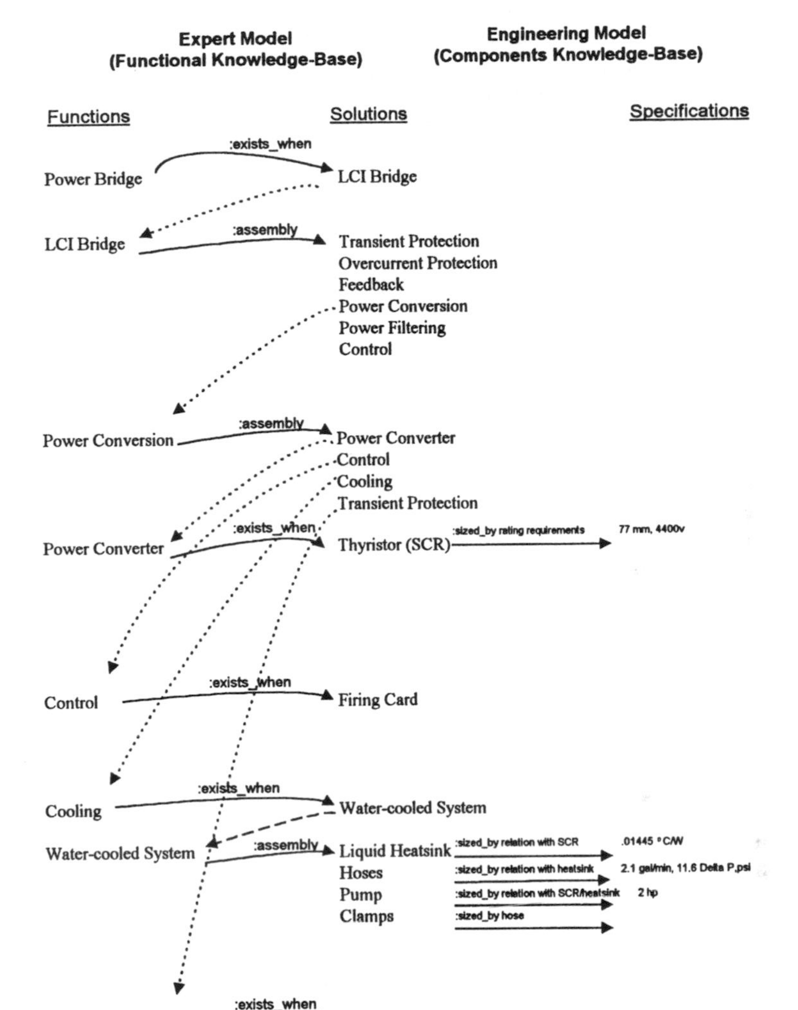

Figure 6. Synthesis and functional decomposition

Designing a new bridge topology requires the expertise of experienced designers, analysis and simulation programs, and a variety of information sources and company databases. In addition, outside vendors may participate in this type of development effort if requirements indicate the need for new components. This is common in power conversion applications where feasibility constraints may require custom heatsinks for cooling, or lead to requirements for new semiconductors to achieve new performance or reliability standards. A description of how the system accommodates this need for a bottom-up approach follows.

As in the previous top-down example, the designer uses the function categories defined in the system to locate design functions desired. For this example, there is no power bridge solution capable of fulfilling the power range. A bottom-up approach must be taken. Figure 7 shows an in-process model with one leg of a power bridge that is under development. As shown, several lower-level function building-blocks have been selected, placed, and interconnected in the modeling work area. To the left of the work area, Figure 7 shows the function category template for semiconductors. Again, rather than choosing a specific type of semiconductor, the general building-block ("Semiconductor" in this case) has been selected and placed in the model.

The designer fills in requirement specification information, additional selection criteria, highlights the semiconductor function in the model, and requests a configuration. The system responds as described previously; using exists_when conditions and rules for sizing and selection. Suppose that for this example, several semiconductor solution types are capable of fulfilling the rating requirements. The designer selects the thyristor (SCR) and then resubmits the configuration request. This time, inferencing invokes the rules of the Engineering Model that size and select the appropriate SCR from the company database.

Next, the designer selects and requests configuration for the heatsink function building-block of the conceptual model. The system returns with the specifications required for the heatsink, but finds no parts available in the company database capable of satisfying this requirement. Continuing to investigate, the designer submits the semiconductor and heatsink component specifications derived by the system to thermal analysis programs. Following several iterations in analysis, the designer is satisfied that specifications for the heatsink will satisfy cooling requirements. Vendors of heatsinks can now be contacted with specifications.

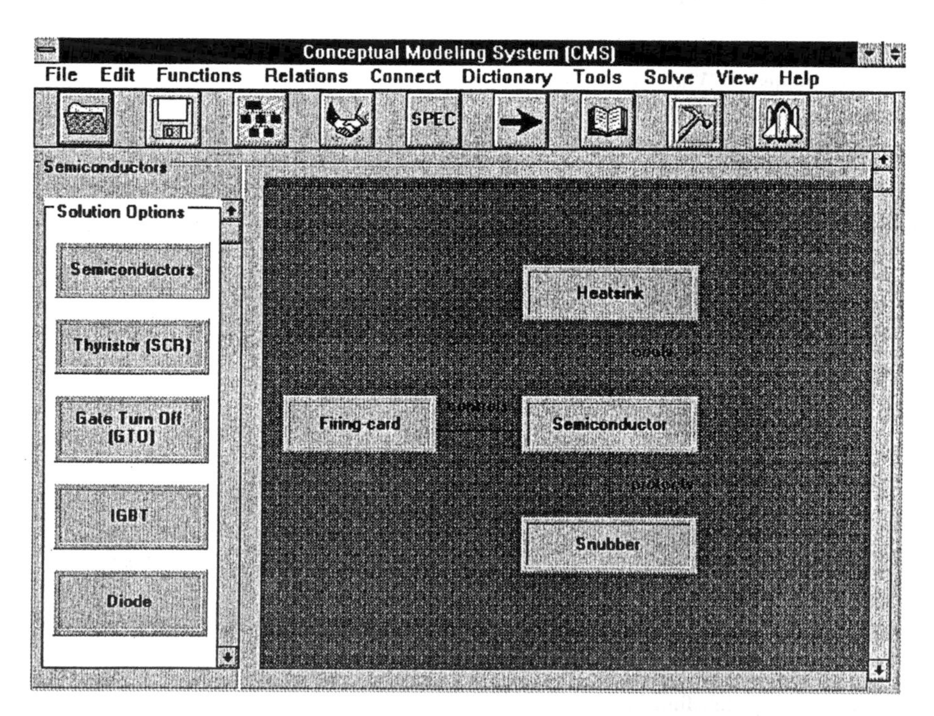

Figure 7. Bottom-up functional modeling

Next, transient protection for the bridge leg (shown as "Snubber" in Figure 7) is considered. In a similar manner to previous iterations, the building-block is selected, constrained, submitted for configuration, and then analyzed. In this case however, snubber is a building-block one level above the component (atomic) level. Thus, the system first determines the appropriate solution for snubber using exists_when conditions. Next, the system considers the capacitor and resistor blocks defining the snubber function. Again, configuration using the system provides the initial specifications of these components. Electrical simulations (circuit analysis) performed outside of the system are then used to validate and refine component specifications. In the end, this iterative bottom-up process of specification, analysis, and evaluation results in a verified solution. The designer then uses the system to capture this model as a functional building-block for reuse in future applications.

This section has illustrated how the framework accommodates a bottom-up approach. As in the top-down example, the example for bottom-up design was related to the design of a power bridge. As shown, the bottom-up approach is most appropriate when new designs are required that deviate significantly from past experience and knowledge. Although this type of approach is limited in the assistance provided from the Expert Model (functional knowledge), much of the highly iterative process of conceptualization is aided by the integrated framework. The system assists the look-up of lower-level functions (components), sizing and selection tasks (Engineering Model), and access to supporting tools and data from the modeling environment. The next section examines blending top-down and bottom-up approaches to conceptual design.

Blending Top-Down and Bottom-Up Approaches

The previous two examples have addressed the design of power conversion systems from what would appear to be opposite directions. In truth, these examples are limited views of a more complete blended methodology. Consider the bottom-up example. It is likely that a top-down approach was first attempted. When no solution was available for the power bridge based on the rating requirements, a bottom-up approach was indicated. Lower-level function building-blocks were combined as a new solution to the higher-level design problem was sought. The system assisted modeling by coupling the low-level functional building-blocks placed in the model with a components knowledge-base (Engineering Model) to aid the configuration process. Designers were not picking parts from a database using a manual trial-and-error method and blindly experimenting; hoping to meet functional requirements. The modeling and synthesis process was assisted by component specifications that were generated as a function of problem requirements and constraints contained within the model description. Thus, although building-blocks placed in the model were low-level (implying a bottom-up approach), the framework supported solution synthesis with a top-down approach (i.e., a systems approach driving requirements to solution).

The bottom-up example also demonstrated how a variety of functions and levels of abstraction can be blended in a single modeling situation. Heatsink, semiconductor, and firing-card were from differing functions, but all at the lowest level of abstraction. Snubber was one level higher and required decomposition prior to specification. This is but a small example of the variety permitted by the modeling system. The designer could select and combine functional building-blocks in a single model from a wide range of abstraction (high and low levels). There is significant benefit to this capability when the design task moves beyond the power conversion section of the power bridge. As discussed in the top-down example, bridge design must also consider high-level functions such as transient protection, overcurrent protection, feedback, power filtering, and control. In many cases, these are functions where all or a part of the solution is common to other bridge types (topologies). It is therefore likely that functional building-blocks and solutions would already exist in the system for these functions. The blended approach allows the designer to couple high-level functions (promoting reuse) with low-level functions (supporting new design) into a single integrated model.

SUMMARY AND CONCLUSIONS

This paper has described an integrated design environment capable of enabling a blending of top-down and bottom-up approaches. Three primary mechanisms defining the integrated design environment were described including: functional modeling, a components knowledge-base, and an integrated design domain. The benefits of the integrated design environment are numerous. Summarized here, features and associated benefits include:

- A building-block approach provides for knowledge capture and reuse, consistency in design practices, and the ability to visually model with functional objects during conceptualization. Building-blocks embody refined and verified engineering expertise and encourage a systems approach to design.

- The framework is customizable. Though demonstrated in the context of power conversion, the framework holds the potential for defining a rich object vocabulary for modeling in any application domain. With each new application domain explored, lower-level functions provide for freedom and flexibility, while higher-level functions provide the greatest benefit of reuse.

- An Engineering Model that automates the specification and selection of parts eliminates the need for manual trial-and-error practices. This greatly reduces the combinatoric effects often present in bottom-up design.

- An integrated modeling environment permits access to valuable sources of tools and data outside of the system. This supports not only reductions in design cycle-time, but improved performance for products that have considered valuable company data related to other life-cycle processes.

- The decoupling of knowledge and data sources provides for greater generalization, reduced system maintenance, and continuous quality improvement.

- And most significant, the framework enables blending top-down and bottom-up approaches to engineering design. This caters to both the need for abstraction and detail that is so important to early design practices; overcoming the difficulties experienced with the exclusive practice of either a top-down or bottom-up approach.

REFERENCES

1. Pahl, G., and Beitz, W. *Engineering Design - A Systematic Approach*, Springer-Verlag, New York, 1988.
2. Paz-Soldan, J.P. and Rinderle, J.R. The Alternate Use of Abstraction and Refinement in Conceptual Mechanical Design, *Proceedings of the ASME Winter Annual Meeting,* ASME, San Francisco, CA, December 1989.
3. Szykman, S., and Cagan, J. A Computational Framework to Support Design Abstraction, *Proceedings - 4th International Conference on Design Theory and Methodology*, 1992 ASME Design Technical Conferences, Scottsdale, Arizona, Sept. 1992, DE v 42, pp. 27-39.
4. Balkany, A., Birmingham, W.P., and Tommelein, I.D., An Analysis of Several Configuration Design Systems,. *Artificial Intelligence for Engineering Design, Analysis, and Manufacturing (AI EDAM)*, v 7, n 1, 1993, pp. 1-17.
5. Keirouz, W., Pabon, J., and Young, R. Integrating Parametric Geometry, Features, and Variational Modeling for Conceptual Design, *Proceedings - Design Theory and Methodology Conference*, ASME, Chicago, Sept. 1990, DE v 27, pp. 1-9.
6. Jacobson, I. *Object-Oriented Software Engineering, A Use Case Driven Approach*. New York, Addison-Wesley, 1994.
7. Microsoft Corporation. *Microsoft Visual Basic Programming System for Windows*, Version 4.0, United States, 1995.

PLUG AND PLAY FRAMEWORK FOR COMBINATORAL PROBLEM HEURISTICS

Andrew Wooster
Panasonic Technologies
Kyushu Matsushita Electric Research Lab
Research Triangle Park, NC, USA

ABSTRACT: Combinatorial problems such as the traveling salesperson, assignment and set-covering problems often rear their heads in process optimization. Because these problems are NP-hard, it remains unlikely that they will yield global optima at reasonable computational cost. Therefore, we must rely on heuristic techniques which seek good, albeit sub-optimal, solutions at an acceptable cost. To this end, meta-heuristic search strategies are interesting because they require very little problem specific knowledge. What little they do need to know about a problem can be encapsulated in a neighborhood operator which provides a mechanism for seeking local optima. Unfortunately, these operators can be very difficult to implement and in this paper we define a practical interface for enumerating neighbors. Once a neighborhood operator has been defined, dynamic strategy selection becomes possible. This means that meta-heuristic strategies, such as simulated annealing and tabu search, can be plugged into problem contexts at run-time.

INTRODUCTION

The factory automation domain is rife with combinatorial process optimization problems such as the traveling salesperson problem, the assignment problem, and set-covering problem, to name but a few. A manufacturer, already facing the capital expenses of high-tech equipment, can be crippled with inefficient operations or under-utilization if these problems are not addressed. Fortunately, over the past three decades much research has focused on these issues and a number of heuristic optimization strategies have been developed. These heuristics come in many forms; operations researchers tend to concentrate on methods that efficiently solve combinatorial problems, while computer scientists are more interested in techniques that quickly provide good, but sub-optimal solutions.

Unfortunately, while much work has addressed specific process problems, researchers have paid little attention to the robustness of their solutions. Most solutions are very problem specific, either from a strategy or design standpoint. That is, the heuristic strategies employed typically optimize only a single problem, or their designs tend to yield little reuse. Most often researchers fail to separate the problem at hand from its optimization strategy, thereby making it impossible to apply different strategies to the same problem.

Furthermore, there has been a tendency to revisit the same problems, over and over, each time with only slight variation from the previous solution. The traveling salesperson problem, for example, has been revisited a myriad times, each version representing a new set of constraints. Most recently, however, products like Ilog's Solver® have effectively addressed the issue of constraint specification and propagation. Separating constraint specification from problem description makes the reuse of objects representing problems and strategies feasible. By isolating problems, constraints and strategies it becomes possible to imagine an environment in which complex problems can be dynamically built from a library of generic problems, and composite strategies can be constructed from a tool kit of heuristic strategies.

Accordingly, the goal of this paper is to provide a framework that meets the three following objectives:

- Simplify the modeling of manufacturing domain problems.
- Identify a small set of common generic problems.
- Separate combinatorial problems from their optimization strategies.

Our interest in this framework resulted from the discussion of a special class of heuristics called *meta-heuristics* in "Modern Heuristic Techniques for Combinatorial Problems." [4] Members of this class hold the promise of problem independence; they can be applied to any combinatorial problem with only a minimal well-defined interface to the problem context. With this in mind, we set about realizing an environment where multiple strategies could be dynamically plugged into complex problems.

COMBINATORIAL PROBLEMS

Combinatorial problems involve finding optimal solutions to problems which can be expressed as a function of discrete decision variables in the presence of constraints. They can be formulated as follows:

$$\begin{aligned} &\text{Minimize} && f(x) && \\ &\text{such that} && g_i(x) \geq b_i; && i = 1, \ldots, m; \\ & && h_j(x) = c_j; && i = 1, \ldots, n; \end{aligned}$$

Where, x is a vector of discrete decision variables, and $f()$, $g_i()$ and $h_j()$ are general functions.

Consider partitioning the multidimensional solution space of a combinatorial problem; each partitioned subspace possess its own "local" optima. These local optima may also be "global" optima, but this can only be determined by searching the entire solution space. Searching among these "local" optima can eventually lead to a "global" solution and brings us to the introduction of the concept of a *neighborhood.*

A neighborhood $N(\mathbf{x},\sigma)$ of a solution **x** is a set of solutions that can be reached from **x** by a simple operation σ. An operation σ might be the interchange of objects in a solution, such as a route swap in the traveling salesperson problem. If a solution **y** is better than any other solution in its neighborhood $N(\mathbf{y}, \sigma)$, then **y** is the local optimum with respect to its neighborhood. [4]

Problem	**Description**
Allocation	n nodes, m tasks $c_{i,j}$ is cost of node i doing task j x_{ij} indicates {0,1} if task j is assigned node i. Find an allocation $\{ \{ x_{11}, \ldots, x_{1m} \}, \{ \ldots \}, \{x_{n1}, \ldots, x_{nm} \} \}$ where $\Sigma_j c_{ij} = 1$ for all nodes i and which minimizes $MAX_i \; x_{ij} \Sigma c_{ij}$ Here the solution is represented by a table $\{ \{ x_{11}, \ldots, x_{1m} \}, \{ \ldots \}, \{x_{n1}, \ldots, x_{nm} \} \}$ indicating with a 1 at x_{ij} if task j is assigned to node i.
Assignment	n nodes, n tasks c_{ij} is the cost of node i doing task j Find an assignment $\{\pi_1, \ldots, \pi_n\}$ which minimizes $\Sigma \; c_{i\Pi I}$ Here the solution is represented by the permutation $\{\pi_1, \ldots, \pi_n\}$ of the numbers $\{1, \ldots, n\}$.
Set Covering	X is a finite set, F subsets of X (F can be very large i.e. $\|F\| = O(2^{\|X\|})$) $X = \cup S$ where $S \in F$ Find a minimum size subset $C \subseteq F$ whose members cover X: $X = \cup S$ where $S \in C$
Traveling Salesperson	Undirected graph $G = (V,E)$ Cost $c(u,v)$ associated with each edge $(u,v) \in E$ Find a tour (hamiltonian cycle) of G with minimum cost. Furthermore, for total cost $c(A)$, where $A \subseteq E$ $c(A) = \Sigma\, c(u,v)$ where $(u,v) \in A$

Table 1. Combinatorial problems in the manufacturing domain.

Table 1 above lists four common combinatorial problems encountered in the manufacturing domain. In addition to these problems it is possible to imagine *composite problems*, that is, problems that are formed by aggregating together sub-problems. [1] Composite problems offer the benefit of not requiring a neighborhood operator; as long as neighborhood operators exist for each sub-problem.

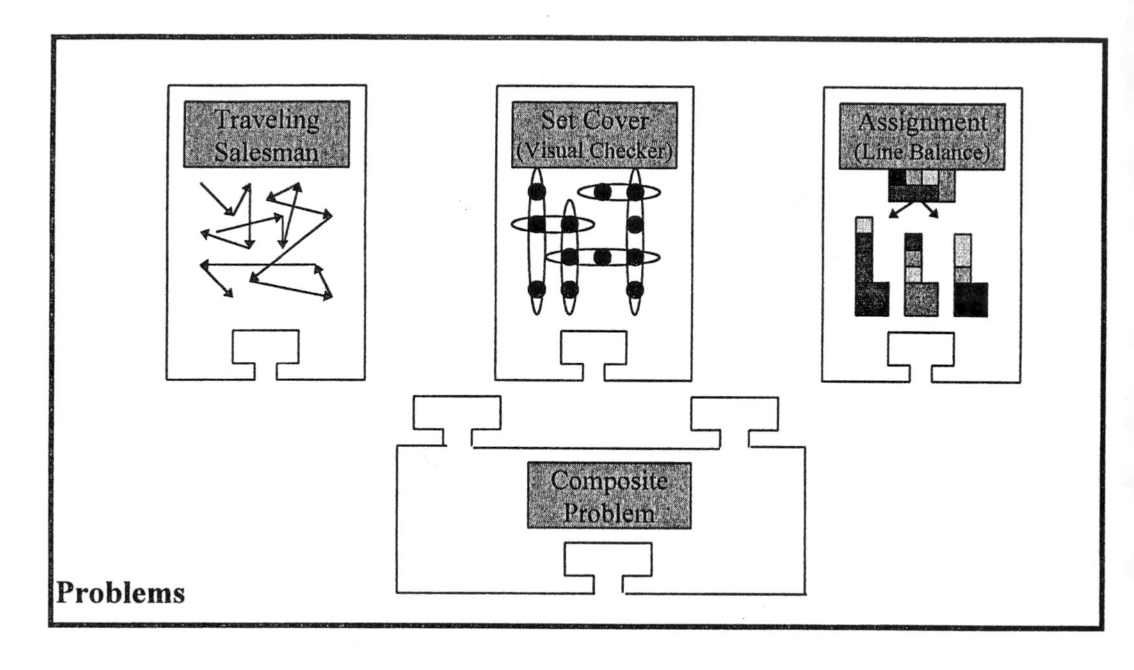

Figure 1. Much like the jigsaw pieces in this figure, composite problems can be built by plugging together sub-problems that have well defined neighborhood operators.

An example of a composite problem is the view minimization problem (VMP). [5] The VMP is common to automated visual inspection machines that have a limited viewing area. In order to optimally perform their inspection, these machines must view all regions of interest on the object being inspected, in minimum time. This is achieved by minimizing the number of views required to cover all regions of interest and minimizing the cost of touring each view. This can be formulated as follows:

Minimize	$a * \lvert V \rvert + \Sigma^{i<\lvert V \rvert} c(V_{i+1}, V_i)$
where	V is the set of views.
	$c(V_1, V_2)$ is returns the cost of traversal between two views.

With regard to problem composition, it is important to realize that the VMP can be decomposed into the more common set-covering and traveling salesperson problems. Minimizing the number of views can be performed by minimizing set-covering, while the cost of touring these views can be optimized via the traveling salesman, thereby obviating the need for a VMP neighborhood operator. Because insight into a neighborhood operator is often non-trivial, problem composition here represents a significant cost savings over defining the VMP from scratch.

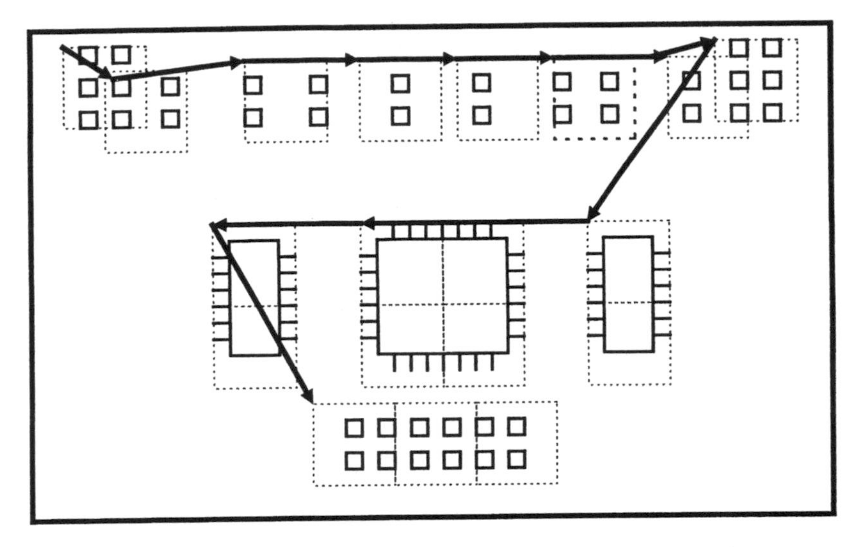

Figure 2. A view minimization problem during PCB inspection.

Like most combinatorial problems, the view minimization problem as well as all of the problems listed in Table 1, have been shown to be *NP-hard* and are therefore intractable. This means that it is unlikely that an optimal solution can be found at reasonable computational cost for anything but the smallest problems. However, it does not mean that good sub-optimal solutions cannot be found after reasonable effort. By applying heuristic strategies it is possible to derive near optimal solutions.

HEURISTIC STRATEGIES

Heuristic optimization strategies fall into two general categories: problem specific heuristics and meta-heuristics. Problem specific strategies involve insight into the problem and often take advantage of some optimal sub-structure. Greedy heuristics, for example, are problem specific heuristics that always make the choice that looks best at the moment. A case in point is the *nearest neighbor* algorithm for the traveling salesperson problem; this algorithm repeatedly finds the node closest to the last node added to the path, and then adds this node to the end of the path. [2] Unfortunately, a heuristic like this applies to a single problem only and cannot typically be used in any other context.

Meta-heuristic strategies, alternatively, can be applied to almost any combinatorial problem and are therefore of great importance to our framework. They differ from problem specific heuristics in that they require no real insight into the problem at hand; rather, they rely on a neighborhood operator interface. As we noted above, neighborhood operators can provide a mechanism for searching a solution subspace for a local optimum. Meta-heuristics typically implement a schedule that seeks a global optimum by pursuing neighborhood optima, but also rely on stochastic perturbations to avoid becoming "trapped" in some local optima. Examples of meta-heuristics include: simulated annealing, tabu search and genetic algorithms.

The problem independence of meta-heuristics offers several advantages. Most notably, they make it possible to dynamically apply multiple strategies to a single problem representation. For example, both the simulated annealing and the tabu search meta-heuristics can be applied to a single specification of the traveling salesperson problem without program code change. Thus we can imagine dynamic graphical user interface in which both problems and strategies are represented by icons. Optimization would be performed by simply dragging a strategy icon and dropping it on a problem icon; results could be monitored in real time, guiding the user's strategy selection.

One drawback, however, of the meta-heuristics is that they tend to require long running times. Because they lack real insight into problems, these strategies must explore a large set of solutions, many of which may be fruitless. Composite strategies described below offer one technique for ameliorating this problem. Nevertheless, it is very important that a user be given some freedom to control the amount of time spent on a problem. When we implemented this framework, we employed preemptive multi-tasking in order to allow the user to specify the duration he or she is willing to give to the problem. Aggressive optimization requires ample time.

Composite strategies, like composite problems, can be assembled by combining sub-strategies. This often results in a efficient high-quality optimization technique that builds on the strengths of its member strategies. One effective composite approach relies on a simple "quick and dirty" heuristic (e.g. nearest insertion) to weed out poor solutions, before turning its results over to a meta-heuristic (e.g. simulated annealing) for final optimization. Such preprocessing not only greatly reduces the amount of time required by a meta-heuristic but, in addition, often yields higher quality results than when applying a single procedure

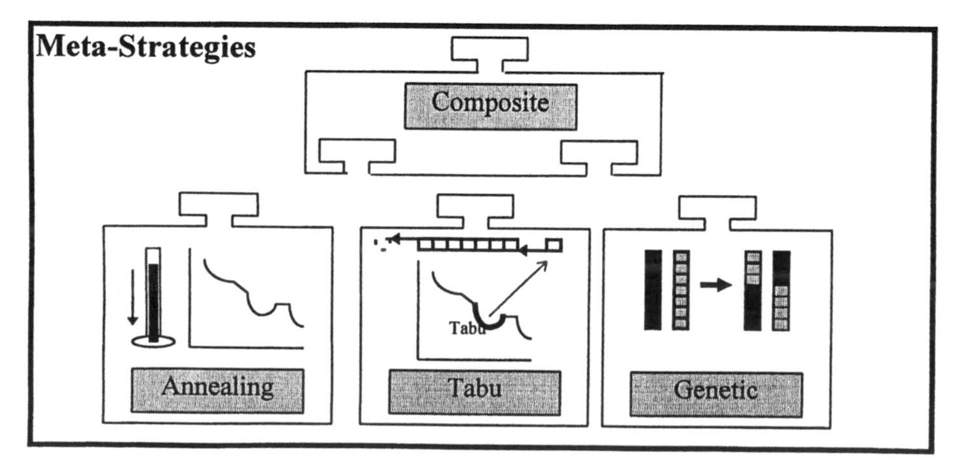

Figure 3. Composite strategies can be built by plugging together sub-strategies.

THE PLUG AND PLAY FRAMEWORK

We use OMT notation [3] below to describe the framework's object-oriented design. There are three main classes of objects: MetaOptSolutions, MetaOptNeighborhoods, and MetaOptStrategies. *MetaOptSolutions* describe problems and internal data structures must represent a solution instance to the problem. At a minimum, a solution must also provide access to a neighborhood operator that allows it to be altered; it may be further augmented with methods that allow for efficient manipulation of its solution. *MetaOptNeighborhoods* implement a minimal interface that allows for neighbor evaluation and selection. Because these operators are used extensively, they must be implemented for maximum efficiency. This means that they are free to take advantage of all OptSolution methods and may, in fact, impose some requirements upon their OptSolution. *OptStrategies* implement general optimization strategies, while the *MetaOptStrategy* subclass implements meta-heuristic strategies that rely on a neighborhood operator.

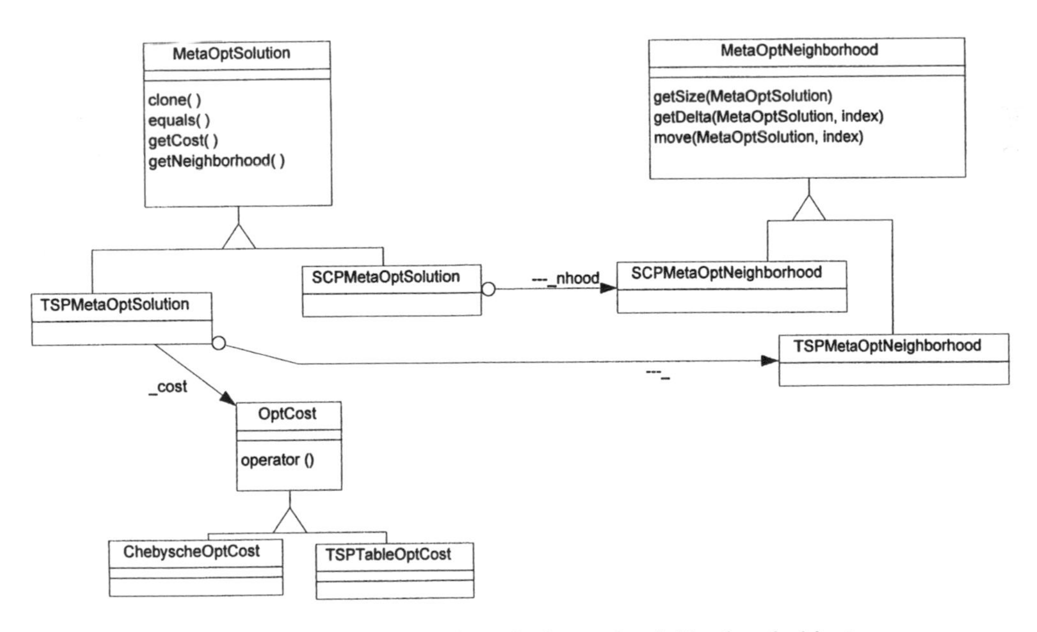

Figure 4. OMT diagram for solution and neighborhood objects.

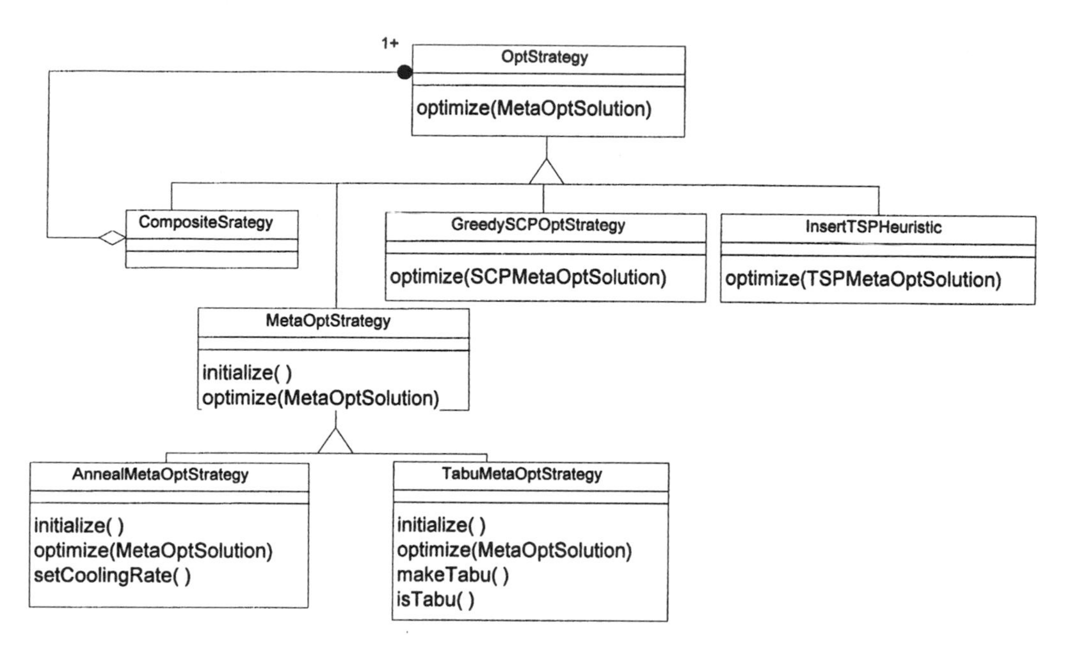

Figure 5. OMT diagram for strategy objects.

The essential connection between Solution and Strategy is provided by the Neighborhood object. Table 2 below lists the three methods that make up the Neighborhood interface, all three of which must be furnished whenever a new neighborhood operator is written. Basically, these methods provide a means for enumerating neighborhood solutions. This enumeration can be randomly traversed using an integer index that ranges from 0 to the neighborhood size. In many cases it is far less computationally intensive to calculate the change in the objective function of moving to a neighboring solution, than it is to recalculate a solution's cost from scratch. It is for this reason, that the interface allows for probing the relative effect of moving to a neighbor with the `getDelta()` method.

Method	**Description**
`size()`	Returns the size of the neighborhood (i.e. the number of neighboring solutions).
`getDelta(int index)`	Returns the change to the objective (cost) function upon moving to a specified neighbor.
`move(int index)`	Moves to the specified neighboring solution.

Table 2. Neighborhood methods.

Neighborhood enumeration can be non-trivial; many problems require some form of *implicit enumeration*. For some sequencing problems, such as the traveling salesperson problem listed below, it is possible to use *a pairwise interchange*. Listing 1 demonstrates this using the *route swapping* technique. Here the neighborhood consists of all solutions that result from reversing the route between any two nodes. Therefore, for a traveling salesperson problem of n nodes this operator enumerates $O(n^2)$ solutions.

```
public final class TSPMetaOptNeighborhood implements MetaOptNeighborhood
{
    public int size ( MetaOptSolution sol )
    {
        TSPMetaOptSolution  tsp  = (TSPMetaOptSolution) sol;
        int                 size = tsp.size();
        return ( size * size );
    }

    public double getDelta ( MetaOptSolution sol, int index )
    {
        double              result = Double.NaN;
        TSPMetaOptSolution  tsp  = (TSPMetaOptSolution) sol;
        int                 size = tsp.size();
        int                 i1, i2, i3, i4;
        double              c1, c2, c3, c4;
        i2 = index/size;
        i3 = index%size;
        if ( tsp.getElemPrev(i2) != tsp.getElem(i3) )
        {
                c1 = tsp.getElemCost( tsp.getElemPrev(i2),tsp.getElem(i2) );
                c2 = tsp.getElemCost( tsp.getElem(i3),tsp.getElemNext(i3) );
                c3 = tsp.getElemCost( tsp.getElem(i2),tsp.getElemNext(i3) );
                c4 = tsp.getElemCost( tsp.getElemPrev(i2),tsp.getElem(i3) );
            result = (c3+c4) - (c1+c2);
        }
        return ( result );
    }

    public void move ( MetaOptSolution sol, int index )
    {
        TSPMetaOptSolution  tsp  = (TSPMetaOptSolution) sol;
        int                 size = tsp.size();
        int i,j;
        i=index/size;
        j=index%size;
        tsp.reverseSubRoute(i,j);
    }
}
```

Listing 1. An example of a traveling salesperson neighborhood operator in Java.

We also list an implementation of the simulated annealing optimization strategy (Listing 2). It is important to note that this class has not problem dependencies; the only interface to the solution is via its neighborhood operator. Essentially, this simulated annealing performs local neighborhood searches, but randomly perturbs the solution based on a cooling schedule. Initially the perturbations are frequent, but they subside once the solution cools and approaches stability.

```
public final class AnnealMetaOptStrategy extends MetaOptStrategy
{
    private Random       _random;
    private double       _cooling_rate;

    protected void optimize ( MetaOptSolution sol )
    {
        MetaOptNeighborhood nhood = sol.getNeighborhood();
        double  temprature;
        double  delta;
        int     neighbor_index, size;
        double  cost, best_cost;
        int     count, count_max;
        int     changes, changes_max;
        size = nhood.size( sol );
        best_cost = cost = sol.getCost();
        temprature=cost/Math.sqrt(size);
        while ( temprature > Double.MIN_VALUE )
        {
                count_max = size;
                changes_max = (int) Math.sqrt(count_max);
                for( count=0, changes=0;
                        (count < count_max) && (changes < changes_max);
                        count++ )
                {
                        neighbor_index = _random.nextInt() % size;
                        delta = nhood.getDelta(sol, neighbor_index);
                        if ( !Double.isNaN( delta ) &&
                             ( (delta < 0.0) ||
                               ( _random.nextDouble() <
                                 Math.exp(-delta / temprature) ) ) )
                        {
                                nhood.move( sol, neighbor_index );
                            size = nhood.size(sol);

                                cost += delta;
                                if (cost < best_cost)
                                {
                          best_cost = cost;
                                        changes++;
                     }
                        }
            }
                temprature *=_cooling_rate;
                fireEvent( new TemperatureEvent( this, temprature ) );
            setSolution( sol, cost );
        }
    }
}
```

Listing 2. The simulated annealing strategy in Java.

As the simulated annealing listing above indicates, strategies make repeated calls to neighborhood operator methods. Neighborhood operators must therefore be carefully crafted for efficiency. Keep in mind that a Solution can be a very large object, containing perhaps thousands of nodes, and completely enumerating its neighbors can be time consuming. It is our eventual hope complex problems can be built from a small set of generic problems with efficient neighborhood operators.

CONCLUSION

The framework described in this paper was implemented in an application designed to optimize the view minimization problem for use with the KME VC32 Visual Checker. Final results were very impressive: 20-40% work reduction over existing solutions with an optimization running time of less than 3 seconds on a 200Mhz Pentium Pro personal computer. The high speed of this optimization is due primarily to the effectiveness of the composite strategies employed.

Problem composition, on the other hand, proved less successful. The output from one problem cannot necessarily be piped directly into the input of another problem. Rather, some "glue," or a adapter interface is required in order to link problems together. For example, the output of the set cover problem is a set of subsets, while the input to the traveling salesperson problem is an undirected graph. Some sort of a adapter will have to provide the desired conversion semantics between these two problems. Thus far we have not found a generic model for these adapters, instead they have been developed on a case by case basis. This approach calls into question the feasibility of problem composition.

Despite the issues of problem composition, the framework proved to be highly interactive. Strategies can be selected and constructed dynamically at run-time using a graphical user interface. In addition, once strategy objects where augmented with an event notification ability, users were able to dynamically monitor a strategy's progress. This kind of interactive environment, one that puts a knowledgeable user at the center of problem solving, realizes our vision of plug and play optimization.

REFERANCES

1. Gamma, E. et al, *Design Patterns: Elements of Reusable Object-Oriented Software*, Addison Wesley, 1995.
2. Golden, B. et al, Approximate Traveling Salesman Algorithms, *Operations Research*, Vol. 28, No. 3, 1980.
3. Rumbaugh, J. et al, *Object-Oriented Modeling and Design*, Prentice Hall, 1991.
4. Reeves, C. R. et al, *Modern Heuristic Techniques for Combinatorial Problems*, Wiley, 1993.
5. Wooster, A. et al, Method for Automated Visual Inspection, U.S. Patent pending, 1996.

A MODEL FOR OPTIMISING THE MASTER PRODUCTION SCHEDULE IN MRP SYSTEMS

Alistair R. Clark
School of Business and Management, University of Teesside,
Middlesbrough, TS1 3BA, England.
a.clark@tees.ac.uk

ABSTRACT: This paper develops a linear programming model to assist in identifying a capacity feasible Master Production Schedule (MPS) in Material Requirements Planning (MRP) systems. It aims to take a more sophisticated approach than *Rough Cut Capacity Planning* and yet avoid the trial and error simulation of *Capacity Requirements Planning*. By assuming lot-for-lot production of components, the model keeps down intermediate stock levels and lead times. In addition, by pre-processing data on the end-items' component structure and the utilization of bottleneck resources over each component's production lead time, the model does not need to explicitly take the MRP explosion of components into account during optimization, thus keeping model size to manageable levels. A simple computational example is tested, showing that the rigidity of the model's lot-for-lot component production can be counterbalanced by the judicious use of overtime as identified by a simple extension of the model.

1. INTRODUCTION

In Material Requirements Planning (MRP) production planning systems, the *Master Production Schedule* (MPS) represents a plan for the production of all end-items over at least the MRP planning horizon. It specifies how much of each end-item will be produced in each planning time period, so that future component production requirements and materials purchases can be calculated using MRP component-explosion logic. As such, the MPS has to be feasible so that components can be produced within the capacity available in each time period.

One approach to ensuring MPS feasibility is Rough Cut Capacity Planning (RCCP) which simply calculates the capacity needs of a proposed MPS on a fairly crude level. If there is not enough capacity, the planner generally has to either supply more or alter the MPS to keep within available capacity. This process must be repeated until a feasible plan is found, generally through experience and judgement coupled with trial and error. Not only can this be time-consuming, but even after finding a viable plan, the MPS might be excessively expensive due to high stocks.

RCCP, being an approximate method, does not preclude the later capacity adjustments which are usually necessary [Vollman, Berry and Whybark, 1992; Correll and Edson, 1990]. Such adjustments are made via Capacity Requirements Planning (CRP) which takes the exploded MRP component production plan as input and calculates in a more precise manner how much capacity is needed.

Again if there is not enough capacity in a period, then the MPS must be adjusted. However MRP systems do not provide any tools beyond simulation to help a planner adjust the MPS to be capacity feasible.

It is clear that there is a role here for a capacity smoothing tool. This paper proposes just such a tool in the form of a linear programming model that also has the aim of keeping work-in-progress under control and finished goods inventory to a minimum. The decision variables are the MPS production quantities. The model aims to minimise the finished goods inventory levels, a consequence of the MPS.

The precise capacity requirements of the MPS depend upon the MRP component production plan which in turn depends on the lot-sizes used in the MRP explosion. However, if we assume lot-for-lot (LFL) component production, then the MPS precisely determines the MRP plan and capacity requirements can be calculated. The model constrains such requirements to be within the available capacity of key resource bottlenecks. The calculations of capacity requirements must be time-phased in order to take production lead-times into account.

The component lead-times used in the model must exclude time spent in work-in-process (WIP) inventory between planning periods that is a result of capacity scarcity. In other words, the lead-times used should be lean and not inflated to include estimates of any significant time (i.e., beyond one planning period) spent waiting for a bottleneck resource to become available [Blackstone, 1989; Hill, 1991]. Such inflation will be eliminated since the proposed model will only schedule production for which capacity is available. Clearly this introduces some rigidity if LFL component production is insisted upon. However, if flexible overtime is available then in many instances not only will a feasible MPS be achieved, but also low WIP inventory and reduced total lead-times due to the LFL production.

2. MODEL SPECIFICATION

Consider the following parameters in an MRP system:

the planning periods, $t = 1,2,3,...$

the MPS end-items, $i = 1,...,M$.

the MRP components $i = M+1,...,N$.

the key production resources $k = 1,...,K$.

$S(i)$ the successor (parent) components of each MRP component i (including purchased components), i.e., at levels higher up in the Bill of Materials.

r_{ij} the number of units of component i in each unit of its successor component $j \in S(i)$

$L(i)$ the production stage lead time of each end item and component i, measured as an integer multiple of planning periods, ensuring that a lot produced in period t is available for consumption only at the beginning of period $t+L(i)$.

T the capacity planning horizon specified in number of planning periods.

From these parameters, we can calculate:

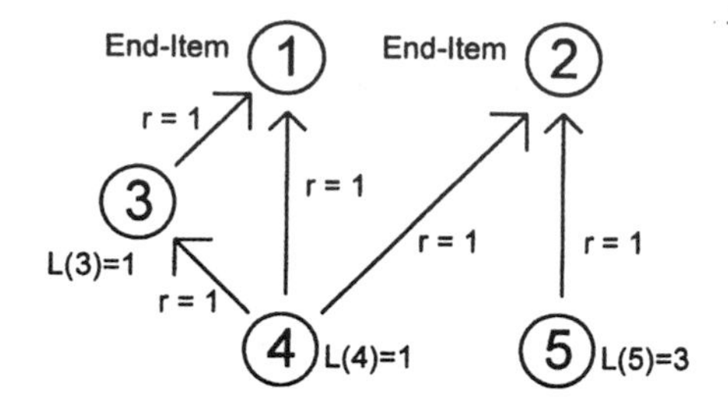

Figure 1. A simple two-product component structure

n_{ij} the number of distinct paths through the MRP bill of materials from component *i* to end-item *j*

$L^n(i,j)$ the sum of the separate component lead times on path *n* from component *i* to end-item *j*, measured in planning periods.

p^n_{ij} the quantity of component *i* used in each unit of end-item *j* on path *n* from *i* to *j*, i.e., the product of the separate r_{ij} on that path.

T(*i*) the sum of the separate component lead-times on the longest lead-time path to component or end-item i from any sub-component of *i* that needs at least one of the key resources *k*.

To better understand the value T(*i*), consider the illustrative two-product component structure of Figure 1

Component 4 is a direct sub-component of component 3 and of end-items 1 and 2. A decision to produce end-item 1 in period *t* causes the LFL production of component 3 in period *t*-1 and hence the production of component 4 in both periods *t*-1 and *t*-2. Thus

$n_{41} = 2$

$L^1(4,1) = 1$ (as a direct component)

$L^2(4,1) = 2$ (as a sub-component of component 3)

$p^1(4,1) = r_{43} \bullet r_{43} = 1$

$p^2(4,1) = r_{41} = 1$

$T(4) = 0$ (it has no sub-components)

In addition

$T(5) = 0 \quad T(3) = 1 \quad T(2) = 3 \quad T(1) = 2$

The value T(*i*) is important as it is the least possible total lead time to produce an end-item *I* from scratch. As such, T(*i*) is the minimum planning horizon for end-item *I* [Mather, 1997; McLeavey and Narasimhan, 1985]. In addition the production of component of end-item *i* before period T(*i*) will be constrained by the inventory availability of some of its sub-components further down in the bill of materials.

The decision variables output by the initial model M1 are

x_{it} the production of each MRP component i in period t

I_{it} the inventory level of component i at the end of period t.

The data inputs to the model M1 are:

d_{it} the independent demand for end-item or component i in period t

h_i the unit per-period inventory holding costs of each component i

$u_{ik\tau}$ the unit requirements by end-item or component i of production resource k, τ periods after beginning production for $\tau = 0,1,2,\ldots L(i)$. A component i may have a lead-time of practically zero in that lot production is very fast and soon available for consumption within the same planning period. In this case, $L(i) = 0$, $\tau = 0$ only and $u_{ikL(i)} = u_{ik0} > 0$. Alternatively, lot production may take a substantial amount of time and be deemed to be available for consumption only at the beginning of the following period. In this case, $L(i) = 1$, $\tau = 0$ and 1, $u_{ik0} > 0$, but $u_{ikL(i)} = u_{ik0} = 0$. For this reason, $\tau = 0,\ldots L(i)$ and not $0,\ldots L(i)$-1.

$b_{kt}^{\min}$ the minimum acceptable utilization of resource k in planning period t.

$b_{kt}^{\max}$ the maximum availability of resource k in planning period t.

x_{it}^{past} for t=1-$L(i)$,...,0, i.e., past production of component or end-item i in period t that is not yet available in inventory.

$I_{i,L(i)}^{known}$ the known projected inventory of component or end-item i based on past production.

Resource setups needs are incorporated into unit requirements to avoid a more complex model and the solution difficulties of combinatorial optimization [Clark and Armentano 1995]. To do so assumes that lot-sizes are predictable and fairly constant. If a component's lot sizes are smaller than expected, then its unit production times will have been understated and model may provide solutions that are capacity infeasible. Similarly, larger lot sizes than expected will result in a production plan that underutilizes capacity. Small variations in lot-sizes should not be a problem, but large ones will need to be corrected with CRP.

Overall the aim is to minimise inventory while remaining within the minimum and maximum levels of capacity. Thus the initial model is:

M1: min $\sum_{i=1}^{N} \sum_{t=1}^{T} h_i I_{it}$

(2.1)

such that

$$I_{i,L(i)+t-1} + x_{it} - I_{i,L(i)+t} = d_{i,L(i)+t} + \sum_{j \in S(i)} r_{ij} x_{j,L(i)+t} \qquad \begin{matrix} i=1,\ldots,N \\ t=1-L(i),\ldots,T+T(i) \end{matrix} \qquad (2.2)$$

$$b_{kt}^{\min} \leq \sum_{i=1}^{N} \sum_{\tau=0}^{L(i)} u_{ik\tau} x_{i,t-\tau} \leq b_{kt}^{\max} \qquad \forall k,\ t=1,\ldots,T \qquad (2.3)$$

$$x_{it} = x_{it}^{\text{past}} \qquad i=1,\ldots,N;\ t=1-L(i),\ldots,0 \qquad (2.4)$$

$$I_{i0} = I_{i0}^{\text{known}} \qquad i=1,\ldots,N \qquad (2.5)$$

$$x_{it} \geq 0; \qquad i=1,\ldots,N;\ t=1-L(i),\ldots,T+T(i) \qquad (2.6)$$

$$I_{it} \geq 0 \qquad i=1,\ldots,N;\ t=0,\ldots,T+T(i)+L(i) \qquad (2.7)$$

In equation (2.2), the production of a component or end-item i is related with:

- its inventory $L(i)$ periods later, since it takes this amount of time to produce the component for entry into stock, be it WIP or spares/finished inventory;
- demand for the component $L(i)$ periods later, be it independent demand $d_{i,L(i)+t}$ or dependent on the production of its successor components $\sum_{j \in S(i)} r_{ij} x_{j,L(i)+t}$.

Expression (2.3) sums the capacity needs of component or end-item i over the distinct periods of its lead time $L(i)$ for each key resource k with the needs of other components and ensures that capacity utilization is kept within acceptable levels.

Model M1 allows lot-sizing during the MRP component explosion, but has at least 2NT variables. In many industrial situations, the model would be very large. Model size can be significantly reduced if we explicitly represent only the MPS end-items, thus lessening the number of variables. To permit this and keep WIP inventory down, lot-for-lot component production is assumed so that the time-phased component production implications are predictable when it is decided to produce a given end-item in a certain period.

Since we assume that the production of a given MRP component is derived on a lot-for-lot basis from the production of its successor components:

$$x_{it} = d_{i,L(i)+t} + \sum_{j \in S(i)} r_{ij} x_{j,L(i)+t} \qquad (2.8)$$

then the production of the MRP components is determined by the production of all the MPS end-items that contain it:

$$x_{it} = d_{i,L(i)+t} + \sum_{j=1}^{M} \sum_{n=j}^{n_{ij}} p_{ij}^{n} x_{j,t+L^{n}(i,j)} \tag{2.9}$$

Further details can be found in [Clark and Armentano, 1993]. Substituting (2.9) into model M1 results in a model being expressed solely in terms of the MPS end-items $j = 1, \ldots, M$:

$$\text{M2:} \quad \min \quad \sum_{j=1}^{M} \sum_{t=1}^{T} h_{jt} I_{jt} \tag{2.10}$$

such that

$$I_{j,L(j)+t-1} + x_{jt} - I_{j,L(j)+t} = d_{j,L(j)+t} \qquad \begin{matrix} j=1,\ldots,M \\ t=1-L(j),\ldots,T+T(j) \end{matrix} \tag{2.11}$$

$$b_{kt}^{\min} \leq \sum_{j=1}^{M} \sum_{\tau=0}^{L(j)} u_{jk\tau} x_{j,t-\tau} + \sum_{i=M+1}^{N} \sum_{\tau=0}^{L(i)} u_{ik\tau} \left[d_{i,t-\tau+L(i)} + \sum_{j=1}^{M} \sum_{n=1}^{n_{ij}} p_{ij}^{n} x_{j,t-\tau+L^{n}(i,j)} \right] \leq b_{kt}^{\max} \qquad \begin{matrix} \forall k \\ t=1,\ldots,T \end{matrix} \tag{2.12}$$

$$x_{jt} = x_{jt}^{\text{past}} \qquad \begin{matrix} j=1,\ldots,M \\ t=1-L(j),\ldots,0 \end{matrix} \tag{2.13}$$

$$I_{j0} = I_{j0}^{\text{known}} \qquad j=1,\ldots,M \tag{2.14}$$

$$x_{jt} \geq 0; \qquad \begin{matrix} j=1,\ldots,M \\ t=1-L(j),\ldots,T+T(j) \end{matrix} \tag{2.15}$$

$$I_{jt} \geq 0; \qquad \begin{matrix} j=1,\ldots,M \\ t=0,\ldots,T+T(j)+L(j) \end{matrix} \tag{2.16}$$

The model M2 is just a starting point and needs to be adapted to the particular reality of the MRP environment being modelled. One possible general disadvantage of the model is that the requirement that production be lot-for-lot could constrain the model so much that the result would be considerable finished stock or, worse, infeasibility. However, most companies do have leeway to alter their capacity, for example, through overtime. This could be reflected in the model as follows:

$$\text{M3:} \quad \min \quad \sum_{j=1}^{M} \sum_{t=1}^{T} h_{jt} I_{jt} + \sum_{k} \sum_{t=1}^{T} c_{kt} o_{kt} \tag{2.17}$$

such that

$$I_{j,L(j)+t-1} + x_{jt} - I_{j,L(j)+t} = d_{j,L(j)+t} \quad \begin{matrix} j=1,\dots,M \\ t=1-L(j),\dots,T+T(j) \end{matrix} \tag{2.18}$$

$$b_{kt}^{\min} \leq \sum_{j=1}^{M} \sum_{\tau=0}^{L(j)} u_{jk\tau} x_{j,t-\tau} + \sum_{i=M+1}^{N} \sum_{\tau=0}^{L(i)} u_{ik\tau} \left[d_{i,t-\tau+L(i)} + \sum_{j=1}^{M} \sum_{n=1}^{n_{ij}} p_{ij}^{n} x_{j,t-\tau+L^{n}(i,j)} \right]$$

$$\leq b_{kt}^{\max} + o_{kt} \quad \begin{matrix} \forall k \\ t=1,\dots,T \end{matrix} \tag{2.19}$$

$$o_{kt} \leq o_{kt}^{\max} \quad \begin{matrix} \forall k \\ t=1,\dots,T \end{matrix} \tag{2.20}$$

$$x_{jt} = x_{jt}^{\text{past}} \quad \begin{matrix} j=1,\dots,M \\ t=1-L(j),\dots,0 \end{matrix} \tag{2.21}$$

$$I_{j0} = I_{j0}^{\text{known}} \quad j=1,\dots,M \tag{2.22}$$

$$x_{jt} \geq 0; \quad \begin{matrix} j=1,\dots,M \\ t=1-L(j),\dots,T+T(j) \end{matrix} \tag{2.23}$$

$$I_{jt} \geq 0; \quad \begin{matrix} j=1,\dots,M \\ t=0,\dots,T+T(j)+L(j) \end{matrix} \tag{2.24}$$

where

o_{kt} is the amount of additional capacity to be supplied through overtime.

$o_{kt}^{\max}$ is the maximum amount of additional capacity that can be supplied through overtime.

c_{kt} is the cost of each additional unit of capacity supplied through overtime.

3. COMPUTATIONAL EXAMPLE

To illustrate the effect of the M2 model, simple tests were carried out using the component structure of Figure 1 over a planning horizon of 12 weeks. The aim was to assess the effect of using the smaller model M2 and its forced LFL component production in comparison with the overtime model M3.

Two key production resources are disputed by all end-items and components. The lead-times for end-items 1 and 2 are 1 and 0 respectively. Given these and the lead times in Figure 1, the data for $u_{ik\tau}$ the unit requirements by end-item or component i of production resource k, τ periods after beginning production are specified in Table 1.

TABLE 1
The Values of the Unit Resource Usage $u_{ik\tau}$

	Resource $k = 1$				Resource $k = 2$			
End-item or Component i	$\tau = 0$	$\tau = 1$	$\tau = 2$	$\tau = 3$	$\tau = 0$	$\tau = 1$	$\tau = 2$	$\tau = 3$
1	1	1	-	-	0	1	-	-
2	1	-	-	-	1	-	-	-
3	0	1	-	-	0	0	-	-
4	1	0	-	-	1	1	-	-
5	1	1	1	1	0	1	0	1

The minimum acceptable utilization of resource k in planning period t, $b_{kt}^{\min}$, is zero for all values of k and t. The maximum availability of resource k in planning period t, $b_{kt}^{\max}$, is 1500 for resource 1 and 700 for resource 2 in all periods. Overtime of up to 100 and 50 units for resources 1 and 2 respectively is available at a cost of 0.5 per unit of overtime.

The past production of end-item 1 in period 0 that is not yet available in inventory, x_{10}^{past}, is 100. The initial inventory of end-items 1 and 2 at the end of period 0 (i.e., at the start of period 1) is 200 and 0 respectively.

The demand for end-item 1 during weeks 1 to 15 oscillates considerably around 100 items/week as follows: 50, 100, 150, 200, 150, 100, 50, 0, 50, 100, 150, 200, 150, 100, 50. The demand for end-item 2 is a steady 100 items/week. Taking the unit echelon stock holdings costs [Clark and Armentano, 1993] of each end-item and component to be 1 per week, the inventory holding cost of end-items 1 and 2 were 4 and 3 respectively.

The optimal inventory levels and capacity usage for both models are shown in Tables 2 and 3. The optimal inventory costs for M2 were 3325 while those for the M3 model were 1850, a 45% reduction. M3's overtime costs were only 400 thus economizing 1075 cost units overall. Thus the judicious use of overtime considerably lowered the levels of finished inventory.

TABLE 2
Optimal Inventory Levels for Models M2 and M3

	Model M2 - Inventory of item:		Model M3 - Inventory of item:	
Week	1	2	1	2
1	150	77.78	150	37.5
2	143.33	46.67	50	33.33
3	33.33	33.33	12.5	12.5
4	0	0	0	0
5	0	0	0	0
6	0	33.33	0	0
7	0	83.33	0	16.66
8	15	120	0	83.33
9	0	150	0	100
10	25	75	0	50
11	0	0	0	0
12	0	0	0	0

TABLE 3
Resource Usage for Models M2 and M3

	Model M2 - Usage of Resource:		Model M3 - Usage of Resource	
Week	1	2	1	2
1	1427	700	1371	717
2	1352	700	1533	750
3	1500	700	1600	750
4	1450	700	1466	750
5	1312	700	1233	667
6	1258	659	1116	567
7	1267	700	1200	617
8	1350	700	1433	750
9	1475	700	1550	750
10	1500	700	1600	750
11	1500	700	1600	750
12	1400	700	1500	750
Mean	1399	697	1433	714

These results, based on just one simple model and data set, merely provide an indication of the usefulness of the basic MPS model and its extensions such as M3. The author intends to test the model in an live production environment with real data and suitable information systems support. Only thus will its merit be established.

4. CONCLUSION

This paper presented a linear programming model for assisting in the challenging task of identifying a capacity-feasible Master Production Schedule in MRP systems. By pre-processing data on the end-items' component structure and the utilization of bottleneck resources over each component's production lead time, the model does not need to explicitly take the MRP explosion of components into account during optimization, thus keeping model size to manageable levels. For many large and complex product structures, the size of the M1 model which does explicitly take the MRP explosion into account would be many times greater than that of M2. A simple computational example showed that the rigidity and high finished inventory levels caused by lot-for-lot production of MRP components can be substantially reduced by selective use of overtime or capacity flexibility. Lot for lot production is also considerably easier to monitor and control on the plant shop floor.

The operational implementation of the models M2 and M3 would require the following elements:

- Pre-processing of data from MRP2 data-bases to calculate n_{ij}, $L^n(i,j)$, p^n_{ij}, and $T(i)$
- A quality fast linear programming optimizer for large MRP systems with many components and end-items
- Graphical presentation of the optimization results, such as the MPS plan, capacity use at key resources, finished goods inventory, inventory, overtime and ideal capacity needs requirements.
- An editor so that the MPS planner can easily alter demand, capacity availability, overtime availability and freeze production when need be.

REFERENCES

Blackstone, J. H., *Capacity Management*, South-Western Publishing Company, Cincinnati, Ohio, 1989.

Clark, A. R. and Armentano, V. A., Echelon Stock Formulations for Multi-Stage Lot-Sizing with Lead Times, *Int. J. Sys. Sci.*, 24, no. 8., 1993.

Clark, A. R. and Armentano, V. A., The Application of Valid Inequalities to the Multi-Stage Lot-Sizing Problem, *Computers and Operations Research*, 22, no. 7, 669-680, 1995.

Correll, J. G. and Edson, N. G., *Gaining Control - Capacity Management and Scheduling*, Oliver Wight Limited Publications, Essex Junction, VT, 1990.

Hill, T., *Production and Operations Management - Text and Cases*, 2nd Edition, Prentice Hall, London, 1991.

Mather, H., *Bill of Materials*, Dow Jones-Irwin, Homewood, Illinois, 1987.

McLeavey, D. W. and Narasimhan, S. L., *Production Planning and Inventory Control*, Allyn and Bacon, Boston, 1985.

Vollmann, T. E., Berry, W. L. and Whybark, D. C., *Manufacturing and Planning Control Systems*, 3rd edition, Irwin, Homewood, Illinois, 1992.

THE DEVELOPMENT OF AN INTEGRATED PRODUCTION AND STRATEGIC PLANNING SYSTEM FOR BUILDING PRODUCTS MANUFACTURING INDUSTRY

Dr Nashwan Dawood and Mr Husam Petros
Division of Civil Engineering & Building
School of Science & Technology
The University of Teesside
Middlesbrough TS1 3BA
UK

ABSTRACT: The Building Products Manufacturing Industry (BPMI) is a major supplier of a wide range of building products to the construction industry. The research project is targeted towards building product manufacturers who supply the market from different factories at different locations. In order to reduce cost of production and maximise profit, products should be made at the most economically viable factories and every region in mainland Britain will be supplied from favoured factories. Management role is, therefore, to plan strategically a production budget for each factory in relation to other factories and the distribution of finished products to regions. The objective of the research is to develop an integrated strategic distribution system that can advise management on strategic and detailed production planning. The system is composed of an integrated data-base module strategic rough-cut capacity planning module, factory module and knowledge rules that direct the decision making process of the system. An industrial case study has been conducted in order to validate the system and expose its potential. It is concluded that the system is beneficial and practical for the BPMI.

Keywords: Physical Distribution, Simulation, Knowledge base, Building materials.

INTRODUCTION

A frequently occurring problem for large manufacturing organisations is the need to rationalise and optimise the utilisation of production facilities and distribution of manufactured products. In the case of similar products being produced at several manufacturing sites (for example, paving blocks), there is a need to optimise production at every factory and satisfy demand regions at very competitive costs. Usually, the production costs at each of the different sites (who produce similar products) involve different levels of fixed and variable production costs. This implies that certain factories might produce products a lot cheaper compared to others. The research project is targeted towards building product manufacturers who supply the market from different factories at different locations. In order to reduce cost of production and maximise profit, products should be made at the most economically viable factories and every county in mainland Britain will be supplied from favoured factories. In the UK, BPMI (Building Products Manufacturing Industry) is a major supplier of a wide range of building materials to the construction industry. It is a capital intensive industry and the instability of the construction market makes it a high-risk business. Dawood (1995) concluded that the rise in the cost of building materials is a major contributory factor to the high bankruptcy rate of contractors. The argument is that while the costs of most manufacturing based products are decreasing, those of building materials keep rising. Emphasis on reducing production and distribution costs has become a priority goal in the BPMI. Almost all major building product suppliers and manufacturers in the UK have several factories that respond to regional demand (Figure 1). In order to reduce the cost of production and increase profit, products should be made at the most economical factories and every region will be supplied from these favoured factories. The role of management, therefore, is split into two parts:

- strategically plan the production budget for each factory in relation to others; and
- plan and monitor the distribution of finished products to regions.

The above points should be achieved in the most economical and efficient way taking into consideration the factories entities in terms of capacity utilisation and production costs.
In this context the prime objective of the research is to develop an integrated strategic distribution system that can advise management on strategic and detailed production planning.

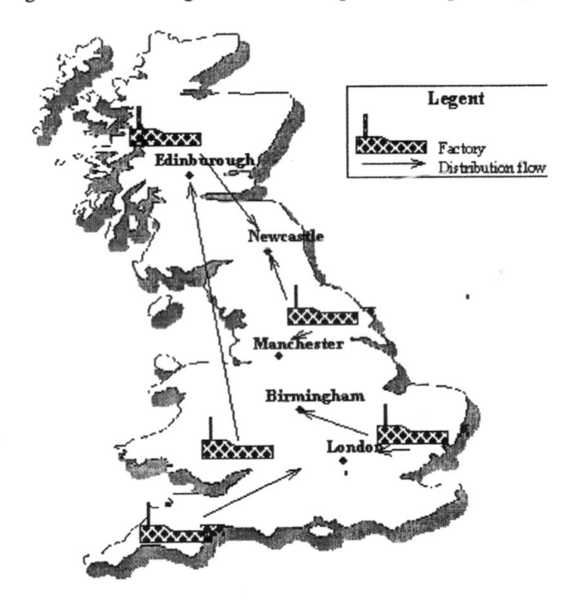

Figure 1. Illustration of physical distribution application

COMPUTER APPLICATION IN THE AREA OF DISTRIBUTION AND STRATEGIC PLANNING: A REVIEW

Bhaskar and Housden (1990) observed that the application of tools to support strategic decision making is still not widespread. Middle and strategic levels of management have always been considered less amenable to automation. The shaded are in Figure 2 shows the relative extent to which computing has been introduced into the different levels of management. This has also been confirmed in a comprehensive study on computer applications in logistic systems by Sijbrands (1993). The utilisation of computers in the transport and physical distribution area has arrived somewhat late relative to other departments in the business firm. There are indications that computers have been used for some period of time as a computational device for routine tasks in distribution. However, the full power of computer applications (Bernard *et al,* 1995) in solving distribution problems and strategic planning systems has not been fully accomplished even at the present time. As physical distribution functions and strategic decisions receive increased attention by top management there is a need to introduce computers at the high levels of decision making process.

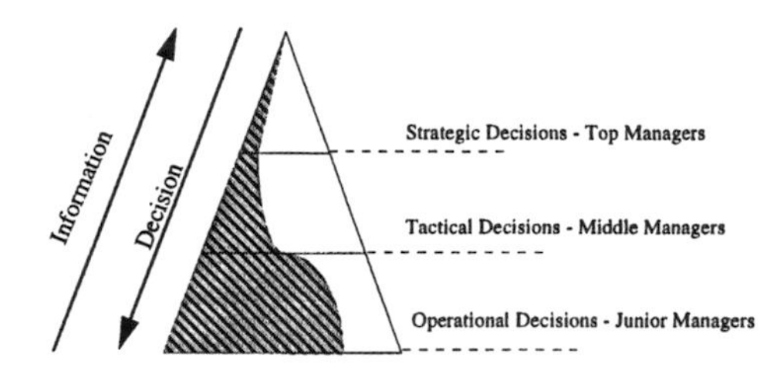

Figure 2. Extent of computerisation in different management levels

RESEARCH OBJECTIVES AND SYSTEM DEVELOPMENT

The prime objective of the research is to develop an integrated strategic distribution system that can advise management on strategic and detailed production planning. The system is composed of an integrated data-base module strategic rough-cut capacity planning module, factory module and knowledge rules that direct the decision making process of the system. In order to achieve the objective, the following tasks have been undertaken:

1. Survey of current industrial practices in order to identify the most critical problems encountered in the production, physical distribution and strategic planning processes in the industry
2. Identify the current research work in the area of strategic planning and distribution systems.
3. Develop system specifications
4. Develop software solution.

The specification of the system is developed for companies (manufacturers) that produce a wide range of standardised building components (in the range of 300 different products) and have a number of factories around Britain. The system is composed of a number of modules: Data-base module, Strategic rough-cut capacity planning module and Factory module. The system also contains knowledge based rules that encapsulate production planning rules which have been developed in the course of this research. These rules direct the planning process of the system. The following sections discuss the modules of the system in greater detail.

The Data base module

The Data base module is developed to encapsulate and calculate the following:

1. Regional demand (RD): This represents the monthly estimated forecasted figures for products for each region in mainland Britain. The method used in this research is to forecast monthly products for 12 months ahead from a national level using the Decomposition method (Dawood, *et al* 1993). The monthly figures will then be split to regional demand using pre-calculated percentages

for each region; a method known as "Budget Regional Split Figures" and is built up from historical information and managerial judgements. As an example, KENT County represents 4% of the national demand for a particular product and if the national demand is 100 tonnes then 4 tonnes goes to KENT County.

2. Haulage Costs (HC): This represents the haulage cost per tonne of delivering the product from a factory to any County, for example it will cost £10.75 to shift one tonne of a particular product from a factory in HULL to KENT. The information is encapsulated in a three dimensional spreadsheet: the X-axes represents the number of factories, the Y-axes represents the regions and the Z-axes represents the number of products. For example, if we have 40 products to be produced in 7 factories that satisfy 70 regions, then 19600 haulage cost figures should be available. These figures are subject to variations and are highly affected by fuel prices, the ages of the drivers, travelling times and maintenance costs of the haulage fleet.
3. Standard Variable Costs (SVC): This represents the cost of production in each factory. It can be differ from one factory to another depending on the efficiency of the plant.
4. Actual Gross Reprisal Price (GRP): This represents the sale price of product groups by region and it can be different from one region to another.
5. Overheads for each factory
6. Profit for each product group per factory.
7. The Total Cost (TC): This represents the cost of producing and shifting a tonne of a product to a particular region.

Identify favourite factories (those with minimum total cost) for each region and theoretically allocate regional demands to those factories. This process should identify the cheapest and ideal solution. The penalty associated with producing products from non-preferred factory will also be calculated.

The Rough-cut capacity planning module

Rough cut capacity planning is an activity that involves the analysis of the master production schedule (MPS). The MPS represents a statement of the plant production level, to determine the implied capacity requirements for manufacturing facilities. Every firm has several critical steps in the production process that need to be carefully monitored as changes occur in the master schedule. At some firms, final assembly or finishing operations represent the critical operations to be closely watched. In others, the critical operations involve the availability of warehouse facilities, equipment and labour forces so that sufficient capacity should be allocated to provide materials when needed.
Having identified the favoured factories (those of a minimum total cost) the next step is to reallocate the production volumes of the factories. The information required in this module interacts with the data provided in the data-base module. Such information indicate the production capabilities in terms of tonnage per product per month and all the information of products being stocked. The output in this stage will indicate the total volume required from each factory to satisfy regions and this is presented as a percentage of the total volume needed for each region as shown in Figure 3. It can be seen in the figure that region 1 will be supplied from factory 1 at 70% and region 3 is supplied from factory 1, 2, and 3 at 30%, 50%, and 20% respectively.

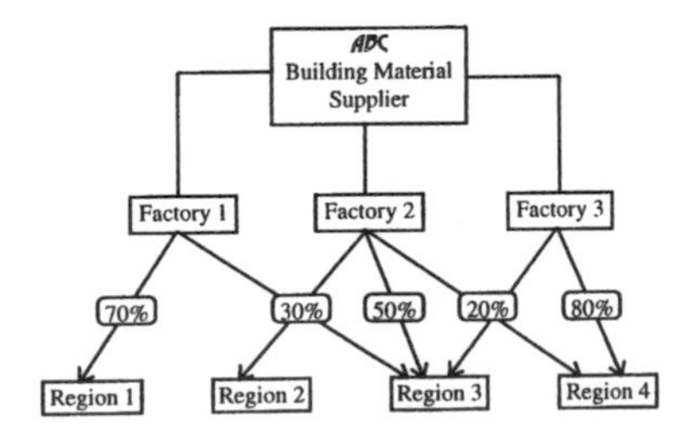

Figure 3. Concept of the Rough-cut capacity planning

Figure 4 represents the steps involved in modelling the process of determining the total costs and production volumes. In order to run the module, window compatible applications is being developed using object-orientation environment.

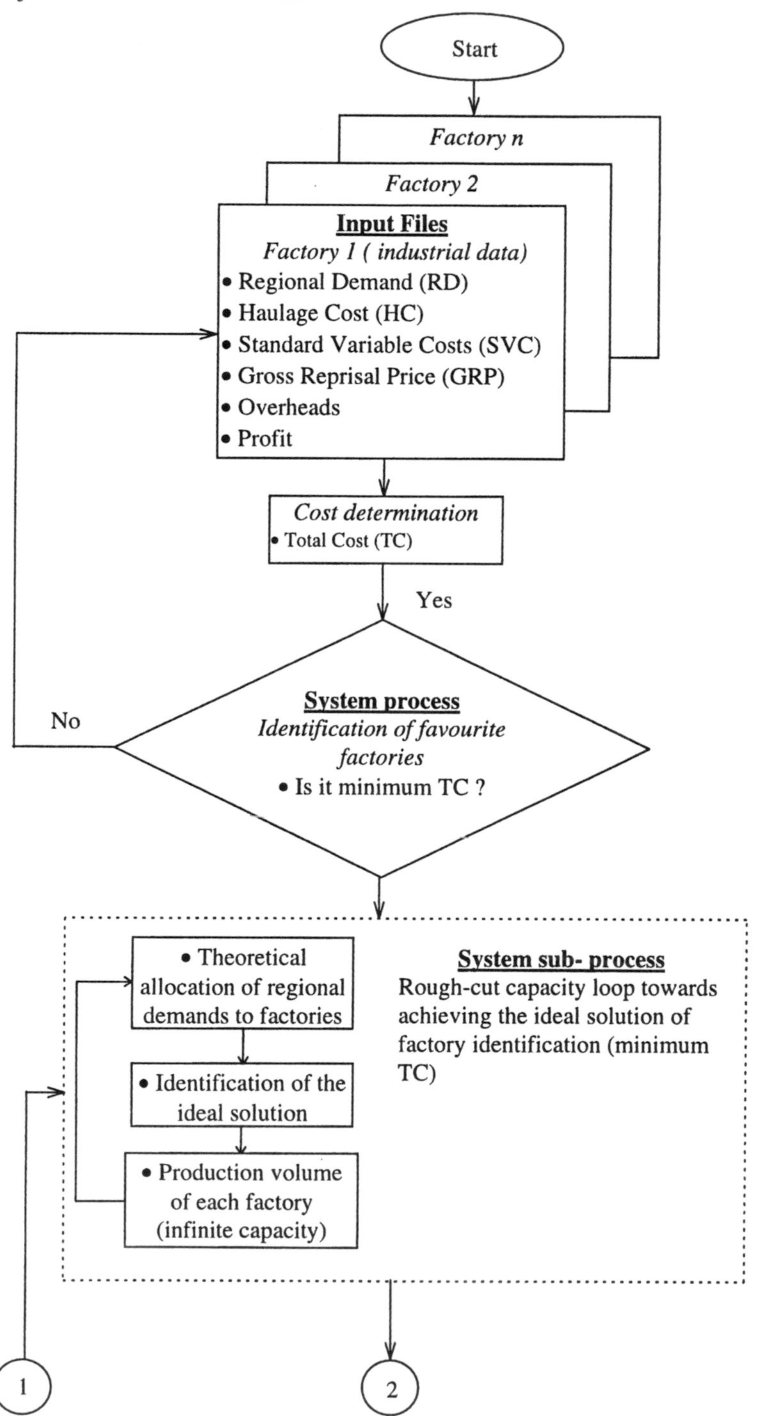

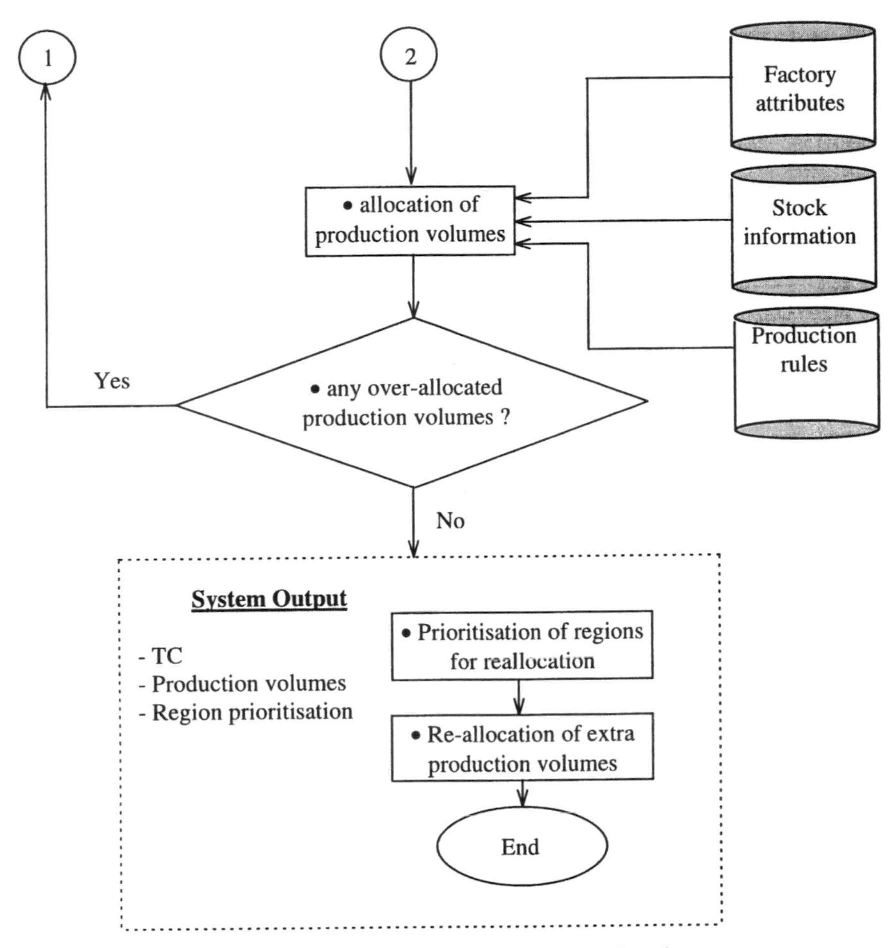

Figure 4. Rough-cut capacity planning flowchart

From the logistical system viewpoint, every firm must have the ability to move materials and finished products between facilities from supply sources and customers. Browersox (1974) identified three factors of importance in establishing the transport capability. These are cost service, speed of service, and consistency of service. The cost of transport results from the actual payment for a movement between two points plus the expense related to having inventory locked in transit. The speed of service relates to the actual time required to complete a transport between two facilities. Consistency of service refers to measured performance of a range of movement between two locations. Two other factors have been added as to the importance of transport capability of such systems and Table 1 shows the performance of transport for each factor. (Slack *et al*, 1995).

Transport capability	Factor of performance
Delivery speed	2
Delivery dependability	2
Quality	2
Transportation cost	3
Rout flexibility	1

Key: 1 = best performance, 5 = worst performance

Table 1. The performance factor of transportation capability

The Factory Module

Once the demand for each factory is determined, the next stage is to produce a production schedule that indicates what, when and how to produce products for each factory in the system. The module is a time-oriented dynamic factory simulator which embodies the knowledge rules and the process of allocating products to the running plant from middle management level for twelve months. The objects of the module are basically entities and attributes. The entities are elements of the system being simulated and they can be individually identified and processed. Plant and products are regarded as entities of the module. Each entity possesses one or more attributes to convey extra information about it. For example certain plants can produce a set of products which others cannot. Another way of using attributes may be to control queue discipline. This priority may be used to select products for processing when there is a choice. The main attributes of the module are:

- The module is a factory simulator which automates the process of planning using attributes and entities of production facilities and production knowledge-base rules to administer the planning process.
- Production facilities are to be modelled as entities so that the model can access and utilise them as required.
- Production plans are to be generated and evaluated automatically without the interference of a human planner so that the manual effort will be reduced.
- Product's and plant's selection rules and allocation rules are main knowledge domain in the model.

Figure 5 shows the specification of the system which is presented by information input, process and information output. The detailed process of the factory model is presented in Dawood (a), (b), 1994. and Dawood, and Neale (1993).

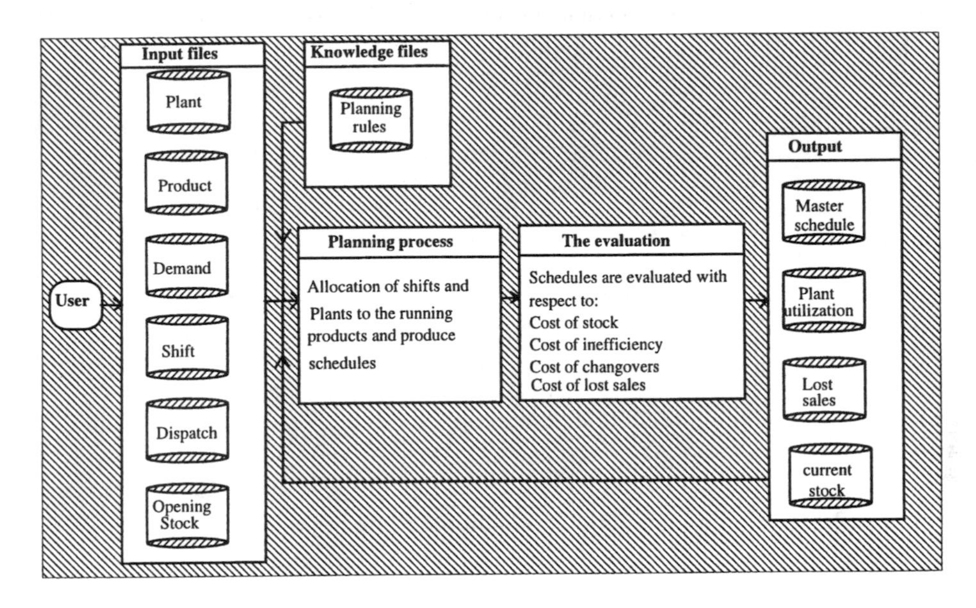

Figure 5. The specification of the system

CONCLUSIONS

The objective of this paper was to present a strategic planning and distribution system for companies who produce a wide range of building products from several factories around UK. The system is a decision support system which integrates a number of methodologies from company level and factory levels. An industrial case study has been conducted and it was concluded that the system is beneficial and can improve strategic plans.

Reference

Bernard, J., Lalonde and Karl, A., A survey of computer applications and practices in transportation and distribution, *International Journal of Physical Distribution and Logistics Management*, Vol. 25, NO. 4, pp 12-21, 1995.

Bhaskar, K. N. and Housden, R. J. *Information Technology Management.* CIMA series, 1990

Bowersox, D. J. *Logistical Management.* A systems Integration of Physical Distribution Management, Material Management and Logistical Co-ordination.,1974.

Dawood, N. N. An integrated knowledge-base/simulation approach to production planning: an application to the precast industry, *Construction Management and Economics Journal*, Vol. 13, 1995.

Dawood, N. N.(a) Applications of "knowledge/heuristic" methods on planning and scheduling in precast manufacturing companies: a case study, *International Journal of Construction Information Technology*, Vol. 2, No 3, 1994.

Dawood, N. N. (b) Developing a production management modelling approach for the precast concrete building products. Paper published in the *Construction Management and Economics Journal*, No. 5, Vol. 12, pp. 393-412, 1994.

Dawood, N. N. and Neale R.H. Forecasting in the precast concrete industry. Paper published in the *Construction Management and Economics Journal*, Vol. 11, pp 81-98, 1993.

Slack, N., Chambers, S., Harland, C., Harrison, A., and Johnston, R., *Operations Management.* Pitman Pulshing, 1995.

Sijbrands, M. J. C., Strategisch en Logistieke Besluitvorming, *PhD thesis*, Tinbergen Institiue Research Series No 35, (English abstract) 1993.

INTERACTIVE METHOD FOR GROUPING OPERATIONS IN ASSEMBLY LINE DESIGN

Fabrice Pellichero[1], Alain Delchambre[1] and Bertrand Mareschal[2]
[1]Department of Applied Mechanics - University of Brussels (ULB)
[2]Institute of Statistics - University of Brussels (ULB)

ABSTRACT: The research work overviewed in this paper aims to develop a method to assist the production line designer for the choice of the best way to feed, to handle, and to insert parts in an assembly line.
They are a lot of criteria influencing his decision, which can be geometrical (symmetry, dimensions,...), mechanical (slipperiness, fitting,...), economical (pay-back period, cycle-time,...),...
It is thus obviously a multicriteria decision-aid problem with three possible solutions : the manual method, the robotized and the automated ones.
The choice of the most significant criteria was based on an extensive literature review and on a very strong collaboration with a Belgian line integrator called FABRICOM which has helped us to quantify the influence of each criteria. In order to implement the decision-aid method in a user-friendly computer tool, we decided to use a statistical method, called PROMETHEE, which has shown in the past his efficiency for multicriteria decision-aid problems.
The paper describes, on the basis of industrial examples, different modules of the computer-aided package, not only for the choice of the assembly method, but also for the analysis of that choice and its archiving.
The link between the decision-aid tool and a CAD method for the layout of assembly lines is also presented.

1. INTRODUCTION

The choice of an assembly method is a complex problem depending on a lot of criteria belonging to various categories as :

- the economical criteria (cycle-time, pay-back period,...)
- the geometry (symmetry, dimensions,...),
- the quality (fragility, slipperiness,...),
- the packaging mode (orientation, position,...),
- the feeding properties of the part,
- the handling properties of the part,
- the insertion properties of the part.

Some researchers, like [Rampersad 95] and [Boothroyd 89], have developed DFA tools, and shown that an assembly operation can be divided in three elementary operations : feeding, handling and inserting. Hence, we decided to decompose our problem into three sub-problems : the choice of the feeding mode, the choice of the handling mode and the choice of the inserting mode.

There are three possible ways to solve these sub-problems :

- the manual method (which includes the semi-automated method and allows the most important flexibility)
- the robotized method (which allows a partial flexibility)
- the full automated or dedicated method (which is the most specific and allows the lowest flexibility)

As these three sub-problems are multicriteria problems, we have decided to use a multicriteria decision-aid method, called PROMETHEE to solve them. This method has been developed in the Institute of Statistics of the University of Brussels and has already shown his efficiency for this kind of problems (see for example [Brans & Maresc 94] & [Briggs et al 90]).
In section 2, we will briefly present what is the global CAD software we are implementing for the design of assembly layout and demonstrate the link with the software presented here.
In section 3, we will describe the PROMETHEE method.
In order to adapt this method to our particular problem, we decided to implement a new user-friendly software, running on Windows 95. This software will be overviewed in section 4.
In section 5 we will draw a general conclusion and give some guidelines for further work.

2. FRAME OF THE PROJECT

The frame of this project is broader than the creation of a software to help the designers in choosing the best assembly methods. Indeed, the ultimate goal of this project is to build a computer-aided tool for the design of an assembly line in the early stage of the product design.

The main characteristics of this software are :

- the suitability to the user's needs (see preference constraints below)
- the user-friendly Man-Machine Interface
- the speed of the algorithm in order to test many variations of the line design problem

Our research work is placed in this context and has to help the designer of a line (the integrator) in proposing a logical layout based on the design of a product and its assembly sequences.
Let's define in more details what we mean by logical layout.
The logical layout gives:

- the number of stations in the line
- the actions done by the stations
- the relative position of those stations

This first layout has to balance the load on the different stations, this is the line-balancing problem.

The line-balancing problem is combinatorial. A software has been developed and described in [Falk & Delch 92] and [Falk. 96].
It uses three kinds of data :

- the assembly sequences
- the duration of the operations

- the cycle-time (time interval between two products exiting the line)
- the preference constraints which are of four types :
    - the associative and dissociative constraints
    - the hard and soft constraints

Associative constraints mean that the designer would like to gather some operations on the same station. The designer has the possibility to specify hard and soft constraints. The difference between them is that the first ones will always have to be satisfied and that the second ones can be violated if the cost of the station is too high.
The dissociative constraints are of the same type but they concern the separation between operations.

Our method of assembly choice will be interfaced with the line-balancing software.
It's at the level of the associative constraints that the link will be done between our problem and the line-balancing problem. The grouping of the operations by assembly method will give us the opportunity to generate associative constraints.
Thus, our project consists in developing the " missing-element " which will allow us to pass quickly from the design of a product to the logical layout of its assembly line.
This will allow us to iterate quickly between the design of the product and its production means. This environment will thus be an excellent support for the introduction of the concurrent engineering concept.

3. THE MULTICRITERIA DECISION-AID METHOD

3.1 Multicriteria decision problems

The multicriteria decision problems can be stated as follows : $\text{Max}\{f_1(a), f_2(a),...,f_j(a),...,f_k(a) \mid a\in A\}$, where A is a set of n possible decisions or alternatives which are evaluated trough k criteria $f_1,...,f_k$. The basic data for such a problem can be presented as shown in Table 1.

Given this table, the dominance relation can be defined as follows ($a,b \in A$): a dominates b (a D b) iff $f_h(a)\geq f_h(b)\quad \forall h=1,...,k$ (with at least one >). The non-dominated alternatives are called efficient solutions. In practice, the dominance relation is often very poor and the number of efficient solutions can be rather large.

Table 1 :
Evaluation Table.

	$f_1(.)$	$f_2(.)$	...	$f_j(.)$	...	$f_k(.)$
a_1	$f_1(a_1)$	$f_2(a_1)$	...	$f_j(a_1)$	...	$f_k(a_1)$
a_2	$f_1(a_2)$	$f_2(a_2)$	...	$f_j(a_2)$	...	$f_k(a_2)$
			...			
a_i	$f_1(a_i)$	$f_2(a_i)$	...	$f_j(a_i)$	...	$f_k(a_i)$
			...			
a_n	$f_1(a_n)$	$f_2(a_n)$	...	$f_j(a_n)$	...	$f_k(a_n)$

Indeed, it is obvious that such data do not generally induce a complete ranking on the set A of alternatives. The problem is not mathematically well-stated and the notion of optimal solution does not exist. However, the problem is most often economically well-stated as it expresses the different and possibly conflicting objectives of the decision maker.

3.2 Principles of the PROMETHEE method

The PROMETHEE method (Preference Ranking Organization METHod for Enrichment Evaluations) includes the three following steps :

Step 1 : Enrichment of the preference structure :
The notion of generalized criteria is introduced in order to take into account the amplitudes of the deviations between the evaluations.
For this purpose we define the preference function P(a,b) giving the degree of preference of "a" over "b" for criterion f. In most cases we can assume that P(a,b) is a function of the deviation
$d = f(a)-f(b)$. After normalization $0 \leq P(a,b) \leq 1$ and :

- $P(a,b) = 0$ if $d \leq 0$, no preference or indifference
- $P(a,b) \approx 0$ if $d > 0$, weak preference
- $P(a,b) \approx 1$ if $d >> 0$, strong preference
- $P(a,b) = 1$ if $d >>> 0$, strict preference.

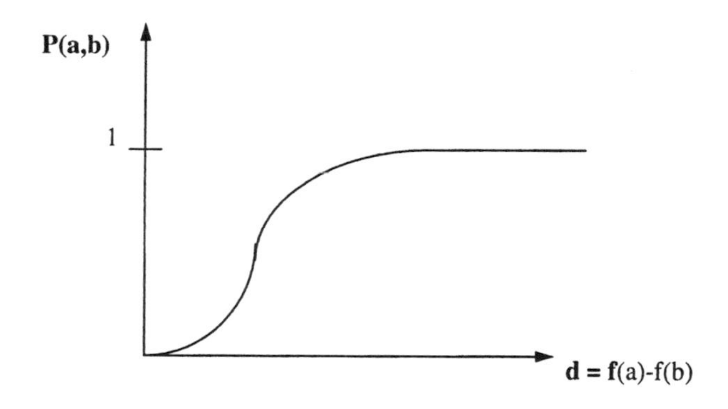

Figure 1. Preference function

It is clear that P has to be a non-decreasing function of d, with a shape similar to that of Figure 1.
The generalized criterion associated to f(.) is then defined by the pair (f(.), P(.,.)).
The PROMETHEE methods request that a generalized criterion is associated to each criterion f_j, $j=1,...,k$.

Step 2 : Enrichment of the dominance relation :
A valued outranking relation is built taking into account all the criteria.
Let us now suppose that a generalized criterion ($f_j(.)$, $P_j(.,.)$) has been associated to each criterion $f_j(.)$.
A multicriteria preference index $\pi(a,b)$ of a over b can then be defined taking into account all the criteria :

$$\pi(a,b) = \sum_{j=1}^{k} w_j P_j(a,b), \quad \left(\sum_{j=1}^{k} w_j = 1 \right) \tag{3.1}$$

where $w_j>0$ (j=1,...,k) are weights associated to each criterion. These weights are positive real numbers that do not depend on the scales of the criteria.
It is interesting to note that if all weights are equal, π(a,b) is simply the arithmetic average of all the $P_j(a,b)$ degrees (j=1,...,k).

π(a,b) expresses how and with which degree a is preferred to b, and π(b,a) how b is preferred to a, over all the criteria.
For each pair of alternatives a,b∈A the values π(a,b) and π(b,a) are computed. In this way, a complete valued outranking relation is constructed on A.

Step 3 : Exploitation for decision aid :
Let us consider how each alternative a∈A is facing the n-1 other ones and therefore define the positive and the negative outranking flows :

$$\phi^+(a) = \frac{1}{n-1} \sum_{b \in A} \pi(a,b) \qquad \phi^-(a) = \frac{1}{n-1} \sum_{b \in A} \pi(b,a) \tag{3.2}$$

The positive outranking flow expresses how much each alternative is outranking all the others. The higher $\phi^+(a)$, the better the alternative. $\phi^+(a)$ represents the power of a, its outranking character.
The negative outranking flow expresses how much each alternative is outranked by all the others. The smaller $\phi^-(a)$, the better the alternative. $\phi^-(a)$ represents the weakness of a, its outranked character.

In the three problems we are faced with, a complete ranking of the alternatives is necessary. For this purpose, we will consider the net outranking flow.

$$\phi(a) = \phi^+(a) - \phi^-(a) \tag{3.3}$$

It is the balance between the positive and the negative outranking flows. The higher the net flow, the better the alternative.

The PROMETHEE II complete ranking is then defined:

$$\begin{cases} aP^{II}b & \text{iff} \quad \phi(a) > \phi(b) \\ aI^{II}b & \text{iff} \quad \phi(a) = \phi(b) \end{cases} \tag{3.4}$$

We will see below that, besides those important results, the software provides powerful tools for the analysis of the results. These tools are fundamental to better understand and control the influence of each parameter.

For more details about the PROMETHEE method, see [Brans & Maresc 94] and [Pellich., Delch. and Maresc., 1996].

4. PRESENTATION OF THE SOFTWARE

4.1 Main characteristics of the software

Let us define the main properties that the software, called PROMWIN, has to be able to satisfy :

- analyse a complete assembly line including 50 stations or more,
- for each station, choose the best method for feeding, handling, and inserting the part,
- give a clear representation of the results for each station,
- give the possibility to see interactively what are the influence of the modifications of the weights associated to the different criteria and also the influence of changing the answer to one question associated to a given criteria,
- give the possibility to analyse the results, for each operation, by families of criteria and thus, to see clearly what are the most important parameters in the analysis of one given operation,
- give the possibility to the user to save the final results or some intermediate ones in order to continue to analyse the line later,
- give the possibility to compare the results of one station with those of one, two, or more other stations.

This software has been implemented in Borland C++ 4.5 and can be run on Windows 3.1 or Windows 95. It was implemented with an Object Oriented approach which will make it more flexible and evolutive for the future.

4.2 The running of the software

The user has two possibilities when he uses the software : create a new file (for the analysis of a new assembly line) or open an existing file. In the former case, he will have to answer to about 30 questions by operations. These questions are classified in three families, corresponding to the three sub-problems associated to each assembly operation, i.e. the feeding, handling and inserting problems.

When the user has answered all the questions, he can run the PROMETHEE II method to see the results. It is also possible to display the weights associated to each criteria and to modify them. When he modifies the weights, or the answers to the questions, the results of PROMETHEE are re-computed in a few seconds (on a PC 486 75MHz) and the display changes automatically. This constitutes a powerful tool to study the influence of each criteria.

When the user wants to change of operation, he has to use the principal window which shows the results obtained for the current operation.

When the user has input the data concerning the assembly operations, he can display a summary of the results obtained for the entire line. Figure 2 shows an example of such a summary. The manual, robotized and dedicated solutions are represented by a "M", a "R" or a "D" respectively. The line displayed on this example is a line including 26 operations. Some operations, like operation 7 or operation 12 don't have any result because they are not, actually, assembly operations, even if they are included in the assembly line. We have chosen to display them to have a general view of the assembly line.

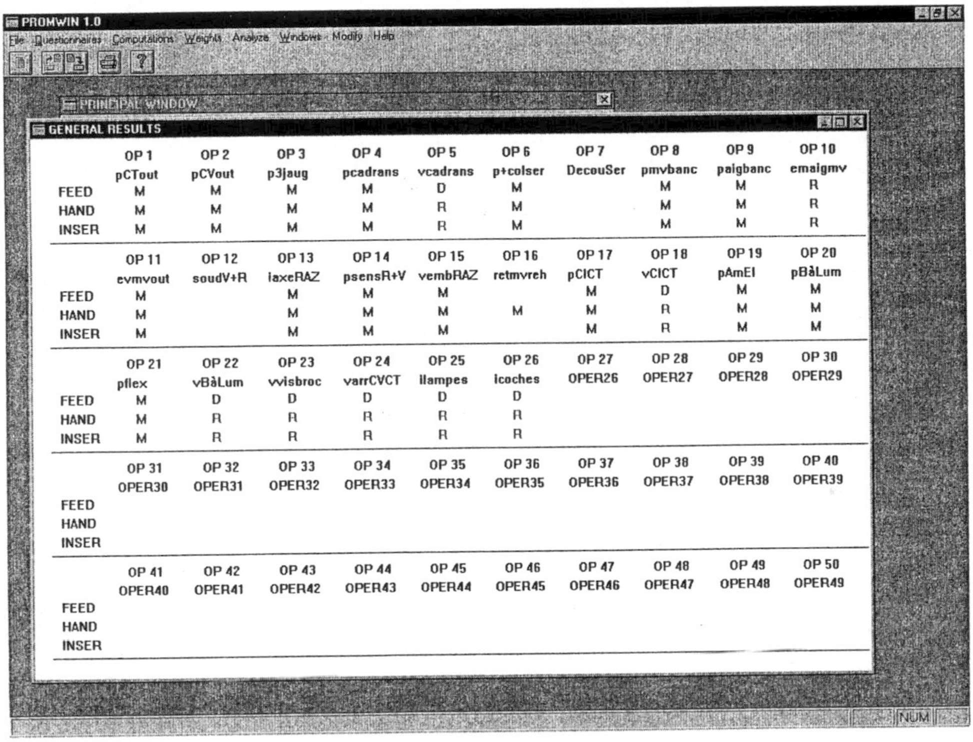

PROMWIN 1.0

File Questionnaires Computations Weights Analyze Windows Modify Help

PRINCIPAL WINDOW

GENERAL RESULTS

	OP 1	OP 2	OP 3	OP 4	OP 5	OP 6	OP 7	OP 8	OP 9	OP 10
	pCTout	pCVout	p3jaug	pcadrans	vcadrans	p+colser	DecouSer	pmvbanc	paigbanc	emaigmv
FEED	M	M	M	M	D	M		M	M	R
HAND	M	M	M	M	R	M		M	M	R
INSER	M	M	M	M	R	M		M	M	R

	OP 11	OP 12	OP 13	OP 14	OP 15	OP 16	OP 17	OP 18	OP 19	OP 20
	evmvout	soudV+R	iaxeRAZ	psensR+V	vembRAZ	retmvreh	pCICT	vCICT	pAmEI	pBàLum
FEED	M		M	M	M		M	D	M	M
HAND	M		M	M	M	M	M	R	M	M
INSER	M		M	M	M		M	R	M	M

	OP 21	OP 22	OP 23	OP 24	OP 25	OP 26	OP 27	OP 28	OP 29	OP 30
	pflex	vBàLum	vvisbroc	varrCVCT	llampes	lcoches	OPER26	OPER27	OPER28	OPER29
FEED	M	D	D	D	D	D				
HAND	M	R	R	R	R	R				
INSER	M	R	R	R	R	R				

	OP 31	OP 32	OP 33	OP 34	OP 35	OP 36	OP 37	OP 38	OP 39	OP 40
	OPER30	OPER31	OPER32	OPER33	OPER34	OPER35	OPER36	OPER37	OPER38	OPER39
FEED										
HAND										
INSER										

	OP 41	OP 42	OP 43	OP 44	OP 45	OP 46	OP 47	OP 48	OP 49	OP 50
	OPER40	OPER41	OPER42	OPER43	OPER44	OPER45	OPER46	OPER47	OPER48	OPER49
FEED										
HAND										
INSER										

NUM

Figure 2 : Summary of the results obtained for an assembly line.

In order to check the results, it was necessary to develop several analysis tools. Their respective role is described in the next section on the basis of an example.

4.3 Overview of the analysis tools through an example

Let's suppose that we have to study the feeding of screws on an assembly line. The screws are standard ones and are thus well adapted to be fed by a vibratory bowl (which is corresponding to the dedicated mode). Moreover, the cycle time is very short and the payback-period is long. All those economical and technical reasons are thus obviously showing that the best solution is the full-automated one. This analysis is clearly illustrated by the analysis tool for feeding and the PROMETHEE result shown in Figure 3. On the left side of that figure, we can see the analysis tool with the grouping of the criteria in different categories and the results of PROMETHEE on the right side.

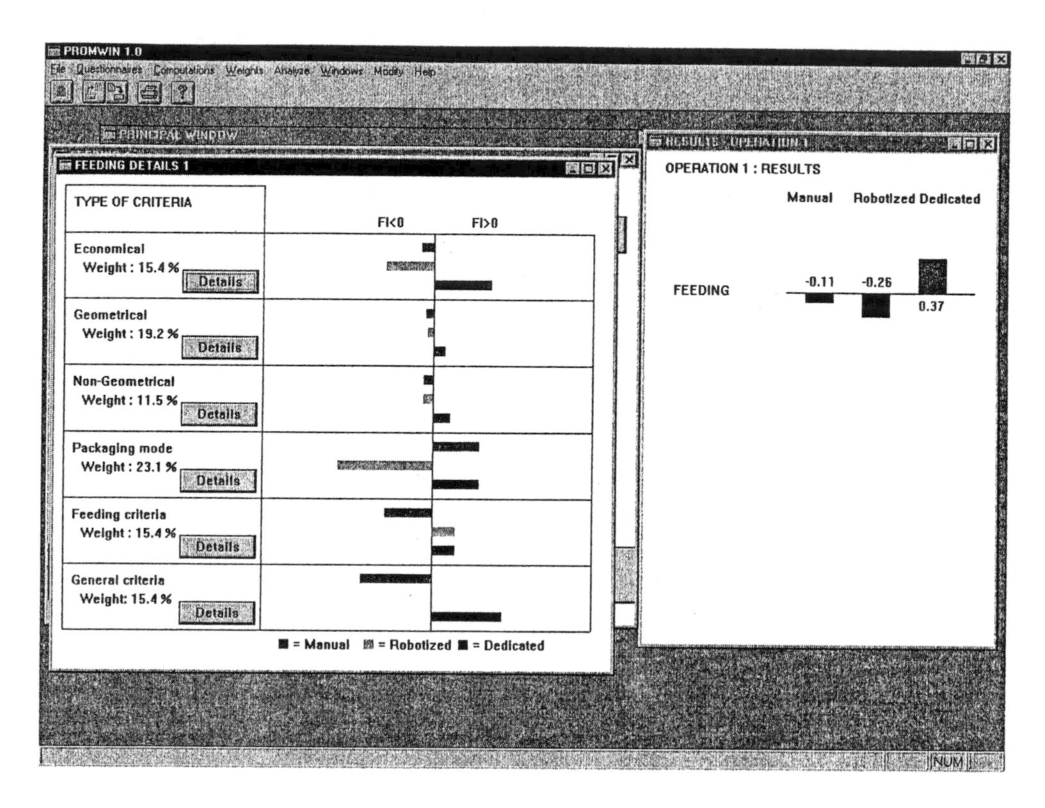

Figure 3 : The results for the feeding of a screw described above.

It shows us that the dedicated mode, is by far, preferred for the economical criteria as for the geometrical, non-geometrical and general criteria.

Let's describe our classification of the criteria for the feeding problem :

- *The economical criteria* are the cycle-time and the pay-back period. Those criteria are general for all the line.
- *The geometrical criteria* are the dimensions of the part, its symmetry, its conicity, the presence of a groove.
- *The non-geometrical criteria* are the fragility of the part, the fact that the part is delicate, sticky, slippery, abrasive, dirty, magnetic, burning, flexible or deformable, and the fact that it has cutting edges.
- *The packaging-mode criteria* are the fact that the parts are packed, their storage method (a bulk, a plan,...), their position and their orientation.
- *The feeding criteria* are related to the risk of tangling, fitting, overlapping, the possibility to handle more than one part, the fact that the orientation is defined by a non-geometrical characteristic.
- *The general criteria of the operation* are the number of repetitions of the operation and the number of variants of the part.

This tool is very useful, because when the user will come back to an analysis done two years ago, he will quickly be able to find the reasons why he had chosen this solution compared to the other ones.

Hence, we can say that this software is useful to help the designer to choose the best solution, but also, to remember him the main aspects of his analysis. Moreover, when the difference between the flows of two solutions is very small, it is up to the designer to take his responsibility and to decide. But, thanks to the analysis tool, he will have a good support to proceed.

For a given operation, it is obvious that each criteria doesn't influence the decision. For example, if a part is not fragile, the fragility criterion will obviously not influence the choice concerning the feeding method. The "Details" buttons present in Figure 3 show to the user the criteria having effectively influenced the decision for the related family of criteria (see Figure 4). After a lot of tests on different lines, those results should allow us to collect statistics concerning the utility and the influence of each criteria.

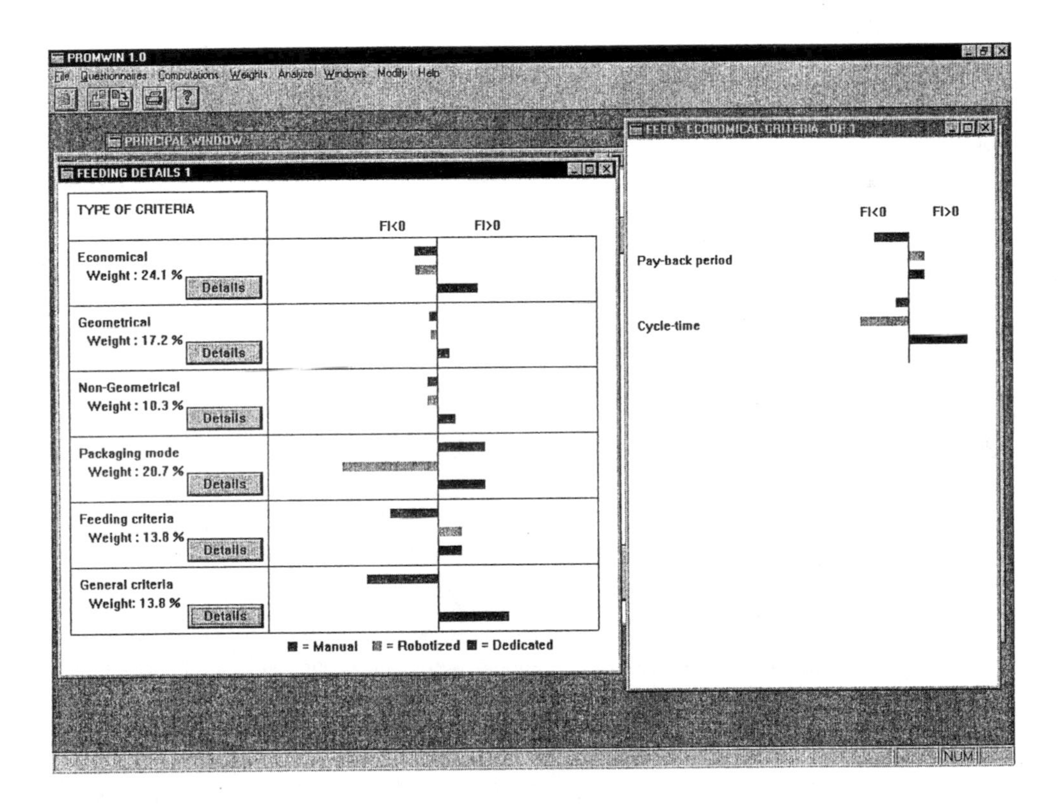

Figure 4 : Result obtained when clicking on the "Details " button for the economical criteria.

The other tool which is helpful for the analysis is the edition of weights.
We can modify those weights and see simultaneously the modification of the PROMETHEE results. This tool is illustrated in Figure 5. On the left side of this figure, we can see the list of the criteria and their corresponding weights. Each weight can be modified. After a modification, the result displayed on the right side will automatically be modified. It is important to notice that, if we modify the weight of a criteria for an operation, this modification will be done in all the other operations. The weights are thus related to the entire line and not to each operation.

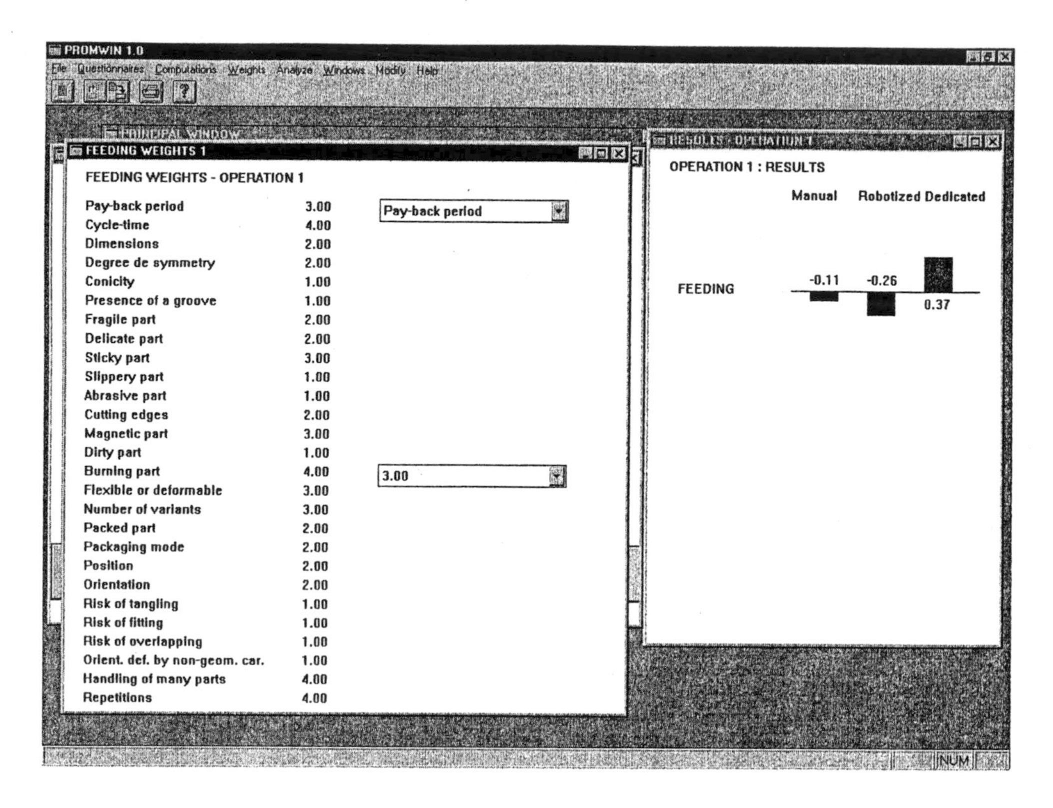

Figure 5 : Weights of the feeding criteria.

The software has already been tested on three different assembly lines designed by FABRICOM, and the results obtained are very encouraging.

It is also important to remind that this method will always have to evolve. It is indeed obvious that the technology will change and that, for example, some operation which can't be realized automatically now, could be possible to realize in the future, thanks to the technological evolution.

5. CONCLUSION AND FURTHER WORK

In this paper, we have presented a method for the choice of the assembly method in a production line. We have seen where the developed method has to be integrated, and its importance in our global project.

The presented method has the following advantages :

- the user-friendliness,
- very short computation times, which gives us the possibility to analyze a lot of data simultaneously and interactively,

- a great adaptability because we can add or suppress some criteria without changing the fundamental structure of the method (this would not have been the case if we had used decision trees for examples),
- the possibility to save the results and to analyze and re-use them later,
- the possibility to analyze a complete assembly line,

The current step of the project, consists in testing the method and analyzing the results with other industrial case studies. It will give us the opportunity to :
- correctly adjust the different weights,
- verify in more details the influence of each criteria and, probably, to suppress some of them and to add some others,

One of the main future improvement to the method, is to reduce the number of questions to be asked. As we said above, it will probably be possible to do it by a statistical analysis of the influence of each criteria after the collection of a lot of results.
Another interesting way of improvement, is to take into account, not only the influence of each criterion, separately with respect to the others, but also the influence of the combination of criteria. Indeed, it is obvious that some criteria have a small influence when considered separately, but when they are associated with some other criteria, they can have a very strong influence.
We think that the current method, and the corresponding computer-package, constitute a very good start for those future improvements.

6. REFERENCES

[Boothroyd 91] G. Boothroyd, " Assembly automation and product design ", Marcel Decker Inc., 1991.

[Briggs et al 90] Th. Briggs, P.L. Kunsch and B. Mareschal, " Nuclear waste management: an application of the multicriteria PROMETHEE methods ", European Journal of Operational Research, vol 44, pg. 1-10, North-Holland, 1990.

[Brans & Maresc 94] J.-P. Brans et B. Mareschal, " The PROMCALC & GAIA decision support system for multicriteria decision aid ", Decision Support Systems, vol. 12, pg. 297-310, North-Holland, 1994.

[Brans & Maresc 94/2] J.-P. Brans et B. Mareschal, " How to decide with PROMETHEE ", ULB, Service de Mathématiques de la Gestion, april 1994.

[Delchambre 96] A. Delchambre (ed.), " CAD method for industrial assembly : Concurrent Design of Products, Equipment and Control Systems ", John Wiley & Sons, London, 1996.

[Falk & Delch 92] Falkenauer E., Delchambre A. - "A Genetic Algorithm for Bin Packing and Line Balancing", IEEE International Conference on Robotics and Automation, May 10-15, 1992, Nice, France. IEEE Computer Society Press, Los Alamitos, CA. pp 1186-1192, 1992.

[Falkenauer 94] Falkenauer E., "The Grouping Genetic Algorithms and their Industrial Applications. ", PhD Thesis, University of Brussels (ULB), 1994.

[Falkenauer 96] Falkenauer E., "A Hybrid Grouping Genetic Algorithm for Bin Packing", Journal of Heuristics, Vol 2, N 1, pp.5-30, Kluwer Academic Publishers, 1996.

[Rampersad 95] Rampersad H. K., "Integrated and Simultaneous Design for Robotic Assembly.", John Wiley &Sons, London, 1994.

[UCL 95] B. Raucent, E. Aguirre, A. Leroy, F. Petit, J. Dejardin, " Conception simultanée d'un produit et de sa ligne d'assemblage ", Formation Continue des Ingénieurs, 1995.

[Vincke 89] P. Vincke, " Multicriteria decision-aid ", John Wiley & Sons, 1989.

GENERIC MODELLING OF CELL CONTROL SYSTEMS AND PRODUCTION RESOURCES

Per Gullander [1], Pär Klingstam [1], Pär Mårtensson [2]
1. Department of Production Engineering, Chalmers University of Technology, SE-412 96 Göteborg, Sweden;
2. Department of Manufacturing Systems, Royal Institute of Technology, SE-100 44 Stockholm, Sweden

ABSTRACT: Frequent changes regarding products, processes, and technologies require the manufacturing systems and the control systems to be inexpensive, flexible, and easy to re-configure. The major aim of the research presented in this paper is to define a generic reference architecture that supports design and implementation of highly flexible control-systems for manufacturing cells. Based on the analyses of two machining cells, a preliminary version of such an architecture has been developed. The main features of this are: (1) modular control structure with one module for each resource in the cell, (2) message-based, generic communication between the system's modules, (3) separation of generic and specific control activities, and (4) separation of the products' operation lists from the description of the resources' capabilities. Among manufacturing cells in the industry, there is a great variety considering system layout, product flow, and the resources' capacity, flexibility, and operation. Therefore, in order to gain a more generic architecture for manufacturing systems, different types of production resources available, processes, manufacturing system layouts, and product flows have been analysed and classified based on information gathered in an industrial survey. Based on this analysis, the applicability of the architecture is discussed. It is concluded that the architecture is directly applicable to most cells, and if some modifications are made in the architecture in order to handle specific features observed, the architecture could be made applicable to all cell types defined except one.

INTRODUCTION

There is an increasing demand on manufacturing systems to be able to be re-configured quickly in order to handle new products, using new machines or processes. In order for the manufacturing system to survive changes in its environment, the control system must be flexible and reusable. Thereby, the system can adapt to changed conditions in its environment, either automatically or with only minor changes needed. This also means that the system can be reused in different environments. Flexibility can be incorporated on different time scales, considering different aspects. On a long time scale, flexibility permits incorporation of new equipment, and on a shorter time scale allows for incorporation of new products. On an even shorter time scale, flexibility concerns the use of alternative operations, product routings, etc.

Because of the complexity of the problem, control software often becomes not only expensive but also inflexible despite the fact that the components of a manufacturing system, e.g. robots and machine tools, are flexible in themselves and can be used in a large variety of applications [13]. Many industrial systems that are considered to be flexible, are only flexible regarding some of the flexibility aspects described above. Instead, in order to lower costs and increase flexibility, we need a truly flexible manufacturing system, FMS, that can be reused without the need to re-program the control software whenever a change in production is introduced.

There are a lot of software tools [4, 6] and communication protocols, e.g. MMS [12], available that make this work easier, but the design and implementation of control systems still are problematic. Even though much work has been done, there is a scarcity of models and other guidelines that support the development of flexible control systems. The models available today often are not sufficiently easy to apply to a specific manufacturing system. They are often too abstract and lack the details necessary for the design and implementation phases.

Research has been conducted at Chalmers University of Technology aimed at supporting the development of truly flexible manufacturing systems capable of absorbing new products and new equipment with a minimum of re-programming of the controlling software [1]. The focus of the research was on the superior control software for machining cells that is responsible for the co-ordination and control of the activities in the subordinate equipment. The approach of this research is to define a generic reference architecture that supports design and implementation of highly flexible control-systems for manufacturing cells. Based on the analyses of two machining cells, a preliminary version of such an architecture has been developed.

Presented in this paper is the results and conclusions from a research project which is based on this architecture, aimed at achieving an architecture for manufacturing systems that are more generic than the present one. In the next section, the architecture is presented. In Section 3, the research aim and method are described (industrial survey). Results from the survey are presented in Section 4, and the applicability of the architecture and ways to adapt the architecture to be applicable to more cell and process types are discussed in Section 5. Finally, conclusions are given in Section 6.

GENERIC ARCHITECTURE FOR CELL CONTROL SYSTEMS

In order to simplify the research work towards developing more flexible systems, the problem has been divided into the following topics:

- Model of the manufactured *products*: A product model is a formal description of an operation list that contains all the necessary operations to be performed on the product [3].
- Model of the manufacturing *resources*: The capabilities, constraints, and the behaviour of the resources are given by the resource models [9].
- The *control algorithm*: The control algorithm synchronises the product models with the resource models, and thereby dedicates a resource to each operation in the operation list [7].
- The *operator interaction*: To achieve better control systems, the operators' roles and interactions with the system are given special consideration [2].

The methodology adopted to reach the goals can be characterised as being bottom-up, i.e. the reference architecture is defined, based on the Production Activity Control model presented by Bauer et al. [5], and gradually refined based on case studies of actual manufacturing control systems. The more cells analysed, the more general the architecture will become. The main features of the architecture are:

- Modular control structure with one module for each resource in the cell;
- Message-based, generic communication between the system's modules;
- Separation of generic and specific control activities;
- Separation of the products' operation lists from the description of the resources' capabilities.

The architecture can be described by the following views:

- *Structural* model, that defines the modules/objects/resources the control system contains, and which activities are carried out in each module;
- *Functional* model, that defines the functions/activities that are carried out, and the information flow between these;

* *Behavioural* model, that defines the temporal relationship between the functions;
* *Information* model, that defines the information entities needed and the relationship between these.

Structural and Functional Description of the Architecture

The control structure defines which modules the controller should be divided into, and the distribution of control activities and responsibilities among these modules. A well-structured modular control-system (see Figure 1), with modules co-operating through the use of generic messages, increases flexibility and reusability since the modules can be reused within different surroundings. Such a system also has the potential of reducing development cost.

The architecture comprises a database that stores both static and dynamic information (see Section 2.3), modules for scheduling and error diagnosis, and modules for control. Based on the product and resource data found in the database, the scheduling modules develop a schedule on-line.

Therefore, the controller comprises, in the general case, several communicating modules. First, the *scheduler* and the *dispatcher* modules that handle the scheduling and control functions, including deadlock handling and optimisation, i.e. decide what to do and gives orders to subordinated modules. Secondly, for each physical device in the manufacturing system there is a corresponding module in the cell controller, called the *internal resource*, that executes resource-specific tasks and keeps track of the current state of the manufacturing resource. Depending on what type of physical resource it controls, the internal resources are sorted into producer, mover, and location modules. The physical resources are classified according to their functionality:

* *Producers* that make the necessary physical or logical changes in product properties, e.g. CNC machines (physical changes) and measuring stations (logical changes).
* *Movers* that transport the products between producers and locations, e.g. AGVs, robots and conveyors.
* *Locations* that only store products and cannot change any properties of the product, e.g. local buffers and storage systems.

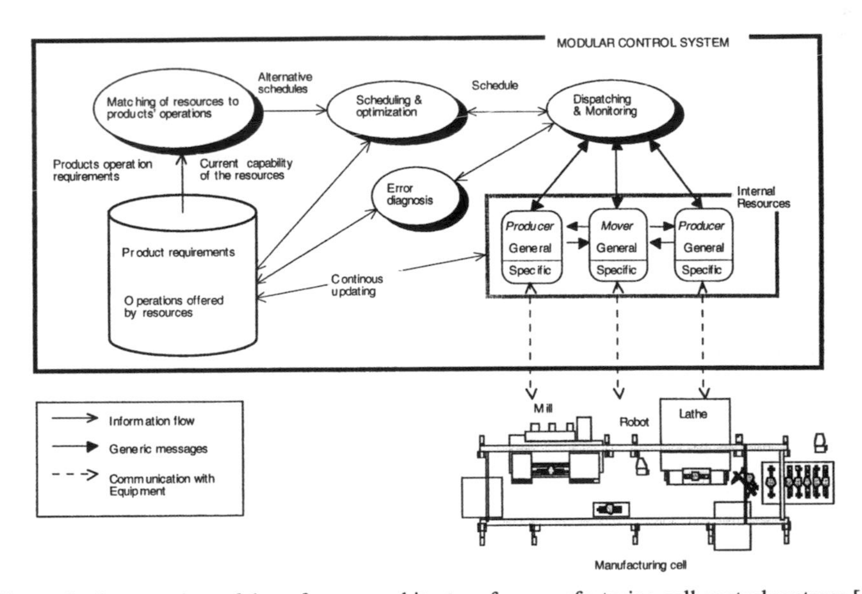

Figure 1. An overview of the reference architecture for manufacturing cell control systems [1].

The controller modules have a client/server relationship and synchronise their activities using high-level messages (see Section 2.5.2). However, between each internal resource and its respective physical device, a message protocol specific for the device is used. To accomplish this and to achieve a homogeneous functional view of the resources despite differences regarding functional capabilities, the internal resources are split in two parts: a *generic*, reusable part providing an abstract view of the resource, and a *specific* part that provides the capability included in the general part but not provided in the physical resource. The dispatcher thus is provided with a homogeneous message-based interface to all physical devices.

Behavioural Description of the Architecture

The behaviour of the cell control system is defined by the dynamic behaviour of modules that are contained in the control system, i.e. the *scheduler*, the *dispatcher*, and the *internal resources*. Below, the behaviour of the internal resources and the behaviour of the whole control system are briefly described.

Generic Resource Models. To support the design and implementation of the internal resources, the implementation of the general parts will be based on the *generic resource models*, i.e. models describing the behaviour and functionality of resources belonging to certain classes.

All resources in a certain class have similar behaviour and can be controlled using the same interface and thus will be modelled in the same way. In order to increase flexibility and standardisation and to support the development of flexible control systems, a library of generic resource models will be created. Because of the variations between various manufacturing resources, it is difficult to generalise their behaviour into generic models. Important differences reside in:

- The number of products processed at a time
- If doors and gripping devices must be controlled by external events
- If the robot also is used for products processing
- The number of products that can be held by the mover

The resources as well as the procedures used to load, process, and unload products, have been studied. Results from this analysis are resource models describing the different states (actions) and transition events (messages) of the internal resource for machine tools and robots [8].

Generic Message-Passing Scheme. A generic model defines the interfaces and interaction between the controller modules has been developed. The generic message-passing scheme is meant to be a support when designing the control software [9]. The model incorporates messages needed for normal operation and human operator intervention. The messages needed for normal operation consist of command messages and handshake messages. The command messages are used by the dispatcher to command the internal resources to move products or to start processing products, and by the internal resources to inform the dispatcher that the operation have finished. The handshake messages are sent between the internal resources to synchronise the loading and unloading of products.

Information View

A lot of information must be collected, stored, processed, and presented within a cell-control system. In order to increase flexibility and openness, the information should be stored in a database. True flexibility can only be achieved when different non-related aspects of a system are kept apart from each other as much as possible, so that a change only affects one part of the manufacturing system. Therefore, the information of the *resources* that are present in the cell and the information of the *products* to be produced, should be kept apart. The linking between these is represented by the matching of product operation requirements with resource operation capabilities. Thus, a new product should not affect anything else but the entering of a new product specification into the system, and likewise, a new resource should only add to the resource descriptions of the system. Only some of the information needed by the control system will be briefly presented here: the product model, the resource model, and the handshake procedure information.

The information in the database can also be divided into *static* and *dynamic* data. Static data is the templates which also exist before the individual products or resources are introduced into the cell. The dynamic data represents the individual products and the resources, that is, using the terminology of object-orientation, the instances of the templates.

Product model. The term "product" is used to refer to the physical object that requires operations. The final product that comes out from the cell can be an assembly of many products. The product model is very important to consider if high flexibility is to be achieved. The product model contains a list of operations that must be performed on the product, and gripping information describing how the product should be handled (gripped) by the mover at different stages. Furthermore, the product model contains information concerning error handling and variables giving information specific for the product type and product individual. For a more detailed description of product models, see [3].

Information about the set of *process operations* that each product requires must be made accessible to the control system. To be able to easily introduce new products (product flexibility), these operation lists are preferably stored in the database instead of hard-coded into the control programs. It is important to note that these operation lists do not identify which specific resources that should be utilised since this is determined at run-time to take full advantage of the current production status.

Apart from these process operations, a number of operations must also be performed on the product to transport it within the cell. However, these *transport operations* are not really required by the product, but are rather requirements put on production by the manufacturing system's layout. Which transport operation to perform depends on the resources involved in the transportation. Since these resources are not identified in the operation lists, the products' transport operations cannot be included in the operation lists, but are stored separately in the database.

In order to fully utilise the separation of product operations from resource capabilities and to achieve a high degree of flexibility, the products' operation lists should not specify a strictly sequential list of operations. Instead, the process planning should leave as much freedom as possible to the control system regarding the sequences of operation steps. In this way, the routing flexibility of the system will increase. Therefore, the operation lists can describe, not only simple sequences of operations, but also complex combinations including conditional, alternative, and parallel (synchronised and asynchronous) operations. A formal language has been defined to specify operations lists [3].

The *handshake procedure* performed when products are loaded and unloaded in/from the procedures varies. Some machines may require complex handshaking including doors, safety systems, and clamping devices, while other devices require no handshaking at all. This depends primarily on the device, but in the general case the correct handshake procedure also depends on the product type that is handled, and on the mover involved.

Resource model. For each manufacturing resource, the following information is normally needed by the control system:

- Resource-*descriptive* information, i.e. information that defines the individual resource's properties. This information is static or close to static.
- Resource *status* information, i.e. dynamic information that describes the current status of the resource (current fixture, current tools available, current product or parts, currently loaded program, error code, and operation mode). This information is retrieved and updated frequently by the control system's modules.
- Resource *error-handling* information, i.e. information that describes the actions to be performed in a given error situation.
- Resource *capability* information, i.e. information that describes which operations a specific resource is capable of performing. This information is divided into a (nearly) static part defining the programmed capabilities, and a dynamically updated part describing the current capabilities (which is a subset of the programmed capabilities). For each such operation, the actual programs to be used in the robots, machine tools etc. must be identified.
- Resource *constraints* information, i.e. the constraints enforced by a resource, e.g. if products only can flow in one direction between two machine tools.

AIM AND METHOD OF SURVEY

Aim of Industrial Survey

The current architecture is mainly based on the study of two machining cells, and therefore not generally applicable to cells found in the industry.

In order to improve the architecture developed so far, more knowledge on the manufacturing cells in the industry was needed. Therefore an industrial survey was conducted covering five large Swedish companies [10]. The main aim was to increase the architecture's applicability, i.e. increase the variety of cell types that the architecture could be used for. The study of the cells were guided by the following general questions:

- Is it possible to use the architecture to design a control system for the cell?
- What specific features in the cell make it difficult to use the architecture?

Method

In the survey more than 100 cells were observed. However, many of these cells were not considered interesting to study in detail, since their layout and operation was of type that made it possible to utilise the architecture as it was. Further study of these cells would not contribute to increasing the generality of the architecture. Therefore, the study was focused on 25 cells, the design and operation of which made it difficult to apply the architecture directly.

In order to simplify the comparison of the cells, they have all been analysed in the same way. Specific interest was directed towards answering the following specific questions:

- Can the resources in the cell be defined according to the classification made in the architecture, i.e. producer, mover and location?
- Are there any handshake signal carried out between the resources that are not considered in the architecture?

RESULTS FROM INDUSTRIAL SURVEY

In this Section, the results from the survey is briefly presented. A more detailed presentation of the results can be found in [10]. The information is summarised in Table 1. For each cell the following information is presented:

- *Company.* The five companies visited are labelled A, B, C, D, and E.
- *Process type* that describes the type of process, i.e. assembly or machining of products.
- *Cell type*, i.e. a classification of the cells based on the resources, layout and the material flow capabilities.
- *Moving Producer.* If a cell contains a producer that also carries out transport operations, the operations it carries out is listed in this column.
- *Producing Mover.* If a cell contains a mover that also carries out processing operations, the operations it carries out is listed in this column.
- Specific *handshake* signals, i.e. handshake signals exchange between the cell's resources and which are not contained in the Generic Message-Passing Scheme (See Section 2.5.2)
- Other features observed in the cell.

Process Types

The main differences between cells with assembly operations and those with machining operations are that:

- Assembly cells have much shorter operation times and therefore the time required for information processing and material transportation is much more important to consider.
- The layout of assembly cells usually differs from the layout of machining cells,

- The product flow in assembly cells is much more complicated since products are also assembled and disassembled.
- The products to be joined together in an operation in a certain producer must be synchronised with each other,
- Assembly operations require the product models and the operation lists to include joining and splitting of products,
- Robots in assembly cells usually carry out operations both for transportation and assembly.
- The control system must be able to create products (disassembly of products) and to delete products (assembly operations)

Cell Types

Among the studied cells, a great variety was identified regarding the number of transporting devices (AGVs, robots, and gantry cranes), the number of machine tools, and, above all, the way these resources were grouped. Based on the classification scheme defined by MacCarthy and Liu [11], a new classification scheme was defined for the types of cells covered by the survey:

- Single Flexible Machine, *SFM*, consisting of a CNC controlled machine tool and possibly also a product storage buffer and a material handling device.
- Flexible manufacturing cells consisting of a number of SFMs and buffers. Two types of FMCs were found in the survey:
  - those with one material handling devices capable of transporting products between all positions in the cell (referred to as *FMC-1*).
  - those with more than one material handling device responsible for the transportation of products in the cell. In the study FMCs with 2 robots were observed (referred to as *FMC-2*)
- Flexible Manufacturing Systems consisting of a multitude of SFMs, buffers, and material handling devices In contrast to FMCs, there are no material handling device capable of transporting products globally in the cell. Instead the transportation between the units is accomplished by the cells having common buffers. All assembly systems covered by the study belonged to this group. (This cell type is referred to as *FMC-3*).
- Multi-cell FMS, *MCFMS*, which consists of FMCs, buffers, and possibly also SFMs. These units are served by a material handling system (AGVs, conveyors, etc.) capable of transporting products between all positions in the cell,

Resources Classification

One of the main interests in the study was whether the resources in the studied cells could be defined according to the classification made in the architecture, i.e. into producers, movers, and locations. The survey showed that, apart from the three types already identified, a fourth type, *hybrid resources*, are very common. Such resources cannot be classified as being a producer or a mover, since they carry out operations that both transport the product (i.e. act as a mover) and carry out some kind of processing operation on the products (i.e. producer). The hybrid resources identified the survey are:

- *Producing movers.* Robots, tripod robots, and gantry cranes that, apart from handling the products (mover), also assemble, act as fixture for, and/or carry out processing operations on products. These processing operations were positioning, washing, greasing, and deburring of products.
- *Moving producers.* Machine tools that are capable of both processing and transporting products (i.e. loading and unloading of products).

Handshake Signalling

Another important question to be answered was whether there were any handshakes carried out between the resources that are not considered in the architecture. The survey showed that approximately 50% of the cells did use handshake signals between the resources that are not included in the reference architecture. The signals identified in the survey are:

- Signals between two movers in order to avoid collisions (referred to as blocking in Table 1),
- Measuring machine that sends correction values to a machine tool,

- Machine tool that sends a signal to the mover that a product is not positioned correctly in the fixture (Error indication),
- Machine tool that sends a signal to the mover before the machine tool is finished, in order for the mover to prepare and be ready when the machine has finished,
- A mover determines the position of a product and sends the co-ordinates to another mover,
- Product identification number is sent from one mover to another.

Table 1

Overview of the Surveyed Cells and their Specific Features. (ass. = assembly operation; fix. = the mover manages the fixturing of the product; transp. = transport operation; proc. = process operation; paral. mach. = machines with similar operation capability; blocking = movers block each other in mutual work space).

Com-pany	Cell #	Process Type	FMS Class	Moving Producer	Producing Mover	Specific Handshakes	Other features in the cell
A	1	machining	FMC-1	-	-	error indication	-
A	2	assembly	FMC-3	-	transp./ass.	blocking	-
A	3	machining	FMC-1	-	transp./fix.	-	-
A	4	assembly	FMC-3	-	transp./ass./ proc.	blocking	product disassembly and assembly
A	5	machining	FMC-3	-	transp./fix.	blocking	-
A	6	machining	MCFMS	-	-	communication of product ID	-
A	7	assembly	FMC-3	-	transp./ass./ proc.	blocking	-
A	8	machining	SFM	proc./transp	-	-	-
A	9	machining	FMC-2	-	-	blocking	-
A	10	assembly	FMC-3	-	transp./ass.	product location data is transferred	co-operation between robots
A	11	machining	FMC-1	-	transp./proc.	signal two minutes before op. ready	-
A	12	assembly	FMC-3	-	transp./ass./ positioning	blocking	-
A	13	machining	FMC-1	-	transp./proc.	-	-
B	1	machining	MCFMS	-	transp./proc.	feedback of measurement data machine	sequence dependent operation
B	2	machining	MCFMS	-	transp./proc.	-	paral. mach.
B	3	machining	FMC-3	-	transp./proc.	-	-
B	4	machining	MCFMS	-	-	-	paral. mach.
B	5	machining	FMC-3	proc./transp.	-	-	-
B	6	machining	FMC-3	-	transp./proc.	-	paral. mach.
C	1	machining	SFM	proc./transp.	-	-	individuals born in cell
C	2	machining	FMC-2	proc./transp.	-	-	fixture transp.
D	1	machining	FMC-1	-	transp./proc.	-	many products handled & processed simultaneously
D	2	machining	FMC-1	-	transp./proc.	-	-
D	3	assembly	FMC-3	-	-	-	very short assembly time
E	1	machining	FMC-1	-	-	-	no tool transp.

Other Features

Other features identified in the cells are:

- Products that are disassembled and then assembled again later in the product flow, requiring synchronisation of product flow.
- Products are created within some cells,
- Operation that positions the products,
- Co-ordination of tools, pallets, and fixtures in the cell.
- Low flexibility because tools cannot be transported between machines automatically.
- Transportation of batches of products between cells.
- Some cells have parallel machines that can carry out the same operations,
- Buffers that are used for cooling products as well as buffering.
- Many product handled & processed simultaneously by robot,
- Separate resources in some cells that identifies products (vision systems, bar-code readers, etc.)

DISCUSSION

Based on the results from the survey, the applicability of the architecture on the surveyed cells is discussed. The discussion mainly considers (i) the applicability of the architecture in general, (ii) the classification and modelling of the resources, (ii) the handshake signalling between the resources, and (iii) the modelling of the products' operational requirements.

Applicability

The aim of study was to increase the applicability of the architecture by analysing if the architecture is applicable to the cells and, if the architecture is difficult to use for a certain cell, analyse the specific features of the cell.

Among the 25 cells selected for further analyses, a great variety was observed, each putting specific requirements and constraints on the cell controller. However, most of the cells in the study was of a type that made it possible to use the architecture directly. The applicability of the architecture depends on (i) the process type in the cell, i.e. assembly or machining of products, and (ii) the physical layout and product flow. These constrain the design of the information system, controller structure, functional de-composition, and the implementation of the controller.

The type of cell strongly affects the applicability of the architecture:

- SFM. This type of cells can easily be controlled according to the architecture concepts. However, the use of the architecture would probably result in a solution that is too complex for such simple task.
- FMC-1. This is the cell type that the architecture originally was built for.
- FMC-2. The architecture can be used for this type of cell, if it is modified to include the specific requirements of using several movers.
- FMC-3. In order to use the architecture for this type of cell it would probably need extensive to be modifications or extensions.
- MCFMS. Considering the hierarchy of these systems, the task of controlling a MCFMS is similar to the task of controlling a single FMC. Therefore, the architecture could be used for controlling MCFMSs, if each subordinate FMC is controlled by a separate control system.

The only observed cell types that the architecture is not directly applicable to, are the FMC-2 and FMC-3 types.

Resource Models

The definition of generic resource models that describe the generic functions and behaviour of the resources, depends on (1) the way the resource interacts with the other resources (i.e. the handshake signals) and (2) on the type of operations it performs. The exact operations carried out by the resources, e.g. milling, lathing, or assembly, must not be considered since all processing operations on an abstract level can be considered to be equal. Furthermore, the actual type of physical resource used to execute the operations is not relevant, only the type of operations carried out, e.g. it does not matter if it is an AGV, a robot or a conveyor system that transport the products, or if the assembly operation is carried out by a robot, a tripod robot, or a special assembly machine.

If the handshake signalling protocol is defined in the database, each general resource model can represent a vast number of manufacturing resources. Each resource then uses the handshake information retrieved from the database, to be able co-operate with the other resource in a correct manner. The correct manner depends on the resources (resource types and resource individuals) and on the product involved.

A manufacturing system can be viewed as a set of resources needed to manufacture products. Based on the survey, the resources are divided into four types: producers, movers, locations, and hybrid resources. Each of these can be described using a generic model.

The resource models must include functions needed for assembly and disassembly operations, e.g. functions for creating and deleting products and their operation lists.

Studies of the surveyed cells, indicate that the use of an additional hierarchical level could simplify the control of the systems, i.e. grouping of certain resources in the cell into stations that to the cell control system can be regarded as one resource, capable of carrying out operations (or transporting products). In this way the design of the control system would be simplified in many cases.

Product Models

The survey made it evident that a much clearer definition of the terms "product" and "operation" are required. The control system of manufacturing cells must also include capability to handle batches or groups of products, e.g. a pallet of products. The architecture defined so far only considers products.

Process operations. A process operation is something that is required by the product. Transporting operations, however, are not considered as something that the product requires, rather it is a requirement based on the layout of the production system.

An operation is something that, in some way, changes any property of the product. Usually, it is the shape of the product that is changed, but it can also be the position, colour, surface, temperature, etc. Moreover, it can be a change in the information regarding the product, e.g. geometry data defining the shape (measuring operation), decision whether the product fulfils quality requirements, or knowledge of the products position. As an example of the different types of operations, there are two ways of determining the location of a product: (1) measuring the position and storing the co-ordinates, or (2) changing the position of the product to a known position.

Operations can be classified into the following three groups:

- Operations that take one product and results in one product (with changed properties).
- Operations in which the number of products decreases, (assembly operations) .
- Operations in which the number of products increases (e.g. disassembly).

One generalisation of the operation concept is that for each operation more than one resource might be required. This must be included in the product models and in the control system logic. Furthermore, it must also be possible to include in the description of the operations that certain fixtures are required.

Assembly operations demands the possibility to describe in the operations which other products or product types, that are required to execute the assembly operation. Disassembly operations must include information on which new product types should be created as a result of the operation.

Transporting operations. The capability of the movers is defined in the information model, i.e. which resources and products that the mover can handle. However, the current architecture does not consider situations where the transport of a product requires the use of several movers that transport the product in sequence.

Handshake

The handshake signalling protocol between the resources as it is defined in the architecture, does not cover all the signals found in the survey. Handshake between two movers in order to transfer products can be treated in the same way as if the mover was to load the product into a machine: the mover holding the product initiates the handshaking, and the handshake procedure is defined in the database. The way the hand-shaking is carried out depends on the product, product state, and on the resources involved.

CONCLUSIONS

This paper presents research on increasing the generality of a reference architecture for cell control systems defined at Chalmers University of Technology. In order for the architecture to be more general, an industrial survey has been conducted to gain knowledge on the variety among cell control systems.

The architecture was found to be directly applicable to many manufacturing cell control systems. However, among the observed cells, 25 cells were selected for further analysis since the layout, process type, product flow or other features in these cells differed from the specific cell type which the architecture is aimed for. These cells were classified into five cell types.

The analysis showed that it is difficult to apply the architecture on control systems for one of these cell types since the layout and product flow differ fundamentally. These cells consist of a multitude of SFMs, buffers, and material handling devices, but in contrast to the other cells, there are no material handling device capable of transporting products globally in the cell. Instead the transportation between the units is accomplished by the cells having common buffers. All assembly systems covered by the study belonged to this group, and, since assembly systems differs also in other aspects, a separate architecture should be defined for such systems, rather than aiming at defining one general architecture.

The analysis also showed that the architecture could be applied to a the other four cell types if the following features are handled in the reference architecture:

- A single operation on a product may require multiple resources to execute,
- Products can be created or deleted within the cell, in order to be able to handle assembly or disassembly operations,
- Multiple movers that must co-operate in order to execute transportation tasks, in contrast to the cells where one transporting device is capable and responsible for transporting the products between all the resources in the cells,
- Specific handshake signals, e.g. signals between movers to avoid collision in mutual work space,
- Hybrid resources, i.e. resources capable of carrying out both transporting and processing operations.
- Product operation lists that include disassembly, assembly operations.

REFERENCES

1 **Adlemo, A., Andréasson, S-A., Fabian M., et al.** "Towards a Truly Flexible Manufacturing System", *Journal of Control Engineering Practice,* Vol. 3, No 4, pp 545-554, April 1995.

2 **Adlemo, A., Gullander, P., Andréasson, S.-A., et al.**, "Operator Control Activities: A Case Study of a Machining Cell", *Symposium on Information Control Problems in Manufacturing Technology, INCOM'95*, Beijing, China, to be presented in August October 1995.

3 **Andréasson, S-A., Adlemo, A., Fabian M., et al.**, "A Machining Cell Level Language for Product Specification", *Symposium on Information Control Problems in Manufacturing Technology, INCOM'95*, Beijing, China, to be presented in August October 1995.

4 **Burgos, A., Sarachaga, M. I., Llorente, J. I., et al.**, "The Role of Integrating Platforms", *First World Congress on Intelligent Manufacturing Processes and Systems*, Puerto Rico, 1995.

5 **Bauer, A., Bowden, R., Browne, J., et al.**, *Shop Floor Control Systems. From Design to Implementation*, London, Chapman & Hall, 1991.

6 **EasyMAP,** *EasyMAP User's Manual* (MAN/6000), Computer Resources International (CRI) Industrial Systems A/S, Birkerød, Denmark, 1992.

7 **Fabian, M., Gullander, P., Lennartson, B., et al.**, "Dynamic Products in Control of an FMS Cell", *Symposium on Information Control Problems in Manufacturing Technology, INCOM'95*, Beijing, China, to be presented in August October 1995.

8 **Gullander, P.,** *Architecture for Flexible Cell Control Systems*, Licentiate Thesis, PTA 96:01, Department of Production Engineering, Chalmers University of Technology, Göteborg, Sweden, 1996.

9 **Gullander, P., Fabian, M., Andréasson, S-A., et al.**, "Generic Resource Models and a Message-Passing Structure of an FMS Controller", IEEE International Conference on Robotics and Automation, ICRA'95, pp 1447-1454, Nagoya, Japan, May 1995.

10 **Klingstam P., Mårtensson, P., Gullander, P.,** "Survey on FMS-cells in Swedish Manufacturing – Evaluation of Reference Architecture for Cell Control Systems " (in Swedish), External Report, PTE 97:01, ISSN: 1102-2817, Department of Production Engineering, Chalmers University of Technology, Göteborg, Sweden, 1997.

11 **MacCarthy, B., Liu, J.,** "A New Classification Scheme for Flexible Manufacturing Systems," *Int. J. of Production Research*, Vol. 31, No. 2, pp. 299-309, 1993.

12 **MMS,** *Industrial Automation Systems – Manufacturing Message Specification – Part 1: Service Definition*, International Standard, ISO/IEC 9506-1, First ed., 1990.

13 **Sargent, P.,** "Inherently Flexible Cell Communications: A Review", *Journal of Computer Integrated Manufacturing*, Vol. 6, No. 4, pp 244-259, November 1993.

PRE-TREATMENT OF DATA FOR IMPROVED ARTIFICIAL NEURAL NETWORK MODEL GENERATION

Parvin S. D., Peel D., Leck M.
Division of Chemical Engineering
School of Science and Technology, University of Teesside, Middlesbrough, TS1 3BA

e-mail : S. Parvin@tees.ac.uk
: D. Peel@tees.ac.uk

ABSTRACT

Neural networks are now being used extensively in many areas of control. Model predictive control schemes and several other techniques have adapted rapidly to encompass the non-linear predictive capabilities of neural networks and many papers have been written looking at the feasibility of these systems. Much work has also been carried out in the area of neural network learning algorithms and architecture.
This paper deals solely with the pre-treatment of real data produced by a binary methanol/water distillation column to produce a prediction of top composition using variables other than top temperature. The predictor used is a Feed Forward Artificial Neural Network with a standard back error propagation learning algorithm. The work covered shows that with additional pretreatment of the data the neural network model produced is an improvement over a neural network model produced using the untreated data, over the column operating range. The pretreatment techniques used are easily adaptable to other data sets and systems. Techniques used involve data smoothing, frequency content analysis and a trial and error based computational technique.

1. INTRODUCTION

Neural networks have been show to be very good at mapping complex continuous non-linear functions.[1] It is for this reason that they are becoming more widely used in advanced control schemes and for process modelling.[2-6]

Generally process models can be used in one of two main ways either in conjunction with a sensor for control purposes or by itself as a soft-sensor. A soft-sensor being an on-line estimator which would produce accurate estimates of the outputs from a process to be used in a control scheme, instead of a measured value. Industrial applications of soft-sensors have been successful in the past and papers have been published to support this.[7]

The generation of a model as a step ahead predictor for control purposes does not deal with as many constraints as for some cases of soft-sensing. For slow dynamic systems a process which has the last measured variable available and only requires a small step ahead predictor, a relatively good model can be produced easily by placing the majority of the weighting of the new variable on the last measured variable. As y(t-1) is available errors are not compounded. In the case of soft-sensing errors are compounded as the measured variable is unavailable and only the previously predicted output y*(t-1), the model must be more accurate to prevent the compounding of errors.

To produce a good model for purposes of control the historical data that is used should be carefully selected to be appropriate to the operating conditions over which the model will be used. An initial consideration is whether the model is required to capture either the steady-state relationships only or the dynamic responses of the process also. If a dynamic model is required then some way to introduce dynamics into the system must be determined. This can be achieved either by incorporating dynamics in the network structure or within the data itself.

Current good practice requires that a 'good spread' of time domain data be presented to the modelling technique, but pays little heed to the weighting of the data or the frequency domain content. This study suggests some techniques, that may be useful in model generation and then applies those techniques to a real distillation column process.

As with time domain a good spread of low frequency domain information seems useful also. This is why a step response is often used in linear system identification. A step response being made up of a good range of frequencies. With this in mind the initial training set selected should contain a reasonable spread of low frequency data.

Frequency content analysis has been used previously for a more detailed analysis of neural network learning and frequency reproducibility.[8]

2. THE DISTILLATION COLUMN

The data used for model generation and validation was generated on a full-scale methanol/water distillation column. The data was sampled once every minute with readings taken for Top, Bottom and Feed temperature, Feed Flowrate, Bottoms take-off, Reflux ratio, Steam to reboilers, Bottom level and Feed Composition. The column is a full scale automated distillation column, used for research, teaching and business collaboration. A more detailed description can be found in a further reference.[9]

3. THE NEURAL NETWORK

The neural networks used for the prediction of top composition of a methanol/water distillation column for use in an advanced control algorithm, as a step ahead predictor, and to replace the estimate produced from a linear algorithm in conjunction with at top temperature sensor, as a means of soft-sensing. The neural network for soft sensing was a three layer feed forward recurrent network with 7 input neurons and 5 neurons in the hidden layer, as determined by examining the learning capability of several networks and following some simple guidelines. The training algorithm used was a standard back error propagation algorithm, with a sigmoid squashing function and the inputs scaled between 0.1 and 0.9. Several good texts exist dealing with neural networks and the back error propagation method of training.[10]

The inputs to the neural network were Feed Temp, Reflux Ratio, Reboiler Steam, Feed Flow, Feed Composition and previous Top Composition. The value of the previous top composition used for training was the actual previous top composition as measured off-line. However for the validation set the last predicted value of the top composition was used, known as a recurrent net.

The data set used for training the model contained extreme points of data for each of the training variables, so that the neural network model need only interpolate. The validation set consisted of data that was deemed as a typical operation for the column.

4. DATA PRETREATMENT

Data pre-treatment should begin by outlier identification and removal, correlation analysis and phase shifting analysis to determine time delays. The best way to begin this is to plot the time series data. This will give an initial indication of outliers, which variables are correlated with which and of time delays. This becomes more difficult to do the greater the number of variables and the more dynamic the data.

The required time delays should then be introduced to the data by time stepping the variable a number of sample periods forward or backwards as required. The techniques described do not allow for processes with variable time delays.

4.1 Repetitive Analysis Sorting

This program was designed to help determine the optimum training set for a process, given a defined operating range, in this case a representative validation set. Care must be taken to ensure that the validation is not too similar to the training set as this would lead to the process reselecting the validation set from the training set. Each training run must include the maximum and minimum operating ranges for each variable.

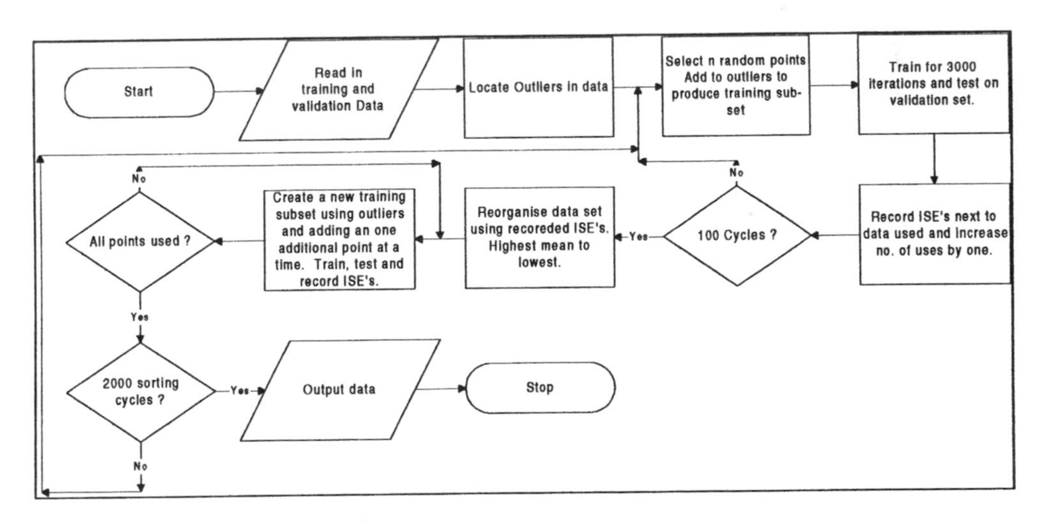

Figure 1. Data Sorting Flowchart.

The sorting program (Fig 1.) works by selecting a fixed no. of random points and compiling a sub-set of the full training set using these points and the extreme points. The sub-set is then used to train the network over a fixed number of iterations. The model is tested on a validation set and the ISE recorded. The ISE is added to a total ISE for the data points used and the used variable is increased by one. This process is repeated 100 times and the mean ISE's for each set of data points is calculated and recorded. After the 100 repetitions the data is then reorganised from highest to lowest mean ISE values.

A series of further runs are then carried out using a second sub-set containing the extreme points and adding one point at a time until the full data set had been included. These runs gave a minimum ISE at a certain number of points. The minimum ISEs and the number of points were recorded and the run continued to produce a final set of results (Fig 2.).

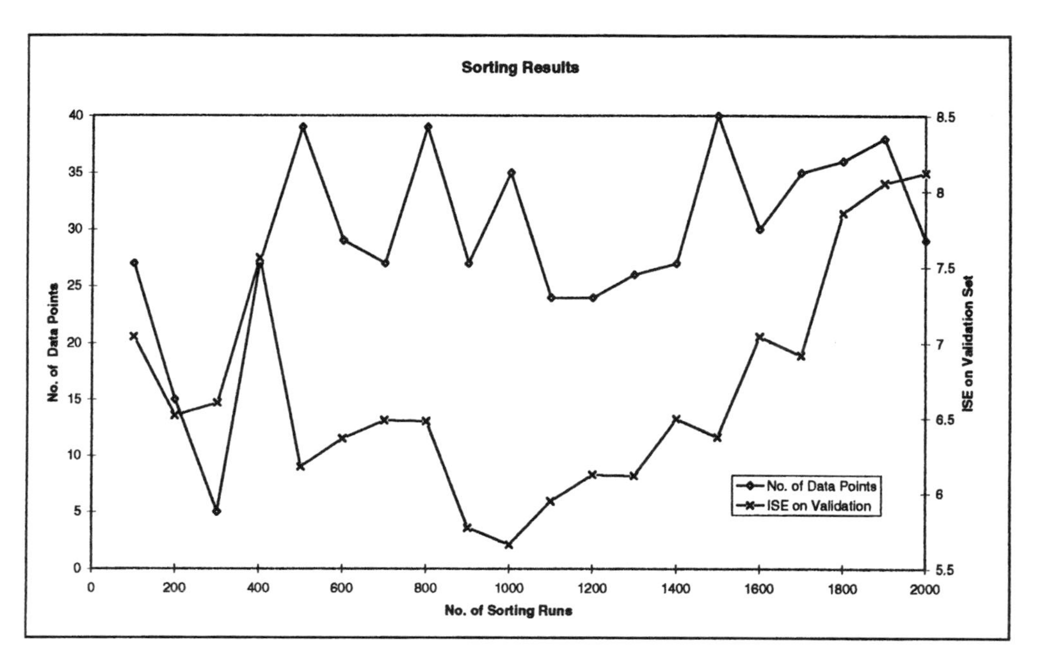

Figure 2. Sorting Technique Results 1

Figure 2 shows that the sorting technique produces a minimum ISE of around 5.71 after some 1000 cycles of the sorting program. This minimum ISE is produced with the best 25 rows in addition to 9 containing the outliers.

Although the sorted data uses fewer data points it does show a good improvement in ISE. The lower ISE results from the model focusing it's attention on points which give a clear relationship and focus on the operating range presented by the validation set.

The graph Figure 3. shows similar trends in the model performance with the unsorted data having a few quite bad predictions. It is these bad predictions introduced into the model by erroneous points in the training data that the sorting technique has removed. In each run through of the sorting technique weights against any data which does not comply to the relationships contained in the validation set.

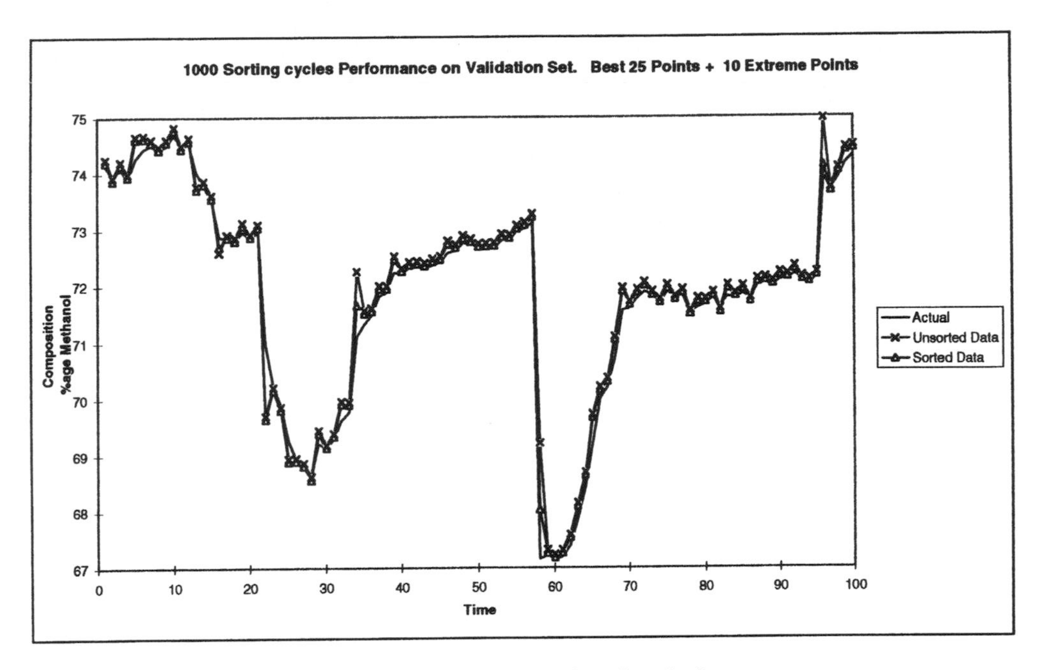

Figure 3. Sorting Technique Results 2

TABLE 1
Sorting Technique Results

Data Set	ISE on Validation Set
Unsorted	11.01
Sorted	5.71

4.2 Data Smoothing

It has been shown in previous work that the input data density for model generation does have a noticeable effect on the performance of the model over the range.[11] Therefore if a model is desired to be equally accurate over a range then it follows that an equal distribution of data should be used to produce the model.

This procedure works best for steady state models or with data that has had dynamics introduced into it, by using past output and input values, as the dynamics could be lost by the removal of points within the set. In this case the model is assumed to be using data which has had previous values of the output inserted into the data set.

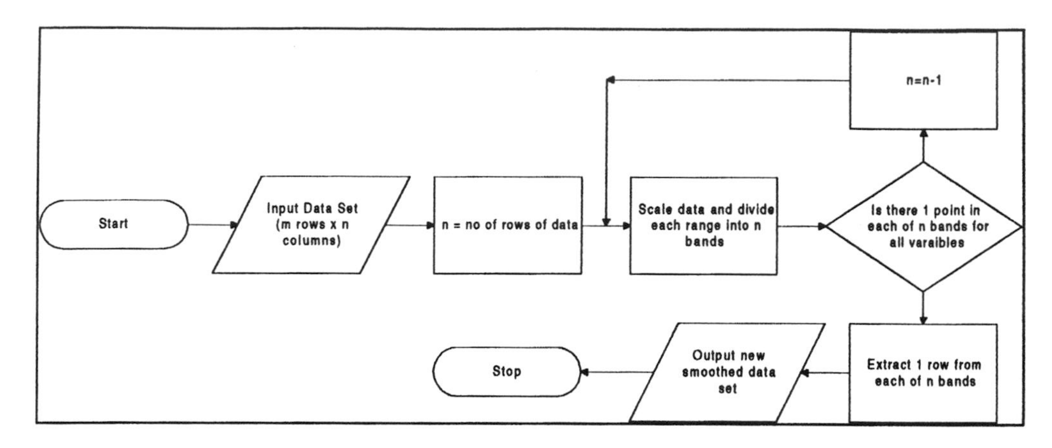

Figure 4. Data Smoothing Flowchart

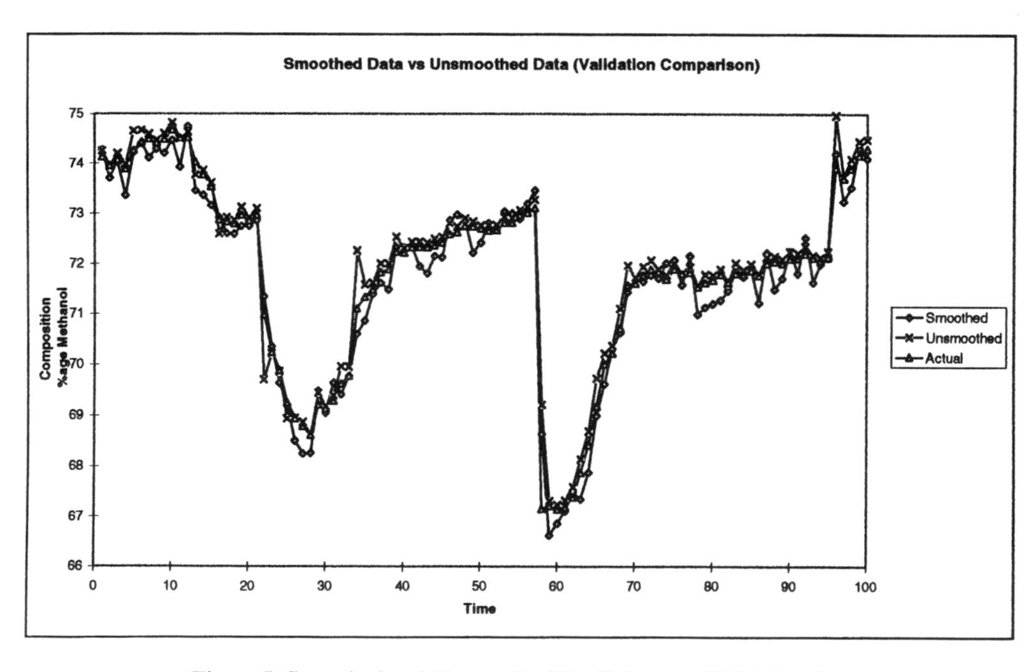

Figure 5. Smoothed and Unsmoothed Predictions on Validation Set

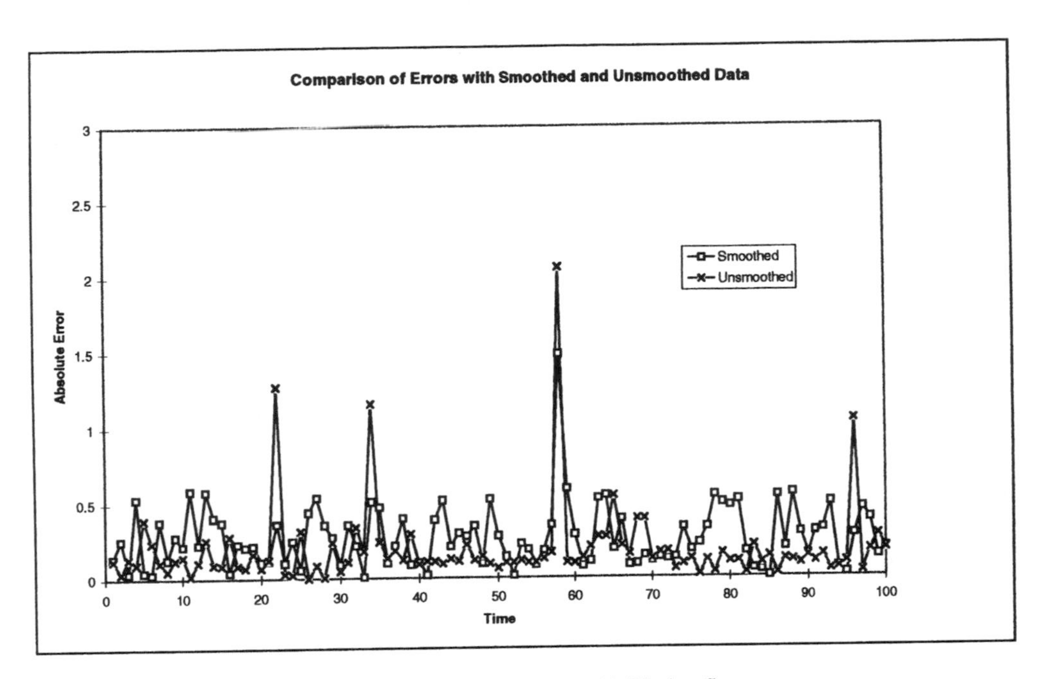

Figure 6. Absolute Errors on Validation Set

TABLE 2.
Smoothing Technique Results

Data Set	Mean Abs Error	SD	ISE
Smoothed	0.27	0.21	11.62
Unsmoothed	0.19	0.27	11.01

Although the smoothed data produces a higher ISE on the validation set than the unsmoothed the maximum absolute error is only 1.5 as opposed to around 2.1. This could prove beneficial in processes with sensitive dynamics. It also shows that the smoothing of the data points produces a more even model of the process.

4.3 Frequency Analysis of the Training Data Set

A Base 2 Fast Fourier Transform is identical to the Discrete Fourier Transform with the algorithms being handled in a slightly different manner to facilitate faster calculation of the frequency domain representation. The matrices are factorised and rearranged to reduce the number of complex additions and multiplications which must be carried out.[12-13]

The end result from the FFT is the conversion of the input signal into it's component frequencies, both real and imaginary parts.

The data set is converted into the frequency domain using an algorithm based on the transform :

$$H\left(\frac{n}{NT}\right) = T\sum_{k=0}^{N-1}\left[e^{-kT}\right]e^{-j2\pi nk/N} \qquad (n=0, ..., N-1)$$

(2)

This being the discrete time version of the Fourier Transform.

One large problem with the use of a FFT to transform from the time domain to the frequency domain is that of leakage which can cause small ghost points to appear at frequencies close to a frequency of large amplitude. This means that in such a complex frequency domain as a data set the margin for error from the actual signals would be fairly large. In this case however as comparisons are being drawn between two sets of frequency domain data which have both been inverted using the same method then the ghost signals caused by leakage would appear in both and the same baseline is used.

Another problem with analysis of a data set's frequency components, results from the Nyquist sample theory. This states that the highest frequency that can be extracted from a continuous series of data is equal to half the sample period. In this case 1/(60*2) = 0.00833 Hz. This means that noise with a frequency higher than this can not be isolated from the data if it exists. If noise did appear in a data sets frequency composition then the noise could be isolated and removed and the original signal reconstructed.

Also non-linear systems cause new frequencies to appear in the output, so direct comparison of complex input and output frequency content is almost impossible.

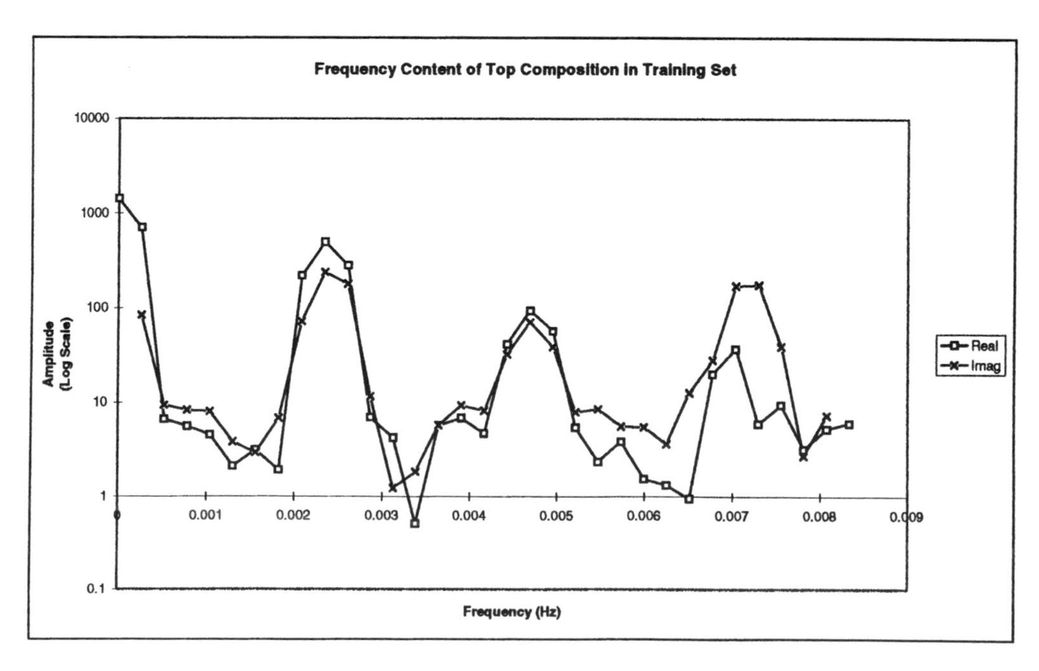

Figure 7. Frequency content of Top Temperature in Training Set.

Examination of the frequency content show similar characteristics in both the real and imaginary components, with a good spread across the frequency range in question. Had the training data set contained only very low frequency information 0.001Hz or less then a separate data set should be selected, with a more even spread of frequency information.

5. CONCLUSIONS

The additional pre-treatment of the data in this case has generated a significally better set of models the smoothed model produced a predicted output with a smaller SD of the error for a small increase in ISE. This means that although the model overall is slightly more inaccurate, it does produce a more even model for the full range and reduces the maximum absolute error produced by the model.

The sorting technique produces a far better model in ISE terms than the untreated data set. This is achieved by the elimination of inaccurately recorded data and possibly incorporates some data smoothing also.

6. FUTURE WORK

As the correct pre-treatment of data is essential for the generation of a good model, neural or otherwise there exists a need for further development in the areas of data pretreatment. Frequency based analysis of data sets can provide insight into the dynamic information in a data set, which is invaluable for model generation purposes.

Many useful statistical techniques currently used for fault diagnosis eg. PCA could also be used to examine data sets to identify erroneous points, which would hinder model generation.[14]

PCA is also a useful tool for data reduction, by reducing the dimensionality of a problem. Work using Principal Components instead of standard input data, for the generation of models is ongoing.

7. REFERENCES

[1] **Narendra, K. S. and Parthasarathy, K.**, Identification and Control of Dynamical Systems Using Neural Networks, IEEE Transactions on Neural Neural Networks, Vol. 1, No. 1, March 1990, pp4-26

[2] **Morris, A. J., Montague, G. A., Willis, M. J.**, Artificial Neural Networks Studies in Process Modelling and Control, IChemE Chemical Engineering Research and Design, Special Edition on Process Control, Jan 1994, sec 2.1-2.16.

[3] **Hunt, K. J., Sbarbaro, D., Zbikowski, R. et al.**, Neural Networks for Control Systems : A Survey., Automatica, Vol 28, No. 6, pp1083-1112.

[4] **Margaglio, E. and Uria, M.**, Identification of Multivariable Industrial Process using Neural Networks : An Application, IEEE Proceedings on the Intl Conference on Systems Man and Cybernetics, 1994.

[5] **Willis, M. J., Montague, G. A. and Morris, A. J.**, Modelling of Industrial Processes using Artificial Neural Networks, Computing and Control Engineering Journal, May 1992, pp113-117.

[6] **Turner, P., Montague, G. A., Morris, A. J.**, Neural Networks in Process Plant Modelling and Control, Computing and Control Engineering Journal, June 1994, pp131-134.

[7] **Tham, M., Morris, A. J. and Montague, G. A.**, Soft-sensing : A Solution to the Problem of Measurement Delays, Chemical Engineering Research and Design, Vol. 67, Nov 1989, pp547-554

[8] **Parvin, S. D. and Peel, D.**,Frequency Content Analysis of Predicted Model, Conference Proceedings SICICA '97, June 9-11. (To be published)

[9] **Leck, M. and Peel, D.**, Advanced Control of a Distillation Column, Conference Proceedings, FAIM '97, June 1997.

[10] **Wasserman, P. D.**, Neural Computing Theory and Practice, Van Nostrand Reinhold. New York.

[11] **Williams, C. K. I., Qazaz, C., Bishop, C. M. et al.**, On the relationship between Bayesian Error Bars and the Input Data Density, Artificial Neural Networks, 26-28 June 1995, Conference Publication No. 409, IEE.

[12] **Brigham, E. O.**, The Fast Fourier Transform, Prentice-Hall, 1974.

[13] **Elliott, D. F. and Rao, K. R.**, Fast Transforms : Algorithms, Analyses, Applications. Academic Press, 1982.

[14] **Zhang, J., Martin, B. E. and Morris, A. J.**, Non-linear Statistical Process Monitoring, Advances in Fault Diagnosis in Process Control 3, York, 22 April 1996.

ADVANCED CONTROL OF A DISTILLATION COLUMN

Leck, M., Peel, D.
Division of Chemical Engineering, University of Teesside

ABSTRACT: A pilot-scale methanol-water distillation column has been purpose built in the Chemical Engineering Department as a test bed for advanced control algorithms. A computer is interfaced to the column via a serial link to two Eurotherm controllers, making implementation of algorithms simple and quick. Simple P.I.D. control has been implemented as a benchmark with Generalized Predictive Control implemented as an example of an advanced linear adaptive control algorithm. A simple, error driven, modification to the classical Generalized Predictive Control algorithm has been proposed and tested on the distillation column and gives improvement over its predecessor. Furthermore, a simple non-linear neural net based controller has also been tested on the distillation column and found to give better performance than fixed parameter GPC.

Process Description

A pilot-scale methanol-water distillation column exists as a research test bed within the Division of Chemical Engineering at the University of Teesside. The process description (Figure 1) is simplified and shows only the basic layout and information relevant to this paper. The column is 9" in diameter and contains 18 sieve trays with a tray spacing of 10". Overall the distillation column is 22 feet high and spans 3 floors. The column and associated pipework are made almost exclusively of glass making it a unique opportunity for observation of hydrodynamic effects. Pumps are not shown on the diagram but are all made of glass and can pump approximately 100 gal/h against a head of about 5m. There are 3 reboilers grouped as one pair and one individual and a feed preheater all supplied by a common steam main. There is a double pipe condenser, a bottom product cooler and a top product sub-cooler.

Reflux Ratio is controlled via a mechanical divider underneath the condenser which allows either condensate to flow down the top product line or to flow back onto the top plate.

Top product quality is inferred from the top plate temperature, calibrated via a Gas Chromatograph.

The circles included in the diagram indicate internal P.I.D. loops supervised by two Eurotherm controllers which are directly connected to plant signals. These P.I.D. loops are tuned and function independently of any advanced control algorithms being implemented. The P.I.D. loops indicated are as follows :

1. Feed Flowrate into the distillation column
2. Feed Temperature
3. Feed Concentration of Methanol
4. Bottom Product Flowrate
5. Water Flowrate to Mixing Tank
6. Methanol Flowrate to Mixing Tank

The last two P.I.D. loops mentioned above are under ratio control to mix new batches of feed to the correct concentration before transfer to the feed tank where it will be fed to the column.

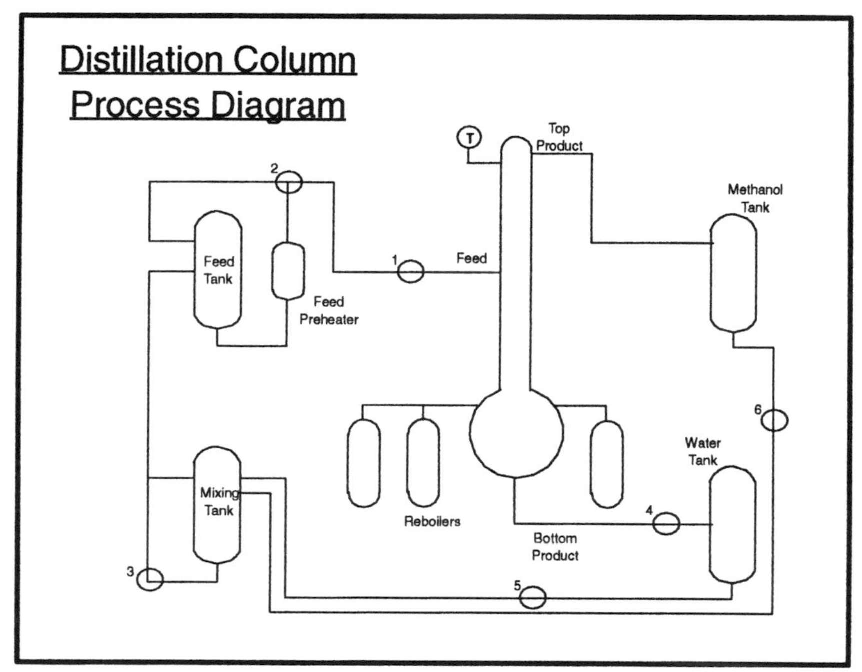

Figure 1. Distillation Column Process Diagram

The hardware configuration makes implementation of advanced control algorithms a simple matter of programming hence a library of control methods can quickly be generated.

System Identification

For control purposes it is necessary to examine the system in terms of it's response to changes in input. Distillation columns, typically, have non-linear and direction dependent dynamics which have a more pronounced effect at the top end of the operating range. If this is true then linear controllers will not give the optimum control performance and a non-linear controller should be proposed.

There are 5 basic inputs to any distillation column, all of which could have a significant effect on the quality of top product :

1. Feed Flowrate
2. Feed Concentration
3. Feed Temperature
4. Reflux Ratio
5. Reboiler Steam Flowrate

On examination of the influence of each of these inputs it can be easily concluded that the manipulated variable for controlling top product quality should be reflux ratio. This is the standard conclusion in industry[1]. This means that we can design a controller based on reflux ratio as the sole manipulated varialbe with inferred top product quality control as the control objective. The other 4 inputs can be used as pseudo-unmeasurable disturbances, that is a change can be introduced deliberately at a predefined time but will not be accounted for by the controller. This would only be done to examine controller performance prior to including these disturbances as feed forward signals.

Step tests in reflux ratio (Figure 2) will establish a great deal about the dynamic behaviour of the distillation column. Column data was sampled on a minutely basis and all other variables held at constant values or set points.

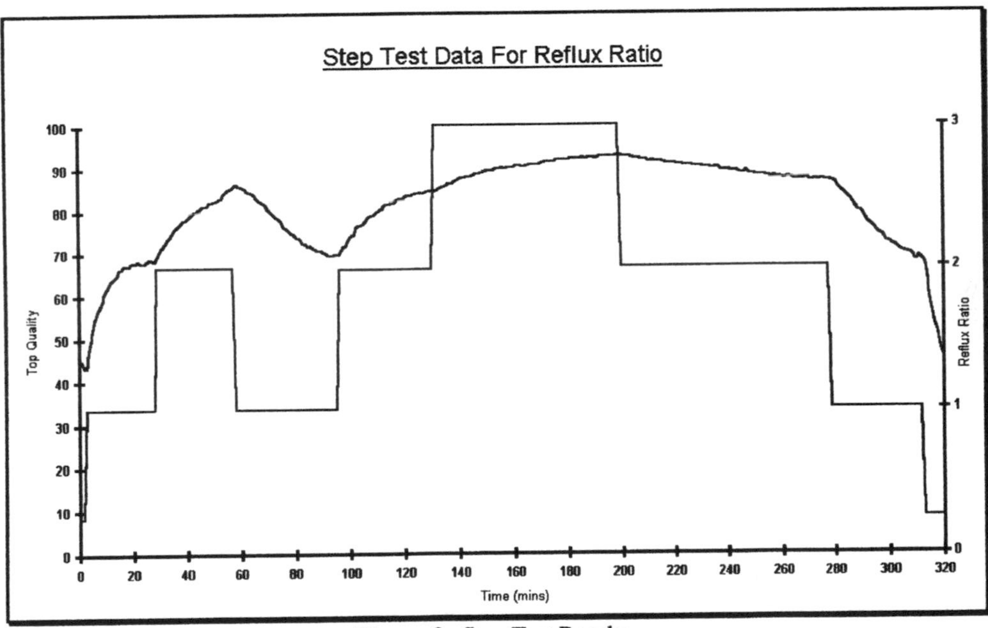

Figure 2. Step Test Results

The reflux ratio is limited in practice to upper and lower limits of 0.25 and 3.0 respectively. From Figure 2 it can be seen that this variation give us almost 50% variation in top product content of methanol

It is quite obvious from a visual inspection of Figure 2 that the column is positionally non-linear in both gain and time constant. There is no evidence of direction dependence or hysteresis. There is, however, a significant change in the gain and time constant over the range considered which could be problematic for a linear controller. There appears to be only a one sample time delay in the process. This can be ascribed to the digital nature of the sampling.

The system is adequately modeled by ARMA equation of the following form :

$$y_t = a . y_{t-1} + b . u_{t-1} + e \quad (2.1)$$

The exact parameter values 'a', and 'b' converge on different solutions for different operating ranges. This will be demonstrated later in this paper.

Generalized Predictive Control

Generalized Predictive Control (GPC)[2] has been in existence for a decade and is still one of the most advanced linear algorithms. Its capacity for adaption to varying processes and it's multi-step-ahead prediction capabilities make it ideally suited for application to the distillation column as a benchmark for the state-of-the-art in linear adaptive control.

At the time of it's creation GPC was the next step in the development of realistic advanced adaptive controllers. Until that point several control algorithms did exist which could control adequately, but each of them had limitations in the types of plant they could model or control. GPC was an attempt to encompass all of the good features of the previous algorithms and create one single algorithm that could successfully control all typical types of plant.

The 4 criteria laid down in the original development[1] of GPC, that GPC must be able to cope with are as follows :

1) Nonminimum-phase plant : most continuous time transfer functions tend to exhibit discrete time zeros outside the unit circle when sampled at a fast enough rate[6].

2) Open loop unstable plant or plant with badly damped poles[1].

3) Plant with variable or unknown dead time. Some methods (e.g. minimum variance self tuners) are highly sensitive to the assumptions made about the dead time and approaches which attempt to estimate the dead time using operating data tend to be complex and lack robustness[1].

4) Plant with unknown order : Methods such as pole placement or LQG self tuners perform badly if the order of the plant is overestimated because of pole/zero cancellations in the identified model, unless special precautions are taken[5].

GPC overcomes all of these problems. The controller can cope with plant with variable parameters, variable dead time and with model order which changes instantaneously (providing input/output data is sufficiently rich). The algorithm can cope with overparameterization without special precautions being taken. Nonminimum phase and open loop unstable plant are also accommodated[1].

Abridged Derivation

'The basic ARMA model used is :

$$A(z-1).y(t) = B(z-1).u(t-1) \quad (3.1)$$

The noise term has been omitted here. (from here the (z^{-1}) will be dropped)

The Diophantine Equation is defined as :

$1 = E.A.\Delta + z\text{-}j.F$ (3.2)

The polynomials 'E' and 'F' are uniquely defined given 'j' and the order of 'A'. Multiplying (1.1) by '$E.\Delta.z^{j}$' and substitute the result into (3.2) and rearrange we see the following :

$$y(t+j) = E.B.u(t+j-1) + F.y(t) \quad (3.3)$$

$$y(t+j) = G.u(t+j-1) + F.y(t) \quad (3.4)$$

It can be shown that one way to calculate **G** is simply to calculate the plant's incremental step response to incremental unit impulse and take the first 'j' terms as required. Hence, although recursion of the diophantine equation is possible it is unnecessary. The currently estimated parameters of the plant model can be used to directly generate the **G** polynomial.

y(t+j) can be said to contain 3 distinct components :

1) One depending on future control action yet to be determined.
2) One depending on past, known controls together with filtered measured variables.
3) One depending on future noise signals (not considered here).

The free response of the process is then introduced and defined as containing that part of the output which is composed of signals known at time 't'. Hence :

$$f(t+1) = [G_1 - g_0].\Delta.u(t) + F_1.y(t) \quad (3.5)$$
$$f(t+2) = [G_2 - z^{-1}.g_1 - g_0].\Delta.u(t) + F_2.y(t) \quad (3.6)$$

This procedure is basically separating the 'G' matrix from all of the terms which are caused by inputs known up until time 't'. In practice the control horizon is usually set to '1' and the " matrix becomes a polynomial in the backward shift operator. The free response can be simply calculated from the current recursively estimated parameters.

Hence the equation which describes the process over the next 'j' time periods is as follows :

$$y = G.\Delta u + f \quad (3.7)$$

The Control Law

The control law considered for GPC is the following :

$$J(N_1, N_2) = \sum_{j=N_1}^{N_2} (y(t+j) - w(t+j))^2 + \sum_{j=1}^{N_2} \lambda(j)(\Delta u(t+j-1))^2 \quad (3.8)$$

Here a new polynomial has been introduced; w(t+j) is simply a vector of set points used to generate a set of errors upon which to control. The second new parameter 'λ' is a control weighting sequence.

Minimisation of this cost function gives the following control law :

$$\Delta u = (G^T.G + \lambda I)^{-1}.G^T.(w-f) \quad (3.9)$$

This clearly demonstrates that if the control weighting sequence, 'λ' is zero then the control action is simply the error from set point divided by the 'G' polynomial, hence scaling the forced response to fit the required approach to set point. Addition of a non zero λ value serves to smooth the control action over a longer period of time.'[1]

A Simple Extension

One of the main reasons that Minimum Variance Control is not favoured on industrial systems is that it tends to make the control action too active in trying to drive the output to set point during the next sample period. The lack of projection into the future also prevents more intelligent control action based upon a projected response. The closer to set point the process actually is could be used to influence the control action taken. For instance, if the output is far from set point then an extreme control action may be necessary, closer to set point only small adjustments may be necessary.

It is already known than the control actions implemented by GPC tend to be slower and less extreme as 'j' increases. Hence if we make 'j' a function of error then we can tailor GPC's response a function of the magnitude of the error from set point.

Hence the following is proposed as a simple extension to the GPC algorithm :

$$j \propto \max(j) - (R * (\mathrm{abs}(error)^{-1})) \quad (3.10)$$

The upper and lower boundary of 'j' can be set to influence performance. The minimum 'j' would be unity, assuming no time delay. The horizon reduction parameter, 'R', describes how 'j' increases as the output approaches set point, i.e. at what point does 'j' become it's minimum value and how does 'j' increase as output approaches set point. In this study 'j' decreases linearly as the absolute magnitude of error increases (R=0.5).

An Optimising Neural Controller

Neural networks have gained in popularity for control purposes over the last decade. The model that a neural network produces after training has been proved to be able to represent, to an arbitrary degree of accuracy, any continuous non-linear function with only a single hidden layer[7]. This property is invaluable in providing models of inherently non-linear systems which, up until now, could not be satisfactorily controlled by linear controllers. References [9][10][11][12] provide a good background to neural networks and their applications for control purposes.

Neural networks have been successfully employed in control architectures for a wide variety of purposes. However the number of applications in simulation[13][14][15][16] far outweigh the number of applications on actual processes. Examples of neural network application to a distillation column simulations are not unknown[17][18] but examples of advanced control schemes implemented on real plant are few and far between.

The main difficulty with fixed linear models is the decrease in accuracy with time because of process variation. Such variation can occur due to fouling or catalyst degradation or if, as in our case, the plant in question is non-linear in nature then special care must be taken to operate in a linearised region. Adaptive linear models gleaned from adding a recursive estimation technique around the model relieve this problem to a certain degree but the adaption may be too slow to converge on the correct parameters in sufficient time to have little impact on control performance.

A simple multilayer perceptron[3] network trained off-line using the backerror propagation algorithm. Step test data along with steady state data can be used to train the neueral network to model the dynamics and steady states of the distillation column to a very high degree of accuracy. After data sorting and pretreatment[4] the data can be presented to the network and the network trained on in-house software over 70,000 training iterations taking 20 minutes.

This fixed model of the plant can then be used to predict the free response of the process from its present condition given any control action across the range of available control actions. Each free response will have an error associated with it and based upon the assumption that, the error surface is uncomplex (i.e. contains a single minimum), we can select a control action that minimises the error. This selection takes place via a simple optimisation technique. Overall computation of the appropriate control action takes a fraction of a second.

Experimental Procedure

Firstly, GPC was tested against normal P.I. & P.I.D. controllers to evaluate performance. An optimum value of the prediction horizon was selected via experimental determination and the performances compared.

The distillation column was subjected to a series of set point changes and disturbances in all other process variables with the exception of feed temperature. The relative performances of the control methods was compared by the sum of the squared errors for each sample (ISE) for each experiment.

In a second series of experiments the optimum performance GPC with fixed parameters was compared against the optimised neural controller over a series of set point changes only.

Conclusions

From Figures 3 - 9 the following information can be found :

Table 1
Comparison of Errors for Linear Controllers

CONTROL	ISE
PID	2377
GPC N=10	1974
GPC N=5	1115
GPC N α 1/error	1011

Table 2
Comparison of Errors for Linear Vs Non-Linear Fixed Model Controllers

CONTROL	ISE
Fixed Model GPC, N=5	4115
Optimised Neural Controller	3931

ISE = sum of squared errors for every sample period

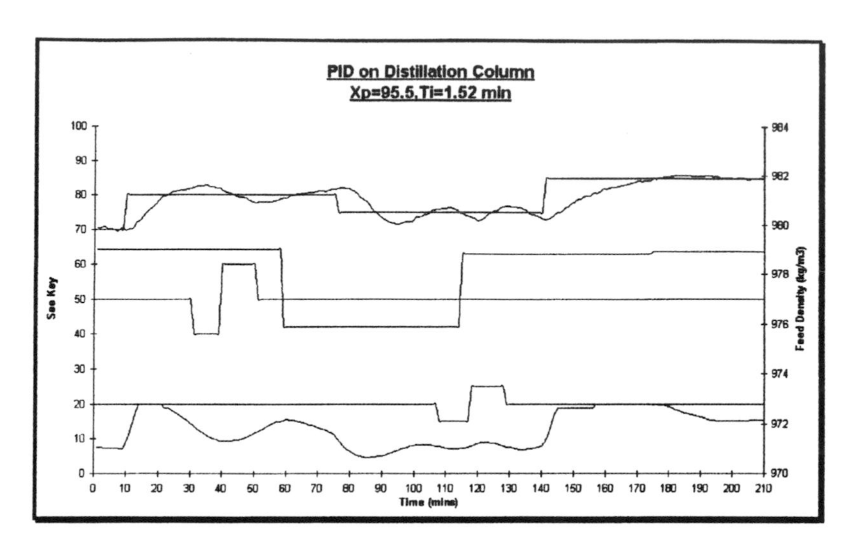

Figure 3. P.I.D. Control of the Distillation Column

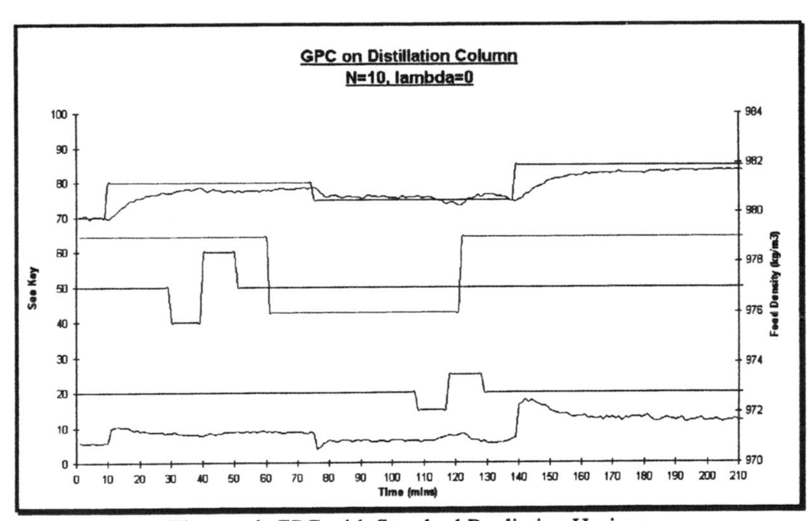

Figure. 4 GPC with Standard Prediction Horizon

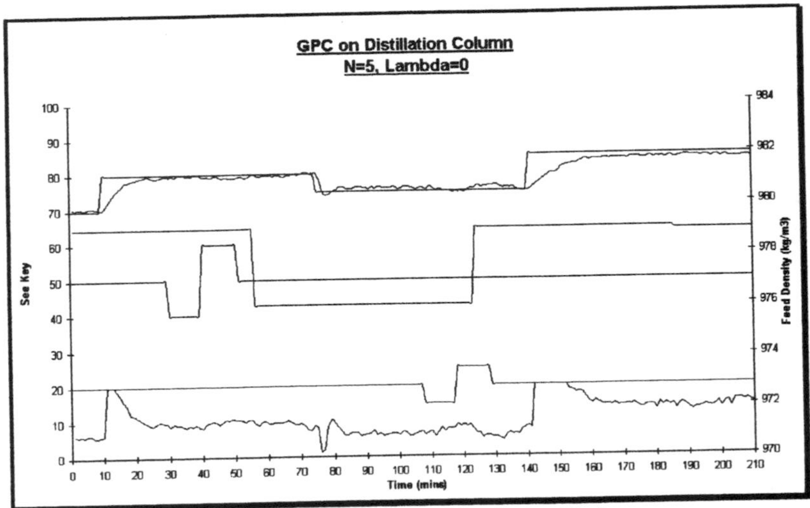

Figure 5. GPC with Tuned Prediction Horizon

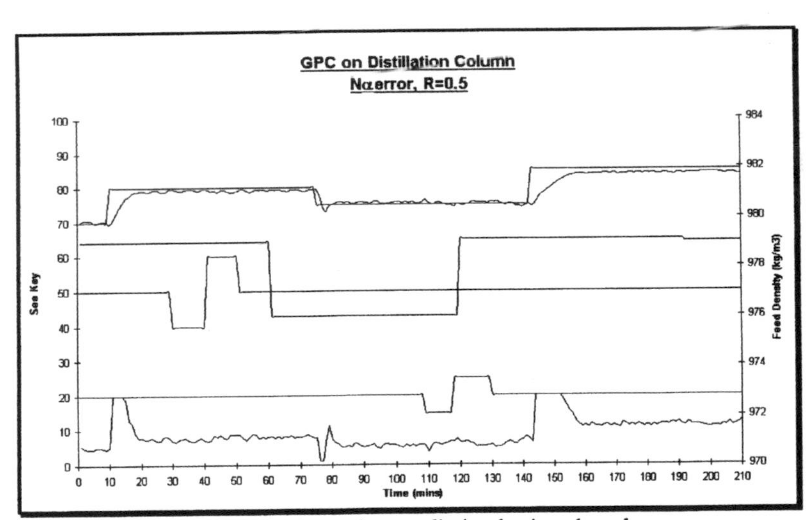

Figure 6. GPC with varying prediction horizon based on error

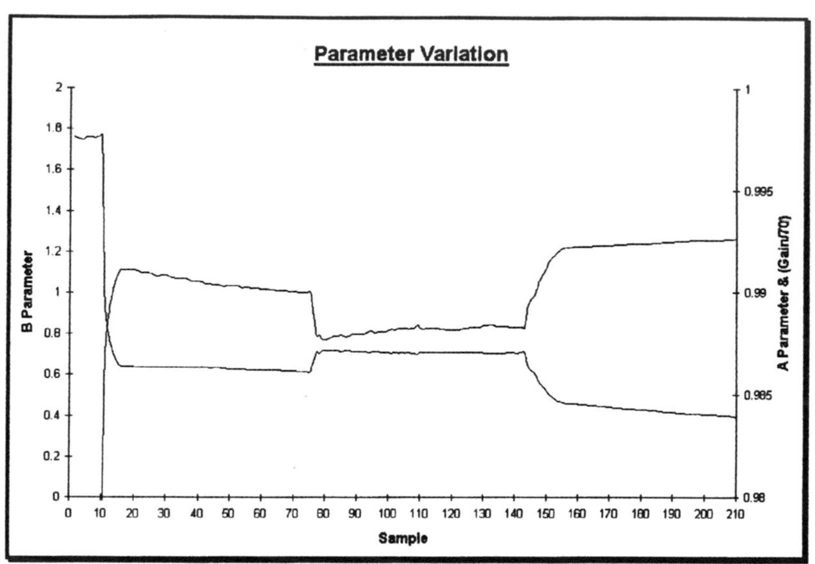

Figure 7. Typical adaptive parameter variation over a test run

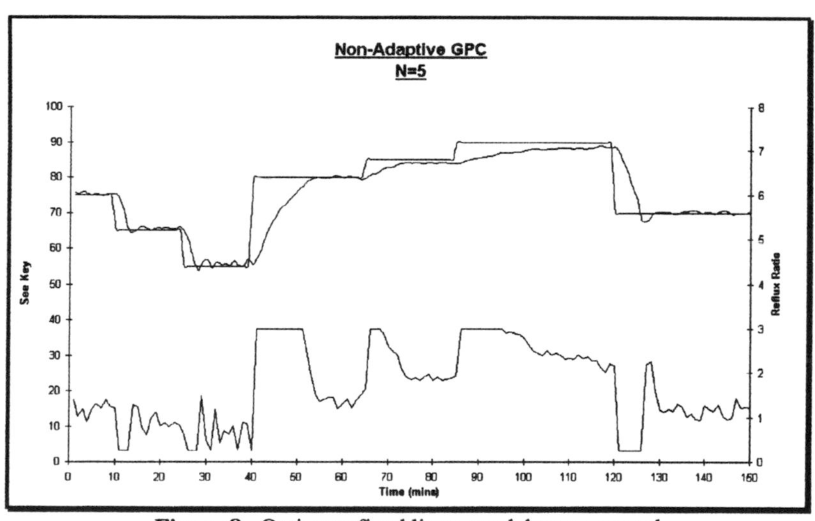

Figure 8. Optimum fixed linear model servo control

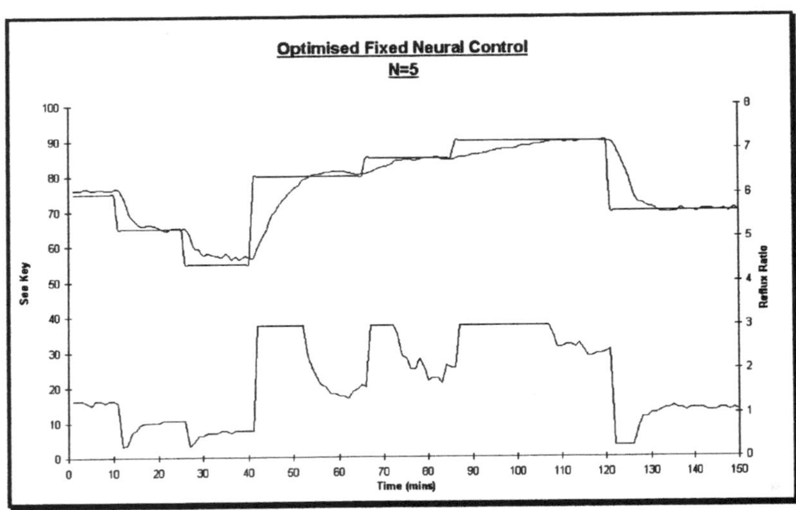

Figure 9. Optimised Fixed Neural Model Control

The P.I.D. performance is inadequate and requires more analysis and better tuning Standard GPC performs much better, as expected, but the adaption of parameters is slow and leads to slow removal of offset which will inevitably get slower as learning progresses due to covariance matrix degradation in the recursive least squares algorithm used to estimate parameter values. GPC with a shorter prediction horizon performs much better due to faster reactions to offset because of the shorter prediction horizon. There is an extra improvement of approximately 9% in performance when utilising the varying horizon approach proposed in this paper as opposed to the best of the previous GPC tests.

When comparing the best fixed model linear controller with the proposed optimised fixed neural model controller there is a visible improvement in performance born out by the ISE measure of error. In fact the neural controller was the only controller which managed to reach set point at the extreme end of the range due to the fact that the controller does not have to linearise or adapt.

Hence we can conclude that a fixed model neural controller embedded within a least squares optimiser performs better than GPC with a fixed linear model on this non-linear system. The extension of this technique to other systems will depend highly upon the type of non-linearity observed and the accuracy to which the neural model is trained.

References

1. **Buckley, P. S., Luyben, W. L., Shunta, J. P.,** Design of Distillation Column Control Systems, 'Book', 1885

2. **Clarke, D. W., Mohtadi, C., Tuffs, P. S.,** Generalized Predictive Control-part I: The Basic Algorithm., Automatica, 23(2), p55-74, 1987

3. **Wasserman, P. D.,** Neural Computing Theory and Proctice., Van Nostrand Reinhold Press, 1989

4. **Parvin, S., Peel, D., Leck, M.,** Pretreatment of Data for Improved Artificial Neural Network Model Generation, FAIM '97

5. **Astrom, K. J., Wittenmark, B.,** Self-tuning controllers based on pole-zero placement., Proc. IEE Pt D, vol. 127, No. 3, p 120-130, 1980

6. **Clarke, D. W.,** Self-tuning control of non-minimum phase systems., Automatica, vol. 20, No. 5, p501-517, 1984.

7. **Cybenko, G.,** Approximation by superpositions of a sigmoidal function., Math. Control, Sig. Syst., 2, p303-314.

8. **Turner, P., Montague, G., Morris, J.,** Neural Networks in Dynamic Process State Estimation and Non-Linear Predictive Control., Artificial Neural Networks, 26-28 , Conference Publication No. 409, IEE, 1995

9. **IEEE** (1988, 1989, 1990). Special Issues on Neural Networks, Control Systems Magazines - nos. 8, 9, 10

10. **Hunt, K., J., Sbarbaro, D., Zbikowski, R., and Gawthrop, P., J.,** Neural Networks for Control Systems - A Survey., Automatica, vol 28, No. 6, p1083-1112, 1992

11. **Thibault, J., Grandjean, B., P., A.,** Neural Networks in Process Control - A Survey., IFAC Conference on Advanced Control of Chemical Processes, Toulouse France, p251-260, 1991.

12. **Morris, A., J., Montague, G., A., Willis, M., J.,** Artificial Neural Networks: Studies in Process Modelling and Control., Trans IChemE Part A, vol. 72, p3-19.

13. **Eliezer Colina Morles,** Nonlinear Adaptive Control Design Using On-Line Trained Neural Networks, Proc. IEEE Conf on Sys, Man and Cyb. vol. 2, p1453-1456, 1994

14. **Jin, L., Nikiforuk, P., N., Gupta, M., M.,** Neural Networks for Modelling and Control of Discrete-Time Nonlinear Systems., Proc. IEEE Conf on Sys, Man and Cyb. vol. 2, 1122-1127, 1994

15. **Owens, A., J.,** Process Modelling and Control Using A Single Neural Network., Proc. IEEE Conf on Sys, Man and Cyb. vol. 2, p1475-1480, 1994

16. **Tan, Y., De Keyser, R.,** Neural Network Based Predictive Control For Nonlinear Processes with Time-Delay., IEEE, 1116-1121, 1994

17. **Luo, R., F., Shao, H., H., Zhang, Z., J.,** Fuzzy-Neural-Net-Based Inferential Control For A High Purity Distillation Column., Control Eng. Practice, vol. 3, No. 1, p31-40, 1995

18 **Ramchandran, S., Rhinehart, R., R.,** A very simple structure for neural network control of distillation., J. Proc. Cont., vol. 5, No. 2, 115-128, 1995

ADAPTIVE CONTROLLERS FOR THE REGULATION OF THE GEOMETRICAL AND THERMAL CHARACTERISTICS OF GMA WELDING

S.G. Tzafestas, E.J. Kyriannakis and G.G. Rigatos

Intelligent Robotics and Automation Laboratory
Department of Electrical and Computer Engineering
National Technical University of Athens
15773, Zografou Campus , Athens , Greece

ABSTRACT : The application of automatic control to arc welding processes is necessary for minimizing the defects caused by the improper regulation of parameters like arc voltage and current, or travel speed of the torch.Among the various controllers available, the adaptive ones seem to be more appropriate and successful, producing weldments of high quality and strength. This paper investigates and compares the effectiveness of three different classes of adaptive controllers when applied to GMA welding, namely (i) the Lyapunov model reference adaptive controller, (ii) the adaptive pseudogradient controller, and (iii) the multivariable self-tuning controller.

1. INTRODUCTION

GMA welding is a complex physical process that needs careful control for minimizing the defects caused by the improper regulation of parameters like arc voltage and current, or travel speed of the torch [1-9]. The main defects of a weldment are the following [5,6] :

- Defects related to the geometrical characteristics. This means that the weldment's width, height or depth differs from the desired one.
- Thermally induced discontinuities in the weldment's structure such as porosity, incomplete fusion, inadequate penetration, undercutting and cracking.

The structure of the paper is as follows. Section 2 presents the standard linearized models for the geometrical and thermal characteristics of the arc welding process. Sections 3.1 and 3.2 present the pseudogradient and the Lyapunov model reference adaptive controllers, and show how they are applied to the welding process. Section 3.3 treats the multivariable case presenting a 2x2 model of the welding process and a one-step ahead self-tuning controller. Finally, section 4 gives representative results of the methods, Section 5 provides a comparative evaluation of the methods, and Section 6 gives some concluding remarks.

2. MODELLING THE GEOMETRICAL AND THERMAL CHARACTERISTICS OF ARC-WELDING

2.1. Arc-Welding-Geometry Model

The geometry of the weld is described by the width W, the height H, and the penetration depth D. The experiment has shown that in all cases the dynamics of W, H, D can be described by a time constant transfer function

$$K/(\tau s+1) \qquad (2.1)$$

where both the gain K and the time constant τ depend on the conditions of the welding and the amplitude of the welding's inputs. This means that the welding process which is highly nonlinear can only be locally approximated by its linear equivalent .

The linearized first-order models for W,H and D were determined by fitting the experimental data as :

$$\frac{W(s)}{U(s)}=\frac{K_w}{\tau_w s+1} \quad , \quad \frac{H(s)}{U(s)}=\frac{K_h}{\tau_h s+1} \quad , \quad \frac{D(s)}{U(s)}=\frac{K_d}{\tau_d s+1} \qquad (2.2)$$

where the values of K and τ are given in Table 1 [8] :

TABLE 1
Linearized Model of the GMA Geometrical Characteristics

INPUT	From	To	Width W		Depth D		Height H	
			τ_w	K_w	τ_d	K_d	τ_h	K_h
Power	2.5 Kw	3.0 Kw	0.75	2.12	0.52	1.67	0.82	-0.22
Power	2.5 Kw	2.0 Kw	0.78	1.83	0.57	1.74	0.69	-0.17
Velocity	6 mm/s	10 mm/s	0.60	-0.75	0.40	-0.30	0.80	-0.15
Velocity	4 mm/s	6 mm/s	1.40	-1.70	1.20	-0.32	0.70	-0.24
Wirefeed	600 in/min	800 in/min	0.95	0.026	0.65	0.0094	0.40	0.0042

2.2. Arc-Welding Thermal Model

The final microstructure and material properties of the joint can be described by the following welding outputs [5] :

(a) *The weld nugget cross section* NS defined by the solidus isotherm Tm.

(b) *The heat affected zone* HZ defined by an enveloping isotherm T_h. This may indicate the extent of weak zones, such as the recovery, recrystallization and grain growth areas.

(c) *The centerline cooling rate* CR defined at the critical temperature T_c. This may determine the crystallization of undesirable, kinetically favoured phases, or supply a measure of the cracking tendency of the weldment caused by thermal stresses

$$(CR=\partial T/\partial t \mid T=T_c) \qquad (2.3)$$

The welding outputs are pictorially illustrated in Fig.1

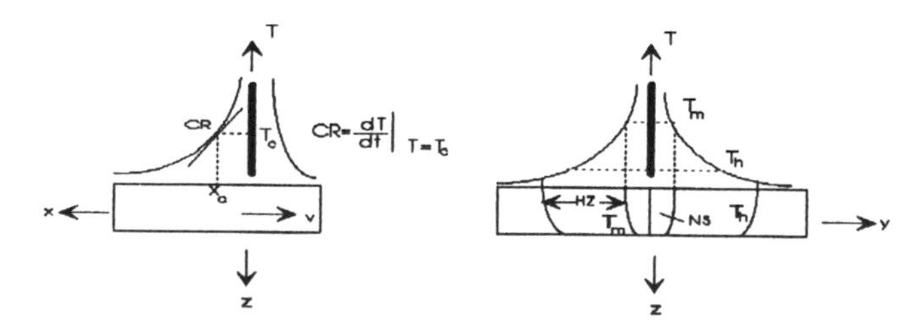

Fig.1 Definition of the welding outputs NS, HZ, CR

The response of the bead cross section area NS to step inputs (either Q_1 ,V or Q_2) has the following form :

$$\frac{NS}{Q_1}(s) = \frac{K_a}{\tau_a s+1}, \quad \frac{NS}{V}(s) = \frac{K_a'}{\tau_a' s+1}, \quad \frac{\mathrm{NS}}{\mathrm{Q}_2}(\mathrm{s}) = 0 \tag{2.4}$$

The response of the heat affected zone HZ to steps in either Q_1 or V has a nonminimum phase second- order behavior while its sensitivity to the third input Q_2 is almost insignificant .Thus :

$$\frac{HZ}{Q_1}(s) = \frac{K_b(\tau_b s+1)}{(\tau_1 s+1)(\tau_2 s+1)}, \frac{HZ}{V}(s) = \frac{K_b'(\tau_b' s+1)}{(\tau_1' s+1)(\tau_2' s+1)}, \frac{HZ}{Q_2}(s) = \frac{K_b''}{\tau_2'' s+1} \approx 0 \tag{2.5}$$

Finally the response of the centerline cooling rate CR to steps in either Q_1, V and Q_2 may be approximately described by the following transfer functions :

$$\frac{CR}{Q_1}(s) = \frac{K_c}{(\tau_a s+1)(\tau_b s+1)}, \frac{CR}{V}(s) = \frac{K_c'}{(\tau_a' s+1)(\tau_b' s+1)}, \frac{CR}{Q_2}(s) = \frac{K_c''}{\tau_a'' s+1} \tag{2.6}$$

The calculated values of the gains and the time constants of the linearized model are presented in Table 2 [5].

TABLE 2
Linearized Model of the Thermal Characteristics

INPUT	From	To	NS (mm^2)		HAZ (mm)				CR(K°/s)		
			K_a	τ_a	K_b	τ_h	τ_1	τ_2	K_c	τ_a	τ_b
Q_1	2.5 Kw	3.0 Kw	0.00574 mm^2/W	1.57 s	0.00356 mm/W	-1.56 s	5.42 s	0.50 s	0.0036 K°/J	5.33 s	0.50 s
Q_1	2.5 Kw	2.0 Kw	0.00434 mm^2/W	0.63 s	0.00221 mm/W	-1.78 s	2.75 s	0.33 s	0.0632 K°/J	4.43 s	0.61 s
V	5 mm/s	6 mm/s	-0.91 mm/s	1.78 s	-1.21 s	-1.39 s	3.04 s	0.52 s	-27.9 K°/mm	4.61 s	0.45 s
V	5 mm/s	4 mm/s	-1.12 mm/s	4.95 s	-3.40 s	-1.16 s	11.0 s	0.65 s	-25.5 K°/mm	6.02 s	0.41 s
Q_2	0 Kw	0.25 Kw	0 mm^2/W	0 s	0.00051 mm/W	0 s	2.65 s	0 s	0.0976 K°/J	2.40 s	0 s

3. ADAPTIVE CONTROL OF ARC-WELDING

This method is based on the work of Henderson et. al. [9]. Here the following adaptive control techniques will be reviewed :

(i) Pseudogradient adaptive control
(ii) Lyapunov adaptive control
(iii)Multivariable adaptive control

3.1. Pseudogradient Control

In the pseudogradient algorithm the number of the controller parameters or its structure does not tie to the order of the plant transfer function. The stability of the method has been established in [14].
The sensitivity functions are given by the following equation [8] :

$$\frac{\partial y}{\partial \theta_i} = \frac{y(k,\theta_i + \delta\theta_i) - y(k,\theta_i)}{\delta\theta_i} \tag{3.1}$$

The sensitivity function can be generated by the scheme shown in Fig. 2.

The transfer function $H_{yv}(z,\theta_i)$ derived cannot be considered to be accurate because it is related to the H(z) transfer function of the plant which is not absolutely known. Thus the filter's output is called *pseudosensitivity function* and the resulting adaptive control law is called *pseudogradient control law.*

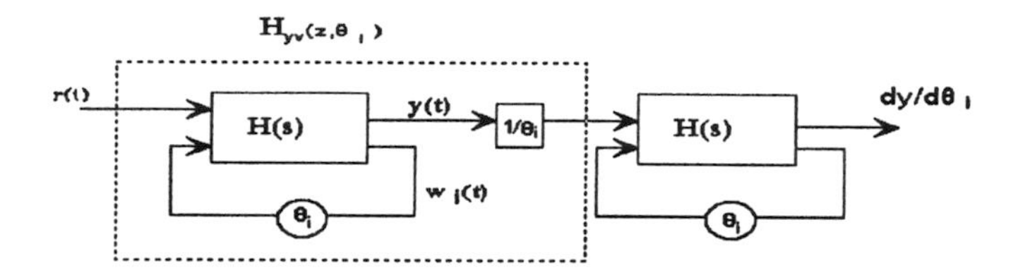

Fig.2 Method for generating the sensitivity function

The pseudogradient adaptive controller (PAC) : The goal of PAC is to minimize the cost criterion

$$J=\frac{1}{2}e^2(k) \tag{3.2}$$

where $e(k)=y(k)-y_m(k)$ (see Fig. 3).

The adaptation for the parameters K_p and K_i of the PI controller is

$$K_p(k+1)=K_p(k)-\eta Te(k)\frac{\partial y_p}{\partial K_p}(k) \tag{3.3}$$

$$K_{i(k+1)}=K_i(k)-\eta Te(k)\frac{\partial y_p}{\partial K_i}(k) \tag{3.4}$$

The transfer function $H_{yv}(z)$ of the sensitivity filter is given by

$$H_{yv}(z)=H_{yr}(z)\frac{1}{C(z;K_p{}^n;K_i{}^n)} \tag{3.5}$$

where $H_{yr}(z)$ is the nominal model of the plant, and $C(z;K_p^n;K_i^n)$ is the transfer function of the PI controller

$$C(z)=K_p+\frac{T}{(z-1)}K_i \text{ (2.10)} \tag{3.6}$$

which is calculated for the nominal values of K_p and K_i.

3.2. Lyapunov Adaptive Control

Controller structure :

The following linear approximation of the plant is assumed to be available :

$$\dot{x}_p(t) = A_p x_p(t) + B_p u(t) \tag{3.7}$$

The reference model is assumed to be :

$$\dot{x}_m(t) = A_m x_m(t) + B_m r(t) \tag{3.8}$$

Suppose the plant's order is n. The control law is

$$u(t) = Q(t)(r(t) + F(t)x_p(t)) \tag{3.9}$$

where $Q(t) \varepsilon R^{mxp}$ and $F(t) \varepsilon R^{mxn}$ are the time varying gain matrices of the controller [11]. The state space equations of the closed-loop are :

$$\dot{x}_p(t) = (A_p + B_p Q(t)F(t))x_p(t) + B_p Q(t)r(t) \tag{3.10}$$

The error vector between the plant state and the state of the reference model is $e(t) = x_m(t) - x_p(t)$.The algorithms for the adaptation of the gain matrices **F** and **G** are given by the following equations :

$$\dot{Q}(t) = Q(t)G_2(t)B_m^T Pe(r + Fx_p)^T Q^T Q \quad , \quad \dot{F}(t) = G_1(t)B_m P^T e x_p \tag{3.11}$$

where $P \varepsilon R^{nxn}$ is a positive definite and symmetric matrix, and $G_1 \varepsilon R^{mxm}$, $G_2 \varepsilon R^{pxm}$ are the gain matrices chosen for speed of convergence [11,12].

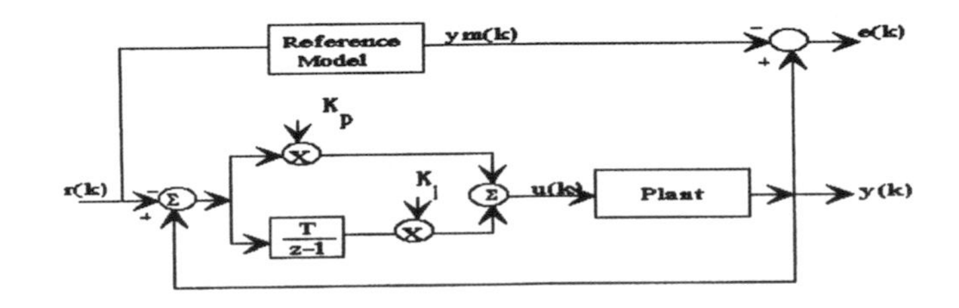

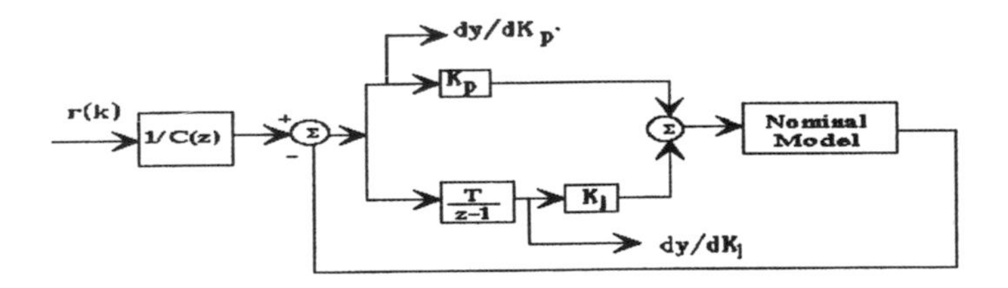

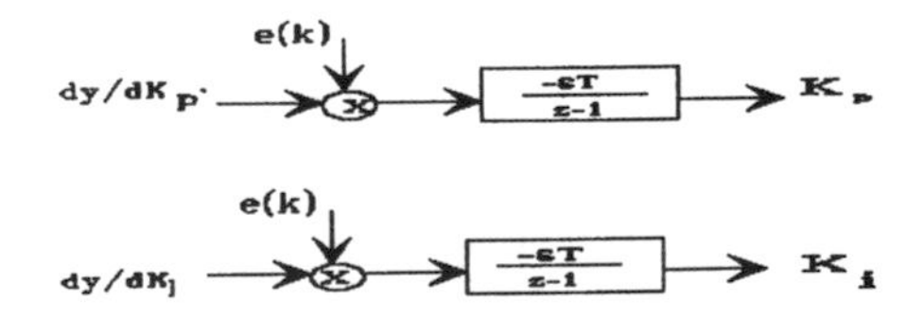

Fig.3 Block diagram of the adaptive control system

Lyapunov adaptive control of the welding process : In the welding process the plant state vector is frequently not fully known. What one usually knows is the plant output which is a geometrical or thermal characteristic that can be measured.

From Tables 1 and 2, one can see that many reference models of the process are of first order. In these cases we make the assumption that the corresponding plant models are of first order as well. This implies that the plant can be described by only one state variable which is identical to its measurable output. Consequently one can assume that the Lyapunov's method of adaptive control is applicable to the above mentioned plants.

3.3. Multivariable Adaptive Control of the Welding Geometrical Characteristics

Up to now the control of both the geometrical and thermal characteristics of the welding process has been pursued by splitting the process in a number of SISO sub-systems. However the outputs of the welding process are highly coupled. Hence for a more correct treatment of the control problem, it is necessary to regulate all outputs simultaneously and not sequentially. This implies that one should handle the welding process as a multi-input multi-output (MIMO) system.

A satisfactory solution to this problem springs from the "one step ahead" multivariable adaptive control algorithm which is based on the self-tuning control technique.

MIMO modelling of the welding process : A 2x2 system is adequate to describe the welding process. As inputs we select the wirefeed rate F and the inverse V^{-1} of the velocity of the torch. On the other hand among many possible outputs, the width W and the penetration D are selected.

Since the process is nonlinear, a locally linearized process model around the operating point is employed. Based on the experiments, the following second-order, discrete time transfer function matrix (TFM) has been developed [15] :

$$\begin{bmatrix} y_1(z) \\ y_2(z) \end{bmatrix} = \begin{bmatrix} \dfrac{(0.18+0.09z^{-1})z^{-2}}{1-1.10z^{-1}+0.40z^{-2}} & \dfrac{(2.48-1.29z^{-1})z^{-2}}{1-1.10z^{-1}+0.40z^{-2}} \\ \dfrac{(0.035-0.01z^{-1})z^{-1}}{1-0.99z^{-1}+0.32z^{-2}} & \dfrac{(0.80-0.44z^{-1})z^{-1}}{1-0.99z^{-1}+0.32z^{-2}} \end{bmatrix} \begin{bmatrix} u_1(z) \\ u_2(z) \end{bmatrix} \qquad (3.12)$$

Although the dynamics are assumed to be the same for each output, the gains (b's) are different for each input-output pair.

General principles of the "one step ahead" control : A MIMO system with m inputs and r outputs can be described by the equation :

$$\mathbf{A}(q^{-1})\mathbf{y}(k) = \mathbf{B}(q^{-1})\mathbf{u}(k) \qquad (3.13)$$

where q^{-1} is the backward shift operator, **y**(k) is the output vector and **u**(k) is the control vector. In general, time delays appear between the inputs j and the outputs i. Since each delay usually differs from the others the delay of the i-th output is defined as : $d_i = \min\{d_{ij}\}$, $i = 1,2,\ldots,r$.

As far as the above mentioned 2x2 model is concerned we have :

d_1 = delay of the first output = $\min\{d_{11}, d_{12}\} = 2$

d_2 = delay of the second output = $\min\{d_{21}, d_{22}\} = 1$

Introducing the matrix $\mathbf{D}(q) = diag(q^{d_1},\ldots,q^{d_r})$, then the matrices **A**, **B** can be represented as follows :

$$\mathbf{A}(q^{-1}) = \mathbf{I} + \mathbf{A}_1 q^{-1} + \mathbf{A}_2 q^{-2} + \ldots + \mathbf{A}_n q^{-n}$$

$$\mathbf{B}(q^{-1}) = \mathbf{D}(q^{-1})(\mathbf{B}_0 + \mathbf{B}_1 q^{-1} + \mathbf{B}_2 q^{-2} + \ldots + \mathbf{B}_m q^{-m}) = \mathbf{D}(q^{-1})\mathbf{B}'(q^{-1}) \qquad (3.14)$$

The future output vector is

$$\hat{y}(k) = [\hat{y}_1(k+d_1),....,\hat{y}_r(k+d_r)] \tag{3.15}$$

Hence the predictor is described by the equation

$$\hat{y}(k) = \alpha(q^{-1})y(k) + D(q^{-1})u(k) \tag{3.16}$$

where

$$\alpha(q^{-1}) = \alpha_0 + \alpha_1 q^{-1} + \alpha_2 q^{-2} +.... \tag{3.17}$$

$$\beta(q^{-1}) = \beta_0 + \beta_1 q^{-1} + \beta_2 q^{-2} +.... = \beta_0 + q^{-1}\beta'(q^{-1}) \tag{3.18}$$

and expresses the future outputs by using the input and the output terms up to the current time k. Now consider the following performance index :

$$\hat{J}(k) = [\hat{y}(k) - \hat{y}^*(k)]^T[\hat{y}(k) - \hat{y}^*(k)] + [u(k) - u(k-1)]^T R [u(k) - u(k-1)] \tag{3.19}$$

where **R** is a positive definite control weighting matrix with dimensionality rxr. In the present welding problem the matrix **R** is taken to be diagonal :

$$\mathbf{R} = \begin{bmatrix} r_1 & 0 \\ 0 & r_2 \end{bmatrix} \tag{3.20}$$

and the terms r_1 and r_2 define how much each input influences the corresponding output. We choose $r_1=r_2$. The first term in the cost criterion represents the divergence of the plant's predicted output from the desired output. With the first term alone the control system can achieve the desired output in just one step but it may produce an excessively large control signal which sometimes causes instabillity. The control weighting matrix **R** in the second term has been introduced to overcome this problem.The difference **u(k)-u(k-1)** penalizes the variations of the control law and prevents us from using a very large control input that could drive the whole system to instabillity.

By minimizing the above performance index (cost function) the following generalized "one step ahead" control law is obtained:

$$\mathbf{u}(k) = [\beta_0 + \mathbf{R}^{-1}]\beta_0[\hat{\mathbf{y}}^* - \alpha(q^{-1})\mathbf{y}(k) - \beta'(q^{-1})\mathbf{u}(k-1)] + \mathbf{R}\mathbf{u}(k-1) \tag{3.21}$$

Parameter estimation : The parameters of the welding process a_{ij} and b_{ij} change depending on both the operating conditions and the thermal history of the metal. Thus the parameters cannot be determined accurately, and an on-line estimation is required to achieve consistent control.This is done by the implementation of a recursive least squares (RLS) algorithm. [16].

The "one-step-ahead" control law in the welding process : The predictor of the outputs is given by the equations [15]:

$$\begin{aligned} y_1(k+2) &= a_{11}y_1(k-1) + a_{12}y_1(k-2) + b_{11}u_1(k) + b_{12}u_1(k-1) + b_{13}u_1(k-2) \\ &+ b_{14}u_1(k-3) + b_{15}u_2(k) + b_{16}u_2(k-1) + b_{17}u_2(k-2) + b_{18}u_2(k-3) \end{aligned} \tag{3.22}$$

$$\begin{aligned} y_2(k+1) &= a_{21}y_2(k-1) + a_{22}y_2(k-2) + b_{21}u_1(k) + b_{22}u_1(k-1) + b_{23}u_1(k-2) \\ &+ b_{24}u_2(k) + b_{25}u_2(k-1) + b_{26}u_2(k-2) \end{aligned} \tag{3.23}$$

where $y_1(k)$ is the width and $y_2(k)$ the penetration of the welding, while $u_1(k)$ is the wire feed rate and $u_2(k)$ the inverse of the velocity of the torch. The control signal **u(k)** is

$$u(k) = [\beta_0^2 + R^{-1}][\beta_0[\hat{y}^*(k) - a(q^{-1})y(k) - \beta'(q^{-1})u(k-1) + Ru(k-1)] \tag{3.24}$$

where

$$\mathbf{u(k)} = \begin{bmatrix} u_1(k) \\ u_2(k) \end{bmatrix}, \quad \beta_0 = \begin{bmatrix} \beta_{11} & \beta_{15} \\ \beta_{21} & \beta_{24} \end{bmatrix} \quad \mathbf{R} = \begin{bmatrix} r_1 & 0 \\ 0 & r_2 \end{bmatrix} \quad \hat{\mathbf{y}}^{*} = \begin{bmatrix} y_1^{*}(k+2) \\ y_2^{*}(k+1) \end{bmatrix} \tag{3.25}$$

$$\alpha(q^{-1})\mathbf{y}(k) = \begin{bmatrix} a_{11}y_1(k-1) + a_{12}y_1(k-2) \\ a_{21}y_2(k-1) + a_{22}y_2(k-2) \end{bmatrix}, \beta'(q^{-1})\mathbf{u}(k-1) = \begin{bmatrix} b_{12}u_1(k-1) + \ldots + b_{18}u_2(k-3) \\ b_{22}u_1(k-1) + \ldots + b_{26}u_2(k-2) \end{bmatrix} \tag{3.26}$$

Another drawback of the "one-step-ahead " control method is that the plant outputs are highly coupled to each other. A consequence of this coupling is that there is a limited number of control inputs for which we can reach good results. That's why we can't reach an arbitrarily large width of the welding and simultaneously an extremely small penetration depth.

The "one-step-ahead" control method can also be applied for regulating the thermal charactreristics of the welding (HAZ,CR,NS).

4. SIMULATION RESULTS

In this section some representative results are provided which have been obtained through simulation using the above methods. In general, these results are very encouraging, with the methods having relative advantages and disadvantages.

4.1.Lyapunov and Pseudogradient Control

i) Regulation of the width W using as control input the thermal power Q [2.5-3.0 Kw]
The results obtained are shown in Figures 4 and 5.

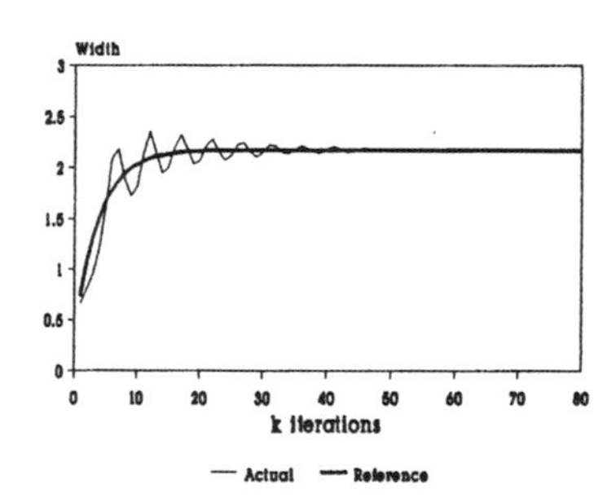

Fig. 4 : Lyapunov method

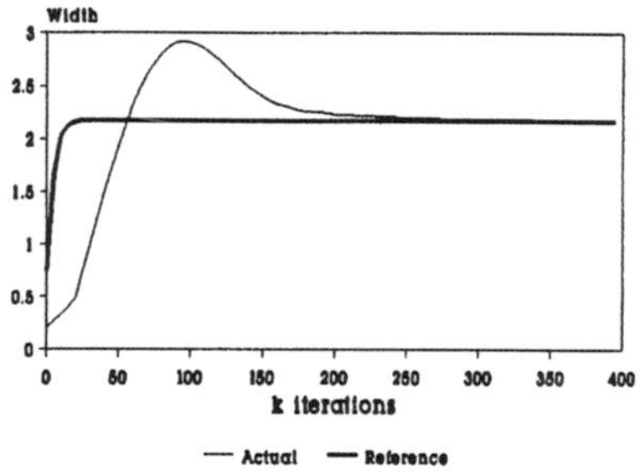

Fig. 5 : Pseudogradient method

ii) Regulation of the nugget section NS using as control input the thermal power of the main torch Q1 [2.5-3.0 kw]. The results are depicted in Figures 6 and 7.

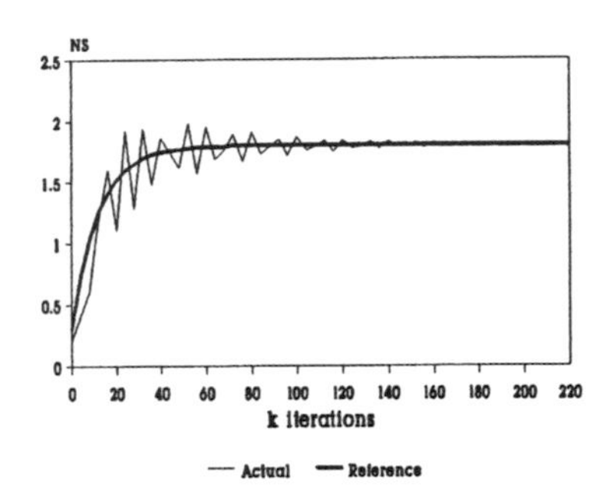

Fig. 6 : Lyapunov method

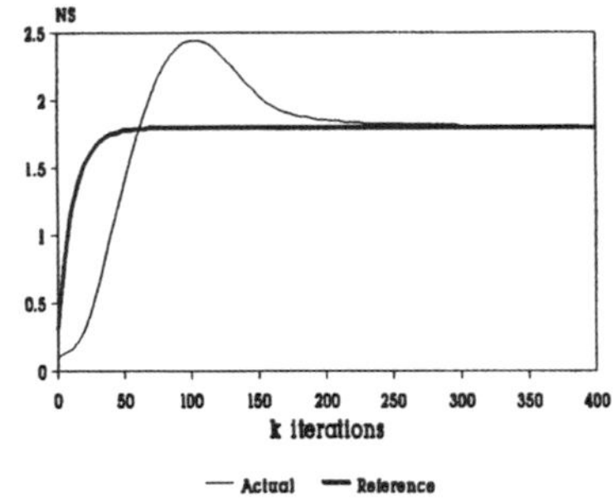

Fig. 7 : Pseudogradient method

iii) **Regulation of the heat affected zone HAZ using as control input the velocity of the torch V [5-6 mm/s].** The results obtained are shown in Figures 8 and 9 :

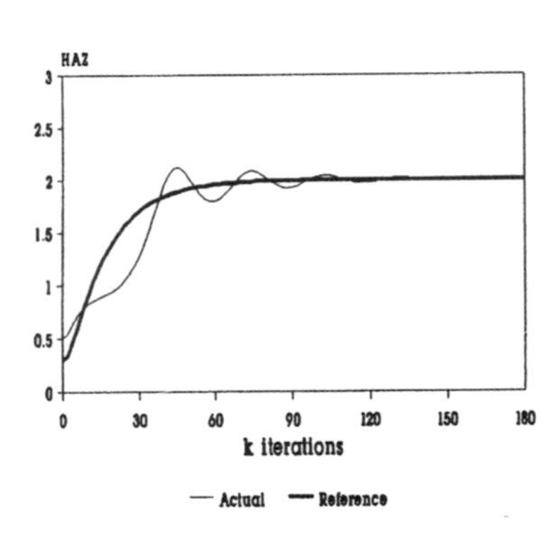

Fig. 8 : Lyapunov method

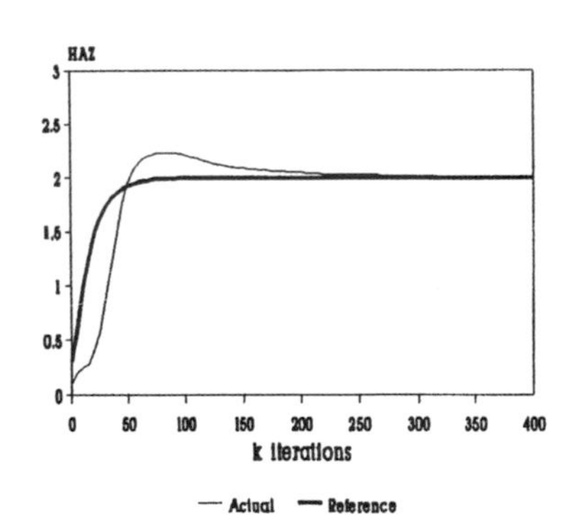

Fig. 9 : Pseudogradient method

iv) **Regulation of the cooling rate CR using as control input the velocity of the torch V [4-5 mm/s].** The results are shown in Figures 10 and 11.

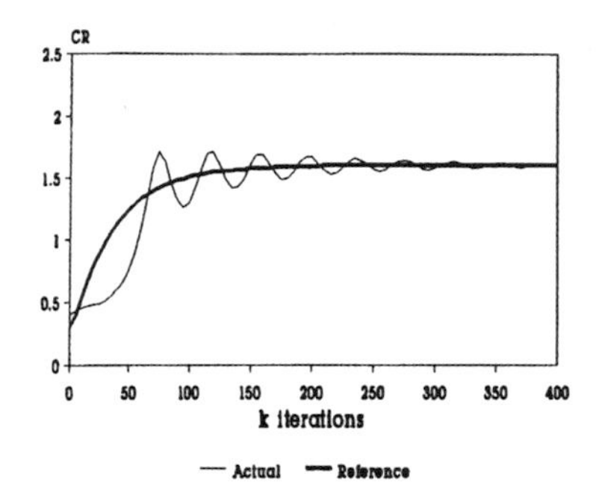

Fig. 10 : Lyapunov method

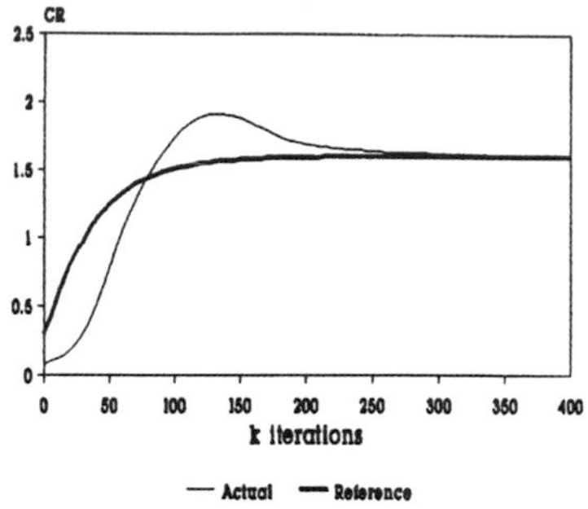

Fig. 11 : Pseudogradient method

5.2. Multivariable adaptive control

i) **Regulation of the width W and the depth D using as control inputs the inverse V^{-1} of the torch velocity and the wirefeed rate f.** The results obtained are depicted in Figures 12 and 13.

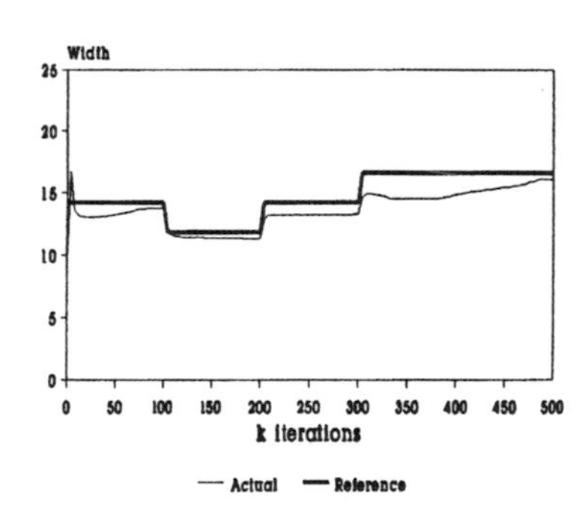

Fig. 12 : Inputs V^{-1},f - Output W

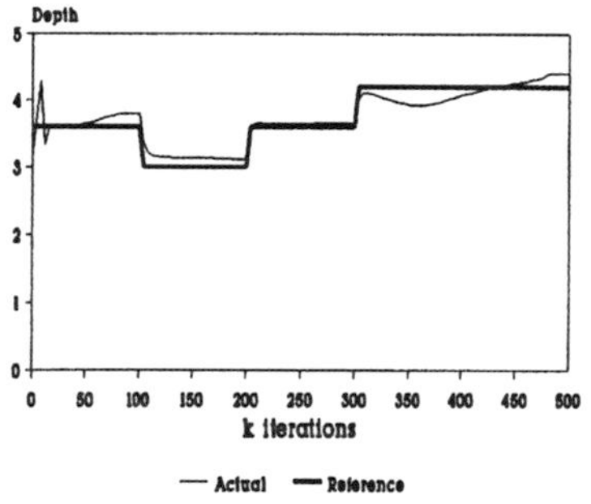

Fig. 13 : Inputs V^{-1},f - Output D

5. COMPARATIVE DISCUSSION

The desired closed-loop specification in all cases was to achieve zero steady state error while obtaining as fast transient response as possible. As already mentioned there are two different approaches to the welding control problem. The first is to control separately each one of the SISO systems in which the welding procedure can be analysed. This strategy was adopted for the Lyapunov and Pseudogradient techniques and is capable to produce a weld of high quality in some welding situations. The second approach is to control the process using the MIMO model (Self-Tuning technique).

The SISO case : The Lyapunov adaptive control law requires the state vector of the plant to be fully known. However this is not always feasible. Consequently it might be non practical to apply the Lyapunov MRAC method to systems that describe HAZ and CR of the welding. In the light of this fact, the pseudogradient method seems to be better, because it does not use a description of the plant in the state space but the only information that needs is the output of the plant.

Regarding the pseudogradient adaptive algorithm , we point out that in contrast to other adaptive methods it does not tie the number of parameters or controller structure to the order of the plant transfer function.Even though in some simulation experiments an extended overshooting was recorded the output converged to the desirable response in less than 400 iterations. The pseudogradient algorithm is actually less complicated but much slower than the equivalent Lyapunov algorithm.

In all simulation results for the SISO case, after 400 iterations, the steady- state of the output did not deviate from the target setpoint more than 0.01-0.1 %.

Unlike the MIMO self-tuning control approach, the Lyapunov, and the pseudogradient designs make no attempt to deal with the delay between the control inputs and the geometrical or thermal outputs.

The MIMO case : In the SISO case the system shows a good disturbance rejection ability, but it has limited success in the MIMO case due to inherent lack of output decoupling in the welding process. If the disturbance affects one output much more than the other then it is difficult to reject the disturbance.

The simulation results of the multivariable self-tuning control law were quite satisfactory. The main problem was to achieve simultaneous convergence of the outputs to the desired set-points. It was because of the highly coupled nature of the outputs that when the first output (e.g. width) reached the desired value, the other output (depth) diverged from its corresponding set-point and vice-versa. In all cases this deviation , after 200 iterations, did not exceed 5%.

6. CONCLUSIONS

This paper has presented some results of the application of three major control techniques for the regulation of the geometrical and thermal characteristics of the GMA welding process. After a short discussion of the basic linearized models of the geometrical and thermal characteristics, the following conventional controllers were considered and applied : (i) pseudogradient adaptive controller, (ii) Lyapunov adaptive controller, and (iii) multivariable self-tuning adaptive controller. The pseudogradient controller was firstly applied by Henderson and coworkers [9] for the regulation of the geometric characteristics with input only the speed of the torch using a different model than the one employed here. The same technique has also been applied by Nishar et al. [8] for the control of the centerline cooling rate also using different models than ours.

All these techniques were applied here for both the geometrical and the thermal characteristics in a unified way using the models of Doumanidis [5,6] who has applied the MIMO control technique. The Lyapunov control method is applied to the welding process for the first time in the present paper. The MIMO controller applied here is based on the work of Song and Hardt [14].

The simulation results for all of these controllers are very encouraging and their implementation in actual robotic welding systems is expected to be beneficial.

REFERENCES

1. **Hunt ,V.D.** , *Industrial Robotics Handbook*, Industrial Press Inc., New York (1983).
2. **Tzafestas , S.G.** (ed.), *Intelligent Robotic Systems*, Marcel Dekker, New York (1991).
3. **Tzafestas , S.G.** (ed.), *Applied Control : Current Trends and Modern Methodologies*, Marcel

Dekker, New York (1993)
4. **Tzafestas , S.G. , Verbruggen , H.B.** (eds.), Artificial Intelligence in Industrial Decision Making, *Control and Automation*, Kluwer Academic Publishers, Dordrecht/Boston (1995).
5. **Doumanidis ,C. , Hardt ,D.E.** , A model for In-Process Control of Thermal Properties During Welding , *ASME J. Dynamic Syst. Meas. and Control* , vol. 111, pp. 40-50 (1989).
6. **Doumanidis ,C. , Hardt ,D.E.** , Multivariable Adaptive Control of Thermal Properties During Welding , *ASME J. Dynamic Syst. Meas. and Control*, vol. 113, pp. 82-92 (1991)
7. **Doumanidis ,C.** , Multiplexed and Distributed Control of Automated Welding , *IEEE Control Systems Magaz.*, pp. 13-24, Aug (1994).
8. **Nishar ,D.V., Schiano ,J.L. , Perkins ,W.R., Weber ,R.A.** , Adaptive Control of Temperature in Arc Welding, *IEEE Control Systems Magaz.*, pp. 4-12 , Aug (1994).
9. **Henderson ,D.E. , Kokotovich ,P.V. , Schiano ,J.L. , Rhode ,D.S.** , Adaptive Control of an Arc Welding Process , *Proc. 1991 American Control Conference*, vol.1 , pp. 723-728 (1991).
10. **Rhode ,D.S. , Kokotovitch ,P.V.** , Parameter Convergence Conditions Independent of Plant Order , *Proc. 1989 American Control Conference*, vol. 1, pp. 981-986 (1989).
11. **Slotine ,J., Welping ,L.**, *Applied Nonlinear Control* , Prentice Hall , New Jersey (1991).
12. **Narendra, K.S., Lin ,Y.**, Stable Adaptive Control, *IEEE Trans. Autom.Control*, Vol. AC-25 No.3 ,June (1980)
13. **Suzuki ,A., Hardt ,D.E., Valavani, L.**, Application of Adaptive Control Theory to GTA Weld Geometry Regulation, *ASME J. Dynamic Syst. Meas. and Control* , vol . 113 , pp. 93-103 (1991).
14. **Song ,J.B. , Hardt ,D.E.**, Dynamic Modelling and Adaptive Control of the Gas Metal Arc Welding Process , *ASME J. Dynamic Syst. Meas. and Control*, vol. 116, pp. 405- 413 (1994).
15. **Haykin ,S.** , *Adaptive Filter Theory* , Prentice Hall , New Jersey (1991)
16. **Tzafestas ,S.G.**, Digital PID and Self -Tuning Control, In : *Applied Digital Control* (S.G. Tzafestas ,ed.) North Holland , Amsterdam , pp. 1-49 (1985).

HIERARCHICAL PETRI NET MODELING FOR SYSTEM DYNAMICS AND CONTROL OF MANUFACTURING SYSTEMS

Liu, C. S.[(1)] **, Ma, Y. H.,**[(1)] **and Odrey, N. G.**[(2)]

[(1)] Research Assistant ; [(2)] Professor and Director of Automation and Robotics Laboratories
Department of Industrial & Manufacturing Systems Engineering, Lehigh University, Bethlehem, PA, USA

ABSTRACT: The objective of this paper is to introduce a Petri net-based modeling structure with a state space representation for manufacturing systems within the context of hierarchical control. A Flexible Manufacturing Cell (FMC) having parallel processing capability and a material handling system with limited capability is investigated. A machine-oriented timed colored Petri net (TCPN) modeling technique is introduced to represent both sequential and parallel processes characteristics of the cell. Based on a TCPN model at the cell level, a workstation level model is configured through a top-down decomposition approach resulting in a hierarchical multilevel PN model. The dynamical system representation of the TCPN model at the cell level is transformed into a mathematically based state space model. The control for the FMC is formulated within a discrete-time optimal control framework with varying time intervals and a time-variant state equation. A scheduling heuristic solution and associated performance evaluation results are discussed. The state space representation at the workstation level is also discussed to demonstrate the continuity of the proposed modeling structure.

INTRODUCTION

In a discrete-part production industry, flexible manufacturing systems (FMSs) have been adopted in an effort to increase productivity and to shorten system response time to product variation. The particular class of flexible manufacturing cells considered here is a non-dedicated flexible machining system (NFMS) as classified by Browne et al. [1] A NFMS is similar to a job shop by having the same multi-directional material movements but differs in that it considers both parallel machining and a material handling system (MHS). These features increases the difficulty of NFMS cell modeling and scheduling.

The focus of this paper is on a machine-oriented timed colored Petri net (TCPN) modeling technique to represent the system behavior of a NFMS cell. The TCPN cell model considered here is dependent on the current production workload and the system capacity of a cell . The production workload determines the number of job tokens and a coloring scheme to differentiate jobs in the TCPN cell model and consists of the number of job types, the processing times, and the processing routings of jobs. The system capacity dictates the configuration of a cell model and includes the number of different types of workstations in a cell, the number of parallel resources in each workstation type, and its load/unload stations or material handling systems. Based on the abstract representation of a workstation in the TCPN cell model, the detailed workstation level model is configured through a top-down decomposition approach resulting in a hierarchical multilevel Petri net model. The application of the TCPN cell model for solving scheduling problems is presented through a mathematically based state space model and an optimal control problem formulation. A state space representation of the workstation level model is also discussed to demonstrate the continuity of the proposed modeling structure. Fig. 1 indicates the information flow between a cell and a workstation level and the corresponding Petri net modeling framework.

MACHINE ORIENTED PETRI NETS

Timed Petri nets (TPNs) and colored Petri nets (CPNs) have been used extensively in system modeling and performance evaluation for flexible manufacturing systems.[2-5] TPNs have been used primarily for the performance estimation of steady-state behavior. Multiple colors of tokens in a CPN model can be used to differentiate job types with different processing routings and gives a much more compact system representation for systems with multiple job types. For solving scheduling problems in a FMC, both time delays and multiple job types with different routings are essential in cell modeling. This has led to the development here of a timed, colored, Petri net (TCPN) to represent a non-dedicated flexible machining cell.

A timed, colored Petri net is defined as a six-tuple, TCPN=(T, P, A, M, C, τ) which contains the representations for token color (C), time delays (τ) and the basic structure of a Petri net, PN=(T, P, A, M) where T defines the set of transitions, P represents the set of places, A is the set of arcs and M the set of tokens. In this paper, a transition (represented by a bar '|') represents the initiation or the termination of an activity, and a place (represented by a circle) is a queue (called a resource place) or an activity (called an operation place) that cause time delays. An activity may be a processing operation or a transport operation. There are two types of tokens: tokens representing physical entities and tokens for information flow. A physical token refers to either a job, a buffer space, a machine or a material handling device. A message token represents a request of transportation from a workstation to the material handling system. In such a TCPN model, to release a job token from an operation place requires the elapse of a processing time associated with the particular type of job in the operation place.

Three basic TCPN constructs for cell modeling, as shown in Fig. 2(i), are used as the modular graphical elements. A resource assignment decision is modeled by Fig. 2(i)-a which depicts the job-machine assignment. In case of parallel availability (simultaneous availability of tokens in input places), parallel assignments are allowed to represent potential parallel processing in a cell. Fig. 2(i)-b represents the token branching decision based on a token color which differs for different job types. With the process routing information embedded in token colors, the branching decision is simply to find a matching place on the routing list of a job from among the output places of the parallel transitions. This branching construct serves as an intermediate step from a resource assignment decision to a set of parallel decision-free constructs. A decision free construct, given in Fig. 2(i)-c, represents processing operations which have time delays. Such a construct is called decision free because the transition between two places can be fired immediate after the release of a token from the input place (i.e., an operation place).

The TCPN cell model here follows a machine-oriented modeling approach which emphasizes the representation of machine activities in the net structure. A typical TCPN cell model is composed of independent closed-loop subnets; one subnet for each entity is defined for the FMC. An entity is either a processing entity, e.g., a processing workstation or a load/unload station, or a transport entity such as an Automated Guided Vehicles (AGV). A closed-loop structure allows repetitive renewal of system resources after the completion of an operation in an entity subnet (e.g. a process workstation).

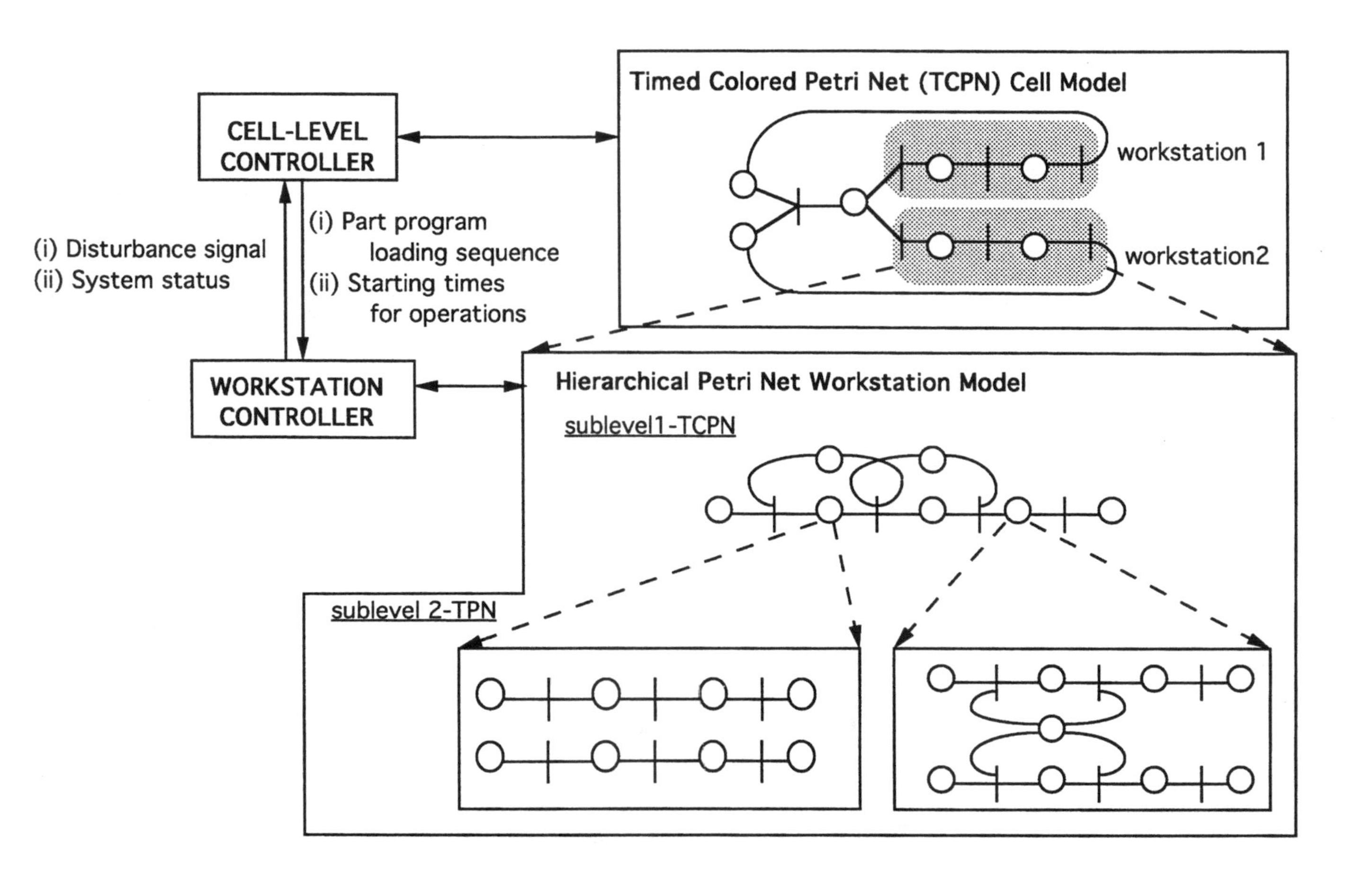

Figure 1. Petri net based modeling structure within the context of hierarchical control

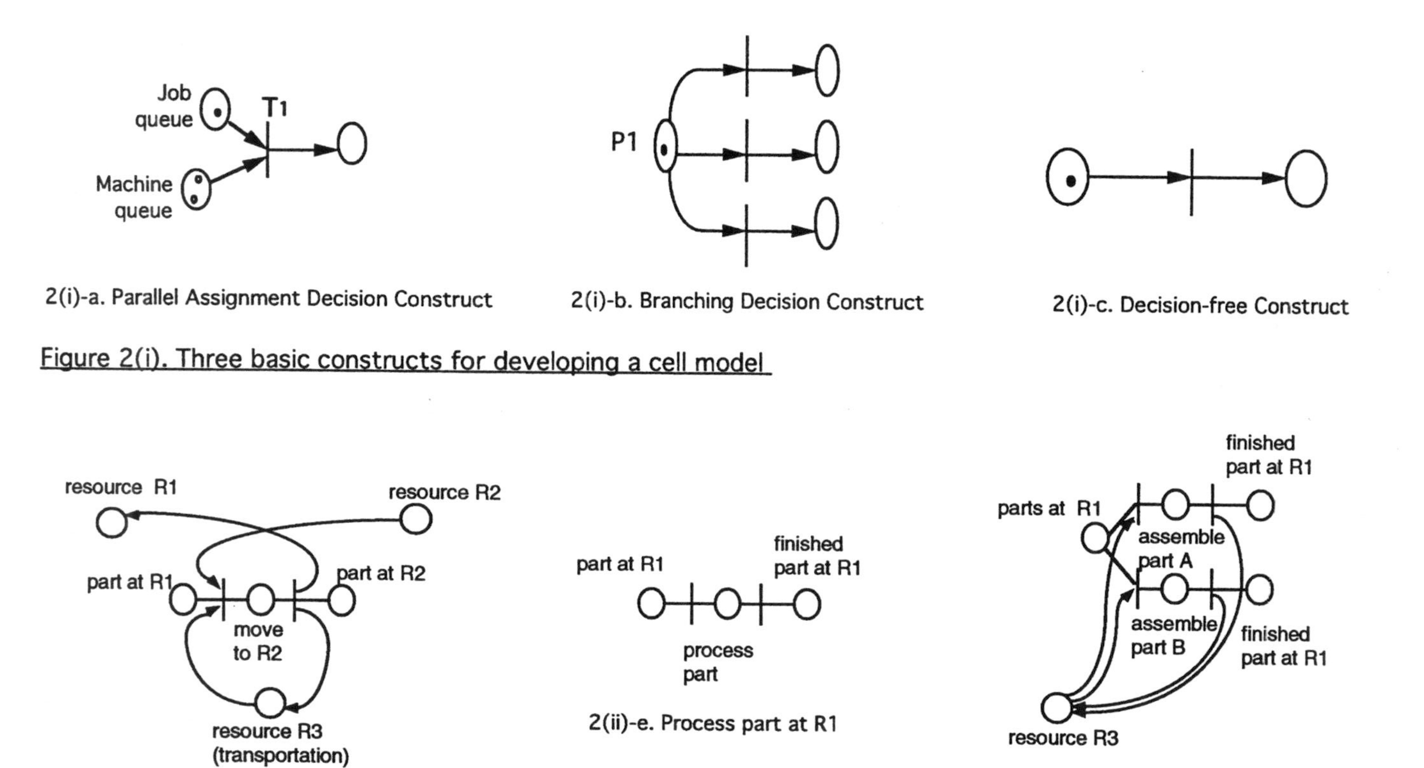

2(i)-a. Parallel Assignment Decision Construct

2(i)-b. Branching Decision Construct

2(i)-c. Decision-free Construct

Figure 2(i). Three basic constructs for developing a cell model

2(ii)-d. Move part from R1 to R2 via R3

2(ii)-e. Process part at R1

2(ii)-f. Assemble part A&B using R3

Figure 2(ii). Basic constructs for sublevel 1 workstation model

Figure 2. Modeling constructs for manufacturing systems

Each operation is an activity performed by the renewable entity resource. The interactions between a material handling system and workstations are depicted by the parallel connections of the processing entity subnets with the transport entity subnet. These parallel connections of processing entity subnets allow the representation of asynchronous part movement which is typical in a NFMS.

HIERARCHICAL PETRI NET MODELING

Within the cell-workstation hierarchy considered in this paper, a TCPN is developed at the cell level which serves as a graphical model and basis for a dynamic scheduling problem formulation. This cell level TCPN model serves as an abstract representation for further detailing at the workstation level. The workstation model presented here consists of two sublevels which result from a more detailed task decomposition. The first sublevel is an expanded TCPN representation of the abstract TCPN model at the cell level and contains more detailed routing information within a specific workstation. The second sublevel consists of a TPN model where different parts (represented by different colored tokens of the sublevel 1 TCPN workstation model) are modelled separately. Further sublevel decomposition is possible for more detailed analysis.

Task Decomposition

Tasks are decomposed into smaller subtasks (or subactivities) which is commensurate with the downward flow of the command structure in hierarchical control. Higher levels deal with longer range and more general operations, and lower levels deal with short term and specific commands. In this research, the first workstation sublevel of the decomposition consists of general operations such "machining", "assembling", and a sequence of movements among processing entities (e.g. an input/output buffer, a machining center, an assembly table). The modeling objective at the first sublevel is to capture the interface with the cell model and the basic routing of a part within a workstation, along with the time and resource constraints. The second sublevel decomposes these general processing and movement commands into more detailed commands such as "setup", "bore", "reach", "pick up", etc. Alternative process plans and different control strategies (e.g. collision avoidance, selection of resources), if any, are incorporated at this sublevel. Further decomposition proceeds in a similar way.

Multilevel Workstation Modeling

Based on the task decomposition approach, a multilevel Petri net model for a workstation is developed where both TCPN and TPN representation are included. One requirement for workstation modeling is the ability to model decision alternatives (e.g. among two robots). As mentioned, the approach taken here is to construct a TCPN as the first sublevel of a workstation model. The inclusion of color tokens are used to distinguish different parts and resources in a compact way and to provide possible alternatives for resource selection. The time factor is used to monitor workstation operations and provide performance information.

The construction of the first sublevel TCPN workstation model is based on the abstract workstation representation in the TCPN cell model. A modular concept using basic constructs is extended from a cell model to a workstation model. The basic constructs for the sublevel-1 workstation model are identified as "move", "process", and "assemble" (see Fig. 2(ii)). Precondition, postcondition, and resource constraints are embedded in each construct. In particular, the "move" construct imposes a constraint on releasing resource R at the end of movement. This constraint prevents other operations from entering other parts into resource R until the current part has been removed completely.

PETRI NET MODELING EXAMPLES

Example for Cell Level Petri Net Modeling

Using the cell level TCPN constructs and the machine-oriented modeling approach, one can construct flexible manufacturing cells having any configuration. Fig. 3 depicts a TCPN cell model which consists of a MHS subnet, two workstation subnets (W1 and W2) and a L/UL station subnet. The two workstation subnets and the load/unload station subnet are connected in parallel through the MHS subnet. The interface between cell entities are the two sets of places {P4, P7, and P15} and {P27, P25, and P26} which represent the input queues and output queues to the L/UL station and workstations W1 and W2, respectively.

The number of tokens in each closed-loop subnet represents the total availability of a particular resource in a cell entity. For example, two tokens in place P9 represent two identical machine resources in workstation W1, whereas a single token in places P6 and P12 represent a single space for the input and output buffer of workstation W1, respectively. In a TCPN cell model, token colors are useful for both visual identification and mathematical representation. In Fig. 3, two job types identified by their different token colors (black and white patterns) are depicted in place P1. In the case of parallel resources, i.e., two parallel machine tokens in P9, distinctive colors would be used for individual resource identification.

In this modeling approach, a three-attribute coloring scheme (part number, workstation number, resource number), is used to differentiate token colors. Part number (pt#) represents the job number; Workstation number (wks#) indicates the workstation where a part is currently being processed or is to be processed; Resource number refers to either a buffer number (b#) or a machine number (m#) in a particular workstation, an equipment number (e#) in a load/unload station, or a device number (d#) for material handling systems. These resource attributes provide a tracking record for the resource assignment decisions. Hence, a token color, (i, j, m), indicates that the token is the ith job which uses the mth resource in the jth workstation. The coloring scheme is embedded in the matrix representation of the TCPN cell model used in the system dynamic equations (see the next section for details).

Example for Workstation Level Petri Net Modeling

Consider a workstation with two input buffers, two machining centers, one assembly table, and one output buffer. Fig. 4(a) depicts a sublevel 1 TCPN model for such a workstation which is based on the workstation level constructs and the following routing : input buffer -> machining center -> assembly table -> output buffer. A control place is added to ensure that two parts moved to and assembled at the assembly table are in sequence. In addition, this ensures that only after the finished product is moved to the output buffer can another assembly operation be performed. The coloring scheme is similar to the cell model, except that only two attributes are necessary, i.e. part number (pt#), and resource number (b#, m#, d#, or e#).

The operations at the assembly table can be modeled in further detail at a sublevel 2 using a TPN as shown in Fig. 4(b). The "move to Ba/Bb" and "in Ba/Bb" places at the sublevel 1 are split into two paths to represent different parts (colored tokens). To construct a sublevel 2 TPN, the "move to" operation at the sublevel 1 model is decomposed into "reach-pickup-transfer-putdown" operations. In addition, the following control strategies are implemented: (a) assemble part A (produced by machine 1) then part B (produced by machine 2), (b) use robot 1 (robot 2) to assemble part A (part B), and (c) use robot 2 to move finished products to the output buffer.

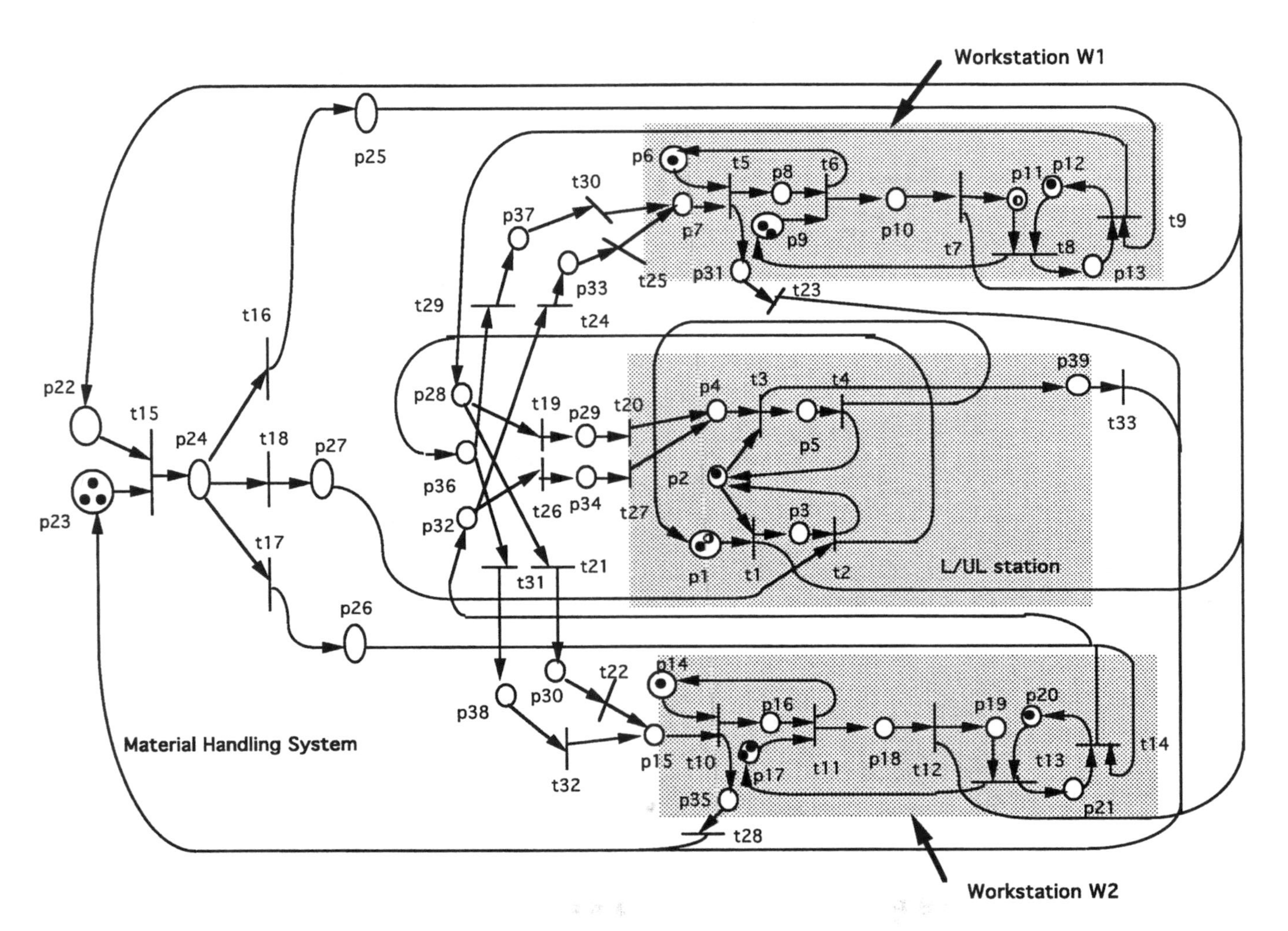

Figure 3. A timed colored Petri net cell model with one L/UL station, two workstations and a MHS

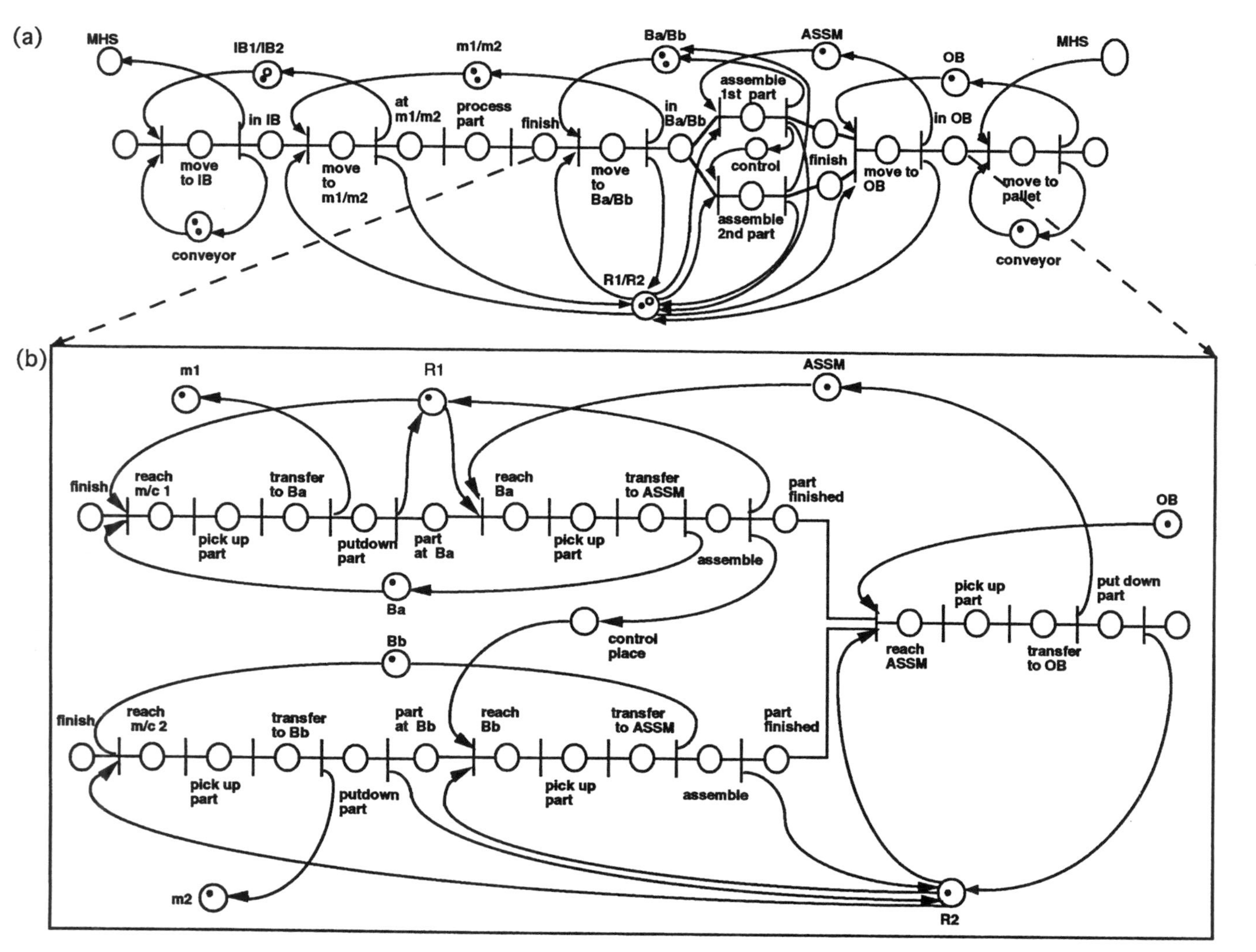

Figure 4. Workstation model : (a) sublevel 1 TCPN model, (b) sublevel 2 TPN model

STATE SPACE REPRESENTATION OF SYSTEM DYNAMICS

A system state of a TCPN model for the cell level is defined in this research by the combination of current token positions (i.e. system marking) and the temporal status of job tokens in operations (i.e. remaining processing times). A system marking is used to indicate the distribution of tokens within a net where a token type may consist of a job token, a machine token, or a combined job-machine token in an operation place. The remaining processing time depicts how long until a specific job-machine token in an operation place can be released (i.e., an operation is completed). The state equations corresponding to the system marking and the remaining processing time are given by the following (terms in state equations are defined in the subsequent subsections) :

system marking :

$$\overline{M}^p(k+1) = \overline{M}^p(k) + \overline{L}\bullet\bar{u}(k) \quad (1)$$

remaining processing time :

$$\overline{M}^r(k+1) = \overline{M}^r(k) - \tau(k)\bullet\overline{P}\bullet\overline{M}^p(k) + \overline{W}\bullet\overline{L}^+\bullet\bar{u}(k) \quad (2)$$

Hence, the system state of the TCPN cell model can be expressed as follows :

$$\overline{X}(k+1) = \begin{bmatrix} \overline{M}^p(k+1) \\ \overline{M}^r(k+1) \end{bmatrix} = \begin{bmatrix} \overline{I} & \bar{0} \\ -\tau(k)\bullet\overline{P} & \overline{I} \end{bmatrix} \bullet \begin{bmatrix} \overline{M}^p(k) \\ \overline{M}^r(k) \end{bmatrix} + \begin{bmatrix} \overline{L} \\ \overline{W}\bullet\overline{L}^+ \end{bmatrix} \bullet\bar{u}(k) \quad (3)$$

$\overline{X}$(k+1), the vector of state variables at stages k+1, is a $(2n\delta\times1)$ column vector consisting of a $(n\delta\times1)$ system marking vector $\overline{M}^p$(k+1) and a $(n\delta\times1)$ remaining time vector $\overline{M}^r$(k+1) where n is the number of jobs. δ is the sum of all colors on tokens associated with places, i.e., $\delta=\Sigma\delta e$ for e=1,..,p. p is the total number of places in a TCPN, and δe is the number of token colors that may appear in a place Pe.

The control vector $\bar{u}$(k) is a $(n\gamma\times1)$ vector with entries equal to ones or zeros where γ is the sum of all colors that can go through transitions, i.e., $\gamma=\Sigma\gamma_s$ for s=1,..t, t is the total number of transitions in a TCPN, and γ_s is the number of token colors that may go through a transition T_s. A scalar, λ, is defined

$$\lambda = (j-1)\times\gamma + \sum_{s=1}^{i-1}\gamma_s + m$$

with the purpose being identify a particular entry (control variable) in the control vector. An entry u_λ (k) equal to one indicates that at the kth stage the ith transition is fired (i.e., a particular operation on a job is being initiated) with a job j's token which has a color m. In this sense, the system evolves through the selective firings of transitions.

System Marking State Equation

The state equation for the system marking has a distribution matrix $\overline{L}$ to account for the evolution of the renewable resource tokens (e.g., tokens for machines, buffers and MHS devices). The distribution matrix $\overline{L}$, a $(n\delta\times n\gamma)$ matrix, acts to determine which places in a TCPN model will be affected by a particular transition firing (i.e., a non-zero entry in the control vector). In our model, parallel machine resources may be available to multiple jobs for assignment. The number of parallel resource tokens available at any particular time instance should be consistent for all jobs in the queue in front of a workstation. As a result, the distribution matrix $\bar{\mathit{L}}$ is partitioned into a set of submatrices with the diagonal consisting of incidence submatrices, $\bar{C}^{(i)}$, and identical off-diagonal resource transition submatrices, $\bar{Q}$.

Job routing information and technological precedence is embedded in the incidence submatrices by establishing a routing (color matching) matrix for each individual transition in the net which represents the completion of a particular job processed in a workstation. The resource transition matrix $\bar{Q}$ depicts specific input/output relations between transitions and the resource queuing places. Detailed construction of matrix $\bar{\mathit{L}}$ for the cell model depicted in Fig. 3 can be found in [6].

Remaining Processing Time State Equations

In a TCPN model, an enabled transition requires not only the condition for token availability as in a PN model, but also that time delays associated with the input places of the enabled transition must all be elapsed. Two situations are considered here which may result from the execution of a transition firing vector $\bar{u}(k)$: (1) some downstream transitions are enabled without any time delay, and (2) no downstream transitions are ready to fire. In the first situation, enabled transitions can continue to fire until no downstream transitions are enabled. The second case suggests that all tokens are either in operation places with time delay or waiting in queuing places for resources to arrive.

When a control action $\bar{u}(k)$ is taken, the remaining processing times (see equation 2) associated with job tokens in operation places are updated with two dynamic terms: (1) an across-the-board or universal reduction caused by the time elapsed between two consecutive stages (represented by the term $\tau(k)\cdot\bar{P}\cdot\bar{M}^{p}(k)$) and (2) processing times added because of new firings, i.e., $\bar{W}\cdot\bar{L}^{+}\cdot\bar{u}(k)$. Here, $\tau(k)$ is a scalar representing the elapsed time between the kth and the (k+1)th transition firing. $\bar{P}$ is a (np×np) square matrix of submatrices that identifies operation places for all job types. Each submatrix $\bar{P}_{ij}$ is either a zero matrix (if $i \neq j$ and place j is not an operation place) or an identity matrix. The dimension of submatrix $\bar{P}_{ii}$ is $(\delta_i \times \delta_i)$ where δ_i is the number of colors for place p_i. $\bar{W}$ is a (np×np) square and diagonal matrix of submatrices. A submatrix $\bar{w}_{ii}$ of $\bar{W}$ represents the processing time of an operation represented by an operation place p_i in a timed colored Petri Net model. $\bar{L}^{+}$ has been defined as the $(n\delta \times n\gamma)$ input matrix for matrix $\bar{L}$.

Since a firing decision needs to be made whenever there are enabled transitions, a natural decision point for transition firings is when the remaining processing time associated with a job token in an operation place is exhausted and its output transitions are enabled. The immediate decision point is hence determined by the shortest time (the smallest entry) in the remaining processing time vector, $\bar{M}^{r}(k)$. Note that the shortest time may be different from stage to stage and not known a priori. The use of unknown sampling intervals between stages for a dynamic system representation has been reported by Tabak.[7] The approach taken here is similar to Tabak in that we introduce a time-variant coefficient $\tau(k)$ to incorporate varying-length time intervals. See reference [6] for details.

System Dynamics of Workstation Models

The system state of the first sublevel TCPN workstation model can be derived using equations similar to the TCPN cell model. The only difference is that the dimension of the system state vector may be smaller than $(2n\delta \times 1)$. The reduction in the dimension of the state equation is due to the incorporation of assembly operations where multiple parts are combined together into one final product. As a result, some token colors are absorbed by the assembly process. To obtain the state dimension, δ is partitioned into two distinct sets : (1)δa, representing the sum of all colors on tokens associated with the process of handling assembled final product, and (2)δb, the sum of all colors on tokens associated with the rest of the places. The dimension of the state equation is then determined by $(2(\delta a+n\delta b)\times 1)$. For the given example in Fig. 4(a), $\delta=48$, $n=2$, $\delta a=5$, and $\delta b=43$, the dimension of the state vector would be only 182 as opposed to 192. The state space representation of the second sublevel TPN is obtained by partitioning the system state of the first sublevel TCPN model into several subsets. Each subset represents the operations at a particular location (e.g. input buffers, machining center, or assembly table). The operation places within each subset are then augmented to incorporate detailed process information. Detailed equations at this level are still under investigation.

APPLICATION OF PETRI NET BASED MODELING

FMC Scheduling Problem Formulation

Given the dynamic equations of the TCPN cell model, one can formulate a FMC scheduling problem with features similar to an optimal control problem formulation. For example, the FMC scheduling problem with a minimum squared due date deviation objective can be modeled as follows:

Problem P_t: $$MIN \quad C = \overline{X}^T(N)\bullet\overline{H}\bullet\overline{X}(N)$$

Subject to :
Dynamic Systems :

$$\overline{X}(k+1) = \begin{bmatrix}\overline{M}^p(k+1)\\ \overline{M}^r(k+1)\\ \overline{M}^d(k+1)\end{bmatrix} = \begin{bmatrix}\overline{I} & \overline{0} & \overline{0}\\ -\tau(k)\bullet\overline{P} & \overline{I} & \overline{0}\\ -\tau(k)\bullet\overline{R1} & \overline{0} & \overline{I}\end{bmatrix}\bullet\begin{bmatrix}\overline{M}^p(k)\\ \overline{M}^r(k)\\ \overline{M}^d(k)\end{bmatrix} + \begin{bmatrix}\overline{L}\\ \overline{W}\bullet\overline{L}^+\\ \tau(k)\bullet\overline{R2}\end{bmatrix}\bullet\bar{u}(k)$$

Boundary Conditions : $$\overline{M}^p(0),\ \overline{M}^d(0),\ \overline{M}^r(0)$$

System Input Constraints :

$$\bar{u}(k) \geq [0]$$
$$\overline{M}^p(k) \geq \overline{L}^-\bullet\bar{u}(k)$$
$$\overline{M}^r(k) \geq \tau(k)\bullet\overline{P}\bullet\overline{L}^-\bullet\bar{u}(k)$$

The $\overline{H}$ matrix in the objective function is a $(3n\delta\times 1)$ distribution matrix which has entries (i.e., weights) of all 1's and 0's signifying that the objective function is based on time deviations alone. Note that the system state variable X(k) is augmented to include the remaining slack time vector, $\overline{M}^d$(k), that is specific for the representation of a due-date based FMC scheduling problem. The remaining slack time state equation is expressed as follows:

$$\overline{M}^d(k+1) = \overline{M}^d(k) - \tau(k)\bullet\overline{R1}\bullet\overline{M}^p(k) + \tau(k)\bullet\overline{R2}\bullet\bar{u}(k)$$

where, $\overline{R1}$ is a $(n\delta\times n\delta)$ matrix and $\overline{R2}$ is a $(n\delta\times n\gamma)$ matrix with '1' entries indicating the remaining routing for places and transitions respectively. The first term, $\tau(k)\bullet\overline{R1}\bullet\overline{M}^p(k)$, represents an across-the-board effect of time advancement on all job tokens currently in a TCPN model. The second term $\tau(k)\bullet\overline{R2}\bullet\bar{u}(k)$ is employed to compensate for the excess slack reductions.

Cell Level Scheduling Results

A heuristic solution of problem Pt was obtained. The approach was based on forward dynamic programming coupled with an estimate of future idle times via an average conflict measure approach. Performance evaluation was carried out by comparing the response time and the due date deviation measures between the developed due-date based scheduling heuristic (DSH), a minimal remaining slack time (MRST) dispatching rule and the Mean Value Analysis of Queues (MVAQ) technique.[8] A set of 18 ten-by-ten jobshop problems were tested on a Gateway 486/33 personal computer. The test problems were chosen owing to the availability of their optimal makespan schedules and were supplied by Applegate.[9] Significant improvements in due-date deviation reduction for both loose and tight due-date situations were documented. For tight due dates, the average improvement ratio for the DSH approach compared to the MRST dispatching rule was 54.9%. For loose due dates, the average improvement ratio was 59.8%. For the steady-state scheduling procedure developed, the computation time in solving these 10-by-10 problems was less than two minutes for all problems. Table 1 gives the test results for a set of representative examples.

Table 1 Scheduling performance results for cell level controller

	Test	Performance	Computation Times(sec)			Ratio (%)	
	Problem	(dev/due)%	DSH	MRST	MVAQ	(DSH/MVAQ)	(DSH/MRST)
Tight Due-Dates	OBR1	12.8	79.75	30.71	464.78	35.2	62.8
	OBR6	5.16	69.64	28.72	856.67	81.8	89
	OBR5	15	79.47	22.35	462.53	22.4	23.7
	OBR10	12.8	86.29	21.58	654.27	41	59.6
	OBR2	15.2	113.2	41.47	601.05	5.3	32.8
	OBR3	10.2	63.50	16.59	510.42	60.8	76.4
Loose Due-Date	OBR1	7.79	74.64	28.07	325.21	22.5	58.9
	OBR6	5.29	69.21	26.86	490.98	64.1	76.2
	OBR5	2.62	63.22	21.09	571.07	78.6	48.5
	OBR10	2.47	86.29	19.72	385.09	57.1	72.3
	OBR2	3.08	72.40	38.38	555.85	72.5	12.2
	OBR3	5.67	77.17	15.26	581.99	65.6	71.2

Workstation Control

The control function at the workstation level is partitioned into a monitoring function, diagnostics, and recovery plan generation. Petri net based workstation modeling provides a basic representation in performing task decomposition and gives necessary information for control purpose. A general framework for a Petri net based error recovery approach has been presented previously.[10] The time factor in the Petri net representation serves as a watchdog in monitoring the workstation operations. To-date, the state space representation has been integrated with a Neural network structure for diagnostic purpose.[11] The proposed approach used Petri Neural networks which combines a T-gate threshold logic concept with a state space representation to facilitate the process of determining proper preliminary actions based on a sublevel-3 TPN model. The hierarchical Petri net modeling structure can also be a useful representation for solving error recovery problems at the workstation level. For example, for errors that need to use different resources and a longer time span to recover (e.g. a machine breakdown), a recovery plan generated from a higher sublevel workstation model may be appropriate. Lower sublevel models can be used to generate recovery plans for small scale errors such as positioning errors or collision errors.

CONCLUSIONS

A Petri net-based modeling structure with a state space representation for manufacturing systems within the context of hierarchical control has been presented. A modular, machine-oriented timed colored Petri net (TCPN) modeling technique was introduced for a non-dedicated FMC. A TCPN cell model consisting of a material handling system, a load/unload station, and two workstations is given. The advantage of such a cell model is its ability to incorporate the job re-entrant situation where a job may repeatedly use a particular device during the processing duration of the job. Detailed representation of workstation operations was configured through a top-down decomposition approach resulting in a hierarchical multilevel PN model. The system state of a TCPN cell is defined through a system marking state equation and a remaining processing time state equation. The construction of the state space representation of a workstation model was discussed.

The state space representation was used to solve scheduling problems at the cell level and to aid the control of manufacturing operations at the workstation level. Performance evaluation of a developed dynamic scheduling heuristic scheduling approach was conducted for a static scheduling environment. The scheduling heuristic is capable of solving a typical jobshop problem and has the added capability to perform job-machine assignments for the case of parallel machines in a cell. The average improvement ratio was in the range of 50% to 60%.

Future work will consider fully automated flexible manufacturing systems subject to configuration changes, demand variations, machine breakdown, or operational failures. Mathematically, the occurrence of system disturbances can be introduced to the FMC scheduling problem by changing the boundary conditions of the cell control problem. Operational failures can be incorporated into the workstation model using the basic modular constructs developed in this paper without altering the overall model.

REFERENCES

1. Browne, J., Dubois, D., Rathmill, K., Sethi, S. P., and Stecke, K. E., Classification of Flexible Manufacturing Systems, *FMS Magazine*, vol. 2, no. 2, pp. 114-117, 1984.

2. Balbo, G., Chiola. G., Franceschinis, G., and Molinar Roet, G., Generalized Stochastic Petri Nets for the Performance Evaluation of Flexible Manufacturing Systems, *IEEE International Conference on Robotics and Automation*, pp.1013-1018, 1987.

3. Kamath, M., and Viswanadaham, N. Applications of Petri Net Based Models in the Modeling and Analysis of Flexible Manufacturing Systems, *IEEE International Conference on Robotics and Automation,* pp.312-317, 1986.

4. Valavanis, K., On the Hierarchical Modeling Analysis and Simulation of FMSs with Extended Petri Nets, *IEEE Transactions on System, Man and Cybernetics*, vol. 20, no. 1, pp.94-110, Jan./Feb. 1990.

5. Viswanadham, N., and Narahari, Y., Stochastic Petri Nets Models for Performance Evaluation of Automated Manufacturing Systems, *Information and Decision Technologies*, 14, Elsevier Science, pp.125-142, 1988.

6. Liu, C.-S., Real-Time Control of Flexible Manufacturing Cells With Alternative Routing Strategies, Ph. D. Dissertation, Department of Industrial Engineering, Lehigh University, 1993.

7. Tabak, D., and Kuo, B. C., Optimal Control by Mathematical Programming, Englewood Cliffs, New Jersey, Prentice-Hall, 7.4: Discrete-time Controlled Systems with Unknown Sampling Periods, pp.150-158, 1971.

8. Suri, R., and Hildebrant, R. R., Modeling Flexible Manufacturing Systems Using Mean-Value Analysis, *Journal of Manufacturing Systems*, vol. 3, no.1, pp.27-37, 1984.

9. Applegate, D., Personal Communication, 1992.

10. Odrey, N. G., and Ma, Y. H., Intelligent Workstation Control: An Approach to Error Recovery in Manufacturing Operations, *Proceedings of 5th International FAIM Conference*, pp. 124-141, June 28-30, 1995.

11. Ma, Y. H., and Odrey, N. G., On the Application of Neural Networks to a Petri Net Based Intelligent Workstation Controller for Manufacturing, *Proceedings of the ANNIE'96 Conference*, pp. 829-836, November 10-13, 1996.

PROGRAMMABLE CONTROLLERS PROGRAM GENERATION USING A TEXTUAL INTERFACE

Dr. Orlando Durán
duran@upf.tche.br
FEAR Universidade de Passo Fundo
C.P. 566 or 567 CEP 99001-970
FAX 55-054-3111307
Passo Fundo (RS) Brazil

Dr. Antonio Batocchio
batocchi@fem.unicamp.br
DEF / FEM UNICAMP
FAX 55-019-2393722 C.P. 6122
CEP 13083970
Campinas SP Brazil

ABSTRACT: This paper reports the implementation of an automatic system for programming programmable logic controllers (PLC) using as an input scripts that describe the control logic of automated manufacturing/assembly cells. The systems makes use of the object oriented paradigm for representing all the manufacturing entities that participate in a given application and their relationships. Though the use of the object oriented paradigm, issues such as consistence and traceability are guaranteed. The implementation of the code generation module is made by using the definite clause grammars methodology, that allows construction of compilers, interpreters and code generators in a readable manner. The user can define once the control logic for an application, and generate source code for a selected PLC many times as he/she needs giving reusability to all the programs generated. Through the use of this kind of systems, the programming task, that is an error prone activity, is enhanced and can be performed attaining high productivity and reliability levels. The systems was written in Arity Prolog and runs under Windows on a personal computer.

INTRODUCTION

Programmable logic controllers (PLCs) are widely used for sequential operations control, specifically in the discrete manufacturing industry. This fact is due to the advantages obtained when compared with fixed logic systems and relays-based systems. These improvements are refereed to flexibility, safes and low maintenance costs, and the reduction in start up and operation times (Cox, 1986, Witzerman and Nof, 1995).

The growing up success of this equipment brought a high diversity of PLCs into the shop floor. Each day more and more manufacturers enter to the market, offering new solutions, with a wide range of possibilities and features. The main features that the manufacturers outline for attaining the high competitions' levels are: throughput, input/outputs points, programming facilities and languages, etc.

When a company purchases PLC from different vendors, each one of these devices incorporating a proprietary programming language, the user faces some difficulties. Such difficulties are: no chance for programs reusability, no program sharing among solutions and so on. A solution for this problem is to have a programming specialist for each one of the PLCs used into the shop floor (Halang and Krämer, 1992). Other solution is to create a neutral input format for describing the control logic and the further translation into a PLC source code. Bhatnagar and Linn (1990) have shown that a program written in a logical syntax can be translated directly into the task code for a particular model program and download it to a controller in the same manner that is currently used for robot control programs.

Another trouble related to PLCs software development is the extended life cycle. It is very difficult pass from one phase to another in a smoothly and natural manner, mainly because of the traditional analysis approaches do not have been realized for manufacturing software, but for data intense applications. Manufacturing engineers realize the manufacturing domain as composed by different entities, such as machine tools, fixtures, pieces, material handling devices, sensors, actuators, etc. Each one of these entities has associated a set of attributes and capabilities. Thus, these entities can be considered as abstract constructions called objects. Through an abstraction process real world objects may be represented as abstract objects belonging to an object - oriented model. Using this type of models, a control logic definition task is simplified and it can be made in a very natural manner.

The object oriented paradigm, that is being widely used in software engineering, is used here for automated program generation for a PLC. This paper presents an approach for developing PLCs software using an object oriented language called object oriented specification technique (OOST).

The automatic generation process is performed from a textual and semi-structured description (written in OOST) of the control logic for a manufacturing cell (Durán and Batocchio, 1993; Durán and Batocchio, 1995a, 1995b).

BACKGROUND

There are some researches in the literature aiming at the development of manufacturing systems modeling and control software specification. Menga and Morisio (1989) defined a specification and prototyping language for manufacturing systems. The language is based on high-level Petri-nets and is object-oriented. The approach uses graphical and textual tools for defining the components of the system and their behaviors. The language integrates two formalisms with the object-oriented development paradigm. The first formalism is the hierarchical box and arrow graphical formalism. The second one is the Petri-net graphical formalism for the detailed specification of the control flow in objects. According the authors, the use of this language allows the simulation of manufacturing systems, hierarchically structuring them and enriching the library with the new objects.

Another initiative (Boucher and Jarafi, 1992) addresses the definition of an interface between manufacturing system design and controller design. The high-level design methodology enhances communications between manufacturing system designers and controller designers. It also allows the automatic control code generation from that design. The high-level design methodology is based in the IDEF0 methodology, created under the development of USAF, and generates a Petri Net representation of the control logic through the use of a rule-based interpreter.

A similar approach is given by Roberts and Beaumariage (1993) that present a methodology to design and validate supervisory control software specification. This methodology is based on a network representation schema. The networks are composed by nodes and arcs. Messages are passed between the nodes, resembling the object-message paradigm. A version of this control specification system have been coded on a Texas Instruments Explorer and other platforms.

A textual specification technique for PLC software generation was reported by Bhatnagar and Linn (1990). The generator uses a user-defined control process specification and initially converts it into a standard control logic specification and finally generates an executable program for the PLC specified. The bi-directionality of the code is the main virtue of this system, providing total portability of programs from a PLC to another.

Halang and Krämer (1992) presented an interactive system with a graphical interface for constructing and validating PLC software. It combines the function block diagram with a graphical language. The approach emphasizes the description of composite PLC software from a library of reusable components and may be considered as an object-oriented approach.

Finally, Joshi, Mettala and Wysk (1992) created a systematic approach to automate the development of control software for flexible manufacturing cells called CIMGEN. The specification is based on context free grammars (CFGs) providing a formal basis for control strategies descriptions. CIMGEN generates automatically control software for workstation and cell control levels. This automated code generation is not totally satisfactory, since the generated code requires hand manipulation for completion.

There are two other papers that present object-oriented approaches for modeling manufacturing systems. Mize, Bhuskute, Pratt and Kamath (1992) showed the results obtained in exploring alternative approaches to the modeling and simulation of complex manufacturing systems. These results argue that is necessary a paradigm shift in developing models for manufacturing systems. Through this paradigm shift, the system planner can now define models through the use of building blocks, called objects. The authors assert that this approach is an strategic opportunity for the fields of industrial engineering, operations research/management science and manufacturing systems engineering. Joannis and Krieger (1992) reported an object-oriented methodology for specifying manufacturing systems. The specifications are made by building successive models, each containing more details than the previous one. The desired behavior of the system is described using a set of concurrent co-operating objects and the behavior of each object is defined through the use of communicating finite state machines (CFSM). According to the authors, the CFSM allow the execution of the specifications. The technique has a text format.

MANUFACTURING CELLS CONTROL

Manufacturing systems are sets of different subsystems, all working together and co-ordinately to get the expected results. The design of these manufacturing systems and the development of manufacturing systems controllers has become more closely linked as the manufacturing environment has become more automated (Boucher and Jarafi, 1992). Controller design addresses issues of communication, control logic, tasks sequencing, error handling and programming of programmable devices.

Usually the control of a manufacturing systems is made up in different levels of abstractions. These levels of abstractions have different time scales, event types and decision kinds. Menga and Morisio (1989) identified these levels:

Plant Level: the decisions here are long term scheduling, weekly or at least daily, inventory control, orders control and processing. The timing of the events at this level can take from hours to months or years.
Shop Level: A shop is composed of cells and a material handling system. The events are the beginning and the ending of an operation or a set of operations, a breakdown situation or a scheduled stop for maintenance. The timing at this level is from hours to days.
Cell Level: A set of machines or stations linked by a material handling equipment. The events that can occur at this level are transfer of batches or workpieces among workstations, and the scheduled stops for maintenance and tool changes. The timing at this level is in minutes or seconds.
Machine Level: At this level the elements are simple machines, inspections devices and manipulators. The events are the beginning and ending of an operation in a machine, the completion of movement by the manipulator or of a measure by the inspection equipment.

The scope of this paper is the manufacturing cell control and the definition of the cell controller. The cell controller is responsible for the sequencing of activities within a cell. It is necessary to define the functions related to the main operations performed for each one of the elements that make up the cell. The control at the cell level is normally implemented using a PLC or combinations of PLCs and a cell host computer or a factory floor computer. Programming a PLC is an error prone task, where the developed programs are very difficult to read, understand and maintain. Hence, there are many efforts to define a high level programming standard to simplify the program definition and the total PLC software life cycle. In the next section some programming methods for PLCs are analyzed.

PLCS' PROGRAMMING PARADIGMS

There are different approaches for programming a PLC. Each manufacturer incorporates one or more types of programming methods to its equipment. The most common methods for programming a PLC are:

- Ladder Diagrams.
- Instructions Lists.
- Structured Texts (STs).
- Sequential Function Charts (SFCs).
- Function Block Diagrams.

Figure 1 shows a very simple example of the three graphical programming languages defined above.

how they interoperate are the major concerns. These are lumped into the application layer in OSI and TCP/IP models and should be distinguished clearly in interoperability framework.

We choose the term "level" instead of "layer" in this paper because it has a connotation more closely related to the effort required to accomplish certain objectives. We also use the term "component" to represent an intelligent controller, a host or a piece of control software for interoperability qualification. We use the term "partner" to refer to a component that is to interoperate with another component for qualification.

2.1. The Levels of Interoperability in Process Control

In our definition, the interoperability is ranked in four levels shown in Figure 2.1.

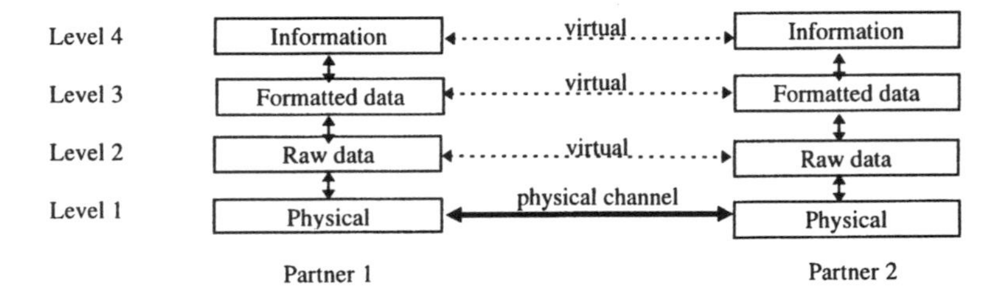

Figure 2.1 Interoperability framework

A minimum level for a component to share data or exchange services with its partner is a common physical link. The physical interface should define the mechanical and electrical characteristics of the connection as defined in OSI or TCP/IP physical layer. This can be an RS-232 port, an Ethernet card, etc. A system integrator is often delighted to see a simple existence of a common communication port. Such port may provide many interoperation possibilities with the programming. The physical link is a big step towards interoperation.

A better connection between two components will provide access to raw data. This level is related to OSI data link layer through session layer in OSI model and through part of the application layer in TCP/IP model. Examples for possessing a data level interoperability would be SECS-I in SEMI standard, streams in TCP/IP, FTP application in TCP. For a better description of SECS-1, please refer to the descriptions in Case II in Section 3. At this level, the components should be able to exchange raw data with its partner without error. Since this level of interoperability only supports raw data without meanings or formats, data parsings and usages must be performed by user applications.

At the third level, formatted data can be transferred. A formatted data such as recipe files or commands can be accepted into applications in the partner. No tokenizing, parsing or categorizing are required. The formatting eliminates tedious and duplicate software developments in applications. With the third level support, users can focus on the storage, retrieval and use of data rather than manipulating data. The standard recipe file format by the SMEAMA is intended to be protocol at such a level. The SECS-II message definitions are also examples of the formatted data level interoperability support.

At the information level, the formatted data in the component is meaningful to its partner for actions or services. For example, the observations of statistical process control data are formatted data (can be unformatted too). However, the mean and variation, and their

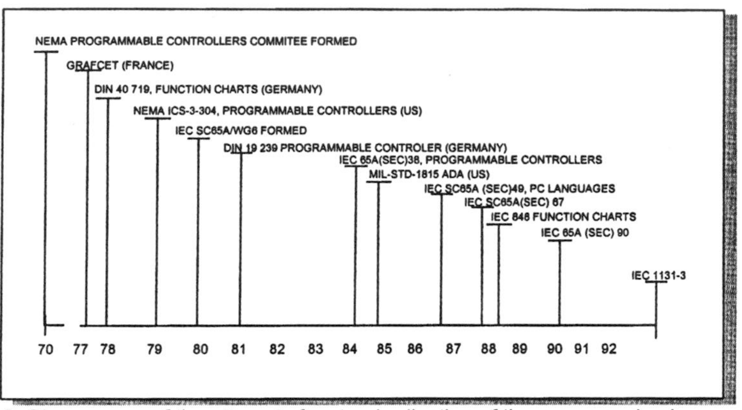

Figure 2. Chronogram of the attempts for standardization of the programming languages for PLCs.

THE APPROACH

The system that was implemented allows the programmer define a new application control program through the use of an object oriented specification technique (OOST). This specification technique may be used from the requirement description phase to the implementation phase (Booch, 1986), and it may be used as an specification input language for an automatic PLC software generator. The development of PLC software using the object oriented specification technique can shorten the life cycle phases, even some phases are no more needed, because they are contained within the previous phases (Hodge and Mock, 1992)

The methodology when applied in the requirement analysis phase allows the natural contact between the programmer/analyst and the final user of the system. Hence, the manufacturing engineer that is in charge of project the automated system can communicate his needs and requirements with the programmer/analyst using a natural common language, based on abstract constructions (objects), avoiding misunderstandings and excessive documentation.

To make up the specification technique, an exploratory survey among various PLC programming languages was made. This survey aimed at defining a sufficient collection of basic operations supported by any PLC performing sequential control tasks, table 3 shows the set of basic operations considered to define the OOST.

Besides the instructions showed by table 3, there is a set of instructions for data manipulation and arithmetic operations. These instructions were also considered by our research and OOST supports them. OOST does not support complex instructions, such as PID functions and Fuzzy Logic-based operations, because they are out of the scope of our research. The basic operations were grouped into five categories, considering types of devices state changes. These five categories are:

- Momentary state changes caused by the beginning or ending of an event.
- State changes that last as long a given event exists.
- State changes caused by the existence of an event during a given time.
- State changes caused by the existence of an event a given number of times.
- State changes caused by a given event and maintained until another event begins.

Next, the structures needed to describe each one of the categories was analyzed. Thus a set of syntactical structures representing each one of the mentioned categories was defined. This set of structures constitutes the description language, that we called OOST. Next each syntactical structure was mapped into a set of rules making up a free context grammar. In these grammars each rule consists of a left-hand side and a right-side, separated by a "-->" symbol. At the left-side, there is a non-terminal symbol that may be decomposed in a series of syntactic entities, namely, reserved word or other non terminal symbols. Each non-terminal symbol needs, at least, one rule lead its decomposition. As Szpakowicz (1987) mentioned in his paper, logic grammars are a succinct and powerful notation for language specification and processing. Finally, this set of rules was translated into Prolog code. As Szpakowicz (1987) outlined with Prolog one can write compilers and interpreters of programming languages, as well as natural language interfaces in a readable manner.

The Prolog code can be considered as a rule-based interpreter. It performs lexical and syntactical analysis of the specifications written by the user and provide an intermediate code of the control logic (see figure 3). This rule-based interpreter also supports the object-oriented paradigm. Through this facility the objects can be hierarchically structured using the concept of superclasses and classes. Objects representing actual manufacturing devices may be defined and incorporated as terminals symbols to the grammar. Grammatically, these symbols are classified as nouns and the methods associated to the classes are incorporated as verbs to the same grammar. Through that, a new object class or object instance can inherit attributes and capabilities of its superclass. It is the main feature of the paradigm, that allows models reusability. One sentence of the OOST is composed by two phrases linked by a reserved word. These two phrases are called subordinated phrases, because the first one completes its meaning with the meaning of the second one. This kind of sentences are very suitable for representing either the causal relation (causal sentences), the timing of an event (temporal sentences) or a conditional relation between a series of events and actions (conditional sentences).

The intermediate code, generated by the interpreter, is used as an input for an automatic PLC program generation module. The general structure of this systems is shown in figure 3.

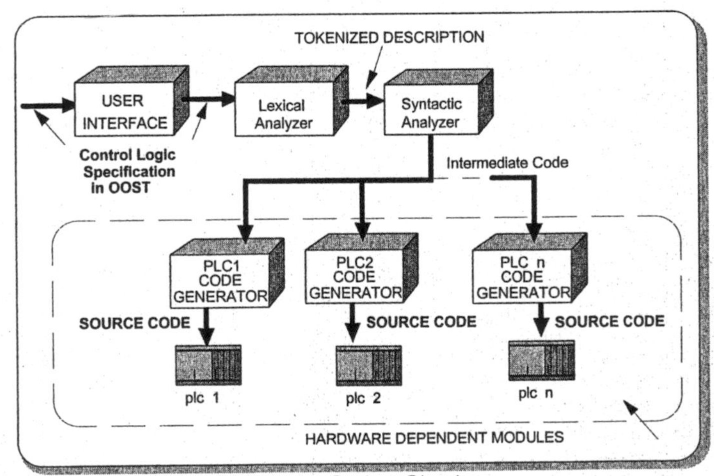

Figure 3. System Structure

As Ram and Lai (1992) pointed out, a core instruction set is common to all manufacturers of PLCs. However, the data addressing requirements vary among vendors. Thus, the translation of code into a PLC program involves an specific assignment of valid data addresses to each device connected to the PLC. The system has a module, called the *wiring module*, that acts as a link between the OOST and the code generator in the sense that it allows the user to define the address associated for each one of the devices (represented by nouns in the specification). A sample view of the system is shown in figure 4.Through this definition the system captures the location and translate it into the correct code in the source PLC program. The windows-based user-interface was written in Visual Basic.

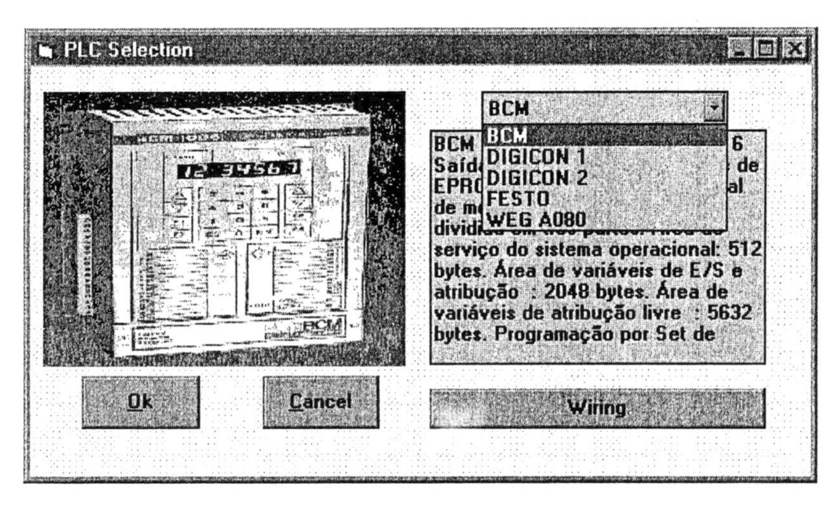

Figure 4. Sample view of the system

CONCLUSIONS

An automatic programmable controller software generator was proposed. The user can define the control logic in a natural manner writing a textual description of the states changes and the conditions associated of an actual manufacturing system. This textual definition, called object oriented specification technique (OOST) is based on the object oriented paradigm. Thus, the devices and machines are represented by objects, and the actions to be taken are represented as the methods associated to these objects or to their superclasses. This specification technique may be considered as an efficient means for performing an informal requirement description, and from it, to obtain in an automated manner the source code for an specific PLC, ready for downloading within PLC memory.

The system was developed in Arity Prolog and the user interface was implemented in MS Visual Basic.

The approach allows to create object library for accelerating new applications developments and fomenting software reutilization. This fact leads to important improvements in productivity and reliability of PLC programming tasks.

REFERENCES

Bhatnagar, S. and R.L.Linn, 1990, Automatic Programmable Logic Controller Program Generator with Standard Interface, *Manufacturing Review*, Vol. 3, No. 2, 98-105.

Booch, G.,1986, Object Oriented Development, *IEEE Transactions on Software Engineering*, Vol. SE-12, No. 2, 211-221.

Boucher T.O. and M.A.Jarafi,1992, Design of a Factory Floor Sequence Controller from a High-Level System Specification, *Journal of Manufacturing Systems*, Vol.11, No. 6, 401-417.

Cox, R.A., 1989, *Technician's Guide to Programmable Controllers*, (Delmar Publishers Inc.), Albanny, New York, USA.

Durán, O. and A. Batocchio,1993, A high-level object-oriented programmable controller program interface, *ISIE'94, Proceedings of the IEEE International Symposium on Industrial Electronics*, Santiago, Chile.

Durán, O. and A. Batocchio,1995, Object Oriented Development Methodology for PLC software, *Proceedings of the IEEE/ECLA/IFIP International Conference on Architectures and Design Methods for Balanced Automation Systems*, Vitoria (ES), Brazil.

Durán, O. and A. Batocchio, 1995, Generating Manufacturing Code with an Object-Oriented Development Methodology, *Proceedings do LCA'95, 5th. Simposium of Low Cost Automation*, Buenos Aires, Argentina.

Halang, W.A. and B.Krämer, 1992, Achieving high integrity of process control software by graphical design and formal verification, *Software Engineering Journal*,January, 53-64.

Hodge, L.R. and M.T.Mock, 1992, A proposed object-oriented development methodology, *Software Engineering Journal*, March, 119-129.

Joannis, R. and M. Krieger, 1992, Object-oriented approach to the specification of manufacturing Systems, *Computers Integrated Manufacturing*, Vol.5, No. 2, 133-145.

Joshi, S.B., E.G.Mettala and R.A.Wysk, 1992, CIMGEN- A Computer Aided Software Engineering Tool for Development of FMS Control Software, *IIE Transactions*, Vol.24, No.3, 84-97.

Menga, g. and M.Morisio, 1989, Prototyping Discrete Part Manufacturing Systems, *Information and Software Technology*, Vol. 31, No.8, 429-437.

Mize, J.H., H.C.Bhuskute, D.B. Pratt and M.Kamath, 1992, Modeling of Integrated Manufacturing Systems using an Object-Oriented Approach, *IIE Transactions*, Vol. 24, No. 3, 14-26.

Ram, B. and S.H.-Y.Lai, 1992, The development of a program generator for Programmable Logic Controllers, *Computers and Industrial Engineering*, Vol. 23, Nos. 1-4, 335-339.

Roberts, C.A. and T.G.Beaumariage,1993, A Specification Technique for Generating and Simulating Supervisory Control, *Computers and Industrial Engineering*, Vol. 25, Nos. 1-4, 515-518.

Szpakowicz, S., 1987, Logic Grammars, *Byte*, August, 185-195.

Witzerman, J.P. and S.Y.Nof, 1995, Integration of simulation and emulation with graphical design for the development of cell control programs, *Intenational Journal of Production Research*, vol.33, n. 11, 3193-3206.

GLOBAL BUSINESS PROCESS RE-ENGINEERING: A SYSTEM DYNAMICS BASED APPROACH

J. Ashayeri and R. Keij
Department of Econometrics
Tilburg University
Tilburg, The Netherlands

A. Bröker
AWB Business Development B.V.
Eindhoven, The Netherlands

ABSTRACT

This paper presents a structured approach for Business Process Re-engineering (BPR) in a large (global) corporation. The paper defines BPR, explores the principles and assumptions behind Re-engineering, looks for common factors behind its successes or failures. In order to overcome the potential failures, we propose to view a global business as a system and use the System Dynamics methodology (Forrester (1961)) in conjunction with Analytical Hierarchy Network Process (Saaty (1996)). This approach will enable corporations to understand the relationships among the components of a global business and link different sub-systems properly when a Re-engineering process is taking place. The purpose of this approach is to increase the effectiveness of the Re-engineering process.

INTRODUCTION

Recent developments

Large (global) corporations are caught up in a massive restructuring process in order to address their customer requirements (excellence with respect to criteria like cost, quality, delivery and service) taking into account new technological solutions (e.g. Information Technology). This dramatic change is being described as a paradigm shift that requires a new context for leadership and management practice in all sectors of the economy. This new paradigm can be described as an integrated approach for restructuring processes in all functions considering customer values. The integrated approach involves major strategic decisions to be made. Therefore, the complexity of strategic decision problems has increased in two ways. First, there is an increase in the number of factors that influence decisions about organizational change, i.e. *detail complexity*. Second, it has become more important, especially in global businesses, to take system-dynamic relationships among business processes into account, i.e. *dynamic complexity*. The intention of this paper is to put forward a conceptual framework of a decision support system, assisting management in handling these dimensions of complexity.

What is BPR?

In order to restructure and improve processes in all functional areas in a corporation, three alternative approaches are suggested in the literature.

- *Continuous Process Improvement* that reduces variation in the quality of output products and services and incrementally improves the flow of work within a functional activity.
- *Business Process Redesign* that removes non-value added activities from processes, improves our cycle-time response capability, and lowers process costs.
- *Business Process Re-engineering* that fundamentally or radically redesigns processes (through the application of enabling technology) to gain drastic improvements in critical contemporary measures of performance, inspired from a new mission, such as cost, process efficiency, effectiveness, productivity, and quality.

The first two approaches can be associated with the Process Of On Going Improvement (POOGI) or Deming loop of Total Quality Management (TQM) discipline. The third approach, Business Process Re-engineering (BPR), was originally defined by Hammer (1990) and is the major concern of this paper. BPR is often undertaken in response to dramatic changes in the external environment (a paradigm shift, for instance) that apply considerable pressure on the ability of the organization to fulfill its mission, improve its competitive positioning, or to even survive as an entity. BPR actions are radical in transforming all functions within the organization. We believe that the first approach will enable organizations to improve *team* oriented activities (an operational change) while the second approach will help organizations to get more in line with their existing mission and can be considered as a tactical change. However, the third approach demands an innovative look at the mission and may result not only in drastic process changes but also in mission changes, which is more a strategic change.

Sage (1995) distinguishes three levels of Re-engineering: product Re-engineering, process Re-engineering and systems management Re-engineering. *Product Re-engineering* does not apply directly to this article. With *process Re-engineering* any actual process is compared with the ideal process and on this basis activities are eliminated and improvements are made. *Systems management Re-engineering* mainly focuses on interrelations between business processes. Therefore, a company is seen as a system of business processes that continuously influence each other. Here not only direct effects are taken into account, but also the indirect and longer term effects are considered as well. This highest level of Re-engineering applies BPR where it really contributes most to the business success.

According to Berman (1994) BPR is often used by companies on the brink of disaster to cut costs and return to profitability. The danger is that during this process the company may slash important processes needed for future growth. To reap lasting benefits, companies must be willing to examine how strategy and Re-engineering complement each other -- by learning to quantify the strategy (in terms of cost, milestones, timetables); by accepting ownership of the strategy throughout the organization; by assessing the organizations current capabilities, and processes realistically; and by linking strategy to the budgeting process. Otherwise BPR is only a short term efficiency exercise.

Berman (1994) concludes that BPR can achieve a radical improvement, but is also very risky. Therefore, it is crucial to do some high-level planning to appoint where BPR will result in a high contribution to the business success. BPR should be a tool that facilitates organizational change. This means that it should always be used as a part of a strategic top-down planning process, based on a clear and innovative mission for the organization. Our conceptual framework will address this major deficiency in the application of BPR theories.

Although the practice of BPR has started since early 90's, the literature on BPR is extremely rich. Hundreds of papers, books, and Web sites are dedicated to this subject. For an overview of BPR literature, we refer to "Decisions About Re-engineering" written by O'Brien (1995). In this book a comprehensive discussion on BPR issues based on a list of major books and journal articles can be found.

BPR in relation to global management and systems thinking: the big picture

The world's problems have not lessened much, if any, since 1961 when Jay Forrester wrote the first book on Industrial Dynamics (Forrester (1961)). In fact, there is an increasing trend towards globalization. Many corporations have found globalization of their business as a competitive advantage. The global economy and the advanced technological solutions have allowed production and distribution to become increasingly transnational. One of the major questions in globalization is the degree of (de-)centralization. Here, there is clear trend towards local production. While in the past large corporations used to produce locally and distribute globally, nowadays they locate their manufacturing operations closer to customers in order to be more flexible towards the customer requirement changes (Kutschker, 1995).

In order to operate successfully in a global business environment, processes must be restructured such that products and services can be standardized and in the meanwhile they should be flexible enough to meet the customer requirements in different market segments. In another word, world-wide operating companies need to render "*global processes, but for local customers*" Johansson et al. (1993). This way, a global business can take the advantage of synergy, while the feeling for local customers remains in tact. The biggest benefit of synergy effects in a global business can only be obtained when all different (sub)systems within the business are viewed as a system.

BPR in such situations considers the world-wide activities as primary processes, while the other activities at local level are considered as supporting processes. However, the effects of strategic decisions regarding process changes and their corresponding synergy effects cannot be easily quantified since no tool or theory has been offered in the BPR literature, other than standard simulation. Therefore, it is essential to conduct a rigorous System Dynamics-driven analysis of the proposed restructuring models and perform simulations, based on these models.

A CONCEPTUAL FRAMEWORK

Strategic decision making calls for process modeling to find out what is the right alternative for organizational change. Given two dimensions of complexity (detail and dynamic), decision problems, mainly of global businesses, are too complex to be worked out manually, without the help of a decision support system. Many strategic feedback processes make it difficult to predict the outcome of decisions. Therefore, a conceptual framework is proposed to support decision making process about organizational changes. This framework can be used to prioritize projects and to make a long term strategic plan in a constructive and structured way.

According to what was discussed earlier in the introduction of this paper, we can distinguish four conditions for a good decision support system.

1. It should be customer-oriented, in order to guarantee competitive advantage.
2. It should establish a clear link between criteria that are important to customers (external criteria) and performance measures for internal usage (internal criteria), in order to quantify the customers' requirement and preferences.
3. It should be based on a Systems Thinking approach, in order to manage system dynamics relations among business processes.
4. It should allow scenario analysis from a systems perspective.

Literature on the application of decision support tools for Re-engineering is limited to process modeling such as Role Activity Diagram (RAD), hierarchy diagram, matrix diagram and traditional simulation software (see O'Brien (1995)). Only Hahm and Lee (1994) have implemented a quantitative method, namely Analytic Hierarchy Process (AHP) in their proposed evaluation framework. In our approach we implement Analytical Network Process (ANP), System Dynamics concept, and ITHINK simulation software (High Performance Systems Inc., 1994). In order to present our conceptual framework, first we will brief AHP based which ANP is developed and System Dynamics concept.

Analytic Hierarchy Process

The Analytic Hierarchy Process (AHP) was developed by Saaty (1980) as a general concept for multi-criteria decision problems. Decision problems are structured by translating the goals, or at a higher level the vision of the decision maker, into measurable criteria, which are in turn related to alternative actions. Large problems can be broken down in a hierarchical way into small sub-problems, which are easy to solve just by giving judgmental weight for every compared pairs of elements. The result is a priority number for each element at each level of the hierarchy, under the restriction that the sum of those priority numbers is equal to one. Important criteria get high priority numbers and for each criterion the alternatives are evaluated and compared against each other. Then the priorities of the alternatives are weighted against those of the criteria, so that the eventual importance of the alternatives related to the goals are quantified. For a detailed explanation of AHP we refer to Saaty (1980) and Saaty and Kearns (1985).

The Analytic Hierarchy Process was used for BPR-planning by Hahm and Lee (1994), as a part of a decision support system. The AHP was used as a selection procedure for business processes that are candidates for Re-engineering. In other words, which BPR-projects should be executed first.

The approach of Hahm and Lee is quite useful, as it provides a structured way of dealing with change problems. Besides, it suffices mostly to the first two conditions of a good decision support system. One drawback is that the evaluation of business activities is not done in a systems thinking manner. Feedback relations are taken into account, but only within processes to appoint the order of redesign-activities. They have no influence on choosing the right processes for change. This means that indirect effects of redesigning a process are not taken into account. Another drawback is the impact-analysis, which is done by eliminating an activity and then estimating the change in performance of the following activity in the chain. This is quite a short-sighted way of providing critical tasks.

System dynamics

The system dynamics methodology is introduced by Forrester (1961). It is a "rigorous method for qualitative description, exploration of complex systems in terms of their processes, information, organizational boundaries and strategies; which facilitates quantitative simulation modeling and analysis for the design of system structure and control" (Wolstenholme et al., 1993).

The biggest advantage of system dynamics simulation is that it facilitates experimentation with business systems. This is common use in operational models, but not in strategic situations. Because system dynamics models do not require detailed information nor exact data about relations, a strategic system can easily be modeled and simulated. The reason that no exact data are required, is that system dynamics focuses on the dynamic behavior of the combination of feedback-loops. This means automatically that only structural, longer term behavior is of interest and therefore system dynamics is a useful tool for change management and BPR.

System dynamics is strongly related to Systems Thinking (Senge, 1990), who states that structure determines behavior. So changing the business system's structure means changing the behavior of the system and thus changing the future of a company. By simulating scenarios for organizational change, their effects can be observed immediately and a strategy can be developed that shapes the future of the company according to its vision.

A new conceptual framework for decision support system

The approaches currently known in the literature contain good aspects, but are incomplete with respect to the four conditions of a good decision support system mentioned earlier. Therefore, a new concept is developed that satisfies all conditions (see Figure 1). The decision support system concept is focused on supporting two important and difficult issues: detail complexity and dynamic complexity.

Detail complexity relates to the enormous number of variables and factors that concerns strategic decision problems. The before mentioned Analytic Hierarchy Process deals with this by structuring problems using a top-down way of planning. We, however, suggest the use of Analytic Network Process (ANP), a more recent method of Saaty (1996). The reason for this is that ANP takes interaction effects between business processes into account. Organizational goals can be linked to a network of business processes, activities, and the influences among them, taking into account all criteria that are important in today's competition, such as cost, quality, service, delivery and innovation. Furthermore, a clear relationship is established between these external criteria and internal performance measures. Processes and activities are then weighted to the criteria in a systems-oriented way, according to the management insights. The result is a reliable, quantitative outline of the importance of business system components, which is used by management to reduce the large number of variables (detail complexity).

Dynamic complexity relates to the interaction effects and the delays that exist in a system. For this the system dynamics methodology (Forrester, 1961) is proposed and the simulation package ITHINK in particular. It is a way of describing systems in terms of feedback loops and relationships among system components, used for evaluating the dynamic behavior of the systems. System dynamics is characterized by quick model building, insensitivity for parameter choice, which means that no exact data are required, and the possibility to include 'soft' or hard quantifiable information. ITHINK facilitates easy implementation and evaluation and interaction with the model during the simulation. The drawback that no exact results are generated, is of no concern, because the emphasis in strategic situation lies on the long term dynamic behavior of a system. This is especially important for organizational change, because a good control of the dynamics can result in the most radical improvements in performance. ITHINK is a useful tool that can help management to gain insight in the dynamic complexity of business systems. A BPR strategy can be chosen in a reliable and well-considered manner.

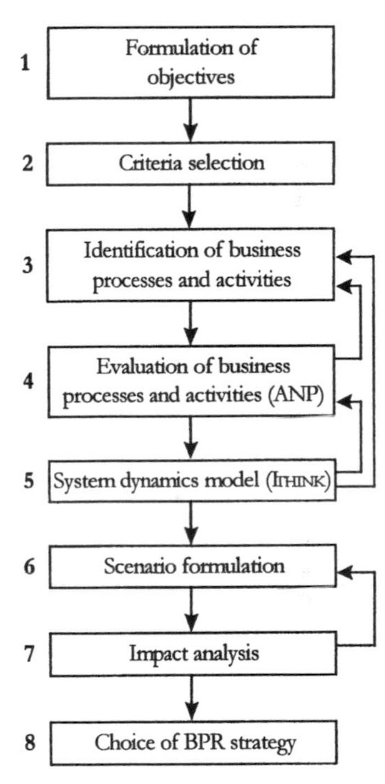

Figure 1: The conceptual Framework

The Analytic Network Process can assist in the so-called modular approach to system dynamics model building, which uses building-blocks for model construction. Important business system components, or building-blocks, can be deduced from general business objectives. This makes sure that improvements with respect to those components will result in a high contribution to the objectives, and thus to the mission of the company.

The combination of ANP and ITHINK, guided by the steps in the decision support system, satisfies all conditions for a good concept. Lane (1994) points out that soft-OR methods, like ANP, and system dynamics can supplement each other. It should however be mentioned that the eventual BPR strategies result from the ITHINK part. ANP is just a tool for top-down model construction, which means that the use of it is not necessary when Critical Success Factors (CSF) and critical processes are known and the scope of the system dynamics model is evident.

EDISCO: A CASE STUDY

The above decision support system concept was applied to Edisco Ltd. Although the study is not a profound re-engineering study, it shows the application of our approach in practice. Edisco is the European distribution company of Abbott Laboratories USA.

The Analytic Network Process at Edisco

For Edisco, the Analytic Network Process was used to determine the business processes that are crucial for the long term success of the company. Because Edisco is a distribution company, the emphasis is placed on logistics processes. Four criteria are considered. In order of their importance these are: *quality* (with weight 0.39), *flexibility* (0.24), *cost* (0.20) and *lead-time* (0.17). Each impact-relation between business processes is quantified, which results in a cross-impact matrix for each criterion. The limit of powers of such a matrix results in a vector which contains the priority weights of the processes. These are a quantification of the importance with respect to the matching criterion. The vectors are then weighted against the criteria to determine the eventual priorities. The results are summarized in Table 1. For detailed examples of ANP see Saaty (1996).

Business process	*Cost*	*Quality*	*Lead-time*	*Flexibility*	*Priorities*	*Rank*
Weight:	***0.20***	***0.39***	***0.17***	***0.24***		
DRP-planning	0.02	0.07	0	0.08	**0.05**	
Production	0.01	0.04	0	0.07	**0.03**	
Distribution	0.01	0.03	0	0.06	**0.03**	
Distribution (Affiliate)	0.01	0.04	0	0.06	**0.03**	
Recruitment/quits	0	0.01	0.05	0.02	**0.02**	
Training + development	0.01	0.07	0	0.02	**0.03**	
Appraisal + rewarding	0	0.03	0	0.01	**0.01**	
Financial management	0	0.05	0	0.10	**0.05**	
Management accounting	0.04	0.04	0	0.04	**0.03**	
Quality control	0.11	0.03	0	0	**0.03**	
Process improvement	0.44	0.13	0.22	0.11	**0.20**	*1*
Quality projects	0.02	0.12	0	0	**0.05**	
Information management	0.05	0.19	0.27	0.15	**0.16**	*2*
External relations	0.25	0.13	0.11	0.14	**0.15**	*3*
Capacity management	0.01	0.03	0.27	0.08	**0.08**	*4*
Subcontracting	0.02	0	0.09	0.05	**0.03**	

Table 1: Priorities Determined Using ANP

ITHINK at Edisco

The ITHINK simulation software package was used in order to construct the system dynamics relationships among the selected processes. Figure 2 illustrates the conceptual system dynamics model of Edisco logistics activities.

Model construction

The structure of the ITHINK-model is as follows (see Figure 2): from the *market demand* a forecast is derived and inserted in the DRP-system. The *net requirements* per month is calculated as the *forecast* minus available stock at affiliates plus safety stock. Once a month the *net requirements* are sent to the sources as a *planned order*. *Forecasts* are consumed by real *disposals* and the difference at the end of a month results in an *exception message* based on which the Edisco planners decide whether or not to make an *order adjustment*. The received amount of *planned orders* are compared with the production *capacity*, which results in an availability report. What cannot be produced is removed from the system at the end of the month. A *production schedule* is made and after production the finished goods are shipped by a *shipping schedule*.

Cause and effect relations

1. The number of *exception messages* influences the *production efficiency*, because the production schedule is adjusted during the run, which causes an increase in change-over times and decrease in efficiency. As a result customers will perceive higher lead-times and, for safety, they will give forecasts that are higher than necessary, which in turn results in more exception messages at the end of each month.

2. Due to scarce capacity, the number of exception messages influences production flexibility and thus delivery reliability. This effects the forecast quality, which finds expression in higher forecasts and safety stocks.
3. Out of stock at affiliates results in a decrease of total market demand, due to loss of customers.

The model is validated by confirmation of every element of it by the management of Edisco. This has turned out to be a confident validation method for system dynamics (Forrester, 1961; Forrester and Senge 1980; Wolstenholme et al., 1993).

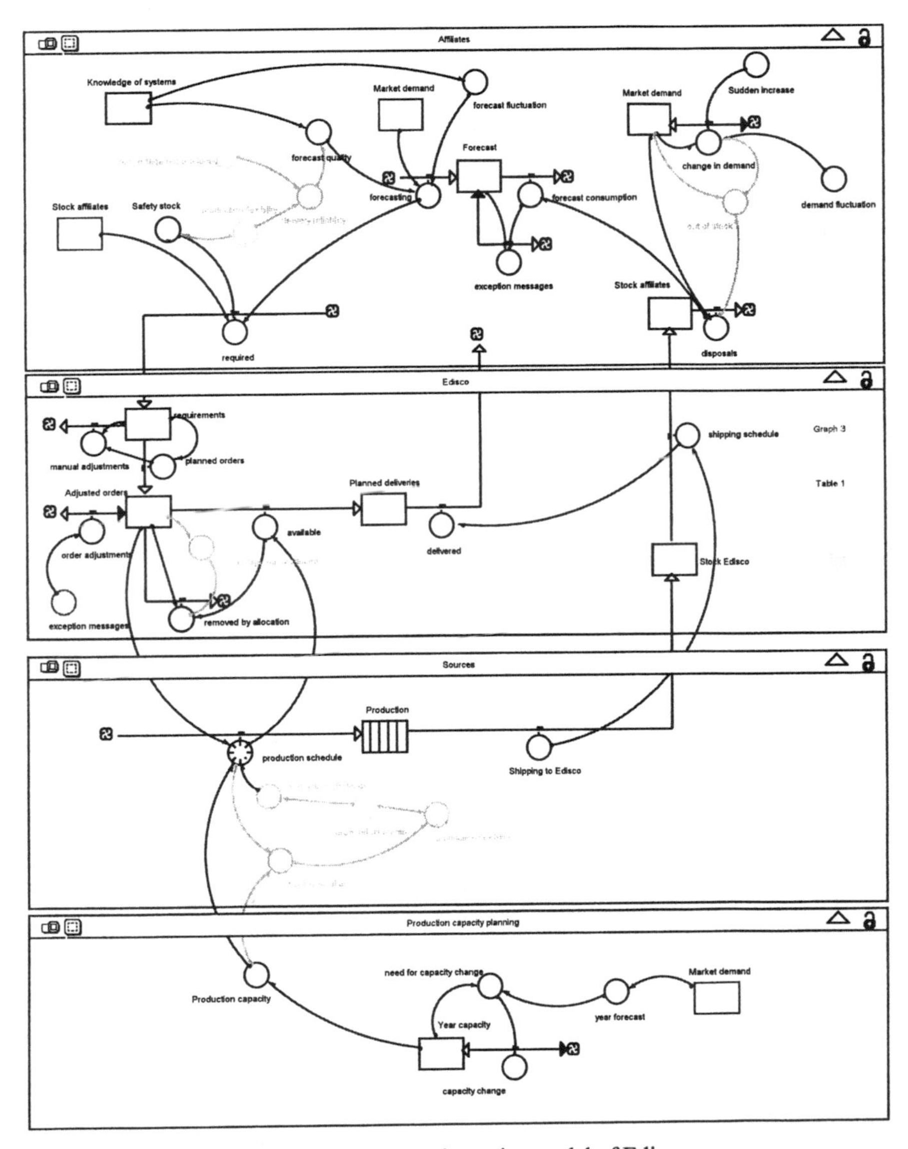

Figure 2: The system dynamics model of Edisco

Experiments with the model

To test the behavior of the model structure, some experiments are conducted (see Table 2).

Experiment	*External situation*	*Internal situation*
Experiment 0:	Stable situation	
Experiment 1:	Sudden increase in market demand	with infinite capacity
Experiment 2:	Forecasts too high	with infinite capacity
Experiment 3:	Sudden increase in market demand	with capacity 100% scheduled
Experiment 4:	Forecasts too high	with capacity 100% scheduled
Experiment 5:	Sudden increase in market demand	with infinite capacity and lead-time higher than 1 month

Table 2: Conducted Experiments

Experiment 0 is the stable situation of the model. Before other experiments are performed, the model must be made stable, i.e. all variables retain their initial value during the simulation run. This is necessary because otherwise when policies are applied, it is not clear which behavior is caused by instability of the original system and which behavior is the effect of the policy changes. The next step is to perform some sensitivity analysis on important variables. Experiments 1 and 2 are based on the assumption that production capacity is never a bottleneck, so increase in demand at the sources can always be processed. Both with the increase in market demand and an increase in forecast peaks, a dynamic process is initiated that converges slowly to a new equilibrium. When capacity is 100% scheduled (experiments 3 and 4) the increase in demand or forecast cannot be managed by production. The effect is that back-orders increase and delivery reliability decreases. Because of the loss of customers, as a result of out-of-stocks, the demand declines and the orders can be processed again. This is illustrated in Figure 3.

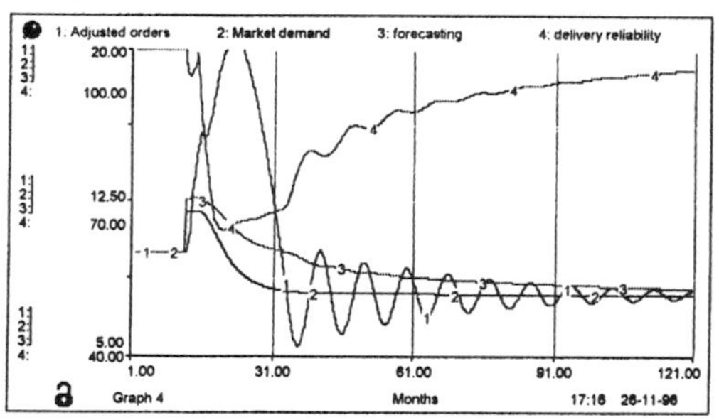

Figure 3: Experiment 3, increasing demand with limited capacity

Conclusions from the experiments

1. Increase of production capacity (a structural change) does not guarantee a stable supply chain. It appears that when the system is brought out of balance, a dynamic process is initiated, caused by delayed feedback-processes. With uncertain demand patterns or unreliable forecasts, a new impulse is given continuously and the system remains unstable.
2. The higher the total lead-time, the bigger the degree of instability. This can even lead to divergence until manual intervention.
3. Although sufficient production capacity does not guarantee a stable supply chain, full scheduling of capacity is disastrous. Small changes in demand have an enormous impact. Net requirements that cannot be produced, shift to the next month. Delivery reliability is declining.
4. A structural too high forecast causes differences with the real disposals at each month. The production schedule has to be adjusted during the production run, which results in a permanent lower delivery reliability.

Scenarios for redesign

The fact that there are many manual adjustments, means that a dynamic process is playing a role that is not well understood. These are short-term solutions that do not resolve the problems. Therefore, some logistic redesign alternatives are evaluated:

Scenario 0: Current situation (as a reference scenario)
Scenario 1: Increase production capacity at the sources
Scenario 2: Improve the support to affiliate-planners
Scenario 3: Combination of scenarios 2 and 3

Impact-analysis

Scenario 0: Current situation (as a reference scenario)

In the current situation production capacity is 100% scheduled. This results in an inflexible production process. When there is a fluctuating demand or poor forecast quality, immediately a dynamic process is initiated, through no margin in capacity. The average delivery reliability is 80.1%. (See Figure 4)

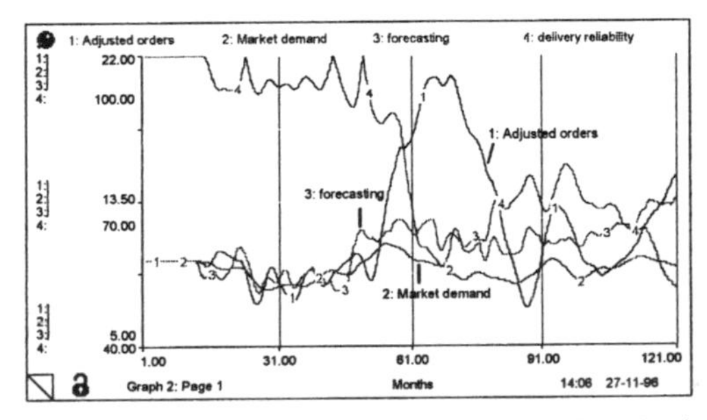

Figure 4: Current situation, limited capacity and fluctuations in demand forecast

Scenario 1: Increase production capacity at the sources

If there is a margin in production scheduling, the worst peaks of the dynamics can be handled and order backlog can easier be processed. Average delivery reliability is 94.0%.

Scenario 2: Improve the support to affiliate-planners

As follows from the experiments, the chain is brought out of balance through fluctuation in demand and through poor forecasts of affiliates. Little can be done to the demand pattern, but the quality of forecasts can be improved. Introduction of a better forecasting system and training of affiliate-planners is a way to reach this. As a result the system-input is better and less adjustments are necessary. From the simulation it turns out that there is less dynamic behavior and that the forecasts follow the demand pattern more accurately. This results in a better performance with respect to the delivery reliability (on average 95.9%). Figure 5 shows however that escalation of the order quantity, caused by a small changed in forecast quality, is still possible, due to the inflexible production process.

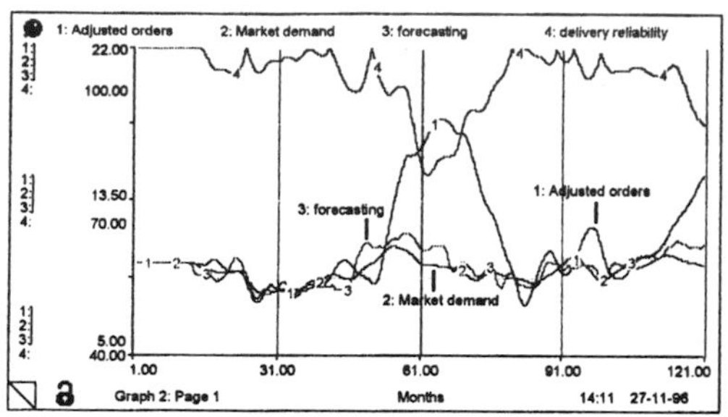

Figure 5: Scenario 2, improve the support to affiliate-planners

Scenario 3: A combination of scenarios 1 and 2

If the forecast quality is improved, while in the meantime the capacity is not a bottleneck, no escalation of order quantities exists anymore. The average delivery reliability is high (97.6%) and that means that theoretically this is the best alternative.

Choice of BPR-strategy

Although scenario 3 seems to be the most suitable strategy, there are some drawbacks. Expansion of production capacity is expensive and Edisco has very little influence on it. Besides, the biggest improvements can be realized by executing scenario 2. Delivery reliability improves from 80.1% to 95.9%. The major reason for choosing scenario 2 as redesign-strategy is that is causes significant improvements with very little investments. Choosing scenario 3 would only add marginal improvements, involved with high investments.

CONCLUSIONS

The proposed conceptual framework for conducting a structured BPR takes the advantages of the Analytic Network Process and Systems Dynamics concepts. ANP allows management to structure problems in a top-down way and breaks them down in elementary sub-problems. A clear relationship is established between the criteria that are important to the customers and internal performance measures. The added value of the ANP against the AHP is that indirect influence and feedback-processes are taken into account. A drawback is that the method requires very much insight data. Therefore, the Analytic Network Process should only be used in case of a general approach of complex world-wide change. The problem can then be reduced significantly.

Advantages of ITHINK, or system dynamics in general, are insensitivity for parameter choice so that exact data are not required, and the fact that hard quantifiable variables can be included. ITHINK cannot generate precise outcomes, but this is not necessary in strategic decision problems. What counts is the long-term behavior pattern, which is mostly the result of dynamic feedback processes. The possibility to influence these dynamics can be crucial for a company's future. Therefore, ITHINK is very useful for the planning of change management.

The fact that system dynamics model can be build in a modular way, brings possibilities for the Analytic Network Process, which can make a selection for important business system components. This results in a system dynamics model that then can be simulated in order to find out which BPR strategies will result in the highest performance improvements and help a company to change towards its vision.

REFERENCES

1. Berman, S., (1994), Strategic Direction: Don't Reengineer Without It; Scanning the Horizon for Turbulence, *Planning Review*, November, pp. 18.

2. Forrester, J.W. and P.M. Senge (1980), Tests for Building Confidence in System Dynamics Models. In: Legasto, A.A., J.W. Forrester and J.M. Lyneis (1980), *System Dynamics: TIMS Studies in Management Science*, Volume 14, Amsterdam: North-Holland.
3. Forrester, J.W. (1961), *Industrial Dynamics*. Cambridge: Massachusetts Institute of Technology.
4. Hahm, J. and M.W. Lee (1994), A Systematic Approach to Business Process Re-engineering. *Computers and Industrial Engineering*, 27, No. 1-4, pp. 327-330.
5. Hammer, M. (1990), Re-engineering Work: Don't Automate, Obliterate. *Harvard Business Review*, 68, July, pp. 104-112.
6. High Performance Systems Inc., (1994), ITHINK Simulation Software, 45 Lyme Road, Hanover NH 03755, Germany.
7. Johansson, H.J., P. McHugh, A.J. Pendlebury and W.A. Wheeler III (1993), *Business Process Re-engineering: Breakpoint Strategies for Market Dominance*, Chichester: John Wiley & Sons.
8. Kutschker, M. (1995). Re-engineering of Business Processes in Multinational Corporations. *Working Paper No. 95-4*, Katholische Universität Eichstatt, Germany.
9. Lane, D.C. (1994), With a Little Help From Our Friends: How System Dynamics and Soft OR Can Learn From Each Other, *System Dynamics Review*, 10, No. 2-3, pp. 101-134.
10. O'Brien B., (1995), *Decisions About Re-engineering: Briefings on Issues and Options*, London: Chapman & Hall
11. Saaty, T.L. (1996), *Decision Making with Dependence and Feedback: The Analytic Network Process*, Pittsburgh: RWS Publications
12. Saaty, T.L., (1980), *The Analytic Hierarchy Process : Planning, Priority Setting, Resource Allocation*, New York: McGraw-Hill.
13. Saaty, T.L. and K.P. Kearns (1985), *Analytical Planning: The Organization of Systems*, Oxford: Pergamon Press.
14. Sage, A.P. (1995), Systems Engineering and Systems Management for Re-engineering. *Journal of Systems Software*, 30, pp. 3-25.
15. Senge, P.M. (1990), *The Fifth Discipline: The Art & Practice of The Learning Organization*, New York: Currency Doubleday.
16. Wolstenholme, E. F., S. Henderson and A. Gavine (1993), *The Evaluation of Management Information Systems*, Chichester: John Wiley & Sons.

ENHANCING LOSS PREVENTION THROUGH EFFECTIVE MANAGEMENT AND EXCHANGE OF SAFETY CASE INFORMATION

L. P. McAlinden, P. W. Norman, P. J. Sitoh
Engineering Design Centre, University of Newcastle upon Tyne, England

ABSTRACT: Major hazards in the chemical and nuclear industries over the last two decades have led to regulations being developed which require operators to demonstrate the safety of their installation. In particular, under UK regulations for potentially hazardous process plant, the operators are required to produce a safety portfolio. This document or 'Safety Case' as it has come to be known, has to be submitted to the Health & Safety Executive for approval. The Safety Case provides detailed information on the operations and chemistry of the installation, along with predictive criteria which establishes the likelihood of major disasters occurring.

Conforming to these requirements calls for an in-depth analysis of the regulations. Moreover, gathering the necessary information and managing it in a structured way can prove an arduous task. At Newcastle University's Engineering Design Centre a research project has been investigating these and related issues. Key aspects of this project have included a detailed analysis of the regulatory requirements, and the subsequent development of an object-oriented information model. This model is now being implemented into a computerised electronic data management and exchange system for Safety Case information.

This paper discusses these developments, which are being carried out within the framework of ISO 10303, the emerging international electronic data interchange standard. The benefits of computerisation, data life-cycle management, and information sharing for concurrent engineering, such as design for safety, are explained in the context of their contribution to enhancing current loss prevention procedures.

INTRODUCTION

Major disasters such as Piper Alpha in 1987 [Cullen, 1993], Chernobyl in 1986, and Bhopal in 1984, have given impetus to legislation world-wide regarding safety for land-based and offshore process plant. For example, in the UK the European Union's (EU) 'Seveso' Directive is implemented through the Control of Industrial Major Accident Hazards (CIMAH) regulations [Health & Safety Executive, 1990]. These require the operators of potentially hazardous chemical installations to submit a legally binding safety report or Safety Case to the Government controlled Health & Safety Executive (HSE) for approval.

To produce a Safety Case, information has to be gathered, structured, and exchanged, throughout many disparate activities and locations of the controlling operator. Satisfying these requirements demands effective communication channels through which information can be exchanged. Such channels would not only facilitate production of the Safety Case, but would also allow other interested parties, such as designers, plant operators, and regulatory authorities to access this important information. For example, in the case of designers, access to safety information will complement safety concepts such as 'design for safety' and 'inherently safer design'. These rely on designers having knowledge of the safety characteristics of an installation at key stages of the design process. This sharing of safety information will also promote greater safety awareness and complement other loss prevention procedures. In today's dynamic engineering world the sharing and exchange of information is best served electronically, allowing large volumes of information to be exchanged quickly, at low cost, over great distances.

Traditionally the information exchange process has been complex due to incompatible interchange formats and translation problems. The emergence of ISO 10303, also known as 'STEP', has overcome this complexity by introducing a standard interfacing methodology allied with the concept of product data models. STEP is an acronym for 'STandard for the Exchange of Product model data', and has been an on-going activity of ISO Technical Committee 184 (TC184) for more than ten years under the title of '*Industrial automation systems and integration - Product data representation and exchange*'. The standard provides mechanisms by which 'neutral' data can be exchanged, e.g., via a neutral file format. Information accessed in this way will mean that considerably fewer conversion and translation programs will be needed than previously [McAlinden and Norman 1996]. The product data model, defined using the EXPRESS data modelling language [Schenck and Wilson 1994], provides a definition of the information structure and content of the modelled domain. This information model is also representative of a product throughout its life-cycle from design to decommissioning.

This paper describes the development of a computerised STEP compliant Safety Case information management system. The system contributes to current and emerging loss prevention procedures by making safety information available throughout the process plant life-cycle. Although no formal safety information exchange standard for the process industries has yet been developed within the STEP standard, this system demonstrates how such a standard could be accommodated. Furthermore, the Safety Case information models have stimulated the development of a generic model for process safety in the process industries. This is currently being offered as a proposal to interested European consortia for the development of such a standard.

LOSS PREVENTION IN THE PROCESS INDUSTRIES

Changes in the process industries in recent years have meant that inventory levels as well as process conditions, such as temperature and pressure, have become more severe [Lees 1993]. This has increased the likelihood of hazards occurring, thus increasing the potential for loss in both human and economic terms. Despite this, the industry has a relatively good record, which can be attributed, in part, to stern safety legislation imposed on plant operators by regulatory authorities.

The main safety legislation in the UK comes under the 1974 Health and Safety at Work Act (HSWA). The result of consultations by the Robens Committee between 1970 and 1972, the HSWA [Dewis and Shanks 1988] places emphasis on employers and employees collectively to resolve health and safety problems. It also created new powers of inspection and enforcement. The HSWA enforces many regulations which are applicable to the process industries, such as the Notification of Installations Handling Hazardous Substances (NIHHS), Control Of Substances Hazardous to Health (COSHH), and CIMAH regulations.

Elsewhere in Europe safety legislation is increasingly becoming similar as the European Commission attempts to harmonise legal systems between Member States. In particular, the Seveso Directive which supports the submission to a regulating authority of a Safety Case by operators of certain types of chemical installations, is a requirement throughout many EU Member States.

It is difficult to make direct comparisons between the UK regulations and those elsewhere in Europe. Aspects of each country's regulations tend to be stronger in certain areas than those of other countries, and vice-versa. For example, the Dutch safety regulations have a stronger bias towards Quantitative Risk Assessment (QRA) than those of Germany, with the British approach somewhere in between [Lees and Ang 1989].

New Developments In Safety

The process industries have been adopting new approaches to safety in recent years. In particular, methodologies such as 'design for safety' and 'inherently safer design' have emerged. The former is a concept which minimizes hazards through a five phase approach:

1. Problem Definition
2. Risk Identification
3. Risk Estimation
4. Risk Evaluation
5. Design Review

One design for safety methodology [EDC 1992] uses these five phases iteratively to satisfy design goals highlighted in the initial problem definition. This process identifies modes of failure which may otherwise have been overlooked. Inherently safer design [Kletz 1991] on the other hand involves the following concepts:

- Intensification - Using small quantities of hazardous substances
- Substitution - Replacing a material with a less hazardous substance
- Attenuation - Using less hazardous conditions or a less hazardous form of material
- Limitation of Effects - Designing facilities that minimise the impact of a release of hazardous material or energy
- Simplification - Designing facilities that make operating errors less likely, and that are forgiving of errors that are made

Furthermore, safety has been introduced as an intrinsic aspect of management in the form of Safety Management Systems (SMS). These have been outlined by the Institution of Chemical Engineers [1994] and have the following key elements:

- Policy - Safety policy
- Organisation - Resources to implement the safety policy
- Management practices and procedures - Life-cycle management
- Monitoring and auditing - Consistent performance monitoring
- Management review - Implementing corrective action based on the monitoring results

Although an SMS is formally documented, it should remain flexible to achieve widespread applicability regarding different hazard types. Line management is generally responsible for the effectiveness of the system. The role of regulatory authorities in relation to an SMS is to verify its adequacy and to check that the monitoring and follow-up process is consistent with the hazards and risks of the installation.

These emerging concepts and methodologies, of which the aforementioned are only a few, are proactive in nature, and are contributing to a safer industry. Since the late 1980's the HSE has focused on the necessity for inherently safer process plant. Prior to this there was a tendency for operators to take extrinsic safety measures, such as adding protective equipment to an already agreed design, and subsequently trying to justify it. Unfortunately these measures proved inadequate in many cases, and accidents resulted. The concept of inherent safety underlines the need for 'avoidance' rather than 'control'. It can be looked upon as taking a more proactive rather than reactive approach to safety hazards.

With the growing amount of legislation and developments regarding safety in process engineering, we can expect the amount of safety information in the industry to increase yet further. The control of this information will become an issue of crucial importance throughout the industry.

THE 'SAFETY CASE' CONCEPT

A Safety Case is a large safety report which the operators of potentially hazardous installations are required to submit to a regulating authority. They are a UK HSE requirement under the European Union's 'Seveso' *Directive 82/501*. This Directive has no legal status, but is an instruction to EU Member States to align their own legislation with it. The Seveso Directive was not completed until 1982, and was not implemented in the UK until 1984. It was stimulated by the UK Flixborough accident in 1974 and more particularly by the 1976 accident at Seveso in Italy, hence the term 'Seveso Directive'. The Directive is implemented in the UK through the Control of Industrial Major Accident Hazards (CIMAH) regulations. A similar regime now exists for the offshore industry in the UK which has to comply with the 1992 Offshore Regulations. The CIMAH Regulations require operators to submit information in the form of a Safety Case on the following:

- Dangerous substances used on the installation.
- The installation and its greater geographical area.
- The management system for the installation.
- Potential major hazards that could occur at the installation.

The offshore regulations require basically the same information with additional detail required on particular safety aspects which are inherently characteristic of the offshore industry. The offshore requirements are beyond the scope of this paper.

Safety Cases can be very large with each section containing a considerable body of information. They require many man-hours to produce and maintain, with information having to be gathered from many disparate sources within an engineering enterprise. Moreover, the logistics of this gathering exercise, and the structuring of the information is substantial. Regulatory requirements dictate that the Safety Case should then be updated regularly to reflect modifications made by the operator, as well as periodic updates to take into account new technical knowledge that may have arisen in the field.

The Safety Case is an ideal source of information regarding the life-cycle safety data for a process plant. It provides an excellent basis for the development of a life-cycle information model which will be discussed. Moreover, the wide variety of information contained in the Safety Case ensures that the data models developed have links to mainstream process engineering information such as process design schematics, plant layout, and major equipment specifications. Data exchange standards are currently being developed under STEP for these process engineering activities.

INFORMATION EXCHANGE IN THE PROCESS INDUSTRIES

Information exchange has not been an issue to the fore within the process industries until recent years [McAlinden, Sitoh and Norman 1997]. Traditionally the domain of the automotive and aerospace industries, the process industries have only become seriously active in setting and complying with information exchange standards since the beginning of the 1990's. Various national and international projects have been created to focus the attention of the industry on emerging standards, and most notably on ISO STEP. This standard has been developed for the exchange of technical information in the architectural, engineering, and construction (AEC) industries. The key aim of the standard is to remove dependencies on proprietary data exchange standards and software. These imposed vendor specific data formats and software on systems which needed to exchange information, particularly in the field of CAD/CAM. This led to the emergence of non-neutral data formats which were supported by the larger software vendors. The resulting translators and interpreters for these formats were often erroneous, inefficient, and prone to losing data.

International standards prior to this have existed and are discussed by Fowler [1995]. One of the key concepts behind STEP is that it has been designed to be 'technology-proof' in that data archived today should (in theory) be accessible in fifty years time, if required. Similarly, the exchange mechanisms the standard provides are unique with respect to previous EDI standards, by employing standard data interfacing methods and product information models.

The standard has been growing since its inception in 1984 and includes many part numbers. Convenient overviews of STEP are provided by Owen [1993] who highlights its structure and development, and Bjork and Wix [1993] who give a brief introduction and highlight some of the key factors which will help the standard to succeed. Three key components of the standard in relation to this paper are now discussed.

Information Modelling And Product Data Models

Schenck and Wilson [1994] describe an information model as '*a formal description of types of ideas, facts and processes which together form a model of a portion of interest of the real world and which provides an explicit set of interpretation rules*'. Information models can be either lexical or graphical, or a hybrid of both. The key aspect is that they are in a formally specified format.

Information modelling as it is known today was originally conceived in the database community in the 1970's as a design tool to allow information to be properly structured in relational databases. In particular, the Entity-Relationship (ER) modelling technique developed by Chen [1976], was one of the first popular techniques for representing information. ER models are graphical, easily understood, and readily translated into database implementations.

Because models were such a useful design aid for producing databases, they gradually became more prevalent in conventional software design. Today object-orientation is superseding the structured paradigm [Booch 1991] and there is a myriad of information modelling methodologies to select from when considering object-oriented analysis and design (OOAD) for software projects. The main benefit of the object-oriented paradigm is its support for the re-use of computer code. This allows more robust and reliable computer systems to be built, in less time, and at a lower cost. These formal methodologies are computer processible and supported by Computer-Aided-Software-Engineering (CASE) tools. These CASE tools have sophisticated graphical editors allowing the model(s) to be drawn and represented on-screen, thereafter their corresponding computer language code can be automatically generated at the touch of a button.

Figure 1 shows the high-level Safety Case information model which has been developed by McAlinden and Norman [1997]. The model has been constructed using the object-oriented notation known as Object Modelling Technique (OMT) [Rumbaugh *et. al.*, 1991]. Lower-level detailed models have been developed from these high-level 'classes' of information. A complete glossary of terms provide definitions for each of the information classes. Along with this a detailed logical classification of each information class, from a framework of generic types, has been performed. This latter task is important from the perspective of building robust and extensible models. These qualities will be reflected in the resulting computer system.

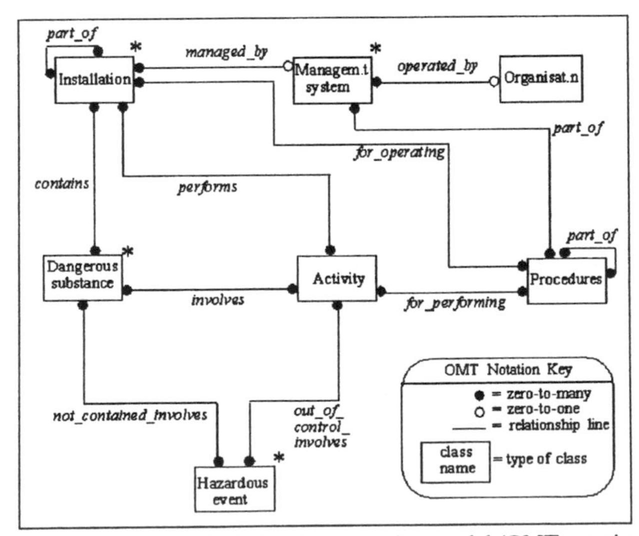

Figure 1. Safety Case high-level information model (OMT notation)

A key aspect of the STEP standard is the definition of a product using formal information modelling methodologies. In particular, product data models, as they are referred to in the STEP standard, are constructed using the EXPRESS information modelling language [ISO 1991]. The language is based on initial work undertaken within the McDonnell Douglas Information Systems company in the USA, and later became known as 'EXPRESS'. It is a formally specified lexical language with basic constructs of entities, attributes, and relationships. This provides features which enable concepts such as generalisation and constraint specification. EXPRESS includes some 'object-oriented' features, but is not a programming language, as there is no executable model. A significant feature of EXPRESS is the fact that it is computer processable, which enables STEP compliant exchange mechanisms to be rapidly developed.

The OMT Safety Case models were subsequently mapped into their corresponding EXPRESS form. These EXPRESS models facilitated the development of a standard data access mechanism to information repositories. A computerised exchange system for Safety Case information has been developed using commercial software which has an implementation of STEP's Standard Data Access Interface (SDAI) methodology.

In relation to this project, formal information modelling methodologies provide an efficient and robust mechanism for modelling safety information. This is especially important in the context of the computerised demonstrator for Safety Case information that has been developed.

Extensible Information Modelling

Information models capture the information requirements for the system they were designed for. As time goes by it may be required that the current system be extended, or integrated to another system. More often than not this can present problems as the information to be shared in each system has a different meaning depending on the viewpoints from which the systems were designed. Moreover, inconsistencies in information models between systems can arise because the models were not developed from a common basis. Indeed, this lack of a definitive basis for classifying objects is a feature of most modelling languages and techniques, including OMT [Fayad *et. al.* 1994].

This problem has been addressed by the EPISTLE group which is actively promoting STEP based information exchange in the process industries. EPISTLE, an acronym for the European Process Industries STEP Technical Liaison Executive, is a group of companies affiliated to the process industries who are working to influence the emerging STEP standard.

It has developed a framework of generic entity types, partially shown in figure 2, which provides a consistent basis for deriving information models. The base entity of the generic framework, hereafter referred to as the 'Framework', is simply known as 'THING'. A THING refers to anything that exists, tangible or intangible. By virtue of this definition any derived entity type is a subtype of THING. We derive new information models by building new subtypes from the Framework. This encourages a consistent basis for any derived model. The entity types in the Framework are presented as orthogonal subtypes, in the sense that any derived entity can only be an instance of one Framework subtype. For example, depending on what is being referred to, an object such as a pump can be an instance of either the generic

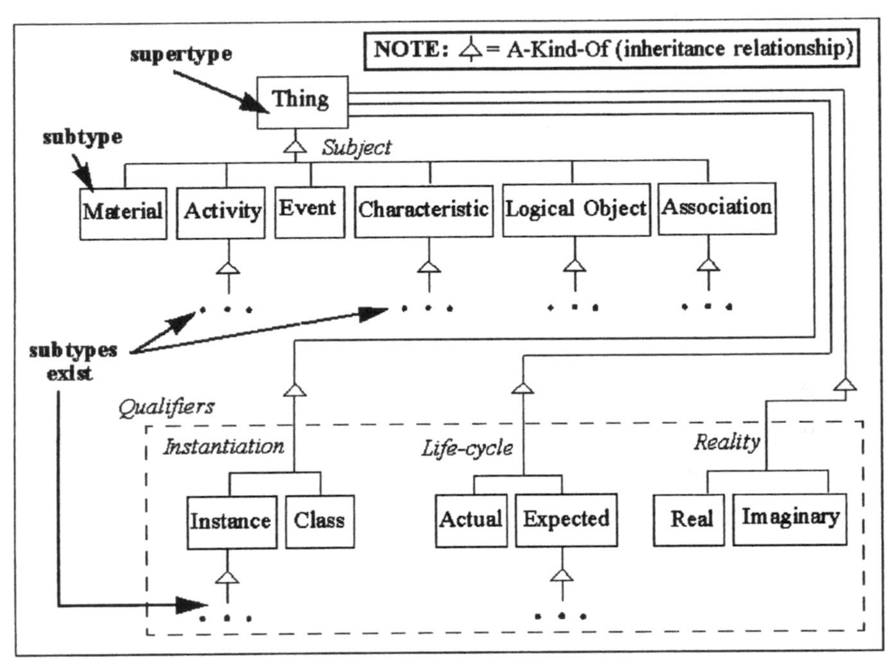

Figure 2. The EPISTLE Framework, Version 1.01

entity type MATERIAL or of FACILITY (the latter being a subtype of the generic entity type LOGICAL OBJECT). When referring to the substance of a pump, a reference is being made to an instance of MATERIAL. However, when referring to the functionality of a pump, reference is being made to a FACILITY. The rationale behind orthogonality is that each entity type in the Framework refers to distinct aspects of the underlying nature of things. Hence the term 'generic entity types'. Applying the Framework in this way means derived information models will have a consistent meaning as well as a common inheritance hierarchy.

This entity derivation process is illustrated by way of an example from the Safety Case models. Consider the 'Hazard potential to humans' class of information from the dangerous substances section of a Safety Case. This is subtyped from the Framework as illustrated in figure 3. The 'Hazard potential to humans' class, shown at the bottom of the figure, is a subtype of the subject type 'activity'. The class contains 'typical' information regarding a set of 'actual' hazard potentials to humans. The information in this class is also 'imaginary' in the sense that these are potential hazards and are not factual until they actually occur.

The entire information regarding this derived entity can be captured in the class name by utilising a naming convention provided by the Framework [EPISTLE 1995]. Thus the 'Hazard_potential_to_humans' class could, after subtyping, be called an 'Imaginary_Actual_Typical_Hazard_potential_to_humans' class of information.

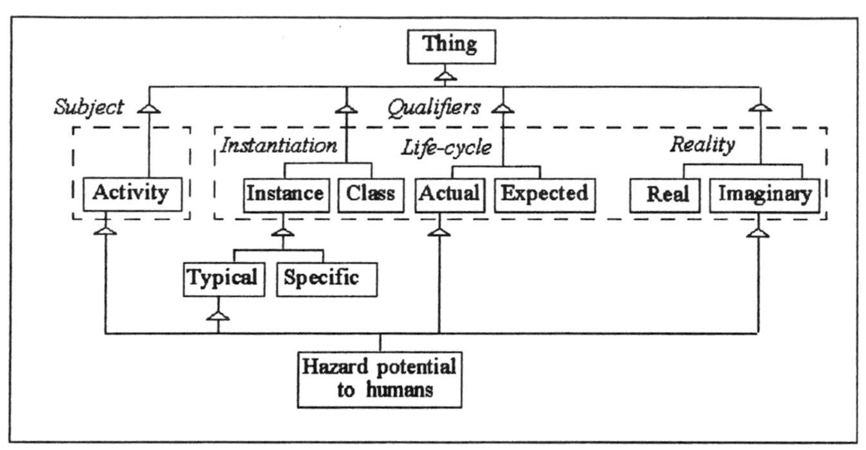

Figure 3. Subtyping of new entities from the EPISTLE Framework

The Standard Data Access Interface (SDAI)

The Standard Data Access Interface (SDAI) specification is given in Part-22 of the STEP standard. This specifies a language-independent access interface to neutral data. Part-23 and Part-24 of the standard specify the 'language bindings' for the C++ and C high-level programming languages respectively. Figure 4 shows the main components of a typical SDAI. The figure shows that an information model created in EXPRESS is mapped into a Data-Dictionary (DD), which is then used in conjunction with SDAI C++ classes and an Application Programming Interface (API). This provides the application writer with a series

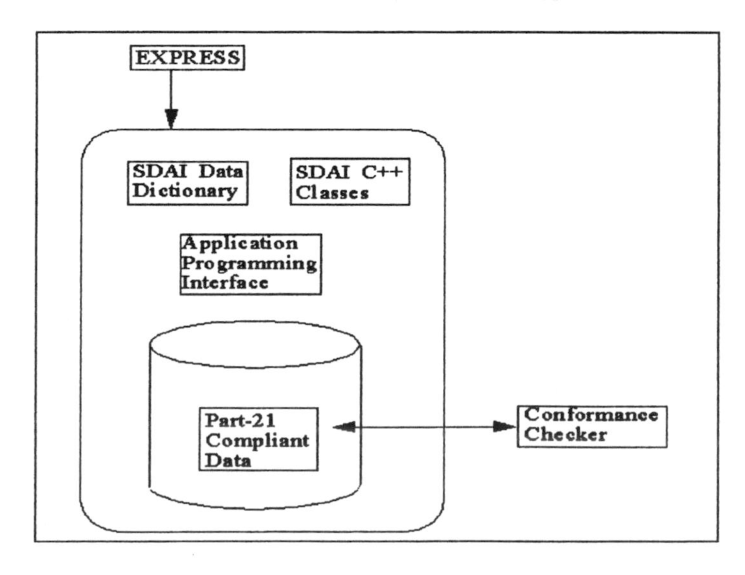

Figure 4. Main components of a SDAI

of functions and classes to manipulate neutral data in the desired manner for the application. The neutral data resides in an information repository which is typically a file or database, as shown in figure 4. The neutral data in the information repository can then be checked for integrity and compliance with the EXPRESS model by a conformance checker.

SIMEX - A SAFETY CASE INFORMATION MANAGEMENT AND EXCHANGE SYSTEM

Figure 5 shows a schematic of the overall system architecture that has been developed for the electronic Safety Case. A request for report information, such as a HAZOP, is sent to the main safety database from a front-end Safety Case browser (SIMEX). If the information exists in the database, then it will be retrieved and displayed on-screen as an electronic report. If the information does not exist, the database system will attempt to generate or retrieve the information from application packages and other data repositories. Access to database information is made consistent through the SDAI which mediates all information exchanges in and out of the database. When users eventually get the information and display it on-screen, they will be given the option of either viewing or editing the information, which will subsequently be saved to the database again. The architecture contains many links to heterogeneous information repositories and application packages, thus demonstrating how disparately located information can be exchanged and managed.

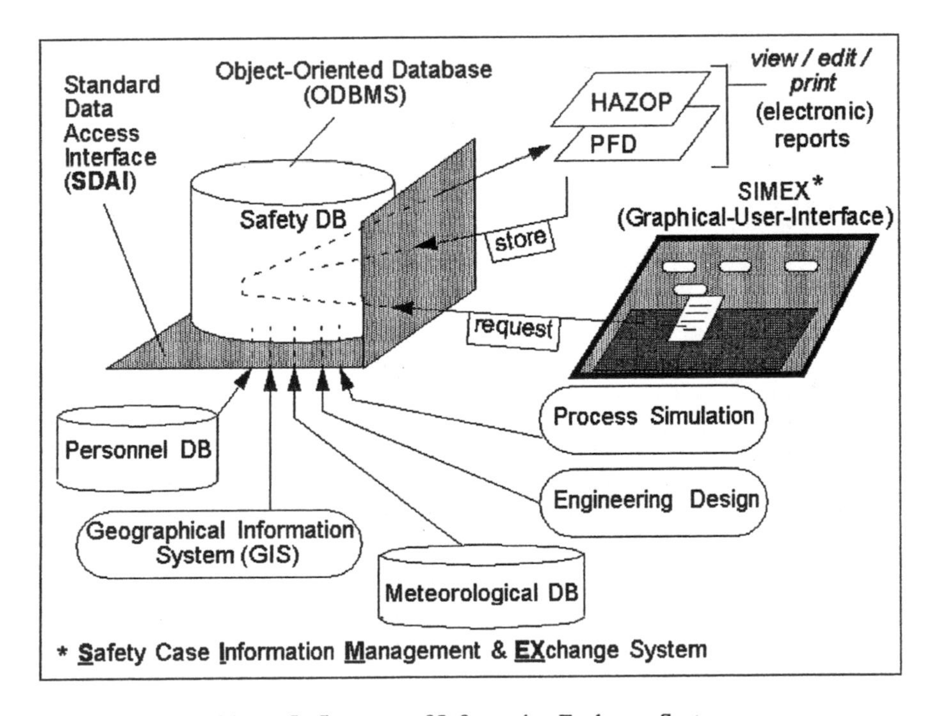

Figure 5. Structure of Information Exchange System

The SIMEX software, supporting an on-line Safety Case, enables safety information to be kept up-to-date and available to all parties throughout a project lifetime, and beyond. SIMEX presents the Safety Case in sections and assembles the required data on demand. The Safety Case is therefore always current, or 'living'. At particular points in the project lifetime, data can be archived so that Safety Cases at particular instances in the life-cycle can be addressed. This not only has the potential to lead to safer design, but will also ensure that safety is a key feature of all an operator's activities, since all operations are accessible to the electronic Safety Case.

CONCLUSIONS

A prerequisite for greater safety in the process industries is the need to share information throughout all aspects of the plant operator's activities. This exchange of information has traditionally been complex, inefficient, and error prone. Information modelling can resolve data exchange problems by structuring the information content. Using formal methods of information modelling such as OMT and EXPRESS means that information content can be rigorously defined into a computer processable format, facilitating the rapid development of a computerised information exchange system. A key concept behind all the models developed is that they should be extensible and able to integrate with other models of interest in the process domain. This can be accomplished by deriving all entities from a consistent basis. The EPISTLE Framework with its generic entity types provides one such basis.

A description of how an integrated data management and exchange mechanism for safety information can be both designed and implemented has been given. Of prime importance in this work has been the need for compliance with the emerging international EDI standard ISO 10303 (STEP). This supports the development of exchange mechanisms that are efficient, error free, and can be supported throughout the entire life-cycle of a product such as process plant.

A key aim of this project is to demonstrate the ability to share safety related information within the process plant life-cycle. To date, this has not been undertaken by the ISO committees working with STEP. Benefits highlighted by West [1995] would be augmented by a safety related standard for process engineering.

The computer system which has been developed demonstrates how information can be shared more efficiently. This will mean that safety information will be more readily accessible to interested parties such as designers, plant operators, and regulatory authorities. The approach taken will also facilitate in safety information exchange throughout the entire life-cycle of the installation, and not just for a transient period. This will enhance a safety awareness culture throughout the enterprise which will complement other loss prevention procedures of the plant operator.

The analysis from this research has also led to the development of a generic process engineering safety model. This is proposed as the basis for the development of an international safety information exchange standard for the process industries, based on the ISO 10303 information exchange methodology.

REFERENCES

Bjork B. and Wix J., An Introduction to STEP, *VTT Technical Research Centre of Finland and Wix McLelland Ltd,* England, November 1993.

Booch, G., Object-oriented design with applications, *The Benjamin/Cummings Company, Inc.,* 1991.

Chen P., The entity-relationship model-toward a unified view of data, *ACM Transactions on Database System 1,* March 1976.

Cullen, W. D., The Public Inquiry into the Piper Alpha Disaster, *Volumes 1 & 2, Dept. of Energy, HMSO,* Reprint 1993.

Dewis M. and Shanks J., Trolley's Health and Safety at Work Handbook, 2nd Edition, *Trolley Publishing Company Ltd,* 1988.

EDC, Design for Safety Project: Design Methodology, *Engineering Design Centre Publication,* University of Newcastle upon Tyne, EDCN/SAFE/RESC/12/1, 1992.

EPISTLE, EPISTLE FRAMEWORK V2.0 (Issue 1.01), *Angus, C.,* Angus Associates, Cumbria CA8 2HH, England, 1995.

Fayad M. E., Tsai W. T., Anthony R. L., and Fulghum M. L., Object Modelling Technique (OMT): Experience Report, *The Journal of Object-Oriented Programming,* Vol 7, No 7, pp. 46-58, 1994.

Fowler J., STEP for Data Management Exchange and Sharing, *Technology Appraisals,* Twickenham, 1995.

Health & Safety Executive HS(R)21(Rev), *A guide to the Control of Industrial Major Accident Hazards Regulations 1984 - Guidance on Regulations, HSE,* 1990.

The Institute of Chemical Engineers (IChemE), Safety Management Systems, *IChemE,* Davis Building, 165-189 Railway Terrace, Rubgy, Warwickshire CV21 3HQ, UK, 1994.

ISO TC184/SC4/WG5, ISO 10303:Part 11 - EXPRESS Language Reference Manual, *ISO,* Release Draft, April 1991.

Kletz T., Plant Design For Safety: A User Friendly Approach, *Hemisphere Publishing Corporation,* 1991.

Lees F.P., Loss Prevention in the Process Industries (2nd Edition), *Volume 1, Butterworth & Co (Publishers) Ltd,* 1996.

Lees F.P. and Ang M.L., Safety Cases within the Control of Industrial Major Accident Hazards (CIMAH), *London Butterworth Scientific,* 1989.

McAlinden L.P. and Norman P.W., Integrated Safety Case Data Management Using ISO 10303 STEP, *The 1996 IChemE Research Event, Conference Proceedings,* University of Leeds, England, 2-3 April 1996.

McAlinden L.P., Sitoh J.P., and Norman P.W., Integrated Information Modelling Strategies For Safe Design In The Process Industries, *To be published in PSE-7/ESCAPE Conference Proceedings,* University of Trondheim, Norway, 25-29 May 1997.

McAlinden L.P. and Norman P.W., OMT Safety Case Models & Data Dictionary (Alpha Release 4.0), *Newcastle University Engineering Design Centre Publication,* Jan 1997.

Owen J., STEP An Introduction, *Information Geometers Ltd,* December 1993.

Rumbaugh J., Blaha M., Premerlani W., Eddy F., Lorensen W., Object-Oriented Modelling and Design, *Prentice-Hall International,* 1991.

Schenck, D. A. and Wilson, P. R., Information Modeling: The EXPRESS Way, *Oxford University Press,* 1994.

West M. (Shell International Petroleum Company), STEP: The Future of Engineering Information, *Shell International Petroleum Company Report No. IC94-063,* September 1995.

Design for de-manufacturing and generation of disassembly plans: The current state of knowledge

George Ioannou and Subhash C. Sarin
Department of Industrial and Systems Engineering
Virginia Polytechnic Institute and State University, Blacksburg, VA 24061-0118

Abstract

In this paper we critically review methodologies for evaluating end-of-life alternatives for products that have completed their initial life-cycle. We consider final assemblies incorporating several discrete parts and subassemblies, which are coupled through appropriate fasteners. Our survey concentrates on issues pertaining to the determination of cost effective disassembly plans that would allow the recycling or reuse of sub-assemblies, parts, or materials. We examine product data representations that are most suitable for devising such plans, and describe the economics of disassembly, recycling, re-manufacturing, and reuse. We also review existing approaches for generating disassembly sequences from a network or other type of product representation, and discuss software tools that can assist this process. Finally, we present an overview of a novel method for determining cost effective disassembly plans through practical and easy to implement steps. The method employs data that are readily available in databases of modern manufacturing companies.

1 Introduction

It is a common observation that the life cycle of modern consumer products in the hands of end-users continuously decreases due to market trends or technological obsolescence as well as to the very high cost of service [11]. This shrinkage of a product's useful life coupled with a constant increment in market demand, boosts the disposal rates and the associated costs to record heights [31]. The limited earth's inventory enters the waste stream faster and a large portion of US landfill sites are expected to reach capacity within the next few years [6]. To bring about a change in this status quo, there is a need to develop procedures and guidelines to aid manufacturers in adopting environmental standards in product designs and processes [24]. Furthermore, to end the constant waist stream and to preserve energy and natural resources, some form of product recycling or reuse is necessary. To achieve this, manufacturers have to select from a set of potential end-of-life options for used products. These options include: materials recycling, and part recovery, reuse, re-manufacture, and re-engineering. The latter four options, which all require the dismantling of used products, are feasible due to the high quality final assemblies, both at the micro (tolerances, compliance to specs) and the macro (operational performance and long product life with minimal service requirements) levels. This fact, in conjunction with shorter product life-cycles, leads to the conclusion that the majority of items within customer-discarded products are in near-perfect condition and can be reused in a cost-efficient manner for the manufacture of new assemblies or the remanufacture of used ones [32].

The motivation for this paper stems from the interrelationship between environmental considerations and benefits of reusing and recycling products. We consider products consisting of discrete parts, i.e., items and/or subassemblies which are held together with appropriate fasteners (joints). At the end of their life cycles, products are returned to the manufacturers for de-manufacturing, which includes any of the end-of-life, disassembly-requiring options mentioned above. Disassembly is the *backwards process* of dismantling final products through successive steps that may or may not follow the reverse order of assembly operations; it is acknowledged as the focal coupling point of environmental consciousness and profitable materials recycling or part reusing. The key decision-making problem associated with disassembly is the identification of the components that should be recovered from a final product and the depth of disassembly, i.e., the product level at which the reverse process stops; it is clear that disassembly may proceed down to each individual part (e.g., ram chip), or terminate at higher level subassemblies (e.g., power supplies and integrated cards, for the example of a personal computer). The problem highly depends upon the tradeoff between the benefits from recovered parts or subassemblies, and the costs incurred during the disassembly and disposal processes.

This paper presents a thorough survey of existing approaches for modeling product data and generating disassembly plans, and outlines a methodology to enable part recovery. Once such a methodology is in place, it can be applied to products that are already in use or to the design of new products (design for disassembly). The remainder of the paper is organized as follows: Section 2 summarizes the state of knowledge in the area of disassembly. Section 3 develops a network-based model for the conceptual representation of product data necessary in end-of-life-cycle evaluation procedures. Section 4 overviews our overall method for determining cost effective disassembly plans through practical and easy to implement steps. The paper concludes in Section 5.

2 A survey of related work in the literature

The study of the life-cycle of a product after it has completed its conventional useful life has been addressed by several authors during the last few years. The main research themes evolve along three axes. The first axis deals with product data modeling suitable for disassembly sequencing and material or part recovery. The second axis addresses economic issues pertaining to disassembly problems, including cost of disassembly, value of recycled or remanufactured components, and profitability of materials recycling. The final axis concerns optimization formulations and solution approaches to aid the processes of: i) evaluating alternatives at the end of the conventional life-cycle of products, and ii) incorporating such decision-support modules at the product design phase for environmentability analysis. In the following, we survey current research efforts in all these interrelated themes.

2.1 Representation of product data for disassembly

One of the basic ways to capture geometrical and relational product data for disassembly studies, and generate disassembly trees from the basic part mating relationships is through the AND/OR graphs (Homem de Mello and Sanderson [12]). AND/OR graphs offer an elegant representation of disassembly operations. However, the nodes of such graphs do not always refer to actual subassemblies and part numbers included in the bill-of-materials of final assemblies. As a result, AND/OR graphs may not be appealing to industrial problems. Zussman *et al.* [43] extended the AND/OR representation of [12] by attaching a decision tree of feasible recycling options to each node. The resulting augmented graph is a *recovery graph*, and is used in evaluating all technically feasible product disassembly sequences. However, the authors do not address the restrictions to the number and part composition of each node from design and materials management indexing. Furthermore, such a complex representation of decision trees within the nodes of a directed graph, may not aid the development of rigorous mathematical models.

Laperriere and ElMaraghy [19] proposed a directed graph to model the components of final assemblies and their mating relationships. The disassembly tree follows all possible arc elimination sequences and explodes with the number of components and relationships; thus, its application to realistic problems is limited. Furthermore, the data required to generate such a graph may not be readily available, and the scheme cannot effectively capture complex relationships between parts. Pu and Purvis [30] and Spath [37] developed relational product models similar to the one in [19]. The applicability of these models is limited to simple assemblies that can be dismantled to their individual components at a single level.

Molloy *et al.* [23] used a combination of B-rep CAD data structures and simple mating relationships such as *against, fits, contacts*, etc., to model final assemblies and generate relationship graphs. The latter exhibit the characteristics of the graphs in [19]. Nevertheless, this work identified the hierarchical interaction between parts and subassemblies, and proposed a method to generate this hierarchy from the part relationships. Ishii *et al.* [15] developed a systematic scheme (LINKER) to model the structure of mechanical final assemblies and capture data necessary for disassembly evaluation. The model does not follow the hierarchical nature of the bill-of-materials, and does not have the rigorous structure of AND/OR graphs. However, it is one of the two product models in the literature that separate fasteners (or joints) from parts and subassemblies. Vujosevic *et al.* [42] proposed the other model of this kind. The graph generated from the model of [42] includes bi-directional arcs that reflect qualitative relationships (e.g., covering, attached to, connected to, engaged with).

Subramani and Dewhurst [40] developed a comprehensive model for the representation of parts, contacts, attachments, and relationships in mechanical assemblies. This model is particularly useful for the analysis of final assemblies, but may not be appropriate for high level decisions such as the recovery of parts and subassemblies. Navin-Chandra [26] proposed a relationship table based on the graph of [40] to describe assemblies. The models of [26] and [40] do not reflect the hierarchical structure of the bill-of-materials, and the associated graphs are cluttered with a vast number of arcs. As a result, the potential of these models for practical applications may be limited. Chen *et al.* [8] proposed mating graphs to model the relationships between parts in a two dimensional space only. Shin and Cho [34] developed a graphical representation of mating parts using *contact level graphs* that was employed by Shin *et al.* [35] to derive hierarchical disassembly diagrams that account for part separability and instability. The models of [8, 35] are mostly applicable to the generation of assembly sequences of simple finished goods, and their application to industrial problems may not be feasible.

2.2 The economics of disassembly and end-of-life-cycle alternatives

The long-range view of a product incorporating components with multiple useful life-cycles is the theme of some recent research efforts. The main finding of the work to-date is that used products contain items of significant value; this, coupled with high disposal penalties, may render disassembly a value-added function [5]. However, end-of-life-cycle options have to be carefully analyzed before cost and benefit studies are to be initiated [18]. Materials recycling is the simplest form of value recovery, and has been employed for years in the production of steel, aluminum and other metals [1]. This form of product recycling, as well as the remaining end-of-life options introduced in Section 1 are interrelated through the disassembly problem. The costs and benefits associated with disassembly, disposal and reuse for various products and materials are critical. Research efforts considering these issues are examined next; the findings are critiqued with respect to their suitability for disassembly optimization studies.

Boothroyd and Alting [5] presented a design perspective of end-of-life alternatives. Their work devised rules and guidelines that should be followed at the design stage of a product when targeting multi-life-cycle elements within complex subassemblies. Navin-Chandra [26] introduced the concepts of stepped obsolescence, sustainable leasing, and materials mortgage. In his scheme, materials are to be returned to the producers for recycling, and products to the manufacturers for remanufacture and reuse. He puts forth the thesis that raw material and product manufacturing costs will be substantially increased by design to sustain multiple uses. However, the additional cost will not be directly transferred to the customer, since multiple life cycles of the product will compensate for such increases. The author developed no rigorous cost estimates, and his findings cannot be directly employed in disassembly studies.

Lund [21] discussed remanufacturing practices in the automotive, industrial equipment, consumer, and residential product sectors. This work was among the first to point out the need for developing remanufacturing technologies and investing in the study of the disassembly process. Holt [13] outlined the status of materials and components recycling in the automotive industry, and two articles in the same issue of *Automotive Engineering* [1] reported on the costs and benefits of plastics and aluminum recycling, all emphasizing the important role of disassembly in determining the effectiveness of recycling. Jovane *et al.* [17] and Tipnis [41] described the tradeoffs between disassembly costs and value of recycled materials and recovered components without attempting to estimate these costs. Boks *et al.* [4] presented the results of a case study for the disassembly of television sets, but did not provide any information about the value of the recovered components and subassemblies, while Steinhilper [38] reported some estimates of the contribution of disposal expenses in the total life-cycle costs of products. Corbet [10] proposed an activity-based life cycle perspective in estimating manufacturing costs, but did not provide any activity sets or procedures for translating accounting data into useful cost functions for disassembly. Stuart *et al.* [39] used an activity-based costing approach to calculate the net present worth of recyclable materials revenues and costs of disassembly. Low *et al.* [20] proposed several formulae for allocating overhead on top of direct labor costs for resale, remanufacture, upgrade, recycling and disposing used products. The results of this study may be useful only if the disassembly process is established and the need for overhead allocations arises.

Dewhurst [11] proposed generic parametric equations to approximate the costs and benefits of disassembly. The author proposed a recycling efficiency index, which attains maximal value for a certain disassembly time. Zussman *et al.* [43] employed separate estimates (through utility functions) for the various costs and benefits in the recovery process, but did not provide any indications on how the complex functions can be used in practical cases. Ishii *et al.* [15] developed a *system disassembly cost* formula that incorporates times to remove components and fasteners from an assembly. However, this formula may not capture cases where multiple parts are kept together by a single joint, or tool and direction changes are necessary. Chen *et al.* [7], Johnson and Wang [16], McGlothlin and Kroll [22], and Shu and Flowers [36] defined recycling cost functions that are linear with respect to the disassembly time. However, none of these expressions does capture the interrelationships between disassembly costs and benefits, and disassembly sequences and depth. The formula of Chen *et al.* [7] includes the costs of shredding, material recovery, waste dumping, and labor, while that of Shu and Flowers [36] provides a convenient representation of remanufacturing costs that, if appropriately adjusted, can be employed in disassembly optimization methodologies. The cost function of Johnson and Wang [16] does not account for tools changes and orientation set-up costs.

Navin-Chandra [26] devised a profit function associated with each node of the disassembly tree that includes part/subassembly reuse value and materials recycling revenue. However, this formula depends on the path that is followed by disassembly operations, and may not be easily used in disassembly optimization methods that require fixed disassembly costs to aid decision making concerning disassembly sequences and disassembly depth. Finally, Barkan and

Hinckley [3] considered the problem of determining product defect probabilities. This problem is particularly important during disassembly, which may involve destructive processes.

2.3 Disassembly optimization

Once the various costs associated with disassembly operations and disposal, as well as the benefits from materials recycling, part/subassembly remanufacturing and reuse are established, the main decisions concerning the depth of disassembly and the order of dismantling operations have to be addressed. A limited number of research studies have considered the disassembly optimization (sometimes referred to as *recovery*) problem. Navin-Chandra [26] modeled this problem as a decision tree. At each decision node, this model considers the following options: dismantle further, send to shredder for material separation, sell, remanufacture, and check for hazardous materials. The decision tree is explored in a greedy fashion (i.e., exploring nodes of maximal recovery value first) through a complete enumeration scheme. However, since the model does not account for disassembly cost and sequencing relationships, the algorithm does not guarantee optimality. Furthermore, the proposed model cannot be formulated as a rigorous mathematical optimization problem and analyzed accordingly.

Navin-Chandra and Bansal [27] formulated the disassembly and recovery problems as a restricted prize collecting traveling salesman problem (TSP) [2]. This formulation requires *a priori* knowledge of the cost of extracting parts from a final assembly, which is not possible without specifying the disassembly sequences. Furthermore, algorithms that efficiently solve the unrestricted prize collecting TSP cannot be employed in this case, since no constraint on the number of cities visited can be imposed. Johnson and Wang [16] developed a search scheme similar to the one described in [26]. This scheme is myopic and does not consider opportunities for material and part recovery that may arise from a global perspective of the disassembly operations. Penev and de Ron [28] proposed the use of AND/OR graphs coupled with an exhaustive search scheme to find optimal disassembly steps and determine groups of parts that have to be recovered to maximize the net value of disassembly. Their method is limited to final assemblies that incorporate a small number of components, and does not account for the interrelationship between disassembly sequences and costs. Furthermore, the use of AND/OR graphs restricts the development of a comprehensive mathematical model. Finally, Zussman *et al.* [43] proposed an explicit enumeration scheme based on the recovery hypergraphs that shares the same limitations with the one presented in [28].

3 A network representation of final assemblies

As discussed in Section 2, existing models do not offer the most convenient, for disassembly optimization, representation of product data. In the following, we present a new comprehensive product data model that is directly related to material management and assembly planning modules. We consider products assembled in a discrete parts manufacturing facility. The end items or final assemblies comprise of several make items and/or subassemblies, which are coupled by appropriate fasteners. Figure 1 depicts a roller assembly comprising six components as an example of such a product. Parts L1 and L3 are welded on part L2 through joints (welds) W1 and W2, respectively. The resulting subassembly is identified as part R1. Two through holes on parts H1 and H2 support R1. H1 and H2 are threaded on the base P1 by joints (screws) S2, S3 (H1) and S1 (H2). The roller assembly is a modified version of the one presented in [40].

Given the multilevel, tree structure, bill-of-materials (BOM) and the assembly plans (AP), extracted from the appropriate MRP or other databases, as well as the set of joints or fasteners (such as threading, welding, adhesive, snap, force fit, etc.) that hold parts together, we devise

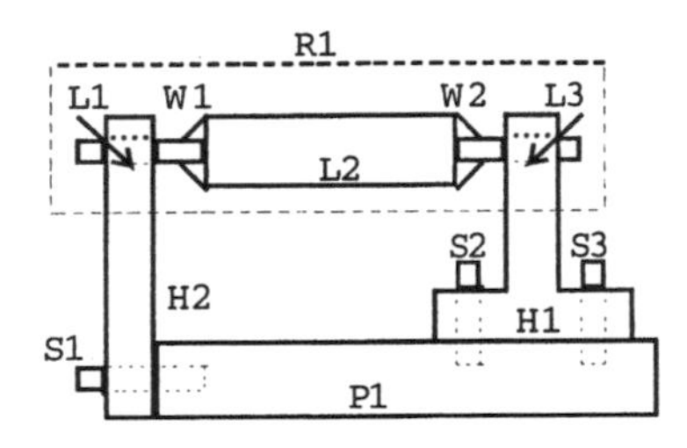

Figure 1: Example of a roller assembly

a convenient network representation of the product data with attached disassembly process information. To this end, we make the following assumptions: a) Only parts or subassemblies with pre-assigned identification numbers are considered as nodes of our network; these part ID's are developed during either the design of the finished products or the stage that immediately precedes actual manufacturing. b) Joints or fasteners couple items that are either at the same BOM level or exhibit a parent-child relationship in the BOM tree. c) The elimination of a set of joints from a subassembly results in the recovery of some or all of the children (which may be either individual parts or other subassemblies) of the subassembly. d) Only parts within the same BOM level may have hierarchical assembly or disassembly relationships. These relationships are necessary for parts that have no preceding joints associated with them.

The network representation of the disassembly operations starts from the final assembly and explodes the part BOMs in a tree-like fashion until the leafs are reached. In between the various BOM levels, we use information from the assembly plans to identify the fasteners that hold the components together. The fastener-part/subassembly relationships define the necessary arcs of the network. The part design is represented by a directed graph G, comprising of three sets of nodes: a) nodes that represent parts or subassemblies, b) nodes that represent joints or fasteners, and c) *dummy* nodes that model terminal disassembly points. Two types of arcs are required: a) arcs that connect parts or subassemblies with terminal disassembly nodes, or with the appropriate joints or fasteners and vice-versa, and b) arcs that couple joints, and model the possible sequencing of the disassembly process. The first type of arcs define the set of joints that have to be broken in order to recover a part or subassembly. Furthermore, these arcs model the relationship between parts or subassemblies that lie in consecutive levels of the BOM.

The network representation of the roller assembly is shown in Figure 2. The labels at the far right of Figure 2 refer to the BOM level. The remaining labels refer to the type of nodes that each row represents, i.e., joints, terminal nodes, parts or subassemblies. The terminal nodes are denoted by the capital letter T followed by the name of the part or subassembly they are associated with; e.g., TR1 for the terminal node of subassembly R1. The directed arcs of the network, denoted by solid lines in Figure 2, connect each part to the joints it comprises, and each joint to the parts that it holds in place. Furthermore, the dashed directed arcs (e.g., the arcs between H1, H2 and R1) connect parts or subassemblies with no immediate joint predecessors to other nodes that have to be removed before this part or subassembly can be recovered. Note that dashed arcs represent OR (disjunctive) relationships, i.e., only one of them has to be traversed in order to recover the target part or subassembly. The inter-joint arcs of the network are not shown in Figure 2.

From the network representation of any product design, we can define the following sets [14]: a) The set I of parts/subassemblies that the end item comprises. The elements of this set are indexed by $i, j = 1, ..., |I|$, where $|I|$ is the cardinality of I. b) The set K of joints included in

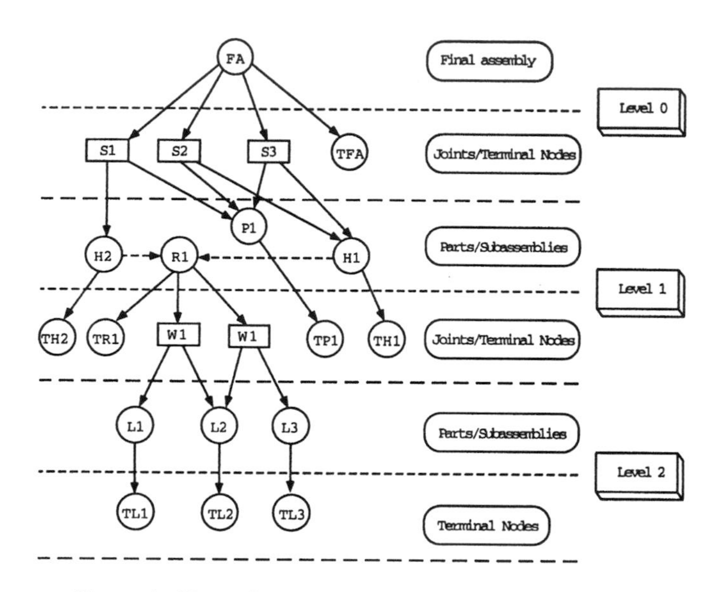

Figure 2: Network representation of the roller assembly

the final assembly. The elements of K are indexed by $k = 1, ..., |K|$. We also use index n to refer to the order in which joint $k \in K$ is broken in the sequence that joints are eliminated from the final assembly. c) The set A of directed arcs $(k, j), (j, k), (i, j), (i, T_i)$ in graph G. d) For each part $i \in I$, the sets of preceding and succeeding joints, as well as the parts that obstruct its recovery: $K_i^p = \{k \in K : (k, i) \in A\}$, i.e., the set of joints immediately preceding part $i \in I$ that have to be broken in order to recover this part; $K_i^s = \{k \in K : (i, k) \in A\}$, i.e., the set of joints that are within part $i \in I$; $P_i = \{j \in I : (j, i) \in A\}$, i.e., the set of parts j at the BOM level of i that have to be removed before part i can be recovered. In case $K_i^p = \emptyset$, then $P_i \neq \emptyset$, while if $K_i^p \neq \emptyset$, then P_i may or may not be empty. The above sets can be constructed from the BOMs and the assembly planning data.

4 Evaluating end-of-life alternatives

Disassembly is a required step in every de-manufacturing option, and any decision concerning the disassembly plans depends upon the tradeoff between the benefits of recovered parts or subassemblies and the costs of breaking joints and disposing items. In case a product, part, or subassembly is simply reused or remanufactured (e.g., defected parts are repaired or replaced), the benefits of recovery can be readily estimated based on the past use of this product. Also, the value of remanufactured components usually constitutes a significant percentage of new components (ranging from 40% up to 75% [21]). However, if the product, part, or subassembly is reengineered, then the process of devising such estimates of the recovered value may be much complicated. Nevertheless, knowledge about realistic benefit figures is central to the accuracy of the solutions obtained using mathematical models. We are developing [14] a Quality Function Deployment (QFD)-based method (see [9] for an extensive presentation of QFD) to obtain the value of products, parts, or subassemblies that are reengineered. This method encompasses possible redesign specifications, expected customer needs, logistic implications, life-cycle costs, and other significant parameters, and devises accurate potential benefits.

The net benefits from the disassembly/recovery process can be expressed as the difference between the cost of disassembly and disposal, and the recovered value of the components that can be reused or recycled as materials. Let c_i be the disposal cost related to component $i \in I$, and b_i be the composite recovered value. Cost element c_i incorporates the per unit cost of land-filling batches of items of type i, as well as the associated transportation and other expenses. On the other hand, b_i includes the benefits from all possible alternatives that can follow disassembly, weighted with appropriate probabilities [14]. These alternatives include direct reuse of the component as a make item in an identical or similar product, remanufacture for upgrade or reuse, and finally shredding and material recovery. Also, let $\bar{b}$ and $\bar{c}$ denote the cost and benefit vectors, with $|I|$ elements b_i and c_i, respectively.

Additional costs arise from the actual dismantling operations. The braking of joints within a final assembly is a complex task, that requires both human labor as well as automated processes. As is the case with assembly, the sequence of operations highly influences the total cost of disassembly. It is evident that a change in the product's orientation, different mechanisms for holding the part in place, and different tools that are required by each disassembly operation (which corresponds to the removal of a joint) result in different costs of a disassembly sequence. For example, if two threaded joints and a weld are to be removed, the total costs of removing the threaded ones first and then the weld is obviously smaller than the cost of removing the weld in between the two threaded contacts. The reason is that the tools required to remove the threaded contacts are different from those required to break the weld. This property is similar to the sequence dependent set-up times encountered in scheduling problems [33], and severely complicates the problem of determining the optimal sequence of breaking the joints of a final assembly. We can capture the effect of the sequence dependent set-up by imposing artificial arcs on the network that represents the final assembly. These arcs interconnect the nodes that model joints, and form a complete graph $G_k = (K, A_k)$, where K the set of joints and A_k the set of artificial arcs. Let f_k be the cost of braking joint k and s_{kl} the set-up cost when disassembly operations related to joints $k, l \in K$ are performed in sequence. It is clear that if both disassembly operations are of the same type (e.g., snap fits) then the set-up cost is negligible. However, the latter cost may be high if the disassembly operations have significantly different backwards processing requirements (e.g., removing adhesives and welds). Also, we denote by $\bar{f}$ the vector of the joint braking costs and by S the $|K| \times |K|$ matrix of set-up costs.

The network structure of a final assembly can be automatically generated based on the information provided by the bill-of-materials and the assembly plans. This information is known, given that the final assembly under consideration has been produced by the manufacturer who will perform the disassembly, or by one of its partners in an agile manufacturing environment that allows standardized information exchange. Therefore, if we access these data from the appropriate materials requirement planning (MRP) or other databases of the manufacturing enterprise, we can create the appropriate application protocol for this information exchange. We are developing [14] a special purpose procedure for generating the graph from MRP data. The procedural steps are: i) retrieve final assembly and generate root node; ii) retrieve children and identify fasteners or joints; iii) create one arc between the current node and all the joints it contains; iv) retrieve assembly plan and identify relationships between the current node and its children; v) create one arc between a joint (node) and all the parts that require its removal to be extracted; vi) set current node equal to one child node, and repeat steps (ii)-(v) for all children; vii) continue until the BOM is exhausted.

It is important to note that not all part and subassembly identification numbers are relevant to assembly plans; special care must be taken to distinguish between such *inactive* elements, and parts or subassemblies that participate in the assembly process. We are devising an application specific knowledge-based system (i.e., the system differs from final assembly to final assembly) to enable this data separation. This module is a distinct aspect of the overall graph generation procedure, and requests information from typical process and assembly planning databases. The graph constructed until this point does not include terminal nodes, disjunctive relationships, or inter-joint arcs. Such elements are included to the graph *a posterior*.

The final task is the development of a decision support system (DSS) to automatically assess final assemblies. Such a DSS should incorporate all the individual modules described above, and should be accompanied with a suitable user interface. The DSS could serve as the basis of the analysis of the disassemblability of finished goods and the recyclability or reusability of their components. Figure 3 illustrates the preliminary architecture of our DSS. Note the interaction between different databases and the flow through data modeling and optimization algorithms. The proposed DSS is not intended to be an open-end assessment tool. Although

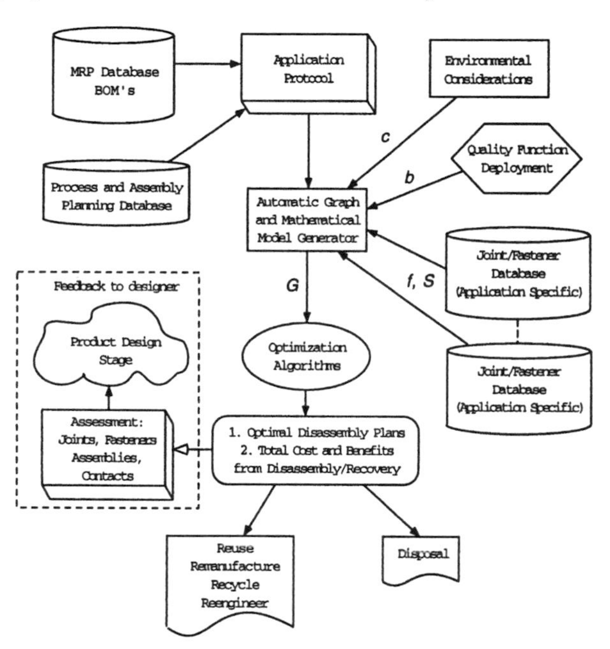

Figure 3: Decision support system architecture

our research concentrates on the evaluation of end-of-life options and the *backwards* disassembly process, the long term goal is the integration of this DSS with product design work stations. The necessary interface between end-of-life assessment and product design will evaluate the results of the disassembly/recovery optimization and will identify design changes that could improve environmental indices. The work of Boothroyd and Alting [5] and Navin-Chandra [25] provide some key pointers in this direction. The heart of the DSS is a mathematical model for disassembly optimization; the model and the solution algorithms are presented in [14].

5 Conclusions

The value of recoverable materials from a majority of products at the end of their life-cycles has traditionally been considered insignificant by the users and the manufacturers. However, environmental consciousness is fast changing this attitude. Issues related to the environment that used to be completely overlooked are now becoming competitive advantages, need to be complied with, constitute social responsibility, and give rise to new opportunities to improve products and processes. The work presented in this paper is a distinct contribution in this vain. The proposed methodology is novel in its approach to the disassembly problem. It is easy to implement, since it can be directly linked to typical MRP databases, and is generic in nature and not application specific and product or process bound. Consequently, it can be implemented in a variety of product, process, and enterprise scenarios. Adoption of this methodology would create cost-related advantages for companies that would effectively retrieve useful parts or subassemblies from customer discarded products. Furthermore, the proposed decision support system could be employed in the early design stage of new products, to provide valuable information on the environmentability and disassembability of the design, and to identify specifications that need to be reconsidered and product/process features that have to be redesigned. Finally, the theoretical foundation of the proposed methodology would allows the quantitative analysis of the disassembly process and the associated costs and benefits, in an integrated manner.

References

[1] *Automotive Engineering*, 1993. Design for aluminum recycling, October issue, pp. 65-68; Modeling costs of plastics recycling, October issue, pp. 53-57.

[2] **Balas, E.**, 1989. The prize collecting traveling salesman problem, *Networks*, vol. 19, no. 6, pp. 621-636.

[3] **Barkan, P. and Hinckley, C.M.**, 1993. The benefits and limitations of structured design methodologies, *Manufacturing Review*, vol. 6, no. 3, pp. 211-220.

[4] **Boks, C.B., Brouwers, W.C.J., Kroll, E. and Stevels, A.L.N.**, 1996. Disassembly modeling: Two applications to a Philips 21 inch television set, *Proceedings of the 1996 IEEE International Symposium on Electronics and the Environment*, pp. 81-86.

[5] **Boothroyd, G. and Alting, L.**, 1992. Design for assembly and disassembly, *Keynote Paper, Annals of CIRP*, vol. 41, no. 2, pp. 625-636.

[6] **Brooke, L.**, 1991. Plastics recyclability update, *Automotive industries*, April issue, pp. 40-44.

[7] **Chen, R.W., Navin-Chandra, D. and Prinz, F.B.**, 1993. Product design for recyclability: A cost-benefit analysis model and its application, *Proceedings of the 1993 IEEE International Symposium on Electronics and the Environment*, pp. 81-86.

[8] **Chen, S.-F., Oliver, J.H., Chou, S.-Y., and Chen, L.-L.**, 1994. Parallel disassembly by onion peeling, *Proceedings of the 1994 ASME Design Automation Conference*, pp. 1-9.

[9] **Cohen, L.**, 1995. *Quality function deployment: How to make QFD work for you*, Addison Wesley Publishing Co., New York, NY.

[10] **Corbet, K.S.**, 1996. Design for value maximization: Putting a business lens on environmental activities, *Proceedings of the 1996 IEEE International Symposium on Electronics and the Environment*, pp. 81-86.

[11] **Dewhurst, P.**, 1993. Product design for manufacture: design for disassembly, *Industrial Engineering*, vol. 25, no. 9, pp. 26-28.

[12] **Homem de Mello, L.S. and Sanderson, A.C.**, 1991. AND/OR graph representation of assembly plans, *Robotics*, pp. 1113-1119.

[13] **Holt, D.J.**, 1993. Recycling and the automobile, *Automotive Engineering*, October issue, pp. 42-50.

[14] **Ioannou, G. and Sarin S.C.**, 1997. Disassembly optimization for environmentally conscious decision making at the end of a product's useful life. *Working Paper*, Department of Industrial and Systems Engineering, Virginia Tech.

[15] **Ishii, K., Eubanks, C.F. and Di Marco, P.**, 1994. Design for product retirement and material life-cycle, *Materials and Design*, vol. 15, no. 4, pp. 225-233.

[16] **Johnson, M.R. and Wang, M.H.**, 1995. Planning product disassembly for material recovery, *International Journal of Production Research*, vol. 33, no. 11, pp. 3119-3142.

[17] **Jovane, F., Alting, L., Armillotta, A., Eversheim, W., Feldmann, K., Seliger, G. and Roth, N.**, 1993. A key issue in product life cycle: disassembly, *Annals of CIRP* vol. 42, no. 2, pp. 651-658.

[18] **Kriwet, A., Zussman, E. and Seliger, G.**, 1995. Systematic integration of design-for-recycling into product design, *International Journal of Production Economics*, vol. 38, no. 1, pp. 15-22.

[19] **Laperriere, L. and ElMaraghy, H.A.**, 1992. Planning of products assembly and disassembly, *Annals of the CIRP*, vol. 41, no. 1, pp. 5-9.

[20] **Low, M.K., Williams, D. and Dixon, C.**, 1996. Choice of end-of-life product management strategy: A case study in alternative telephone concepts, *Proceedings of the 1996 IEEE International Symposium on Electronics and the Environment*, pp. 112-117.

[21] **Lund, R.T.**, 1984. Remanufacturing, *Technology Review*, February/March issue, pp. 19-29.

[22] **McGlothlin, S. and Kroll, E.**, 1995. Systematic estimation of disassembly difficulties: Application to computer monitors, *Proceedings of the 1995 IEEE International Symposium on Electronics and the Environment*, pp. 83-87.

[23] **Molloy, E., Yang, H. and Brown, J.**, 1991. Design for assembly within concurrent engineering, *Annals of the CIRP*, vol. 40, no. 1, pp. 107-110.

[24] **Moore, A.H. and Anhalt, K.N.**, 1995. Manufacturing for reuse, *Fortune*, February 6 issue, pp. 102-112.

[25] **Navin-Chandra, D.**, 1993. ReStar: A design tool for environmental recovery analysis, *Proceedings of the International Conference of Engineering Design, ICED 93*, the Hague.

[26] **Navin-Chandra, D.**, 1994. The recovery problem in product design, *Journal of Engineering Design*, vol. 5, no. 1, pp. 65-86.

[27] **Navin-Chandra, D. and Bansal, V.**, 1994. The recovery problem, *International Journal of Environmentally Conscious Design & Manufacturing*, vol. 3, no. 2, pp. 65-71.

[28] **Penev, K.D., and de Ron, A.J.**, 1996. Determination of a disassembly strategy, *International Journal of Production Research*, vol. 34, no. 2, pp. 495-506.

[29] **Plummer, T.**, 1992. Automotive parts recycling: A case study of alternator remanufacturing, Technical Report, Carnegie Mellon University, Pittsburgh, PA.

[30] **Pu, P. and Purvis, L.**, 1995. Assembly planning using case adaptation methods, *Proceedings of the IEEE International Conference on Robotics and Automation*, pp. 982-987.

[31] **Renich, C.R.Jr.**, 1991. First recyclable appliance, *Appliance Manufacturer* June issue.

[32] **Scheuring, J.F., Bras, B. and Lee, K.-M.**, 1994. Significance of design for disassembly in integrated disassembly and assembly processes, *International Journal of Environmentally Conscious Design & Manufacturing*, vol. 3, no. 2, pp. 21-33.

[33] **Sherali, H.D., Sarin, S.C. and Kodialam, M.**, 1990. A mathematical programming approach for a two-stage production process: Applications, models and algorithms, *International Journal of Production Planning and Control*, vol. 1, no. 1, pp. 27-39.

[34] **Shin, C.K., and Cho, H.S.**, 1994. On the generation of robotic assembly sequences based on separability and assembly motion study, *Robotica*, vol. 12, pp. 7-15.

[35] **Shin, C.K., Hong, D.S. and Cho, H.S.**, 1995. Disassemblability analysis for generating robotic sequences, *Proceedings of the IEEE International Conference on Robotics and Automation*, pp. 1284-1289.

[36] **Shu, L.H. and Flowers, W.**, 1995. Considering remanufacture and other end-of-life options in selection of fastening and joining methods, *Proceedings of the 1995 IEEE International Symposium on Electronics and the Environment*, pp. 75-80.

[37] **Spath, D.**, 1994. The utilization of hypermedia-based information systems for developing recyclable products and for disassembly planning, *Annals of the CIRP*, vol. 43, no. 1, pp. 153-156.

[38] **Steinhilper, R.**, 1994. Design for recycling and remanufacturing of mechatronic and electronic products: Challenges, solutions, and practical examples from the European viewpoint, *Design for Manufacturability*, DE-vol. 67, ASME, pp. 65-76.

[39] **Stuart, J.A., Ammons, J.C., Turbini, L.J., Saunders, F.M. and Saminathan, M.**, 1995. Evaluation approach for environmental impact and yield tradeoffs for electronics manufacturing product and process alternatives, *Proceedings of the 1995 IEEE International Symposium on Electronics and the Environment*, pp. 166-170.

[40] **Subramani, A.K. and Dewhurst, P.**, 1991. Automatic generation of product disassembly sequence, *Annals of CIRP*, vol. 40, no. 1., pp. 115-118.

[41] **Tipnis, V.A.**, 1994. Challenges in product strategy, product planning, and technology development for product life-cycle design, *Annals of the CIRP*, vol. 43, no. 1, pp. 157-162.

[42] **Vujosevic, R., Raskar, R., Yetukuri, N.V., Jothishankar, M.C. and Juang, S.-H.**, 1995. Simulation, animation, and analysis of design disassembly for maintainability analysis, *International Journal of Production Research*, vol. 33, no. 11, pp. 2999-3022.

[43] **Zussman, E., Kriwet, A. and Seliger, G.**, 1994. Disassembly-oriented assessment methodology to support design for recycling, *Annals of the CIRP*, vol. 43, no. 1, pp. 9-14.

THE IMPORTANCE OF CLIENT REQUIREMENTS PROCESSING IN CONCURRENT ENGINEERING

Mr. J. M. Kamara[1], Dr. C. J. Anumba[1], Dr. N. F. O. Evbuomwan[2]
[1]Construction Research Unit (SST), University of Teesside, Middlesbrough, TS1 3BA;
[2]Department of Civil Engineering, University of Newcastle, Newcastle upon Tyne, NE1 7RU.

ABSTRACT: The application of concurrent engineering (CE) is becoming widespread in product development. Many manufacturing organizations are reported to have adopted its principles to shorten time to market and improve product quality. The different definitions and approaches to understanding concurrent engineering and its implementation, reflect the different interpretations of the concept. However, CE is basically a client-oriented product development process and the requirements of the client are therefore central to its concepts. In this paper, it is argued that the processing of these requirements is both essential to satisfy clients, as well as facilitate concurrency. In particular, requirements processing provides for compliance checking at every stage of the design and construction process and ensures the traceability of design decisions to explicit and implicit client requirements. The approach being adopted in the development of a Clients' Requirements Processing Model (CRPM) for construction clients is also described.

INTRODUCTION

The use of concurrent engineering is becoming widespread in the manufacturing industry. This is in direct response to the need for faster development of high quality products at lower costs, and the associated need for companies to remain competitive by becoming dynamic and lean. Syan [1994] suggested that the use of CE will soon become imperative for most companies in the near future. The popularity in the use of CE is, no doubt, a result of the associated benefits in adopting its principles. These include: a reduction in product development time and time to market, overall cost savings, products that precisely match customers' needs, assured quality, low service cost throughout the life of the product and earlier break-even point [Madan, 1993; Dowlatshahi, 1994; Frank, 1994; Carter, 1994; Constable, 1994; Thamhain, 1994; Nicholas, 1994; Evbuomwan et al, 1994; Smith et al, 1995; Prasad, 1996]. However, as a result of its growing use in industry, many definitions and interpretations of concurrent engineering have emerged and expectations vary as to its benefits. Prasad [1996] comments that, "expectations range from modest productivity improvements to complete push-button type automation, depending on the views expressed". Research and development activities in CE therefore tend to focus on the methodologies, technologies and tools for achieving 'concurrency'. This focus however, may disguise the fact that CE is basically a client-oriented product development process which seeks to fully satisfy the requirements of clients. In this paper, the importance of client requirements processing within a concurrent engineering framework, is reiterated. It is argued that a rigorous up-front processing of these requirements provides a sound basis for fully satisfying the customer and enhances concurrent life-cycle design and construction. In particular, requirements processing facilitates early consideration, and subsequent management, of all life-cycle issues affecting a product by enabling compliance checking at every stage of the design and construction process. It also enables traceability of design decisions to explicit and implicit client requirements. Within the context of this paper, 'client requirements processing' refers to the identification, structuring, analysis, rationalization and translation of explicit and implicit client requirements into solution-neutral specifications for design purposes. The use of the word 'client' is also taken to be synonymous with 'customer' but the former is used for convenience. Following a discussion of, the principles of concurrent engineering, the nature and classification of client requirements, the relationship between client requirements processing and concurrent life-cycle design and construction, the paper concludes with a description of a methodology for processing construction clients' requirements which is embodied in a client requirements processing model (CRPM).

CONCURRENT ENGINEERING

Concurrent Engineering is a concept which embodies several other methodologies such as multi-disciplinary teams, parallel scheduling of activities and cross-functional problem solving. It is a logical and structured framework which facilitates up-front integration of the myriad aspects of the product development process. These include: members of the product development team, technologies, tools, knowledge, resources, experience, concepts, techniques, information and data [Kamara et al 1997]. A consideration of one or some of the different methodologies embodied in concurrent engineering, results in the various approaches adopted in its implementation [Dowlatshahi, 1994, Kuan, 1995]. These are also reflected in the many ways in which CE is defined [Winner et al, 1988; Monroy, 1992; Koskela, 1992; Barkley, 1993; Madan, 1993; Bayliss et al, 1994; Carter, 1994; Prasad, 1995A, 1996; Harding and Popplewell, 1996]. However, the general goals and the fundamental principles which describe the various aspects of the CE process, provide broad guidelines for its implementation.

Goals and Principles

The goals and principles of concurrent engineering according to: [Madan, 1993; Ranky, 1994; Evbuomwan et al, 1994; Nicholas, 1994; Bayliss et al, 1994; Carter, 1994; Handfield, 1994; Prasad, 1995B; Evbuomwan et al 1995; Harding and Popplewell, 1996; Prasad, 1996] are as follows:

Goals: These include: (1) the reduction of the product development time; (2) getting rid of waste; (3) reducing cost; (4) increasing quality and value; (5) design-it-right-first-time; (6) simultaneously satisfying the requirements for functionality, produceability and marketability; and (7) fully satisfying the customer.

Principles: CE principles include: (1) the need for organizational support to implement business process changes that will facilitate concurrent working practices; (2) the use of multi-disciplinary teams involving all the parties in the product development process; (3) early or up-front consideration of all life-cycle issues affecting the product which determines life-cycle costing, and provides for effective utilization of resources - this facilitates early problem discovery and early decision making; (4) concurrent or parallel processing wherever possible as a result of work structuring; (5) information management to facilitate the flow of timely, relevant and accurate information, within and between teams, and across the stages of the product development process; (6) integration of all the technologies and tools that are used to enable concurrent product development by simultaneous product and process design; (7) continuous process improvement by incorporating lessons learned;

and (8) continuous focus on the requirements of the customer. These principles describe the various aspects of the CE process and they serve as guidelines for its implementation. They also provide a basis for defining the requirements for a concurrent engineering system.

Requirements for a Concurrent Engineering System

In defining the requirements for a concurrent engineering system , there is need to focus on enabling technologies, tools and techniques which need to be integrated into a CE framework as well as the principles which introduce cultural, human and organizational changes within the enterprise [Evbuomwan and Sivaloganathan, 1994; Molina et al, 1995]. These will be considered under goals, processes or interactions, level of operation and supporting infrastructure.

Goals of a CE system: A CE system should, (1) enable the development of quality products at affordable prices; (2) enable the reduction of product development lead times; (3) enable the requirements and expectations of customers to be fully satisfied; (4) provide for continuous improvement of the design and product development process; and (5), enable the integration of all related life-cycle issues throughout the product development process [Evbuomwan, 1994; Evbuomwan et al, 1995].

Processes and interactions: The requirements for the processes and interactions in a concurrent engineering system include, (1) Product and process classifications (product parameters and function interdependence; concurrent design process independence, product design consistency); (2) life-cycle interactions which would require the development of an integrated information architecture and the capture of knowledge that includes both upstream and downstream information; (3) information modelling to capture and represent different forms of information; (4) teaming and sharing to define teaming and sharing agents; (5) planning and scheduling to identify and distribute independent product design activities; (6) networking and distribution to exchange data among and between designers, agents, tools, and systems; (7) reasoning and negotiation to define negotiation and conflict resolution strategies at different levels of granularity; (8) collaborative decision making to capture design intent knowledge, rationale, heuristics, and decision processes; (9) organization and management to set up various interfaces and multi-disciplinary multigroups [Prasad et al, 1993].

Level of operation: Support for a concurrent engineering system is also required at the organizational, team and individual levels. At the organizational level, support involves the provision for interactions and information exchanges between different design teams, or between individual team members and other members of their particular discipline group. It also includes provision of information to support senior management in strategic decision making. At the team level, the requirements are imposed by team-working methods (activities which will assist the team to work as a single, effective entity). This includes promotion and maintenance of a common view of the team's objectives and the encouragement of exchange of knowledge between team members. At the individual level, there must be sufficient flexibility to provide assistance for all members of the design team, irrespective of their role within the team. The diverse disciplines of all the individual members of the team should therefore be supported [Harding and Popplewell, 1996]. The satisfaction of these requirements at the three levels described above can place diverse and even contradictory demands on a computer aided engineering (CAE) system. They should therefore be assessed with respect to the need for distribution, heterogeneity and autonomy. This support requirement matrix is illustrated in Table 1.

Supporting infrastructure: A supporting infrastructure is required to facilitate a concurrent engineering approach to product development. Such an infrastructure should therefore, (1) identify, coordinate and communicate between the different perspectives involved in CE; (2) provide information on knowledge sources that are able to represent evolving expertise and product designs, that can be readily modified, and that are easily accessible; (3) monitor the history of the decision process so as to enable future design procedures to capture best practice and to maintain accountability; (4) control and configure the various system elements in a way that is transparent to the user and ensures system integration; (5) provide an interactive, multimedia interface for the system user and (6) provide integrating strategies for integration of the project team, the design evolution process, life-cycle activities, and integration of textual (non-graphical) and geometric (graphical) project design data [Eversheim et al. 1995; Molina et al., 1995; Evbuomwan and Anumba, 1996]

Table 1

Support Requirement Matrix for Concurrent Engineering

		LEVELS		
		Organizational	**Team**	**Individual**
DIMENSIONS	**Distribution**	Move information between multiple sites.	Reduce remoteness and promote exchange of information between team members at different physical locations.	Make information available to individuals.
	Heterogeneity	Support organizations to achieve different missions.	Support Project Teams to achieve different goals.	Support Individuals to perform different jobs.
	Autonomy	Discourage multiple individual stores of information.	Support team members to work as individuals, or as a group, and transitions between these two types of working.	Support individual's preferred manner of working.

[Source: Harding and Popplewell, 1996]

CLIENT REQUIREMENTS AND CONCURRENT ENGINEERING

From the foregoing discussion, it is seen that concurrent engineering is a client-oriented product development process. The primary focus or ultimate goal is the complete satisfaction of the client by directly responding to their explicit and implicit requirements as understood and interpreted by the product development team. For example, the focus on quality which Kuan, [1995] identifies as a key approach to concurrent engineering is client focused since quality is defined with respect to meeting the expectations of a client. All the methodologies, technologies or infrastructures highlighted above are enablers which are used to facilitate a process for meeting the requirements of the client. These requirements are therefore central to concurrent engineering. Although companies which implement concurrent engineering benefit as a result of their increased competitiveness and profitability, it is because they effectively responded to client expectations that they became profitable and competitive. Thus, the need to fully satisfy client requirements is paramount in concurrent engineering, and understanding those requirements from the client's point of view is essential, especially as there are different categories of client requirements.

The Nature of Clients Requirements

The satisfaction of client requirements requires that their 'voice' is appropriately incorporated in the product development process. This is because, there are different categories of client requirements. According to [Griffin and Hauser, 1991; Mallon and Mulligan, 1993], they could be classified as, basic or expected needs, articulated or demanded needs, and exciting needs. Basic needs are those which are not voiced but are assumed to be present in a product. The fulfillment of basic needs would not excite the client, but their omission will reduce the satisfaction of the client. Articulated needs are those which are voiced or demanded. Exciting needs are those which, although not voiced, will pleasantly surprise and delight the client if fulfilled. To fully satisfy clients, all three categories of needs (requirements) must be fulfilled. Client requirements can also be categorized into primary or needs, secondary or tactical needs, and tertiary or operational needs. Primary needs are described as strategic needs and they relate to the overall business objectives of the client. Secondary needs are elaborations of primary needs. Operational needs are further elaborations of tactical needs and they constitute the details on which engineering solutions are based. This decomposition is very important in understanding client requirements and it is useful in their prioritization. It is also important to note that, although the ultimate objective is the satisfaction of client requirements, the requirements of the product development team are also important. Other factors also have to be considered in developing client requirements as illustrated in Figure 1 [Prasad, 1996]. These different sources of data to develop client requirements and the categories within those requirements therefore necessitate that they are effectively processed.

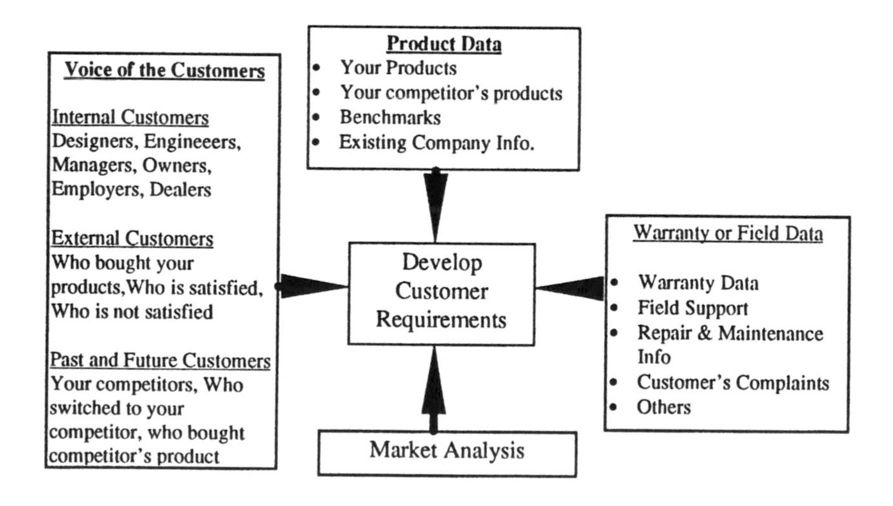

Figure 1. Source of Data to Develop Customer Requirements
(Source: Prasad, 1996)

The Need for Requirements Processing

Encapsulating the 'voice of the customer' requires that the customer needs are identified, structured and prioritized [Griffin and Hauser, 1991]. However, requirements processing, as used in this context, goes beyond prioritization and involves the translation of explicit and implicit client requirements into solution-neutral specifications for design purposes. Within a concurrent engineering context, the need for requirements processing basically centres around the need to focus on the customer and provide traceability of requirements in life-cycle considerations. Processing also enhances concurrent working as described below.

Focus on the client: The fact that concurrent engineering is client-oriented, means that the needs of clients must be precisely defined. This requires the evaluation and rationalization (processing) of competing factors (Figure 1) from the standpoint of the client.

Facilitating teamwork: The use of multi-functional teams, within a supportive organizational framework is considered to be a key aspect of concurrent engineering. Although the principles of teamwork do facilitate the working of teams, the processing of requirements is very important to the effective working of different teams. If the requirements have not been properly defined, there is the tendency that parallel teams will give different interpretations to ill-defined requirements, creating confusion and waste of time, thereby negating a key benefit of concurrent engineering in reducing product development time.

Facilitating parallel or concurrent working: To enable concurrent working, tasks have to be decomposed and those which can be independently executed can be done concurrently or in parallel. This therefore requires that requirements are unambiguous to facilitate the decomposition of product development tasks into sub-tasks which can be scheduled to run concurrently or in series.

Information sharing: Concurrency requires the sharing of partial or incomplete information about designs, and other processes, to enable parallel processing. Therefore, to enable this to be effectively implemented, formal mechanisms need to be set in place at every stage of the product development process. Adequately processed requirements will provide a continuous point of reference for stages in the product development process which are executed with partial information from dependent tasks upstream.

Life-cycle management: To effectively consider all life-cycle issues that affect a product, it is very important to relate client requirements to every stage in the entire process. This can only be done if the requirements are adequately processed by evaluating upstream and downstream actions against the original requirements. Furthermore, support for concurrent engineering should include the traceability and correlation of customer requirements throughout the product development cycle [Carter, 1994].

Continuous process improvement: To ensure that there is continuous process improvement that fully satisfies the client, the requirements should be in a form that can be easily altered or modified with reference to the original expectations of the client. Adequate processing is therefore essential to facilitate this process.

Change monitoring: Some client requirements (for example in construction) are often dynamic and time-varying and design solutions may need to be modified to take account of these. It is therefore important that client requirements are effectively processed to facilitate the monitoring of such changes in both requirements and design solutions. Disputes and claims are bound to arise when changes in client requirements are not explicitly monitored.

CLIENT REQUIREMENTS PROCESSING

It has been established that client requirements processing is very important in concurrent engineering, not only to encapsulate the 'voice of the customer', but also to enhance the process (concurrent working) which translates that 'voice' into a product that fully satisfies the client. What follows is a brief description of a methodology that is being developed for the processing of client requirements in a concurrent life-cycle design and construction (CLDC) context. It is based on structured techniques such as QFD, DFD (design function deployment [Evbuomwan, 1994]), and requirements engineering. Prior to describing the approach being adopted, it is useful first to discuss the requirements for requirements processing in concurrent engineering.

The Requirements for Requirements Processing in CE

The requirements for requirements processing in concurrent engineering derive from the requirements for a CE system discussed earlier. Specifically, they include the following:

Formal methodology: Requirements processing in CE should be based on formal methodologies. This is because, in a CE system, the need to apply principles which introduce cultural, human and organizational changes in an enterprise requires the use of formal methodologies (such as QFD) which are (in some cases) supported by information technology [Molina et al, 1995].

Precision and clarity: A framework for requirements processing in CE should provide effective procedures and techniques for the precise establishment of client/user requirements and desires. It should also allow for the articulation of these requirements in such a way that makes them easy to implement [Evbuomwan et al, 1995].

Traceability and correlation: To ensure that requirements are traceable and easily correlated during the entire product development process, they must be, (1) complete with no omission; (2) sufficiently decomposed so that their relationship to higher level requirements is possible; (3) adequately justified so that the rationale for those requirements are clearly understood; (4) properly documented with respect to the source of requirement and their dependencies; (5) unambiguous so that, as much as possible, only one interpretation can be given to a requirement statement; and (6) consistent. [Kott and Peasant, 1995].

Analysis and prioritization: Because there are many inputs in determining the client requirements, there are bound to be conflicts in requirements. A requirements processing framework should therefore enable the analysis and prioritization of requirements against the identified constraints (for example, resources, company policy, government regulations).

Facilitating design creativity and integration into product development process: To facilitate design creativity, requirements should be stated in a solution-neutral format so that the range of possibilities for meeting the need are not restricted. This should make it possible for their integration into the product development process.

Client Requirements Processing Model (CRPM)

A client requirements processing model (CRPM) is being developed for the processing of client requirements within a CLDC framework. This model is a three stage process (Figure 2) which involves requirements identification, requirements analysis and prioritization, and requirements translation. This constitutes the outline of the model which will be developed in full using structured techniques such as QFD and DFD. The translation of the prioritized requirements into solution-neutral specifications should facilitate design creativity.

Requirements identification: This is the first step in the processing of clients requirements [Anumba and Evbuomwan, 1996]. At this stage (Figure 1), all data required for the processing of client requirements are listed. The client should be encouraged to describe in his/her own words the benefits s/he desires from the facility (or product). This clear description should facilitate the categorization of requirements into basic, articulated and excitement needs in the next stage of the model.

Requirements analysis and prioritization: It is important in processing client requirements that these are analyzed and prioritized. The analysis will result in the structuring of the identified needs into appropriate categories. The specific categories will depend on the nature of the envisaged product. Prioritization of requirements is essential to facilitate balancing the cost of fulfilling a requirement with the benefit to the client and the enterprise. The use of QFD and similar techniques in this stage will be helpful.

Requirements translation: The prime objective in the processing of client requirements is the production of a solution-neutral requirements specification which will inform the design process. The requirements translation stage, therefore, involves reviewing the prioritized client requirements with a view to developing specifications that fully satisfy them, as well as removing all unnecessary constraints to design creativity.

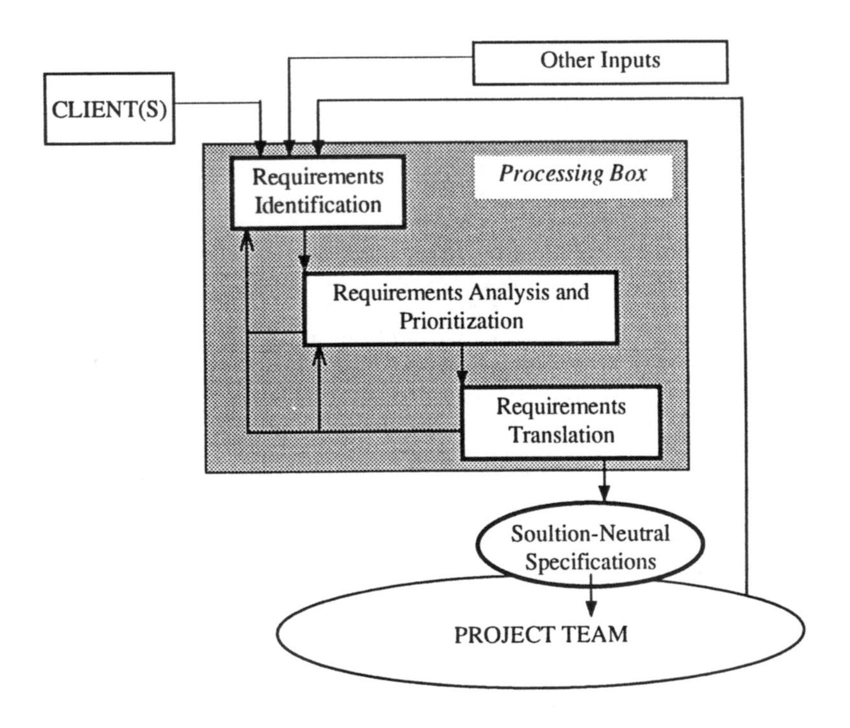

Figure 2. Stages in Client Requirements Processing

Benefits of the model: The potential benefits of this model include, (1) a clear definition of the client requirements; (2) thorough analysis and prioritization with respect to all the competing factors in the product development process; and (3) the opportunity to fulfill the requirements for requirements processing in a concurrent engineering context.

CONCLUSION

In this paper, the importance of client requirements processing in a concurrent engineering context has been discussed. It has been shown that the processing of client requirements, enables the project team to focus on the customer by facilitating teamwork, parallel or concurrent working, information sharing, life-cycle management, continuous process improvement and change monitoring. The methodology that is being adopted in the development of a clients' requirements processing model (CRPM) within a concurrent life-cycle design and construction (CLDC) framework has also been described. This model, which provides for a three-stage process (identification, analysis and prioritization, and translation) should satisfy the requirements for requirements processing in concurrent engineering. These include: the need for a formal methodology, precision and clarity, traceability and correlation of requirements, analysis and prioritization, and the need to facilitate design creativity and integration into the product development process. Further work involves the development of the details of each stage, implementation, and evaluation of the model using some construction projects as case studies.

REFERENCES

Anumba, C. J. and Evbuomwan, N. F. O., A Conceptual Model for Construction Clients' Requirements Processing, *Computing in Civil Engineering: Proceedings of the third ASCE Congress*, Anaheim, California, June 17-19, 1996, 431-437.

Barkley, J., Applications of Concurrent Engineering, *Autotestcon 93: Enhancing Mission Effectiveness: Proceedings of the IEEE Systems Readiness Technology Conference*, 1993, 451-457.

Bayliss, D.; Akueson, R. Knight, J. and Parkin, R., Implementing Concurrent Engineering using Intelligent Systems, *Proceedings of the 2nd. International Conference on Concurrent Engineering and Electronic Design Automation (CEEDA '94)*, 7-8 April 1994, 75-80.

Carter, D. E., Concurrent Engineering, *Proceedings of the 2nd. International Conference on Concurrent Engineering and Electronic Design Automation (CEEDA '94)*, 7-8 April, 1994, 5-7.

Constable, G., Concurrent Engineering - Its Procedures and Pitfalls, *Measurement and Control*, 27:8, 245-247, October, 1994.

Dowlatshahi, S., A Comparison of Approaches to Concurrent Engineering, *International Journal of Advanced Manufacturing Technology*, 9, 106-113, 1994.

Evbuomwan, N. F. O., Design Function Deployment: A Concurrent Engineering Design System, *Ph.D. Thesis*, City University, London, 1994.

Evbuomwan, N. F. O.; Sivaloganathan, S. and Jebb, A., A State of the Art Report on Concurrent Engineering, *Proceedings of Concurrent Engineering: Research and Applications, 1994 Conference*, Pittsburgh, Pennsylvania, August 29-31, 1994, 35-44.

Evbuomwan, N. F. O. and Sivaloganathan, S., The Nature, Classification and Management of Tools and Resources for Concurrent Engineering, *Proceedings of: Concurrent Engineering: Research and Applications, 1994 Conference*, Pittsburgh, Pennsylvania, August 29-31, 1994, 119-128.

Evbuomwan, N. F. O.; Sivaloganathan, S. and King, A. M., Requirements for a Concurrent Engineering Design System, *Proceedings of: Concurrent Engineering 1995 Conference*, McLean, Virginia, August 23-25, 1995, 565-572.

Evbuomwan N. F. O. & Anumba, C. J., A Software Architecture for Concurrent Life-Cycle Design and Construction, *Computing in Civil Engineering: Proceedings of the Third ASCE Congress*, Anaheim, California, June 17-19, 1996, 424- 430.

Eversheim, W.; Rozenfeld, H.; Bochtler, W. and Graessler, R., Methodology for an Integrated Design and Process Planning based on a Concurrent Engineering Reference Model, *CIRP Annals - Manufacturing Technology*, 44:1, 403-406, 1995.

Frank, D. N., Concurrent Engineering: A Building Block for TQM, *Annual International Conference Proceedings - American Production and Inventory Control Society (APICS),* Falls Church, VA, USA, 1994, 132-134.

Griffin, A. and Hauser, J. R., The Voice of the Customer, *Working Paper*, Sloan School of Management, MIT, Cambridge, MA 02139, 1991.

Handfield, R. B., Effects of Concurrent Engineering on Make-to-Order Products, *IEEE Transactions on Engineering Management*, 41: 4, 384-393, November, 1994.

Harding, J. A. and Popplewell, K., Driving Concurrency in a distributed concurrent engineering project team: a specification for an Engineering Moderator, *International Journal of Production Research*, 36: 3, 841-861, 1996.

Kamara, J. M., Anumba. C. J. and Evbuomwan, N. F. O., The Principles and Applications of Concurrent Engineering, *Technical Report*, Construction Research Unit, School of Science and Technology, University of Teesside, Middlesbrough, UK, January, 1997.

Koskela, L., Application of the New Production Philosophy to Construction, *Technical Report No. 72*, Center for Integrated Facility Engineering (CIFE), Stanford University, CA, USA, September, 1992.

Kott, A. and Peasant, J. L., Representation and Management of Requirements: The RAPID-WS Project, *Concurrent Engineering: Research and Applications*, 3: 2, 93-106, 1995.

Kuan, K-K., Facilitating Conceptual Design in Concurrent Engineering, *Proceedings of: Concurrent Engineering 1995 Conference,* McLean, Virginia, August 23-25, 1995, 223-229.

Madan, P., Concurrent Engineering and its Application in Turnkey Projects Management, *IEEE International Management Conference*, 1993, 7-17.

Mallon, J. C. and Mulligan, D. E., Quality Function Deployment - A System for Meeting Customers' Needs, *Journal of Construction Engineering and Management*, 119: 3, 516-531, 1993.

Molina, A.; Al-Ashaab, A. H.; Ellis, T. I. A.; et al., A Review of Computer-Aided Simultaneous Engineering Systems, *Research in Engineering Design*, 7: 1, 38-63, 1995.

Monroy, J., Concurrent Engineering, *APEC 92: Seventh Annual Applied Power Electronics Conference and Exposition*, 1992, 311-314.

Nicholas, J. M., Concurrent Engineering: Overcoming Obstacles to Teamwork, *Production and Inventory Management*, 35: 3, 18-22, 1994.

Prasad, B., On influencing Agents of CE (Editorial), *Concurrent Engineering: Research and Applications*, 3: 2, 78-80, 1995 (A).

Prasad, B., Sequential versus Concurrent Engineering - An Analogy (Editorial), *Concurrent Engineering: Research and Applications*, 3: 4, 250-255, 1995 (B).

Prasad, B., Morenc, R. S. and Rangan, R. M., Information Management for Concurrent Engineering: Research Issues, *Concurrent Engineering: Research and Applications*, 1: 1, 3-20, 1993.

Prasad, B., *Concurrent Engineering Fundamentals, Volume 1: Integrated Products and Process Organization*, New Jersey: Prentice Hall PTR, 1996.

Ranky, P. G., *Concurrent/Simultaneous Engineering (Methods, Tools & Case Studies),* CIMware Limited: England, 1994.

Smith, J.; Tomasek, R. Jin, M. G and Wang, P. K. U., Integrated System Simulation, *Printed Circuit Design*, 12: 2, pp. 19-20, 22,63, Feb. 1995.

Syan, C. S., Concurrent Engineering: A Survey of Practice in the UK, *Short Paper Proceedings: 2nd International Conference on Concurrent Engineering and Electronic Design Automation*, April 7-8, 1994, 3-8.

Thamhain, H. J., Concurrent Engineering: Criteria for Effective Implementation, *Industrial Management*, 36: 6, 29-32, 1994.

Winner, R. I. et al, The Role of Concurrent Engineering in Weapons Systems Acquisition, *IDA Report R-388*, Institute of Defense Analysis, Alexandria, USA, 1988.

A COMPUTER-AIDED PLANNING SYSTEM FOR THE PRE-CAST CONCRETE INDUSTRY

N. Dawood, B. Hobbs, A Kalian
Division of Civil Engineering and Building
School of Science and Technology
The University of Teesside
Middelsbrough, TS1 3BA
UK.

ABSTRACT: The UK construction industry is going through a major re-appraisal with the objective of reducing construction cost by at least 30% by the end of this millennium. Prefabication and off-site construction are set to play a major role in improving construction productivity, reducing cost and improving working conditions. In a survey of current practices of the prefabrication industry by the authors, it was concluded that the industry is far behind other manufacturing-based industries. It is suggested that some form of systematic approach to presenting and processing information is needed, using a computer-aided approach. The objective of this research is to develop an integrated intelligent computer-based manufacturing information system for the precast concrete industry. The system should facilitate; the integration of design and manufacturing operations; automation of production schedules directly from design data and factory attributes and generation of erection schedules from site information, factory attributes and design data. It is hypothised that the introduction of such a system would reduce the total cost of precasting by 10% and encourage clients to choose the precast solution.

INTRODUCTION

Precasting is the process of producing building products and components, mainly concrete products, in a factory-based environment. It offers the greatest potential for radical improvement of productivity and quality in building.
A factory production environment usually provides higher technological efficiency, better working conditions and more rigorously enforced quality control (Retik, et al 1995). A building component which can be prefabricated and then assembled on site promises a better performance and a more economic use of resources than when constructed on site. It has been argued that activities on site should be reduced to a bare minimum and off-site manufacturing should be increased (Powell 1995). In a manufacturing environment, the application of IT has had a substantial impact on almost all activities in a factory. Often, the introduction of computers in the work place has

changed the organizational structure of a factory and made necessary the adaptation of a completely different and new management structure.

In a survey of current managerial practices by the authors, it was concluded that:

- Managerial practices and in particular planning are fairly basic, manually driven and may be inaccurate.
- Communications inside a factory are often primitive and some form of IT application has the potential to bring the industry into the 21st century.
- Detailed design, capacity planning, production scheduling and erection schedules are fragmented and manually driven. This is causing extra financial burden on the industry and depriving it from substantial growth. The need for a coherent manufacturing information system is quite apparent.
- Designers may have no appreciation of downstream operations which include; production, transportation and erection. This has caused extra burden on the industry resulting from inefficient utilization of moulds.

Figure 1 sets out potential advantages to be gained from the proposed development of an integrated methodology for the precast business process which has been identified from the survey.

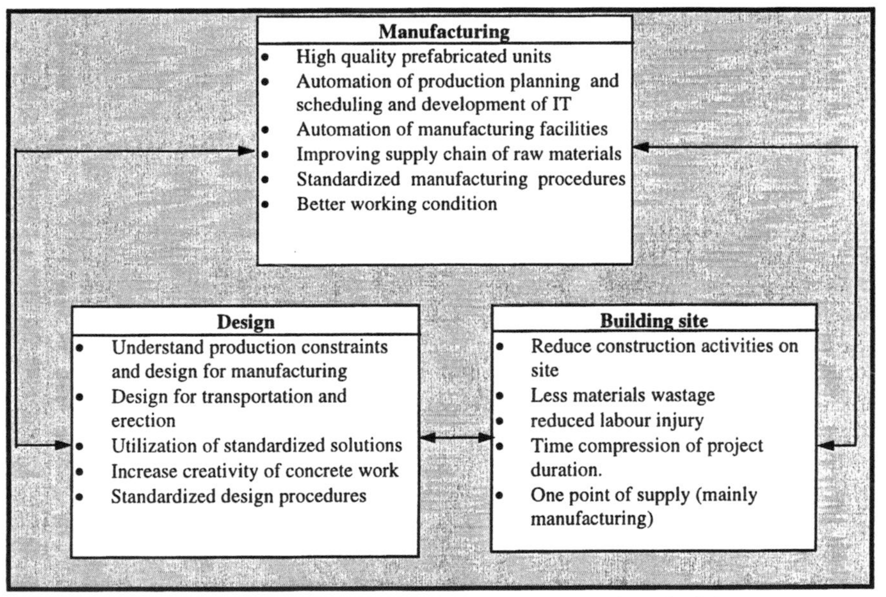

Figure 1. A potential advantages of the proposed integrated methodology

The objective of this research is to develop an integrated intelligent computer-based manufacturing information system for the precast industry. The system should facilitate; the integration of design and manufacturing operations; automation of production schedules directly from design data and factory attributes and generation of erection schedules from site information, factory attributes and design data. It is hypothised that the introduction of such a system would reduce the total cost of precasting by 10% and encourage clients to choose the precast solution.

In order to achieve the objective of the research, the following tasks have been undertaken:

- Development of the theory of the model.
- Designing the specification and the concept of the proposed model. This is presented in terms of product and process models.
- Re-engineering business process of precasting.
- Development of the computerisation aspects and validation of the model.

Due to the restriction of the size of this paper, the computerisation of the model is not given.

THE THEORY OF THE MODEL

The model developed in this paper is theoretically based on the CIM (Computer Integrated Manufacturing), and CIC (Computer Integrated Construction) models (Rembold, 1993). It consists of both the use of information technology in different phases and tasks of the manufacturing process (including design) and the integration of these phases and tasks through the use of digitally stored data and data transfer. Among the elements needed to create a computer integrated environment are (Bjork, 1992):

- The extensive use of data bases for storing design and project management data
- Intelligent user interface
- An infrastructure of data transfer networks
- Data structuring and transfer standards
- Data and knowledge bases with general construction/manufacturing information.

The modelling approach adopted in this research is the conceptual modelling technique which has been particularly popular in product model research. The approach focuses on modelling the structure of the information describing the products, processes, resources and other elements of the precasting process. A conceptual model specifies the categories of information used in a specific domain or database. In a conceptual model only the information itself is modelled, not the format in which the information is stored or presented.

A significant volume of construction-related product modelling research has been developed and reported by a tremendous amount of literature and amongst the models being developed are: RATAS (computer-aided design of building) , COMBINE (energy-conscious builiding design), CIMSTEEL (construction steel work), ATLAS (large scale engineering), Process Base (process plants) and CAESAR Offshore (offshore project). The intention of this research is utilise the wealth of theory developed in such projects.

The basic concept used in almost all data models is the object or entity. An object is a set of closely interrelated data about "something" in the modelling domain, see Bjork (1992). "Something" can be a physical object, but it could be an equation system, or any kind of abstract object. Similar concepts to objects are frames in knowledge-base systems, abstract data types in programmes and the "objects" of object oriented programming. The overall concept of the research is to standardise the way in which engineering information, relating to a type of object (precast units, moulds, etc.) is held to facilitate the transfer of such information between process models (applications software). The following section explains and discusses the structure and specifications of the proposed model.

THE STRUCTURE AND SPECIFICATION OF THE MODEL

The model is composed of a central "product model repository" where physical objects are modelled and stored. Such objects are: precast units and their attributes and factory attributes. The objects are designed to interact with application software for the creation of application objects specifically tailored to a particular job, processing of information to produce plans and schedules. Figure 2 shows the structure of the model in terms of product model repository and applications software.
The data structure of the "product model repository" is designed to facilitate the transfer of information to different applications in a standardized way. Figures 3 and 4 show the structure for modelling information for the main objects in the model, namely building and factory respectively. As can be seen in Figure 3, the attributes for each sub-object are presented in terms of: type of component, dimensions, bill of quantities, type of joint, mould type, structural dependency, weight, man-hr. needed for steel fabrication, man-hr. needed for mould preparation, proposed erection time and proposed casting time. Some of the information is provided directly by users and some is a product of information processing by applications software. For example, dimensions and type of joint will be developed through CAD while Bill of Materials and Man-hr. needed for mould preparation and steel fabrication can be provided by the estimation software.
Figure 4 shows the structure for modeling production facilities. A factory is composed of moulds attributes, casting man-hr. available and transportation facilities. The information describing the entities is given by production managers for each given period.

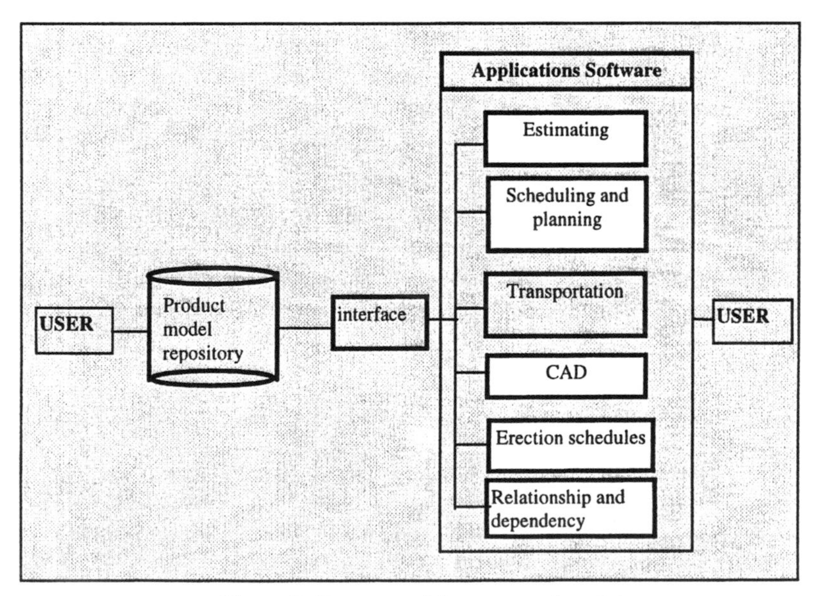

Figure 2. Structure of the proposed model

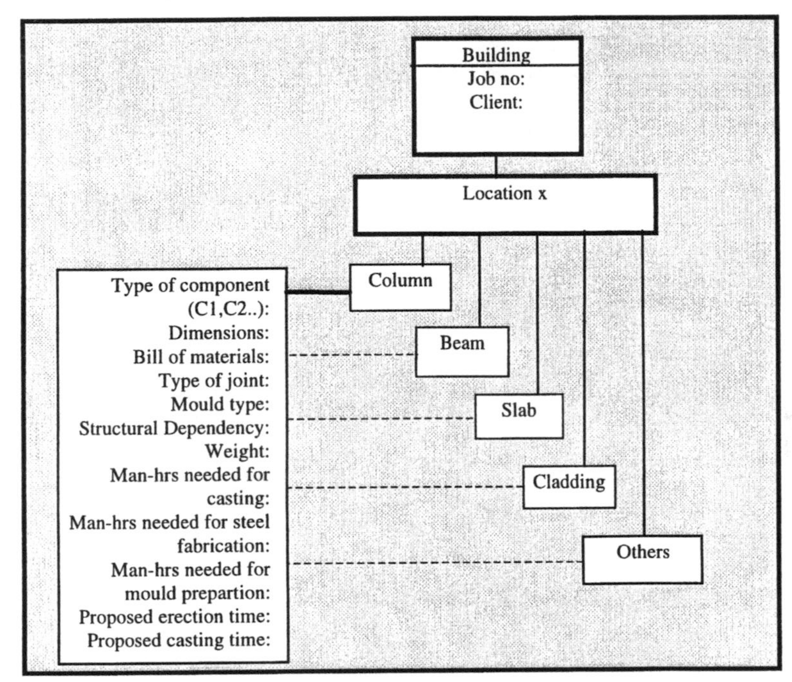

Figure 3. Structure for modelling information for a given job

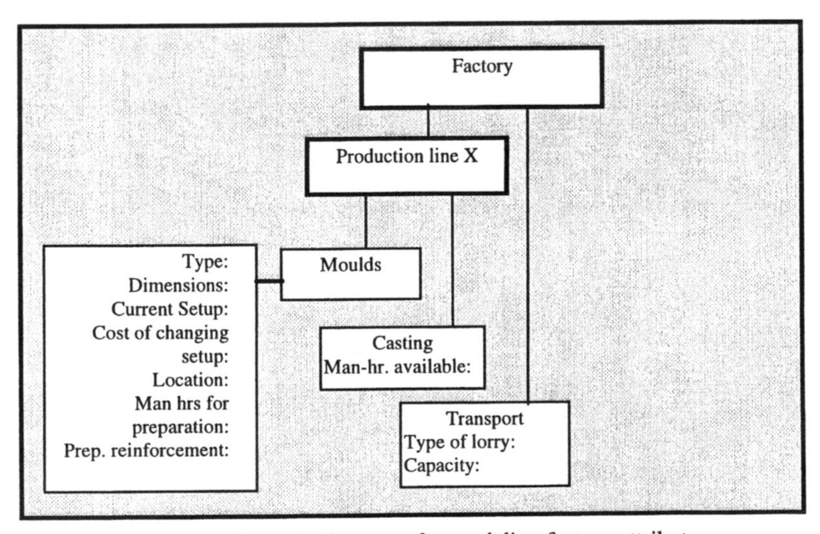

Figure 4. Structure for modeling factory attributes

The Processes of the model

The processes of the model are the operations which use the objects discussed above, process their information and produce: relationship and dependency between units, erection schedule, and finally production planning and scheduling information. The processes are developed using integrated methodologies of KBS and heuristic rules which are being developed from elicitation of knowledge from precast production mangaers. The following introduces briefly the above processes.

Relationship and dependency of units

From previous literature (Kähkonen, 1993), it was concluded that factors affecting the sequence of project locations are contracts, site condition and working practice. The planner's decisions on the sequencing of individual precast units are affected by structural factors, but safety, production technology and site conditions can also be important. Additionally, when defining the overall erection programme, the decisions are affected by resources, work area and safety. Typical factors of dependency are presented in Table 1.

	Decision	Affecting Factors
1	Sequence of project locations	Contractual, site conditions
2	Sequence of precast units	Structural, safety, erection technology and site condition
3	Overall erection programme	Erection equipment and resources, work area and safety

Table 1: Typical factors which have an influence on dependency

It is proposed in this research that structural relationship and dependency between precast units in a given building location is established in the first instance. For example, columns to floor *i-1* precede beams to floor *i*. The information is presented in a hierarchical form or a network or dependency bar chart. The structural dependency will be then revised with respect to the availability of erection resources which in turn reflect site condition and erection technology. The output of this process is the erection schedule. Planners should be able to interact with the process and make modifications to suit contractual arrangements if needed. The procedure for automating the dependency (sequencing) process is as follows:

I. The project will be split into locations. A location dependency schedule will be produced which is based on contractual and site conditions. For example, a multi-storey building might be split into floor locations. This process is achieved manually and a code will be attached to each location. Each location will be known as a high level object.

II. Each location will then be decomposed into precast units which are known as sub-objects. The relationship between the units is generated automatically using a knowledge-base and constraints available in the model. This is achieved as follows:

a) For each location, the structural engineer should identify the structural relationships between precast units and store them in the code system of each unit. The relationships will be presented in a dependency hierarchical form as shown in figure 5. The figure shows a location in a building which is composed of four precast columns and four precast beams. The structural dependency states that the four columns can be erected at the same time and beam code B1 is dependent on columns code C1 and C2, Beam code B2 is dependent on columns code C2 and C3 and so on.

b) Once the 3D model of the whole building is developed and locations are identified, the structural dependency of precast units should be established and embedded in the unit's universal coding system. A knowledge-base is developed to generate dependency charts from the codes of individual units. The output information of this process is a logic dependency network or bar chart.

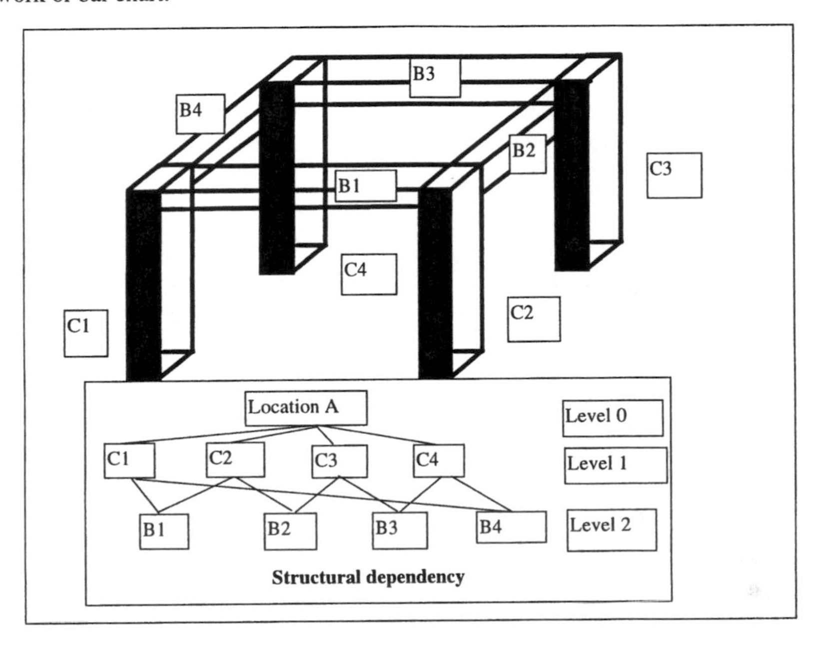

Figure 5. Example of the sequencing process

Erection shedules

These schedules are generated to reflect realistically the erection of precast units on site. Once the dependency sequencing process is accomplished the output schedule will be refined and altered with respect to the following factors:

I. Erection equipment on site
II. Erection gang on site
III. Site conditions and working space which should be reflected in points I and II above.

In order for a precast unit to be erected on site, resources are needed to accomplish this within a specified time. Planners should decide on this based on their experience. The total amount of resources available on site will be kept in a central pool and units will be competing for such resources. If resources are insufficient to accomplish the dependency network generated in the above section, certain units might be delayed until resources become available. In this case, priority rules to be used are:

I. Units with high erection time have high priority compared to other units.

II. Large units should have priority over small units.

In the example given in Figure 5, the four columns can be erected at the same time. However, if one crane is available on site, the four columns cannot be erected at the same time and a revised schedule will be produced which should indicate that the columns will be erected sequentially and not concurrently.
Once an erection schedule is developed, the next stage in the process is to produce factory production plans.

Production planning and scheduling

Once the erection schedule has been developed, the next process is to produce the precast units. Each factory might supply several construction sites and the planning process in the system is accomplished as follows:

I. The erection schedule of precast units, for a given building, is transformed into ALAPCS (As Late As Possible Casting Schedule) assuming unlimited capacity is available. This is achieved by calculating the minimum lead time needed for casting, curing and transporting of precast units. The latest production time will be (erection time for unit(i)- (casting,curing and transporting of unit(i)).

II. The ALAPCS schedule is incorporated in a mould schedule. The system searches for available moulds in the shop and compares them with the required moulds for units. If the available moulds are sufficient, then the ALAPCS schedule will be converted to a mould schedule after subtracting the mould set-up time and other requirements that are needed before casting. Otherwise, the ALAPCS will be altered using the backward scheduling technique (Dawood, 1996). In this case, certain units might be produced earlier than they should and therefore provision for stock holding should be considered. The main criterion used in the evalution of different moulds allocation is to minimise mould changeovers. A mould schedule is considered to be an important operation in a precasting shop and cost of production can be increased rapidly if moulds are not utilized efficiently and effectively. Setup changes should be minimised in order to reduce cost. In a study of the effect of mould changeovers and repetition of casting in a precasting company, it was concluded that the cost of a precast unit produced in a special mould which is used to produce 2 or 3 casts is five times greater than a precast unit produced from a mould which can produce 30 casts of the same units, see Figure 6 (the information in figure 6 has been complied from one of the leading precast companies in the UK).

III. Once a mould schedule is produced, the system automatically produces steel fabrication schedules for the moulds.

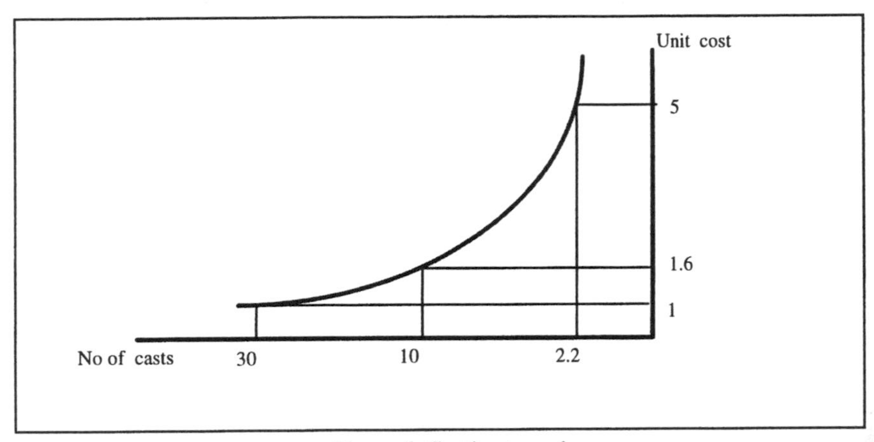

Figure 6. Cost/cast graph

Scheduling knowledge rules are developed and encapsulated into the scheduling process. The rules are developed to mimic the decisions needed to accomplish the scheduling process (Dawood, 1993, 1996).
The knowledge scheduling rules are developed to satisfy the following criteria:

- efficient utilization of resources by minimising changeovers of moulds
- satisfy erection programmes
- minimise disturbances in the shop floor

The planning and scheduling process is meant to be interactive and schedules should be able to be modified and altered as work progresses.

Re-engineering of the business process

For the proposed approach to work, certain aspects of the business process should be re-engineered. Scheme and detailed designers should be aware of certain manufacturing and erection processes. The design process of a prefabricated building should be developed collaboratively and manufacturers should play a focal point in the design and construction processes. Figure 7 shows an outline of the proposed integration of design, manufactureing and construction process.

CONCLUSION

The objective of this paper was to introduce and discuss the specifications of an integrated intelligent computer-based information system for the prefabrication industry. The theory of the system has developed using the "conceptual modeling technique". Product and process modelling procedures have been developed and discussed. An outline of the proposed business process approach has been introduced and discussed.

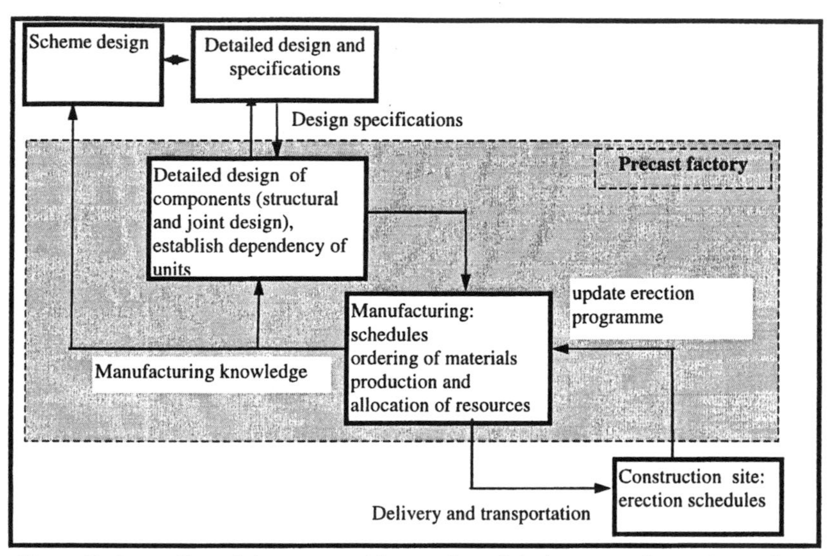

Figure 7. An outline of the proposed integration of design, manufactureing and construction process

REFERENCES

Bjork, B. C. A unified approach for modelling construction information, *Building and Environment*, 27, 2, 173-194, 1992a.

Bjork, B.C A conceptual model for spaces, space boundries and enclosing structure, *Automation in construction*, 1, 3, 193-214, 1992.

Dawood, N N and Neale, R. H., A Capacity Planning Model for Precast Concrete Building Products, *Construction Management and Economics*, 28, 1, 81-95, 1993.

Dawood, N. N. and Smith, M., Validation of an automated capacity planning in the concrete building materials industry using an industrial case study, *Built and Environment*, 31, 2, 129-144, 1996.

Kähkönen K. Modelling Activity Dependencies for Building Construction Project Scheduling, *VTT Publications*, Finland, ESP00,1993.

Powell, J, IT Research for Construction as Manufacturing Process, an EPSRC View, Colloquium on Knowledge Based Approaches to Automation in Construction, *The Institute of Electrical Engineers*, Digest No 1995/129, 1995.

Rembold, U., Nnaji, B.O. and Storr, A., Computer integrated Manufacturing and Engineering, *Addison-Wesley* Publishing Company Inc., 1993.

Retik, A. and Warszawski, A., Automated Design of Prefabricated Building, *Building and Environment*, Vol. 29, No. 4. pp 421-436, 1995.

A METHOD FOR INTEGRATING FORM ERRORS INTO TOLERANCE ANALYSIS

Scott Pierce and David Rosen
Systems Realization Laboratory
George W. Woodruff School of Mechanical Engineering
Georgia Institute of Technology
Atlanta, GA 30332-0405

The design of mechanical components is traditionally carried out under the assumption that components have perfect geometry. However, actual manufacturing processes produce components whose form and dimension deviate from the nominal design. Mechanical tolerances serve as a common ground between the need to produce reliable products which conform to the nominal design and the ability of manufacturing processes to produce parts in a cost-effective manner. Selection of geometric tolerances for an assembly or mechanism can be a complex problem, involving analysis of mating conditions between several components. This complexity forces designers to make simplifying assumptions, resulting in overly conservative tolerances and high manufacturing cost. The purpose of this research is to investigate a computer-aided approach to tolerance analysis for assemblies and mechanisms. A CAD system which can generate freeform surfaces is used to produce imperfect-form, variant models. A method is developed for simulating the mating of these imperfect-form components by formulating surface mating relationships as a nonlinear programming problem. Using this approach, tolerance analysis can be performed by generating a series of typical component variants, simulating the positioning of these imperfect-form components then measuring geometric attributes of the assembly or mechanism which correspond to functionality. This approach to tolerance analysis is demonstrated using the slider/guideway components of a slider-crank mechanism.

INTRODUCTION

Traditionally, the design of mechanical components has been carried out under the assumption that a manufactured component has exactly the form and dimension specified by the design engineer. Manufacturing processes are, however, inherently imperfect and produce parts which vary from the nominal design model. In order to control these variations, mechanical tolerances are appended to the model. In general, the design of a highly reliable product calls for tight tolerances, so that manufactured parts are as close as possible to perfect dimension and form. Unfortunately, the cost of producing such near perfect parts can be quite high, as manufacturing costs usually increase with a decrease in allowable tolerance. Tolerance specification can serve as a common ground between the need to design reliable products and the need to produce these products at a competitive cost.

An important step in tolerance selection is tolerance analysis. Tolerance analysis is concerned with defining the relationships between tolerances and product performance. If the product is an assembly or mechanism which has several attributes which are critical to functionality, tolerance analysis can become a complex problem. As an example, consider the mechanism shown in Figure 1. The components shown in this drawing comprise part of a high-speed stapling mechanism which is being used as a motivating example in this research. The mechanism consists of four prismatic components which, when assembled, form three prismatic sliding joints. A good measure of functionality of the mechanism is how well the driver, bender and bonnet components are aligned as the mechanism moves through the bottom of its stroke.

Suppose that the designers of this product know from experience with their manufacturing processes that there is typically some curvature in the (nominally straight) bonnet and that the (nominally parallel) side faces of the driver are typically not quite parallel. How can this information be

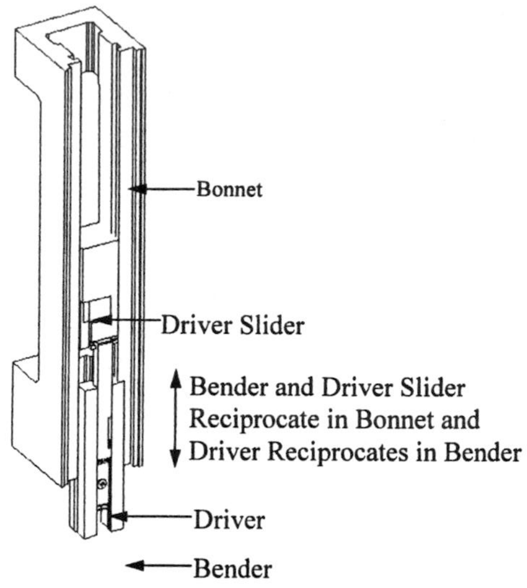

Figure 1. A High-Speed Stapling Mechanism

incorporated into the design process and used to specify tolerances which insure that misalignment between components stays within acceptable limits? This tolerance analysis problem requires the designer to determine mating conditions between several different faces. Furthermore, this analysis must be repeated over a range of variant magnitudes, for a number of different sets of variants, and at a series of positions through the stapling cycle. Clearly, this analysis is too complex to be performed manually.

This paper describes the development of computer-aided tolerance analysis capabilities in the CODA system. CODA (COnfiguration Design of Assemblies) is a feature-based design and virtual prototyping environment which is being developed in the Systems Realization Laboratory at Georgia Tech.[1] Tolerance analysis capabilities within CODA are incorporated into the Variational Modeling Module (VMM). The VMM is intended to allow a designer to: 1) Model components of imperfect form and size (component variants) which might occur as the result of manufacturing errors, 2) Simulate positioning of different combinations of component variants (assembly variants), 3) Assess geometric measures of functionality of assembly variants.[2]

The remainder of this paper is organized as follows: A brief review of relevant work in the fields of variational modeling and simulation of component mating is provided. Next, a detailed description of the Variational Modeling Module is provided. Finally, the work to date is summarized and the direction of future work is discussed.

BACKGROUND AND LITERATURE REVIEW

Variational Solid Modeling

Variational modeling is the process of applying variations in form or dimension to a solid model of a part. Research in this field has focused on two areas: 1) The mathematically rigorous definition of geometric tolerances,[3-5] 2) Methods which can be used to manipulate the data base of a solid modeler in order to permit variational modeling.[6-8] Work in the first area has lead to a formal mathematical definition of the geometric tolerancing standards.[9] Most of the research in the second area concentrates on methods for allowing variations in model variables. Relatively little research has been done on the problem of generating solid models of particular geometric variants.

Tolerancing is, by definition, concerned with variations in the boundaries of a solid. Thus, a boundary representation (B-rep) is a natural choice for the modeling core of a tolerance analysis system. Ideally, a B-rep which is to be used for variational modeling should be capable of representing a broad spectrum of surfaces, including free-form surfaces and surfaces where the variation from nominal form is localized to only part of the surface. In many cases, a variant surface will be of higher order than the nominal surface. There are two approaches which can be used to handle this problem: 1) The nominal surface representation can be replaced by a higher-order surface, 2) The nominal surface can be divided into a mesh of smaller surfaces.

Pandit and Starkey develop a variational modeling scheme in which variant surfaces are modeled by meshing a nominal surface into planar facets.[10] In order to generate variant surfaces, each planar facet is divided into a series of planar, triangular patches. Variant surfaces are then generated by perturbing the vertices of the patches in a direction normal to the nominal surface.

Gupta and Turner extend this "triangularization" approach to use non-planar patches and to generate particular surface variants.[11] First, the nominal surface is triangularized into planar patches in the manner described above. The edges of each triangle are then bisected and mid-edge vertices are defined. Each triangular face is then used to form a triangular, quadratic Bezier patch. For each patch, six control vertices are defined (the three corners and three mid-edge vertices). Interior vertices are moved only in the direction normal to the nominal surface. Moving a particular vertex changes only the patches which are defined using that vertex. Thus, this approach can be used to model localized surface variations. The minimum size of the variation which can be modeled is determined by the mesh density.

Modeling the Mating of Imperfect-Form Components

While commercially available CAD software has been developed which allows modeling of a limited set of mating relations, very little work has been done in modeling the mating of actual components which do not have perfect form. For such components there may be several sets of positions which are feasible. The final set of positions will depend on the form of component errors, the sequence in which contacts are established and on the forces transmitted between components. A system for modeling the positioning of component variants must provide some mechanism for choosing from amongst the set of feasible final positions.

Early work in modeling the assembly process focused on the definition of mating relations as constraints on possible component coordinate transformations. Lee, Andrews and Rocheleau describe a strategy for generating coordinate transformations directly from component geometry and mating conditions.[12-13] In this work the mating conditions "against" and "fits" are formulated as a set of vector equalities. Use of equality relations means that only perfect-form components which mate exactly can be modeled. Mullineux proposes a modeling strategy which is a combination of equality and inequality constraints.[14] He points out that the existence of clearances between mating parts means that there is often a range of positions over which mating conditions can be satisfied. Thus, inequality constraints often provide a more accurate model of the requirements which must be satisfied in order for parts to assemble. Kim and Lee propose that components in an assembly can be broken into groups and that each group can be formulated as a separate mathematical programming problem.[15] They point out that solving the positioning problem for each group in sequence is more computationally efficient than positioning the entire assembly simultaneously.

Turner asserts that this sequential positioning approach best represents the manner in which assemblies are usually built.[16] The author models the construction of an assembly as a series of linear programming problems. Each mate between a pair of elements is formulated as a separate problem. The convex hulls of mating surfaces are identified and these convex hulls are substituted for the (possibly non-convex) variant geometry. The mating relations between these convex surfaces are then linearized and solved using a linear programming algorithm. When the mating position is found between a pair of surfaces they are fixed by expressing their relative position as an equality constraint in subsequent positioning operations. An important aspect of Turner's work is the manner in which the positioning problem is formulated. The objective function is always formulated as "minimize the maximum distance from ideal fit." The requirement of non-interference is then embodied as a set of inequality constraints. By posing the problem in this fashion, there does not have to be an exact fit between mating surfaces in order for a feasible solution to exist. Thus, this formulation can be used to model the mating of parts which have imperfect form.

Inoue and Okano demonstrate the use of the linear programming approach to the mating of components which have randomly generated form errors.[17] Deviation from nominal fit is defined by measuring the differences in position and orientation between points on mating components which are coincident in a perfect assembly. Forces and moments at each point are calculated using pre-specified loading conditions. These values are then used to calculate the potential energy associated with a particular set of component positions. The objective of the positioning algorithm is then to minimize the potential energy of the system.

VARIATIONAL MODELING FOR TOLERANCE ANALYSIS IN CODA

The "Generate and Test" Approach

As discussed in Section 1, the goal of tolerance analysis is to establish relationships between a proposed set of tolerances and measures of product functionality. In order to determine whether a proposed tolerance set is sufficient to ensure functionality, a designer must consider the form and magnitude of variations which are allowed (the variational class). The designer must analyze the manner in which component variants will mate and assess the functionality of different groupings of component variants. In order to perform such an analysis the designer must: 1) specify variations from nominal size and form which are likely to occur for each component, 2) generate models of imperfect form component variants, 3) analyze the mating relationships between variants, 4) assess the functionality of different sets of component variants. We refer to this approach to tolerance analysis as the "generate and test" method (a term first used by Pandit and Starkey[10]).

Figure 2 is a schematic diagram of the Variational Modeling Module. The VMM will allow a designer to perform the tasks listed above using a boundary-representation solid modeler. The module uses the CODA front-end running on top of the ACIS geometry engine. Using the CODA interface, a designer can generate nominal solid models of the components of an assembly or mechanism. The VMM then allows the generation of "as manufactured" component models by specifying sets of points which lie on the surfaces of a variant component. The reasoning behind this approach is that information about probable manufacturing errors for a particular class of component often exists in the form of discrete point measurements. Such a point set might be the result of measuring similar or prototype components using a coordinate measuring machine (CMM). Note that this is a significant departure from previous approaches to tolerance analysis in which random form errors are generated using a technique such as Monte Carlo simulation. Our method allows the designer to incorporate existing manufacturing process information into tolerance analysis procedures.

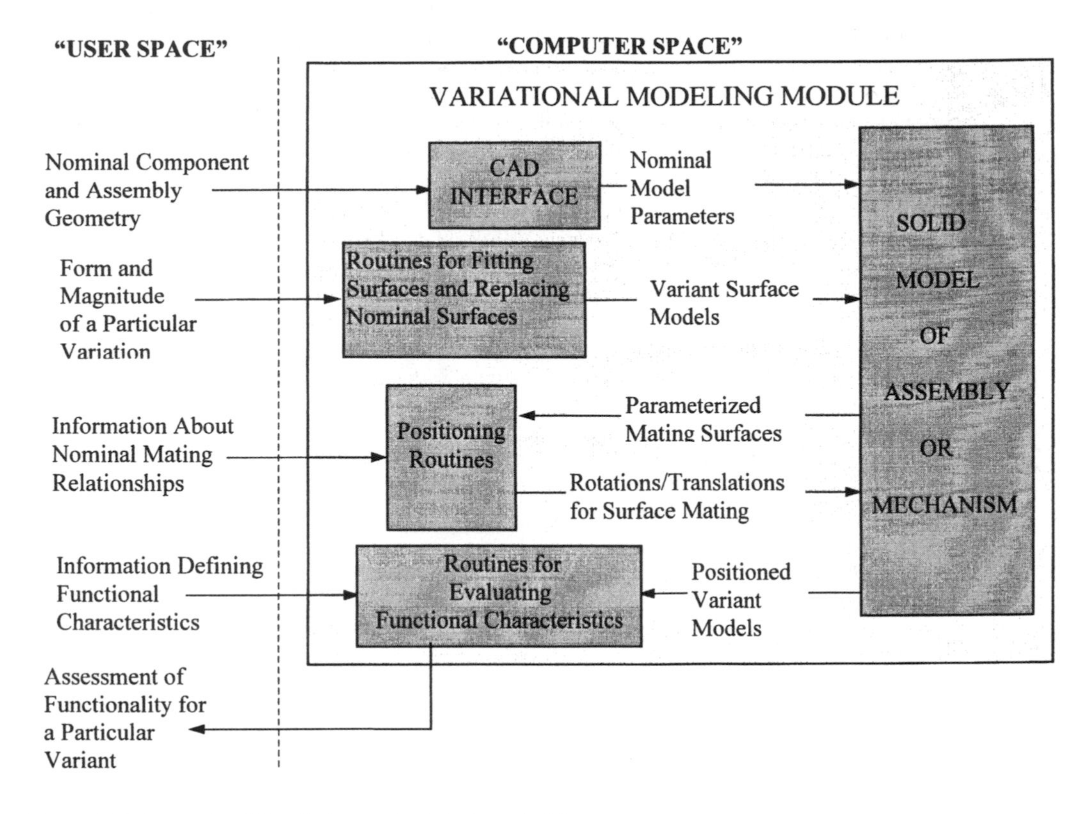

Figure 2. Schematic of the Variational Modeling Module

Point measurements which define component variants are used to fit freeform surfaces. These freeform surfaces are then used to replace the corresponding nominal surfaces in the component solid model. Once a set of variant components has been constructed in this manner, mating between components can be simulated by formulating mating relations as a nonlinear programming problem. Geometric attributes of the assembly or mechanism which correspond to functionality can then be measured using this model. In the case of a mechanism, each set of variant components can be tested in a series of different points in the mechanism cycle. By repeatedly generating and testing variants which are allowed under a proposed set of tolerances, a designer can test the effect of particular tolerances on the overall function of the assembly or mechanism.

Development of the VMM is an ongoing project. Currently, the capabilities to generate variant solid models have been implemented in the modeling system. The methods which are to be used for assembly simulation have been developed and we are in the process of implementing the positioning algorithms as the next step in the development of the tolerance analysis system.

Generation of Variant Component Models

As discussed above, the ACIS geometric modeler is being used as the modeling engine for the CODA system. ACIS is a b-rep solid modeler which incorporates non-uniform, rational, b-spline surfaces (NURBS). These freeform surfaces can be used to model a wide array of variant shapes, including form errors which are localized to a small area of the surface. This flexibility eliminates the need to mesh a nominal surface into many smaller surfaces in order to model a variant.

Figure 3 illustrates the process of generating a variant component using point measurements. In this case, the variant component is the slider component of a slider-crank mechanism (which is a simplification of the high-speed stapling head). Figure 3b shows a pointset which represents a manufacturing error which can occur in end milling operations as a result of cutter deflection combined with a chucking error.[18] The sample data set was generated by sampling an analytical approximation to the cutter deflection data presented in reference 18. This approximation has the form of an offset sinusoidal surface, as the cutter deflection error is approximately sinusoidal in form. In order to simulate the effects of surface roughness and measurement error, each data point was randomly perturbed by multiplying it by a scaled, normally distributed random number. Note that the error magnitude has been multiplied by a factor of 10 in this figure so that the error is visible.

Figure 3c shows the bicubic NURBS surface which is fitted to this pointset. In order to account for the possibility that the width of the slider may be larger than the nominal dimension, extra points are generated outside of the nominal surface by extrapolating the form of the original, measured points. The surface is then fit to the dataset and incorporated into the solid model of the slider by first extending the nominal face of the model so that the interior of the surface is embedded within the slider, then slicing the model into two pieces with the surface and discarding the extraneous piece (Figure 3d). Figure 3e shows the slider after the same set of operations has been performed on a second face using a pointset which represents a milling error caused by spindle tilt.[19]

Simulation of Mechanism Component Mating

As discussed in Section 2, previous research has utilized a linear programming algorithm in order to solve for the mating positions between surfaces. While such an approach is computationally efficient, it requires excessive simplification of the part model. In particular, these approaches consider only the convex hull of a surface or positions of extreme points on mating surfaces. However, as-manufactured components may have several maximum and minimum points across their surfaces. Consideration of only the extreme points of a surface ignores the interaction between these local surface variations. The existence of these local extrema may cause a traditional linear programming algorithm to get stuck in a local minimum which is not the global minimum solution. This possibility is particularly likely when simultaneous mating between multiple faces is considered.

A major goal of this research is to solve the component mating relationship problem without simplifying the shape of the variant surfaces. Towards this end, a nonlinear programming algorithm is used to solve the positioning problem. The problem will be formulated as:

Minimize: Z = Total Distance Between Mating Surfaces

Subject To: Non-Interference Between Components

This problem will be implemented by normalizing the total distance and the interference terms, then incorporating the non-interference constraint into the objective function as a weighted penalty term. Thus, the final problem formulation will be:

Minimize: Z = Normalized Total Distance Between Mating Surfaces +
(Weighting Constant) * Normalized Interference Between Components

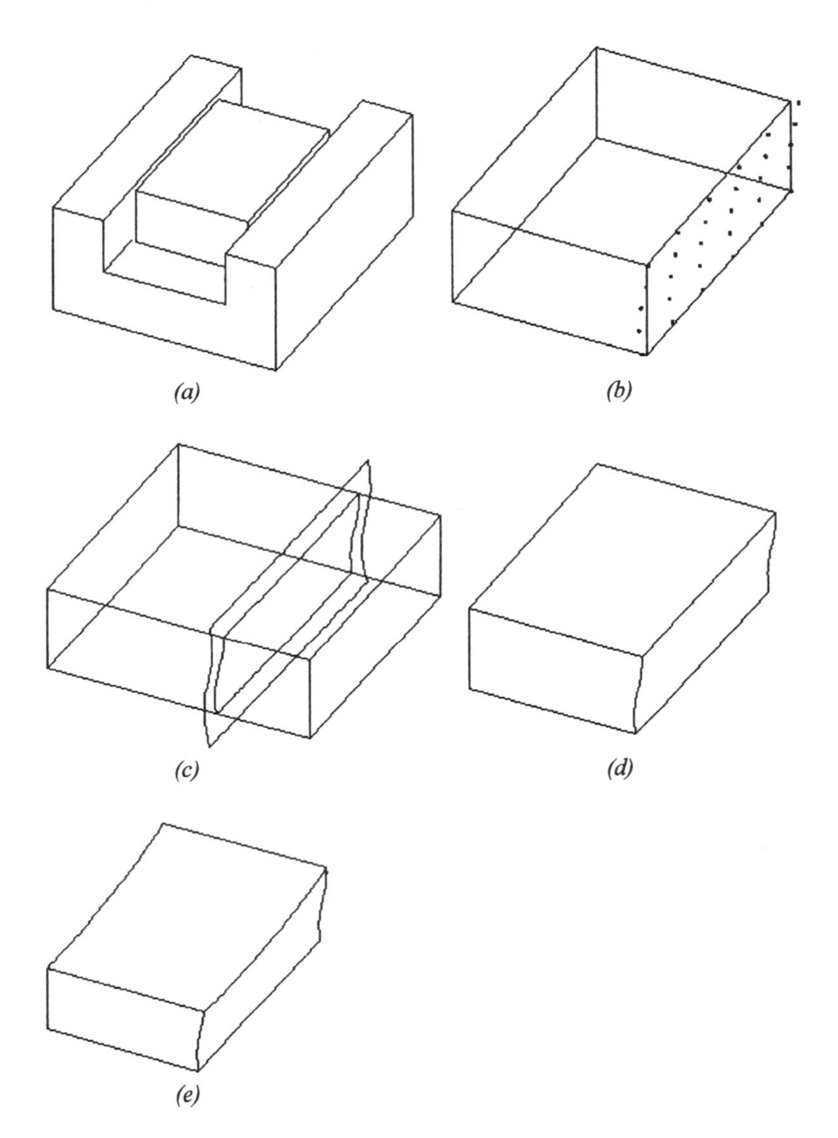

Figure 3. Constructing an "as manufactured" variant of a slider. *(a)* Nominal slider/guideway. *(b)* Pointset which defines a manufacturing error on the side face of the slider due to cutter deflection during milling. *(c)* NURBS surface fitted to the pointset with the face of the slider extended to cover the surface. *(d)* Slider with the variant face replacing the nominal side face. *(e)* Slider with both the top and side faces replaced with variant surfaces. The top face reflects an error due to spindle tilt during milling.

There are two major reasons that the penalty formulation is used instead of formulating non-interference as a "hard" constraint: 1) If the nominal component positions do not yield a non-interfering initial position, the optimization algorithm (which is discussed at the end of this section) can be used to search for such a position by minimizing the interference term. If no non-interfering position exists the algorithm will at least find the position of minimum interference. 2) The penalty method allows the optimization algorithm to move the search through a position where there is slight interference in order to find the optimum position.

Figure 4 illustrates the simulation of component mating for the slider as it is being pushed into the corner of the guideway. To begin the simulation, a base component (in this case the guideway) is fixed in model space. The next component in the assembly is then "coarse positioned" so that there is no interference between components. Coarse-positioning is accomplished by generating bounding boxes for each component, then moving the component being positioned so that the faces of the bounding boxes are coincident. Clearly, it is possible that coarse positioning in this manner may not

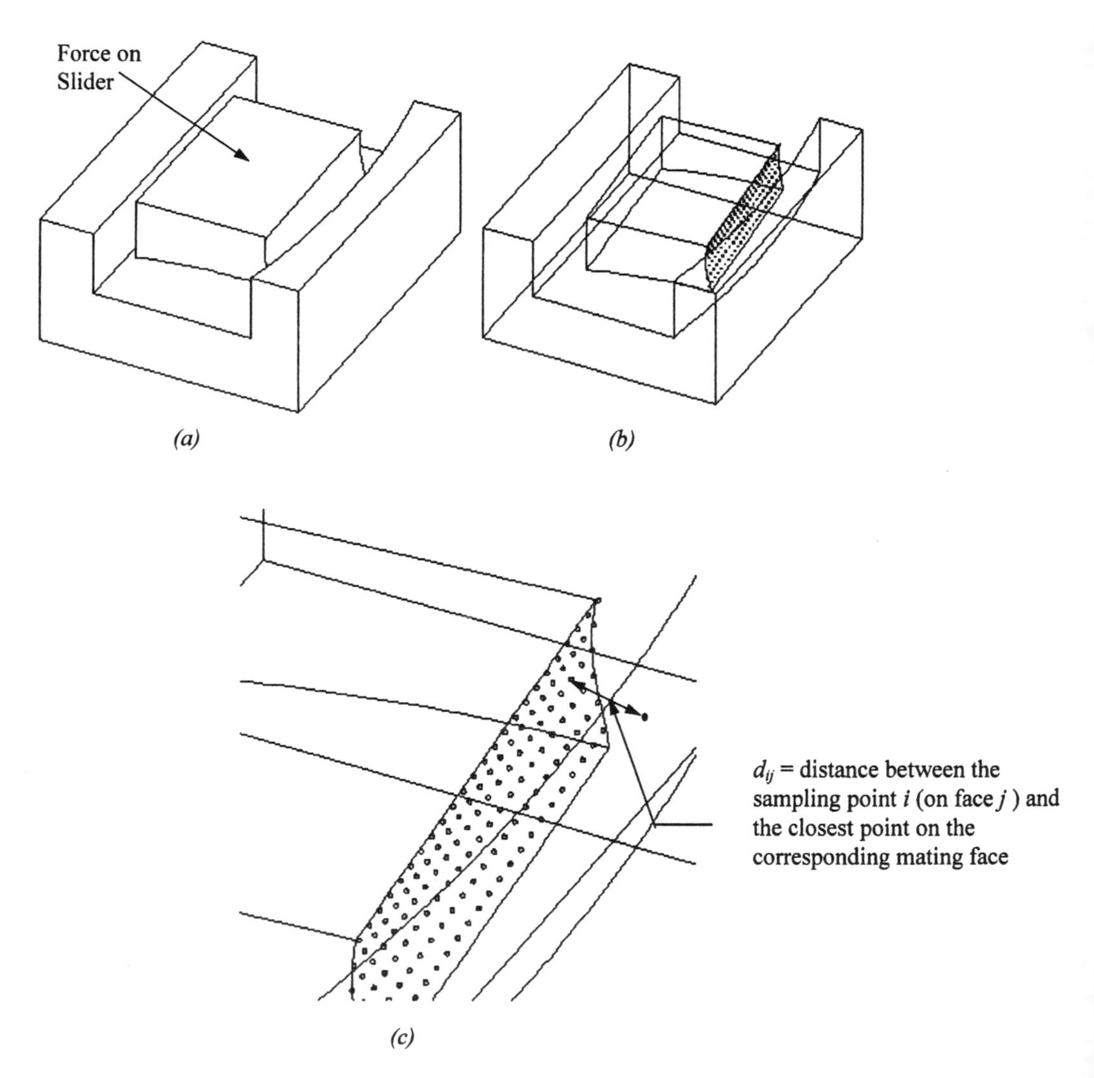

Figure 4. Simulation of mating between two components of imperfect-form. *(a)* Components in their nominal positions. *(b)* Slider being tested in a candidate mating position with distance sampling points shown on the side face. *(c)* Measurement of the distance between sampling point d_{ij} and the corresponding mating face.

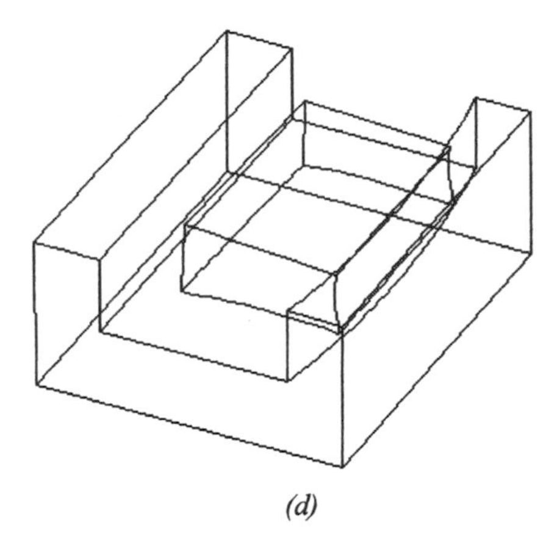

(d)

Figure 4 (cont.). *(d)* Final position of the slider in the guideway.

yield a non-interfering position. In this case, positioning is started from the nominal configuration and the positioning algorithm is used to search for a feasible position.

Once the component is coarse positioned, the optimization algorithm begins generating candidate sets of position and orientation variables (Figure 4b). Evaluation of a candidate variable set begins by moving the component model into that position, then evaluating the interference between the component being positioned and the other components in the assembly or mechanism. If the bounding boxes interfere, a Boolean intersection between the two bodies is performed. If the Boolean intersection returns a non-NULL result, the interference volume is calculated. The interference volume is then normalized by dividing by the total volume of the body being positioned. The normalized total interference over all mating components is then multiplied by the penalty term, W. This term is chosen to be large enough to overwhelm the other term in the objective function, so that a position with any significant interference has a high objective value.

The next step in the evaluation of the objective function is to measure the distance between mating faces. For each mating face, a grid of sampling points is generated on the component being positioned (Figure 4b). For a particular grid point (point i) on mating face j, the closest point on the corresponding face of the stationary component is found and the distance between the points (d_{ij}) is measured (Figure 4c). If a negative distance is found (i.e. the parts interfere) d_{ij} is defined to be zero at that point. The sum of these distances over all of the component mating faces is calculated and this sum is normalized by dividing by the sum of the maximum possible values for each of the d_{ij}.

Using the normalized intersection volume and normalized distances, the value of the objective function is calculated. This is then compared against a minimum value which is specified as a stopping criterion. If the stopping criterion is not met the optimization algorithm generates a new set of candidate position/orientation variables and a new iteration is started. If the stopping criterion is not met after a prescribed number of iterations the algorithm is stopped and the position corresponding to the minimum value of the objective function is reported.

In summary, the steps involved in positioning a component variant are:

- Starting with the component in its nominal position, generate bounding boxes for all mating faces.

- Use the bounding boxes to move the component into a non-interfering initial position (if possible).

- If no position is found where bounding boxes do not interfere, start the positioning algorithm from the nominal position.

- Use the optimization algorithm to generate a set of candidate position and orientation variables. Move the component to the candidate position/orientation.

- Generate bounding boxes for the component and all potentially interfering mating faces. Test for interference between these bounding boxes.

- If interference is detected between bounding boxes, perform a Boolean intersection between the potentially interfering components. Measure the total interference volume for all components and normalize this value.

- Generate a sampling grid over each of the component mating faces. For each sampling point measure the distance between the point and the closest point on the mating face. Normalize this distance and sum the normalized distances over all mating faces.

- Calculate the value of the objective function.

- Check the objective function value against the stopping criterion value. If the objective function value is less than or equal to the stopping criterion value, stop the algorithm and report the position. Otherwise, iterate.

- If the maximum number of iterations is reached, stop the algorithm and report the position which gave the minimum objective function value.

Note that instead of establishing one mating pair at a time, our approach considers all of the mating faces of a component simultaneously during the positioning process. We feel that this is a more accurate representation of the component mating process than sequentially mating a single face at a time. This formulation can easily be extended to simultaneous positioning of multiple components in cases where such a formulation more accurately represents the assembly process.

As previously stated, development of the assembly simulation component of the VMM has not yet been completed. The algorithm which will initially be tested for generation of candidate positions will be based on the method of simulated annealing.[20] The reason for this choice is the ability of simulated annealing algorithms to find a global minimum in the presence of several local minima. As discussed above, when mating between multiple variant faces is considered there is a strong possibility that local minima in the objective function will exist. By occasionally making a search move in an "uphill" direction, the algorithm can avoid getting stuck in a local minimum and will search through the local minima until the global minimum is found. The main disadvantage of the simulated annealing approach is that it is computationally inefficient. Testing is needed to determine whether this method can be used to simulate positioning of multiple components in a reasonable amount of CPU time.

CONCLUSIONS AND FUTURE WORK

This paper has discussed progress toward the development of an environment for computer-aided tolerance analysis. This environment is intended to allow a designer to test the efficacy of a proposed set of geometric tolerances by generating models of as-manufactured component variants and simulating the positioning of these variants in an assembly or mechanism. Using this computer-aided approach, a designer can perform a detailed tolerance analysis and can incorporate existing information about manufacturing process capabilities into the tolerance selection process. This allows the selection of functional tolerances without the use of conservative simplifications which increase manufacturing cost.

The strengths of our approach to tolerance analysis are:

- Information about the probable form of manufacturing errors is included in the analysis.
- It is possible to model a broad range of dimensional and form variants.
- It is not necessary to simplify the shape of component variants in order to simulate positioning.
- The positioning algorithm is robust and can work when many local minima exist.

Limitations of our approach are:

- It does not take into account the transmission of forces and moments between mating components.
- It does not explicitly consider the statistical distribution of manufacturing errors.
- Evaluation of interference volume in the objective function and use of the simulated annealing algorithm are computationally expensive.

Future work will focus on the implementation and evaluation of the positioning algorithm. Progressively more complex mechanisms will be simulated and mating conditions will be developed for non-planar contacts. In the long term, we plan to investigate the use of design optimization techniques along with the VMM in order to move towards computer-aided selection of geometric tolerances.

ACKNOWLEDGMENT

The authors would like to thank the National Science Foundation, which sponsors this work under NSF grants DMI-9420405 and DMI-9414715.

Author Contact Information:
Primary Contact: Dr. David Rosen, (404) 894-9668, david.rosen@me.gatech.edu
Secondary Contact: Scott Pierce, (404) 894-8528, gt8122a@prism.gatech.edu

Bibliography:

1) Rosen, D., "CODA - Configuration Design of Assemblies Prototypical CAD Systems," Systems Realization Laboratory, Woodruff School of Mechanical Engineering, Georgia Institute of Technology, Atlanta, Ga., 9/20/94.

2) Pierce, R.S. and Rosen, D., "Freeform Surface Modeling as a Tool for the Analysis and Selection of Assembly Tolerances," *Proceedings of the 29th International Symposium on Automotive Technology and Automation,"* June, 1996.

3) Turner, J.U., " The M-Space Theory of Tolerances," *Proceedings of the 1990 ASME Design Automation Conference*, Chicago, Ill., September, 1990.

4) Rossignac J.R. and Requicha A.A.G., "Offsetting Operations in Solid Modeling," *Computer Aided Geometric Design,* Vol. 3, 1986, pp.129-148.

5) Requicha, A.A.G., "Toward a Theory of Geometric Tolerances," *International Journal of Robotics Research*, Vol.2, No. 4, 1983, pp. 45-60.

6) Aldefeld, B., "Variation of Geometries Based on a Geometric-Reasoning Method," *Computer-Aided Design*, Vol. 20, No. 3, April, 1988, pp. 117-126.

7) Light, R. and Gossard, D., "Modification of Geometric Models through Variational Geometry," *Computer-Aided Design*, Vol. 14, No. 4, July 1982, pp.209-214.

8) Guilford, J. and Turner, J., "Representational Primitives for Geometric Tolerancing," *Computer-Aided Design*, Vol 25, No. 9, Sept., 1993, pp. 577-586.

9) Walker, R.K. and Srinivasan, V., "Creation and Evolution of the ASME Y14.5.1 Standard," *ASME Manufacturing Review*, Vol. 7, No. 1, 1994.

10) Pandit, V. and Starkey, J.M., "Mechanical Tolerance Analysis Using a Statistical Generate-and-Test Procedure," *ASME International Computers in Engineering Conference*, 1988, pp. 29-34.

11) Gupta, S. and Turner, J.U., "Variational Solid Modeling for Tolerance Analysis," *IEEE Computer Graphics and Applications*, May, 1993, pp.64-74.

12) Lee, K. and Andrews, G., "Inference of the Positions of Components in an Assembly: Part 2," *Computer-Aided Design*, Vol. 17, No.1, February, 1985, pp. 20-24.

13) Rocheleau, D.N. and Lee, K., "System for Interactive Assembly Modeling," *Computer-Aided Design,* Vol. 19, No. 2, March, 1987, pp. 65-72.

14) Mullineux, G., "Optimization Scheme for Assembling Components," *Computer-Aided Design,* Vol. 19, No. 1, Jan., 1987, pp. 35-40.

15) Kim, S.H. and Lee, K., "An Assembly Modeling System for Dynamic and Kinematic Analysis," *Computer-Aided Design*, Vol. 21, No. 1, 1989, pp. 2-12.

16) Turner, J.U., "Relative Positioning of Parts in Assemblies Using Mathematical Programming," *Computer-Aided Design*, Vol. 22, No. 7, September, 1990, pp. 394-400.

17) Inoue, K. and Okano, A., "Variant Shape Model Applicable to Combined Geometric Tolerances," *Proceedings of the 1996 ASME Design Engineering Technical Conference and Computers in Engineering Conference*, 96-DETC/DAC-1062.

18) Kline, W.A., DeVor, R.E., Shareef, I.A., "The Prediction of Surface Accuracy in End Milling", *ASME Journal of Engineering for Industry*, v. 104, August, 1982, pp.272-278.

19) Gu, Melkote, Kapoor and DeVor, "A Model for Prediction of Surface Flatness in Face Milling," The Physics of Machining Processes , vol. II, Stephenson and Stephenson.

20) Press, W.H., Vetterling, W.T., Teukolsky, S.A., Flannery, B.P., Numerical Recipes in C, 2nd ed., Cambridge University Press, New York, N.Y., 1995, pp. 444-455.

AN OPEN OPTIMIZATION SYSTEM FOR CONTROLLING OF MANUFACTURING PROCESSES

W. Sauer, G. Weigert, D. Hampel
Dresden University of Technology
Electronics Technology Laboratory

ABSTRACT: A lot of scheduling problems of flexible manufacturing processes can be described as combinatorial optimization problems. The sequence optimization problem, which asks for an optimal sequence of products, is a well known example. Although many different scheduling techniques were developed in the past, the application of them is very difficult. In our opinion the optimization of scheduling only can be solved satisfactorily in combination with methods of simulation which guarantee the required high flexibility of the optimization system. We describe in this paper an approach for an open optimization system, where the optimization cycle consists of two independent parts - the simulation and the search algorithm. The easy replacement of these parts is the advantage of the open architecture. So it is possible to combine several simulation systems and optimization algorithms in the same system and to use it as a experimental environment for examining new optimization strategies. Some optimization algorithms were already examined e.g. Genetic Search Algorithms and Tabu Search. The practical test will be carried out in the field of electronic production.

1 INTRODUCTION

The economic viability of an enterprise is fundamentally determined by the scheduling of manufacturing processes. Because of that the optimal scheduling has got a significant importance. In this case economic viability is equated with time based objectives, such as minimal makespan, deadline keeping, maximal machine utilization, minimal stocks etc.
On the other hand there exist control quantities, their variation also allows to achieve these objectives. Methods of influencing the manufacturing process are for instance input job sequences, setup strategies, making batches a.o. All the different optimization problems and their methods of solution have a common basis: they are mainly combinatorial optimization problems which are by a majority part of the class of NP-complete problems. Although Operations Research developed a lot of statements of solution in the past there are problems in application of them in a practical environment (see [1]). The reasons for this are the following:

- Real conditions in manufacturing and their constraints are very complex and unclear that mathematical description with enough accuracy is often impossible.
- Most of the developed statements of solution are too specific for the problem. The adaption to deviate conditions needs either a high expenditure or is impossible at all.
- Objectives and control quantities are nor only specific for the enterprise but also variate in connection with time inside the same enterprise.

For the optimization algorithms we can observe a trend which goes away from the more mathematical based procedures (e.g. Dynamic or Linear Programming) but goes to heuristic based procedures. An important source for inspiration will be found in nature. In the last years some promising methods were developed by simulating nature more or less (PPSN - Parallel Problem Solving by Nature). Genetic Search Algorithm [2], Evolution Strategy [3], Neuronal Networks, Simulated Annealing etc. are parts of this group of procedures. In addition to that new heuristic procedures are developed continuously, e.g. Tabu Search. The common feature of this methods is that a well known connection between objective quantities and control quantities is not necessary, in opposite to the gradient method for example, where you need a mathematical formula for this. The PPSN-methods and the other heuristic methods are widely independent from the kind of problem modeling. In any case a simulation system can take on the task of transferring the control quantities to the objectives. But instead of a simulation model the real process can be used too. That is why these optimization methods are particularly adaptable and they are principally suitable for application in modern flexible manufacturing.
Development of optimization systems with high flexibility and adaptability should be the main target. Such an optimization system should be adaptable to different problems like a modern flexible manufacturing system is able to be adapted to the e.g. changing conditions in the market. Excluding the user from the optimization process is certainly an illusion. Because of that the optimization system must have an interface for both visualization of the results and the optimization process itself. In order to examine a lot of different procedures for their suitability for solving special optimization problems and in order to compare each other, it is necessary to develop a program system which can be used as an experimental system. The inclusion of several optimization algorithms but also of several simulation systems should be possible, also for the case the programs were not developed by ourselves and/or their sources are not available. Optimization and simulation exchange their data with the help of problem specific interface modules. The adaptation of the general optimization system to a special problem (e.g. Traveling Salesman or Job Sequencing) is a final step. So it is possible to have an unified and problem independent controlling and monitoring of optimization process.

2 THE OPEN OPTIMIZATION SYSTEM

The open optimization system consists of four main components (Figure 1):

- the optimization component,
- the simulation component,
- the controller and
- the interface.

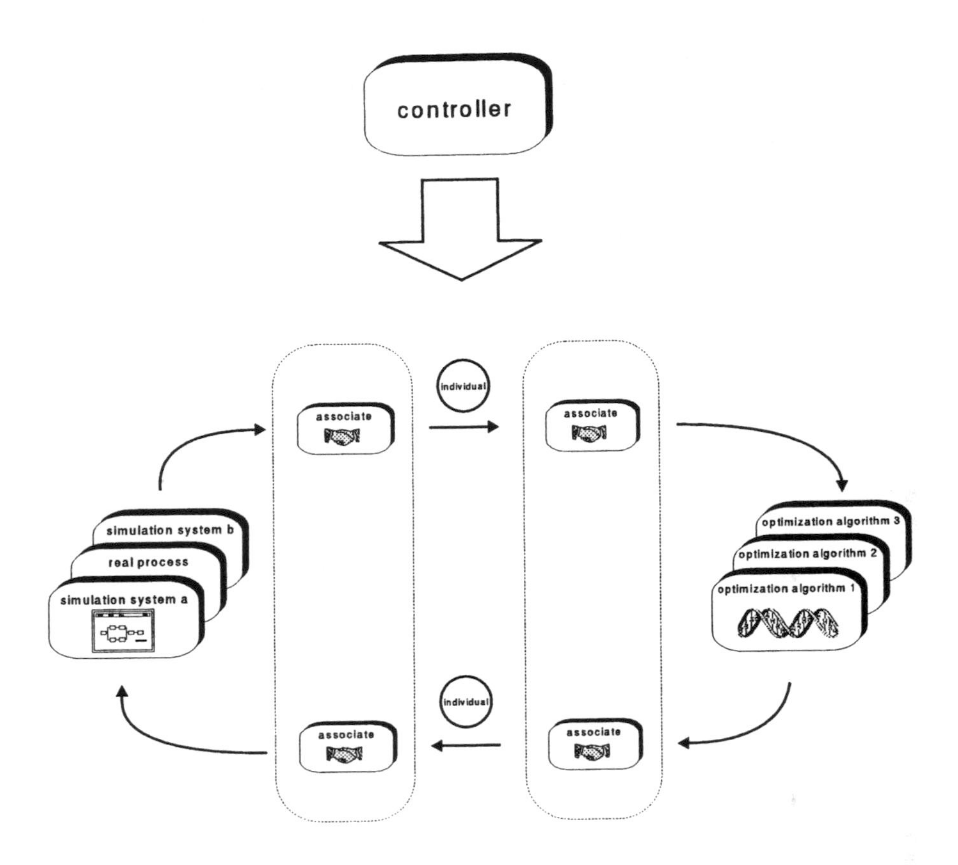

Figure 1. Optimization Data Flow

The data flow in the optimization system is controlled by a program written in the script language Tcl/Tk. This main tool is the controller (Figure 2). It is possible to influence the following characteristics of the optimization system:

- choice of the optimization algorithm and the optimization operators, adjustment of the necessary parameters
- choice of the simulation system
- choice of the appropriate interface, definition of the interface
- supplying of necessary data (e.g. definition of a final criterion)
- representation and analysis of the optimization results
- controlling of the data flow in the optimization system

Objects which are to be optimized can be job sequences, machine arrangements, queue job sequences and so on. The description of the optimization objects is completely independent of

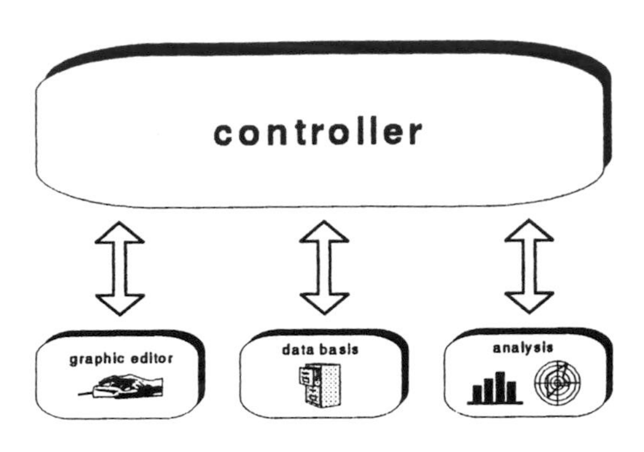

Figure 2. Controller

the optimization problem. They are described on an abstract level. In order to structure the problem we use some terms of the genetics. These terms got their popularity by the development of genetic algorithms.
A single solution (e.g. a job sequence) is described as an individual (Figure 3) consisting of one or more chromosomes. An individual for a job sequence problem in the field of electronic manufacturing can consist of an input job sequence and a queue job sequence. The input job sequence is encoded in chromosome 1 and the queue job sequence in chromosome 2. The chromosomes are defined by genes, and not the jobs are stored in the genes, but the sequence relations between the jobs. So the order of the genes in the chromosome defines a fixed input job sequence (chromosome 1) and a fixed queue job sequence (chromosome 2) without describing the problem as a job sequence problem. We don't have to tell the optimization algorithm, that he processes jobs. The algorithm is only interested in the knowledge, that there are elements having a fixed sequence relation. Several simultaneously existing individuals (several possible job sequences) form a population. It is imaginable, that a chromosome consists of only one gene

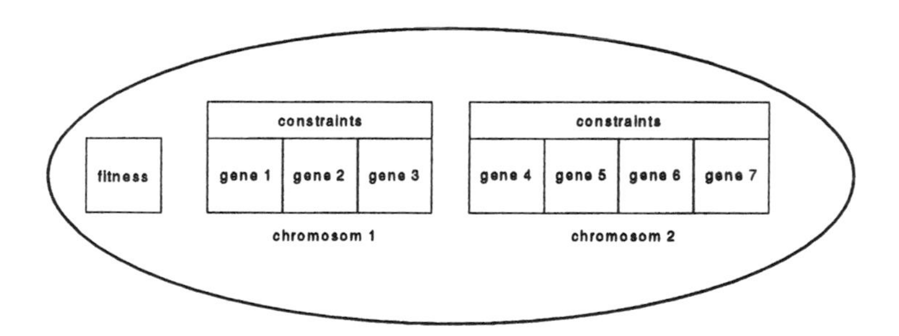

Figure 3. Example of an individual with chromosomes consisting of several genes

(Figure 4), which contains a real number. This real number can be changed by an evolution strategy operator. The open optimization system should be available to the solution of non-combinatorial problems. Optimization of technological processes is a part of these problems, for instance the optimization of laser cutting or the printing of solder paste before placing surface mounted devices in the field of electronic manufacturing. As the result the individual consists of several chromosomes with only one gene, each contains various parameter adjustments, which can have different results carrying out this technological process. The decision on the number

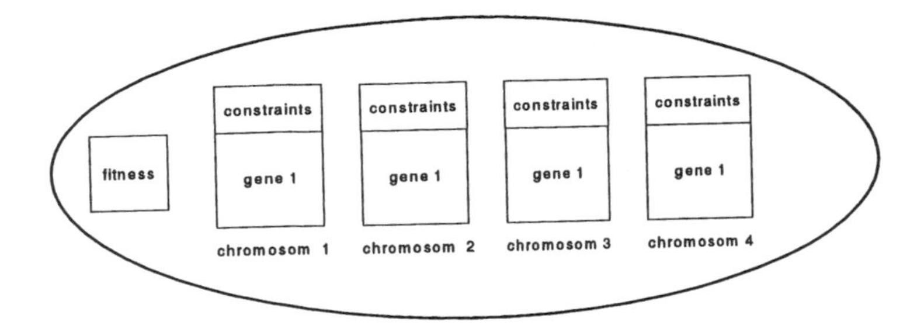

Figure 4. Example of an individual with chromosomes consisting of only one gene

of chromosomes depends on the problem, but important is in any case the application of the right optimization operator to the particular chromosome.
Besides, every individual contains a data field, in which one or more fitness values can be written after a simulation or can be read during an optimization. In addition every chromosome within an individual has a data field for constraints. Constraints are necessary, if some elements in a sequence have to hold together, or, in the case of technological process, to define limits of a parameter (real number). In this way we try to describe any problem (e.g. input job sequence, queue job sequence, machine arrangement, technological process) in the same manner.
The association between the single components (simulation, optimization, controller) and the conversion of the individuals in their original meaning is the task of the interface component (Figure 5). The creation of an interface to adapt different optimization algorithms or simulation

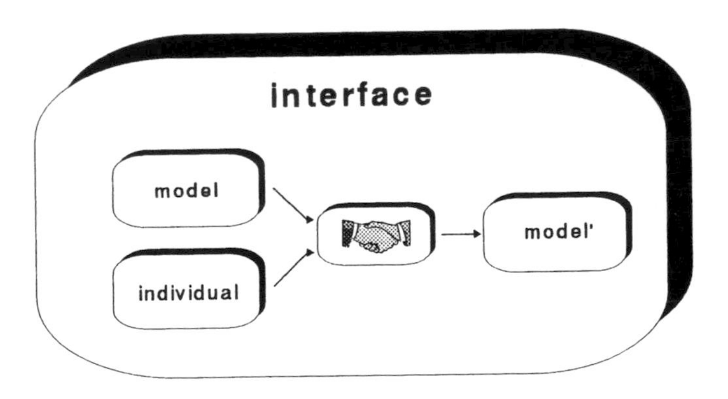

Figure 5. Interface

systems to the abstract description of optimization objects is the main part of the work, because every simulation system, every optimization algorithm or analysis tool has an own data format. But it is only necessary to program the interface once for every component. The advantage is the adaptation of all different components to the same abstract description. This is a guarantee for the simple and quick replacement of components. Once with an interface adapted simulation systems or optimization algorithms can be replaced within a component. We gained first experiences with the adaptation of the simulation system XROSI [4]. The simulation system has an own data representation in form of an ASCII-file. The jobs, the machines, queues and their associations are presented in this file. If we want to optimize the makespan by means of changing the input job sequence, we have to encode possible job sequences as abstract individuals.

In the genes the sequence relations, not the jobs, were stored. If we want to simulate concrete job sequences, we have to assign the encoded information of the individuals to their original meaning, in our case the job sequence. The individuals are associated with an existing model (Figure 6). In this manner a new updated XROSI file is created and simulated to determine the makespan as the fitness value. The individuals get their specific fitness value. Afterwards the controller transmits the individuals alone or in the population to the optimization component. If the optimization algorithm is adapted to the abstract description, there is no conversion of the individuals to an algorithm specific form necessary. If we use a commercial or a free available algorithm, we have to adapt this algorithm with an interface to the description of the individuals. The controller transmits during the optimization changed or new created individuals to the simulation component for the purpose of determination of the fitness. This cycle goes on till the user stops the data flow or a final criterion is reached. It is possible to visualize determined fitness values with suitable analysis tools.

An other example of a combinatorial problem in the field of electronic production, which can be solved with the open optimization system, is the placement of surface mounted devices (SMD) on printed circuit boards (PCB). Different placement ways and times result from different placement sequences. The possible sequences are described as individuals. The placement sequence is encoded in a chromosome, the genes store the sequence relations between the SMD's, not the devices themselves. So this description of the placement problem is equal to the description of the job sequences. It is possible to use the same optimization algorithms for the solution of both problems. The original meaning of the sequence is assigned to the individuals after the association of individual and the layout of printed circuit board (with the coordinates of the devices) to determine the fitness value, in our case the placement way or time. To do the association we have to program a specific interface (Figure 7).

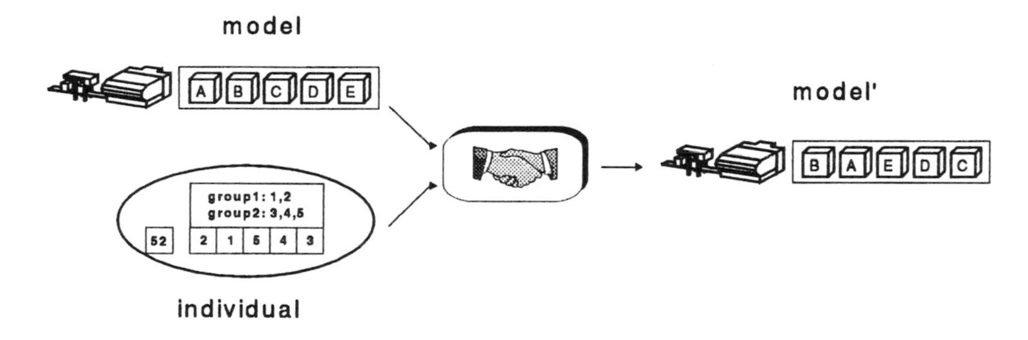

Figure 6. Job sequence - example of the interface

3 OPTIMIZATION OF JOB SEQUENCES - AN EXAMPLE

A prototype for an open optimization system was developed in our laboratory. [5], [6] For the simulation of manufacturing processes at first XROSI, a self made simulation program, was used. [4], [7] The optimization algorithms also were developed by ourselves at first. Both the simulation system XROSI and the optimization programs run as independent processes on UNIX machines. A simple Tcl/Tk-script controls and visuals the total optimization process. The different modules exchange their data by special ASCII-files.

Two optimization algorithms were implemented: Genetic Search Algorithm and Tabu Search. In

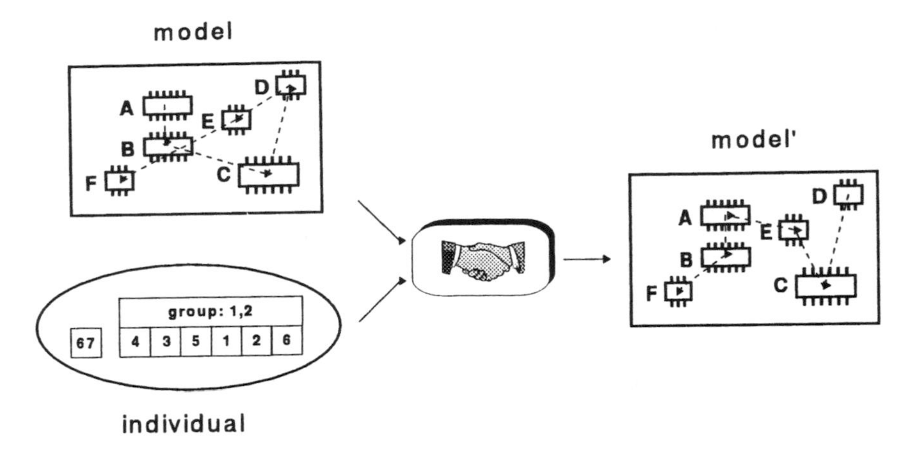

Figure 7. Placement problem - example of the interface

addition a simple Hill Climbing algorithm for job sequencing is included in the simulation system XROSI itself. These three algorithms were used for optimization of job sequences of several manufacturing systems, which were modeled by XROSI. Figure 8 shows the improvement of the makespan in dependence of the optimization cycles for all the three optimization algorithms for the well known 10 x 10 model from Muth & Thompson. [8] The variated quantity is the job sequence. Some more practically based models were examined too. We got the best optimization

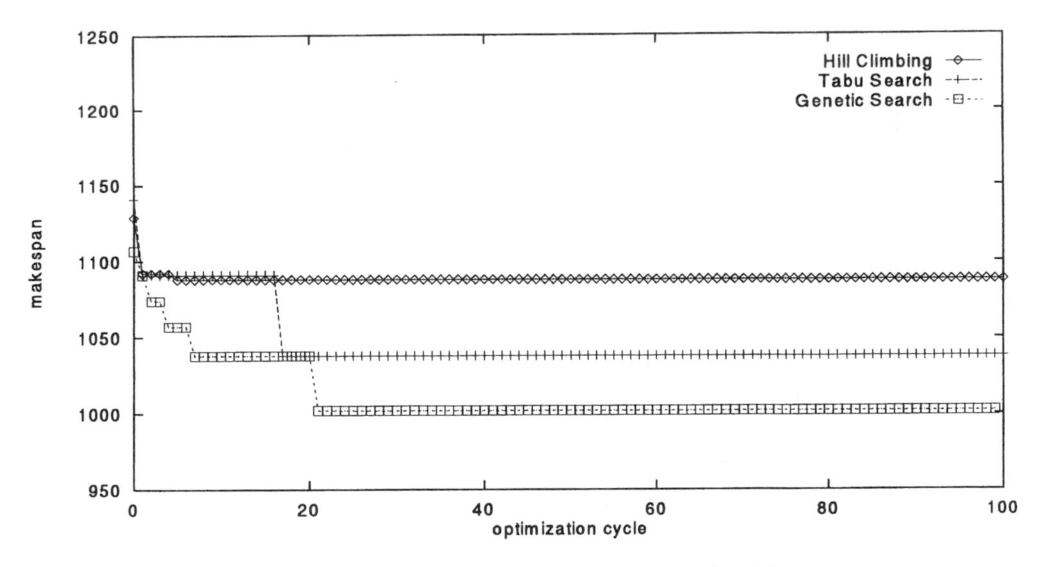

Figure 8. Comparison of three optimization algorithms

results with the Genetic Search Algorithm. In our opinion this algorithm includes the biggest potential of optimization especially for the solving of complex and NP-complete problems. That is why we use terms of genetic for describing our open optimization system. But of course it is possible to include in this system also other algorithms.

4 OUTLOOK

At present we prepare the application of algorithms like simulated annealing and evolutionary strategy. Besides we want to realize the integration of two simulation systems, Arena (Systems Modeling) and Simple++ (Aesop GmbH). In addition we want to create tools for presentation and analysis of optimization results and we think about the application of commercial analysis tools. In future we will examine problems with a number of objectives, but the description should go beyond a simple evaluation of the objectives.
At first we examined combinatorial problems, like job sequences. In future we will solve problems of technological processes. Often we can't model these problems, because they are very complex and have got too many parameters, therefore it is impossible to use the simulation as a tool to determine the fitness. That's why we determine the fitness in a real process by using the parameters getting from the optimization. The parameters are changed in the process. The quality of the process is converted into the fitness and is the basis for further optimizations.
First of all the open optimization system is an experimental system. We use it to find appropriate optimization algorithms and simulation systems to solve special problems or classes of problems, especially in the field of electronic manufacturing. After an experimental stage it is possible to adapt the open optimization system to the needs of a real user. On the basis of the optimization results the user can make his decisions quickly and well-founded to control his manufacturing system. In addition the optimization system will be open for the integration of further algorithms, simulation systems and analysis tools. This openness is necessary to react to new problems, especially in the field of electronic manufacturing.

References

[1] M. Pinedo. *Scheduling.* Prentice Hall International Series in Industrial and Systems Engineering. Prentice-Hall, Englewood Cliffs, New Jersey, 1995.

[2] J. Heistermann. *Genetische Algorithmen.* Teubner-Texte zur Informatik, Band 9. B. G. Teubner Verlagsgesellschaft, Stuttgart, Leipzig, 1994.

[3] J. H. Holland. *Adaption in natural and artifical systems.* Ann Arbor, MI: The University of Michigan Press, Michigan, 1975.

[4] W. Sauer, G. Weigert and P. Goerigk. Real time optimization of manufacturing processes by synchronized simulation. In *5th International Conference FAIM '95, Flexible Automation & Intelligent Manufacturing,* pages 271–282, Stuttgart, June 1995.

[5] R. Schulze. Kombinatorische optimierung in der elektronikproduktion. Master's thesis, Dresden University of Technology, Department of Electrical Engineering, 1996.

[6] G. Weigert and R. Schulze. Genetic algorithms schedule todays manufacturing process. In *2nd International Mendel Conference on Genetic Algorithms,* pages 192–196, Brno, June 1996. Technical University of Brno.

[7] W. Sauer, G. Weigert and P. Goerigk. A learnable simulation system for scheduling fms. In *6th International Conference FAIM '96, Flexible Automation & Intelligent Manufacturing,* Atlanta, May 1996.

[8] J. Muth and G. Thompson. *Industrial Scheduling.* Prentice-Hall, Englewood Cliffs, 1963.

AN UP-TO-DATE SHOP FLOOR MODEL FOR SHORT TERM PRODUCTION CONTROL

Prof. Dr.-Ing. Dr.-Ing. E.h. Hans Kurt Tönshoff,
Dipl.-Ing. Marc Sielemann

Institute for Production Engineering and Machine Tools,
Schloßwender Straße 5,
30159 Hannover,
Germany

ABSTRACT: Short range shop floor control is dependent on up-to-date information from the shop floor. In order to gain a detailed picture of the shop floor situation, a great effort is required in collecting this information. This is a result of the complexity of the shop floor system. The system elements such as machines, tools, fixtures etc. constantly change their state and their relationships to each other. The collection of all relevant events not only causes a great effort, it also requires a large investment in hard- and software components for data collection. But even if the technical prerequisites are given, incorrect or incomplete information can occur because of missing data collection discipline [1], [2]. In this paper, a system will be presented which guarantees that collected data is correct, complete, up-to-date and that the data collection is economic.

INTRODUCTION

Information on the present shop floor state is of elementary importance for shop floor control. On the one hand this information concerns resources such as machines and plants, tools and fixtures, on the other hand the progress of manufacturing orders. Up-to-date information is a prerequisite for an effective order submittance, for introduction of corrective steps in case of disturbances, as well as for disposition of tools. Availability of this data presupposes a functioning shop floor data collection system [3]. Because of the complexity of the shop floor environment, not only a great deal of effort during data collection has to be accomplished, but also the data quality can suffer due to poor collection discipline. This was the reason for initiating the project „an up-to-date shop floor model" which is funded by the German Research Council and carried out at the Institute for Production Engineering and Machine Tools, University of Hannover. The aim of the project is the development of methods which makes it possible to gain a complete and accurate picture of the shop floor situation. The present picture is to be provided by a computer supported system. The basis of this system is a model of the shop floor.

THE CONCEPT FOR GAINING UP-TO-DATE INFORMATION

Changes taking place on the shop floor have to be forwarded to the computer model. These changes effect the values of object attributes as well as the relationship between objects, e.g. tools are in storage at one point in time and used by a machine at another, machines are set, work on parts or are out of order.

If one analyses the production flow, different events and phases can be identified (figure 1).

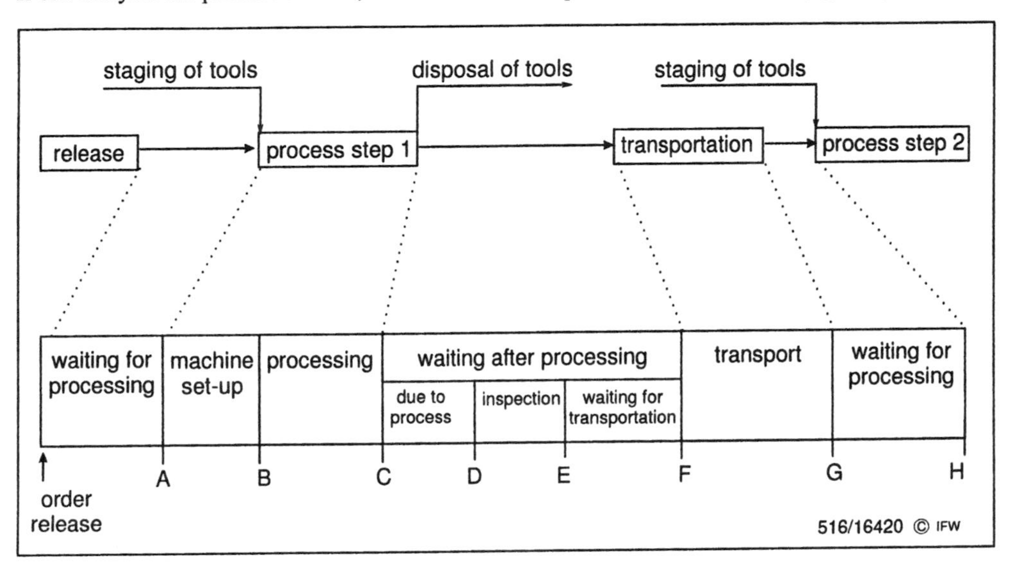

Figure 1. The course of events during manufacturing

The beginning and end of each phase, marked in the picture by a letter, is equivalent to a change of the shop floor state. For example, the end of processing (marked by a C) means that the machine and the used tools are available again. If the beginning and the end of each phase were to be registrated via shop floor data collection, it would be easy to determine the present state of the shop floor.

In order to guarantee a minimal collection effort, possibilities have to be found to gain further information on the shop floor environment. The two following possibilities can be used. The one possibility is simulation and the other is to use causal connections within the shop floor. The causalities can be formulated as meta-rules which can be used to derive further implicit information, based on feedback messages. The present shop floor state can be determined with the help of a computer system which applies both of these methods stepwise [4] (figure 2).

Within the first step, the feedback messages which have been collected since the last determination of the shop floor state, are evaluated. These feedback message are checked with respect to their plausibility in order to filter incorrect data. This is done by comparison with the data of similar orders processed in the past and by juxtaposition of feedback data with one another. Based on the remaining feedback data an approximate picture of the present shop floor situation is established. Further information implicitly contained in the feedback messages, is derived by applying meta-rules. For example, a feedback message which states that the third process step of a certain order has been finished, implicitly means that the first two steps have also be accomplished. Another example is, if someone notifies that a special tool has been stored, one can conclude that the corresponding process step has probably been completed (figure 3). In some cases, the progress of a order can be deduced when the progress of another order is known (figure 3). In this figure, order 1 as well as order 2 are to be processed on machine 5. Order 2 is an express order. The completion of the first process steps are notified from machine 1 and 4. If machine 6 notifies that order 1 has been completed, then it can be concluded that order 2 has been processed and is now at machine 3.

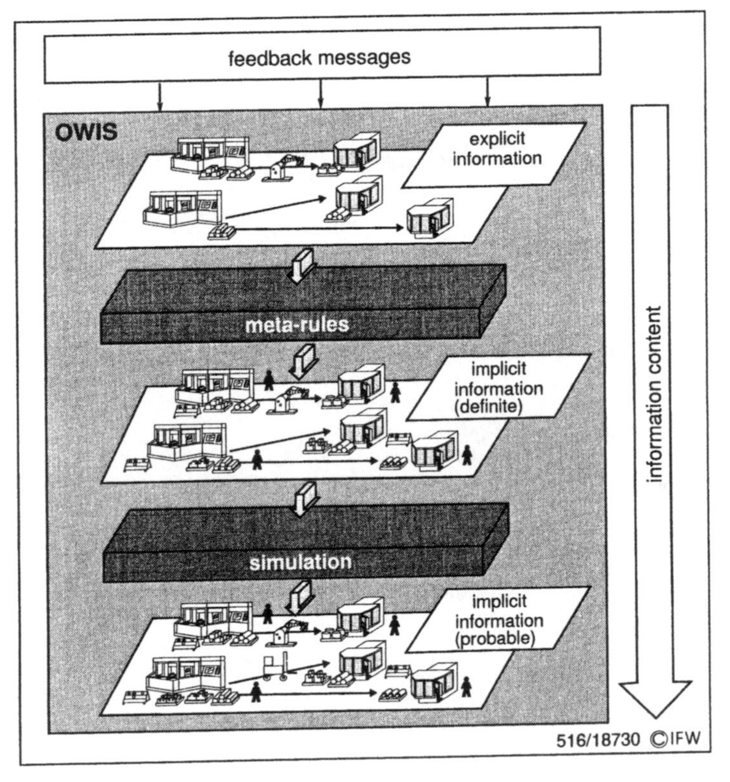

Figure 2. The approach for determining the present shop floor state

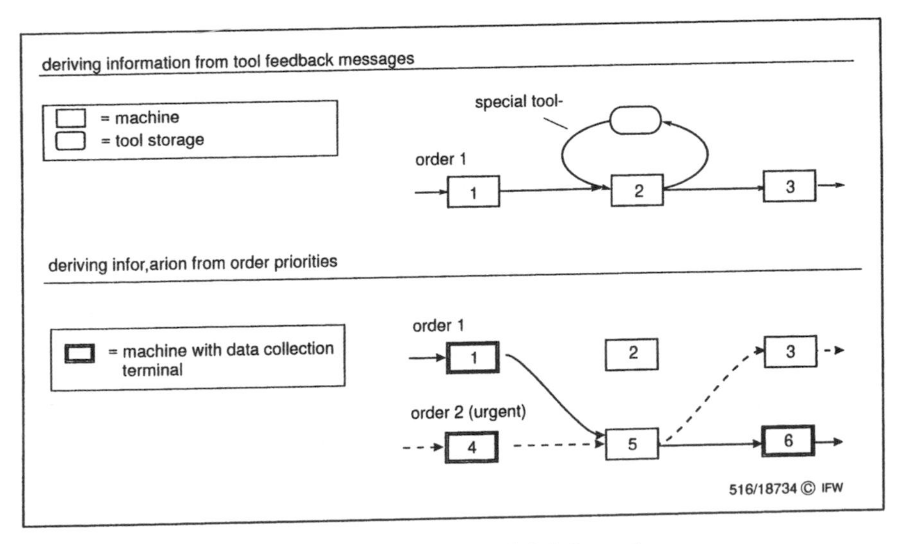

Figure 3. Deriving of implicit information

Further general meta-rules which can be applied to determine the shop floor state, can be formulated.

After the internal computer model of the shop floor state has been updated, as far as possible, with the help of the meta-rules, more information is gained by means of simulation. The simulation makes use of three information sources:

- the shop floor feedback data,
- the object states and relationships stored in the shop floor model,
- the orders which are being processed or are planned to be processed.

The first two sources provide constraints which have to be met during the simulation run. In contrast to the aim of most simulations which simulate into the future, an afterward reconstruction of the past production flow is attempted, with the aim of determining the present shop floor state. Because of the simulated reconstruction of the production flow missing feedback messages due to poor discipline are of minor importance.

As customary for simulation of the production flow, the work times for each of the process steps, specified within the orders, build the planning basis [5], [6]. Furthermore, the control strategies at the resources (first-in-first-out, shortest operation time, shortest set-up time etc.) as well as the transport strategies and times are considered. If the given times and control strategies were followed exactly and if the production flows were totally without disturbances, then the simulation run would depict the shop floor environment precisely. Reality is however obviously characterised by several disturbances. Operations are done slower or quicker than planned times. Manufacturing personnel make their own planning decisions. The simulation control has to consider this. The shop floor model and the collected feedback data serve as constraints. Every simulation step has to be checked for discrepancies between simulation results and the given constraints. If this is the case, the simulation has to jump back and the run has to be continued with a different alternative. This jumping back and renewed simulation is equivalent to the back-tracking mechanism known in rule processing. An example is shown in figure 4.

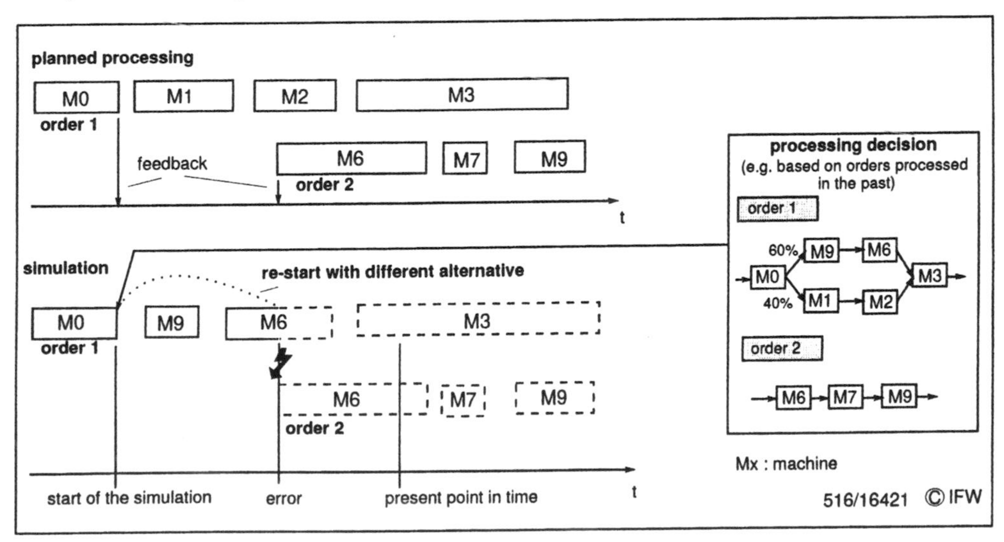

Figure 4. The back-tracking mechanism of simulation control

Looking at the figure, the planned processing of two orders is shown. In the bottom half the simulated processing is indicated. After work has been completed on machine 0, the simulation has to decide whether order 1 was manufactured on machine 9 or machine 1. Because of the past experience (see box „processing decision") it is first assumed that machine 9 was used. The order then goes to machine 6. Here the simulation violates a feedback message, which was submitted from machine 2. Therefore, the simulation must go one step back and continue with the alternative, machine 1. This procedure is repeaied until a consistent result is achieved. The result is characterised with probabilities in order to indicate its reliability.

An updating of the model is called for whenever planning or scheduling is to be done (figure 5). Between two points of updating, the state of the shop floor changes. The gap between the real shop floor state and the assumed one which is stored in computer model steadily gets larger. If the system is requested to give information on the present state of the shop floor, the feedback messages collected since the last update, are evaluated. The starting point is the last known state. If required, an update can even be carried out each time a feedback message is received.

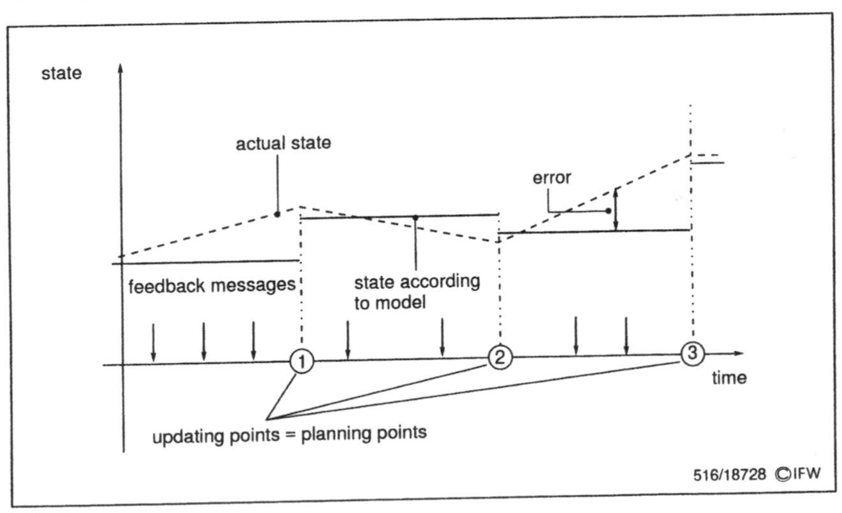

Figure 5. Points in time for updating

DETERMINATION OF FEEDBACK POINTS

With the presented concept, the prerequisites for reduction of the collection-effort has been established. However, the quality of the computer model information depends on the arrangement of the feedback points. They must be arranged so that the required collection effort and the quality of the derived picture of the present shop floor state are in an optimal ratio. At the moment, an adequate procedure for determining these points is being developed. In the following, the approach will be described, as far as it has been elaborated.

On determination of the feedback points, company-specific constraints have to be taken into consideration. Depending on these constraints, different strategies for arranging the points can be imagined. Companies with a constant product spectrum and throughput could install fixed feedback points. In the case of a varying production program, it can be better to determine new points for each production cycle and to use mobile feedback devices. Also, the combination of fixed and varying feedback points is a possible variant.

In order to guarantee a suitable collection effort, various points for feedback can be taken into account. These are independent of the installation of fixed or varying feedback points. For example, it makes sense to arrange them along the main flow of orders. A main order flow represents machine sequences, which often occur during order processing. By this means, a large amount of orders, on the shop floor, can be covered. Furthermore, those machines where orders cross, are favourable feedback points. Here the state of the preciding and of the following machines can be deduced. Independent of these denoted possibilities, the state of bottle-neck resources should always be known, because at these locations throughput times and stock can be strongly influenced [7].

For determination of main order flows and junctions, an analysis of certain representative orders and their underlying process plans, is necessary. The process plans not only contain machining and set-up times, but also the required machines. Because of the amount of data to be managed, a computer aided analysis is preferable. Most companies have process plans in electronic form. In a first step the plans are read and a list of the used machines is constructed. Based on this list, a transport matrix is generated. After this, the orders are analysed process step by process step. While doing this, the discovered transport processes are marked in the matrix.

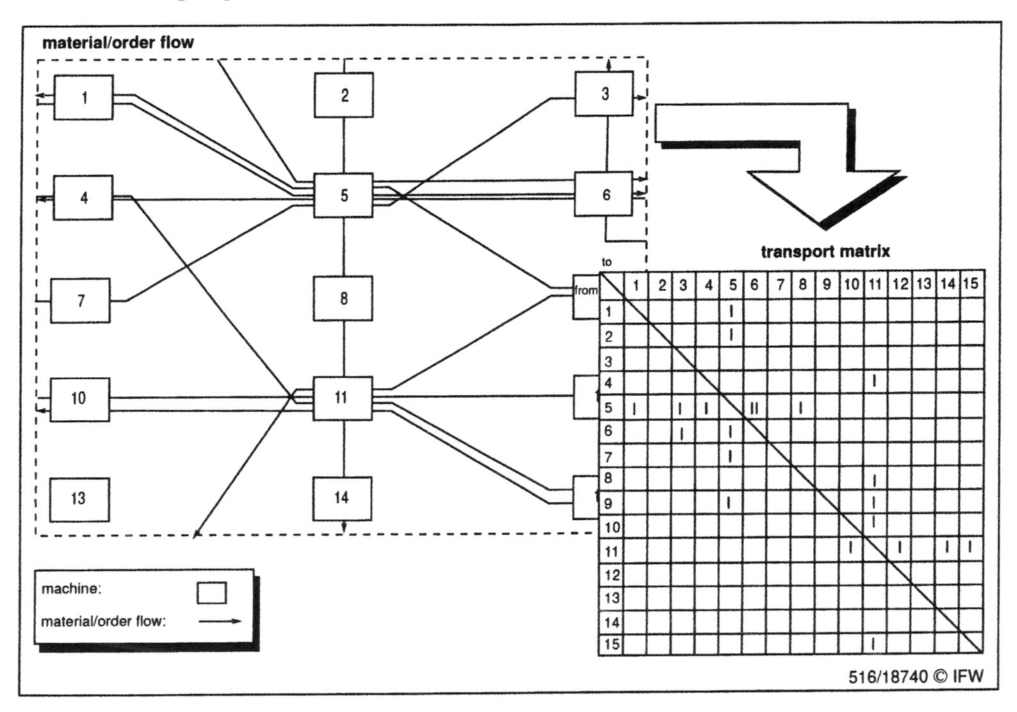

Figure 6. The determination of possible feedback points

The matrix fields, which contain a high amount of entries, mark preferable feedback points. Machines which are located at junction points are recognised due to the fact that the rows and columns are strongly occupied. Two cases have to be differentiated, when determining bottle-neck machines. In the first case, bottle-neck machines do not change. These machines are known from experience. The second case concerns varying bottle necks. Here simulation must be used. The basis is the same computer model which provides information on the shop floor. Beginning at the present state the future processing of orders is simulated. Planned orders are used as input. The machines which have a high load over a time period are normally the bottle necks. At all these machines mobile feedback devices can be set up.

The determined possible feedback points (or a subset of them) can be grouped to one or more arrangements, which are to be accessed. The assessment can be used to determine the most suitable variant, where there are cases of several arrangements. The arrangements are compared with respect to the collection effort, the fault vulnerability, the accuracy and the completeness of information. For this a procedure comprising several steps is foreseen. Within the first step a simulation is carried out.

The order data used to determine the feedback points, is used as input. Beginning at a known initial shop floor state, a simulation is run for a certain period of time. Every event which occurs during the simulation is recorded. These events simultaneously represent all theoretically possible feedback messages. For assessment of a certain arrangement, it is assumed that only the messages which would have been recorded at the planned feedback points are available. With help of the elaborated updating procedure (figure 2), the rest of the shop floor state is attempted to be determined. The completeness and the accuracy of the constructed picture, as well as the economy of collection, can be quantified for each point in time by comparison with the result of the simulation. The fault vulnerability can be determined by well-aimed fading out of individual feedback points and renewed determination of completeness and accuracy. The procedure is repeated for each arrangement so that the most suitable one can be determined. This effort is only advisable if fixed feedback points are installed. Due to the fact that varying points are determined periodically, the assessment and comparison of various arrangements is too costly.

TECHNICAL REALISATION

The whole concept needs to be implemented in a software system. Several demands are therefore put to the software system. The most important demand,is that the system must provide all possible information on the present shop floor state at any time. Prerequsite for this are possibilities to build a company-specific model of the shop floor. This model is catered for by means of shop floor data collection in connection with the developed methods. Therefore, an interface to the shop floor data collection is needed. Furthermore, the updating methods have to be implemented. For these methods, mechanisms for inclusion and interpretation of meta-rules as well as for simulation including an intelligent control have to be realised. Additionally, information on the planned orders (lot size, starting time, process steps etc.) is needed, so that an interface to MRP/Leitstandsystem must be available.

The realised system which is called „OWIS“, runs under the graphic user-interface WINDOWS, on a PC. Due to the experience made in the past, an object-oriented environment for implementing the system was chosen [8]. Data storage is based the object-oriented database system POET, programming was done with C++.

The OWIS system has a layer-like structure (figure 7).

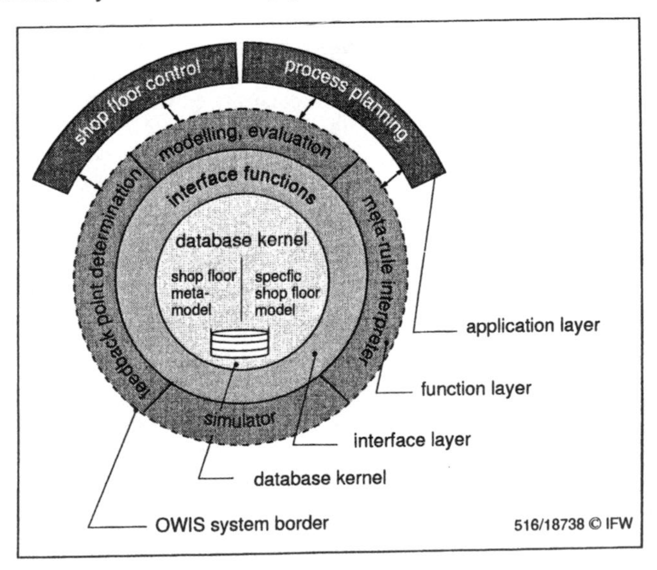

Figure 7. The structure of the software system

The database in which a shop floor meta-model and the company specific model is stored represents the system kernel. The meta-model contains the abstract structures and relationships of all possible physical and logical elements of reality to be modelled - i.e. the shop floor [9]. Within the object-oriented paradigm it is equivalent to the class schema. In this schema for example, the attributes which are used for the description of a machine or the principle relationships between a machine and a tool are specified. It is not possible to design a lasting all comprising meta-model. For this reason, a basic structure which can be modified by the user was defined. The classes, their behaviour, their attributes and relationships to each other, are on the one hand derived by analysing the tasks of shop floor control, on the other hand by examining the course of events during order processing. The shop floor meta-model is the basis for creation of a company-specific shop floor model. This means, the model of a certain shop floor is created by instantiating the classes of the meta-model. This way, the design effort is reduced to a minimum.

The stored model data can only be manipulated via a defined interface layer, so that its consistency is guaranteed. The communication with external systems can also be done by means of this interface.

In the function layer, function units are arranged. These are units which serve for model design and evaluation, updating, as well as for planning of the feedback point arrangement.

The design and evaluation unit allows the interactive graphic design of the meta-model and of the company-specific model of a concrete shop floor (figure 8). The classes of the meta-model are represented by grey shaded rectangles, the relationships between classes by the connecting lines. Every class can be assigned as many attributes as needed for characterisation. Three different types of relationships between classes can be differentiated: „is-a"-relationships, „is-part-of"- relationships and associations. With these three relationship types, all imaginable relationships which exist in reality can be expressed. The „is-a" relationship enables the expression of hierarchical relationships between classes. By means of this relationship, one can for example express that the class „turning machine" is subordinate to the class „machine". Subordinate classes are also called derived classes. The hierarchical structure built by this kind of relationship leads to a structured data stock. Easy searching of stored data is facilitated. The principle of inheritance is directly coupled with the „is-a"-relationship. For example, every attribute which belongs to the class „machine" is inherited by the class „turning machine" and can be used for description. „Is-part-of" relationships serve for depicting physical or logical relationship between classes. This kind of relationship is suitable for expressing that a machine consists of several components, like tool changer, body, table etc. This type of relationship also conveys a hierarchy. „Is-part-of"-relationships are inherited to derived classes. Associations describe loose relationships between classes. In contrast to the other relationship types, this relationship does not express a hierarchy. It connects equal valued classes. Therefore, this relationship is the weakest one. Nevertheless, the expressive possibilities are very powerful. With this kind of relationship, one could express that certain tools can be used on a certain machine. Associations, the same as „is-part-of" relationships, are inherited to derived classes. Apart from this, both can be characterised with cardinalities.

During design of the company-specific shop floor model, objects of the classes are created and relationships between objects are established. The attributes and relationships are specified according to the present shop floor state. All further changes of the shop floor are traced in the model by means of shop floor data collection and the implemented updating mechanisms.

In order to receive the present shop floor state, the updating procedure is called for. Updating effects the values of the object attributes, as well as the relationships between objects. For this, the feedback messages collected since the last model update are evaluated. In a first step, the messages are checked with respect to their plausibility. This is done by juxtaposition with same or similar orders, processed in the past. Suspicious messages are sorted out. In order to filter out further faulty messages, they are compared with one another. For example, feedback messages which are incorrect due to their technologically required sequence of process steps, are identified. For processing of the remaining messages the meta-rule interpreter is run and afterwards the simulation is started.

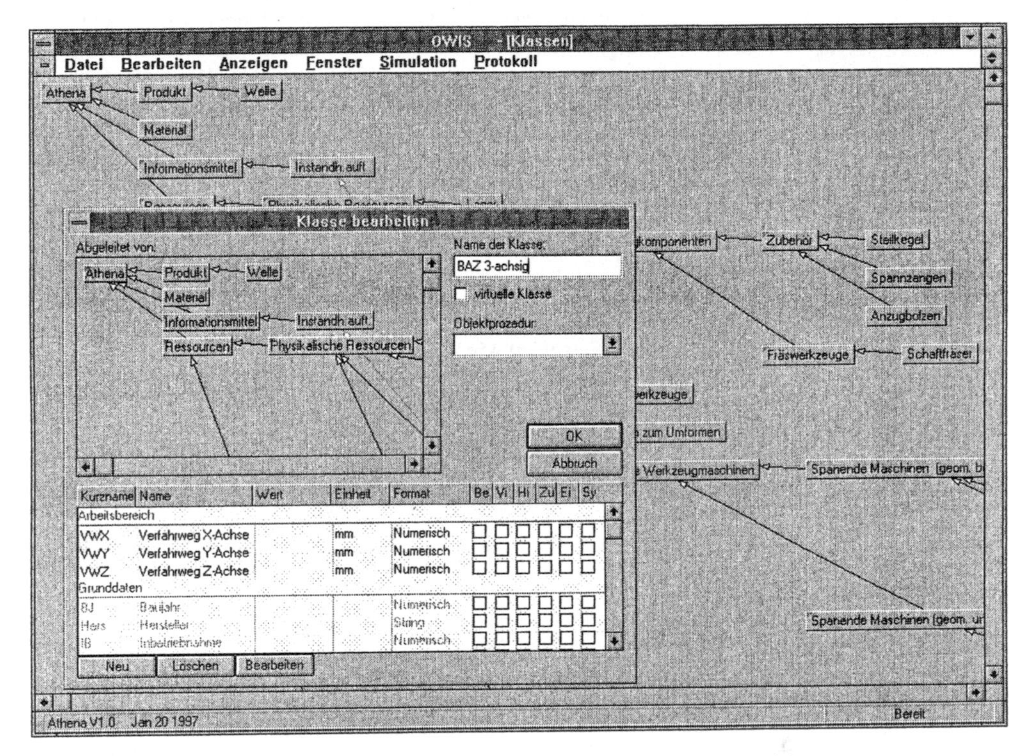

Figure 8. Graphic interface for design of the meta-model

Meta-rule interpreter and simulation where realised by giving the objects stored in the database, the possibility to receive messages and to send messages themselves. In order to facilitate this mechanisms for transfer of messages were established. Additionally, the concept of object procedures was introduced. The object procedures which have to be coded in C++, can handle one ore more messages. The way an object reacts, is controlled by the message code. For each code, a corresponding routine within the object procedure must be foreseen. Here object attributes and relationships can be changed and new messages can be generated. The procedures must be designed as Dynamic-Link-Libraries according to a specific format and be kept in the working directory of OWIS. When the system is started, the working directory is searched for procedures. They can be assigned to classes and therefore, to all objects of the class, for description of their behaviour during design of the meta-model. This shows that the strict distinction between simulation and meta-rule interpreter, as described in figure 8, does not really exist, because the functionality is embedded in the object procedures.

The call for a model update, first leads to a chronological processing of the feedback data collected since the last update by the meta-rule interpreter. Feedback messages are structured so that they possess a emitter, a receiver and a feedback message code. According to the feedback code and meta-rule respectively, the corresponding routine is executed. This can lead to a change of object attributes and relationships, as well as to the sending of messages. Feedback messages have to be delivered by the data collection system. OWIS can either be directly connected to the system, via the interface layer, or read feedback data from an ASCII file. The feedback messages which have been accumulated since the last update are then processed and stored in a special class within the OWIS-system.

Similar to the execution of meta-rules, the simulation is done by the objects stored in the database. The data of the orders which have been released since the last model update, are needed as the basis for simulation. Feedback messages and the data derived, with help of the meta-rules, serve as constraints. The simulation begins by sending an initialising message to all objects. These can react to this by changing object attributes and relationships, or by sending further messages. The simulation then runs almost autonomous, due to the initiated message exchange. An intervention only takes place, if the shop floor state, determined by the simulation, violates the given constraints. In this case the simulation control must go back one or more steps and continue with a different alternative. The result is an up-to-date model of the shop floor. The state of the objects can be called with help of the graphic modeller.

Furthermore, there is a unit for determination of suitable feedback points. It does the analysis of order data and allows the assessment of feedback point arrangements.

FUTURE PROSPECTS AND CONCLUSIONS

Substantial benefits result from correct and complete shop floor data collection. [10], [11]. With the elaborated approach and the realised computer system, the prequisites for guaranteeing an economic and exact shop floor data collection, for short time production control, are established. Feedback messages are checked with regard to their plausibility. In the next step further implicit information is derived by means of meta-rules and simulation. The result is an up-to-date model of the shop floor. The present shop floor state is stored in a database and can be viewed with help of the developed software system. The quality of the stored data, representing the picture of the shop floor state, as well as the collection effort, depend on the arrangement of feedback points. For this reason a concept for determination of suitable points and for assessment of different arrangements was developed.

The aim of future work will be to analyse further aspects with respect to determination of feedback points. Examples are the amount of centrally stored tools or the existence of DNC. In case of a large amount of centrally stored tools it can be suitable to use the tool storage as a feedback point. The identity number of the tool and the process plan can be used to deduce the manufacturing progress. If DNC is used, the loading of NC-programmes can simultaneously serve as a feedback message for the beginning of machining.

REFERENCES

1. **Schmager, B.,** Betriebsdatenerfassung und -verarbeitung als Instrument der betrieblichen Informationsgewinnung, *Disseration, Technische Universität Hamburg-Harburg*, 1989.

2. **N.N.,** Automation the key to higher accuracy and lower costs, *Manufacturing Computer Solutions*, Vol. 1, No. 7, September 1995, p. 48-49.

3. **Roschmann, K.,** Trends und aktuelle Entwicklungen bei BDE-Systemen. *CIM Management*, 10 (1994), Heft 4, S. 6-10.

4. **Tönshoff, H.K. and Sielemann, M.,** Reduzierung des Rückmeldeaufwands in der Fertigung, *Zeitschrift für wirtschaftlichen Fabrikbetrieb,* Jahrgang 92, Ausgabe 1-2, 1997, S. 42-44.

5. **Tönshoff, H.K. and Eblenkamp, M.,** Simulation and Optimization in Production Planning, *Production Engineering*, Vol. 1/2, 1994, No. 1, p. 165-168.

6. **Novels, M. D. and Wichmann, K. E.,** Simulation applied to production scheduling, *Second International Conference on Factory 2001 - Integrating Information and Material Flow Conference*, Cambridge, July 10-12, 1990, Publ. by IEE, Michael Faraday House, Stevenage, p. 130-134, 1990.

7. **Goldrat, E.M. and Fox, R.,** The Race, *North River Press, Inc*, 1986.

8. **Tönshoff, H.K. and Dittmer, H.,** Object- instead of function-oriented data management for tool management as an example application, *Robotics & Computer Integrated Manufacturing*, 1990, Vol. 7, No. 1/2, p. 133-141.

9. **Tönshoff, H.K. and Witte, H.H.,** Produkt- und Werkstattmodell für die Planung und den Betrieb von Produktionsanlagen, *Modellbasiertes Planen und Steuern reaktionsschneller Produktionssysteme,* Tagungsbericht, 15.-16.10.1991, München: gfmt.

10. **Kahan, S.,** Implementing product monitoring and control, *Manufacturing Systems*, Vol. 8, No. 3, March 1990, p. 20-22.

11. **Keene, A.,** Automatic data, *Manufacturing Engineer*, Vol. 74, No. 3, June 1995, p. 131-133.

INTEROPERABILITY AND ITS IMPACT ON RAPID DEVELOPMENT OF PROCESS CONTROL SYSTEMS

Hua X. Du, Jun Wei and Chen Zhou
School of Industrial and Systems Engineering
Georgia Institute of Technology
Atlanta, GA 30332-0205
USA

ABSTRACT: The manufacturing control systems are complex. In order to develop manufacturing process control applications rapidly, practitioners rely on the existing commercial enable software such as manufacturing execution systems, supervisory control and data acquisition systems. The enable software has evolved to provide powerful application development tools such as drag-and-drop graphical programming environment. However, the enable software must be interfaced with equipment controllers and other applications since single product is rarely sufficient to support all the functionality required by a process control application. Therefore, the ability of equipment controllers and enable software to interoperate with other components determines the effort required for system integration. This research is to address this ability, also referred to as interoperability. The ability is defined in general terms of service levels and is related to the effort required to achieve categories of process control applications. The methods for qualitative analysis of interoperability will be presented. The method can be used to estimate effort required in system integration in a systematic manner, to provide reference in the development of new process control systems, and to develop reusable software modules. The methods are illustrated with two case studies at the Manufacturing Research Center in Georgia Institute of Technology.

1. INTRODUCTION

In advanced manufacturing systems, the cost of integration is commonly higher than the cost of hardware and software combined [MESA, 1996]. Due to proprietary nature of equipment interface, software interface, and the lack of an adopted open communication standard for process equipment [Nguyen, 1993], a large portion of the effort in process control applications is in the development of special interfaces and controller dependent softwares in integrations. The special interface and software result in four major problems. The first problem is related to the high costs and long lead times in initial system launch. The second problem is the poor flexibility for products, processes or control changes. The third problem is the difficulty in the system maintenance [Smith, 1994] because of the complexity and poor documentation often associated with application dependent solutions. The fourth problem is that the resultant application is not commonly reusable, portable or scaleable and results in repeated development of modules of the identical functionality.

Some of the approaches in the rapid development of process control systems are:

- standardization
- open architecture system design
- object oriented technology
- the development of general purpose application enable software

In standardization, the Manufacturing Message Specification (MMS) in Open System Interconnection (OSI) reference model, Semi Equipment Communications Standard (SECS) and Generic Equipment Models (GEM) from SEMI International, Object Linking and Embedding (OLE) and Dynamic Date Exchange (DDE) from Microsoft, and various fieldbus standards are some of the emerging standards that can potentially enhance interoperability of equipment and software components. Standardization can allow the development of reusable, scaleable and general purpose applications because of known behaviors of the interface partners, and can accelerate the development of process control systems. SEMATECH estimates that standard equipment interfaces will reduce equipment integration time by a factor of 5 and reduce equipment integration costs by a factor of 4 [Dinnel, 1994]. Motorola has estimated that the use of GEM standard can port over 90% of the cell control software to other applications (Hawthorne, 1995). Standardization may also pose interface restrictions, excess overhead and restrict new technological advancement.

The equipment users and software users have long voiced desires for open architectures. Although the open architecture may mean different things for different people [Robotics World, 1996], many vendors have started to provide some level of access to their product under the pressure from the user community [Owen, 1996]. Currently, a consortium with National Institute of Standard and Technology for Enhanced Machine Controller (EMC) represents one such effort in developing open architecture controllers for the machinery.

The object oriented technology has been an important driving force in rapid development of process controls. The objects in object oriented systems are loosely coupled through the message passing. Such scheme provides better modularity and reusability of resulting control softwares and therefore accelerate the development of process control systems.

The various software enablers provide tools or platforms for the rapid development of process control systems at a conceptual level. Two categories of such enablers are manufacturing execution systems (MES) and supervisory control and data acquisition (SCADA) systems. These enablers provide most general purpose services such as statistical process control, man-machine interface (MMI), data manipulation and data base management. More importantly, users of these enablers can tailor the system to a specific application with simple drag-and-drop manipulations and simple data replacements. Normally, MES is not connected to manufacturing equipments directly, partially because the interface of equipment controllers is often proprietary. SCADA, on the other hand, often supports various drivers for different equipments. One can guess that the driver set can never be inclusive. It is highly likely that a driver is not available for an equipment one wants to integrate.

A review of the work done so far on the interoperability indicated that the previous researches focus on a specific communication protocol or individual software testing and analysis. The interoperability testing verifies that products do actually interwork. Two products may conform to standards, but still not be able to effectively interwork, and two products may interwork but not conform to standards. Bonnes describes the testing of OSI communication protocol [Bonnes, 1991]. Dssouli and Fournier present a new architecture to measure the communication software interoperability [Dssouli and Fournier, 1991]. In this research, we will investigate the meaning of interoperability in the development of process control systems. Specifically, we will study the effect of SECS/GEM standards and the use of commercial enablers.

2. INTEROPERABILITY

Today, users of controllers demand "plug-and-play" connectivity. The term interoperability is often associated with such demand. In Computer Dictionary [Spencer, 1993], *Interoperability* is defined as "the ability of equipment from a variety of vendors to share data and communicate effectively". In real world controllers, the communication effectiveness can vary. For example, a controller can support a standard interface port, such as RS-232 or Ethernet. What one can operate and the amount of effort required for one to operate through the port may be rather limited and may require significant efforts. On the opposite end of the scale, a controller can provide an interface that works with many SCADA software products to accomplish monitoring or manipulation with minimum requirements from the user. So, interoperability should also include services that can be exchanged between connecting components.

It is desirable to qualify the interoperability in more concrete terms that can be used to assess the effort required in system integrations. The qualification should be generic in the sense that it is not restricted to a specific controller or system. The qualification should also be closely associated with process control in manufacturing rather than internetworking or software engineering. The qualification should be well defined so that connecting components can be qualified without ambiguity.

Modern manufacturing equipment controllers are often referred as "embedded systems" or "intelligent systems" because of built in computers in the controller. The interoperability from another component in the system (a host or another intelligent equipment) refers to the ability to exchange data and/or to exchange services. The content of exchange can be raw data that has no meaning without further processing. It can also be data that are readily processed because of known formats. It can even be services that involve the manipulation of process and data from another component. Therefore, interoperability is a complex issue that can not be concrete with simple definitions.

Open system interconnection (OSI) reference model and TCP/IP are well adopted models in network connectivity. OSI and TCP/IP specify the internetworks in a layered structure. The layered structure provides a clear partition of functionality and abstraction of complex internetworks. Since interoperability is also a complex issue it should also benefit from a layered framework. However, the concerns of a system integrator differ from that of an internetwork designer. Whether two components in a manufacturing system can interoperate has less to do with complex internetworking because the components are often connected in point-to-point communication or a simple local area network (LAN). Therefore, the middle layers in the OSI layers associated with the control in complex internet is often of little concern to system integrators. On the other hand, what two components can interoperate and how they interoperate are the major concerns. These are lumped into the application layer in OSI and TCP/IP models and should be distinguished clearly in interoperability framework.

We choose the term "level" instead of "layer" in this paper because it has a connotation more closely related to the effort required to accomplish certain objectives. We also use the term "component" to represent an intelligent controller, a host or a piece of control software for interoperability qualification. We use the term "partner" to refer to a component that is to interoperate with another component for qualification.

2.1. The Levels of Interoperability in Process Control

In our definition, the interoperability is ranked in four levels shown in Figure 2.1.

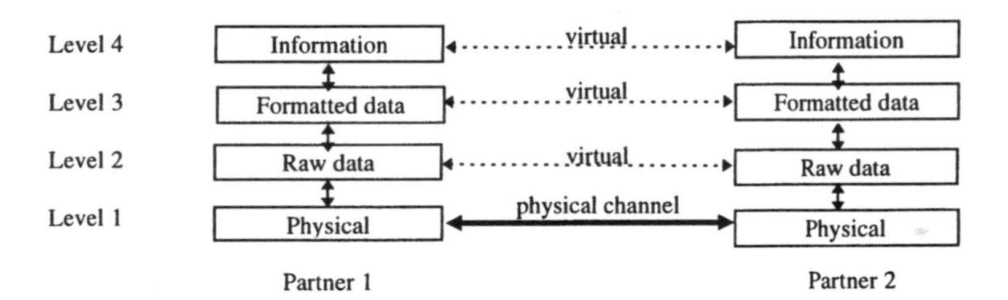

Figure 2.1 Interoperability framework

A minimum level for a component to share data or exchange services with its partner is a common physical link. The physical interface should define the mechanical and electrical characteristics of the connection as defined in OSI or TCP/IP physical layer. This can be an RS-232 port, an Ethernet card, etc. A system integrator is often delighted to see a simple existence of a common communication port. Such port may provide many interoperation possibilities with the programming. The physical link is a big step towards interoperation.

A better connection between two components will provide access to raw data. This level is related to OSI data link layer through session layer in OSI model and through part of the application layer in TCP/IP model. Examples for possessing a data level interoperability would be SECS-I in SEMI standard, streams in TCP/IP, FTP application in TCP. For a better description of SECS-1, please refer to the descriptions in Case II in Section 3. At this level, the components should be able to exchange raw data with its partner without error. Since this level of interoperability only supports raw data without meanings or formats, data parsings and usages must be performed by user applications.

At the third level, formatted data can be transferred. A formatted data such as recipe files or commands can be accepted into applications in the partner. No tokenizing, parsing or categorizing are required. The formatting eliminates tedious and duplicate software developments in applications. With the third level support, users can focus on the storage, retrieval and use of data rather than manipulating data. The standard recipe file format by the SMEAMA is intended to be protocol at such a level. The SECS-II message definitions are also examples of the formatted data level interoperability support.

At the information level, the formatted data in the component is meaningful to its partner for actions or services. For example, the observations of statistical process control data are formatted data (can be unformatted too). However, the mean and variation, and their relationships with control limits are information because they can be used for automatic or supervisory control. Another example will be is the SECS-II messages associated with GEM models [SEMI, 1995] versus a user defined SECS-II message. GEM defines a number of process behaviors of semiconductor manufacturing equipment in state models [Harel 1987]. For more detailed presentation of GEM model, please refer to Case II in Section 3. The fixed messages in manufacturing message specifications (MMS) can also be considered as the information because of its definite meaning in process controls. Another example on information level is automatic statistical process control (SPC). In SPC, the process data are processed and compared to control limits for process regulations.

This definition of interoperability reflects the level at which components can interoperate. It is reasonably generic, concrete and can be linked to the ability for the data sharing and service exchange between components in a manufacturing system.

2.2 Interoperability Requirements and Effort Requirements in Applications

In this section, we will first develop a general model for estimating the effort required for a control system. Then, we will identify the required level of interoperability in various applications and modify the general model to relate to real applications better.

In a process control system, there exist many hardware and software components that require interoperation. The hardware components can be an equipment controller or a host, the software component can be recipe files (programs running in the equipment controller), a data base management system, an application enable system (such as MES or SCADA), or a user program in various languages. The interoperations are connected with the shared data. Here, the shared data can be files, objects, strings, or a simple on/off signal. Let,

d: $d = 1, 2, ..., D$ be dth shared data set out of a total of D that require interoperability level improvement to achieve application goals.

c: $c = 1, 2, ..., C_d$ be cth component in the dth data set.

i: $i = 1, 2, 3, 4$ be ith level of interoperability.

$E(i, c, d)$: Effort function representing the effort to achieve ith level of interoperability for a component when $i - 1$ *th* level of interoperability is available, where $i = 2, 3, 4$.

$a_{(c, d)}$: $a \in \{1, 2, 3, 4\}$ the levels of available interoperability for component c in data set d.

$r_{(c, d)}$: $r \in \{1, 2, 3, 4\}$ the level of required interoperability for component c in data set d.

E: total efforts required to achieve the required level of interoperability.

$$E = \sum_{d=1}^{D} \sum_{c=1}^{C_d} \sum_{i=a_{(c,d)}}^{r_{(c,d)}} E(i,c,d)$$

The definition of d implies $a_{c,d} < r_{c,d}$. If $a_{c,d} \geq r_{c,d}$, no effort is required for that set of data in system integrations.

A specific application may require different levels of interoperability. We identified four categories of process control applications that have different requirements on the level of interoperability:

1. simple process monitoring
2. computer assisted process monitoring
3. process monitoring with manual controls
4. process monitoring with automatic controls (interlock)

In the simple process monitoring, the monitoring partner simply gets raw data from its interoperating partner and presents it to the operator in its raw form. An example of the simple process monitoring is host monitoring of nozzle vacuum pressure of a placement machine. The raw data of all nozzles coming from the equipment roll up in a text window on the host. It is up to the operator to find any abnormalities in fast scrolling data and make decisions. In the computer assisted process monitoring, the monitoring partner formats the data and presents it to the operator in a formatted (a table for example) or even graphical form. It is much easier for the operator to monitor the process in this way compared to the simple process monitoring. In the same example with computer assisted process monitoring, the nozzle pressure for different nozzles can be depicted in different locations. In this way, the operator can easily associate pressure with individual nozzles. In the process monitoring with manual controls, not only the monitoring component gets the data, it will take the decision from the operator and feed it back to the partner to be monitored for manipulation. In the process monitoring with automatic controls, the feedback loops in the process monitoring with manual controls is closed automatically. The interoperability requirements for each of these application categories are presented in Figure 2.2.

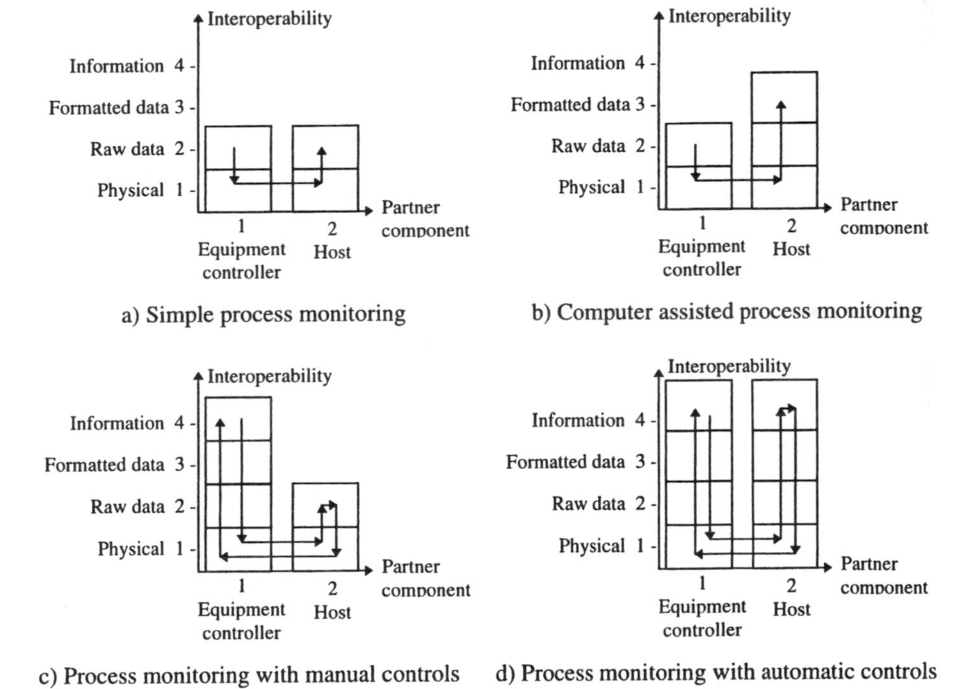

Figure 2.2 Required level of interoperability of four process control applications

The interoperating pairs in Figure 2.2 are special cases between an equipment controller and a host. In reality, the interoperating pairs can be other components in a system, such as MES, Visual Basic user application software or a data base management system. An important factor affecting the ability and convenience in integration is the programming platform. Assembler is powerful. It can be used to program a physical link to get the raw data. However, it requires significant efforts. However, the use of the assembler to create a graphical man machine interface (MMI) becomes too great of a task. On the other end of the scale, the use of MES or SCADA system to create MMI is very convenient involving mostly drag-and-drop and some prompted parameter entering. However, it is almost impossible to program a physical link to get the raw data in this platform. For an equipment, the interoperability can be programmable from process control program or from the environment. It is commonly easier for integrator to improve interoperability through process control program (recipe). Therefore, we categorize the partner platforms into the following six groups.

1. equipment: environment
2. equipment: user program
3. host: assembler
4. host: general purpose programming language
5. Host: fourth generation programming language (4GL)
6. host: MES or SCADA systems

There are other issues related to the effort concerning process controls and interoperability. First, there is the issue of the real-time control versus the supervisory control. In Oxford Dictionary of Computing, real time control requires "The lag from input time to output time must be sufficiently small for acceptable timeliness". Young (Young, 1982) defines the real time system that "has to respond to externally-generated input stimuli within a finite and specified period". The key concept in real-time control is specified and quick timeliness response. The second issue is the on-line control versus off-line control. The on-line process control is process control when the process is in progress. The fourth issue is the data sharing protocols: who is the initiator, event driven versus polling, etc. We leave these details in the actual analysis rather than categorizing them at this time.

The existing and required interoperability in one process control function (such as monitoring of one parameter) can be viewed as a function of two communicating partners in Figure 2.3.

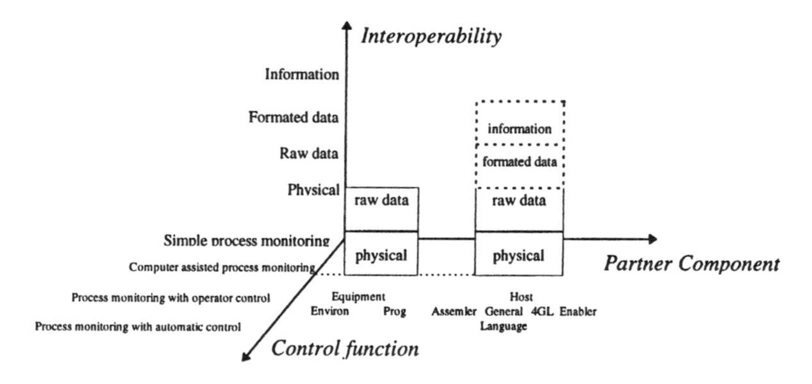

Figure 2.3 Parameters and effort requirements in process control applications

Figure 2.3 shows the operator assisted supervisory control. The solid boxes represent the available interoperability between an equipment and a host running Visual Basic. Two dashed boxes represent the required effort in operator assisted supervisory control.

3. IMPACT OF INTEROPERABILITY ON INTEGRATION

We present two case studies to illustrate the impact of interoperability on the effort required in system integrations. In both cases, the intent was to rapid prototype the control application for the system in laboratory environment. The first case involves a robotic machining cell. The cell consists of educational manufacturing equipment with controllers that resemble the state-of-the-industry control characteristics. The information sharing is minimum for our very basic control objectives. The second case involves a printed circuit board assembly line using state-of-the-art industrial equipment. The information sharing is more substantial for more sophisticated control objectives.

3.1 Case I. Process Control in a Robotic Machining Cell

The robotic machining cell contains a Light Machine SPECTRA mill, a SPECTRA lathe, a SCORBOT ER-V robot and a SPARC SUN workstation as a cell controller. The schematic of the cell is shown in Figure 3.1. The cell is able to cut various patterns on two types of raw materials coming in from gravity feeders. The control objective is to sequence a known set of jobs to minimize cycle times or idle times in a finite horizon on line. In order to find the sequence and execute the scheduled sequence, the cell controller must monitor the progress in the cell. The minimum information requirements include part lists, part descriptions, the completion times of the robot and machines.

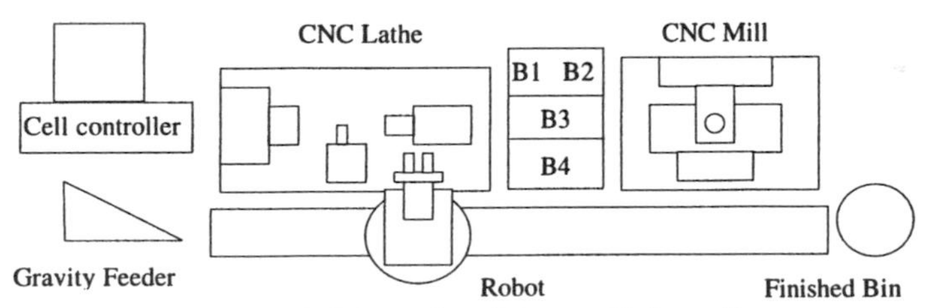

Figure 3.1 Schematics of the machining cell

The cell controller runs a SCADA enabler CELLworks from FASTech Inc. The user program in the cell controller can be partitioned into three modules:

- monitor
- executor
- scheduler

Part of the data sets, interoperating partners and data channels are described in Table 3.1.

Table 3.1 Data set, interoperating partners and data channels in the machining cell

Data sets	Partner 1	Partner 2	Data channels
Robot Completion	Robot Program	Cell controller Monitor	Robot program sends messages to monitor through RS-232 interface
Robot Instruction	Robot Environment	Cell controller Executor	Executor sends message to robot environment in the form of "run program_x"
CNC Completion	CNC Program	Cell controller Monitor	CNC sets a high signal through one of its DO to robot, robot then sends a message to monitor
CNC Instruction	CNC Environment	Cell controller Executor	Executor downloads a NC program to CNC, CNC starts execution after downloading (DNC)
Part Information	Cell controller Scheduler	Database System	Scheduler accesses a database system established based on CELLworks

The detailed description of these modules can be found in Jun and Zhou (1996). In the process of monitoring robot operations, the available interoperability at both sides is at the raw data level because the meaning of the message must be defined and interpreted by the integrator. The available level of interoperability for monitoring the CNC is indirect but at the same level. In the process of executing instructions, the executor simply sends unformated data to the receiver and the receiver can fully understand the predefined commands or NC codes. Therefore, the sender is at the raw data level while the receiver is at the information level.

The scheduler is internal to the cell controller. It accesses the developed database system for the jobs information, the status of the machines and robot collected by the monitor module, and passes its decisions to the executor module for executing. The links between components in CELLworks and other developed servers are through a proprietary message bus. The links provide the raw data interoperability. It is up to the users of CELLworks to parse and interpret the data. Therefore, $E(3)$ and $E(4)$ level enhancements are required for the scheduler to digest the information.

Figure 3.2 illustrates the available interoperability (solid boxes) and the required level of interoperability (dashed line boxes) for the monitoring and execution between the cell controller and the robot.

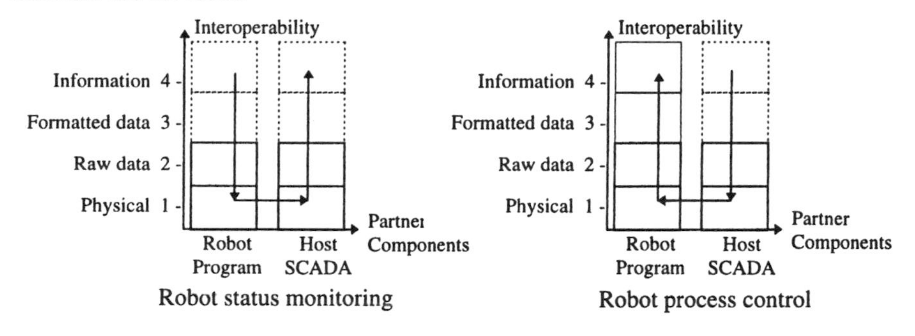

Figure 3.2 The available and required interoperability in robot controller and cell controller.

The effort required for monitoring, scheduling and executing the robot is

$$\begin{aligned} E_{\text{Robot}} &= [E(3,\ \text{Robot Prog},\ \text{Monitor}) + E(4,\ \text{Robot Prog},\ \text{Monitor})] \\ &+ [E(3,\ \text{Host SCADA},\ \text{Monitor}) + E(4,\ \text{Host SCADA},\ \text{Monitor})] \\ &+ [E(3,\ \text{Host SCADA},\ \text{Executor}) + E(4,\ \text{Host SCADA},\ \text{Executor})] \end{aligned}$$

For brevity, we did not include the terms for other data sets and components. One can perform the same analysis for E_{CNC} and E_{Schedule}.

3.2 Case II. Process Control in a Circuit Board Assembly Line

In this case, we perform supervisory control of a circuit board assembly line with a MPM screen printer, two Siemens placement machines and a BTU reflow oven. The system is controlled by a single host. The host is a Pentium computer running Windows NT, WinSECS and SPCWin from FASTech. The important characteristics of this circuit board assembly line are that all equipment controllers are compliant to SECS/GEM standards, and the host controller is compliant to SECS standard. To understand the interoperability for this system, one must have some understanding of SECS/GEM standard. In the following, we provide a brief description for the SECS/GEM standard.

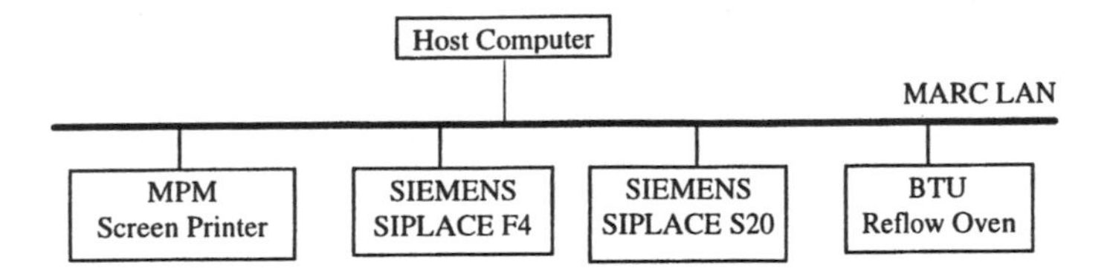

Figure 3.3 Configuration of the Circuit Board Assembly Line

SECS-I restricts the options exists in RS-232-C and expands the single character frame into blocks to allow addressing and more sophisticated handshaking. Therefore, the SECS-I provides plug-and-play physical interoperability and robust raw data transfer.

SECS-II defines the named stream (Sx) and function (Fy). Function defines the data structure, data type and some times the fixed data values in the SECS-I blocks. The data structure and type elevate the interoperability to formatted data. For example S1, F3 is (stream 1 function 3) is named Selected Equipment Status Report. It defines the structure and data type of the status variables communicating partners will share. In the case when the data values are fixed, the message has clear meaning to the receiver and should be considered at information level of interoperability. For example, a S1, F1 (Are You There?) consists of a null block for status checking. Modern manufacturing systems are often connected through a local area network. A companion standard, High-Speed Message Services (HSMS), is also available. It is the network counterpart of SECS standard based on TCP/IP protocol.

GEM defines a set of equipment behaviors that are related to a subset of SECS-II messages and are modeled with state models. The behaviors defined include a limited number of state and state transitions in: Communication, Processing, Control, Alarms, Spooling, Limits Monitoring, Process Programs, Material Movement, Trace Reports and Terminal Services. Compliance to GEM indicates the "plug-and-play" functionality and its associated messages should be considered at information level. For example, the control state model in GEM is defined in Figure 3.4.

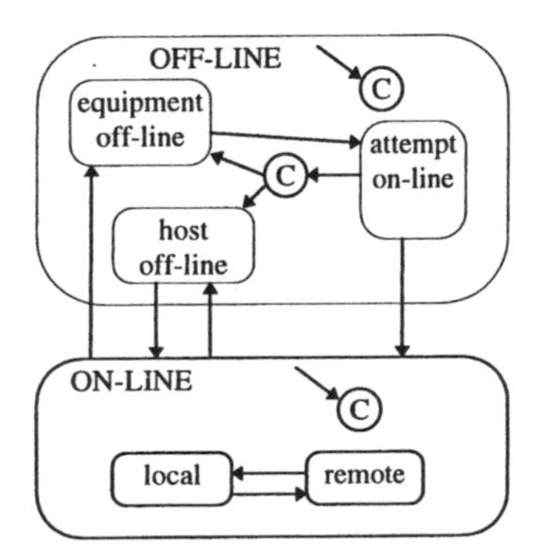

Figure 3.4. An control state model in GEM

The state transitions, represented by the arrows, are triggered by designated SECS-II messages. Such SECS-II messages are understood by a machine that is compliant to GEM standard in a plug-and-play fashion, and are therefore considered information. This can be generalized to all SECS-II messages that are associated with transitions in some defined GEM models.

In this board assembly line, we need to perform several supervisory control functions:

- process monitoring and graphical presentation
- remote control (Start, stop and recipe selection)
- process pogrom (recipe) management (uploading and downloading)

For brevity, we will only present the process monitoring and graphical presentation. In parameter monitoring, we used polling or event reporting for different process parameters on different machines. The machine name, process parameter name, sample frequency are tabulated in the first three columns in Table 1. The numbers in the fourth column are required level of interoperability (*a*) versus the available level of interoperability (*r*). The difference between the two numbers *r* - *a* represents the effort required for the particular data set on the equipment side and is listed in fifth column Effort. The numbers related to the host are listed in sixth and seventh column.

Table 1. Process monitoring in different equipment

Machine	Variable	Frequency	Equipment		Host	
			r / *a*	Effort	*r* / *a*	Effort
MPM printer	Idle Time between successive boards	after each print	2/2	0	4/2	3 4
SIPLA C80-F4	Placement Pressure of a component	in each placement	3/3	0	4/3	3 4
BTU oven	Controller Input Temperature in zone 1	every second	3/3	0	4/3	4

It is clear from this table, in order to monitor the process through this set of parameters, we must perform *E*(3) and *E*(4) functions on the host. WinSECS provides excellent utility in performing this functions partially in drag-and-drop programming and partially in programming in Visual Basic. The graphical presentation with associated control parameters using SPCWin is also easy.

3.2 Summary of Two Cases

One may notice that for every data set in both cases, the available level of interoperability is at least at raw data level. This is because one of our objective is to integrate the system rapidly. The rapid system development is commonly possible only with available raw data level interoperability, For example, if we wanted to get robot current location, we will have the physical connection. However, the information is not accessible from the robot program. We will need to enhance the interoperability. This will involve low level and proprietary programming in the robot controller. This is true for getting placement head locations on a placement machine too. Another example would be if we wanted to get CNC status directly, we will not even have the physical link because RS-232 between the cell controller and the CNC is simplex. To make that available can be even more challenging. This is true for instructing the placement machine to go at a different offset on-line. Since we only work with SCADA software, we were able to accomplish the control goals in four months for the first case and two months for the second case. From two cases, we can see the power of standardization and enabler softwares. Another benefit of standardization is the portability or scalability of applications. When we finished recipe management and remote control functions for one machine, it is rather easy to complete the second, third and fourth. However, it will be very difficult to accomplish physical or raw data level interoperability with enabler software.

4. CONCLUDING REMARKS

In this paper, we presented a qualitative framework for the estimation of the effort required to achieve certain control functions. The framework is based on an abstraction of interoperability. With the known data sharing requirements, one can use the framework to identify the qualitative and intuitive measures for the effort to achieve the control goals. We finally provided two case studies to show the use of the framework and to illustrate the usefulness in estimating the effort required for integration.

We would like to acknowledge National Science Foundation with its ARPA TRP project. We would also like to acknowledge Siemens, BTU, MPM and FASTech for their support.

REFERENCES

1. Bonnes, G. (1991), IMB OSI Interoperability Services, 1996
2. Dssouli, R. and Fournier, R. (1991), Communication Software Testability, 1995
3. Dinnel, T., Implementation GEM on process equipment, 227-236, 1994
4. Drusinsky, D. and Harel, D. (1989). Using statecharts for hardware description and synthesis, IEEE transactions on computer aided design, Vol. 8, No. 7, July 1989, 798-806.
5. Harel D. (1988). On visual formalisms. Communications of the ACM, Vol 31, No. 5, May 1988, 514-529
6. Harel D. (1987). Statecharts: a visual formalism for complex systems. Science of computer programming 8, North - Holland, 231-274
7. Hawthorne, J. et. al., CIM integration of SECS and GEM compliant equipment, 243-246, 1995
8. Hooman, J.M., Ramesh, S. and Roever, W. (1992). A compositional axiomatization of statecharts. Theoretical computer science 101, Elsevier, 289-335
9. MESA International, "MES Software Evaluation / Selection," White Paper No. 4, MESA International.
10. Nguyen, K. (1993), The development and implementation of a cell controller frame, 1993 IEEE/SEMI international semiconductor manufacturing science symposium, 54-47
11. Robotics World Editorial, "Open Architecture: Myth or Pipe Dream," , Spring, 1996, pp 18 - 20.
12. Smith, J.S. and Joshi, S.B. (1994), A shop floor controller class for computer integrated manufacturing, Working paper, Texas A&M university, College Station, Texas
13. Spencer, D. (1993), Computer dictionary, fourth edition, Camelot Publishing Company, Ormond Beach, Florida.
14. Owen, J. "Open up the Plant Floor," Manufacturing Engineering, Vol. 117, No. 5, Nov. 1996, pp 46 - 58.
15. Wei, J. and Zhou, C. "Cell Control Rapid Prototyping using Object Oriented Technology and SEMI Standards", Proceedings of Six International FAIM Conferences, 1996, pp263-273
16. Young, S. J. *Real Time Languages: Design and Development.* Chichester: Ellis Horweed.

MODELLING AND CONTROL OF A PLASTICATING EXTRUDER

C. K. S. Ho*, I. G. French, C. S. Cox*, A. Jones*****
*Control Systems Centre, University of Sunderland, U K
**EPICC, University of Teesside, U K
***ICI Films Ltd., U K

ABSTRACT: Extrusion is a very common process in the plastic production industry. Accurate extruder barrel temperature control has proved to be essential as it is closely related to the final product quality. However, it is widely understood that the process, although very straightforward, imposes a lot of difficulties in control. This paper begins with a general introduction to the modelling of an extruder using SIMULINK. Then the paper will proceed to discuss the fundamental difficulties in extruder temperature control. The particular problem of Stop-Start events will be highlighted. Finally, a strategy based upon the scheduling of the set-point is proposed. Simulation studies show that both settling time and temperature deviations caused by the change of screw speed can be reduced using this strategy.

INTRODUCTION

Plastic product has been used in a wide variety of areas. These areas include packaging material for foodstuffs, industrial use such as capacitor, motor and cables insulation, or for the storage and display of information, typically as floppy discs, audio and video tapes, photographic films and overhead projection slides. Some of these products require a very high quality consistency, hence the requirement for good process control is stingent.

In plastic product manufacturing, the extrusion process plays a key role. Consequently, the quality of the final product is greatly affected by the performance of the extrusion control system. One of the control problems commonly associated with extrusion control is the accurate control of the temperature of the extruder barrels. This is important because the quality of the finished product depends heavily on material properties such as viscosity and elasticity, and these properties vary with temperature.

In this paper, the operation of the extruder is first introduced. The paper then proceeds to illustrate an approach to modelling the temperature control process using heat balance equations. Finally, difficulties in temperature control will be discussed, especially during the stop-start event.

MODELLING OF A PLASTICATING EXTRUDER

Extruder Operation

An extruder is a machine with a hollow barrel which has a screw feed in the centre. Polymer chip together with reclaimed material is fed into the extruder where it is melted. The extrusion process involves the softening of the solid plastic material through heat and mechanical work to form continuous plastic products. The 'melt' is subsequently forced out of the extruder through a die. Variables that can be controlled include feed rate, screw speed, temperature, pressure and output rate .

Heat is produced by the action of the screw mixing and compressing the extrudate. In general, the amount of heat produced by an extruder depends on the plastic material being used and increases with the speed of the screw. However, the frictional heat produced may be excessive or not enough to maintain the temperature of the melt at a constant desired temperature, due to diverse reasons such as unplanned variations in screw speed or material property changes. Hence, an extra heating/cooling media is needed. Thermocouples are widely used to measure the melt temperature. A typical single-screw extruder is shown in Fig. 1.

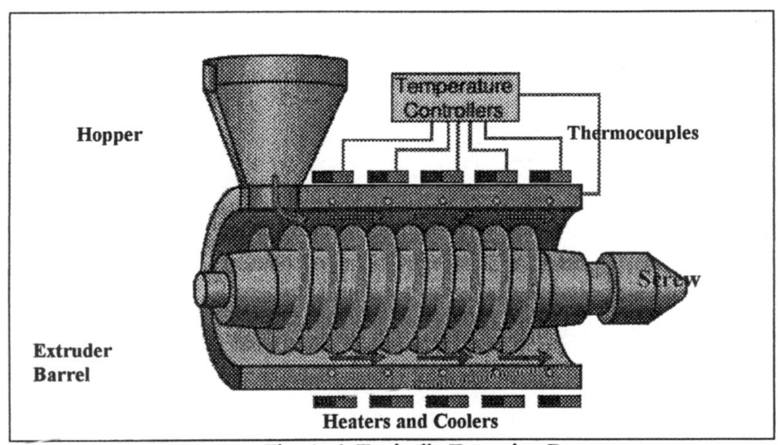

Fig. 1 A Typically Extrusion Process

There are a number of heaters and coolers surrounding the barrel which are used to control the temperature of the melt. The barrel is divided into different heating/cooling zones and each of these zones is usually controlled separately. Usually electrical heaters are used for heating whilst cooling is performed by solenoid controlled shots of water evaporating on the barrel wall. To achieve heat balance, heat is conducted either to or from the melt through the barrel. To save energy as well as to minimise the amount of external disturbance from the environment, the barrel is usually 'jacketed'. In practice, 'jacketing' the barrel can also help provide acceptable working conditions for the shop floor personnel.

Modelling of the Extruder

Over the years, many different extruder models have been developed by various researchers [1,2,3]. However, most reported dynamic models used for control synthesis are still based upon a first order plus time delay transfer function and using such formulations most work has concentrated on the thermal lag of the barrel wall and the heater/cooler dynamics. However, as will be shown in the next section, the effect of screw speed is also very important.

In this study, a simplified model of an extruder is developed using SIMULINK. The model is formulated based upon a heat balance equation. Assuming that the temperature of the inner barrel wall is sufficiently close to the actual melt temperature, then the heat balance equation can be written as follow :

Heat Gain = Energy from Heater/Cooler + Frictional work from screw
- Heat absorbed by the plastic melt

(1)

The corresponding SIMULINK model is shown in Fig. 2. The model comprises five mains elements; the heater dynamics, the barrel wall dynamics, heat loss to the surroundings, frictional work from the screw and the heat loss to the plastic melt. It should be noted that a dead-zone block has been added to the screw speed section, this is because at low screw speeds the frictional heat produced by the screw is insignificant. The model, which is shown in Fig. 2 has been validated using industrial data.

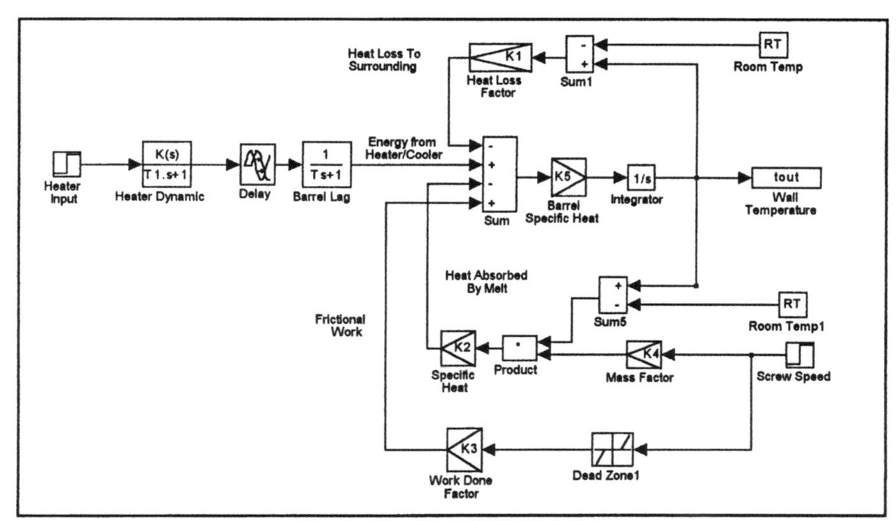

Fig. 2 The SIMULINK Model

PROBLEMS IN TEMPERATURE CONTROL

The General Problem

From the description given in Section 2, the extrusion process would appear to be relatively simple. However, there are several practical features which cause difficulty during operation.

Firstly, the thick barrel wall (which can be characterised by thermal lags) introduces significant time delay resulting in the inability to achieve true equilibrium instantaneously, a problem which of course is more acute with the larger systems. For example, an 10" extruder can take up to 2 hours to settle to the point where end melt temperature and pressure are sufficiently stable to meet product quality requirements.

Secondly, the dynamics of heating and cooling are completely different, a fact which is further complicated because the cooling process is itself nonlinear [4]. Thus, the simulation model presented in Section 2 can be further refined by the incorporation of a variable gain in the heater/cooler block.

Another major source of nonlinearity arises as a function of the screw speed. Typically, the heating/cooling time constant at high speed is only about one-tenth of that at low speed. This is because the amount of material being fed into the system, which also dictates the system load, depends on the screw speed. The shear work produced is a nonlinear function of the screw speed and the material properties vary with temperature. Thus in general, the system gain is a nonlinear function of screw speed. More importantly, the screw speed has a much bigger impact on the barrel temperature than the heating/cooling actuators. Ideally, screw speed should be used to control the barrel temperature, however, this is impossible in practise as the process throughput has to be maintained constant.

Given the apparent level of nonlinearity, it is perhaps supprising to learn that conventional PI/PID type controllers, with fixed controller settings, are still employed almost universally within industry for extruder temperature control. Especially because, it has been shown in various applications that while conventional PI/PID control strategies provide acceptable performance in steady state, they fail to provide satisfactory response when the extruder is subject to screw speed changes, either during a planned shut down or during an unplanned process interruption. This problem is discussed in the next section.

Stop-Start Event

One of the major areas of concern for extruder temperature control the stop-start event. Stop-Start events, refer to situations when the screw was stopped and is then subsequently restarted. The transients which occur during a typical stop-start event are shown in Fig. 3

From Fig. 3, it can be seen that the stoppage of the screw causes a rapid reduction in temperature due to the absence of shear from the screw. At this point, the temperature control system will increase the output to the heater in an attempt to maintain the barrel zone temperature.

When the screw speed is gradually restored (note that rapid screw speed restoration is usually prohibited in practise as it may cause problems in material feeding), there is an initial decrease in temperature. This may be somewhat unexpected as in general, a higher screw speed will generate more heat. However, the fall in temperature is due to the feeding in of 'cold' materials (sudden load change) and at low screw speed the shear work produced is insignificant. At this point, the temperature controller will again increase its output to try to restore the temperature. When the extruder screw speed is increased to a level where the shear work produced is significant, the barrel temperature rises very quickly and, as the effect of the screw speed on the melt temperature is faster than that of the heating/cooling due to the barrel wall lag, excessive and undesirable overshoot in temperature is obtained. This overshoot may last in excess of 2 hours.

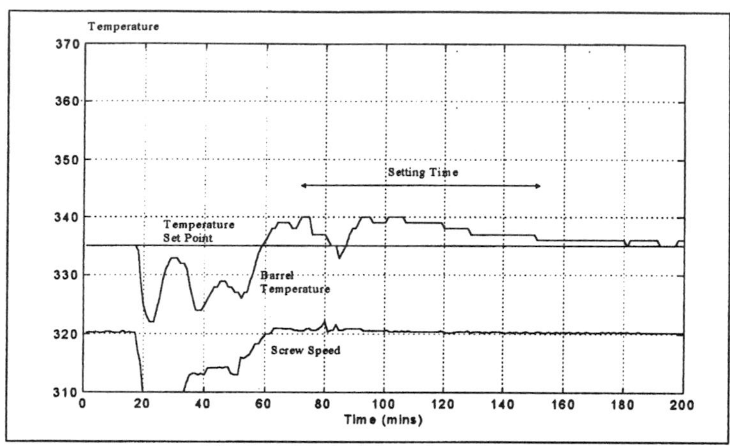

Fig. 3 A Typical Stop-Start Event

The big deviation in temperature is partly due to the inadequacy of the fixed term controller. This is because, system behaviour during the idle condition is very much different from that when the screw is running. Further, since it is often the case that the loop gain is set to a value to meet the normal running requirement, then the use of this gain results in oscillation under start-up conditions. To illustrate the above point, a step test was carried out under standby conditions, that is when the screw is stopped. In the test, the heating power is increase by 5% from an equilibrium condition. The result is shown in Fig. 4.

It can be seen from an inspection of Fig. 4 that the system has not settled even after 5 hours. Actually, the system behaviour is almost that of an integration process. This is because almost all the extra heat supplied by the heater to the system was used to raise the temperature of the extruder barrel since, because of the insulated jacket, the energy losses to surroundings are negligible.

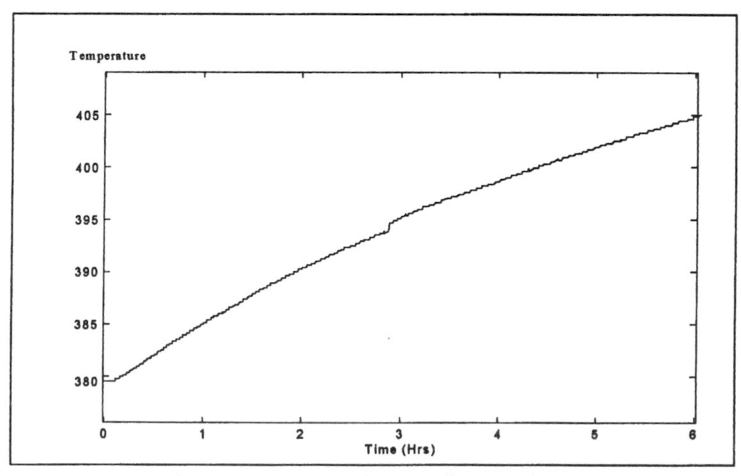

Fig. 4 Step Test Under Standby Conditions

THE PROPOSED CONTROL STRATEGY

In tackling the problems of extruder temperature control, many authors [5,6,7] have suggested schemes to try to improve the control performance. For example, Kramer [5] suggested the use of adaptive reset time to shorten the time required to restore the temperature after a screw speed change or a disruptive upset. Also, several studies have suggested that feedforward control should be used to compensate for disturbances in the screw speed. And indeed, these approaches can successfully reduce the settling time after a change in screw speed. The problem, however, is that finding a good adaptive mechanism or feedforward controller is not an easy task, since in general it requires a good process model.

Therefore, having gained a better understanding of the causes of the problem of large temperature deviation during the stop-start event, it may be possible to improve the controller performance by taking some more simple corrective actions.

For example, as it is clear that temperature will fall when the screw is stopped, it is desirable to increase the heating power beforehand. One simple approach to achieve this is to increase the set point to provide the initial kick. An example of this approach is shown in Fig. 5 below.

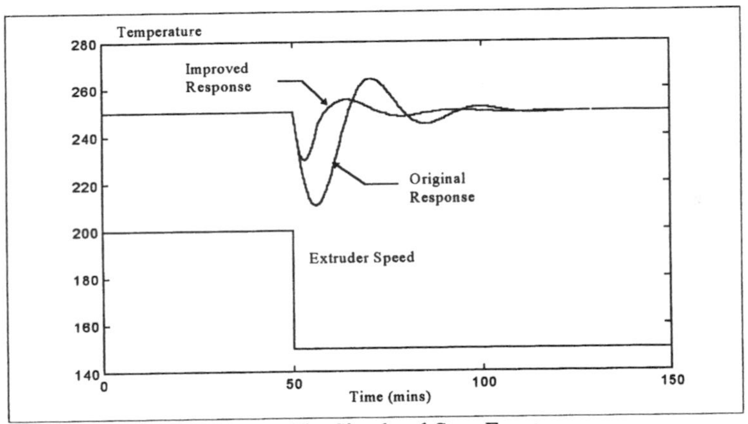

Fig. 5. The Simulated Stop Event

The responses shown in Fig. 5 are simulated responses obtained under closed-loop PI control of the model presented in Section 2. The responses show the typical barrel temperature deviation during a stop event. In our scheduling strategy, at time t=50 mins, the set-point was increased from 250 to 290. The set point was restored when the temperature rose to a value close to the original set point of 250. It can be seen that the maximum temperature deviation has been reduced by more than half.

If the stoppage of the screw is a planned event, the increase in set point can be carried out a few minutes before the screw is stopped to compensate for the system thermal lag and time delay. This can further reduce the temperature deviation and settling time. This is shown in Fig. 6.

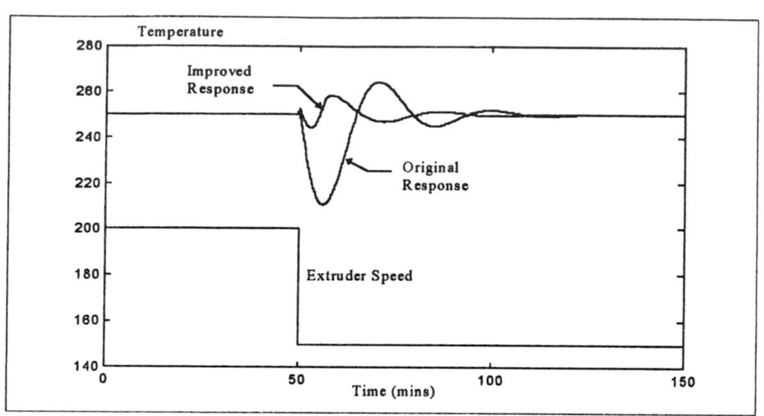

Fig. 6 The Simulated Stop Event (with predictive action)

At can be seen that the temperature fall has been dramatically reduced.

The rescheduling approach can also be applied to the scenario when the screw speed is restored (i.e. the Start event). In these cases, the temperature set point is decreased initially as a rapid increase in temperature is expected. The simulated responses are shown in Fig. 7. Again, the settling time and temperature deviation have been dramatically reduced.

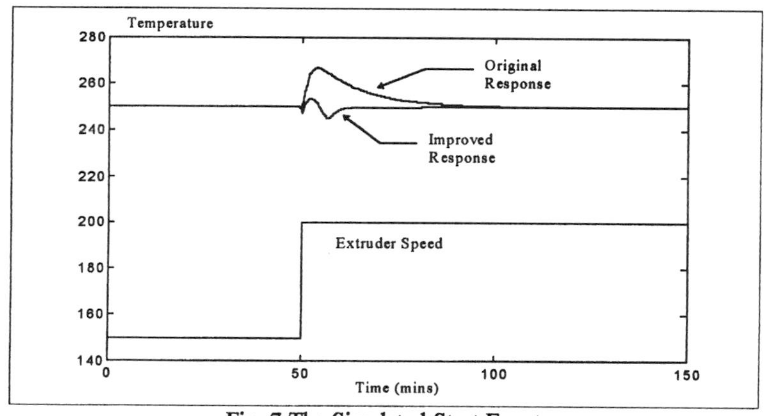

Fig. 7 The Simulated Start Event

The proposed approach is very similar to a feedforward approach except that the only tuning parameter is the amount of temperature setpoint deviation which can easily be determined empirically. In addition, for a planned event, predictive action can easily be applied to further improve the performance.

DISCUSSION

One of the major components of plastic extrusion is the use of temperature control to ensure the variation in material properties have minimum effect on the finished product. The physical construction of the thick extruder barrel wall introduces significant time delay which has an adverse effect on the control strategy. The control problem is further complicated by the process nonlinearity caused by the difference in the characteristics of heating and cooling. These factors have led to problems in control especially when the screw speed is changed significantly. Typically, these problems manifest themselves as prolonged settling time of the extruder barrel temperature, thus reducing first pass efficiency. A simple strategy is proposed which utilises scheduling of the set-point to reduce the settling time as well as the temperature deviations caused by the change of screw speed. The approach requires no modification of the PI/PID controller settings which is particularly attractive. In practise, the setpoint scheduling scheme proposed here can easily be implemented as an automatic event driven strategy in the SCADA and monitoring system.

REFERENCES

[1] D. Chan, R. W. Nelson and L. J. Lee : 'Dynamic Behaviour of a Single Plasticating Extruder - Dynamic Modelling', Polymer Engineering and Science, Vol. 26, No. 2. Jan, 1986.

[2] G. K. McMillan : 'Effect of Cascade Control on Loop Performance', Proc. Automatic Control Conference, 1982, p363-368.

[3] P. R. Krishnaswamy and P. R. Rangaiah, : 'Role of secondary integral action in cascade control', Trans. I Chem E, Vol 70, Part A 1992, p149-152.

[4] G. W. Howes, Eurotherm Application Notes : AN-PLAS-3: 'Water Cooling of Plastics-Extruder Barrels', 1981

[5] Eurotherm Application Notes AN118 : 'Temperature Control on Machinery for Extrusion of Plastics and Rubber' 1977

[6] W. A. Kramer : 'Extruder Temperature Control with Adaptive Reset', Tappi Journal, Vol.79, No. 2, 1995, pp247-251.

[7] B. Yang & L. J. Lee : "Process Control of Polymer Extrusion", Polymer Engineering and Science, Vol 26, No. 3, 1986.

[8] P. Harriott : 'Process Control', McGraw Hill, New Your, 1964, p.157

[9] F. G. Shinskey : 'Controlling Temperature in Batch Reactors' , Intech, p69-72.

IMPROVED CONTROL OF MIMO PROCESSES USING GENETIC ALGORITHMS

I. G. French*, C. K. S. Ho, C. S. Cox**, K.C.S. Ng*****
*EPICC, University of Teesside, U K
**Control Systems Centre, University of Sunderland, U K
***Eurotherm Controls Ltd., U K

INTRODUCTION

Multivariable systems frequently occur in industry. In many cases there is considerable interaction between the different manipulated and controlled variables. The use of single-input single-output strategies in these situations should be avoided since the cross-coupling can often result in non-optimal sub-standard performance. Thus, modern industry requires advanced control solutions based around multivariable system formulations.

One such formulation is the MIMO PIP strategy introduced by Billington et al. [1] and subsequently developed by Young and his co-workers [2]. This method relies only on input-output data gathering so no state estimation is required. However, the accuracy of the final control strategy is strongly dependent on the quality of the mathematical model obtained to characterise the process. This introduces two problems, namely, the efficient identification of the structure, order and parameters of the MIMO discrete-time transfer description which forms the basis for the controller design philosophy, which for MIMO plant this can be a very time consuming operation, and the related problem of quickly establishing the control 'pairings' between the appropriate manipulated variables and the specified outputs.

This paper suggests a framework to help solve both problems by the development of automated search procedures based on genetic algorithms.

SYSTEM IDENTIFICATION

Methods for Single-Input Single-Output (SISO) model identification are now well accepted but as stated by Young [3], it is well known that the identification of multivariable model structure and order presents a major practical challenge and that the failure to properly define such a model structure can lead to many practical problems.

To place the problem into perspective, consider a 2-input 2-output system characterised by the Left Matrix Fraction description

$$\begin{bmatrix} y_1 \\ y_2 \end{bmatrix} = \begin{bmatrix} A_1(z^{-1}) & 0 \\ 0 & A_2(z^{-1}) \end{bmatrix}^{-1} \begin{bmatrix} B_{11}(z^{-1}) & B_{12}(z^{-1}) \\ B_{21}(z^{-1}) & B_{22}(z^{-1}) \end{bmatrix} \begin{bmatrix} u_1 \\ u_2 \end{bmatrix} \tag{1}$$

where each element $A_*(z^{-1})$ and $B_{**}(z^{-1})$ have the structures

$$A_* = 1 + a_1 z^{-1} + \cdots + a_p z^{-p}; \qquad B_{**} = (b_1 z^{-1} + \cdots + b_q z^{-q}) z^{-(d-1)} \tag{2}$$

and highlighting that the individual a, b, p, q & d values will, in general, be different for each element.

In this case an 'exhaustive search' for p=1 to 3, q=1 to 3, d=1 to 5 will yield 1350 different models. However, for $p = 1$ *to* 5, $q = 1$ *to* 5, $d = 1$ *to* 9 there will be 20,250 different models to be considered and for a system with 5 inputs and 4 outputs, and with the original search range for p, q & d, the result would be 9,000,000 different models.

The objective of the MIMO system identification phase is to 'pick' that single structure from the large number of candidate models which best 'explains' the collected experimental data. It can be seen, therefore, that finding the optimum structure for MIMO systems by trying out all the combinations is inefficient and unwise. Therefore, an alternative search approach is needed.

One possible solution to the above problem is to implement a new approach which relies on the attributes established [4] for genetic algorithms (GAs). The strategy proposed here, has proved to be very successful, and is formulated as follows:-

1. Establish a chromosome which embodies the structure of the system to be identified. This may take the form of a string made up of integers which represent the number of 'a' terms, the number of 'b' terms for each input and the number of delays before the first non-zero element of each input. For a 2-input-1-output system, the structure of this string may take the following form

$$\text{Struct} = \begin{matrix} n_a & n_{b1} & n_{b2} & d_1 & d_2 \\ 1 & 1 & 1 & 2 & 4 \end{matrix}$$

2. Define a search range for each element of the string. Say 1 to 3 for the 'a' terms, 1 to 3 for the 'b' terms and 1 to 5 for the delays.

3. Generate and code an initial population, say 20 randomly chosen individuals from within the search range.

4. Using least squares or an alternative, such as Instrumental Variables (IV), estimate the coefficients of each model defined within the population.

   The quality of each model is tested using the a fitness function based on a Young's Information Criterion (YIC). Hence each member of the population is ranked according to their YIC.

5. Terminate the algorithm if any model within the population contains only parameters that satisfy the 90% significance bound

   $$|\hat{\theta}_i| > 1.95\sqrt{\text{Variance of } \theta_i}$$

and has a fit of greater than a given threshold, say 80 %.

6. Based on the fitness define a population for mating. This population consisting of multiple copies of the most highly fit individuals (to total 20) based on the relationship

$$\textit{Expected copies of individual} = \frac{\textit{Fitness of Individual}}{\sum \textit{Fitness of Individuals}} \times \textit{Population}$$

7. Mate the population by randomly exchanging information between pairs of strings. Pairings are chosen at random. Also the number and location of exchange sites within the string are randomly selected.

8. To avoid convergence to local minima. Mutations are introduced by changing randomly selected elements from within the population to other values within the search range. A typical mutation rate is 1 every 200 information exchanges.

9. Goto 4

CONTROL STRUCTURE SELECTION

Given a sufficiently accurate model, the controller design process can be undertaken. For MIMO control systems, the selection of an appropriate set of manipulated variables to control a set of specified outputs via feedback is a major component of the control system design. By selecting alternative sets of manipulated variables the input structure and therefore the system subject to feedback control is modified. Thus changes in the set of manipulated variables can be considered simply as a special case of the design modifications undertaken to improve dynamic performance. Consequently, failure to find the most appropriate input-output pairings will at best reduce the efficiency of any ensuing control system design and at worst result in a system for which a realisable controller design is not possible.

Perhaps the most popular tool, used by designers, to aid in the selection of the most appropriate input-output pairings is the Relative Gain Array (RGA) [5]. In the 30 or so years since its origin numerous extensions [6,7] have been proposed to the basic technique, yet, in whatever form the RGA is defined the primary design requirements remain consistent. A matrix of numbers (the RGA), each of which represents the relative influence of a given input on a defined output, is manipulated, by moving complete rows and columns, until it most closely approximates to the ideal solution, represented by the identity matrix.

For small systems, the RGA manipulations are straightforward. However, as systems become larger such manipulations become more complex and some form of automated solution is advisable. In this paper it is proposed to use a GA for this task. The basic structure of this algorithm is very similar to that outlined for model structure selection. Here, therefore, we will be primarily concerned with those features which are application specific, namely, the structure of the strings, the mechanisms for crossover and the selection of an appropriate fitness function.

String Selection

To utilise the GA to search for the optimal input-output pairings, the relative placement of the rows/columns of the RGA must be encoded. The manipulation of the RGA to find the 'best' input-output assignment is, of course, an example of the classical permutation encoding problem, the most famous example of which is the travelling salesman problem. In the case of the RGA, to represent a candidate configuration, each individual comprises two permutation strings, one for the rows and another for the columns. The two strings contain mutually exclusive integer numbers in the range one to row/column length. These integers identifying the relative placement of the row/column of the original RGA in the arrays representing the candidate configurations.

For example if a column string were

col_string_n = [5 3 4 1 2]

this would indicate that column one of the candidate configuration was column five of the original RGA, that column two of the candidate configuration was column three of the original RGA and so on.

Recombination Mechanism

For the structure selection problem it is not feasible to use simple crossover. For example consider the two parent strings P_{S1} and P_{S2} with crossover point '|'

P_{S1} = [5 3 4 | 1 2] O_1 = [5 3 4 3 5]

↕↕

P_{S2} = [2 1 4 | 3 5] O_2 = [2 1 4 1 2]

The offspring O_1 and O_2 are clearly not legal since each contains repeated elements. It is obvious, therefore, that simple crossover will never work satisfactorily and that some alternative mechanism must be found.

The mechanism used here takes the following form :

1. The number of crossovers for a particular pairing is chosen at random in the range one to string length.
2. For 1 to number of crossovers.
3. Generate a random integer in the range one to string length. e.g. 3.
4. Identify the location of the crossover site within each parent

↓

P_{S1} = [5 3 4 1 2]

P_{S2} = [2 4 1 3 5]

↑

5. rotate the elements in the two identified sites

O_1 = [5 1 4 3 2]

O_2 = [2 3 1 4 5]

6. Repeat for the required number of crossovers.

Fitness Functions

The fitness function is evaluated as a function of the sum of the absolute differences between the candidate solutions and a template matrix. The template matrix being chosen as follows for a a 4x4 example.

$$\begin{bmatrix} 1 & 0.75 & 0.5 & 0.25 \\ 0.75 & 1 & 0.75 & 0.5 \\ 0.5 & 0.75 & 1 & 0.75 \\ 0.25 & 0.5 & 0.75 & 1 \end{bmatrix}$$

The philosophy being to attract values closest to unity to the leading diagonal, with decreasing values being positioned further away. For non-square problems additional zero rows (or columns) are appended to the template matrix.

CONTROLLER DESIGN - PROPORTIONAL-INTEGRAL-PLUS CONTROLLER

PIP control was first proposed by Young [8] as a general purpose algorithm. It can handle, without modification, non-minimum phase systems, unstable systems and those with appreciable time delay. The design is based around a non-minimal state-space description of the process dynamics using discrete time data trains defined to be the present and past values of the process output and past values of the process input. An 'integral of error' state is also included to provide offset free performance in practical applications. Through the use of full linear state-variable feedback, the PIP controller permits arbitrary closed-loop pole assignment, however, more recently and optimal PIP [1] algorithm has been proposed which has proved to be particularly robust for a range of real control problems.

Consider a n-output, m-input, discrete-time system written in the left MFD form,

$$Y_k = A^{-1}(z^{-1})B(z^{-1})U_k \tag{3}$$

or

$$A(z^{-1})Y_k = B(z^{-1})U_k$$

where

$$A(z^{-1}) = I_n + A_1 z^{-1} + ... + A_p z^{-p} \quad \text{and} \quad B(z^{-1}) = B_1 z^{-1} + ... + B_q z^{-q}$$

in which A_i (i=1,..p) are nxn matrices with A_p not null, and B_j (j=1,..., q) are nxm matrices with B_q not null. However, some of the initial B_j's can take null values to accommodate pure time delays.

In a similar manner to the SISO analysis [8], the NMSS description of the system can be defined in terms of present and past Y_k, past U_k and a delay free integration M_k.

$$X_k = [Y_k \quad Y_{k-1} \quad \cdots \quad Y_{k-p+1} \quad U_{k-1} \quad U_{k-2} \quad \cdots \quad U_{k-q+1} \quad M_k \;]^T \tag{4}$$

The following NMSS representation can then be obtained straightforwardly,

$$X_{k+1} = FX_k + GU_k + EY_{dk}$$
$$Y_k = H\ X_k \tag{5}$$

where

$$F = \begin{bmatrix} -A_1 & -A_2 & \cdots & -A_{p-1} & -A_p & B_2 & \cdots & B_{q-1} & B_q & 0 \\ I_n & 0 & \cdots & 0 & 0 & 0 & \cdots & 0 & 0 & 0 \\ 0 & I_n & \cdots & 0 & 0 & 0 & \cdots & 0 & 0 & 0 \\ \vdots & \vdots & & \vdots & \vdots & \vdots & & \vdots & \vdots & \vdots \\ 0 & 0 & \cdots & I_n & 0 & 0 & \cdots & 0 & 0 & 0 \\ 0 & 0 & \cdots & 0 & 0 & 0 & \cdots & 0 & 0 & 0 \\ 0 & 0 & \cdots & 0 & 0 & I_m & \cdots & 0 & 0 & 0 \\ \vdots & \vdots & & \vdots & \vdots & \vdots & & \vdots & \vdots & \vdots \\ 0 & 0 & \cdots & 0 & 0 & 0 & \cdots & I_m & 0 & 0 \\ A_1 & A_2 & \cdots & A_{p-1} & A_p & -B_2 & \cdots & -B_{q-1} & -B_q & I_n \end{bmatrix}$$

and

$$H = [I_n \quad 0 \quad \cdots \quad 0 \quad 0 \quad 0 \quad \cdots \quad 0 \quad 0]$$
$$E^{\mathrm{T}} = [0 \quad 0 \quad \cdots \quad 0 \quad 0 \quad 0 \quad I_n]$$
$$G^{\mathrm{T}} = [B_1^T \quad 0 \quad \cdots \quad 0 \qquad I_m \quad 0 \quad \cdots \quad 0 \qquad 0 \quad -B_1^T]$$

The multivariable optimal control law is now simply obtained by minimising the following quadratic performance index :

$$\mathrm{J} = \sum_{k=1}^{k=\infty} \mathrm{X}_k^{\mathrm{T}} \mathrm{Q} \mathrm{X}_k + \mathrm{U}_k^{\mathrm{T}} \mathrm{R} \mathrm{U}_k \tag{6}$$

where Q and R are, respectively, symmetric positive semi-definite and positive definite weighting matrices.

To yield the state variable feedback control law :

$$\mathrm{U}_k = -\mathrm{K}\mathrm{X}_k \tag{7}$$

SCRUBBER PROCESS EXAMPLE

In the process industries, recent years have seen a rapid growth in legislation related to environmental protection standards. Classic examples can be found in rubber processing, where waste from natural and synthetic rubbers, oils, waxes and sulphur compounds are prone to agglomerate. Removing this waste residual material from an industrial process is usually straightforward. The choice of scrubber will be determined by several factors, such as energy requirements and desired abatement efficient, together with the properties of the particle laden gas stream. Generally, there is a direct functional relationship between efficiency, pressure drop and power consumption, and consequently running costs, of a gas scrubber. Improving capture efficiency demands a proportionate increase in both the other factors.

The scrubber process to be considered here is shown schematically in Figure 5.1. Efficient operation requires tight control of three key variables; namely output flow (F6), inlet pressure (P1 = 0.9 bar A) to prevent discharge of contaminated material via the temperature valve which allows cooling air to be drawn from atmosphere and, to optimize operation and prevent poisoning of the scrubber, the inlet temperature (T1), which must be maintained at 523°K.

The existing control system, which consists of three SISO PID controllers. Unfortunately this controller configuration results in a large amount of interaction between the pressure and flow controllers. The system either goes into oscillation or control is lost depending on the controller parameters, see Figure 5.2. Practically this is overcome by switching the pressure controller to manual and regulating only flow and temperature. This provides stable control and prevents poisoning of the scrubber, but, offers little protection against discharge of contaminated material to atmosphere, via the temperature control valve. Therefore, it is intended, here, to investigate the possibility of bringing all three loops under closed-loop control using a multivariable type control scheme.

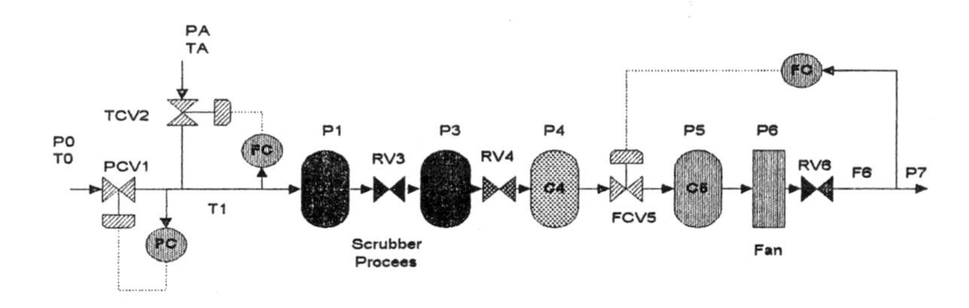

Figure 5.1 Schematic representation of the scrubber process

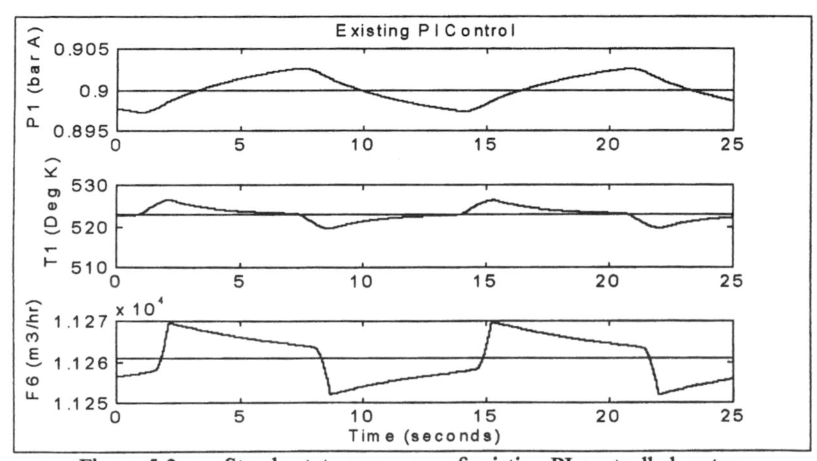

Figure 5.2 Steady state responses of existing PI controlled system

System Identification

System identification tests were carried out to obtain a linear discrete time model of the scrubber process. Three uncorrelated PRBS signals were applied to the process inputs and the output data was recorded. The sampling time of 1 second, was selected for each of the tests after a study of the open-loop step responses of the process. The sampling time is selected so that 5-15 samples covers each process rise time. In each test 100 samples were recorded. The amplitude of the PRBS excitation for the pressure and flow loops is 10% whilst 5% is used for the temperature loop. This being sufficiently large to overcome valve friction yet small enough to maintain operation close to the set point values, in order to obtain a good linear approximation. Prior to the model estimation, the input-output data was preprocessed. This involved the removal of steady state levels by differencing, and the noise by applying a first order digital filter with filter constant equal to 0.7.

Three, three-input-single-output, transfer functions were subsequently identified using the genetic algorithm outline in Section 2. The most efficient parameterizations for the transfer functions were determined to be

		Input 1	Input 1	Input 2	Input 2	Input 3	Input 3	YIC
	na	nb	nd	nb	nd	nb	nd	
P1	2	2	1	2	1	1	1	54.7143
T1	1	1	1	2	1	1	1	54.6273
F6	2	2	1	3	1	3	1	53.0452

Figure 5.3 shows a comprison of modelled and actual process outputs based on this parameterization.

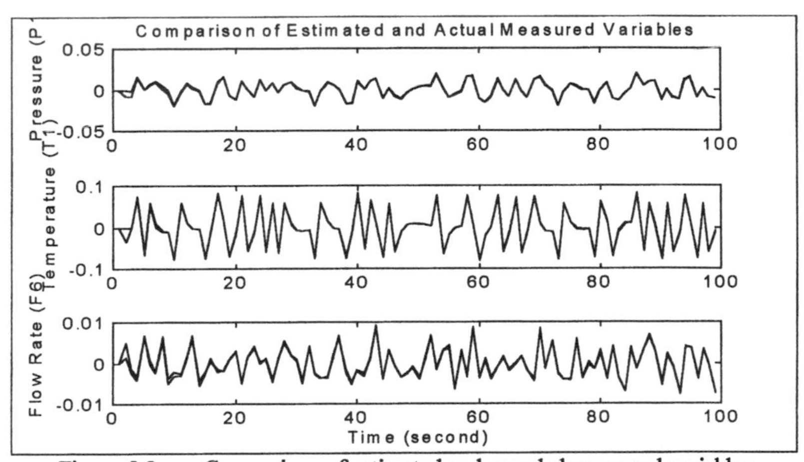

Figure 5.3 Comparison of estimated and sampled measured variables

The estimated transfer function matrix was:

$$\begin{bmatrix} P1_k \\ T1_k \\ F6_k \end{bmatrix} = A^{-1}(z^{-1})B(z^{-1})\begin{bmatrix} PV1_k \\ TV1_k \\ FV6_k \end{bmatrix}$$

where

$$A^{-1}(z^{-1})B(z^{-1}) = \begin{bmatrix} \frac{0.2643z^{-1}+0.1216z^{-2}}{1-0.7725z^{-1}+0.1294z^{-2}} & \frac{0.0036z^{-1}}{1-0.7725z^{-1}+0.1294z^{-2}} & \frac{-0.0052z^{-1}-0.013z^{-2}}{1-0.7725z^{-1}+0.1294z^{-2}} \\ \frac{1.157z^{-1}}{1-0.3591z^{-1}} & \frac{-0.0111z^{-1}-0.034z^{-2}}{1-0.3591z^{-1}} & \frac{-0.0081z^{-1}}{1-0.3591z^{-1}} \\ \frac{0.0289z^{-1}+0.0456z^{-2}}{1-1.0027z^{-1}+0.325z^{-2}} & \frac{-0.0034z^{-1}+0.0031z^{-2}-0.0075z^{-3}}{1-1.0027z^{-1}+0.325z^{-2}} & \frac{0.0719z^{-1}-0.0647z^{-2}+0.0102z^{-3}}{1-1.0027z^{-1}+0.325z^{-2}} \end{bmatrix}$$

Multivariable Control Structure Design

The first stage of the control structure selection, based upon the search algorithm developed in Section 4 is to evaluate the RGA based on relationship

$$\Gamma = G^{*}.(G^{-1})^{T}\Big|_{\omega=0} \tag{8}$$

to give

$$\begin{bmatrix} P1_k \\ T1_k \\ F6_k \end{bmatrix} = \begin{bmatrix} 1.5065 & -0.7911 & 0.2847 \\ -0.4046 & 1.4057 & -0.0011 \\ -0.1019 & 0.3855 & 0.7165 \end{bmatrix} \begin{bmatrix} PV1_k \\ TV1_k \\ FV6_k \end{bmatrix}$$

Applying the optimal search algorithm returns the modified RGA shown below:

$$\begin{bmatrix} T1_k \\ P1_k \\ F6_k \end{bmatrix} = \begin{bmatrix} 1.4057 & -0.4046 & -0.0011 \\ -0.7911 & 1.5065 & 0.2847 \\ 0.3855 & -0.1019 & 0.7165 \end{bmatrix} \begin{bmatrix} TV1_k \\ PV1_k \\ FV6_k \end{bmatrix}$$

This is shown graphically in Figure 5.4.

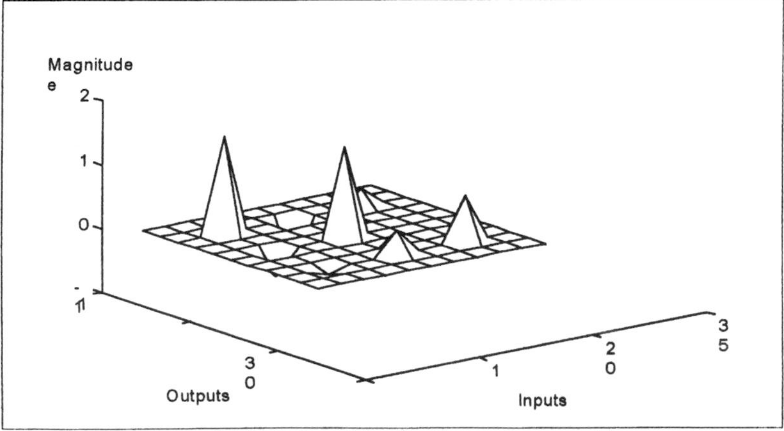

Figure 5.4 Relative Gain after Input-Output pairings

Inspection of the figure indicates the presence of significant off diagonal elements in the modified RGA. This, in turn, suggests a requirement for a full 3x3 multivariable controller design.

Multivariable Control Design

To design an optimal multivariable PIP control scheme, the NMSS description is formed, giving:

$$F = \begin{bmatrix} 0.7725 & 0 & 0 & -0.1294 & 0 & 0 & 0.1216 & 0 & -0.013 & 0 & 0 & 0 & 0 & 0 & 0 \\ 0 & 0.03591 & 0 & 0 & 0 & 0 & 0 & -0.034 & 0 & 0 & 0 & 0 & 0 & 0 & 0 \\ 0 & 0 & -0.2959 & 0 & 0 & 0 & 0.0456 & -0.0031 & 0.0647 & 0 & -0.0075 & 0.0102 & 0 & 0 & 0 \\ 1 & 0 & 0 & 0 & 0 & 0 & 0 & 0 & 0 & 0 & 0 & 0 & 0 & 0 & 0 \\ 0 & 1 & 0 & 0 & 0 & 0 & 0 & 0 & 0 & 0 & 0 & 0 & 0 & 0 & 0 \\ 0 & 0 & 1 & 0 & 0 & 0 & 0 & 0 & 0 & 0 & 0 & 0 & 0 & 0 & 0 \\ 0 & 0 & 0 & 0 & 0 & 0 & 0 & 0 & 0 & 0 & 0 & 0 & 0 & 0 & 0 \\ 0 & 0 & 0 & 0 & 0 & 0 & 0 & 0 & 0 & 0 & 0 & 0 & 0 & 0 & 0 \\ 0 & 0 & 0 & 0 & 0 & 0 & 0 & 0 & 0 & 0 & 0 & 0 & 0 & 0 & 0 \\ 0 & 0 & 0 & 0 & 0 & 0 & 1 & 0 & 0 & 0 & 0 & 0 & 0 & 0 & 0 \\ 0 & 0 & 0 & 0 & 0 & 0 & 0 & 1 & 0 & 0 & 0 & 0 & 0 & 0 & 0 \\ 0 & 0 & 0 & 0 & 0 & 0 & 0 & 0 & 1 & 0 & 0 & 0 & 0 & 0 & 0 \\ -0.7725 & 0 & 0 & 0.1294 & 0 & 0 & -0.1216 & 0 & 0.013 & 0 & 0 & 0 & 1 & 0 & 0 \\ 0 & -0.03591 & 0 & 0 & 0 & 0 & 0 & 0.034 & 0 & 0 & 0 & 0 & 0 & 1 & 0 \\ 0 & 0 & 0.2959 & 0 & 0 & 0 & -0.0456 & 0.0031 & -0.0647 & 0 & 0.0075 & -0.0102 & 0 & 0 & 1 \end{bmatrix}$$

$$G = \begin{bmatrix} 0.2643 & 0.0036 & -0.0052 \\ 1.1517 & -0.0111 & -0.0081 \\ -0.0289 & -0.0034 & 0.0719 \\ 0 & 0 & 0 \\ 0 & 0 & 0 \\ 0 & 0 & 0 \\ 1 & 0 & 0 \\ 0 & 1 & 0 \\ 0 & 0 & 1 \\ 0 & 0 & 0 \\ 0 & 0 & 0 \\ 0 & 0 & 0 \\ -0.2643 & -0.0036 & 0.0052 \\ -1.1517 & 0.0111 & 0.0081 \\ 0.0289 & 0.0034 & -0.0719 \end{bmatrix}$$

If the weights are chosen as

$$Q = \text{Diagonal}[50\ 50\ 50\ 50\ 50\ 50\ 50\ 50\ 50\ 100\ 100\ 100\ 50\ 50\ 50]$$
$$R = \text{Diagonal}[50\ 50\ 50]$$

The feedback gain matrix evaluated by solving the Matrix Riccati Equation is

$$K = \begin{bmatrix} 0.2465 & 0.1704 & 0.0015 & -0.0465 & 0 & 0 & 0.0435 & -0.0163 & -0.0041 & 0 & 0 & -0.0001 & -0.1463 & -0.2460 & -0.0053 \\ 0.5898 & -0.1123 & -0.0056 & -0.1184 & 0 & 0 & 0.1121 & 0.0105 & -0.0129 & 0 & -0.0001 & 0.0002 & -0.3284 & 0.1988 & -0.0276 \\ -0.4449 & 0.08 & -0.0608 & 0.0893 & 0 & 0 & -0.0746 & -0.0076 & -0.0034 & 0 & -0.0015 & 0.0021 & 0.2477 & -0.1417 & -0.1969 \end{bmatrix}$$

The performance of the optimal multivariable PIP control scheme applied to the non-linear scrubber process is illustrated in Figures 5.47 to 5.49. Figure 5.47 shows the response when the pressure (P1) setpoint is changed from 0.9 to 0.92 bar A, Figure 5.48 shows the response when the temperature (T1) setpoint is changed from 523 to 543 °K and Figure 5.49 shows the response when the flow (F6) setpoint is changed from 11260.9 to 11560.9 m^3/hr.

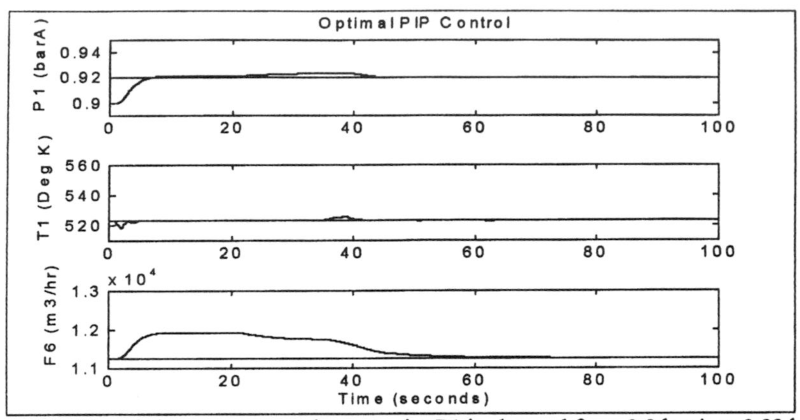

Figure 5.5 Closed -loop response when setpoint P1 is changed from 0.9 barA to 0.92 barA

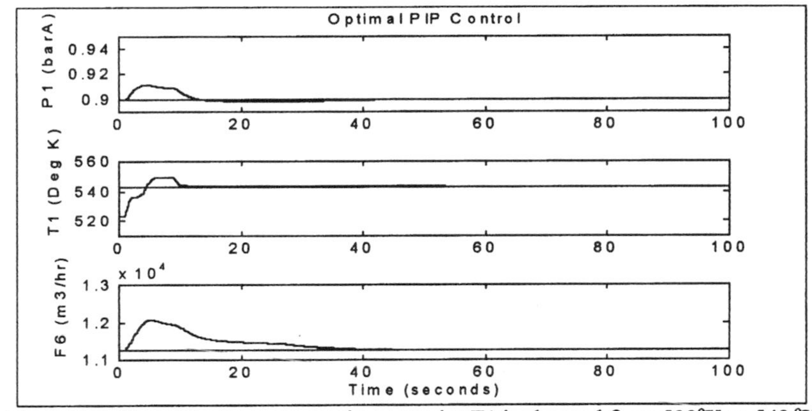

Figure 5.6 Closed-loop response when setpoint T1 is changed from 523°K to 543 °K

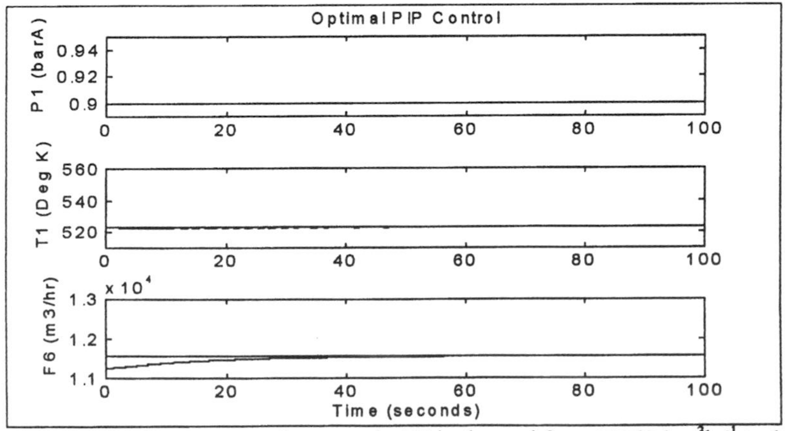

Figure 5.7 Closed-loop response when setpoint F6 is changed from 11260.9 m^3hr^{-1} to 11560.9 m^3hr^{-1}

As can be seen, the optimal multivariable PIP control scheme provides satisfactory dynamic and steady state performance with acceptable levels of interaction.

CONCLUSIONS

The paper has outlined, briefly, two instances where a GA can be used as an effective tool to help the control engineer in the design of MIMO control schemes. The ultimate success of these tools can, however, only be judged against the success or otherwise of any ensuing controller policy. In the example presented this requirement for effective control would appear to have been met. However, it must be noted that the tools themselves are still in a relatively early development phase and further refinement and testing is still required. In addition, there is a need for further exploitation on a wide range of industrial examples and perhaps an investigation into the use of alternative controller design approaches, such as GPC.

REFERENCES

1. Billington, A.J., Boucher, A. R., Cox, C. S. 'Optimal PIP Control of Scalar and Multivariable Processesl', Control' 91.
2. Young, P.C., Lees, M., Chotai, A., Tych, W., & Chalabi, Z.S., 'Modelling and PIP control of a glasshouse micro-climate', Control Engineering Practive, Vol. 2, No.4, 591-604, 1994.
3. Young, P. C. and Wang, C, 'Identification and Estimation of Multivariable Dynamic Systems.' Chapter 15, Multivariable Control for Industrial Applications, Ed J. O'Reilly, Peter Peregrinus Ltd. 1987
4. Goldberg D.E., 'Genetic Algorithms in search, optimisation and machine learning', Addison-Wesley Publishing Co. ISBN 0-201-15767-5
5. Bristol E.H., 'On a new measure of interaction for multivariable process control', IEEE trans. Automatic Control. Vol. 11, 1966.
6. Maciejowski, J. M., 'Multivariable Feedback Design', Addison-Wesley, 1989
7. Chang J.W. and Yu C.C., 'The relative gain for non-square multivariable systems' J. Chem. Eng. Sc. vol 45 No. 5
8. Young, P. C., Behzadi, M.A., Wang, C.L. and Chotai, W.A., 'Direct Digit and Adaptive Control by Input Output State Variable Feedback Pole Assignment', Int. J of Control, vol. 46, No. 6, 1987.

A COLLABORATIVE MANUFACTURING ENGINEERING EDUCATION PROGRAM IN OREGON

G. A. Spolek[1], R. D. Dryden[2], S. J. T. Owen (deceased)[3]

[1]Chair of Mechanical Engineering, Portland State University, Portland, Oregon, USA
[2]Dean of Engineering, Portland State University, Portland, Oregon, USA
[3]Dean of Engineering, Oregon State University, Corvallis, Oregon, USA

ABSTRACT: This paper discusses the establishment of a joint Master of Manufacturing program offered by Portland State University (PSU) and Oregon State University (OSU). Departments of Mechanical Engineering and Engineering Management at PSU, along with Mechanical Engineering and Industrial and Manufacturing Engineering at OSU, enrich the program with their unique strengths. A curriculum was developed by a task force comprised of industries representing a wide range of products.

The curriculum has been offered in the classical live lecture format as well as with several distance learning methodologies. These include two way audio-one way video format on satellite and by microwave, as well as delivered by VCR tapes supplemented by an on-site tutor and an on-call teacher located on campus. All types of formats appear to be well received by the students with diverse needs based upon their employment situations and routine.

Future outlook for the program appears promising and enrollment continues to increase. Experimentation continues in delivery mechanisms as well as timing, location, and course sequences.

INTRODUCTION

As concerns grow regarding the USA's competitiveness in a global environment, the importance of manufacturing engineering has concomitantly raised. A rapidly changing economic world demands products of high quality that are available immediately.

Competitiveness, to a large extent, does not simply rely on the ability of companies to better utilize the available technology; working in teams to focus on the wishes of the customer requires today's engineers to blend both traditional analytic and experimental skills with interpersonal communication and cooperation skills. As an example, Philip Condit, Chairman and CEO of the Boeing Company, listed these core competencies of Boeing: (1) Understanding the customer, (2) large-scale integration of complex systems, and (3) lean and efficient design and production systems. [1]

In the state of Oregon, the need for efficient manufacturing has been complicated by the changing nature of its primary industries. Historically, Oregon relied on its natural resources for economic stability. Forest products (lumber, pulp, paper), fishing, and agriculture dominated the resource-based manufacturing scene. With time, though, dominance by those industries was replaced by "high technology" industries who emphasized adding value to their products. Building on Tektronix's success in manufacturing electronic test equipment, several other electronic, semiconductor, and computer-related industries grew to thrive in the region surrounding Portland, the state's largest metropolitan area. Additionally, specialty metals (aluminum, titanium, zirconium) and metal processing industries developed. A cadre of companies producing equipment in optics, transportation, robotics, material handling, flexible packaging, and biotechnology also arose. In short, Oregon, and especially Portland, evolved within thirty years into a major manufacturing center in markets that were rapidly changing and expanding. The need for highly qualified manufacturing engineers to sustain that growth was desperate.

Like many states, Oregon established state universities in key locations during its resource-based manufacturing years. The land grant college was placed in Corvallis, in the center of the rich Willamette Valley agricultural region, to support agricultural, mining, railroading, and the like. That institution, Oregon State University (OSU), was commissioned to offer engineering. Subsequently, as Portland outgrew the rest of the population centers to dominate the economic needs of the state, Portland State University (PSU) was founded and also chartered to offer engineering programs. However, the larger and more established OSU had already developed its programs in industrial and manufacturing engineering, so PSU was not allowed to expand into that arena by rules established by the governing board of state higher education. Hence, the needs of the Portland metropolitan area and its thriving manufacturing base were not being served by the engineering education structure of the state.

Recognizing the special needs of the manufacturing industries of the Portland metropolitan area, especially for graduate education and advanced research, the state legislature created a special branch of higher education called the Oregon Joint Graduate Schools of Engineering (OJGSE). Its mission was to draw on the combined resources of PSU and OSU to provide the graduate education needs of industry and its working professionals, in their workplaces throughout the state. Five specific areas were identified as key to the state's economic health; one was manufacturing. Several mechanisms to support manufacturing were investigated and initiated, including targeted research, short course and seminar series, and graduate education. The project discussed in this paper is the development of the graduate education program in manufacturing engineering in Oregon.

PROGRAM DEVELOPMENT

From the onset, it was clear that any education program being developed specifically to address the needs of professionals working in industry must involve representatives of that industry. A Technical Advisory Board (TAB) was formed drawing from a wide range of products, as represented by the following companies:

Boeing (Aerospace)
Blount (Forestry harvest equipment)
Gunderson (Heavy manufacturing)
Harris Group (Facility design)
Hewlett-Packard (Computer systems and peripherals)
Intel (Semiconductors, computer systems)
James River (Paper products)
Precision Castparts (Metal casting)
Sequent Computers (Large computer systems)
Tektronix (Electronic measurement equipment)

Faculty representatives from PSU, OSU, and OCATE (Oregon Center for Advanced Technology Education) met with the TAB members regularly to develop all aspects of the program and curriculum. Topics of greatest interest for inclusion in the curriculum varied with industry, but a focus group study revealed [2] those of greatest common value. The resulting curriculum included the following areas of emphasis:

Analysis	Statistics, Numerical Methods, Statistical Process Control, Design of Experiments.
Manufacturing Management	Systems Engineering, Systems Management, Project Management, Communication and Team Building, Concurrent Engineering
Management	Strategic Planning, Human Resources, Cost Accounting
Technical Electives	CAD/CAM, Industrial Safety, Human Factors, Mechanical Tolerancing, Total Quality Management, Design for Manufacturability, etc.

With the curriculum in place, program administration and delivery were developed. The key objective was to deliver the program to the engineers at convenient locations and times, drawing on existing resources. Neither PSU nor OSU had the faculty to deliver all courses to divergent sites, and still maintain ongoing programs. The solution arose by combining the faculty of two departments from each campus: Mechanical Engineering and Engineering Management departments of PSU; Industrial and Manufacturing Engineering and Mechanical Engineering departments of OSU. The core and specialty courses could and would be taught by faculty located at either campus. Alternate campus sites, as well as those stationed within corporate sites or at institutional sites around the region, would be serviced through interactive television. Students would be allowed to register through either PSU or OSU, and all credits would accumulate toward a Master of Engineering in Manufacturing Engineering degree granted jointly by both universities. Admissions, transcription, and graduate standings would be transparent across institutional lines. An academic committee was assembled to administer and adjudicate programmatic matters.

The solution, as envisioned and developed, would solve the problem of providing pertinent graduate Manufacturing Engineering education to working professionals in the Portland metropolitan area by using these unique features:

- Faculty and facilities from both institutions would be utilized.
- The curriculum was developed by a Technical Advisory Board to respond to current needs.
- The program emphasized practice-based, rather than research-based, outcomes.
- Extensive use of interactive television allowed course delivery to local sites.
- Course schedules were arranged to accommodate the constraints of working professionals, and taped backup was available for students unable to meet regularly scheduled times.

RESULTS

The program began on a pilot basis in 1993, and admitted its first students in 1994. Each quarter, four courses were televised, nominally two each originating from PSU and OSU; therefore participating departments each taught about three courses every term. On-campus students took the classes as well, so normally the courses were covered in-load by the departments. Hence, the only additional load was that associated with preparing and delivering the televised course, and the overhead of servicing the needs of remote students.

Courses were scheduled with a frequency to allow students to graduate within two years of regular study; very few did graduate that rapidly. Of the twelve courses offered each year, typically eight covered core material and four, elective material. Surveys taken after two years of operation revealed that many students sought broader offerings of electives, especially in advanced design and manufacturing topics. Core courses appeared to lose popularity, even though required for the degree. It is unknown whether the demand was being saturated or more students were taking targeted courses for the material rather than pursuit of a degree. In any case, topical offerings were adjusted to meet the demands of the students (customers).

A large aerospace company, Boeing, entered the program on a trial basis. Several Boeing engineers registered for the program even though the program's center of operations was 200 miles away. Boeing chose to forego use of interactive television, opting instead for a tape/tutor format. In this mode, the live lectures were recorded on tape and sent to Boeing headquarters. There, tapes were viewed by students on a schedule of their convenience, while being supported on site by a trained tutor. Some students regretted the loss of live interaction with the instructor, but appreciated the flexibility of schedule. Interaction between the instructor and students was facilitated through regularly scheduled telephone office hours, e-mail exchange, and quarterly visits by the instructor to the Boeing site.

To date 31 students have been admitted to the program; 6 have graduated. The demand has not met initial expectations, but seems to be increasing as the program reputation spreads.

CONCLUSION

A practice-based Master of Engineering in Manufacturing Engineering program has been developed in Oregon, USA, by drawing on the combined resources of four departments in two universities. The typical student is the working professional engineer who attends class part-time at interactive television sites conveniently located. The program blends technology and management aspects of manufacturing engineering in a curriculum designed in close consultation with representatives of industry. Preliminary evaluations indicate that the program is achieving some success in improving the manufacturing engineering abilities for local industries.

REFERENCES

1. **Condit, Philip,** "Opening Remarks," *Boeing/University Workshop*, Seattle, WA, February, 1997.
2. **West, T., E. McDowell, and G. Reistad,** "Development of a Joint University Graduate Program in Manufacturing Engineering," *Fifth International Symposium on Robotics and Manufacturing - ISRAM '94*, Maui, HI, USA, August, 1994.

MODELLING OF SOCIAL SYSTEMS FOR ENHANCING THEIR PERFORMANCE

Petia Sice
European Process Industries Competitiveness Centre
University of Teesside

ABSTRACT: Most of the work in modelling of social systems focuses on the actual codification of such models thus taking for granted the underlying assumptions and the modelling process itself. This paper describes an innovative conceptual framework for modelling combining the opposing traditions of subjectivism and objectivism, materialism and idealism, chaos and equilibrium, absolutism and relativism, ethics and pragmatism . It is based on perspectives opened by cognitive science, chaos theory, systems thinking, requisite variety and hermeneutics. The enquiry introduces an observer and his cognising capability and addresses human factors' concerns in modelling of social systems. References to practical management issues have been emphasised where possible. The authors suggest that research into *human factors*, and specifically how people interact among themeselves and with the world and how they commit themselves to action and acquire new behaviours should be enhanced. They refer to the importance of models that include generation of organisational competencies and knowledge creation.

DEFINITIONS AND ASSUMPTIONS

A social system is here defined as any system constituted of humans that remains distinct through continuously regenerating its interactions over time. Thus an entertainment party is not a social system while an enterprise or a nation are. What we are saying is that social systems actively maintain identity.

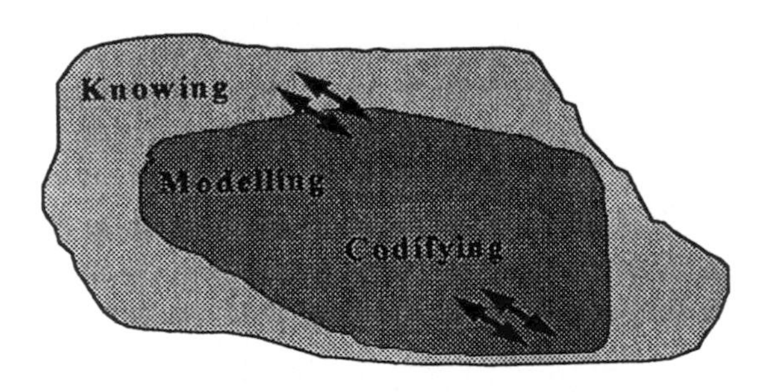

Figure 1. Knowing-Modelling-Codifying Tripartite

It is important to emphasise that we are here referring to modelling as knowing about a system. Thus our models are identified as our full (full to the extent of our cognising capability) knowledge about a system. Thus our models are messy accounts, including our implicit and explicit assumptions, our conceptual framework, our feelings and awareness about the subject. Our models breathe and develop with us. The double-pointed arrows emphasise our understanding of modelling as knowing and modelling as creating the situation. The modeller through his models and relevant interactions is also the creator of the world he/she lives in. This echoes in the writings of Ferguson [1980]: 'If we believe the universe and ourselves to be mechanical , we live mechanically....If we imagine that we are isolated beings, we ....will lead different lives than if we know a universe of unbroken wholeness. ' And as the sculptor molds the clay so do our models 'mold' our lives. *Thus modelling is contemplation and creation.*

The act of creating explicit representations of models we are here addressing as codifying. [Boisot, 1994] Codifying our models allows us to structure our experiences and communicate them. Codification methodologies and tools already offer an implicit organisation of representing a model and this can influence back the content of the model and thus the modelling perspective as well. Codifying deserves accurate attention as otherwise it can restrict the modelling perspective. Constructive issues about choosing the level of resolution in codifying our models is given by Beer in 'Diagnosing the System for Organizations'. [Beer, 1990] Different levels of codification are required for different issues of concern. Thus the vision statement is a codified model of what the social system is about at a very high codification level, while the operation plan is an example of a low level of codification of the system's model exhibiting a lot of detail.

FRAMEWORK OF ENQUIRY IN MODELLING OF SOCIAL SYSTEMS

An innovative framework of enquiry has been developed based on combining the alternative traditions of subjectivism and objectivism, chaos and equilibrium, absolutism and relativism. The main issues of concern are summarised below :

Models are specifications

A *model is a specification of a system rather than an objective account of it*.

The understanding of the above comes from the paradigm of enactive cognitive science. A well noted observation reveals that when a stick is illuminated by white light from one side and a red light from the other, it casts two shadows one of which appears to be red, and the other green (both against a pink background). However, 'objectively' there is no light within the spectrum of wavelength called 'green' in the above experiment, but only shades of white, red and pink. The features of the entities we (see) recognise are not determined by the entities themselves only. *Instead, in our interactions we specify a world rather than recognise one and this world constitutes our knowledge*. Our activity brings forth our world. The word activity is here used in the meaning of all the dimensions of our daily life - thus including action as well as reflection

The Modeller is the Ultimate Point of Reference

The main operation performed by the modeller is the operation of distinction - conscious or subconscious distinguishing of entities, relations, feelings, intuitions, etc. from a background and

from him/herself. Thus the modeller acts as an observer. The modeller is thus the ultimate point of reference in knowing a system.

Thus the members of a social system are acting as observers and creators of it at the same time. The above will have further implications on the system since culturally different men live in different cognitive realities that are specified through their living in them. Consequently, a model of a situation is limited by the cognitive capacity of the modeller. What is needed in better management is power based on competence, distributing the power and responsibility to people with most relevant cognitive capacity in the context under concern. The realisation of the above requires fundamental cultural change.

The modelling of social systems as knowledge-oriented and knowledge-distributed systems will allow better performance and will help communicate cultural change.

System, Organisation and Structure are epistemological qualities

It is important to emphasise that an entity is brought forward in our description in an act of distinction (the main cognitive operation of the observer) and it is this operation of distinction that defines it and makes it possible and specifies its properties as a 'simple' unity. If the observer applies the operation of distinction recursively and thus distinguish the components within the unity he redefines it as a 'composite' unity (system - a unity with structure). [Maturana & Varela, 1980].

The concept of a system (a set of elements) and its organisation and structure are epistemological qualities that the modeller (observer) recognises performing an operation of distinction and uses them in his/her descriptions. They are not claims of how phenomena objectively occur.

The relations between components that define a system as a composite unity of a particular kind, constitute its organization. Thus the organisation of the social systems constitutes of the processes of regeneration of the processes of communication between its members .
Structure includes the components and relations that actually constitute a particular system and makes its organisation real. The definition of structure and organisation are given by an observer and depend on the level of recursion in distinction.

The phenomenal domains (all actions and interactions) generated by the operation of a social system as a unity is different from the phenomenal domain in which its components operate. This makes phenomenal reductionism (and thus explanatory reductionism) in modelling impossible.

Applying Systems Thinking to the modelling process is the only way to avoid ambiguities in mixing different phenomenal domains.

Environment and system are mutual sources of perturbations and not of instructions

Every organisation can be realised through different structures. It is the structure of the system that determines its interactions in the world it lives in. As observers we distinguish a unity from its background, thus distinguishing two structures : of the system and of the environment. The interactions between the system and the environment can bring about a change in the system but do not determine it. What happens to the system is determined by its organisation and its structure. The same holds for the environment. Thus environment and system are mutual sources of perturbations and not of instructions. This process Maturana and Varela call *structural coupling.*

The ontogeny (history of structural change without loss of identity) of a social system takes place with conservation of organisation and structural coupling in its environment.

It is not the variations in the environment that the observer might see that determine the structural changes of the system, but the conservation of structural coupling of the system in its own environment (niche, medium).

The niche *(medium)* is defined by the classes of interactions into which a system can enter. The environment is defined by the classes of interactions into which the observer can enter and which he treats as a context for his interactions with the observed entity. The observer beholds the system and environment simultaneously and he considers as the niche of the organism that part of the environment which he observes to lie in its domain of interactions. Accordingly, as for the observer *the niche appears as part of the environment*, for the observed system the niche constitutes its entire domain of interactions (its cognitive domain), and *as such it cannot be part* of the environment that lies exclusively in the cognitive domain of the observer. [Maturana, 1970]

Models of Social Systems' Behaviour Cannot Claim Complete Predictability

Niche, and environment intersect only to the extent that the observer (including instruments) and subject have comparable organisations, but even then there are always parts of the environment that lie beyond any possibility of intersection with the domain of interactions of the system, and there are parts of the niche that lie beyond any possibility of intersection with the domain of the interactions of the observer. Thus the modeller knows that under each situation observed there are deterministic processes but in order to model what happens he needs a descriptive account that is practically inaccessible.

The above has an impact of our understanding of predictability. If we acknowldegde that the prediction reveals what we as observers expect to happen to the observed system, we cannot affirm that the structural determination of the system denotes a complete predictability. This is so because we may not be in a position as observes to know what we need to know about the operation of the system when we cannot identify all the variables involved or when the system changes when under observation. and thus can go unnoticed. Nobody can observe accurately the slightest variations in the interactions of human beings, the variations in their moods and motivation.

Models of social systems cannot claim complete predictability. They can possibly predict the class of phenomena that may occur, but no event in particular.

Dynamical Models are most appropriate for social systems

Dynamical models as they allow evolution and change are most appropriate for social systems. What matters is not to formulate the model once and forever and then choose from alternative behaviours. What is important is to explore the situation and continuously broaden the horizon of the issues we are addressing. Dynamical models.presuppose continuous learning. Static models (models with rigid logic) threaten society with fossilization.

In a dynamic system (like living beings, social systems, etc.) the structure of the system changes continuously and so does its domain of possible interactions although at any moment they remain determined by the present structure. 'All attempts to prescribe or predict from outside are doomed to failure. The system has got its history (ontogeny) and the future is path dependent. [Prigogine, 1989] *Time is a creative dimension in the ontogeny of the system :*

> 'If a new activity is launched at a certain time, it may well grow and stabilise....However if the same activity is launched at a different time, it need not succeed; it may regress to zero and represent a total loss. It illustrates the dangers of ... planning by direct extrapolation Such static methods threaten society with fossilization, or in the long term with collapse. *..the adaptive possibility of societies is the main source allowing them to survive in the long term, to innovate of themselves, and to produce originality.*' [Prigogine, 1989].

Social Systems are Autonomous Systems

Social systems and their components are autonomous systems and cannot be explained through treating them as non-autonomous systems and this knowledge should be taken into consideration while modelling/explaining social systems.

A formal definition of autonomy states that autonomous systems are :' ...defined as a composite unity by a network of interactions of components that (I) through their interactions recursively regenerate the network of interactions that produced them, and (ii) realize the network as a unity in the space in which the components exist by constituting and specifying the unity's boundaries as a cleavage from a background...'. [Varela, 1987]

Notions as purpose, regulation , transmission of information do not enter in the realization of an autonomous system because they do not refer to actual processes in it. They are cognitive notions which represent the interactions of the observer but do not reveal any feature of the autonomous organisation. However they reveal the consistency in operation of the observed system within the domain of observation.

Thus treating (modelling, understanding) social systems as non-autonomous does not reveal their particular way of being and does not give insight into their operation. The same applies to treating the components of the social system as non-autonomous systems. Thus human beings and social systems cannot change over night when 'instructed to do so' as this 'instructions' is a mere perturbations that trigger changes in them but these changes depend on their structure and are subservient to their autonomy. The 'instruction' might even be outside of the systems domain of interactions.

Modelling of social systems should include modelling of a learning capability

Acquiring knowledge is an irreversible process. Knowledge itself is an ever changing stance, there is no absolute truth as such. We need to free ourselves of the image of knowledge as a stock that merely accumulates with time adding quantity to itself. It is the very quality of knowledge that changes : cognition is our engagement with the world and as such reshapes, breaths, exists, develops with the cognitive ability of the observer. It is this cognitive ability of the observer in the enquiry that we need to address. As reality it is revealed to us only through the active construction in which we participate we need to develop a learning capability in human enterprises to accommodate the variety of the individual and the self awareness of the whole. The fundamental role of the learning capability is to continuously question the pre-understood assumptions of enquiry and open it to evolution and thus accommodate the awareness of the unknown and the insight and blindness [Winograd & Flores,1986] introduced by our prejudices.

Language is a Venue for Action

An organism (in our case the human being) can enter into structural coupling with other organisms and if the interacting organisms as dynamic systems have continuously changing structures, and if they reciprocally select in each other their respective paths of ontogenic structural changes through their interaction without loss of autopoiesis, then they generate a domain of communicative interactions. [Maturana&Varela, 1980] The individual ontogenies of the participating organsims occur as part of the network of co-ontogenies that they bring about in constituting higher order unities (social systems).

Communication consists of coordinated behaviours mutually triggered among the members of the social system.

This consensual (linguistic) domain of communicative interactions is non-informative, although the observer may describe it as if it were so. Communication depends not entirely on what is transmitted, but what happens to the person who receives it. [Maturana, 1970]

Communication thus is a matter of mutual orientation- 'primarily with respect to each others behaviour, and secondarily (only via the primary orientation) with respect to some subject.' Whitaker [Internet Notes on Autopoiesis].

In the traditional approach the interaction between humans is described as semantic coupling - a process by which each of the participants computes its appropriate response state from some informative input of the other. As we have shown above this is not the case and the notion of information is valid only in the descriptive domain of the observer.

The key feature is that language enables those who operate in it to describe themselves and their circumstances through linguistic distinctions of linguistic distinctions. *Word are tokens for linguistic coordination of action. Linguistic interaction is a venue for action, coupling the cognitive domains of two or more actors.* 'Language was never invented just to take in an outside world. Therefore it cannot be used as a tool to reveal that world. ' ...it is by languaging that the act of knowing, in the behavioural coordination, which is language, brings forth a world. *We work out our lives in a mutual linguistic coupling, not because language permits us to reveal ourselves but because we are constituted in language in a continuous becoming that we bring forth with others.'*. [Varela, 1987]

The unity of the human society is generated through the network of conversations which language generates and which through its closure generate this network of conversation.

Social Systems are Cognitive Systems

The author's theoretical and empirical [Sice P., Elliott M.] research have led to the development of the codification of modelling social systems' ontogeny given in Fig. 1. The proposed framework has been developed on the ground of the issues discussed above. It represents a generic conceptual construct that recognises that a social system is a cognitive system. [1] *Its cognitive domain is the entire domain of its interactions. It is realised through the interplay of the capabilities of its components.* The cognising capability of the social system is dependent on its capacity to interact, i.e. to distinguish an environment, to identify its medium of recognisable interactions. (Fig. 1.) The variety of the medium is determined by the social system's coupling with the environment, that in turn is changing with the emergence of new structures and thus new interactions.

The conceptual framework (Fig. 1) allows space for introducing modelling as both contemplation and creation. The order loop deals with observing the system from a meta-level. The disorder loop promotes creativity through partially codified micro-events models.

Knowledge creation happens through development of new interactions with the internal and external environment of the social system.

Management models for change of performance, organisatinal learning , knowledge creation should explore not what instructions need to be introduced into the system but what interactions should be designed to trigger appropriate changes. This applies to both the internal (the network of components) and external environment of the social system.

CONCLUSION

As a human activity modelling is an act of knowing, thus our models are not and can never be 'true' or 'complete' in the absolute sense of the objectivist tradition. In interaction with our environment we do not acquire a direct mapping (representation) of it, rather we acquire relative and temporary specifications that develop with our ability to distinguish while at the same time enhancing this ability by continually bringing forth a world.

We can model a social system in different domains, depending on the distinctions we make. On one hand we can consider a system in the domain where its components operate, in the domain of its internal states and considered from the internal dynamics of the system, the environment appears irrelevant. On the other hand we can consider a system that also interacts with its environment and describe its history of interactions with it. From this perspective of the observer, the internal dynamics is irrelevant.

Both of the above descriptions are necessary to complete our understanding of a social system. It is a fact that in our current models of social systems the operational description has been neglected. Thus models of market driven organisations are presented as an up-to-date insight into good management. We should though, notice the danger of 'emptying' the organisations of their content as we do not

---

A cognitive system is a system whose organisation defines a domain of interactions in which it can act with relevance to the maintenance of itself, and the process of cognition is the actual (inductive) acting or behaving in this domain. [Maturana, 1970]

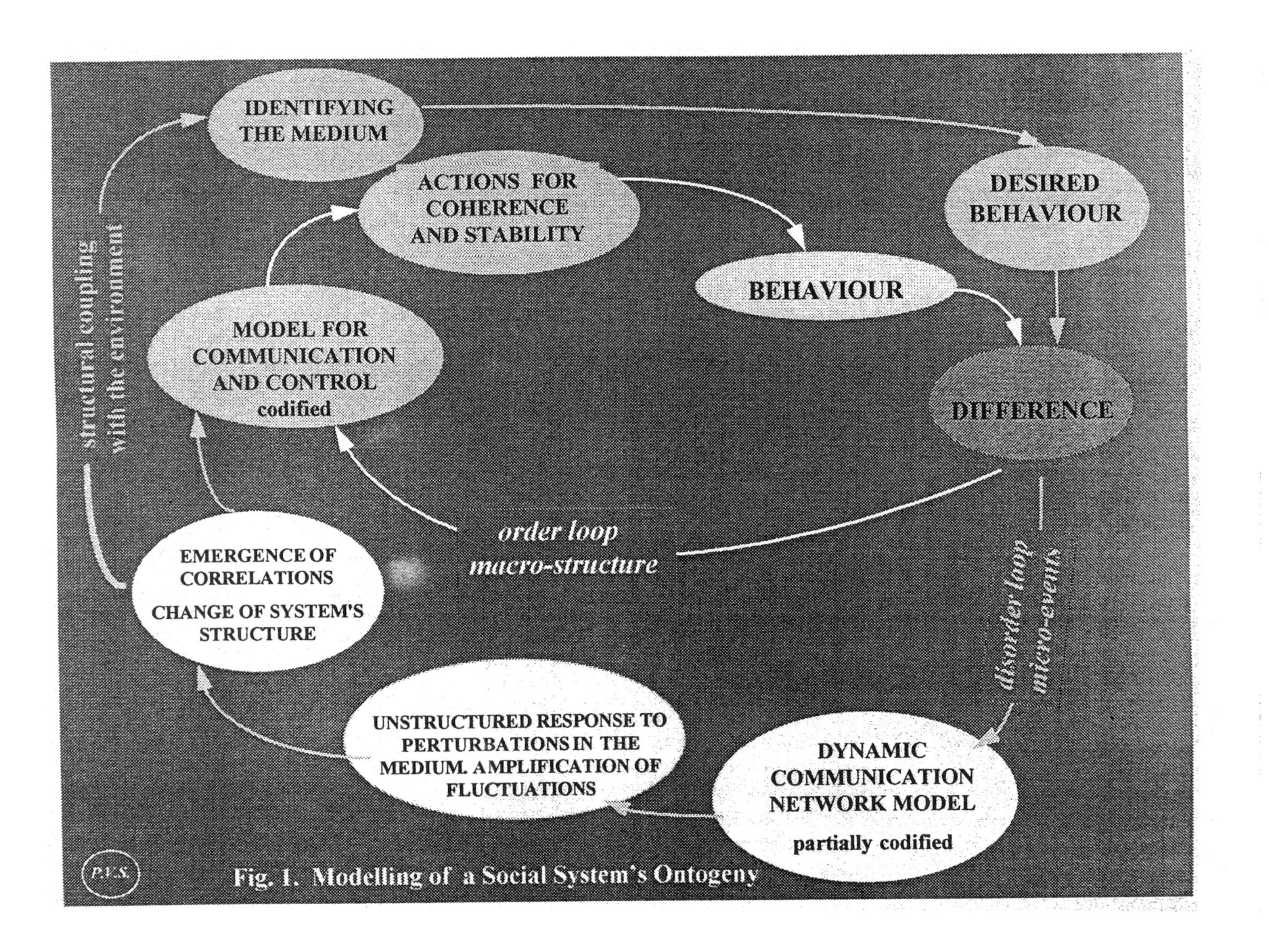

Fig. 1. Modelling of a Social System's Ontogeny

pay enough attention to the generation of behaviour from inside. On the other hand authors such as Nonaka and Hall, realising the above, are bringing to research the healthy issue concerning the development of organisational competencies and the creation of knowledge within the organisation.

REFERENCES

Ashby, W. R., Intorduction to Cybernetics, Chapman and Hall, 1956.

Beer, S., Diagnosing the System: for Organizations, John Wiley and Sons, 1993.

Boisot, M., Information and Organizations, Harper Collins, London, 1994.

Hall, R., The Strategic Analysis of Intengible Resources, Strategic Management Journal, 13, 1992.

Flores, F., Winograd, T., Understanding Computers and Cognition, Ablex Publishing, 1986.

Maturana , Biology of Cognition, North Holland, 1970

Maturana , Varela, F., Autopoiesis - the Organization of the Living, North Holland, 1980

Nonaka, I. , The Knowledge Creating Company, Harvard Business Review, 1991.

Ferguson M., The Aquarian Conspiracy: Personal and Social Transformation in 1980's, , J.P. Tarcher, Los Angeles.], 1980.

Progogine I., Stengers, I., Order out of Chaos, Heinemann, London, 1989

Senge, P., The Fifth Discipline: the Theory of the Learning Organization, John Wiley& Sons, 1990.

Sice, P., Complexity and Knowledge. Questioning the Assumptions in the System of Enquiry about Management, International Conference in Knowledge Transfer, 1996.

Sice P., Elliott M, Using Systems Dynamics Approach for Improving Manager's Imagination Models, International System Dynamics Conference, Stirling, 1994.

Varela F., The Tree of Knowledge, New Science Library, Boston, Massachusettes, 1987.

Whitaker R., Internet notes on Autopoiesis, 1997.

ERGONOMICS OF MATERIAL HANDLING

Karl H. E. Kroemer, Dr. Ing.
Professor and Director, Industrial Ergonomics Laboratory
Human Factors Engineering Center
Industrial and Systems Engineering Department, Virginia Tech (VPI&SU)
Blacksburg, Virginia 24061-0118, USA

ABSTRACT: The goal of Human Factors/Ergonomics is "humanization" of working and living conditions. This goal can be symbolized by the "Ergonomic Double E" of Ease and Efficiency for which technological systems, and all their elements, should be designed. This requires knowledge of characteristics of the people involved, particularly of their dimensions, capabilities, and limitations; and the conscientious engineering and managerial application of this knowledge. Altogether, there are seven key issues that must be balanced to achieve ease and efficiency in manual load handling: facility layout, job design, equipment design and selection, consideration of people involved, training, screening, and the design of work place and work task.

Ergonomic information on many of these aspects used to be scattered throughout the literature; a systematic approach to combine all knowledge into one set of comprehensive guidelines was needed. This has been achieved now for each of the seven keys of efficient and safe material handling.

In principle, ergonomics can be applied both in the original design of material handling systems and in the modification of existing ones. Application of ergonomic knowledge ensures prudent use of human capabilities and abilities, and safeguards people from overexertion and undue strain.

Ergonomics can be used in two major strategies:

- Fitting the person to the job. This means selection of individuals for their ability to perform certain tasks, and training of these persons to perform their tasks better, more safely.
- Fitting the job to the person. Here, the task, equipment, and work organization are adjusted to fit human capabilities, limitations, and preferences.

Especially in material handling, both approaches can be used at the same time to supplement each other. However, fitting the job to the person has highest priority.

KEY #1: FACILITY LAYOUT

It is the purpose of purposeful facility layout, or facility improvement, to select the most economical, efficient, and the safest design of building, department, and workstation. Of course, specific details depend on the overall process, but in general a facility with well laid out material flow has short and few transportation lines. Transportation is always costly in terms of space, machinery and energy; it does not add value to the object being moved about but it is full of hazards to people. In many existing facilities, reduction and simplification of material movement can lower the expense of material transport considerably which often amounts to 30 to 75 percent of total operating cost. Of course, even more important from a human point of view is the possibility of reducing the chances of overexertion and injury to workers by re-designing, improving or eliminating transport lines. For this reason it is very important to include an ergonomist in the team that is planning a new facility.

There are two major design strategies: "process layout" and "product layout." In the first case, all machines or processes of the same type are grouped together, such as all heat treating in one room, all production machines in another section, and all assembly work in a different division. The major advantage of process layout design is that quite different products or parts may flow through the same work stations, keeping machines busy. But much floor space is needed, and there are no fixed flow paths. Process layout requires relatively much material handling. In contrast, in product layout all machines, processes, and activities needed for the work on the same product are grouped together. This results in short throughput lines, and relatively little floor space is needed. However, the layout suits only the specific product, and breakdown of any single machine or of special transport equipment may stop everything. Altogether, product layout is advantageous for material handling, because routes of material flow can be predetermined and planned well in advance.

It is rather easy to describe events and activities with simple sketches, symbols, and words. The *flow diagram* is a picture or a sketch of the activities and events. It indicates their sequence, and where they take place. The *flow chart* is a listing or table of the same activities and events. It indicates their duration and provides detailed information on related facts or conditions. Flow charts can be modified to reflect special conditions in specialized industries, or to point out specific concerns such a repetitive manipulations. Both recording and analysis techniques can be done in the traditional paper-and-pencil fashion or with the help of computerized tools. Either way, they provide comprehensive, easily comprehended, and convincing information on the features of existing facility layouts and on possible newer and better solutions.

Ergonomic solutions *are needed, and largely available, for the planning of new facilities, and the redesign of existing ones, for a minimum of material handling. Modern technology allows flexible recording, and then analysis, of material flow by using computer templates and software. However, management must be convinced that it is necessary and efficient to have ergonomic participation in the planning and review of facility layout.*

KEY #2: JOB DESIGN

This is one of the most important decision points in the planning and design of a new facility, or in the improvement and re-design of an existing plant: which material movement should be assigned to machinery? Or what activities need to be performed by people?

"Task allocation" primarily falls into three categories:

- Automation of the material movement, with no people directly involved.
- Mechanization of the material movement, with people as controllers and operators.
- Manual material movement, with persons "doing the job" at risk of injury and overexertion.

People function best as supervisors and controllers: they can deal with unexpected events, can make decisions even when only incomplete information is available, and are able to function under overload conditions. Machines are much superior to people in speed, power, and their ability to perform routine, repetitive, and precise actions. This speaks for as much automation, or mechanization, as technologically and economically feasible.

Traditional "principles of material movement" can help in task allocation and job design, but one must be cautious in applying some of the old axioms. The "simplification principle," for example, is part of classic industrial methods engineering of the middle 1900s but now has acquired special meaning: with the recent recognition of the dangers associated with repetitive motions, specialization and simplification at the job should not be carried to the extend that a person must perform the same body motions and exertions thousands of times, over and over.

The one-time sudden overexertion leading to an injury is only one type, and not the most frequent type, of trauma related to material handling. Cumulative trauma disorders (CTDs, RSIs, OMDs, etc.) occur when an activity is repeated so often that is overloads the body parts involved. Such overuse disorders have been described in the literature since the early 1700s, under various names. The major job factors in CTDs are rapid, often repeated movements, forceful exertions and movements, static muscle loading including maintenance of odd posture, and vibrations, especially in a cold environment. All of these can occur in load handling. Their negative effects are likely to be aggravated if several components are combined, such as numerous repetitions with high force exerted in awkward posture.

High *repetitiveness* has been defined as a cycle time of less than 30 seconds, or as more than 50% of the cycle time spent performing the same fundamental motion. High *force* exerted with the hand, e.g. more than 45 N, may be a causative factor by itself. A strong maintained isometric contraction of muscles needed to keep the body or its part in position, or to hold a hand tool, or to carry an object, is often associated with a CTD condition. Also, inward or outward twisting of the forearm with a bent wrist, a strong deviation of the wrist from the neutral position, and the pinch grip can be stressful.

Jobs should be analyzed for their posture, movement, and force requirements using e.g. the well established industrial engineering procedure of "motion and time study." Each element of the work should be screened for factors that can contribute to CTDs. A convenient tool to do so is the "Flow Chart - Flow Diagram" technique discussed in Key 1. This procedure can be applied on a mini-scale to the minute details of a job as well as for macro-evaluation of a whole facility. The analysis may show that material handling activities may in fact be quite different from expected. For example, an evaluation of the load handling activities in the distribution center of a large transport company (Baril-Gingras and Lortrie, 1995) showed that of 3217 handling of objects other than boxes,

- 23% were sliding,
- 22% raising,
- 19% pivoting,
- 9% turning,
- 8% lowering,
- 7% carrying,
- 6% voluntary dropping,
- 3% rolling,
- 2% overcoming a resistance (when objects got caught or jammed).

In 57% of the handlings, the object remained supported in some way on a surface; in 42% it left the surface (was lifted, lowered, carried, etc.); in 04% it fell unintendedly. Such a break-down indicates, not only the details of the job, and hence how the task might be re-designed to make it easier and less hazardous; it also provides guidance on how to provide equipment such as roller conveyors that make the material movement least strenuous and fastest (see Key 3). Finally, the analysis also shows which specific body activities should be avoided or preferred for health and effort reasons, as discussed in Key 5. After the job analysis has been completed, workstation and equipment can be ergonomically engineered and work procedures organized to avoid unnecessary stress on the operator. Of course, the considerations also include the work environment regarding lighting, sound, climate, and housekeeping.

Ergonomic solutions *can be derived from the recognition of "seven sins" that specifically need to be avoided in job design:*

1. *Job activities with many repetitions.*
2. *Work that requires prolonged or repetitive exertion of more than about one-third of the operator's static muscular strength available for that activity.*
3. *Putting body segments in extreme positions*
4. *Work that makes a person maintain the same body posture for long periods of time*
5. *Pressure from tools or work equipment on tissues (skin, muscles, tendons), nerves, or blood vessels .*
6. *Work in which a tool vibrates the body, or part of the body.*
7. *Exposure of working body segments to cold.*

KEY #3: EQUIPMENT

One can distinguish between equipment that provides assistance to the material handler at the workplace, and equipment that provides for in-process movement between work stations.

Equipment for assistance at the workplace includes

* Lift tables, hoists
* Ball transfer tables, turn tables
* Loading/unloading devices
* Non-powered trucks, walkies, and dollies.

Equipment mostly used for in-process movement includes

* Powered walkies, rider trucks, tractors
* Cranes.

Obviously, several of these can be used both at the workplace and for in-process movement, such as hoists, conveyors, and trucks. Finally, there is a group of material movement equipment primarily *used at receiving and in warehousing*. This includes

* Stackers
* Reach trucks
* Lift trucks
* Cranes
* Automated storage and retrieval systems.

There is specialized industry that supplies and installs automated and highly mechanized machinery that makes it unnecessary for humans to be directly involved. The (US) Material Handling Institute is one source of related information. In this text we concentrate on the ability of equipment to relieve persons from strainful load handling, with emphasis on the safety and ease of use of such equipment for the operator. Often, simple carts and dollies can be used to transport and lift objects to the correct working height. Electric, hydraulic and vacuum hoists or small cranes can avoid that a person has to lift or lower, load or unload manually. Many conveyors are available on which one can move objects easily from one workstation to the next.

There are many kinds of often inexpensive equipment that can do the holding, turning, carrying, pushing, pulling, lowering, and lifting of loads which would otherwise be performed by persons. However, whether this will indeed be done "by machine" depends, besides economical considerations, on the layout of the workstation and organization of the work itself. For example: will a lift table be installed next to an assembly workstation if this means removal or relocation of other workstations in order to make sufficient room? Will an operator use a hoist to lift a 20 kg-object if this is awkward and consumes more time than to simply grasp it and pull it up - even at the risk of some back pain?

Obviously, facility layout (Key 1) as well as work design (Key 2) must be suitable for the use of equipment. Furthermore, the operator must be convinced (trained) that it is worthwhile to go through the effort of using a hoist instead of heaving the material by hand. Equipment must not only be selected to be able to perform the material handling job, but it must also "fit" the human operator Key 4). Unfortunately, some material movement equipment such as cranes, hoists, powered and hand trucks, and particularly fork lift trucks have shown an alarming lack of consideration of human factors and safety principles in their design. When selecting equipment one needs to consider that it must be safe and easy to operate. The old-type forklift truck was neither: the load obscured the operator's forward view and, therefore, much of the driving was done rearward, with the driver's body contorted and the controls awkward to operate. The driver's space was cramped and the seat hard, directly transmitting shocks and vibrations to the trunk.

***Ergonomic solutions** regarding human factors aspects of equipment design and selection are at hand. For example, usability aspects of simple trucks and trolleys were reviewed in industry and hospitals (Mack, Haslegrave and Gray, 1995). From the great variety of observations and comments made by the users of equipment, the following primary areas were identified:*

- *Required operating force, especially for getting the aid into motion, controlling when in motion, and stopping it.*
- *Stability, mostly with two-wheeled and tall aids (made "tall" either due to device design or by putting on a tall load).*
- *Steerability, mostly related to the number, location, size, and type of wheels; quality of wheel bearings; and the size ("wheelbase") of the aid.*
- *Handles, often too low, too small, too short, inappropriately bent and badly located*
- *Starting, particularly with heavy loads and small wheels on the hand truck, and when the device must be tilted before starting to move.*
- *Stopping, particularly on down-slopes, when cornering and on wet or slippery floors.*
- *Loading and unloading, especially with pallets that have too low an opening space, making insertion of the fork difficult; and with "sack trucks," where the front edge of the "shoe" may be difficult to slip under the load.*
- *Security of the load, especially when going over bumps or around corners with aids that by design have no provision to keep the load in place, or with aids equipped with insufficient or awkward-to-use securing devices.*

KEY #4: PEOPLE

People come in all sizes, weak and strong, fit and untrained, female and male, young and old. Ergonomic knowledge of human characteristics stems from several disciplines, such anthropometry (body dimensions, motion capabilities, muscles); work physiology (capabilities for physical work); biomechanics (mechanical structure and strain behavior of the body); industrial psychology and sociology (attitudes and behaviors, perception within society).

Throughout history, attempts have to be made to establish simple rules that describe body dimensions. Among these is the idea that one could define an "average person" which might be used as a standard design template. This approach was and is wrong: there are no persons who are average in many or all respects. Even people with an average height may not have an average weight, have no average arm length, are not of average strength, and do not want to be treated as average. Therefore, equipment, workstations, and material to be handled must be designed and controlled to *suit and fit the whole range of body dimensions, and capabilities, of the people who handle material.* Handle sizes must fit small and large hands. Working surfaces must be designed for short as well as tall people. Even relatively weak people must be able to move material, while the strongest should not break it.

The human body must maintain an energy balance between external demands, caused by the work and the work environment, and the capacity of the internal body functions to produce that energy. The body is an "energy factory," converting chemical energy derived from nutrients (protein, carbohydrate, fat and alcohol) into externally useful energy. Final stages of this process take place at skeletal muscles. The energy conversion needs oxygen transported from the lungs by the blood. Also, the blood removes byproducts generated in the energy conversion, such as carbon dioxide, water, and heat which are dissipated in the lungs where, of course, oxygen is absorbed into the blood. Heat is also expelled through the skin, much of by sweat. The blood circulation is powered by the heart. Thus, the pulmonary system (lungs), the circulatory system (heart and blood vessels), and the metabolic system (energy conversion) establish *central limitations* of a person's ability to perform strenuous work.
A person's capability for labor is limited also by muscular strength, by the ability for movement in body joints (e.g. the knees) or by the spinal column (e.g. "weak back"). These are *local limitations* for the force or work that a person can exert. Local limitations establish very often the upper limits for performance capability. For example, one may simply lack the strength to lift an object because the hands are too far extended in front of the body. The mechanical advantages at which muscles must work often determine one's

While handling material, the force exerted with the hands must be transmitted through the whole body, that is via wrists, elbows, shoulder, trunk, hips, knees, ankles and feet to the floor. In this chain of force vectors, the weakest link determines the capability of the whole person to do the job. If muscles are weak, or if they have to work at mechanical disadvantages, the handling force is reduced. Often the weakest link in this chain of forces is the spinal column, particularly at the low back. Muscular or ligament strain, painful displacements of the vertebrae and/or of the inter vertebral disks in the spine, may limit a person's ability for material handling.

Ergonomic solutions *can be derived from models that describe body size and functional capabilities as well as limitations. Simplified for convenience, they may be grouped as follows:*

- *Physiological models provide information on energy conversion and expenditure (measured in kilocalories) and on the loading of the circulatory system (e.g. heart beats per minute).*
- *Muscle strength models provide information on the ability to exert force to an outside object (usually measured in N). Up to now, these measurements were usually done with the muscle kept at a given length ("isometric strength"), i.e. with involved body segments in "frozen positions". Isokinematic ("isokinetic") and isoinertial techniques are applied to assess dynamic muscle strength, that is, while the body moves.*
- *Biomechanical models provide information on strain tolerance and performance capabilities of the body, particularly of the spinal column and its disks, and on the effect of body positions on available muscle strength.*
- *Psychophysical models combine physical measures with subjective assessments of the perceived strain. They provide synergistic judgments of material handling capabilities and limitations.*

KEY #5: TRAINING

Numerous attempts have been made to train material handlers to do their work (particularly "lifting") in a safe manner. Unfortunately, hopes for significant and lasting reductions of overexertion injuries after training in "safe lifting" have been generally disappointed. There are several reasons for disappointing outcomes of training attempts:

- If the job requirements are stressful, "doctoring the symptoms" such as behavior modification will not eliminate the inherent risk. Designing a safe job is basically better than training people to behave safely.
- People tend to revert to previous habits and customs if practices trained to replace previous ones are not reinforced and refreshed periodically.
- Emergency situations, the unusual case, the sudden quick movement, an increased body weight, or reduced physical well-being may overly strain the body since training did not include these conditions.

Thus, unfortunately, training for safe material handling (not limited to the popular "lifting") cannot be expected to really solve the problem -- although, if properly applied and periodically reinforced, training should help to alleviate some aspects. Yet, even well-intended training programs are ineffective unless the trainee becomes personally convinced that safe material handling techniques are available, feasible, and that it is truly in one's own interest to learn and follow them.

Clearly, the material handler is the person who shall be trained to handle objects and their own body prudently to avoid overexertion, injury and accidents. But the supervisor, foreman, manager, and engineer direct, assign, and design the tasks and procedures, objects and work places; therefore, they need similar training so that they understand what may or may not be expected in handling loads. Furthermore, they are the persons who mostly set the stage for attitudes and behavior in the shop or construction site in terms of “safe efficiency” or reckless “get the job done.” Making employees aware of authority concern is an underlying theme. There is enough anecdotal evidence to support the common sense opinion that training the boss is about as important as training the worker.

Low back disabilities (due to pain and/or injury) are a major health problem among industrial workers. They are the most frequent and most expensive musco-skeletal disorders in the United States, accounting for at least 25% of all compensable overexertion injuries. Back pain has been related to weak trunk musculature, muscular fatigue, degenerative diseases, and to improper posture and inappropriate lifting techniques; however, the majority of industrial back injuries has not been associated with objective pathological findings, and about every second back pain episode cannot be linked to a specific incident.

One of the problems in training for safe load handling is that broad generalizations concerning material handling tasks across all work environments may not be possible. Task characteristics and requirements differ much among industries (e.g. tire-making, mining, nursing) as well as within one industry, depending on the specific job, handling aids and equipment available, successful implementation of worker selection, and ergonomic job design.

Even the group characteristics of material handlers in different industries might be important in designing a training program; for example, hospital workers might have higher educational skills than heavy-industry workers. Female employees are predominant in certain industries or occupations; this can influence training because, on the whole, women are about two-thirds as strong as men.

Recent findings indicate not only that transfer of the same training techniques from one facility to the next may be inappropriate, but also that a job analysis (see Key 2) might be advisable because it can indicate that changes in work details and equipment (see Key 3) are warranted before the training can commence.

Ergonomic solutions *are needed, and available, for these topics: Who shall be the trainer? What exactly to teach? Individual or group training? Voluntary or enforced participation? Which format of the course? How many repetitions? When to train: before first activity, after some time on the job? What method is most effective? Where should the sessions be held? What is the retention of learned information by the trainees after training? How to judge the effectiveness of training?*

KEY #6: SCREENING MATERIAL HANDLERS

Screening of material handlers, that is the selection of persons suitable for stressful material handling, is an important and difficult task. It is important because its success determines whether or not a person can expect to perform load handling safely and in good health, or whether this individual is at risk of an overexertion injury. It is difficult because techniques and procedures are still being developed in research and experimentation, and because care must be taken to avoid discriminatory implications. Matching job demands with a person's capability requires that both

- individual capabilities or limitations and
- the related job demands are quantitatively known.

Screening procedures are meant to place the right worker to the right job. For this, they must fulfill several requirements: Of course, they must be *safe* which is not always easy to garnet, because the testing is meant to detect the upper limit of a person's performance capability. Tests must be *objective* as opposed to subjective: the outcome does not depend on the whims of the tester. They must be *repeatable* by bringing about the same results when done again. They must be *valid, accurate* and *sensitive* , and *specific* .

In the past, selection procedures and techniques have been applied which were primarily based on one or several of the following: *The epidemiological approach* analyzes the circumstances and characteristics that were present in previously observed incidents. While the epidemiological approach provides general information regarding accident probabilities, it does not yield information suitable for individual selection. The *medical examination* tries to identify an individual's impairments which would make this person vulnerable to overexertion injury risks or, conversely, to identify persons who are healthy and able to perform material handling. While obvious abnormalities (for example, of the spinal column or of the musculature) are likely to be identified in a "physical," more subtle impairments often remain undetected.

Furthermore, the medical examination is usually not specific to the job requirements. The *physiological approach* (often part of the "physical") tries to identify physiological capabilities, or limitations, regarding load handling. As discussed earlier, "central limitations" are unlikely to be the limiting factors in today's material handling, while "local limitations" are often not specific enough to be detected in the physiological exam.

The *muscle strength-measuring approach* attempts to assess a person's muscular strength capability with respect to the work demands. Most previous attempts to measure muscle strength were limited to isometric (static) muscle efforts, which is not the same as many exertions actually required in load handling. Measurements of dynamic muscle strength are becoming feasible now, and may alleviate that shortcoming. The*biomechanical approach* considers the human body simply as a system that can be understood and measured in mechanical terms. Mathematical and computerized modeling of the mechanical properties of the body is often used. Inputs into current models are mostly body position details, muscular strength information, and assumptions about spinal column compression capabilities. The *psychophysical method* is based on the concept that human capabilities are synergistically determined by bodily, perceptual and judgmental capabilities. The psychophysical approach relies on a person's ability to "rate the perceived strain" stemming from the physical work performed, and to judge whether or not the strain is personally acceptable for safe performance of that task.

Ergonomic solutions *are being sought for several grave and largely unresolved problems with the screening of applicants for material handling jobs: Why is the task so demanding? Couldn't the requirement be lowered so that anybody of normal health and strength can do the job? How to identify the critical job demands, and how to quantify them? What exactly makes a job so hard? If we don't know the exact job demands, we cannot test applicants appropriately. This generates another set of problems, namely that the examining physician or nurse too often must resort to testing only generic health aspects and performance capabilities of a job applicant, while job-specific evaluations would be more meaningful and protective.*

KEY #7: ERGONOMIC DESIGN OF WORKPLACE AND TASK

We all "handle material" daily. We push, pull, lift, lower, hold, and carry while moving, packing, storing objects, while loading and unloading machinery. The objects may be soft or solid, bulky or small, smooth or with corners and edges. We may handle one object, or perhaps several at a time, and they may be by themselves or come in bags, boxes, or containers, with or without handles. We may handle material occasionally or repeatedly, often as part of our jobs, but also during leisure.

Handling material can involve risks of injury or health hazards, particularly if the objects are too heavy, too bulky, cannot be grasped securely, or if they must be handled too often, in awkward postures and body motions. Manipulation of even light and small objects can strain us because we have to stretch, move, bend, or straighten body parts, using fingers, arms, trunk and legs. Heavy loads pose additional strain on the body owing to their weight or bulk, or lack of handles. Exerting force on an object with the hands strains wrists, elbows, shoulders, the trunk and particularly its lower back, and the legs. The directions and magnitudes of the internal forces are different depending on body posture, the location of the object, and the direction of force that must be exerted on it. The primary area of physiological and biomechanical concern has been the low back, particularly the disks of the lumbar spine.

If the body generates force in motion, the acceleration determines the amount of musculoskeletal force needed, according to Newton's Second Law: force = mass x acceleration. Thus, simply setting a "weight limit" (such as in kg or lb) does not do enough to prevent overexertion injuries. The location of the object to be handled with respect to the body (see Keys 1, 2, and 3) is a major determiner of the needed musculoskeletal effort, as are the magnitude and direction of required force. The load must be within close and easy reach of the hands in front of the trunk. Bending and stooping, particularly sideways twisting of the body while handling materials, are likely to result in injuries.

Pushing and pulling are preferable to lifting and lowering objects; if done at proper height, rolling or sliding an object strains the body less. Use of pushing and pulling is particularly preferred to lifting and lowering an object if it is heavy, difficult or fragile to handle, and handling must be done often. To initiate the movement of an object, a larger force must be generated than to sustain the movement. Both the initial push or pull, and the sustained effort, are facilitated if the object is put on a surface that allows easy gliding, better rolling -- see Key 3. Snook and Ciriello (1991) have developed tables of acceptable push and pull forces for American males and females.

Loads may be *carried* in many different ways. The appropriate manner depends on many variables: the weight of the load, its shape and size, its rigidity or pliability, the provision of handholds or points of attachment, and the bulkiness or compactness of the load. What is the best technique also depends on the distance of carry, on whether the path is straight or curvy, flat or inclined, with or without obstacles, whether one can walk freely or must duck under obstacles or move around them. Kroemer, Kroemer, and Kroemer-Elbert (1994) have listed different ways to carry loads and indicate their advantages and draw-backs. Carrying a heavy load in one hand is especially fatiguing and stressful, particularly for the muscles of the hand, shoulder, and back. Nevertheless, it is often done because of the convenience of quickly grasping an object. For carrying loads in the shop or factory, Snook and Ciriello (1991) published a table that indicates suitable weights to be carried by American males and females.

An existing need for *lifting and lowering* often indicates poor job design. In a shop or factory, the object should have been delivered to the work place at proper height and location -- (see Keys 2 and 3) or lifted or lowered there by equipment to a height where a human can push and pull it easily. Keep in mind that lifting or lowering any object also means lifting and lowering of body parts, often not only of hands and arms, but also involving shoulders, head, neck and torso - easily 25 kg .

In 1981, the US National Institute for Occupational Safety and Health (NIOSH) published its "Work Practices Guide for Manual Lifting." This document contained distinct recommendations for acceptable masses that must be lifted. These recommendations considered location and length of the lift path, and the frequency of lifting. In 1991, NIOSH revised the guidelines and established a new recommended weight limit of, in the best case, 23 kg (Waters, et al., 1993). It represents the maximal mass of a load that may be lifted or lowered by ninety percent of US industrial workers, male or female, who are physically fit and accustomed to physical labor.

In 1991, Snook and Ciriello also published tables of maximally acceptable lift and lower forces which they had derived in psychophysical experiments for US workers. Other recommendations and guidelines have been published nationally and internationally and/or are available as computer programs. Some of these agree fairly well with each other, at least in part and under certain conditions, but one may also find major discrepancies. In Europe, for example, recommendations are rather different from those in the USA. In the case of differing recommendations it is prudent to follow that set which protects the load handler best.

Ergonomic solutions *are up to the "ergonomist" (engineer, manager, designer, health care provider) who can determine the suitable conditions of material handling from existing information (Kroemer 1997, Kroemer et al. 1994). As a rule, the job must be done, but not at the expense of the material handlers. Nature and details of the job are to be determined with the welfare of people foremost in mind. Therefore the ergonomist will select those conditions that are the easiest, least stressful, and safest.*

SUMMARY and CONCLUSION

Of course, guidelines and recommendations such as those discussed above do not absolutely assure that a material handler will be satisfied with the work load, nor that he or she may not be overexerted or even injured when working. People and conditions differ from each other, from day to day and place to place. General guidelines and recommendations have many assumptions, clearly stated or often hidden in fine print. They simplify and normalize, for example by assuming that only one type of work task is done (such as lifting) or that one type of strain (such as spinal disk compression) loads the material handler. Furthermore, different experts and authorities (such as researchers, legislators, judges, government or international agencies) in various regions of the earth have differing opinions about what is tolerable or desirable in handling loads. What is acceptable in, say, Europe may not be applicable in Asia or North America, for various reasons.

As a general rule, material handling is unproductive, costly, and hazardous. Therefore,

- *reduce material transport to the absolute minimum; and/or,*
- *make material transport a mechanical process.*

These are the PRIMARY measures to "take the hand out of material handling." If it cannot be avoided that people handle material, ergonomic considerations for the design of facility, job, equipment, and of the material to be handled become of utmost importance, because now "people are the kingpins of the work."

This results in SECONDARY measures:

- *Fit the work system to the people.*
- *Use mechanical aids to eliminate hand labor.*
- *Design the work for a minimum of manual handling.*

TERTIARY measures are:

- *Assessment of existing physical job demands for*
- *Selection of individuals capable of doing the job, and for*
- *Training of persons in proper material handling practices.*

Elimination of all hazardous or overly strenuous load handling through proper work design is the strategy preferred over worker selection and training. An existing need for worker selection and training indicates that working conditions persist which are too strenuous and strainful. Yet, sufficient information has been developed and published internationally to establish guidelines in the seven distinct areas of manual movement of loads. Ergonomic design of material handling systems assures ease and efficiency.

REFERENCES

Baril-Gingras, G, and Lortie, M. (1995). The Handling of Objects Other Than Boxes: Univariate Analysis of Handling Techniques in a Large Transport Company. *Ergonomics, 38,* 905-925.

Kroemer, K. H. E. (1997). *Ergonomic Design of Material Handling Systems.* New York, NY: CRC Press/Lewis.

Kroemer, K. H. E., Kroemer, H., B., and Kroemer-Elbert, K. E. (1994). *Ergonomics. How to Design for Ease and Efficiency.* Englewood Cliffs, NJ: Prentice Hall.

Kroemer, K. H. E., Kroemer, H, J., and Kroemer-Elbert, K. E. (1997). *Engineering Physiology: Bases of Human Factors/Ergonomics* (3rd ed.). New York, NY: Van Nostrand Reinhold.

Mack, K., Haslegrave, C.M. and Gray, M.I. (1995). Usability of Manual Handling Aids for Transporting Materials. *Applied Ergonomics, 26,* 353-364.

Snook, S.H. and Ciriello, V.M. (1991). The Design of Material Handling Tasks: Revised Tables of Maximum Acceptable Weights and Forces. *Ergonomics, 34,* 1197-1213.

Waters, T.R., Putz-Anderson, V., Garg, A. and Fine L.J. (1993). Revised NIOSH Equation for the Design and Evaluation of Manual Lifting Tasks. *Ergonomics, 36,* 749-776.

OPERATOR HEARING IN INDUSTRIAL NOISE: SIGNAL DETECTION, SPEECH COMMUNICATIONS, AND CONSIDERATIONS IN SELECTING HEARING PROTECTION

John G. Casali, Ph.D., CPE
Grado Professor and Department Head
Human Factors Engineering Center
Industrial and Systems Engineering Department
Virginia Polytechnic Institute and State University
Blacksburg, Virginia 24061-0118

ABSTRACT: To combat the threat of noise-induced hearing loss posed by intense noise environments, hearing protection devices (HPDs) have been utilized in U.S. industry since the 1950s, although their use was not widespread until promulgation of the Occupational Health and Safety Act of 1971. When properly selected and used, conventional HPDs offer an effective means of reducing the incidence of occupational hearing loss and noise annoyance, although they are not a panacea. In many industrial situations where speech communications and signal detection are necessary, workers complain that HPDs compromise their abilities to hear desired sounds. Recent litigation of a few industrial cases has implicated HPDs in accidents in which workers were injured when they did not hear a warning sound. This paper provides guidance for the proper application of HPDs in industry and reviews their effects on a worker's hearing in industrial noise. Also covered are several new technologies in HPD design (active noise cancellation, active sound transmission, frequency selectivity, adjustable attenuation, amplitude sensitivity, and uniform attenuation) which are aimed at improving the worker's ability to hear while providing adequate protection.

NOISE IN INDUSTRY

The din of noise pervades many a workplace, its effects on workers ranging from minor annoyance and interference with work tasks to major risk of hearing damage. Unfortunately, at least within the current limits of technology, noise is a by-product of many industries, especially those involving manufacturing and certain services, such as air transport, construction, and farming. Workers complain about the negative effects of noise on their ability to communicate, hear warning and other signals, and concentrate on tasks at hand. However, the safety-related effect which has historically been of most concern to industry has been permanent *noise-induced hearing loss,* or NIHL. In addition to posing as a threat to hearing loss, noise and the resultant application of personal hearing protection devices (HPDs) have both been associated with safety problems through their effects on the audibility of warning signals, speech communications, and other important workplace sounds.

Scope of Occupational Hearing Loss in the U.S.

Noise-induced hearing loss is one of the most widespread occupational maladies in the U.S., if not the world. In the early 1980's, it was estimated that over 9 million workers were exposed to noise levels averaging over 85 dBA for an 8-hour workday (EPA, 1981). Today, this number is likely to be higher because the control of noise sources, both in type and number, has not kept pace with the proliferation of industry, construction, and service development. Due in part to the fact that before 1971 there were no U.S. federal regulations governing noise exposure in general industry, many workers over 50 years of age now exhibit hearing loss that relates to the effects of occupational noise. Of course, the total noise exposure from both occupational and non-occupational sources determines the NIHL that a victim experiences. Of the estimated 28 million Americans who exhibit significant hearing loss due to a variety of etiologies, such as pathology of the ear, ototoxic drugs, and hereditary tendencies, over 10 million have losses which are directly attributable to noise exposure (NIH, 1990). Therefore, the noise-related losses are *preventable* in nearly all cases. The majority of losses are probably due to on-the-job exposures, but leisure noise sources do contribute a significant amount of energy to the total noise exposure of some individuals.

Prevention of *noise-induced permanent threshold shift* (NIPTS), for which there is no possibility of recovery, is the primary aim of the industrial hearing conservationist. NIPTS can manifest suddenly as a result of acoustic trauma; however, industrial noise problems that cause NIPTS most typically constitute exposures that are repeated over a long period of time and have a cumulative effect. In fact, the losses are often quite insidious in that they occur in small steps spanning a number of years of overexposure and the worker is not aware of the loss of hearing sensitivity until it is too late. This type of exposure produces permanent neural damage, and although there are some individual differences as to magnitude of loss and audiometric frequencies affected, the typical pattern for NIPTS is a prominent elevation of threshold at or near the 4,000 Hz audiometric frequency (sometimes called the "4 KHz notch"), followed by a spreading of loss to adjacent frequencies of 3,000 and 6,000 Hz. As noise exposure continues over time, the hearing loss will spread over a wider frequency bandwidth inclusive of midrange and high frequencies, encompassing the range of most auditory warning signals as well as the critical speech bandwidth. In some cases, the hearing loss renders it unsafe or unproductive for the victim to work in occupational settings where the hearing of certain signals are requisite to the job.

OSHA Noise Exposure Limits to Prevent NIPTS

In combating hearing loss, the Occupational Safety and Health Act (OSHA, 1983) requires that if the noise dose exceeds the OSHA action level of 50%, which corresponds to an 85 dBA time-weighted average (TWA) referenced to an 8-hour day, the employer must institute a *hearing conservation program* (HCP). Facets of an HCP must include: 1) noise exposure monitoring, 2) employee notification, 3) audiometric testing, 4) hearing protection, 5) training program for employees, and 6) recordkeeping. If the criterion level of 100% dose is exceeded (which corresponds to the Permissible Exposure Level [PEL] of 90 dBA TWA for an 8-hour day), the regulations specifically state that steps must be taken to reduce the employee's exposure to the PEL or below via administrative work scheduling (e.g., rotation of workers on and off noisy machines) and/or the use of engineering controls (e.g., absorptive materials or barriers in the noise path, vibration isolation or enclosure at the noise source). It is specifically stated that HPDs shall be provided if administrative and/or engineering controls fail to reduce the noise to the PEL, must be worn by all employees exposed to greater than the PEL, and must be supplied to all workers exposed to the action (85 dBA TWA) level or above. Therefore, in applying the letter of the U.S. law, HPDs are only intended to be relied upon when administrative or engineering controls are infeasible or ineffective. However, due to cost and implementation problems associated with other countermeasures, HPDs are by far the most popular method used in U.S. industry. The final OSHA noise level requirement pertains to impulsive or impact noise, which is not to exceed a true peak sound pressure level limit of 140 dB.

Noise Effects on Signal Detection and Communications

In regard to industrial noise hazards, the most prominent concern (and the motivation behind the OSHA regulations) is with the long term effects on the worker's audiogram; however, another prominent hazard is the effect of noise on hearing critical signals in the workplace.

Interference and the Signal-to-Noise Ratio

One of the most noticeable effects of noise is its interference with speech communications and the detection of nonverbal signals. Workers often complain that they must shout to be heard and that they cannot hear others. Likewise, noise interferes with the detection of workplace signals such as alarms on equipment, those for general area evacuation and warnings, and machine-related sounds which are relied upon for feedback. The ratio (actually the algebraic difference) of the speech or signal level to the noise level, termed the *signal-to-noise ratio* (S/N) is the most critical parameter in determining whether speech or signals will be heard in noise. A S/N of 5 dB means that the signal is 5 dB greater than the noise, while a S/N of -5 dB means that the signal is 5 dB lower than the noise.

Masking

Masking is technically defined as the tendency for the threshold of a desired signal or speech *(the masked sound)* to be raised in the presence of an interfering sound *(the masker)*. As an example, in the presence of a hair dryer near the user's ear, the ringer volume for a telephone must often be increased to enable detection, whereas a lower volume will be more comfortable while affording audibility when there is masking noise present. The *masked threshold* is defined as the SPL required for 75% correct detection of a signal when that signal is presented in a two-interval task wherein, on a random basis, one of the two intervals of each task trial contains the signal and the noise and the other contains only noise. In a controlled laboratory test scenario, a signal that is about 6 dB above the masked threshold will result in near perfect detection performance (Sorkin, 1987), while in workplace settings the S/N should be at least 15 dB to provide reliable detection in noise.

Analytical prediction (as opposed to actual experimentation with human subjects) of the interfering effects of noise on speech communications may be conducted using the Articulation Index (AI) technique defined in ANSI S3.5-1969 (R1986). Essentially, this technique utilizes a weighted sum of the speech-to-noise ratios in specified frequency bands to compute an AI score ranging between 0.0 and 1.0, with higher scores indicative of greater predicted speech intelligibility. Nonverbal signal detectability predictions can also be made analytically, with the most comprehensive, standardized computational technique being based on a spectral analysis of the noise and appearing in ISO 7731-1986. While a full discussion of these analytical procedures is beyond the scope of this paper, the reader is referred to the individual ANSI and ISO standards for detail, and Robinson and Casali (in press) for examples. The AI and masked threshold computational techniques provide better resolution and accuracy for speech intelligibility and signal detectability predictions than a simple evaluation of broadband S/N ratios because the techniques incorporate the frequency-specific information that simple S/N ratios do not reflect.

Indicators of the Need for Attention to Industrial Workplace Noise

The need for management, or perhaps more appropriately, abatement of industrial noise is indicated when: 1) noise creates sufficient intrusion and operator distraction such that job performance (and sometimes even job satisfaction) are compromised, 2) noise creates interference with important communications and signals, such as inter-operator communications, machine- or process-related aural cues, and/or alerting/emergency signals, and 3) noise exposures constitute an NIHL hazard. A full discussion of noise management strategies and techniques appears in Casali and Robinson (in press). The remainder of this paper concentrates on the selection and application of hearing protection devices, which are most typically aimed at problem 3, which is governed by OSHA federal regulations in general industry (OSHA, 1983) and MSHA (Mine and Safety Health Administration) regulations in mining.

CONVENTIONAL HEARING PROTECTION DEVICES (HPDs)

Categories of HPDs

In industrial settings, the vast majority of HPDs are of the conventional passive type, that is, devices which produce attenuation of noise strictly by passive means without the use of electronic circuitry, and result in a noise reduction between the environmental sound level and the sound level under the protector. Passive attenuation is accomplished through one or more avenues, including the use of construction materials with high sound transmission loss properties, liner materials which absorb and dissipate sound, trapped air volumes which provide acoustical impedance, and compliant materials which establish an acoustical seal against the skin. There are three basic styles of conventional HPDs that are used for industrial purposes. *Earplugs* consist of vinyl, silicone, spun fiberglass, cotton/wax combinations, and closed-cell foam products that are inserted into the ear canal to form a noise-blocking seal. Proper fit to the user's ears and training in insertion procedures are critical to the success of earplugs. A related device is the *semi-insert* or *ear canal cap* which consists of earplug-like pods that are positioned at the rim of the ear canal and held in place by a lightweight headband. The headband is useful for storing the device around the neck when the user moves out of the noise. *Earmuffs* consist of earcups, usually of a rigid plastic material with an absorptive liner, that completely enclose the outer ear and seal around it with foam- or fluid-filled cushions. A headband connects the earcups, and on some models this band is adjustable so that it can be worn over-the-head, behind-the-neck, or under-the-chin, depending upon the presence of other headgear, such as a welder's mask.

With regard to general attenuation performance, as a group, earplugs often provide better attenuation than earmuffs below about 500 Hz and equivalent or greater protection above 2,000 Hz. At intermediate frequencies, earmuffs typically have the advantage in attenuation. Earmuffs are generally more easily fit by the user than earplugs or canal caps, and as such, their rated attenuation values typically correspond more closely to actual protection achieved in the field than do those of earplugs or canal caps. Canal caps generally offer less attenuation and comfort than earplugs or earmuffs, but because they are readily storable around the neck, they are convenient for those workers who frequently move in and out of noise. A thorough review of HPDs and their application may be found in Berger and Casali (1997). Recent new technologies in hearing protection have emerged, including electronic devices offering active noise cancellation, communications capabilities, and noise-level-dependent attenuation, as well as passive, mechanical HPDs which offer level-dependent attenuation and near flat or uniform attenuation spectra; these devices are briefly reviewed herein and more thoroughly covered in Casali and Berger (1996).

Regardless of its general type, HPD effectiveness depends heavily on the proper fitting and use of the devices (Park and Casali, 1991). Therefore, the employer is required to provide training in the fitting, care, and use of HPDs to all affected employees (OSHA, 1983). Hearing protector use becomes mandatory when the worker has not undergone a baseline audiogram, has experienced an OSHA standard threshold shift (STS), or has a TWA exposure which meets or exceeds 90 dBA. In the case of the worker with an STS, the HPD must attenuate the noise to 85 dBA TWA or below. Otherwise, the HPD must reduce the noise to at least 90 dBA TWA.

Determining HPD Adequacy for a Noise Exposure Situation

The protective effectiveness or adequacy of an HPD for a given noise exposure must be determined by applying the attenuation data required by the EPA (1979) on protector packaging. These data are obtained from psychophysical threshold tests at nine 1/3 octave bands with centers from 125 to 8,000 Hz that are performed on human subjects, and the difference between the thresholds with the HPD on and without it constitutes the attenuation at a given frequency. Spectral attenuation statistics (means and standard deviations) and the single number noise reduction rating (NRR), which is computed therefrom, are provided. The ratings are the primary means by which end-users compare different HPDs on a common basis and make determinations of whether adequate protection and OSHA compliance will be attained for a given noise environment.

The most accurate method of determining HPD adequacy is to use octave band measurements of the noise and the spectral mean and standard deviation attenuation data to determine the *"protected exposure level"* under the HPD. This is called the *"NIOSH long method"* or the *"octave band"* method. Computational procedures appear in NIOSH (1975). Because this method requires octave band measurements of the noise, preferably with each noise band's data in TWA form, the data requirements are large and the method is not widely applied in industry. However, because the noise spectrum is compared against the attenuation spectrum of the HPD, a "matching" of exposure to protector can be obtained; therefore, the method is considered to be the most accurate available.

The NRR represents a means of collapsing the spectral attenuation data into one broadband attenuation estimate that can easily be applied against broadband dBC (or less accurately, dBA) TWA noise exposure measurements. In the calculation of the NRR, the mean attenuation is reduced by two standard deviations; this translates into an estimate of protection theoretically achievable by 98 percent of the population (EPA, 1979). The NRR is primarily intended to be subtracted from the dBC exposure TWA to estimate the protected exposure level in dBA, as via the following equation:

$$\text{Workplace TWA in dBC} - \text{NRR} = \text{Protected TWA in dBA}$$

Unfortunately, because OSHA regulations require that noise exposure monitoring be performed in dBA, the dBC values may not be readily available to the hearing conservationist, although some dosimeters do provide both weightings simultaneously. In the case where the TWA values are in dBA, the NRR can still be applied, albeit with some loss of accuracy. With dBA data, a 7 dB "safety" correction is applied to the NRR to account for the largest typical differences between C- and A-weighted measurements of industrial noise, and the equation is as follows:

$$\text{Workplace TWA in dBA} - (\text{NRR} - 7) = \text{Protected TWA in dBA}$$

While the above methods are promulgated by OSHA (1983) for determining HPD adequacy for a given noise situation, a word of caution is needed. The data appearing on HPD packaging are obtained under optimal laboratory conditions with properly fitted protectors and trained human subjects. In no way does the "experimenter-fit" protocol and other aspects of the current test procedure (ANSI S3.19-1974) represent the conditions under which HPDs are selected, fit, and used in the workplace (Park and Casali, 1991). Therefore, the attenuation data used in the octave band or NRR formulae shown above are inflated and cannot be assumed as representative of the protection that will be achieved in the field. Recent efforts by ANSI Working Group S12/WG11 have focused on the development of a testing standard which utilizes subject (not experimenter) fitting of the HPD and relatively naive (not trained) subjects to yield attenuation data that are more representative of those achievable under workplace conditions wherein a HCP is operated (described in Royster et al., 1996). However, at press time this draft standard, while receiving a positive ballot from ANSI constituents, had not been promulgated into law for use in producing the data to be utilized in labeling HPD performance.

Real-world NRR data from a recent review of studies in which manufacturers' on-package NRRs were compared against NRRs computed from actual subjects taken from field settings with their HPDs make it apparent that the laboratory NRRs for earplugs overestimate the field NRRs by an average of about 75%, while the laboratory NRRs for earmuffs overestimate the field NRRs by an average of only about 40% (Berger et al., 1996). These data would argue for the "derating," or reduction of the on-package NRRs, on a basis that differs by device type, as opposed to a constant "derating", such as the 50% OSHA recommendation. But in any case, the use of derating factors or other modifications of the NRR to adjust it for field applications is tenuous at best and is a practice that should not be expected of the end user. The best solution is to establish a testing standard (and attenuation rating therefrom) which accurately predicts workplace protection achieved by HPDs, and this is the ANSI standard work described in Royster et al., (1996).

Other Factors Bearing on HPD Suitability

While the adequacy of an HPD's attenuation for a specific noise situation is a critical factor in device selection, there are other influences which should be considered. Compatibility with the environment of use is essential. For instance, in environments which comprise a shock hazard, HPDs should be made exclusively of dielectric materials; on the other hand, in food processing environments, it is helpful if HPDs do have metallic parts so that they can be located via electronic sensors in the event they find their way into the food product. Earmuffs may be more uncomfortable than earplugs in a hot environment due to heat and humidity buildup underneath the cups, but they constitute a welcome insulator in cold environments. Compatibility with other headgear is also important. Earplugs do not interfere with welding shields or hard-hats, and certain styles of earmuffs, such those with neckbands, are designed to interface with these types of safety gear. When workers must move in and out of noise frequently, it is important that they be able to store their HPDs on their person. Earplugs connected by a lanyard as well as canal caps offer the advantage that they can be stored around the neck, while some earmuffs are designed to be worn on a belt clip.

HPDs must also be compatible with the type of work being performed. For example, porous foam or spun fiberglass earplugs which can harbor bacteria are not the best choice for dirty environments, food processing areas, or those characterized by metal filings and grit, especially where workers must doff and then reinstall HPDs with soiled hands. Ease of size selection and fit are also important factors. For instance, if workers do not have the advantage of close supervision and training in the use of HPDs, it is not a good strategy to supply multi-sized earplugs from which they must select their personal size. In situations where spot checks of worker compliance is essential, it is important that the HPDs be conspicuous, pointing to the advantage afforded by earmuffs or brightly-colored earplugs.

Finally, there is the influence of the need for hearing warning sounds and speech in the workplace on HPD selection. The most basic guidance in this regard is to insure that the HPD provides adequate protection, according to the aforementioned HCP regulations (OSHA, 1983) for the noise at hand with a margin of safety, but does not provide such high levels of attenuation that hearing thresholds for signals and speech are overly compromised. For normal hearing workers who experience noise above the OSHA action level of 85 dBA TWA, this balance is usually achievable, but it is a more difficult task for the hearing-impaired worker. The remainder of this paper addresses the issue of hearing protector effects on in-workplace communications and signal detection, and provides guidance concerning non-conventional HPDs which are designed to facilitate operator hearing in noise.

Effects of Conventional HPDs on Workers' Hearing in Noise

The research evidence on subjects of normal hearing abilities suggests that conventional passive HPDs have little or no degrading effect on the wearer's understanding of airborne speech sounds, in ambient noise levels above about 80 dBA, but cause increased misunderstanding over unoccluded conditions in lower sound levels. Although hearing protection is not required at such low levels, it may be desired for reduction of annoyance, or worn for convenience so that when at some later time an intermittent sound increases in magnitude, the wearer will already have his/her HPDs in place. In the latter case, the use of conventional protectors in the quiet periods of intermittent noise can be problematic.

At ambient noise levels greater than about 85 dBA, most studies have reported slight improvements in intelligibility with certain HPDs (e.g., Casali and Horylev, 1987, Suter, 1989), while others attempting to simulate on-the-job conditions have reported small decrements, especially when the talker is also wearing protection which causes a reduction of voice output (Hormann et al., 1984). Noise- and age-induced hearing losses generally occur in the high-frequency regions first, and for those so impaired, the effects of HPDs on speech perception are not clear-cut. These persons are certainly at disadvantage due to the fact that their already elevated thresholds for mid-to-high frequency speech sounds are further raised by HPDs, which typically have greater attenuation at higher frequencies than at low. Though there is not consensus among studies, it does appear that sufficiently hearing-impaired individuals will usually experience reduced communications abilities with HPDs worn in noise (Suter, 1989).

Conventional HPDs do not selectively pass speech (or nonverbal signal) versus noise energy at a given frequency; therefore, the devices do not improve the S/N ratio, which is the most important factor for achieving reliable intelligibility. In fact, nearly all conventional devices attenuate high-frequency sound more than low-frequency sound, thereby reducing the power of consonant sounds that are important for phoneme discrimination and also allowing low-frequency noise through, thus setting the stage for an upward spread of masking. While increased attenuation as a function of increasing frequency comprises the general spectral profile of conventional HPDs, it should be noted that inter- and intra-HPD category differences do exist. As discussed above, earmuffs as a category generally exhibit slightly higher attenuation in the mid-frequencies compared to earplugs, while the reverse is true at low frequencies.

How then do protectors sometimes afford intelligibility improvements in certain high-noise situations? The accepted theoretical explanation is that by lowering the total incident energy of both speech and noise, HPDs alleviate cochlear distortion that occurs at high sound levels. "Acoustic glare" is thereby reduced and the sensori-neural system operates under more favorable conditions. The situation is somewhat analogous to the reduction in visual glare and enhanced vision that results from the use of sunglasses on a bright day. However, it must be kept in mind that prediction of the effects of protectors on speech intelligibility in noise is a complex issue that depends on a host of factors, including the listener's hearing abilities, whether or not the talker is occluded and/or in noise, HPD attenuation, speech and noise levels, reverberation time of the environment, facial expressions and lip movements, and content/complexity of the message to be interpreted. Differences in testing protocols with respect to these factors contribute to the variance in reported results across studies.

The same HPD influence on S/N ratio and the theoretical basis for reducing cochlear distortion apply to the detection and recognition of nonverbal signals, such as warning horns or sirens, annunciators, and machinery sounds. The high-frequency bias in attenuation of conventional HPDs, coupled with the typically elevated high-frequency thresholds of those suffering from NIHL and the upward spread of masking from low-frequency noise, render warning signals and sounds of above about 2,000 Hz as those most likely to be missed. However, warning signal parameters, such as frequency, intensity, and temporal profile may be adjusted to help alleviate detection problems.

Also due to their increased attenuation with frequency, conventional HPDs create an imbalance in the listener's experience of the relative amplitudes of different pitches and cause broadband acoustic signals to be heard as spectrally different from normal, in that they take on a muffled, sometimes bassy tone. However, while signal *interpretation* may be affected, the bulk of empirical studies, with noise levels ranging from 75 to 120 dB, indicate that signal *detection* will not be compromised by HPDs for normal hearing individuals (Wilkins and Martin, 1987). While the evidence is less extensive for hearing-impaired listeners, they can be expected to experience detection and recognition difficulty, depending on their hearing loss profile, the particular signal, ambient noise, and hearing protector worn.

Since some of the high-frequency binaural cues (especially above about 4,000 Hz) that depend on the pinnae are altered by HPDs, judgments of sound direction and distance may be compromised. Earmuffs, which completely obscure the pinnae, interfere with localization in the vertical plane and also tend to cause horizontal plane errors in both contralateral (left-right) and ipsilateral (front-back) judgments (Suter, 1989). Earplugs may result in some ipsilateral judgment errors, but generally cause fewer localization problems than muffs, with a few studies demonstrating exceptions.

AUGMENTED HEARING PROTECTION DEVICES (HPDs)

As a result of the undesirable auditory effects of conventional HPDs previously described, especially for those with existing hearing impairments, special consideration is required when specifying S/N ratios and other design parameters for communications and auditory warning systems. Furthermore, safety issues must be contemplated when selecting HPDs for work areas where speech and signals are important information sources, such as in the vicinity of in-plant vehicles which rely on back-up alarms for conspicuity. In addition, the HPD may further compromise the situational awareness of the hearing-impaired employee, who is already at a sensory disadvantage. One opportunity for a solution, in addition or as an alternative to adjusting the communications/signal parameters, is that the HPD itself offers an avenue for change, with special technological augmentation that will enhance, or at least maintain hearing with the protected ear. Of course, it is a fundamental premise that the individuals' auditory sensitivity be preserved via the proper use of an adequate protector, but if the device can yield both acceptable attenuation as well as reasonable auditory perception, it will more likely be worn and may offer a benefit to workplace safety as well. For this reason, new HPD designs have been developed to improve communication and signal reception for the noise-exposed wearer. Of these, technologies which incorporate electronics to achieve such features as noise cancellation, signal transmission, or DC-powered communications capabilities are typically termed "active," while those that rely strictly on mechanical means to provide various qualities such as amplitude-sensitivity or uniform attenuation are termed "passive."

Active Hearing Protectors

Active Noise Cancellation (ANC) HPDs and Communications Headsets

Many environments, including tracked vehicle and helicopter interiors, ship engine rooms, and some industrial processes such as wood chipping have noise levels that test the limits of conventional communications headsets and hearing protectors. For this reason, much effort has been recently devoted to perfecting ANC noise control technologies for use in HPDs and headsets with the promise of providing a better and more wearable product due to improvements in attenuation, auditory perception, speech communication, and user comfort. In ANC, the basic premise is that superposition of two sound pressure waves of equal amplitudes, but with an exactly out-of-phase relationship, will result in destructive wave cancellation. The original noise is not masked in a psychoacoustic sense, but is actually reduced in amplitude, or "canceled," in a physical sense.

ANC has been incorporated into two types of personal systems: 1) those designed solely for hearing protection, and 2) those designed for communications, having the associated required boom- or throat-mounted microphone and earphone components, commonly referred to as ANC headsets. Both types are further dichotomized into open-back (or supra-aural) and closed-back (or circumaural earmuff) variations. In the former, a lightweight headband connects ANC microphone/earphone assemblies which are surrounded by foam pads that rest on the pinnae. In that there are no earmuffs to afford passive protection, the open-back devices provide only active noise reduction, and if there is electronic failure, no protection is provided by the device. Closed-back devices, which represent most ANC-based HPDs to date, are typically based on a passive noise-attenuating earmuff which houses the ANC transducers, and in some cases, the ANC-processing electronics. The ANC electronics and/or power supply may also be located on a belt-mounted pack and connected via cable to the headset. If backup attenuation must be provided by the device in the event of electronic failure of the ANC circuit, the closed-back HPD is advantageous due to the passive attenuation afforded by its earmuff.

The majority of ANC products currently marketed rely on analog electronics to capture, phase-invert, and condition the original noise, and then reintroduce it in the earphone of the muff or earplug worn by the user. A sensing microphone transduces the noise which has penetrated the passive barrier posed by the earmuff. The microphone signal is then fed back through a phase compensation filter which reverses the phase, to an amplifier which provides the necessary gain on the signal, and finally is output through an earphone loudspeaker to effect cancellation in the occluded sound field. Other contemporary systems use digital components and are based on either feed-forward and feedback loops; a complete discussion of the technologies appear in Casali and Berger (1996).

An example of the performance of an ANC headset in terms of its inherent passive attenuation compared to its total attenuation (i.e., the ANC-on mode) is provided in Figure 1. The computed active attenuation, that is, the difference between the total attenuation and the passive attenuation, evidences the range in which the ANC circuitry functions most effectively. The gains in low-frequency attenuation, and losses in the midrange frequencies are clearly illustrated. The low frequency (up to about 1,000 Hz) effectiveness demonstrated in analog ANC tests to date is particularly fortuitous for earmuff design in that ANC can potentially bolster the low frequency attenuation of conventional passive earmuffs, which tend to be most protective in the frequencies above 1,000 Hz. However, there are sizable incremental cost and weight tradeoffs associated with the addition of ANC components to a passive muff; therefore, it is important that the gain in low frequency attenuation be significant over that afforded by the muff alone. Furthermore, the ANC earphone/microphone components partially fill the occluded volume under a passive earmuff; the concomitant reduction in the acoustical impedance afforded by the trapped air mass decreases the passive attenuation afforded, especially at the low and middle frequencies. ANC devices are useful for certain intense noise situations; however, as shown in Figure 1, a properly-fitted passive earplug worn under a passive earmuff can exceed the performance provided by some ANC devices, with the added advantage of much lower cost. Battery maintenance and electrical circuit reliability are other considerations in the selection of ANC devices.

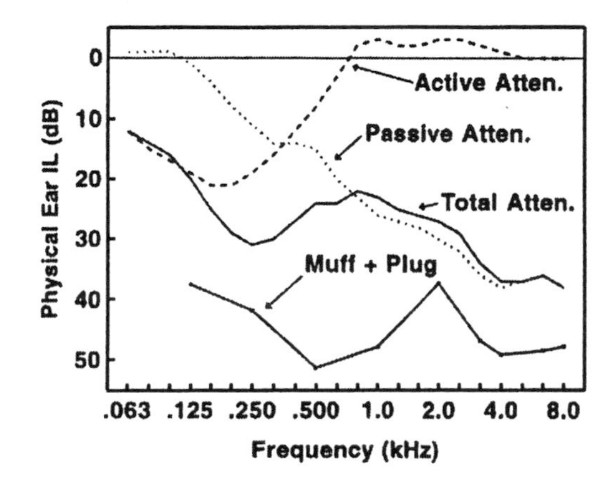

Figure 1. Microphone-in-real-ear insertion loss (IL) values for a high-performance closed-back ANC headset in the passive-only, active-only, and total (ANC on, i.e. active + passive) modes. For comparison, the real-ear attenuation of a dual hearing protector, i.e., well-fitted foam earplug plus earmuff, are shown. (Adapted from Casali and Berger, 1996).

Typical ANC HPDs will primarily be beneficial in noisy areas with a strong bias toward low frequency sound spectra. If the (dBC minus dBA) value of a noise is large (that is, above about 4 dB), a low frequency energy bias is evident in the noise and the environment may constitute a good candidate for ANC. If one can be assured that the noise spectrum is exclusively low frequency (or for certain specially-tuned ANC HPDs, narrow band), open-back (supra-aural) ANC devices may suffice, but it must be realized and stressed to users that no protection is afforded if the ANC circuitry fails. Open-back devices are a particularly attractive alternative in hot environments and when acoustic signals and/or voice must be heard from outside the headset.

Amplitude-Sensitive Sound Transmission (ASST) HPDs

Amplitude-sensitivity, or synonymously, level-dependence, refers to an augmentation technique wherein an HPD's attenuation increases as incident noise increases. By electronically amplifying a pre-selected frequency range of sound and allowing it to pass through the protector to the ear, *active* ASST protectors are aimed at improving speech intelligibility when high-intensity noise is not present. They consist of modified conventional earmuffs or earplugs that house microphone and amplifier systems to transmit external sounds to tiny loudspeakers (earphones) mounted inside the earcups or earplugs. The electronics can be designed to transmit and boost only those sounds within a desired passband, usually the critical speech band. Typically, the limiting amplifier maintains a predetermined (in some cases user-adjustable) earphone level, often at about 85 dBA, unless the ambient noise reaches a cutoff level of 115 to 120 dBA, at which the electronics cease function (Maxwell et al., 1987). In these high levels, the earmuff simply continues to provide the level of passive attenuation established by its earcups.

Typical and ideal performance as a function of incident sound level for an ASST HPD are depicted in Figure 2. While an example gain of 6 dB is shown in the figure, the gain for these systems may typically be set over a wide range from negative to high positive values. With some variations among designs, "amplitude-sensitive" properties of the devices are achieved by: 1) using a clipping circuit to effect instantaneous reduction of intense sound transmission, and/or 2) incorporating amplifier shutoff at a certain sound threshold value, such that the earmuff provides its highest attenuation (i.e., the passive attenuation of its earcups) when incident sounds are at and above the predetermined cutoff level. At levels below cutoff, the HPD has less attenuation since the sound transmission circuit passes and amplifies a bandwidth of frequencies. Sound transmission HPDs may afford better hearing in quiet and in moderate noise because of the amplification they provide in the design bandwidth; this may be particularly beneficial for the high-frequency hearing-impaired since the passive attenuation of the HPD is generally strongest at higher frequencies. Also, in intermittent noises, especially those with impulse-type (e.g., gunfire) and brief on-segments, the ASST HPD can be beneficial. For this reason, certain active sound transmission HPDs have been marketed as "hunter's ears," with claims of improved hearing in quiet periods and suddenly available ear defense during a gunshot. However, caution should be exercised when selecting a sound transmission HPD for use in noises of a continuous high-level character, because of the earphone distortion that may occur. As gain settings increase and the amplifier is pushed into clipping, some ASST devices will produce distortion products and spurious earphone noises sometimes characterized as popping, raspy, and static in nature. These distortion products will likely cause annoyance and mask desired signals that would otherwise be detectable.

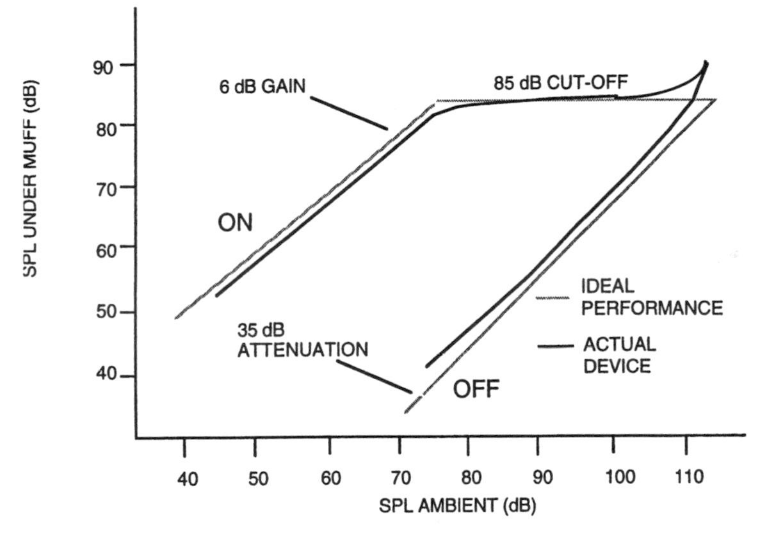

Figure 2. Typical and ideal performance for an active amplitude-sensitive, sound transmission earmuff as a function of incident sound level. (Adapted and modified from Maxwell et al., 1987)

A paucity of empirical studies have addressed the performance of ASST HPDs in improving signal detection and speech communications. Casali and Wright (1995) determined that subjects' masked thresholds for detecting a vehicle back-up alarm were not significantly improved by an electronic ASST HPD (as compared to a conventional passive HPD condition) in broadband noises of 75, 85, and 95 dBA (Figure 3). However, under certain other signal, noise, or subject hearing ability conditions, it is likely that the ASST HPD will result in improvements. Dynamic noise attenuation and user hearing acuity under sound transmission earmuffs depend on a host of electronic system design factors such as: cutoff sound level and sharpness of attenuation transition at this level, system response time to sharp onsets, frequency response and bandwidth, distortion and residual electronic noise, S/N ratio at sound levels below the cutoff, sensitivity to wind effects, and battery condition. There is variance among available products with respect to these factors. Another design issue is that of microphone configuration. It may be *diotic*, wherein a single microphone in one earcup feeds both earphones, or *dichotic*, in which each earcup has an independent microphone to afford interaural triangulation of signals. The latter approach should provide better localization performance where wearers must ascertain the source and direction of environmental sounds (Noble et al., 1990).

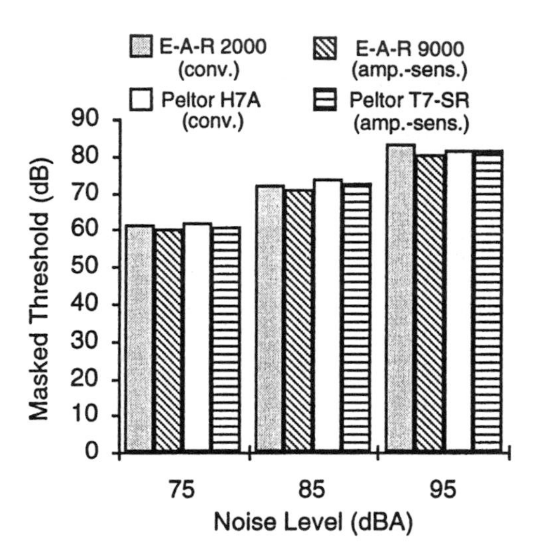

Figure 3. Masked backup alarm threshold means in dB for active (Peltor T7-SR) and passive E-A-R 9000 sound transmission HPDs compared to their conventional earmuff counterparts. Within a single noise level there are no muff differences at $p < 0.05$. (Adapted from Casali and Wright, 1995).

Passive Hearing Protectors

In contrast to active devices which rely on electronics for their dynamic influence on sound, passive HPDs incorporate structural features and mechanical elements including orifices, slits, ducts, diaphragms, valves, ports, resonators, and other designs to effect sound modification for better aural perception. In general, passive devices offer the advantages of being less expensive and requiring less maintenance. However, they are more limited in the performance features that can be provided through non-electronic techniques.

Frequency-Sensitive HPDs

An example of technically straightforward efforts to improve communications under earplugs involves the use of apertures or channels through an earplug body. One very early technique incorporated an air-filled cavity encapsulated by the walls of a premolded earplug (Zwislocki, 1957). In such a design, the cavity is vented to the outside air and also to the ear canal via a tiny port on either end, and the acoustical impedance of the configuration increases with the air cavity's volume and decreases with the size of the ports. The net effect is that the passive earplug functions similarly to a two-section low-pass filter, and it can be designed to provide an attenuation transition from negligible attenuation levels below about 1000 Hz up to about 35 dB at 8,000 Hz. Because most of the speech frequencies critical to intelligibility lie in the 1,000 to 4,000 Hz range, the communications benefit potential of the low-pass feature may be relatively small depending on the situation, especially in noisy environments which have considerable low-frequency energy that causes a spread of masking upward into the critical speech band.

An even more simple approach, common today in custom-molded earplugs, is to drill a small (about 0.5 mm) port or stepped-size vent longitudinally through the plug. The resultant air leak typically reduces attenuation in the low frequencies to a greater extent than in the high, roughly providing a low-pass characteristic. However, the attenuation achieved is insufficient for many industrial noise environments.

Adjustable-Attenuation HPDs

To help overcome the problem of "overprotection" in moderate noise environments, earplugs which afford some level of control over the amount of attenuation have recently appeared. These devices incorporate a leakage path that is adjustable (currently performed by the manufacturer) via the setting of a valve which obstructs a channel through the body of the plug, or alternatively, through the use of interchangeable filters or dampers which interface with the plug. An example of the former is the "Varifone," a Dutch product, that is constructed from an acrylic impression of the user's ear canal. According to the manufacturer's data, below 500 Hz the attenuation adjustment range is approximately 20 to 25 dB, with a maximum attenuation of about 30 dB at 500 Hz. At higher frequencies, the range of adjustment decreases, while the maximum attenuation attainable increases slightly. At any valve setting, the Varifone provides attenuation which increases with frequency.

Passive, Amplitude-Sensitive HPDs

Hearing ability under a conventional HPD is compromised during the quiet periods of intermittent sound exposures because the device yields constant attenuation regardless of ambient noise level. Amplitude-sensitive (level-dependent) HPDs reduce this problem by providing diminished attenuation at low sound levels but increased protection at high levels of steady-state and impulsive noise. A dynamically-functional valve or sharp-edged, round- or slit-shaped orifice that opens into a duct in the protector constitutes the nonlinear element that changes attenuation.

The valve-type devices incorporate a flutter valve which is said to close off the duct when activated by high sound pressures. However, given the very sharp rise time profiles of impulsive gun blasts and explosive detonations, a potential problem is that the mass and friction of the valve will inhibit its closing in time to effect full protection in impulses. On the other hand, the orifice-duct technique takes advantage of the "nonlinear" acoustical behavior that develops when sounds of high pressure levels, above about 120 dB, are incident upon a small opening. Low-intensity sound waves predominantly exhibit laminar airflow and pass relatively unimpeded through the aperture, whereas high-intensity waves involve more turbulent flow and are attenuated due to the sharply increased acoustic resistance of the aperture (ratio of acoustic pressure across a material to the particle velocity through it). This technique has been applied with success in earplugs, such as the "Gunfender" which has been used by British military for twenty years (Mosko and Fletcher, 1971), and in earmuffs, such as the Cabot Safety "Ultra 9000" (Allen and Berger, 1990).

For passive, amplitude-sensitive devices, a critical performance parameter is the transition sound level at which insertion loss becomes significant. As illustrated in Figure 4, at sound levels beyond the function's knee of 110 to 120 dB, insertion loss increases at a rate up to about half the increase in sound level, up to a point where the measured insertion loss tends to asymptote where the HPD behaves much like the orifice is sealed shut.

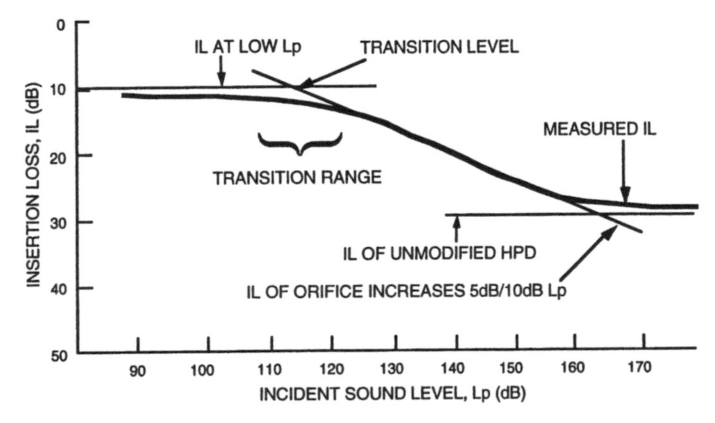

Figure 4. Representative insertion loss (IL) and illustration of the transition level at a single frequency, as a function of incident sound level, for an amplitude-sensitive hearing protector with a nonlinear orifice. The transition range indicates the sound levels at which nonlinearity, due to turbulent flow in the orifice, begins to increase the device's attenuation. (Adapted from Allen and Berger, 1990).

At lower, but still potentially hazardous sound levels, most amplitude-sensitive devices exhibit behavior similar to that of a leaky or vented earplug, affording very weak attenuation below 1,000 Hz and highly frequency-dependent attenuation above. An exception is an orifice-type earmuff, the Cabot Safety Ultra 9000 that provides roughly 25 dB attenuation from 400 to 8,000 Hz (Allen and Berger, 1990]). Because most passive amplitude-sensitive protectors require such high sound levels to yield significant attenuation, they are best suited for intermittent impulsive blasts and gunfire exposures, especially outdoors. The orifice-based Cabot Ultra 9000 earmuff lends itself for these purposes, and as demonstrated by the Casali and Wright (1995) data in Figure 3, it should *not* be relied upon to provide improved detection of warning signals in continuous broadband noises of 75 to 95 dBA.

Uniform-Attenuation HPDs

Conventional earplugs, muffs, and canal caps exhibit a nonlinear attenuation profile, with an increase in attenuation as frequency increases (Figure 5). Not only are sounds reduced in level, they are also imbalanced in a spectral sense. Since many auditory cues depend on frequency coding for information content, conventional HPDs compromise these cues. For instance, machine tool operators complain that auditory feedback from a cutting tool is distorted, aircraft pilots and tank operators indicate that important signals cannot be discerned, and musicians report pitch perception problems under conventional HPDs. To counter these effects, flat or uniform attenuation HPDs, which impose attenuation that is nearly linear from about 100 to 8,000 Hz, were developed in the late 1980's. The attenuation for two such products is shown in Figure 5.

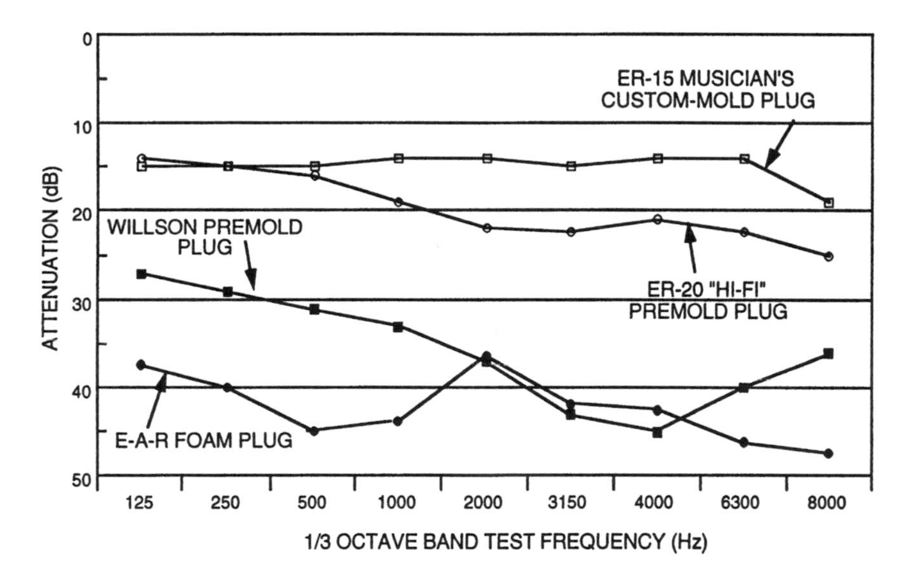

Figure 5. Attenuation by frequency for two conventional earplugs (user-molded E-A-R Classic foam and premolded Willson EP-100) and two uniform-attenuation earplugs (custom molded ER-15 and premolded ER-20). Manufacturers' data are illustrated.

Although empirical proving studies are as yet lacking on uniform HPDs, better hearing perception and adequate protection should be achievable in low to moderate noise exposures of about 90 dBA or less. Those with sensori-neural hearing losses and professional musicians may find them particularly beneficial, due to the reduction in high-frequency attenuation and uniform effect on the sound spectrum, respectively. However, for noises having substantial high-frequency energy, uniform attenuation earplugs generally offer less protection than conventional earplugs, as shown in Figure 5.

Conclusions

Hearing protection devices are an important countermeasure against noise-induced hearing loss in industry. Especially in situations where speech communication, aural signal detection, and/or sound interpretation is an issue, or where the user has a hearing impairment, the industrial hygienist or safety engineer should consider the potential disadvantages posed by conventional passive HPDs and investigate alternative devices which incorporate special features to augment hearing acuity while providing adequate protection. Further research is needed to determine both the real-world attenuation as well as speech intelligibility/signal detection performance with these augmented HPDs. It is important to recognize that augmented HPDs are more specialized in application than are conventional HPDs, and therefore they must be selected after careful consideration of the workplace scenario noise, signal, and listener characteristics. Furthermore, augmented HPDs may require more maintenance and user training than conventional devices, thus necessitating additional attention in a hearing conservation program.

REFERENCES

Allen, C. H. and Berger, E. H. (1990). Development of a unique passive hearing protector with level-dependent and flat attenuation characteristics. *Noise Control Engineering Journal, 34*:3:97-105.

ANSI S3.5-1969 (R1986). *Methods for the calculation of the articulation index.* New York: American National Standards Institute, Inc.

ANSI S3.19-1974 (1974). *Method for the measurement of real-ear protection of hearing protectors and physical attenuation of earmuffs.* New York: American National Standards Institute, Inc.

Berger, E. H. and Casali, J. G. (1997). Hearing protection devices. In M. J. Crocker (Ed.) *Encyclopedia of Acoustics.* New York: Wiley.

Berger, E. H., Franks, J. R., and Lindgren, F. (1996). International review of field studies of hearing protector attenuation. In A. Axelsson, H. Borchgrevink, R. P. Hamernik, P. Hellstrom, D. Henderson, and R. J. Salvi (Eds.), *Scientific Basis of Noise-Induced Hearing Loss.* (pp. 361-377), New York: Thieme Medical Publishers, Inc.

Casali, J. G. and Berger, E. H. (1996). Technology advancements in hearing protection: Active noise reduction, frequency/amplitude-sensitivity, and uniform attenuation. *American Industrial Hygiene Association Journal, 57,* 175-185.

Casali, J. G. and Horylev, M. J. (1987). Speech discrimination in noise: The influence of hearing protection. In *Proceedings of the Human Factors Society—31st Annual Meeting.* New York, NY: Human Factors Society, pp. 1246-1250.

Casali, J. G. and Robinson, G. S. (in press). Noise in industry: Auditory effects, measurement, regulations, and management. In Karwowski, W. and Marras, W. (Eds.) *Handbook of Occupational Ergonomics,* Boca Raton, Florida: CRC Press. (manuscript available from authors).

Casali, J. G. and Wright, W. H. (1995). Do amplitude-sensitive hearing protectors improve detectability of vehicle backup alarms in noise? *Proceedings of the 1995 Human Factors and Ergonomics Society 39th Annual Conference,* San Diego, California, 1994-998.

EPA (1979). 40CFR211, Noise labeling requirements for hearing protectors. Environmental Protection Agency, *Federal Register, 44*(190), 56130-56147.

EPA (1981). *Noise in America: the extent of the noise problem.* Environmental Protection Agency Report No. 550/9-81-101. Washington, D.C.: EPA.

Hormann, H., Lazarus-Mainka, G., Schubeius, M. and Lazarus, H. (1984). The effect of noise and the wearing of ear protectors on verbal communication. *Noise Control Engineering Journal 23*:2:69-77.

ISO 7731-1986 (E) (1986). *Danger signals for work places- auditory danger signals.* Geneva, Switzerland: International Organization for Standardization.

Maxwell, D. W., Williams, C. E., Robertson, R. M. and Thomas, G. B. (1987). Performance characteristics of active hearing protection devices. *Sound and Vibration, 21*:5:14-18.

Mosko, J. D. and Fletcher, J. L. (1971). Evaluation of the gunfender earplug: Temporary threshold shift and speech intelligibility. *Journal of the Acoustical Society of America, 49*:1732-1733.

National Institutes of Health (NIH) Consensus Development Panel. (1990). Noise and hearing loss. *Journal of the American Medical Association, 263*(23), 3185-3190.

NIOSH (1975). *List of personal hearing protectors and attenuation data.* National Institute for Occupational Safety and Health-HEW Publication No. 76-120, 21-37. Washington, D. C.:

Noble, W. G., Murray, N. and Waugh, R. (1990). The effect of various hearing protectors on sound localization in the horizontal and vertical planes. *American Industrial Hygiene Association Journal. 51:*7:370-377

OSHA (1983), 29CFR1910.95. *Occupational noise exposure; Hearing conservation amendment; Final rule.* Occupational Safety and Health Administration. *Code of Federal Regulations*, Title 29, Chapter XVII, Part 1910, Subpart G, 48 FR 9776-9785. Washington, DC: Federal Register.

Park, M. Y. and Casali, J. G. (1991). A controlled investigation of in-field attenuation performance of selected insert, earmuff, and canal cap hearing protectors. *Human Factors, 33*(6), 693-714.

Robinson, G. S. and Casali, J. G. (in press). Audibility of reverse alarms under hearing protectors and its prediction for normal and hearing-impaired listeners. In N. Stanton, (Ed.) *Auditory Warnings.* London: Avebury Technical (manuscript available from authors).

Royster, J. D., Berger, E. H., Merry, C. J., Nixon, C. W, Franks, J. R., Behar, A., Casali, J. G., Dixon-Ernst, C., Kieper, R. W., Mozo, B. T., Ohlin, D., and Royster, L. H. (1996). Development of a new standard laboratory protocol for estimating the field attenuation of hearing protection devices. Part I. Research of Working Group 11, Accredited Standards Committee S12, Noise. *Journal of the Acoustical Society of America, 99*(3), 1506-1526.

Sorkin, R. D. (1987). Design of auditory and tactile displays. In Salvendy, G. (Ed.), *Handbook of Human Factors,* (pp. 549-576). New York: McGraw-Hill.

Suter, A. H. (1989). *The effects of hearing protectors on speech communication and the perception of warning signals* (AMCMS Code 611102.74A0011). Aberdeen Proving Ground, MD: U.S. Army Human Engineering Laboratory.

Wilkins, P. and Martin, A. M. (1987). Hearing protection and warning sounds in industry — a review. *Applied Acoustics 21*:267-293.

Zwislocki, J. (1957). Ear protectors. In C. M. Harris (Ed.),*Handbook of Noise Control,* (8-1 to 8-27). New York, NY: McGraw-Hill.

MACROERGONOMICS TO IMPROVE HUMAN AND TECHNICAL PERFORMANCE IN AGILE MANUFACTURING SYSTEMS

Brian M. Kleiner
Department of Industrial and Systems Engineering
Virginia Polytechnic Institute and State University
Blacksburg, Virginia 24061-0118 USA

ABSTRACT: The value of industrial ergonomics in traditional manufacturing environments is well established [1,2]. Agile manufacturing systems operate in complex, dynamic environments and require maximum flexibility of human operators. This paper provides a better understanding of agile environments through discussion of an empirical study of how time is best allotted to the personnel and technological subsystems. The paper also contributes to the design of agile systems by reporting on the empirical results of a laboratory study which investigated computer-supported collaborative work in concurrent engineering with differing personnel configurations.

INTRODUCTION

The Challenges of Agile Manufacturing

A study from Lehigh University's Iacocca Institute concluded the manufacturing strategy of the 21st Century must be to adopt agile manufacturing in order to survive and prosper [3]. Although its exact definition remains elusive, agility refers to the capability to quickly go from a set of novel customer requirements to a quality, finished product [4]. An agile company will be relatively flat with decentralized management, will have a multi-skilled and flexible work force, will have flexible supplier strategies, will master lean production, will have flexible tooling and automation, will deeply understand the customer and will utilize real-time market data [5]. Among its many characteristics, agile competition involves the integration of information and communication technologies to coordinate geographically distributed resources regardless of location and achieve powerful competitive advantages. These advantages have created organizational environments characterized by very broad product ranges, short product life cycles, arbitrary lot sizes, and highly individualized and customized products [6].

Virtuality will play a central role in these agile environments by allowing organizations to combine resources (e.g. information, expertise, production capacity) regardless of geographical location to optimize organizational outcomes. The Virtual Corporation, a temporary network of companies that come together quickly to exploit fast-changing opportunities, exemplifies this exploitation of information technology [7].

Macroergonomics

New approaches to competitiveness such as Agile Competition create new challenges for the design of the work system, such as the optimal use of information technology. Macroergonomics is concerned with the optimization of work system design through consideration of relevant personnel, technological, and environmental variables and their interactions [8]. The central focus of macroergonomics is to optimize work system design. This is what distinguishes it from both micro-ergonomics and organizational psychology and makes it a valuable contribution to manufacturing. It is through a systematic consideration of the relevant sociotechnical system variables in the ergonomics analysis, design, implementation, evaluation, and control process, that this is accomplished. These variables relate to three basic, empirically-identified sociotechnical system components, the technical (technological) subsystem, social (personnel) subsystem, and the external environment. A fourth sociotechnical system element, job (and task) design, falls within the purview of microergonomics, but is influenced by, and interacts with, design of the work system. In reality, macroergonomic design of the work system helps determine many of the characteristics which should be designed into individual jobs for joint optimization of the total system. Work system design, thus, includes both the structure and related processes of the entire work system.

The term "sociotechnical system" was first coined by Emery and Trist to better convey the nature of complex human-machine systems [10]. Sociotechnical systems theory proposes a top-down view of a system as a transformation process permeable to the external environment. Two major components interact in this transformation process, the social or personnel subsystem and the technical or technological subsystem [11]. The social subsystem consists of those who comprise the work system. The technical subsystem focuses on the methods and tools used to transform inputs to outputs in the work system. Sociotechnical systems theory emphasizes the concept of joint causation which dictates that both subsystems react together to causal events in the environment. Since both systems react jointly to changes in the external environment the only way to avoid suboptimization is to avoid optimizing one subsystem and then fitting the other [11]. A revised STS model was recently proposed for agile systems where the two subsystems are viewed as components of a highly integrated system [12].

Historical perspective. Macroergonomics is a recognized field of specialization within the International Ergonomics Association (IEA) and the Human Factors and Ergonomics Society (HFES). The HFES Macroergonomics Technical Group. At the initial organizing meeting held during the 1981 HFES Annual Meeting, the name "Organizational Design and Management" was chosen for the new technical group by popular vote. At that time, the entire concept of macroergonomics did not exist. Today, macroergonomics has rapidly been developing a perspective, approach, and methodology of its own. Further, there are now well-defined applications of macroergonomics and constituent technical areas. Therefore, in January of 1997 by popular vote and by ratification of the HFES Executive Council, the Organizational Design and Management Technical Group changed its name to the Macroergonomics Technical Group of the HFES [9].

In summary, an appreciation of the human factors inherent to agile product development is pivotal to the successful integration of agility-enabling technologies, as well as the coordination of personnel working within a concurrent engineering environment [4]. Since the goal of macroergonomics is to improve the whole system through a top-down, joint design of all involved subsystems, it provides an ideal perspective to better understand and design agile environments and systems. This paper provides a deeper understanding of agile environments through its discussion of how time is best allotted to the personnel and technological subsystems. The paper also contributes to the design of agile systems by reporting on the empirical results of a laboratory study which investigated computer-supported collaborative work in concurrent engineering with differing personnel configurations.

METHODOLOGY

Field Research Study

Macroergonomics can help improve work force capabilities in lean and agile environments [13]. A study was designed to gain a better understanding of time allotment in such modern manufacturing environments through the precepts and principles of sociotechnical systems theory. Table 1 illustrates the major research questions and hypotheses in this study.

Participants in a comprehensive field study included 92 full-time, first-level managers (85 male and seven female). Typically, these managers were foremen or supervisors and their job duties included such titles as production lab supervisor, quality control engineer, maintenance supervisor, manufacturing supervisor, warehouse or logistics supervisor. The youngest manager was 25 years old and the oldest was 61 years old and the mean age of the participants was 42 years olds. First level managers had as little job

TABLE 1
Field Study Research Questions

Research Question	Research Hypotheses
Research Question 1: How do reports of departmental joint optimization compare between different types of first-level managers?	1a: *Upstream, transformation, and downstream first-level managers across different manufacturing firms report different levels of joint optimization in their departments.* 1b: *Upstream, transformation, and downstream first-level managers across different manufacturing firms spend different proportions of time on the social and technical subsystems of their departments.*
Research Question 2: What balance of time between the social and technical subsystems do first-level managers report for jointly-optimized departments?	2a: *Departments rating high in joint optimization will have first-level managers who spend equal amounts of time addressing the technical subsystem and the social subsystems. Departments with lower ratings in joint optimization will have managers who spend more time on the technical subsystem.* 2b: *Departments rating high in joint optimization will rate high in overall performance.*
Research Question 3: What relationship do time dimension of work variables have on departmental joint optimization and how managers allot their time to the social and technical subsystems?	3a: *Time dimension variables affect the association between joint optimization and the difference of time spent on technical and social activities.* 3b: *Time dimension variables predict department performance.*
Research Question 4: What proportion of time do first-level managers spend on the social and technical subsystems in departments with perceived high performance ratings?	*4: Department with first-level managers who spend equal proportions of time on the social and technical subsystems are higher performing departments*

experience as four months to the most tenured supervisor with 34 years experience. The average department tenure was 7.3 years and the average organization tenure was 16.3 years. The number of direct reports for each manager ranged from zero to 130 operators where the average number was 23 direct reports in each department. Twenty-seven first-level managers reported no level of change in their departments. Sixteen departments reported department restructuring, 13 reported staff retraining, 25 reported production advancement, 27 reported information technology or improvement, and 12 departments reported other projects including plant expansion, centralization of lab location, and job safety analysis. Many departments reported more than one on-going project (13 departments: two projects, four departments; three projects, and two departments: four projects). Eleven of the organizations classified their primary function to be production and only one company classified its primary function to be distribution. The number of managers from the companies were American facilities (n = 4) and Canadian facilities (n=8).

Laboratory Research Study

In agile environments resources can be directed at technical alternatives such as the engineering process methodology used or the use of computer-support technology. Alternatively, managers can direct resources at social alternatives such as increasing the number of personnel on the project. Sociotechnical systems theory suggests that there is an optimum combination of social and technical alternatives for performance. This research investigated the relationship between the important design process, personnel, and technological issues that must be considered by those responsible for organizations performing the design of complex work systems. This research considered two technical considerations: (1) the overall design process methodology - concurrent engineering versus sequential engineering, and (2) whether to use or not use computer-supported collaborative work technology, and one social consideration, since agility emphasizes a collaborative design process whereby engineers are integral participants in making decisions [4]: (1) whether large teams of six persons would be more effective and efficient than small teams of three persons.

Research Question 1 asked how design performance is affected by engineering methodology, group size, or computer support. Research Question 2 asked how process time is affected by engineering methodology, group size, or computer support. Research Question 3 asked how process cost is affected by engineering methodology, group size, or computer support. Research Question 4 asked how the satisfaction of group members toward the engineering design decision process is affected by engineering methodology, group size, or computer support. Research Question 5 asked if there was a optimum combination of engineering methodology, group size, and computer support that creates the greatest technical and social outcome.

The experimental design of this research was a 2 x 2 x 2 factorial, between subjects design. Large teams, consisting of six engineering students, and small teams, consisting of three engineering students were given a set of requirements to design a transportation system that moved a payload from one point to another. Each team was asked to: (1) develop a design concept, (2) develop a detail design in the form of engineering drawings, (3) manufacture the system based on their design products (e.g., drawings and specifications) using plastic LEGOS, and (4) test the system to determine if it met the design requirements.

RESULTS

Field Research Study

Research hypotheses 1a and 1b focused on the level of joint optimization and the proportion of time allotted to the personnel and technological subsystem compared across manager type. The results of hypothesis 1a were significant at the a = .05 level. Downstream managers reported higher levels of joint optimization than transformation managers (see Table 2).

TABLE 2
One way ANOVA of Joint Optimization using Manager Type as the Factor

Source	SS	df	MS	F	Sig.
Between Mgr_Type	.631	1	.631	4.418	.038*
Within Mgr_Type	12.701	89	.143		
Total	13.332	90			

* Significant at α = .05

The results of hypothesis 1b showed there were no significant differences between the percentage of time spent on the personnel and technological subsystems for each level manager type.

Hypothesis 2a focused on the relationship between the amount of time first-level managers spent on both the personnel and technological subsystem and the reported level of joint optimization. The hypothesis predicted a positive relationship between department performance and time allotment for managers who balanced their time equally between both subsystems. The hypothesis was tested using Pearson's correlation coefficient as well as ANOVA testing. This hypothesis was not supported The correlation test supported a significant relationship between the balance of time variable (DIFF) and joint optimization. However, the ANOVA test did not support a relationship between level of joint optimization (low or high) and the amount of time spent in either subsystem. Hypothesis 2b tested the relationship between joint optimization and department performance. The research hypothesis was supported by a strong positive correlation between department performance and joint optimization where r = .607 was significant at the α = .05 level. Figure 1 illustrates this relationship.

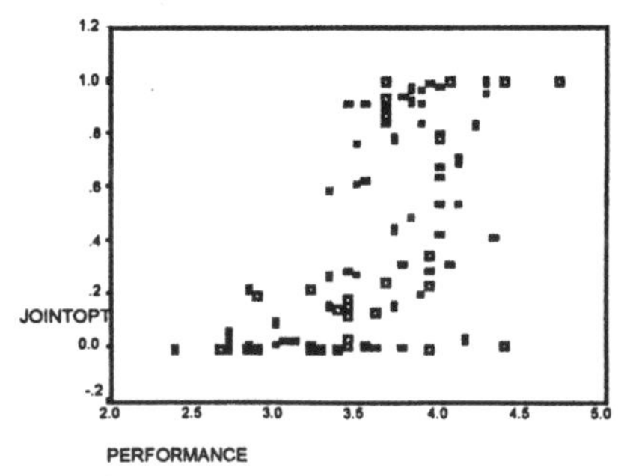

Figure 1. - Joint Optimization vs. Department Performance

Hypothesis 3a tested whether the time dimension variable affected the association between joint optimization and the difference of time spent in the personnel and technological subsystems. This hypothesis was supported. The results of the set of regression tests showed that scheduling, future orientation, awareness of time use each had significant affect on the relationship between joint optimization and time allotment between the subsystems at $p < .001$ as seen in Table 3. Autonomy of time use and synchronization also affected the relationship at $p < .01$ and $p <. 05$ respectively.

TABLE 3
Summary of Regression of Joint Optimization on DIFF and Time Dimension Variables

TDW Variable	b_1	t	sig	b_2	t	sig	R^2
Scheduling	.246	2.509	.011	-.406	-4.288	.000	.197
Future	.169	1.811	.074	.444	-4.75	.000	.244
Allocation	.212	2.054	.043	.147	1.430	.156	.072
Awareness	.279	3.006	.003	.451	4.857	.000	.251
Autonomy	.217	2.181	.032	.279	2.799	.006	.128
Synchronization	.204	2.003	.048	-.220	-2.168	.033	.099.

Hypothesis 3b was tested by using linear regression to determine which time dimensions of work variables predict department performance. The research hypothesis was supported. Awareness of time use, autonomy of time use, and scheduling achieved significant regression weights. However, both DIFF and synchronization variables contributed to the final adjusted $R^2 = .378$ value of the regression model where $F(5,85) = 12.81$, $p < .001$ (see Table 4).

TABLE 4
Regression of Performance on the Time Dimensions

Model	b	Std. Error	t	sig
Awareness of Time Use	.224	.064	3.490	.001
Autonomy of Time Use	.140	.046	3.062	.003
Scheduling	-.301	.090	-3.327	.001
DIFF	.032	.018	1.796	.076
Synchronization	.113	.073	1.545	.126

Hypothesis 4 tested whether the equal proportions of time spent on the personnel and technological subsystems resulted in higher performance. The hypothesis was not supported by the results of both correlation tests and scatterplots. The correlation coefficient for DIFF (balance of time spent on both subsystems) and performance was .085 which indicates that there is little relationship between the two variables. The scatterplot of Performance versus DIFF and Performance versus PCTDIFF revealed that there was some variation with respect to performance scores for each possible DIFF or

PCTDIFF score. However, the majority of managers reported their time was allotted in one of four proportions to the technical and social subsystems - 20/80 or 80/20, 30/70 or 70/30, 40/60 or 60/40, and 50/50. Managers who reported spending 80% percent of their time on the technological subsystem scored higher on performance and joint optimization than managers who reported spending 80% of their time on the personnel subsystem. Managers who split their time 40/60, 50/50, or 60/40 between the technological and personnel subsystems tended to score higher on both joint optimization and department performance. Those high performing managers who had a 50/50 time allotment also tended to have consistent performance scores from their evaluating managers with their own evaluation of departmental performance.

Laboratory Research Study

There were two levels of the engineering methodology variable: sequential and concurrent engineering. An ANOVA showed no significant difference between the effect of concurrent engineering and sequential engineering on design performance. However, the average means of those groups using concurrent engineering exceeded those groups using sequential engineering. There were two levels of the group size variable: small groups consisting of three people and large groups consisting of six people. The design performance of small groups was significantly greater, $F(1,32) = 13.14$, $p<0.001$, than that of large groups as shown in Table 4.

TABLE 4

Design Performance ANOVA Table

Source	df	SS	MS	F
EM	1	0.0714	0.0714	0.20
C	1	0.3404	0.3404	0.96
GS	1	4.6717	4.6717	13.14*
EM*C	1	0.7102	0.7102	2.00
EM*GS	1	0.7981	0.7981	2.24
C*GS	1	0.0004	0.0004	0.00
EM*C*GS	1	0.2176	0.2176	0.61
Error	32	11.3808	0.3557	
Total	39	18.1906		

* Significant at $\alpha = 0.001$

There were two levels of the computer-supported cooperative work variable: groups which used groupware in conceptual design and those that did not. An ANOVA showed no significant difference between the effect of using or not using groupware. The average means of those groups using groupware was less than those that did not use groupware.

Given that design performance is the ratio of system effectiveness and life-cycle cost, these variables were also analyzed. While there were no main effects for the variable with respect to system effectiveness, there was an interaction among engineering methodology, group size, and the use of computer-supported cooperative work, $F(1,32)=4.51$, $p<0.01$, as shown in Table 5.

TABLE 5

System Effectiveness ANOVA Table

Source	df	SS	MS	F
EM	1	18966	18966	0.42
C	1	6528	6528	0.14
GS	1	75951	75951	1.67
EM*C	1	116964	116964	2.57
EM*GS	1	13506	13506	0.30
C*GS	1	65529	65529	1.44
EM*C*GS	1	205492	205492	4.51*
Error	32	1456586	1456586	
Total	39	1959522		

* Significant at $\alpha = 0.01$

The interaction of these variables is shown in Figure 2.

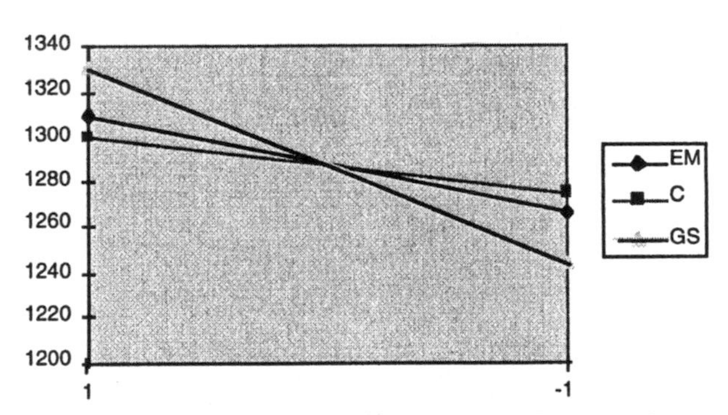

Figure 2. System Effectiveness Interaction

For life cycle cost, Group size was significant, $F(1,32)=12.44$, $p<0.001$, as shown in Table 6.

There were no significant effects of engineering methodology on process time. However, the mean process time of sequential groups was greater than that of concurrent groups. There were no significant effects of group size on process time. However, the mean process time of large groups was greater than that of small groups. There were no significant effects of computer-supported cooperative work on process time. However, the mean process

TABLE 6
Life-Cycle Cost ANOVA Table

Source	df	S S	M S	F
EM	1	44	44	0.00
C	1	44489	44489	0.60
GS	1	918090	918090	12.44*
EM*C	1	180634	180634	2.45
EM*GS	1	55801	55801	0.76
C*GS	1	21437	21437	0.29
Error	32	2360781	73774	
Total	39	3581275		

* Significant at $\alpha = 0.001$

time of computer-supported groups was greater than non-computer-supported groups. There were no significant differences between the process costs of groups using sequential or concurrent engineering. However, the mean process cost of concurrent engineering was less than sequential engineering. Cost for large groups was significantly higher than small groups, $F(1,32)=128.70$, $p< 0.001$, as shown in Table 7.

TABLE 7

Process Cost ANOVA Table

Source	df	S S	M S	F
GS	1	1408501	1408501	128.70*
C	1	149818	149818	13.69*
EM	1	6656	6656	0.61
EM*C	1	37088	37088	3.39
EM*GS	1	7290	7290	.67
C*GS	1	27878	27878	2.55
EM*C*GS	1	9181	9181	.84
Error	32	350201	10944	
Total	39	1996613		

* Significant at $\alpha = 0.001$

The process cost of computer-supported groups significantly exceeded the cost of non-computer-supported groups $(F1,32)=13.69$, $p<0.001$. Finally, member satisfaction differences failed to reach significant levels.

CONCLUSION

In the field study, first-level managers' increased focus on time-oriented constructs such as time allotment to skill development and compensation, and customer needs and strategic planning, in conjunction with the department's value of awareness of time use, autonomy of time use, and future orientation were significant predictors of the level of joint optimization as well as performance. Increased focus on time-oriented constructs such as schedules and deadlines and future orientation had negative relationships with department performance and the level of joint optimization. The manager's ability to maximize STS characteristics in the technical subsystem was a significant predictor of the level of joint optimization. These managers also tended to have higher joint optimization scores. Managers involved with an information technology project had significantly higher scores which is relevant to agile environments. More of these projects resided with downstream managers and these managers had higher levels of joint optimization than transformation managers. The overall relationship between department performance and the level of joint optimization was positive and strong. Skill development and compensation, awareness of time use and autonomy of time use were significant predictors of both variables. Joint optimization was also predicted by the department's future orientation. Performance was also predicted by customer needs and schedules/deadlines. The strong correlation between department performance and level of joint optimization and the overlapping set of time-oriented predictor variables support the notion that time allotment can be used to operationalize joint optimization. Managers should therefore balance the needs of the people with the needs of technology. Agile environments can become technology-driven if a human centered approach is not taken.

In the laboratory study, no condition emerged as statistically significant, although non-computer-supported, small, concurrent engineering groups exceeded the design performance of all other combinations of variables. This negates the "myth" behind technology which says something like "what can be automated should be automated". Hendrick [14] cited three dysfunctional work system design practices: technology centered design, a left-over approach to function and task allocation, and a failure to integrate the organization's sociotechnical system characteristics into its work system design. Designing for agile manufacturing environments can easily be performed according to these dysfunctional practices. Rather, a macroergonomic approach, driven by a deeper understanding of subsystems and their interactions, will lead to human-centered function and task allocation [15]. Empirical macroergonomic research can result in what Hendrick called a "fully harmonized work system". Such a system assumes joint optimization, not maximization of the technological or personnel subsystem and can result in higher levels of both social performance (e.g. quality of work life) and technical performance (e.g. productivity).

REFERENCES

1. **Kleiner, B. M. and Drury, C. G.** "Implementation of Industrial Ergonomics in an Aerospace Company," *Success Factors for*

Implementing Change: A Manufacturing Viewpoint, Society of Manufacturing Engineers, Dearborn, 1988.

2. **Drury, C. G. and Kleiner, B. M.** "Human Factors Support to Modern Manufacturing," *Proceedings of 10th IEA Congress,* 1989, pp. 175-177.
3. **Hormozi, A.M.** "Agile Manufacturing", *37th International Conference Proceedings of the American Production and Inventory Control Society,* 1994, 216-218.
4. **Forsythe, C.** "Human Factors in Agile Manufacturing" *Human Factors and Ergonomics in Manufacturing Special Issue: Human Factors Contributions to Agile Manufacturing,* Vol.7, No.1, 1997, pp. 3-10.
5. **Brooke, L.** "Is Agile the Answer?" *Automotive Industries,* August 1993, pp. 26-28.
6. **Goldman, S.L., Nagel, R.N. and Preiss, K.** "Agile Competitors and Virtual Organizations", *Manufacturing Review,* Vol.8 No.1, 1995.
7. **Byrne, J.A.** "The Virtual Corporation" *Business Week,* February 1993, pp. 99-103.
8. **Brown, O. Jr., Imada, A., Hendrick, H. and Kleiner, B.M.** *Macroergonomics Bulletin,* The Human Factors and Ergonomics Society, Santa Monica, 1997.
9. **Brown, O. Jr.** "The Human Factors and Ergonomics Society". *Human Factors in Organizational Design and Management,* Elsevier, Amsterdam, 1996, pp. 635-641.
10. **Emery, F. and Trist, E.,** "The causal nature of organizational environments", *Human Relations,* 1965, 18, 21-32.
11. **Kleiner, B.M.** "Macroergonomics Lessons Learned from Large Scale Change Efforts in Industry, Government, and Academia". *Human Factors in Organizational Design and Management,* Elsevier, Amsterdam, 1996, pp. 483-489.
12. **Baba, M.L. and Mejabi, O.** "Advances in Sociotechnical Systems Integration: Object-Oriented Simulation Modeling for Joint Optimization of Social and Technical Subsystems" *Human Factors and Ergonomics in Manufacturing Special Issue: Human Factors Contributions to Agile Manufacturing,* Vol.7, No.1, 1997, pp. 37-61.
13. **Plonka, F.E.** "Developing a Lean and Agil Work Force" *Human Factors and Ergonomics in Manufacturing Special Issue: Human Factors Contributions to Agile Manufacturing,* Vol.7, No.1, 1997, pp. 11-20.
14. **Hendrick, H.W.** "Humanizing Re-engineering for True Organizational Effectiveness: A Macroergonomic Approach" *Proceedings of the Human Factors and Ergonomics Society 39th Annual Meeting* (San Diego, California), 1995, pp. 761-765.
15. **Kleiner, B. M. and Drury, C. G.** "The Use of Verbal Protocols to Understand and Design Skill-based Tasks". *(in press) International Journal of Human Factors in Manufacturing,* 1997.

THE USE OF THE MONTE CARLO METHOD FOR TOOL SELECTION IN THE SHEET METAL INDUSTRY

E. Summad, E. Appleton
School of Engineering, University of Durham, Durham DH1 3LE, England

ABSTRACT: Using modern numerically controlled fabrication methods, a single punch tool can be used to punch holes in more than one product design and it is also possible to punch complex shapes by using a combination of tools which are available from a turret. Therefore, numerically controlled methods have the potential to overcome many of the tool inflexibility drawbacks of traditional methods, thereby significantly increasing overall flexibility in production and design.
Since reducing or eliminating set-ups is a primary advantage of numerically controlled methods, implementing them in an inefficient fashion, that requires unnecessarily frequent set-ups, is obviously an ineffective strategy. Thus, selecting appropriate tooling in numerically controlled manufacture is an important economic issue.

In the present work, the approach has been to develop a generic algorithm to select the best set of tools for producing any given sheet metal part, based upon searching a huge explosive decision tree which, in a practical sense is impossible to solve analytically. This paper will show how the well known Monte Carlo simulation technique can be used in the decision tree search and hence in the process of selecting appropriate tooling.

INTRODUCTION

The main effort in generating a production plan for punching operations lies in the process of tool selection . Few attempts have been made to solve the tool selection optimisation problem. Raggenbass, A, and Reissner, J.[1-3] suggested that sheet metal features, could be divided into edges and holes and also into common shapes (circles, squares, rectangles, long holes) or uncommon shapes. Then, for all inner features with common shapes, rule-based programming seeks a tool with which the feature contour can be punched out in one blow, or if necessary two blows. Next, for those inner features and outer contour features that remain unpunched, tools are subsequently selected, again using rule-based programming, such that the individual feature can be manufactured with a minimum number of blows.

Liebers, A., et. al.,[4] proposed a system with the objective of minimising the processing time, in which the basis for tool selection was the use of the largest possible commonality between tools. A " method selection " module, using AI techniques, based upon rules is used which matches features to a nibbling or punching process. For each feature a predetermined punching or nibbling tool is determined. The tool that can be used for the largest number of features is the one selected.

Choong, N. F. , et. al., [5,6] suggested that profiles of the part usually belong to one of the following: general external profile, general internal profile, square hole, rectangular hole, triangular hole, oblong hole, arc- slot hole, or circular hole. The proposed algorithm divides the process of tool selection into two stages. In stage one tools are selected for each profile based on its shape and geometry. In stage two an algorithm rationalises those independently selected tools in the primary stage and plans for a single tool set-up to produce the complete part. Two steps are involved in the secondary tool selection. Step one reduces the total number of primary selections to the number of tool stations that can be accommodated. Based on the tools selected in step one, step two selects only one solution from the few primary selections obtained for every profile.

Cho, K H, and Lee, K.,[7] developed an algorithm through which one can find the successive matching curves between two curve lists, one for the punching tools and the other for the boundaries of the sheet metal part design. This is very compatible with the way shape data is stored in a CAD package such as AutoCAD.

As can be seen from the above discussion most of the previous work has involved restricting the shapes of the holes to be considered, or focusing on selecting the best tools for one particular feature or boundary edge. Such approaches will normally result in one of two extremes, that is, either more than one tool set-up is required and / or free slots are left on the machine turret causing extra unnecessary blows. In practice, one tool set-up is much preferred. Therefore implementing a tool selection strategy which will require unnecessarily frequent set-ups is obviously an ineffective strategy.

BACKGROUND INFORMATION

One of the interesting applications in the area of computer vision is the recognition of partially occluded objects and their location in a overlapping scene. The idea behind this is to find a correspondence between the boundary entities of the complex scene and those of the single objects. Adopting such ideas into the process of tool selection for sheet metal components shows promising results. Cho, K H, and Lee, K.,[7] have introduced what they called the shape-index-set , which uniquely describes each boundary entity. In their approach the focus was on selecting the best tool for a particular curve or a sequence of curves; and the first interference free matching tool is selected. Such a policy does not take into account either the expensive tool set-up problem, nor the capacity of the tool rack.

To design a system which will result in the selection of the best set of tools to produce any given sheet metal design, without any restriction on the geometry of the features, it was necessary to consider all the feasible tools and not to limit the search to the first interference free tool. In doing this, and especially in cases where boundary curves can be partially punched out by a particular tool, a huge unpredictable size decision tree emerged. In this work it was considered not practical to optimise the selection procedure using an analytical method and so a numerical method based upon a Monte Carlo approach was considered.

To solve such an optimisation problem without violating the objective of the research, a simulation approach has been adopted, where a random tool is selected for punching a randomly selected feature. The aim was to find the best match between one tool curve and only one feature curve at a time, providing that such a match will not cause any tool-feature interference. Another tool is then randomly selected for another randomly selected feature and so on until the part is completely punched out. Carrying out such a simulation will result in the selection of a set of tools for producing the whole part. Running such a simulation for a number of times will result in the generation of a large number of different sets of tools. Of course, most of these solutions will be obviously unacceptably inefficient. A cost-function value has been introduced to evaluate each solution and the set of tools with the minimum cost value will be selected. Subsequently, an interesting point arose, regarding how long this simulation should run in order to have an acceptable sets of tools to choose from and how far this selected set of tool is from an optimum answer.

The rest of this paper will show with more detail the data structure used in the proposed system, the definition of curve-signature, the matching process and finally the application of the Monte Carlo simulation.

THE DATA STRUCTURE

In AutoCAD, a popular and easily available CAD software package, design data is stored in the form of a list which contains entities (straight lines and / or circular curves). For example, a square will be saved as four separate straight lines. A straight line is defined by the co-ordinates of its starting (x_1, y_1, z_1) and ending (x_2, y_2, z_2) points while a circular curve is defined by the co-ordinates of the centre point (x_0, y_0, z_0) and the radius of curvature (R) and the starting (A_S) and finishing (A_F) angles from the centre point. If the entity is a circle, it will be stored by the co-ordinates of the centre point and the radius of the circle.

CURVE SIGNATURE

In the present work a curve signature, similar to the shape-index-set introduced by Cho, K H, and Lee, K.,[7] has been introduced in which each feature / tool entity is given a unique description. Each curve signature is composed of four elements: length of entity, radius of curvature (if the entity is a straight line the radius of curvature is made equal to 999), the angle with the preceding entity and angle with the proceeding entity. Using the data stored in AutoCAD, each entity is allocated its unique curve signature which can be used directly in the tool-feature matching process.

MATCHING PROCESS

The aim of this matching process is to find the best match between a randomly selected tool and a randomly selected feature. Therefore, every tool curve signature is compared with every feature curve signature and given a "priority-rank" value. The match with the highest priority-rank value will be selected providing there is no tool-feature interference. The priority-rank value is based upon considering the curve-signature elements and calculated as follows:

$$R^{nm}_{ij} = R1^{nm}_{ij} + R2^{nm}_{ij} + R3^{nm}_{ij} + R4^{nm}_{ij}$$

where,

i : feature number,
j : tool number,
n : feature curve number,
m: tool curve number,
R^{nm}_{ij} : priority-rank value of matching the nth curve on the ith feature with the mth curve of the jth tool,
$R1^{nm}_{ij}$: Comparing feature curve length with tool curve length,
$R2^{nm}_{ij}$: Comparing feature curve radius of curvature with tool curve radius of curvature,
$R3^{nm}_{ij}$: Comparing feature curve angle with preceding entity with tool curve angle with preceding entity, and
$R4^{nm}_{ij}$:Comparing feature curve angle with proceeding entity with tool curve angle with proceeding entity.

The values of $R1^{nm}_{ij}$, $R2^{nm}_{ij}$, $R3^{nm}_{ij}$, and $R4^{nm}_{ij}$ are calculated as follows:

$R1^{nm}_{ij} = 2$	if	$R1^{m}_{j} / R1^{n}_{i} = 1$
$R1^{nm}_{ij} = R1^{m}_{j} / R1^{n}_{i}$	if	$R1^{m}_{j} / R1^{n}_{i} < 1$
$R1^{nm}_{ij} = -99$	if	$R1^{m}_{j} / R1^{n}_{i} > 1$

$R2^{nm}_{ij} = 2$	if	$R2^{m}_{j} / R2^{n}_{i} = 1$
$R2^{nm}_{ij} = 1$	if	$R2^{m}_{j} / R2^{n}_{i} < 1$
$R2^{nm}_{ij} = -99$	if	$R2^{m}_{j} / R2^{n}_{i} > 1$

$R3^{nm}_{ij} = 2$	if	$R3^{m}_{j} / R3^{n}_{i} = 1$
$R3^{nm}_{ij} = 1$	if	$R3^{m}_{j} / R3^{n}_{i} < 1$
$R3^{nm}_{ij} = -99$	if	$R3^{m}_{j} / R3^{n}_{i} > 1$

$R4^{nm}_{ij} = 2$	if	$R4^{m}_{j} / R4^{n}_{i} = 1$
$R4^{nm}_{ij} = 1$	if	$R4^{m}_{j} / R4^{n}_{i} < 1$
$R4^{nm}_{ij} = -99$	if	$R4^{m}_{j} / R4^{n}_{i} > 1$

MONTE CARLO SIMULATION

In many applications of Monte Carlo, the physical process is simulated, and there is no need to derive equations that describe the behaviour of the system. A description of the way Monte Carlo simulation was used in the process of tool selection will appear in the following case study.

CASE STUDY

Figure 1 shows the drawing of a sample sheet metal part and the tools used. The part consists of one boundary feature and four inner features.

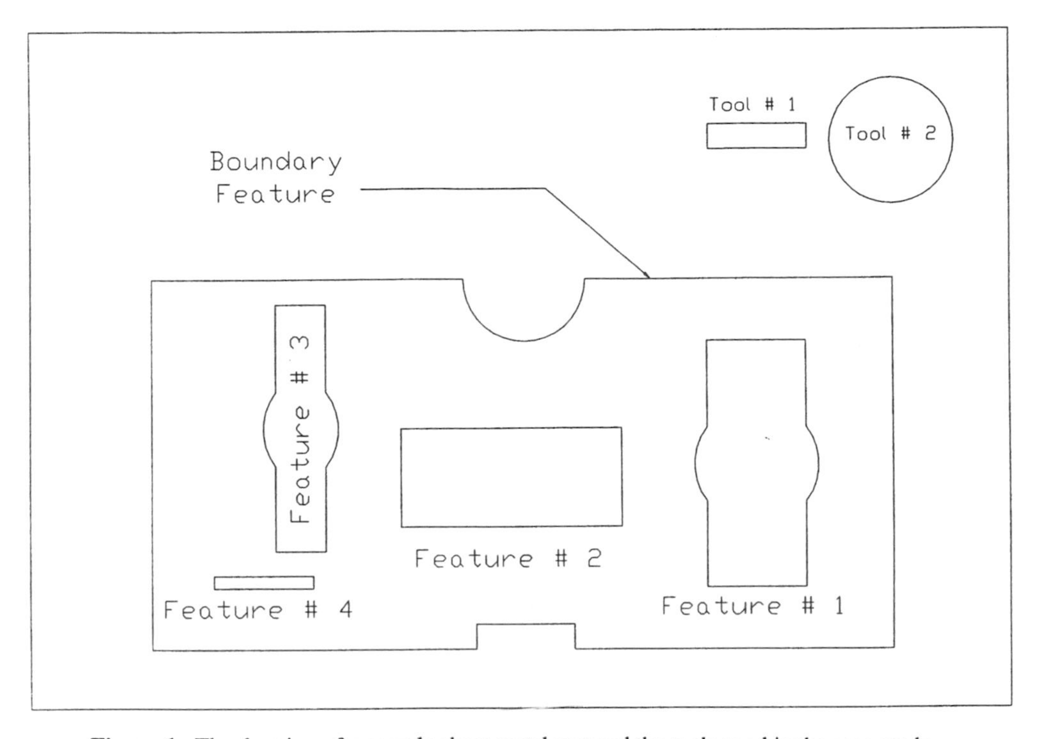

Figure 1. The drawing of a sample sheet metal part and the tools used in the case study.

To assist in interchanging drawings between AutoCAD and other programs, a drawing interchange file format (DXF) has been defined [8]. All implementations of AutoCAD accept this format and are able to convert it to and from their internal file representation. Table 1 shows the format of the entities which belong to the part shown in Figure 1.

Reading these data, each entity can be given a curve signature. For example the 18th entity i.e. Entity 178 will have a curve signature [40, 999, 90, 90] where 40 is the length of the entity, 999 is the radius - straight line - of curvature, the first 90 is the angle with the preceding entity i.e. Entity 179, and the second 90 is the angle with proceeding entity i.e. Entity 18F.

A " long hand " simulation approach has been used, as a start, to gain more insight into the issues involved in the process, as well as avoiding the confusion of mixing algorithmic inconsistencies with programming / coding errors. In this " long hand " study, to insure a random selection of features and tools, seven penny coins have been used where each represented a tool or a feature.

Table 1
The format of the entities which belong to the sheet metal part shown in Figure 1.

S. No	Entity	Description	x_0	y_0	z_0	R	A_S	A_F	x_1	y_1	z_1	x_2	y_2	z_2
1	164	Line							55	30	0	185	30	0
2	16F	Line							185	30	0	185	40	0
3	16E	Line							185	40	0	225	40	0
4	16D	Line							225	40	0	225	30	0
5	170	Line							225	30	0	355	30	0
6	165	Line							355	30	0	355	180	0
7	166	Line							355	180	0	230	180	0
8	171	Arc	205	180	0	25	180	0						
9	172	Line							180	180	0	55	180	0
10	167	Line							55	180	0	55	30	0
11	17F	Line							280	55	0	280	90	0
12	17E	Arc	300	105	0	25	216.869898	143.130102						
13	17B	Line							280	120	0	280	155	0
14	17A	Line							280	155	0	320	155	0
15	180	Line							320	155	0	320	120	0
16	181	Arc	300	105	0	25	36.869898	323.130102						
17	179	Line							320	90	0	320	55	0
18	178	Line							320	55	0	280	55	0
19	18F	Line							155	80	0	155	120	0
20	18E	Line							155	120	0	245	120	0
21	18D	Line							245	120	0	245	80	0
22	18C	Line							245	80	0	155	80	0
23	184	Line							105	70	0	105	105	0
24	185	Arc	125	120	0	25	216.869898	143.130102						
25	186	Line							105	135	0	105	170	0
26	18A	Line							105	170	0	125	170	0
27	183	Line							125	170	0	125	135	0
28	182	Arc	105	120	0	25	36.869898	323.130102						
29	188	Line							125	105	0	125	70	0
30	18B	Line							125	70	0	105	70	0
31	16B	Line							80	55	0	80	60	0
32	16A	Line							80	60	0	120	60	0
33	169	Line							120	60	0	120	55	0
34	168	Line							120	55	0	80	55	0

The simulation started by rattling five numbered coins and randomly selecting one. By using the coins to represent features or tools random features and tools were selected to match the entities of each randomly selected tool with the entities of each randomly selected feature. Each match was given a priority-rank value. The match with the highest priority-rank was selected. The tool was then laid on the top of the feature after the necessary tool rotation and transform. A check for any tool-feature interference was carried out in order to make sure that the tool will not break into any of the sheet metal part feature profiles. A check for the punched curves was then necessary in order to update the feature details. If the tool failed the interference test, the simulation started again i.e., a new random feature and a new random tool were selected. This simulation is continued until all the features are completely punched out. Table 2 shows a complete example simulation.

Table 2

An example of a complete simulation.

Trial Number	Feature Number	Tool Number	Feature Curve	Tool Curve	Interference Test	Trial Number	Feature Number	Tool Number	Feature Curve	Tool Curve	Interference Test
1	2	1	1	2	OK	45	B	2	5	Circle	OK
2	3	1	8	1	OK	46	1	2	1	Circle	Fail
3	1	1	4	2	OK	47	1	2	1	Circle	Fail
4	3	2	2	Circle	Fail	48	4	1	2	2	Fail
5	1	1	8	4	OK	49	3	2	2	Circle	Fail
6	4	2	2	Circle	Fail	50	1	2	1	Circle	Fail
7	4	1	2	2	Fail	51	B	1	6	2	OK
8	4	2	1	Circle	Fail	52	3	1	4	1	OK
9	B	1	3	2	OK	53	1	1	3	1	OK
10	4	2	3	Circle	Fail	54	2	2	4	Circle	Fail
11	B	1	1	2	OK	55	1	2	1	Circle	Fail
12	4	2	4	Circle	Fail	56	2	1	4	2	OK
13	1	2	2	Circle	OK	57	1	2	1	Circle	Fail
14	2	1	3	2	OK	58	1	2	1	Circle	Fail
15	B	1	1	2	OK	59	3	2	2	Circle	Fail
16	2	1	3	2	OK	60	3	2	2	Circle	Fail
17	4	2	2	Circle	OK	61	1	1	3	1	OK
18	3	1	8	1	OK	62	4	1	2	2	Fail
19	4	1	2	2	Fail	63	2	2	2	Circle	Fail
20	B	2	8	Circle	OK	64	B	1	6	2	OK
21	1	2	1	Circle	Fail	65	2	2	2	Circle	Fail
22	2	1	2	2	OK	66	3	2	2	Circle	Fail
23	B	1	1	4	OK	67	2	1	2	3	OK
24	2	2	2	Circle	Fail	68	1	2	1	Circle	Fail
25	B	1	1	1	OK	69	2	1	4	1	OK
26	4	1	2	2	Fail	70	1	1	1	1	OK
27	2	2	2	Circle	Fail	71	3	1	3	1	OK
28	4	1	2	2	Fail	72	2	1	4	1	OK
29	3	2	2	Circle	Fail	73	2	2	2	Circle	Fail
30	1	2	1	Circle	Fail	74	2	2	2	Circle	Fail
31	3	2	2	Circle	Fail	75	B	2	6	Circle	OK
32	1	1	3	1	OK	76	2	1	4	1	OK
33	2	2	2	Circle	Fail	77	2	1	2	1	OK
34	4	2	1	Circle	Fail	78	B	1	7	2	OK
35	2	2	4	Circle	Fail	79	1	1	1	1	OK
36	1	2	1	Circle	Fail	80	2	1	2	1	OK
37	B	1	5	2	OK	81	1	2	1	Circle	Fail
38	3	2	2	Circle	Fail	82	B	2	7	Circle	OK
39	1	2	1	Circle	Fail	83	B	1	9	2	OK
40	3	2	2	Circle	Fail	84	B	1	9	2	OK
41	2	2	2	Circle	Fail	85	B	2	9	Circle	OK
42	1	2	1	Circle	Fail	86	1	2	1	Circle	Fail
43	B	1	5	2	OK	86	B	2	10	Circle	OK
44	4	1	2	2	Fail	88	1	1	1	1	OK

The results of this simulation is shown in Figure 2.

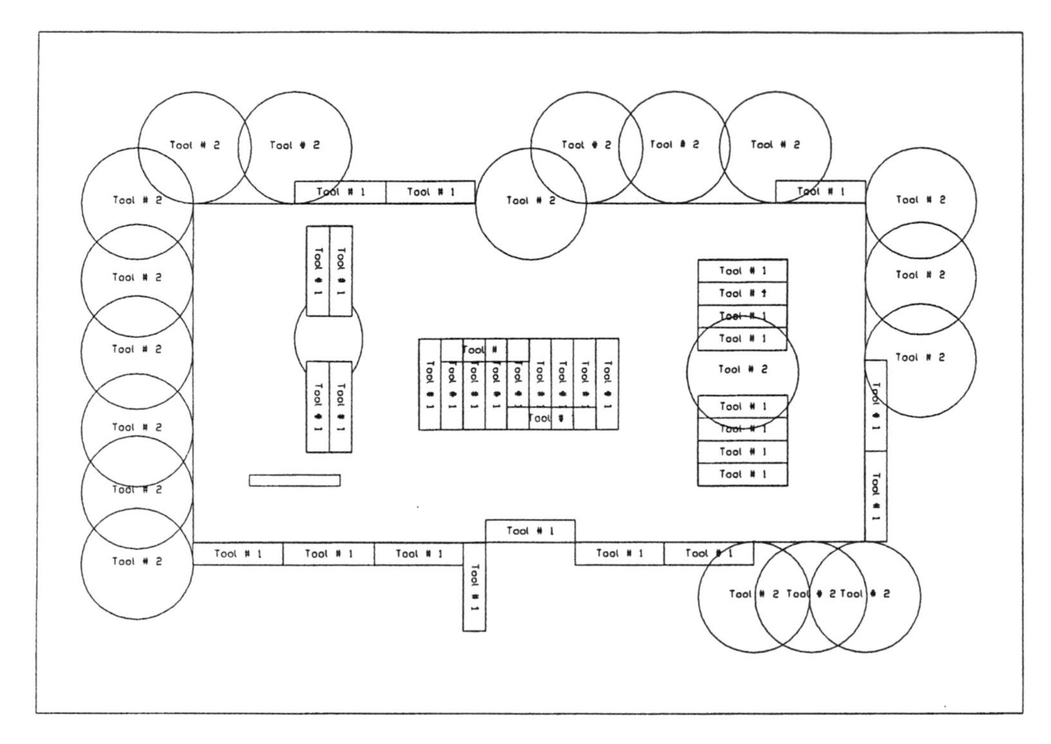

Figure 2. The solution of the example simulation.

As can be clearly seen, in Figure 2, a great number of the blows are unnecessary. A cost function value which takes into consideration the number of blows, the number of tools, the tool set-up, tool loading, and material utilisation was devised in order to evaluate each simulation result. Running such a simulation a number of times will result in different sets of tools, each with its own cost function value. The set of tools with the minimum cost function value is the one to be selected. In order to randomly reach an optimal answer, see Figure 3, a large number of simulation trails would be needed.

In some cases, see Figure 3, some features or parts of features may be impossible to punch with the existing tools.

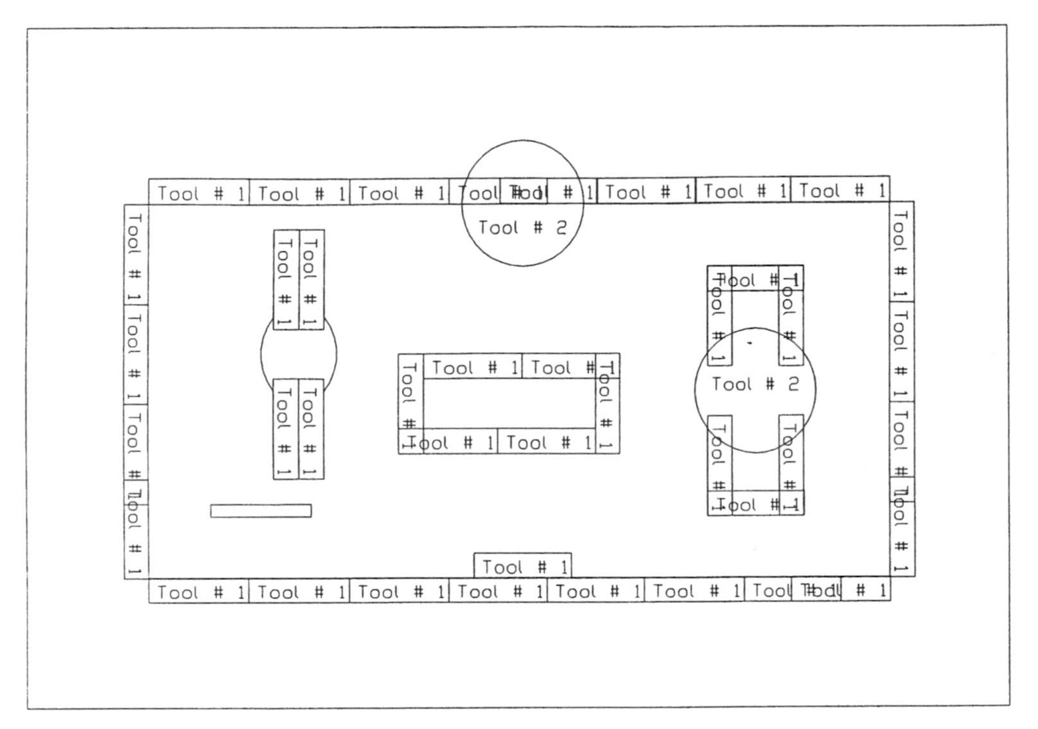

Figure 3. Optimal answer.

The next step of this work will be computerising the simulation and the use of Monte Carlo analysis in order to be able to determine how long the simulation needs to run in order to arrive at a number of good tool sets. It will also be important to determine how close this selected set of tools is from the optimum answer.

A detailed description of the use of the computer based Monte Carlo method in the process of tool selection optimisation will appear in a separate paper, by the same authors, in the 32nd International MATADOR Conference, 10th-11th July 1997, at the University of Manchester Institute of Science and Technology (UMIST), England.

CONCLUSIONS

1. The AutoCAD data storage structure is a convenient starting point for automatic tool selection.
2. Finding the optimum set of tools involves searching an explosively large decision tree.
3. A Monte Carlo method using a cost function evaluation has been demonstrated to be feasible.
4. Further investigation are required to explore the relationship between number of simulation runs and closeness to an optimum tool selection.

References

1. **Raggenbass, A. and Reissner, J.**, Stamping - laser combination is sheet processing, Annals of the CIRP, 38, 291-294, 1889.

2. **Raggenbass, A. and Reissner, J.**, An expert system as a link between computer aided design and combined stamping - laser manufacture, Proc Instn Mech Engrs, 205, 25-34, 1991.

3. **Raggenbass, A. and Reissner, J.**, Automatic generation of NC production plans in stamping and laser cutting, Annals of the CIRP, 40, 247-250, 1991.

4. **Liebers. A., Streppel, de Vin, L. J., de Vries, J., and Kals, H. J. J.**, Toolselection and toolpath generation for nibbling, Proceedings of the IMC-10 Conf., 1993, 37-449.

5. **Choong, N. F., Nee, A. Y. C., and Loh, H. T.**, The implementation of an automatic tool selection system for CNC nibbling, Computers in Industry, 23, 205-222, 1993.

6. **Loh, H. T. Nee, A. Y. C., and Choong, N. F.**, A PC-based software package for automatic punch-selection and tool-path generation for a CNC nibbling machine, Journal of Materials Processing Technology, 23, 107-119, 1990.

7. **Cho, K. H., and Lee, K.**, Automatic tool selection for NC turret operation in sheet metal stamping, Journal of Engineering for Industry, 116, 239-246, 1994.

8. **Drawing interchange and file formats** - AutoCAD Release 12, Autodesk, part no. TSPSD-R12.Custom.Man-1, July 1993.

MATERIAL TRANSPORT SCHEDULING IN FLEXIBLE MANUFACTURING SYSTEMS

Bala Ram[1], Sanjay Joshi[2]
[1]North Carolina A&T State University, Greensboro, NC 27411
[2]The Pennsylvania State University, University Park, PA 16812

ABSTRACT:
Material transport activities assume special importance in the content of automated manufacturing environments such as Flexible Manufacturing Systems (FMSs). The performance of the total system in terms of throughout and timeliness of order completions, and utilization of material transport and handling equipment depends on scheduling of material transport and handling activities. "Material transport", in the context of this paper, refers to inter-workstation movement, while "material handling" refers to handling activities within a workstation. We address the issue of scheduling material transport activities in the context of any FMS.

In order to address material transport in an FMS, in an integrated way, we need a unified generic scheme that provides appropriate links between subsystems in an FMS. Such a representation must also be based on some assumed architecture for an FMS; this is to ensure that algorithms and procedures developed are compatible with all other aspects of making an FMS operational. For example, the material transport schedules developed must be at a level of detail that interactions with processing stations can be evaluated. The architecture we assume here is that proposed in Smith, Hoberecht and Joshi, and consists of three levels: shop, workstation and equipment.

In this paper, we present a representation for a Material Transport System in an FMS. Based on the representation, we identify a few scheduling subproblems. We present some mathematical models and example problems for some of the scheduling subproblems.

INTRODUCTION

Flexible Manufacturing System (FMS) came about as an answer to efficient manufacturing of products in medium quantities with the added requirement of frequent changes in the products to be manufactured. Such a manufacturing system has the stated dual objectives of "productivity" through automation of processes and materials handling, and "flexibility" through the ability to quickly reprogram the system. Several FMSs are in operation throughout the world. However, the cost of such systems is still prohibitive, primarily due to the high cost of the software that has the task of accomplishing integration. Merchant suggests that it has proven much more difficult than expected to integrate the so-called "islands of automation" into a fully integrated system. [Merchant, 1988] Bjorke attributes this difficulty to the lack of a theoretical basis by which the individual functions of the manufacturing systems can be analyzed. [Bjorke, 1988]

Implied in the observation of Bjorke is the need for the explicit statement of an efficient control architecture for FMSs. [Bjorke, 1988] Smith, Hoberecht and Joshi [1992] present such an architecture. [Smith, Hoberecht and Joshi, 1992] The proposed architecture is a generalized version of that proposed by Joshi, Wysk and Jones; it consists of three levels: shop, workstation and equipment. [Joshi, Wysk and Jones, 1990] The possible classes of machines in a production system under this architecture are: material processing, material handling, material transport, and automated storage. The class of material handling machines is essentially made up of robots, indexing devices, and other devices capable of moving parts from one location to another in a specified orientation. The class of material transport machines is made up of AGV's, conveyors, fork trucks, and other manual or automated transport machines. The primary function of these machines is to transport parts to various locations throughout the factory. The distinction between material handling and material transport machines is that material handling machines perform intra-workstation part movements functions and material transport machines perform inter-workstation part movement functions. Smith, Hobercht and Joshi propose that control activities be partitioned into planning, scheduling and execution classes. [Smith, Hobercht and Joshi,1992] Planning refers to selection from among possible processing routes; scheduling involves setting start and finish times for processing, material handling and material transport activities; and execution is concerned with performing tasks.

Material transport and materials handling are important integration activities in a FMS. We address the issue of scheduling material transport and materials handling activities. Often, reactive scheduling is adopted for material transport and handling activities. We present a representation framework and some mathematical models for proactive scheduling, which is likely to be more efficient than a reactive approach.

BACKGROUND

We review the literature in the following areas relevant to our discussion here: modeling for material handling systems, and scheduling in the real-time operational environment.

Matson and White present a survey of operations research applications in material handling. [Matson and White, 1982] The survey points out that very little basic research has been devoted to integrated material handling systems. Research has addressed various aspects of material handling such as: robotics, conveyor theory, transfer lines, flexible manufacturing systems, equipment selection, storage alternatives, automated storage and retrieval systems, vehicle routing, and facilities layout and location. While analytical models that have been developed tend to be focused on a narrow aspect of material handling, simulation studies have addressed the design of integrated materials handling systems.

Smith, Graves and Kerbache have developed and applied an analytical, open-queuing network model for examining topology routing and resource-planning problems in the material handling component of a manufacturing environment. [Smith, Graves and Kerbache, 1986]

Kouvelis and Lee present a graph-theoretic modeling framework of material handling system design. A network representation is used to propose solutions to the equipment selection and topological design problem. [Kouvelis and Lee, 1990]

In view of the volume of scheduling literature, it is instructive to get a broad perspective of past work. Solberg provided an excellent overview of approaches taken to handle scheduling problems in manufacturing; he reveals three separate paradigms that have driven research in scheduling over the past few decades; the optimization paradigm the date processing paradigm, and the control paradigm. The optimization paradigm views planning and scheduling problems as primarily associated with optimization. The data processing relating to scheduling; the focus here is on computer-based methods to organize and manipulate data relating to scheduling. The control paradigm focuses on the commands and feedback loops necessary to keep a production system operating as intended. While in implementing scheduling all three paradigms assume importance, both research and practice in scheduling use only any one of the three paradigms. [Solberg, 1989]

Parunak provides a good insight into the characteristics of the scheduling problem; the framework presented addresses issues that are important in automated manufacturing environments. Five parameters that need to be addressed in a general scheduling problem are discussed. These parameters also provide a framework to understand currently known approaches to scheduling. [Parunak [1991]

The area of material transport and material scheduling, which is the subject in this paper, has received very little attention in the literature. Most reported research in this area has addressed the scheduling of AGVs; such research has primarily taken an isolated view of the AGV network, in context to the integrated material handling system. This research seeks to address material transport and material handling scheduling in an integrated manner, in the context of automated manufacturing environments as in FMSs.

AN INEGRATED REPRESENTATION OF MATERIAL TRANSPORT

In order to address material transport, in an FMS, in an integrated way, we need a unified generic scheme that provides appropriate links between subsystems. Such a representation must also be based on some assumed architecture for an FMS; this is to ensure that algorithms and procedures developed are compatible with all other aspects of making an FMS operational. For example, the material transport schedules developed must be at a level of detail that interactions with processing stations can be evaluated.

The architecture we assume here is that proposed by Smith, Hoberecht and Joshi. A feature of the architecture relevant to our discussion here, is that the FMS is assumed to consist of three levels as mentioned in the previous section: shop, workstation and equipment. [Smith, Hoberecht and Joshi, [1992].

It must be mentioned here that "material transport" in the context of the Smith, Hoberecht and Joshi work refers to inter-workstation movement only and use the term with the same meaning in the paper.

Kouvelis and Lee, as mentioned in the previous section, propose a network representation in their work on design of material handling systems. [Kouvelis and Lee, 1990] We also propose a directed-graph representation that is augmented to capture additional characteristics relevant to the operation of a material transport system.

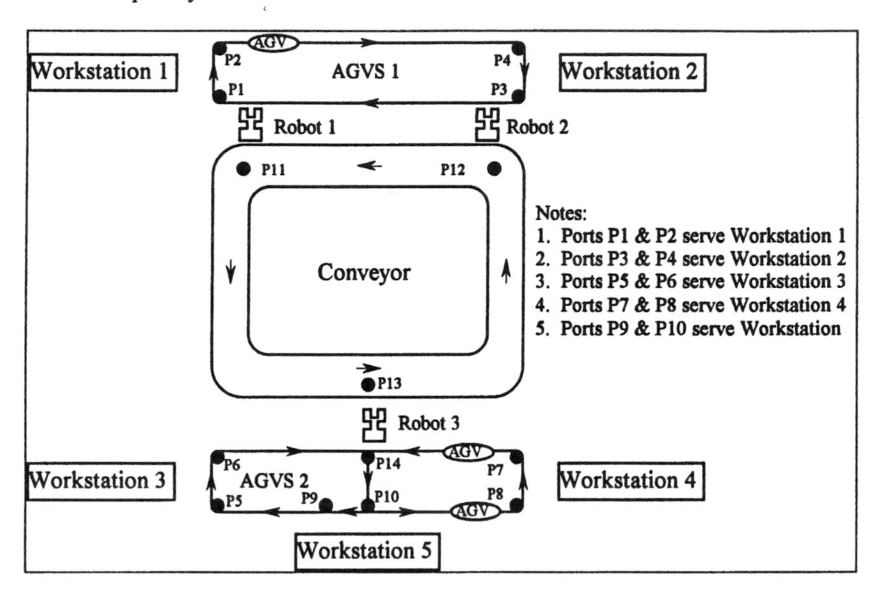

Figure 1. Example Shop with Material Transport System

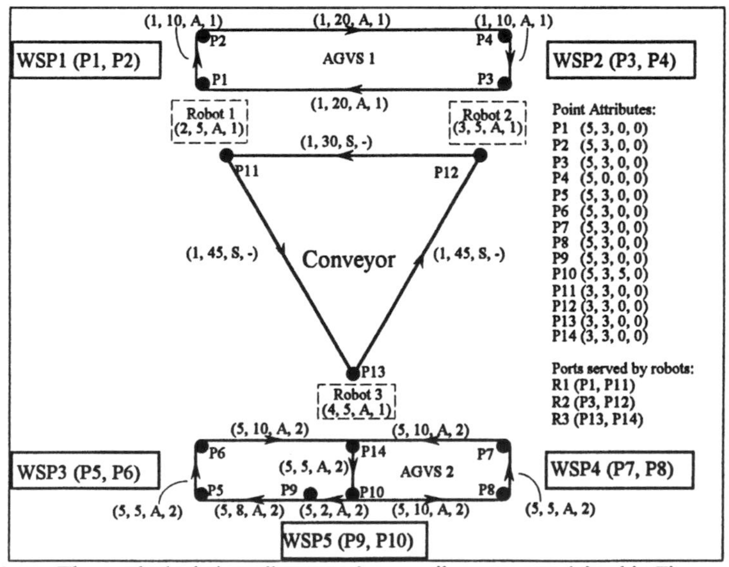

Note: The node (point) attributes and arc attributes are explained in Figure 4.

Figure 2. Representation of Material Transport System in Shop of Figure 1.

Figure 1 shows and example of a shop consisting of five workstations. Material movement between the workstations is made possible by a conveyor, two AGV networks and three robots; the configuration of this set of material transport equipment is shown in Figure 1`. P1, P4, P5, P8, and P10 serve as input ports and P2, P3, P6, P7 and P9 serve as output ports for workstations 1, 2, 3, 4, and 5 respectively. P11, P12, P13 and P14 serve as interface points between the material transport subsystems in this example.

The representation scheme proposed here is a directed graph shown in Figure 2. The ports and interface points in Figure 1 are nodes in the network. The transport segments are represented by directed arcs. Each node and arc has a four-component vector that stores their characteristics. Figure 3 shows the subsytems of Figure 2. Figure 4 presents notations that summarize the meanings of the components of the vectors for nodes and arcs. Additional notations for transport requests to the material transport systems and direct address transport units are also shown in Figure 4.

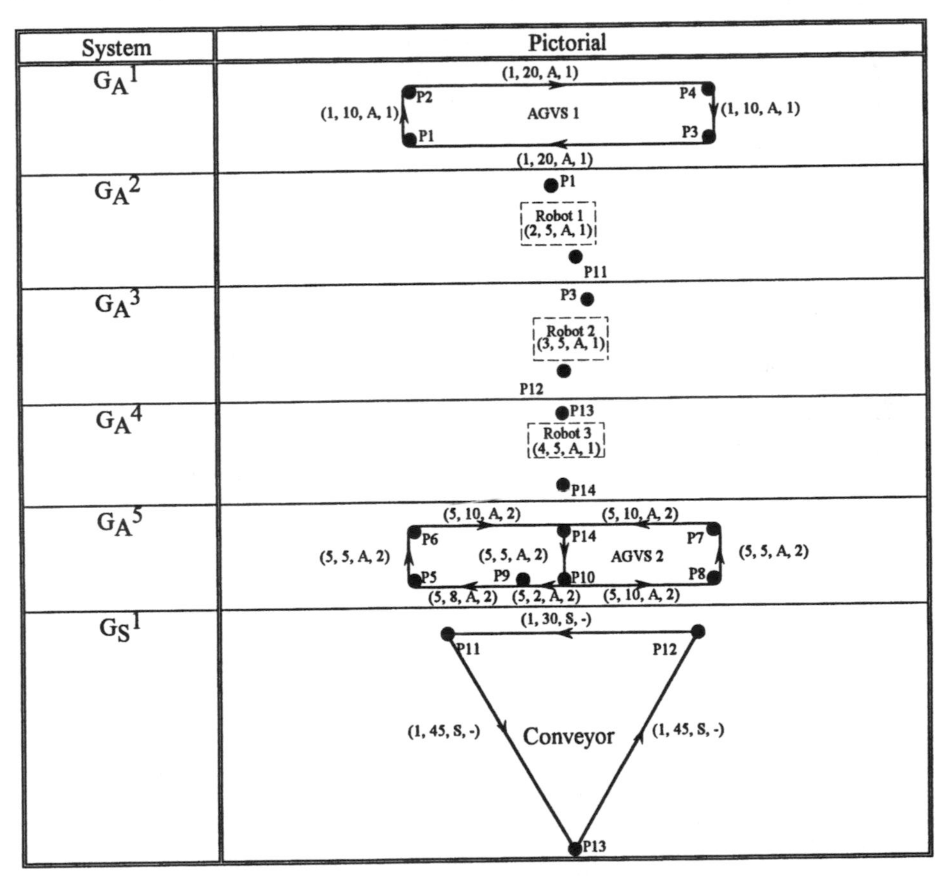

Figure 3. Representation of Material Transport Subsystems in Shop of Figure 1.

Nodes.

n = node index

N = set of nodes

A = set of nodes for asynchronous transport devices

S = set of nodes for synchronous transport devices

A_a = set of nodes for asynchronous transport system "a"

S_s = set of nodes for synchronous transport system "s"

t_l^n = load time for node n

t_u^n = unload time for node n

t_t^n = transfer time for node n

b^n = buffer size for node n

parameter vector for nodes = $(t_l^n, t_u^n, t_t^n, b^n)$

Graphs and arcs.

G_A^a = augmented directed graph for asynchronous transport system "a"

G_S^s = augmented directed graph for synchronous transport system "s"

G = augmented directed graph for entire material transport system

G_i = set of all sub-graphs designating possible routes for request i

g_i^k = a sub-graph designating k^{th} route for request i

g_i = selected route for request i

parameter vector for arcs = (a/s index, time, asynchrononus(a) or synchornous(s), number of directed units)

Transport requests.

i = transport request index

I = set of transport requests

R = number of transport requests

d_i = destination node for request i

t_i = time of request i

c_i = required completion time for request i

Direct address transport units. (automatic guided vehicles , pallets on conveyors, etc.)

V_a= set of direct address transport units in asynchronous transport system "a"

Figure 4. Notations for Representation of Material Transport Systems.

The nodes in the network, which essentially represent transfer points are described by a load time, an unload time, possibly a transfer time and a buffer size. Load time is the time to place material from a node; transfer, refers to moving material off a node by diverting, instead of unload followed by load operation. For example, the attributes of P3 (see Figure 2) are (5,3,0,0) where the load time is 5 time units, unload time is 3 units, and both transfer time and buffer size are 0.

Transport devices are classified as asynchronization or synchronous. A synchronous device is always available for a transport action; for example, any conveyor. A synchronous device must be checked for availability at any time, at any point along its path; for example, any automatic guided vehicles (AGVs) or robots. Direct address transport units are a special case of asynchronous transport; AGVs and pallets on conveyors are examples of such devices.

Arcs are characterised by the index of the device, transport time along the arc, an indicator for whether asynchronous or synchronous devices are available along it, and for arc with asynchronous devices, the number of units that travel along the arc. For example, attributes of Robot 3 (see Figure 2) are (4,5,A,1) where 4 is an index, 5 time units is the transportation time, "A" denotes an asynchronous device, and 1 denotes the number of such devices corresponding to index 4. As another example, the attributes of one of the conveyor segments (see Figure 2) are (1,45,S,-) where 1 is an index, 45 time units is the tranportation time, and "S" denotes a synchronous device.

SOME MATERIAL TRANSPORT SCHEDULING SUBPROBLEMS

Any underlying material transport system in an FMS can be represented in the form of directed graphs as presented in the previous section. Once such a representation is developed, we can state two subproblems that will enable the formulation of schedules for material transport activities for the FMS. While the scheduling problem can be stated as a single problem addressing both the route selection and scheduling, it is likely that the resulting formulation will be intractable. Hence, we take the approach of formulating two subproblems.

The first subproblem deals with the selection of a route for servicing a transport request. Such a request will have a origin and a destination node in the directed graph, with multiple possibilities for path from the origin to the destination node. We call this the Material Transport Routing problem. In the next section we formally state the problem and propose three approaches to solve this problem.

The second subproblem addresses the issue of scheduling these transport activities, or making a decision on the sequence in which these transport activities will be undertaken at each node. We call this the Material Transport Scheduling problem. This problem is also formally stated and we propose three approaches to solve this problem.

MATERIAL TRANSPORT ROUTING PROBLEM

Problem Statement:

Given a set of transport requests select a good route g_i for each request i.

Additional Notations:

M_i = matrix of "node use" times for transport request i (see more explanation under "Approach")

P_i = priority of request i (larger the number, the higher the priority)

L_i^K = estimated lateness for request i if route g_i^k is used

E_i^K = estimated transport time for request i if route g_i^k is used

r_i = vector of route selection decisions for request i (Here, r_i^K = k^{th} component of r_i; if k^{th} route g_i^k is selected, $r_i^K = 1$ and 0 otherwise

a" = $|N|$ dimensional vector of units on available times at nodes during the planning horizon

Approaches:

MTR-1.

Under this approach the objective is to minimize the total weighted lateness of all requests that are to be processed; the weighting is based on priorities assigned to each of requests. The constraints in the formulation are based on the idea that the total times spent by the transported parts at individual nodes (due to load, unload and transfer activities) have to limited by the availability of the device or resource represented by that node.

Step 1:

Construct all possible routes G_i for each transport request i

Step 2:

Construct node utilization matrix M_i for each transport request i.

M_i is a matrix of size $|N| X |G_i|$; the entries represent the times to be spent at the node by the transported unit, due to a load, unload or transfer action. The times are appropriately taken from t_l^n , t_u^n , and t_t^n .

Step 3:

The following mathematical model is solved to select a route in G_i for each request i.

Minimize: $\sum_i \sum_k P_i L_i^k r_i^k$

Such that: $\sum_i M_i r_i \leq a$ (1)

$\sum_i r_i^k = 1$, for each i (2)

where: $L_i^k = E_i^k - (c_i - t_i) \quad if \quad E_i^k > L_i - t_i$

$= 0 \; otherwise$

The following modification is suggested to tighten the node availability constraints (1).

Define $f_n(g_i^k)$, f_n, a" and a' as follows:

$f_n(g_i^k)$ = time to first transition into node n for k^{th} route for request i

f_n = Minimum $f_n(g_i^k)$

= earliest possible first transition time into node n

a" = vector of times to earliest possible for transition into each node; f_n is the nth component of this vector

a' = a - a"

= adjusted $|N|$ dimensional vector of available times at nodes during the planning horizon

Replace a in (1) by a'.

MTR-2.

This is a heuristic approach based on looking at nodes with high "node use" times and deleting, where possible, alternate routes that contribute to high node use.

Additional Notations:

$t_n(M_i)$ = total node utilization for node n derived from matrix M_i

N_t = {n: arranged in decreasing order of $t_n(M_i)$}

$N_t^{[m]}$ = m^{th} node in N_t

$j(G_i, N_t^{[m]})$ = number of routes in G_i that is selected

Step 1:

Set i = 1, m = 1, $\bar{n} = N_t^{[m]}$

Step 2:

- If $j(G_i, \overline{n}) > 0$ and $|G_i| > 1$ then set $G_i = G_i - g_i^q$ and revise M_i, $t_n(M_i)$, N_t, and $\overline{n}$.

  where,

  $g_i^q = \{g_i^k$: Maximum (node utilization for node $\overline{n}$},
- If $|G_i| = 1$ for all i, then go to step 4.

Step 3:

- Set $i = i + 1$.
- If $i > |I|$ then set $i = 1$ and $m = m + 1$.
- If $m > |N|$ then set $i = 1$.
- Go to step 2.

Step 4:

Set $\overline{g_i} = G_i$ for each i.

MTR-3.

Approaches MTR-1 and MTR-2 do not consider the effect of scheduling on choice of routes for each transport request.

The mathematical model in MTR-1 is expanded to include constraints pertaining to scheduling. These constraints are based on a model based on discretizing time. [Sherali et al, 1990].

Additional Notations:

m_i^k = number of transitions in route k for request i

p_{ikl} = node use time for transition l of alternative route k for request i; this is derived from M_i

x_{iklt} = 1, if l^{th} node of alternative route k for request i is used during time interval t
= 0, otherwise

Minimize: $\sum_{i=1}^{R}\sum_{k=1}^{m_i^k} P_i L_i^k r_i^k$

Such that: $\sum_{i=1}^{R} M_i r_i \le a$ (3)

$\sum_{i=1}^{R} r_i^k = 1$, for each i (4)

$\sum_{t=1}^{T} x_{iklt} = P_{ikl} \times r_i^k$, for l = 1, 2, ..., m_i^k each $g_i^k \in G_i$ and i =1, 2, ..., R (5)

$\sum_{\alpha=1}^{l=1}\sum_{\beta=1}^{T} x_{ik\alpha\beta} + \sum_{\beta=t+P_{ikl}}^{T} x_{ilk\beta} \le (r_i^k - x_{iklt})\sum_{\alpha=1}^{l} P_{ik\alpha}$

for l =1, 2, ..., m_i^k, each $g_i^k \in G_i$, i =1, 2, ... R; and t = 1, 2, ..., T (6)

The typical problem size will be about 3000 binary variables and about 3000 constraints.

Number of binary variables = $\sum_{i=1}^{R}|G_i| + T\sum_{i=1}^{R}|G_i| m_i^k$

Number of constraints = $|N| + R + \sum_{i=1}^{R}|G_i| m_i^k + T\sum_{i=1}^{R}|G_i| m_i^k$

For R = 10, average $|G_i|$ =3, average m_i^k = 10, T = 10, and $|N|$ = 15:

Number of binary variables	=	10 x 3 + 10 x 10 x 3 x 10
	=	3030
Number of constraints	=	15 + 10 + 10 x 3 x 10 + 10 x 10 x 3 x 10
	=	3325

MATERIAL TRANSPORT SCHEDULING PROBLEM

Problem Statement:

Given a set of transport requests and having selected routes to service the requests, to find the for each:

(i) a sequence of processing to requests at each node (synchronous transport segments), and

(ii) to form "trips" for asynchronous transport segments that have direct address transport units.

Approaches for solving (i) will have prefix MTS1.

Approaches for solving (ii) will have prefix MTS2.

Approaches:

MTS1-1.

The mathematical formulation in MTR-3 is applicable as is to provide answers to both the routing and scheduling problem; the resulting problem size if very large even for moderate values of number of requests, average number of alternate routes per request, average number of nodes per request, time window size (discretized) and total number of nodes (see MTR-3 above).

Also, the MTR-3 formulation could be modified to assume a routing decision; however, the problem size will still be one-third, if on an average there are alternate routes per request.

MTS1-2.

A heuristic approach is proposed based on the premise that delays at nodes are the primary cause for tardiness of transport requests.

Additional Notations:

$d_i^{\overline{n}}$ = latest time at which request must leave node $\overline{n}$, based on times for nodes and arcs but assuming no wait times

$I_{\overline{n}}$ = set of indices of requests that have node $\overline{n}$ in their selected route

I_S = set of indices of requests that have been scheduled

$S_{\overline{n}}$ = selected sequence of requests in $I_{\overline{n}}$ for action at node $\overline{n}$

$t_i^{\overline{n}}$ = earliest time at which request i will reach node $\overline{n}$, based on times for nodes and arcs but assuming no wait times

t_i^r = time at which request i must be released at its initial node

Step 1: Initialize

Set $m = 1$, $\overline{n} = N_t^{[m]}$, and $I_S = 0$.

Step 2: Sequence bottleneck nodes

- Set $I_{\overline{n}} = I_{\overline{n}} - I_S$.

- For requests $I_{\overline{n}}$, solve a one-machine sequencing problem with objective of minimizing absolute lateness based on $t_i^{\overline{n}}$, $d_i^{\overline{n}}$, and node use times for node $\overline{n}$ from matrix M_i to get $S_{\overline{n}}$. The heuristic algorithm proposed by Holsenback and Russell can be adapted for this problem. [Holsenback and Russell, 1992]

Step 3: Compute release times

- Use $S_{\overline{n}}$ to compute t_i^r via backward scheduling, assuming no wait times.
- Set $I_s = I_s + I_{\overline{n}}$

Step 4: Next bottleneck node or stop

- Set $M = m + 1$ and $\overline{n} = N_t^{[m]}$.
- If $I_s = I$ then stop ; all requests have been scheduled.
- Otherwise go to Step 2.

MTS2-1.

For segments with direct address transport units a relevant problem is to form trips for units in an attempt to service the transport requests. A mathematical formulation is presented for the "trip formation" problem for transport sub-systems $G_A{}^a$.

Additional Notations:

I_i^k = set of indices of requests that are incompatible with using vehicle k for request i on the same trip

t_i = time at which request i will be ready for transport action

$t_{ki\alpha}$ = time at which vehicle k will visit initial node for request i on trip α

$v_{ki\alpha}$ = 1, if vehicle k is assigned to request i on trip α
= 0, otherwise

Minimize: $\Sigma_k \Sigma_i \Sigma_\alpha d_{ki\alpha} v_{ki\alpha}$

Such that: $\Sigma_k \Sigma_\alpha v_{ki\alpha} = 1$ for each i (7)

$t_{ki\alpha} + M(1 - v_{ki\alpha}) \geq t_i$ for each i, k, α (8)

$v_{ki\alpha} + \Sigma_{j \subset I_i^k} v_{ki\alpha} \leq 1$ for each i, k, α (9)

where M is a large positive number

CONCLUDING REMARKS

In this paper, we have introduced a representation for the underlying material transport system in any automated manufacturing system such as an FMS. In the context of such a representation, the issue of scheduling material transport activities in a proactive manner is addressed. Two subproblems are stated in an attempt to develop schedule for material transport activities. Three approaches are proposed to address each of the subproblems.

ACKNOWLEDGEMENTS

The authors acknowledge the partial support of this work by a National Science Foundation grant.

REFERENCES

Bjorke, O., Towards a Manufacturing Systems Theory, *Robotics and Computer Integrated Manufacturing*, 4, 625-632, 1988.

Joshi, S.B., Wysk, R.A., Jones, A., A Scalable Architecture for CIM Shop Floor Control, *Proceedings of CIMCOM '90*, A. Jones (Ed.), National Institute for Standards and Technology, 21-33, 1990.

Kouvelis, P. and Lee, H.L., The Material Handling Design of Integrated Manufacturing Systems, *Annals of Operations Research*, 26, 379-396, 1990.

Matson, J.O. and White, J.A., Opertional Research and Material Handling, *European Journal of Operational Research*, 11, 309-318, 1982.

Merchant, M.E., The Precepts and Sciences of Manufacturing, *Robotics and Computer Integrated Manufacturing*, 4, 1-6, 1988.

Parunak, V.D.H., Characterizing the Manufacturing Scheduling Problem, *Journal of Manufacturing Systems*, 10, 241-259, 1991.

Sherali, H.D., Sarin, S.C., and Desai, R., Models and Algorithms for Job Selection, Routing, and Scheduling in a Flexible Manufacturing System, *Annals of Operations Research*, 26, 433-453, 1990.

Smith, J.S., Hoberecht, W.C., and Joshi, S.B., A Shop Floor Control Architecture for Computer Integrated Manufacturing, *IIE Transactions*, 28, 783-794, 1996.

Smith, J.M., Graves, R.J., and Kerbache, L., QNET: An Open Queueing Network Model for Material Handling System Analysis, *Material Flow*, 3, 225-242, 1986.

Solberg, J. J., Production Planning and Scheduling in CIM, *Proceedings of the 11th World Computer Congress : IFIP*, 919-925, 1989.

FRONTIER BASED SCHEDULING IN FMS

Gregory B. Jones[1], Manbir S. Sodhi[2] and Victor Fay Wolfe[3]
[1]Naval Undersea Warfare Center, Newport, Rhode Island.
[2]Industrial and Manufacturing Eng., University of Rhode Island.
[3]Computer Science and Statistics Dept., University of Rhode Island.

ABSTRACT: In this paper the problem of scheduling to meet two performance criteria is considered. The first criteria relates to timeliness while the second relates to precedence (or technological) constraints. An ideal schedule is one that yeilds the best measure of timeliness without violating the precedence constraints. However in some systems, such as flexible manufacturing systems and real time scheduling systems, it is sometimes possible that all precedence constraints need not be met. Given deadlines by which all tasks must be completed, it may be desirable to perform an alternate set of tasks, perform tasks out of order or even omit some tasks given these deadlines. In this paper the case where it is permissible to violate some precedence constraints for a commensurate improvement in meeting timing constraints is considered. Thus, the precedence constraints are no longer inviolate constraints, but can be represented by another measure of performance. The problem of preparing task schedules is formulated such that a trade-off between timing and precedence constraints is made when both constraints cannot be met and procedures are developed to obtain solutions for arbitrary deadlines.

INTRODUCTION

In practical scheduling applications there are often many performance measures that must be considered simultaneously. Conflict among these performance measures makes an optimal solution hard to find. This paper considers one such scenario of bi-criteria scheduling. Of the two performance measures to be considered here, one relates to timeliness while the other relates to precedence constraints. Timeliness considerations can include due date criteria, tardiness measures etc. Precedence considerations arise from technological considerations requiring the completion of specific tasks before inception of others. In scheduling problems it is usually assumed that all precedence constraints must be met. Within the set of schedules that meet all precedence constraints, the goal is to find a schedule that yeilds the best measure of completion time. Here we consider the case where it is permissible to violate some precedence constraints for a commensurate improvement in meeting deadlines. Thus, the precedence constraints are no longer inviolate constraints, but may be accounted for as another measure of performance. There are several real-world systems where such compromizes are permissible. One example is the real time control of distributed systems, represented in Figure 1. Each sub-system provides some input to a central decision making system, which determines the action to be taken based on the data provided by each of the sub-systems. This central system gathers the input according to some pre-programmed order. The order in which the sub-system data is collected is important for the interpretation of the data. Asynchronous data requests are also considered possible by the central processor. If a sub-system cannot respond to the central control system's demand for a status reading, the central system is forced to make a decision without the sub-system data. This results in a penalty for not being able to maintain the precedence constraints scheduled into the system.

Another example, detailed later in this paper, is from flexible manufacturing. A salient feature of flexible manufacturing systems is that each machine is capable of a variety of tasks, and control is distributed so that many of the task processing decisions are determined at the local (machine) level without central intervention. In such as scenario, it is forseeable that the task of maintaining the schedule is delegated to the central control system, while the role of local processors is expanded to the determination of processing rates and tasks so as to meet deadlines issued by the central controller. Thus, in response to a deadline issued by the central scheduling system, the local processing system may respond by merging two operations (such as a single pass of metal removal with a greater depth of cut instead of two separate, lighter passes) at a higher cost to the system (increased tool and machine wear).

In the next section the scheduling model is discussed in detail, followed by the mathematical representation and the solution procedure for generating points on the performance frontier. An illustration of this is detailed next. A more general formulation of the case where individual tasks have deadlines is then constructed, followed by conclusions and future research directions.

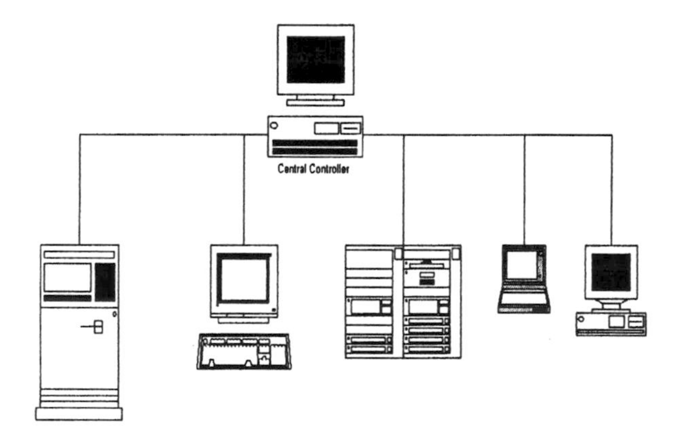

Figure 1: Distributed Real Time Control Architecture

SCHEDULING MODEL

The single machine flow shop problem is closely related to the problem discussed here, and a number of algorithms useful for this problem are described in French [4]. For example, the EDD (earliest deadline due) algorithm minimizes maximum lateness, and is a generally accepted method of creating an optimal schedule with respect to timing constraints [7]. EDD, however useful for meeting deadlines, does not consider precedence constraints. Lawler's algorithm [6], by contrast minimizes maximum lateness (or tardiness) while preserving precedence constraints. The notion of a frontier in relation to efficient schedules for the twin measures of average flow time and maximum tardiness ($Tmax$) is introduced in [4]. However, the discussion is limited to timing constraints only and does not consider precedence constraints. Gangadharan and Rajendran [5] use simulated annealing, to minimize makespan and total flow time in a flowshop. They consider the case with more than one machine, and show improved computational times over more conventional techniques. However, they too do not consider precedence constraints. Similarly, Chen and Bulfin [1] discuss the bi-criteria problem, and suggest a branch and bound solution which attempts to find a non-dominated schedule for makespan and total flow time minimization, without considering the re-ordering of precedence constraints. Chang and Hsu [2] consider scheduling tasks on both a uniprocessor and multiprocessor system. They propose using an algorithm based on the A* algorithm to minimize the total completion time such that all precedence constraints are met. However, they do not consider schedules where precedence constraints are violated. Other references to precedence constrained or deadline bound scheduling can be found in [9] and [3]. However, the issues of precedence constraints and deadlines are not jointly considered in either.

When two performance measures are simultaneously considered, as discussed earlier, it is possible to construct an x-y scatter plot of all possible sequences, where each (x, y) point represents a schedule with a value of x for one measure and y for the other. Figure 2 is one such plot.

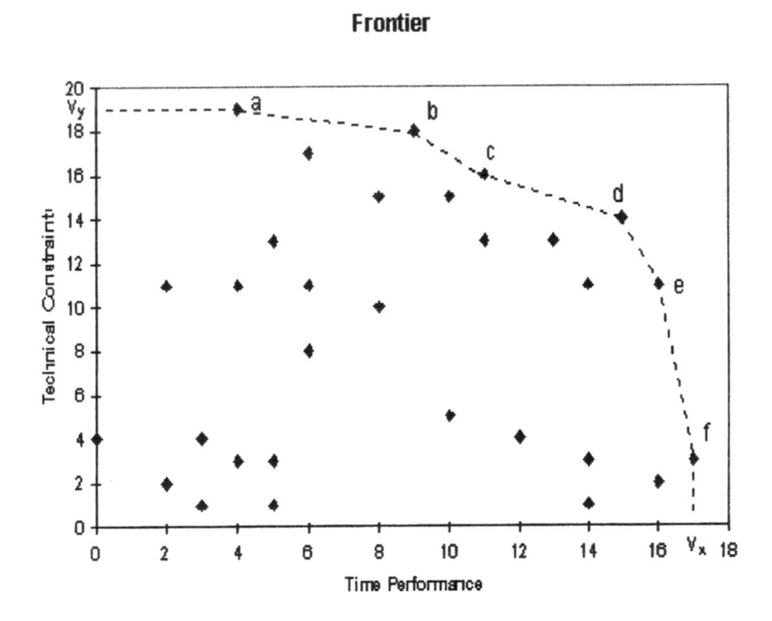

Figure 2: Scatter Plot of Schedule Objectives

In this plot, the x and y axis are shown such that greater performance in time or constraints is to the right (for time) and up (for constraints), which would be the negative of a regular performance measure. The line defined by points [a, b, c, d, e, f] is the frontier. A schedule is on the frontier if there does not exist a schedule that is at least as good in both measures and better in one. This also defines the set of optimal schedules. Any schedule, which does not lie on the frontier, can be no better in both measures, and therefore cannot be an optimal schedule. If the frontier can be determined, then the best schedule can easily be chosen from the set of optimal schedules using one of several objectives. This is especially useful when the central controller is given deadlines by an external agency. In this case, any idle time available prior to the delivery of the deadline can be used to compute the frontier. This information is then stored. When the deadline is issued, the optimal schedule (i.e. one that meets the deadline with the least penalty) is easily identifiable from the set of points defining the frontier. This sequence is then executed adn the timing constraints are satisfied with the least possible penalty.

In the next section we focus upon the problem of generating a frontier for a common deadline for all tasks. This is then illustrated with a small example from flexible manufacturing. We conclude with a formulation for an extension of this problem where tasks are given individual deadlines.

Model 1: Makespan Reduction

Consider a partially ordered sequence of tasks, $1, 2, \ldots, N$. For each task i , there is a (possibly empty) set of successor tasks $\{S_i\}$ such that task i precedes task j for all $j \in \{S_i\}$. Furthermore, for each successor task there is a penalty $p_{ij} \geq 0$ if task i is scheduled after task j. Each task has a known processing time, $t_i > 0$. Furthermore, for some tasks, a time advantage $t_{ij} \geq 0$ is deducted from the execution time for task i if a successor j of this task is scheduled before it. We assume that $\sum_{j \in \{S_i\}} t_{ij} \leq t_i$. The maximum time required to complete all tasks is $\sum_{i=1}^{i=N} t_i$, and the minimum processing time required is

$$\sum_{i=1}^{i=N} \left(t_i - \sum_{j \in \{S_i\}} t_{ij} \right).$$

Also, the maximum penalty possible is

$$\sum_{i=1}^{i=N} \sum_{j \in \{S_i\}} p_{ij},$$

and the minimum penalty is zero.

The problem can then be stated as follows: given a common deadline T by which all tasks must be completed, it is necessary to determine the set of precedences that will have to be violated to meet this timing constraint. This can be formulated as

$$\min \sum_{i=1}^{i=N} \sum_{j \in \{S_i\}} I_{ij} P_{ij} \tag{2.1}$$

subject to:

$$\sum_{i=1}^{i=N} \left(t_i - \sum_{j \in \{S_i\}} (1 - I_{ij}) t_{ij} \right) \leq T \tag{2.2}$$

$$I_{ij} \in \{0, 1\} \tag{2.3}$$

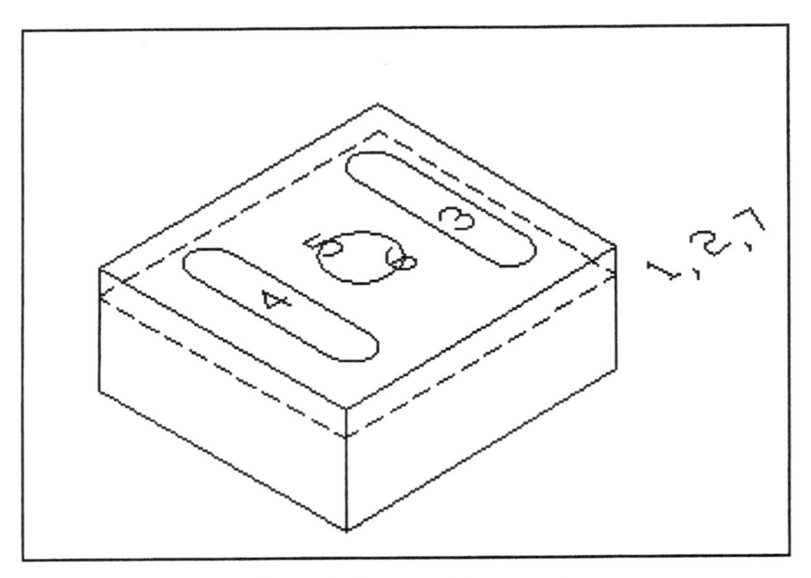

Figure 3. Part used for example

The binary variable I_{ij}, if positive, indicates task i is executed before task j.

This problem is a knapsack problem, and a variety of solution methodologies can be found in the literature [8]. However, for generating the frontier, we need to solve this problem for all knapsacks of size (i.e. T) 0 to $\sum_{i=1}^{i=N} t_{ij}$. A particular approach, using dynamic programming, for solving the knapsack problem with the largest possible knapsack will result in the generation of the entire frontier. Define $P(t)$ as the minimum penalty that results from a deadline of $\sum_{i=1}^{i=N} t_i - t$, and t_{ij} is the benefit earned by violating the precedence ij. Clearly, $P(0) = 0$. Then,

$$P(t) = \min_j \min_i \{p_{ij} + P(t - t_{ij})\}.$$

Thus, the process of computing $P(\sum_{i=1}^{i=N} t_i - t)$ can be used to efficiently generate all points on the frontier.

EXAMPLE

A simple example is used to illustrate the technique developed. Consider a sequence of machining operations on a part. This part, illustrated in Figure 3, consists of a metal blank which is machined in several steps. The required operations are partially ordered, that is, certain operations should preferably occur before others. The remaining operations can be performed in any order. Figure 4 shows the relationship of operations to be performed on the part. Table 1 lists the required tasks along with the time required for each task. The precedences, which can be derived from Figure 4, are listed in Table 2 along with the penalties that are applied for each violation.

TABLE 1. Data for Test Problem

Task	Description	Time
1	Rough Mill Face	2
2	Finish Mill Face	4
3	Mill Pocket (1)	2
4	Mill Pocket (2)	3
5	Bore Hole	2
6	Finish Hole	2
7	Polish and Deburr Face	3

In accordance with our model, if a task is scheduled to execute after all successor tasks, then that task has zero processing time. This is a logical assumption, since it is often pointless to perform an operation that is required to precede another operation. This also enables the tradeoff between meeting deadlines and meeting precedence constraints. By scheduling a displaced task after all its successor tasks, the task is given an execution time of zero, resulting in a reduction in the makespan. While at first the violation of precedence constraints may appear unacceptable, this is often the method by which deadlines are met in real life. As can be seen in Table 2, each precedence constraint is associated with a penalty. The penalty is a measure of the importance of the constraint. As an example illustrating this, consider operations 1 and 2 on this part. The first operation is a roughing operation, while the second is a finishing operation. Naturally the roughing operation should be performed before the finishing operation. If, however, due to time constraints, the makespan for completing this part must be reduced, operation 1 can be omitted and a single operation combining operations 1 and 2 (within technological limits) is performed. While this single operation results in a reduction in time, the total cost to the system for this operation may be higher, either due to increased tool wear or wear on the machine, than operations 1 and 2 separately.

TABLE 2. Violation Penalties for Tasks

Task	2	3	4	7	7	6	7	5	7
Predecessor	1	2	2	3	4	5	6	1	2
Violation Penalty	2	5	5	12	12	3	10	3	10

Using the method detailed earlier, the optimal set of solutions which define the frontier are obtained. For this example, the frontier is shown in Figure 5. A solution which ensures that all

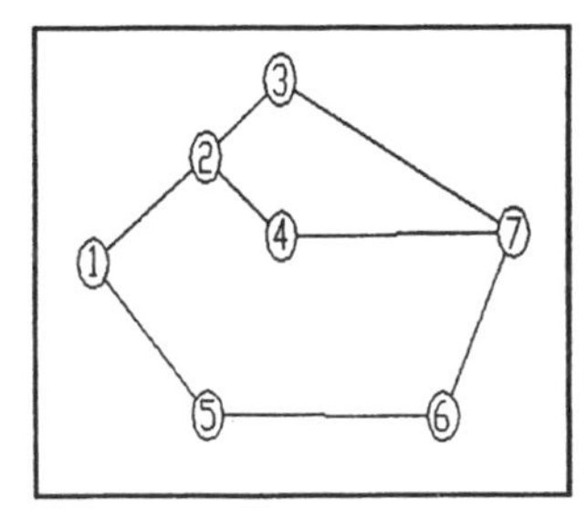

Figure 3: Operation Precedences for Test Part

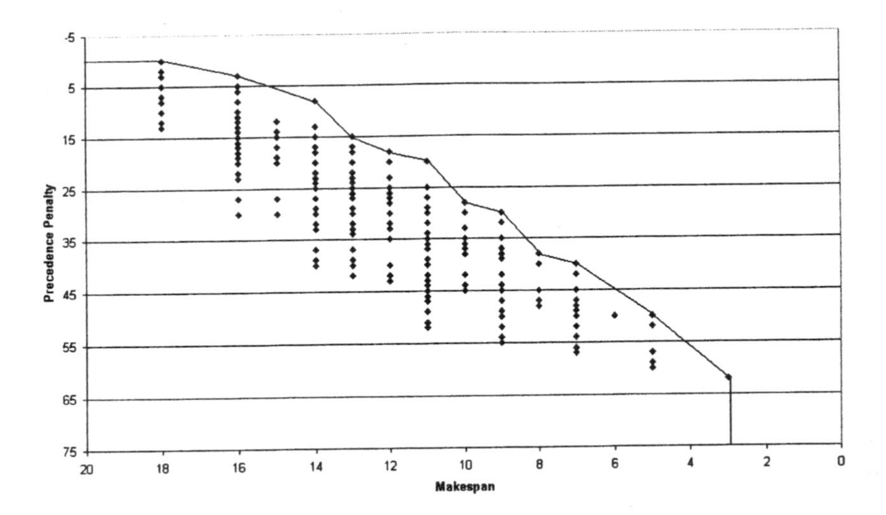

Figure 5: Performance Frontier for Test Problem

precedence constraints are met, for example: (1, 2, 3, 4, 5, 6, 7), has a makespan of 18 and no penalty. This point is easily computable using Lawlers method. If all precedence constraints are violated, for example: (7, 6, 5, 4, 3, 2, 1), requires a makespan of only 3; however the penalty incurred is 62. These two points represent the extremes, other points on the frontier express a trade-off such that makespan is between 3 and 18 with penalties ranging between 62 and 0. For each point on the frontier there are a number of schedules that produce that point (for this example, many frontier points have over 20 associated schedules).

TABLE 3. Frontier Points For Sample Problem

Task Sequence	Makespan	Penalty
(1 2 3 4 5 6 7)	18	0
(1 2 3 4 6 5 7)	16	3
(2 3 4 6 5 1 7)	14	8
(1 2 3 6 5 7 4)	13	15
(2 3 4 7 6 5 1)	12	18
(2 3 6 5 1 7 4)	11	20
(3 4 6 5 7 2 1)	10	28
(2 3 7 4 6 5 1)	9	30
(3 4 7 2 6 5 1)	8	38
(3 6 5 7 4 2 1)	7	40
(3 7 4 2 6 5 1)	5	50
(7 3 4 2 6 5 1)	3	62

MODEL 2: INCORPORATION OF TASK DEADLINES

The inclusion of task deadlines complicates the problem substantially. Extending the notation used earlier, let d_i be the deadline of task i. Let $P(i)$ be the set of predecessors of task i. Because of partial ordering, the deadline of task i is also carried over to each task $j \in P(i)$. Thus, if d_j is an externally specified deadline for task j, the actual deadline for each task $j \in P(i)$ is $d_j = min\{d_j, d_i - t_i\}$. Now, let us define the set $E(i)$ as the set of tasks that have deadlines earlier than task i. Given a schedule completion deadline T, and the individual deadlines $d)i$, the problem now is the determination of a sequence, and consequently the precedences that will have to be violated, to meet these times. This problem can be formulated as:

$$\min \sum_{i=1}^{i=N} I_{ij} P_{ij} \tag{4.1}$$

subject to:

$$(1 - L_i)t_i + \sum_{j \in E(i)} I_{ji} \left(t_i - \sum_{k \in E(j)} (1 - I_{kj})\, t_{kj} \right) - \sum_{j \in P(i)} (1 - I_{ji}) t_{ji} \leq d_i \ \forall i \tag{4.3}$$

$$\sum_{i=1}^{i=N} \left(t_i - \sum_{j \in \{S_i\}} (1 - I_{ij}) t_{ij} \right) \leq T \tag{4.4}$$

$$I_{ij} \in \{0, 1\} \tag{4.5}$$

In this formulation, $L_i > 0$ indicates task i is late, i.e. is completed after its deadline and I_{ij} indicates task i is completed before task j. The first set of constraints, 4.3, accounts for the total time scheduled before each deadline, while 4.4 accounts for the total time scheduled before the schedule deadline. This model is a non-linear integer programming formulation, and suitable heuristics are being investigated for efficient generation of the frontier for this problem.

CONCLUSION

In this paper models for scheduling with precedence constraints have been identified, where the precedence constraints can be violated. Examples of such systems can be found in flexible manufacturing systems and in real time control systems. A procedure for generating schedules that can be used to meet given deadlines by compromising precedence constraints has been discussed. A model where individual tasks have deadlines has also been formulated. Future research is focused upon the solution of this formulation, as well as several additional cases of this class of scheduling problems as well as their application to real life problems.

REFERENCES

[1] Chen, C, Bulfin, R., (1994), Scheduling a Single machine to Minimize Two Criteria: Maximum Tardiness and Number of Tardy Jobs, IEE Transactions, v 26, n5, pp 76-84.

[2] Chang J. and Hsu, C., (1995), Task Scheduling with Precedence Constraints to Minimize the Total Completion Time, International Journal of Systems Science, v 26, n11, pp.2203-2217.

[3] Scheduling Dependent Real-Time Activities, Unpublished Ph. D Dissertation, *Carnegie Mellon University.*

[4] French, S.,(1982),Sequencing and Scheduling, An Introduction to the Mathematics of the Job-Shop, *John Wiley and Sons,* New York.

[5] Gangadharan, R and Rajendran, C., (1992), A Simulated Annealing Heuristic for Scheduling in a Flowshop with Bicriteria, *Computers and Industrial Engineering,* v27, n1, pp. 473-476

[6] Lawler, E.L., (1973), Optimal Sequencing of a Single Machine Subject to Precedence Constraints, *Management Science,* v. 19, pp. 544- 546.

[7] Locke, D., (1986), Best Effort Decision Making for Real-time Scheduling, *Unpublished Ph.D Thesis - CW-CS-86-134,* Carnegie Mellon University.

[8] Martello, S. and Toth, P., (1990), Knapsack problems : Algorithms and Computer Implementations, *J. Wiley & Sons, New York.*

[9] Tei-Wei K. and Mok, A., (1992), Application Semantics and Concurrency Control of Real-Time Data-Intensive Applications, *Real-Time Systems Symposium,* San Fransisco.

ERROR BUDGET CONTROL IN COORDINATE MEASURING SYSTEMS

M.M. Sfantsikopoulos [1], S.C.Diplaris[1], V.S. Issopoulos[1] and D.A. Mills[2]
[1] National Tech. Univ. of Athens ; [2] JADE Networks Ltd.

ABSTRACT : Precision measurements that are based on the measurement of point coordinates, strongly depend on the positioning and orientation of the measured dimensions inside the workspace of the measurement system. The statistical volumetric error for atypical coordinate measurement machine can be readily calculated; this not the case when two or more independent measuring systems are integrated to make such measurements.. The analysis presented in this paper allows for a suitable dimension or systems' components positioning and orientation to be determined. The pursued accuracy level may thus be achieved by trying alternative measurement layouts. Results produced by means of the developed algorithm are also given for a system that measures the dimensions of a machined workpiece held on the machine tool worktable. The measuring system does not use the cutting machine to make any measurement.

INTRODUCTION

A mechanical part is manufactured on the basis of toleranced and untoleranced dimensions. The latter generally have allowances which are permitted in the relevant standards (e.g. ISO 2768) for dimensions without tolerance indication. Because of the relatively large values of these allowances, the measurement of untoleranced dimensions does not present any particular difficulty. The measurement errors however of the toleranced dimensions, which represent the critical dimensions for the functionality of a part, should hold a special relationship with the relevant tolerances, otherwise the measurements will be ineffective. The zone width of these tolerances is very narrow and usually corresponds to the ISO IT5-IT8 tolerance grades. Taking into account the measurement error (e) -dimensional tolerance (T) universally accepted constraint, [1-6],

$$e \leq k \cdot T, \qquad k \leq \frac{1}{5} \tag{1}$$

it is apparent that before preparing any measurement plan, it is necessary to verify the measurement effectiveness according to this relationship, for all the toleranced dimensions that have to be measured.

In coordinate measuring systems, dimensions are indirectly obtained usually either through the coordinates of their ends (if they are taken as distances between two points), or as distances between two parallel planes (which are determined via the coordinates of a cloud of points). The accuracy of the measured dimensions consequently depends on the measurement errors of the point coordinates and on the analytical relationships between coordinates and dimensions. The position and orientation of a measured dimension within the measurement system's workspace, determine the measurement error.

Coordinate measurement errors belong to two main categories :

a. Sensor (tactile or optical) errors and

b. Errors having to do with the establishment of the sensor position with reference to the absolute Cartesian coordinate system.

The measurement error for the typical Coordinate Measuring Machine (CMM) is quoted as 1D-accuracy (i.e. measurement along its x,y,z axes), as 2D-accuracy (i.e. measurement on a plane parallel to the (x,y), (y,z), (z,x) machine reference planes and as 3D-volumetric accuracy for a dimension randomly positioned inside its workspace. Given that this error decreases from the 3D- to 1D-accuracy, there is apparently a potential to improve the measurement accuracy of dimensions with very strict tolerances by positioning them for 1D- or 2D- measurement, provided of course, that this is possible. For large workpieces whose overall size exceeds the available CMM workspace, special transportable CMMs may be, on the other hand, employed (e.g. FARO, [2]). An additional measurement system may then be required in order to obtain the CMM's multiple positions with respect to the absolute coordinate system. This measurement approach is also applicable for the measurement of dimensions of workpieces that are not readily measured on a conventional CMM. An example is the independent measurement of the dimensions of a machined workpiece while this is still mounted on its worktable.

For such a case the relative positions between the workpiece under measurement, the sensor and the measurement system of the sensor's coordinates have to be adapted to the needs of the considered layout - machine tool configuration, sensor's geometry, measurement system for the sensor's coordinates, shape and size of the workpiece. Obviously there do not exist here any prespecified 1D- , 2D-, or 3D- accuracy data for the envisaged measurements, similar to those of a CMM. The only available data are the coordinate measurement accuracy of the sensor's system and the measurement accuracy of the sensor's coordinates. The resulting measurement error is therefore the calculated output of a complex function of numerous parameters. The estimation of this error is nevertheless a prerequisite for the control of the effectiveness of such a measurement. It leads to a sensor positioning and system layout that will provide for acceptable errors.

In this paper a methodology is presented for the error budget control of coordinate measuring system layouts as these described above. The maximum possible measurement error of a given dimension is calculated and through its comparison with the specified dimensional tolerance the effectiveness of the planned measurement is assessed. It becomes possible to determine the most suitable sensor positioning and the most effective measurement system layout for the particular application.

MEASUREMENT SYSTEM ANALYSIS

In principle two typical states may be considered for the measurement system which was described in the previous section :

i. Dimensions to be measured lie within the workspace of the measurement sensor and if it is not therefore required to move the sensor,

ii. dimensions to be measured require, because of their size, the measurement sensor to be moved from a first to subsequent positions. .

One position of the measurement sensor, Fig.1

For this case a dimension D_{12}, represented by the distance (A_1A_2) between the points A_1,A_2 on the workpiece surface, is derived as ,

$$D_{12} = \sqrt{\left({}^m x_2 - {}^m x_1\right)^2 + \left({}^m y_2 - {}^m y_1\right)^2 + \left({}^m z_2 - {}^m z_1\right)^2} \tag{2}$$

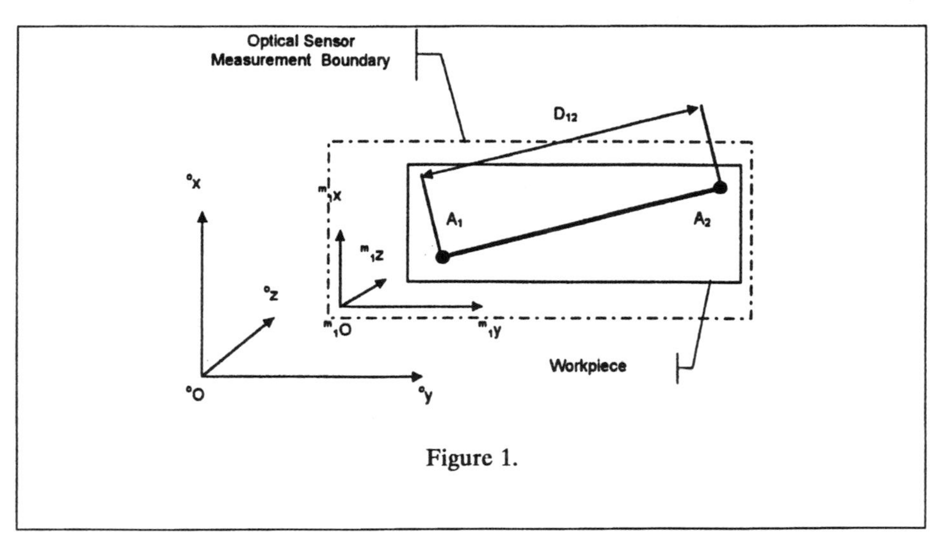

Figure 1.

where $({}^m x_1, {}^m y_1, {}^m z_1), ({}^m x_2, {}^m y_2, {}^m z_2)$ are the measured coordinates of the points A_1,A_2 with reference to the Cartesian coordinate system ${}^m O {}^m x {}^m y {}^m z$ of the sensor. Because of the equation (2), the anticipated measurement error e_{12} of the dimension D_{12} will be equal to,

$$\begin{aligned} e_{12} &= d\left(D_{12}\right) \\ &= \frac{1}{D_{12}} \cdot \left[\left({}^m x_2 - {}^m x_1\right) \cdot \left(d^m x_2 - d^m x_1\right) + \left({}^m y_2 - {}^m y_1\right) \cdot \left(d^m x_2 - d^m x_1\right) + \right. \\ &\quad \left. \left({}^m z_2 - {}^m z_1\right) \cdot \left(d^m z_2 - d^m z_1\right)\right] \end{aligned} \tag{3}$$

with a maximum attainable value,

$$e_{12,\max}=\left|d(D_{12})\right|$$
$$=\frac{1}{D_{12}}\cdot\left[\left|{}^{m}x_2-{}^{m}x_1\right|\cdot\left(\left|d^{m}x_{2,\max}\right|+\left|d^{m}x_{1,\max}\right|\right)+\left|{}^{m}y_2-{}^{m}y_1\right|\cdot\left(\left|d^{m}y_{2,\max}\right|+\left|d^{m}y_{1,\max}\right|\right)+\right.$$
$$\left.+\left|{}^{m}z_2-{}^{m}z_1\right|\cdot\left(\left|d^{m}z_{2,\max}\right|+\left|d^{m}z_{1,\max}\right|\right)\right] \quad (4)$$

In the relationships (3) and (4), $d^{m}x_j, d^{m}x_j, d^{m}z_j, j=1,2$ are the measurement errors of the coordinates x_j, y_j, z_j, $\pm d^{m}x_{j,\max}, \pm d^{m}y_{j,\max}, \pm d^{m}z_{j,\max}$ their respective maximum possible values and D_{12} is the nominal size of the dimension under measurement.

Two positions of the measurement sensor Fig.2

The sensor bound Cartesian coordinate system has to be referred to the absolute Cartesian coordinate system °O°x°y°z, which is determined by the Measurement System of the

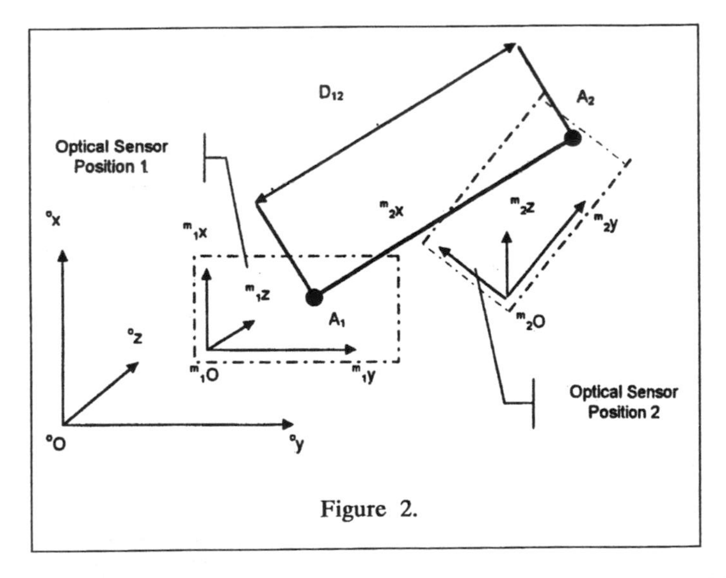

Figure 2.

Sensor's Coordinates for two successive measurements 1,2 of the coordinates of the points A_1 and A_2. If ${}^{m}_{j}\lambda_i, {}^{m}_{j}\mu_i, {}^{m}_{j}\nu_i$ $(i=1,2,3,\ j=1,2)$ are the direction cosines of the ${}^{m}_{j}x, {}^{m}_{j}y, {}^{m}_{j}z$, of the sensor bound Cartesian coordinate systems for the two measurement positions 1 and 2 and $h_j, k_j, l_j, (j=1,2)$ the coordinates of the origins ${}^{m}_{1}O, {}^{m}_{2}O$, all referenced to the absolute coordinate system °O°x°y°z, then the absolute coordinates of the points A_1, A_2 are obtained as

$$
\begin{aligned}
{}^{o}x_j &= {}^{m_j}\lambda_1 \cdot {}^{m_j}x_i + {}^{m_j}\mu_1 \cdot {}^{m_j}y_i + {}^{m_j}\nu_1 \cdot {}^{m_j}z_i + h_j \\
{}^{o}y_j &= {}^{m_j}\lambda_2 \cdot {}^{m_j}x_i + {}^{m_j}\mu_2 \cdot {}^{m_j}y_i + {}^{m_j}\nu_2 \cdot {}^{m_j}z_i + k_j \\
{}^{o}z_j &= {}^{m_j}\lambda_3 \cdot {}^{m_j}x_i + {}^{m_j}\mu_3 \cdot {}^{m_j}y_i + {}^{m_j}\nu_3 \cdot {}^{m_j}z_i + l_j
\end{aligned} \tag{5}
$$

In equations (5) the quantities ${}^{m}_{j}\lambda_i, {}^{m}_{j}\mu_i, {}^{m}_{j}\nu_i \;\; h_j, k_j, l_j, (j=1,2,\; i=1,2,3)$ have to be calculated on the basis of the measured data produced with the help of the sensor, the measurement system of the sensor's coordinates and three reference points on the sensor's fixture. Let B, C, D be these points with $\left({}^{m}x_B, {}^{m}y_B, {}^{m}z_B\right)\left({}^{m}x_C, {}^{m}y_C, {}^{m}z_C\right), \left({}^{m}x_D, {}^{m}y_D, {}^{m}z_D\right)$ coordinates related to the sensor bound Cartesian system. By applying equations (5) their absolute Cartesian coordinates will be,

$$
\begin{aligned}
{}^{o}x_{B_j} &= {}^{m_j}\lambda_1 \cdot {}^{m}x_B + {}^{m_j}\mu_1 \cdot {}^{m}y_B + {}^{m_j}\nu_1 \cdot {}^{m}z_B + h_j \\
{}^{o}y_{B_j} &= {}^{m_j}\lambda_2 \cdot {}^{m}x_B + {}^{m_j}\mu_2 \cdot {}^{m}y_B + {}^{m_j}\nu_2 \cdot {}^{m}z_B + k_j \\
{}^{o}z_{B_j} &= {}^{m_j}\lambda_3 \cdot {}^{m}x_B + {}^{m_j}\mu_3 \cdot {}^{m}y_B + {}^{m_j}\nu_3 \cdot {}^{m}z_B + l_j \\
{}^{o}x_{C_j} &= {}^{m_j}\lambda_1 \cdot {}^{m}x_C + {}^{m_j}\mu_1 \cdot {}^{m}y_C + {}^{m_j}\nu_1 \cdot {}^{m}z_C + h_j \\
{}^{o}y_{C_j} &= {}^{m_j}\lambda_2 \cdot {}^{m}x_C + {}^{m_j}\mu_2 \cdot {}^{m}y_C + {}^{m_j}\nu_2 \cdot {}^{m}z_C + k_j \\
{}^{o}z_{C_j} &= {}^{m_j}\lambda_3 \cdot {}^{m}x_C + {}^{m_j}\mu_3 \cdot {}^{m}y_C + {}^{m_j}\nu_3 \cdot {}^{m}z_C + l_j \\
{}^{o}x_{D_j} &= {}^{m_j}\lambda_1 \cdot {}^{m}x_D + {}^{m_j}\mu_1 \cdot {}^{m}y_D + {}^{m_j}\nu_1 \cdot {}^{m}z_D + h_j \\
{}^{o}y_{D_j} &= {}^{m_j}\lambda_2 \cdot {}^{m}x_D + {}^{m_j}\mu_2 \cdot {}^{m}y_D + {}^{m_j}\nu_2 \cdot {}^{m}z_D + k_j \\
{}^{o}z_{D_j} &= {}^{m_j}\lambda_3 \cdot {}^{m}x_D + {}^{m_j}\mu_3 \cdot {}^{m}y_D + {}^{m_j}\nu_3 \cdot {}^{m}z_D + l_j
\end{aligned} \tag{6}
$$

whereas it is known that,

$$
\sum_{i=1}^{3} {}^{m_j}\lambda_i^2 = \sum_{i=1}^{3} {}^{m_j}\mu_i^2 = \sum_{i=1}^{3} {}^{m_j}\nu_i^2 = 1 \tag{7}
$$

for $j=1,2$.

The Measurement System of the Sensor's Coordinates may now produce either Cartesian coordinates (${}^{o}x, {}^{o}y, {}^{o}z$) or polar coordinates ($\varrho$, ϑ, ϕ), Fig.3,

For the Cartesian coordinates, the relationships (6) and (7) constitute a system of 24 equations with 24 unknown parameters. Solution of this system determines the two positions and orientations of the sensor's Cartesian system regarding the absolute system ${}^{o}O{}^{o}x{}^{o}y{}^{o}z$. When polar coordinates are used, their Cartesian counterparts are calculated as,

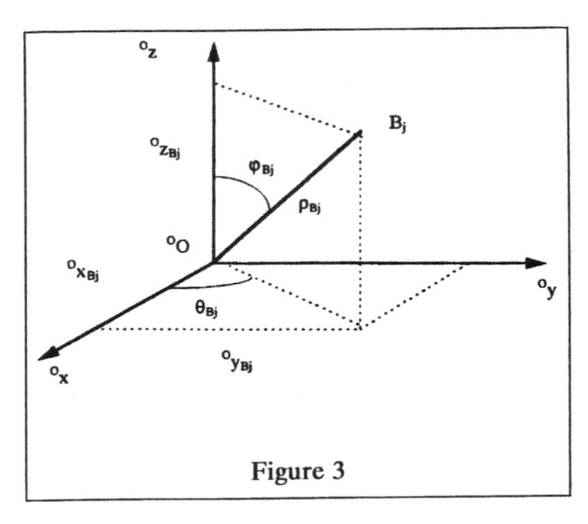

Figure 3

$$
\begin{aligned}
{}^{o}x_{B_j} &= -\rho_{B_j} \cdot \sin\varphi_{B_j} \cdot \cos\vartheta_{B_j} \\
{}^{o}y_{B_j} &= -\rho_{B_j} \cdot \sin\varphi_{B_j} \cdot \sin\vartheta_{B_j} \\
{}^{o}z_{B_j} &= -\rho_{B_j} \cdot \cos\varphi_{B_j} \\
{}^{o}x_{C_j} &= -\rho_{C_j} \cdot \sin\varphi_{C_j} \cdot \cos\vartheta_{C_j} \\
{}^{o}y_{C_j} &= -\rho_{C_j} \cdot \sin\varphi_{C_j} \cdot \sin\vartheta_{C_j} \\
{}^{o}z_{C_j} &= -\rho_{C_j} \cdot \cos\varphi_{C_j} \\
{}^{o}x_{D_j} &= -\rho_{D_j} \cdot \sin\varphi_{D_j} \cdot \cos\vartheta_{D_j} \\
{}^{o}y_{D_j} &= -\rho_{D_j} \cdot \sin\varphi_{D_j} \cdot \sin\vartheta_{D_j} \\
{}^{o}z_{Dj} &= -\rho_{Dj} \cdot \cos\varphi_{Dj}
\end{aligned}
\tag{8}
$$

and the system will then consist of 33 equations with 33 unknown parameters.

From equations (5),(6),(7),(8), the size of dimension D_{12} is derived,

$$
\begin{aligned}
D_{12} &= \sqrt{\left({}^{o}x_2 - {}^{o}x_1\right)^2 + \left({}^{o}y_2 - {}^{o}y_1\right)^2 + \left({}^{o}z_2 - {}^{o}z_1\right)^2} = \\
& f\left({}^{m_j}x_i, {}^{m_j}y_i, {}^{m_j}z_i, {}^{m}x_B, {}^{m}y_B, {}^{m}z_B, {}^{m}x_C, {}^{m}y_C, {}^{m}z_C, {}^{m}x_D, {}^{m}y_D, {}^{m}z_D, \right. \\
& \left. {}^{o}x_{B_j}, {}^{o}y_{B_j}, {}^{o}z_{B_j}, {}^{o}x_{C_j}, {}^{o}y_{C_j}, {}^{o}z_{C_j}, {}^{o}x_{D_j}, {}^{o}y_{D_j}, {}^{o}z_{D_j} \right) = \\
& f\left({}^{m_j}x_i, {}^{m_j}y_i, {}^{m_j}z_i, {}^{m}x_B, {}^{m}y_B, {}^{m}z_B, {}^{m}x_C, {}^{m}y_C, {}^{m}z_C, {}^{m}x_D, {}^{m}y_D, {}^{m}z_D, \right. \\
& \left. {}^{o}\rho_{B_j}, {}^{o}\vartheta_{B_j}, {}^{o}\varphi_{B_j}, {}^{o}\rho_{C_j}, {}^{o}\vartheta_{C_j}, {}^{o}\varphi_{C_j}, {}^{o}\rho_{D_j}, {}^{o}\vartheta_{D_j}, {}^{o}\varphi_{D_j} \right)
\end{aligned}
\tag{9}
$$

which corresponds to an anticipated measurement error,

$$
\begin{aligned}
e_{12} = d(D_{12}) &= \frac{\partial f}{\partial^{m_1} x_1} \cdot d^{m_1} x_1 + \frac{\partial f}{\partial^{m_1} y_1} \cdot d^{m_1} y_1 + \ldots + \frac{\partial f}{\partial^{o} z_{D_2}} \cdot d^{o} z_{D_2} = \\
&= \frac{\partial f}{\partial^{m_1} x_1} \cdot d^{m_1} x_1 + \frac{\partial f}{\partial^{m_1} y_1} \cdot d^{m_1} y_1 + \ldots + \frac{\partial f}{\partial^{o} \phi_{D_2}} \cdot d^{o} \phi_{D_2}
\end{aligned}
\tag{10}
$$

with a maximum value

$$
\begin{aligned}
e_{12} = d(D_{12})_{\max} &= \left|\frac{\partial f}{\partial^{m_1} x_1}\right| \cdot \left|d^{m_1} x_1\right|_{\max} + \left|\frac{\partial f}{\partial^{m_1} y_1}\right| \cdot \left|d^{m_1} y_1\right|_{\max} + \ldots + \left|\frac{\partial f}{\partial^{o} z_{D_2}}\right| \cdot \left|d^{o} z_{D_2}\right|_{\max} = \\
= \qquad &= \left|\frac{\partial f}{\partial^{m_1} x_1}\right| \cdot \left|d^{m_1} x_1\right|_{\max} + \left|\frac{\partial f}{\partial^{m_1} y_1}\right| \cdot \left|d^{m_1} y_1\right|_{\max} + \ldots + \left|\frac{\partial f}{\partial^{o} \varphi_{D_2}}\right| \cdot \left|d^{o} \varphi_{D_2}\right|_{\max}
\end{aligned}
\tag{11}
$$

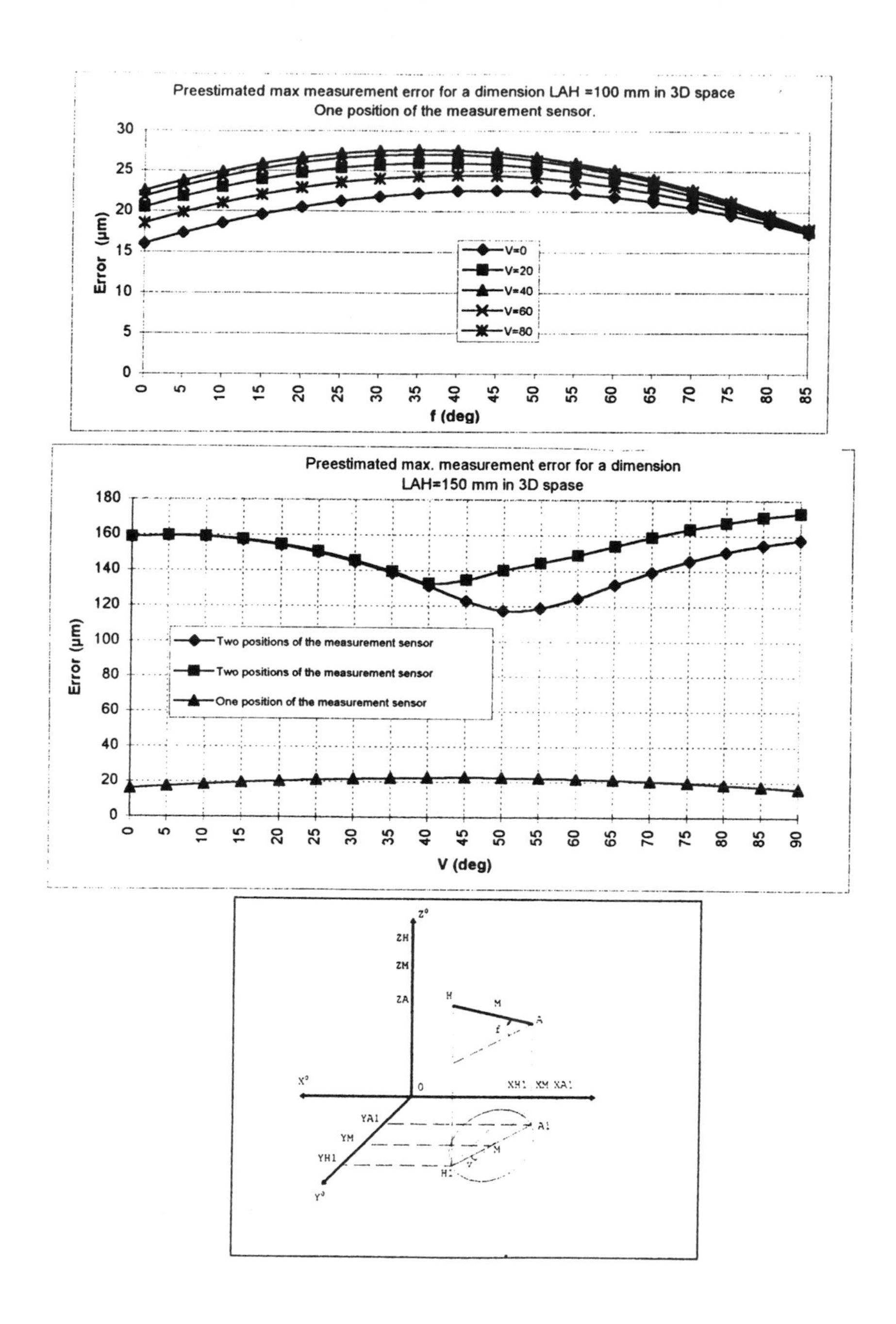

Fig. 4 Max. Possible measurement error for different cases of application (symbols refer to the bottom figure)

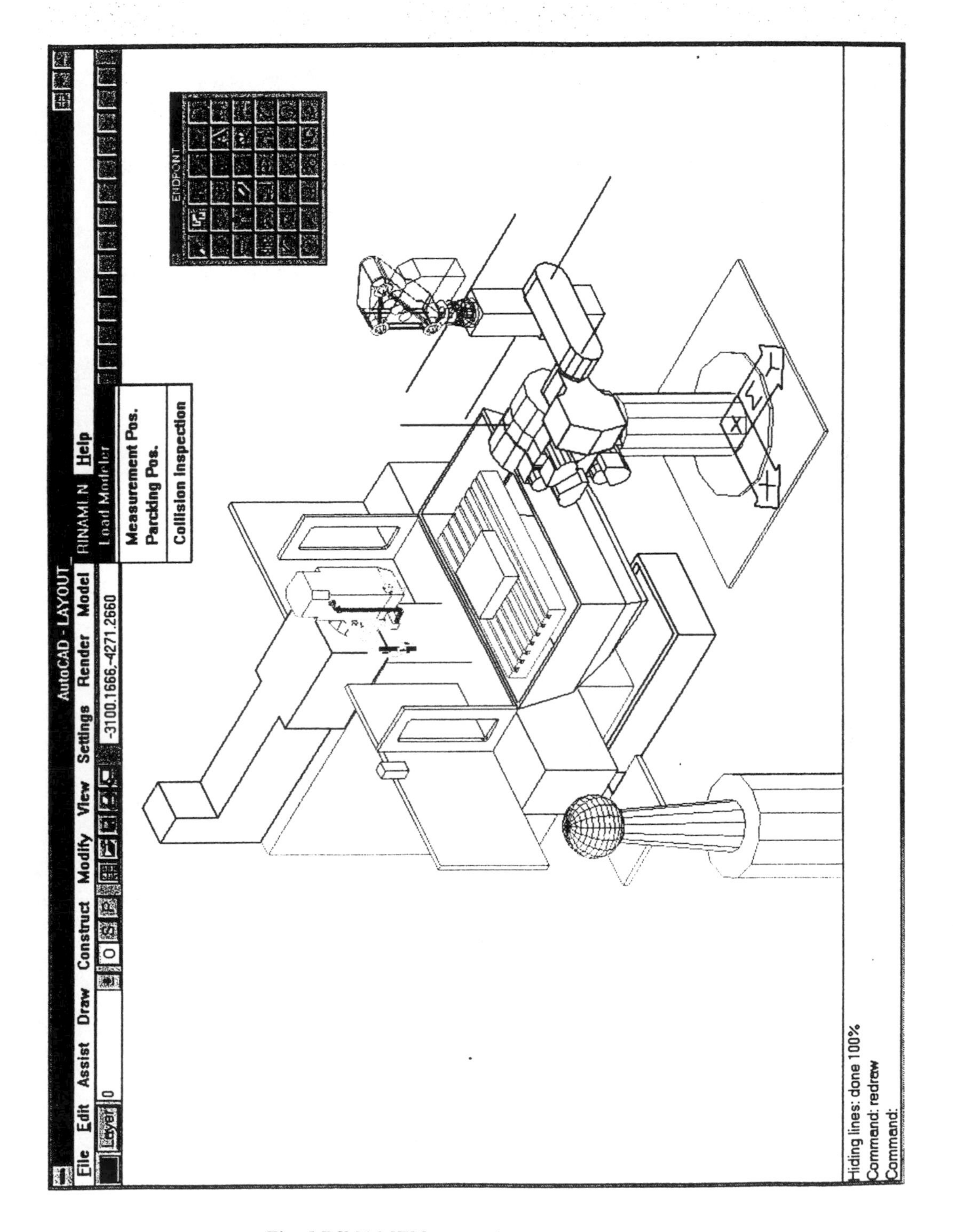

Fig. 5 RINAMEN system layout.

SYSTEM APPLICATION

On the basis of the proceeded analysis a computer code was developed, which for a given measurement layout and known measurement accuracies of the system components, calculates the maximum possible measurement error for every one dimension under consideration.

In Fig.4 results are shown from the application of the algorithm for the measurement system of Fig.5.

The diagrams demonstrate the contribution of the size of the measured dimension and, as well as, of its orientation inside the absolute Cartesian Coordinate System, to the maximum possible measurement error.

The measurement system of Fig.5 was developed within the BRITE - EURAM BE5593 Project (RINAMEN) for the non-contact measurement of dimensions of workpieces mounted on the machine-tool table independently from the machine tool own axes. It comprises an Optoelectronic Sensor ,[8],which produces the coordinates of the points of the workpiece surface with an accuracy of 10μm, an Absolute Distance Laser Interferometer,[7,8,9], which gives the polar coordinates of the sensor with an accuracy of 5μm / 5μrad and a Robot Manipulator for the sensor positioning. The position and orientation of the sensor's system of Cartesian coordinates are in fact determined through the polar coordinates of three mirror reflectors which are suitably attached on its robot fixture.

CONCLUSIONS

The measurement accuracy of a dimension through the measurement of the coordinates of its two ends strongly depends on its positioning and orientation inside the workspace of the measurement system. Whereas, for the typical CMM the statistical volumetric error may permit the positioning and orientation problem to be overcome, this is not the case when more than one measurement systems are cooperate for a particular application. As it was demonstrated in the paper for this type of measurement layout, in order to achieve a given accuracy level, suitable dimension and/or measurement system components positioning and orientation should be used. Taking into account the available accuracies, this task requires the solution of a complex system of equations with numerous parameters. The developed algorithm works towards this direction. It allows for alternative measurements to be directly and quickly assessed and the best suited one for the case to be chosen. The algorithm's contribution becomes, on the other hand, increasingly necessary as the number of the under measurement dimensions with narrow tolerances, and increased therefore accuracy requirements, increases.

REFERENCES

1. V.S.Korsakov, *Fundamental of Manufacturing Engineering,* Mir Publishers, Moscow, (1979).
2. M.M. Sfantsikopoulos and S.C. Diplaris, *Coordinate Tolerancing in design and manufacturing,* Int.J. Of Robotics and Computer-Integrated Manufacturing 8(4), 219-222, (1991).
3. M.M. Sfantsikopoulos, A *cost-tolerance analytical approach for design and manufacturing,* Int. J. Adv. Manufacturing Technology 5,126-134, (1990).
4. M.M. Sfantsikopoulos, *Compatibility of Tolerancing,* Int. J. Adv. Manufacturing Technology 8, 25-28, (1993).
5. M.M. Sfantsikopoulos, S.C. Diplaris and P.N. Papazoglou, An *accuracy analysis of the peg-and-hole assembly problem,* Int. J. Machine Tools Manufacturing, vol. *34,* No 5, 617-623, (1994).
6. M.M. Sfantsikopoulos, S.C. Diplaris and P.N. Papazoglou, *Concurrent dimensioning for accuracy and cost,* Int. J. Adv. Manufacturing Technology (1995) 10 : 263-268.
7. RINAMEN, ZEISS, BRITE-EURAM BE5593, Interim Report, Jena October 1993
8. BRITE-EURAM BE5593, Final Report, July 1996.
9. SMART 310, Product Information, LEICA LTD, Unterentelden, Switzerland, 1992.
10. Hillyard and I.C. Braid, *Analysis of dimensions and tolerances in computer-aided mechanical design,* Computer Aided Design,10(3),161-166, (1978).
11. Warnecke and W. Dutschke, *Fertignngsmesstechnik,* Springer-Verlag, (1984).
12. T. Pfeifer, *Koordinaten-nresstechnik für die qualitatssichernng,* VDI VERLAG.
13. Bambach, M.; Fürst, A; Pfeifer, *T: Ermittlung der Messunsicher-beit von 3D taststsstemen* ,Technisches Messentm, Helt 4, S.161-169 (1979).
14. Neumann, H.l.: *Koordinatenmesstechnik-Technology und Anwen-dung,* Bibliothek der Technik, Bd. 41, Verlag Moderne Industrie, Landsberg /Lech, (1990).

LASER RANGE FINDER AND VISION SENSORS FOR ROBOTIC APPLICATIONS

Hossein Golnabi
Institute of Water and Energy, Sharif University of Technology
P. O. Box 11365 - 8639, Tehran, Iran

ABSTRACT: This article deals with the vision sensors, and, in particular describes the operation and properties of laser vision sensors for robotic applications. Geometrical analysis for usual triangulation method, synchronized scanning geometry, and bidimensional scanning is presented. Variations of the range and depth of field as a function of the baseline are reported. The position of image point on the photodetector as a function of scanning angle for the case of usual triangulation and synchronised geometries is investigated. Autosynchronized scanners based on using polygonal or pyramidal mirrors are also described. A few examples are presented to show the factors affecting vision sensor performance. Typical examples are given to show potential applications of such systems in robot - mounted systems.

INTRODUCTION

During the past ten years considerable effort has been made in the development of new techniques and sensing devices to provide inputs from environment for robotic manipulators [1]. To gather such information, tactile sensors, scene analysers, proximity sensors and range - finders have been investigated and tested. Range - finders are able to provide sensory information required for functions such as object grasping, obstacle avoidance, and moving parallel to prodution table.

A great deal of attention has been given to the development of remote sensing and non-contact techniques. Optical, acoustic, and magnetic sensors have found many advantages, in particular for distance gauging. More recently, the importance of three-dimensional vision sensing in robotic was recognized, and research activities in this field are growing. Considering the importance of new laser vision sensors, the goal of this article is to describe different aspects of the method in terms of its technical merits and applications [2].

Different geometrical techniques in laser range-finders based on synchronized scanners are presented. Synchronized and autosynchronized scanners based on polygonal and pyramidal mirrors are described. Various components of a typical laser vision system are explained.

Problems such as scene illumination, factrors involved in mechanical scanning arrangement, electro-optical imaging and other parameters that may lead to a better overall performances are discussed.

Finally, typical examples of collecting real-time, high resolution imagery data to provide a high quality 3-D information about the object under study are presented. Future applications of such devices in robot-mounted systems are also described [3,4].

VISION SENSORS

Scene illumination, coordinate transformation, electro - optical imaging, and image processing are the main concerns in a vision system. In scene illumination, the need for controlled illumination is well understood but so often the problem of feature extraction from perceived images is unnecessarily complicated by the lack of contrast in image. Backlighting has ofen been used as a solution to enhance the contrast and this problem is solved by using parallel projection optics [5]. It has been shown that whenever it is possible the vision sensor should look into a source of illumination of sufficient magnitude to swamp any ambient lighting. Another solution to the problem of scene illumination has been tackled by the use of structured light [6]. The exploitation of structured light is relaised as a single light stripe projected on the surface of the work area.

One of the most elegant solution to the problem of scene illumination has been developed by using a high intensity laser beam. The concept of illuminatig the work area using a laser beam coupled with a mechanical scanning arrangement provides the best way to recover 3-D information about the objects.

Coordinate transformation between vision and robot axes is required whenever vision sensors are incorporated. In the case that these two are separated in space, the problem of coordinate transformation is considerable. In practice, placing the vision sensor on the end-effector reduces the problem of coordinate transformation. However, it must be pointed out that at this point in evolution laser scanning system can not be incorporated within the end-effector of a robot. consequently, it is a common practice to use laser scanning system as a free-standing sensor mounted above the work area. The problem of coordinte transformation is usually avoided by using laser sensor in conjunction with another end-effector mounted sensor. This combination then promises for a complete sensory understanding of the environment which is probabily the best available [7].

Vision formation includes two separate but related functions that considers electro-optical imaging and image processing. The purpose of the first process is to convert optical radiation to an appropriate electronic signal for the computer. The aim of image processing is to extract and interpret useful information from the electronic image provided by the sensor.

RANGE FINDING SYSTEMS

In single - spot laser arrangement a laser is mounted vertical and the camera adjacent to it but at angle to the laser axis. Fig. 1 shows this arrangement where the range factor is given by

$$r = P/\tan\alpha \tag{1}$$

Where p is the camera/laser offset, r range, and α is the camera/laser angle. This angle is set according to the range and depth of field that is required in measurements [8].

If we consider the camera viewing angle to be θ as shown in Fig. 1, the depth of the field is defined by

$$d = \frac{P}{\tan(\alpha - \theta)} - \frac{P}{\tan(\alpha + \theta)} \tag{2}$$

For a typical laser/camera angle of $\alpha = 10^o$, the variation of the range factor, r, with the baseline, P, is shown in Fig. 2. Variation of the depth of field, d, for a camera viewing angle of $\theta = 5^o$ as a function of the baseline is also displayed in Fig. 2. As can be seen in Fig. 2, both of range and field of view increase linearly with increasing the value of laser/camera offset distance.

The single - spot laser probe system has found many applications for range findings.It can be

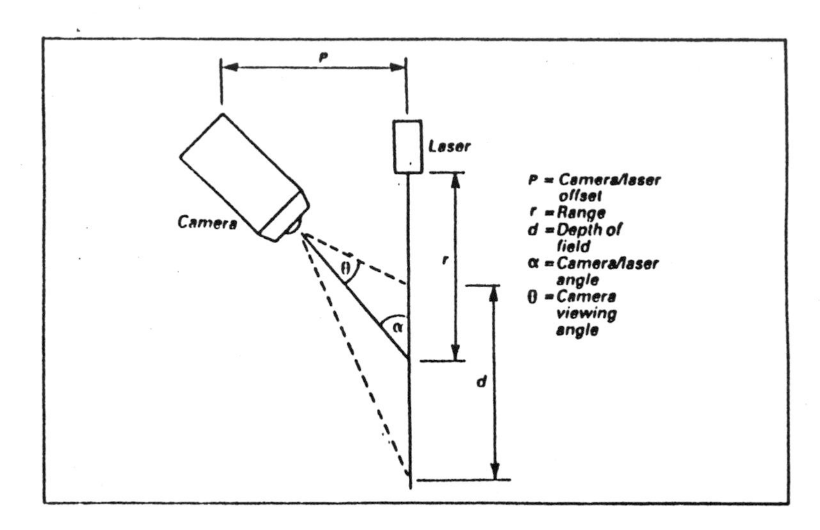

Figure 1. A single - spot laser range finding system [8]

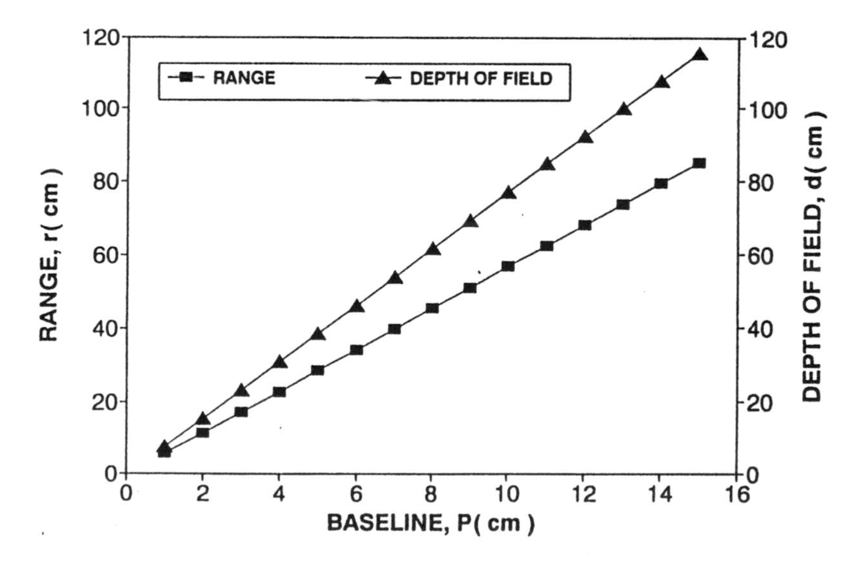

Figure 2. Variations of the range and depth of field as a function of baseline

used for the contour following and height sensing. The object location and its dimension also can be checked by this method of laser probing.

An improved method of laser probe is considered to be the use of a line of light instead of a single spot light .This technique has found many applications in the area of object edge following and object location in robotic systems.

LASER SCANNING SYSTEMS

A laser - based scanning range finder consists of: a laser light source, a transmitting mirror, a receiving mirror, a lens, and a photodetector. The transmitting mirror, rotating at a constant speed refelects the reference and the target - refelected beams through a focusing lens to a position sensitive photodetector. The receiving mirror rotates synchronously with the transmission mirror and serves to sweep a laser beam in a plane.

A triangle geometry is figured by the transmitting mirror, target, and the receiving mirror. The triangulation angle is extracted from the time difference between the passage of the reference and the target reflected beams across the center part of the detector. Time to angle to range conversion is usully performed by a microprocessor and resolution is a function of range, baseline length and scanning speed.

In usual triangulation geometry consider a beam of light orginating from position d along X axis as shown in Fig 3. The light beam is projecting at an angle θ_0 provides a reference point (d/2,l) that is used for calibration purposes. At the origin there is a lens of focal length f used to focus the light on a position sensitive detector aligned parallel to the axis and in focus at - fl/(1 - f) along Z axis.

Under rotation of scanner, the light beam rotates to a new angular position ($\theta_0 + \theta$) and the spot light on the detector moves from df/2 (1 - f) to a new location p as shown in Fig. 3. By trigonometry we can find the relation between the object coordinate (X,Z) and the other parameters as

$$X = dp \left[p + \frac{fl\,(2\,l \tan\theta + d)}{(l - f)\,(d \tan\theta - 2l)}\right]^{-1} \qquad (3)$$

$$Z = -d \left[\frac{p\,(l - f)}{fl} + \frac{2l \tan\theta + d}{d \tan\theta - 2l}\right]^{-1} \qquad (4)$$

where we assame that d,l, and f are known.

A. Synchronised Scanning Geometry

Now consider a synchronised scanning system as shown in Fig. 4. Here we have added a scanner which has it axis of rotation set at (0,0) and moves synchronously with the one at point (d,0), the projection scanner. This geometry has the effect of cancelling the angular movements.

As can be seen in Fig. 4 the result is to bring the point on the detector p′ closer to the reference point. Point p is the position that would be without synchronisation and p′ is related to p as

$$P' = \left[p - \frac{fl \tan\theta}{l - f}\right] \left[1 + \frac{p(l - f) \tan\theta}{fl}\right]^{-1} \qquad (5)$$

The variation of the position of nonsynchronised image point on the detector, p, as a function of scanning angle θ in XZ- plane is presented in Fig. 5. Considering the results shown in Fig. 5 it is noticed that the position of this point with respect to the reference point is increasing by increasing the θ value. In Fig. 5 the position of synchronised image point, p′, is also indicated which shows to be a constant and independent of θ. This feature shows one of the advantages

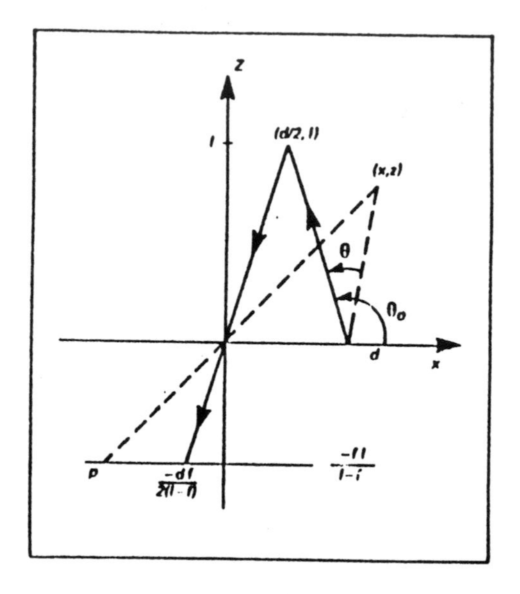

Figure 3. Simple diagram for a conventional triangulation geometry [7]

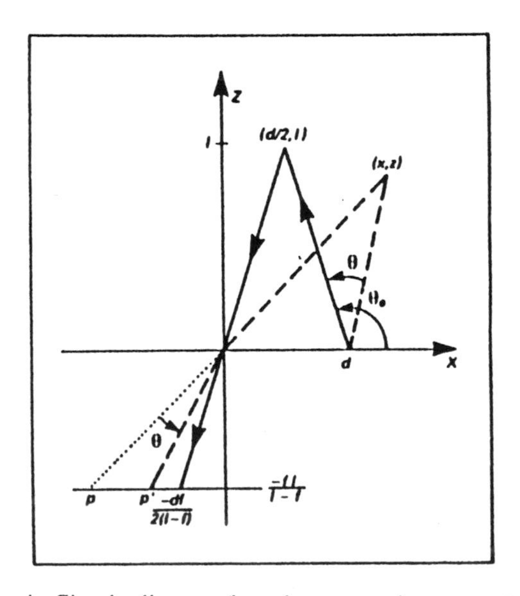

Figure 4. Simple diagram for a laser scanning geometry [7]

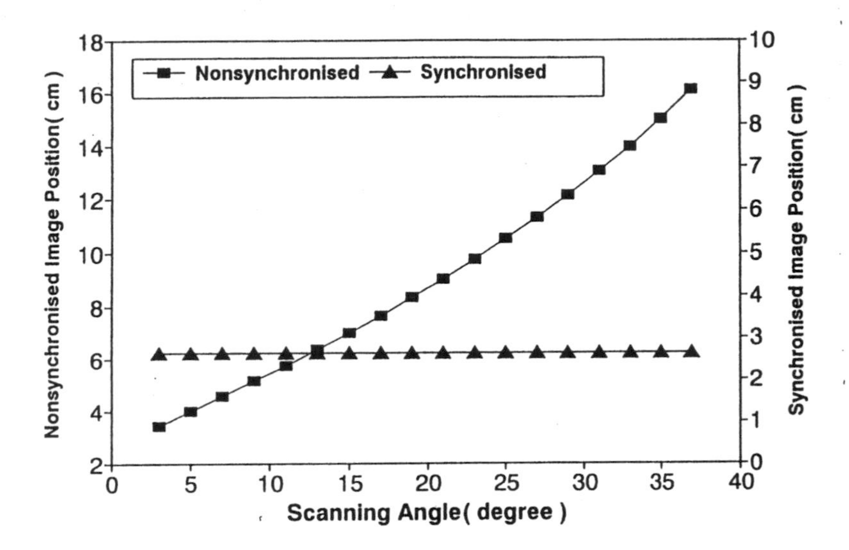

Figure 5. Variations of the image position on the phtodetector as a function of scanning angle for the case of usual triangulation and synchronised geometries

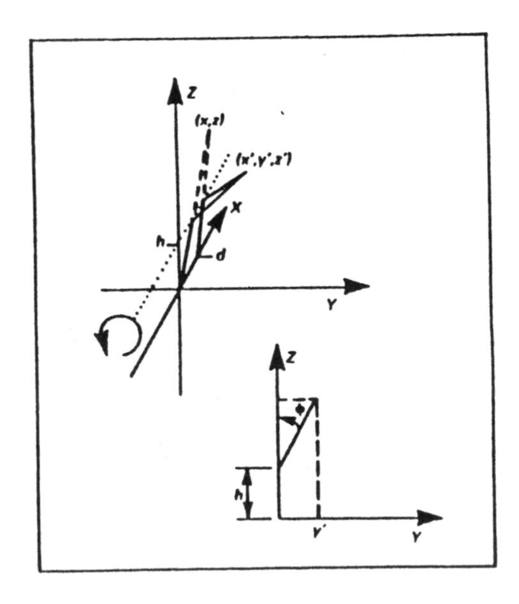

Figure 6. Simple geometry for a dual - axis synchronised scanning approach [7]

of the synchronised geometry in comparison with the usual triangulation method.

It is noticed that a change in the position of the light along Z axis produces an equivalent angular shift in both usual and synchronised geometries. However, any change along X axis produces smaller angular shift in the case of synchronised arrangement. In another word, with the same position sensitive detector used we can increase the focal length of the lens and to obtain an increased resolution in range without reduction of the field of view with synchronisation technique.

B. Bidimensional Scanning

The geometry here is similar to the one described in Fig. 4. but as shown in Fig. 6. a defelecting mirror is added which has its rotation axis parallel to X axis and it located at a distance h along the Z axis from the synchronised line scanners. A dual - axis synchronised scanning geometry is arranged in which θ is the scanning angle in the X Z - plane and Φ is the scanning angle in the YZ - plane.

By trigonometry we have found the relation between the coordinates (X',Y',Z') and the parameters of the geometries as

$$X' = dp\left[p + \frac{fl}{(l - f)\,\alpha}\right]^{-1} \tag{6}$$

$$Y' = [h - (X' - d)\,\alpha]\sin\Phi \tag{7}$$

$$Z' = h + [-h + (X' - d)\,\alpha]\cos\Phi \tag{8}$$

where α is defined as

$$\alpha = \frac{d\tan\theta - 2l}{2l\tan\theta + d} \tag{9}$$

and Φ is twice the rotation angle of the defelecting mirror which is assumed to be known. In this arrangement the three - dimensional coordinates of a surface can be measured by the system.

In finding these coordinates we have found that there is a good agreement between our result for Y' with the equation (15) of reference [7], but our results for X' and Z' are different from those of reference [7]. We believe that our results are correct and it seems to be some typing mistakes in Eqs. (14) and (16) of reference [7] for X' and Z'.

C. Autosynchronised Scanners

There are two mechanisms for autosynchronised scanners, namely the pyramidal and the polygonal geometries [7]. Fig. 7 shows a typical arrangement for the pyramidal scanner which consists of a light source, pyramidal mirror including of six facets, fixed mirrors, a deflecting mirror and a detector.

A beam of light is deflected by a facet of scanner to mirror M1 and second deflection occurs on mirror M3. The scattered light by the object after reflection through M3 is reflected to the oposite facet of the mirror scanner. The reflected light is then focused by a lens onto a detector. The pyramidal mirror produces the scanning in θ direction while deflecting mirror provides scanning in Φ direction.

Although two separate scanning mechanisms can be used to provide rotation of projection and detection axes, but this autosynchronised method provides a more stable system with less adjustment requirements. The usual triangulation technique shows severe shadow effect and the geometry of Fig. 7 still shows some shadow effects. The way to overcome this problem is either to rotate object or translate the camera head.

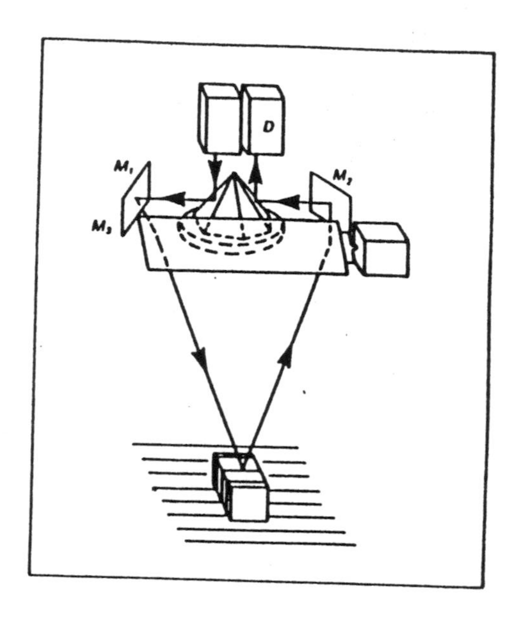

Figure 7. A typical geometry for the autosynchronised pyramidal miror [7]

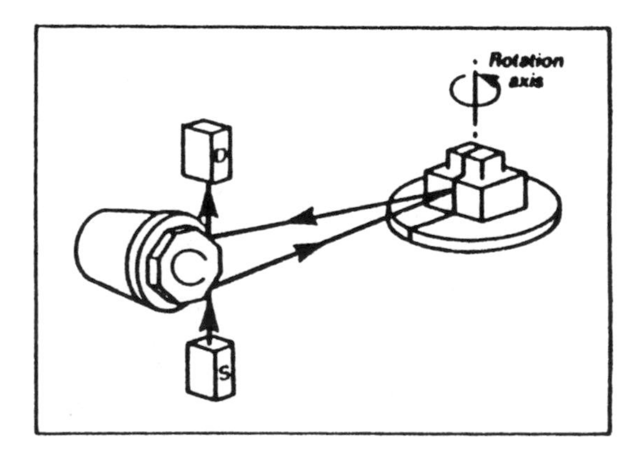

Figure 8. A typical geometry for the autosynchronised polygonal scanning system [7]

The use of a rotating object system is presented in Fig. 8. Here because of object rotation there is no need for deflection mirror (M3), and the system can be simple and compact by using a polygonal scanner. The camera head can also be rotated for the case of a fixed object. Such an arrangement can be useful in robot guidance, where 360° scanning is required.

In comparing two arrangements, pyramidal mirror has the advantages of (1) having a very large fields of view without compromising resolution along range axis, (2) large collecting area for small intertia (3) angular coverage is a factor of two smaller than a polygon system. However, the polygonal geometry of Fig. 8 has less shadow effect and the system can be made very simple and more compact.

APPLICATIONS

Vision sensors can be used in inspections and part identifications [8,9]. The camera system should be suitable for use with most computer systems sharing interface capable of accepting serial data at a high bit rates (1 MHz). Another important parameter of any vision system is the picture resolution.

A practical vision system for use with bowl feeders has been described [10]. The microprocessor at the heart of the vision system is able to recognise different orientations of components and can also detect faulty components. A vision controlled robot system for transferring parts from belt conveyors has been reported as CONSIGHT - I [11]. This system can determine the position and orientation of a wide class of manufactured parts.

Visual guidance techniques for robot arc welding has also been reported [12]. This system uses a structured - light technique. The complete system consists of two CCD area cameras and four laser diode sources which are packed around the welding torch.

As a good example of laser range finders we can mention a laser - based scanning range finder reported in Ref. [13]. A compact (7×7×8 cm) scanning range - finder for robotic applications has been described. The modified triangulation scanning method has been reported in which the angle is extracted from the time difference between the passage of the reference beam and target reflected beam. A typical precision of 0.1% of range at 30 mm and 0.07% of range at 1 m has been reported.

In another report [6] three - dimensional locating of industrial parts has been described by using a structured - light range - finding and data processing technique.

In summary, scanning geometries described in this report could lead to the realisation of compact low - cost three - dimensional cameras [14 - 17]. Beside robotic applications, inspection of manufactured parts, as well medical applications such as prostheses fabrication and diagnostic instrumentation are possible.

CONCLUSIONS

Different geometrical arrangements for laser range finding and vision systems have been described. Analysis for usual triangulation, synchronised and autosynchronised scauners has been presented. Usual triangulation method shows severe shadow effect problems associated with the requirement of a large separation between projection and detection axes in order to get a reasonable resolution in range axis.

In contrast to the usual triangulation geometry, in the synchronised approach a single position on the position sensor defines a surface in the object space. This surface is a circle inscribing the reference point and both scanner's axes of rotations. This feature can be used advantageously when profile measurements of spherically shaped objects are to be made.

Another advantage is that for usual triangulation method a dual - axis sensor is required to do surface profile measurements while in the scaning approach reported a linear position sensor is adequate.

The availability of cw diode lasers, position sensitive photodectors, easily interfaceable microcomputers and the use of reported scanning method enables the design and implementation of compact and accurate optical range finders for robotic applications.

REFERENCES

1. Golnabi H, and Ashrafi A: The role of sensor systems in automated manufacturing. *Proc. of the 4th Int. Conf. On Flexible Automation and Integrated Manufacturing*. FAIM 94 (Begell House Inc., New York, 1994), PP 645 - 654.
2. Pugh A: *Robot sensors* vol.1 - vision, IFS publication Ltd, UK, 1986.
3. Cromwell RL: Sensor and processor enable robot to see and understand. *Laser Focus World*, PP 67 - 78, March, 1993.
4. Gerhardt LA, Mistetta WJ: The use of preconditioned structured light for 3 - D vision based parts inspection. *Proceedings of Manufacturing International 90*. NY, ASME, 1990, PP 97 - 102.
5. Saraga P and Jones BM: Parallel projection optics in simple assembly. *Proc. lst Int. Conf. On Robot Vision and Sensory Controls*, (IFS publication Ltd, Bedford UK, 1981) PP 99 - 111.
6. Schroeder E: Practical illumination concept and techniques for machine vision. *Robot sensors* vol. 1- vision Ed. by Pugh A. IFS publication Ltd, UK, 1986. PP 229 - 244.
7. Rioux M: Laser ramge - finder based on synchronised scanners. *Robot sensors* vol. 1- vision Ed. by Pugh A. IFS publication Ltd, UK, 1986. PP 175 - 190.
8. Loughlin C, and Morris J: Line, edge and contour following with eye - in - hand vision system. *Robot sensors* vol. 1- vision Ed. by Pugh A. IFS publication Ltd, UK 1986. PP 95 - 102
9. Whitehead DG, Mitchell I, and Mellor PV: A low - resolution vision sensor. *Robot sensor* vol. 1 - vision Ed. by Pugh A. IFS publication Ltd, UK, 1986. PP 67 - 74.
10. Cronshaw AJ, Heginbotham WB and pugh A: Apracticl vision system for use with bowl feeders. *Robot sensor* vol. 1 - vision Ed. by Pugh A. IFS publication Ltd, UK, 1986, PP 147 - 156.
11. Holland SW, Rossol L, and Ward MR: Consight - I, a vision - controlled robot system for transfering parts from belt conveyors. *Robot sensors* vol. 1 - vision Ed. By Pugh A. IFS publication Ltd, UK, 1986. PP 213 - 228.
12. Morgan CG, Bromley JSE, Davey PG and Vidler AR: Visual guidance techniques for robot are welding. *Robot sensors* vol. 1 - vision Ed. by Pugh A. IFS publication Ltd, UK, 1986. PP 255 - 266.
13. Nimrod N. Margalith A, and Mergler H: A laser based scanning range - finder for robotic applications. *Robot sensors* vol. 1- vision Ed. by Pugh A. IFS publication Ltd, UK, 1986. PP 159 - 173.
14. Ryan A, Gerhardt LA: Application of laser scanner to 3 - D visual sensing and enhancement tasks. *Proceedings of SPIE*, 1657, Bellingham, WA, 1992, PP 213 - 224.
15. Kennth M, McCalannon P, whelan F, McCordell C: Integrating machine vision into process control. *3 rd International Conference on Flexible Autonation and Integrated Manufacturing*, FAIM 93, PP 703 - 714.
16. Jones BE: Optical fibre sensors and systems for industry. *J. Phys. E: Sci. Instrum.* 18, PP 770 - 782, 1985.
17. Jordan GR: Sensor technologies of the future. *J. phys. E: Sci. Instrum.* 18, PP 729 - 735, 1985.

PERFORMANCE TESTING OF HAND HELD SCANNERS FOR TWO DIMENSIONAL BAR CODE SYMBOLOGIES

Richard E. Billo, Bopaya Bidanda, Martin Adickes
University of Pittsburgh

ABSTRACT: In this paper, a suite of standard tests are presented to measure the decode time of hand held scanners. These tests were applied to a PDF417 two-dimensional bar code symbol, and used to compare the performance of five newly developed hand-held scanners. The tests focused on the following factors that typically affect decode time: lighting conditions, reading distance, orientation angles of the scanner, and use of the scanners in a typical freehand application. Results show the tests to be a valid means for comparing performance of differing hand-held scanner technologies typically used for two-dimensional symbologies.

INTRODUCTION

Two-dimensional (2D) symbologies are becoming increasingly popular in industrial, commercial and retail applications. This popularity is largely due to the ability of such symbologies to contain high data capacities on the symbol itself as compared to traditional linear symbologies which require an external database to retrieve comparable information.[1] In addition, 2D symbologies have excellent error detection and correction capabilities, and can be accurately decoded even when significantly damaged.

One of the most widely accepted 2D symbologies that has recently been accepted as an American National Standard is PDF417 (see Figure 1) developed by Symbol Technologies. This symbology can encode up to 1,850 alphanumeric characters, has eight error correction levels, and can be read using a variety of hand-held and overhead scanning technologies.[2] PDF417 is currently being used for a wide variety of applications, a few of which include small package transport, military identification, drivers license renewal, access control, and warehouse management.

Figure 1. PDF417 Bar Code Symbol

Parallel to the increase in popularity of PDF417 has been developments and advancements in the technologies capable of decoding it. In the early 1990's hand-held bar code scanners primarily utilized laser technology to decode PDF417.[3] With this technology, a laser beam is physically moved across the entire symbol through the usage of an electro-mechanical system consisting primarily of rotating mirrors.[4]

Now, however, through increases in computing power and advances in semiconductor design, there has been a surge in the development of portable *image-based* scanning systems. Image-based bar code scanners, as their name implies, capture an image of the symbol and uses it as the basis for decoding. The symbol is flooded with light and the resultant image is optically transferred to an image detector. The detector converts the photons it receives into electronic signals, organizes them into binary representations of pixels, and sends this data to a host system for decoding.[3] The image detector can contain a single row of photodiodes, in which case the image must be captured row by row, or can have multiple rows of photodiodes, in which case the entire image is captured at once. Two current technologies dominate image based reading devices: charge coupled devices (CCD) and complimentary metal oxide semiconductor (CMOS) devices. CCD technology is the technology most often used for such devices as video cameras and for machine vision applications. CMOS technology uses semiconductor technology, and has a lower power consumption, lower price and is easier to manufacture than CCD.[5]

RESEARCH PROBLEM

Many corporations and agencies that have made the decision to adopt a 2D symbology such as PDF417 often face the problem of selecting the best scanning technology for their particular application. Although all manufacturers of 2D bar code scanners provide technical specifications on the performance of their products, it is difficult to make an objective comparison. Tests performed by different manufacturers typically use different performance metrics, factors, and test conditions. Relying on a manufacturer's specification as a basis for comparing performance of the new 2D scanner technology makes objective comparisons among different product specifications difficult if not impossible.

The purpose of this research was to design a standard test plan and metrics that can be used to assess the performance of 2D symbology-based hand-held scanners. The research focused on hand-held scanning technology as this is the most widely used technology for 2D bar code applications and thus represents the greatest interest at this point in time. The tests were designed to allow a potential user of a 2D symbology the ability to make direct comparisons of scanner performance using not only different models of the same scanning technology but also across competing scanner technologies.

Four tests were developed to provide an objective comparison of hand-held scanners used for reading 2D symbologies. These tests assessed 1) controlled versus freehand reading, 2) reading distance, 3) lighting conditions, and, 4) angle of the symbol (tilt, pitch, skew) with respect to the scanner.

The dependent measure selected for each test was the mean decode time of a scanner. Decode time was defined as the elapsed time from the initial pull of the scanner trigger to the audible indication of the successful decode of the symbol.

The tests were validated through a case study in which they were applied to five different models of scanners representing four different types of state-of-the-art scanning technologies developed for reading 2D symbologies. These included 1) *Linear CCD* scanners which utilize a single row of photodiodes to capture the 2D symbol image, 2) *Area CCD* scanners which have multiple rows of photodiodes to capture the image, 3) CMOS technology, and 4) raster laser scanners. PDF417 was selected as the symbology to be scanned in the case study.

HYPOTHESIS

The mean decode time (u) was compared in all four tests for the five scanners described in the case study. In each test, an analysis of variance was conducted to detect significant differences in mean decode time that may have occurred among scanners. Hypotheses were as follows:

$$H_0: u_1 = u_2 = u_3 = u_4 = u_5$$
$$H_a: u_1 \neq u_2 \neq u_3 \neq u_4 \neq u_5$$

TEST EQUIPMENT AND SYMBOLS

Testing was done using a PDF417 bar code symbol. The data content consisted of 183 ASCII characters. Although the symbol conforms to the specified data format, the actual data contained on the symbol was random. The symbol was printed onto 4" by 6" label stock using direct thermal printing technology, and was printed using an X dimension of 15 mils, and a Y dimension of 45 mils. This resulted in a symbol with overall dimensions of approximately 6 cm. wide by 5.5 cm. high.

To assess the full impact of each test, it was necessary to develop a fixture in which a scanner could be mounted that would eliminate the effect of human handling and would allow testing to be conducted in a consistent and repetitive manner. Without such a fixture, it would be difficult to control such factors as scanner distance, pitch, skew and tilt. A scanner to be tested is mounted into the fixture, and a PDF417 symbol placed on the target panel. The scanner mount was designed to be fully adjustable over the range of most commercial scanners, thus allowing testing of various reading distances. The target panel can be simultaneously rotated in any combination along the x, y, or z axes, and raised or lowered as a unit to allow precise alignment to the scanner. The fixture can also be reconfigured with a small sliding platform to accommodate linear scanners which require the user to systematically move down the symbol to acquire an image.

TEST PROTOCOL

Four major types of tests were developed. In summary, these tests consisted of a controlled versus a freehand test which compared optimally controlled conditions with uncontrolled real life conditions; a reading distance test which assessed the impact of distance from scanner to symbol; the impact of lighting on decode time; and the impact of orientation angle (tilt, pitch, skew) on decode time. For each test, a variable of interest was selected, with all other associated variables being held fixed at the appropriate manufacturers' recommended settings. Each test is described below.

Controlled versus Freehand Scanning

This test was designed to compare decode time for scanners under both optimally controlled conditions and real life conditions. This test was based on the assumption that differences in decode time will occur due to the natural human error that occurs through physical handling of the hand-held scanner during the reading process. For both tests, symbols were oriented 0° relative to the scanner, and were scanned in normal room light (10 ft. candles). In the controlled decode time experiment, the scanner was mounted into the test fixture and the reading distance set to the manufacturer's recommendation. In the freehand experiment, a user physically held the scanner in her preferred hand and alternately scanned two symbols in a back and forth manner until the required sample size was reached. In the freehand test, the two symbols were separated 12" horizontally and 6" vertically.

Scanner Distance

Distance between scanner and symbol can also have an effect on the performance of a bar code scanner. Although product specifications clearly state optimal reading distances, users typically want to know the sensitivity of the products with respect to these specifications. This test measured mean decode time when the scanner was placed at varying distances from the symbol. Scanners were mounted into the test fixture and the symbol oriented 0° relative to the scanner. Selections for reading ranges and optimal reading distances were made based on the manufacturer's recommended settings. For each scanner, the scanner was placed at the closest recommended distance to the symbol, five reads were made, and the scanner moved in 1-inch increments from the symbol.

Lighting

Because bar code reading technologies are primarily optical reading technologies, the lighting test was designed to assess whether there are any differences in decode time based on the reading environment's lighting conditions. In this test, decode time for each scanner was evaluated under three types of lighting conditions: bright lighting conditions (175 ft. candles) simulated with incandescent flood lamps, normal lighting conditions (10 ft. candles) simulated with fluorescent lamps, and low lighting conditions (1 ft. candle) simulated with a fewer number of fluorescent lamps placed far enough from the symbol to attain the 1 foot candle requirement. Scanners were mounted in the fixture and placed at factory recommended distances from the symbol. The PDF417 symbol was oriented 0° relative to the position of the scanner. A Simpson Model 408-2 Illumination Meter incorporating a 1.77 inch diameter selenium photocell corrected to the spectral sensitivity of the human eye was utilized to measure the illumination for the three lighting conditions.

Orientation Angle

A bar code scanner can yield highly different results depending on various orientation angles with which it is positioned relative to the 2D symbol. There are three types of orientation angles that need to be considered (see Figure 2): the tilt angle which can be represented by rotation of the symbol along the z-axis on the Cartesian coordinate system; the pitch angle represented by rotation along the x-axis; and skew angle created by rotation along the y-axis.

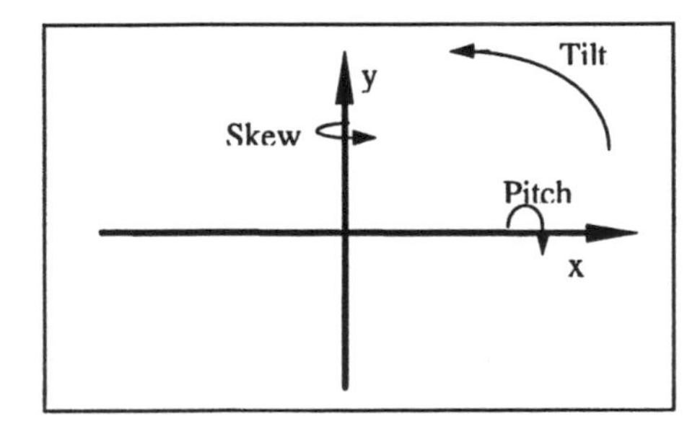

Figure 2. Cartesian Coordinate System illustrating Orientation Angles

For all three tests, scanners were mounted into the scanning fixture and the scanning distance set to the manufacturer's recommendation. Two variations of the tilt test were conducted. For those scanners that claimed to be omnidirectional (360° reading capability), a full 360 rotational tilt test was done. For those scanners that were not omnidirectional, a limited range tilt test was done that varied between ±5°. For the full 360° rotational test, the PDF417 symbol was rotated and scanned at 10° increments through a 360° circle. For the limited range test, each symbol was rotated and scanned at 1° increments through the ±5° range. Although both the pitch and skew tests were done independent of each other, both tests were conducted in a similar manner. Each scanner was tested between it's maximum recommended skew and pitch angle and 0° orientation relative to the scanner. Beginning with 0° orientation, angle of the scanner was changed in 10° increments.

CASE STUDY

To help validate the usefulness of the tests in providing clear and unambiguous comparisons of performance capabilities of 2D symbology hand-held scanners, the tests were applied to five different scanners representing four different hand-held reading technologies. Two scanners incorporated CCD technology, the remaining three scanners incorporated raster laser, CMOS, and Linear CCD technology. Scanner technology as well as any details describing any particular scanner that took part in the case study have been deliberately omitted. Approximately 37,000 scans were conducted over the course of the testing. All tests were conducted using a single PDF417 symbol. Because all scanners were non-contact scanners, there was no damage or degradation in the symbol over time.

Controlled versus Freehand Results

Figure 3 shows the mean decode times for the various scanners under optimally controlled conditions and freehand use. All scanners performed significantly better under controlled conditions when the scanner was affixed in the fixture at optimal distances and angles than in the freehand usage condition. With some scanners, these differences were quite substantial such as the case with Scanner B, with mean decode times as much as twice as high in the freehand condition. Under controlled conditions, Scanner D performed best, followed closely by Scanner C and Scanner E. However, when the scanners were removed from the fixture and used as they would be in a normal situation, Scanner E performed best.

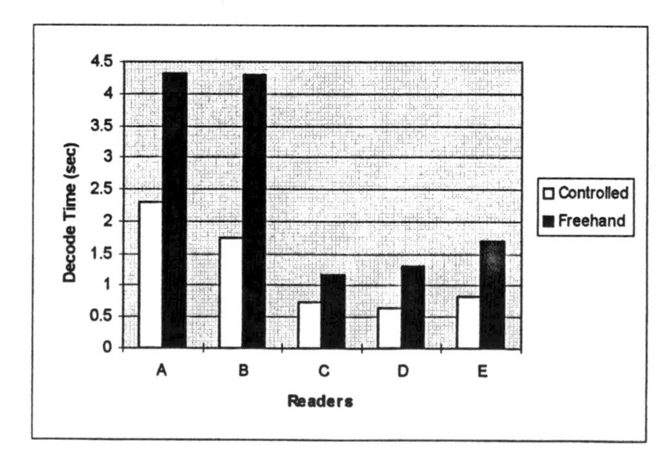

Figure 3. Mean Reader Decode Time under Controlled vs. Freehand Conditions

Scanner Distance Results

Figure 4 shows the effects of scanner distance on mean decode time for each of the scanners tested. Decode times tended to increase as scanners were moved closer to the symbol. In addition, it appeared that Scanners A, B and E were more sensitive to changes in reading distance than Scanners C and D.

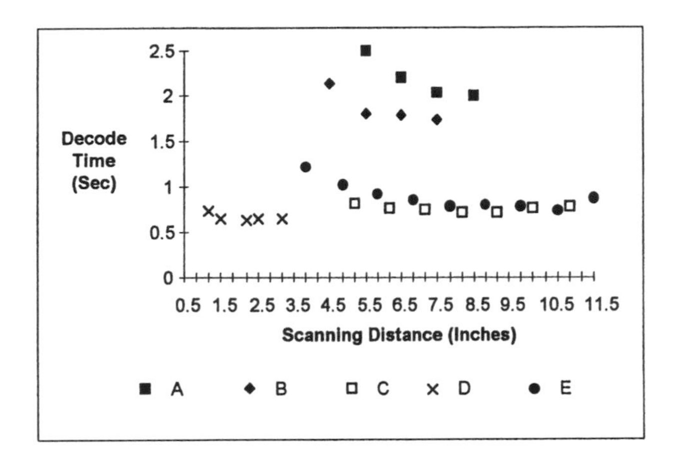

Figure 4. Reader Distance Test Results

Lighting Results

Different results were found among the scanners concerning the effects of lighting on scanner performance. As Figure 5 shows, illumination did not appear to have a very large effect on either Scanner C or E. Scanner D, on the other hand, demonstrated the most sensitivity to lighting, with the overall decode time more than doubling under bright light conditions. In general, all scanners showed consistent performance patterns when lighting conditions were varied, performing worse under extreme lighting conditions, and best under normal lighting conditions.

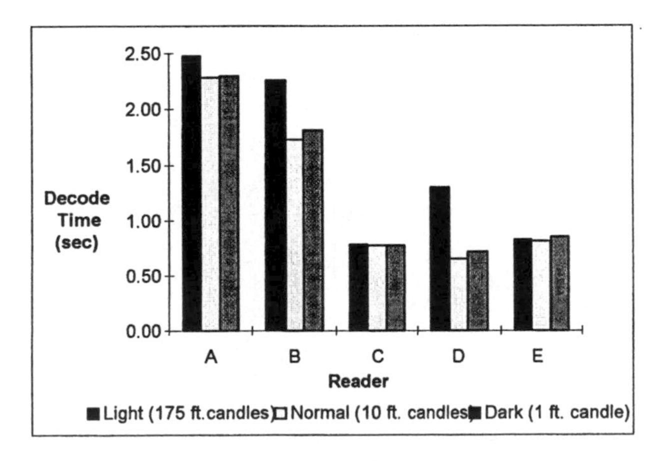

Figure 5. Lighting Condition Test Results

Orientation Results

Tilt test results for scanners that had omnidirectional decoding capability are displayed in Figure 6. Each of the scanners in Figure 6 exhibited periodicity with respect to decode time, although this periodicity was minimized with Scanner E. Scanner A was not able to decode the PDF417 symbol in all of the rotational positions as some angles caused the symbol to fall outside the scanner's field-of-view. Scanner B exhibited periodicity that was reverse of what would be normally expected: decode times that moved from low to high, and back to low as the symbol moved through each quadrant of the circle.

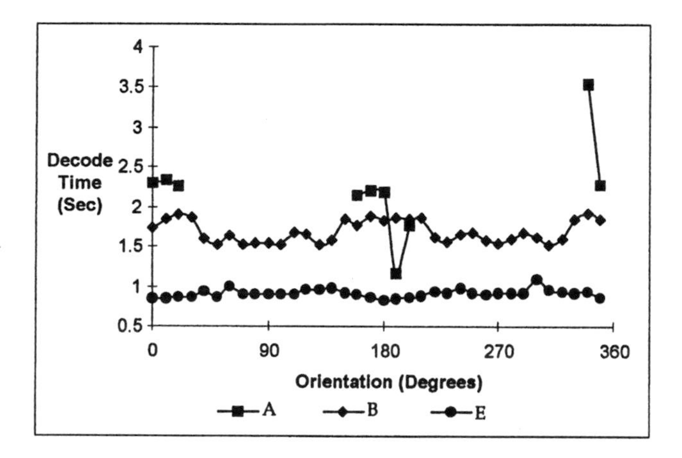

Figure 6. 360° Rotational Tilt Test Results

Scanners not capable of omnidirectional scanning were tested within the range of angles at which they could successfully decode the test symbol. Results are summarized in Figure 9. As expected, decode times symmetrically increased as the symbol angle deviated from optimal 0° orientation relative to the scanner.

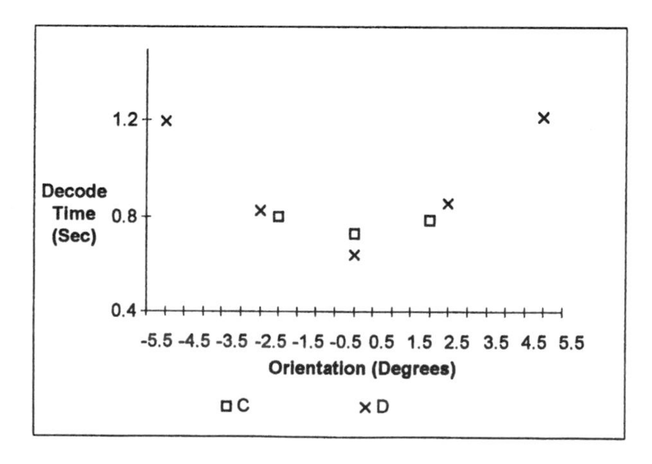

Figure 7. Limited Rotation Test Results

Although not shown, both the pitch and skew tests yielded similar results for all scanners. Best decode times occurred when the symbol was oriented 0° relative to the scanner.

CONCLUSIONS

Figure 8 summarizes individual scanner performance characteristics of five new scanners that have been recently introduced into the marketplace or are currently under development. Results indicate that at the current time, Scanner C performs the fastest and most consistently. Scanner D performs almost as well, but is hindered by a limited depth of field. The other scanners, despite their omnidirectional capabilities, are considerably slower at decoding PDF417 bar code symbols and exhibit periodicity that even further degrades their performance. These results provided feedback to the manufacturers of these scanners that further research is needed into such areas as development of accurate and user-friendly targeting systems, and quicker data capture and decoding algorithms.

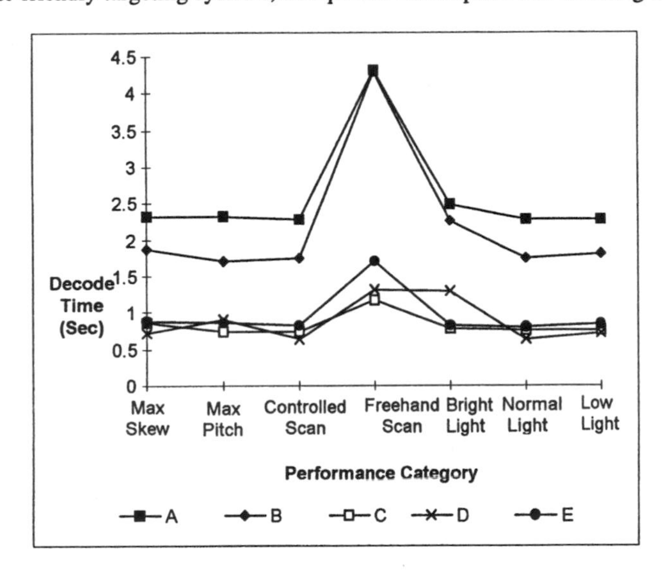

Figure 8. Overall Performance Comparison

The performance tests have been reviewed by several major manufacturers and users of hand-held scanners and accepted as reasonable and valid tests to assess performance of these products for 2D symbology applications.

REFERENCES

1. Pavlidis, T. and Wang Y.P., Two dimensional bar codes, *Proceedings of Industrial Automation Conference*, 10-3 - 10-8, 1990.
2. Itkin, S. and Martel, J., *A PDF417 Primer: A Guide to Understanding Second Generation Bar Codes and Portable Data Files, Monograph 8*, Symbol Technologies, Inc. Bohemia NY. 1992.
3. Palmer, R.C., *The Bar Code Book*, Helmer's Publishing, Inc. New Hampshire, 1992.
4. Ford, W., Hand-held scanning, *Scan Tech 88 Proceedings*, AIM^{USA}, Pittsburgh, PA, 1988.
5. Wilson, R., CMOS goes where CCD has treaded., *Electronic Engineering Times*, April 29, 40-42, 1990.

DESIGN INTEGRATION AND DATA INTERCHANGE IN A PRODUCT DEVELOPMENT ENVIRONMENT

P.W.Norman[1]
R.Jamieson[2]
1. Department of Chemical and Process Engineering
2. Regional Centre for Innovation in Engineering Design
University of Newcastle upon Tyne

ABSTRACT: This paper describes a methodology for identification of opportunities for enabling integration of the design process within Small to Medium Enterprises (SMEs), supported by electronic data interchange. The work is based on case studies carried out in a number of SMEs. Analysis of the potential for electronic integration is used as the basis for specifying the architecture of a demonstrator for managing product data. The demonstrator is based on the adoption of robust electronic data management technologies which are affordable and capable of being handled by a typical SME.

INTRODUCTION

In today's business climate there is a need to reduce the time taken to develop a product from concept through to a production ready version. Recent developments in CAD and other computer based systems have made powerful techniques available to designers. Migration of simple 2-D CAD representations of engineering products to representation on 3-D CAD database systems represents a significant enhancement of the capability of any business but also a major step up in the requirement for enabling technology. To be used effectively, the associated database system must be able to exchange data seamlessly with all the peripheral applications and possibly other databases. There are therefore significant issues of interoperabilty to be considered. The arguments for reliable electronic data interchange (EDI) are well rehearsed and include issues of data accuracy and integrity, security and speed of access compared with manual methods of exchange and representation [1]. In addition, there are issues of integration and reworking of business processes within a company and also between business partners.

To avoid problems of being locked into proprietary systems (or locked **out** of future software developments) it is necessary to evaluate interchange mechanisms and standards that are appropriate to the sector in which a business operates and to adopt an implementation policy that strictly adheres to the chosen standards. It is also essential that potential partners, clients and supply chain companies are also committed to them.

A typical product development process involves multi-functional teams involving experts in marketing and sales through to product design and engineering. Systems need to facilitate communication between the team and their respective departments as well as across the interface with other companies and partners. The aim should be to foster creative thinking and rapid response.

In this paper we describe work that is in progress in the Newcastle Regional Centre for Innovation in Engineering Design (RCID) aimed at developing a methodology which can be used by and between companies to facilitate rapid product development environments based on emerging information technologies. An architecture for a demonstrator has been defined and the rationale for this is described. The work has been carried out in collaboration with some of the RCID member companies. RCID is a collaborative research and development venture based at Newcastle University and supported by ten local Small to Medium Enterprises (SMEs) as well as other universities in the region and the European Regional Development Fund.

TODAY'S MANUFACTURING ENVIRONMENT

Getting new products to market quickly, at the right quality and at the lowest cost are key issues. Over recent years companies have invested heavily in technology tools to improve performance and productivity including: CAD/CAM, MRP and CNC tools, etc. These technology aids have generally improved individual and departmental activities but their role in improving business competitiveness as a whole has often proved disappointing.

Much management time has gone into projects to achieve BS5750 (ISO9000) certification and the discipline introduced will lead to more structure and organisation in manufacturing and engineering processes. Yet, despite this investment and the associated benefits, managers remain frustrated by the time and cost involved in getting new products to market. Even simple engineering changes can take an inordinate amount of time and management effort to carry out. Additionally, today's businesses increasingly deal with project partners, clients and sub-contractors, both nationally and internationally, which just compounds delays and errors.

The core of the problem is that solutions are often employed based upon traditional working practices. The new technologies act to speed up the ***appearance*** of problems rather than ***preventing*** problems occurring in the first place. The need to change is recognised but the solution is elusive. A quantum improvement in performance is possible but only through a fundamental change in working practices.

MANUFACTURING CHALLENGE

The key to that change is the availability of timely, accurate information throughout the product or project life cycle. There needs to be the ability to collect, qualify and distribute the information necessary to speed development, manufacture and support.

This sounds simple but is in fact extremely complex. Information exists in many diverse forms and locations. Examples include: paper specifications, electrical and mechanical CAD data, assembly instructions, bills of materials, accounting information, production analyses, part catalogues and so on. Harnessing this information and making it globally available upon demand is a major undertaking and forms the basis of Product Data Management (PDM).

DATA INTEGRATION, CONTROL AND MANAGEMENT

No company should consider change on a large scale to accommodate EDI/PDM without first putting in place solid foundations. The following are pre-requisites before company-wide integration through data management can be achieved:

1. ISO 9000 accreditation for process structure and control purposes,
2. IT infrastructure and support,
3. A pilot project example to expand into a company wide process,
4. A 'project champion' at senior level,
5. Experience of the benefits of tools such as CAD, CAM, MRP.

Technology on its own must not be the driver in the product development process. If a company has simple products and few designers then EDI/PDM may not provide a cost effective solution. However, historical case studies show that through a Product Data Integration, Control and Management philosophy companies can [2] :

Cut costs by 10 %,
Reduce engineering costs by 15 %,
Cut time to market by 25 %,
Reduce the number of changes by 40 %.

INFORMATION FLOW IN THE DESIGN PROCESS

There have been many attempts to draw maps or models of the design process. Some of these models simply describe the sequence of activities, others models attempt to prescribe a better or more appropriate pattern of working. For example French's model has the following activities [3]:

Analysis of Problem	- specification of problem
Conceptual design	- broad based solutions
Embodiment of schemes	- GA drawings and final choice of design
Detailing	- Large number of small design decisions

There is, of course, a great deal of input and data exchange between the many individuals, groups, departments, clients and contractors in all four stages of the product development process in this model.

As an example of best ***practice***, the German professional engineers' body VDA has produced a number of guidelines for a *'Systematic Approach to the Design of Technical Systems and Products'* [4]. These guidelines suggest an approach similar to ISO 9000 Quality Assurance procedures already implemented in two of the companies participating in this study. The requirements for accreditation are such that the design stages have to be documented and proven to be working. These requirements have meant focusing on management procedures and the control of the information/data flows, storage and accountability within the design process.

The net effect of such requirements is that the hidden 9/10 of the iceberg of data integration is starting to become evident. Whereas large companies have the resources to purchase and implement full scale Product Data Management Systems costing hundreds of thousands of pounds, SMEs are not so fortunate. Nevertheless, SMEs can ensure that they also benefit from data integration by first understanding the potential and pitfalls. They can then put in place the ***appropriate*** procedures to ensure integration, implementation and training to achieve the benefits within a short time span without the wholesale change and cost of full scale commercial PDM systems.

THE NEEDS OF THE RCID COMPANIES

Each company member within the RCID has its own perceived needs regarding the electronic management of information within its product development process. Some companies operate 'islands' of advanced technology while others are just beginning to move from simple 2-D CAD drawing systems. Few have significant integrated electronic facilities. In order to gain an impression of possible needs and concerns three member companies who have indicated significant commitment to the idea of PDM were interviewed. Some common threads were discerned:

- all three companies saw engineering design as a priority area where they would like to implement EDM. However they were all cautious about being overambitious. It was recognised that progress towards EDM would be best served by starting with basic technology such as emaii and then moving in small steps towards eventual full EDM. This was seen by them as a confidence building strategy.
- Two of the companies valued the social interaction and direct contact of traditional design procedures and would not want distributed EDM to interfere with this.
- Two companies saw internal communication as a priority and were not particularly concerned about electronic partnerships. Security was seen as an issue that might inhibit EDM across company boundaries.
- The companies generally wished to deal with preferred suppliers and did not enthusiastically welcome the ability that EDM might provide for unsolicited enquiries.

Having obtained an overview of the environment into which PDM might be placed in typical SME companies, two of the companies were selected for a detailed study of the potential for PDM. Both had already gone some way towards introducing advanced IT systems including 3-D CAD. The purpose of the study was to identify the existing information flows within the product development process in the companies to reveal opportunities for integration of the process through PDM and electronic data interchange (EDI). The task was to identify current activities and data flows within the two surveyed companies and to identify systems that would support these activities. The study did not, though, look at how the processes themselves might be changed given the availability of PDM.

One of the companies designs, markets and manufactures a range of high speed electro-mechanical machinery and exports are 90% of its production. It currently employs 400 people at a single site. Due to the nature of the product there is a range of electrical, mechanical, control and software components within the majority of designs. The company has both Electronic CAD and Mechanical CAD seats with approximately 40% of design currently carried out with 3D-CAD. This is expected to rise to 90% by mid 1997. 3D-CAD work is centered on the Pro-Engineer (Pro-E) system supplied by Parametric Technology Corporation (PTC)

The second SME comprises several different companies employing about 140 people at its sites. The focus of the study was on the group's product design company. Pro Engineer and AutoCad are used for design data that is to be exported to precision engineering manufacturers. The latter, including associated group companies, specialise in high precision CNC machining for the manufacture of injection mould tools, press tools and other special purpose components. The company has full ISO 9001 accreditation and a well developed IT infrastructure. Electronically based designs are currently sent for approval via modem/ISDN links to the sister companies as well as to external clients. At this point in time one third of designs are created in 3D with Pro Engineer. It is intended that this will be 60% by mid 1997.

THE POTENTIAL FOR PRODUCT DATA MANAGEMENT

The starting point for the two case studies was a ***design process diagram***. In order to achieve quality accreditation a company will already have this diagram or an equivalent written procedure in place. For the purposes of defining a strategy for adopting Electronic Data Interchange (EDI) and Product Data Management (PDM) the design process diagram was modified to show the types of information that were generated or used at particular points in the process. This was accomplished through one or more interviews with senior managers. Structured questions were asked to elicit for each stage the type, quantity and format of the information, as well as how the information flow was controlled. A systematic procedure was adopted for recording the information in the form of interview sheets.

This information was then analysed in order to derive an ***activity model***. This diagram shows the ***flow*** of information between activities. Activities are initially identified as major events in the design process. Each activity contains sub-activities listed as imperatives such as *'produce customer requirements specification'*. It would be possible, depending on the degree of differentiation between information flows that was required, to show the links between each subactivity in further sets of diagrams. In the case studies reported here, enough information about information flows was available from the top level diagram.

Interestingly, analysis of the information from both case studies showed that a common activity model (at least at the top level) was appropriate to the design processes of both of the participating companies. The sample was too small to generalise about whether the form of the activity model was typical of all design processes in the RCID companies, but anecdotal experience would suggest that a single model was appropriate. This is shown in Figure 1. Major activity blocks identified were:

Feasibility
Conceptual design
Detailed design (engineering)
Detailed design (manufacturing)

These closely mirror basic elements of design models such as that developed by French, which also suggests the appropriateness of the derived activity model [3].

Inspection of the model shows the types of information flowing between activities. From this it is then possible to start thinking about candidates for inclusion in an EDI/PDM strategy. For example, little would be gained from producing a customer requirements specification (CRS) in electronic form (a typical output from the feasibility activity) if this could not be used by information processes in conceptual design. The only benefit would perhaps be to improve the efficiency of CRS production. The benefit of electronic integration with CAD generation would be missed. The value of the activity model is in providing a structured representation of information flow for development of the EDI/PDM strategy. Some of the data types that were uncovered by developing the activity model are illustrated in Table 1.

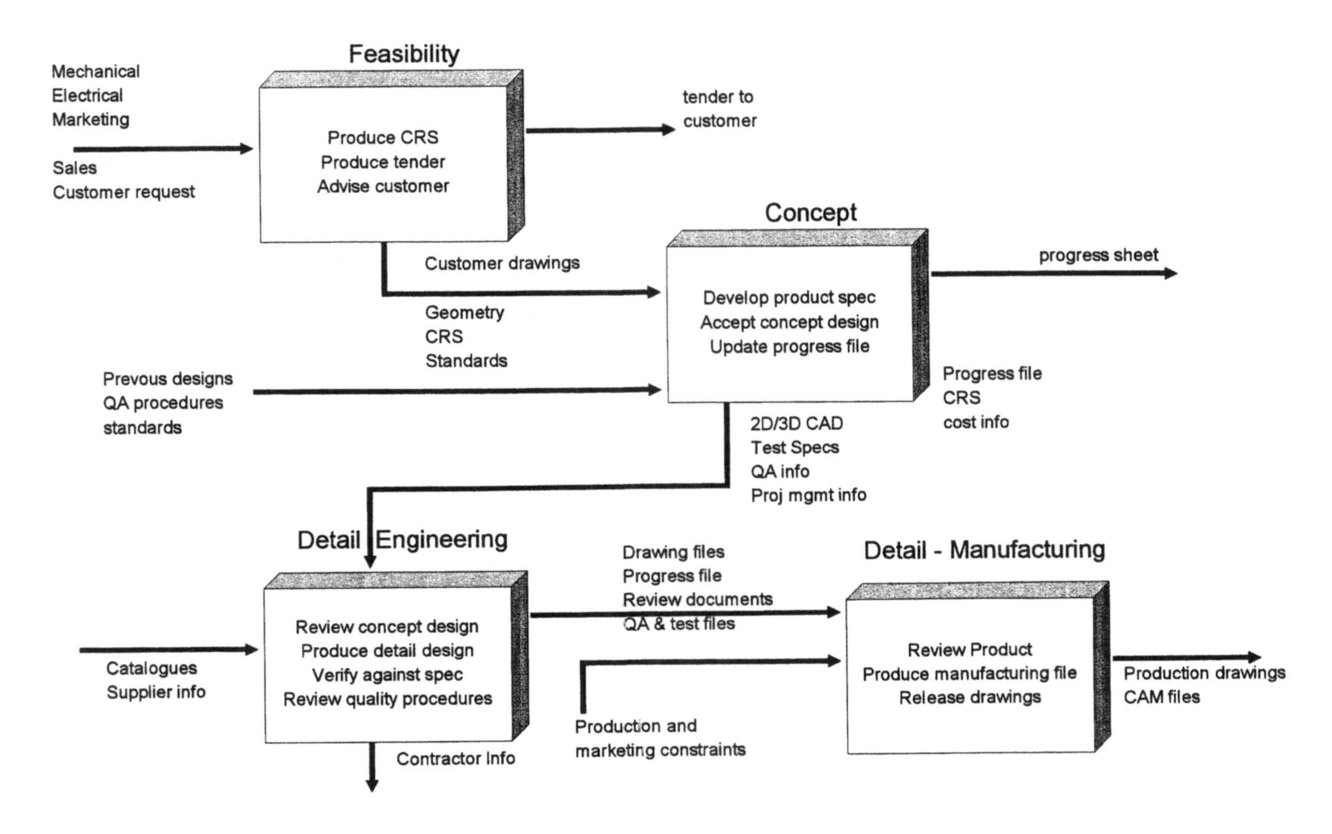

Figure 1 Activity Model

TABLE 1
Types of Data used in the Product Development Process

Data type	Example
Unstructured text	Minutes, Review documents, email
Structured text	Customer requirements, Progress file
Paper drawings	Production drawings, Conceptual sketches
Electronic drawings	3-D geometry
Information resources	Catalogues, standards

The next step was to map the information flows to the functional units that generated or used the data. This brings an organisational viewpoint to a PDM strategy. For example, an electronic CRS may be desired by the design team at the concept stage but if those making an input to its generation, such as Sales and Marketing, are not equipped to participate electronically, then the strategy cannot be implemented. A typical ***Project Information Flow Map*** derived for this case study is shown in Fig.2.

ANALYSIS OF THE SURVEY RESULTS

The activity model for each of the case studies showed that that the design process was very much a paper based system even though there were distinct 'islands' of electronic data production. Data was generated and used in electronic form in the 'obvious' places, eg. complex technical applications such as detail design, where 'off-the-shelf' computer based solutions were available. Generally though, IT infrastructure did not support the electronic communications between all members of the design team, suppliers and partners which is required to enable full use of EDI/PDM.

Although the data exchanged between individuals and departments was almost always paper based it was often electronically generated. Therefore this was where the greatest immediate potential for integration and electronic data exchange exists. EDI would enable accurate up to date information to be available to all individuals involved in the product development process and also to those at the higher management level. There was especially scope for producing the documentary information electronically. This tended to occur more at the feasibility and pre-production stages. Where such documents could be ***structured*** use could be made of word-processing and, particularly, spreadsheets and databases.

However the impact of this is reduced unless the electronic infrastructure and training is available at every desktop of members involved in the product design process, from project coordinators and managers to project team members. As an example, feedback from production and later design stages and mark-up/revision of key documents could be improved if electronic data exchange was implemented but this has to be weighed against the inconvenience and cost of having a VDU on every desk to display the information. Working away from a suitably equipped desk or room would be difficult although this could be overcome by using dialup access from suitably equipped portable PCs using a modem link.

There is clearly a need for training as well as equipment and software procurement and this adds significantly to costs. There are also cultural issues involved. At the senior management level, there must be full commitment to the strategy both in support and financial provision. In other words, management must ***want*** an EDI/PDM strategy to work. At the other end, there must be a cultural change for all employees. An integrated electronic design system must be seen to work effectively and provide real benefits in the workplace. These might be improved creativity, reduction of routine tasks and a perception that the strategy really is improving the performance of the company. As an example,

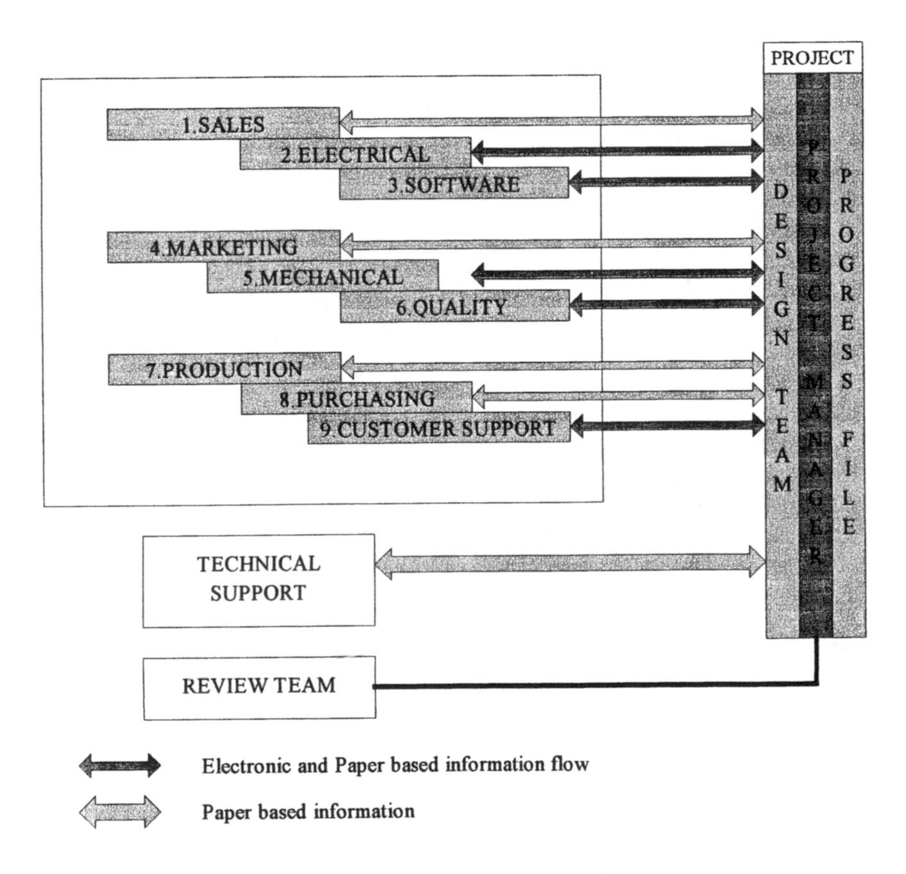

Figure 2 Functional Information Flow map

provision of email facilities will not work if people do not routinely check their mailboxes for messages.

A further problem with an EDI/PDM strategy is that the complexity of the different types of data and information used in the development of a product does not lend itself to an easy implementation. Documents and information are prepared and represented in many different forms. In considering EDI/PDM it is essential to review the available and developing ***standards*** that can be used for information exchange. Some examples are given below together with comments, to illustrate the sort of consideration that must be given to the issue:

> **DXF files**: 2-D CAD *de facto* standard, originating from software vendor AutoDesk. Widely used and accepted but controlled ultimately by AutoDesk. What happens if AutoDesk re-organises its products or is taken over? Is it committed to supporting interchange universally?
>
> **Pro-Engineer files:** Proprietary 3-D CAD file. Could become widely accepted but is not a standard. Pro-E can, in principle, read standard IGES files but an IGES based CAD system probably cannot read a Pro-E file. PTC have built limited STEP support (for 3-D geometry) into PRO-E
>
> **IGES:** the current standard for 3-D geometry. Not suited (and not designed for) handling other types of data. Development frozen and is being subsumed into STEP.
>
> **STEP:** the (slowly) emerging international standard for data exchange. Designed to handle all types of information. Some *parts*, eg 3-D Geometry already specified but much work needed on other parts of the standard exchange specification.
>
> **TRADACOMMS:** the UK standard for exchange of 'document' type information such as orders and invoices. Suitable for commercial departments but no real links to technical data. Will be subsumed into EDIFACT
>
> **EDIFACT:** the international standard for commercial and administration data transfer. Links to technical data exchange are being developed (especially to STEP).

From the above, it is clear that even if the candidate paths for electronic data exchange are identified, a lot of effort must also be put into adopting standards which are robust, will cover the needs and which are stable. Wherever possible, international (ISO) standards should be used but the nature of the standards process is such that this desirable state takes a long time to evolve. The components for a standards-based strategy may thus not be readily available when needed.

DEVELOPMENT OF A DEMONSTRATOR

Work is currently under way to develop a demonstrator for exchanging product development-related data. The aim is that this could be adopted by SMEs with relatively low budgets and limited technical expertise. The architecture of the demonstrator is shown in Figure 3 and is based on a concept that embraces two scenarios:

- An original equipment manufacturer releasing a requirements specification as a set of documents to a number of supply chain companies. The companies will respond with proposals for fulfilling the requirement. This in turn will initiate a subset of the product development process within the supply chain companies.

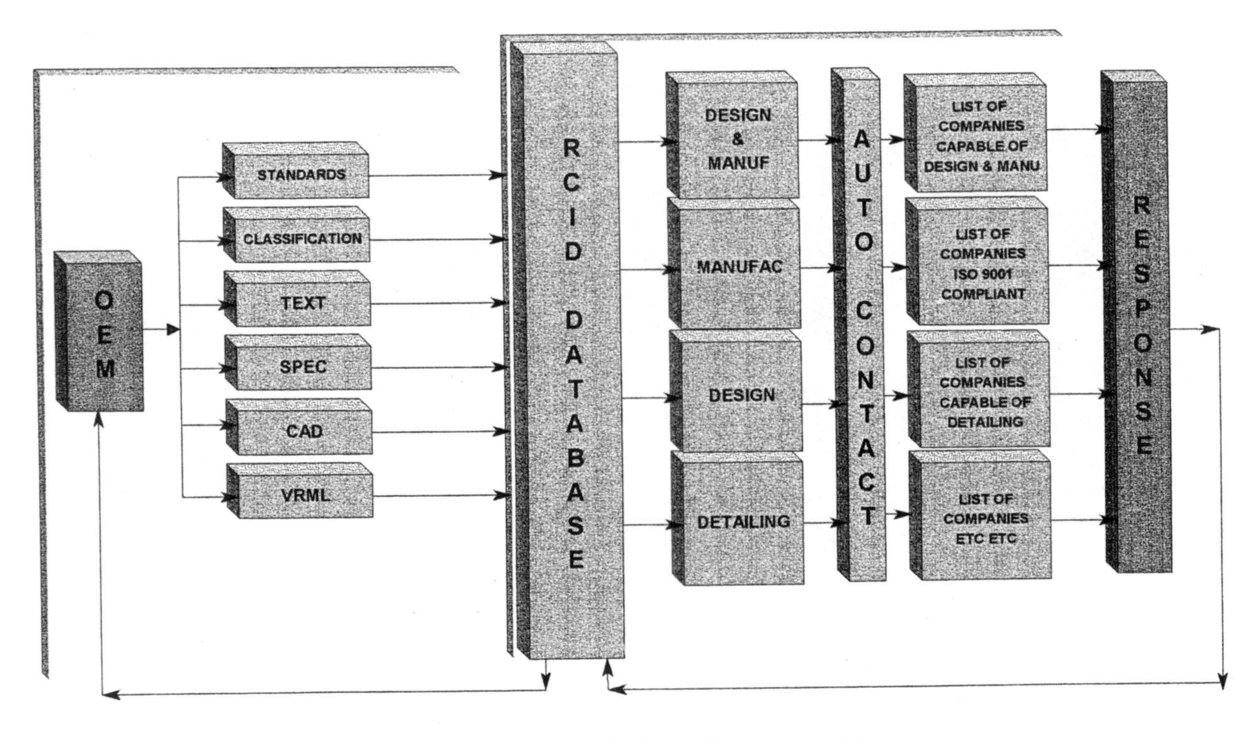

Figure 3 Architecture for Product Develoment Integration

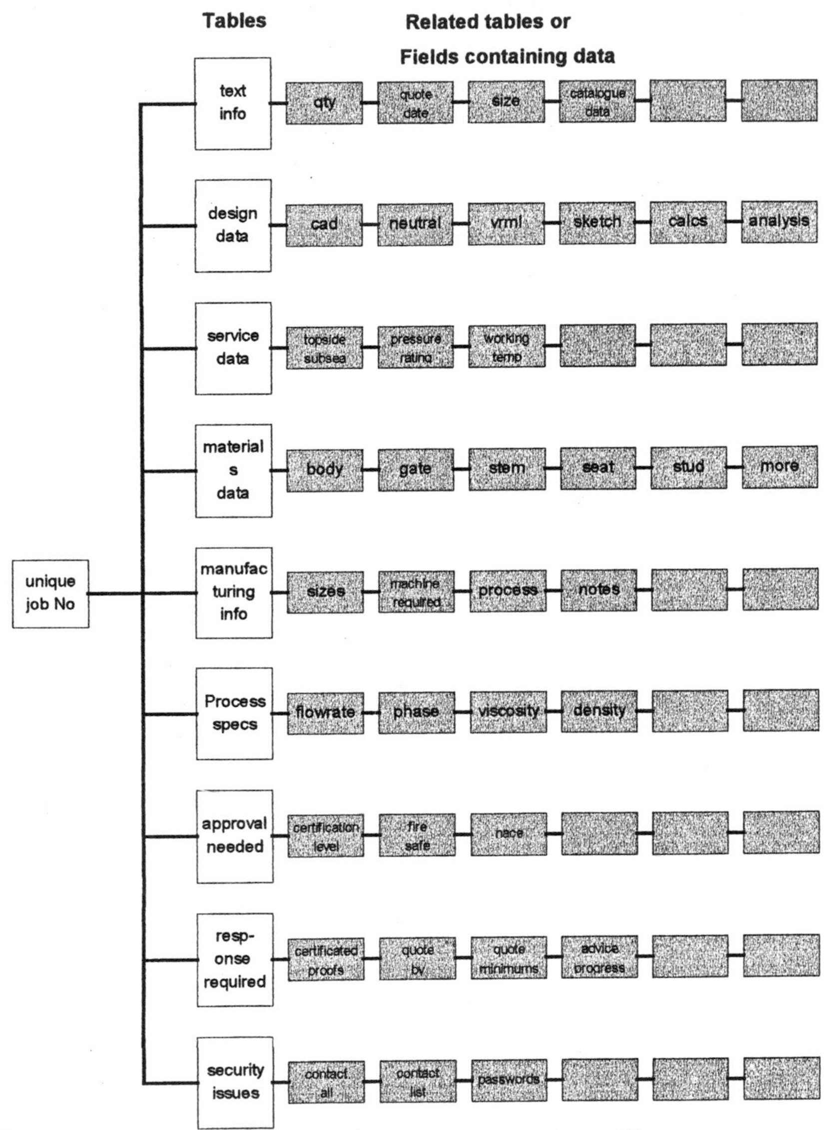

Figure 4 Data Organisation and Types

- A similar scenario could exist in-house with different departments or groups taking the place of supply chain companies.

Given the results of the initial survey of three of the RCID companies, it is likely that the second scenario is more attractive and consequently development of the demonstrator is being given priority in the direction of support for the product development processes internally.

The participating groups, generically called functions (eg. Sales and Marketing, Product Design, etc), are connected to a hub database through any suitable connection ranging from a simple modem to a full LAN connection. The hub database will hold all data relevant to a particular job in a structured form. This might include CAD models, STEP neutral file exchange data, text documents and so on. The data types and database structure are indicated in Figure 4. The notification mechanism is based on email and an email management system is to be developed. This will allow functions which have a need to be activated at particular points of the development cycle to be notified automatically. These functions will then be able to access the information that is specific to their role in a form that is immediately useable. This could be to carry out some specific work or simple to review a document and respond to it. The managed email system will co-ordinate the responses and notify other functions as appropriate. The aim is to retain simplicity and to use technology that is robust and which a typical SME can handle.

CONCLUSIONS

In this paper we have outlined a methodology for establishing ***opportunities*** for integration of the product development process, ie: Structured interview with senior management, Design process diagram, Activity model, Functional information flow map

For the two companies studied, it was possible to define a common top level activity model and this is likely to be common to most of the RCID companies. The methodology establishes the information flow within individual companies. The activity model and functional map, together with information about current support systems, allows companies to identify candidate areas for EDI/PDM. With a clear view of the opportunities highlighted by the analysis, a company can also consider what practices and procedures might be changed in order to maximise the benefits of electronic integration. This, of course, will require iteration of the methodology adopted in this study as an EDI strategy evolves.

Ultimately though, factors such as EDI standards to be adopted, IT infrastructure, training and cultural issues, as well as costs and benefits all influence the priorities in implementing EDI/PDM.

Integration within a company leaves open the question of communication channels with external bodies and may itself represent a bottleneck for improvement of business processes. Inclusion of external bodies in the analysis opens up the potential of improved two-way information flow and the benefits of co-operative working. However, this may require new business practices and an element of mutual trust as partners become linked into a network based on communication. On balance, the rapid advances in technology such as the Internet will act as a powerful driving force for co-operative working.

In order to test the realisation of potential benefits of EDI/PDM in SMEs, a demonstrator system is being built that could be implemented in a typical SME. Work is currently in progress to take the architecture which has been defined into one or more of the RCID companies. The requirements for a particular instance of such a system are being derived from the results of the study reported here.

ACKNOWLEDGEMENTS

The authors would like to acknowledge the contribution to this work that was made by Dr.P.Hackney of Northumbria University and John Richardson of Express Engineering.

REFERENCES

1. **Bytheway, A.,** EDI:Managing the Costs and Benefits, *EDI-91, Proceedings of the National Conference on Paperless Trading,* Birmingham, UK, 29-31st October, 1991, 60-71

2. *The Product Data Workbook,* UK, Published by the Department of Trade and Industry, 1994

3. **French, M.J.** *Conceptual Design for Engineers,* Designe Council/Springer-Verlag, London, 1985

4 *Systematic Approach to the Design of Technical Systems and Products,* Verband der Deutschen Automobilindustrie, Franfurt, Germany.

TOTAL INTEGRATED MANAGEMENT SYSTEM
A NEW APPROACH TO ORGANIZATION IT PLANNING

Mohammed Kazem Farhang Kermani
Industial Engineering Department,Amirkabir Univeristy of Technology,Tehran-Iran
Fax (9821)641-3025

ABSTRACT: The challenge to todays' managmenment is to achieve success in a changing world using precious information.This understanding fits to the western countries, where computer has been used since 1950s. Most IT planning models make certain assumptions not applicapable to the third world countries where undocumented manual systems are prevalent. However, many managers now realize the pressure toward integration via inforamtion.

An approach, Total Integrated Mangement System (TIMS), is conceptualized to change the picture. TIMS provides management supports in three management dimensions called **roles, functions,** and **operational tasks**. This support is done through identifying management functions which should be performed on *procuring, producing, maintaining,* and *distributing* the resources used in the organization

TIMS scrutinizes the manual systems to segregate reengineered manual and computerized activities. This paper presents a definition of TIMS, the planning process using BPR principles, supporting systems, and justification.

INTRODUCTION

There are definitely a global pressure toward integration via information. Even though the pressures may seem to be different in nature, some where along the long the lack of information as the root of the problem can be recognized. The pressures can be analyzed through three main sources named trends in business, world new order, and organization & management changes. A brief description of These sources will point out that information is the key to the survival.

Trends in Business

The challenge to management of organizations today is to achieve success in a world that is changing daily. There are 11 dominant trends identified by Senn [1] and in all of them the importance of information and the way information systems are changing organizations are quite clear. These trends are:

- Blurring of Industry Boundaries
- Deregulation of Industries
- Faster pace of business
- Increasing Foreign Competition
- Global Business Community
- An Information Society

- Increasing Complexity of Management
- Interdependence of Organization Units
- Improvement of Productivity
- Availability of Computers for End-Users
- Recognition of Information as a Resource

Well established organizations with already working systems are able to better analyze the business requirements and come up with sound solutions to cope the pressures. On the other hand, majority of organizations in the third world countries still work with the production rather market oriented philosophy. This with the new requirement set forth with World New Order compounds the management complexities of these organizations.

World new order

To plan, there are two issues that must be considered. The first one is that, planning assumes a ***relative stability*** on the area that planning is being done for. The second issue is that, planning requires ***information*** as a prerequisite. To bring the above two issues to practical reality, *communism*, as the major obstacle to stability, has been removed from the face of the earth. And the trends show that various activities such as Middle East Peace Process, Balkan Issue, lessening Apartide problems, and so on, are under way or they have already started, in order to politically ***Unify the World***.

A term called Global Market or World Market has been designated to handle the business and commerce. If a country wants to do business in this market, she has to obey the rules and regulations set forth for the benefits of every one. The means to achieving this objective has also been provided. An idea which started almost fifty years ago, now has flourished and it has become a major business forum - World Trade Organization. Customers need to become sure of the quality of a product in this market. Therefore, a standard practice of doing business was needed. This requirement is now satisfied by ***ISO Certificates,*** *a document which ensures a customer how an organization takes quality standards into its daily management practice.*

Information is another aspect which has already been tackled. The creation of Internet, and Information Super highways will provide a means to collect, process and distribute information and interact with others regardless of the geographic boundaries. And this is what the new world order is supposed to be. The adjustment needed, to become regulated in this environment, is the added pressure that organizations of the third world are severely touched.

Organization and Management Changes

Changes in the business environment and in technology induce organizations to change the manner in which they operate. It has long been recognized that there are strong relationships among the environment, technology, organizational structure, people in the organization, organizational strategy, and management processes. Significant changes in the environment are likely to change the equilibrium in internal parts of organizations. There is already evidence of some of these changes in many organizations. Changes in one company may impact other companies, creating more business pressures. The major factors that contribute to the pressures are as follows[2]:

- Business alliances.
- Decreased budgets of public organizations
- New management concepts.
- Time-to-market
- Empowerment of employees and collaborative work
- Customer focused approach

The above factors also add another dimension to the pressures that management has to cope with

Survival Key

To survive in this competitive world market, An organization needs to have an acceptable product or services with a reasonably accepted quality and offer that with an acceptable price to customers while maintaining high productivity. In order to achieve these conflicting objectives, organizations have to be equipped with the best information available to them. This information encompasses environments that the organization is competing in, such as *Customers, Suppliers, Competitors, and Government agencies.* Collecting data, processing them and producing precious information on a timely manner is the key to survival of the companies in upcoming years. This Goal can not be achieved unless information technology takes a top organizational or even national priority.

TIMS CONCEPTUALIZATION

The pressures mentioned earlier have certainly changed the management focal point from *production* to *market* orientation. This refoucsing needs understanding about the customer, its needs, and actions required by management to satisfy the needs. In order to conceptualize TIMS, a brief description of quality oriented customer, the business process reengineering needed to change to quality attitude, and then the fundamental management functions needed to achieve the integration are described.

Quality Oriented Customer

Today's market is known to be a customer market. The slogan of *"customer is king"* can be found all over. In this section the customer needs and its related issues as a driving force for integration is explored.

Future Perfect. Due to no competition, and be as a sole public or private organizations in the country, management in the third world organizations has never paid any attention to customer before. On the other hand Stanley Davis [3], in his book titled future perfect cites a simple yet powerful concept. The basic tent of future-perfect vision is that technology is getting better, and better, which means that business process can get better and better. Taken to this logical conclusion, technology will get perfect. Therefore, companies, should develop a business of perfection. What does "perfection" look like? Customers can get what they want ***any time***, ***anyplace***, ***anyway*** they want it, and with the ***quality*** they desire. Quality, or lack of it, has become a major issue worldwide in virtually all industries. Other than increasing productivity, perhaps no other general management issue has received so much attention. And, as many quality experts point out, poor quality is a major cause of low productivity. Thus, this is a new goal that the management should attain. It is definitely a new challenge, since it must be achieved in an environment that the customer concerns have never been taken into account before. Thus a new management philosophy using technologically available tools must be developed.

Mass customization. One of the most innovative concepts of the Industrial Revolution was mass production. The basic idea is to manufacture a large quantity of goods which are placed in inventory and then sold to unknown customers. A major Change in manufacturing developed with the rise of automobile, which resulted in waiting time of several weeks or even months. Today's customers are not willing to wait so long. Therefore, Mass Customization may be essential to the survival of companies in the 1990s. The basic idea is to enable a company to produce large volume (mass), yet to customize the product to specifications of individual customers. Mass customization enables a company to provide flexible and quick responsiveness to customer's needs, at a low cost and with high quality. Mass customization is made possible by allowing fast and inexpensive production changes, by reducing the ordering and sales process, by shortening the production time, and by using prefabricated parts. All of these functions require new methods and information which must be developed soon, and this is another challenge for management of the third world countries..

Business Process Reengineering

The pressures, described earlier, plus new customer concerns create a business environment in which many companies as well as public organizations have difficulty operating or even surviving. The reason is that both the magnitude and pace of the changes around the business are much stronger and faster than ever before. The method that hammer and champy proposed to execute such change was BPR, which means starting all over, starting from scratch. The major pressures are summarized by Hammer and champy [3] as three Cs : ***customers, competition, and change.*** To cope the Cs, they provide the following definitions which are used in TIMS process identification:

- **Reengineering** is the fundamental rethinking and radical redesign of business processes to achieve dramatic improvements in critical contemporary measures of performance such as cost, quality, service, and speed .
- **Business process** is a collection of activities that takes one or more kinds of inputs and creates an output that is of value to the customer. For example, accepting an application for a loan, processing it, and approving (or rejecting) it, is a process in a bank

Thus, when a system is designed , especially when a BPR is taking place, we need to now the followings for every step of the process or operation:

- What needs to be done
- Who has to do it
- Where is it being done
- Why is it being done this way
- Which sequence of steps is needed
- How is it being done
- Input information needed and from where it is coming from and by what means
- Output information produced and to where it is going to and by what means

The information required or produced can vary from an item number from a file , or specific reports to standards, quality objectives or even a complete manual. And at the same time, The source and sink of the information can vary from a manual file cabinet, to a computerized file handling system, to specific department in organization or it can be an agency outside of the organization.

To implement BPR. information systems become essential to organizations and they must be integrated. The failure to approach information systems in an integrated fashion through Reengineering process results in high speed, automated versions of existing manual systems which will collapse soon. Integration approach suggests that all informational and operational systems be viewed as one total system. However, the dimension of this system is not defined uniquely by the authorities. The next section examines how TIMS can help the management to reach at total integrated solution.

Management Integration

Management is a process by which certain goals (outputs) are achieved through the use of resources (inputs) such as people, money, energy, materials, space, and time. The degree of a manager's success is often measured by the ratio between outputs and inputs for which she is responsible. This ratio is an indication of the organization's ***productivity.*** To understand how computers support managers in achieving their goals, it is necessary to first describe what managers do. One of the classical studies was done by Mintzberg [5], who divided the manager's roles into three categories .

1. Interpersonal roles : figurehead, leader, liaison .
2. Informational roles : monitor, disseminator, spokesperson.
3. Decisional roles : entrepreneur, disturbance handler, resource allocator, negotiator .

Early information systems mainly supported the informational roles ; in contrast, the purpose of recent information systems is to support all three roles. From a different perspective, managers traditionally are responsible for the functions of *Planning, Organizing, Staffing, Controlling,* and *Communicating.* These functions are executed through some operational tasks. The tasks, which should be performed on the resources used in the organization, are called *Procuring, Producing, Maintaining, and Distributing.* For examples, management should *plan* for *maintaining* a specific resource such as *material*. Therefore, the idea of Total Integrated Management System (TIMS) is to provide management supports in three dimensions as :

1. *Management Role*
2. *Management Functions*
3. *Management Operational Tasks*

As it is clear, IT plays a major role in this integration process which can be understood and designed using Reengineering principles.

Tims Definition. In this concept, we assume that whatever managers do, is in the direction of an organization goals through the best utilization of resources. This can be accomplished by assuming that managers are functioning toward executing activities of procuring, producing, maintaining, and distributing of resources to satisfy customer needs. This means that a total management system, if needs to be integrated, must take the followings into account:

- ***Management functions.*** They can be grouped into 4 areas of: Planning, Actioning, Evaluating, and Correcting.
- ***Operating functions.*** These are Procurement, Production, Maintenance, Distribution.
- ***Resources.*** Management function through operating functions is supposed to use resources efficiently. These resources in general can be grouped into 4M-EI namely as :Man, Machinery, Material, Money, Energy, and Information
- ***Information type.*** On the other hand as it was mentioned earlier, management in order to do its job or to make decision, needs various information and reports which can be categorized as: Comfort information, Status or Progress information, Warning information, Planning information, Performance information, External intelligence, Externally distributed information.
- ***Management roles.*** The above information must be suited for different management roles of interpersonal, informational, and decisional roles.

The integration of the above five dimensions are graphically depicted as shown in figure 1.

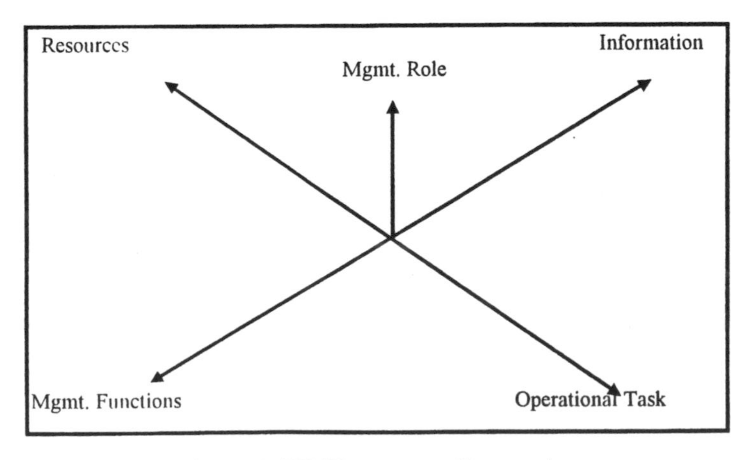

Figure 1- TIMS conceptual integration

TIMS Support System. When it comes to TIMS, it is hard to understands completely how far the integration is necessary. Developed countries, who have been using computers for at least over three or four decades, have already established the basis for computer technology applications. In an organizations, the following activities occur:

1. Occurrence of an event called Transaction or process
2. Collecting of data related to transactions
3. Processing information or transaction
4. Distributing information
5. Acquiring a proper and selective reports or information for decision making
6. Making decisions based on the obtained information
7. Communicating the result or impact of the decision

To bring the TIMS concept to practical reality, 5 major tools or support systems, which complement each others, as shown in figure 2 are required for integration.

- Transaction Processing Systems (TPS)
- Management Information or Reporting Systems (MIS),and (EIS) called MEIS
- Technical Support System (TSS) such as CAD, or different computer-base analytical tools
- Management Decision Support Systems (MDSS) which can include DSS, ES, ANN
- Communication Support Systems or Group Support System (GSS)

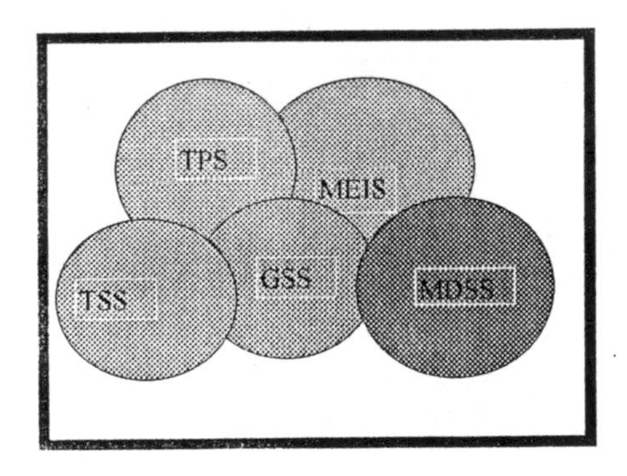

Figure 2- TIMS supporting system

Now the question is, do we want to, or really need to integrate the above 7 activities in order to have a TIMS? In other words, When do we say that we have TIMS? The answer depends on the characteristics of the organization as follows:

- Current available IT
- Information and system culture
- Management knowledge of information-based-organization
- Management desire to achieve integration

But in general, an organization can progress toward total system integration according to the time horizon shown in figure 3.

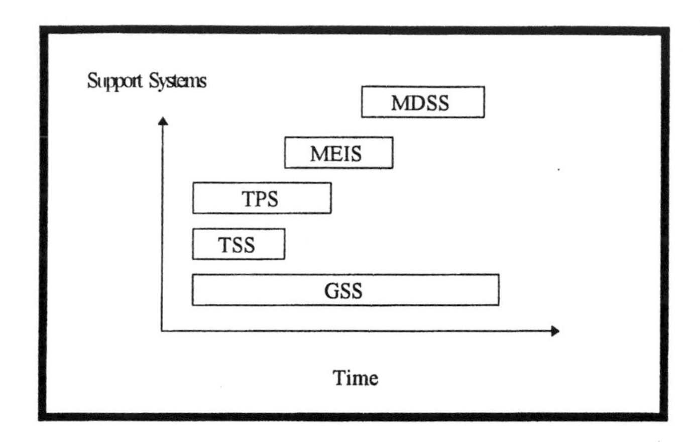

Figure 3- Systems support development horizon

TIMS Prerequisite. In order to achieve goals set for TIMS, commitment is the main prerequisite for successful implementation. A lot of top executives always talk about systems and lack of proper information for running the company. But when the actual process begins, implementation of IT applications takes a back seat, even when the top management says she is committed to using IT. Why is this so? What do we mean by commitment. In fact, Commitment exists when the following can be acquired from the management.

1. Management knowledge. Management must be aware of the benefits of IT to the success of the organization. He must ***feel*** that if the implementation does not occur, the future of the company may be in jeopardy.
2. Time consuming. Management must realized that system implementation takes time. For example 2-3 years project, is not uncommon for a system development project.
3. Money appropriation. Management must realize that system implementation requires a lot of resources in terms of human, software, and hardware resources. All of these will cost the company money, and the money must be available at the time it is needed most.

Even though, the above three factors may not be applicable to developed countries, but it is still a major obstacle in third world countries where information usability culture is still at the growing stage.

TIMS PLANNING PROCESS

A generic information systems planning model is called the **four-stage model of planning.** It consists of four major, generic activities; strategic planning, requirement analysis, resource allocation, and project planning. This separate planning procedure for the managers of the third world organizations where they have just started to feel global pressure toward integration via information can not work. IT planning in these organizations where they have problems even on setting business strategic planning, is quite a challenge. Therefore, these four phases of the 4-stage model are done simultaneously.

TIMS Analysis Methodology

To scrutinizes the manual systems and obtain the current and future organization information needs, a survey analysis consists of the following steps is proposed to be conducted by a team of analyst to segregate ***reengineered*** manual and computerized activities.

Step 1. Identify functional departments of the organization by evaluating the organization chart.

Step 2. For each functional area identify managerial level consists of Vice president, directors, and section manager.

Step 3. Schedule an interview with each of the managers.

Step 4. Prepare for the interview with the intention of gathering the following information.

.1-*Top management.*

- List of major customer concerns.
- List of concerns related to the environment of the organization
- List of major decision making activities and its related information.
- Identify the degree of automation.
- List major control and coordination needed and their type.
- Identify type of internal group communication and external communication.

2-Directoral level

- List of major means of evaluating department performance.
- Identify timeliness measure of information.

3-Managerial level

- List of major activities done in each department.
- List of major data and its sources required to do the activity
- List of major data/information produced in each activity and the destination of that data.
- List of potential problems related to each activity (system, operation, information, staff,...)
- List of potential functions or processes, reports, and controls needed .
- List of major statistics
- Sample of all forms, report and major departmental standard operating procedures
- General process flow chart of activities.

Step 5. Compile and analyze the data in terms of the following:

- Accuracy of information and resolving conflicts.
- Customer identification
- Customer concerns

Step 6. Regroup the processes through Reengineering

- Regroup the processes to form a complete cycle of inputs to outputs
- Eliminate the processes which can be handled by IT using Reengineering principles

Step 7. Build a Creation/Usage matrix

- Identify which group of data is created by which processes
- Identify which group of data is used by which processes
- Create Process-data matrix and determine the score of each data group called DS as follows; During the interview, managers are asked about both the importance and the current availability of different type of information. Responses for both importance and availability are recorded as high, medium, or low. A score is computed for each category of information, according to the following formula:

Score = Importance * Availability

Table 1
Importance/ availability table

	Low	Medium	High
Importance	1	2	3
Availability	3	2	1

Step 8. Build priority table

- List data group
- Identify predecessors and successors of each data group
- Count number of predecessors of each data group
- Assign number 3 for each predecessor. Therefore, if a data group has 3 predecessors, its score will be 3 * 3 = 9 (called PS)

Step 9. Identify systems

- Regroup the data group to form logical system process according to the definition of process mentioned in BPR
- Calculate System Score (called SS) using the following formula , where i's indicate the system number

$$SS = \sum_i DS_i * PS_i$$

- Prioritize the system according to their SS.
- Determine the categorization of each Identified systems such as TPSs &TSSs &..

Step 10. Prepare planning documents

Planning Generated Document for TIMS

To properly document the process of system planning, the following documents are advised to be generated:

Organization Survey Report (OSR). In this report , all the related information regarding the physical activities of each department will be documented. The following topics should be covered in this report:

Departmental documentation including DFD's from department context down to 2 levels; list of inputs, its origination, and its type; list of outputs, its destinations, and its type; list of problems cited, list of requirements cited, list of databases; **Organization machinery** including list of equipment, locations, and general capabilities and **List of hardware & software inventory; General Statistics** including number of different transactions, number of customers, number of jobs, ...

Survey Analysis Report (SAR). In this report, the information gathered during the survey period will be analyzed. This analysis will be used for determining the IT plans and systems. The following topic should be covered in this report:
List of analyzed problems classified in terms of operational, informational, systems, organizational, and personnel; **List of analyzed requirements** classified in terms of operational, informational, systems, organizational, and personnel; **Functional analysis of activities** based on the matrix of management functions, resources used, and operational activities.; **Functional processes identification** focusing on customer requirement life cycle.; **Processes' data identification** including the data needed and generated for each processes.

General System Specification Report (GSSR). Based on the analysis described in SAR, general system specification with the following topic should be generated:
System classification indicating TPSs TSSs,...; **System context diagrams** including organizational and functional systems at two levels..; **Systems detail specifications** including the DFDs of subsystems and functions, and major manual and computerized databases, inputs and outputs; **System linkage** including the list of functions performed manually and automated; **System alternatives** including the list of alternatives in terms of degree of automation or type of IT architecture.; **Alternative financial analysis** including the estimated cost and benefit for each alternative; **System Organization Matrix** indicating which department will contribute to each functions.; **System plan** including the gantt chart of system implementation and phases

Computerized Systems Specification Report (CSSR). In this report based on the selected alternative in GSSR, the specification of computerized section with the following topics should be outlined:
List of systems including the various TIMS supporting systems; **Systems priorities**
System DFDs including the DFDs at the subsystems, and functions levels.; **I/O list** including the list of inputs and outputs of each systems.; **Organizational databases** including the list of databases which will be shared throughout the organization; **System architecture** including general hardware and software required; **Buy /build analysis** including the cost and benefit of buying off-the-shelf packages; **System financial analysis** including the cost and benefit in more detail than in GSSR.; **System implementation plan** including gantt chart and financial expenditure plan; **System-Organization Matrix** indicating which part of each system is created or used by which department.

TIMS JUSTIFICATION PROCESS

Although there are various IT justification models available in the literature, TIMS can be justified based on two components: the cost savings through ***internal cost reduction***, and ***competitive advantage***. The first component can be calculated based on the savings through use of IT in the area of continuous and investment cost of **management functions** (of planning, actioning, evaluating, and correcting) on executing **management operational task** (of procuring, producing, maintaining, and distributing) on **resources** used in each department. The second component, competitive advantage, can bring additional savings to organization through the use of IT. The Savings come from two elements of **capability of remaining competitive** through increasing quality, increasing units sold; and **salvaging new opportunities** through identifying new local and international market.

CONCLUSION

This paper presented challenges that organizations of third world countries face today due to changes coming from different sources. To cope this challenge, a concept called Total Integrated Management System-TIMS was introduced. The components of TIMS described, and the planning process to achieve the integration proposed by TIMS including the required documents was presented. In addition, the cost savings needed to justify implementing TIMS were explored.

REFERENCES

1. **James A. Senn**, Information Systems in Management, 3-8 , 1990
2. **Turban.McLean. Wetherbe**, Information Technology for Management-Improving Quality and Productivity, 17-19,459-460, 637,1996
3. **Stanley S. Davis** , Future Perfect, 1987
4. **M. Hammer and J.Champy**, Reengineering the Corporation, NewYork:Harper Business, 1993
5. **H. Mintzberg**, The Nature of Managerial Work, 1973

DESIGN SOFTWARE FOR SPECIFICATION AND ANALYSIS OF DISCRETE EVENT MANUFACTURING SYSTEMS

T.O. Boucher, M.A. Jafari, R.C. Wurl, A. Yalcin
Department of Industrial Engineering
Rutgers University
Piscataway, NJ 08855-0909

ABSTRACT: This paper describes software for the specification and analysis of discrete event manufacturing systems. The specification is done using the IDEF0 modeling formalism. The specification is a static model of the system. Dynamic system analysis is accomplished by converting the IDEF0 representation to a Petri net and then simulating the behavior of the system. The contribution of this work is a methodology and its implementation in software.

INTRODUCTION

We are concerned with the design of automated production systems and their controllers. We propose three integrated levels of design activity: specification, performance analysis, and control logic design. For specification we use the modeling methodology of Structured Analysis and Design Technique (SADT) [4] or its variant, IDEF0. For performance analysis, we use Petri nets [2]. For control logic design we use a GRAFCET like procedure for controller specification, with automatic code generation from the specification diagram [1,3]. In our work we have brought these design functions together within the same software tool, called IDEF / System Dynamics Software. In this paper we will describe the process of system specification and analysis as implemented in this software.

BUILDING THE IDEF0 MODEL

The building blocks for an IDEF0 model, shown in Figure 1, are the *activity box*, *input arcs*, *output arcs*, *mechanisms* and *controls*. The *activity box* defines a specific activity, or function, that is being modeled. The activity may be a decision-making or information-converting activity or a material converting activity. *Inputs* to the activity are shown as arcs entering from the left of the activity box. Inputs are items (material, information) that are transformed by the activity. *Outputs* of the activity are shown as arcs exiting at the right of the activity box. Outputs are a result of the activity acting on the inputs. *Controls* are shown as arcs entering the activity box from the top. A control is a condition that governs the performance of the activity. For example, a control may be a set of rules governing the activity or a condition that must exist before the activity can be done. *Mechanisms* are shown as arcs entering the activity box from the bottom. A mechanism is the means by which an activity is realized. For example, the mechanism may be a machine or a worker.

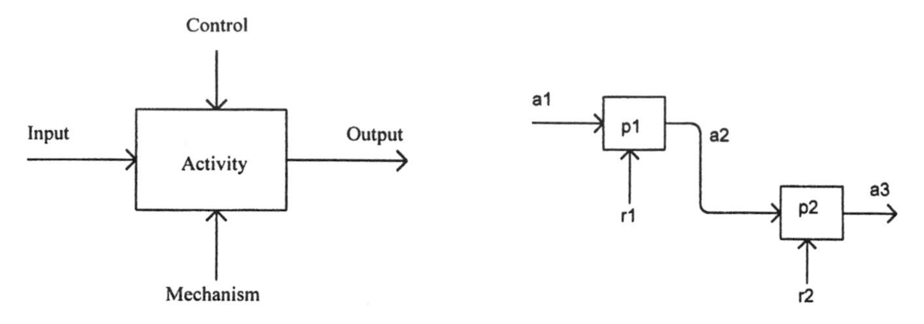

Figure 1. IDEF0 Modeling Components

Figure 2. "TEST" IDEF0 Model

The activity box and the four entities of Figure 1 provide a concise expression: An *input* is transformed into an *output* by an *activity* performed by a *mechanism* and governed by a *control.* The specific activity, its inputs, outputs, mechanisms and controls must be defined for the situation being modeled. Activity boxes represent actions being performed and are labeled with verb phrases. Inputs, outputs, controls, and mechanisms are things and are labeled with noun phrases.

A sequence of related activities can be described by an IDEF0 diagram, such as that shown in Figure 2. Here, two activities are related in sequence. The output of activity 1 becomes the input to activity 2. Any number of activities can be related to each other in order to describe a complete system. We will use Figure 2 as an illustration. We have adopted the following default labeling convention for activities(p), arcs(a), mechanisms (or resources) (r), and controls(s).

In order to illustrate the input conventions used with the software input screen, we will refer to the diagram of Figure 2. The input data that represents that diagram will be shown in Figure 3. The following describes the process by which the model of Figure 2 is input as data in Figure 3. The user begins with the "Activities" input box, Which is located in the upper center of Figure 3. By clicking on the "New" command button, a dialog box appears to ask for the number of activities. When the user responds, the activities are labeled consecutively from p1 to pn and appear in the activities box as shown. The user can "Add" and "Remove" from the activities list at any time by using the "Add" and "Remove" command buttons of the activities box. For the IDEF0 model of Figure 2, there are 2 activities. To the right of the activity is the "capacity" row. This is used to place a bound on the number of inputs that can utilize an activity simultaneously. For example, if the activity is a machining operation and the machine tool is capable of machining only one part at a time, it is necessary to enter a "1" in the capacity cell next to the activity. If the activity is a transportation activity and the transport device is capable of holding only 2 units at a time, a "2" is entered in the capacity cell next to the activity.

The user then moves to the "Arcs" input box, shown at the upper right of Figure 3. In a manner similar to entering the number of activities, the number of arcs are input. In the IDEF0 model of Figure 2 there are 3 arcs. Arc labels from a1 to am are automatically provided. The user must then enter the relationship between activities and arcs in the arcs input box. If an arc is an *input* to an activity, the user must put a "1" in the cell representing that arc/activity combination. If an arc is an *output* of an activity, the user must put a "-1" in the cell representing that arc/activity combination. If there is no input/output relationship between an arc and an activity, the user leaves the cell blank, which is a value "0".

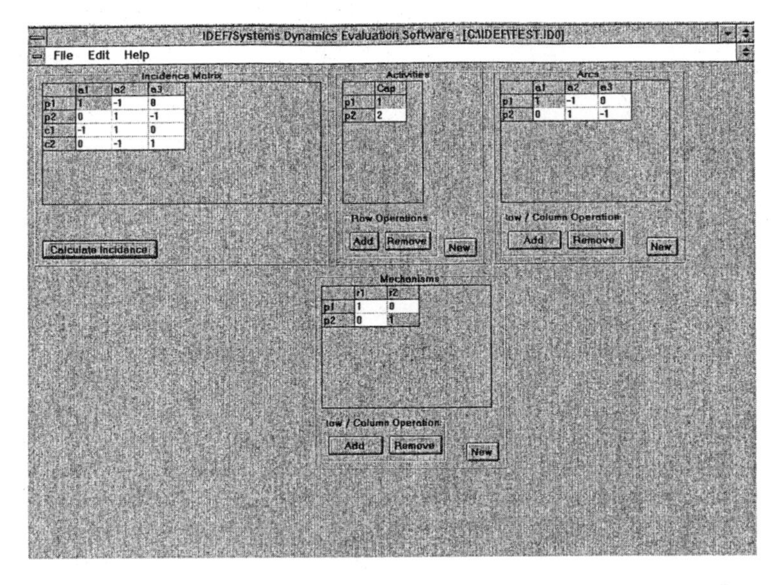

Figure 3 Input for "TEST" IDEF0 Model

The user next moves to the "Mechanisms" input box to input the number of mechanisms in the same manner as the activities and arcs input box. In the case of our example problem, there are 2 mechanisms. If a mechanism services an activity, the user puts a "1" in the cell that shows the relationship between a particular mechanism and activity.

The controls structure of the IDEF0 model is added using "signals" input boxes, to be discussed later. The "TEST" IDEF0 model of Figure 2 does not have controls. Even if controls were indicated, the control structures are unnecessary for the simulation model, which is used to show the movement of input and output flows as well as the utilization of resources. The control, or signal structure, is appropriate to the development of a system controller, to be discussed later.

The user now moves to the incidence matrix, which shows the relationship between the entities of the IDEF0 model. When the user clicks on the "Calculate Incidence" command button, the arc labels appear across the top of the box and the activities labels and the mechanisms labels appear down the rows of the box. The cells are automatically filled with data that is derived from the information previously entered by the user. This "incidence matrix" is a data structure representation of the complete IDEF0 model.

The user will note an additional symbol in the incidence matrix. The symbol "c" is a capacity holder, which is derived from the information entered by the user in the "Activities" box. If the user declares that an activity will have a finite capacity associated with it, a capacity holder will be assigned to the activity. It's purpose will be apparent when we discuss the simulation model.

THE SIMULATION MODEL

The simulation screen, shown in Figure 4, has several input boxes and information displays as follows: "Incidence Matrix", "Initial Marking", "Previous/Current Marking", "Run Control", "Timing Vector", "Database Insert", and "Firing Vector".

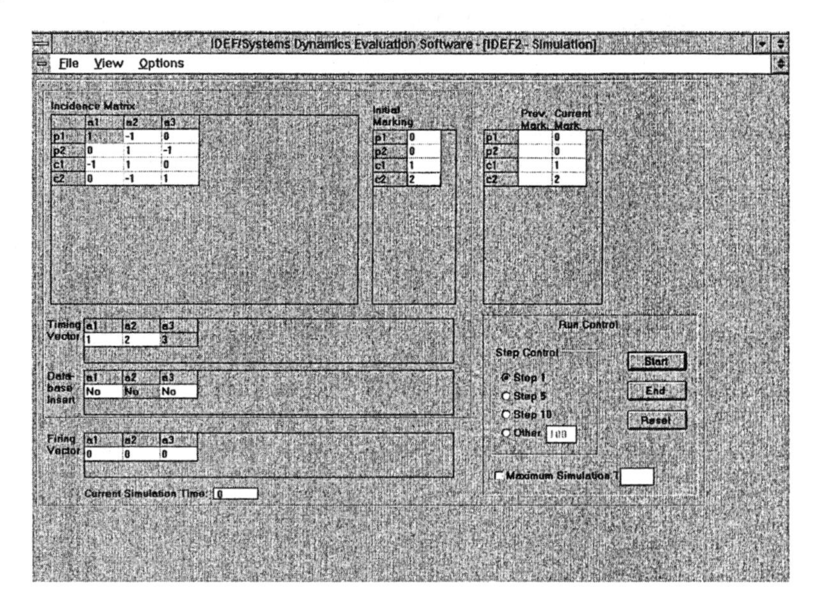

Figure 4 IDEF/System Dynamics Simulation Screen

The "Incidence Matrix" is the data structure of the simulation model. It is derived from the "Incidence Matrix" (data structure) of the IDEF0 model of Figure 3, but it is not necessarily the same. For the "TEST" example, the IDEF0 incidence matrix and the simulation incidence matrix are identical. The simulation model uses the modeling formalism known as Petri nets. A Petri net is a bipartite graph having two types of nodes, known as "places" and "transitions". These nodes are connected by arcs. A graphical model of a Petri net uses circles to represent places and bars to represent transitions. When the IDEF0 model of Figure 3 is converted to a Petri net, the resulting model structure is as shown in Figure 5. Here you can see that the activities of the IDEF0 model have become the places of the Petri net and the arcs of the IDEF0 model have become transitions of the Petri net. In the incidence matrix of the simulation model, the input places of a transition are labeled with a "-1" and the output places of a transition are labeled with a "1". The capacity holder for an activity is shown as a complementary place across the place that represents the activity. The significance of this will be illustrated shortly.

Before a simulation can be executed, it is necessary to initialize the model. The model is initialized by marking the active places. A place is marked by putting a token in it. For example, In Figure 5 there are two capacity places, c1 and c2. Available capacity is indicated by marking these places with tokens that equal their capacities. Presumably these capacities must be present for the system to be initialized. In the "Initial Marking" input box of Figure 4, we place a "1" next to c1 and a "2" next to c2.

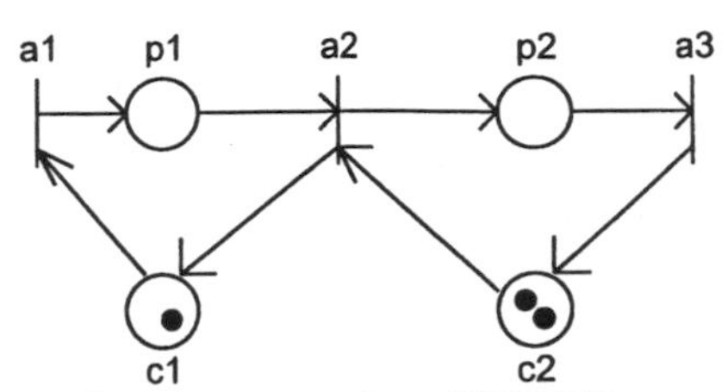

Figure 5 Petri net Representation of "TEST"

Before a simulation can execute, it is necessary to provide timing information on how long it will take to perform an activity. There are two ways to do this in the Petri net modeling methodology. The time can be directly associated with a place or it can be associated with the output transition of a place. For example, if a time of 10 time units is associated with activity p1 of Figure 5, it would indicate that that activity takes 10 time units. Alternatively, if a time of 10 time units is associated with transition a2, it would indicate that activity p1 must be marked for 10 time units before it is completed. Associating time with places or with transitions can result in the same interpretation. In the "Timing Vector" input box of Figure 4, it can be seen that time is being associated with transitions. The user must enter this timing information by placing the number of time units in the "Timing Vector" input box.

We can now describe the dynamics of the simulation model. From the given initial state, a Petri net changes state by firing one or more of its transitions. A transition fires when it is enabled AND its timing vector has expired. A transition is said to be enabled when all of its input places are marked. When the transition is enabled, the timing clock associated with that transition begins to run. When the clock expires, the transition fires. When the transition fires, it removes the tokens from its input places and puts tokens into its output places. Thus, the new state of the system is shown by the new marking of the places.

In order to run the simulation, the user enters the "Run Control" box. There are two parameters to enter. Under "Step Control", the user selects the number of changes of state that should occur before the simulation pauses. The current selection will appear as a black dot in the circle. In Figure 4, "Step 1" is selected. If "Step 5" is chosen, the simulation will run through 5 changes of state (transition firings) before it pauses. The user can enter any desired number of steps in the input box next to "Other". Under "Maximum Simulation", the user enters the maximum number of steps the simulation should run before it is finished. If this option is not selected, the simulation will run until it is ended by clicking the "End" button. At any time during the simulation, the simulation can be "Reset". Clicking the "Reset" button will reset all simulation counters to zero. After choosing the "Step Control" and "Maximum Simulation" options, the user begins the simulation by clicking "Start".

As the simulation runs, the simulation statistics appear on the simulation screen. The "Current Simulation Time" block at the lower left of Figure 4 is a running total of the amount of simulation time that has elapsed since the simulation began. The "Previous Marking / Current Marking" block at the upper right of Figure 4 shows each place (activity) that is marked (in process) at the last step and current step of the simulation. Since Figure 4 is the state of the simulation at current simulation time t=0, there is no "Previous Marking". The "Firing Vector" shows the number of times that transitions fire during the simulation run.

In Figures 4 and 5, at time t=0, the system is in its initial state as shown by the initial marking. At time t=1, the timing vector of transition a1 expires and transition a1 fires. The system moves to the state shown in Figure 6a and 6b.. In this state, activity p1 has begun and capacity c1 has been consumed into activity p1. Since p1 and c2 are the input places of p2 and both are marked at t=1, the clock for transition a2 can now begin timing. Transition a2 times out after 2 time units; i.e., at t=3. At that point transition a2 fires, removing tokens from p1 and c2 and placing tokens in p2 and c1. The new state of the system, Figures 7a and 7b, shows that capacity c1 is now available and that activity p2 is in progress. Transition a1 begins timing again and, after 1 time unit has elapsed, transition a1 fires again at t=4. At that time, activity p1 begins again. At this point both p1 and p2 are in progress. The circulation of tokens and the marking of activities indicates the state of the system at various points in the simulation. This brief introduction to Petri nets will give the user some insight as to how the simulation executes. The results of a simulation run must be interpreted in terms of the origional IDEF0 model of a physical system.

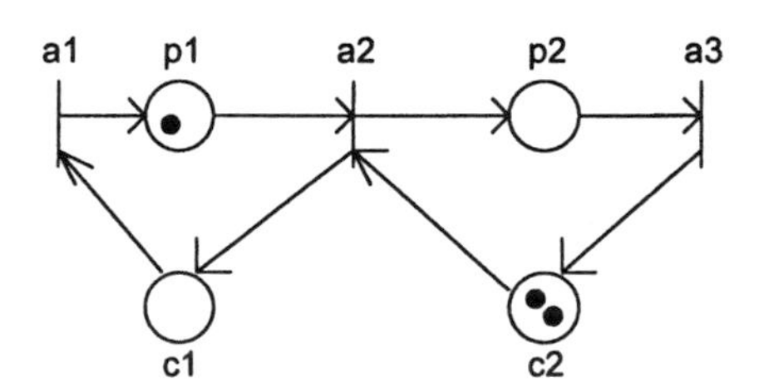

Figure 6a

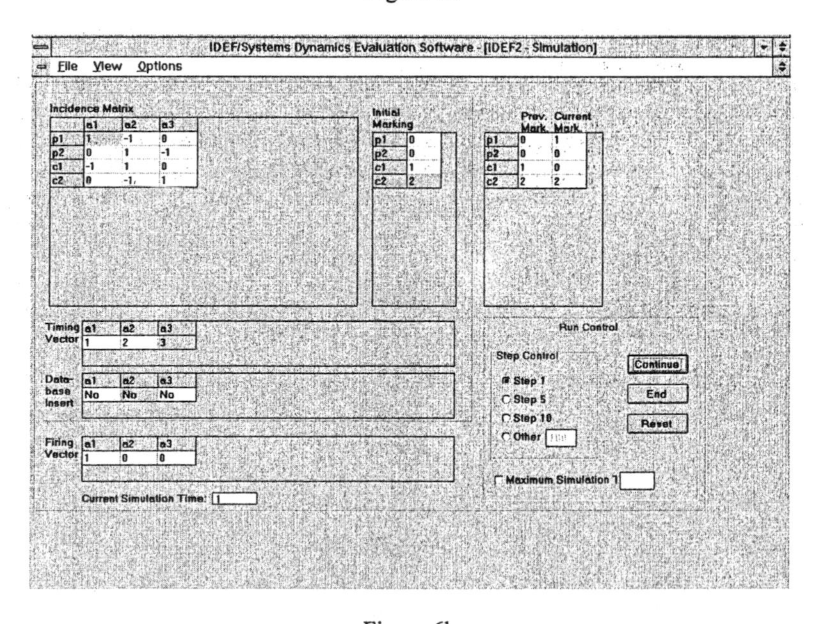

Figure 6b

Figure 6 State of the Simulation at Time t=1

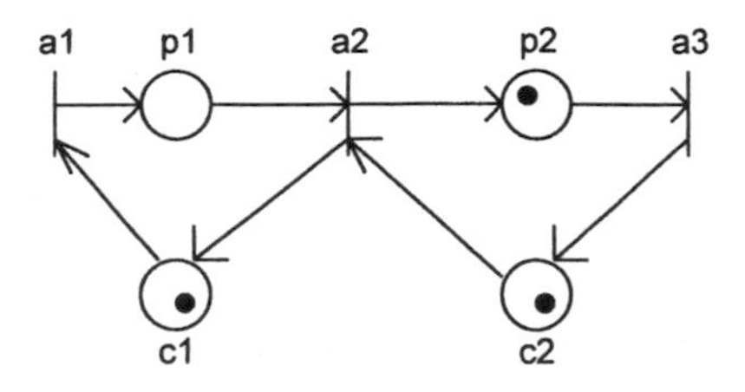

Figure 7a

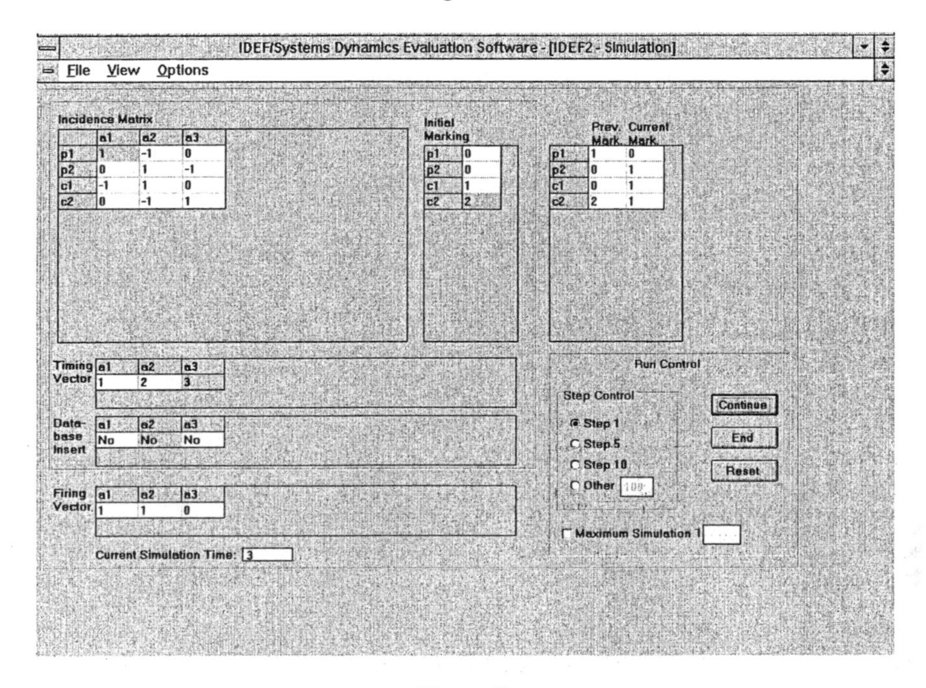

Figure 7b

Figure 7 State of the Simulation at t=3

SIMULATION POLICY OPTIONS

Refer to Figure 8. The menu item labeled *Options* allows the user to select a policy for the firing of transitions. Click on the *Options* menu and click on the *Policy* submenu. There are two policies to choose from: 1) Priority on event time and 2) Priority on enable. These can explained with reference to Figure 9.

There are cases where a conflict can occur in a simulation and that conflict must be resolved by a firing policy. *Priority on event time* settles conflicts by allowing the transition that times out first to fire first. Assume in Figure 9 that p1 obtains a token at t=0. Since the transition a1 requires only a token in p1, the transition a1 is scheduled to fire in 4 time units. Let us suppose that, at t=2, a token arrives in p2. Since a2 requires a token in p1 and p2 to fire, it is now scheduled to fire in 1 time unit; i.e., at t=3. Since both transitions require the token from p1, only one of the transitions will be allowed to fire. The user selected policy determines which token will fire. *Priority on event time* allows a2 to fire first because it will time out first. *Priority on enable* states that the transition that is first enabled captures the token at that time and that transition will fire after its timing vector times out, regardless of whether or not another transition is enabled later and requires the same token. This policy would have a1 fire first because it was enabled first by the arrival of a token in p1.

The appropriate policy is a function of the interpretation of the situation being modeled. When a conflict situation exists in a simulation, the user must decide which policy is the best interpretation for that situation.

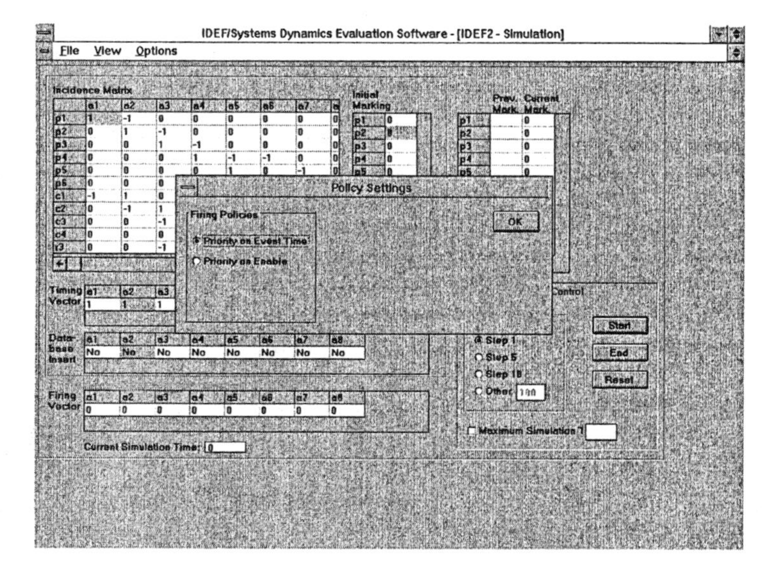

Figure 8 Policy Option Submenu

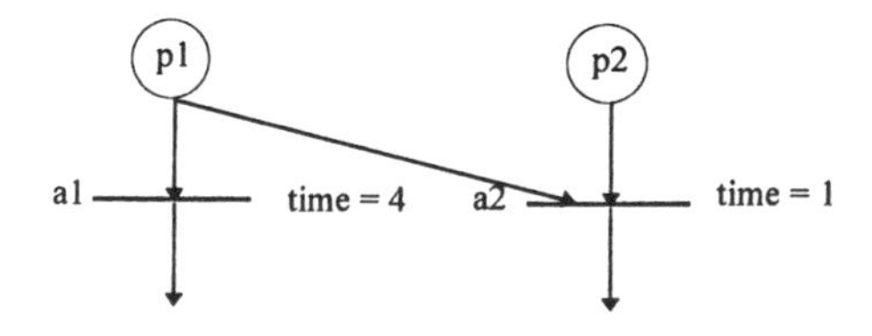

Figure 9 Example of a Conflict Requiring a Firing Policy

CAPTURING SIMULATION DATA

In order to further analyze the system simulated, the data generated by the simulation is stored into an Access database file. For any IDEF0 model, as soon as *Simulation* option is chosen, a database with the IDEF0 model name is created. The database includes a table for each row of the incidence matrix and in each table there are three fields to store information on the number of tokens and simulation time. This is done automatically without any input from the user.

As soon as the initial markings are entered and the *Start* button is clicked, the initial markings are inserted into the database. The tables are updated every time there is a change in the number of tokens in a place. This continues until the user ends the simulation. The database is cleared by clicking the *Reset* button.

When a new project is created using a previously simulated IDEF0 model name or the same project is reopened, the program reopens the existing database for that IDEF0 model. If there is any existing data in the tables, a *Warning* window opens. This window allows the user to either clear the database for another simulation run, or to append information to the previous tables or to get out of the program to save the existing information under another name.

GENERATING A DISCRETE EVENT CONTROL PROGRAM

IDEF/System Dynamics software can also illustrate how an IDEF0 / Petri net incidence matrix structure can be used to generate discrete control logic. In the following sections we will introduce an illustrative example and guide the user through the process of entering the model.

Consider the situation shown in Figure 10. The activities along this production line are controlled by a supervisory controller. We are interested in modeling the problem and the controls from the point-of-view of the supervisory controller.

The package is moving along a conveyor until it reaches sensor 1. The conveyor is stopped momentarily by the supervisory controller while the filler dispenses product into the container. When complete, the filler sends a "filling done" signal to the supervisor and the conveyor is restarted. The package passes to the weighing station, where the conveyor is momentarily stopped while the package is weighed. The scale provides two signals to the supervisor, a "weighing done" signal and a "reject" or "accept" signal. Based on these signals the package is either sent through the accept lane or diverted down the reject lane. For simplicity we assume that only one package is allowed through this portion of the line at a time. Continuous operation could also be assumed, but it requires a different model than the one that will be described in this discussion.

A Petri net model that represents this situation is shown in Figure 11. The packages exiting at p5 and p6 give control back to the front of the line for the reentry of another package. This is indicated by the arcs reentering at p1. Thus, the model is showing the control logic.

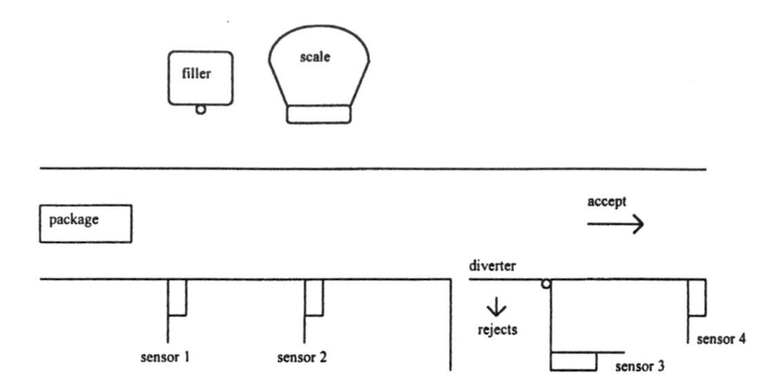

Figure 10 Filling Line Example Situation

A Petri net model that represents this situation is shown in Figure 11. The packages exiting at p5 and p6 give control back to the front of the line for the reentry of another package. This is indicated by the arcs reentering at p1. Thus, the model is showing the control logic.

The transitions between states occur as a result of the input signals from the sensors along the line. The transitions of Figure 11 are labeled with the appropriate signal. The activities along the line must actuate the outputs to control the overall operation of the line. The approriate output for an activity is shown next to the activity. It is left to the reader to examine the control model and confirm that it is appropriate to the situation.

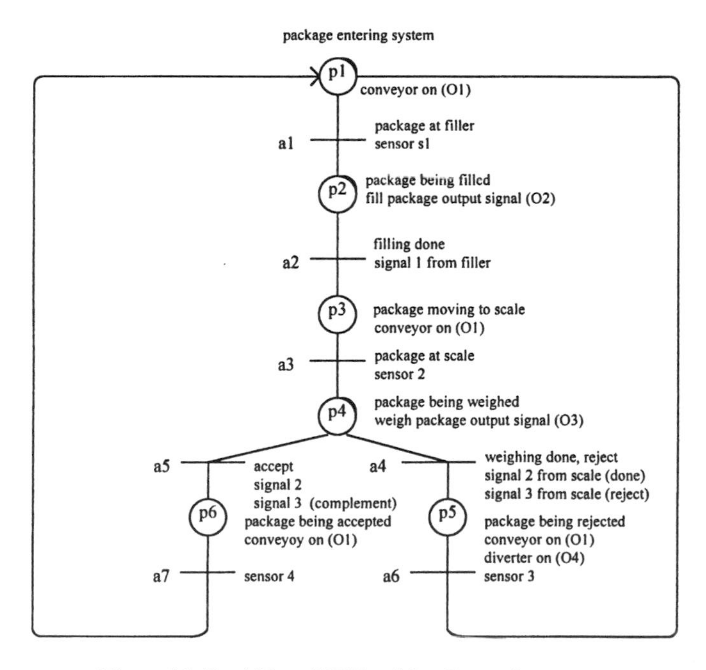

Figure 11 Petri Net of Filling Line Example

The incidence matrix can be entered in the usual way, as previously shown. The incidence matrix for this example is as follows:

	$a1$	$a2$	$a3$	$a4$	$a5$	$a6$	$a7$
$p1$	-1					1	1
$p2$	1	-1					
$p3$		1	-1				
$p4$			1	-1	-1		
$p5$				1		-1	
$p6$					1		-1

In using the incidence screen for modeling controller problems, it is not necessary to enter mechanism data. It is assumed that all inputs and outputs are going to be entered with respect to the supervisory controller.

Once the model is in active memory, the user should exit the incidence matrix screen and click on the *Edit* menu item and the *Input/Output* submenu item. The user will be presented with a screen having the format of Figure 12. The input box is where the user enters the signaling and sensor information. The output box is where the user enters the supervisory controller output actions. A description of the inputs and outputs can be entered in the descriptions panels.

Exit the Input/Ouput Screen and return to the project screen. Click on the menu item *Make* and the submenu item *Ladder Logic*, as shown in Figure 13. The software will compute the control logic and represent it in ladder logic form. There will be multiple pages of the logic code shown on the screen. The user can review earlier pages by minimizing or closing each page. The logic underlying this ladder is created in a particular manner. The interested reader is referred to reference [1].

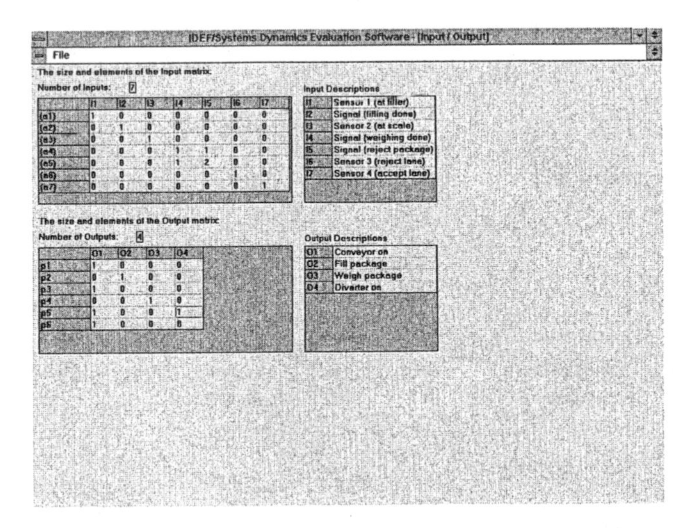

Figure 12 Input/Output Data Entry

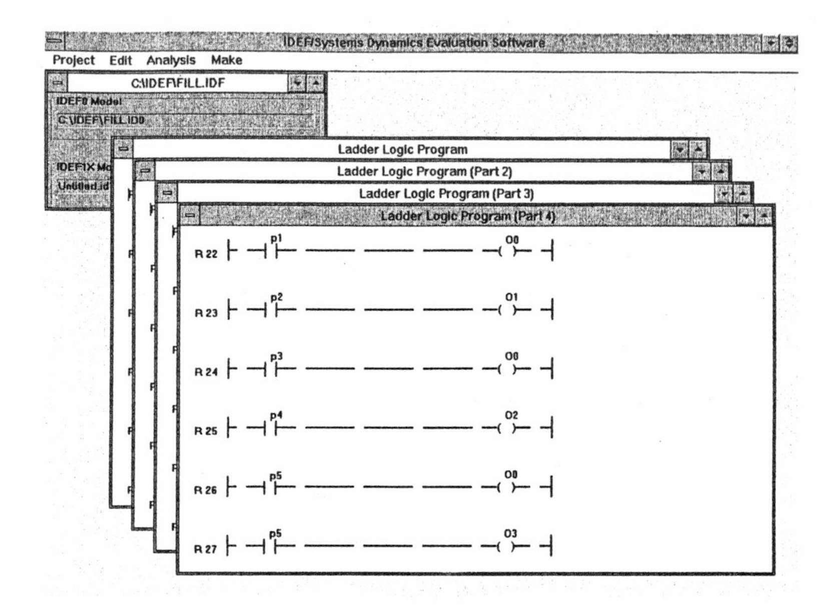

Figure 13 Ladder Logic Output

SUMMARY

We have described the use of the IDEF0 modeling formalism in combination with Petri nets for the specification and analysis of discrete event systems. It has been demonstrated how Petri nets, in conjunction with input/output specifications, can be used to create controller programs. These functions have been implemented in software that is part of a continuing research project at Rutgers University for bringing together specification, dynamic analysis and controller development at the design stage of a discrete event system [5].

REFERENCES

[1] **Boucher, T.O.,** 1996, *Computer Automation in Manufacturing*, London: Chapman & Hall.

[2] **Desrochers, A. A. and R.Y. Al-Jaar,** 1995, *Applications of Petri Nets in Manufacturing Systems*, Piscataway, New Jersey: IEEE Press.

[3] **Jafari, M.A. and T.O. Boucher,** A rule based system for generating a ladder logic control program from a high level system model, *Journal of Intelligent Manufacturing*, Vol. 5, No. 2, 1994.

[4] **Marca, D.A. and C.L. McGowan,** 1988, *Structured Analysis and Design Technique*, New York: McGraw-Hill.

[5] **Wurl, R.C., T.O. Boucher, M.A. Jafari, W. Zhao and G. Alpan,** *User Manual: IDEF/System Dynamics Software, Version 1.0a*, Department of Industrial Engineering, Rutgers University, 1996.

FUNCTIONAL REQUIREMENTS FOR METHODS TO SUPPORT EARLY-STAGE SPACE MISSION DESIGN

Richard C. Anderson[1],
Michael R. Duffey[1],
Kevin W. Lyons[2]
1 The George Washington University;
2 National Institute of Standards and Technology

ABSTRACT: This paper considers a body of requirements for integrated product-process modeling methodologies intended to enhance the early-stage design capabilities of Integrated Product and Process Development (IPPD) teams. In an IPPD environment, the mission systems engineer and the team leader are often one in the same person, making him/her responsible for ensuring that design decisions at all times properly reflect the customers' (both internal and external) fundamental requirements for acceptable levels of life cycle cost, performance, schedule, and risk. In order to do that effectively, the systems engineer must be able to rapidly develop appropriate mission analysis models to estimate the relative life-cycle consequences of design alternatives with respect to those top-level mission parameters. For complex products and processes, this can become exceedingly difficult. This paper explores related modeling requirements for government-sponsored scientific remote-sensing satellite missions. However, requirements in this environment are applicable to many design/manufacturing domains in both the public and private sectors.

INTRODUCTION

The aerospace industry has traditionally used the term *mission* (often interchangeably with either *system* or *project)* to represent the collection of products, processes, and resources required to carry out a specific program objective, such as to land a man on the Moon and bring him safely back to earth, collect and analyze soil samples on the surface of Mars, or monitor changes in the earth's polar ice sheets from orbit. In the days before Concurrent Engineering, or what is now often called Integrated Product and Process Development (IPPD), early-stage space mission design concerned itself primarily with the operational aspects of the engineering systems required to carry out a mission's "user" objectives. In those days, *technical performance* was the primary criterion for mission success. Issues such as cost and process efficiency were given low priority if seriously considered at all.

In recent years, however, the traditionally performance-driven approach to space mission design has been evolving into to a more integrated, concurrent, or holistic one. Aerospace engineering organizations in both the public and private sectors, facing increasing levels of competition and cost constraint in a post-Cold War world, recognize the need to become more *efficient* as well as effective. Many of them, like growing numbers of firms in other technology intensive industries, have begun to reorganize and implement various forms of IPPD, where a multidisciplinary Integrated Product Team (IPT) works together to insure that each early-stage design decision is made with the best possible understanding of its end-to-end ramifications across all functional disciplines and throughout the entire mission life cycle, including manufacture, testing, deployment, operations, and disposal.

While there are many possible ways to structure and operate an Integrated Product Team for doing IPPD [Sheard & Margolis, 95], there can be little doubt that the systems engineer (or *mission engineer* if you prefer) plays a critically important role. Systems Engineering is defined in [EIA, 94] as "an interdisciplinary approach encompassing the entire technical effort to evolve and verify an integrated and life-cycle balanced set of system people, product, and process solutions that satisfy customer needs." As such, Kuhn and Sampson contend that "it is impossible to separate Concurrent Engineering from Systems Engineering, since the latter is required to accomplish the former." [Kuhn & Sampson, 93]

In an IPPD environment, systems engineering is performed in concert with system management, with the systems engineer providing critical information needed by the manager to make sound top-level design decisions [Shishko, 95]. In many cases, the systems engineer and the team leader are one in the same person, making him/her responsible for ensuring that design decisions at all times properly reflect the customers' (both internal and external) fundamental requirements for acceptable levels of life cycle cost, performance, schedule, and risk. In order to do that effectively, the systems engineer must be able to rapidly develop appropriate mission analysis models to estimate the relative life-cycle consequences of design alternatives with respect to those top-level mission parameters. For complex products and processes, such as those encountered in space missions, this aspect of systems engineer's job can become exceedingly difficult.

This then begs the question: What kinds of methodologies and tools do mission systems engineers/managers need to be able to effectively model, evaluate, and communicate design information from a mission life cycle perspective? The authors believe that an important first step is to develop a body of top-level functional *requirements* that captures those needs. In order to do that properly, it is necessary to carefully reexamine how mission products and process are represented.

Space Mission Products and Processes

It is well documented that each and every man-made product is subject to a life cycle process through which it is conceived, designed, acquired, delivered, operated, and ultimately disposed. This is true for products at *any* level of abstraction – whether a complete system (e.g., automobile or satellite) or any of its subsystems, assemblies, or the smallest of its component parts – they all have their own life cycles. It is also true for complex systems of systems, like those frequently encountered in space missions. The *NASA Systems Engineering Handbook* [Shishko, 95] depicts the space mission life cycle as a series of five phases containing a total of ten developmental stages stretching from concept development through disposal, with the progressions between stages determined by a variety of "go, no-go" transitional reviews called control gates. A listing of these life cycle process components is shown in Table 1.

TABLE 1
Major Elements of the Space Mission Life Cycle

Mission Phase	Stage	Control Gate
Pre-A: Advanced Studies	Mission Feasibility	Mission Concept Review
A: Preliminary Analysis	Mission Definition	Mission Definition Review
B: Definition	System Definition	System Definition Review
	Preliminary Design	Preliminary Design Review
C: Design	Final Design	Final Design Review
D: Development	Fabrication & Integration	System Acceptance Review
	Preparation for Deployment	Flight Readiness Review
	Deployment & Operational Verification	Operational Readiness Review
E: Operations	Mission Operations	Decommissioning Review
	Disposal	

Each stage in the space mission life cycle results in a specific intermediate product or products that can serve as resources for subsequent, or "downstream" stages. For example, the designs produced in the early stages are used as resources to guide the fabrication and integration of systems. Those systems are in turn deployed and then used as resources to carry out mission operations, such as the acquisition of scientific data, in ultimate fulfillment of the customer's objectives. When taken collectively, all of these intermediate processes and products constitute a complete mission.

Products have always had life cycles, and so have space missions – that is nothing new. However, before the advent of concurrent engineering, few design engineers had the necessary perspective, knowledge, or incentive to attempt to optimize their product designs with respect to all of the life cycle stages. Early-stage space mission designers (i.e., those working within phases Pre-A and A) have typically been preoccupied with satisfying the end-user's performance requirements for mission operations, without giving due consideration to all of their other downstream *internal* customers, i.e., those responsible for doing the detailed design, manufacture, testing, deployment, and disposal of the mission systems they design. This has often proved to be a rather inefficient arrangement, prone to extensive rework and cost and schedule overruns. In today's more competitive and cost-constrained environment, mission feasibility can no longer be determined solely on the basis of the operational performance of its physical systems. End-to-end mission efficiency must also be designed-in from the beginning of the life cycle (by the end of Phase A) so that the entire mission process is as cost-effective as possible. The mission systems engineer is primarily responsible for ensuring that this happens.

The Mission Systems Engineering Process

The realization of a mission over its life cycle is the result of a succession of decisions among alternative courses of action. The most consequential decisions are usually made very early on in the design process, particularly within the Mission Definition stage (Phase A). The goals for this stage are to establish the top-level requirements, architecture, and implementation plan for the entire mission life cycle, and to estimate overall mission performance, risk, schedule, and total cost, all to a level of detail sufficient for the generation of proposals for the performance (or contracting) of work in downstream phases. This usually requires the specification of mission products and processes all the way down to the component level. However, the purpose of Phase A is to make reasonable program estimates as quickly as possible, based on credible, feasible, but not necessarily optimal, mission designs. What is most important is to insure that 1) at least one solution exists that can perform the mission within the stated resource constraints, and 2) that no major factors or concerns have been omitted from consideration. Work to refine all of the implementation details of the mission is left to subsequent phases, based on the requirements that are generated in Phase A.

From the mission systems engineer's perspective, each and every design decision in effect transforms a requirement (or set of requirements) into a specification of a product or process solution that will hopefully satisfy it in light of known conditions or constraints. His/her primary role in this context is to define and understand mission requirements and their implementation alternatives (i.e., candidate product and process solutions) well enough to support their selection. This is accomplished by performing ***requirements analyses*** and ***trade studies*** in accordance with the "Doctrine of Successive Refinement" [Shishko, 95], as illustrated in Figure 1. This approach is in sharp contrast to the more traditional and risk-prone "waterfall" design practice where much of implementation design would wait until detailed requirements were developed.

Requirements analysis is performed at each level in the process of Figure 1 in order to fully understand what needs to be accomplished. At first, the highest-level, fundamental mission goals are

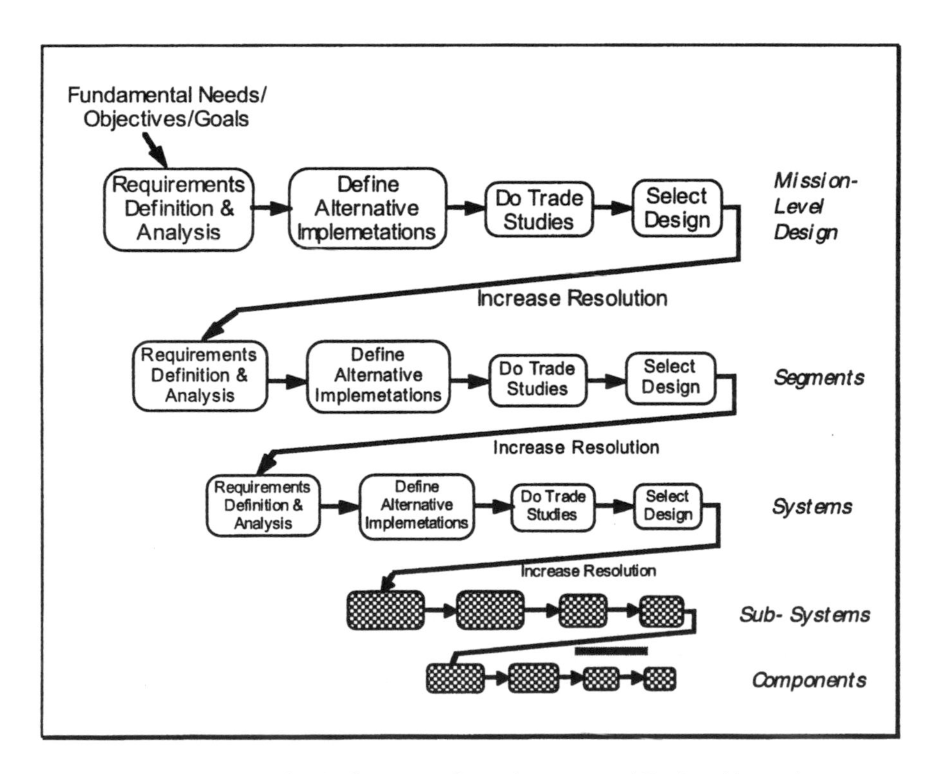

Figure 1. Successive Refinement of Requirements and Design Alternatives

identified and then successively refined, decomposed, and translated into functional requirements (what must be done) and performance requirements (how well to do it) in concert with the design process[Grady, 93]. This process continues iteratively, alternating between requirements discovery and implementation design selection, until a desired level of refinement is achieved. Traditionally, representations and tools such as Functional Flow Block Diagrams (FFBDs), N-squared Charts, and Time Line Diagrams [Grady, 93; Shishko, 95] have been used for requirements analysis.

Trade studies are a critical link in the transformation of requirements into effective design solutions. As a mission design matures, many decisions will be made about functionality, architecture, and implementation. Trade studies are done to ensure that those decisions move the design toward optimality [Shishko, 95; NASA/JSC, 94; NASA/EMC, 92; Larson & Wertz, 92]. They can be used to support design decisions at many levels of detail, ranging from the mission and system levels all the way down to the selection of individual components. Higher level trades are usually the focus of system engineers and project managers, while more detailed trades are performed within individual engineering disciplines.

Within an IPPD environment, a mission level trade model requires the synthesis of lower-level (i.e., segment, system, segment, subsystem, component) design products of the individual engineering disciplines (e.g., Mechanical, Thermal, Power, Propulsion, Communications, Flight Dynamics, etc.) into a higher level, aggregate model of the *entire* mission that would enable its effective evaluation relative to competing alternatives. Note that these evaluations are performed in a complete mission life cycle context, so that a mission design must not only satisfy technical objectives, but must also be the most cost-effective over the entire span of manufacturing, integrating, testing, deploying, operating, and disposing the mission. Four basic "outcome parameters" or criteria are used in NASA trade studies (and in just about any such evaluation elsewhere for that matter): performance, cost, schedule, and risk. Each mission design alternative must be evaluated according to estimates of the life cycle aggregate values for each of those four parameters as provided by a mission level trade model. The results of the analysis and selection process can be used to track design progress, support proposal writing, and feed back information and concerns to the engineering disciplines from the mission systems perspective.

Discussion of Functional Requirements for Methods to Support Integrated Mission Design

Given the above examination of the problems and challenges facing mission systems engineers as they move into IPPD, one can begin to develop a body of requirements for the next-generation tools and methodologies they will need in order to represent and analyze mission processes, products, and requirements concurrently from a complete life cycle perspective. The following discussion is offered as a first step in that direction, and is not necessarily a complete or comprehensive prescription of hard and fast requirements. In keeping with the principles of successive refinement discussed above, requirements will undoubtedly be discovered and revised as possible modeling approaches emerge and are evaluated as to their ability to satisfy objectives. At present, we have three main classifications for requirements: (1) mission-level trade study support, (2) mission design communication and coordination, and (3) integrated mission product-process model representation.

Mission-Level Trade Study Support

Mission systems engineers and managers require methodologies and tools that will allow them to see as clearly and as quickly as possible the many cause and effect relationships between individual design decisions (amongst alternative requirements or their implementations) and their consequences in terms of the overall life cycle mission cost, performance, schedule, and risk. This would take into account the cause and effect relationships that exist between functional disciplines within a particular development stage, as well as those that exist across life cycle phases.

For example, consider a situation where a team is working to develop an earth-orbiting satellite mission. At issue is a decision about whether to employ human operators to control the spacecraft from the ground or allow it to be fully automated with its own on-board computer. Looking at things strictly from an operational viewpoint (within Phase E of the life cycle), one could see that that this decision about the implementation of the command and data handling subsystem will have consequences for many if not all of the engineering disciplines that are involved in designing the other subsystems in both the ground and flight segments of the mission. For instance, an on-board computer could reduce the need for ground support but increase the power and mass properties of the flight segment. It could impose tighter requirements for thermal control or radiation shielding. It might also call for changes in the operation of the communications subsystem, as less contact with the ground would be required if the space craft is made more autonomous. Fully automated computer control might also leave the satellite unable to deal with unexpected contingencies when they arise.

As one can imagine, a trade study to determine which course of action is best (even when considering performance alone) can quickly become rather complicated even for a simple example such as this. Currently, such trade studies are largely confined to a single mission phase (e.g., operations), and they typically take a long time to complete. Without sophisticated tools that allow for at least partial automation of the process, including a means through which detailed discipline information can be aggregated into a mission-level trade model, it is unlikely that they can be conducted quickly and well at the same time. There is simply too much information and too many relationships to deal with "by hand" even within the operations phase. Beyond that, matters are further complicated if we attempt to consider the effect of this spacecraft autonomy decision in terms of its full effects throughout the entire mission life cycle. In doing so, we would need to ask such things as:

(1) Which of the two options will lead to better overall mission performance? One option may be clearly attractive in that it better satisfies the customer's requirements during operations, but what about performance-related issues in other life cycle phases? Do any of the proposed options make the mission systems more complicated so their acquisition, integration, and testing is made more difficult? Will the choice of launch vehicle or the processes by which the satellite is launched, deployed, and verified be affected technically? Will the mission systems now be more difficult to dispose because of any change in the nature of the operations of the command and data handling subsystem?

(2) What are the full implications for each alternative in terms of life cycle cost? Probably an autonomous satellite would result in lower cost in the operations phase since fewer human operators on the ground would need to be trained and put on duty. However, what additional relative cost is incurred by having to acquire or develop the required on-board

controller? What cost effects might be realized as a result of required changes in the products and processes in all of the life cycle phases, including those implied in (1) above?

(3) How could this one decision about the level of spacecraft autonomy possibly affect the overall schedule for the mission? Each alternative implementation clearly has different effects on the scheduling of operations and the operational behavior of mission systems. Will any of the issues discussed in either items (1) or (2) have any important effects on the development schedule? Could such changes make the mission systems more difficult to develop so that there is an increased chance of missing a critical launch window? Does either option significantly affect the mission cash flow distribution so that delays occur when waiting for funds to be released?

(4) How might the alternatives affect the overall mission risk? This can take on the dimensions of performance (technical) risk, cost risk, and schedule risk. Answering this question involves a careful treatment of probabilistic uncertainties to determine not only the likelihood of an alternative hitting its performance, cost, and schedule targets, but also the very reliability of the estimates themselves.

The exact decision issues, design trades, and specific parameter values will undoubtedly vary from mission to mission. Because of this, the supporting modeling methodology needs to provide a sound fundamental structure or framework for capturing and analyzing the complex causal relationships that exist between disciplines and across life cycle phases. It should be flexible enough to allow for tailoring to accommodate a variety of mission possibilities.

Mission Design Communication & Coordination

An advanced product and process modeling tool should also help a mission systems engineer justify any decisions and effectively communicate the results of trade studies to higher level management, external customers, and internal customers such as the members of the IPPD team. Any such model must also provide a convenient framework for conveying and sharing product and process information in such a way that all parties can be involved in the mission design process as early and as completely as possible, thereby increasing the possibility that team members at all levels will remain fully informed and generate better designs from a life cycle perspective. Analytical models used by the systems engineer for mission trade studies can be made more understandable by a variety of audiences (i.e., administrators, customers, contractors, and design team members).

For example, one very important audience that needs to be in close contact with the mission systems engineer, especially in the earliest stages of design, is the external customer. In the case of a satellite mission to conduct scientific observations, the customer is usually a principal investigator (PI) who represents the interests of a team of scientists who need to acquire some data about a particular scientific phenomenon. As in all cases, the customer's requirements are the driving motivation behind the mission and are ultimately its very reason for being. Any judgments about the relative worth of design decisions must ultimately be traceable back to whether they respond properly to customer requirements, subject to practical implementation constraints. One frequent problem that needs to be addressed is that PI's often know what they want to accomplish, but they tend not to understand all of the technical consequences of their specifications. However, these same PI's are being made increasingly responsible for watching how much their missions cost. It would be of significant value if a mission systems engineer could use the output of a mission trade study model to quickly develop a presentation that could educate the PI as to the relative life cycle cost of a particular requirement. In addition, and equally importantly, the PI could be shown *why* the requirement is estimated to be so expensive to implement.

It is likely that some manner of graphical representation will be needed in order to show the relative consequences of design decisions most effectively to a variety of audiences. This might involve the use of different representation schemas, each tailored for the particular needs and concerns of the recipient.

Integrated Mission Product-Process Model Representation

Full realization of the aforementioned goals will require an integrated representation of mission requirements, processes, and products that captures the essential cause and effect relationships between them throughout the entire life cycle. As described earlier in this paper, each and every design decision works to either transform a requirement into some set of intermediate product and process solutions (implementation), or to refine or modify a requirement based on updated information from implementation design efforts. Requirements are most often modified because the products and processes they imply are deemed to be either infeasible or undesirable. Implementations at all levels must always be judged in terms of their ability to fulfill stated requirements in the presence of constraints. If we are to evaluate design decisions with respect to their life cycle consequences, we need a modeling methodology that can explicitly link requirements with their alternative product and process implementations.

An effective mission product-process representation methodology should account for the fact that requirements analysis and process analysis are very closely related. Basically, a functional requirement, no matter where it is encountered in the mission life cycle or how detailed it is, specifies a certain subset of all the things that need to be done in order to fully realize mission objectives. Examples include such things as "fabricate solar panels," "integrate spacecraft and instrument," "prepare for launch," "conduct mission operations," "transmit data," and "execute burn for deorbit." Design decisions attempt to determine the exact processes by which these requirements should be realized, in keeping with stated performance requirements and imposed constraints. Given that, it can be argued that a functional requirement can be represented as an analog of some physical process [Grady, 93]. This correspondence between functional requirements and process activities can be observed in Figure 2. For example, consider the hypothetical function named "Do b" on the left side of the figure. It specifies that some element of work "b" must be done as part of the requirement to complete "Stage 2" of the mission. The corresponding process to perform "b" in the Implementation Domain appears as a "sub-process" of Stage 2 on the right side of the figure. When alternative processes for performing a given requirement are nominated, they need to be further defined in terms of the resources that will be required to implement them (e.g., cost, time, personnel, equipment, material, and so on).

The relative evaluation and selection of such alternatives cannot be done effectively from a life cycle mission perspective without also factoring product information into the mix. Currently available process models do not explicitly include product information to a level of detail sufficient for supporting the requirements outlined in this paper. Historically, process modelers have concerned themselves primarily with capturing the precedence relationships between activities (i.e., reporting

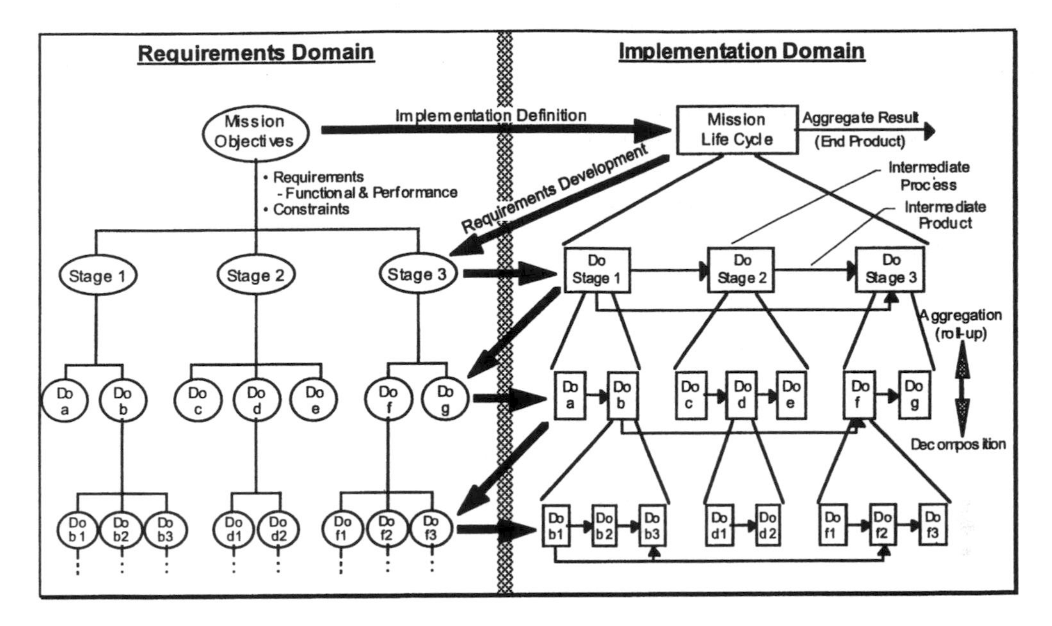

Figure 2. Relationships Between Requirements, Processes, and Intermediate Products

that "this happens, then that happens"). While this provides insight into sequential or time-based relationships between activities, it ignores part of a larger, more complete definition of process as a progression of transformations of *intermediate products* (Figure 2) by activities to achieve defined objectives. This connection is vitally important if we wish to understand how the effects of early-stage design decisions propagate throughout the subsequent implementation stages of fabrication and integration, preparation for deployment, deployment and operational verification, mission operations, and disposal. The success or failure of each element of the life cycle process must be contingent upon the quality of the intermediate products that are exchanged between them as well as the efficiency of the processes by which they are produced.

Concluding Remarks

The emerging practice of concurrent engineering for mission design, whether for aerospace or other technology domains, requires a fundamental reexamination of supporting methods and tools. While it is tempting to proceed directly to creating such tools, we believe that a careful assessment of functional requirements is a necessary but frequently overlooked first step.

Acknowledgments

This on-going research project has received support from the NASA Goddard Space Flight Center and the National Institute of Standards and Technology.

References

Electronic Industries Association (EIA), "Systems Engineering," Publication EIA/IS-632, 2001 Pennsylvania Avenue NW, Washington, DC, 20006, December 1994.

Grady, Jeffrey O., System Requirements Analysis, McGraw-Hill, Inc., New York, 1993.

Kuhn, Dorothy A. and Sampson, Mark E., "A survey of Systems Engineering Design Automation Tools," *Proceedings of the Third Annual International Symposium of the National Council on Systems Engineering (NCOSE)*, July 22-26, 1993.

Larson, Wiley J. and Wertz, James R. (ed.), Space Mission Analysis and Design, Kluwer Academic Publishers, Boston, 1992.

National Aeronautics and Space Administration (NASA), "NASA Systems Engineering Process for Programs and Projects," (NASA Publication JSC 49040), NASA Lyndon B. Johnson Space Center, Houston, Texas, October 1994.

National Aeronautics and Space Administration (NASA), "The NASA Mission Design Process: An Engineering Guide to the Conceptual Design, Mission Analysis, and Definition Phases," The NASA Engineering Management Council, December 22, 1992.

Sheard, Sarah A. and Margolis, M. Elliot, "Team Structures for Systems Engineering in an IPT Environment," *Proceedings of the Fifth Annual International Symposium of the National Council on Systems Engineering (NCOSE)*, St. Louis, Missouri, July 22-26, 1995.

Shishko, Robert, ed., NASA Systems Engineering Handbook, (NASA Publication SP-6105), NASA Program/Project Management Initiative, June 1995.

Analysis of k-Stage Automated Production Flow Lines with Recirculating Conveyor Buffer Storage

William E. Biles, John S. Usher, Jacqueline M. Sprinkle
Department of Industrial Engineering
University of Louisville
Louisville, Kentucky 40292
USA

ABSTRACT: This paper examines the production rate, expected buffer storage level, and part transit time for k-stage automated production flow lines with closed-loop, recirculating conveyor buffers. Analytical models are developed for the performance of the recirculating conveyor buffer. Computer simulation results are presented to validate these analytical models and to compare the operational performance of these recirculating conveyor systems with that of k-stage systems having classical static buffer storage.

INTRODUCTION

An automated production flow line consists of a series of workstations interconnected along a line or around a circle. The workstations are typically untended machines performing such manufacturing operations as turning, milling, drilling, boring, reaming and grinding, as well as assembly and inspection. The in-line production flow lines, also called *automatic transfer lines*, may be laid out in a straight line or have one or more 90° turns. In-line systems may have a large number of workstations and are typically used to process large parts, such as engine blocks and transmission casings. Circular transfer systems use rotary tables for indexing parts at machines arranged at equal angles around the circumference. These latter systems typically have few workstations and are used primarily with small parts, such as pistons and pipe valve bodies.

The operation of automated production flow line systems consists of parts flowing through workstations and storage areas called *buffers*. The time that parts spend in workstations is random. This randomness is due to random processing times, random workstation failure and repair events, or both. If the flow line is rigidly connected, with no buffer storage, the failure of one workstation results in shutting down the entire line until the failed machine is repaired, resulting in a significant degradation of production performance. The installation of buffer storage at strategic positions along the flow line has the effect of decoupling the line, significantly mitigating the effects of workstation failures.

This paper describes a particularly useful form of buffer storage, the closed-loop recirculating conveyor, and shows how to calculate the production rate, mean buffer storage, and part transit time for such systems. A computer simulation study is used to compare the operational performance of k-stage production flow lines with recirculating conveyor buffer storage to that of the classical static buffer systems. This comparison is based on the assumption of a repair crew being assigned to each stage, so that repair on a failed workstation begins immediately upon failure and there is no waiting for a repair crew to complete work on another workstation.

Although the focus of this paper is on the performance of the recirculating conveyor buffer, some discussion is presented for production flow lines with static buffer storage to provide a baseline for comparison. Dallery and Gershwin (1992) provide an excellent review of analytical approaches to automated production flow line modeling.

ANALYSIS OF PRODUCTION FLOW LINES WITHOUT BUFFERS

To begin this analysis, certain basic characteristics of the automated production flow line must be set forth. First, the line consists of *m* machines. Parts are introduced at machine M_1, and are processed and transported at regular intervals to succeeding stations. Transfer between stations is synchronous, with the cycle time T_c corresponding to the longest station processing time. Actually, the cycle time T_c is made up of three components: (1) processing time, (2) idle time (for those stations for which processing time is less than that of the longest station), and (3) transfer time.

Because of machine breakdowns, the actual average production time T_p is greater than the cycle time. If the average downtime per breakdown is T_d, the average production time per part is

$$T_p = T_c + FT_d \tag{1}$$

where

$$F = \sum_{j=1}^{m} F_j \tag{2}$$

is the mean breakdown rate per cycle, and F_j, j = 1, ..., m is the breakdown rate for each machine M_j, j = 1, ..., m. The line production rate is therefore

$$R_p = \frac{1}{T_p} \tag{3}$$

Line efficiency E is the proportion of time the line is up and operating, and is computed as

$$E = \frac{T_c}{T_d} \tag{4}$$

AUTOMATED PRODUCTION FLOW LINES WITH STORAGE BUFFERS

One of the ways automated production flow lines can be made to operate more efficiently is to add one or more part storage buffers between workstations in the line. If one buffer storage is used, the line is divided into two stages. If two buffers are used, the result is a three-stage line. The upper limit on the number of buffers that can be placed in the flow line is *m* - 1. A *k* - stage line would have *k* - 1 buffer storages placed strategically throughout the line. Each stage effectively acts like a single machine, with its associated production rate, failure rate and repair rate.

The focus of this paper is on *k*-stage automated production flow lines. The following section reviews results for two-stage lines, which will then be extended to *k*-stage systems.

Two-Stage Lines with Buffer

Under this category of models, a line having *m* machines is divided into two stages by placing a single buffer storage facility between machines *l* and *l* + 1. Thus, stage one has *l* machines and stage two has *m-l* machines. Several models have been reported in the literature for the two-stage case.

Random processing times

This category of models assumes, in addition to random machine failure and repair rates, random production rates for each stage. Buzacott (1967) and Gershwin, et al. (1980) have described solutions for this case. It should be noted that random processing times are not applicable to transfer lines without internal storage.

Constant processing times

This category of models assumes a constant production rate for each stage, dictated by the longest processing time for machines within the stage, and can be further categorized as follows:

Exponential up and down times with finite buffer, including Levin and Pasko's (1969) model of an asynchronous line with two repair crews.

Infinite buffer models with exponential up and down times.

Geometric up and down times, such as with Buzacott's model (1967).

Stage Failures and Repairs for Two-Stage Lines

When a line is divided into two stages using a storage facility, each stage may contain more than one machine. For example, the *m*-machine line in Figure 1 below is divided into two stages by placing a storage facility between the machines *l* and *l*+ 1.

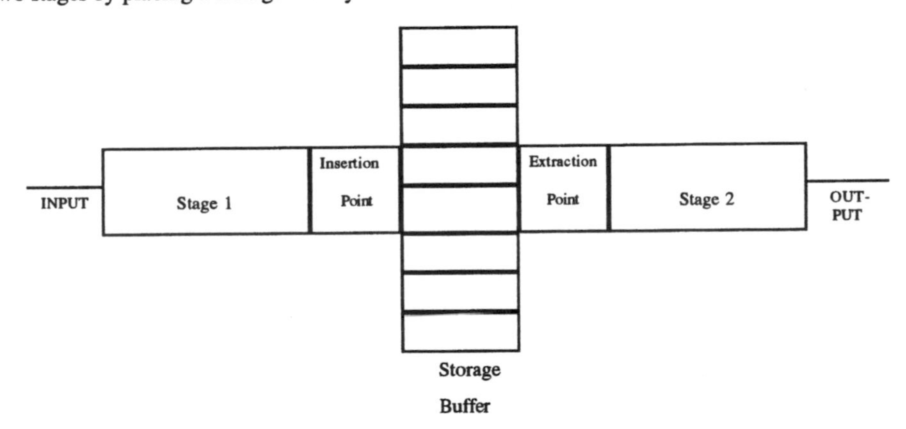

Figure 1. A Two-Stage Line With a Storage.

In such cases, as illustrated in Figure 1, a series of machines make up a stage. The first stage contains *l* machines in series and the second stage *m* - *l* machines in series. In the analytical models for two-state lines, each stage is assumed to be operating as a single machine with a particular production rate, failure rate and repair rate. Therefore, to make these models more useful, we have to find stage failure and repair rates for such lines. If we assume exponential time to failure with failure rate λ_i for each machine, then the failure time distribution is

$$f_i(t) = \lambda_i e^{-\lambda_i t} \quad i = 1, \ldots, m \tag{5}$$

The probability that machine i will fail by time t is

$$F_i(t) = 1 - e^{-\lambda_i t} \quad i = 1, \ldots, m \tag{6}$$

Let R_i (t) = reliability of machine i at time t; that is, the probability that machine i will operate without failure until time t. The reliability of machine i at time t is therefore

$$R_i(t) = 1 - F_i(t) = e^{-\lambda_i t} \quad i = 1, \ldots.., m \tag{7}$$

The probability that the first stage, consisting of l machines, does not fail by time t is

$$R_I(t) = R_1(t) \cdot \ldots \cdot R_l(t) \tag{8}$$

or

$$R_I(t) = \prod_{i=1}^{l} R_i(t) \tag{8a}$$

Then

$$R_I(t) = \prod_{i=1}^{l} e^{-\lambda_i t} = e^{-\sum_{i=1}^{l} \lambda t} \tag{9}$$

The probability that the first stage will fail by time t is given by

$$F_I(t) = 1 - R_I(t) = 1 - e^{\sum_{i=1}^{l} \lambda_i t} \tag{10}$$

and the failure time distribution, $f_I(t)$, is

$$f_I(t) = \frac{dF_I(t)}{dt} = \left(\sum_{i=1}^{l} \lambda_i\right) e^{-\left(\sum_{i=1}^{l} \lambda_i\right) \cdot t} \tag{11}$$

Similarly, the failure time distribution for the second stage, consisting of machines $l + 1, \ldots,$ m is

$$f_{II}(t) = \left(\sum_{i=l+1}^{m} \lambda_i\right) e^{-\left(\sum_{i=l+1}^{m} \lambda_i\right) \cdot t} \tag{12}$$

Therefore, each stage has an exponential time to failure with mean rate $\sum \lambda_i$

Two-Stage Lines with Finite Buffer

There have been several studies which sought to determine the effect of a finite buffer of capacity N on the efficiency of a production flow line. Gershwin and Berman (1980) developed a system of equations which modeled the so-called "boundary conditions," in which the first stage is "blocked" if the buffer is full and the second stage is "starved" if the buffer is empty, as well as the so-called "internal conditions" which hold when the buffer is neither full nor empty. Solution of this system of equations yields the marginal probabilities of the production line being in any of its 4(N+1) possible states, and allows one to compute the line production rate q_r and the expected number of units in the buffer E(n).

A key assumption of the Gershwin and Berman approach (1980) is that there are two repair crews, one for each stage, so that there is no waiting when a stage fails. Savsar and Biles (1984) considered the case in which only one repair crew was available, and developed a system of differential equations which modeled the production line. A key assumption of their approach was that the buffer could be modeled as a continuous system akin to the level of liquid in a tank. Again, their approach gives the line production rate q_r and the expected buffer storage E(n).

Both the Gershwin and Berman (1980) and the Savsar and Biles (1984) approaches assumed exponential failure and repair times in much the same way as was described in the previous sections of this paper.

A Closed-Loop Recirculating Conveyor Buffer

Both the Gershwin and Berman [3] and the Savsar and Biles [5] approaches considered that the finite buffer was a static facility that featured instantaneous insertion and removal of parts, given that at least one space was available in the case of insertion or at least one part available in case of removal. A static storage facility is not always the preferred method for buffer storage, however. Often a closed-loop recirculating conveyor, usually a chain-driven overhead monorail, is the system of choice. Such a system would perform quite differently than one with a static buffer, owing to the random delay that would be incurred for both part insertion and extraction.

The ensuing discussion describes an approach to computing the line production rate q_r and the expected number of units in storage, E(n), for a buffer consisting of a closed-loop recirculating conveyor of capacity N, that is, the conveyor is of length L with N carriers, so that the carrier spacing is

$$d = L/N \qquad \text{ft/carrier} \tag{13}$$

If conveyor speed is v ft/sec, then the rate at which carriers arrive at insertion and extraction station is

$$\eta = \frac{1}{v} \text{ carriers/sec.} \tag{14}$$

Let

$$\delta = E(n) \tag{15}$$

be the expected number of parts in a buffer of capacity N, as determined by either of the methods mentioned previously ([4] or [5]).

Let us first consider the case in which stage 1 has just completed a production cycle and is attempting to insert a part into the closed-loop buffer. If the arrival of a carrier is synchronized to coincide exactly with the machine cycle, then the issue is whether the carrier is empty (as is desired) or loaded. Let p represent the probability that the carrier is empty (success); then, q = 1 - p is the probability that the carrier is loaded. This process can be viewed as a Bernoulli trial with parameter p. Then the distribution of the number of arrivals *x* until the first empty carrier is geometric with probability

$$p(x) = pq^{x-1} \quad x = 1, 2, \ldots \tag{16}$$

where the estimate of p is obtained from

$$p = \frac{N-\delta}{N} \tag{17}$$

The mean and variance of this process are

$$\mu_x = \frac{1}{p} = \frac{N}{N-\delta} \tag{18}$$

and

$$\sigma_x^2 = \frac{q}{p^2} = \frac{\delta / N}{[(N-\delta)/N]^2} = \frac{N\delta}{(N-\delta)^2} \tag{19}$$

The expected blockage delay due to waiting for an empty carrier is

$$\tau_I = \frac{\mu_x}{\eta} = \frac{N/(N-\delta)}{\eta} \text{ sec.} \tag{20}$$

Therefore, the production cycle for stage 1 is τ_I seconds longer than that for the infinite buffer, or

$$t_I = \frac{1}{q_{rI}} + \tau_I = \frac{1 + q_{rI}\tau_I}{q_{rI}} \tag{21}$$

and the adjusted production rate for stage 1 is $1/t_I$ or where q_{rI} is the

$$q_I = q_{rI}[N/\eta(N-\delta)] \tag{22}$$

In the case of stage 2, we are interested in determining the time delay until the first loaded carrier arrives so that a part can be extracted. Through a similar analysis, the mean number of arrivals *y* until the first loaded carrier is

$$\mu_y = \frac{1}{p_y} = \frac{N}{\delta} \tag{23}$$

with variance

$$\sigma_y^2 = \frac{q_y}{p_y^2} = \frac{N(N-\delta)}{\delta^2} \tag{24}$$

The mean delay due to starving is therefore

$$\tau_{II} = \frac{\mu_y}{\eta} = \frac{N/\delta}{\eta} \text{ sec.} \tag{25}$$

Then, the adjusted production rate for stage 2 is

$$q_{II} = \frac{q_{r2}}{1 + q_{r2}[N/\eta\delta]} \tag{26}$$

Finally, the adjusted line production rate is

$$q = \min\{q_I, q_{II}\} \tag{27}$$

Determining the Size of the Buffer
Suppose we have used the buffer size N to compute the expected storage E(n) by either of the methods [3] or [5]. Then we would want the mean delay for starving to equal the mean delay for blockage, so the adjusted production rates remain in the same relationship to one another as for the infinite buffer case. Therefore, from (18) and (23)

$$\mu_x = \mu_y$$

$$\frac{N}{N-\delta} = \frac{N}{\delta}$$

$$\delta = N - \delta$$

$$2\delta = N$$

$$\delta = N/2 \tag{28}$$

This is an interesting result, in that it says we want the expected number of parts stored on the closed-loop conveyor to be one-half the buffer capacity. This is an intuitively realistic result, since it is clear that a larger E(n) favors extraction by stage 2, while a smaller E(n) favors insertion by stage 1. This result suggests the following procedure for sizing a closed-loop buffer.

- Assume a buffer capacity N.
- Using the Gershwin and Berman approach [3] or the Savsar and Biles approach [5], determine the expected buffer storage E(n).
- Set N'=2E(x) and compute the expected delays due to blockage (τ_1) and starving (τ_2) from (20) and (25).
- Using the stage production rates q_{r1} and q_{r2} from either the Gershwin and Berman approach [3] or the Savsar and Biles [5] approach, compute the adjusted stage rates q_I and q_{II} from (22) and (26).
- Compute the adjusted line production rate from (27).

Although not exact, this procedure yields realistic estimates of the production rate of a two-stage automatic transfer line with a closed-loop conveyor buffer.

k - Stage Lines with Recirculation Conveyor Buffers

The results presented above for two-stage lines extend easily to k-stage lines if the following assumptions are made:

- All stages have the same number of machines.
- All stages have the same production rate p, failure rate λ, and repair rate μ.
- The first stage has an uninterrupted supply of parts (cannot be starved), and the *k*-th stage has an infinite output capacity (cannot be blocked).
- The interior stages 2 through *k*-1 can be blocked and starved. Blocking occurs if the succeeding buffer is full, or if the arriving carrier at the part insertion device is loaded. Starving occurs if the preceding buffer is empty, or if the carrier arriving at the part extraction device is empty.

Under these assumptions, the performance of the recirculating conveyor buffer is the same as that shown in equations (13) through (28). If the stages have different production rates p, failure rates λ, or repair rates μ, the stage having the minimum effective production rate paces the entire production line.

A SIMULATION STUDY

A simulation study was carried out to test and validate the analytical results reported in the previous section for a closed-loop recirculating conveyor buffer, and to compare its performance to that of a conventional static buffer. A simulation model was developed for this purpose using the Arena/SIMAN [6, 7] language. The automated production flow line consisted of 12 machines, each with a constant cycle time of 1.0 minutes. The failure rate for each machine is 0.025 failures per cycle (minute), and the repair rate is 0.125 repairs per minute. The mean time to failure is, therefore, 40 minutes and the mean repair time is 8 minutes.

The experimental design for this simulation study is as shown in Table 1. There were 3 replications of 24 treatments, for a total of 72 simulation trials. Each trial was run for 2880 minutes, with the first 480 minutes discarded as "warm-up" time, so that each simulation trial modeled a 40-hour week of production. The warm-up period mitigates the effects of starting the simulation run in an "empty and idle" state, resulting in a 2400-minute steady-state production period in each trial.

Table 1. Experimental Design for Production Flow Line Simulation Study

Factor	Number of Levels	Values
No. of Stages	4	2, 3, 4, 6
No. of Carriers per Buffer	3	20, 40, 60
Type of Buffer	2	Static, Recirculating
No. of Replications	3	--

Table 2 gives the results of this experimental design for the performance measure "production efficiency." The performance measure "production efficiency" is similar to the quantity expressed in equation (4), except that it also includes "time blocked" and "time starved" in the denominator.

Table 2. Simulation Results for Line Efficiency

No. of Stages	Type of Buffer: Static			Recirculating		
	Buffer Capacity					
	20	40	60	20	40	60
2	0.074	0.076	0.076	0.074	0.076	0.076
	0.074	0.074	0.074	0.074	0.074	0.074
	0.072	0.072	0.072	0.072	0.072	0.072
3	0.136	0.137	0.137	0.136	0.137	0.137
	0.126	0.133	0.133	0.126	0.133	0.133
	0.124	0.126	0.126	0.124	0.126	0.126
4	0.198	0.198	0.198	0.198	0.198	0.202
	0.176	0.176	0.176	0.175	0.176	0.176
	0.185	0.189	0.189	0.184	0.189	0.189
6	0.305	0.315	0.307	0.302	0.308	0.308
	0.294	0.303	0.303	0.292	0.303	0.303
	0.327	0.327	0.327	0.326	0.327	0.327

Table 3 gives the analysis of variance results for production efficiency. The F-value for the factor "Number of Stages" is statistically significant at any α- level one might choose, where as "Buffer Capacity" and "Type of Buffer" are not at all significant. Reexamining the results in Table 2, we see that production efficiency increases with the number of stages. Actually, the production efficiency for an unbuffered 12-machine flow line is

$$E=\frac{T_c}{T_c+FT_d}=\frac{1.0}{1.0+(\frac{12}{40})(8.0)}=\frac{1}{3.4}=0.294$$

The theoretical efficiency for a stage having k machines is

$$E=\frac{T_c}{T_c+\frac{k}{12}T_d}=\frac{1.0}{1.0+0.667k}$$

If $k = 2$ for a production line having six stages, then

$$E=\frac{1.0}{1.0+1.33}=\frac{1.0}{2.33}=0.429$$

For this 12-machine line, the greatest number of stages one could have is 12, so that $k = 1$. Therefore, the greatest line efficiency one could achieve for this line is

$$E=\frac{1.0}{1.0+(\frac{1}{12})(8.0)}=\frac{1.0}{1+0.667}=\frac{1}{1.667}=0.6$$

This efficiency assumes there is zero production time lost due to blocking and starving.

Table 3. Analysis of Variance for Line Efficiency

Source	DF	SS	MS	F	P
Stages	3	0.556701	0.185567	2653.92	0.000
Buffer Size	2	0.000133	0.000066	0.95	0.393
Buffer Type	1	0.000001	0.000001	0.02	0.888
Error	65	0.004545	0.00007		
Total	71	0.56138			

Table 4. Simulation Results for Time Part Spent in System

No. of Stages	Type of Buffer: Static			Recirculating		
	Buffer Capacity					
	20	40	60	20	40	60
2	298.95	317.64	317.64	294.90	317.66	317.67
	238.20	238.20	238.20	233.47	233.53	234.67
	244.13	244.13	244.13	240.09	236.58	237.04
3	247.92	247.73	247.73	247.11	244.87	346.43
	222.81	318.32	336.44	217.25	312.76	336.39
	210.75	282.28	282.28	205.57	275.47	272.72
4	191.08	226.66	228.01	186.54	216.23	216.85
	195.45	263.60	307.49	191.83	256.45	294.95
	195.43	232.39	235.56	190.91	225.37	226.02
6	196.45	226.87	230.78	72.59	82.13	81.60
	212.69	344.34	393.67	60.39	75.84	76.83
	199.16	264.02	294.37	66.07	90.19	95.95

Table 5. Analysis of Variance for Time Part Spent in System

Source	DF	SS	MS	F	P
Stages	3	111659	37220	12.41	0
Buffer Size	2	34152	17076	5.69	0.005
Buffer Type	1	40356	40356	13.46	0
Error	65	194928	2999		
Total	71	381096			

Tables 4 and 5 give the simulation results and analysis of variance for the performance measure "time-in-system." This is the mean time a part spends in the production flow line from the instant it is placed in machine 1 to the instant it leaves machine 12. "Time-in-system" is comparable for static buffers and recirculating conveyors for 2, 3 and 4-stage lines, but recirculating conveyors showed greatly reduced time-in-system for 6-stage systems. The analysis of variance results given in Table 5 show that all three factors exert a significant influence on "time-in-system" at an α level of 0.05, with "Number of Stages" and "Type of Buffer" being particularly influential.

Summary and Conclusions

This paper has described a novel procedure for analyzing a k-stage production flow line in which the buffer consists of a closed-loop recirculating conveyor system, instead of the typical static storage facility assumed by other approaches. Equations are given for adjusting the stage production rates for a k-stage production line with infinite buffer by estimating the blockage delay imposed upon stage l and the starving delay for stage $l+1$ caused by having those stages wait for the arrival of a part carrier that is either empty (stage l) or loaded (stage $l+1$) so that part insertion or extraction can take place.

References

Dallery, Y. and Gershwin, S. B., Manufacturing flow line systems: a review of models and analytical results, *Queueing Systems*, 12, 3-94, 1992.

Bugacott, J. A., Automatic transfer lines with buffer stocks, *International Journal of Production Research*, 5(3), 183-200, 1967.

Gershwin, S. B. and Berman, O., Analysis of transfer lines consisting of two unreliable machines with random processing times and a finite buffer storage, *AIIE Transactions*, 12(4), 2-11, 1980.

Levin, A. A. and Pasko, N. I., Calculating the output of transfer lines, *Machines and Tooling*, 40(8), 12-16, 1969.

Savsar, M. and Biles, W. E., Two-stage production lines with a single repair crew, *International Journal of Production Research*, 22(3), 499-514, 1984.

Kelton, W. D., Sadowski, D. A., and Sadowski, R. P., *Simulation with Arena*, Systems Modeling Corp., Sewickley, PA, 1996.

Pegden, C. D., Shannon, R. E., and Sadowski, R. P., *Introduction to Simulation Using Siman*, 2nd Ed. McGraw-Hill, New York, 1995.

Banks, J., Carson, J. S., II, and Nelson, B. L., *Discrete-Event System Simulation*, 2nd Ed., Prentice Hall, Upper Saddle River, NJ, 1996.

A MANUFACTURING CELL CONTROLLER ARCHITECTURE

***António Quintas* [1] *, Paulo Leitão* [2]**

[1] *Professor Associado do DEEC da Faculdade de Engenharia da Universidade do Porto, Rua dos Bragas, 4099 Porto Codex; Tel: 351.2.2041844; email: aquintas@fe.up.pt*
[2] *Assistente da ESTG do Instituto Politécnico de Bragança, Quinta Sta Apolónia, Apartado 134, 5300 Bragança; Tel:351.073.3303083; email: pleitao@ipb.pt*

ABSTRACT : Worldwide competition among enterprises has lead to new needs in the area of manufacturing to answer for price, quality and delivery time. The improvement of productivity and flexibility in manufacturing systems by the introduction of new concepts and technologies, and by the appropriate integration of the different resources, may constitute a key factor for the solution towards the success.
This paper describes the specification and implementation of a Manufacturing Cell Controller integrated into the demonstration pilot of an ESPRIT Project.
Special attention is given to the control structure, communication system, information system and the way the integration was done using the MMS (Manufacturing Message Specification) communication standard. Be also important, an analysis of the performance and costs of the solution is discussed together with alternative development platforms and technologies to achieve a better one.

1. INTRODUCTION

Worldwide competition among enterprises lead to a need for new control systems, which perform the control and the supervision of the manufacturing through the integration of the resources. The implementation of these technologies is the answer to the improvement of productivity and the quality, and the decrease of the price and the delivery time.
The control systems used to integrate and to control the resources of a manufacturing or assembly cell, is called by Cell Controller [1]. Mainly, the Cell Controller must allow the following functions [1,2]:

- the integration of the cell resources;
- the manufacturing control and supervision;
- real time scheduling of cell production, based upon the real time cell capacities;
- the storage and the download of the NC and the RC programs;
- the reaction to fault conditions.

The design and the implementation of a Cell Controller is a complex task, involving real time control restrictions and the manipulation of different operating systems, different

communication protocols and machines supplied from different vendors. The specification of a Cell Controller architecture results into the definition of two main points: control structure and communication system.
The control structure is a key factor for the final performance of the Cell Controller and it can be performed with some basic control architectures: centralized, hierarchical and heterarchical [3]. With the powerful PCs, inexpensive and widely available, the architectures evolued from centralized architecture to the distributed architectures, allowing the improvement of the Cell Controller performance.
The communication system is crucial to implement the integration of the cell resources, as to transfer to them the NC and RC programs and also to control these devices. There are several protocols available to implement the communication system: MAP (Manufacturing Automation Protocol) /MMS, FieldBus, serial link, TCP/IP. The choice of a communication protocol depends upon the costs, the standardization, and the control degree available.
Following the paper, a Cell Controller architecture solution is described, with the analysis of his performance. Then, other platforms and technologies for the Cell Controller are discussed and a comparative performance evaluation is presented.

2. MANUFACTURING CELL CONTROLLER

2.1 Flexible Manufacturing Cell

The manufacturing cell has two CNC machines and an anthropomorphic robot for the load/unload of the machines. One of these machines is a turning center *Lealde TCN10*, with a SIEMENS *Sinumerik 880T* controller; the other machine is a milling center *Kondia B500* model, with a FANUC *16MA* numerical control. The robot is a KUKA *IR163/30.1* with a SIEMENS *RC3051* controller. The manufacturing cell has two transfer tables for the containers loading and unloading. These containers bring the material to be operated into the cell and take away the pieces produced [4].

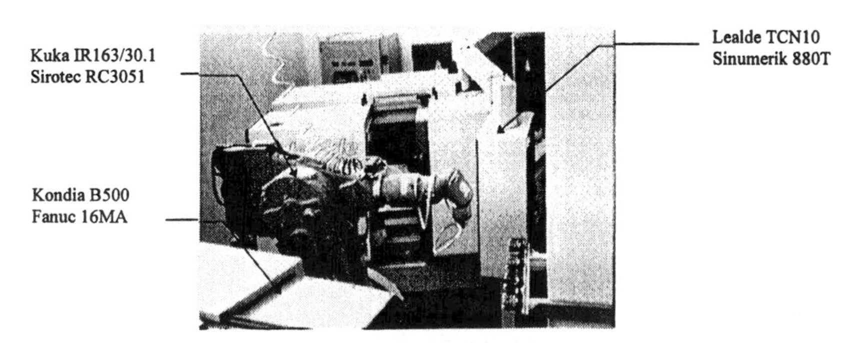

Figure 1 *- Manufacturing cell layout*

The Cell Controller was developed and implemented in a Sun SparcStation 10 workstation with Solaris 2.4 Operating System. This workstation has a network card with two stacks:

- **TCP/IP stack,** for the communication with the Shop Floor Controller and the Project Department. This network allows the transmission of NC and RC programs from fileserver to the industrial machines;
- **OSI MAP stack,** for the communication with industrial machines, like NC machines and robots.

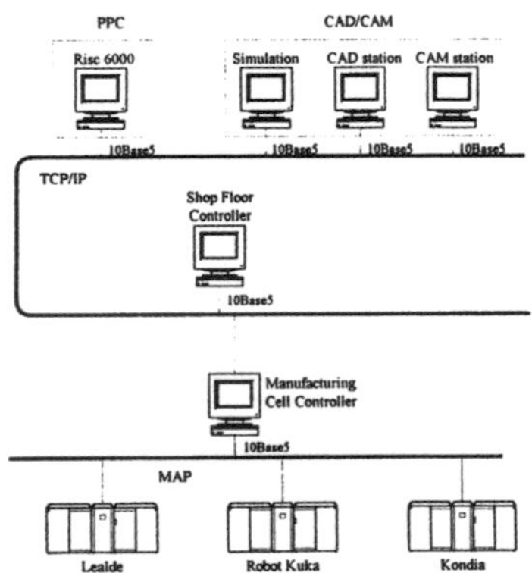

Figure 2 - Shop Floor communication infrastruture

The manufacturing cell is connected to the controlling room by a LAN with a bus structure topology, based on a base band transfer media (10Mb/s). The LLC protocol used is 802.3 (Ethernet CSMA/CD). All the machines have MAP interface boards. These interfaces are: **CP 1476 MAP** for Siemens Sinumerik 880T machine controller, **CP 1475 MAP** for Siemens Sirotec robot controller and **GE FANUC OSI-Ethernet Interface** for GE Fanuc 16MA numerical controller.

2.2. Manufacturing Cell Controller architecture

The Cell Controller architecture implemented in the manufacturing cell is a set of several modules, that share information stored in a local database [4]. The definition of control structure for Cell Controller was done keeping in mind two important aspects:

- **modular structure**, which allows the future expansion of the control system, for instance, if the number of machines grows up;
- **real time requirements**, to guarantee that control system is able to execute all required functions performing the time restrictions, for example the cooperation between machines.

The specified architecture for this Cell Controller is based in the hierarchical architecture; the initial control system is decomposed into individual subsystems, and each one of these subsystems can be updated as necessary without affecting the rest of the system.
This structure has three hierarchical levels, being each one responsible for the execution of control functions. The first level, the Manager Module, is the brain of the Cell Controller, and it is responsible for the control and the supervision of the production process of the manufacturing cell and also for the management of cell resources. The main functions of this module are:

- Start the Cell Controller, verifying if it is a normal or abnormal start;

- Receive and process the messages from the Shop Floor Controller;
- Receive and process the results of services executed by the Device Controllers;
- Determine for each order, the next operation to be executed and dispatch these orders to the Device Controllers;
- Notify the Shop Floor about the evolution of the orders and whenever an alarm occurs;

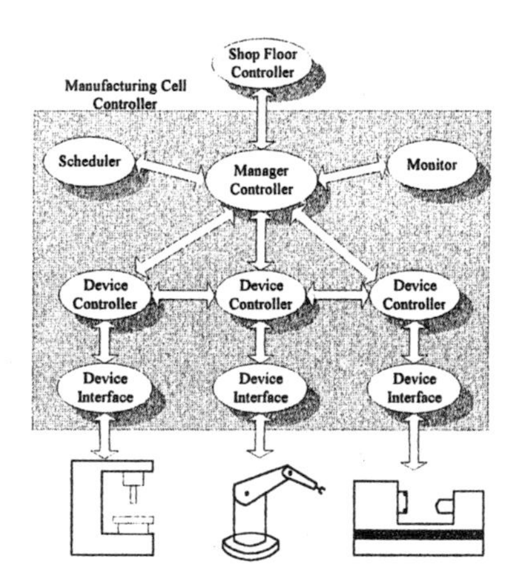

Figure 3 - Cell Controller structure

The management of the manufacturing cell is based upon information stored in the cell database. This database contains information about orders, resources, cell buffers and tools stored in the industrial machines of the cell. In the second level, there are several modules to control industrial machines. Each of these modules, designated by Device Controller, is customized to the industrial machine and it has the responsibility for the execution of the jobs in the machines. Finally, the last level, the Device Interface, contains the interface between the Cell Controller and each of the industrial machines, implemented with the MMS protocol.
All Cell Controller code is written in C language and the communication between modules, inside the Cell Controller, is implemented with a Unix Operating System functionality, called pipes, which is made up of files to exchange the messages.

2.3 MMS, A Standard Protocol

Flexibility and open systems concept lead us to the application of an open systems communication standard protocol at the manufacturing process control level. MMS is a standardized message system for exchanging real-time data and supervisory control information between networked devices and/or computer applications, in such a manner that it is independent from the application function to be performed and from the developer of the device or application. MMS is the international standard ISO 9506 [5], based upon the OSI (Open Systems Interconnection) networking model - it's a protocol of the application layer of the OSI model. It runs on the top of an OSI stack providing a set of 80 services distributed for ten MMS functional classes.
The MMS protocol implements the interface communication between the Cell Controller and the industrial devices. The key features of MMS are:

- **the "Virtual Manufacturing Device" (VMD) model**. The VMD model specifies the objects contained in the server, and with MMS services it is possible to access and manipulate these objects;
- **the client/server model**. The client is a device or application that requests data from the server; the server responds with data requested by the client.

There is a distinction between a real device (e.g. PLC, CNC, Robot) and the real objects contained in it (e.g. variables, programs), and the virtual device and objects defined by the VMD model. Each developer of a MMS server device or MMS server application is responsible for "hiding" the details of their real devices and objects. The executive function translates the real devices and objects into the virtual ones defined by the VMD model.

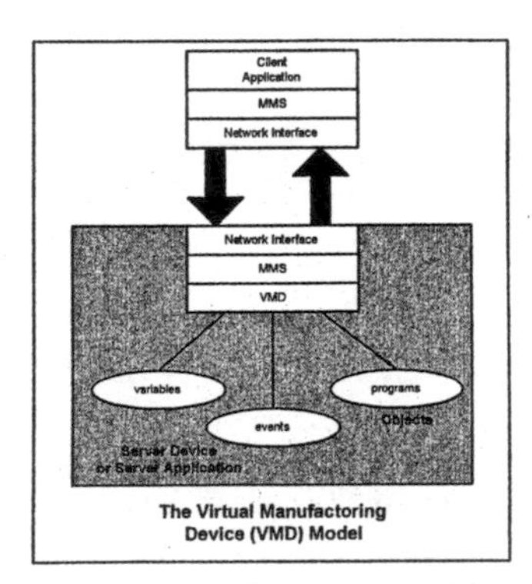

Figure 4 - The VMD model

In many cases, the relationship between the real objects and the virtual ones can be standardized for a particular class of device or application (PLC, CNC, Robots, PCs). Developers and users of these real devices may decide to define more precisely how MMS is applied to a particular class of device or application - the result is a Companion Standard.
It is possible to note the missing of some implementations that would be useful from the control application point of view. There is a gap between the MMS functionalities specified in ISO 9506 [6] and the objects and services provided by the MMS server applications for the controllers we work with. As an example the end of execution of a NC/RC program can be asynchronously reported by the server application by means of an event notification unconfirmed MMS service. This feature is only implemented in the MMS server application for the *Sinumerik* NC controller. The MMS server application for the *GE FANUC* NC controller reports this occurrence with an unsolicited status unconfirmed MMS service. In the case of the robot MMS server application there is no mechanism for the reporting of this occurrence which implied the development of a polling function with the associated lost of efficiency.
It has also been possible to note different solutions in the modeling of some real objects to MMS objects for the MMS server applications related to the two NC controllers. We expect that in the future this problem is solved with the application of MMS Companion Standards.

3. PERFORMANCE ANALYSIS

3.1 Data Transmission rates

The execution of a program in a industrial device requires the existence of this program inside the device memory. Due to the memory capacity limitation, it is necessary to transfer the program to the device whenever it will be necessary. This operation is performed by the Cell Controller with the MMS Download service. During the Cell Controller implementation tests, and with a network analyzer, it was possible to work out the transfer speed: 10 kbits/s. This speed is very slow and origines the Cell Controller lost of efficiency. For example, the Cell Controller spends approximately 30 seconds to download a NC program with 50 kbytes! This time is not profitable in the processing tasks, and causes the lost of productivity in the manufacturing cell.

With this slow transfer speed, it was necessary to know where the problem was located. After several tests, it could be concluded that the problem was not associated with MMS protocol, because two MMS applications, running at SUN worksation, could communicate with 10 Mbits/s transfer speed; thus, the problem is related to the internal machine problems. In fact, when a machine receives a program, this is analyzed inside the machine, line by line, to detect errors inside the program. This procedure causes the reduction of data transmission from 10 Mbits/s to 10 kbits/s.

For increasing the Cell Controller efficiency, it is necessary to do the management of the NC programs inside the machines memory and download the needed NC programs into the Cell Controller, making use of it´s dead times. This is a complex solution to be implemented, and requires the introduction of Artificial Intelligence techniques inside the system control.

5.2 Costs of the solution

The cost is an important factor that will be decisive for the viability of a solution. It is possible to quantify the cost associated with this Cell Controller architecture solution, by analyzing the costs of the several platforms and interface boards used to implement the Cell Controller. These costs can be divided into three main areas: the communication software, the interface boards and the developing platform.

The cost of the MAP/MMS interface boards is around $6000 (US dolars) each one. In this application, the number of MAP/MMS interface boards is 4 (3 servers/machines and 1 client), so it is necessary to spend $24000 to provide all cell resources with MAP interface boards. The price of the MMS software (MMSEASE from SISCO in this case) is around $6700. The cost of the developing platform is $15340, due to the cost of the SUN SparcStation 10 and the developing software SUN SparcWorks.

For a better perception of this cost value, it is possible to make a comparation with the cost associated with the resources of the cell (two NC machines and one robot). The cost of these three machines is around $333300. So, the communication system cost value is approximately 9,2% of the costs associated with the resources of the cell, and the cost of Cell Controller is approximately 13,8%. However, taking into consideration the cost related with the developing phase, it can be assumed the reduction of this percentage (9,2%), due to the less effort allocated to this phase when the MMS solution is adopted.

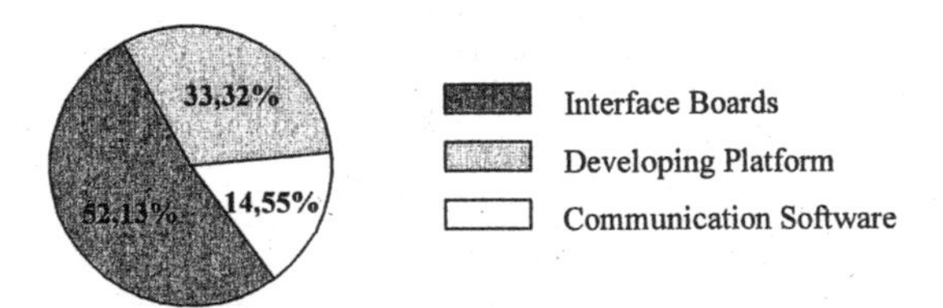

Figure 5 - Cost of the Cell Controller architecture

Even so, with this scenario, it can be concluded that it is expensive to use a standardized communication protocol like MAP/MMS, basically because of the price of the interface boards quite high. In alternative, other communications protocols, such as the FieldBus, could realise the communication system. The need also for new platforms like Windows NT suggests even higher costs not supportable by the generality of SMEs (Small and Medium Enterprises)

4. NEW TRENDS OF THE CELL CONTROLLER ARCHITECTURES

Actually, the questions around the design and the implementation of the Cell Controller solutions are focused in the new technologies and methodologies, to increase his performance and to decrease the developing time of the solution. What is the best communication protocol? What is the adequate developing platform? What is the best control system to be implemented? The answers to these questions require the analysis of the technologies and the methodologies available, together with the requirements of the specific problem.
The trends of the Cell Controllers can be characterized in the following points:

- distributed and user-friendly developing platforms, which allow a quick, easy and also a cheap solution;
- communication protocols, like MMS and Fieldbus, which must allow the easy integration of the resources supplied by different vendors;
- new programming and control paradigms, such as object oriented and Artificial Intelligent methods, which allow the improvement of the solution performance;
- new generation of manufacturing systems, such as holonic, biological and fractal systems, which allow the rapid development and the flexibility needed for the manufacturing systems.

The following points will dispute the application of these technologies and methodologies, with the special attention to the manufacturing systems requirements.

4.1 Developing Platforms

Newish PC operating systems, such as Windows NT, offer the same features of the Unix systems, with the advantages of low setup, administration effort and cost. These operating systems present a platform windows compatible, a good market penetration, an inexpensive platform and a wide range of development tools available.
An important problem with the Unix platforms is the monitoring or graphical interface is very complex to implement. The use of Visual C++, for instance, allows the reduction of development costs and a more quick implementation of the solution for the desired application, using all the potentialities of these programming tools. The Windows NT platform allows the

reduction of hardware costs and the use of more user-friendly programming tools, like Visual Basic or Visual C++.

4.2 New Programming and Control Paradigms

Due to the complexity and diversity of the problems to be solved, it is necessary to use new programming paradigms, such as Object Oriented, instead of traditional programming. The traditional programming should be used only if combined with artificial intelligence methods, such as knowledge based systems.

The Object Oriented paradigm is a new approach, which increases the potentiality of the programming tools. Object oriented programming is a paradigm for a software design and implementation, which is organized as a collection of objects, that operate by exchanging each other, messages of services request. [7]. An object is simply something that makes sense in an application context, and it is characterized by encapsulating his data, designated by attributes, and offering a set of services, designated by methods, to access the data.

Another ongoing research area in the Cell Controller design and implementation is the artificial intelligence methods. The application of the artificial intelligence methods allows the resolution of complex control tasks, such as the faults recover and the resources control. There are several AI methods available to be implemented in the Cell Controllers solutions, such as Expert Systems, Neural Networks and Fuzzy Logic.

The knowledge-based systems use the human and the specialist experience to solve the problems. The knowledge is stored in a database, which can be updated through the system learning, and a so-called inference engine can make decisions with a higher efficiency.

4.3 Communication Protocols

The integration of the industrial resources, using a Cell Controller, interfers with the lower levels of the ISO model: cell, equipment and sensors.

The implementation of the RS232 or RS485 protocol is very cheap, but it requires the development of an API (Application Program Interface) for each industrial resource, customized to the resource funcionalities. For a higher number of resources, this task is complex due to the different functionality of each resource. Thus, the RS232 is acceptable for applications with low degree of control and when it is only necessary to transfer the NC and RC programs to the industrial resources. Applications to control a lot of resources (supplied from different vendors) and with some control and supervision tasks, need a standard communication protocol, such as MAP or Fieldbus.

The MAP architecture is based upon the OSI reference model, specifying an ISO protocol for each layer. In the application layer, two important protocols are specified: MMS for the remote control and monitoring of the industrial resources, and FTAM (File Transfer Access Management) for the files remote access. The advantages of this architecture are the suppliers independence, the decrease of developing cost and the decrease of maintenance cost.

The CNMA project [5], developed under the ESPRIT programme, is compatible with the MAP architecture, and additional specification of the RDA (Remote Database Access) protocol in the application layer. These architectures are a good solution for applications with a large number of different resources (from different suppliers); nevertheless, in situations, which require temporal restrictions, these architectures are to heavy. Other additional disadvantage is the cost of the interface boards, which dissuades the expansion of this communication protocol in the industrial market.

The Fieldbus networks are traditionally used to interconnect sensor and actuator devices, localized in a small local area. These industrial networks only implement three layers: physical, data link and application layers. The physical layer uses the RS485 protocol, adequate to the industrial communications due to the good noise immunity, and also it has the possibility to implement the MMS protocol in the application layer. These networks are characterized by the good adaptation to adverse ambients, an easy update of new resources and the low cost of the interface boards and the maintenance.
The limitative factor to the expansion of the Fieldbus networks is the lack of the standardization in this architecture. In fact there are several standards for the Fieldbus [8]:

- **FIP (Factory Instrumentation Protocol)**, supported by the french manufacturers;
- **PROFIBUS (Process Fieldbus)**, supported by Siemens, Bosh and Kloeckmer-Moeller;

The FiCIM project [9] was an effort to create an unique fieldbus standard, but the aim of the project wasn't reached.

4.4 New Generation of Manufacturing Systems

Manufacturing industry comprises the manufacture of a variety of products in small sized lots, with a small delivery time. These requirements imply the development of new manufacturing control systems with more autonomy and more intelligence, able to handle to the changes and disturbances much better than the actual control systems. There are several theories for the manufacturing systems, which involve a decentralized control system, characterized by intelligent and cooperate nodes, such as holonic, biological, fractal and genetic manufacturing systems. These theories have some similar concepts, but they can be distinguished by their origin: mathematics for the fractal, nature for biological and social organization for holonic.
The Holonic control system is based upon a structure with several holons, each one being autonomous and able to cooperate with the other holons. An holon is a particle or module, and it can represent a physical or logical activity, such as a machine, an order or an operator. The holonic paradigm consists in broking complex tasks into a several sub-tasks, which in turn could be broken into further sub-tasks, which allows the reduction of the problem complexity [11].

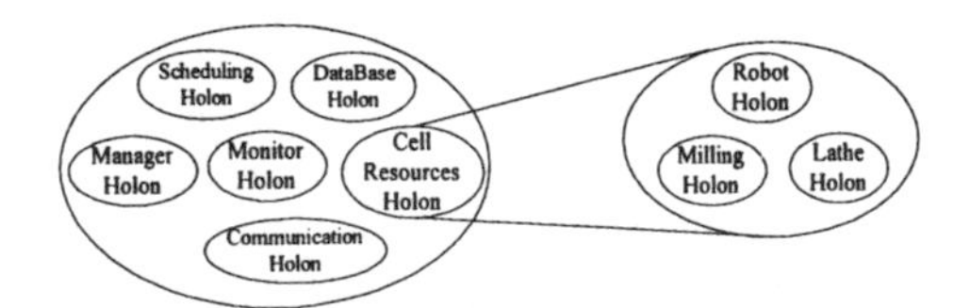

Figure 6 - Holonic system architecture

This structure presents the following benefits:

- fast reconfiguration in response to the system strategical exchange;
- increase of the system flexibility, because it is possible to consider the operator as a holon;
- increase the capability for the expansion of the system, because it is easier to do.

One approach which derives from the Distributed Artificial Intelligence is the multi-agents systems concept. The multi-agents architecture could be defined as a set of nodes, designated by

agents, that represent the manufacturing system objects, such as resources and tasks. The agents are autonomous and intelligent, and they can communicate together to perform the required tasks. In this architecture, the negotiation between different agents is one of the most important problem to solve. This negotiation can be implemented using the Contract Net Protocol [11].
The dissemination of these theories will allow the future implementation of modular, flexible and intelligent Cell Controller architectures, with the advantage of flexibility and the improvement of the manufacturing systems intelligence.

5. CONCLUSIONS

The cell controller architecture presented in this paper is now working well. The two major aspects of a Cell Controller specification were focused: the control structure and the communication system. In the control structure, the use of decentralized architectures is growing up, mainly due to the increase of the computing system capacities and the reduction of the prices. The control structure presented in the Cell Controller solution is an hierarchical architecture, allowing the flexibility and the modularity of the application. The communication system is implemented with the MAP/MMS protocol due, basically, to it's standardization and the simplicity of it's implementation.
With the analysis of the MMS implementation, it was possible to conclude that the costs of the MMS interface boards may be the limiting factor to the implementation of MMS protocol in the integrated applications. Nevertheless, the MMS protocol is a good solution for applications which require a total control of resources, or when the integration involves a great number of resources with high degree of complexity.
The trends of the Cell Controller technologies and methodologies, look for Cell Controller architectures with low cost and with short developing time, which can be afford by SMEs. The cell controllers for the next future are mainly based upon four points: developing platforms, new generation of manufacturing systems, communication protocols, new programming and control paradigms. The development and implementation of these new concepts will allow the increase of the Cell Controllers performance, operationality, and also the decrease of their cost.

6. REFERENCES

[1] **Groover, M. P.**, Automation, Production Systems and CIM, *Prentice-Hall*, 1987

[2] **Rembold, U., B.O. Nnaji, B.O.**, Computer Integrated Manufacturing and Engineering, *Addison-Wesley*, 1993

[3] **Diltis, D., Boyd, N., Whorms, H.**, The evolution of control architectures for automated manufacturing systems, *Journal of Manufacturing Systems*, Vol 10 N°1, pp 63-79, 1991

[4] **Quintas, A., Leitão, P.**, A Cell Controller Architecture Solution: Description and Analysis of Performance and Costs, *Proceedings of Integrated and Sustainable Industrial Productions*, Lisboa, May ,1997

[5] **CCE-CNMA 2**, MMS: A Communication Language for Manufacturing, *ESPRIT Project CCE-CNMA 7096*, Volume 2, Springer, 1995

[6] **ISO/IEC 9506-1**, Industrial Automation Systems - Manufacturing Message Specification, Part 1 - Service Definition, 1992

[7] **Rumbaugh, J., Blaha M., Premelani, W., *et al***, Object-Oriented Modelling and Design, *Prentice-Hall*, 1991

[8] **Pimentel, J.**, Communication Networks for Manufacturing, *Prentice Hall*, 1990

[9] **ESPRIT Project 5206**, FiCIM (Fieldbus Integration into CIM)-Documents, 1992
[10] **Cantamessa, M.**, Agent-based Modelling of Manufacturing Systems, *Advanced Summer Institute 95*, pp 125-131, 1995
[11] **Smith, R.G.**, The Contract Net Ptrotocol:High-Level Communication and Control in a Distributed Solver, *IEEE Transactions on Computers*, Vol C-29, N°12, pp 1104-1113, 1980

DESIGN AND REALIZATION OF INTELLIGENT AUTOMATED WORK-CELLS

Heinz Ulrich, Willi Dürig

ETH Zurich
Institute for Operations Research
CH - 8092 Zurich
Switzerland

E-Mail: [ulrich, duerig]@ifor.math.ethz.ch
WWW: http://www.ifor.math.ethz.ch/

ABSTRACT: We describe a prototypical solution of a process planning and control system for a work-cell based on the idea of dynamic adaptation and self-control which is already partly realized in practice. The key concept relies on what we call "agent-centered" design and control. Instead of allocating the planning intelligence to one authority we distribute planning tasks to agents in real time and provide information access when asked for. Each agent possesses a variety of problem solving strategies to cope with different situations. One particular focus in this project was the development of adequate man-machine interfaces to take advantage of the human potential in knowledge and experience as well. Finally we try to outline a methodology for the design of such process planning and control systems.

1. INTRODUCTION

In today's manufacturing industry automation is one important mean to improve efficiency. An automated manufacturing process is based on a network of individual work-cells. The planning and control task has therefore to include the management of automated processes. The legacy systems in operation however often lack the capabilities to deal with "extra-ordinary" situations: Confined to a stable set of inputs process planning consists mainly in designing rules to manage optimally the coordination of shared resources. As long as the environment of the system is stable such concepts are quite successful. In case of unforeseen situations such as a partial breakdown of certain capacities, unexpected workloads etc. the process planning system may become more a burden than a support.

2. REQUIREMENTS

To determine today's requirements for planning and control of automated manufacturing processes we have to consider two aspects:

First the dramatic change in economical conditions is to mention. The development towards a global economy with global markets and global competition is for most manufacturing companies a tough challenge for survival. The resulting requirements for manufacturing industry in logistics management can be summarized as follows:

- *High productivity*
- *High flexibility*
- *Fast reactions*

On the other hand modern information and communication technology offers promising new features to realize planning and control systems. The Electronic Superhighway realized until now by the Internet indicates the beginning of the information age. A worldwide network for information exchange is established herewith. Modern computer technology offers distributed systems for data processing and storage. As a consequence we can state:

Information and computer performance becomes globally available.

Logistics management is particularly addressed by the above mentioned requirements for manufacturing industry. To take full advantage of modern information and communication technology we suggest a new design for planning and control processes:

Planning and control as *intelligent adaptive process.*

The planning and control process in manufacturing industry can not be automated totally, Human resources will always be involved, be it as operator with unavoidable manual tasks, be it as manager with management responsibility. In the design of the planning and control process beside the potential for automation human resources have to be taken into consideration from the beginning. Therefore we define the work-cell as socio-technical system

3. WORK-CELL AS SOCIO-TECHNICAL SYSTEM

In this socio-technical system we have to deal with an automated production process and the directly involved human resources. The success of the design does not only depend on independent abilities of human and technical resources but also on the quality of their mutual interaction. The task allocation between human and technical subsystem is therefore a crucial question in the system design. Frequently used criteria like economic and technical feasibility and allocation to the presumably better performer have proven to be insufficient as to optimal interaction between operator and technical system. We suggest a *complementary* approach taking into account differences in strengths and weaknesses of human and technical subsystems. Their adequate interaction provides a performance impossible to human or technical systems alone [1] [2] [3].

4. INTELLIGENT AUTOMATION

An automation of a manufacturing process comprises the automation of the technical and the logistical process. We assume a highly automated technical process requiring only manual human interventions for set-up, emergency and management operations and focus on the automation of the logistical process. This automation we intend to design as efficient as possible which implies capabilities we define as intelligent. This intelligence consists in a capability for problem solving. For our purposes we distinguish two levels of intelligence, a reactive and a cognitive level, a distinction which is well known in the community of artificial intelligence.

Intelligence	Reactive	Cognitive
Base	Knowledge	Knowledge , Methods
Activity	Search process	Problem solving procedure

Reactive intelligence bases on knowledge which is represented in rules or in decision trees. The solution process provides an adequate reaction to the actual situation and consists of a search procedure on the individual knowledge base. Reactive intelligence comprises all the well known conventional automation where an apparatus reacts to input signals according to well-defined rules The possible outcome is selected among a number of predefined actions (e.g. finite state automata).

Cognitive intelligence includes reactive intelligence but in addition more demanding problem solving procedures are at hand. A base of methods offers a selection of methods ready for use. These methods can be applied in a problem solving procedure where solutions are generated in view of explicit goals, valued according to relevant criteria to choose finally the best one for realization.

An intelligent automation of manufacturing processes has to deal with several distributed interacting processes running simultaneously, centralized solutions hardly can meet today's requirements as to flexibility and reactivity therefore. Distributed problem solving concepts are asked, one possible approach we will be presented in the following.

The artificial intelligence community is discussing agent based concepts which may be useful in our context.

The agent concept

Definition

For our purposes we use the term agent in its weak notion according to M. Wooldrige and N.R. Jennings [4]:

An agent is an entity in a software-based computer system with the following properties:

- *autonomy:* agents operate without the direct intervention of humans or others, and have some kind of control over their actions and internal state;
- *social ability:* agents interact with other agents (and possibly humans) via some kind of agent-communication language;
- *reactivity:* agents perceive their environment (which may be the physical world or the model), and respond in a timely fashion to changes that occur in it;

- *proactivity:* agents do not simply act in response to their environment, they are able to exhibit goal-directed behavior by taking the initiative.

Agent types

According to the above definition agents have to be equipped with intelligence. Similar to the distinction of two degrees of intelligence, we distinguish corresponding to their degree of intelligence two types of agents.

The reactive agent
Reactive agents are provided with an intelligence of reactive quality, that is they act using a stimulus/response type of behavior and are able to respond to the present state of the environment in which they are embedded. They do not have a representation of their environment in their knowledge base.

The cognitive agent
Cognitive agents are provided with an intelligence of cognitive quality, that is they are driven by intentions, i.e. by explicit goals that conduct their behavior and make them able to choose between possible actions. They have a symbolic and explicit representation of their environment in their knowledge base on which they can reason and from which they can predict future events.

Problem solving engine

Each agent is able to solve problems independently. The proceeding for their problem solving is called engine. A problem solving engine is designed according to a general problem solving proceeding in management science e.g. Systems Engineering [5]. The following cycle for problem solving is proposed:

1. *Problem definition:* Analysis of the present state in search for problems which need to be solved.

2. *Search for goals:* Among the given goal or goal hierarchy choosing the most adequate operational goal to be addressed in the above identified problem.

3. *Search for solutions:* Based on the available potential on knowledge and methods possible solutions are generated

4. *Final decision:* The generated solutions are valued according to criteria corresponding the operational goal determined in step 1.

In case of the lower intelligence level of the reactive type, the last two steps are reduced to one single step, a simple search procedure.

5. THE MULTI-AGENT SYSTEM

To realize distributed problem solving for automation of logistical processes an integration of several agents in a system which is then called multi-agent system [6] is proposed. This community of interacting agents has to be organized. We suggest a structure following human management organization as guidelines as close as possible.

As multi-agent system we define therefore a collection of agents of reactive and/or cognitive intelligence type with internal organization in respect of communication and hierarchy in management level. All agents have their specific planning and control task which they fulfill together in a well defined and cooperative way. Each agent is built up according to a similar internal structure.

Internal structure of an agent

In agent's design we propose a general internal structure for each agent containing the following elements:

Characterization
The characterization comprises individual name and localization within the system.

Intelligence degree
The competence of an agent as to its intelligence is indicated by his association to one of the two agent types (cognitive or reactive).

Position
Each agent has a well-defined position in the management hierarchy level within the agent community and the communication links with other agents are determined.

Information base
An agent has an individual information base which can be extended by the defined communication links. The information base contains knowledge, methods and information about the system state within the running process.

Task
Each agent has well-defined responsibilities and herewith related specific tasks he has to complete alone or in cooperation with other agents.

Goals
Related to his responsibilities an agent has his given goals he is able to adapt to operational goals for his specific tasks. The coordination of the individual goals of the agents is performed within the management process and is a crucial problem of its design.

Capabilities
For problem solving each agent has his individual problem solving engine designed to solve his individual tasks. His capabilities are in correspondence with his available degree of intelligence. In case of a cooperative problem solving each agent may use his engine in a problem solving cycle several times, until the common solution promises satisfactory results.

System Control

The dynamic flow of a process is usually represented in a simulation model. For management processes in particular discrete event simulation is the most appropriate method. The process control is performed by an event list containing events and their due date. The events are sorted after their due dates and executed in the determined time sequence. In a multi-agent system this simple procedure is not sufficient. Events are not only generated as a consequence of other events, agents with there proactive behavior are independent actors in the system generating additional events.

System control in a multi-agent system is therefore a more demanding task which has to include the synchronization of agent's activities as to simultaneousness and clearing up of contradictions in actions.

We suggest the following approach for system control in a multi-agent system:

As in discrete-event simulation we have a sequential processing of the event list. Each agent enters his next proactive action in the event list. An event in the event list is processed as follows:

- Consolidation of the present system state (updating)
- Cyclic call to all the agents for reaction and release of the responding actions.
- Clearing up of conflicts resulting out of the initiated actions beforehand.
- Consolidation of the new system state (unique definition of the result)
- Proceeding to the next time in the event list and repeating the same procedure again

Multi-agent system for work-cells

A work-cell is a socio-technical system, as to logistics, human resources and automated systems have to solve similar management problems. In the design of a automated systems we therefore try to take advantage of the available experience with management by human resources. The internal organization of a multi-agent system for work-cells we conceive in analogy to human organizations with hierarchical levels and increasing abilities and competence towards the higher levels as it is shown in figure 1.

Special design problems

Organizational structure with local autonomy: One key motivation for an agent approach is the reduction of system complexity. In human management organizations concepts of local autonomy and self-regulation are discussed. As these efforts tend also in the direction of complexity reduction, local autonomy is an important issue in multi-agent system design as well. We therefore experiment with systems consisting of locally autonomous agents with their own tasks and goals and master agent having only coordinating and supervising functions. In literature this concept is discussed in context with the so called eco problem solving, eco-agents are defined as actors (see [6]).

Coordination mechanisms: Autonomous agents have to solve problems partly in communities in a competitive way. Without master control automated coordination mechanisms are required. Possible approaches are clearing systems as existing in stock exchange, market mechanisms or simple allocation rules.

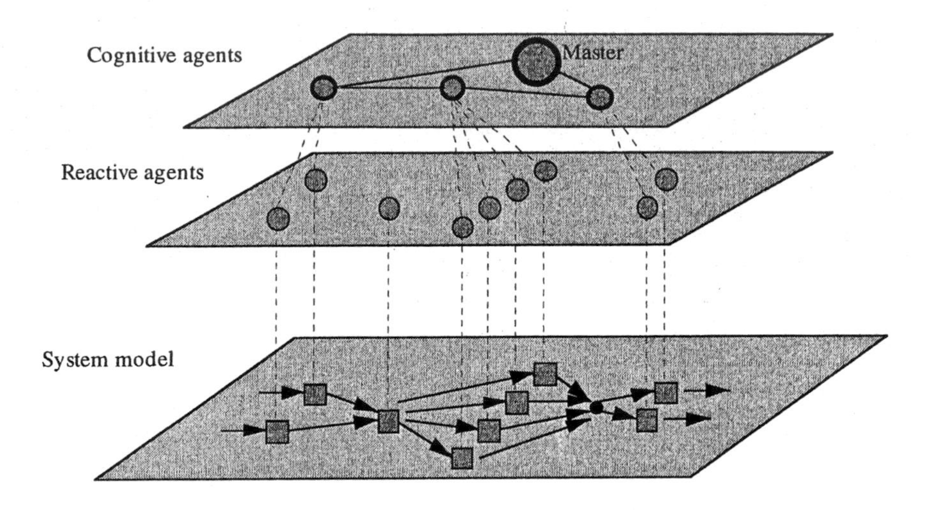

Figure 1. Organization structure of the multi-agent system

Proactivity of agents: A community of proactive agents can not be triggered centrally by an event list as easily as in discrete event simulation. A mechanism has to be added to initiate decentrally the agent's proactive activities.

Simulation for forecast: Some agents may have a global representation of their environment. This representation could be the system model itself. In the decision making process solutions have to be taken in view of their future consequences. This implies a forecast of events initiated by the decisions in question. Potential solutions have to be valued according the relevant criteria. Simulation used to generate future scenarios starting at present time provides an estimation of probable future system development. The outcome can be measured directly in view of the envisioned goals.

Problem identification and goal search: In the problem solving engine intelligence is not only asked for solution search, problem identification and determination of adequate goals are tasks demanding intelligence as well. The base of knowledge and solution methods has therefore to include proceedings in these fields too.

6. METHODOLOGY FOR DESIGN OF SOCIO-TECHNICAL SYSTEMS

Complementary design requires tools for modeling and prospective design of both operator jobs and technical systems. Allocation decisions have to take into account technical, organizational, and human considerations in an integral manner (cf. e.g. Ulich, [7]). A (re-)design of a production system implies its description for which an aggregate model is helpful. The set of variables represented by a model as well as its interaction is defined by both the modeler and the modeling technique. Comparing different design solutions by computer simulation only those variables represented by the model are considered in deriving design scenarios. The effect of variables that are not represented by the model (e.g. the human operators motivation) are neglected. Criteria of complementary system design must therefore be included into design processes and modeling concepts in order to reach satisfactory solutions.

To test the efficiency of potential management support concepts before their implementation, we use the following basic model design principles for the real world system:

1. Distinction between physical, logical and organizational level.

In the real world system we have a clear distinction between physical process, the information flow in relation with planning and control logistics and the management level. For our process model we introduce therefore three separate levels, a *physical* , a *logical* and an *organizational* level In the physical level we model the "hardware" elements of the process, basically the material flow. In the logical level, we represent the information flow related to the planning and control process and the organizational level is provided for the management process. There, the main process tasks as well as their interdependencies are modeled. According to design criteria for socio-technical systems we allocate the tasks to human and technical subsystems These decisions fundamentally influence the quality of the workers task and the requirements for the man-machine-interaction. The model of the technical subsystem on the organizational level contains only the automated part of the process as well as the necessary interfaces to the human subsystem.

Whereas in real world human resources directly interact with the automated process, in the model situation we have to model human resources as well. This can be done by introducing directly in the automated subsystem simple assumptions concerning the human part or by developing a separate model representing human management.

2. Model and real world with same interface to logical level

The above mentioned separation between physical and logical level in the process model allows to introduce a similar interface between real world and planning and control as between logical and physical level (see Fig. 2). In this way an elaborated solution for planning and control can be tested in the model and afterwards be implemented directly into the real world system.

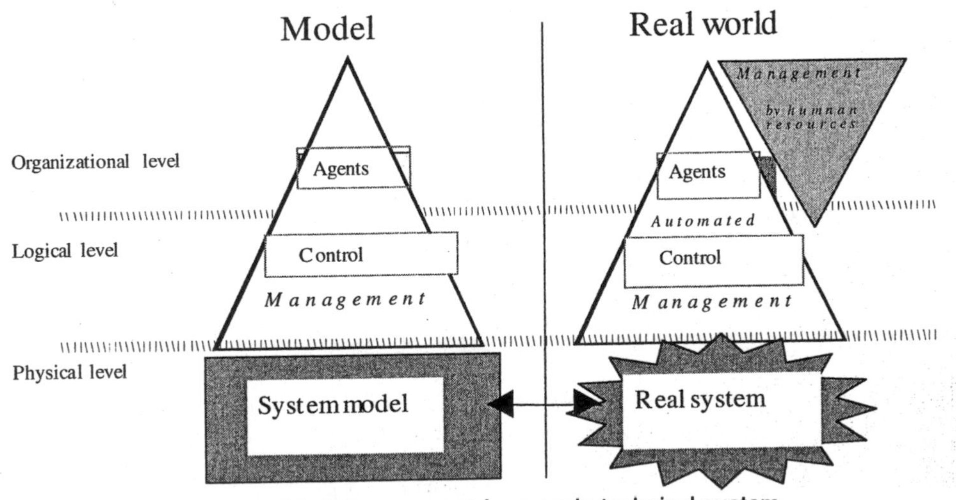

Figure 2. Modeling concept for a socio-technical system

Simulation is the method to analyze the dynamic behavior of the system. At our institute we developed our own simulation kernel HIDES (Highly Interactive Discrete Event

Simulator). to model and simulate logistic systems. A HIDES model is represented as a directed graph with two types of nodes: stations (locations in the graph where specific activities can be performed) and queues. A third important model element are resources that are introduced as abstract elements that can be allocated dynamically and contain information about their availability and their behavior. HIDES allows the building of hierarchical models and therefore supports top-down modeling. A more detailed description of HIDES and the general concepts it is based on can be found in [8].

For the automated planning and control task a distributed problem solving approach was indicated. Our approach is based on the multi-agent concept using our simulation kernel HIDES as modelling base. The appearing planning and control tasks are distributed to several agents. The internal organisation is built up as described in chapter 6. Each of these agents supports a clearly discernible task and has its own specific problem solving strategies. Agents have information about the problem area they are responsible for (e.g. a group of equipment) and they can request services of other agents. In the work-cell the simulation is used by the agents to determine the future consequences of their decisions. In operation events from the work-cell are permanently forwarded to the agents. The agents compare the actual state of the work-cell with the planning data and take appropriate actions if necessary. [9]

7. A COMPLEMENTARY DESIGN APPROACH

As design methodology we base on the KOMPASS (complementary analysis and design of production tasks in socio-technical systems) [10] approach developed at the Work and Organizational Psychology Institute (IfAP) at the Swiss Federal Institute of Technology (ETH). The design heuristic KOMPASS aims at supporting complementary design by providing design criteria and a moderation aid leading a design team through a discussion of design philosophy, design criteria, and the actual development and evaluation of design options. [11] It provides guidelines for the analysis and assessment of existing systems and a heuristic for system design. A participatory approach is envisioned, such that future system operators are included in the design decisions as early as possible; although, participation and the use of expert design criteria should be balanced. The KOMPASS criteria concern job design on three levels: socio-technical system, individual work task and human-machine-interaction. Design goals aimed at are, for example, local control of variances and disturbances (e.g. through independence of organisational units) and intrinsically motivating job design (e.g. through complete tasks).

In order to provide designers with an efficient methodology for complementary system design we use as modeling base the model representation of the simulation kernel HIDES according to our modeling concept described in chapter 6. The main focus is the integration of the KOMPASS approach into the modeling process of the work-cell. based on our simulation kernel.

8. PRACTICAL EXAMPLE

Our practical example was a highly automated work-cell for micro-chip assembling in semi-conductor industry. The work-cell supports three process steps: on the die-bonder the electrical devices are picked from a wafer and glued on a metallic strip (the leadframe). A typical strip carries up to 15 devices. That semi-finished product goes to an oven where the glue is dried. Afterwards the electrical connections (gold wires) between the device and the leadframe are made on the wire-bonder. On the die-bonder the strips are filled into magazines. The transport of these magazines within the work-cell is automated and is performed by a robot. Due to the different performances of the equipment a work-cell normally contains one die-bonder, several oven chambers and several wire-bonders. The scheduling problem to determine the sequence of lots to be started on the die-bonder and the distribution of magazines to the wire-bonders is more complicated as it appears at first sight. Depending on the user of the work-cell different optimization criteria and several different restrictions may be addressed in addition. For this work-cell we conceived an intelligent automation of the planning and control process based on the multi-agent concept. Until now the system is only partly realized, we hope to finish a first version soon.

References

[1] Bailey, R.W. 1989, *Human performance engineering*, 2nd edn, (Prentice-Hall International. London).

[2] Bainbridge, L. 1982, Ironies of automation. In G. Johannsen & J.E. Rijnsdorp (eds.), *Analysis, design and evaluation of man-machine systems*, (Pergamon Press, Oxford) 129-135.

[3] Older, M.T., Waterson, P.E., Clegg, C.W. et al., 1995, *Task allocation: A critical assessment of task allocation methods and their applicability.* Internal Report, Institute of Work Psychology, University of Sheffield.

[4] Wooldrige, M., Jennings,N.R., 1995 Intelligent Agents: Theory and Practice *Knowledge Engineering Review.*

[5] Daenzer, W.F., (Hrsg), 1976, Systems Engineering, Verlag Industrielle Organisation, Zürich.

[6] Ferber, J., .Drogoul, A., 1992, Using Reactive Multi-Agent Systems in Simulation and Problem Solving. In *Distributed Artificial Intelligence*, pages 53-80. Kluwer Academic Publishers: Boston, MA.

[7] Ulich, E., 1994, *Arbeitspsychologie*, 3rd edn, Verlag der Fachvereine, Zürich; Poeschel, Stuttgart.

[8] Graber, A., Ulrich, H., Schweizer, D et al.. 1993. A Highly Interactive Discrete Event Simulator designed for Systems in Logistics. In A. Verbraeck and E. Kerckhoffs, editors. *European Simulation Symposium Proceedings*, Delft.

[9] Ulrich, H., Dürig, W. , 1994, The Contribution of Simulation for the Management of an Automated Work-Cell. In J. Halin, W. Karplus and R. Rimane, editors, *CISS - First Joint Conference of International Simulation Societies Proceedings*, Zurich.

[10] Grote, G., Weik, S. & Wäfler, T., 1996, KOMPASS: Complementary allocation of production tasks in sociotechnical systems. In S.A. Robertson (Ed.), *Contemporary Ergonomics,* 306 - 311. London: Taylor & Francis.

[11] Grote, G., Weik, S., Wäfler, T. et al., 1995, Criteria for the complementary allocation of functions in automated work systems and their use in simultaneous engineering projects. *International Journal of Industrial Ergonomics, 16*, 367-382.

INTER-CELL LAYOUT DESIGNS IN A CELLULAR MANUFACTURING ENVIRONMENT- A CASE STUDY

Massoud Bazargan-Lari
Faculty of Science and Technology, University of Western Sydney - Nepean,
P O Box 10, Kingswood, NSW 2747, Australia.

ABSTRACT: Cellular Manufacturing (CM) represents a rewarding strategy, finding its application in a variety of manufacturing environments. A key element in exploiting the benefits of CM is efficient layout designs. A poor physical layout may negate some or all of the benefits expected from CM.

This paper presents the application of recent proposed methods by the author to a dynamic food products and packaging company in Australia. The large volume of shop floor material movements, confusion over production planning, losing customers due to long lead times and high overheads were among the problems expressed by the company because of poor shop floor layout. Furthermore the company was deeply concerned about the increasing number of accidents and injuries on the shop floor caused by poor layout of machinery and the lack of proper aisles for movement of the lift-trucks.

This paper shows that unlike many existing layout models, the travelling cost is not the only major dominant force in generating the layout designs. This paper presents the process of developing the final inter-cell layout designs by providing the management with multiple layout configurations and showing the impact of each design on the material handling cost at each stage. These solutions not only provide a safer shop floor but also significant reductions in material handling cost, waste, need for large capital investment and the number of lift-trucks needed on the shop floor.

1. INTRODUCTION

Cellular Manufacturing (CM), an application of Group Technology (GT), utilises the concept of divide and conquer and involves the grouping of machines, processes and people into cells responsible for manufacturing or assembly of similar parts or products.[Ham and Hitomi, 1985]

A recent survey conducted by a joint academia/industry research team on the impact of CM on 209 Australian manufacturing companies show that 52% are already using or are in process of-implementing CM with a further 28% indicating their future plans to introduce CM. Of the companies already using CM, more than 70% reported improvements in one or more aspects such as lead times, lot sizes, labour productivity, set up times, on time delivery, labour flexibility and quality.

The main reason that manufacturing companies are attracted towards implementing CM is that the benefits of CM can normally be realised with relatively low capital investment by relocating and possibly duplicating certain machines as opposed to other automated strategies.

The design for cellular manufacturing involves three stages (i) grouping of parts and production equipment into cells, (ii) allocation of the machine cells to areas within the shop-floor (inter-cell or facility layout), and (iii) layout of the machines within each cell (intra-cell or machine layout).[Jajodia *et al.*, 1992]

The first sub-problem concerning part/family formation and machine grouping has been the subject of research by many scholars [see for example Heragu, 1994 for references]. A key element in exploiting the benefits of CM is, however, efficient layout designs. A poor physical layout may negate some or all of the benefits expected from CM. It is therefore absolutely essential to have the means and knowledge for generating efficient and intelligent machine/cell layout designs. Accordingly these two phases require and deserve more dedicated attention and research in order to provide a better platform for the companies planning to introduce CM.

Although the concept of general layout problem has been studied for decades [see Kusiak and Heragu, 1987 or Welgama and Gibson, 1995 for a comprehensive review of such models], certain specific and practical constraints normally present in a real CM environment restrict the implementation of general-purpose layout models.

A major difficulty with the existing layout models is that they normally generate a single take it or leave it layout design. These models consider the travelling cost as the sole criterion and the layout with the minimum cost is presented to the decision maker, ignoring many real-world constraints. Certainly designing an industrial layout involves multiple criteria and objectives of which reducing the travelling cost represents one of them. Because of these limitations many industrial layout designs are being developed by the companies based on the their in-house expertise.

Bazargan-Lari and Kaebernick, 1997,1996 introduced a new approach to solving the machine layout problem and how the three phases of the design for CM can be integrated. These methods are multi-objective continual plane approach which use a combination of non-linear goal programming and simulated annealing to generate efficient and feasible solutions. These algorithms also include features addressing the problems of implementing cellular manufacturing in an existing traditional manufacturing operation.

The present paper demonstrates the capability of this model in generating alternative inter-cell layout designs for a food manufacturing and packaging company. This paper shows the need to involve the decision maker in arriving at the solutions at each stage. In this paper it is shown that although travelling cost is important to the decision maker, it is not the dominant force in deriving the final designs. Furthermore, the capability of the model in capturing realistic constraints such as closeness relationships, location restrictions/preferences, orientation, aisles and irregular site shapes is presented.

Section 2 provides a brief description of the mathematical model and the solution methodology presented in Bazargan-Lari and Kaebernick, 1997. Section 3 presents an introduction to the company, work-centres, shop floor plan and the flow of materials. Section 4 describes the exiting layout together with the process of generating the preliminary, interim and final inter-cell layout designs. It also shows how the feedback from the company was used to arrive at a new set of configurations. Finally section 5 concludes this paper.

2. MATHEMATICAL MODEL

This section briefly explains the mathematical model and the solution methodology presented in Bazargan-Lari and Kaebernick, 1997,1996. For more detailed explanations, the interested reader is referred to these publications.

In this model the machines/work-centres are considered to be rectangular blocks with known dimensions. Three variables are associated with each block namely the coordinates of the bottom left hand corner (x_i, y_i), and the aspect ratio o_i. The orientation of a machine is considered horizontal if o_i is less than 1 and vertical otherwise (see Figure 1). Therefore

$$w_i(\text{width block})_i = \left(\frac{a_i b_i}{o_i}\right)^{\frac{1}{2}} \tag{2.1}$$

$$h_i(\text{height block})_i = \left(a_i b_i o_i\right)^{\frac{1}{2}} \tag{2.2}$$

Where a_i, b_i are the dimensions, and o_i is the aspect ratio of block i. The task of this model is to evaluate the coordinates (x, y) and aspect ratio (orientation) of each machine in the continuum plane by achieving and satisfying certain goals and constraints.

2.1 Non-Overlapping Conditions

The two blocks B_i and B_j are not overlapping if and only if they are at least x-projection or y-projection non-overlapping (Tam and Li 1991), i.e.,

$$\max \begin{pmatrix} \left(x_i - \left(x_j + w_j\right)\right)\left(\left(x_i + w_i\right) - x_j\right), \\ \left(y_i - \left(y_j + h_j\right)\right)\left(\left(y_i + h_i\right) - y_j\right) \end{pmatrix} \geq 0 \tag{2.3}$$

2.2 Shop Floor Boundaries

Each machine must be fully located inside the associated boundary. Taking the upper x and y boundaries of the shop floor as max_x and max_y we should have:

$$\begin{aligned} x_i + w_i &\leq \max\_x & \forall\, i = 1,2,\ldots,n \\ y_i + h_i &\leq \max\_y & \forall\, i = 1,2,\ldots,n \\ x_i, y_i &\geq 0 & \forall\, i = 1,2,\ldots,n \end{aligned} \tag{2.4}$$

where w_i and h_i are width and height of block i respectively as defined in (2.1) and (2.2) above.

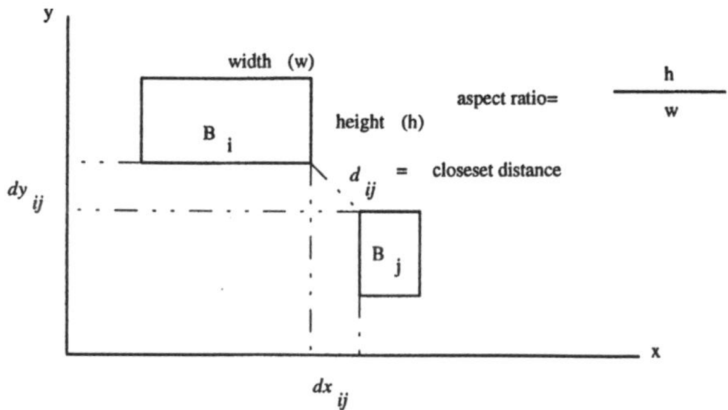

Figure 1. Definition of co-ordinates and distances.

2.3 Closeness Relationships

The distance between two machines (blocks) is measured by the nearest pair of points, each on the perimeter of one block. Referring to Figure 1, the closest distance between the two blocks B_i and B_j is given by:

$$d_{ij} = \left(dx_{ij}^2 + dy_{ij}^2 \right)^{\frac{1}{2}} \tag{2.5}$$

2.4 Location Restrictions/Preferences

Certain restrictions or preferences might exist for the physical location of a machine or work-centre. For example, consider a case where a machine is already installed and its relocation would be economically unjustified, subsequently the machine should retain its position in the new layout configuration. Let (α,β) be the co-ordinates of the bottom left-hand corner with γ_x and γ_y as the dimensions of such an area (see Figure 2). If we now desire that this area be restricted for possible location of machine i then we should ensure that these two blocks do not overlap. i.e.,

$$\max \begin{pmatrix} \left(x_i - \left(\alpha + \gamma_x\right)\right)\left(\left(x_i + w_i\right) - \alpha\right), \\ \left(y_i - \left(\beta + \gamma_y\right)\right)\left(\left(y_i + h_i\right) - \beta\right) \end{pmatrix} \geq 0 \tag{2.6}$$

On the other hand if it is desired that this machine be located inside such an area, then the following should hold:

$$\begin{aligned} x_i &\geq \alpha \\ y_i &\geq \beta \\ x_i + w_i &\leq \alpha + \gamma_x \\ y_i + h_i &\leq \beta + \gamma_y \end{aligned} \tag{2.7}$$

Location restrictions can be used to address aisles and irregular shape sites (see Figure 2).

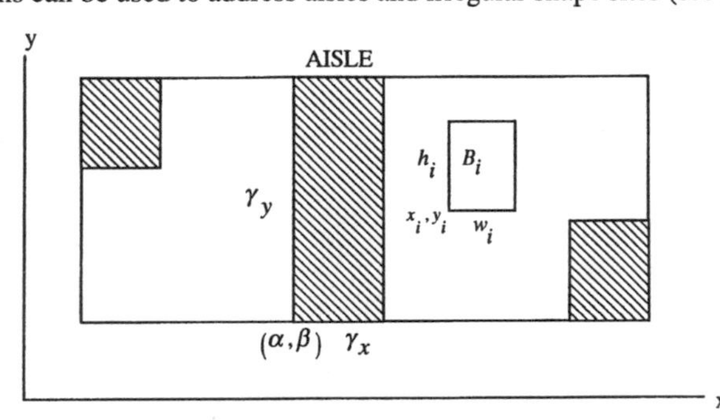

Figure 2. Utilising location restrictions to address aisles and irregular shape sites (shaded areas).

2.5 Travelling Cost

In this model rectilinear distances between centroids of the blocks have been used to evaluate the travelling distances. In most machine layouts, material is transported cell by a carrier. It is therefore appropriate to consider frequencies of trips made by these carriers to account for the material flow between machines (Heragu and Kusiak, 1988). Let f_{ij} denote the total frequencies of trips for all parts between machines i and j over a period of time. Then, the total rectilinear distances travelled by all parts and between all machines in the same period of time is given by:

$$\sum_{i=1}^{n-1} \sum_{j=i+1}^{n} f_{ij}\left(\left|x_{ci}-x_{cj}\right|+\left|y_{ci}-y_{cj}\right|\right) \tag{2.8}$$

where n is the total number of machines and $x_{ci}, x_{cj}, y_{ci}, y_{cj}$ are the x and y co-ordinates of the centroids of blocks i and j respectively.

2.6 Solution Methodology

Nijkamp *et al.*, 1990, proposed that the main reason for increasing influence of multiple criteria models is to avoid the decision analysis to end up with a single and "forced" solution dictated by a researcher, but with a spectrum of feasible solutions from which a choice can be made.

Solving the non-linear mathematical programming model presented above for problems with realistic constraints and numbers of machines (5 machines or more) using conventional non-linear single objective packages often leads to infeasible solutions. By infeasible, it is meant a solution that violates constraint(s) which should absolutely be satisfied such as non-overlapping constraint.

Similar difficulties were also reported by other researchers in solving the non-linear optimisation problems for layout in continuum plane such as the penalty methods in Heragu and Kusiak 1991, Van Camp et al 1992 and the mixed integer programming in Lacksonen 1994.

Infeasibility of the solution occurs because of the assumption that all constraints are equally important and, the existing non-linear algorithms try to satisfy all constraints simultaneously. However, in practice not all of the constraints/objectives are of equal importance. For example, a feasible solution with a slightly higher travelling (cost) is preferred to an unfeasible but lower cost configuration.

Two optimisation methodologies were found to be most useful in generating multiple solutions and overcoming the above difficulties, namely non-linear goal programming in the class of multiple-objective models and Simulated annealing in the class of random search methods.

The space limitation does not allow the descriptions and procedure of merge between these two methodologies in this paper. The interested reader is again refereed to Bazargan-Lari and Kaebernick 1997, 1996.

3. CASE STUDY

This section presents a profile of the company together with the required information to run the inter-cell layout designs.

3.1 Company's Profile

Steric Pty. Ltd, a dynamic and a major producer of a diverse range of food products, operating for more than 30 years, is situated in south west of Sydney - Australia with 4 more plants in other states of Australia and another one to be constructed soon.

Steric specialises in a wide range of food products including Cordials, Toppings, Sauces, Packet Soups, Cooking Oils, Spices, Oats, Cocoa and Coconut. As well as the food products, the company is a major producer of Birdseeds and cleaning products such as Dishwashing and Laundry Detergents, Disinfectant, Baby shampoos and Fabric-softeners. Furthermore the company has a set of 12 blow Moulding machines which supply almost all the PVC, PE bottles, Canisters, Caps, etc., needed for bottling the finished products. The company has extended its operation from private label/ housebrand/ generic area to its own branded proprietary ranges of products.

3.2 Background

The growth of operations and increase in the number of work-centres over the last 30 years has resulted in an inefficient shop floor layout with a large number of unnecessary material movements. This difficulty arises almost with every growing company and worsens over time. These companies normally start with one or two work-centres and as they grow, they introduce new work-centres on the shop floor without considering the impact of material movements or other layout considerations such as ease of material handling on the total operation of the company. In many cases the location of a new work-centre is dictated by other work-centres which already exist on the shop floor. The following represents some of the major difficulties experienced by the company:

- Increase in the number of accidents and injuries on the shop floor caused by poor layout of machinery and the lack of proper aisles for movement of the lift-trucks,
- Growing cost due to unnecessary large volumes of material handling,
- Huge amounts of work in process and finished products,
- Confusion over production planning,
- Losing customers due to long lead times and high overheads,
- Increase in the waste of finished products.

The management of the company was interested to explore the benefits of cellular manufacturing and possibility of utilising the Layout Design Model [Bazargan-Lari and Kaebernick 1997] in order to overcome these difficulties. In particular, the company was keen to evaluate perfomance measures such as travelling cost and safety of the existing layout configuration and compare such measures with the possible solutions generated by the layout design model.

3.3 Layout Data

In the first few visits to the company the severity of the problems expressed by the management was confirmed. Efforts were then focused at collecting the information required by the CM design model.

Table 1 presents the description, sizes, flow of materials and constraints imposed for each work-centre.

The flow presented in the table represents the number of pallets per year from batching station to packing and from packing work-centre to the warehouse. The lift-trucks handle one palette at a time and each pallet consists of a number of boxes/cartons which in turn every box contains a certain number of packed products according to the customers specifications.

The company requires that certain work-centres must not be relocated as special operational environment were in use for these stations. Any relocation of such work-centres implied that new and costly environments must be re-installed. Furthermore, some of the work-centres also have to be close to each other as they are using the same facilities such as mixing tanks. The restrictions column in Table 1 signifies the constraints imposed on such work-centres.

Almost all the products are a one-off operation, i.e. the batching work-centres produce the products and then they are pumped/sent to the packing station for bottling/packing. This implies that there is virtually no inter-cellular movements between unrelated work-centres, representing perfect autonomous cells and a very special inter-cell layout problem. The cell-partitioning model also confirmed the company's groupings.

Description	Length (m)	Width (m)	Flow Pal.	Restrictions
CORDIAL BATCHING	9.0	3.2	3645	-
CORDIAL PACKING	21.1	4.8	3645	-
TOPPING BATCHING	8.0	3.8	4452	-
TOPPING PACKING	17.0	3.1	4452	-
PISTON FILLER BATCHING	6.0	4.0	4974	close to asset batching
PISTON FILLER PACKING	15.0	4.8	4974	-
OIL PACKING 2LT	16.7	5.2	3477	-
OIL PACKING 4LTR	2.0	.9	1401	close to oil packing 2LT
ASSET BATCHING	6.0	1.5	636	-
ASSET PACKING	15.0	3.2	636	close to piston filler batching
DETERGENT BATCHING	11.2	9.5	10287	-
DETERGENT PACKING	13.1	11.7	10287	-
HIGGINS BATCHING	4.3	4.2	1326	-
HIGGINS PACKING	8.0	2.8	1326	-
SOUP BATCHING	5.9	4.7	1953	fixed
SOUP PACKING	10.5	6.8	1953	fixed
SALT PACKING	11.5	5.9	1353	fixed
SPICE BATCHING+PACKING	3.9	3.0	243	fixed
BLOW MOULDING + INJECTION MOULDING	50.0	11.0	19767	fixed
DETERGENTS 4LTR BATCHING	2.1	1.6	360	close to det. Batching
CHQ BATCHING	5.9	5.5	0	fixed
FRONT MACHINE/ ROVEMA	5.6	3.5	1000	-

Table 1

Description, sizes and limitation for each work-centre.

The company has the potential of producing more than 600 different products distinguished by brand, size, weight, volume, etc. However, if we classify them according to their generic names this number reduces to 120 distinct products. For example lemon cordial and orange cordial are counted as two distinct generic products, while each one branches into 10 different products based on the brand and size. Table 2 presents the grouping of the products into 14 autonomous cells.

Having established the parts/machines grouping the main task was to generate alternative intra and inter-cell layout designs. However, As 90% of the work-centres have only one machine, the problem is then reduced to determining the inter-cell layout designs satisfying certain constraints and achieving the company's objectives.

Description	Number of Products brands/sizes
CORDIALS	53
TOPPINGS	62
ROVEMA	24
PISTON FILLER	49
OIL	7
OIL-4LTR	7
ASSET	10
DETERGENTS	76
HIGGINS	36
SOUPS	32
SALT	10
SPICE PACKING	28
BIRDSEED	112
BLOW MOULDERS	48

Table 2
Partitioning of products into autonomous cells.

3.4 Shop Floor Plan

Figure 3 shows the shop floor plan together with the corresponding dimensions. This plan basically consists of two separate adjacent buildings with two doors of 5m width connecting them.

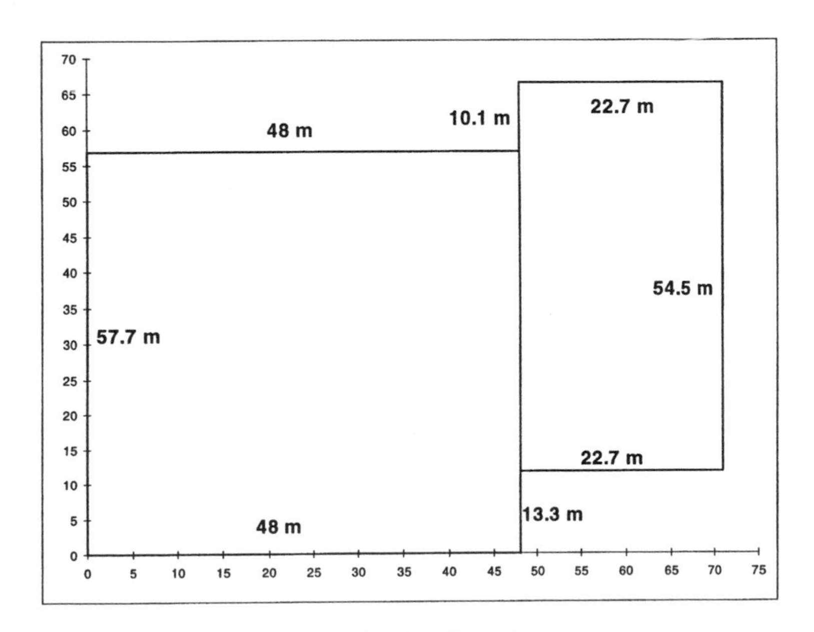

Figure 3. Shop floor plan

4 EXISTING AND PROPOSED LAYOUT DESIGNS

Figure 4 presents the current location of the work-centres on the shop floor. Presently the shop floor not only accommodates the machinery but also serves as the raw material and finished products stores.

The travelling costs reported in this paper include all the rectilinear distances for transferring the materials/pallets to the batching then to the packing work-centres and finally to the warehouse's loading/unloading point located at the bottom right hand side of the plan. The exact co-ordinates of this loading/unloading point is (71.2,0).

As the figure suggests there is no established and organised aisle system for the movements of the 7 lift-trucks operating on the shop floor. The travelling distances reported for present layout represents a very modest figure as the pallets of raw materials and finished products are scattered all over the vacant spaces on the shop floor causing a great deal of difficulties for the lift-trucks to manoeuvre between the work-centres. This limitation forces the lift-trucks to travel much more than needed, at a very slow speed and at a high risk to the safety of shop floor employees.

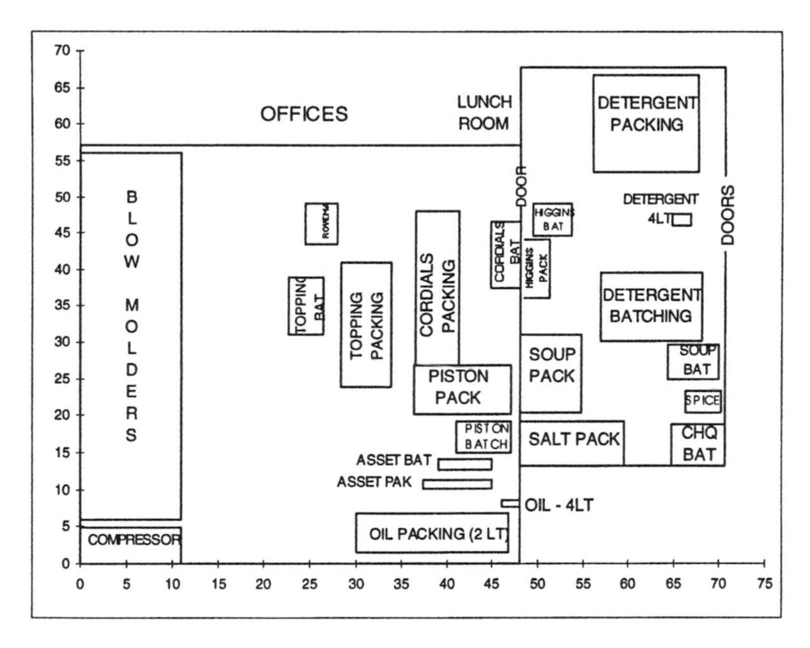

Figure 4. Present layout design with 8700 km total travelling distances/year.

4.1 Preliminary Layout Designs

The layout model was run with the company's data to generate some preliminary results. The CPU time for generating each solution reported in this paper was 4-5 minutes running under Digital Unix on DEC Alpha. Figure 5 shows one of the layout design among some of the early results submitted to the company for evaluation. This design not only provides a structured aisle system but also reduces the travelling cost by more than 38%.

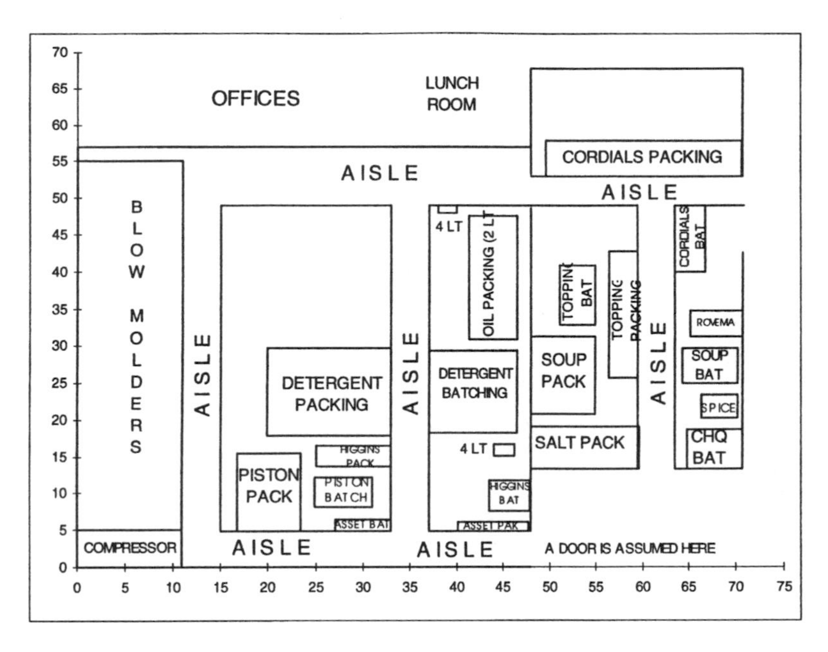

Figure 5. Preliminary layout design with 5335 km total travelling distances/year.

4.2 Interim Layout Designs

The preliminary layout designs were found to be useful by the management to get familiar with the type of the solutions the model can generate. Especially the aisle structures were found to be impressive. Having observed the preliminary solutions, the company imposed the following restrictions on the new set of layout designs:

- It is preferred to have the food products work-centres to be located in the larger building and non-food products work-centres such as detergents in the smaller one,
- Company's customers requested that the boxes and cartons be bar-coded. To satisfy this restriction, the management planned to purchase the bar-coding machinery which each cost about \$50,000. Considering the number of work-centres, the company had to spend over \$600,000 to meet this requirement. It is therefore very desirable if the work-centres are arranged in a manner so that the company achieves this objective with minimum number of bar-coding machines.
- It is preferred that the Detergent batching and packing work-centres moved to the top of the smaller building.
- At least 2.5m space required between packing work-centres to accommodate two pallets of WIP.

To address the minimum number of required bar-coding machines, it was decided and agreed with the management that all the cartons/boxes from various work-centres be transferred by a conveyor to a carousel where the company installs just one single variable bar-coding machine.

This new set of constraints was introduced to the inter-cell layout model. The model generated many feasible solutions(over 50 designs) satisfying the above and previous requirements. Figures 6 and 7 present two of the layout designs presented to the company for further evaluation and/or refinements. In this figure a conveyor is assumed on the aisle between the batching and packing work-centres transferring the boxes/cartons to the carousel where a variable bar-coding machine is located.

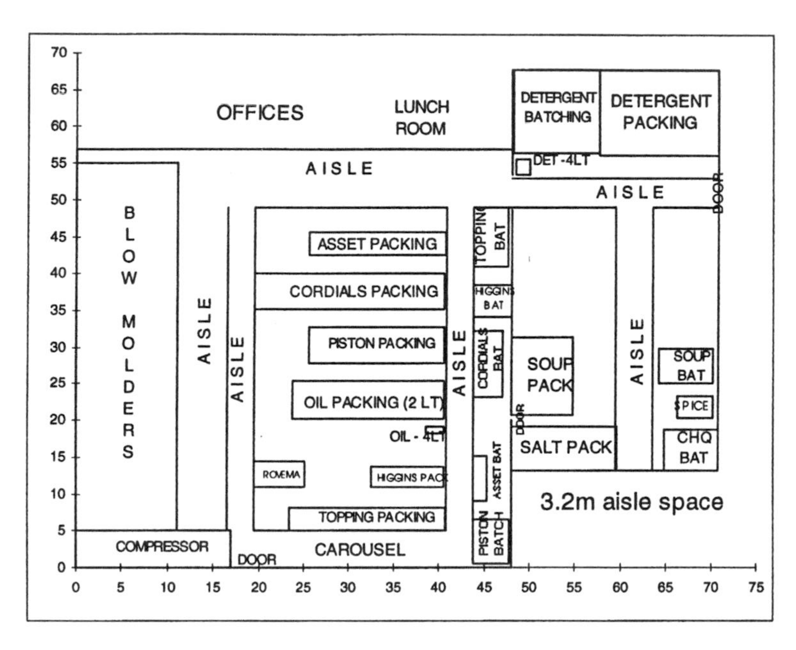

Figure 6. Interim layout design 1 with 5710 km total travelling distances/year.

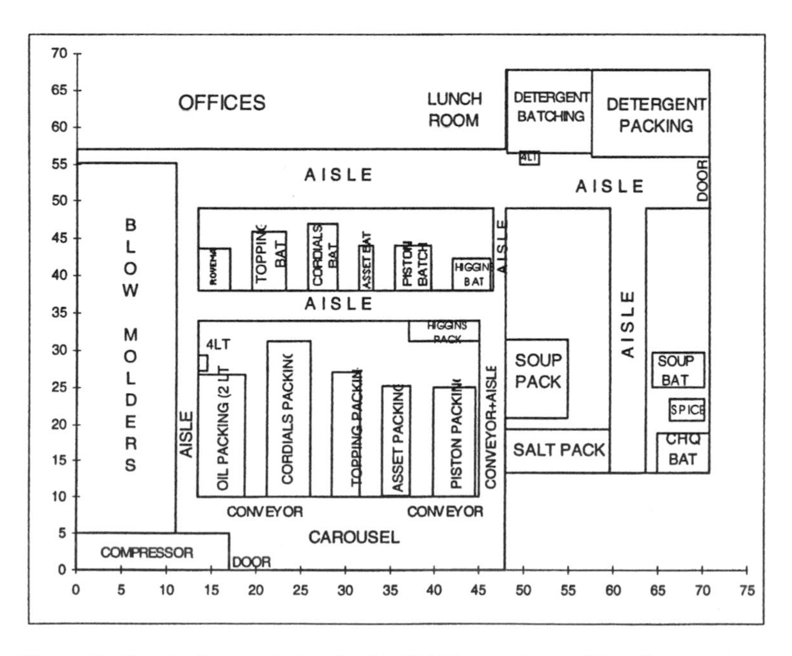

Figure 7. Interim layout design 2 with 6244 km total travelling distances/year.

4.3 Final Layout Designs

Observing the interim layout designs, the company requested for the final touches and refinements as follows:

- The width of the aisle between food batching and food packing work-centres should be enough to ensure the safe and easy movement of two lift-trucks in parallel as well as .6m space for the conveyor. An isle width of 4.2m was considered to be adequate to meet this requirement.
- The company's preferred sequence of the batching work-centres is (from the top) Cordials, Topping, Piston, Asset and Higgins batching work-centres.
- The preferred sequence of the packing work-centres is (from the top) Cordials, Topping, Piston, Oil, Asset, Higgins and Rovema work-centres.
- The detergent packing work-centre should be relocated.
- A new work-centre, namely juicy filler will be introduced which has to be located close to the asset packing station.
- More space is needed around the carousel area.

The revised layouts based on the above requirements were generated. Figure 8 presents one of the final layout designs meeting all the constraints. This configuration is selected by the company for implementation.

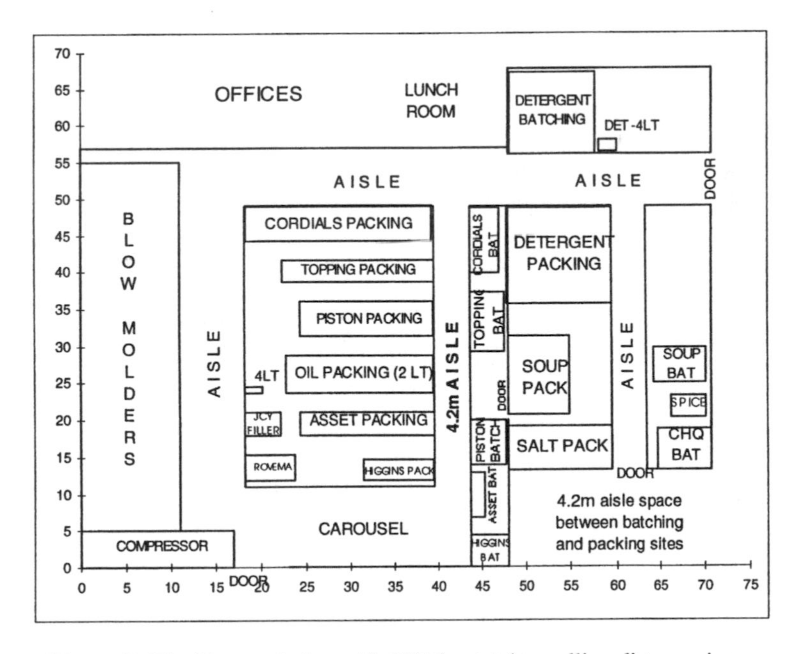

Figure 8. Final layout design with 6081 km total travelling distances/year.

Some of the expected benefits through the final designs are:

- Safe working environment,
- 30% reduced material handling cost,
- Half a million dollars reduction in purchasing bar-coding machines,
- Reduced number of lift-trucks needed,

- Increased employees efficiency,
- Reduced waste.

5. CONCLUSION

In this paper, the application of a recently developed model for the design of CM to a food manufacturing and packaging company was demonstrated. The capability of the model in addressing a number of issues related to the practical implementation of cellular manufacturing structures such as closeness relationships, location restrictions/preferences, orientation, aisles and irregular site shapes was presented.

This paper demonstrated the process of developing the final inter-cell layout designs by providing the management with multiple layout configurations and showing the impact of each design on the material handling cost at each stage. The performance of the model designs became more fine-tuned to the company's requirements as the layout design model was adjusted to the feedback from the company. This process reconfirms that the travelling cost is not the sole criterion to generate the layout designs. For a layout configuration to be accepted and implemented, the decision maker must be involved and present in all phases of the design.

REFERENCES

Bazargan-Lari, M., Kaebernick, H., (1996), An Approach to the Machine Layout Problem in a Cellular Manufacturing Environment, *Production Planning & Control*, Vol 8, No. 1, 1997.

Ham J., Hitomi K., Yoshida T., Group Technology, Boston, MA: Kluwer-Nijhoff, 1985

Heragu, S.S., Group Technology and Cellular Manufacturing, *IEEE Transactions on Systems, Man, and Cybernetics*, Vol. 24, No. 2.,1994.

Heragu, S.S., Kusiak, A., Machine layout problem in flexible manufacturing systems, *Operations Research*, Vol. 36 (2), pp. 258-268, 1988.

Jajodia, S. , Minis I. , Harhalakis, G., Proth J.M., CLASS: Computerised layout solution using simulated annealing, *International Journal of Production Research*, Vol. 30 (1), pp. 95-108, 1992.

Kaebernick, H., Bazargan-Lari, M., An Integrated Approach to the Design of Cellular Manufacturing, *CIRP ANNALS*, pp 421-425, Vol 45/1, 1996.

Kusiak, A., Heragu, S.S., The facility layout problem - Invited Review, *European Journal of Operational Research*, Vol. 29, 229-251, 1987.

Lacksonen, T.A., Static and Dynamic layout problems with varying areas, *Journal of Operational Research Society*, 45, 59-69, 1994.

Nijkamp, P., Rietveld, P., Voogd, H., Multi-criteria evaluation in Physical planning, Elsevier Science Pub.,1990.

Tam, K.Y., Li, S.G., A hierarchical approach to the facility layout problem, *International Journal of Production Research*, Vol. 29, No. 1, 165-184,1991.

Van Camp, D.J., Carter, M.W., Vanneli, A., A non-linear optimization approach for solving facility layout problems, *European Journal of Operational Research*, Vol. 57, pp. 174-189, 1992.

Welgama, P.S., Gibson, P.R., Computer-aided facility layout-a status report, *International Journal of Advanced Manufacturing Technology*, Vol. 10, pp. 66-77., 1995.

A VISION BASED INSPECTION SYSTEM OF PISTONS

Dezhong Hong, Thompson Sarkodie-Gyan, Andrew W. Campbell

Laboratory for Intelligent Systems Technology
School of Science & Technology
University of Teesside, UK

Email: d.hong@tees.ac.uk

ABSTRACT: In this paper, we describe a range measuring method to inspect various different pistons by projecting a laser line onto the crowns. There exist a number of similar piston types with very slight differences but with the same overall diameter such that it is readily possible to fit the incorrect piston to an engine. Differences between a number of families of similar pistons are found in obvious features such as gudgeon pin diameter and position, etc. Within a family, the differences are more subtle and may be simply a change in the shape of the bowl in the crown of the piston. A vision based measurement system is designed to confirm the identity of the piston just after it has been fitted to the engine assembly.

INTRODUCTION

Many of today's flexible manufacturing systems (FMS) contain fully integrated 'contact probing' inspection devices. Yet still in the small batch manufacturing environment, such devices have been found to be expensive, requiring long set-up times and large amounts of manual input. Work has been carried out to develop non-contact inspection techniques to replace contact probing. It has been found that many optical systems are faster than their more dedicated counterparts, and this allows components to be inspected at production speeds.

The main objective of this project is to measure the depth of the bowl of the piston since this is an important parameter to determine the compression ratio of the engine. It is thus possible to differentiate between pistons for the correct classifications. A problem has arisen at the assembly stage of diesel engines, involving the selection of the correct pistons. There exist a number of similar piston types with very slight differences but with the same overall diameter such that it is readily possible to fit the incorrect pistons. Yet still the engine with the wrongly fitted piston can work but at the expense of performance and life. The measurement system is developed to confirm the identity of the piston just before or just after it is fitted in the engine assembly.

Based on the mechanical data provided by the manufacturer it appears that differences between families are found in obvious features such as gudgeon pin diameter and position, piston rubber ring positions etc., which are easy to find by human eye. Within a family the differences are more subtle and may be simply a change in the depth of the bowl in the crown of the piston. The depth of the bowl is a very important parameter as it determines the compression ratio of the engine. The minimum difference value might be as little as 0.2 mm. The variations involved in detecting the similarities or dissimilarities are beyond the capabilities of the human eye. Hence, the measurement system is designed to reveal the 3D features of the crown of the piston as accurately as possible.

It is not the intention to reconstruct the 3D structure of the crown of a piston, since this will involve huge times to describe an object in 3D, and even more time at the pattern matching stage. For the industrial purpose, a real-time piston detecting system is needed. Since the range differences of depth of the crown among the piston types may be as little as 0.2 mm, a structured laser line is used to obtain more accurate 3D information on the crown of a piston. Related work using structured laser lines or structured light stripes to obtain the range image can be found in [1]~[7].

The similar topics were studied by the authors[1] and a vision system with a single laser line projector was designed to extract essential features for piston identification. In order to obtain stable pattern recognition result, fuzzy sets theory was employed in the scheme[2]. In this approach, a completely new probe is used to make the system more adaptable and more reliable. The range data of the piston crown is obtained using a laser line generator with three parallel laser lines projecting onto the surface inspected at the same instance. The object (piston) feature extracted from the main beam can be compensated or verified with the results from side beams.

Because of the symmetry of the piston, a single laser line projected onto the crown of a piston is enough to extract the piston features required for the recognition of the various piston types. We use the same method to describe the features of pistons at the learning stage, so that the recognition of the piston by matching the sample features to the standard one stored in the computer is easier and costs less time.

At the low level image processing stage, a line detector is developed to locate the centre of the laser line from the grey image captured with a CCD camera, since the laser line disperses on the surface in terms of Gaussian distribution. We design a one dimensional line detector to reduce the processing time. The line detector concept was introduced by Ziou [8], which was based on Canny's criteria [9].

Some line data detected at specific points are less reliable than at the other points, because there exists some shining or other noise on the crown of the piston when a laser line is projected onto the crown, also the laser line reflected by the crown is very dim when the angle of the slope is over 60°,. According to the cutting tolerances of the components provided, some parts of the components are more important and more accurate than other parts. In that case, we choose some key points on the crown surface to obtain the features of the component detected in order to obtain more reliable results. The system has successfully been used in identifying a number of pistons from different families. Some of the experimental results are presented in this paper.

SYSTEM SETTING-UP

As shown in Fig. 1, the setting up is based on the specifications for the future work where a prototype will be produced to measured the piston. The CCD camera is mounted on the bar and the laser is clamped at an angle so that it can generates the laser lines directly to the piston crown without using the mirror, and this has also made the system design become more simple. With this experimental design, the CCD camera can be moved up and down (in Z axis) by turning the lead screw clockwise or anticlockwise, whereas the laser could also be moved up and down and rotated at any angle which is most suitable. The piston is located on a thin plate where the plate has a hole with diameter about the diameter of the piston. This is to ensure that the piston will not be able to move while taking the measurement of the piston.

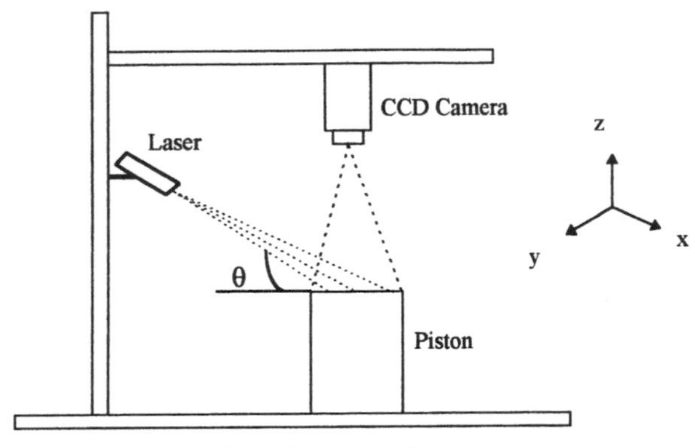

Fig. 1 System setting-up

The laser lines are projected onto the surface of a piston with fixed angles, and the camera is used to capture the image from another direction. The curvature of the laser lines reflected from the crown of a piston will be measured. Since the laser line is projected with a fixed angle, by calculating the curvature of the curve detected, we can determine the depth of the points on the crown at which the laser line meets the crown. Because of the symmetry of the piston, we do not need all the structural information on the piston crown. The three fixed laser lines are enough to extract the piston features required for the recognition of the various piston types.

LINE DETECTION

Since it is impossible to get perfect laser line, when laser line is projected onto the surface of an object, it will disperse. Generally, the dispersion can be approximated by the Gaussian distribution function. Our task is to locate the centre of the line. According to Canny's edge detection criteria, an optimal filter is required to detect the centre of a line in the bath of noise with good location and single response. The filter must be symmetrical, so that the located centre is the point where the filter is applied. It should produce zero when it is applied to invariant image even in the presence of white noise. A proper size of the filter should be chosen based on the average width of the laser line.

Since the Gaussian matches the line dispersion fairly well, an approximate filter is constructed by means of Gaussian. Because there is only one laser line on the crown, and the line does not overlap along vertical direction, it is possible to use 1D line detector to obtain satisfactory result. For the industrial application, the efficiency of the recognition processes is important. It only takes one fourth as much time as a 2D line detector.

The digital line detector is given by:

$$R(n) = \sum_{k=0}^{M-1} i(n-k)G''(k) \tag{1}$$

Generally, the laser line lies across 4 to 10 pixels in the image. The central hill part of $G''(k)$ should cover most of the pixels on the line in order to produce the biggest output for the line centre pixel. So a digital line detector based on the second derivative of Gaussian is suggested as below:

$$(-1, -2, 0, 1, 4, 1, 0, -2, -1) \tag{2}$$

Obviously, it produces zero on the constant image. If the line covers less than 3 pixels, a few maximum values will be produced around the centre of the line. In this case, we can define the centre of the line as the middle of the several maximum value. Since shining intensity is lower than line intensity, and the shape of shinning is different from the line, it rarely produces false responses.

From the image it may be found that the intensity of the laser line at some points is higher than at other points. At the bottom of the image, there is a relatively strong shinning. The whole image is in the bath of white noise. Fortunately, the line detector can deal with this phenomenon effectively. One of the experimental result of line detection is presented in Fig. 2.

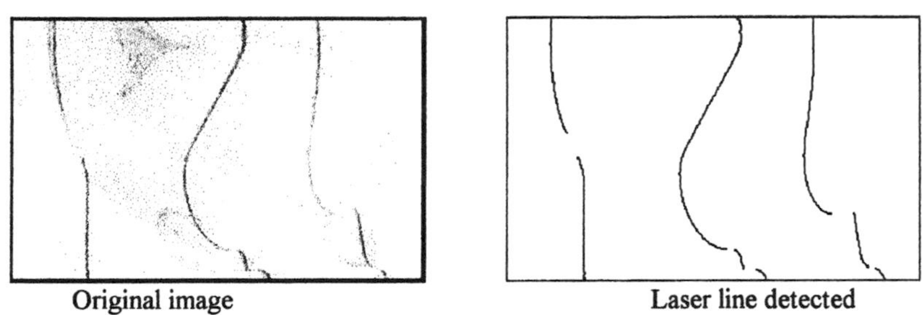

Original image Laser line detected

Fig. 2 Laser line detection

FEATURE EXTRACTION AND PATTERN RECOGNITION

When a laser line is projected onto the piston crown, a profile of the piston crown from the image captured will be produced as shown in Fig. 3.

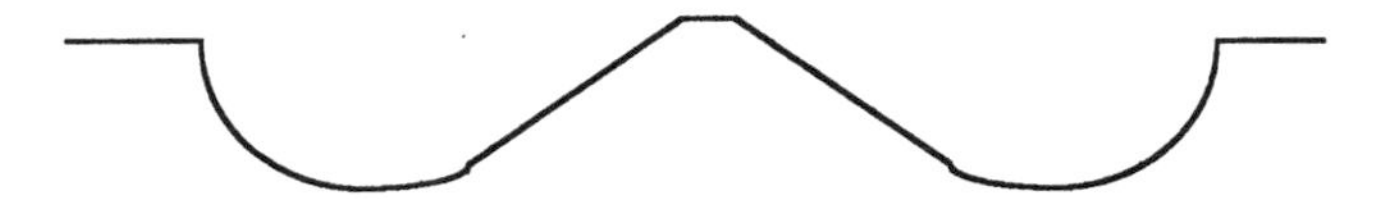

Fig. 3 The sectional profile of piston crown from a laser line

In this project, multiple laser lines are used, therefore more information of the piston crown can be obtained. From the image in Fig. 2, it is easy to find that some pixels in the curve are bright, other pixels are very dim. When the slope of the surface is over 60°, it will be very difficult to locate the curve accurately. Some of the result of line detection are reliable and some are not. The key points chosen are all from the reliable and repeatable segments of laser curves and they are significant in identifying the types of pistons. Ten different reference points are chosen and the specific locations of the points are shown in Fig. 4.

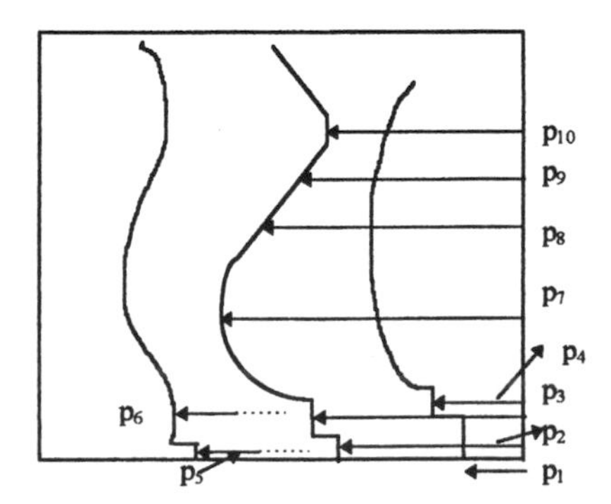

Fig. 4 The key reference points for piston features

We combine the model component feature learning and the sample feature matching with the model feature in the measurement system, so that it is easy for the user to extend the range of detected piston types. At both the learning stage and the matching stage, we employ the same feature extraction and description methods. Thus, piston recognition is simply carried out by finding the minimum value of the Euclidean distances between sample feature vector and the model feature vectors in the database. The Euclidean distance can be calculated by:

$$E = \sqrt{\sum_k ((x_{c_k} + p_{c_k}) - (x_{m_k} + p_{m_k}))^2} \tag{3}$$

The best match of a model will be picked up as a candidate. If the Euclidean distance obtained from (3) is less than a threshold, we say the piston is recognised as this particular type. Otherwise, the test piston will be consider as an incorrect one.

EXPERIMENTAL RESULTS

The model database is built using the same descriptions of piston features. In order to make model more adaptable, two images are captured for each piston from different placing orientation. The key reference points will be chosen from the laser curves from the both images. In this paper, we present eight different piston models and the reference profiles generated for all the models are also displayed. To generate a model for a piston, two images are captured. A sample of the results is illustrated in Fig. 5. The original images captured by the CCD camera are displayed in the left column and the results of laser line detection are displayed on the right. A set of feature values are sampled for each piston model, which are composed of the measurements of the ten key reference points. The feature values are listed in Table 1 in pixel.

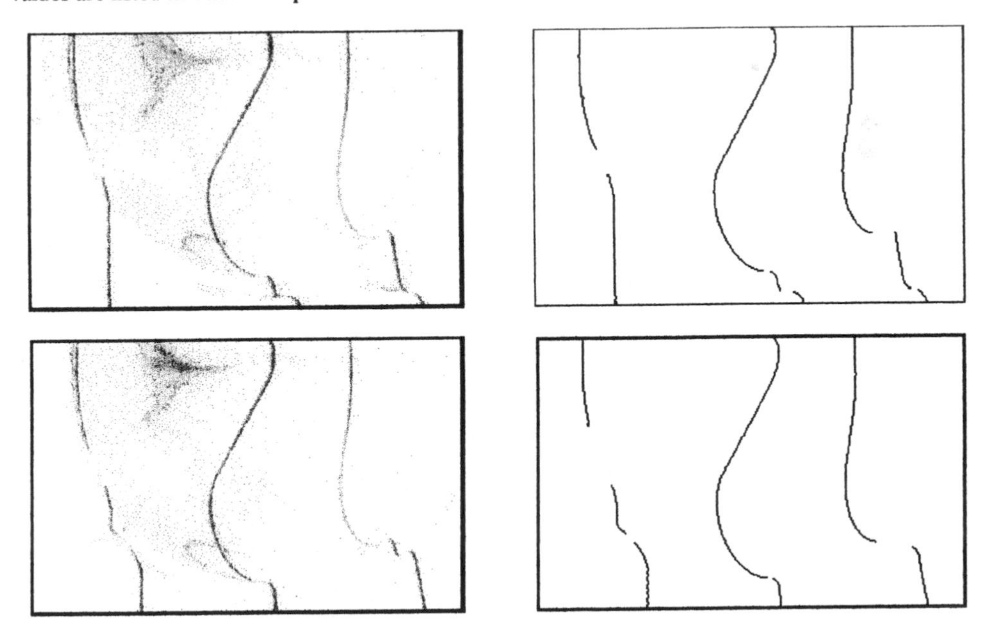

Fig. 5 Piston 1

Table 1 The piston features extracted from key reference points

Piston Model Key Points	1	2	3	4	5	6	7	8
1	118	122	124	119	119	119	119	119
2	120	121	120	121	120	121	121	120
3	138	139	129	144	138	140	145	134
4	138	138	132	144	140	139	144	135
5	118	118	118	118	118	118	119	118
6	140	139	129	149	141	140	148	135
7	186	185	175	181	180	180	184	174
8	178	173	171	171	173	175	177	167
9	148	148	152	146	147	147	145	151
10	139	139	146	139	139	139	139	146

The feature profiles generated for the piston models are shown in Fig. 6. The eight piston types can be distinguished according to the profiles. When a test piston is inspected by the system, only one image is captured for the piston and a set of feature values are extracted to compare with the model profiles. The closest one will be the winner.

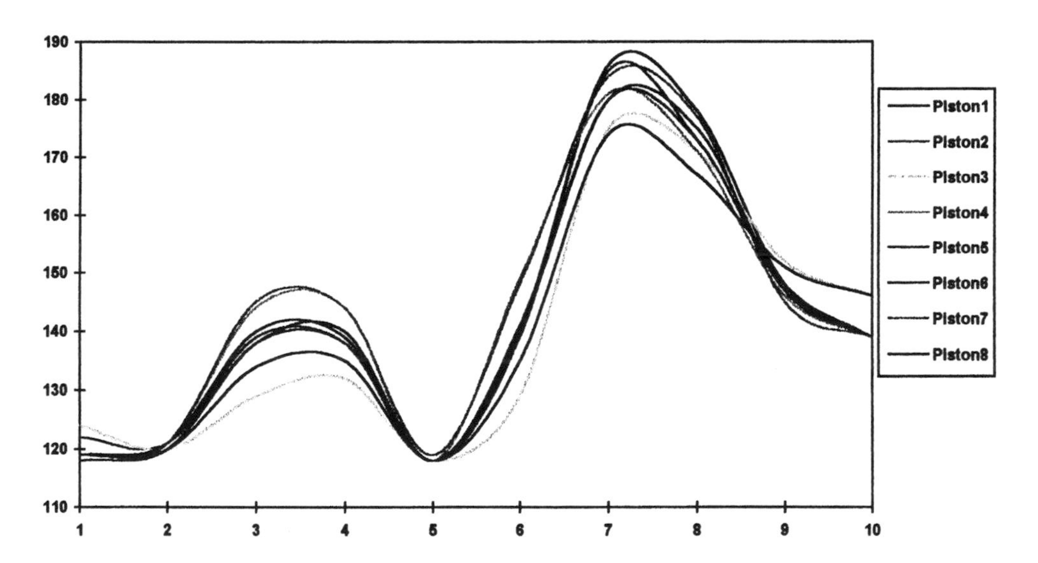

Fig. 6 The feature profiles for piston models

CONCLUSION

In this paper, we present a unique method to extract and describe features of a piston. A vision based measurement system is designed to confirm the identity of the piston just after it has been fitted to the engine assembly. This enables us to distinguish between the components among a considerable number of models with small tolerances. Structured laser light line is employed in the measurement system to obtain the information of the depths on the crown of a piston. A kind of calibration has to be done before it can be used to measure the objects. Because the tolerance of the measurement is as little as 0.20 mm or even less, a special compensation is designed in our system to reduce the measurement error caused by camera nonlinearity. The system has successfully been used in identifying a number of pistons from different families. The experiments has provide the evidence that the vision system is reliable and shows potential abilities on a wide range of applications.

REFERENCES

[1] D. Hong, A. W. Campbell, Y. Yan, T. Sarkodie-Gyan, "Pseudo-three-dimensional vision based measurement system of pistons", Proceeding of 28th ISATA Conference, Stuttgart, Germany, pp. 323-328, 1995

[2] D. Hong, T. Sarkodie-Gyan, A. W. Campbell, Y. Yan, "Three-dimensional feature description of pistons using the fuzzy sets theory", Proc. of ACCV'95, Singapore, III 554-558, 1995

[3] K. L. Boyer, A. C. Kak, "Color-encoded structured light for rapid active ranging", IEEE Transactions on Pattern Analysis and Machine Intelligence, Vol. PAMI -9, No. 1, pp.14-28, 1987

[4] M. Maruyama, S. Abe, "Range sensing by projecting multiple slits with random cuts", IEEE Transactions on Pattern Analysis and Machine Intelligence, Vol. 15, No. 6, pp. 674-651, 1993

[5] A. Blake, D. McCowen, H. R. Lo, P. J. Lindsey, "Trinocular active range-sensing", IEEE Transactions on Pattern Analysis and Machine Intelligence, Vol. 15, No. 5, pp. 477-483, 1993

[6] R. A. Jarvis, "A perspective on range finding techniques for computer vision", IEEE Transactions on Pattern Analysis and machine Intelligence, Vol. PAMI -5, No. 2, pp. 122-139, 1983

[7] C. H. Wong, "3-D vision using structured lighting", MPhil Thesis, University of Newcastle upon Tyne, UK, 1994

[8] D. Ziou, "Line detection using an optimal IIR filter", Pattern Recognition, vol. 24, no. 6, pp. 465-478, 1991

[9] J. Canny, "A computational approach to edge detection", ", IEEE Transactions on Pattern Analysis and Machine Intelligence, Vol. PAMI-8, No. 6, pp. 679-698, 1986

[10] A. Huertas, G. Medioni, "Detection of intensity changes with subpixel accuracy using Laplacian-Gaussian masks", IEEE Transactions on Pattern Analysis and Machine Intelligence, vol. PAMI-8, no. 5, pp. 651-664, 1986

[11]I. Pitas, "Digital image processing algorithms", Prentice Hall International (UK) Ltd., 1993

Flexible Automation and Intelligent Manufacturing Conference
University of Teeside; Middlesbrough, U.K.
(June 25th-27th, 1997)

Use And Implementation Of Digital Close-Range Photogrammetry To Enhance Competitiveness: Use, Application, And Cost Justification

George W. Johnson, Bath Iron Works Quality Engineer (Photogrammetry); Student, School of Industrial Technology , University of Southern Maine; Gorham, Maine

H. Fred Walker Ph.D., Faculty Member, Departmentl of Industrial Technology, University of Southern Maine; Gorham, Maine

ABSTRACT

Photogrammetry, as its name implies, is the science of obtaining precise coordinate measurements from photographs. Until recently, photogrammetry used film photographs taken with specially designed high-accuracy cameras. With the development of high-resolution solid-state video imaging sensors and the emergence of Digital Cameras into the open market a new era in precise three dimensional non-contact measurement has arrived. Digital Close-Range Photogrammetry (DCRP) as it is often called is a powerful enabling technology that not only performs many current measurement tasks faster and more efficiently than related technologies, but also, now makes feasible many types of measurements which previously were not practical or possible. The capability for quick, accurate, and reliable in-place measurements of static or moving objects in vibrating or unstable environments is a powerful combination of features all in one system.

This paper will provide information pertaining to DCRP theory, describe measurement application, and finally present results using both implemented manual and verified automated DCRP methodology from work performed at the Bath Iron Works Corporation (BIW), located in Bath, Maine. Our approach stresses overall process improvement through Total Quality Management (TQM) techniques as well as complimenting the use of 'traditional' optical tooling equipment (theodolites, transits, optical micrometers, right angle prisms/optical squares, etc.).

INTRODUCTION

To remain competitive in dynamic markets, companies continue to seek ways to improve product quality with limited resources. Central to this issue of improving product quality with limited resources is developing innovative technologies such as DCRP and integrating this technology with widely used tools and technologies such as Computer Aided Design (CAD), Statistical Process Control (SPC), and Information Management Software (IMS). DCRP then is needed as a cost effective means for improving quality by providing a means of measuring three dimensional (3-D) geometry at various points in manufacturing processes.

BACKGROUND

The Shipbuilding Industry has many diverse measurement applications all performed under a wide range of operating conditions. Traditionally theodolites and optical micrometers have been the standard tools. This equipment is accurate and reliable, but one dimensional. The emergence of three dimensional measurement systems and their diversity is having a significant impact on the way ships are manufactured. In 1992 Bath Iron Works (BIW) assembled a highly skilled and diverse group of employees representing the various divisions of the company and initiated a Continuous Process Improvement "CPI" project to determine the cost effectiveness/ feasibility of utilizing photogrammetry for shipyard processes requiring precision measurement. 1993 brought about a year of reviews and extensive testing of many different types of existing measurement systems (Johnson 1993) which can be categorized as follows:

Laser Based Thoedolites are probably the most widely known 3-D measurement systems since they closely resemble the traditional surveying instruments they were named after. These instruments, in simplest form, measure distances and angles from the reflection of a laser beam between the theodolite sensor and a desired reflective surface. Single measurements are fast and relative accuracies are acceptable for many jobs. Results are electronically output in standard formats and thus are readily imported into CAD and other softwares for visualization or analysis. These features are significant improvements in measurement capability over existing one dimensional measurement methods. Some key limitations of this equipment are that measurement need to be performed on the at the job site and the time needed for large numbers of points is a major impediment to Production personnel since the Theodolite and object being measured need to remain motionless. The shipyard environment is a constant flow of people, equipment, and heavy components, the interruption of a measurement setup is typical and can leading to lost time and possible errors. Adaptations of initial Laser Theodolites have been made specifically for the shipbuilding industry making this tool very nice for specialized jobs. However; the gains made in functionality have been very much at the expense of the portability, hurting the overall usability of the product.

Coordinate Measurement Machines (CMM) have track records in accuracy, reliability, and functionality within the high precision machinist industry that will ensure their use for a long time. The CMM (in simplest form) consists of a solid foundation, base machine with 3-D readout area, extension arms, and contact probe. Wherever the probe touches the object being measured 3-D coordinates may be obtained in real time. For shipyard use a CMM could see use in the Machine Shop where precision parts/components are either made, machined, or drilled off. But, since the quality of Machine Shop outputs at BIW consistently meets or exceeds current customer expectations the need for increased dimensional control is limited. Add to this the high costs for a measurement system that is fixed in one place (with limited reach) and you have the reason that BIW does not have a CMM at a time when significant break-throughs in performance can be attained by improving measurement capabilities in other areas.

Film Based Photogrammetry is of the most intriguing technologies, offering both stereo and convergent techniques for specific applications over a wide range of industries and environments (Fraser and Brown 1986). Stereo photogrammetry relies on manual extraction of 3-D coordinates from overlapping (stereo) pairs of images and is most widely used in mapping. Convergent

Photogrammetry is based on the principle of triangulation and enables the user to economically measure large numbers of targeted points or features in an off line manner and is most widely used in industry. By taking pictures from at least two different locations and measuring the points of interest in each photograph, one can develop lines of sight from each camera location to the points of interest on the object. The intersection of these pairs of lines of sight can then be triangulated to produce the 3-dimensional coordinate of the point on the object. Some key features of this technology is that impact to production workers is minimal, the object being measured can be moving/vibrating (although it must remain rigid), and the systems are extremely portable. Cost effectiveness can be reduced from measurements with film media due to such problems as difficulties in setting up photographic equipment in manufacturing environments, lost time spent handling and developing film, and excessive costs of system components.

DCRP THEORY AND BASICS

During this same period of time (1992 ... 1993) research of using charged coupling device (CCD) cameras instead of the existing film based types was being pursued in a cooperation with Prof. Armin Gruen and researchers at the Swiss Federal Institute of Technology (ETH) in Zurich, Switzerland. At issue was the testing of ordinary photography cameras as tools for 3-D measurement in an industrial environment. The target camera tested was a KODAK DCS200mi black and white digital camera which was designed for Photographers and Journalists. Since the images are 'captured' as electronic signals by the CCD chip (instead of as a negative or positive on film) they could be directly downloaded to a PC or Laptop computer via SCSI terminal then transmitted over telephone lines via modem to customers from remote locations. These features enhance the quality and quantity of images provided in some very competitive businesses. It was hoped that these same benefits, as well as higher levels of measurement automation would also be possible for the extraction of 3-D geometry within industry.

For control purposes this camera was compared to digitized measurements from a JVC camcorder utilizing a frame grabber and Leica 32mm film camera with a desktop scanner. The results (Maas and Kersten 1994) proved beyond a doubt that measurement accuracies from the KODAK DCS200mi camera and off the shelf components were well within tolerances required in the shipbuilding industry for a wide range of applications. And in 1994 an ongoing dialogue with Geodetic Services Inc. resulted in the development, acceptance, and subsequent implementation of the first DCRP Measurement System in shipbuilding. This event changed the way BIW approaches Dimensional Control for many reasons but two stand above the rest. First and foremost is the technological ability to measure rigid bodies quicker, more accurately, and with less impact to surrounding workers than ever before. Secondly is the versatility of the software in handling data. Digital cameras with solid state imaging chips (CCD sensors) are now available with sufficient dimensional integrity and resolution to provide accuracies suitable for many measuring tasks. DCRP systems based on these digital cameras and are very accurate, portable, reliable and can be purchased for about 1/3 the cost of the film based systems. The typical components of a basic DCRP system consist of the following:

Digital Camera: The initial KODAK DCS series of cameras used at BIW consists of a standard NIKON 8008S body, KODAK CCD sensor, attached to an 80 megabyte hard drive (1992 & 1993 models). Later models have been improved significantly by the use of portable PCMCIA hard drives for image storage and increased sensor size for increased accuracy. These cameras are light, readily available from KODAK, and relatively inexpensive compared to similar measurement cameras.

Computer: DCRP systems are used across a variety of platforms, Windows and Unix being the most popular at this time. For Windows based systems a typical configuration could be anything from a 486dx 66hz with 12 megabytes of ram and a one gigabyte hard drive with PCMCIA reader upwards to Pentium Pro 200hz models with 64 Megabytes of ram and 2+ gigabytes of hard drive space. On the Unix side a basic system would be configured around a Sparc 5 with 64 megabytes of ram and 2+ gigabytes of hard drive space. It should be noted here that these configurations are based on each vendors software needs and their own unique measurement processes and development plans. Both platforms efficiently utilize laptop computers for off-site applications requiring maximum portability.

Software: The software used throughout industry for measurement of targeted points is conceptually the same. In simplest and abbreviated form one needs to have a set of known coordinates in each image, which are used to calculate the camera location, which is in turn used to approximate the x and y coordinates for each target by image. A least square bundle adjustment then combines the data from each image into 3-D coordinates calculated to the least amount of residual error. Incorporated into this data are corrections for and lens distortion encountered.

Automation: Development is focused towards increased automation of the measurement process leading to results in near real-time (minutes/seconds for DCRP as opposed to days/hours for traditional theodolite). From measurements to date, automation through the use of coded targets offers the most significant advantages in measurement time savings. This is achieved by using the coded targets to automatically orient, then measure all the images in one process (although single image measurement is also readily available to the operator). A blunder detection process is then run on all the images to eliminate bad points or targets. Finally, the bundle adjustment is used to calculate the X-Y-Z coordinates for each measured point while also compensating for any lens distortion encountered. The use of coded targets is a feature being researched, tested, and implemented at a feverish rate and will lead to some very promising advancements in the near future.

SHIPYARD IMPLEMENTATION

The significant and multiple benefits derived from DCRP features has proven to be both economical and social in nature. Simply stated, one can typically obtain three dimensional data (versus one dimensional from theodolite) in about one-half for manual measurement to one tenth for automated measurements the time for selected jobs requiring thirty or more points to be measured. Savings are maximized when the number of points to be measured is increased. At BIW this is readily being achieved by re-focusing and combining many of the quality inspection and production service checks. Socially (yet still very positively linked to the economy) is the improvements in shop floor productivity through enhanced visualization of measurement results utilizing CAD. The old adage "one picture is worth a thousand words" is quite applicable here. Another reason is the pride associated with using the first standard DCRP system and experiencing a true 'change of pace' from years of traditional optical tooling practice. Safety, which is of high social and economic importance, has also been improved. In two years of implementation there has yet to be an on the job injury. And finally the integration of Information Management Software with the final coordinate outputs of measurements from the shop floor will enable the shipyard to utilize data for SPC analysis that in the past has been neglected due to the high costs of manual input. This is a very economical effort because the data output is a standard ascii delimited text file it is easily imported and shared between softwares. It is very satisfying to see measurement data be used for multiple purposes. Not only is data used for production service; but for process control and capability determination then monitoring as indicators of quality levels in the quest for continuous improvement.

SHIPYARD APPLICATION

Neat Cut Unit Erection

Figure 1. Erection Unit ready for imaging with retro-targets in place

With the acceptance of modularized construction in shipbuilding a common practice is to build interfacing steel edges with excess material on one end to be trimmed during the fabrication, assembly, pre-outfit, and/or erection process. Initial manufacturing practices positioned the module in the ship's plane, at an approximate location on the ship, and traced the existing ship's shape on the erected unit so that when the excess steel was burned off the unit should fit into its proper place. Significant gains in productivity were realized by the trimming of this steel prior to erection so that the unit could be mated together neat in one short lift of the cranes. However; the traditional 'neat cutting' process was a theodolite based labor intensive process for the Surveyors performed as part of a large and intensive series of quality checks and measurements performed on the modular unit before it is allowed to leave the production shop floor. Implementation of DCRP techniques has many cost effective benefits; of primary importance to the shipyard worker is that it facilitates the combining of many different types of measurements (as mentioned above) into one, all performed on the modular unit at a time and place within the construction/storage sequence where it is more cost effective and less disruptive to manufacturing personnel. A description of this transition follows:

The traditional approach of neat cutting unit erections utilizes two to three Surveyors with a theodolite and rods/tapes to measure the stock edge of a unit after it is assembled. The theodolite is set up in the production area, referenced into the local coordinate system, and each point is measured for length, width, and height manually from visual readings to a rule or rod placed at the required spots one point at a time until the complete modular unit has been measured. This involves multiple instrument setups (often times 5+) repeated for each of between three to five edges (forward edge, aft edge, port side, starboard side, and the top deck. This is an expensive, time consuming, and stressful process due to obstructions (staging, welding leads, sucker tubes, etc..), people working around the instrument, job site noise levels, and pressure to get the job done right, quickly (it is not uncommon for assembly units to be shipped to the next construction stage immediately after Surveyor readings are completed). The matching data to derive the neat cut and final fitup is obtained by repeating this process for the interfacing edge by taking readings of the actual condition of the ship on the erection ways. The output (neat cut data) from these two measurements is manually computed for each point and noted on a table provided to the structural team. Accompanying this list is a hand sketch depicting the location of each point. The results for this process as a whole are fully dependent on a stable working platforms. Preferably the units are positioned gravity level in assembly or on ships declivity out on the ways. This process is very subjective to surrounding working conditions, schedule, and operator skills but is cost effective, none the less. For many three dimensional measurement systems and companies the job begins and ends as described above; a request for service is forwarded to the Surveyors/Optical Tooling group and the data is provided in the form of a measurement file to the shop floor mechanic. BIW's approach utilizes CAD software for enhanced graphical visualization of a job and its results as well as Information Management Software to analyze and breakdown the data for manufacturing process control and capability purposes. This information is 'on-line' and shared with all divisions within BIW via the Corporate Local Area Network. Thus not only is the measurement performed for the requesting trade but in-depth quality and engineering data is available for process improvement initiatives.

From the beginning of the measurement process "Planning" DCRP techniques offer a much different approach because the measurements are not dependent on special positioning or stable platforms, only that the object being measured remains structurally rigid regardless of its position. The first step towards measurement minimizes the impact to production schedules by replacing the measurement of the complete edge conditions in the assembly areas by establishing predetermined dimensional references or control points for scale at the extremes of selected girths, seams, and/or decks to be measured. After the units are shipped out, they are tracked so that when they are the most accessible and the surrounding work is at a minimum the complete edge conditions can then be obtained in a very safe, fast and accurate manner. For the DDG 51 class of ships, this is very often done in storage areas (Figure 1) because most of the units are structurally rigid. Units that twist or rack are typically measured in Pre-outfit areas where they are leveled as part of their normal work cycle.

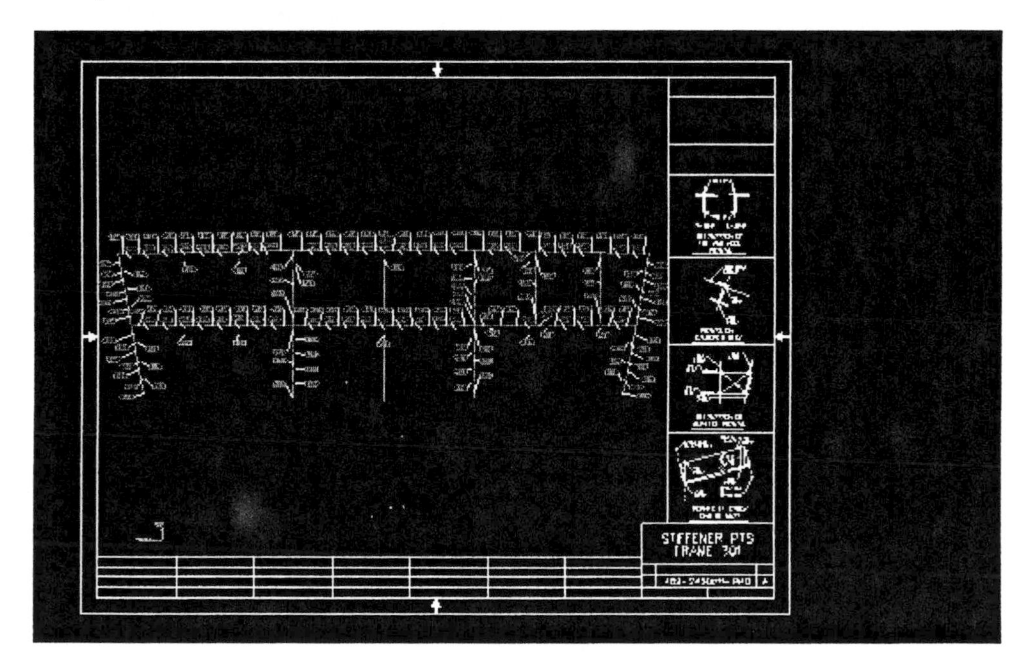

Figure 2. ACAD target layout sketch

The measurement process starts off by placement of the initial orientation bar, coded orientation targets, and then measurement targets by placing magnetic retro-reflective targets at all key structural locations for all edges to be measured as requested by the structural fitting team. This task if usually performed with the use of a scissors lift or condo lift and really has few hindrances provided the targets are clean and in good shape. Target placement for one of three edges is depicted on an AUTOCAD sketch (Figure 2). This sketch is directly tied into the Engineering AUTOKON database for accuracy. Customized LISP routines enable the Surveyor to extract design XYZ dimensions for each targeted point automatically. Next, images (typically 20 to 60) are taken of the targeted unit at many converging angles from the same lift used in the targeting. Camera settings are usually quite constant at ISO 200, aperture setting of F16, and a shutter speed of 250. Image measuring is initiated by downloading the images into the Sparc station, PC or laptop, entering the orientation bar coordinates, control point data, scale distances, and importing the design (or comparison) data for the target locations. Each image may then brought up on the computer screen, measured individually, and saved. This can be automated to a significant degree using the coded targets and measure all feature within the software. After many or all of the images are measured the data is merged together and XYZ coordinates are calculated through a least squares, self-calibrating bundle adjustment. Measurement is completed when all targets are within imaging and dimensional requirements and the resulting XYZ dimensions in object space have been transformed to the units local coordinate system. Quality control features enable the operator to view estimated accuracies for each target point measured and to edit images to obtain extremely consistent accuracy levels.

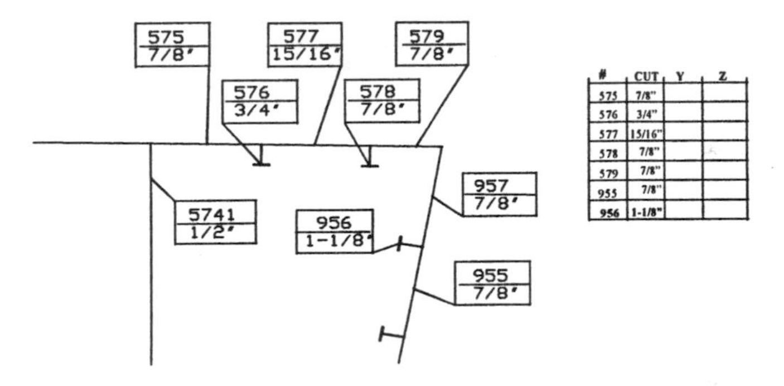

#	CUT	Y	Z
575	7/8"		
576	3/4"		
577	15/16"		
578	7/8"		
579	7/8"		
955	7/8"		
956	1-1/8"		

Figure 3. ACAD Neat Cut sketch

The data for each edge is stored until the measurement process is completed for the interfacing edge of the mating unit. Once the stock edge and neat edge conditions are obtained, a comparison program is then run to calculate the amount of excess steel. The output to the structural trades is in the form of the original AUTOCAD sketch (Figure 3 shows a close up of the upper right corner) being modified to provide the target identification number and the amount of excess steel to be cut at each targeted point measured

The 3 dimensional data from the digital photogrammetry neat cut process which is stored as a standard ASCII text file is readily imported into BIW's Information Management System for Statistical Analysis. Process control, process capability, and out of spec. measurement information is reviewed for each stage of construction contributing to the Neat Cut process. This provides all BIW employees with access to the corporate LAN with on-line feedback of Process and Product Quality levels. The analysis starts with the Main Selection Menu screen (Figure 4) which is easily accessed from the corporate LAN by the customization of a unique icon for BIW Quality within the IMS software. Once here; any one of three construction stages may be analyzed, performance to corporate strategic objectives reviewed, or basic data entry options can be accessed.

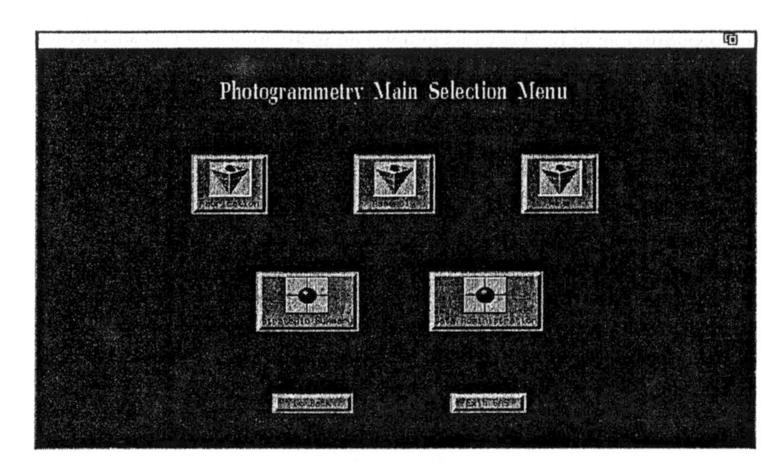

Figure 4. Main Selection System

Figures 5 through 8 are representative of the versatility and functionality of software involved as they depict a Quality Control system and its SPC outputs. For this example a sample of units from various sections were selected for analysis of their neat edges (edges that are supposed to be built exactly to design lengths). In Figure 5 a box and whisker control chart shows that although the process is out of control the bulk of measurements for each unit are within the limits. Also note that the units are uniformly distributed around the mean. From this one could surmise some levels of manufacturing stability. By examining the outliers we hope to identify areas of concern for more in-depth analysis and root cause identification. Another user friendly way to look at the measurement data is to lump all the readings into one big histogram that specifies the construction tolerances needed for neat construction levels of performance (Figure 6).

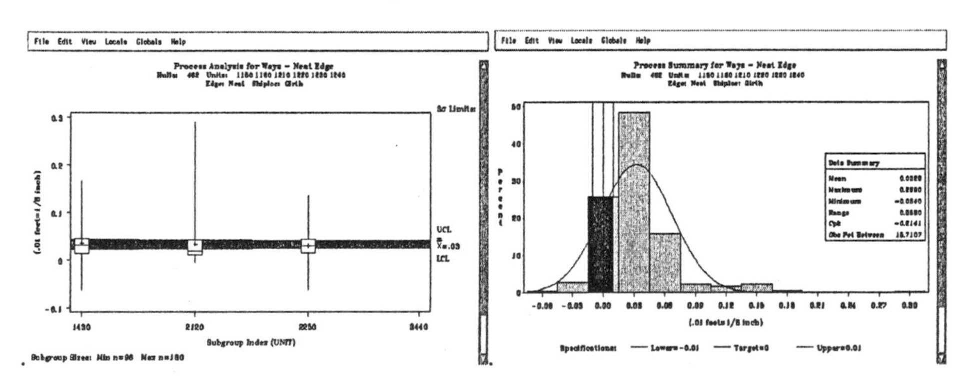

Figure 5. Process Control Chart Figure 6. Process Summary Histogram

From this view of the data (which has purposely left out the control limits for simplicity) one can see that there are few points within specs and there are many that are very close which may also indicate some stability as in the previous graph. But there are a few measurements that are way out of range and these points need to be identified and reviewed for immediate corrective action.

Long term results can be obtained through analysis of a drill down bar chart of the non-conforming (or out of spec.) measurements. This type of "Exceptions Report" is very important because we can "drill down" through the data to look for groupings of bad points by product or process giving a much more in-depth look at things. For our example (Figure 7) we can see that of the points out of spec. unit 3440 has approximately three times as many bad points as unit 1430 and almost twice as many as unit 2120 which is a similar unit. This could be for many reasons and this is where the drill down feature helps us because by selecting a unit with the mouse we can see the breakdown of bad points by subassembly within that unit.

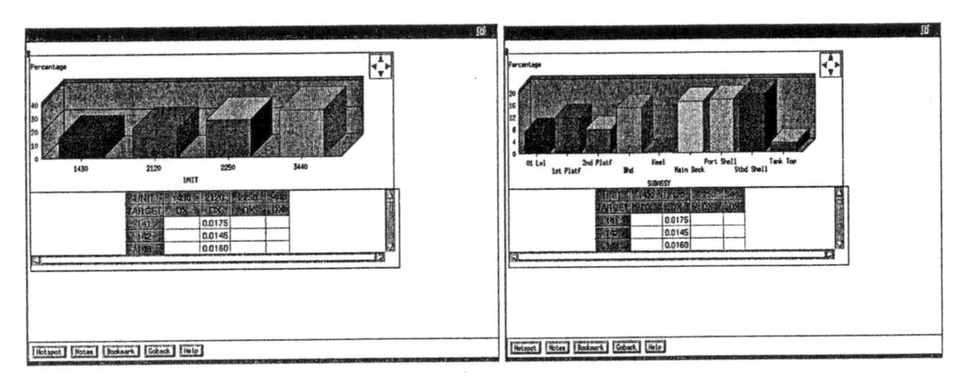

Figure 7. Overall Drill Down Graph

Figure 8. Unit Specific Drill Down Graph

Figure 8 shows the dispersion of bad points within unit 2120. Here we can see that the keel and tank top subassemblies appear to be quite good and the out of spec. points are somewhat evenly dispersed across the other subassemblies. Initial investigation could begin in many directions such as 'drilling down' into the subassemblies to see if there is anything that stands out. However; another interesting concept would be some initial investigation to properly interpret the graph. The question to be investigated would be ...(is the error really evenly spread across the other assemblies or are the keel and tank top areas the problem areas with the others somewhat constant?). For this type of unit this could very well be the case because they are the points of origin for the master reference controls for measurements. An error here could indeed ' throw' the rest of the data out of spec. and lead to a few dead ends with regards to investigation.

When complete Shop floor mechanics and Engineers from Structural Design, Quality Control, Welding areas will have current information available pertaining to items like the amount of excess steel (stock) being cut, misalignment of structural members for height and half-breadth, and structural alignment data at key shapes intersections (bulkheads, decks, shell plating). The examples above are provided to depict the need to visualize results in many different but user friendly ways. By analyzing this data in conjunction with the process control, process capability, and exception functionality then applying sound shipbuilding methodology, BIW plans to achieve a level of robustness that will shorten the transition time to neat construction (Johnson 1996). Making it easier for the structural trades to maintain higher levels of fitup quality throughout the entire construction process.

COSTS AND BENEFITS ANALYSIS

	Surveying hours per unit (3 edges)	Accuracy	Neat cut data points per Unit (3 edges)	Predominant data gathering sites	Units erected per month	Surveyors req'd. for neat cutting	Job rel. injuries avg per quarter
Theodolite	75+	+/- 1/8"	150 - 225	Ways & assembly	9	4 - 7	2
DCRP Manual	32 +/-		450 - 900	Storage & PreOutfit	13+	2 - 3	0 over 10 quarters
DCRP Automated	10 +/-	+/- 1/16"	450 - 900	Storage & PreOutfit	13+	2	0 over 10 quarters

Table 1. Neat Cut process comparison

Currently digital photogrammetry has neat cut over 300 structural units and the results clearly show its measurement performances are far superior for this application (Table 1). The bottom line to Shipbuilding is that the shop floor mechanic is receiving three dimensional data, graphical views, and quality information in a fraction of the time and much more safely than past "measurements". A summarized breakdown of benefit/financial impact to the overall corporation follows:

A. *Improved Response time:* Perhaps the first improvement experienced is the improved response time that occurs when implementing a DCRP system. Since the number of man-hours and people needed for Unit Erection decreased significantly, their time can be used elsewhere for existing work and/or new measurement applications (each saving the company time and money for in their own right). At BIW the Surveyor Department has reduced the man-hours per unit to what is depicted in Table 1 and has shown up in a 30% reduction in response time for other measurement tasks. And these savings were realized even though the departmen. has seen a reduction of 6 surveyors (of 24 total) from the work force due to retirement, promotion, and lost time injuries (non-DCRP related). A factor that is beginning to contribute additional savings to this area is the implementation of the Automated and shared quality data. By utilizing this information for each unit, Surveyor visits to each unit are reduced (combined) or their efforts maximized on important and identified issues. This reduces the Mechanic's waiting time for service and leads to standardization of day to day work.

B. *Improved Quality of Structural fitup*: The next area of improvement noticed should be in the improved quality of unit/structural fitup. This is a result of increased measurement accuracy and increasing the number of data points measured so that a better neat cut line is measured on. Large savings in welding costs are realized when fitups are such that weld gaps are uniform and correct. Note: a 3-5mm error in weld gap root opening can double the total welding costs as well as reduce weld quality for each unit. With such tight tolerances on such large and heavy objects this factor is used as a prime indicator of erection success.

C. *Reduced Schedule time*: One of the most significant benefits to be associated with DCRP implementation is the overall reduction in schedule time for the manufacturing of ships. In Assembly when the theodolite work would have had to occur, a one to two day reduction in time for

Surveyor measurements is realized. Checks of this nature (quality checks for final heights, final half breadths, unit controls, and neat cut data) are performed elsewhere in production when manufacturing is at a lull and access is optimum. Another immediate impact item is that the number of units that can be erected per month has increased from 9 units to more than 13. This was always a 'bottleneck' for production due to surveyor availability and response time. At the current manning levels and work load DCRP has made this process easy to support for the Surveyors. The limiting factor now is the handling and welding of the units by manufacturing personnel, not Surveyor support! This has reduced ways erection time by 25% (55 weeks+/- to 45 weeks +/-) for each ship, a significant money saver since this is a milestone monitored by the customer in performance reviews.

D: *Reduced Waste and Unplanned Labor*: Long range goals are encompassed by the effort to build ship components and assemblies completely free of the excess stock lofted into joining edges for construction errors. This is a company wide effort that when realized will yield reductions of between 10% to 20% percent of overall construction costs. The ability to burn steel the right size, assemble it together with minimum distortion, and deliver it to the next phase of construction within quality levels that enable similar methodology 'down the line' is a major effort being undertaken by most all of the successful shipyards world wide. Essential to achieving this level of manufacturing robustness is a consistent and thorough quality control plan that focuses on dimensional accuracy. The utilization of neat cut DCRP data for visualization, analysis, and quality improvements is cost effective, accurate, and easily shared. thus; making it an excellent base for current as well as futuristic Corporate initiatives.

SUMMARY

Digital Close-Range Photogrammetry is a powerful and emerging technology. For BIW it is functioning quite well in the demanding shipyard environment on daily basis. Seasonal variations in temperature have been of little consequence for measurements to date provided the camera has been acclimated to environmental surroundings. Over 450 measurements for neat cut, quality control checks, and production support have been obtained in temperatures ranging from -10 to +100 degrees Fahrenheit over the last three years. The speed of obtaining images and measuring off line is well appreciated at these extremes. Also important is the portability and ability to work on unstable platforms (ensuring the least possible impact to surrounding production work) is a decided advantage over existing measurement systems.

The use of DCRP methodology to compliment traditional equipment and techniques has led to faster data turn around, improved accuracy, higher productivity, and increased safety as provided in the application discussions. The use of other digital components and software combinations (Gruen 1994) like video cameras, frame grabbers, surface reconstruction, etc. are very attractive for measurement and documentary purposes. To date not much of this aspect of digital technology has been tested in the Shipbuilding Industry.

The future promises to be exciting for BIW as this overall effort becomes refined and focused. More applications will be found while new equipment and technologies are developed then brought on-line. The current costs of a DCRP system range from $120,000 to $150,000 U.S. and has easily be paid back to BIW within one year using only the $$ saved from this one application (manual or automated). The cost justification is made more lucrative as other applications are undertaken. In a shipyard there are many diverse applications ranging from the installation and fitup of foundry castings, to pre-determined liner/shim sizes for advanced installation of critical equipment, to verification of battery alignment (ordnance) lines and foundations that are worked on a daily basis with similar savings for an even greater return on investment. Also to be considered is the safety issue. The costs savings associated with reduced lost time injuries and related workers compensation and medical costs would also pay for a system on its own merit. DCRP technology has positively impacted modern shipbuilding as well as other manufacturing industries (automotive and aerospace to name just a few) more than any other measurement technique to be developed over the last 20 years. There is great potential for even further productivity gains and cost savings as automated measurement processes and in some cases machine vision applications (Hagren 1986) mature, provided the research and development keeps the functionality of DCRP systems at an implementation level that encompasses a wide range of users.

REFERENCES

Hagren, Henrik, 1986. Real-time Photogrammetry As Used For Machine Vision Applications, International Archives Photogrammetry And Remote Sensing, 26, 5, pp. 374-382.

Maas, Hans-Gerd & Kersten, Thomas, 1994. "Digital Close Range Photogrammetry for Dimensional Control in a Shipyard" Video Metrics III

Fraser, C.S., & Brown, D.C. 1986. "Industrial Photogrammetry: New Developments and Recent Applications" The Photogrammetric Record,

Gruen, Armin, 1994. "Digital Close-Range Photogrammetry - Progress Through Automation" ISPRS Commission V Symposium.

Johnson, George W., 1993. "Digital Photogrammetry a New Measurement Tool" Corporate Summary of Three Dimensional Measurement Technology Comparisons for Unit Erection.

Johnson, George W., 1996. "Practical Integration of Vision Metrology and CAD in Shipbuilding" Invited paper Commission V, Working Group V/3 ISPRS 1997 Congress, Vienna, Austria.

ENGINEERING 2D BAR CODE SYMBOLOGIES FOR MANUFACTURING INFORMATION SYSTEM MANAGEMENT

Duane D. Dunlap, Soumya K. De
Purdue University

ABSTRACT: Two-dimensional (2D) bar codes literally add another dimension to engineering information management systems when compared to the common linear bar code symbol. The power of the 2D symbol is in its' design and application use. First, a 2D bar code symbology allows a single, complete information record (up to millions of characters) to move with the product, process, or shipment. The second significant attribute involving 2D symbologies is the use of error correction. Simply put, the ability of a bar code symbology to use error correction allows the symbology in question to be partially destroyed or damaged and still be completely readable by an input device. This paper will address three of the most commonly used 2D bar code symbologies based on usage, standards, public access, and application use. The three 2D bar code symbologies that will be discussed include PDF417, DataMatrix, and MaxiCode. Mathematical descriptions and applications as they related to manufacturing environments will be the focus.

Figure 1. PDF417 Symbol

Why 2D Symbologies?

Two-dimensional (2D) bar code symbologies add a new manufacturing dimension in data communication and information management. Portable Data File 417 (PDF417) symbology and other 2D symbologies like DataMatrix and MaxiCode answer the immediate need to capture, store and transfer large amounts of data fast and inexpensively. Two-dimensional symbologies can exchange complete data files (such as text, numerics or binary data) and encode production control, statistical process control, shipping manifests, electronic data interchange (EDI) messages, equipment calibration instructions, CAD drawings, hazardous materials data sheets and much more for use in the manufacturing environment. Two-dimensional symbologies provide a powerful communications capability-without the need to access an external database. For almost no increase in cost, you can add 2D bar code symbols to manufacturing documents and labels you are already printing.

Think of 2D symbologies as distributed databases with complete freedom of data movement, traveling together with a person or an item, object, package, form, document, card or label. These graphical symbologies accomplish what wired networks cannot: it allows one to immediately access their data regardless of location. Data encryption is available as an advantage when additional data security is required.

Figure 2. DataMatrix

Moreover, since 2D symbologies are a machine-readable method of transporting data, they eliminate time-consuming and error-prone manual data entry. The 2D symbol is like a write once, read many (WORM) diskette or CD-ROM. Additionally, 2D symbologies behave as a universal machine language, which communicates with all host computer operating systems. Two-dimensional symbologies can encode full ASCII, numeric, or binary data using sophisticated error correction algorithms to keep intact 100 percent of the data even when as much as half of the symbol is damaged. Moreover, many of the 2D symbologies are self-verifying, so data errors can be detected and then corrected maintaining data integrity.

What a Difference a Dimension Makes!

One-dimensional (1D) bar codes contain an access code or license plate number that serves as real-time key for opening or pointing to a record a database. In today's world-class, information hungry manufacturing environments, a 15-20 character bar code license plate message just isn't enough. There are numerous manufacturing applications where large amounts of data need to be encoded in an efficient manner and shared immediately. This often cannot be accomplished by printing 1D bar codes repeatedly. The physical area required to accommodate bar code symbols sometimes is just too small and the inconvenience of the built-in information retrieval process may not just be where or when you need it. These reasons have limited the use of 1D bar code symbols to manufacturing applications demanding more than a traditional string of encoded license plate characters. The benefits of 2D symbols are ideal for manufacturing applications that are limited by the natural constraints of 1D bar codes. Two-dimensional symbologies are designed to encode large amounts of data in an efficient manner. Some 2D symbologies can encode up to almost of 100MB worth of information or 3,000 characters into space the size of a postage stamp. A single scan of a 2D symbol can easily replace thousands of multiple 1D bar code scans which result in faster delivery of more information to make world-class manufacturing decisions.

In addition, 2D symbols were developed to survive in rugged manufacturing and industrial environments. Robust by design, these symbols use error correction algorithms like Reed-Solomon to correct lost or missing data for damaged symbols. Up to half of the symbol can be destroyed or missing and yet the symbol can still be read and decoded.

Figure 3. CodeOne

Which 2D Symbology?

Two-dimensional bar code symbologies can be broken into two discrete families, stacked or matrix. A stacked symbology, has multiple rows of bars and spaces. The most predominant stacked symbology in use based on standard's implementation is PDF417 (Figure 1). A matrix symbology is an arrangement of regular polygon shaped cells where the center to center distance of adjacent elements is uniform. Depending upon the matrix symbology in question, they may include recognition patterns which do not follow the same rule as other elements within the symbol. Some examples of the most predominant matrix symbols in use based upon AIM International standards include DataMatrix, CodeOne, and MaxiCode (Figures 2-4).

Emerging manufacturing applications require a symbology providing both high information density and high information capacity. To ensure cost effective and cost sensitive solutions, the symbology in question must be compatible with existing state-of-the-art bar code printing and scanning technologies.

Figure 4. MaxiCode

International standards are shaping the automatic data collection (ADC) industry. A major push by the U.S. Technical Advisory Group (TAG) for the Joint Technical Committee 1 of ISO (International Standards Organization) and IEC (International Electrotechnical Council) are supporting the engineering and development of 2D application standards. Standards that have been developed or are near completion include electronic data interchange (EDI), product shipping manifest and labels, ID badges, maritime container IDs, rail and truck rolling stock ID, and all discrete electronic product manufacturing processes.

Information Error Detection and Correction

One physical construction characteristic of a linear bar code symbology is that encoded information is repeated over the height of the symbol. Often linear scans at alternate positions can successfully read a damaged 1D bar code. In cases where the damage is so severe or the print quality so poor, or the symbol is totally unreadable, the end-user typically has the option of manually keying in the character string from the printed, human-readable data. Although elimination of manual key entry is the optimal goal, the potential for introducing errors into the host computer system by this method was the original reason for implementing bar code scanning. However manual data entry is still a backup solution when there is a totally unreadable linear bar code. All one dimensional bar codes have error detection. However, all 1D symbologies in use for manufacturing applications do not have any automatic error correction capabilities. Two-dimensional bar codes have both error detection and correction. Two-dimensional bar code information can be scanned by a raster, CCD, or another input device like a vision system providing 20,000 times faster input than linear bar code capture.

2D Bar Code Symbology Construction

The first 2D symbology we will describe is PDF417. PDF417 stands for Portable Data File 417 which is referred to in the literature as a stacked symbology. Portable Data File 417 symbology, has become of the most prominent of all stacked symbologies. Symbol Technologies first released PDF417 in 1989.

PDF417 is composed of three to ninety rows of data or information. Each row contains three to thirty-two code words (two of which are left and right row indicators). Each code word is formed by four bars and spaces that total seventeen modules, with the largest bar or space totaling six modules in width. A module can be the "X" dimension, or minimum bar/space width (see Figure 6). To encode information in a PDF417 symbol, the end-user has the flexibility to choose the bar height as well as the minimum bar width of the symbol expressed in mils. He or she can also decide on the level of error correction needed and the symbol's aspect ratio to determine the size and shape of the printed symbol. An aspect ratio is defined as the printed symbol height related to symbol width.

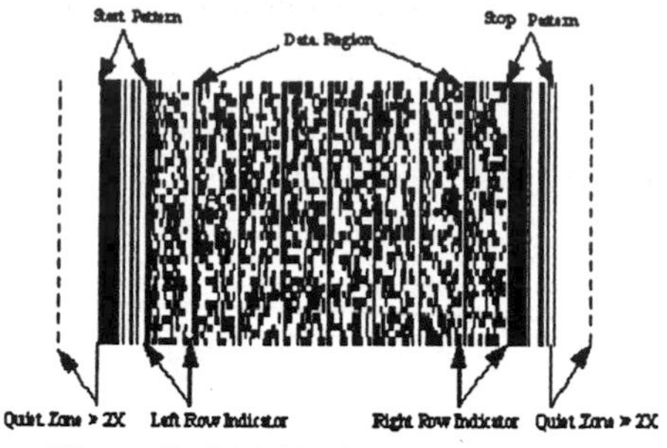

Figure 5. PDF417 Symbol Structure

Quiet zones must surround the PDF417 symbol. These zones need to be areas that are free of any other printing marks and precede the start symbol character and follows the stop symbol character as shown in Figure 5. Unique start and stop patterns are defined for PDF417. The start symbol pattern is made up of a unique 8 bar-space sequence while the stop pattern is defined by a 9 bar-space sequence. These patterns allow a scanner in part, to know when to begin reading and when to finish reading the symbol. Left and right row indicators in the PDF417 symbol allows each row to contain a unique row identification number. In PDF417, local row discrimination is achieved by using a different symbology for adjacent rows. This enables a feature called block decode, where the right half and the left half can be decoded in separate scans and then "stitched" together to define the contents of the entire symbol. Similarly, PDF417 has three distinct subsets of bar/space patterns to represent the 929 PDF417 codewords. Each unique subset is called a "cluster". Each row uses only one of the three clusters to encode data. Each cluster repeats sequentially every third row. Because any two adjacent rows use different clusters, the decoder can distinguish a codeword decoded in one line from one in the next on a single linear scan by identifying the respective bar/space patterns or clusters for each. By using this composition, the PDF417 symbol is vertically synchronized to detect row-to-row transitions.

A codeword when used with PDF417, is the basic data unit or minimum segment containing interpretable data. Figure 6 below shows a PDF417 codeword. Every codeword in the symbol is the exact same physical length and each codeword can be divided into seventeen equal modules. A codeword set in a PDF417 symbol can contain up to 929 different, defined codewords. Symbol characters used in a PDF417 symbol consists of seventeen modules arranged into four bars and four spaces. Each codeword is associated with a numeric value from 0 to 928. Thus, the PDF417 symbology defines 929 distinct codewords, each represented by a particular bar/space pattern, and each associated with a value between 0 and 928. The entire set of symbol characters is divided into three mutually exclusive encodation sets called clusters. The Figure 6 below graphically describes a PDF character structure.

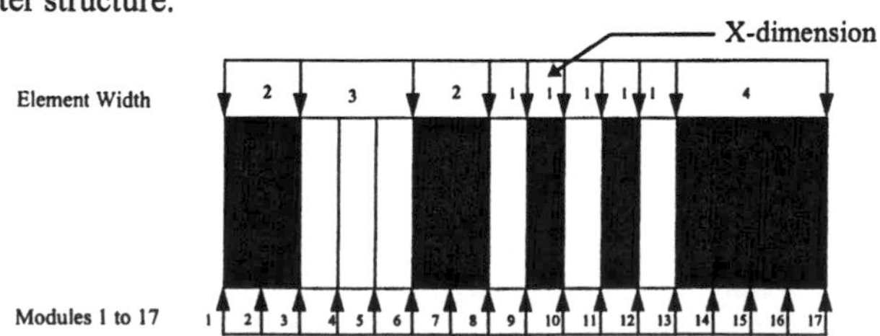

Figure 6. PDF417 Codeword Character Structure

The encodable character set for a PDF417 symbol includes ASCII characters (0-127 values in accordance with ANSI X3.4) and extended ASCII data values (128-255 values in accordance with ISO8859)[2]. PDF417 is capable of encoding 8-bit binary data and up to 811,800 different character sets and/or interpretations. Data compaction modes in PDF417 efficiently pack a series of bytes into a stream of codewords. Each mode defines a conversion or mapping scheme between codeword sequences and byte sequences. The three modes that can be used by a PDF417 symbol include Extended Alpha Numeric Compaction Mode, Binary/ASCII Plus Mode, and Numeric Mode. *Mode Latch* and *Mode Shift* codewords are used for switching between modes. The interpretation of the byte sequences encoded by a compaction mode is determined by the Global Label Identifier (GLI). A GLI is a codeword sequence that activates a set of interpretations, assigning meaning to the stream of bytes encoded by the compaction modes.

Every PDF417 symbol contains two error detection codewords that are used like the check digit in linear bar code symbology. The purpose of this is to detect decode errors and verify that all data has been read and decoded accurately. Additionally, PDF417 provides error correction in the event that portions of the symbol have been damaged, destroyed, or are illegible. There are nine levels (0-8) of security or error correction available when generating a PDF417 bar code as indicated by Table 1. The security level determines the number of error correction codewords added to the symbol. The higher the level, the greater the symbol damage that can occur while maintaining 100% integrity of the data contained in the symbol.

Security Level	Number of Error Correction Codewords	Maximum Number of Unreadable Codewords and/or Misdecoded Codewords
0	2	0
1	4	2
2	8	6
3	16	14
4	32	30
5	64	62
6	128	126
7	256	254
8	512	510

Table 1. Error Correction Levels in PDF417

Error correcting codewords are generated using Reed-Solomon algorithm and appended to the encodation data stream. The theory that associates a certain group G of substitutions on the roots of a given algebraic equation[1] was initiated by E. Galois, who was killed in a duel in 1832 at the age of 21.[3] The coefficients of the polynomial defined by the expansion of:

$$g(x) = (x-3^1)(x-3^2)\dots(x-3^k) \text{ ---------------------------(1)}$$

where k = number of error correction codewords or check characters required. The coefficients of the polynomial defined by the expansion of (1) are integral coefficients. These coefficients then become the data codewords for the message in question. The Reed-Solomon error correction symbol character values are the remainder after dividing the data symbol character polynomial, multiplied by x^k, given below:

$$d(x) = d_{n-1} x^{n-1} + d_{n-2} x^{n-2} + \dots + d_0, \text{-----------------(2)}$$

Macro PDF417 is another version of PDF417 that provides a means for encoding data files too large to be represented by a single PDF417 symbol. The original data file is broken down into segments which are then encoded in Macro PDF symbols with control characters that enable the reconstruction of the original file irrespective of the order in which the multiple Macro PDF symbols may be scanned.

Matrix Symbologies

The Data Matrix symbology was originally developed by International ID Matrix and placed in public domain in 1994. Data Matrix is made up of square modules arranged within a perimeter finder pattern. Data Matrix is designed to pack a lot of information into a very small space. A Data Matrix symbol can store between one and 500 characters in a single symbol. The symbol is also scaleable between a 1-mil square to a 14-inch square. That means that a Data Matrix symbol has a maximum theoretical density of 500 million characters to the inch! The practical density will, of course, be limited by the resolution of the printing and reading or input technology used.

Data Matrix has several other interesting features worth noting. Since the information is encoded by an absolute dot position rather than a relative dot position, it is not as susceptible to printing defects as is a traditional bar code. The coding scheme has a high level of redundancy because the data is "scattered" throughout the symbol. According to International ID Matrix, the scattering of data allows the symbol to be read correctly even if part of it is missing. Each Data Matrix symbol has two adjacent sides printed as solid bars, while the remaining adjacent sides are printed as a series of equally spaced square dots. These patterns are used to indicate both orientation and printing density of the symbol.

The Data Matrix symbology comes in different types. One type of Data Matrix, the ECC000-140, involves convolutional coding for error checking and correcting (ECC). This mathematical technique comes uses only addition and subtraction [4]. Data Matrix ECC 200 symbology uses Reed-Solomon error correction and is in the public domain recommended by AIMI Technical Specifications - Data Matrix. The following section will introduce the Data Matrix symbology and define some of its basic characteristics.

Each Data Matrix ECC 200 symbol consists of data regions composed of square shaped modules set out in a regular array. In larger ECC 200 Data Matrix symbols, there are two or more data regions. Each data region is bounded by alignment patterns. The data region is surrounded by a finder pattern on its left and lower sides and a quiet zone equal to one module wide on all four sides. Figure 7 shows some the different characteristics of a Data Matrix symbol. The ECC 200 comes in twenty-four square and six rectangular configurations. The size and shape of the symbol may be chosen to suit the requirements of the application.

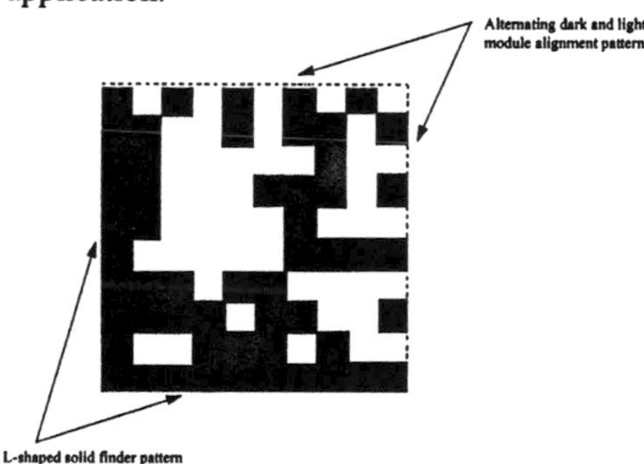

Figure 7. Data Matrix Symbol Characteristics

A single cell in a matrix symbology is used to encode one bit of data/ information. Each module represents a binary bit. A dark module represents a one while a light module represents a zero. In Data Matrix, a module is one square in the symbol as shown in Figure 8. A codeword is a symbol character value in DataMatrix. It is an intermediate level of coding between source data and the graphical encodation in the symbol. The newer (1996) USS-Data Matrix specification includes expanded encoding capabilities that include all 128 ASCII characters (0-127 values in accordance with ANSI X3.4), and extended ASCII data values (128-255 values in accordance with ISO8859) [5]. In addition ECC 200 Data Matrix symbols have special symbology control characters, having particular significance to its unique encoding scheme.

A DataMatrix symbol character consists of one or more codeword, each consisting of eight square shaped modules, arranged in three rows. Eight modules are positioned from left to right and top to bottom to form a symbol character. Within each symbol character the most significant bit is the lowest numbered module and the least significant bit is the highest numbered module. A typical encoded symbol character for Data Matrix is shown in Figure 8.

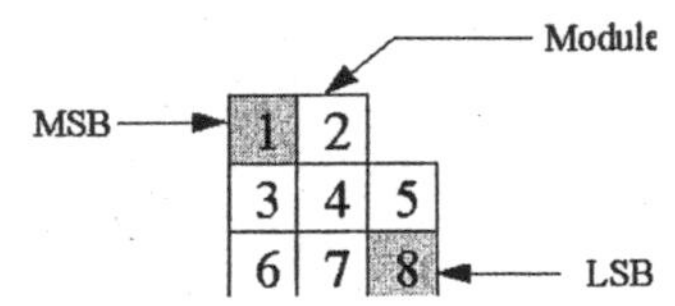

Figure 8. A Data Matrix Symbol Character

The finder pattern in a Data Matrix symbol is typically one module wide made of solid dark lines that forms a L shaped boundary on the symbol's left and lower sides. The finder pattern is used to determine the symbol's physical size, orientation, and symbol distortion. Alternating dark and light modules make up the two opposite sides of a Data Matrix symbol. These modules primarily define the cell structure of the symbol while assisting in determining its physical size and distortion. See Figure 7. The following flow chart shows the steps involved in encoding a message in a Data Matrix EEC 200 symbol. Later each step will be discussed briefly to introduce the main ideas behind each step and the terminology used.

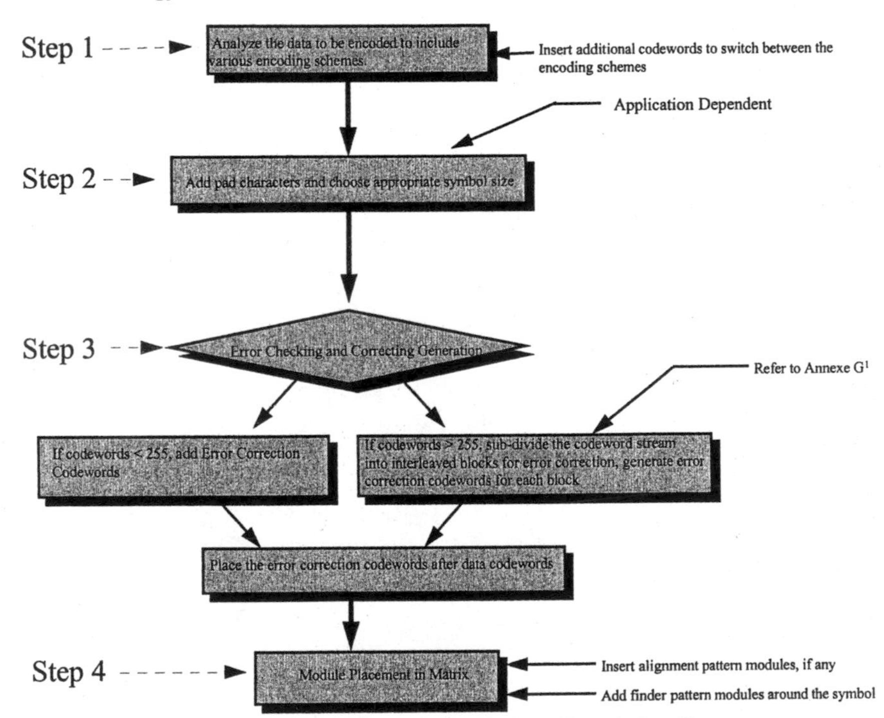

Figure 9. USS Data Matrix Symbology Encodation Steps

[1] AIM International Technical Specification: International Symbology Specification - Data Matrix

Step 1 analyzes the data by identifying the different symbol characters to be encoded. The lowest of the six encodation schemes capable of encoding the data is selected to ensure maximum compaction efficiency. The ASCII encodation scheme is the basic scheme. In choosing the most optimum encoding scheme, one should not limit his/her choice to fewest bits per data character. The aim should always be to attain maximum compaction efficiency by taking into account switching between code sets and between encodation schemes. In this step we also choose the symbol size satisfying the real estate restrictions and other demands of the application in question.

Step 2 is the error checking and correcting algorithm generation stage. Error correcting codewords are generated using Reed-Solomon algorithm and appended to the encodation data stream. The theory that associates a certain group G of substitutions on the roots of a given algebraic equation[1] was initiated by E. Galois. The polynomial arithmetic for ECC 200 is calculated using bit-wise modulo 2 arithmetic and byte-wise modulo 100101101 arithmetic. This is a Galois Field of 2^8, since each codeword in the case of Data Matrix is made up of 8 modules. The coefficients of the polynomial defined by the expansion of:

$$g(x) = (x-2^1)\,(x-2^2)\ldots(x-2^k) \text{ ----------------------------(1)}$$

where k = number of error correction codewords or check characters required. The coefficients of the polynomial defined by the expansion of (1) are integral coefficients. These coefficients then become the data codewords for the message in question. The Reed-Solomon error correction symbol character values are the remainder after dividing the data symbol character polynomial, where each symbol character is the coefficient of the dividend polynomial in descending power, by the generator polynomial g(x) of degree k. The payoff of the Reed-Solomon approach was a coding system based on groups of bits-such as bytes-rather than individual 0s and 1s. That feature makes Reed-Solomon codes particularly good at dealing with "bursts" of errors: Six consecutive bit errors, for example, can affect at most two bytes. Thus, even a double-error-correction version of a Reed-Solomon code can provide a comfortable safety factor. Mathematically, Reed-Solomon codes are based on the arithmetic of finite fields.

In Step 3, the final codewords from Step 2 are placed in the binary matrix as symbol characters. The ECC 200 Data Matrix symbol is constructed by placing the codeword modules in mapping matrices of different shapes and sizes (depending on symbol shapes and sizes) described in detail in Annexe M.2 and illustrated in Annexe M.3 in AIMI - Data Matrix pages 71-81[5]. The alignment pattern modules, if any, are then placed. The symbol character placement in any ECC 200 Data Matrix symbol starts at the top left corner and follows the path shown in Figure 10 ending in the symbol's bottom right corner.

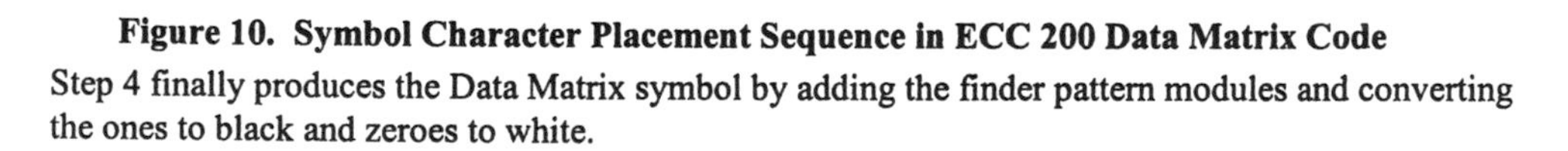

Figure 10. Symbol Character Placement Sequence in ECC 200 Data Matrix Code

Step 4 finally produces the Data Matrix symbol by adding the finder pattern modules and converting the ones to black and zeroes to white.

The structured append feature allows files of data to be represented in up to 16 Data Matrix symbols. The original data can be retrieved regardless of the sequence in which the symbols are scanned. If a symbol is part of a Structured Append, the first symbol character position is occupied by a symbol sequence indicator, which is followed by three structured append codewords providing the file identification.

MaxiCode

MaxiCode is a fixed-size capacity 2D matrix bar code symbology especially designed for the high-speed scanning application of package sorting and tracking. Introduced by the United Parcel Service, Inc. (UPS) in 1992 (with underlying development dating from the late 1980s) and has recently been published in an AIM USA "Uniform Symbology Specification - MaxiCode"[6]. MaxiCode can be scanned in any orientation as long as the symbol is perpendicular to the input device.

MaxiCode was designed with a number of features that uniquely suit it to the environment in which most parcel carriers function. The central finder pattern allows accurate omni-directional location of the symbols while they are traveling at high speeds. The fixed element size allows highly reliable reading of labels on packages of varying heights on wide conveyor belts. The nested hexagonal or honeycomb pattern permits reading on warped or angled surfaces. Extensive error correction capability means extremely reliable performance.

The following information will introduce Maxicode and define its characteristics. The effort here is to make the reader aware of the terminology, the processes, and different mathematical tools used in Maxicode symbol encodation. A single cell in matrix symbology used to encode one bit of data/ information. In Maxicode a module is one regular hexagon in the symbol as shown in Figure 11.

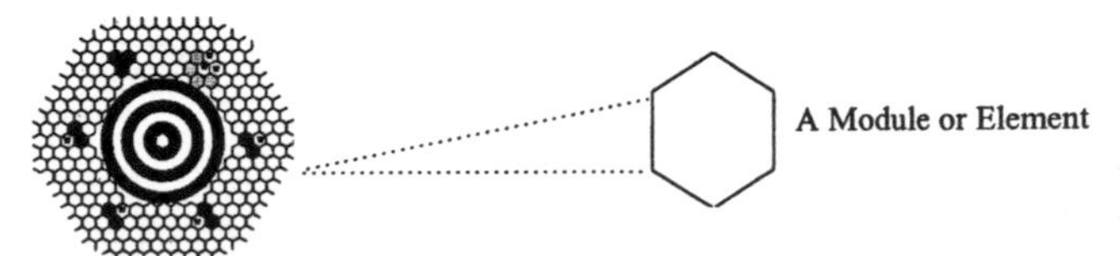

Figure 11. A Module in a MaxiCode Symbol

A codeword is a symbol character value. It is an intermediate level of coding between source data and the graphical encodation in the symbol. A codeword set consists of 2^6 (=64) values ranging from 0-63; 000000 to 111111 in binary notation. The codewords form the basis for error correction. The newer USS-Maxicode specification includes expanded encoding capabilities of all 256 International characters including all 128 ASCII characters (0-127 values in accordance with ANSI X3.4, and 128-255 values in accordance with ISO8859) [6]. Furthermore, numeric compaction allowing nine digits to be compacted in six codewords is also available. Besides, various other symbology control characters for code switching and other control purposes have been included. A Symbol Character consists of one or more codeword, each consisting of six modules, arranged in three rows of two modules each. They are placed from upper right to lower left. Within each symbol character the most significant bit is the lowest numbered module and the least significant bit is the highest numbered module. A typical symbol character is shown in Figure 12.

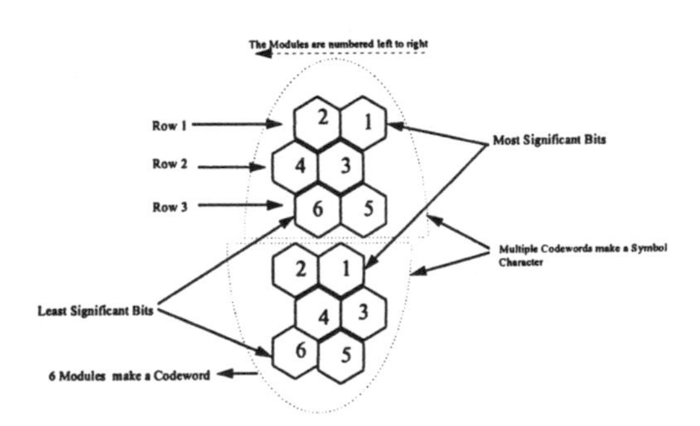

Figure 12. A Codeword and a Symbol Character in MaxiCode

The finder pattern noted in Figure 13, is made up of three concentric rings and three included light areas, centered on the "virtual hexagon". The finder pattern may remind some people of a bullseye.

The central finder pattern allows accurate omni-directional location of the symbols while they are traveling at high speeds.

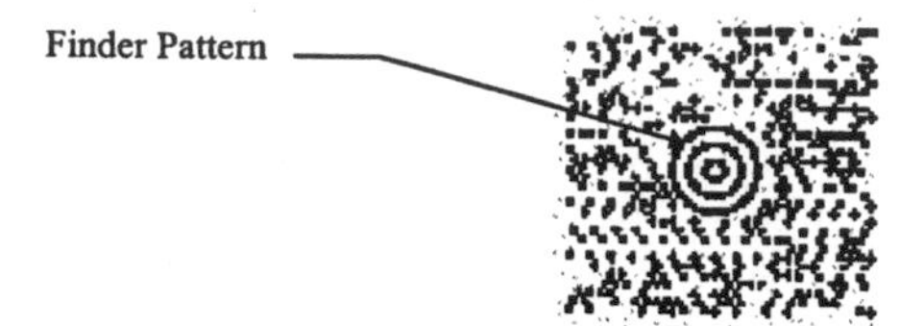

Figure 13. The Finder Pattern in MaxiCode

There are six orientation patterns surrounding the central finder pattern. The orientation information is provided by six patterns of three modules each. The orientation pattern are ordered from 30 degrees, clockwise around the symbol. Figure 14 shows the position of the orientation patterns in a Maxicode symbol.

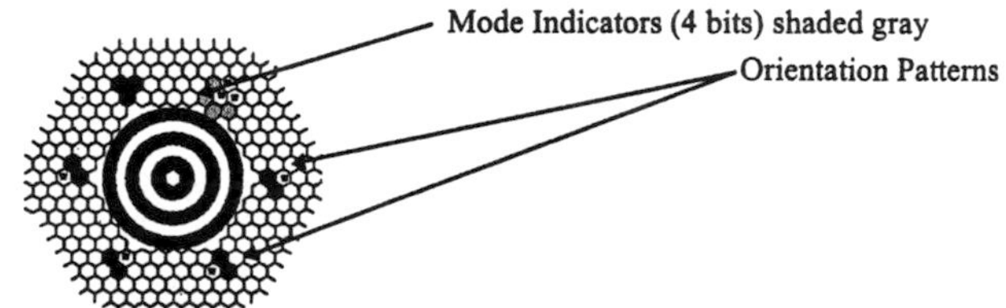

Figure 14. Orientation Patterns and their Positions

Modes in a Maxicode symbol are used for defining the data structure and error correction level in a symbol. The mode indicator is a group of modules that is encoded as part of the primary message to define the data structure (see Figure 14). The following flow chart shows the steps involved in encoding a message in a Maxicode symbol. Later each step will be discussed briefly to introduce the main ideas behind each step and the terminology used.

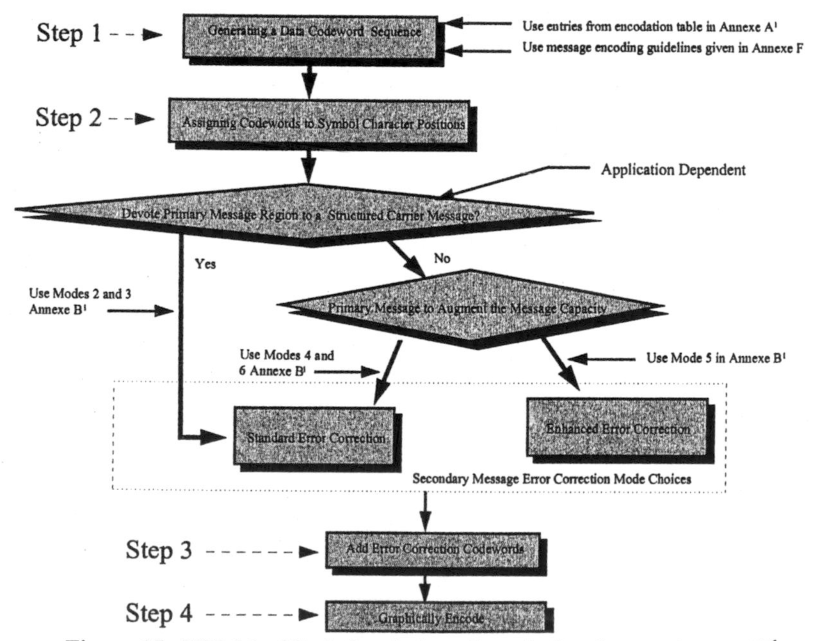

Figure 15. USS MaxiCode Symbology Encodation Steps: Annexe H[1]

AIM International Technical Specification: International Symbology Specification - Maxicode

Step 1 in the above figure describes the process involved encoding information in a Maxicode symbol. The encodable 256 characters noted in the AIM USS-Maxicode has defined five code sets. This step also takes into account the symbology control functions required to be encoded. Maxicode has fifteen symbology control characters with no ASCII equivalents.

Step 2 determines how many data codewords fit in a Maxicode symbol and where they are placed. A Maxicode symbol message is divided into two categories, primary message, and secondary message. Modules around the finder pattern in a MaxiCode constitute the Primary message. It contains sixty bits of data plus sixty bits of Reed-Solomon error correction. Its close proximity to the center and high redundancy makes it the most secure region of the symbol. It serves two very important purpose:

- It contains the Mode Indicators (4 modules), that indicate the "mode" the rest of the symbol is encoded in.
- In the rest of the 56 data bits, the information needed for package sorting and tracking is encoded.
- In high volume environments the overhead scanner on the conveyor then just needs to read and decode the information in the primary message and not the secondary message.

A secondary message often contains additional source/destination information useful for manual handling of the package. The symbol characters in this case are divided into two error corrected sub-messages which are fully interleaved, the odd-numbered characters comprising one Reed-Solomon field and the even numbered characters comprising the other. These are laid into the symbol in a boustrophedonic pattern, shown in the figure below, distributing both the sub-messages evenly throughout the secondary message region [6].

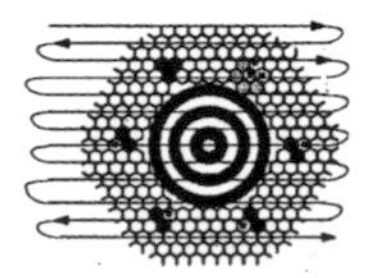

Figure 16. Secondary Message Layout in MaxiCode Symbol

Each of these message categories contain data and error correction codewords. If one decides to use the primary message region to a Structured Carrier Message (Modes 2 and 3), then only the secondary message is available for Standard Error Correction (SEC). Otherwise if one elects to augment the message capacity, he/she may do so by electing Modes 4 and 6 (standard error correction) or Mode 5 (enhanced error correction).

Step 3 involves employing Reed-Solomon error correction algorithm. There are two possible levels of error correction available in Maxicode. They are Standard Error Correction (SEC) and Enhanced Error Correction (EEC).

Since a codeword set consists of 2^6 (=64) values ranging from 0-63; 000000 to 111111 in binary notation. The codewords form the basis for error correction. The theory that associates a certain group G of substitutions on the roots of a given algebraic equation[1] was initiated by E. Galois, who was killed in a duel in 1832 at the age of 21. The coefficients of the polynomial defined by the expansion of:

$$g(x) = (x-2^1)(x-2^2)\ldots(x-2^k) \text{ ---------------------------(1)}$$

where k = number of error correction codewords required. For e.g. for EEC in primary message, k =10, EEC in secondary message, k = 28, and SEC in secondary message, k= 20. The coefficients of the polynomial defined by the expansion of (1) are integral coefficients. These coefficients then becomes the data codewords for the message in question. The Reed-Solomon error codewords for the message are the remainder after dividing the n data codewords by the generator polynomial g(x) of degree k. The payoff was a coding system based on groups of bits-such as bytes-rather than individual 0s and 1s. That feature makes Reed-Solomon codes particularly good at dealing with "bursts" of errors: Six consecutive bit errors, for example, can affect at most two bytes. Thus, even a double-error-correction version of a Reed-Solomon code can provide a comfortable safety factor. The mathematical encoding in step 3 results in values for all the 144 symbol characters in the Maxicode symbol.

Step 4 involves the use of the Maxicode symbol character sequence, as given in the USS-Maxicode specifications, to establish where the 6 bits for each codeword actually is positioned in the Maxicode symbol.

A Structured Append feature has also been added to the message encoding protocol that allows larger data messages to be spread across up to 8 different MaxiCode symbols. Thus messages up to 728 "characters" long can be encoded in multiple MaxiCode symbols and reconstructed no matter what order the symbols are scanned in. Alternately, just the Secondary messages in up to eight Mode 2 or 3 symbols can be strung together, with the Structured Carrier messages in all those symbols encoding the same destination data.

References

[1] AIMI International, Inc. 11860 Sunrise Valley Drive, Suite 100, Reston, Va.. 22091 USA.

[2] AIM International Inc. *AIM International Technical Specification - International Symbology Specification - PDF417*, July 1994, v1.00.

[3] Dickinson, Leonard E, (1926), *Modern Algebraic Theories*, Benjamin H. Sanborn & Co., Chicago, IL.

[4] Sharp, K. R. "Symbology Developments at Scan-Tech" ID Systems, Vol 17, No 1, 1997, pp 84 & 70

[5] AIM International Inc. *AIM International Technical Specification - International Symbology Specification - Data Matrix*, December 1996, v1.01.

[6] AIM International Inc. *AIM International Technical Specification - International Symbology Specification - MaxiCode*, November 1996, v1.01.

United Parcel Service of America. Inc. "Maxicode Briefing Document." [cited 10 February 1997]. Available http://www.maxicode.com/doc_brif.com

Longacre, Andy. 1996. "Maxicode Overview." [cited 10 February 1997]. Available http://www.maxicode.com/doc_ovr.com

"Data Matrix: A New Dimension in Automatic Identification." [cited 16 February 1997] Available http://www.mit.edu:8001/afs/athena.mit.edu/user/k/a/kacandes/www/TwoDsyms/Datamatrix/

Automatic Data Capture - http//:www.tech.purdue.edu/it/adc1

ACORN 1479

AN INTEGRATED TOOLKIT FOR THE DESIGN OF FIELDBUS APPLICATIONS

Iain G Brownlie

Eutech Engineering Solutions Ltd

Belasis Hall Technology Park, Billingham, Cleveland, UK

ABSTRACT: Fieldbus technology has been under development for many years and is approaching maturity. This will have widespread implications for the way in which industrial process plants are designed, operated and maintained. This paper reviews the benefits which the new generation of highly distributed process control systems promise to deliver to end users, and identifies that the introduction of this new technology carries considerable risk. It is proposed that the use of appropriate engineering tools will be an important factor in reducing the risks to acceptable levels. The principle user requirements for fieldbus engineering tools are discussed. The ACORN 1479 collaborative research project is implementing such a design toolkit which will be validated in the context of pilot applications in the Chemical and Power sectors. This paper presents the key findings of the project at its mid term including an overview of the toolkit architecture and the pilot applications.

INTRODUCTION

International efforts to develop a single digital fieldbus communication standard have been underway for much of the last decade. In the meantime some 40 or more different architectures have come to the market with various degrees of success. End users have largely remained sceptical of the technology and have held back from large scale implementation. Significant developments over the last twelve months including the publication of the EN50170 European fieldbus standard, and the launch of the Fieldbus Foundation fieldbus products indicate that key concerns may soon be resolved.

This new technology will lead to a considerable change in the information requirements to specify the design of control applications. A new generation of computer aided software engineering tools will be required to support the design process. These tools must be able to describe application functionality, manage the device hardware descriptions in a vendor independent manner, define the network of devices to be installed, and map the application functionality to these devices. A variety of new design decisions relating to integrity, safety and performance will need to be addressed through a variety of techniques including simulation.

The ACORN 1479 collaborative project to develop a design toolkit started in February 1996 and is due to complete in July 1998. This paper will discuss the key issues which need to be addressed by the design tools, the consolidated user requirements, the resulting toolkit architecture, and the pilot applications where the end users seek to validate the project results.

PROJECT OVERVIEW

The ACORN 1479 project receives financial support from the Commission of the European Communities under the Brite EuRam III Workprogram for collaborative research and development in the area of Industrial and Materials Technologies.

ACORN 1479 Objectives

The objective of the project is to develop a software toolkit to support the design and implementation of the emerging new generation of open, highly distributed control systems, based on fieldbus technologies and to validate these tools in the context of real process applications.

ACORN 1479 Consortium

The project brings together a group of end users, system integration and research organisations with considerable knowledge and experience of fieldbus technologies.

***End Users. Eutech Engineering Solutions* is the overall Project Co-ordinator. *ICI* are responsible to the project for defining the end user needs within the Chemical Sector. *EDP* and *ENEL*, the Portuguese and Italian Power Generators, are addressing the needs of the Power Sector. These end users will also evaluate the toolkit in the context of a number of pilot applications.**

***Developers.* The consortium has five technology providers. *Siemens,* the major system integrator, are responsible for the Application Design Tool. The *ifak* research institute are implementing the Resource Management Tool. *CISE,* the engineering subsidiary of *ENEL,* are responsible for developing the Configuration Management Tool. The *Fraunhofer IITB* research institute are implementing the Simulation Tool. *Intrasoft,* a software house, have the responsibility for providing the framework for integrating the individual tools into a coherent toolkit.**

ACORN 1479 Dimensions

The ACORN 1479 project started on 1st February 1996 and is due to complete at the end of July 1988. Some 40 man years of research and development effort will be spent by the consortium over the 30 month duration of the project. The consortium would like to thank the European Commission for their generous financial support towards the 5.4 MECU budget.

ACORN 1479 Current Status

At the time of the mid-term review held in May 1997 the following activities had been completed

- **Production of the consolidated User Requirements for a Design Toolkit**
- **Generation of the Field Architecture Reference Model**
- **Implementation of Prototype versions of the ACORN 1479 tools**
- **Formal Evaluation of the Prototypes**
- **System Architectural Design Document**
- **Design Specifications for each of the Individual Tools**
- **Specification of the Pilot Applications**
- **First Dissemination Workshop held at BIAS '96 in Milan**

ACORN 1479 Ongoing Development

The project partners are now progressing the implementation of the toolkit and the design of the pilot applications. Further information about our experiences validating the toolkit in the context of the pilots applications will be disseminated at formal workshops to be held later in the project.

SOURCES OF INDUSTRIAL BENEFIT AND ASSOCIATED RISKS

Business Drivers

It should be self evident that any innovative technology will only succeed if it delivers substantial benefits to the user without introducing unacceptable risk.

It is our considered view that the tangible and intangible benefits for fieldbus are real and will ultimately lead to a substantial change in process control system architectures. However, we should not lose sight of the very real and significant additional risks involved.

Those risks have been compounded by the competing architectures and the lack of practical solutions to solve the whole range of end user problems. There have been some significant developments over the last year, notably the launch of the Fieldbus Foundation products, and the introduction of the EN50170 standard which suggest that some of these concerns may now be approaching resolution.

Our belief is that end users in the Process Industries will begin to experiment with some relatively small scale fieldbus applications in the near future, but may postpone major investment decisions until the implications of the technology are better understood.

Benefits

There are three principle areas where fieldbus technology can deliver financial benefits to end users.

Direct Cost Savings Much is made of the direct cost savings resulting from reduced plant wiring. These savings are derived from the ability to multi-drop as many as 32 devices on a single "home run" cable. In practice, we expect users to seek to maintain "single loop integrity" on a bus segment, limiting the average number of devices to perhaps 3 or 4, thus considerably reducing the potential wiring savings. The practical requirement to provide device isolation junction boxes in the field could add to cost of installation. Direct cost benefits will generally only apply to green field sites, and very little savings can be expected from upgrading existing installations.

While reduction in capital cost can have a significant impact on the financial viability of a proposed investment, many end users do not believe that direct cost savings will be sufficiently large to influence early investment decisions in fieldbus.

Functionality The increased functionality of smart instrumentation can offer immediate user benefits through enhanced accuracy, easy calibration and flexibility derived from software configuration. We are beginning to see considerable uptake of smart HART devices which exploit existing 4-20mA media. Key factors in the success of HART has been the relatively low incremental cost over traditional devices, and negligible risk of deployment.

However, the real potential of smart devices lies in the increased amount of information available. This can take the form of secondary measurements, enhanced diagnostics of the device and process conditions. These benefits are particularly difficult to quantify, but the potential for reduced maintenance costs, increased plant reliability and efficiency, and the corresponding reduction in plant downtime could be very significant. This potential for increased device functionality is seen by many end users as the prime justification for introducing fieldbus systems, and should enable richer knowledge based software applications in the areas of control, maintenance and plant scheduling.

Performance and Safety The introduction of a digital communication protocol gives an immediate benefit on accuracy and reliability. This should ensure that the risk of acting on bad data is substantially reduced giving an important safety benefit. There is a strong perception that any communication system introduces new failure modes and complexity which is not present in current control architectures. In practice, reliability may increase as less emphasis is placed on the central DCS controller. Systematising the design process based on industry best practice should also give benefit. Time will be required to build up end user confidence levels.

Risks

Fieldbus Architecture The protracted and continuing saga of the international fieldbus standardisation effort have undoubtedly delayed the introduction of fieldbus by many end users. The goal of a single international standard remains elusive. Indeed the needs of different industry groups imply that several fieldbus architectures are required. End users are seeking truly open systems. This means that the specifications must be complete, available and widely implemented. The absence of a practical user layer in EN50170 means that this standard is not complete.

Interoperability and Interchangeability Most of the existing fieldbus architectures support some degree of device interworking, that is to say, a special application program can be written to read and write information to remote devices across the network. For interoperability, we expect to be able to easily distribute application functionality across whichever devices we decide to use (plug and play). This requires "device profiles" and "standard function blocks" to define the user layer interface to the common set of parameters and functions supported by all vendors. For interchangeability, devices must not only provide the same functionality, but also perform identically.

In practice, the various fieldbus architectures offer interoperability to a greater or lesser extent. However, it is clear that interoperability may be incompatible with using the smart "added value" features that vendors will introduce to differentiate their product. The clever diagnostic features available in a particular control valve may require special configuration and analysis software provided by the vendor.

Integration. Even if primary continuous control is delegated to the field devices, the central DCS systems will still be required to provide advanced control, sequential batch control and operator interface functions. A practical installation will require integration of the DCS system with a bus topology involving a number of fieldbus segments. End users need to see a range of fieldbus products being available for their preferred DCS systems before committing to fieldbus. While PC SCADA systems may be sufficient to experiment with the technology on a small scale, a new generation of DCS systems may be required to truly exploit fieldbus on full scale applications. At the device level a wide selection of instrumentation must be available from several suppliers.

Design Complexity The amount of information required to design modern control systems is increasing rapidly as the devices and control strategies become more complex. The design of the fieldbus topology requires a significant number of new decisions regarding the cable routing and device allocation strategies. As with partitioning of I/O in a traditional DCS system, these decisions can have a serious impact on the performance of the control system. Furthermore, any errors in this design process will be difficult to rectify, and may not be discovered until very late in the plant commissioning process. A set of guidelines for the application of fieldbus needs to be developed to help end users through this complexity. Computer Assisted Software Engineering (CASE) tools offer considerable potential for reducing the risk through structured information management. The choice of which set of development tools are used in the design process may yet prove to be the most significant factor governing the success of a fieldbus project.

Configuration Distributed applications need to be configured at the system level. Configuration of single instruments using handheld configuration tools may still be useful in some cases, but a system wide configuration tool will be necessary in the general case.

Commissioning The distribution of control functions into the field devices means that it will no longer be practical to test control software in a "factory acceptance test". Simulation facilities which allow early testing of the control system design will be important to minimise commissioning delays.

Maintenance Fieldbus devices promise to provide much more performance related data, but real benefit will not be achieved unless this new information is analysed and presented in a timely manner to enable corrective actions to be implemented. Fieldbus provides the necessary communication mechanism to enable the data to be delivered to the point of use, but this is not sufficient. The need for management procedures and integrated application software must not be overlooked.

REQUIREMENTS

Integrate the Control and Instrumentation View

Process, Control and Instrumentation engineering have traditionally been separate functions in the plant design organisation. The needs of Operations and Maintenance have often been neglected in a project environment strongly focused towards delivery within time and cost constraints. Smart devices offer a number of enhanced features including multiple sensors, the capability of running control algorithms in the field and maintenance diagnostic information. The selection of the most appropriate instrument for a specific duty must balance all these aspects against cost. There is a pressing need for an integrated approach considering asset management over the total plant lifecycle.

The ACORN 1479 toolkit focuses on the design of the process control system. In a traditional DCS environment the control engineer can operate with generic Inputs and Outputs, safe in the knowledge that the 4-20mA standard will allow any sensor and actuator technology to be used. In the fieldbus environment, the control engineer must be aware of the capabilities of the instrumentation in order to properly distribute the control functionality. The dynamic performance of the control system will be critically determined by the fieldbus topology so tuning the network design must be considered.

Support Vendor Independent Design conforming to Open System Standards

New Design Processes The toolkit should support the overall design process. It should be possible to formally describe the control application functionality much earlier in the design process. There should be a close relationship between the process design and control system design. The toolkit should separate the functional requirements from the physical implementation, to allow portable designs to be developed which can be implemented on different target architectures.

Vendor Independence The toolkit should support a framework for describing the capabilities of specific devices in a generic manner (c.f. HART device description language).

Standards Compliance The toolkit should be designed with due regard to the relevant standards. This applies both to application standards (e.g. EN50170/IEC 1158 - Fieldbus, ISA S88.01 - Batch Control and IEC 1131-3 - Programming Languages) and standards which influence the design of the toolkit itself (e.g. ISO 10303 - Standard for the Exchange of Product Model Data, Client Server architecture with Windows GUI).

Support Configuration of Highly Distributed Control across Flexible Network Topologies

Logical Application Design The toolkit should allow control applications to be hierarchically structured according to the physical plant design. The instrumentation requirements for measurement and actuation should be identified from the process design, and continuous control functionality should be specified in terms of function blocks. Specific integrity requirements for trip, control, alarms, certification, redundancy, and performance must be captured.

Configuration The toolkit should allow selection of appropriate instrumentation subject to the specific instrumentation requirements, individual device capabilities, and additional criteria (e.g. preferred supplier). Control function blocks must be mapped to devices and devices should be allocated to fieldbus segments to form the control network.

Reduce Risks through Increased Quality of Design

An important requirement of the toolkit should be to improve the quality of the design process. This should be achieved through Completeness, Consistency and Design Integrity Checking. Performance of the design should be assessed by simulation. The toolkit could also provide early indication of cost.

TOOLKIT OVERVIEW

ACORN Framework

Figure 1 illustrates how the individual tools in the ACORN 1479 toolkit communicate with each other to design a process control application.

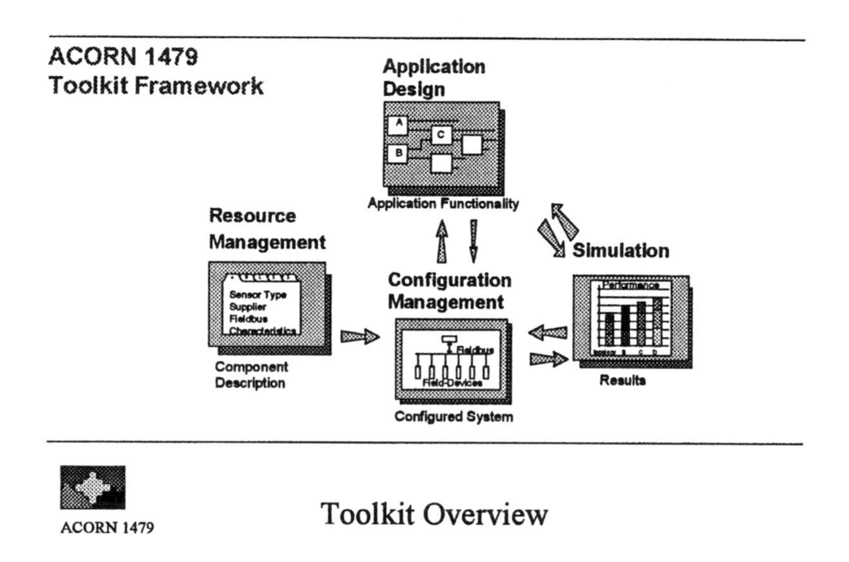

Figure 1. ACORN 1479 Toolkit Framework

Application Design This tool supports an object oriented approach to application specification based on the Function Block paradigm. An appropriate level of abstraction is required to enable specification of the important aspects of object behaviour, interaction and dynamics in a vendor independent manner.

Resource Management This tool is based on the formal description of hardware, software and network capabilities of physical control and network devices. Although the focus of the project is to prove the concepts in the context of fieldbus, it should be possible to describe the capabilities of traditional control equipment, thus providing a migration path from current technologies.

Configuration Management This tool allows systems to be configured by selecting the required components from the available resource library, and then allocating application software components from the design model into these hardware devices, subject to the capabilities of the specified devices.

Simulation This tool provides facilities to help assess run time behaviour of the target architecture early in the design process. Areas of specific interest include the simulation of network loading and the simulating the dynamic behaviour of applications to facilitate early validation of the design.

Application Design Tool

Many of the potential benefits from modern information technologies are only achieved with full integration between functionally separate parts of an organisation. In particular, the design of a fieldbus based process control system should be developed in parallel to the process and instrumentation engineering, and should take due regard with the longer term requirements for operation and maintenance. The close relationship between process and control system design is illustrated in Figure 2.

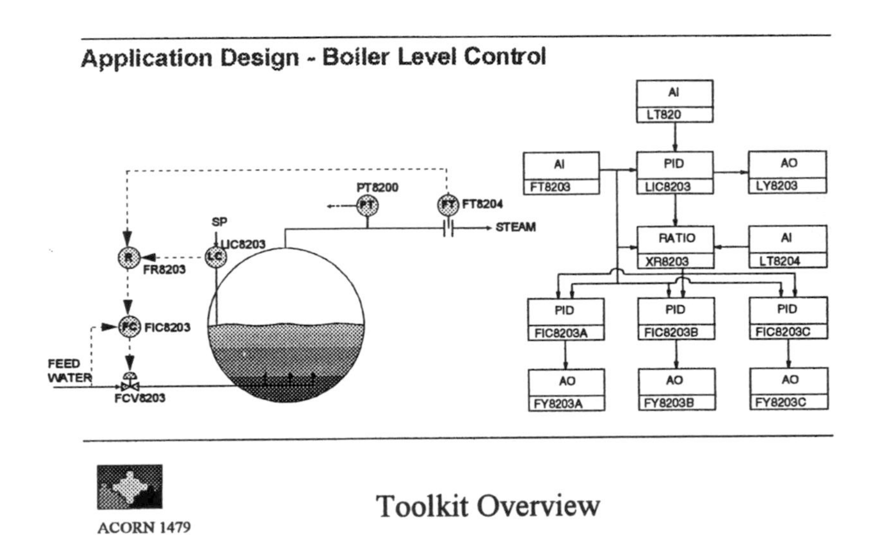

Figure 2. Integrated Process and Control System Design

The Application Design Tool (ADT) captures the functional requirements of the continuous control system independent of the implementation architecture. An ADT logical design can be implemented on a traditional DCS or totally distributed into the field devices, leading to portable designs.

Hierarchical Design This is supported by using the Physical Plant Model from ISA S88.1 as a basis for structuring applications closely mimicking the physical plant layout as defined in the related Piping and Instrumentation process drawings.

Instrumentation Requirements The fundamental measurement and actuation requirements can be readily identified from the process design. These are used to specify constraints on the selection of appropriate instruments during configuration. The class of instrument is mandatory (e.g. Flow), but the designer can define additional requirements (e.g. certification requirements, sensor technology, fieldbus type).

Control Loops The implementation of a control loop is defined as a network of function blocks. The toolkit will support a generic set of commonly available basic function block types (e.g. AI, PID, AO etc.), and a "function block editor" will allow "special" function block types to be created by the user. The tool will capture a number of requirements which will constrain the implementation of the loop. These will include timing, redundancy and certification constraints which will be validated within the Configuration Management Tool.

The ADT will support good design practice by implementing various consistency, completeness and design integrity checks. For example, the concept of loop "duty" is introduced to validate the separation of trip and control functionality.

Resource Management Tool

The Resource Management Tool (RMT) is used to define the characteristics of real devices using a formal device description data model.

Identification Information about manufacturer and model, device class and version

Offered Functionality Information about which function blocks are supported in the device

Fieldbus Interface Information about the ports available for connecting to bus segments

Characteristics Information about certification, device power, housing

The data model is shared with the Configuration Management Tool and represents both the capabilities available and the facilities used in a specific device configuration

Program Invocation Information about function block used

Linkages Information about communication links

Scheduling Information about device scheduling

Configuration Management Tool

The Configuration Management Tool (CMT) is used to specify how the application functionality is to be distributed across a specific hardware architecture. The resulting configuration of the boiler control example is illustrated in Figure 3.

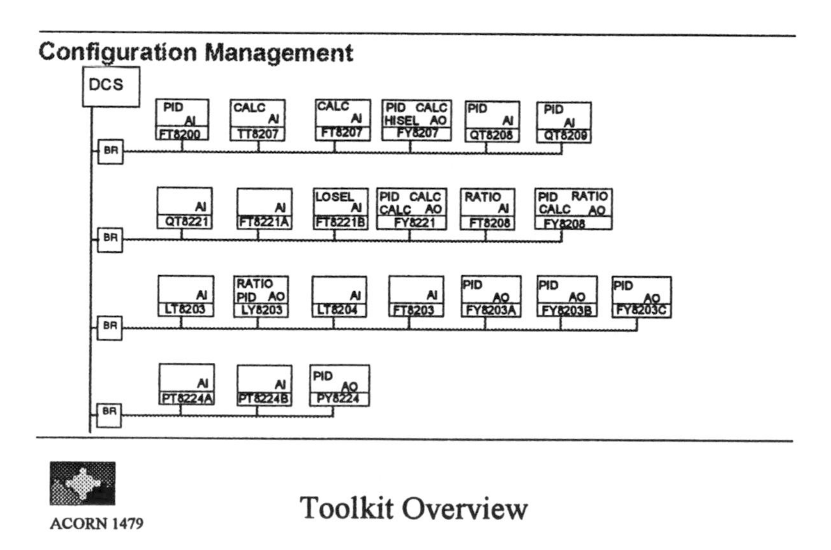

Figure 3. Configuration of the Boiler Control Example

Device Selection The instrumentation requirements defined in the application design identify the basic requirements for measurement and actuation devices. Further constraints may be defined during the configuration process including device power constraints, certification requirements, fieldbus technology, and preferred supplier. The user can then choose from the set of compatible devices defined in the RMT device library. In addition to the instrumentation, the user may define other devices required to implement the control scheme (e.g. DCS or PLC controllers).

Network Topology Design The user can connect the devices to form networks. These are formed by attaching devices to bus-segments according to the available interface ports and bus capacity. Gateways devices supporting multiple interfaces allow different network technologies to be combined in a single application.

Function Block Allocation In order to complete the application the user must map the function blocks onto the devices which will implement them. This process needs to consider the capabilities of the devices, and the scheduling of device communications and function block processing.

The CMT will implement a variety of consistency, completeness and design integrity checks. For example, the tool will validate whether "single loop integrity" constraints are satisfied by ensuring that all devices involved in a control loop are connected to the same fieldbus.

Simulation Tool

The Simulation Tool (SMT) aims to predict the dynamic behaviour of configurations. The information from the ADT and CMT tools is used to characterise a set of mathematical models defined in a number of layers.

Process Layer The control system monitors the plant status by reading measurement information, and manipulates the plant actuators. The consequential effect on the measurements should be modelled by a dynamic process model. These are application specific, so a very simple process model is provided with the toolkit but the architecture should support integration of rigorous process models.

Process Control Layer The toolkit will simulate the execution of function blocks by rigorous emulation of the control algorithms. This will allow the basic control strategy to be tested.

Communication Layer The toolkit will implement a semi-rigorous stochastic modelling of the communication of the fieldbus networks. This will allow the effect of different network topologies and different function block allocation strategies on network loading to be determined.

PILOT APPLICATIONS

ICI Pilot

The ICI pilot application is designed to monitor product flows between various production plants and storage facilities to create a mass balance for an integrated petrochemical site.

This application will be implemented using a Profibus fieldbus with instrumentation being provided from several suppliers. The main requirement is to bring information into the works records system from measurements which are widely distributed across the 5 kilometres length of the site, much of which is classified as a Zone 2 hazardous area. The proposed architecture is illustrated in Figure 4 and involves running a high speed RS485 "DP" backbone with repeaters between the main plant control rooms. From each control room, bus couplers will provide the links to a number of low speed IEC 1158-2 "PA" segments which will run to the field devices. The IEC physical layer standard carries both power and signals on a single twisted pair, and offers an intrinsically safe option.

Chemical Pilot - Site Monitoring System

- Monitoring Application
- Hazardous area
- Long cable runs
- 30-40 instruments mainly Flow, Temperature, Pressure
- Multivendor instrumentation
- High and Low speed fieldbus
- Profibus fieldbus

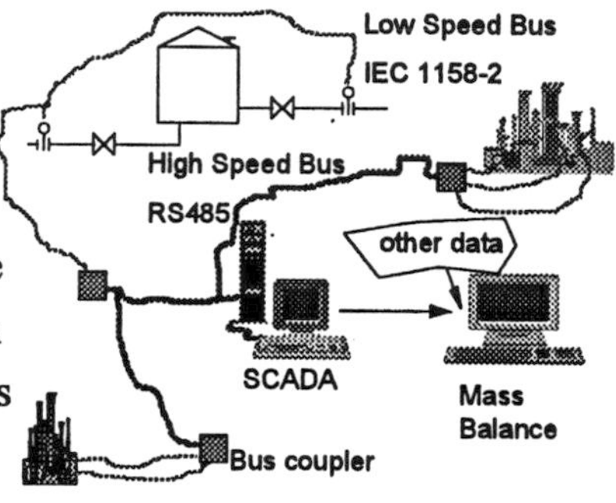

ACORN 1479

Pilot Applications

Figure 4. Chemical Pilot Application

The mass balance application requires a number of compensation calculations to calculate mass flow from the measured volumetric flow together with temperature, pressure and density measurements. This requires the raw measurement data to be combined with information from other sources including the laboratory test results. This is a further illustration of the need to integrate the fieldbus systems with higher level information systems.

EDP Pilot

The EDP Power Pilot Application will focus on an application on the Carregado Power Station. This heavy fuel oil power station is to be completely revamped and will receive new burners to allow operation with natural gas as a second type of fuel. While the main project will be implemented using a traditional DCS, a parallel study will be undertaken to assess the impact of using fieldbus technology to control part of the boiler systems. The ACORN toolkit will be used to design and simulate this application.

The main objective of the study is to validate the ACORN toolkit, through the implementation of the application for Carregado, and to assess the suitability of the toolkit to fieldbus applications in the power industry. Some of the key benefits which the toolkit is expected to deliver include the following

- Project design time saving
- Commissioning savings due to simulation
- Increased project quality due to consistency, completeness and design integrity checks
- Increased robustness and performance

The study will seek to derive quantitative "measurements" of the impact of fieldbus and the ACORN tools in comparison with the traditional approach.

ENEL Pilot

The control system for the steam generation cycle of the ENEL experimental power station at Santa Gilla in Sardinia was equipped with a FIP fieldbus network in an earlier research project. It is planned to update this installation using WorldFIP devices and the ACORN 1479 toolkit will be used to design the new configuration.

Power Pilot - ENEL Santa Gilla Power Station

- Process Control System
- 5 Level Sensors
- 5 Temperature Transmitters
- 9 Pressure Transmitters
- 5 Flow Transmitters
- 6 Actuators
- 1 On Off Valve
- WorldFIP fieldbus

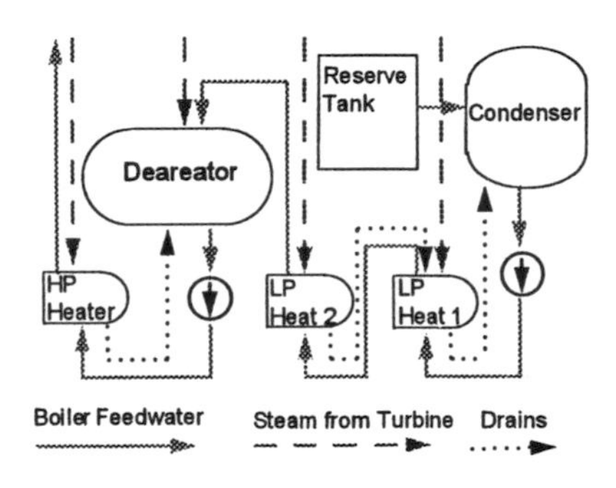

ACORN 1479

Pilot Applications

Figure 5. Power Pilot Application

Condensate is extracted from the condenser by two pumps (one in standby) at a rate of 125 te/hr, and sent via two low pressure heaters to the deareator. Three further pumps (one in standby) pump the feedwater to the boiler via a third high pressure heating stage.

The control system monitors the temperature and pressure of the feedwater exit each of the five main process vessels (the condenser, deareator and each of the three heaters). The pressure of the exhaust steam from the turbines is monitored. The level in each of the three heaters is controlled by manipulating the backward flows via the drains. The level in the deareator is controlled by manipulating the flow into the vessel from the second heating stage. The level in the condenser is controlled by manipulating the flow from the reserve condensate tank.

VALIDATION

The ACORN 1479 project will formally validate the toolkit in the context of the pilot applications in order to establish quantitative and qualitative measurements of the impact of fieldbus and the software tools on the design process.

The results of this validation exercise will be presented at future information dissemination events which are planned to be held later in the project. Please contact the author if you wish to be added to our mailing list so that you may be kept informed of project progress and results.

CONCLUSIONS

The main conclusions to be drawn from this paper are the following

- Fieldbus technologies are maturing rapidly
- The technology will give benefits but there are associated risks
- End users need to establish a strategy for introducing fieldbus
- The strategy needs to consider *system* and *lifecycle* implications of the technology
- Small scale applications can be effective in building confidence and understanding
- Software tools have a role to play in reducing risk
- Vendor and technology selection should be based on the ease of integration
- The ACORN 1479 toolkit will offer some ideas and solutions
- The industry as a whole must respond to the integration challenge

The ACORN 1479 project would welcome any comments on the information presented in this paper.

INDUSTRIAL ENGINEERING RESEARCH AND AUTOMATION IN ELECTRONIC PRODUCT MANUFACTURING

John T. Tester
Department of Industrial and Systems Engineering
Virginia Polytechnic Institute and State University
Blacksburg, Virginia 24061-0118 USA

ABSTRACT: This paper addresses the industrial engineering research directions taken in the electronic product manufacturing industry. "Electronic product" is considered to be an assembly which is composed of a printed circuit card subassembly, but may contain other major components as well. Examples cited from literature lead one to the conclusion that the research is primarily concerned with the printed circuit card assembly alone. The manufacturing processes necessary to produce the final, electronic product are not as highly targeted for research investigation. A proposed transition from manual to robotic assembly will highlight the need to include capital investment in evaluating any approach's cost-effectiveness.

INTRODUCTION

This paper examines some general directions of industrial engineering (IE) research in the field of electronic product manufacturing (EPM) in the 1990's. The need for such research was well documented by Wilhelm and Fowler in a workshop specifically devoted to this purpose.[1] By using the term "electronic products," this article refers to consumer goods in which a printed circuit card assembly (PCCA) is a major component of the product. The PCCA may not necessarily be the primary component; a VCR, for example, contains a majority of major components and subsystems in addition to the PCCA. The final, electronic product has also been referred to as the "third-level packaging" when discussed within the electronics packaging context.[2] In such a context, second-level packaging is associated with the PCCA. First-level packaging, i.e., integrated circuit manufacturing, is not a part of this paper.

Section 2 examines the general trend of published IE research in the EPM field. A search of literary publications shows three general trends: Component placement problems, robotic workstation problems, and production planning and control (PPC) issues. Section 3 discusses the apparent absence of manufacturing research in the realm of the final assembly of the third-level packaging. Another gap in current research concerns the introduction of robotic automation into previously manual assembly processes. An advantage of using capital investment analyses, applied to these areas, will also be discussed. Section 4 presents the conclusions of this paper.

LITERATURE REVIEW

In the review of EPM published research there is an almost exclusive focus on PCCA issues. The printed circuit card, and its assembly, by nature, are a natural target for research. A personal computer motherboard, for example, can have hundreds of components per unit; the placement of one component can constitute a single step within hundreds of nearly-identical steps in a serial, assembly process. That serial process is the placement of the components upon the board. Though other important process steps are researched as part of PCCA manufacture, the component placement problem is a dominant research topic.

Component Placement/Insertion

The latest machine advancement in PCCA manufacturing is generally considered to be in the surface mount technology (SMT) area.[3] SMT machines are highly specialized machines which place specialized components on only one side of a board at a time at a very high rate of speed. Though very fast, these machines are also very expensive; one machine could require an investment of over one million dollars. With such a large investment in a single workstation, a natural thrust of IE research is to reduce the SMT machine process time as much as possible. Many different types of SMT machines are available in the market place, so many different approaches are taken to minimize this processing time. The publications in this field have several common characteristics:

- Integer (or mixed integer) programming is usually required for strictly addressing the problem formulation.
- The problems are computationally intractable.
- Lower bounds on the solutions are obtainable by computationally tractable means.
- Reliability of the surface-mounted placement activity is not generally considered.

A general example of an SMT problem is illustrated in Figure 1. In this machine, the components tray has a degree of freedom in the x direction. The components tray is the location at which various components can be picked by the placement head of the machine. The printed circuit board (PCB) has two degrees of freedom, in both the x and y directions. The combined three degrees of freedom of the PCB and the component tray allow simultaneous motion of these machine subsystems during placement, thus allowing a faster rate of placement than if they were all fixed in place. Many articles dealing with this problem have been published (for example, see McGinnis, Ammons and Tovey [4]). It should be noted that this area of research is still ongoing, since the problem formulation is very machine-dependent.

The older method for placing electronic components is by through-hole insertion. The components for PCB placement have wire leads which are inserted into the board, through pre-milled holes. This insertion process technology (IPT) has process time minimization objectives, which are similar to the SMT component placement problem mentioned above. However, when automated IPT machines are considered, the main differences between the two technologies are the speed (SMT is faster) and the capital expense (IPT is historically cheaper).[3] Boards with both SMT and IPT components are considered "hybrid" products. The hybrid boards are usually produced by mounting the SMT components in one dedicated machine/workstation and the IPT components on a second machine/workstation.

There are placement/insertion problems associated with hybrid PCCA, an example of which is presented by Kumar and Li.[5] A single PCB is to be populated by assorted SMT and IPT components from the tray at the bottom, as shown in Figure 2. In the specific machine under study, only the placement head had any degree of freedom; the component tray and the PCB are fixed in

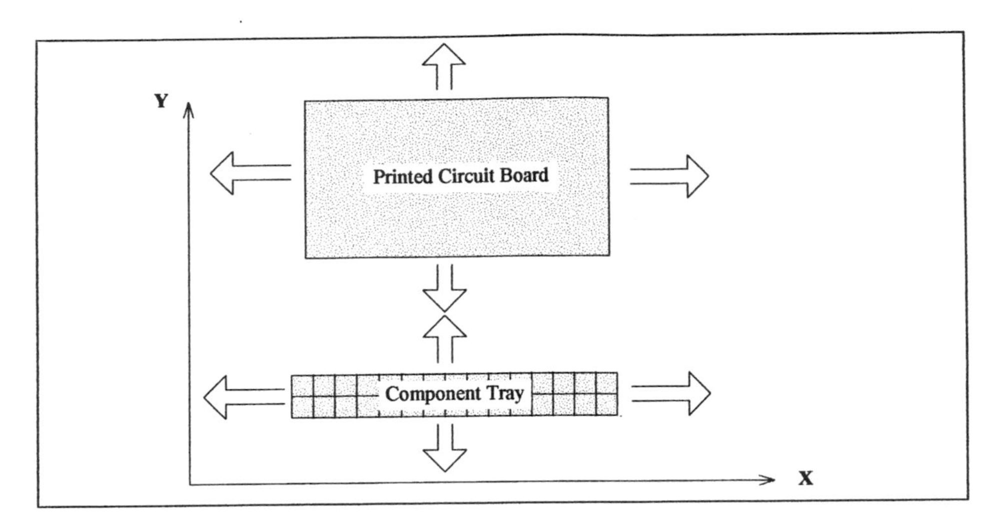

Figure 1. Example of an SMT component problem

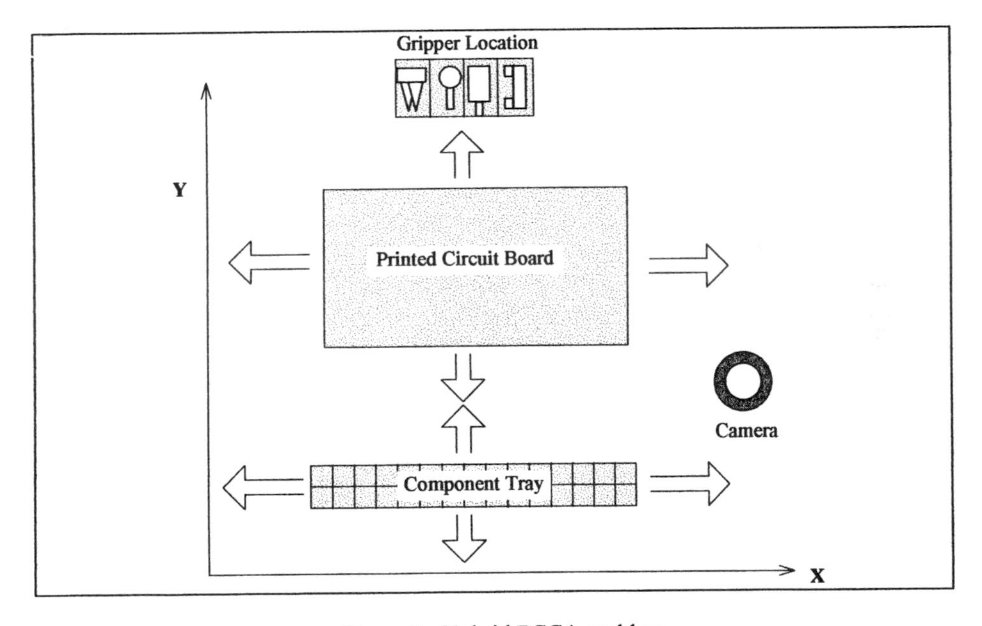

Figure 2. Hybrid PCCA problem

place. Due to the difference in shapes of SMT verses IPT components, multiple grippers are available for the placement head. Some components require special inspection by a camera before insertion or placement; this additional requirement would add time to the placement of the component. The problem is to minimize the time required to accomplish all the required motions necessary to populate the PCB; these motions include component pick-up, inspection, and placement. Though optimization is not feasible, Kumar and Li establish an upper bound by treating the situation as a traveling salesman problem with weighted factors for the assignment of the pickup slots.

Robotic Placement

Some PCB layouts require components which do not lend themselves towards either SMT or IPT. These components must either be mounted manually or via specialized robots. Manual electronic assembly, though delicate, has assembly techniques similar to that of non-electronic products.[6] Assembly robots, though not as specialized as SMT machines, are still customized for the particular model for which they assemble. Both component insertion optimization and robot workstation layout have become recent research topics of interest.

Figure 1, used for illustration of the SMT placement problem, can be reference for the robotic placement problem as well. The key difference between the SMT and the robotic placement problem was illustrated by Su et. al.[7] The main problem is that there exists a significant wait time for the robot at the pick-up and placement points. This wait time is generally ignored in SMT placement problems, but cannot be dismissed for the slower, robotic placement machines. Su et. al. overcome some of the wait time by creating an algorithm which allows for "dynamic pick-and-place" calculation. That is, the points for pick-and-place are not restricted to specific locations but are allowed to be anywhere within the PCB or component tray's range of motion. The main thrust of the research was to show how, for some robots, not all degrees of freedom are necessary for the PCB motion in order to have near optimal solutions.

A variation of the above research deals with the layout and implementation of two robots to place components on a single board.[8] The purpose of using two robots is to increase the placement cycle time per board. Two layout options were studied. The first option dealt with concurrent placement of the components by both robots on the same PCB. The robots, on opposite sides of the PCB, had obvious interference problems, which were addressed by the authors. The second layout option was in a sequential placement arrangement of the robots. The robots were placed very close to each other, but their workspace was not common (i.e., no interference). The PCB was shuttled in one, linear direction between the two robots for sequential placement. In other words, the PCB would shift to the first robot for placement while the second robot moves to pick up the next component; the appearance of the workstation motion is reminiscent of a tug-of-war between the robots, with the PCB in the middle. The most obvious results of the study was that either of the two-robot layouts significantly decreased the cycle time over that of a one-robot layout. Very little difference in performance was observed between either of the two-robot layouts.

Layout of the manufacturing workstations affects the manufacturing process, overall. Significant work in the process planning and control aspects of EPM has been published.

Process Planning and Control (PPC)

An excellent overview of PCCA PPC and related issues was developed by McGinnis et al.[9] The reader is referred to the detailed descriptions of the essential PPC elements required for PCCA which are presented in that article. In terms of grouping, allocation and component sequencing, the IE problems considered by a PCCA facility can be considered in an hierarchical manner, as presented in Figure 3.[10] PPC decision-making categories can be related to the increasing complexities of the products and the manufacturing systems which produce the assemblies. For example, the placement and sequencing problems discussed above are directly associated with the arrangement/sequencing process of placing a finite number of components on a single PCB. The next-higher level of PCCA complexity can be the consideration of component allocation. When several SMT/IPT machines are used over multiple PCB models, one must consider the loading and allocation of the different components distributed across a set of these machines to minimize the required setups and to balance the line. The component allocation may be improved (i.e., number of setups reduced) by the grouping of the product families, which may be considered as the higher-level PPC analyses conducted in PCCA manufacturing.

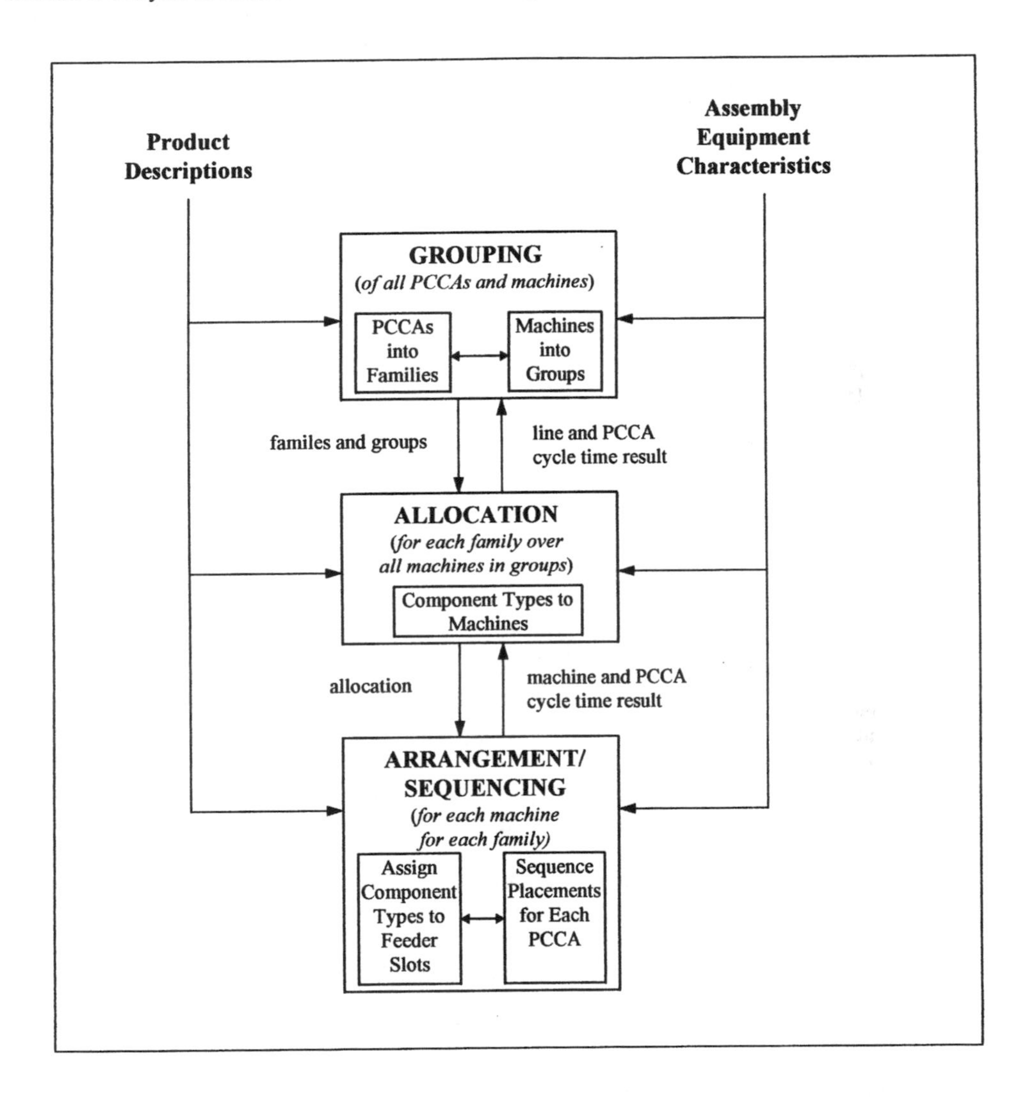

Figure 3. PCCA hierarchy of PPC problems, with respect to grouping, allocation and arrangement issues [10]

Research which addressed a single family, component allocation, across multiple machines, was addressed by Ammons et al.[10] A two-part objective function was defined for minimization. The first part of the algorithm considered the allocation problem, in order to balance the multiple machines; the second part of the method focused on reducing the time necessary to place all the components on a PCB. The machines under consideration were considered dissimilar and constraints on component feeder placement for the machines could also be considered in the model. These last two characteristics of the research problem were considered necessary to the research in order to apply the algorithm to an existing, PCCA, manufacturing site.

In addition to the placement time and component allocation problem, Ellis included setup strategy considerations in the PCCA manufacturing line.[11] An approach was developed by which the machine-level PCCA placement time was estimated and applied to the higher-level problems associated with product family grouping and component allocation across the SMT machines. The model formulation process results in a mixed-integer programming model; branch-and-bound techniques are applied to solve the problem.

Another component allocation and machine setup problem was addressed in a unique manner at Hewlett-Packard.[12] Fuzzy logic methods were applied for a set of PCCA products in order to group the products into families. The effort of developing families on product similarities contributed to the larger goal of reducing the machine setup times. Additionally, this effort was integrated into the problem of balancing the line, specifically across the three SMT machines required for its production. The fuzzy method solution resulted for no setup changes for 95% of the products, after the families were grouped.

A final note to group technology issues: The electronics industry has applied these methods to their existing product lines with some success.[11-14] Not only are quantitative performance measures such as component placement and setup time improved, but qualitative measures are improved as well. Such measures can include reduced complexity in record-keeping as well as a structured means by which to evaluate new product manufacturability.

Group technology is usually applied to improve the flexibility of the manufacturing system. Just what "flexibility" means is a subject of debate; however, Suarez et al attempted to define the term in relation to PCCA manufacturing systems.[15] They conducted a survey of PCCA manufacturers in Europe, Japan and the United States on the strategic use and implementation of flexible manufacturing. One of their findings suggest that the PCCA industry has less variation in their manufacturing processes than other assembly-oriented industries. This rationale is primarily due to the fact that the 31 facilities under study all used the same, automated equipment vendors. Other findings are detailed in the authors' research; they give a rare insight to a broad spectrum of PCCA manufacturing issues.

Forecasting and inventory issues can take on unique characteristics when studied in the EPM environment. For example, inventory levels should not be considered as a relatively static measure over time, nor should the forecasts for their demand. Electronic products have a short life cycle in the market place, and correspondingly in the manufacturing environment as well. Kurawarwala and Matsuo combined a forecasting analysis with an inventory model, in order to assist the production planning of a personal computer manufacturer.[16] The personal computer manufacturing life cycle is typically only twelve months; the optimization of this model produced an inventory policy for the lifetime of the product.

DISCUSSION

As described above, there is considerable research related to the EPM industry. However, much of this research concentrates on the PCCA manufacturing process. Figure 4 illustrates a schematic of a generic, electronic product manufacturing process. The product is composed of both printed circuit board and non-electronic components used in final assembly. For example, a telephone might be composed of the main PCCA, the keypad, the display, the enclosure and associated connecting components. The final assembly of all these components comes together in the last stage of the manufacturing system. This manufacturing concept is common fpr EPM, but research is rare on the topic of an integrated IE methodology for EPM. Generally, much research is restricted to the outlined area in Figure 4; this area contains the manufacturing processes devoted solely to PCCA issues.

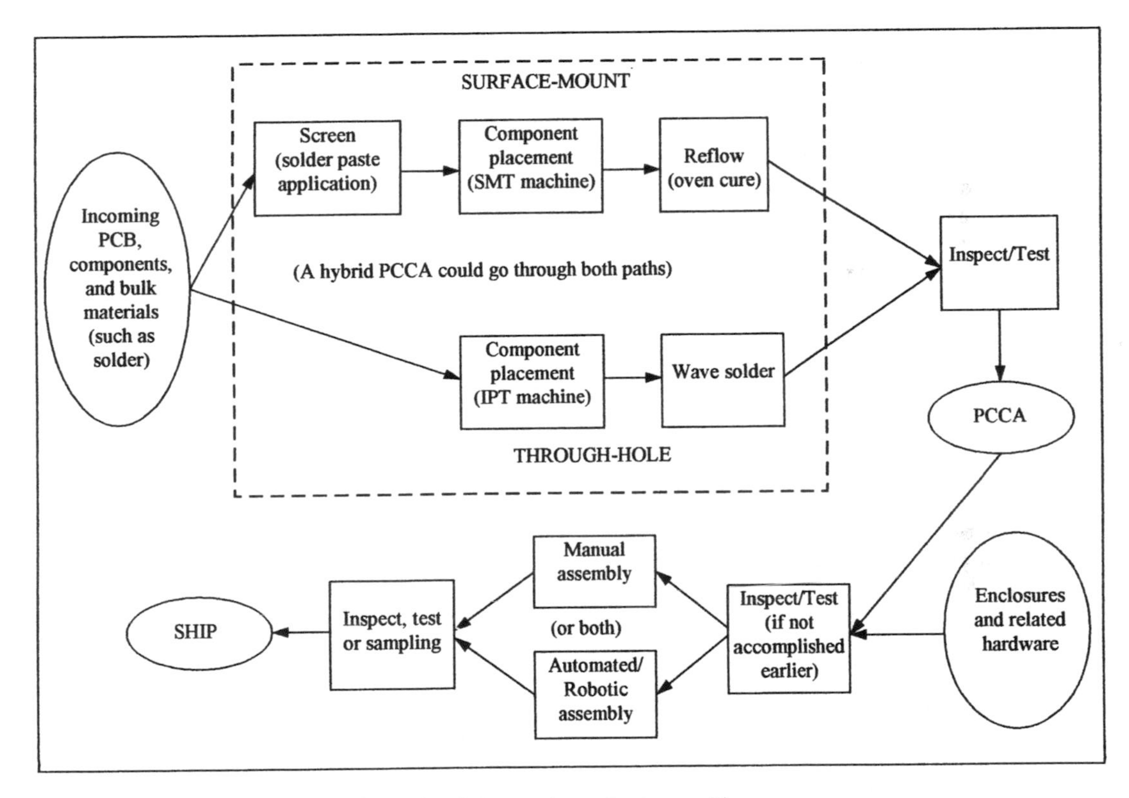

Figure 4. A schematic of electronic product assembly processes.

Component placement research is concerned only with the machine-level problem of SMT or IPT machines. Component allocation research deals with multiple machines and product grouping, so this research may be considered an integrated approach to EPM problems. On closer analysis, however, one realizes that the analyses are strictly concerned with the PCCA and not the end product. This approach is very appropriate for "captive" industries, where the end product of a manufacturing facility is the PCCA itself; the PCCA are then used by the parent company to assemble the final product.[15] For a facility which makes the entire product (for example, a VCR) under one roof, creating part families based upon the PCCA may be only the first part of planning the products entire manufacturing process. Other processing activities, downstream of the PCCA manufacturing process, could possibly be improved by either production planning techniques or automation implementation.

The concentration on PCCA research is understandable; these subassemblies have common characteristics that fit well into IE applications and techniques. The PCCA assembly process involves many, discrete, subcomponents which are installed in a designed layout for a given model. Even when one compares radically different products, such as personal computers and VCRs, the PCCA manufacturing process is essentially the same and uses very similar components (resistors, capacitors, and so on). This similarity allows IE investigators to apply similar, mathematical techniques across a broad spectrum of product industries. However, the third-level packaging of these products is often quite different. A personal computer has few moving parts, whereas the VCR has many; such differences in the products' features can require significantly different methods of final assembly and testing.

The final assembly of the electronic products, discussed above, has been dominated by manual assembly methods in the past. Automation, including robotic automation, has been introduced this past decade to reduce the variability of the cycle times in final assembly.[6] The primary reasons for this trend are to either increase the throughput or reduce the work-in-process (WIP), since high variability means high WIP.
For any electronic product manufacturing firm which produces the PCCA along with the final electronic product: A manual process downstream of a highly automated process will require a significant WIP in its buffer, in order to account for the manual process' variability. Compounding this problem is the likelihood that the WIP subassemblies have a significant portion of their value added strictly from the upstream, highly automated process. In other words, the largest accumulation of WIP in the total manufacturing line very likely contains the highest-valued subassemblies in that line. Therefore, automating that manual section of the process should be given significant priority.

Specialized automation machines, such as those used in dedicated flow lines, have very low cycle variability, but are not flexible to changes in the models produced on the line. Robotic workstations, placed in sections of the line where model variations require changes in the manufacturing process at that point, can increase the flow line's flexibility and still have the potential to keep the pace of product flow near to that of the fully automated system. The expense of the transition for either of this alternatives will warrant investigation. Capital expense is usually the primary consideration, but training and operating expenses of an automation decision are possible drivers as well.

Automating the assembly process, in the context discussed so far, may be related to implementing computer-integrated manufacturing (CIM) into manufacturing facilities. It has been shown that companies require an higher, CIM capital investment payback for implementation than that of other industry investments.[17] This higher standard may be understandable, since CIM is not always necessary to produce a product; it merely enables the manufacturer to produce with greater throughput, less variability and improved yields. However, it can also be argued that CIM enables companies to remain competitive in the global, technology-oriented, consumer-goods marketplace. A similar argument can be made for automating the manual EPM processes which occur downstream of the PCCA manufacturing processes.

Incremental introduction of automated assembly, downstream of the PCCA process, may be the true manner in which full automation is gradually implemented. This concept seems to demand robotic assembly stations; the research into their workspace layout and placement algorithms has already been illustrated.[7,8] However, the actual integration of the robotic workstation into the existing product flow is less studied. Some research, targeted toward robotic integration into *manual* assembly lines, has already been accomplished.[18] The research investigates an incremental introduction of robotic assembly on a manual assembly flow line; it is primarily concerned with this introduction in the less-developed countries, where manual assembly is the norm. Though not associated with electronic products, insights into this research can provide a conceptual seed for growing the research in EPM robotic applications.

CONCLUSIONS

Industrial engineering research, in the area of EPM, has been mostly concentrated around PCCA manufacturing. Within this area, some categories can be defined: SMT and IPT component placement, robotic workstation layout, and PPC. The PPC research discussed above can be broadly refined into product family definition and the component allocation problem.

Significant research can be accomplished for the cost-effective integration of automation downstream from the PCCA processes. This research must be combined with capital investment analysis to enable such an integration more appealing in commercial industry.

REFERENCES

1. **Wilhelm, W. E. and Fowler, J.**, Research directions in electronics manufacturing, *IIE Transactions*, v. 24, n. 4, 6-17, September 1992.
2. **Dally, James W.**, Packaging of Electronic Systems: A Mechanical Engineering Approach, McGraw-Hill, Inc., New York, 1990.
3. **Manko, H. H.**, Soldering Handbook, for Printed Circuits and Surface Mounting, 2nd Ed, Van Nostrand Reinhold, New York, 1995.
4. **McGinnis, L. F., Ammons, J. C. and Tovey, C. A.**, Circuit card assembly process planning, *Proceedings ASME Winter Annual Meeting*, New Orleans, LA, 51-56, November 29-December 1, 1993.
5. **Kumar, R. and Li, H.**, Integer Programming approach to printed circuit board assembly time optimization, *IEEE Transactions on Components, Packaging, and manufacturing Technology-Part B*, v. 8, n. 4, 720-727, November 1995.
6. **Noble, P. J.**, Printed Circuit Board Assembly, Halsted Press/John Wiley & Sons, Inc, New York, 1989.
7. **Su, Y., Wang, C., Egbelu, P. J. and Cannon, D. J.**, A dynamic point specification approach to sequencing robot moves for PCB assembly, *International Journal for Computer Integrated Manufacturing*, v. 8, n. 6, 448-456, 1995.
8. **Lin, H., Egbelu, P. J. and Wu, C.**, A two-robot printed circuit board assembly system, , *International Journal for Computer Integrated Manufacturing*, v. 8, n. 1, 21-31, 1995.
9. **McGinnis, L. F., Ammons, J. C., Carlyle, M. W., Cranmer, L., DePuy, G. W., Ellis, K. P., Tovey, C. A. and Xu, H.**, Automated process planning for printed circuit card assembly, *IIE Transactions*, v. 24, n.4, 18-30, September 1992.
10. **Ammons, J. C., Carlyle, M., Cranmer, L., DePuy, G., Ellis, K., McGinnis, L. F., Tovey, C. A. and Xu, H.**, Component allocation to balance workload in printed circuit card assembly systems, Accepted for publication in *IIE Transactions*, 1995.
11. **Ellis, K. P.**, Analysis of Manufacturing Setup Strategies in Electronic Assembly Systems, Ph. D. Dissertation, Georgia Institute of Technology, Atlanta, GA, 1996.
12. **Krucky, J.**, Fuzzy family setup assignment and machine balancing, *Hewlett-Packard Journal*, v. 45, n. 3, 51-65, June 1994.
13. **Harhalakis, G. and Minis, I.**, A group technology-based manufacturability evaluation system for a class of electronic products, *ASME Intelligent Concurrent Design: Fundamentals, Methodology, Modeling and Practice*, v. 66, 105-117, 1993.

14. **Burns, G., Rajgopal, J. and Bidanda, B.**, Integrating group technology and TSP for scheduling operations with sequence dependent setup times, *IIE 2nd Industrial Engineering Research Conference Proceedings*, 837-841, 1993.
15. **Suarez, F. F., Cusumano, M. A. and Fine, C. H.**, An empirical study of manufacturing flexibility in printed circuit board assembly, *Operations Research*, v. 44, n. 1., 223-240, 1996.
16. **Kurawarwala, A. A., and Matsuo, H.**, Forecasting and inventory management of short life-cycle products, *Operations Research*, v. 44, n. 1, 131-150, January-February 1996.
17. **Slagmulder, R., Bruggeman and W., van Wassenhove, L.**, An empirical study of capital budgeting practices for strategic investments in CIM technologies, *International Journal for Computer Integrated Manufacturing*, v. 40, 121-152, 1995.
18. **Wu, P. S., Tam, H. Y. and Venuvinod, P. K.**, Hybrid assembly: A strategy for expanding the role of "advanced" assembly technology, *Computers and Electrical Engineering*, v. 22, n. 2, 109-122, 1996.

SETUP MANAGEMENT ISSUES IN PRINTED CIRCUIT CARD ASSEMBLY SYSTEMS

Kimberly P. Ellis,[1] Leon F. McGinnis,[2] and Jane C. Ammons[2]

[1] Industrial and Systems Engineering
Virginia Polytechnic Institute and State University
Blacksburg, VA 24061-0118

[2] Industrial and Systems Engineering
Georgia Institute of Technology
Atlanta, GA 30332-0205

ABSTRACT: The importance of process planning for improving productivity in printed circuit card assembly has been identified by researchers and practitioners. This paper briefly describes process planning in printed circuit card assembly systems, outlines the major issues in setup management, summarizes the current research in this area, and illustrates the impact of setup management decisions through case study results.

INTRODUCTION

The electronics industry continues to rank as an important industry as the twenty-first century approaches. In the United States, annual sales of electronic products exceed $329 billion [9], and the electronics industry is also the largest industrial employer and the fastest growing manufacturing industry [17]. In order to maintain this trend, manufacturers strive to provide electronics products with advanced technology while meeting demands for increased productivity, customer responsiveness, and cost effectiveness.

Printed circuit card assemblies are the critical elements of virtually all electronic products manufactured today. The assembly of these circuit cards requires a large investment in capital equipment as well as significant expenditures for labor and overhead. In order to remain competitive, companies in the electronics industry must utilize these assembly resources efficiently.

Expertise in a variety of areas including production planning and control, labor and facility management, quality assurance, and process planning is required to achieve this efficiency. Of these areas, process planning is particularly important in printed circuit card assembly. Process planning determines how efficiently available resources are used to meet production requirements. The development of a process plan involves translating product design information into machine instructions while considering production requirements and machine capabilities. In the assembly of printed circuit cards, this is a complex function to perform due to the large number of components to assemble, the diversity of product types, and the variety of assembly machine technologies.

The purpose of this paper is to briefly describe process planning in printed circuit card assembly systems, present the issues in setup management, describe the current research in this area, and illustrate the impact of these decisions through case study results.

PROCESS PLANNING DECISIONS

In general, process planning involves addressing two closely related issues: *process optimization* and *setup management* [10]. Process optimization refers to the decisions required to minimize the time to place parts on a particular card for a given arrangement of assembly machines. Specifically, process optimization decisions involve allocating the components required for the circuit cards to the various placement machines and configuring the machines for the assembly operations. Machine configuration involves arranging the components in the feeder slots of the assembly machines and sequencing the placement of these components onto the circuit cards. The primary focus of process optimization is processing time efficiency.

Setup management refers to the planning decisions regarding the organization and allocation of resources and products in the assembly system. Setup management decisions involve determining the appropriate setup strategy for a set of cards, the family of circuit cards to produce together, the machines to group together for production, the assignment of circuit cards to a group of machines, the sequence in which the cards are to be assembled, and the effects of these decisions on the system. The primary focus of setup management is setup time efficiency. The decision problems required for setup management and process optimization are summarized in Table 1.

Table 1. Process Planning Decisions

Process Planning Decisions
Setup Management
Strategy Selection
Card Grouping
Machine Grouping
Line Assignment
Card Sequencing
Process Optimization
Component Allocation
Feeder Assignment
Placement Sequencing

The relationship among these process planning decision problems is illustrated in Figure 1. The process optimization decisions appear near the center with the setup management decisions appearing in the two outer rings.

The machine optimization decisions of feeder assignment and placement sequencing are shown in the center of the circle as these problems determine the detailed configuration of the placement machines. Feeder assignment involves assigning component types to locations on the feeder carriage of an assembly machine, and placement sequencing involves specifying the sequence that the components are placed on a circuit card by the assembly machine. The objectives of these decisions are processing time reduction through proper machine configuration. The form and complexity of these problems are a function of the type of assembly machine under consideration as well as the setup strategy employed. In general, however, heuristic solution approaches are employed due to the complexity of the problems and the amount of information that must be analyzed. McGinnis *et al.* [10] and Peters and Subramanian [13] summarize the recent research for feeder assignment and placement sequencing.

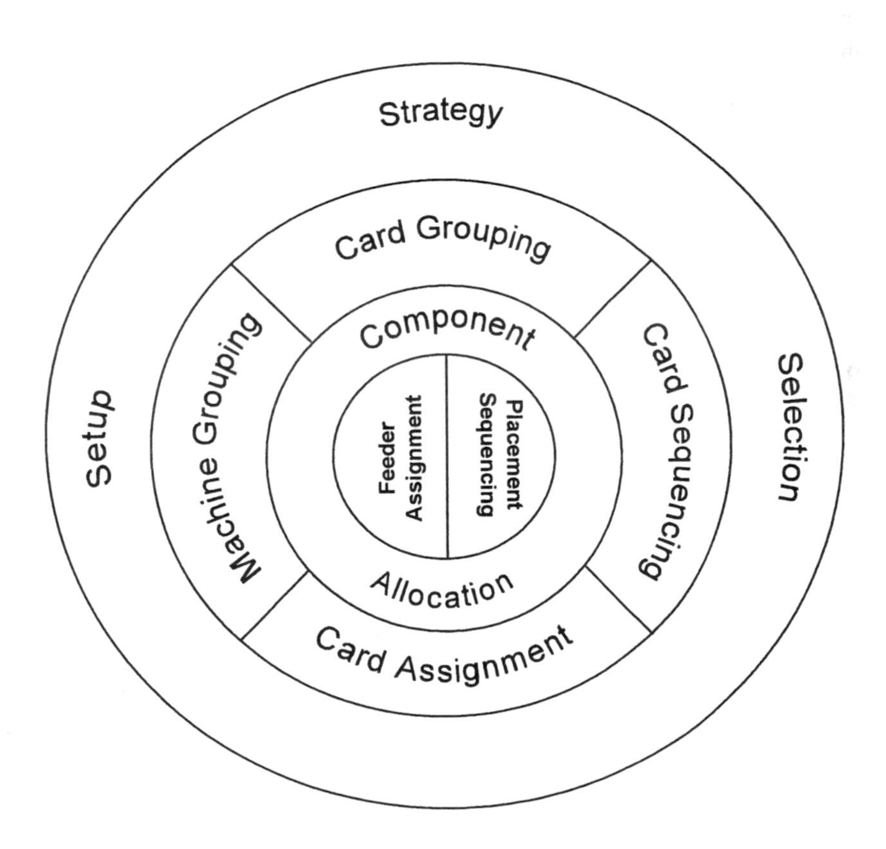

Figure 1. Relationship Among Process Planning Decisions [4]

At the next level is the component allocation problem. The purpose of the component allocation decision for a set of card types is to distribute component types, and the corresponding workload, among available placement machines. The primary objective of component allocation is processing time reduction through workload balancing. A summary of the literature related to the component allocation problem has been presented by several researchers [1, 2, 3, 10]. Specifically, in Ammons *et al.* [1], the component allocation literature is categorized by the setup strategies for which the solution approach is applicable. The solution approaches employed in the literature include both heuristic solution approaches and discrete optimization for finding optimal solutions [1]. Clearly, component allocation is related to the machine optimization decisions.

The next level of decision problems includes card grouping, card sequencing, machine grouping, and line assignment. Card grouping decisions involve the selection of the circuit cards to group together as a family for production [6, 15]. Card sequencing decisions involve sequencing the production of the circuit cards on the selected assembly line [7, 14]. Machine grouping decisions involve the selection of the machines to group together as an assembly line or cell [16]. Line assignment decisions involve the selection of the assembly line or cell for the production of the circuit cards [5]. The objectives of the problems vary for different applications with setup time reduction often used as the primary objective. The problems have generally been solved independently of the other decision problems in process planning. Although illustrated (and often addressed) as separate decision problems, these problems are related to each other.

The outer decision problem is setup strategy selection, which involves selecting a line operating policy for the circuit cards to be produced. The focus of this problem is on the reduction of total assembly time, which includes setup time and processing time. The selected setup strategy often dictates the nature and complexity of the lower level decision problems. The determination of the appropriate strategy, however, is a difficult decision that is interrelated with the lower level decisions [4].

SETUP STRATEGY SELECTION ISSUES

A variety of setup strategies are currently employed by manufacturers in printed circuit card assembly. Several classifications of setup strategies for printed circuit card assembly are provided in the literature. The classification presented by McGinnis *et al.* [10] incorporates the classifications discussed by other authors and is used as the reference classification in this paper. As described by McGinnis *et al.* [10], setup strategies are classified as *single setup strategies* and *multisetup strategies*.

The single setup strategy requires only one setup of the production line to assemble a given setup of card types. If the set contains only a single card type, then the strategy is referred to as a *unique setup strategy*. With a unique setup strategy, the line is set up for a particular product type. After completing the production of that product type, the line is reset for the next product type. If the set contains multiple card types, then the strategy is referred to as a *family setup strategy*. With family setups, the products are combined into product families and the line is set up such that any member of the product family can be assembled without a change in the physical line setup.

The multisetup strategies require multiple setups of the production line to complete the assembly of a given set of card types, often due to the limited staging capacity of the production line. There are many possible multisetup strategies, but two possible strategies are *decompose and sequence* and *partition and repeat* [10]. With a decompose and sequence strategy, a family of card types is divided into subfamilies, the subfamilies are then sequenced to minimize the incremental setups between subsets. With a partition and repeat strategy, the set of components required for the family of card types is partitioned into subsets such that the assembly line has enough staging capacity for each subset of components. The assembly line is then configured to process each subset of components for the family.

Selecting the appropriate setup strategy is a difficult process planning decision. Some setup strategies may be more applicable in certain production environments. For example, a high volume, low variety environment requires that the process planner focus on reducing process time. Process efficiency is the primary issue since a smaller variety of cards is produced in greater volume. Conversely, a low volume, high variety environment requires that the process planner focus on setup time. Since a higher variety of cards is produced, setup efficiency is the primary issue. Often the decision on whether to focus on setup time or process time is not clearly defined when selecting a setup strategy. Also, the strategy selection decision is interdependent on the other setup management decisions as well as the process optimization decisions [4].

For purposes of discussion, assume an assembly line is available for the production of a given set of circuit cards. The planning engineers are trying to decide which type of setup strategy to employ. Assume the choices are limited to the single setup strategies (unique setup strategies and family setup strategies). The advantage of a unique setup is that, at least conceptually, the line may be optimized for each card type and the maximum production rate can be achieved for that product. The disadvantage of a unique setup of that long downtimes may be required to reset the line between card types. The advantage of a family setup is the elimination of individual setups between card types in the family. The disadvantage of a family setup is that the line may be less efficient for assembling a given card type than if a unique setup for that product had been performs. Thus, there is a tradeoff between the increased line availability and the reduced line efficiency for individual card types when choosing among these strategies.

In essence, the unique setup strategy can be viewed as a family setup strategy with a family size of one card type. Thus, for single setup strategies, the objective of the setup strategy problem is to determine the circuit cards to group together as a family in order to minimize the total assembly time for the set of circuit cards. As illustrated in Figure 2 [4], the decision problems to address include card grouping, component allocation, feeder assignment and placement sequencing issues.

The setup strategy selection problem is a complex decision problem that requires the evaluation of several underlying decision problems as well as consideration of processing time and setup time tradeoffs. The research challenge is to develop a solution approach which addresses the primary strategy selection problem and incorporates the related lower level decision problems.

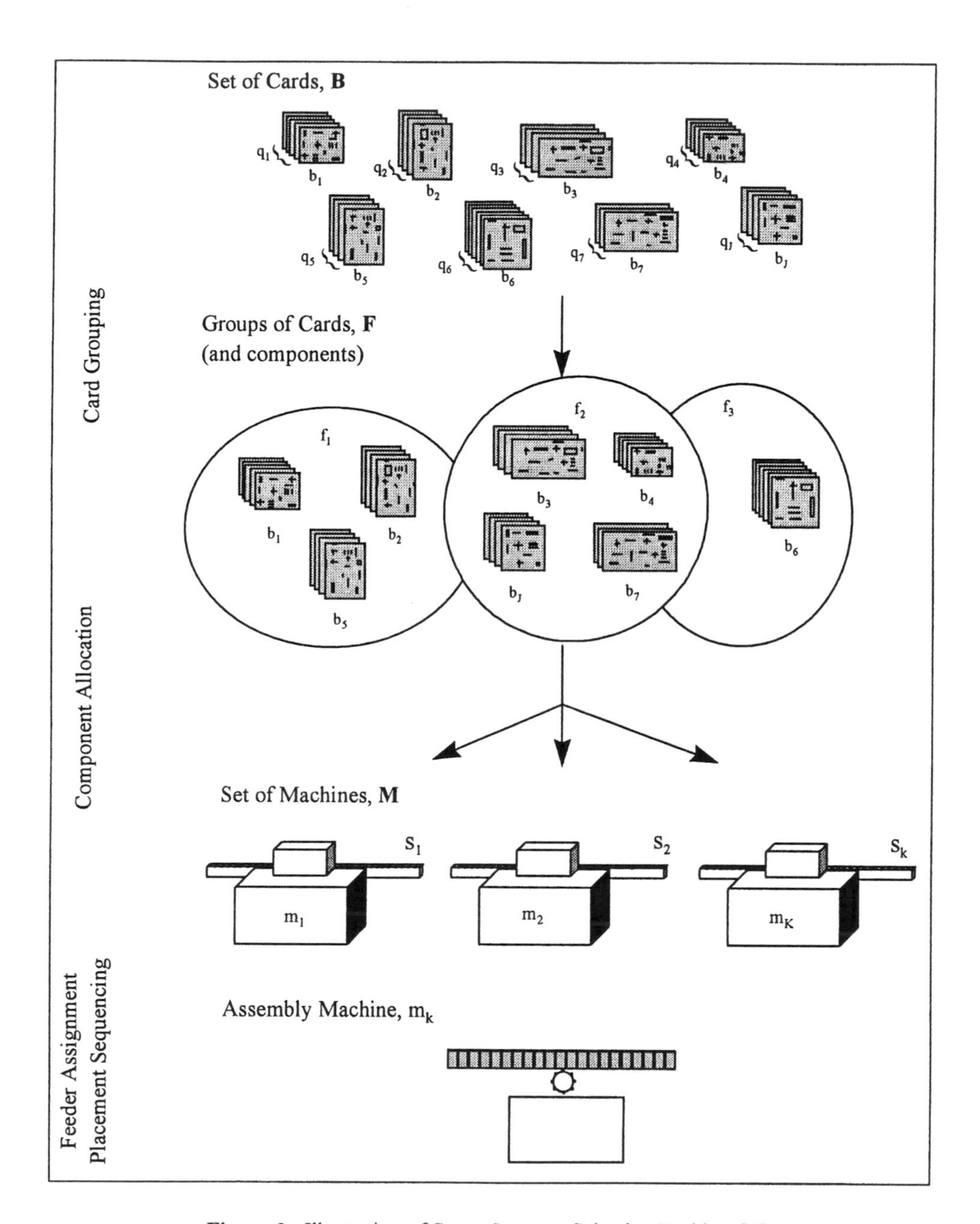

Figure 2. Illustration of Setup Strategy Selection Problem [4]

SOLUTION APPROACH

A solution approach for addressing the setup strategy selection problem for the single setup strategies has been developed and presented by Ellis [4]. The overall approach for addressing the setup strategy selection problem involves capturing the effects of lower level machine configuration decisions through an empirically based processing time estimator function and incorporating this function with the higher level decision problems of component allocation and card grouping.

The processing time estimator function approximates the time required to assemble a circuit card on a specific assembly machine type. This estimator function is developed empirically for a specific placement machine type and machine optimization software. For example, processing time on a Panasonic MV150 surface mount placement machine that is optimized using Panatools Factory Automation Software [12] can be approximated by a linear function of several variables. These variables include:

the total number of components placed on the circuit card by the machine;
the number of unique component types staged on the machine for the card;
the number of component types staged on the machine that are common between the card type and the other card types in the group;
the number of component types staged on the machine for other card types; and
the order that the card is evaluated by the process planning software [4].

The higher level decision problems of card grouping and component allocation together with the empirical estimator function are modeled as a large scale mixed integer programming problem [11]. The setup strategy selection problem is solved using a branch-and-bound algorithm [11], supplemented with specialized techniques for improving the computational times [4]. For typical single machine case studies such as those illustrated below, the solution times are on the order of seconds on a Sun SPARCStation 10 with a model 30 processor. For typical multiple machine case studies, the computational times are longer (on the order of hours) due to the additional complexity of the problem. The solution times are considered reasonable for the planning horizon under consideration (days or weeks) and indicate this approach is feasible as a process planning tool.

ILLUSTRATIVE CASE STUDY

The following case study illustrates the importance of evaluating the underlying decision problems as well as considering processing time and setup time tradeoffs when solving the setup strategy selection problem. For this case study, a data set of four industry representative circuit card types are evaluated. The characteristics of these card types are summarized in Table 2 with additional details provided in [4]. The component setup time used for this case study is 45 seconds, which includes the time to load and unload a component type at a specified feeder carriage location. The setup time for a family of card types is modeled as a function of the number of unique component types that are required for the family. The processing time is modeled based on the Panasonic

MV150 surface mount placement machine that is optimized using Panatools Factory Automation Software [12].

If the objective is to minimize setup time, then the cards will be grouped together in order to minimize the changeover time required between the groups. For this example, all the components required for the card types will fit on the feeder carriage of the assembly machine. Therefore if setup time minimization is the objective, then the cards will be grouped together into a single family containing the four unique card types. Using this setup strategy (SS1), the total assembly time for the required production volumes is approximately 2169 minutes (36.1 hours).

Table 2. Characteristics of Circuit Cards

Card Identifier	Unique Component Types	Total Component Placements	Production Requirements
1	38	146	450
2	31	69	1550
3	28	78	1350
4	7	76	850

As with many placement machines, however, the processing time required to assemble a card type on a Panasonic MV150 is dependent on the machine configuration. If a family setup is used, the machine configuration for processing each individual card type may not be optimized and the resulting processing time may increase. If the objective is to minimize total assembly time (including both setup time and processing time), then the decrease in setup time should be evaluated against the possible increase in processing time.

With the objective of minimizing total assembly time for all the card types, then the optimal solution for this data set is to process the card types individually. Using this setup strategy (SS2), the assembly time for the required production volumes is approximately 2011 minutes (33.5 hours). This results in approximately a 7% decrease in assembly time.

The production schedules for both these strategies are illustrated in Figure 3. The total amount of time reduced when using the setup strategy identified by the solution approach (SS2) is approximately 158 minutes. In the electronics industry, this amount of time can represent significant savings since processing times are relatively short (generally on the order of seconds or minutes). In this 158 minute savings, an additional 479 units of card type 4 or an additional 443 units of card type 3 could be produced.

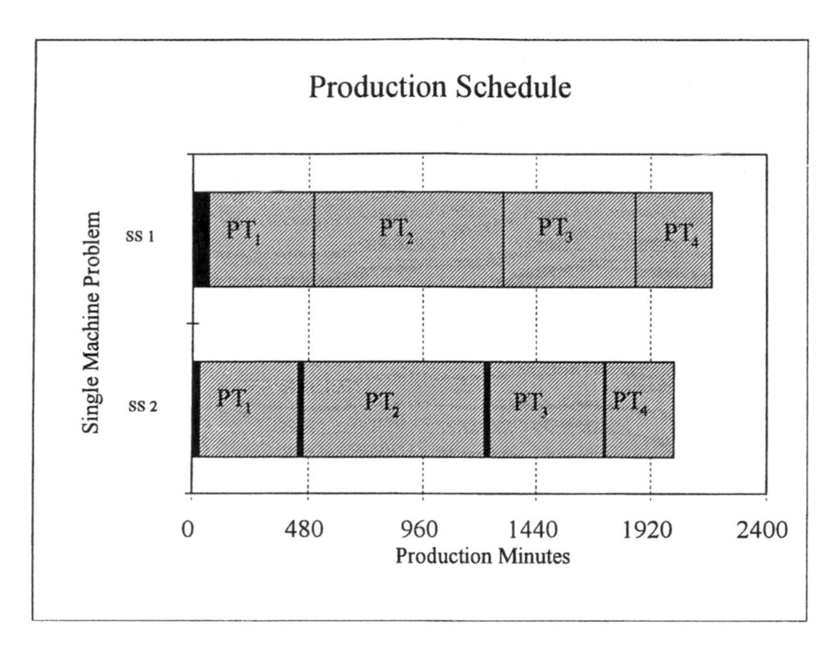

Figure 3. Production Schedule for Card Data Set [4]

CONCLUSIONS

The comprehensive problem of setup strategy selection is a complex decision problem. The important issues related to setup management and setup strategy selection are highlighted in this paper. The research challenge is to develop a solution approach which addresses the primary strategy selection problem and incorporates the related lower level decision problems. The importance of addressing the lower level process optimization problems along with the higher level setup management problems is illustrated through a case study example. The solution approach described here provides a means for addressing these issues for unique and family setup strategies.

ACKNOWLEDGEMENTS

This research has been funded by the National Science Foundation (Grant DDM-9102556) as well as the School of Industrial and Systems Engineering, the Manufacturing Research Center, and the Material Handling Research Center at the Georgia Institute of Technology. This paper includes results previously published in a doctoral dissertation [4].

REFERENCES

[1] J.C. Ammons, M. Carlyle, L. Cranmer, G. W. DePuy, K. P. Ellis, L. F. McGinnis, C. A. Tovey, and H. Xu, "Component Allocation to Balance Workload in Printed Circuit Card Assembly Systems," *IIE Transactions*, accepted for publication.

[2] M. L. Brandeau and A. Billington, "Design of Manufacturing Cells: Operation Assignment in Printed Circuit Board Manufacturing," *Journal of Intelligent Manufacturing*, Vol. 2, pp. 95-106, 1991.

[3] G.W. DePuy, "Component Allocation to Balance Workload in Printed Circuit Card Assembly Systems," Doctoral Dissertation, School of Industrial and Systems Engineering, Georgia Institute of Technology, September 1995.

[4] K.P. Ellis, "Analysis of Setup Management Strategies in Electronic Assembly Systems," Doctoral Dissertation, School of Industrial and Systems Engineering, Georgia Institute of Technology, May 1996.

[5] K. Feldmann, N. Roth, and K.G. Gunther, "Optimization of Set-up Strategies for Operating Automated SMT Assembly Lines," *Annals of the CIRP*, Vol. 40, No. 1, pp. 433-436, 1991.

[6] S. Loh and G.D. Taylor, "An Evaluation of Product Commonality and Group Technology Production Methods in a Predefined Multiple Machine Scenario," *Production Planning and Control*, Vol. 5, No. 6, pp. 552-561, 1994.

[7] S. Jain, M.E. Johnson, and F. Safai, "Implementing Setup Optimization on the Shop Floor," Operations Research, Vol. 43, No. 6, pp. 843-851, 1996.

[8] O. Maimon and A. Shtub, "Grouping Methods for printed Circuit Boards Assembly," *International Journal of Production Research*, Vol. 29, No. 7, pp. 1379-1390, 1991.

[9] P. McCloskey, "Exports Propel the U.S. Electronics Industry to Record Heights," *Electronic Engineering Times*, No. 850, p. 46, May 1995.

[10] L. F. McGinnis, J. C. Ammons, M. Carlyle, L. Cranmer, G. W. DePuy, K. P. Ellis, and C. A. Tovey, "Automated Process Planning for Circuit Card Assembly," *IIE Transactions*, Vol. 24, No. 4, pp. 18-30, 1992.

[11] G.L. Nemhauser and L.A. Wolsey, *Integer and Combinatorial Optimization*, New York: John Wiley and Sons, 1988.

[12] Panasonic Factory Automation, "Panasonic Electronic Component Placement/Insertion Machines: General Catalog for Panasert Series," Matsushita Electric Industrial Co., Ltd,. Doc. No. TI-581-0292, 1992.

[13] B.A. Peters and G.S. Subramanian, "Analysis of Partial Setup Strategies for Solving the Operational Planning Problem in Parallel Machine Electronic Assembly Systems," *International Journal of Production Research*, Vol. 34, pp. 999-1021, 1996.

[14] M. Sadiq, T.L. Landers, and G.D. Taylor, "A Heuristic Algorithm for Minimizing Total Production Time for a Sequence of Jobs on a Surface Mount Placement Machine," *International Journal of Production Research*, Vol. 31, No. 6, pp. 1327-1341, 1993.

[15] A. Shtub and O. Maimon, "Role of Similarity Measures in PCB Grouping Procedures," *International Journal of Production Research*, Vol. 30, No. 5, pp. 973-983, 1992.

[16] U. Wemmerlov and N.L. Hyer, Procedures for the Part Family/Machine Group Identification Problem in Cellular Manufacturing,"" *Journal of Operations Management*, Vol. 6, No. 2, pp. 125-147, 1986.

[17] W. E. Wilhelm and J. Fowler, "Research Directions in Electronics Manufacturing," *IIE Transactions*, Vol. 24, No. 4, pp. 6-17, 1992.

FUZZY LOGIC CONTROL OF MATERIAL FLOW IN FLEXIBLE MANUFACTURING SYSTEMS

Eman Kamel, William E. Biles
Department of Industrial Engineering
University of Louisville
Louisville, Kentucky 40292
USA

ABSTRACT: Automated material flow is essential to the optimal performance of a Flexible Manufacturing System. For those Flexible Manufacturing Systems for which material flow is carried out using conveyors, the control of the conveyor system is the key to managing material delivery to the several machines comprising the system. This paper describes a fuzzy logic-based control model for a Flexible Manufacturing System conveyor system. Fuzzy logic technology is employed to realize Proportional Integral Derivative control in real-time. Conveyor speed on each conveyor segment is controlled using fuzzy logic techniques to regulate the arrival rate of pallets to work stations. In this way, the production throughput of the Flexible Manufacturing System is maximized.

INTRODUCTION

The rapidly converging global economy is forcing radical changes in the industrial manufacturing community. High quality coupled with low cost are pushing advanced manufacturing automation as the dominant factor in global trade competition. The recent history of the automobile and computer industries clearly support the need for quality, economy of scale, and economy of scope. Not even IBM or GM can survive global competition without big changes in design and manufacturing. Expanding customer's demand for a variety of models and choices in a product has led more companies to implement Flexible Manufacturing Systems (FMS) on the plant floor.

The major characteristics of flexible manufacturing systems can be summarized as follows:

- FMS must meet requirements for both production volume and flexibility of attributes.
- FMS must meet the requirements of repetitive manufacturing applications and the need for efficient task-oriented robot or automated machine modules.
- FMS must perform well for non-repetitive applications by providing universal computer control, flexible conveyor systems, a centralized computer interface, and efficient integration of all system resources.
- FMS should provide the same level of utilization and efficiencies of the repetitive manufacturing system while supporting the flexibility of general purpose tools/equipment modules.

Modular design, conveyor systems, and cellular manufacturing modules provide for increased flexibility and allow a plant to handle surges in volume and the number of customized products orders. FMS implementation can reduce machine setup times, shorten the overall manufacturing cycle, reduce inventory, and minimize the overall cost of tools.

Due to manufacturing constraints, parts handling, inspection, assembly, safety and mechanical requirements, the spacing between pallets needs to be regulated. Typically, the spacing is best maintained at a fixed and small value. The transportation/conveyor system speed can be used to regulate pallet spacing under varying conditions in FMS. This can lead to conservation of energy and a more effective production environment.

This paper presents a model and its implementation, for controlling an FMS conveyor system. The developed work uses fuzzy logic techniques to accomplish supervisory, Proportional Integral Derivative control in real time. A summary of the literature pertaining to fuzzy control in industrial automation and process control systems, and an overview of the developed fuzzy controller, are included in this paper. Details of the implementation and experimentation on the test conveyor system are also given.

LITERATURE REVIEW

The field of intelligent control has converged as a combination of control theory, operations research, and artificial intelligence. One of the most popular technologies based on artificial intelligence is fuzzy logic. This is evident in the thousands of papers and hundreds of patents that have treated fuzzy logic in one way or another. The use of fuzzy logic in the field of control systems has resulted in the development of fuzzy controllers.

Fuzzy sets were introduced in 1965 by Lofti Zadeh as a new way to represent vagueness in everyday life.[1] Although fuzzy logic can be considered a more powerful tool than the classical predicate logic, it was established based on the understanding and the past experience of the latter. Fuzzy logic, upon which fuzzy control is based, is much closer to thought processes than the traditional logical system.[2] Basically, it provides an effective means of capturing the approximate, inexact nature of the world. Viewed from this perspective, the essential part of the fuzzy logic controller (FLC) is a set of linguistic control rules related by the dual concepts of fuzzy implication and the compositional rules of inference.

The methodology of the FLC appears very useful when the controlled processes are too complex for analysis by conventional quantitative techniques, or when the available sources of information are interpreted qualitatively, inexactly, or with uncertainty. Thus, fuzzy logic control may be viewed as a step toward a rapprochement between conventional, precise mathematical control and human-like decision making.[3]

Fuzzy control systems are thus rule-based systems in which a set of fuzzy rules represents a control decision mechanism to adjust the effects of certain causes coming from the system.[4] These fuzzy rules are in the form of IF-THEN rules whose antecedents (IF part) and consequences (THEN part) are themselves labels of fuzzy sets.

Typical rules can result from a human operator's knowledge; e.g., "**IF** the Temperature is hot, **THEN** set the heater at low level." The development of control rules depends on whether or not the process can be controlled by a human operator.[5] If the operator's knowledge and experience can be explained in words, then linguistic rules can be written directly. If the operator's skill can keep the process under control, but this skill cannot be expressed immediately in words, then a fuzzy model of these actions must be used to generate the rules.[6-7] The fuzzy model in this case refers to the description of the operators input/output control actions using fuzzy implications. However, when a process is exceedingly complex, it may not even be controllable by a human expert. In this case, a fuzzy model of the process is built and the control rules are derived theoretically.

The basic configuration of a fuzzy logic controller is comprised of four principal components: a fuzzification interface, a knowledge base, decision-making logic, and a defuzzification interface.[2] The five steps are listed below.

Step 1. Identify the controlled process: define the input to the control, the output, the controlled variable, and the controlling variable.
Step 2. Define input and output fuzzy set labels: define the membership functions for the fuzzy sets defined in step 1.
Step 3. Construct the rule base: define the rule-base structure which represents the control policy.
Step 4. Design a fuzzy computation unit: this is the heart of the fuzzy controller. It performs as an inference engine for the fuzzy controller.
Step 5. Design a defuzzifier: this part converts the fuzzy output of the inference engine into a crisp output (actual control value).

Flexible Manufacturing Systems (FMSs) typically have from three to twenty machine tools/robots (modular cells) and use a computer to control the various tasks in the manufacturing process, moving parts handling activities between modular cells, and the regulation of the flow of parts through the system. FMSs are generally suitable for manufacturing situations with moderate production volumes and a large variety of requirements. Thus, FMS automation is normally justified based on economies of scope, not on those of scale.[8-9] Control procedures for FMSs have been extensively investigated by researchers.[10] Reviews of FMS analytic models and scheduling rules have been reported.[11-12]

In industrial automation, fuzzy logic technologies enable the efficient and transparent implementation of human control expertise. Here, the individual control loops of single process variables mostly remain controlled by conventional models such as PIDs. The fuzzy logic system then gives the set values for these controllers based on the process control expertise put in the fuzzy logic rules.

Fuzzy logic was used to save energy at the world's largest oral penicillin production facility in Austria.[13] While keeping single process variables at their command values is relatively easy, the determination of the optional operating point is often a complex multi-variable problem. In most cases, a solution based on a mathematical model of the process is far beyond acceptable complexity. In some cases, the derivation of a mathematical model of the plant consumes considerable effort. Hence, in a large number or plants, operators control the operation point of the process manually. In these cases, fuzzy logic provides an efficient technology to put the operation control strategies into an automated solution with minimum effort.[14-17]

Other examples of fuzzy logic applications in the area of industrial automation and process control include automatic train operation, start-up and shut-down control in a power plant, blast furnace control, neutralizing chemical waste water, control of an arc-welding robot, and AC motor control.[18-21]

SYSTEM DESCRIPTION

The test conveyor system used in this study was successfully configured with the needed computer control software and user interface as the basis for a comprehensive study of an FMS advanced computer control using fuzzy logic. The conveyor system was modified to allow for adaptive control of speed for two separate conveyors. The system PLC was equipped with an analog input module. This input module accepts signals from speed sensors mounted on the conveyor drive motors. A C-face ring tachometer was interfaced to the PLC through a signal isolation and scaling card. Fuzzy logic and heuristic techniques were implemented in an adaptive closed PID control loops. Fuzzy logic membership functions were investigated and a function suitable for PLC programming was developed, implemented, and tested. Allen Bradley PID implementation was used to test the candidate fuzzy logic algorithm.

The test conveyor system is a 7-station/cellular module FMS. Pallets (transporters) run on two independently motorized belts, one providing closed circular motion of pallets between stations while the other provides straight motion. Two transient belts allow for pallet transfer between the two conveyors. The conveyor system was configured to provide the platform flexible manufacturing system for this study.

FUZZY CONTROLLER MODEL

Consider a machine feeding pallets to a section of conveyor belt. The machine cycle time at pallet (i) release is Tc_i and the conveyor speed is n_i. The position of pallet (j), measured from the machine at time t is given by

$$d_j(t) = \sum_{i=j}^{k-1} n_i Tc_i + \left[t - \sum_{i=1}^{k-1} Tc_i \right] n_k \quad (1)$$

$$\sum_{i=1}^{k-1} Tc_i < t \leq \sum_{i=1}^{k} Tc_i \quad (2)$$

The spacing between pallet (j) and pallet (j + 1) is given by

$$\Delta d_j\ (t) = d_j\ (t) - d_{j+1}\ (t) = n_j\ Tc_j \quad (3)$$

The above equations show that pallet spacing changes as machine cycle time and conveyor speed vary. Pallet spacing can be regulated at constant distance through changes in conveyor speed to compensate for changes in cycle times. The energy consumed by the FMS transportation system is proportional to the speed of the conveyor. Increasing the speed of the conveyor does not affect the throughput from a given section.

Due to manufacturing constraints (handling, inspection, assembly, safety, and mechanical requirements), the spacing between pallets must be maintained at a minimum and fixed distance. Thus, at a higher machine cycle time the system can conserve energy by reducing the speed of the conveyor and still maintain control over both throughput and the number of pallets on the conveyor section.

For a long conveyor section, speed adjustment can be achieved through measurement of the spacing between pallets. This task becomes complicated as we consider several linked sections of a transportation system, where each is driven by an independent motor drive. In a typical system, the operator observes, analyzes, and then decides on a conveyor speed set value. The PID configured for the AC drive adjusts the conveyor speed to the desired set value. It is usually very difficult for an operator to provide an accurate speed set value. Obviously, the type of speed change-increase or decrease-is easily defined. In a supervisory control mode, the system keeps a log history of process behavior which is analyzed to produce the desired set values. This technique produces good performance at the expense of complex data acquisition and analysis techniques. In the proposed work, a fuzzy logic technique is used to provide an incremental approach to supervisory control.

The fuzzy logic technique defines fuzzy time parameters, one for each section of the conveyor system. Each fuzzy parameter provides a sense of the spacing between consecutive pallets on one of the conveyor sections. The pallet spacing is categorized as a Low, Medium, High (L, M, H) time value.

Two such fuzzy parameters are assigned to the same, but longer, sections of the conveyor system, T_{11} and T_{12}. The two fuzzy parameters are interrelated using the fuzzy matrix in Table 1. The matrix defines the fuzzy logic rules for combining the fuzzy parameter values into a definite change in conveyor speed as shown below.

The fuzzy conveyor speed is categorized as Very Low, Low, About Right, High, and Very High (VL, L, AR, H, VH). Each fuzzy speed is mapped to a deterministic change in speed through the defuzzification process. Preliminary implementation assumed a binary membership function for the defined fuzzy parameters.

The fuzzy controller, shown in Figure 1, defines nine states and three types of transitions. Each state is a combination of two fuzzy time values, one for the leading and the other for the trailing part of the conveyor section. For example, the state (L,H) defines the fuzzy state having a low pallet interarrival time through part one and a high pallet interarrival time through part two of a conveyor section. The interarrival time variables are categorized into (Low, Medium,
and High) (L, M, H) fuzzy values.

TABLE 1
Fuzzy Logic Rules for Changing Conveyor Speed

		Time Value for Conveyor Section 2 T_{12}		
Time Value for Conveyor Section 1		L	M	H
	L	L	AR	H
T_{11}	M	AR	AR	VH
	H	H	VH	VH

Transitions from one state to another are caused by anticipated uncontrolled disturbances. Planned disturbances are dominated by changes in the machine cycle time. Others include diversion, rejection, or removal of pallets during quality checks in real-time inspections. Uncontrolled disturbances include machine tool wear out, operator change, intermittent failures, pallet jamming, and changes in other associated conveyor systems. All these types of disturbances can be simulated as an equivalent change in the machine cycle time. In our work, only transitions caused by machine cycle times are considered. Transitions used in the our experimentation were caused by insignificant changes, significant increases, and significant decreases in machine cycle time. The three state diagrams (Figures 2, 3, and 4) show the fuzzy controller status under these conditions. Note that, independent of the starting states of the conveyor system, the fuzzy controller will eventually converge to the state (M,M). At this state the controller will not change the speed of the conveyor as long as cycle time changes are insignificant. The final speed will be different from the initial speed after a sequence of transitions.

For a significant increase in machine cycle time, the (M,M) state can be unstable. Transitions can take place converging again at the (M,M) state, but at a considerably lower speed. The same is true for significant decreases in cycle time, but with the final speed considerably higher. Dotted lines are shown in the state diagram as other possibilities for a transition from a given state. The reflexive transition is used to accommodate the fact that the system stays in the same state for some time before moving to another state. The time duration at which the system stays in one state depends on the tuning of the fuzzy logic controller. Tuning involves the selection of parameter ranges, linguistic rules, and the final control action.

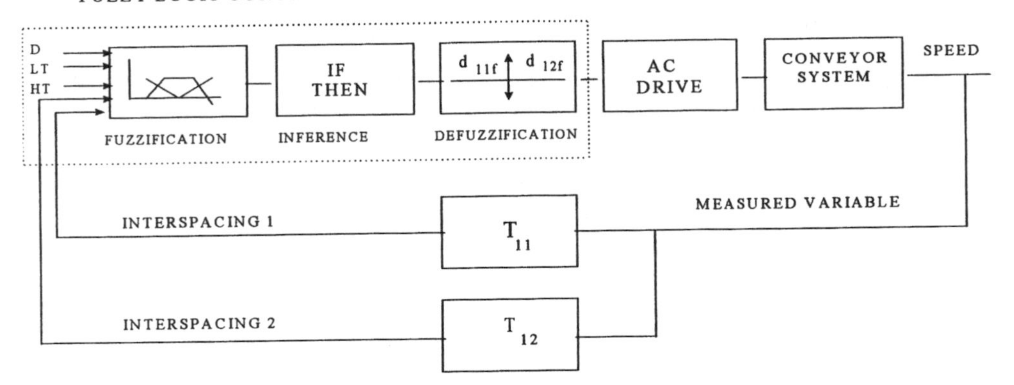

Figure 1. Fuzzy Logic Controller Configuration

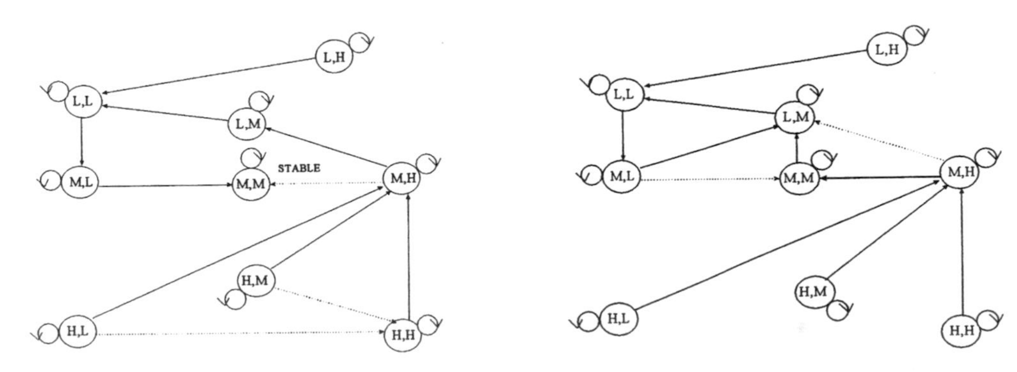

Figure 2. State diagram for insignificant changes in cycle time

Figure 3. State diagram for significant increase in cycle

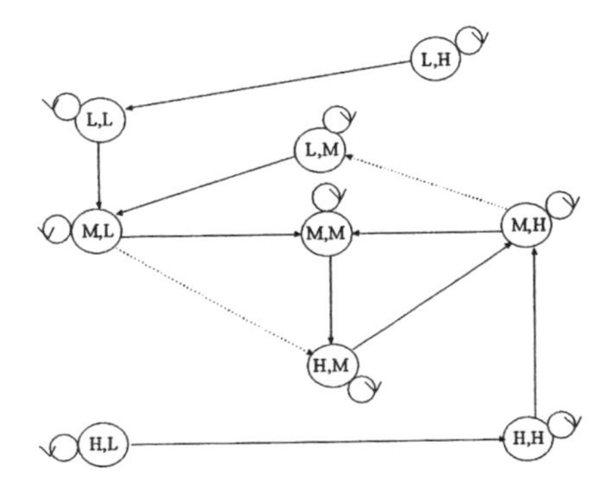

Figure 4. State diagram for significant decrease in cycle time

CONCLUSIONS

This paper has described a model for investigating the development of a fuzzy logic controller for supervisory control of an FMS conveyor. The belief is that such a system can efficiently maintain both highly regulated pallet flow and high throughput, as well as reduce the overall energy requirements used by the motors drive. Preliminary investigation and implementation on the test conveyor system has been discussed in this paper. The unique aspects of this research are the investigation of a new supervisory fuzzy logic controller for FMS scheduling and pallet regulation, and the creation of a universal FMS control system architecture which requires little or no modifications when ported to more advanced manufacturing applications. The fuzzy controller provided much smoother response to disturbances in cycle time than both manual and PID control. It also required much less tuning and resulted in lower energy consumption by the conveyor motor drive system for the same pallet traffic pattern.

REFERENCES

1. **Zadeh, L.,** Fuzzy sets, *Information and Control,* 8, 8-352, 1965.
2. **Lee, C.,** Fuzzy logic in control systems: Fuzzy logic controller--Part I, *IEEE Trans. System, Man, and Cybernetic*, 20, (2), 404-418, 1990.
3. **Gupta, M., and Tsukamoto, Y.,** Fuzzy logic controllers, a perspective, *Automatic Control Conf.,* San Francisco, California, August 1980, pp. FA10-c.
4. **Jamshidi, M., Vadiee, N., and Ross, T. (Eds.),** Fuzzy logic and control, Prentice Hall, Englewood Cliffs, New Jersey, 1993.
5. **Sugeno, M., and Kang, G.,** Fuzzy modeling and control of multi-layer incinerator, *Fuzzy Sets and Systems*, 18, 329-346, 1986.
6. **Matsushima, K., and Sugiyama, H.,** Human operator's fuzzy model in man-machine system with a nonlinear controlled object, in M. Sugeno (Ed.), *Industrial Applications of Fuzzy Control,* North-Holland, Amsterdam, 1985, 175-185.
7. **Takaki, T., and Sugeno, M.,** Fuzzy identification of systems and its application to modeling and control, *IEEE Trans. Syst., Man, and Cybern.*, 15, 116-132, 1985.
8. **Tompkins, J., and White, J.,** Facilities planning, John Wiley and Sons, 1984.
9. **Vollmann, T., Berry, W., and Whybark, D.,** Manufacturing planning and control systems, 3rd Ed., Irwin, Homewood, Illinois, 1992.
10. **Stecke K., and Solberg, J.,** Loading and control procedures for a flexible manufacturing system, *Journal of Production Research,* 19 (5), 481-490, 1981.
11. **Buzacott, J., and Yao, D.,** Flexible manufacturing systems: A review of analytical models," *Management Science,* 32 (7), 1986.
12. **Denzler, D., and Boe, W.,** Experimental investigation of flexible manufacturing system scheduling rules, *International Journal of Production Research,* 25 (7), 979-994, 1987.
13. **Von Altrock, C., Franke, S., and Froese, T.,** Optimization of a water-treatment system with fuzzy logic control, *Computer Design Fuzzy Logic '94 Conference,* San Diego, California, 1994.
14. **Evans, G. W., Karwowski, W., Wilhelm, M. R.,** Application of fuzzy set methodologies in industrial engineering, North-Holland, Amsterdam, 1989.
15. **Roffel, B., and Chin, P.A.,** Fuzzy control of a polymerization reactor, *Hydrocarbon Processing,* 47-50, June 1991.
16. **Tobi, T. and Hanafusa, T.,** A practical application of fuzzy control for an air-conditioning system, *International Journal of Approximate Reasoning* 5, 331-348, 1992.

17. **Yagishita, O., Itoh, O., and Sugeno, M.,** Application of fuzzy reasoning to the water purification process, in Sugeno (Ed.), *Industrial Applications of Fuzzy Control*, North-Holland, Amsterdam, 1985, 19-40.
18. **Yasunobu, P., and Miamoto, P.,** Automatic train operation by predictive fuzzy control in Sugeno (Ed.), *Industrial Applications of Fuzzy Control*, New York, 1985, 1-18.
19. **Bien, Z., Hwang, D. H., Lee, J. H., et. al.,** An automatic start-up and shut-down control of a drum-type boiler using fuzzy logic, *2nd Int'l Conference on Fuzzy Logic and Neural Networks Proceedings*, Iizuka, Japan, 1992, 465-468.
20. **Wegmann, H., and Prehn, E.,** Fuzzy logic mit sps standardkomponenten zur ph-neutralisation chemischer abwasser, *Swiss Fuzzy Logic Conference '92*, Baden, 1992.
21. **Hofmann, W., and Krause, M.,** Fuzzy rules for regulation of torque in ac-drives, *EUFIT' 93--First European Congress on Fuzzy and Intelligent Technologies*, 1993, 1059-1065.

A STUDY OF DYNAMICAL AMPLIFICATION COEFFICIENT UNDER THE DIFFERENCE OF SPECIFIC CUTTING RESISTANCE IN LATHE CHATTER (PART II)

Yoshiaki IWATA

Faculty of Science and Technology, Kinki University

Kowakae 3-4-1, Higashi-Osaka, Osaka 577, Japan,

Tel: +81-6-721-2332, Fax: +81-6-730-1320,

Email iwata @ im. kindai. ac. jp

ABSTRACT: The stabilizing threshold of chatter vibration is obtained from the characteristics of resonance curve in machine tool system, which can be established the dynamical amplification coefficient and spring constant. This study is taken the external cutting of a round bar by right handy blade bit with side cutting edge angle. The bit is fixed tightly in the tool post of lathe. A round bar fitted between the both center is treated as bending vibration. In lathe turning under the vibratory characteristics, the cutting area change between normal cutting and outbreak of chatter vibration after one rotation of work becomes to alter by the size of bit and cutting condition. When the cutting area change is set theoretically, the machine tool characteristics become to influence chatter vibration in lathe turning. In this study, the difference of specific cutting reristance before and after the chatter vibration in normal force direction is taken by $-0.10\,kgf/mm^2$, $0\,kgf/mm^2$, $0.10\,kgf/mm^2$, $0.30\,kgf/mm^2$, and $0.50\,kgf/mm^2$. And dynamical amplification coefficient in machine tool system is analyzed from the formula of stabilizing threshold.

Introduction

When a work fixed in both center of lathe is turned by the right handy blade bit with side cutting edge angle, the change of dynamical amplification coefficient in lathe turning characteristics influence the lathe turning chatter vibration. The dynamical amplification coefficient is given from the resonance curve by transfer function analyzer. The influence of lathe turning chatter vibration is taken from the dynamical amplification coefficient.

A MODEL OF CHATTER VIBRATION IN LATHE TURNING

When a installed round bar in lathe with both center is cut down by a single point tool of right handy blade bit with side cutting edge angle, the relation among tangential force, axial force and normal force under the cutting for lip part of blade is shown in Figure 1.

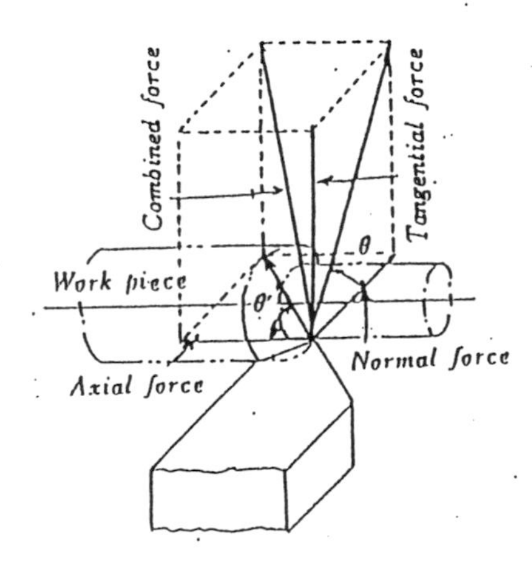

Figure 1. Three component force of cutting

When the resultant of these component forces is projected to each component force direction, the projected angle for normal and axial force direction is θ and θ'. These angle θ and θ' are not changed under the cutting. The direction of vibration for work piece and the direction of cutting force are the same. When the vibratory amplitude puts into x, the component of cutting in normal force direction is put into $x\cos\theta$. The work piece moves approach and go away towards a single point tool from the component of vibration for the normal force.

SETTING OF EACH NUMERICAL VALUE

A. SETTING OF THE SIZE OF BIT

The blade part related to cutting on work piece takes the domain of same size between normal and outbreak of chatter vibration. The size of bit takes inclination angle of cutting edge $0°$, end relief angle $6°$, side relief angle $6°$, normal side rake angle $\alpha_n = 5°$, end cutting edge angle $C_e = 15°$, side cutting edge angle $C_s = 15°$ and nose radius $r_n = 0.5$ mm, 1.0 mm in P 20.

B. SETTING OF SPECIFIC CUTTING RESISTANCE FOR THE NORMAL FORCE DIRECTION

The specific cutting resistance for the normal force is given from the results of cutting experiment. The conditions of cutting experiment by the bit of A are cutting speed

100m/min, depth of cut 1.0 mm and solid round bar (ϕ 15×600 mm) by S 35 C for work piece. The specific cutting resistance k_{s1} is given from the amount of normal force in cutting divided by cutting area.

C. SETTING OF THE ANGLE FACING NORMAL AND AXIAL FORCE DIRECTION IN VIBRATION

The angles facing normal and axial force direction in chatter vibration are set from the result of cutting experiment in cutting speed 100 m /min by the bit from A. The relation among A, B and C is shown in Table 1.

Table 1

The numerical symbols among r_n, ks_1, θ and θ'

r_n (mm)	k_{s1} (kgf/mm^2)	θ (deg.)	θ'(deg.)
0.5	$Exp4.12-0.36 \log_e S_o$	$18.36S_o+61.70$	$20.90S_o+34.71$
1.0	$Exp4.05-0.45 \log_e S_o$	$25.49S_o+59.01$	$19.64S_o+39.60$
1.5	$Exp4.08-0.44 \log_e S_o$	$29.53S_o+55.05$	$9.21S_o+48.46$
2.0	$Exp3.94-0.50 \log_e S_o$	$29.68S_o+59.90$	$24.23S_o+37.26$

D. SETTING OF THE OVERLAP FACTOR

The resultant force between normal force and axial force in the intersection of before and after one rotation of work piece is inclined to θ' facing axial force direction under the lathe turning. Angle θ' shows the direction of vibration for work piece. The resultant force is treated changeable for the dimension in the direction of vibration. The overlap factor μ is set from the condition in Table 1.

E. SETTING OF THE VIBRATORY CHARACTERISTICS OF A MACHINE TOOL CUTTING SYSTEM

In order to set the vibratory characteristics of a machine tool cutting system, the work piece with the radius 30 mm of the central part in Figure 2 is fitted in both center of lathe with the soecification in Table 2.

The resonance curve is gained by transfer function analyzer. The vibratory characteristics of machine tool cutting system is set as follows:

The number of proper vibration $f_0 = 228Hz$

A natural angular frequency $W_0 = 2\pi f_0 = 1431.84$ rad/sec

A spring constant $k = \dfrac{P}{X} = 333.3$ kgf/mm $\quad \dfrac{X}{P} = 3$ (μ/kgf)

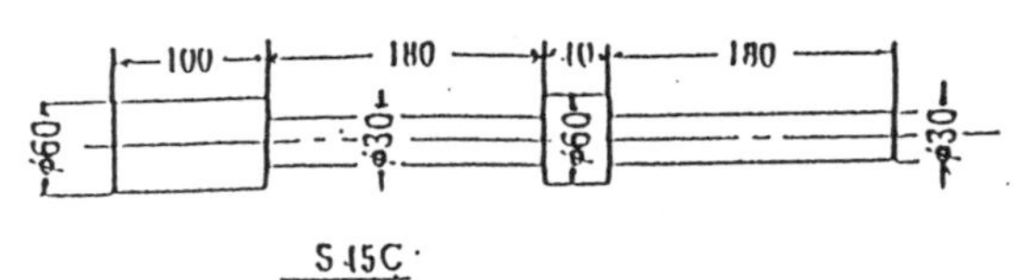

Figure 2 . Dimension of test piece.

Table 2

Specification of a lathe

Swing over compound rest	mm	178
Distance between centers	mm	600
Weight of machine	kg	1660

SETTING OF DYNAMICAL AMPLIFICATION COEFFICIENT IN THE STABILIZING THRESHOLD

In order to set the stabilizing threshold, the blade part of bit related to work cutting takes the domain of same size of bit between normal and outbreak of vibration. The formula of stabilizing threshold is shown in (1), (2) by the dynamical specific cutting resistance for normal force direction k_{1A}, the rotation of work N, angular frequency in chatter vibration ω and $K_A = k_{s3} - k_{1A}$

$$Q = -\frac{\frac{N}{\omega_0}}{H_2 \cos\theta \left(\frac{K_A}{k} + \frac{k_{1A}}{k}\mu\frac{N}{\omega}\sin\frac{\omega}{N}\right)} \qquad (1)$$

$$\left(\frac{\omega}{\omega_n}\right)^2 = 1 + \frac{k_{1A}}{k} H_2 \cos\theta \left(1 - \mu\cos\frac{\omega}{N}\right) \qquad (2)$$

$d \geqq r_n(1-\sin C_S)$, $S_0 \leqq 2 r_n \sin C_e$, $S_0 < 2 r_n \cos C_S$

$$H_2 = \{d - r_n(1-\sin C_S)\}\tan C_S + r_n \cos C_S + \frac{S_0}{2}$$

$$\mu = 1 - S_0 \Big/ \left[[\{d - r_n(1-\sin C_S)\}\tan C_S + r_n\cos C_S + \frac{S_0}{2}] + [d - r_n + \sqrt{r_n^2 - \left(\frac{S_0}{2}\right)^2}\,]\cot\theta' \right]$$

$d \geqq r_n(1-\sin C_S)$, $2 r_n \sin C_e < S_0 < 2 r_n$

$$H_2 = \{d - r_n(1-\sin C_S)\}\tan C_S + r_n\cos C_S + S_0\cos^2 C_e - r_n \sin C_e \cos C_e$$

$$\mu = 1 - S_0 \Big/ \Big[[r_n \sin C_e + \{d - r_n(1-\sin C_S)\}\tan C_S + r_n\cos C_S + S_0\cos^2 C_e - \cos C_e\sqrt{(2 r_n - S_0 \sin C_e) S_0 \sin C_e} - S_0] + [r_n \cos C_e + d - r_n - S_0 \sin C_e \cos C_e + \sin C_e \sqrt{(2 r_n - S_0\sin C_e) S_0 \sin C_e}]\cot\theta' \Big]$$

$$d < r_n(1 - \sin C_s), \quad S_o \leqq 2 r_n \sin C_e$$

$$H_2 = \sqrt{d(2r_n - d)} + \frac{S_o}{2}$$

$$\mu = 1 - S_o \Big/ \left[\left[\sqrt{d(2r_n - d)} + \frac{S_o}{2} \right] + \left[d - r_n + \sqrt{r_n^2 - \left(\frac{S_o}{2}\right)^2} \right] \cot\theta' \right]$$

From (1), (2), Q is shown from N/f_0. Examples of stabilizing chart by the difference of specific cutting resistance K_A are shown in Figure 3 and Figure 4 .

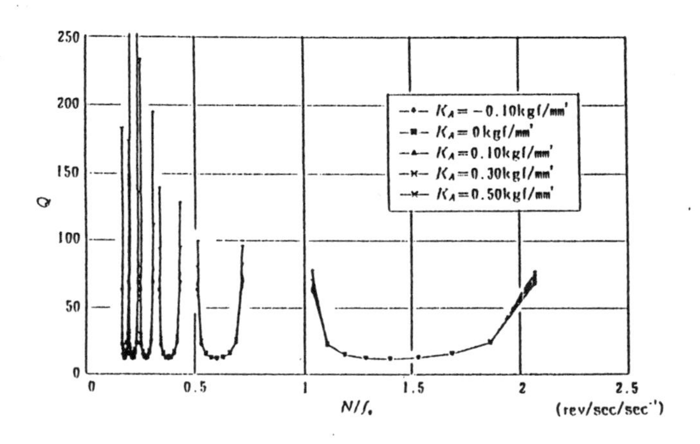

Figure 3. Relation between Q and K_A ($n=1 \sim n=6$, K_A : changeable) ($d=1.0$mm, $C_e=15°$, $r_n=0.5$mm, $C_s=15°$, $S_o=0.2$mm/rev, $k_{s_3}=109.88$kgf/mm², $\theta=65.37°$, $\mu=0.899$, $H_2=0.751$mm constant)

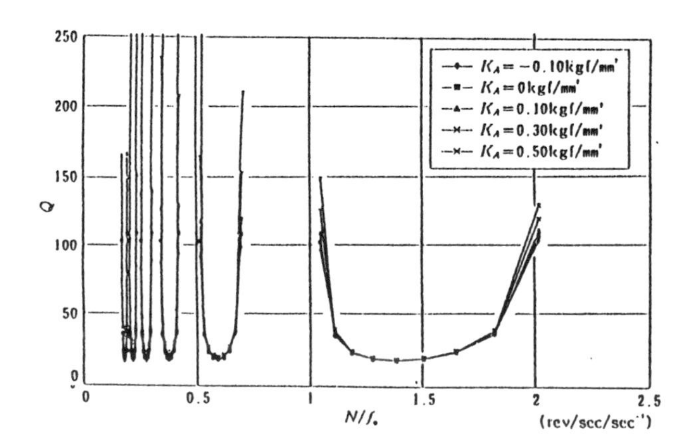

Figure 4. Relation between Q and K_A ($n=1 \sim n=6$, K_A : changeable) ($d=1.0$mm, $C_e=15°$, $r_n=0.5$mm, $C_s=15°$, $S_o=0.5$mm/rev, $k_{s_3}=79.01$kgf/mm², $\theta=70.88°$, $\mu=0.732$, $H_2=0.993$mm constant)

Figure 5, Figure 6 are shown the stabilizing chart in the domain to rotation 256 r.p.m. from 960 r.p.m. . In this case , the relation between the lowest value Q and N/f_n is obtained the range of $\frac{K_A}{k}+\frac{k_{1A}}{k}\mu\frac{N}{\omega}\sin\frac{\omega}{N}<0$.

So, the range of $\frac{\omega}{N}$ is taken in n = 1 $(\frac{\omega}{N}=\pi\sim 2\pi)$, n = 2 $(\frac{\omega}{N}=3\pi\sim 4\pi)$, n = 3 $(\frac{\omega}{N}=5\pi\sim 6\pi)$ and so on.

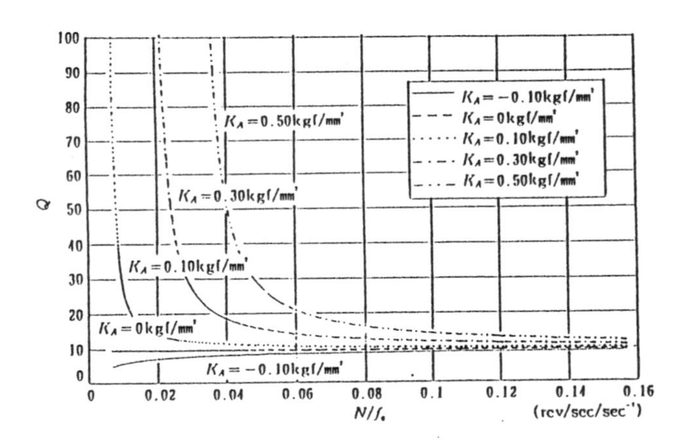

Figure 5. Relation between Q and K_A ($n=20\sim n=60$, K_A : changeable) (d=1.5mm, $C_e=15°$, $r_n=0.5$mm, $C_s=15°$, $S_o=0.2$mm/rev, $k_{s_1}=109.88$kgf/mm², $\theta=65.37°$, $\mu=0.923$, $H_1=0.885$mm constant)

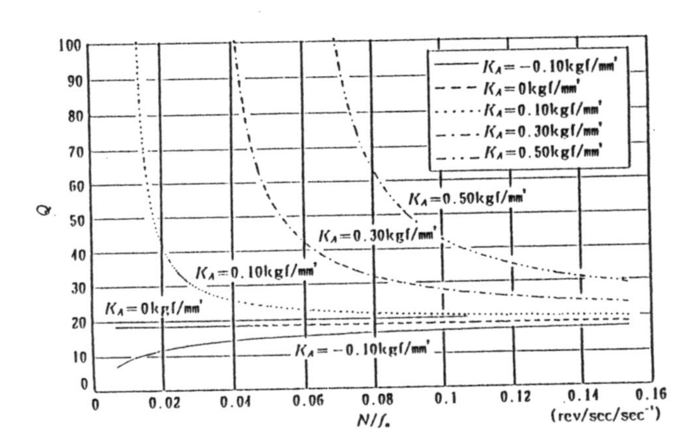

Figure 6. Relation between Q and K_A ($n=20\sim n=60$, K_A : changeable) (d=1.0mm, $C_e=15°$, $r_n=0.5$mm, $C_s=15°$, $S_o=0.5$mm/rev, $k_{s_1}=79.01$kgf/mm², $\theta=70.88°$, $\mu=0.732$, $H_1=0.993$mm constant)

When the depth of cut d , end cutting angle C_e , nose radius r_n , feed S_o, difference of specific cutting resistance K_A are not changed and side cutting edge angle C_s changeable in Figure 7 , Figure 8 , the stabilizing threshold of dynamical amplification coefficient Q decreases with growing the side cutting edge angle C_s .

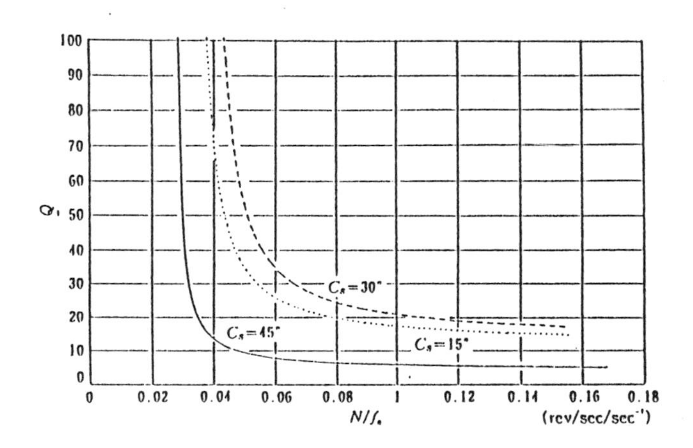

Figure 7. Relation between Q and C_s ($n=20 \sim n=60$, C_s : changeable)
($d=1.0$mm, $C_e=15°$, $r_n=0.5$mm, $S_o=0.2$mm/rev, $K_A=0.5$kgf/mm^2 constant)

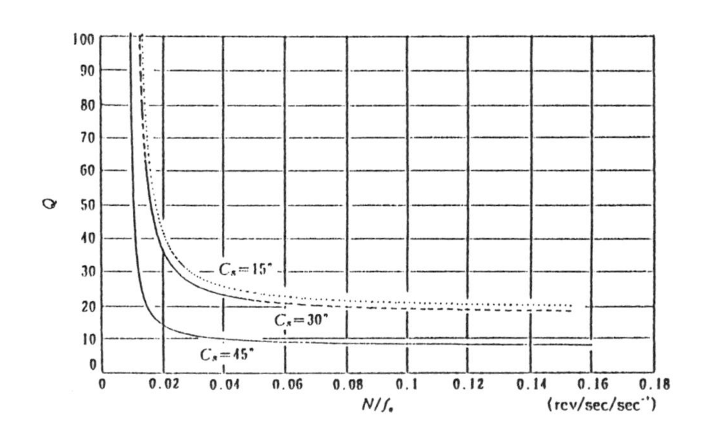

Figure 8. Relation between Q and C_s ($n=20 \sim n=60$, C_s : changeable)
($d=1.0$mm, $C_e=15°$, $r_n=0.5$mm, $S_o=0.5$mm/rev, $K_A=0.1$kgf/mm^2 constant)

When the depth of cut d , end cutting edge angle C_e , side cutting edge angle C_s , feed S_o , difference of specific cutting resistance K_A are not changed and nose radius r_n changeable in Figure 9 and Figure 10 , the stabilizing threshold of dynamical amplification coefficient Q decreases with growing the nose radius r_n .

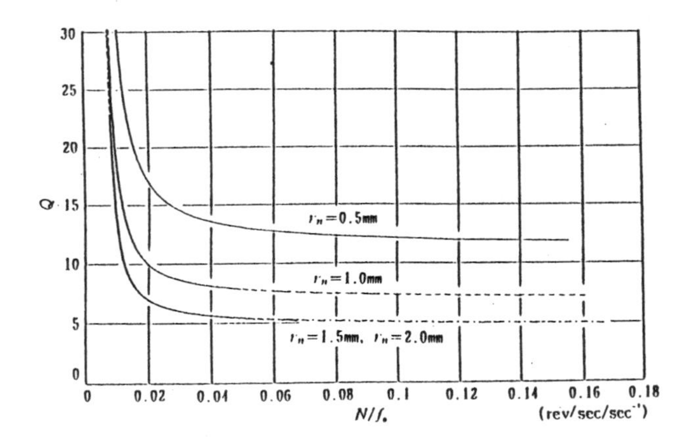

Figure 9. Relation between Q and r_n ($n=20 \sim n=60$, r_n : changeable)
($d=1.0$mm, $C_e=15°$, $C_s=15°$, $S_o=0.2$mm/rev, $K_A=0.1$kgf/mm² constant)

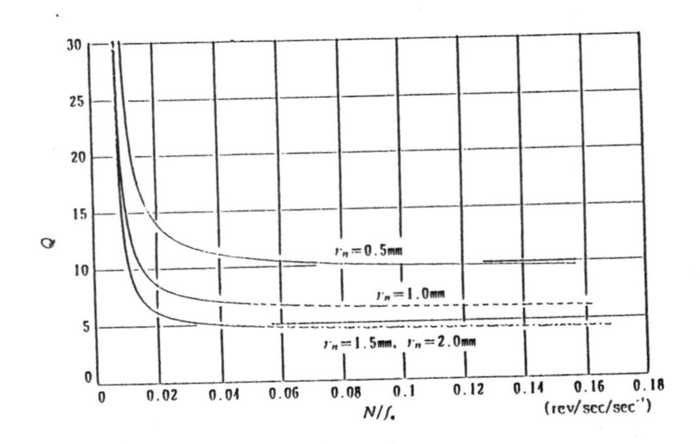

Figure 10. Relation between Q and r_n ($n=20 \sim n=60$, r_n : changeable)
($d=1.5$mm, $C_e=15°$, $C_s=15°$, $S_o=0.2$mm/rev, $K_A=0.1$kgf/mm² constant)

An example of resonance curve by the change of difference of specific cutting resistance in the domain to frequency 10 Hz from1000 Hz is shown in Figure 11 (a) and (b).

CONCLUSION

In this study, a size of bit is taken as the normal side rake angle 5 deg. , inclination of cutting edge 0 deg. , end relief angle 6 deg. , side relief angle 6 deg. and end cutting edge angle 15 deg. are not changed and side cutting edge angle , nose radius changeable. When the work piece is turned in many different cutting conditions, the results as follows:

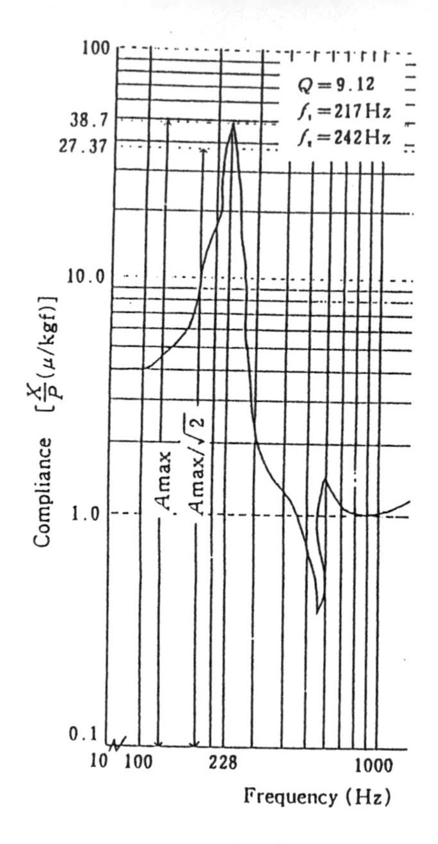

(a) $K_A=-0.10$kgf/mm^2

Figure 11. A resonance curve
(d=1.0mm, $C_e=15°$, $r_n=0.5$mm, $C_s=15°$, $S_o=0.2$mm/rev, $k_{s_3}=109.88$kgf/mm^2, $\theta=65.37°$, $\mu=0.899$, $H_1=0.751$mm, $N=530$r.p.m.)

1. When the nose radius and difference of specific cutting resistance are not changed and side cutting edge angle changeable, the stabilizing threshold of dynamical amplification coefficient decreases with growing the side cutting edge angle.
2. When the side cutting edge angle and difference of specific cutting resistance are not changed and nose radius changeable, the stabilizing threshold of dynamical amplification coefficient decreases with growing the nose radius.
3. When the size of bit and cutting condition are not changed and difference of specific cutting resistance changeable, the stabilizing threshold of dynamical amplification coefficient increases with growing the difference of specific cutting resistance.

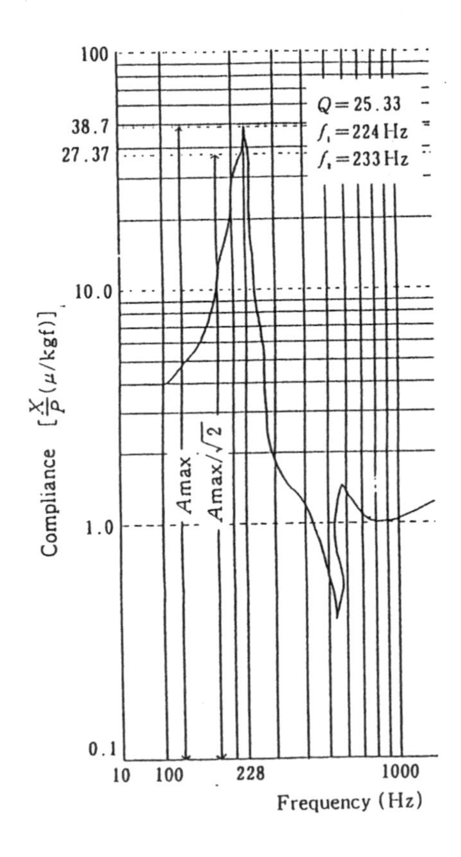

(b) $K_A=0.3$kgf/mm²

Figure 11. A resonance curve
(d=1.0mm, $C_e=15°$, $r_n=0.5$mm, $C_s=15°$, $S_o=0.2$mm/rev, $k_{s_1}=109.88$kgf/mm², $\theta=65.37°$, $\mu=0.899$, $H_1=0.751$mm, $N=530$r.p.m.)

REFERENCES

1. **Iwata, Y, ,** A study of the stabilizing threshold in lathe turning chatter, Proceedings of IXth International Conference on Production Research, 1987, pp 788-794
2. **Iwata, Y, ,** A study of the most suitable cutting condition in lathe chatter, Proceedings of Automation Technology, 1990, pp 387-395
3. **Iwata, Y, ,** A study of machine tool characteristics under the condition of chatter vibration in boring, Proceedings of XIth ICPR, 1991, pp 950-954
4. **Iwata, Y, ,** A study of normal force change under the condition of the variation of normal side rake angle in lathe turning chatter, Proceedings of the second China - Japan International Symposium on Industrial Management, 1993, pp 423-428

SPIN CASTING OF METAL PARTS DIRECTLY FROM RP MASTERS

P M Hackney[1], M Sarwar[1]; S Widdows[2]

[1]Centre for Rapid Product Development
School of Engineering
University of Northumbria at Newcastle
Tel: 0191 2273644/2274589
Fax: 0191 227 2854
http://www.unn.ac.uk/~mfx2

[2]RapidCast (UK) Ltd
Consett
Co Durham
Tel: 01207 500050
Fax: 01207 581617

ABSTRACT: Industry has now become familiar with the ideas of Rapid Prototyping (RP) and emerging Rapid Manufacturing technologies and is using them to reduce product development risks and life cycles (1). A great emphasis has been placed on the production of realistic plastic parts using techniques such as vacuum casting of polyurethane components which have similar properties to injection moulded parts.

Until recently the production of metal parts has concentrated on using traditional techniques, such as sand and investment casting. These techniques are now being replaced by using RP models to produce wooden/metal/wax masters.

The process of spin casting is capable of reproducing the fine detail provided by most RP systems. It offers a high degree of accuracy and repeatability and also minimises post casting operations.

The development of new silicone rubbers which bond at lower operating temperatures allow most RP components to be used as masters to create a mould in which materials such as zinc can be formed.

This paper describes the process of spin casting and how accurate parts can be produced.

INTRODUCTION

No rapid prototype process can produce a component that has the same physical properties of either plastic or metal parts. The purpose of RP machines is to produce one off parts or dies for the creation of production like quality components so that realistic physical, manufacturing and assembly testing may be carried out.

Rapid tooling (soft tooling) is a range of techniques capable of producing low to medium volume production of either plastic, metallic or ceramic parts.

Plastics - vacuum casting, vacuum forming, RIM, Bridge tooling, Keltool and Rapid Tool.

All provide a polymer part or die for production of plastic parts.

As metal parts are typically larger than the economic operation of several RP techniques most development has been focussed on small plastic parts.

Metals - Sand Casting (patterns, matchplates, coreboxes) wax patterns, and investment casting.

These have taken traditional techniques and replaced the traditional masters with RP masters with some process modifications.

Ceramics - slip casting, compaction and isostatic forming.

These use RP masters and initial dies for formation of ceramic inserts.

The process of spin casting takes an RP master made by SLA, SLS, LOM, FDM or traditional processes and creates a cavity in a silicone mould. The mould is then bonded to form a durable mould tool. This tool is then rotated and metal or polyurethane resin spun to reproduce the original master.

This paper describes how the development of spincasting has been used to create:

- Realistic detailed metal components.
- Ability of RP masters to produce rubber moulds capable of withstanding the metal casting process.
- Process optimization to enable the production of reliable and accurate components.

The material being cast is Zinc (ZA1) which is commercially available and widely used for low temperature (400^{0}C) high precision, lightweight components utilised in aerospace, automotive, consumer and toy industries.

SPIN CASTING OVERVIEW

Spin casting is an economical prototyping and medium production process (2) for the production of zinc alloy castings. (95% Zn, 4% Al, 0.9% Cu.) Masters are laid out on a disk of uncured silicone rubber and cavities cut into the silicone to accommodate the part together with cores and pull out sections if required. The mould with the masters in place is loaded into the vulcanising press for curing. Hydraulic pressure clamps the mould shut between the heated platens forcing the silicone into all the crevices and around the details of the master. The heat cross links the uncured silicone and the resulting mould is tough, resilient and dimensionally accurate with good heat and chemical resistance. Undercuts can be accommodated by flexing the mould to remove the part. The masters are removed and gates, runners and air vents cut into the mould. The finished mould is loaded into the spin caster and automatically centred and clamped. When the spin cycle starts, liquid metal is poured into the central feed channel and the centrifugal force pushes the liquid through the mould runner system completely filling each mould cavity. When the metal solidifies, the mould is opened and the parts removed. Gates, runners and air vents can be broken away by hand and the parts usually require no further finishing.

The maximum production rate when casting zinc alloys is around 50 cycles per hour and using a 12 cavity mould this can result in a production rate of 600 parts per hour (3). The average part size possible with this process is approximately 250 x 160 x 90 mm.

Advantages

The process is ideal for prototyping and short run production of die castings. Accuracy is similar to that achieved with vacuum casting (0.1 mm per 25 mm) with excellent surface finish. Tooling and parts can be produced within five days of receiving masters. Large product volumes can be achieved by using silicone rubbers with higher bonding temperatures which require metal masters produced from the prototype moulds. The RP masters required to produce the metal masters need to be scaled "oversized" to increase end product accuracy.

Disadvantages

Rapid prototyping models are not always robust enough to withstand the vulcanising process and an intermediate step of producing more robust masters through spin casting, vacuum casting or plaster moulding is often necessary. This adds to the turn around time of the process and the cost. Post machining is often required to achieve the extreme accuracy often required of castings.

THE PROCESSES

The spin casting route to low and high volume production.

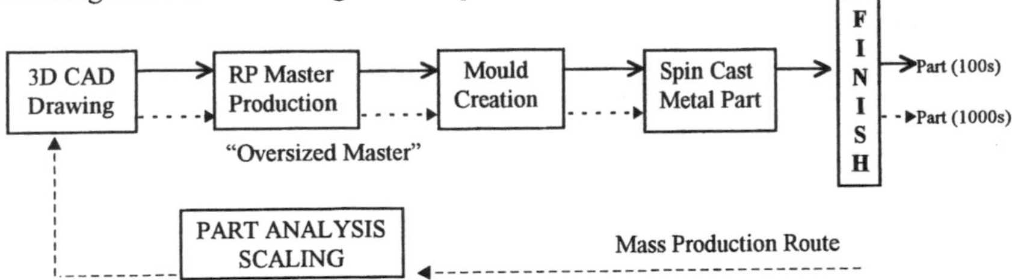

Figure 1. The Processes

Many types of silicon rubbers can be used for production of mould tools. The criteria for selection is the thermal/mechanical durability of the master, the complexity and detail of reproduction and the expected production quantity. The company concerned have collaborated with their suppliers to produce a material capable of bonding at low temperatures and yet having the mechanical properties to produce economic metal parts.

MOULD CREATION

Two methods of mould creation can be used:

Room temperature Vulcanising Rubbers (RTV)

Moulds can be produced using room temperature vulcanising rubber using the following stages (4). Fig 2.

Stage 1. A mould box is made and the RP master suspended inside. The complex parting line is established using modelling clay. Pull cores in RTV rubber or Teflon are made where necessary. When completed, a parting compound is sprayed on the RP models, RTV cores and clay.

Stage 2. After the RTV material has been catalysed and de-gassed, it is poured into the frame to form the top half of the mould and allowed to cure at room temperature. Mould pins placed around the perimeter act like pins of a die for locking thus preventing the mould from shifting.

Stage 3. After the top half of the RTV mould has cured, the clay is completely removed. Parting compound is again spayed on the cured RTV rubber and the models. The other half of the mould is now created by pouring catalysed and de-gassed RTV mould material over the first half of the mould.

Stage 4. After the other half of the mould has cured the RTV is now completed. It is removed from the ring frame and separated at the pre-established parting line. Gates, runners and air-vents are easily cut into the rubber using simple hand tools.

Stage 5. The RTV mould is assembled and slid into the spincasting machine. As the mould spins, 400^0C zinc base metal alloy is poured down the mould's centre sprue section. Centrifugal force pushes the liquid metal through the mould's runner system, completely filling each and every section, corner and detail of each mould cavity.

The disadvantage of RTV rubbers is their low mould life, low rigidity and labour intensive method of production. Heat cured silicone materials exhibit better properties and have been used by RapidCast UK for many years for the production of moulds from hard masters.

Heat cured rubbers:

Moulds produced from heat cured silicone rubbers are created by the following stages: Fig 3.

Stage 1. The heat-cured silicone moulds are composed of several disks of uncured silicone material. In the uncured state this material is soft and pliable and has the consistency of clay. Cavities are cut and shaped by hand to accommodate the parts and to form the complex parting line locations. Pull cores are also shaped at the same time. Both the top and bottom halves of the mould are prepared simultaneously.

Stage 2. Parting compound is sprayed to ensure easy opening of the mould after vulcanisation. Top and bottom halves are placed together, with small anti-shift lock nuts positioned around the perimeter of the mould. The unvulcanised mould is placed inside a vulcanising ring frame, which is slid into the vulcanising press and cured by heat and by hydraulic pressure.

Stage 3. After vulcanisation the top and bottom halves of the mould are easily separated. At this point the mould's gating and runner systems are cut in a similar way as with the RTV moulds.

Stage 4. The mould is then placed together and slid into the spincasting machine. While the mould spins, a high strength zinc alloy (ZAMAK) ZA1 is poured into the mould at 425^0C to produce a large series of precision zinc parts. As the mould material is resistant to high temperatures it can be used for up to 200 cycles.

Stage 5. The mould is removed from the machine and opened. The wheel of material comprising of feed sprue, runners and components is then pressed out.

The sprues and runners are easily removed at the gates and recycled in the melt pot. The mould is coated with a release powder and reused.

EXPERIMENTATION

The experimentation involved the creation of a set of test components which could be scientifically evaluated at different operating parameters.

The test specimens chosen were tensile test pieces, and specimens with varying length to width ratios. These were then positioned around the mould at various radial distances and angles.

The moulds were tested at spin speeds up to 1000 rpm with various clamping pressures and metal temperatures. In total over 600 samples were taken and evaluated.

The specimens were analysed for:

- dimensional accuracy of the parts in three planes in relation to spin direction.

- casting defects such as flashing, incomplete filling of results, porosity etc.
- tensile strength of specimens.
- microscopic analysis of cast structure.
- Hardness of castings.

Theoretically the centrifugal forces causing mould filling are defined as

$$F = mR(2\Pi n)^2$$

where F = Force (N), m= mass (kg), r = radial distance *(m)* = spin speed (rps).

RESULTS

For the range of components, die design and mould materials utilised the following results were established.

Dimensional Accuracy

When the test pieces were analysed it was found that the accuracy increased as the spin speed increased up to an optimum point. After this point the number of casting defects due to flashing and incomplete mould filling rose significantly. This is due to the hardness of the rubber. It was evident that larger casting only required lower spin speeds to achieve the same results as for the test specimens.

It was found that dimensional accuracy was within -0.2 mm per 25 mm. Fig 4. The process was found to have a standard deviation of 0.1 mm proving that if "accurate" patterns are used then accurate parts can be reliably produced.

Casting Defects

The clamping pressure and the spin speed effect the amount of flashing and mould sizes. At speeds above 800 rpm the rubber moulds locally flashed material causing rejects. At low spin speeds incomplete filling occurred. Fig 5 & 6.

Porosity was found on the inner edge of the large test pieces Fig 6. This was caused by the centrifugal action and the gating arrangement to feed the casting.

Tensile Strength

All specimens exhibited a small increase in tensile strength and hardness with increase in spin speed with no maximum point being found. The tensile strength at 625 rpm was 150 N/m^2 and Brunell hardness of 125 (VPN 2.5 kg). Fig 7.

Microscopic Examination

Samples were mounted, etched and examined at various magnifications. The grain structure was found to be homogeneous in all areas except those shown in Fig 6 where porosity was evident.

The castings all had regular, small, equal axial grain structure, proving the homogeneity of the castings.

CONCLUSIONS

Spin casting can produce real metal functional parts from RP masters. However the process and the materials currently being used require experienced casters, moulders and mould makers, and further development.

The tests have found the spin cast process provides:

- Typical error of ± 0.2 mm per 25 mm
- Excellent repeatability with a typical standard deviation of 0.1 mm
- Parts which give excellent property values with fine grain structure and 120 BHN typical giving material specification 20% better than ZA1.
- A process which can operate from solid or 'soft' masters including SLA, SLS, FDM to produce high quality parts "rapidly".
- Industrial quality zinc alloy castings from 1 to 100s in days at economic cost. Larger volumes can also be accommodated.
- Size limitations currently X 240 mm Y 130 mm Z 70 mm (All sizes maximum working parameters).

The process has an excellent repeatability, resulting in the ability to make RP masters "oversized", producing zinc parts for low production numbers (<1000) within a period of days.

Higher production can be achieved with higher temperature rubbers but this would require an "oversized x 2" master to produce metal masters for production mould creation.

The above process provides new opportunities, but requires further development for the full potential of this process to be utilised.

REFERENCES

1. Adaption of Time Compression Technologies into Small and Medium Sized Enterprises, Gill Morey, Materials Information Services, tct '96, Gaydon UK.

2. Rapid Tooling: A Review of the Alternatives, Phil Dickens, University of Nottingham, TCT '96, Gaydon UK

3. Spin Casting Assists Automotive Product Designers in Developing Fulling Function Metal and Plastic Test Parts from RP Models, John Pauwells, Symore, Leuven, Belgium, TCT '96, Gaydon UK.

4. The Fine Art of Perfect Reproduction, Dan Corning, 1988

5. Six Steps to Mould Production, Teckcast.

ACKNOWLEDGEMENTS

Collaboration and assistance provided by RapidCast (UK) Limited, Unit 39, No 1 Industrial Estate, Medomsely Road, Consett, Co Durham, DH8 6TW.

Figure 2. Room Temperature Mould Preparation

A TWO PART BLOCK MOULD

Step 1
Prepare the original.

Step 2
Prepare a moulding box with sides, base and lid of wood, plastic or metal making sure there is a minimum of 15 mm clearance all around the original.

Step 3
Seal the edges of the box with plasticine.

Step 4
Wrap half of the original in aluminium foil (the half that will be at the bottom when the original in positioned in the box) and position it in the box resting on 1 cm high wooden blocks. Pour melted plasticine into the box until it comes half way up the sides (to the level of the aluminium foil wrapping).

Step 5
Insert location pegs into the plasticine.

Step 6
Prepare the moulding material (see page 6) and pour it into the box holding the container as low as possible. Pour slowly to allow the material to fill all the crevices and level off. Continue pouring until the top of the silicone material is a minimum of 10 mm above the highest point of fhe original.

Step 7
Leave to cure for 24 hours at room temperature.

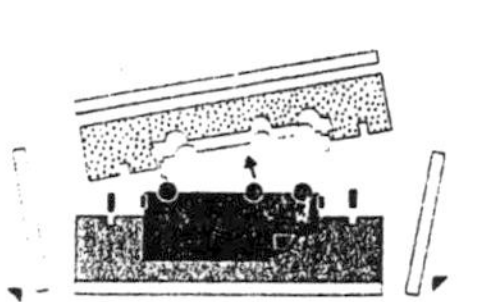

Step 8
Invert the box and disassemble it. Remove and discard the plasticine, the aluminium foil, the 1 cm wooden blocks and the location pegs.

Step 9
Reassemble the box and apply a release agent to the exposed area of silicone.

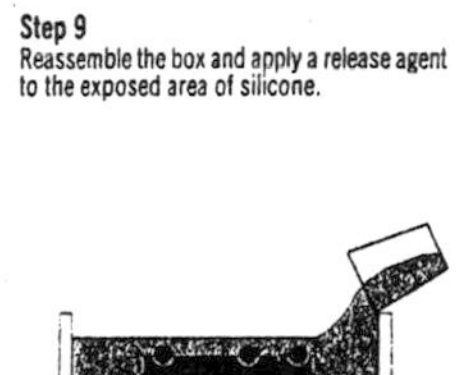

Step 10
Holding the container as close to the box as possible, pour in the silicone material. Pour slowly to allow the material to fill all the crevices and level off. Continue pouring until the top of the silicone material is a minimum of 10 mm above the highest point of the original.

Step 11
Leave to cure for 24 hours at room temperature.

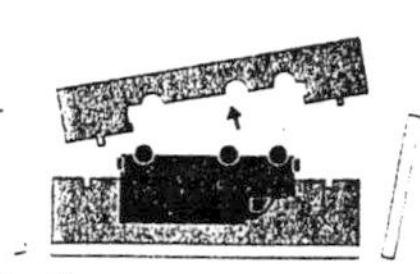

Step 12
Disassemble the box and separate both parts of the mould from the original.

Figure 3. Heat Cured Mould Creation and Production

1. Preparing Mould

2. Vulcanising Mould

3. Gating and Venting

4. Placing Mould in Spin Casting

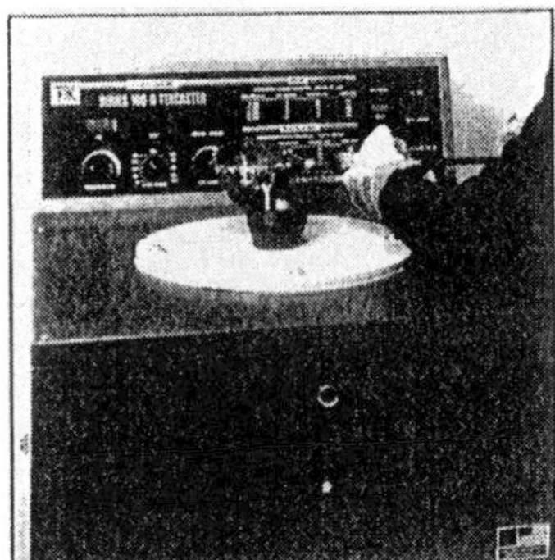

5. Pouring and Spin Casting

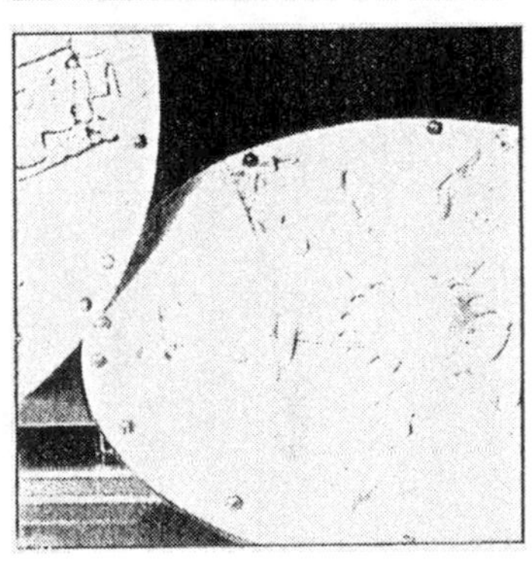

6. Removal of Spin Cast Parts

Figure 4a. Error versus mould position

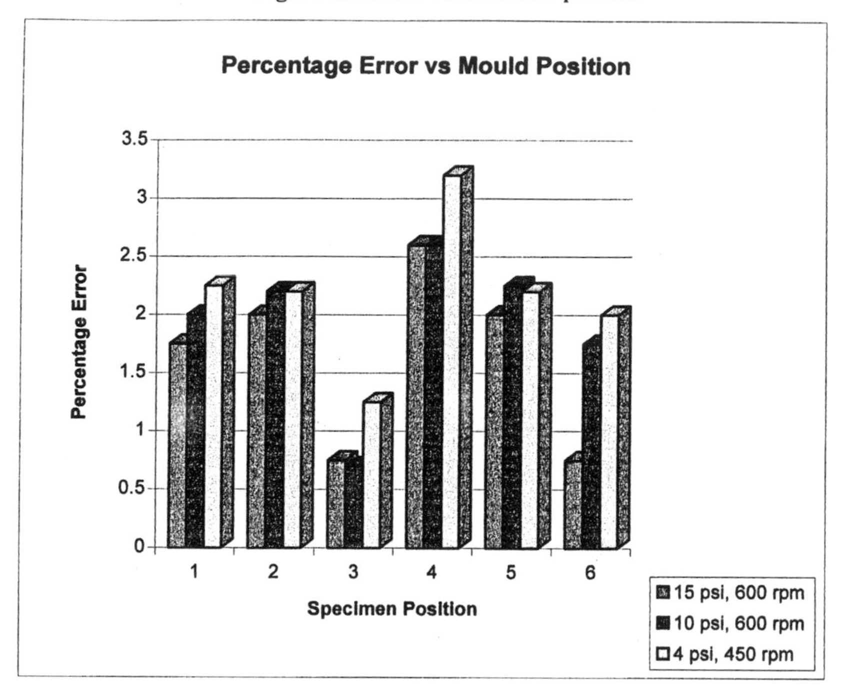

Figure 4b. Error versus mould pressure and spin speed

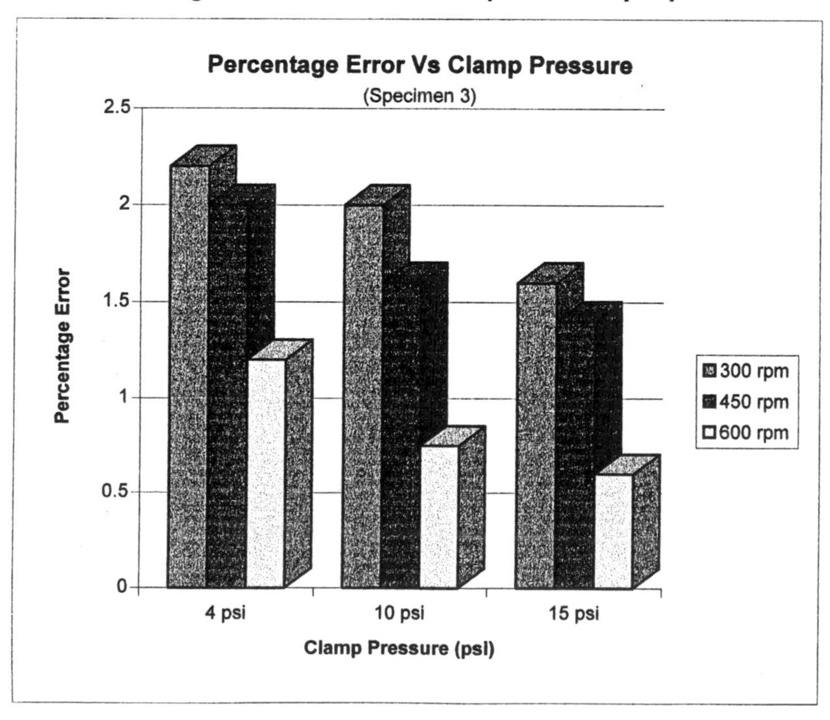

Figure 5. Zinc crystals and mixing alloys filling voids between crystals

Photograph 8
Large specimen at position middle II. Magnification 400x
Shows zinc crystals and mixing alloys filling voids between crystals

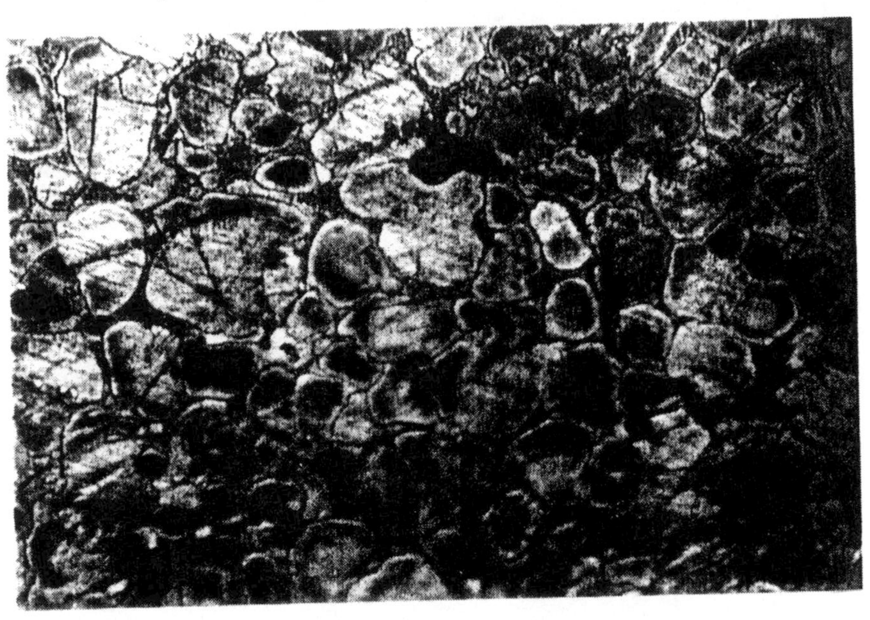

Figure 6. Porosity in Centre of Large Specimen

Photograph 2
Large specimen, position middle I. Magnification 100x.

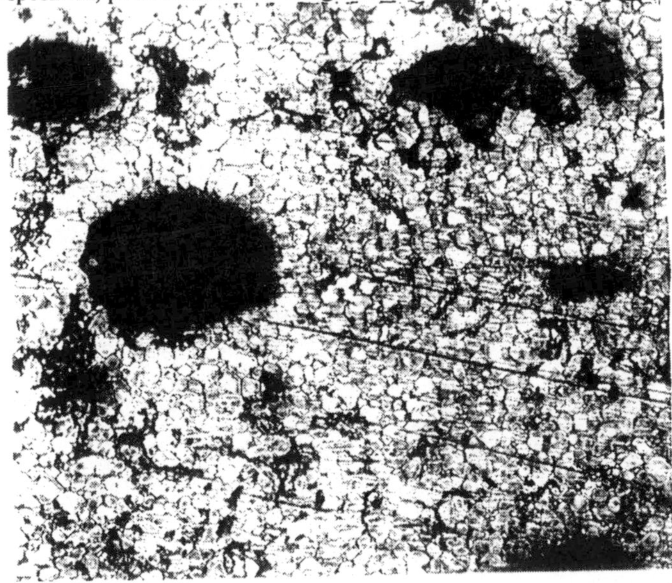

Shows porosity large dark voids, grain structure around voids regular and fine.
Porosity extended for a region of 50% of the middle section.

Figure7. Tensile Strength and Hardness

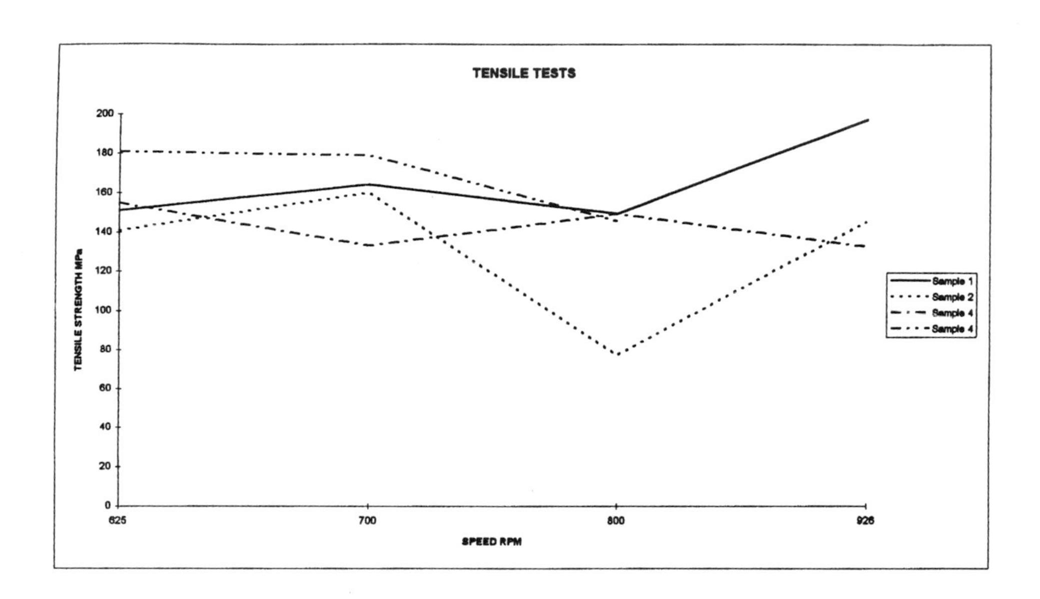

Figure 7 (cont). Tensile Strength and Hardness

Hardness of Blocks (VPN 2.5Kg)

Test Number	Process conditions	Position	Spot	Middle	End	Average
			Position			
1	625 rpm / 20 psi	i	95	60	143	
		ii	145	104	142	123
		iii	138	138	138	
2	700 rpm / 20 psi	i	115	64	69	
		ii	121	137	131	117
		iii	133	142	138	
3	800 rpm / 75 psi	i	93	80	138	
		ii	137	123	143	127
		iii	134	148	143	
4	926 rpm / 75 psi	i	91	49	146	
		ii	135	145	143	126
		iii	143	146	135	
5	small block @ 625 rpm / 20 psi	i	135	140	99	
		ii	135	99	206	136
		iii	126	126	155	

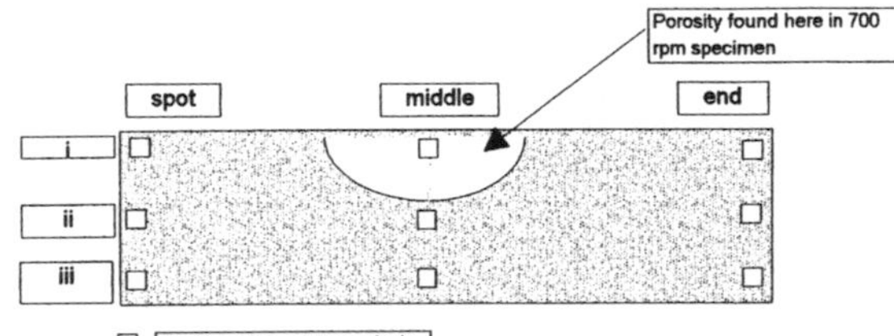

THE PRODUCTION OF OVER 2000 INJECTION MOULD IMPRESSIONS OF AN SL MASTER USING COPPER ELECTROFORMED INSERTS

C.Bocking[1], G.Bennett[2], S.Dover[2]
GEC[1]; Buckinghamshire College[2],

ABSTRACT: A polyurethane model derived from a stereolithographic master was used as a mandrel for copper electroforming. The electroform then acted as inserts for an injection moulding tool. Over 2000 impressions of glass filled polypropylene were made and measurements of their dimensional accuracy are given. A second set of electroformed copper inserts were made directly from a stereolithographic mandrel.

1. INTRODUCTION

Rapid Prototyping is becoming a well established technique. Adapting well to concurrent engineering concepts[1], it is an ideal process for the production of one-off physical prototypes for various processes. A number of downstream processes are available to produce varying numbers of replicas of the SL part with varying lead times, costs and accuracies[2]. A relatively new method was investigated with the intent to quickly produce a prototype injection moulding tool (to give good replication of material properties) that can withstand the production of several thousand impressions, with little manual intervention and at low cost. The process chosen was to electroform copper onto an SL part (or a replica) to produce tool inserts. These were then to be backed up with an aluminium powder filled epoxy resin and bolstered with aluminium.

The part chosen to test the process is shown in figure 1.

2. SHELL FORMATION USING ELECTROFORMING.

Electroforming is a process by which solid structures are produced by electrodepositing thick layers of metal onto a model, known as a mandrel. Deposit thicknesses of the order of millimetres are produced such that when separated from the mandrel, a freeform structure remains. The surface features, geometric structure and dimensions are directly reproduced within the electroform as a reverse of the mandrel. When used to produce injection mould cavity shells, it is necessary to arrange that the electroform is produced in two parts, which comprise a reverse representation of each half of the original model. In addition, a parting plane has to be produced as an integral part of each shell.

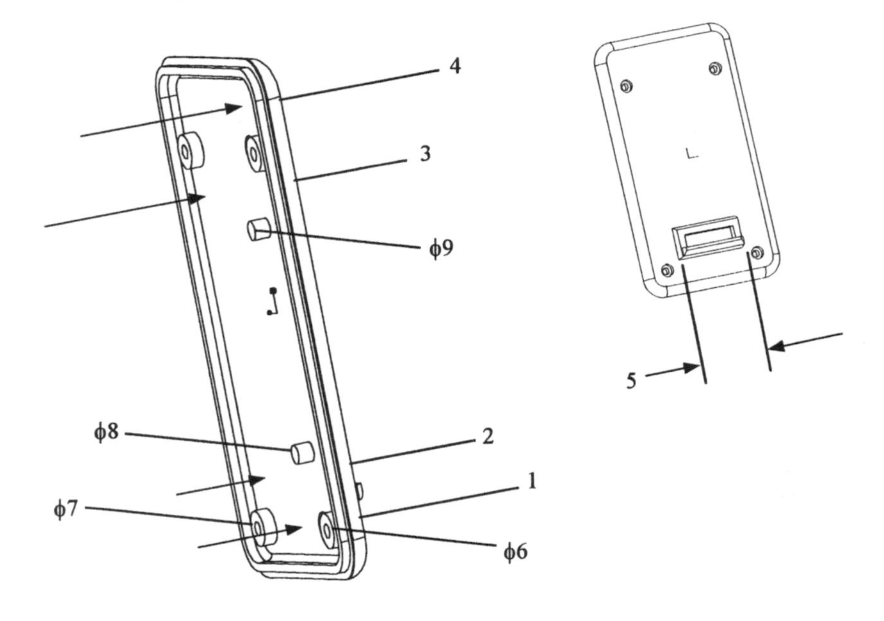

Figure 1. The pattern used as a mandrel for electroforming and the dimensions measured an the subsequent injection moulding impressions.

When using non-conducting polymer mandrels, the surface has to be made conducting. There are several ways in which this can be achieved such as the use of special silver loaded paints, magnetron sputter coating, silver reduction spraying (as used in the production of mirrors) , coating the surface with graphite powder or direct methods. Direct methods include depositing "seed" layers of palladium particles followed by electroless copper or nickel that build up a thin conducting layer. In this work, silver loaded paint was used for the initial metallisation process.

There are a number of ways in which the two shell halves can be produced. The method used will largely dictate the way in which the parting plane is introduced.

3. INSERTION OF THE PARTING PLANE.

In order to produce a satisfactory injection mould cavity, it is important that each cavity half is furnished with a parting plane that is as close a match to the other half, otherwise excessive flash and inadequate mould filling will occur. However, a perfect fit would not allow air to escape from the mould during the injection process. This does allow a certain flexibility in the fabrication process when designing the model for electroforming. There are a number of strategies that may be used to introduce a suitable parting plane and the each controls the methodology used for the electroforming process. The strategy used is also dependent on the design of the model.

1. Each half of the shell can be produced separately from the other. In this case, only one half of the model is metallised and the electroformed shell produced. The parting plane is introduced by fitting the model into a flat sheet of another material such as perspex in such a way that allows the face of the perspex to act as a mandrel for the parting plane. Both the perspex face and model half are metallised and electroforming is carried out until the shell reaches an appropriate thickness. Subsequently, the perspex is separated from the electroform, but leaving the model in place. The remaining half of the model is then metallised and electrodeposition is continued so that the second half of the shell is produced. To prevent this second half of the shell adhering strongly to the first parting face, the face is treated in such a way as to prevent strong adhesion but still remain conductive. During the production of the second half of the shell, the first half can be masked or isolated from the plating electrolyte if further build up of the first half is undesirable although this is not essential.

After the second half of the shell has reached the desired thickness, the completed structure is rinsed and dried and any masking material removed. The model can then be separated from the shell using suitable thermal cycling, which makes use of the difference in thermal expansion between the model and the electroformed shell. this effectively splits the two halves. Alternatively, the shell halves can be bolstered using metal filled epoxy resin as described later and then split. The main advantage with this method is that the parting face on each shell half provides an almost perfect fit to the other. However, by plating one half at a time, any excessive internal stress within the deposit can distort the model, particularly thin sections, so that careful choice and control of the electrolyte is essential when using this method.

2. An alternative approach is to build into the model a suitable parting plane. this requires that the model is built oversize by the thickness of the parting plane. The entire model is then metallised, except for a region around the edge of the parting plane (to facilitate subsequent separation). The entire model is then plated to a suitable thickness. Separation of the model from the electroformed shells is then carried out as before. This method does suffer from the disadvantage that each half of the parting face will be slightly different and thus not give a very good fit. Differences in surface finish on each half of the model parting face (such as "staircasing" from the SL process) will exaggerate the imperfection further. However, the method does offer the advantage that any internal stress that could cause distortion is compensated for by plating both halves at the same time.

3. In a situation where the parting face has to be curved, the use of the above strategies is precluded. It would defeat the object of rapid tooling if a curved parting plane had to be machined from a material such as perspex. The inclusion of a parting plane within the model design as used in (2) would not be helpful as both the parting planes produced would have a slightly different radius of curvature and would not fit well. A curved parting plane requires the use of a modification of method (1) whereby a strip of thin (approx. 0.5 mm) modellers plastic sheet is attached to the model where the parting plane is required. Such a material has the required flexibility to form a curved surface and the electroformed shells are produced as in method (1) with the sheet being removed prior to electroforming the second half. In this way, slight undulations and variations in curvature are compensated for as the second parting face is grown directly on the first.

Whilst each methods 1 and 3 require some manual intervention in the fitting of the parting plane, timescales of production are not significantly increased.

For this work. Method 1 was chosen as it was the most suitable for the design of the part.

4. ELECTROFORMING CONDITIONS.

For this work, copper from the acid sulphate electrolyte was chosen. This was because this electrolyte operates under ambient temperature conditions and by use of suitable additive, the internal stress of the electroforms could be maintained close to zero or slightly compressive. This minimised the possibility of distortion of the electroform during the formation of the first shell. The electrolyte made use of proprietary additives that have been used extensively for the production of distortion free electroforms for use in the aerospace industry. Not all proprietary processes are suitable for this type of electroforming as internal stress can vary significantly during the lifetime of the electrolyte. The system used was the result of many years of research into copper plating additives from many sources and represents the most stable system studied.

Metallisation of the model was carried out using a high silver content thermoplastic paint designed for electroforming applications. In this work, the coating was applied by brush although spray application can be used. The advantage of this method is that during the model removal at elevated temperatures, the thermoplastic melts and acts as a kind of lubricant, aiding the separation of the model from the shell.

Electrodeposition was carried out at a current density of 20 mA cm^{-2} for a period of 80 hours for each half of the model. During deposition , the surface area of the model gradually increases as the electroform builds in thickness and the total applied current was increased to compensate for this. The concentration of additives was maintained by continuous dosing. The rate of consumption of the additives had already been predetermined by previous experience and was based on the number of coulombs of electricity passed.

Each half of the model was plated with an average thickness of 4 mm of copper although the thinnest region was of the order of 2 mm thickness.

5. BOLSTERING OF THE INSERTS

The copper inserts were backed with (vacuum dearated) DEVCON™, an epoxy resin filled almost to saturation with a fine aluminium powder, one at a time as shown in figure 2.

The copper inserts were sealed with an epoxy glue to prevent the DEVCON™ from adhering to the insert tool face. The plastic planes were inserted to separate one DEVCON™ backing from the next due to the possibility of the two backings sticking together. This had the effect that the entire clamping force was applied through the copper parting plane. The backed inserts were then removed from the mould box and the plastic plane, machined and bolstered with aluminium plates.

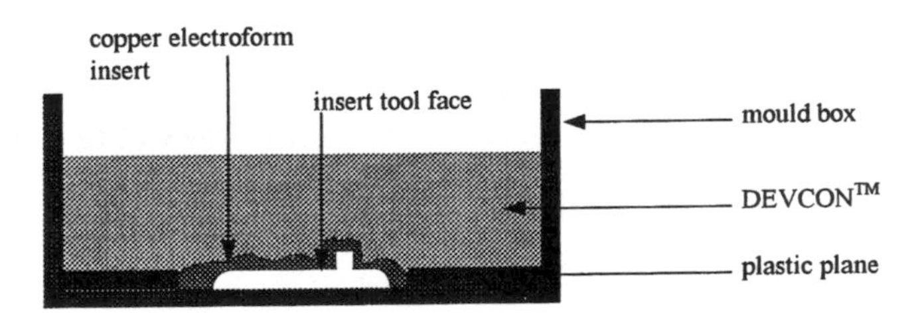

Figure 2. The backing apparatus for the electroformed inserts.

6. ADDITION OF SPRUE AND EJECTOR PINS

A sprue bush and 3 ejector pins were added. Unlike the holes for the ejector pins, the hole for the sprue bush was drilled and reamed from the back of the aluminium plate, through the DEVCON™ and then through the copper. It was noted that the drilling and subsequent reaming operations pushed the copper away from the DEVCON™ in a small area around the hole. This shows that even though the backs of the copper inserts were slightly rough and nodular, sufficient adhesion was not achieved between copper and DEVCON™. This meant that the sprue side of the tool had to be lightly sanded and polished, increasing the lead time and decreasing the accuracy of reproduction of the mandrel. The holes for the ejector pins were drilled and reamed from the other direction and, apart from small burrs, no deformity was noticed.

7. RESULTS OF INJECTION MOULDING

The tool was injected at Henley Mouldings of High Wycombe using a Negri Bossi NB80 machine. Glass filled (>10%) polypropylene, including some reground material, was used with a clamping pressure of 14.4 - 16.0 metric tonnes and an injection pressure of 25 bar at a temperature of 208 - 220°C. The cycle was semi-automatic with a cycle time of approximately 22 - 25 seconds. The first 100 impressions were packaged in groups of 20 with subsequent impressions being batched in groups of 100. Four impressions from selected batches were measured with nine measurements made on each measured impression. These measurements are shown in figure 1 and the resulting graphs are shown in figures 3 to 6.

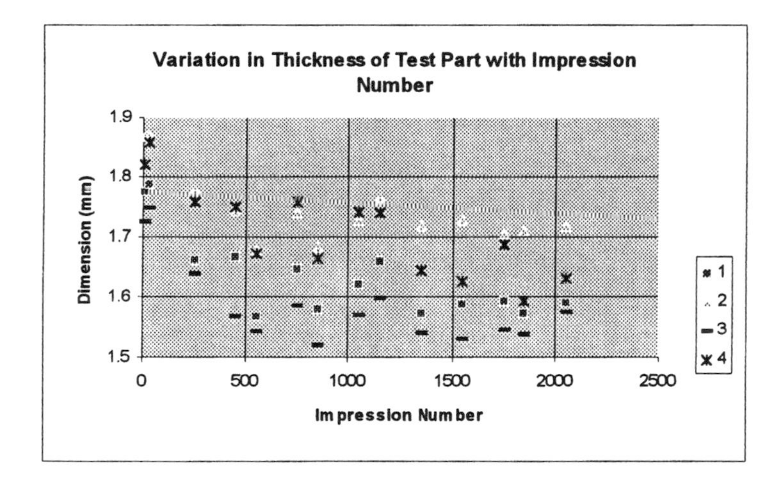

Figure 3. Variation of the first four measurements (as shown in figure 1) with impression number.

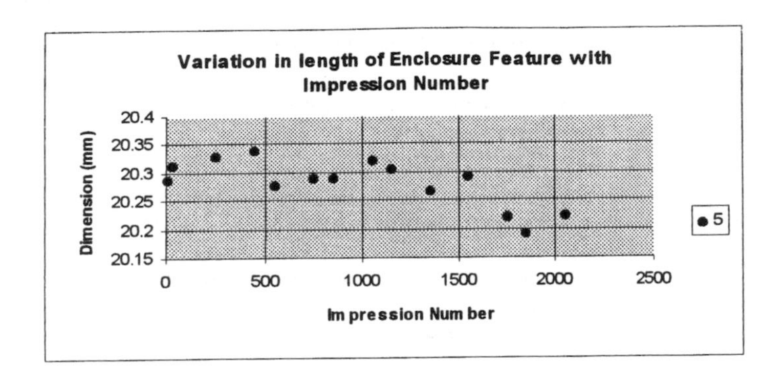

Figure 4. Variation of the fifth measurement (as shown in figure 1) with impression number.

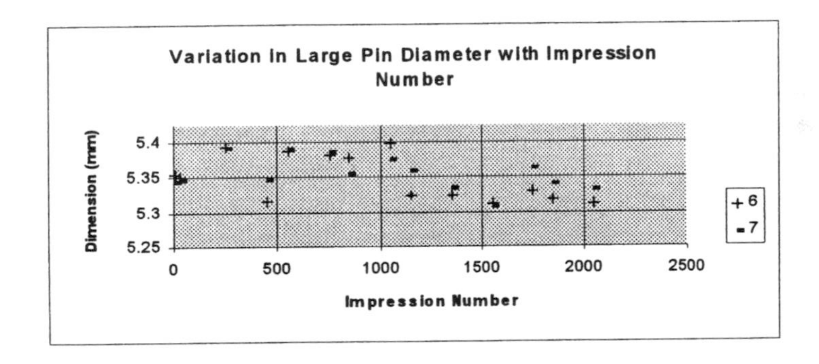

Figure 5. Variation of the sixth and seventh measurements (as shown in figure 1) with impression number.

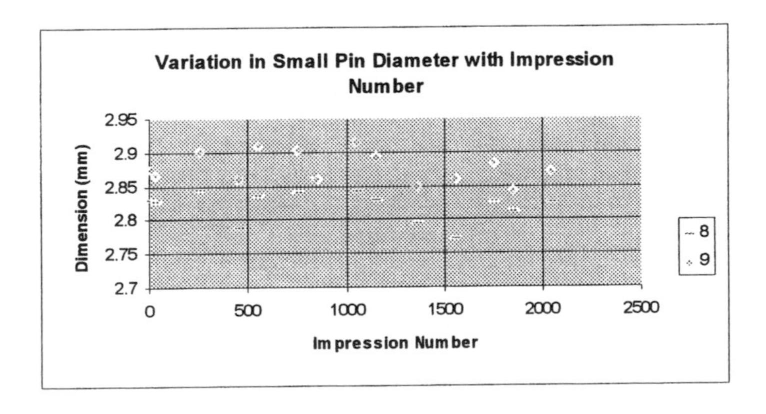

Figure 6. Variation of the eighth and ninth measurements (as shown in figure 1) with impression number.

8. ELECTROFORMING ONTO AN SL PART.

A number of shells have been made using SL models. The procedure is similar to that of the polyurethane model although model removal is more difficult. Polyurethane is less brittle than the epoxy SL model and is removed easily. The SL models tended to fracture during removal. As a consequence, the shell halves were electroformed and the backing of DEVCON™ cast around them prior to splitting the shells. This gave a greater resilience to the shell during the model removal.

One of the shells produced was that of a component that had been used as part of a rapid tooling project conducted by members of the European Action on Rapid Prototyping (EARP). A number of different RP methods were used to produce models for the production of aluminium cast injection mould cavities. None of the methods were able to accurately produce a suitable cavity. However, within 4 days of receiving a working copy of the STL file, an SL model was made, and an electroformed shell insert had been made that faithfully reproduced the structure of the original model.

9. CONCLUSIONS

The measurements of the impressions showed no appreciable tool ware. The only definite trend that can be discerned from the measurements of the impressions is shown in figure 3. The four graphs in figure 3 correspond to four measurements, roughly in each corner of the parts. They show that the thickness decreased to a limiting value. Tool wear (induced by the abrasive glass filled polypropylene) of the tool face would produce a thickening of the impressions. The measured variation was later discovered to be due to a slight misalignment of the tool faces causing one tool face to rest on a small (~1mm wide) ridge which subsequently deformed under the clamping pressure.

The other graphs show no obvious trend other than a seemingly random distribution of approximately ± 0.05mm. This could be attributed to various factors including the varying quality of the feed stock (some of the material was 'reground'). However, the main reason was thought to be the low injection and clamping pressures (necessary due to the nature of the tool) which often lead to variability in dimensions.

The problem of the insert being pushed away from the bolstering material could easily be solved by making the back of the inserts rougher. This could readily be achieved by increasing the electroforming current density at the end of the process.

The results obtained suggest that this method is capable of producing well over 2000 injection moulding impressions from an SL master in an abrasive material at a relatively low cost. Initial cost comparisons with equivalent technologies suggest cost savings of the order of a third to a half the nearest competing rapid tooling technology. In addition, turnaround times of the order of 2 weeks from CAD to final tool would be expected. Furthermore, the electroforming process reproduced the surface of the mandrel to an accuracy much greater than all RP processes can achieve.

Further scope exists to create conformal cooling channels between the copper and the bolster or even embedded within the copper electroform as well as the substitution of copper with nickel electroforming allowing a higher hardness and faster rate of metal deposition.

10. REFERENCES

1. **D.G.Cheshire, M.A.Evans, P.W.Wormald**, "An investigation into the feasibility of rapid prototyping technologies being adopted during industrial design activity," First National Conference on Rapid Prototyping and Tooling Research, 1995, p 1-10.
2. **J.C.Male, G.Bennett, H.Tsang**, "A Time, Cost and Accuracy Comparison of Soft Tooling for Investment Casting Produced Using Stereolithography Techniques," 7th Solid Freeform Fabrication Symposium, Austin, Texas, 1996.

Economic Evaluation of Rapid Prototyping in the Development of New Products

Guangming Zhang, Mark Richardson and Rena Surana
Department of Mechanical Engineering & Institute for Systems Research
University of Maryland, College Park, 20742 U.S.A.

ABSTRACT

This paper presents results obtained from a study of cost structure and cost estimation for applying the rapid prototyping technology in the process of developing new products. The laser-based rapid prototyping technology is creating a unique platform for the production of physical models using the solid free form fabrication technology. However, the high initial equipment investment calls for a careful balance between economic gains and losses. In this study a cost hierarchical structure is proposed using a combined industrial and engineering approach, which decomposes the process of rapid prototyping into four stages: 3D solid modeling, data preparation for free-form fabrication, part building, and quality inspection. A case study utilizing the stereolithography process to build physical prototypes for constructing 3D tactile graphics, is presented to demonstrate the cost structure and to justify the economical feasibility of using rapid prototyping in the process of developing new products.

I. Introduction

In the globally competitive market place, innovative technologies play a unique role in the development of new products. The laser-based rapid prototyping technology provides a platform for the production of physical models using the solid free form fabrication technology. The availability of physical prototype offers great opportunities to evaluate the product design. The enhancement of manufacturability in the early stage of product development improves product quality, shortens the time to market, and signifies great potential to increase profitability.

Traditional methods to produce prototypes include method of wood making fabrication, rubber molding, vacuum casting, etc. However, machining has been one of the major fabrication processes of making prototypes. The versatility of computer numerically controlled (CNC) machine tools in producing geometrically complicated features offers industry a unique tool to fabricate prototypes for design verification and cost evaluation. The accuracy of numerically controlled (NC) machining stands out. The disadvantage associated with the method of NC machining is the lead time. The time needed to complete the machining operation starts from preparing cutting tools and selecting fixtures. In addition, NC machining has limited capability of dealing with geometrical complexity characterized by a wide range of surface curvatures.

Attempts to generate physical objects directly from CAD data without part-specific tooling or human intervention led to the development of solid freeform fabrication in 1980s. First, with the advancement of computers and CAD systems, 2D design drawings have been gradually replaced by solid modeling in three-dimensional space. The designed object is uniquely characterized by fully-closed and water-tight surfaces using STL data representation. With the rapid advancement of laser technology, the layer manufacturing method is introduced. A designated laser traces the slice geometry on a photosensitive liquid polymer or on sheet material. By solidifying each layer, a physical prototype of the designed solid model is built without any additional tooling. As a result, the time needed to produce physical prototypes is thus reduced from weeks and/or months to days. With innovative technologies and expanding applications, rapid prototyping is now revolutionizing the process of product development in almost every sector of industry.

There has been a significant number of success stories concerning applications of rapid prototyping, such as revenue increase due to early introduction of products to market, and the strengthening of concurrent engineering of design and manufacturing. Although rapid prototyping is a fast-growing industry, its acceptance by industry in the product development cycle has been slow. High initial capital investment in equipment raises the issue of affordability for companies to purchase rapid prototyping equipment for in-house applications. Most small and medium sized companies are shied away by high prices charged by engineering service bureaus for rapid prototyping. The uncertainty and high risk on the rate of investment and capital return have built barriers between the technology providers and the technology users. To promote the emerging technology in industrial applications, there is a pressing need to provide a procedure to perform cost estimating and cost analysis of the rapid prototyping technology.

This paper presents a study which aims at enhancing the potential for efficient utilization of the rapid prototyping technology. The complexity of rapid prototyping is characterized by the requirement for special skills, methods, and equipment. Recognizing the interaction between CAD solid modeling and the process of rapid prototyping, the study begins with meeting the requirement of 3D solid modeling. As there are very few texts or high technical treatises on the mathematical and statistical aspects of estimating cost items which are directly associated with the rapid prototyping technology, a logical methodology is employed in this study. First, a top-down approach is used to define the major steps involved in applying the rapid prototyping process. Then, a decomposition approach is used to itemize individual engineering steps involved in each stage. The approach calls for estimating labor-hours, material consumption, equipment depreciation, pricing of other work related to the rapid prototyping process, and the accumulation of the total cost. Integration of these two approaches signifies economic evaluation of applying the rapid prototyping technology from the unique systems engineering perspective.

II. Process Flow of Rapid Prototyping

During the past decades, several formats of the rapid prototyping technology have been developed. Representative methods are stereolithography, laminated object manufacturing, and laser sintering. Although these methods employ different approaches to produce prototypes, there are four fundamental steps in applying the rapid prototyping technology. These fundamental steps are 3D solid modeling, data preparation for solid freeform fabrication, the process of part building, and quality inspection of the built prototype(s). The four key steps and the process flow of rapid prototyping are illustrated in Fig: 1.

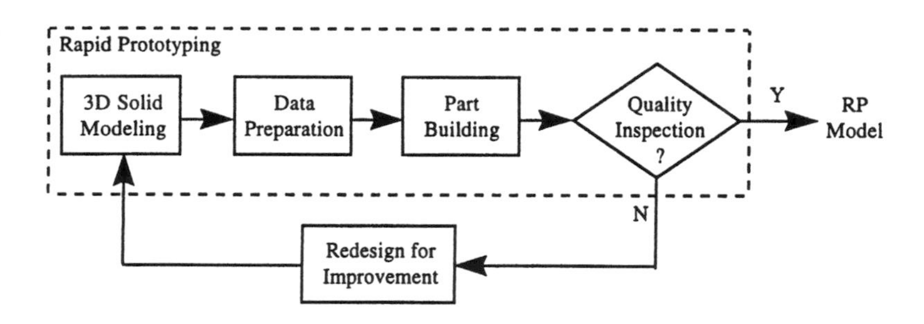

Figure 1 Process Flow of the Rapid Prototyping Application

The architecture of economic evaluation for rapid prototyping applications is built on the process flow. Cost estimation of each of these four fundamental steps is based on the requirement in terms of personnel, equipment and facilities, which are the three main resources that provide companies with the capability of carrying out business activities.

III. Cost Structure and Cost Estimates

Based on the process flow of the rapid prototyping application, an industrial engineering approach to cost estimating is used in this study. Cost elements are identified through the engineering activities involved in each of the four fundamental steps of the rapid prototyping application. Cost analysis is performed to justify the presence of cost elements, and data obtained from an industry survey of fifteen engineering service bureaus are used to calibrate the entire cost structure of the rapid prototyping application.

1. Cost Estimates Related to 3D Solid Modeling

In analyzing the engineering process of 3D solid modeling, three important cost elements are evident. They are the costs of converting 2D drawing to 3D solid models, recreation of cosmetic features, and adjustment of the dimensions associated with tolerances.

(1) Cost Related to Conversion of 2D Drawings to 3D Solid Models

The method of solid free form fabrication requires the contour of a layer to be in a closed format to drive the building process. In this respect, CAD solid modeling plays a critical role when applying rapid prototyping. A significant number of companies have adopted 3D solid modeling in their product development because they recognize the importance of concurrent engineering. Solid modeling allows them to model and prototype a new design - a design which may not even work - long before documenting and implementing it. For those companies, there is no cost involved because engineering designs begin with the process of 3D solid modeling. However, cost related to 3D solid modeling will be added to the new product development for those companies where 2D drawings are the documenting format. The two cost elements for converting 2D drawings to 3D solid models are:

Equipment Costs. They are related to the purchase of computer(s) and CAD software which can be very high and should be entered into the overall cost estimate of a rapid prototyping application. In general, these purchased items are also used for supporting other engineering tasks in the company. It would be appropriate to account these equipment costs as one of the important elements in overhead costs, which also include the expenses related to maintaining and operating these purchased items.

Labor Cost. It has been one of the major cost elements in doing any type of work. Solid modeling requires the special skills of operating computer(s) and utilizing software. Additional expenses to pay for design engineers with high wages are expected.

A survey of 15 engineering service bureaus in the Eastern region of the United States indicates that an average charge rate for converting 2D drawings to 3D solid models is $40 per hour with an overhead rate of 35%. Thus, for 10 hours of work, a total cost of $540 is expected.

(2) Recreation of True Geometric Shape(s) as Replacement of Cosmetic Feature(s)

For those companies where engineering designs begin with 3D solid modeling, there would be no need to convert 2D drawings to 3D solid models. However, cost related to the rapid prototyping application often occurs. A typical cost item would be the cost related to recreating cosmetic features of certain components in the design stage. One of the representative examples would be designs with threads. Based on the ISO 6410 and ANSI standard, threads are drawn in dashed lines and specialized symbols, respectively. It has become a convention in engineering design and has been implemented in most of the commercial CAD systems. However, such a cosmetic feature representation does not fit the requirement of solid modeling for the rapid prototyping application. Replacement of the cosmetic feature representation by true geometric shape(s) of the thread on the designed object is required. Figures 2a and 2b present a pair of two components coupled with external and internal threads. True geometrical shapes of these threads, in terms of thread form, thread pitch, location of starting surface, and the total thread length, have to be created accurately.

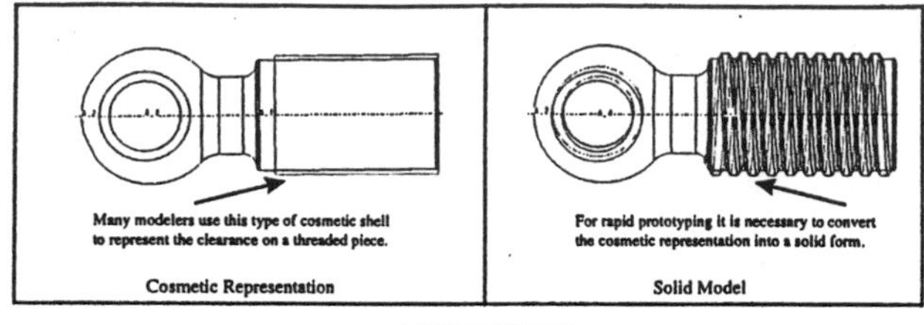

(a) External Threads

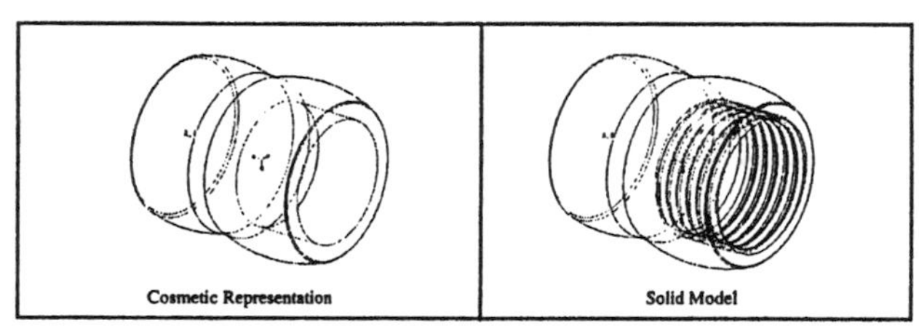

(b) Internal Threads

Figure 2 Cosmetic Features Versus Solid Models

(3) Dimension Adjustment to Accommodate Tolerance Requirement.

Additional costs may also be needed for adjusting dimensions related to assembly. When prototypes of two or more parts are being made, dimensions associated with tolerances have to be adjusted to reflect the design intent. The two components with threads shown in Figs. 2a and 2b serve as an example. Without adjustments of the thread diameters, additional work may be required to ensure the assembling of the two prototypes of these components together.

Equipment costs and labor costs are involved in recreation of true geometrical shape(s) and dimension adjustment. Cost estimating should follow the same procedure as stated in cost estimating for converting 2D drawings to 3D solid models. For 3 hours of replacement work and 2 hours of adjustment work, the cost can be high as much as $270, including the overhead cost involved.

2. Cost Related to Data Preparation for Solid Free Form Fabrication

Solid freeform fabrication is characterized by building a solid part layer by layer. There are several critical requirements in solid freeform fabrication. To minimize the trapped volume of liquid during fabrication, selection of part orientation for the building process is critical. Deflection of a thin, solidified layer without sufficient supports causes distortion of the geometrical shape required to build. To ensure accuracy, a support structure has to be designed and added to the part geometry before the process of building the prototype. A software tool is also needed to generate slicing files of horizontal cross sections for part building. All these requirements call for special skills and, thus, additional labor costs. On the other hand, special and designated software tools are needed to assist the design of support structures and the generation of slicing files. As these software tools come with the rapid prototyping equipment, and they are specialized only in rapid prototyping applications. Relatively high charge fees are enforced by most engineering service bureaus, ranging from $60 to $80 per hour for labor cost with 35% overhead rate. Thus, 3 hours of service in the rapid prototyping application can cost as high as $324.

3. Cost Related to the Process of Part Building

Costs related to the process of part building are determined by the specific rapid prototyping technique used. In this paper, cost estimating of the rapid prototyping technology is focused on applications of stereolithography, a process for building a solid plastic part out of a photosensitive liquid polymer using a directed laser beam to solidify the polymer. Figure 3 illustrates a general setup for the stereolithography process where part fabrication is accomplished as a series of layers in which one layer is added onto the previous layer to gradually build the desired 3D geometry.

The stereolithography process is a slow building process as the rate of the solidification process limits the rate of energy absorption, or the size of the laser spot, to ensure the part accuracy. The initial investment of equipment is relatively high. A machine capable of building a 250 x 250 x 250 mm solid is priced at $200,000. At present, the maintenance cost charged by the equipment manufacturer accounts for 10 - 15% of the equipment purchasing price. The cost of laser replacement adds another important element to the equipment costs. Although a fully automatic and unattended stereolithography operation is claimed by the equipment manufacturer to signify low labor cost, the operational expense is relatively high. Results obtained from the 15 engineering service bureaus indicate an hourly charge rate with a mean of $130 per hour, and the charge rate can be as high as $200 per hour.

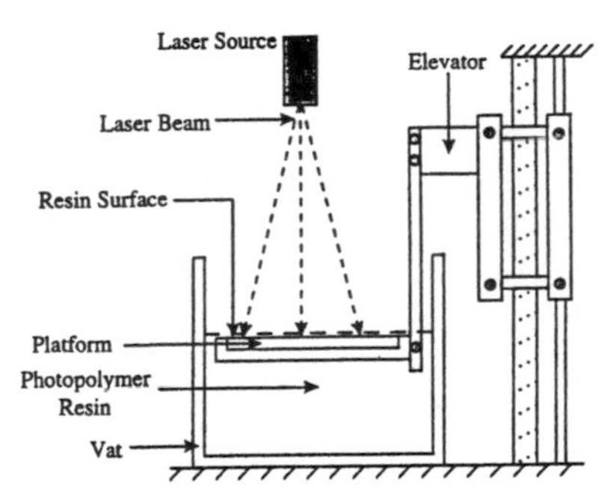

Figure 3 Stereolithography Process

Cost of the photosensitive resin stands for a variable cost item to cover the material consumption not only for the built solid, but also the support structure. The charge fee for a built solid weighing 1000 grams ranges from $100 to $200.

A surcharge fee is generally enforced to cover the labor cost for setting up the machine, cleaning the supports from the built solid, the post-process curing to complete the liquid solidification, and surface smoothing, such as light sanding. Survey data indicate the average surcharge fee is $150 per part. By summing the equipment cost, material cost and labor cost, a part to be built weighing 500 grams and needing 12 building hours may cost as high as $400 + $2,400 + $150 = $2,950.

To summarize the discussion of cost structure of applying the rapid prototyping technology, and to quantify each of the cost estimates, we present an example. It is making a prototype of the fan used in a sander for dust collection. . As illustrated in Fig. 4, the part weighs approximately 200 grams. As the geometry and size of the designed fan imply, building a prototype using the stereolithography process requires 10 hours of work for converting a 2D drawing to a 3D solid model, 4 hours for data preparation, and 10 building hours. The itemized cost elements and the total cost, as expected and without overhead rate charge, are listed below:

2D to 3D conversion:	\$40 x 10	\$400
Data Preparation:	\$70 x 4	\$280
Material Cost:	\$150 x 0.5	\$75
Building Hours:	\$130 x 10	\$1300
Surcharge Fee:	\$150	\$150
	Total:	\$2,205

Figure 4 Isometric View of a Dust Collection Fan

IV. Case Study - Production of a Mold Pattern of Tactile Graphics

To demonstrate the procedure used for cost estimating preparation, a case study is presented in this section. The study deals with the production of a mold pattern for tactile graphics using the stereolithography process. Figure 4 is a picture of human heart. The picture contains features, such as arteries, ventricles, and blood flow, depicting the function of a human heart in the study of human biology science. Note that these features are created in the process of 3D solid modeling. On the picture, Braille characters are embedded. Therefore, the picture is also called tactile graphics, which specifically fits the educational need for blind children.

Tactile graphics learning aids are produced using a thermal forming process where a mold pattern is needed. Figure 6 illustrates the process flow of tactile graphics production. The product starts with the creation of geometrical features, then the production of a mold pattern, and finally the

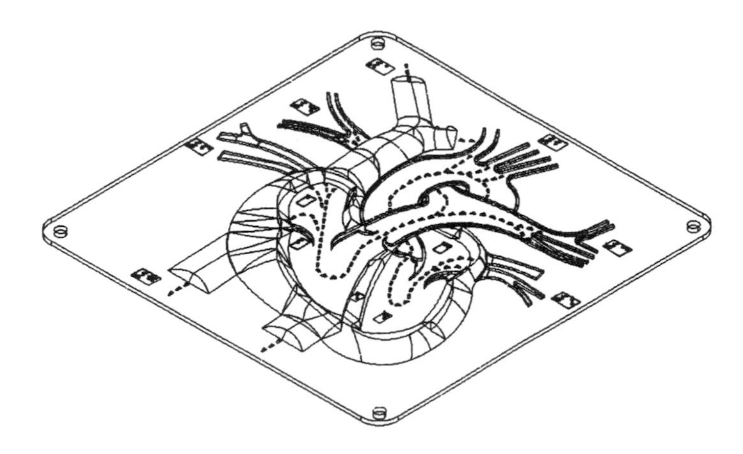

Figure 5 3-Dimensional Representation of the Human Heart

production of duplicates made from the mold pattern using a thermal forming process. Traditionally, mold patterns are made by hand. NC machining has recently been introduced to automate the task of mold creation. However, it also requires a significant amount of time to machine the geometrically complicated surfaces. The associated tooling cost is extremely high to ensure accuracy of the mold pattern. Breakage of Braille dots on the mold pattern often occurs during machining. Quality assurance is a real challenge. Table 1 lists cost estimates of the major operations involved in applying NC machining.

To meet this challenge, the stereolithography process is introduced to carve the complex surfaces covered on the mold pattern as well as to produce the Braille dots descriptions on the mold pattern. The versatility of the layer fabrication method has made the rapid prototyping application of making the mold pattern particularly attractive. For an area of size 125 x 125 mm, 10 building hours are sufficient to produce the mold patterns. The list below illustrates the cost estimates of the major operations involved in applying rapid prototyping. Comparing the two sets of data, economic benefits are evident for a reduced cost ($2,265 Vs. $2,880) and shortened time period (27 hours Vs. 47 hours) to produce the mold pattern using the rapid prototyping technology. Additional advantages gained by using rapid prototyping include the reliability of the building process with no risk of breakage occurring, and easy accommodation of a variety of surface texture patterns to improve the learning efficiency for blind children.

Rapid Prototyping Application		NC Machining Application	
3D Solid Modeling:	$40 x 15	3D Solid Modeling:	$40 x 15
Data Preparation:	$70 x 2	NC Path Preparation:	$70 x 2
Material Cost:	$150 x 0.5	Material Cost:	$40
Building Hours:	$130 x 10	Machining Hours	$60 x 30
Surcharge Fee:	$150	Tooling Cost:	$300
Total Cost:	$2,265	Total Cost:	$2,880

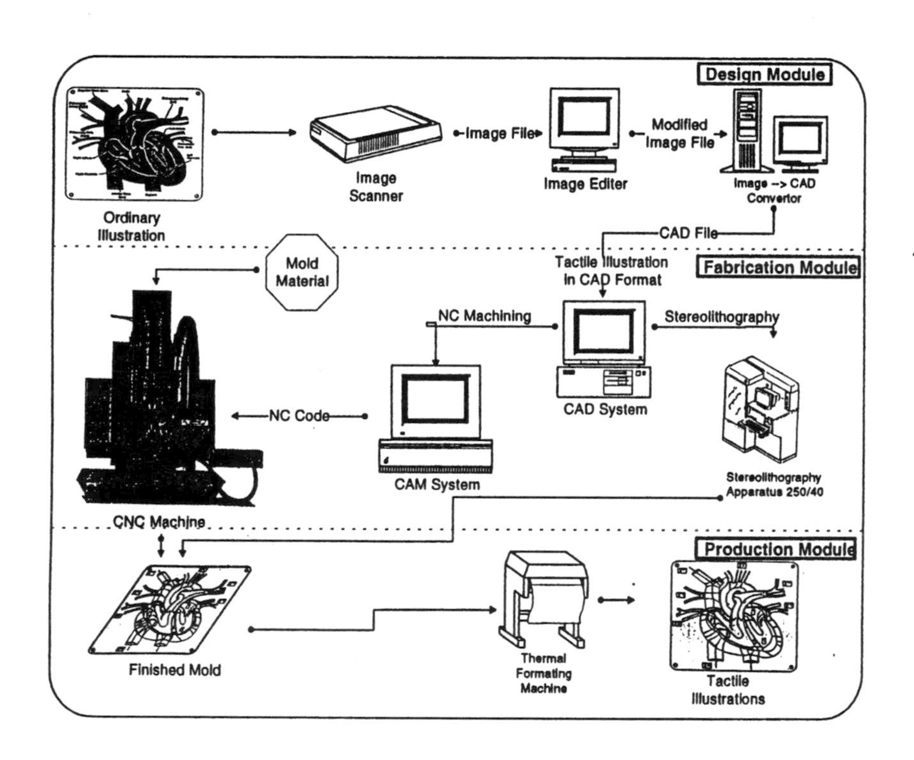

Figure 6 Process Flow of Tactile Graphics Production

V. Discussion of Results

The ultimate objective of a product development process is to produce products with superior quality and competitive prices for the functionality the products can offer. With global competition, companies are becoming leaner and getting their products to market faster by reformulating their business strategy. Rapid prototyping, as a tool in the product development process, is playing a unique role in speeding up product design, and assisting in identifying product improvements in the early design stage. Based on the cost estimating and cost analysis presented in this paper, the relatively high cost associated with rapid prototyping applications is mainly contributed by the following factors:

(1) High Initial Investment of the Rapid Prototyping Equipment. The 1995 market price of a representative stereolithography facility for building capacity 250 x 250 x 250 mm was $200,000 per unit. The rapid prototyping equipment industry is young, and has a history of less than a decade. The accumulated expenditure made by the industry in research and development has been high. Although the market growth rate of rapid prototyping is one of the highest among promising technologies, such as aerospace and computer chip manufacturing, the industry wealth has not accumulated sufficiently enough to offer low equipment prices to its customers. However, a significant price drop can be anticipated in the very near future as the rapid prototyping equipment industry is making tremendous progress to introduce new and innovative technologies with significantly lower costs.

(2) High Reliance on the Technical Support Provided by Engineering Service Bureaus. Based on a survey conducted in 1996, for United States alone, there are about 150 engineering service bureaus located all over the country. At present most of the companies are still in the process of accepting the 3D solid modeling approach in their product development process. As a result, a significant amount of drawings made today are in 2D format. When those companies try rapid prototyping, they have to seek needed technical support from engineering service bureaus. An investment of $200,000 for a five-year return comes up with an annual payment of $52,000 at an interest rate of 10%. An annual maintenance fee is in the order of $15,000 and the cost for laser replacement is about $10,000 for every 2000 hours. The sum of these three cost elements constitutes the major portion of the operational expense besides the labor cost. An hourly rate for building a part using the stereolithography equipment, calculated based on these estimates, may come up with $38 per hour. This number is significantly lower than what is actually charged by most of the engineering service bureaus. Such an inflated service charge puts companies at a disadvantage for maintaining the product development cost at a low level. However, corrections on the service charges are anticipated as the competition among service providers in the industrial neighborhood drives the price down. More important is the fact that most companies are in their final stage of implementing the 3D solid modeling approach in the product design. Due to this, the reliance on engineering service bureaus to provide technical support for rapid prototyping applications becomes less critical, and will eventually be limited to the narrower scope of building prototypes.

(3) Economic Potential of Applying Rapid Prototyping. Product innovation has come to be seen as a fundamental solution for companies to maintain their competitiveness. It has become very critical to develop their new products fast enough to keep up with the turbulent, shifting market. With the rapid prototyping technology, increasing the pace of developing products is not a dream. The benefits of a short development cycle are evident for extending the product's sales

life, strengthening customer loyalty, and gaining an extra time of revenue and profit. If a product profit model is introduced in the case study of producing a mold pattern for tactile graphics, and we assume that a one month earlier completion of producing the required number of learning aids would save one month labor cost for three operators, the expense of $2205 spent on using the rapid prototyping technology, instead of using NC machining, would be well justified. As rapid prototyping applications stretch out, it can be anticipated that companies will gradually create a culture of prototyping to guide the product design because of the compelling economic potential which can be gained from shortening the product development cycle.

VI. Conclusions

As an emerging technology, rapid prototyping is revolutionizing the engineering design approach. This paper presents a comprehensive and detailed coverage of the elements needed to embark on a cost estimating task for applying the rapid prototyping technology in the product development process. Using cost estimates of the work activities involved in rapid prototyping applications, this study points out a critical phenomenon, which has led to a slow acceptance of the rapid prototyping technology by industry. It is the combined effect of high equipment cost and a low volume of work load. Such an effect has led to an uneven distribution of the capital investment in rapid prototyping applications, and has positioned rapid prototyping applications as high-risk activities to gain short-term investment return. Price reduction of rapid prototyping equipment and the creation of a culture of prototyping are the two main challenges facing the design and manufacturing community in promoting the rapid prototyping technology.

Acknowledgments: The authors acknowledge the support from the Engineering Research Center, the Mechanical Engineering Department, and the Institute for Systems Research under NSFD CDR-88003012 grant. Special thanks are due to the Society of Manufacturing Engineers, 3D Systems, Inc., and Mr. Peter Sayki, President of SICAM Corp.

References

1. G. Zhang, M. Richardson, and R. Surana, "Development of a Rapid Prototyping System for Tactile Graphics Production," Proceedings of the 1996 Flexible Automation and Intelligent Manufacturing, Atlanta, Georgia, May 12-15, 1996.
2. P. Jacobs, "Rapid Prototyping and Manufacturing: Fundamentals of StereoLitho-graphy," First Edition, Society of Manufacturing Engineers, 1992.
3. P. Jacobs, "Stereolithography and other RP&M Technologies," ASME Press, 1996.
4. R. Stewart, "Cost Estimating," Second Edition, John Wiley & Sons, Inc., 1991
5. "The Edge: Competitive Advantage Through Rapid Prototyping and Manufacturing," 3D Systems Publications, Vol. V, No. 1, 1996.
6. T. Wohlers, "Rapid Prototyping: Past, Present, Future," RP Directory, Connect Press, Ltd. Publication, 1996.
7. G. Zhang and S. Lu, "An Expert System Framework for Economic Evaluation of Machining Operation Planning," Journal of Operations Research Society, Vol. 41, No. 5, pp. 391-404, 1990.
8. B. Dent, "Principles of Thermatic Map Design," Addison-Wesley Publishing Company, 1985.
9. J. Wiedel, "Summary of Tactile Mapping in the U.S.A.," Third International Symposium on Maps and Graphics for the Visually Handicapped People, 1989.
10. G. Jansson, "Tactile Maps as a Challenge for Perception Research," First International Symposium on Maps and Graphics for the Visually Handicapped, 1983.

INFORMATION SYSTEMS FOR MAKE-TO-ORDER MANUFACTURING

S A Abbas
Merryweather Management Systems Limited
Darlington, UK

ABSTRACT: This paper analyses the information systems needs of make-to-order companies and considers some current approaches addressing these needs. The main distinguishing characteristics of companies in this marketplace are identified. An integrated view for systems application is adopted, viewing the full spectrum of activities from enquiry, through quotation, order receipt, supplier purchase orders, materials control, planning, works scheduling, production monitoring, time and attendance, to goods dispatch. The contrasts with traditional MRP systems are discussed, showing how these approaches have not properly addressed the requirements. An illustration is given of an operational system specifically designed for this marketplace, and a case study of a company utilising an integrated system for its manufacturing operations is described.

THE MARKETPLACE AND TECHNOLOGY AREA

The overall manufacturing arena may be perceived as segmented in many ways. For the purposes of this paper the representation depicted in Figure 1 may be considered as a model. At one end of the spectrum there are the continuous flow process industries and at the other end there are the "make-to-order" manufacturers. Because the term "make-to-order" is used with different connotations, let us first define what we understand by it in the context of this paper.

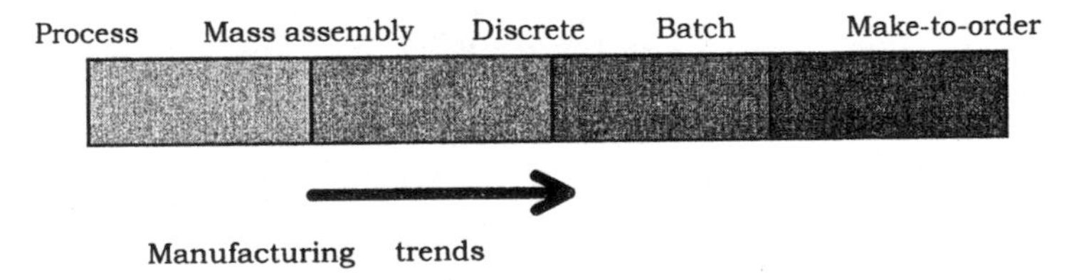

Figure 1. Representation of the manufacturing spectrum

Make-to-order manufacturing has the following characteristics:

- There are few, if any, standard products, although there may be variations
- Orders are received for one-offs or for small batches
- The products are complex and of high value
- Specifications are defined by the customer
- There may be design changes after manufacturing has commenced
- There are usually strict delivery dates and sometimes penalty clauses for lateness
- Purchase orders are placed upon suppliers *after* the customer order is received
- Sub-contractors may have to manufacture some items specifically for the contract
- Sub-contractors' work schedules may have to be monitored and managed
- Many different jobs are running on the shop floor
- Jobs are of widely differing scopes, nature, and timescales
- Stock is minimal, usually being mainly made up of consumables and basic standard items

These characteristics mean that the information systems needs of such companies are fundamentally different from companies which "make-to-stock". Let us compare and contrast these requirements broken down under the main functions of:

- Sales order processing
- Purchasing and materials control
- Production planning and scheduling

which may be considered as the main software modules of an information system.

Sales order processing

MAKE-TO-STOCK - Standard range of products. Prices and deliveries fixed, and delivery times quite short. Invoicing fairly simple.

MAKE-TO-ORDER - Variable products, customised or engineered to order. Prices negotiated, and subject to changes during progress of the job. Delivery times sometimes short and sometimes lengthy, with staged invoices and payments.

Purchasing and materials control

MAKE-TO-STOCK - Fixed bills of materials. Known and standardised parts to be bought in. Quantity orders on small number of suppliers. Infrequent runs to check inventories.

MAKE-TO-ORDER - Configured bills of materials. Engineering changes in most orders, reflected in purchase orders upon suppliers. Small quantity orders upon large number of suppliers. Supplier may need to engineer to order. Frequent runs on materials requirements systems to cope with changing situations.

Production planning and scheduling

MAKE-TO-STOCK - Based on part numbers. Inventory levels crucially dependent upon delivery times. Small numbers of work orders, for large quantities. Production line changes infrequent. Forward planning based upon market forecasts made internally.

MAKE-TO-ORDER - Based upon job numbers. Inventories minimal. Large numbers of work orders, for small quantities. Production resources flexibly utilised. Continual monitoring of each work order and "dynamic scheduling" required to cope with changing situation.

In the above discussion we have categorised manufacturing activity into two distinct forms, make-to-stock and make-to-order. Another category, sometimes referred to as a "hybrid", has arisen due to present trends towards customised manufacture. This involves the use of flexible systems to produce goods based upon standard concepts, with particular variations chosen by the customer. The producer makes the product by:

- assembling the product from standard parts, bought-in or from in-house manufactured stock

or by

- making a small proportion of parts as bespoke to the order and finally fabricating and assembling the end product using standard components and sub-assemblies.

Information systems requirements for this type of manufacturing are not the same as for the type of make-to-order manufacturing characterised above, as should be clear from the detailed features discussed. A definition of make-to-order given by O'Toole, Etheridge and Warrington is succinct, and particularly apt:

"...a make-to-order company is defined as one in which the total manufacturing time is less than the delivery lead time". [1]

THE FAILURE OF MRP

The MRP approach grew out of the requirements for make-to-stock manufacturers and this has been reflected in all MRP systems developed for the market. O'Toole, Etheridge and Warrington have suggested that this is embedded in the "mind set" of the software developer of MRP systems.[1] The previous section has outlined fundamental differences in the requirements for make-to-order, and many industry practitioners have argued that production management software should be developed specifically for the make-to-order market.

The last twenty years or so have seen numerous examples of expensive failures as make-to-order manufacturers have attempted to implement commercially available MRP products modified to suit their company's operations. Singleton refers to a 90% failure rate. [2] Those organisations which had in-house software expertise developed their own systems. The true costs of adopting this approach only became apparent later, as hardware platforms became obsolete, end-users demanded continuing improvements, key software personnel left the organisation, and, overall, maintenance and upkeep became untenable.

Researchers looking into manufacturing trends and the implications for information systems lying at the heart of enterprise operations, have identified the emergence of supply chain strategies as a dominant force. Dailey, of the Gartner Group, concludes that an increasing number of manufacturing enterprises are questioning their manufacturing practices and applications, particularly the use of MRP:

"For more advanced manufacturers, the MRPII concept is already dead, while others have adopted other applications that stretch the life of their existing MRPII applications." [3]

In make-to-order manufacturing the customer will increasingly drive the manufacturer to service his specific order, and the overall production management approach, including materials procurement and planning/scheduling, will need to adapt accordingly. The Production functions will have to be responsive to Sales and Marketing. According to Dailey, manufacturing organisations will not be able to take full advantage of the new integrated technologies unless:

"...they have standardised planning and scheduling concepts across the enterprise and are willing to take control of the manufacturing schedule out of the hands of manufacturing and place it in the hands of order takers." [3]

AN INFORMATION SYSTEM FOR MAKE-TO-ORDER

We shall now illustrate the above arguments by outlining the features and functions of one system that was specially designed for make-to-order organisations. The Whessoe company, based in the north east of England, manufactured large steel plant items and equipment mainly for the industry sectors of oil and gas, power generation, chemicals and petrochemicals. In the 1970s it saw the need for computer-based tools to aid the management of its manufacturing operations. A search of available commercial offerings revealed that there were none suitable for the company, whose business was almost wholly made-to-order. Because there was an active and growing technical computing function within the organisation it was decided to develop a capability in-house. Stage by stage, a system was developed, and gradually put to use in the company's manufacturing operations. A major advantage was the ongoing feedback from users which meant that there was continuous input of practical engineering expertise while the developments were carried out. The problems that arose in time, however, were connected with the issues mentioned previously in this paper, ie the costs of continuous bespoke development, maintenance, and difficulties in retaining staff with the product expertise.

Other companies operating in similar industry sectors, notably Motherwell Bridge Fabricators, of Scotland, became aware of the Whessoe systems, and expressed an interest, observing that they too had investigated the commercial market for a long time and not found a product suitable for their purposes. Subsequently, Whessoe worked with Motherwell Bridge and also other manufacturing organisations, aiming to develop systems which would not be restricted to company-specific capabilities but would have wider applicability. Systems were installed at a number of company sites, and valuable experience was gained as to how individual manufacturing organisations needed to operate their information systems.

Merryweather Management Systems Limited, a company specialising in management information systems whose management and technical staff had long connections with Whessoe, were aware of these activities, and in April 1996 negotiated with Whessoe to acquire the full rights to the software that had been developed together with all related business and key technical staff. Merryweather's objective was to develop fully the potential of the software, now known as MC2000, and to launch and promote it as a product in the commercial marketplace.

A brief overview of the scope and main functions of this system will illustrate how the principles described above in the previous sections may be applied to provide a generalised solution to the requirements of make-to-order manufacturing.

The system comprises the following modules:

- Production planning and scheduling
- Materials control
- Sales order processing
- Time and attendance
- Shop floor data collection
- Job costing
- Financial

These functions are integrated such that data is not input repeatedly and not duplicated in the system. Figure 2 depicts a typical configuration in a manufacturing environment.

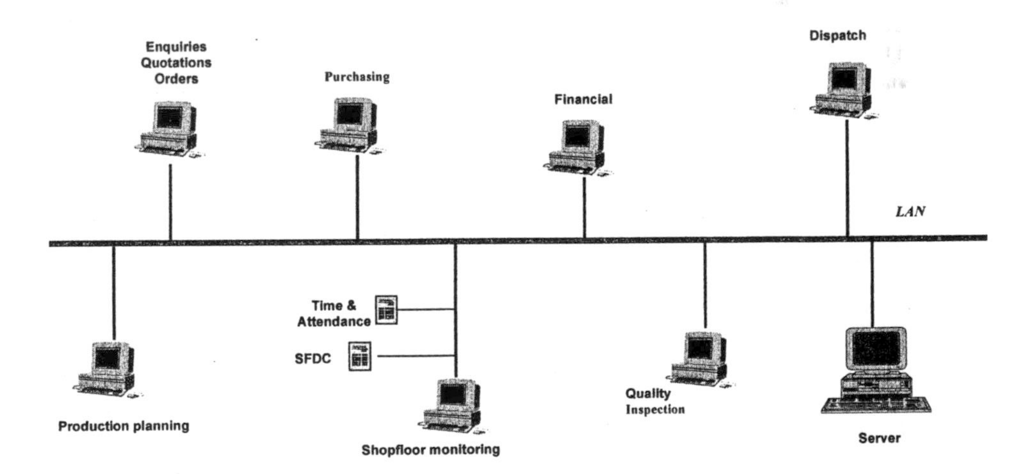

Figure 2. Typical system configuration in a manufacturing environment

The **production planning and scheduling** function operates at three levels. The highest level is intended for senior management control and client reporting and gives overview information. The intermediate level is used for planning anticipated workloads and resource requirements. The third level is most detailed, and used to schedule jobs on the shop floor.

A key point, bearing in mind the earlier discussion on MRP, is that scheduling can be "dynamic", reflecting the volatile environment in the make-to-order shop floor. Delays may occur due to the unique nature of some jobs, perhaps due to unforeseen production problems, changes in design initiated by the client or the manufacturer, or problems with a subcontractor who may be struggling with a job that has unusual characteristics. Electronic "shuffleboards" are available to deal with such situations, and may be updated on a daily basis to control the progress of each job.

"What if" scenarios may be examined to determine the effect on delivery schedules of all affected jobs if certain selected jobs are re-allocated. These features are crucial in this type of manufacture.

The **materials control** function has flexible facilities for generating purchase enquiries and orders to satisfy bought-in items. These include the consolidation of similar items from different requisitions, the splitting of requested items across several suppliers and delivery dates, etc. At the same time links can be defined between purchase order items and activities in the production plan. This allows the latest promised delivery dates to be automatically taken into consideration when scheduling production work. Alternatively, the schedule can calculate the latest required-by dates for inclusion on the purchase orders. These methods of operation are of course not conventional in the standard MRP make-to-stock approaches.

In **sales order processing** the features provided are very different from standard MRP. Enquiries are entered and broken down into detailed items. Sales orders may be linked to previously entered enquiry details. Variations to orders are tracked, and a link with production is maintained. Despatch of the full orders or part-contracts are signaled to invoicing.

Basic **time and attendance** functions apply across all manufacturing, and indeed to all business activities. **Shop floor data collection** (SFDC) is particularly important for make-to-order activities, because it provides details of resource utilisation across diverse contracts in a dynamically changing situation. With the increasing use of specialised terminals in the work place along with swipe cards and bar codes, the potential for management control is enhanced considerably.

Job costing functions pull together information on all the resources - people, machines, and material, highlighting when actual or forecast costs for a job deviate from plan or estimate. More detailed reports allow the determination of exactly where and why such variances have occurred, with breakdowns into resource types, cost centres, etc.

Financial utilities allow the creation of invoices, including staged invoices, and provide interfaces to widely used standard financial packages for consolidation into company accounts.

Figure 3 shows one system screen as an example of the sort of graphical representation which users are demanding today. An important point to note for all system developers is that the majority of users in a manufacturing environment will not be very computer literate; they will be people who have their particular roles to perform in the production process and use the computer based system as tool to help them. Therefore the system must be as natural and convenient to operate as possible.

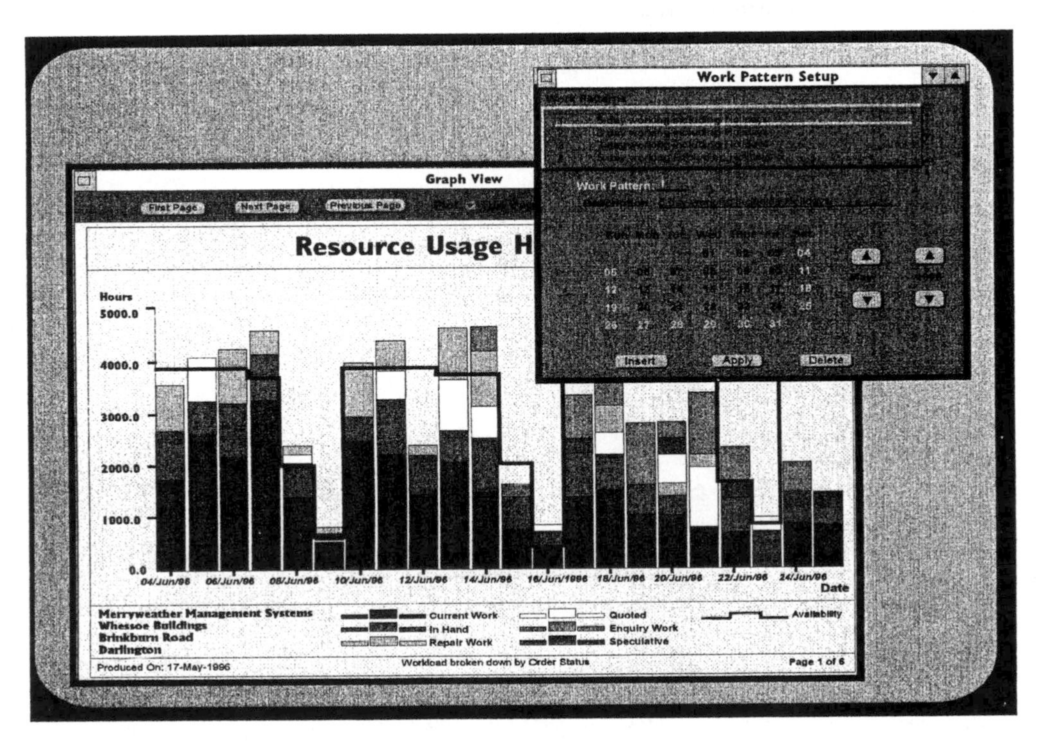

Figure 3. Example of a system screen showing total production workload

A CASE STUDY - METALCRAFT LTD

The Metalcraft company is based in Chatteris, near Cambridge, in England. It employs around 130 people and has a turnover of £8 million per annum. It belongs to a group which has a turnover of £20 million with 280 employees. Its business is subcontract fabrication and machining for the medical, nuclear, defence, petrochemical and process plant sectors. At any one time it has about 150 orders being processed through its works, of varying types, size and scope.

Since the early 1990s Metalcraft had been looking to tighten up its manufacturing management and compete effectively in the market. The company had examined its existing systems and decided that drastic change was needed. It had been using Manman for several years, an MRPII system designed for controlling mass produced commodities, not suitable for a company manufacturing varied products to customers' orders. The company had recognised that it was in the business of selling hours, not product; it needed systems to manage its resources effectively, and to control *projects.*

In searching for new information systems Metalcraft identified certain key issues:

- the system concepts should show an understanding of this type of business
- the system should have capabilities for being contract specific and flexible
- the real and speculative workload could be modeled and actions taken
- current contracts could be monitored "in real time" and necessary adjustments made

After exhaustive examination of the market Metalcraft chose MC2000 as the product which met most of the company's needs. The system was implemented in stages, with training and familiarisation gradually being given to all staff affected by the implementation. Today there are 25 PC workstations available to users, linked to a powerful Digital Unix server, and virtually the whole organisation has access to the system in order to fulfill the various necessary functions.

The following people use the system:

Sales	enquiry tracking, order taking and administration.
Estimating	quotations.
Product engineers	looking at progress, material arrival, etc.
Planning	linking to sales order items.
Engineering	front-end activities: quality plans, weld procedures, generation of requisitions for materials required.
Buying	sending out enquiries, generating purchasing orders from requisitions, expediting.
QA	reporting on "non-conformances" in material and work.
Inspection	passing or failing in-house or sub-contracted work.
Stores	receiving material and issuing to workshops.
Accounts	monitoring hours worked and comparing with hours clocked, invoicing clients and paying suppliers.
Dispatch	generating dispatch notes.
Directors	looking at the contract from many angles.

Furthermore, it is intended to extend and improve the shop floor data collection facilities, enabling all shop floor employees to be brought into the system.

CONCLUSION

It has been argued in this paper that the traditional systems for production management still in widespread use are not appropriate for make-to-order manufacturing. The special characteristics of make-to-order manufacture have been discussed, and a currently available system has been described for illustration. Practical experience of a user company, small in size compared with the large corporations, moving to integrate such a system into its operations, has been outlined.

With market demand and technology push, the make-to-order manufacturing area will continue to show significant growth. Demand for suitable information systems to assist management is increasing in companies of all sizes in varied industry sectors, as manufacturers strive to respond flexibly to customers' requirements. A number of systems products are beginning to emerge in the marketplace designed particularly to address these needs.

ACKNOWLEDGEMENTS

The author expresses his appreciation to Metalcraft Limited for their kind permission to use information regarding the company, and thanks Adrian Seal particularly for providing the case material on the company's manufacturing operations.

REFERENCES

1. **O'Toole, B., Etheridge, R. and Warrington, A.,** Assessing the suitability of manufacturing control software for small make-to-order manufacturing companies, *BPICS Control*, October/November 1993.

2. **Singleton, R.,** MRP: a waste of time, *Consultants' Conspectus*, August 1995.

3. **Dailey, A.,** Redefining the role of manufacturing for SCM efficiency, *Gartner Group Conference on Supply Chain Management*, Amsterdam-Birmingham-Paris, June 1996.

STRATEGIC INFORMATION SYSTEM - INTEGRATION OF THE MANAGEMENT AND ENGINEERING ACTIVITIES

Brane Semolič, Jože Balič
Faculty of Mechanical Engineering, Maribor, Slovenia

ABSTRACT: In the paper we consider the concept of "Strategic Information System" (SIS) which is based on findings of Synnott and other authors. The starting points of SIS were set by C. Wiseman. According to him SIS is an information system designed to support and increase the competitive strength of company. SIS is an information system which is essentially different from the classical definitions of information systems. A critical analysis of modern approach to SIS tells us that these are oriented mainly into searching methodologies. In our work we are trying to connect SIS into engineering system such as Simultaneous Engineering. To manage this we have developed a five level manufacturing model and a five level business model and connect both model together according to strategic information system. These models are developed for small and medium sized companies and is already implemented in tools industry.

1. STRATEGIC INFORMATION SYSTEM (SIS)

1.1 Introduction

The starting points of SIS were set by C. Wiseman, according to whom ***SIS is an information system designed to support and increase the competitive strength of a company*** [1]. SIS is an information system which is essentially different from the classical definitions of information systems, these being mainly based on the paradigm of Anthony's Triangle [2]. Interpreters of Wiseman's concept of SIS, such as D.Remeny, even think that Anthony's paradigm is inconvenient for interpreting the strategic importance of individual parts of SIS for the increasing corporate strategic strength..[3]

1.2 SIS and its application fields

The SIS assuring the corresponding information support to corporate management, which has to provide, at different levels and areas of operation, successful and effective management of corporate business processes.
The findings of contemporary authors dealing with SIS, and the results of our research concerning the characteristics of contemporary corporate management have led us to the conclusion that SIS should be considered from two aspects, i.e.:

- the aspect of defining possible fields of application of SIS, their characteristics and interrelations,
- the aspect of searching the fields of application which would, in a concrete case, help a company

In our paper, we restricting ourselves to the aspect of defining possible fields of application of SIS and to the defining of their main characteristics and interrelations.
We starting from the general needs of an profit oriented company, which business objectives are:

- economical production of existing products/services of the company,
- adaptation of products/services and the company operation to changes in the environment and
- performing and directing of the company's development activities,

and their aim is making profit today and in the future. What we expect from a modern company is therefore:

- OPERATIONAL EFFICIENCY - that is successful solving of the problem how to make profit with existing products and services,

- SUCCESSFUL ADAPTATION - to the requirements and changes in the environment and

- SUCCESSFUL DEVELOPMENT - assuring business success in the future, or making it possible to keep the existing or assure new strategic positions of the company's success.

According to these facts in any company next types of the business processes should be taken into consideration; with regard to their characteristics, these include (Figure 1) [4]:

- operations business processes
- adaptation business processes and
- development business processes.

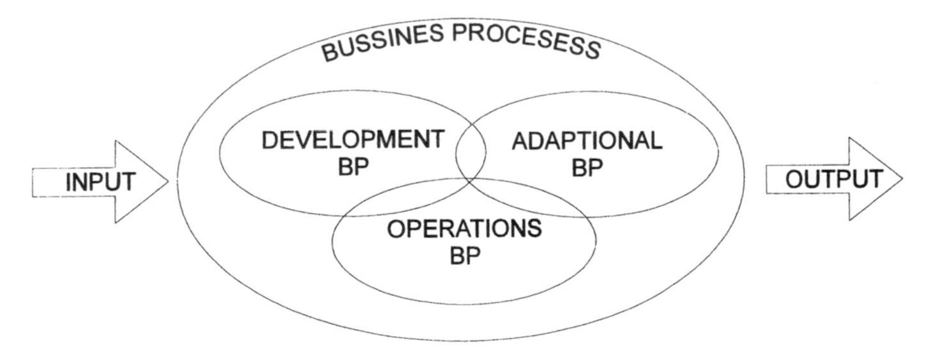

Figure 1. The company business processes

The above division of business processes into operations, adaptation and development business processes serves as a basis for the definition of information systems needed for carrying out and managing these processes. Accordingly, there are two basic fields of application of SIS [4]:

- ***the operations information system (ISOP) and***
- ***the development information system (ISD).***

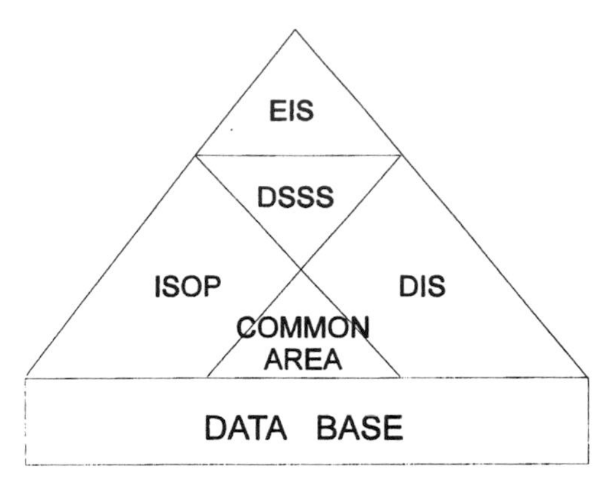

Figure 2. SIS structure

SIS also includes information support to operations and to the management of adaptation business processes, as both these processes involve projects or project processes which are a component part of corporate strategic development.

Figure 2 illustrates the SIS concept and its fields of application. As it is evident from the picture, ISOP and ISD partly coincide, forming, in this way, a common area, where ISOP has to take into account the requirements of ISD, and vice versa. Each of the above mentioned information systems its own partial data base (DBP_o and DBP_d), and for the operation of the common area, a common data base (CDB), which has obligations both towards ISOP and ISD. A group of decision support systems (GDSS) is intended to support different decision making situations that managers at the medium management level and experts preparing expert foundations for the needs of decision making have to face. They have at their disposal different information technologies (ITs). From the point of view of current following of operation and the production of different analyses connected with it, the decision support systems (DSS) are the central information technology of GDSS. For the needs of top management, however, there exist the Executive Information Systems (EIS) which provide an insight into the operation of a company "from the top". Through the links of EIS with external data bases, top management has access to business data and information from the company's environment; SIS also includes all kinds of computer links of the company with its environment. These are various kinds of inter-organizational links with computer data exchange (CDE), use of above mentioned business data bases, and also all other technical data bases needed in carrying out corporate development activities.

2. OPERATIONS INFORMATION SYSTEM (ISOP)

The task of ISOP is to provide all data and information necessary for the management of operations business processes. The operations management includes the management of strategic business units (SBUs) of a company, and inside these, those operations business functions whose presence is essential for making profit by implementing the SBU's or the company's products/services. This includes operations business functions such as purchase, finance, manufacture, sales, and the like. ISOP users are presented in a partial display of a modified Anthony's paradigm (Figure 2), which illustrates the satisfying of informational needs of performers in operations business processes inside the above mentioned organizational units and their management. In accordance with the different characteristics of operations business processes (mass, batch or individual production), the implementation of an adequate ISOP has to be provided which would take into account and support mentioned specifics

3. DEVELOPMENT INFORMATION SYSTEM (ISD)

Figure 2 shows the fields of application of ISD in accordance with the starting points given by the modified Anthony's paradigm. The Development Information System (ISD) should provide all the data and information required by the performers and managers responsible for carrying out the corporate adaptation and development projects. These projects can run on corporate level or within the framework of individual strategic business units (SBUs), or in otherwise defined areas of corporate operations. As the projects are single, i.e. they do not have the characteristics of continuous business processes, it is possible in the majority of cases to set up, in advance, such an information system that will be valid and suitable for all cases. The exception are projects, such as the preparing of corporate annual plan, characterized in that that they are composed of a standard set of activities that have to be repeated every year. The characteristic of the majority of projects one encounters in a company is that they can only be standardized to a certain point, because the contents of a certain part of these projects depend on a concrete case. In order to cover, by ISD, as wide an area as possible, it makes sense to standardize certain parts of the area to the highest possible level still possible with a concrete project. The notion of project standardization comprises:

- contextual classification of corporate projects,
- production of reference models for performing individual standard project types, and
- project organization for performing standard projects.

Besides the structure and the technology of implementing a project, the forms of project organization for individual types of projects should be defined. In this way the structure of potential information users is obtained, which is necessary for establishing standard bases of a project information system.

Of course, every reference model and project organization should in a concrete case be adapted to the specific requirements of a given situation. The adaptation is carried out during the project start-up period (PSU), which lasts from the moment we decide upon a project to the point when the project implementation is launched.

Roughly, ISD should meet the information requirements of:

- development and adaptation projects' management,
- project organization,
- experts (performers), and
- other participants in the process of project management.

The experts (performs) need all the necessary information enabling them to carry out the project activities. These information we divide into:

- program information and
- structural information.

The program information include all data of a plan that are necessary for the implementation of projects. The data of a plan include all the necessary information on:

- the content of an activity,
- the duration of an activity,
- the time parameters of an activity (date of beginning, date of conclusion, time reserve, fixed terms, milestones),
- preceding and following activities,
- resources necessary for the performance of an activity,
- costs of and activity, etc.

During the performance of an activity and at its conclusion, the program information are transformed into control information. These tell us whether an activity was carried out in accordance with the plan (program information), and whether there were any deviations.

The structural information provide all the necessary instructions for the physical implementation of an activity. Such information are:

- data on materials,
- data on working tools,
- internal organizational instructions,
- internal and external technical standards,
- data on licenses and patents,
- blue prints, etc.

The structural information appear both at the input and at the output of an activity, depending on the case. ISD should provide the program and control information necessary for the implementation of a project as a whole, as well as for the implementation of individual activities. It should also provide all structural information necessary for the performance of individual activities of a project. Information, such as data on patents, licenses, external standards and the like, can be obtained only in a company's environment. Therefore the linking of ISD with the information systems in the company's environment is essential. This means above all linking between the corresponding domestic and worldwide data bases. In order to be able to use the same resources and the system, ISD has to be linked and coordinated with the operation of ISOP.

4. DECISION SUPPORT SYSTEMS (DSS)

In an company we can identify three characteristic levels of decision - the executive (top management) level, which is responsible for strategic management of a company, the medium management level, which is responsible for individual SBUs, and the business functions, projects and operations management level, which is responsible for the performance of business activities. Individual levels are interrelated, yet relatively independent in their decisions and planning. As the business processes on the operations level are determined, the decision processes can be structured in advance, too, and they are in this way easy to manage. There is a quite different situation on the remaining two management levels. The more we approach top management, whose task is strategic management of a company, the more the decision processes grow stochastic, or they can be only partly structured or not structured at all. A contemporary manager or expert has at his disposal a range of information technologies which form a Decision Support Systems Group (DSSG). These are:

- Decision Systems (DS),
- Decision Support Systems (DSS),
- Expert Systems (ES),
- Group Decision Support Systems (GDSS).

The central part of DSSG represent the Decision Support Systems (DSS). In practice, the notions of DSS and EIS (Executive Information Systems) are often confused. Thus, sometimes, the first or the second notion is interchangeably used for the same system. Nevertheless, DSS and EIS are two different systems, and they should be treated separately, as each of them is intended for different users, and they also have different purposes and modes of application.

5. EXECUTIVE INFORMATION SYSTEMS (EIS)

In the past, executives primarily used for their decisions the intuition, which was based on their practice and experience. The new circumstances are pushing these intuition-based decisions into the background and forcing executives to take their decisions on the basis of concrete information. Advanced information technology is forcing its way also into this level of decision making. A research made at the MIT (USA) revealed that computer executive support in companies is becoming increasingly important in the nineties, and will have a decisive influence on the success of companies in the turbulent post-industrial environment. [5]
The problem top management is facing is how to extract, from a multitude of data referring to a company's operation and its environment, the information which is vital for further orientation of the company's operations. Some theoreticians, like L. Orman talk about "pollution", which can be considerably limited by an adequate computer support; Ms. S. E. Weber and B. R. Konsynsky talk about tools for the extraction of knowledge, etc.[6,7] Generally speaking, this is about computer tools which enable the extraction of the data or information individual executives wish to follow, and upon which they base their business decisions.
As the Executive Support Systems have their foundation in the Decision Support Systems (DSS), the structure of both systems has many common points and links, even though the Executive Support Systems include a range of new components. The database and the model base, which make up the decision support subsystem, are adapted to the new requirements. The informing subsystem is composed of two parts, one informing on company operations and the other informing on the events in the company environment. In many cases, it also includes the "Electronic Meeting Support System". Accordingly, ESS can be defined as a system designed for executives to support them in:

- making decisions,
- acquiring information on the operation of the company,
- acquiring information on the events in the company's environment,
- communicating with their environment (e.g. meetings, correspondence, presentations, etc.)

In that part of ESS which is intended for seeking information for the support of deciding and informing on the operation of the company, it is, unlike the Decision Support Systems, not possible to "play with numbers" or seek different scenarios, states A. Young; only alpha-numeric and/or graphic displays of chosen data are usually possible here.[8]
Access to these data is organized according to the "Top- Down" principle, whereby a multidimensional data model is used as basis, stemming from the Decision Support System (DSS). This model is usually called "Executive Cube" and it is made up of those dimensions and data on the corporate operation that are relevant for the management of the field which is the responsibility of a certain manager. In this way, all members of a company's top management should have, with regard to their responsibilities, their own "cubes" reflecting the data that refer to their proper fields of operation and for which they are responsible, and the interrelations between these data. Each manager should provide for such a system that would enable him or her to follow:

- the effectiveness of the operations subordinated to him or her,
- how successful is the adaptation of the company or the SBU to the changes inside the company and in its environment,
- how successful and efficient is the implementation of development activities - projects.

The effectiveness of operations can be followed with the use of corresponding indicators referring to budgetary values, various accountancy data, and the like. The success of implemented adaptation projects can be analyzed by means of data on market share and trends, level of exploitation of operations' capacities, and the like. The development activities and the activities of adaptation can be followed by following individual projects (according to the principle "planned - implemented") as well as complete investments in the field of development and adaptation of the company or SBU.

6. CASE STUDY

6.1 Introduction

As an illustration we presenting the case study of a tool factory. The tool production is the unique process – project. In this specific case the property of the ISOP are the equal to the ISD. The hierarchy structure of such project information system within ISOP is show in figure 3.[9,10]

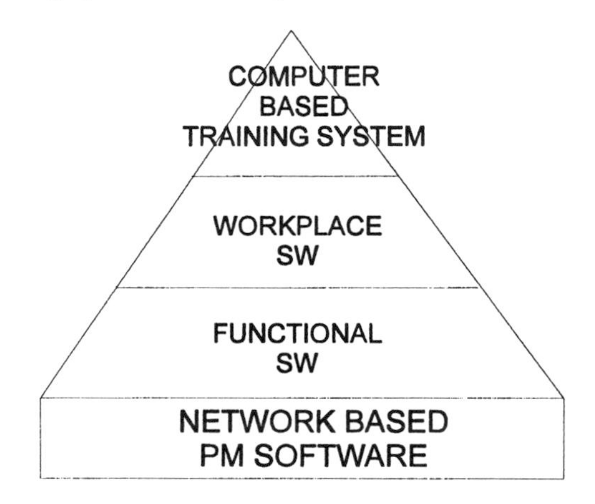

Figure 3. The hierarchy structure of the project information system

The network based SW helping us to plan and control the business processes of the "tool-order projects". Functional SW like configuration management SW, risk management SW, prototyping SW, principles of simultaneous engineering etc. can be tools for solve the economic and technical problems within business activities. Workplace SW represents the well known SW tools like spread sheet SW (Excel, Lotus, etc.,), word processors etc. Computer based training systems (CBT) are expert systems which has the "knowledge base" to help the experts and managers to solve the particular business or technical problem.

The main phases of the business process of "tool-order project" are [11]:

- Acquisition phase,
- Design phase,
- Procurement phase
- Production phase and
- Guarantee phase.

As in the many of other businesses also here the short delivery time, quality and costs are key words when we talking about the problem how to increase the competitiveness. With the proper usage of the modern IT (information technology) we can decrease the tool delivery time, increase the quality and cut down the costs. With the good knowledge and capability of project management, simultaneous engineering and by using the latest IT "tools" we can dramatically decrease the tool delivery time.

6.2 Five-level model of manufacturing system

The process of conceiving, designing and manufacture is represented in our approach as a five-level hierarchical model with the basic data-bus between manufacture on one side and the technological data base on the other side (Figure 4)

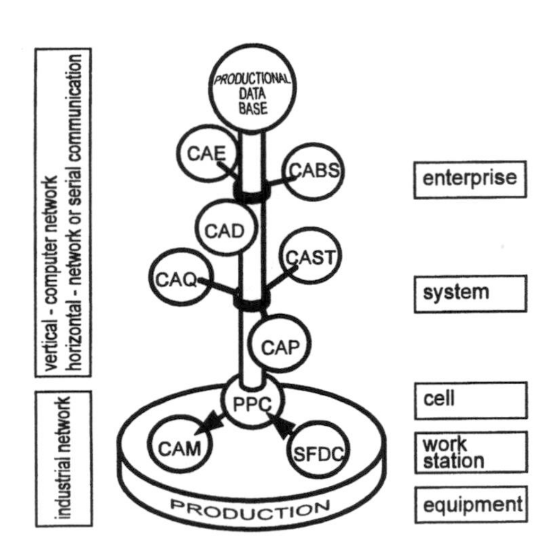

Figure 4. Computer integrated engineering (CIE) system

The top hierarchical level is the company with computer aided integrated engineering system for conceiving and designing, quality assurance and control, for storing and for production process planning. The cell level comprises the system for direct distribution of control information, monitoring of work stations and collection of operation data.

To establish the CIE an effective informational flow have to be provided. Before approaching to the logical solutions, some physical demands have to be fulfilled. The most important are: a computer network establishment on all levels of an enterprise [12] and a production data base realisation. The informational flow in an enterprise can be furthermore divided into a flow of orders among executioners and an information flow between an executioner and the production data base. To provide both flows and to meet the above mentioned demands an idea of a special interface have occurred which was later on developed into a so called informational interface.

6.3 Advanced approach to simultaneous engineering

The principle of simultaneous engineering is one of the most important part of agile manufacturing. The conventional method of solving of engineering problems anticipate successive progress of activities, which also means separable solving of individual problems. The development time is prolonged, the number of defects is increased and the product becomes incompetitive on the market. This result in decreased demand and production, and finally shut-down of the production. The principle of simultaneous engineering (Figure 5) ensures simultaneous approach to conceiving, designing and manufacturing of tools, dies and moulds at several levels at a time.

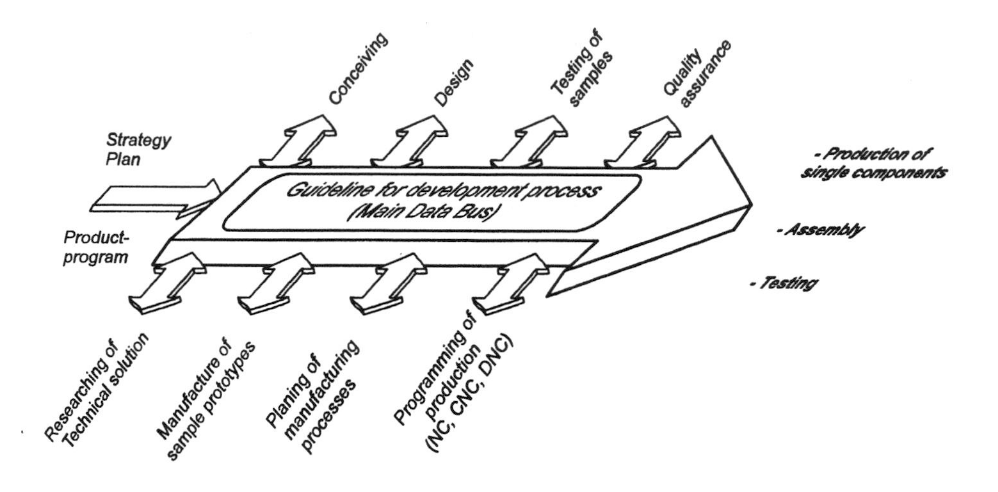

Figure 5. Simultaneous engineering approach [13]

The process of simultaneous engineering in practice isn't so smooth, because of inner and outer disturbances that affect the production process (Figure 6).

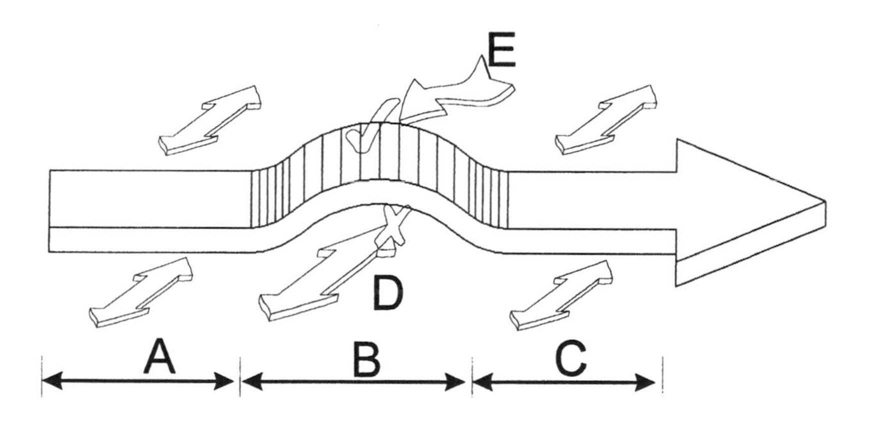

Figure 6. Disturbances in the process of simultaneous engineering

The A area represents a so called "calm development phase". All activities run normally corresponding to the plan of activities. In the B area, a disturbance occur that prevents a normal realisation of an activity in a process. The disturbance has to be perceived early enough to trigger utility activities to adapt the process for new circumstances. Perceiving and triggering are enabled in the strategic informational system included in the CIE strategy. In the E case we were able to adapt the process, and the activities' flow is normal again. In the D case the adaptation was unsuccessful what caused the interruption of the process. In such case methods and principles of so called adaptation or crises management have to be used.

7. REFERENCES

1. Wiseman C., Strategy and Computers Information System as Competitive Weapons, Dowe-Jones, Irwin, 1985,
2. Zeni W., "Blueprint in MIS", Harvard Business Review, 1970,
3. Remeny D., Introducing Strategic Information Systems Planning, NCC Blackwell, 1991,
4. Semolič B., Integrcija poslovnega in projektnega informacijksega sistema, doktorska disertacija, Univerza Maribor, EPF, Maribor, 1993,
5. Sprague R.H., A Framework for Development of Decision Support Systems, Prentice Hall, 1986,
6. Rockart, J., DeLong D.W., Executive Support Systems: The Emergence of Top Management Computer Use, Dow Jones – Irwin, Homewood, 1988,
7. Orman L, Fightin Information Pollution with DSS, Journal of Management Information Systems, 1984,
8. Young A., Implementing Executive Information Systems, Handbook, 1990,
9. Dworatschek S., Theses on the Development of Project Management Software, International Symposium project Management Software "Application – Implementation- Trends", Garmisch-Partenkirchen, 1986,
10. Semolič B., Priprava zagona projekta, Univerza Maribor, EPF, Maribor, 1995,
11. Semolič B., Zlicar A., Kako obvladovati proizvodnjo orodij s pomočjo projektnega pristopa, posvetovanje "Orodjarstvo '96", Postojna, 1996,
12. Drstvenšek, I., Živec, Z., Balič, J.: Informational Flow Assurance In Small And Medium Sized Enterprises, Proceedings of 12th international conference BIAM '94, KoREMA, Zagreb, 1994,
13. Balič J., Čus F., Simultaneous Engineering for Die-Mould Industry – a Practical Approach, FAIM'96, May, 1996, Atlanta, Begell House inc.,

Architecture for low volume AGV controlled FMS

Fergus G. Maughan, and Huw J. Lewis
University of Limerick
Limerick, Ireland

ABSTRACT

As computer integrated manufacture expands the search for lower cost and reliable machine communication links continues. In a typical Flexible Manufacturing System (FMS) there is a wide range of equipment available which may be used such as personal computers, Programmable logical controllers (PLC), CNC machines or robots. Each of these having their own "intelligence" and library of data. Providing a means of communication between a series of these individual controllers and the host controller has traditionally been achieved using a Local Area Network (LAN). The complexity of successful integration of each work station using a LAN results in a relatively high cost system. It is the objective of this project to devise an alternative to a LAN, in a flexible manufacturing cell, for communication of each station to the cells' controller. In the proposed solution an automatically guided vehicle (AGV) will be used as both the materials handling unit and the communications line, linking the work stations to the cell controller which will contain the scheduling algorithms for the materials within the cell.

Communication between the work stations to the host controller via the AGV, will be achieved by infra red data transmission eliminating hard wiring and network protocols. The AGV will operate in both download and upload mode. Downloading the necessary instructions with the machining file and uploading information the current status of the machine such as tooling or set-up constraints effecting material scheduling. The AGV acting as the host will have knowledge of the current status of the manufacturing cell as if it were connected to each work station by a LAN.

Scheduling of materials within the cell will also be performed by the AGV on board computer, being updated each time a work station is visited by the AGV which will in turn inform the host as to the condition of the cell.

Keywords, Flexible Manufacturing Systems, Cell Controller, Automatically Guided Vehicle.

INTRODUCTION

This project is concerned with designing a shop floor control architecture for the manufacture of parts in a batch production environment. The existing layout consists of CNC machines, PLC controlled machines and manually operated machines. Due to the high cost of converting all machines to CNC control and implementing a full scale Flexible Manufacturing System with a expensive complex control system such as, hierarchical control, centralised control, heterarchical control, or hybrid control, and the cost of linking all these devices together it is necessary to develop a low cost control system for implementing a Flexible Manufacturing System to the shop floor.

Chen *et. al.*, describes a Flexible Manufacturing Systems as,

> *"...a class of highly automated systems which consist of (1) numerically controlled (NC, CNC, DNC) machining tools, (2) an automated materials handling system (MHS) that moves parts and sometimes tools through the system, and (3) an overall computer control network that coordinates the machine tools, the material handling, and the parts..."* [1]

However these philosophies, which describe Flexible Manufacturing Systems, neglected to include the human intervention in FMS. Such as a cell of machines containing turret lathes drill presses or milling machines, which are manually controlled by the operator who receives instructions on a print out from the cell controller, which in turn is linked to the CIM system.

In order for small to medium sized Irish firms to adapt to Flexible Manufacturing Systems, stand-alone manually operated machines will have to be incorporate into the flexible manufacturing concept.

These Flexible Manufacturing System will have to consist of a group of machines, not necessarily all CNC control, arranged and controlled by an "intelligent" system which instructs cell controllers or human operators to perform, in a optimal manner, manufacturing operations on a variety of components.

In the proposed solution, the decision function (planing and scheduling) and the execution function (material transportation and program delivery) will be separated. This separation allows the development of both functions independently. Also the only means of communication between the cell controller and the AGV controller, the AGV controller and the FMS host, is through a standard serial infra-red link.

Components of the proposed software

Scheduling and routing

As Wysk *et al.*[2] described, planning has been attributed to the selecting of tasks that the manufacturing system will preform, scheduling as identifying a good sequence for these planned tasks based, on some performance criteria and execution as preforming the scheduled tasks through the direct interfaces with the physical equipment. These functions will be preformed by the AGV on board computer. Information flow between these devices is conducted in a hierarchical fashion as shown in Figure 1. For updating of the hierarchical controllers, information flow is conducted in the opposite fashion.

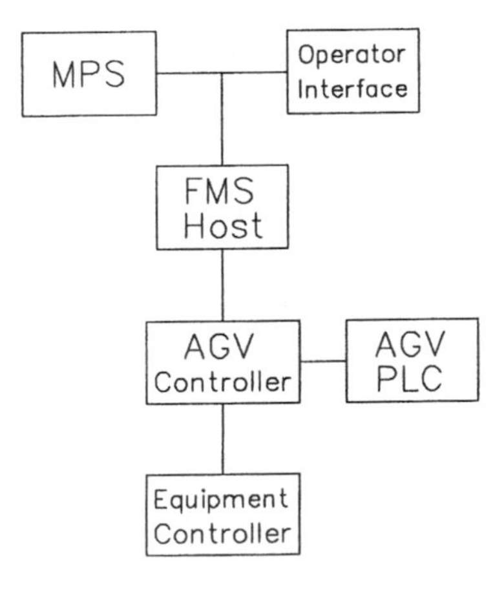

Figure 1: Information Flow

The function of the AGV on board computer (the decision maker) will be to allocate parts thorough the cells and the machines and setting priorities between individual parts at the processing station. In Figure 3, this would relate to the loading and unloading of parts into the Flexible Manufacturing Systems, the loading and unloading of parts into each cell and the prioritising of parts for machining in each cell. These decisions being made by the AGV on board control computer. When the AGVs on board computer receives a production requirement information from the FMS controller, it is the responsibility of the AGV to ensure the parts are delivered to the correct cell. The FMS controller receives its production requirement information for either an operator input at the FMS host or from the *master production schedule* through the plant network. The production requirement information is defined in the *process plan*. The process plan contains the necessary job instruction to process a part, indicating machining precedence, part routing, tools and fixture requirements, etc..

With this structure the decision maker does not concern its self with the operations the AGV performs to execute the decision makers instructions. By separating the decision controller and the transportation decisions, the decision makers functions are subsequentially reduced and simplified. The separation of these functions has been described in detail by Smith *et al.*[3].

The host computer will simply instruct the AGV on board computer to pick up a part at the load/unload station and instruct it as to the sequence of operations it requires, for example a part must be worked on machine 1 then machine 3 at cell 2, before proceeding to cell 1 for work on machine 1. When the AGV on board computer receives this information a part file is generated, which will log the work done on the part as the sequence of operations is executed.

Execution System

The execution system consists of two levels of hierarchical control consisting of the instructor and the transporter. The instructor which determine where a particular part must be delivered to, depending on it's Process Plan, instructs the AGV to begin transportation. Transportation commences when the AGVs PLC receives a signal from the instructor by means of an input signal from the output of the serial port of the instructing PC. A graphical representation is shown in Figure 2.

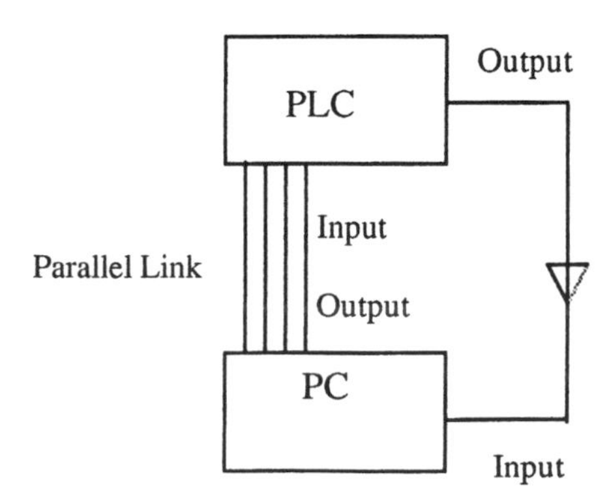

Figure 2 PC link to PLC

When the PC decides what station a part is to be delivered to a signal is sent through the 4 bit bus, allowing 15 (omitting 0000) different instructions or destinations to be instructed to the PLC. The PLC on receiving an instruction drives to the destination. On arrival the PLC signals the PC as to it's location. The PC can then commence downloading programs, printing job sheets and dropping off parts. A schematic layout of the FMS system is shown in Figure 3.

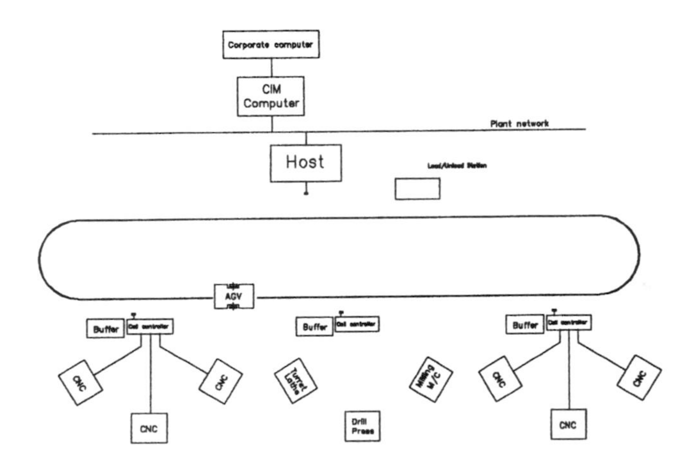

Figure 3 FMS layout

DESCRIPTION OF OPERATION

The transportation of all parts is achieved by use of an Automatically Guided Vehicle (AGV). Parts are carried on a pallet fixture to accommodate transportation by the AGV. Several parts may be fixed to a pallet. In this context, The pallet and the parts contained will be considered to be a part group, which is the smallest unit that can be assigned to a work cell. However within the work cell the parts and the pallet are not considered to be a part group since the operator can remove individual parts for machining. For example, the AGV picks up a pallet of parts from the load/unload station of the FMS and delivers it to a work cell. The AGV on arrival drops off the pallet containing the parts at the cell. The operator can now remove the parts from the pallet, machine each one separately, and replace them back on the pallet where the AGV will treat them as a part group.

There are three classes of primary functions within this architecture, being control, transportation and equipment. Within the control function there are a further three categories, 1) the FMS controller, 2) the AGV on board controller, and 3) the cell controller. The functions of each of these controllers will be described in the following section.

Jones and Saleh [4]describe three possible ways of providing each controller access to various types of process information.

1) Each controller receives its information from it's superior,
2) Each controller has it's own database management system,
3) Each controller must have a interface to a global CIM database management system.

Since it is the objective of these architectures to abstain from using a LAN, the third option in not economically viable, since the use of an Infra Red (IR) or Radio Frequency (RF) wireless network incurs high cost. The system described here, aims to use a combination of the first two alternatives. For example the FMS controller will download CNC programs to the AVG on board computer, which will both store the program for future use and download the program to the necessary CNC machine.

As described by Smith *et al.*[3] the equipment level controller and its subordinate machine are referred to as a single piece of equipment. With this system the machine controller monitors all the machines within the cell, keeps track of part locations, and monitors the operation of the machines under its supervision. The individual machine having their own controller, which may be a CNC controller a Programmable Logic Controller (PLC) or another form of controller standard to the specific machine, effetely operate on a stand alone basis.

FUNCTION OF THE FLEXIBLE MANUFACTURING SYSTEM

Operating procedure

When a process plan has been developed, it is transferred to the AVG on board computer along with the relevant CNC programs. Assuming the system is idle the AGV will firstly pickup a pallet of parts. The first "un ticked" line of the process plan is then scanned and the destination for the pallet either identified or selected by the AGVs task allocation decision rules (SPT, LPT, FCFS etc.). The pallet is the delivered to its destination, a process plan printed and the relevant CNC programs downloaded to the cell controller computer. Should a manufacturing cell which consist wholly of stand alone machines be selected, CNC programs are irreverent. In this scenario the parts are delivered to the cell buffer and process plan printed out. For stand alone machines a more detail job instruction may be required. These job instructions include detailed drawings machining parameters and machining precedence. The AGV also receives updated machine files from the cell controller. Should a machine become off line or is restricted to a particular machining operation, due to a mechanical failure, it is necessary for the AGV controller to be aware of its status, for successful task allocation .

FMS controller

The FMS controller can operate on a "stand alone" basis with process plan developed either at the station or off line and copied on to the controller for processing.

The controller being operated on a stand alone basis allows greater maneuverability to the system, for example the FMS control hardware and software could be relocated in a different section of the plant to accommodate seasonal demands

Principle of operation of the AGV controller

When a process plan for a new batch of parts is received by the AGV a destination for the parts must be identified. This is achieved by the AGV controller scanning the digital process plan and identifying the first process in the process plan.

On identifying the operation a decision is made as to where the pallet is to be transported to. This is achieved by the AGV controller identifying what machines are capable of preforming the required operation for the parts. The AGV uses the machine file data, which is received from the system controller, to perform such a task On identifying the capable machines, a machine must be selected that is, if more than one is capable of preforming the operation. This may be achieved by using one of many basic rules such as Shortest Processing Time (SPT) for the Pallet, Longest Processing Time (LPT), taking into account existing Work in Progress (WIP) for the machine, or the first machine to come available. (Upton[5] describes a similar system where each pallet contained a RF transmitter & receiver to communicate to the AGV host, Shaw[6] describes a network bidding scheme).

When a decision is reached as to the destination of the pallet the AGV must deliver the pallet to the assigned location. If a case should arise that the selected machine or cell is already busy, the pallet of parts is stored in the cell buffer until the machine becomes idle.

Principle of operation of the Cell Controller

The primary function of the cell controller is to monitor the operation of the cell equipment. It is the responsibility of the cell controller to update tool file data with the assistance of the detailed process plan.

For downloading the CNC programs to the CNC machines the operator identifies the necessary CNC program form reading the process plan. When in Download Mode of the user interface, the operator is prompted for the part no., the destination machine, and the CNC program name. This system allows more flexibility for the operator. The operator can decide on the destination machine to achieved an optimal through put time by scheduling at a local level.

CONCLUSION

FMS technology is not a manufacturing practice that would be adopted by Irish firms overnight. It is a slow learning process. One of its major drawbacks is the level of technology present in the software. Software engineering is very specialised and many manufacturing organisations find it difficult to develop high skill in this complex area.

The proposed system would be capable of being implemented without upsetting production in a existing plant. On successful implementation management can get a feel for the low level FMS and see the benefits and effects of integrating FMS into an existing system. Form this small scale system FMS can grow within the plant at a rate which company can control, allowing the management and workers to familiarise themselves with the new manufacturing concept.

This model shows that a Flexible Manufacturing Systems could be implemented without the high cost associated with converting all machines CNC machines, automated handling equipment and centralised computer control by means of a distributed system utilising the intervention of human operators and stand alone manual machines.

REFERENCES

[1] Chen I.J., Chung C.S., *Sequential Modeling of the Planning and Scheduling Problems of Flexible Manufacturing Systems,* Journal of the Operational Research Society (1996) 47, pp 1216-1227

[2] Wysk R.A., Smith J.S. *A Formal Functional Characterization of Shop Floor Control.* Computers in Industrial Engineering, Vol. 28, No. 3, (1995), pp. 631-644.

[3] Smith, J.S., Joshi, S.B. *A Shop Floor Control Architecture for Computer Integrated Manufacturing,* International Journal of Computer Integrated Manufacture, (1994).

[4] Jones, A.T, Saleh, A., *A Multi-level/Multi-layer Architecture for Intelligent Shop Floor Control,* International Journal of Computer Integrated Manufacture, Vol. 3, No. 1,(1990), pp.60-70.

[5] Upton M. David, *A Flexible Control Structure for Computer-Control Manufacturing Systems*: Manufacturing Review Volume 5(1), (1992) pp 58-74

[6] Shaw M., Wiegand G. *Intelligent Information Processing in FMS,* The FMS Magazine, 6(3), 137-140, (1988), IFS Publications.

A TWO-HOLE CLAMPING STRATEGY WITH GEOMETRIC TOLERANCING FOR SETUP PLANNING

Jueng-Shing Hwang, William A. Miller
Industrial and Management System Engineering
University of South Florida
Tampa, FL 33620

ABSTRACT: This paper describes a method for integrating a clamping strategy with geometric tolerancing information to automate setup planning for machining prismatic parts. This procedure, referred to as a two-hole clamping strategy, uses text and graphical forms to provide individual feature and feature-related descriptions on engineering drawings. To provide more alternatives for setup planning, the two-hole clamping strategy is used instead of the traditional vice clamping strategy. Each strategy can assign a variety of priorities for the feature interactions. According to patterns on the part designs, a grid-based algorithm can be applied to find a set of pilot holes for feasible machining sequences. The clamping strategy is addressed in three ways: tolerance effect, sequence of reference surfaces and sequence of features.

INTRODUCTION

In recent years significant research has been done to develop computer software methods for generating satisfactory sets of machining features for automating process planning. However, one aspect of process planning research that has received little attention is clamping and its influence on setup planning.

This paper describes a two-hole clamping strategy for automating setup planning, taking into consideration geometric tolerancing information. Usually, experienced planners have several alternatives for deciding how to clamp a workpiece and develop a logical processing sequence. To implement a multiple alternative manufacturability analysis, process planning software needs more than fragmented heuristic rules. The results of this research provide one new procedure that could be used to model the "multiple alternatives" planning strategy without a complicated fixture design.

The central issue of a clamping strategy is to decide the "appropriateness" of a setup. Workpiece, cutting tools and clamping devices, are three of the most crucial factors for determining appropriateness of a setup. By considering the mutual interactions of workpiece, cutting tools, and clamping devices, the valid sets of machining features can be generated. A two-hole clamping strategy has been implemented to handle the appropriateness of the given case under different levels of feature interactions.

LITERATURE REVIEW

Setup planning is mainly concerned with the grouping of features into setups and the setup sequencing in process planning. Such grouping and sequencing usually need to consider the complication of fixture design [Ong and Nee, 1994]. Ong and Nee proposed a setup system using fuzzy set theory to carry out the setup sequence. Sakurai developed algorithmic and heuristic methods to synthesize and analyze the setup planning for fixture configurations [Sakurai, 1992]. Although there have been many research efforts concerning the automatic generation of process plans for machining, setup planning is one of the least studied areas.

The appropriateness of each setup dominates the consistence between grouping and sequencing machinable features. To group the features into different setups and to sequence the features within each setup is an iterative process. Maintaining the consistence between grouping and sequencing depends on deciding the appropriateness of each setup. Appropriateness of each setup depends on applying an individual strategy to each interpreted characteristic of a given case and combining each useful item of information for decision making with a perspective view. So far, very little research has been done in this area usingf human strategy modeling.

More and more attention is beening given to combining heuristics with algorithms in setup planning research. There are two major focuses. The first focus is to develop certain forms of meta rules where rules need to be categorized and controlled according to the case situations. Instead of capturing human knowledge only using IF-THEN rules, higher abstract representations are needed. The second focus is to formulate more algorithms using subjective value systems such as using fuzzy numbers and de-fuzzifying the number instead of an objective function as in linear programming. These two efforts greatly affect in modeling of the human experts' decision making process. The purpose is to allow more flexible modeling for handling complicated cases.

In present practice, there are multiple alternatives for a human expert to devise a setup plan. Each human expert groups the features according to the process planner's strategy. Then, by taking advantage of the recognized characteristics at each feature, as well as the whole case, the feasible solution can be developed in favor of his/her strategy. The present modeling of a human expert still cannot describe the characteristics of the pattern into a perspective view. Such a perspective view can enable a single strategy to be applicable to various cases. Thus, the complexity in modeling the perspective view has a dynamic property caused by the interactions between summarizing the case and developing detailed interpretation. Such abstract reasoning in strategy decision making is not clearly defined yet.

Besides setup sequencing, tolerance analysis is crucial to process planning. Few CAPP systems are capable of tolerance analysis. At most, tolerance analysis is used to check the dimensional tolerance stack-up rather than to generate feasible setup/process plans [Chen, 1993]. Recently, researchers [Huang and Gu, 1994] have begun to recognize the importance of tolerance analysis in generating appropriate setup/process plans. Although some rules of thumb have been developed for automating setup planning, no systematic approaches are available in the literature. Two missing focuses are semantics in tolerances and the positioning of the clamping devices, which affect the critical decisions when a clamping strategy applies to setup planning.

The motivation for this research was to find a procedure to verify the appropriateness of a setup plan for different strategies under various setup situations. The appropriateness of a setup can include the clamping positions, feature availability of tools and feasible sequence in reference to surfaces and machinable features. Such an appropriateness of each setup plan should be evaluated by different clamping strategies to decide the least compromised alternative for meeting the requirements of reference surfaces, setups and tools.

COMPUTER AIDED SETUP PLANNING

Setup planning usually includes at least two activities: machining of reference surfaces and machining of features. For machining reference surfaces, a machining sequence of reference surfaces is determined through the selection of locating, supporting and clamping surfaces according to the status of clamping devices. Meanwhile, for the machining of features, the features with the same tool accessible directions are grouped into one setup. During the reasoning process, reference surfaces and features that can be machined in the same setup are searched and grouped together. To minimize time and cost, as many of these as possible are grouped together.

When considering more sophisticated scenarios, setup planning includes: (1) grouping features into setups, (2) selecting locating datums for the setups, and (3) sequencing the setups. Theoretically, design datums should be used as locating datums. However, the complexity of a given case can greatly affect the strategy of using datums for sequencing. The selection of locating datums is critical to tolerance control and sequencing in setup planning.

Under the same clamping strategy, three methods can be used to sequence the features when bilateral precedence feature relationships are considered. They are: (1) machining two features in the same setup, (2) using one feature as the locating datum and machining the other later, and (3) using an intermediate locating datum to machine the two features in different setups. The latter is the most complicated method in setup planning.

The decision of determining the appropriateness of a setup plan includes three characteristics: (1) availability of a feature to reference surfaces, clamping devices and the cutting tools; (2) specification of precedence relationships in positive interactions and negative interactions between clamping devices and machine or clamping device and cutting tools, and (3) alternative bonding to connect the features by general guidelines. Clamping strategies allow subjective values to be assigned to all characteristics through sequencing the feature operations.

Setup Problem

Compared with the traditional approach in operation sequencing, solving a clamping strategy here is actually solving a partial ordering problem instead of a classical sequencing problem. The traditional approach sees the problem as a well-defined searching process achievable by one search, where the parameters of each feature are predefined and the boundaries clearly stated. However, human experts evaluate and prioritize features with subjective values according to the characteristics of the patterns. Most of all, human experts devise the answer through connecting the partial solutions.

To consider multiple setups, conflicts among clamping devices, tools and parts need to be resolved before the partial feasible sequence can be generated. To identify the needed elements in a partial sequence, the feasibility of a setup plan must be verified for machines, clamping devices and tools, plus their interactions. However, the verification depends on the status of machines, clamping devices, and tools. Among these three, clamping devices are a critical variable that influence the feasible clamping regions on each part. The availability of each feature referring to the machining condition strongly depends on the positions of clamping devices and the interaction between tool accessible direction and clamping devices.

In this paper, the part assumed to be machined begins as a solid rectangular block with setups on a three-axis vertical milling machine. This research focus is on regions suitable for clamping and interactions of clamping devices with machining operations. In present expert systems, regions suitable for clamping are not considered nor expressed explicitly, which affects the availability of each machinable feature during sequencing a setup or setups. Interactions of clamping devices with machining operations can be considered in two ways: the hidden meanings of feature priorities and the tool accessibility of each feature when the clamping position is specified.

CLAMPING STRATEGIES

Clamping strategies establish the consistency by assigning subjective values to process planning such as selecting the reference surface using the clamping device, determining the sequence for reference surfaces, prioritizing the feature relationships, and connecting the features when no explicit constraints are applicable. To apply a clamping strategy to a set of machinable features is to link the needed operations in a dynamic hierarchical structure. Each clamping strategy provides alternatives including machining work pieces, holding work pieces on the table and identifying suitable features from setup to setup.

Developing a clamping strategy contains at least four steps: pre-process the user input, detect the hole position, sequence the surfaces, and sequence the feature within each surface. To pre-process the given user input, usually the input has two parts, text description and graphic representation. The symbolic meaning is more dependent on the text part, which indicates the shape, size, dimensions, etc. Graphic representation reflects the spatial relationships and critical areas in terms of high tolerance. In graphic representation, tolerance information is crucial in determining the spatial relationships, especially the feature stack-up situation and clustering the feature in bilateral precedence relationships. The existing datum can be used to identify the leading element among bilateral precedence feature relationships. Datums affect the decisions of clustering two or more features to be grouped into the same setup; whenever, these features belong to the same surface and have the same tool accessible directions. Moreover, when two features do not belong to the same setup but use one datum as the locating datum (for example, parallel or perpendicular notation) the other feature can be grouped to later setups to satisfy the bilateral precedence relationship.

A two-hole clamping technique representing the use of two precisely drilled holes, called two auxiliary holes (Figure 1), is illustrated by an example case (Figure 2). The idea of using these two holes is to articulate a plan according to the principle of reducing the number of setups. By using two holes to bolt the part to the workable, all the machinable features can be related to the parallel planes that are perpendicular to the holes or the axis of the holes. This approach is very useful when a part has a characteristic that most of the features are related to the top and bottom reference surfaces.

Two situations are considered for the initial setup. The two pilot holes can be machined in an internal or external style. The internal style assumes that the customer agrees to violate the original design by adding two extra holes to the part. With the external style, the two holes are placed outside the shape envelope, see Figure 3. On the engineering drawing, if hole patterns exist and they can be used, the algorithm will consider this possibility first. If there is no existing hole pattern, then possible hole patterns will be located and an optimal one will be determined by the diagonal length between these two holes. If there is difficulty in convincing the customer, the pilot holes can be machined outside the shape envelope. Before the final finishing operations, these two holes will be eliminated.

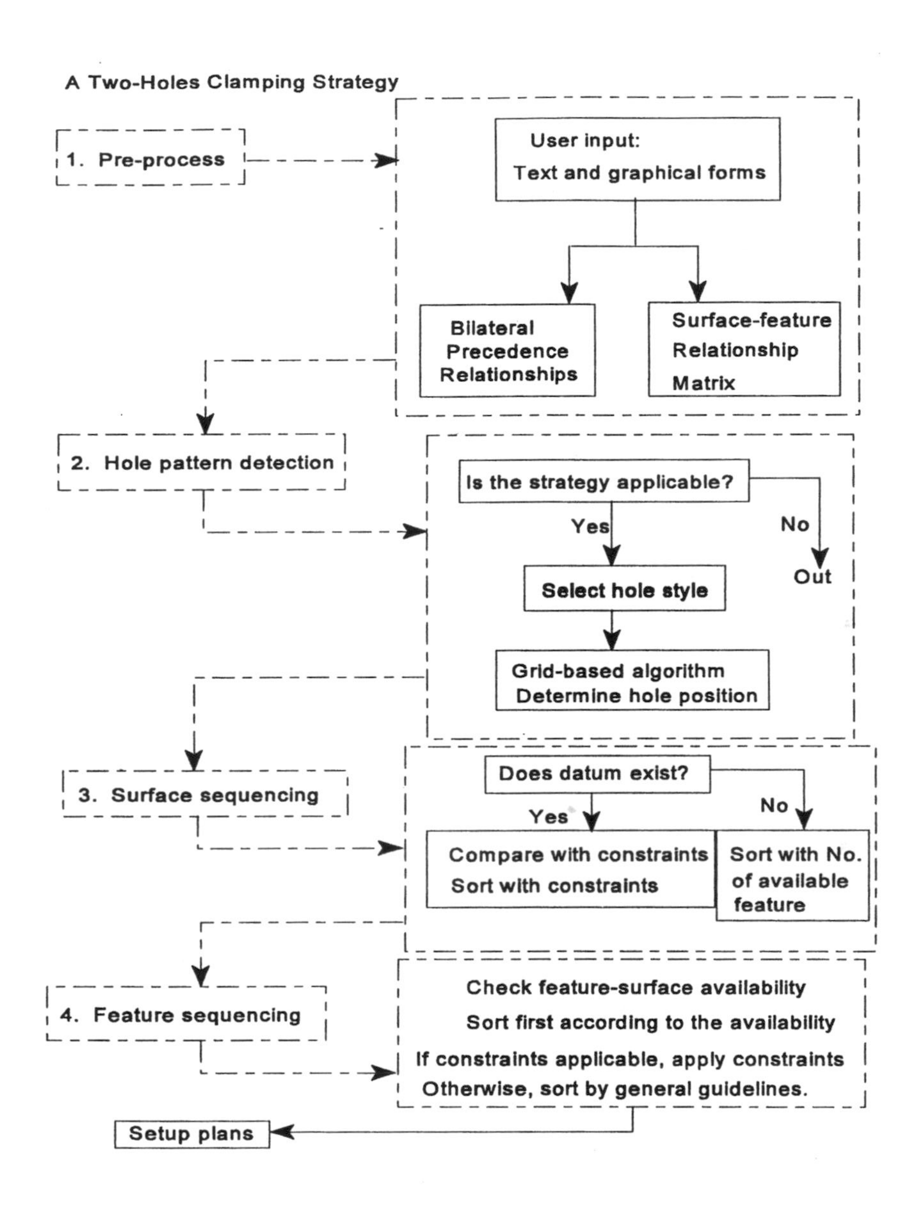

Figure 1. Flowchart for the Two-hole Clamping Strategy

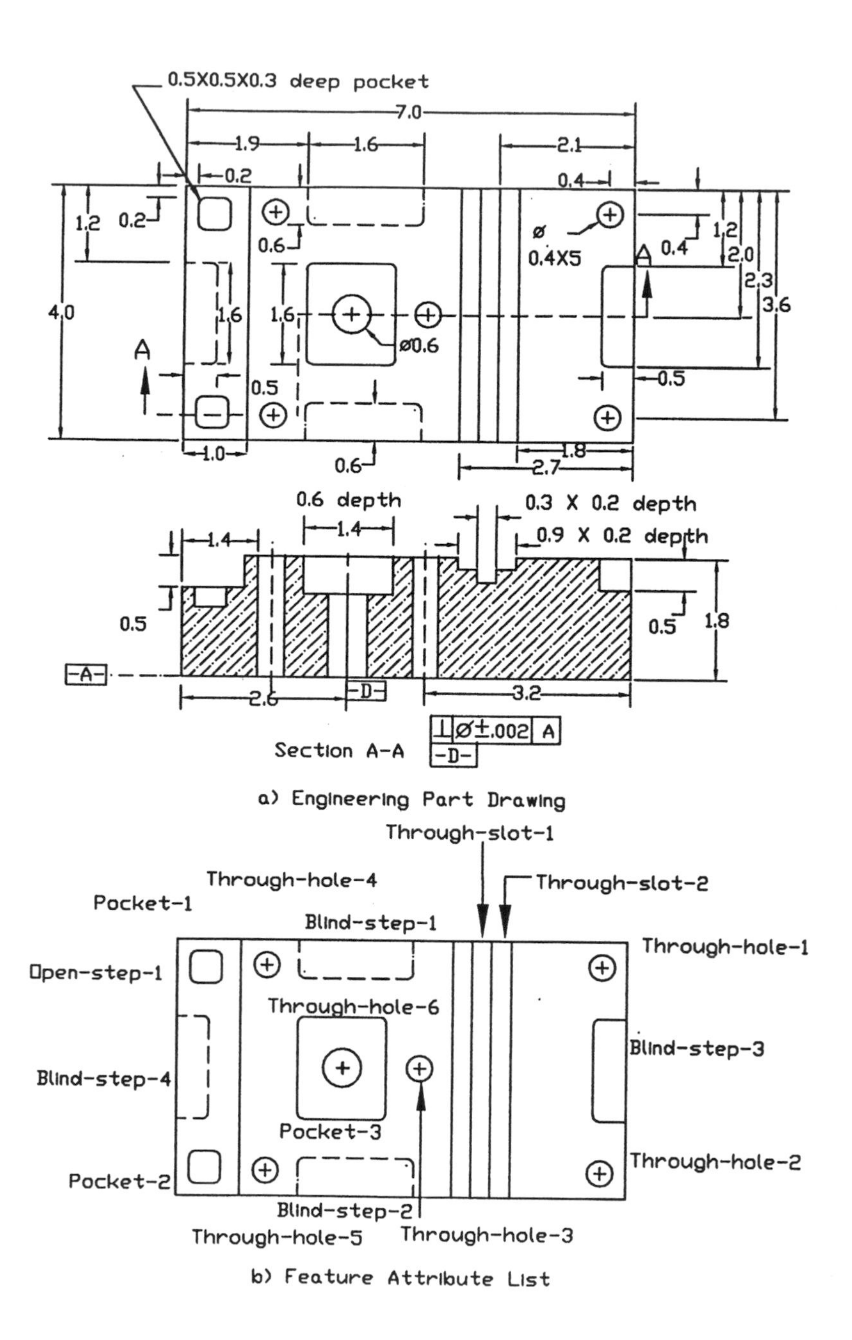

Figure 2. Engineering Part Drawing and List of Features

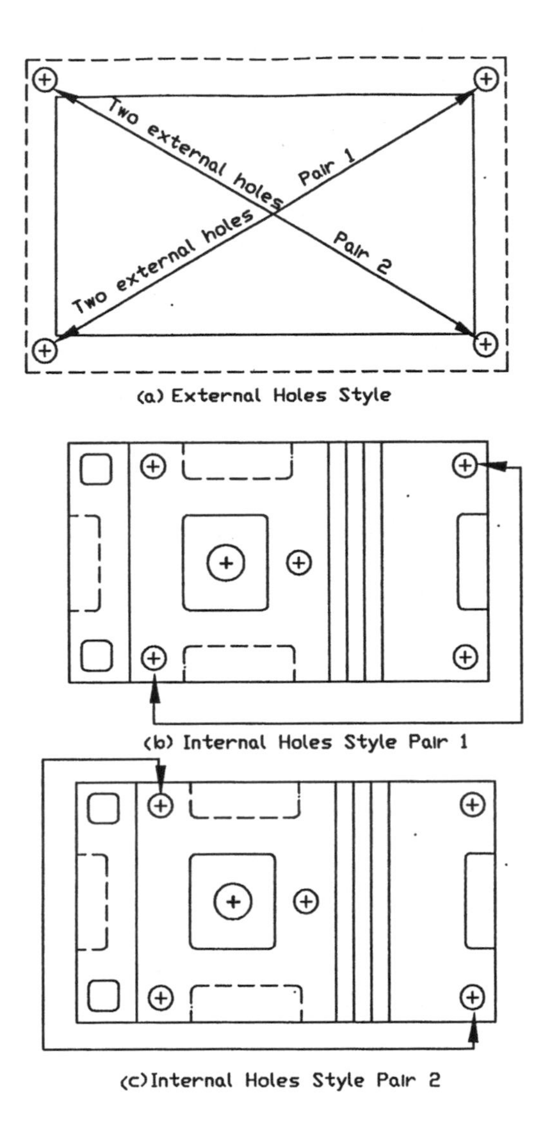

Figure 3. Hole Styles Locations

Geometric Tolerancing and Pre-Processing

The tolerance affects the setup planning in four ways: (1) the individual feature priority for the sequence, (2) the bilateral precedence relationships for critical features, (3) the hidden importance of each reference surface, and (4) the interactions with the spatial relationships under different strategies. Thus, tolerance generates greater complexity in the setup planning decision making processes.

Geometric tolerances are modeled mainly through feature control frames, which is not included in present setup systems or expert systems. A feature control frame usually contains four parts: the geometric characteristic symbol, tolerance value, modifiers, and datum references. The hidden meaning relies essentially on geometrical characteristic symbol and datum reference. Each geometrical characteristic symbol can imply the applicable setup sequences and tools. Besides the characteristic symbol, the datum reference can indicate multiple sets of bilateral precedence relationships as constraints. To convert each hidden meaning in the feature control frame is one of the major steps to pre-process the user input.

After converting the hidden bilateral precedence relationships, a square N by N matrix is used to represent the mutual relationship among N features. Using the upper triangle, the mutual precedence can be expressed in an increasing order from the first row as F1 to F2, F1 to F3, ... F1 to FN. The second row should be F2 to F3, F2 to F4, ... F2 to FN, until the second last row FN-1 to FN. The diagonal line from left top to right bottom should be all zeros to show that there is no precedence relationship for each feature to itself. Three kinds of notations are used to express the precedence, 1 for the row feature to precede the column feature, 0 for no precedence relationship, and -1 for the row feature to follow the column feature. To avoid the deadlock, each given single precedence relationship is checked as a chained reaction. Such as if F1 > F2 and F2 > F3, then the number in F1 to F3 should be filled with 1. After N loops to check the validity of chain reactions, the upper triangle will be transposed to lower triangle to express the bilateral precedence relationship in a decreasing order.

Interpreting the geometric tolerancing means that: (1) the text information must be separated from graphical representation to identify the semantics in tolerance, and (2) the hidden meaning in spatial related feature relationships must be extracted for decision factors according to the degree of interactions. Before the features can be prioritized, the subjective value of each feature and the precedence relationships among them need to be decided according to the strategies. Constraints derived from the given information such as "feature stack-up situations," are one of the crucial hidden feature relationships under investigation. Besides spatial relationships and tolerances, datums can be used as a leading element to structure a sub-list of operations.

Pre-processing contains two major matrices, surface-feature-relationship matrix and bilateral precedence relationship matrix. Surface-feature-relationship matrices depict how each feature attaches to the six outside surfaces of a shape envelope. Each feature should have at least one tool accessible direction and relate to at least one surface. The bilateral precedence relationship matrices describe the hidden meanings in text and graphic representation. As mentioned above, bilateral precedence relationship can use datums as a locating element to group features into different setups. Another important factor, spatial relationship, can determine the bilateral precedence relationship. Usually there are three kinds feature stack-up situations: flat group overlap with flat group feature, cylindrical group overlaps with cylindrical group, or flat group overlaps with cylindrical group. Each stack-up situation limits the positioning regions to the clamping devices and the tool accessibility to features. To determine the feature precedence in a stack-up situation according to the view, two circumstances are important, the removal volume for the feature and the hidden lines of the feature. To calculate the overlapped volume between two features, assuming the cutting tool comes from the top, the bigger removal volume always stacks on top of the smaller volume. If hidden lines exist, then a cross section can be used to see the actual stack-up without interfering with the other features.

Hole Patterns Detection

Before the reference surface can be machined, the position of pilot holes needs to be determined. According to the given case, six surfaces will be evaluated by sorting the maximum number of available features. Determining the hole pattern includes three major decisions: (1) checking the feasibility of using the two-hole strategy, (2) selecting the hole style, and (3) using a grid-based algorithm to determine the actual hole position. According to the given part designs, the feasibility of using the two-hole strategy needs to be verified with the existence of potential hole locations and stability of using two holes for clamping the part. If the hole patterns exists, but cannot satisfy the kinematic analysis, such as a bending moment which causes undesired deformation, the strategy is inapplicable in that situation. By considering the hole patterns, two options can be used to decide the pilot holes: drill two new holes or use the existing holes. Once the analysis suggests that the potential hole patterns exists, a grid-based algorithm can be used to search the hole position. In such a grid-based algorithm, each surface is divided into small pixels, each pixel can be articulated by given resolution in length and width. There are two possible situations: utilize the existing holes as the pilot holes for clamping the part on the table or drill new holes with customer's agreement.

Sequencing Reference Surfaces

To sequence the reference surface, two factors are considered, the existence of the datum and the number of features related to the reference surface according to the tool accessible directions. Tool accessible direction and clamping position indicate the availability of a feature to a reference surface. By assigning the value to those six reference surfaces, the priorities among these six can be sorted decreasingly.

Sequencing the Features

In setup planning for machining the features, the features that can be machined in the same setup are grouped and a machining sequence between these groups of features is determined. These are implemented by comparing feature attributes such as feature type, size, location, tool an accessible direction, geometrical tolerance, and related datums.

The basic access direction of each classified feature is a tool approach direction for machining a feature and it has one or more basic directions. If a feature has two or more basic directions, then a preferential unique direction of this is determined. The direction is determined by comparing tolerance and datum with other feature's and comparing position and size of each surface.

CONCLUSION

The two-hole strategy has three advantages over side clamping with a vice:

(1) Once the holes are drilled, the reference surface can be presented so all related features can be machined and the only need is to consider how to minimize the number of tool changes. When the features are scattered on two major opposite

surfaces, flipping these two surfaces will complete all or most of the work. The clamping effect, such as losing the accuracy or stability of the parts, can thus be simplified.

(2) The two-hole pattern can be used to locate the pilot holes to secure the parts on the table. Using the existing holes for clamping reduces the number of setups and generates the availability of the feature list in a quicker and easier way.

(3) The part can be secured without considering slippage of the part. The vice clamping method needs to consider the effective contacting area that is to prevent the slippage of gripping the parts during the cutting operation or losing the desired accuracy.

The proposed system implements three activities of setup planning: interpreting the hidden meaning of tolerance, sequencing the reference surfaces for machining and sequencing the machinable features. This system integrates these three activities according to the principles of maximizing the number of features available to one setup and reducing the number of tool changes. The tolerance meaning has been converted into bilateral precedence relationships for deciding the feature sequence. For sequencing the reference surfaces, the priorities can be assigned according to the selected setup methods.

This work has been performed using a relatively popular set of clamping strategies to illustrate the concept. This has provided an understanding from which a future industrial applicable setup planning module can be built. More IF-THEN rules in the knowledge-based can be replaced by using more flexible algorithms. Moreover, the independent algorithms can be applied to more comprehensive product data considering their interactions. This offers the prospect of design aids that can provide concurrent interactions between design for function and design for manufacture.

The use of strategy-oriented features has given this research a basis for identifying appropriate features for clamping and locating a workpiece to test the method employed. For more complex parts, a method of recognizing appropriate faces for clamping on different contours should be researched.

REFERENCES

Chen, C.L.P., "Setup Generation and Feature Sequencing Using Unsupervised Learning Algorithm," *Proceedings of the 1993 NSF Design and Manufacturing Systems Conference* 1, (1993): 981-86.

Huang, X. and P. Gu, "Tolerance Analysis in Setup and Fixture Planning for Precision Machining," *Proceedings of the 4th International Conference on Computer Integrated Manufacturing and Automation Technology,* Troy, NY: Rensselaer Polytechnic Institute, (1994): 298-305.

Ong, S.K. and A.Y.C., Nee, "Application of Fuzzy Set Theory to Setup Planning," *Annals of the CIRP* forty-three, no. 1, (1994): 137-44.

Sakurai, H., "Automatic Setup Planning and Fixture Design for Machining," *Journal of Manufacturing Systems* 11, no. 1, (1992): 30-37.

OVERVIEW OF THE DATA MODELS AND THE KNOWLEDGE SOURCES THAT FORM THE BACK-BONE OF AN INTERACTIVE CAPP KERNEL.

G. Van Zeir, J.-P. Kruth, J. Detand
Katholieke Universiteit Leuven,
Department of Mechanical Engineering, Division P.M.A,
Celestijnenlaan 300B, B-3001 Leuven, Belgium.
Tel. (+32) 16 32 24 80, Fax. (+32) 16 32 29 87
e-mail : Geert.VanZeir@mech.kuleuven.ac.be

ABSTRACT : This paper gives an overview of the data models and the knowledge sources that form the back-bone of an interactive Computer Aided Process Planning (CAPP) kernel. The CAPP kernel is based on a blackboard system architecture. The paper describes the information that resides on the blackboard and how this information is instantiated : the blackboard approach provides the human process planner with assisting 'expert modules', each capable of performing a specific automated process planning task. In each process planning step that is performed by such expert module, the human operator has full control over the generation of the process plan and can always change the solution proposed by the expert modules.

INTRODUCTION

The magnitude of the CAPP problem has been consistently underestimated. The very nature of human problem solving has probably been the primary reason for this dilemma. Recent research has shown that human cognition, when addressed to a problem like process planning, incorporates a vast network of concepts and perceptions. Indeed, human beings tend to perform planning in an opportunistic fashion by intermittently postulating goals, generating sub-goals, gathering constraints, reformulating ideas, exploring consequences, etc., until an acceptable solution has been reached. Therefore, process engineers do not adhere to rigid CAPP algorithms for the specification of which step to take, given every possible situation.[1]

To force complete order on the process planning procedure through the development of a rigid process planning algorithm, is to lose most of what makes human process engineers such adaptable problem solvers. It is clear that an effective CAPP system should provide comprehensive means for user interaction. Ultimately, to computerise/automate the process planning problem is to understand what is at the core of human cognition and to transform this into models.

In this paper, a model is presented for an interactive CAPP kernel. This kernel is based on a blackboard system architecture, because this approach represents a good approximation of the way process planners plan in real world.[2] Indeed, they do not always plan in a hierarchically layered manner, but rather, they will often employ an opportunistic reasoning, beginning with difficult features that constrain the plan, making preliminary decisions based on those features, and exploring the ramifications of those decisions, independent of other part features.

In a first section of this paper, the object-oriented taxonomy of the developed CAPP kernel is addressed. A second section specifies the diverse knowledge sources and their relation with the data models is explained.

BLACKBOARD BASED CAPP

Process planning deals with many diverse, specialised applications (call it process planning tasks) that have to be integrated in some way. Some planning tasks can be executed arbitrarily (e.g. selection of a specific tool before determining the machine or vice versa). However, the outcome of each task can (will) depend on the results returned by previously completed planning tasks (e.g. the selected tool can only be mounted on a limited set of machining centres). Furthermore, the knowledge representation in each planning application can be different (e.g. rule-base, Petri-net, neural net, fuzzy logic, table, algorithm,...). Such complex environment can typically be handled by a blackboard system.[3,4]

The architecture of a blackboard system can be seen as a number of people sitting in front of a blackboard. These people are independent specialists, working together to solve a problem, using the blackboard as the workspace for developing the solution. Problem solving begins when the problem statement and initial data are written onto the blackboard. The specialists watch the blackboard, looking for an opportunity to apply their expertise to the developing solution. When a specialist finds sufficient information to make a contribution, he records his contribution on the blackboard, solving a part of the problem and making new information available for other experts. This process of adding contributions onto the blackboard continues until the problem has been solved. A manager, separate from the individual experts, attempts to keep problem solving on track and ensures that all crucial aspects of the problem are receiving attention.

Translating this metaphor into a computerised blackboard system, the distinct specialists should be considered as expert modules and the blackboard as a global database containing input data, partial solutions and other data in various problem solving states. For triggering and controlling these expert modules, three ways of managing the blackboard could be distinguished :[2]

- User driven : The decision of which expert module (or expert) to call is in the hands of the user.
- Automatic triggering : The experts constantly observe the information on the blackboard and add new information as soon as they can. The manager mediates if experts have conflicting goals.
- Scenario driven : A scenario is designed in advance, that determines the behaviour and sequence of calling expert modules. This approach resembles somewhat the Part system, described by van Houten and Jonkers, or the DTM/CAPP system presented by Jasperse.[5-7]

In the presented developments, a user driven approach for managing the blackboard is chosen. It not only promotes the interactiveness or human involvement; it also makes the CAPP system transparent and facilitates the understanding of its structure, behaviour and outcome. The difference with (and advantage over) the PART or DTM/CAPP system is that the different expert modules can be called arbitrarily and more than once. Moreover, the 'automatic triggering' and 'scenario driven' approaches could easily be implemented in a later stage, since full flexibility is being offered.

The next paragraph explains how the blackboard data are organised. It proposes an object-oriented model that allows expert modules as well as the human expert to produce the information.

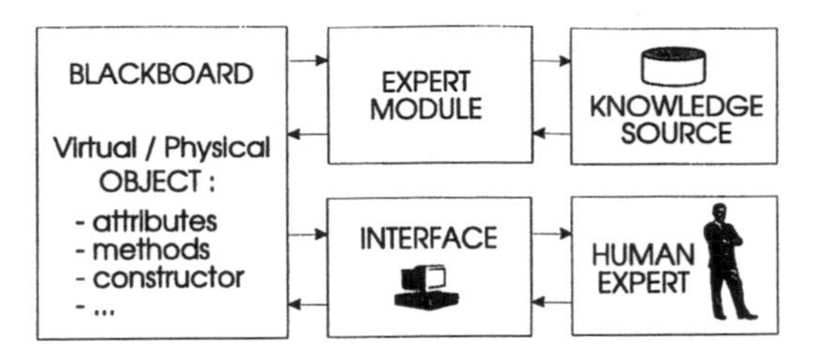

Figure 1. Object-oriented data model for blackboard systems

An object-oriented Blackboard model

Objects are used to represent the blackboard information. Each data-object has attributes (slots that contain information), a set of methods, a constructor, etc. for handling this object (figure 1). The information contained in such object can be supplied by :

- an *expert module* that consults the appropriate knowledge source,
- the *human expert*, by means of an interface that is provided for each type of object.

Consulted expert modules will take into account the information that was added by the human expert (or by other modules). Moreover, the interface to the human expert allows him/her to verify, accept or alter the information generated by an expert module, at any time.

Objects with partial information (empty attributes, attributes that describe parameter intervals or constraints or multiple discrete values rather than a fixed value,...) will be called *virtual objects*, while objects that are unambiguously determined by the information contained in their attributes will be called *physical objects*.

An object-oriented model for the developed Blackboard CAPP system

This paragraph illustrates how manufacturing data and knowledge have been incorporated in the CAPP system. The following data models are distinguished (figure 2) :

The blackboard. The CAPP blackboard contains both part and process planning information in various states during the process plan generation. The part description can be considered as the initial input data (see metaphor). From this input, new blackboard objects are created by expert modules or by the human operator.

A part model. The CAPP kernel is feature-based. Consequently, the part information, which serves as input to the CAPP system, should a/o contain a detailed description of the part's features. This model incorporates a/o company specific feature types.

A resource model. The resource model embeds machine tools, fixtures, tools, and other auxiliary equipment, available in the factory and considered during process planning. It includes all data that are important to inquire about during the process planning task (e.g. power, accuracy, outer dimensions, axis data, etc.).

A process model. This model contains the manufacturing processes that are used in the company (e.g. end-milling, face-turning, welding, laser-cutting, wire-EDM, etc.). Further, it embeds related process parameters (cutting conditions, costs, accuracy, etc.), and associated geometric constraints and technical parameters (roughness,...).

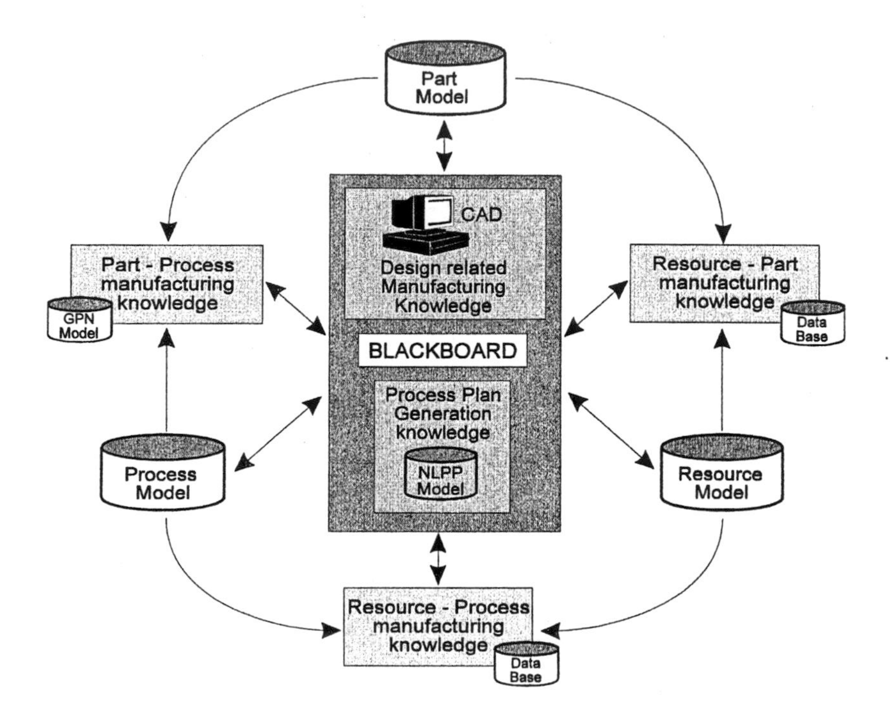

Figure 2. Data models and manufacturing knowledge for CAPP system.

A process plan model. The blackboard CAPP system supports graph-based process plans that allow the modelling of alternative manufacturing sequences. Such process plans with alternatives are called non-linear process plans or NLPP's.[8] All process plan data that are required for further order processing, manufacturing and all administrative data are included in the model. The process plan model contains the newly generated process plans (in NLPP format) of specific parts.

The outlined data classes are not just some isolated data structure but are interdependent and related to one another by some specific constructs. The manufacturing knowledge in a CAPP system holds the following relationships (figure 2) :

- The 'design related manufacturing knowledge' is employed for instance by the 'CAD expert' module, which allows to extract/add process planning information from/to the part design.
- The 'part-process manufacturing knowledge' associates the data content of the part model to the process model; it embodies the 'process selection expert' module which determines the different manufacturing steps to be undertaken on a certain part type or feature type (modelled in a Generic Petri Net or GPN), and the sequencing relationships between those manufacturing steps. The use of GPNs is explained in the next section.
- The 'resource-process manufacturing knowledge' associates the data of available resources to the process model; it embodies for instance the 'machine selection expert' which determines the candidate machine tools for the operations on a certain workpiece. However, this expert module will also have to consult the 'resource-part manufacturing knowledge' (e.g. because there are limitations on part dimensions for a certain machine tool). It can be modelled by means of tables in a data base.
- The 'resource-part manufacturing knowledge' relates the part data with the resource data (e.g. the selection of a tool is influenced by the dimensions of the feature to be processed). This type of

knowledge is consulted by for instance the 'tool selection expert' and the 'machine selection expert' modules, and can be modelled with tables in a data base.

- The 'process plan generation knowledge' encloses the knowledge that brings all other knowledge sources together. In this interactive CAPP kernel, the 'blackboard manager' that triggers the distinct process planning expert modules could be considered as part of this knowledge.

In figure 2, the 'data and knowledge loop' around the blackboard can be considered as the generic data and knowledge that is used to build a process plan for any given part. In contrast, the data residing on the blackboard, always refers to a specific part instance.

Generating a process plan requires the analysis of relevant part information, the selection of the right manufacturing processes and the appropriate resources thus building the objects on the CAPP blackboard. The following section elaborates on how the information of each blackboard data-object is supplied by triggering the different expert modules.

KNOWLEDGE SOURCES FOR BLACKBOARD CAPP

This section elaborates on the distinct knowledge sources (figure 2) and their relation with the different data models. It explains for each knowledge source :

- the *knowledge representation* : its content and how it is stored
- the *knowledge instantiation* : how to supply the information contained in each blackboard data-object (by expert modules or by human interaction)
- the aspect *virtual vs. physical object* : each object created on the blackboard is considered as a virtual object as long as some of its attributes remain unknown; when all attributes are determined, the virtual blackboard objects has evolved towards a physical one.

Design related manufacturing knowledge

This knowledge source provides a link between the CAD model and the manufacturing practices (figure 2). It embodies a/o the 'CAD expert' module. This module allows to extract/add process planning information from/to the part design. Also the 'set-up expert' module uses this knowledge to visualise information like set-ups, operation sequences, etc. on the CAD drawing. These expert modules are explained hereafter and their function is illustrated through some examples.

CAPP oriented part information extraction

Knowledge representation. In this era of CAD systems, it would be advantageous to use the part specification, which is stored in a CAD database, for CAPP purposes. Therefore, the CAPP kernel is provided with a CAD application program that serves as a link to several wire-frame and feature based CAD systems (Applicon Bravo, AutoCad, Unigraphics). This CAD interface can be considered as the 'CAD expert' module. The module supports the human process planner with analysing the geometric and technological information on the part drawing and extracts all part information that is relevant for the feature based process planning system (overall part data, feature data and feature relation data). The design related manufacturing knowledge is thus residing in the CAD drawing itself and in the implementation of the CAD expert module that can interpret this drawing.

Knowledge instantiation. According to the object-oriented blackboard model, explained previously, there are basically two ways to generate the part information (as objects on the blackboard) :

- Manual editing : The complete part description (part, features and feature relations, with corresponding parameters) can be entered by means of appropriate user interfaces, either via the CAD system or directly on the blackboard (e.g. if no CAD system is available).
- Expert consultation : The 'CAD expert' module transfers the part design from CAD to objects on the blackboard. However, some human interaction is required, before the module can generate a complete process planning oriented part description. This interaction consists of identifying the features interactively by selecting their geometry and associating it with the corresponding user defined feature type. When a feature is identified, the module can automatically inquire its geometric and technological parameters from the CAD database. Feature relations must be defined explicitly, as most commercial CAD systems do not explicitly model relations between features in their native database. Form and location tolerances, which also result in feature relations, are automatically recognised if contained in the CAD database.[9]

Figure 3 shows an example of the type of information that the 'CAD expert' module extracts from the CAD model. It illustrates how a part, its features and feature relations are specified. This information is placed in a part description file that can be inspected by the human operator, and which is translated by the CAD expert module into objects on the blackboard. To provide flexibility, the information gathered by the CAD expert module can always be manually changed or completed by the human operator.

```
WORKPIECE D5641256A (CARRIER)
{
 last_update       = "23.06.1996";
 last_upd_author = cad_capp;
 last_upd_module= "CAD EXPERT";
 WORKPIECE_DATA
 {
  LENGTH          = 60.000;
  WIDTH           = 60.000;
  HEIGHT          = 20.000;
 }
 FEATURES
 {
  hole_1          (DEEP_THROUGH_HOLE)
  {FEATURE_DATA
   {DIAMETER              = 10.000;
    DIAM_STRING_TOL  = H7;
    DIAM_UPPER_TOL   = 0.015;
    DIAM_LOWER_TOL   = 0.000;
    LENGTH                 = 60.000;
    ROUGHNESS          = 1.600;
    VOLUME                = 4710.000;}
   LOCATED_FEATURES
   {1
    {POSITION
     {10.000, 0.000, 12.500}
     ORIENTATION
     {0.000, 1.000, 0.000,
      0.000, 0.000, 1.000,
      1.000, 0.000, 0.000}}}}
...
```

```
...
  ext_b_pocket  (EXTERNAL_BLIND_POCKET)
  {FEATURE_DATA
   {LENGTH                      = 40.000;
    WIDTH                       = 32.000;
    WIDTH_UPPER_TOL = 0.100;
    WIDTH_LOWER_TOL= 0.050;
    ...
...
 RELATIONS
 {
  1  (PERPENDICULARITY_TOL)
  {object                        = hole_1;
   object_instance_id       = 1;
   reference                    = ext_b_pocket;
   reference_instance_id  = 13;
   RELATION_DATA
   { TOL_VALUE               = 0.010;
     TOL_ZONE                = LINEAR;}}
  2  (NEIGHBOUR_RELATION)
  {object                        = ext_b_pocket;
  object_instance_id        = 13;
  reference                     = hole_1;
  reference_instance_id   = 1;
  RELATION_DATA
   { WALL_THICKNESS   = 0.500;}}
...
}}}
```

Figure 3. Part information extracted by the 'CAD expert' module

Virtual vs. physical design objects. The part information is transformed into new objects (part, features and feature relations) on the blackboard. During the process of completing the information contained in these objects, virtual objects can be transformed to physical ones. To explain this concept, the case of a feature is considered as an example. A feature will evolve towards a physical object as more characteristics (parameters) of the instance of this type of feature are known. If for instance the corner radius of a pocket may range between 5 mm and 20 mm (e.g. because this parameter is not important for the feature's functionality), a wide range of tools may be valid to produce this feature. When the actual, tool is chosen, the actual radius will be determined.

Selection, visualisation and simulation of possible set-ups

Knowledge representation. The 'set-up expert' module takes 'manufacturing elements' as input. Each manufacturing element holds one or more 'manufacturing direction entities' (MDE). MDEs model the possible orientations of the tool with respect to the manufacturing element. A manufacturing element can be a 'feature' or an 'operation'.

In the first case (feature level), the manufacturing elements are a set of features and a machine tool on which to process the given features. The machine tool can be physical or virtual, depending on the level of information yet available on the blackboard; at least the kinematics of the machine should be known. The search algorithm of the set-up selection expert calculates the most economic number of set-ups, taking into account the machine-tool kinematics, the feature MDEs and possible constraints (e.g. due to feature tolerance relations).

In the second case (operation level), the manufacturing element is an operation; i.e. an aggregation of processes executed on one machine tool. Each process refers to a certain feature and inherits the MDEs from this feature. Again, the set-up expert module will calculate the most economic set-up plan for this group of processes, by means of the same algorithm.

The operation level approach will apply if operations have been selected prior to invoking the set-up module (i.e. selected operations are found on the blackboard). If no or insufficient operation data is available on the blackboard, the set-up selection module will automatically base its decisions on feature information.

Figure 4 summarises both approaches : in this example the MDEs indicate that the through hole can be drilled from +Z and -Z direction, while the step can be milled from -Z and -X direction. When features are used for set-up planning, the actual processes on those features are not yet known. They are however restricted by the MDEs of each feature. For example, the MDEs related to the step in figure 4 allow end-milling form the -X direction or peripheral milling from -Z direction.

Knowledge instantiation. The instantiation deals with the actual creation of a set-up plan for a given set of features or for a set of processes in an operation. Set-up planning can be performed manually or through consulting an expert :

- Manual editing : On the CAD drawing, the operator can indicate the features to be processed in one set-up. Also reference features or clamping surfaces can be interactively identified on the drawing. The appropriate attributes (e.g. referring to a certain set-up id, or indicating that it is a reference feature, etc.) are automatically created and attached to these features.

- Expert consultation : The 'set-up expert' module, reads the type of machines (e.g. 3, 4½, or 5 axis) from the blackboard. From the CAD drawing, the module gathers all MDEs and the position and orientation of the features w.r.t. the part.[10,11] With this information, the expert module calculates the set-up plan. For each set-up in the plan, the expert module indicates (1) which features are to be manufactured in this set-up; (2) the so called "optional features", which can be executed in this set-up or others; and (3) the positioning features (e.g. datum plane that is milled formerly in another set-up). Each set-up is visualised on the CAD screen by giving the different feature groups a specific colour. Further the module draws a symbolic representation of how the part should be clamped on the machine.

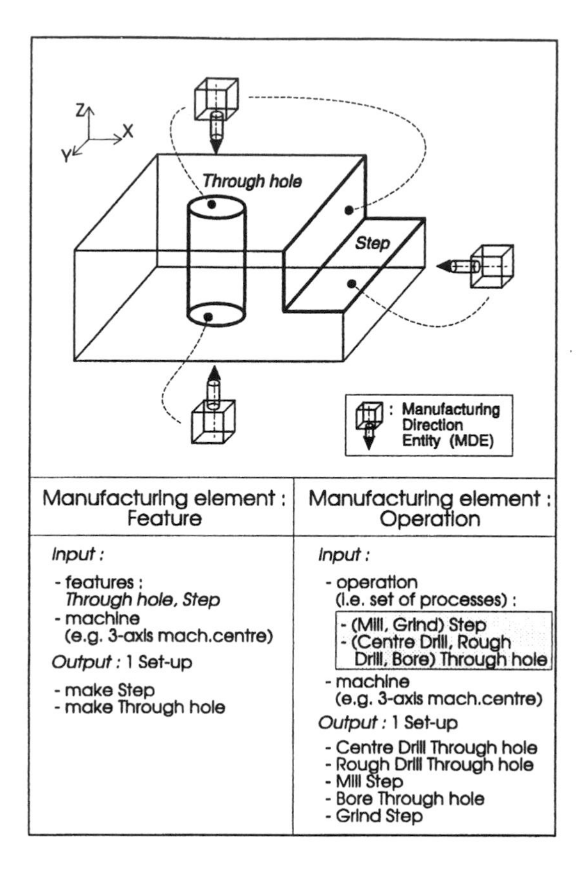

Z
X
Y
Through hole
Step
: Manufacturing
Direction
Entity (MDE)
Manufacturing element :
Feature
Manufacturing element :
Operation
Input :
- features :
Through hole, Step
- machine
(e.g. 3-axis mach.centre)
Output : 1 Set-up
- make Step
- make Through hole
Input :
- operation
(i.e. set of processes) :
- (Mill, Grind) Step
- (Centre Drill, Rough
Drill, Bore) Through hole
- machine
(e.g. 3-axis mach.centre)
Output : 1 Set-up
- Centre Drill Through hole
- Rough Drill Through hole
- Mill Step
- Bore Through hole
- Grind Step

Figure 4. Set-up planning schemes

When the manufacturing element is an operation, the processes are extracted from this operation. The set-up plan for the given collection of processes is calculated using the same algorithm (see also figure 4). Visualisation in the CAD screen is done by colouring the features on which the processes are performed.

Virtual vs. physical set-ups. There are four cases where the set-up can be considered as a virtual object on the blackboard :

- there exist optional features that still can be appointed to another set-up
- the design is still incomplete, and newly added features could be added to this set-up
- the planning is done only on feature level, not on process level.
- the fixture has not yet been designed/selected; note that one fixture can contain multiple (virtual) set-ups.

Part-process manufacturing knowledge

This knowledge relates the data content of the workpiece model to the process model. It is consulted mainly by the 'process selection expert' which determines the different manufacturing steps to be undertaken for a certain workpiece or feature, and the sequencing relationships between those manufacturing steps.

Knowledge representation. The CAPP kernel uses the concept of 'generic Petri nets' (GPNs) as a tool to model the manufacturing knowledge for **part types** (families) and **feature types**, which enables the CAPP kernel to support variant, generative and hybrid planning modes.[2,12] A generic Petri net represents company specific knowledge, structured in a graph, that describes all possible process routings for all conceivable instances of a specific feature type or workpiece type.

A part related GPN models processes that could typically be executed on that specific type of part, like 'saw from stock', 'final inspection', 'paint', 'electrolytic treatment', etc. A feature related GPN outlines the particular processes that could be performed on the feature type at hand (e.g. 'centre drill', 'die sink EDM', 'bore', 'hone', etc.).

In the example of the feature type GPN in figure 5, each rectangular block is a possible manufacturing step that is linked to a process from the process model.

GPNs generally provide manufacturing alternatives by *or-splits* (OS), *or-joins* (OJ), *conditional-splits* (CS) and *conditional-joins* (CJ). While an OS refers to alternatives that are valid under all circumstances, a branch succeeding a CS models a valid alternative only if the condition related to this branch is obeyed.

Knowledge instantiation. Like any other information on the blackboard, the processes to be executed on a given part, can be entered manually or retrieved from an expert module.

- Manual editing : The CAPP kernel offers a graphical editor (figure 6) that allows the operator to define a set (branched graph or sequence) of company specific manufacturing processes to be performed on the part or on its features. Moreover, process graphs or sequences of similar parts or features can be loaded into the editor. These graphs/sequences can still be edited according to the specific instance of the feature or part at hand (e.g. add or delete a specific process or a complete alternative branch).

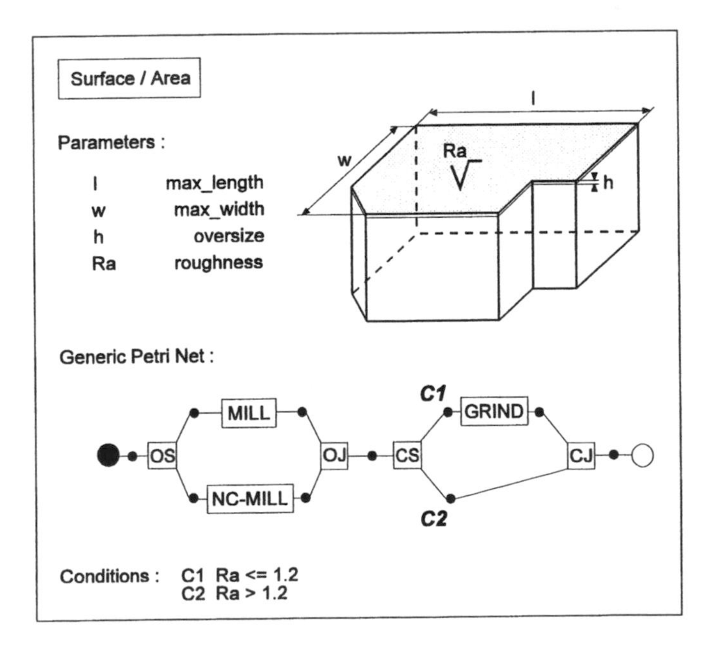

Figure 5. Example Generic Petri Net (GPN) for a 'Surface'

- Expert consultation : The ‘process selection expert’ module analyses the type of part and the associated feature types and retrieves the corresponding part and feature GPNs from the process model database. The module generates the *part-process knowledge* for the part and feature instances at hand, by ‘evaluating’ their GPNs. The process selection expert module does this by removing all branches after a conditional split (CS) for which the conditions do not meet the actual part or feature parameters (e.g. dimensions, roughness, etc.). An evaluated GPN contains no more conditional elements. It is not generic anymore, but only valid for the feature or part instance at hand. At this stage it is simply called a Petri net (PN). The resulting PN can be visualised by a graphical editor (figure 6). If the part-process manufacturing knowledge is incomplete, the user can still edit the result returned by the expert module.

Virtual vs. physical processes. The Petri nets only model the process types and sequences; no resource information is associated to the processes at this stage. Until all process parameters are determined, the process (created as an object on the blackboard) remains virtual. These parameters highly depend on the selected resources (machine, tool and fixture). When the resources are selected, the process parameters (e.g. spindle speed, feed rate, cutting force for milling, laser-power for laser cutting, etc.) can be calculated.

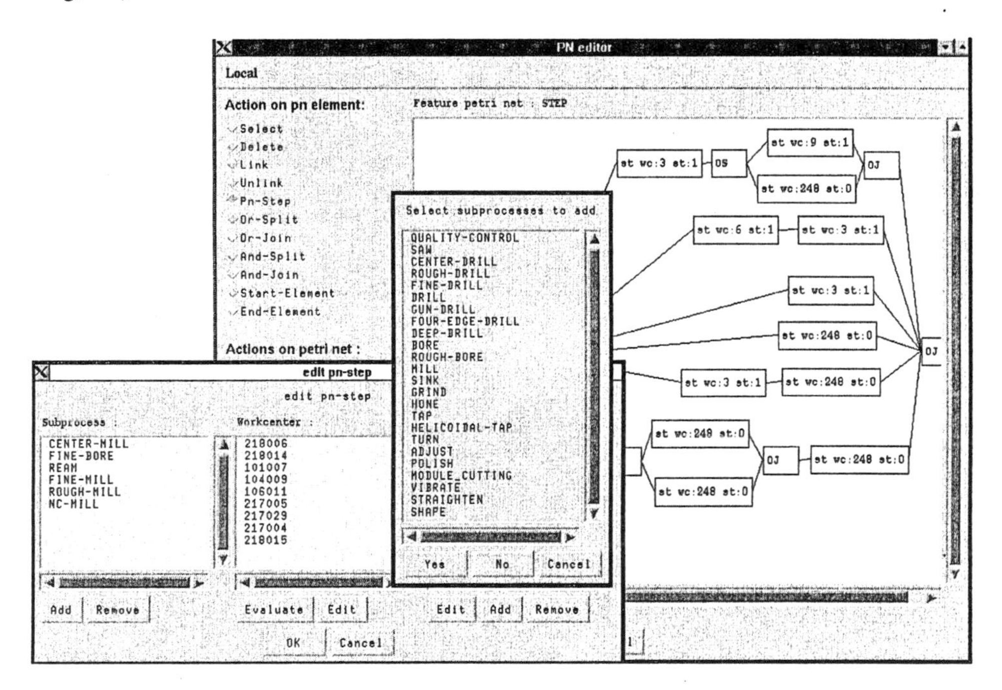

Figure 6. The graphical editor for modelling part-process knowledge

Resource related manufacturing knowledge

Resource related knowledge sources are consulted by different experts : the ‘machine selection expert’, the ‘tool selection expert’, the ‘fixture selection expert’,.... If an expert module tries to find a resource for a given process (e.g. a tool for deep-drilling), it will also have to take the part information into account (e.g. hole diameter, length, roughness,...). Another example : the selection of a resource (e.g. a fixture) for a given part will be influenced by the process parameters (e.g. the cutting force during turning). Therefore, the resource related manufacturing knowledge comprises both the resource-part and the resource-process related knowledge, explained here in this section.

Knowledge representation. Resources are modelled using tables in a relational database. Resource related knowledge is structured through links between these tables and other tables, delineating the capabilities of the resources. A machine tool for instance does not only refer to its parameters, statistical data (e.g. mean lead time, operation time, set-up time, etc.), axis data, etc. but also to feature and workpiece parameters that impose boundary conditions : i.e. a machine tool is capable of executing a process on a feature or part whose parameters (e.g. length, tolerance, roughness, etc.) are within the specified bounds.

Knowledge instantiation (The machine selection expert). The relation between the company specific processes and the company's machine tools is described in tables of the relational database. The machine selection expert module can generate standard SQL for executing the proper queries. It will thus return one or more machines that can perform a certain process. However, during the selection of a machine tool the expert module will also take into account the parameters of the part and its features, where necessary (e.g. a milling operation on a heavy part cannot be done on a tiny milling machine). The 'machining capabilities' table models the possibilities of each machine tool and parameter limits for a specific manufacturing process for manufacturing a specific geometry (e.g. maximum diameter for drilling in steel, maximum clamping force for milling, power, maximum spindle torque, maximum dimensions of the clamped part, kinematic range, etc.).[13]

Knowledge instantiation (The fixture selection expert). The selection of a fixture (modular fixture, standard vice, calibre,...) strongly depends on :

- the type of process : e.g. a chuck for turning or a modular fixture for milling
- the kinematics and dynamics of the machine : controllability, reachability, and accuracy of the distinct axes, the clamping facilities (vice, pallet exchange device, etc.).
- the geometric part data : e.g. positioning and clamping points, geometry to manufacture, machining direction, weight, etc.
- the number of set-ups, returned by the set-up selection module : e.g. a fixture or pallet can comprise two or more set-ups (rotation-table).

For machining of prismatic parts, a fixture selection expert module is developed that uses this information for determining the components of a modular clamping device.[10]

Knowledge instantiation (The tool selection expert). This expert module is incorporated in the commercial P2G system that is linked with the developed CAPP kernel.[14] The module selects tools for a given process that is executed on the part or on one of the part's features. Naturally, geometric and technological part information is of vital importance during this process. Some examples :

- the radius of a face mill must be smaller than the corner radius of the pocket that has to be milled
- if a hole has a roundness tolerance of 0.01 mm, a twist drill will not deliver the required result, instead a reamer or bore will have to be used during the 'hole making' process.

The selection is made from a standard or a customer tool catalogue, taking into account the part and feature parameters. If the tool is chosen, the module calculates the most economic process parameters based on Kienzle's law for cutting forces and Taylor's law for tool life time. Note that new types of processes may require new experts : an in-house developed software for selecting the appropriate process conditions for EDM could be a complementary expert to some commercially available expert system for calculating metal cutting parameters.[15]

Virtual vs. physical resources. A resource remains a virtual blackboard object until it is unambiguously described by its parameters (i.e. only one single tool can be appointed). When a machine tool is described by characteristics like : numerical controlled, 4½ axis, possible processes, etc., a query for a suited machine will probably result in an enumeration of possible candidates.

When one machine is selected from this list, the resource becomes a physical object on the blackboard.

Process plan generation knowledge

Knowledge representation. In the developed CAPP kernel, the process plan generation knowledge consists of a Petri net based search algorithm and the 'blackboard manager' that triggers the distinct process planning expert modules (process selection, machine tool selection, fixture selection, etc.). The ways of managing a blackboard system has been described previously. As stated there, the user driven approach for managing a blackboard is chosen. At any time during the creation of a process plan, the user can invoke an automated module (an expert) for a specific process planning task and inspect or eventually change the returned results.

During the generation of the process plan, the objects (features, processes, machines, fixtures, set-ups, feature relations, etc. residing on the blackboard) become part of the process plan generation knowledge/data.

Knowledge instantiation. By triggering the expert modules or by manual editing, several blackboard objects are created. At some time, these objects will all have to fit into the non-linear process plan that is being generated. Because of the high flexibility of plan generations one can expect many scenarios for finalising the process plan. Again, the concepts of manual editing and expert consultation are important to mention within this context. They are shortly explained hereafter.

- Manual editing : Graphical editors for each planning task (e.g. figure 6) allow for the interactive construction of process plans. Plans can be built up *form scratch* or through *similarity planning* (group technology oriented). In the latter case the resembling plan is loaded from a file or database and manually changed according to the specifications of the part at hand.
- Expert consultation : The 'process sequencing expert' generates a graph of operations (i.e. the NLPP) through performing a search. Part and feature Petri nets (PNs) and the assigned machine tools form the input for the expert module. A number of techniques have been implemented to ensure high performance of the developed search algorithm :
    - ◊ *Combined variant/generative planning*: The algorithm combines the part related operations (part PN) and all feature related operations (feature PNs) to one non-linear process plan (NLPP). During this search, all feature relations (e.g. tolerance relations, interference relations) are taken into account. These relations are evaluated, resulting in constraints (e.g. grouping or sequencing of operations) which will guide the search for the NLPP.[8]
    - ◊ *Constraint based search* : Apart from the constraints that result from feature relations, planning restrictions coming from the scheduling department can be taken into account (e.g. avoid the use of bottle-neck machine X, create an alternative for machine Y, try to use modular fixture Z, etc.). Such constraints can be generated automatically (e.g. via statistical analysis of workshop data) or entered manually by schedulers.[16]
    - ◊ *Cost based search* : Weight factors can be entered for a number of cost criteria (e.g. number of set-ups, number of conventional tools, number of CNC centres, etc.). Specific search algorithms (e.g. A*, best first, branch & bound) take these factors into account and produce solutions very quickly.[12]
- Opportunistic planning expert : Opportunistic process planning is a new concept in the CAPP (research) domain. The idea is fairly simple and resembles the way human process planners think. The opportunistic search algorithm takes an existing non-linear process plan and a feature Petri net as input, and checks whether the feature can be produced on the already determined sequence of

machines (outlined by the NLPP). Opportunistic planning is an effective instrument in many cases :

◊ *Generative planning with large number of features* : When the part consists of many features (>100), this tool becomes very powerful. In this case, the human expert selects the features that will most likely determine the general outlay of the process plan. For these features a NLPP is generated. The other features are added to the NLPP afterwards using the opportunistic search algorithm.[2]

◊ *Similarity planning* : The user retrieves an existing process plan for a part family. Features that make up the specific differences of the part at hand are fitted into the process plan.

◊ *Design For Manufacturing* : A process plan is generated at an early stage of the design. When new features are added to the design, an on-line checking whether the feature can be made on the selected machines, can be performed.

Virtual vs. physical process plan. The NLPP is an aggregation of objects. The NLPP - as one object on its own - is a physical blackboard object if all its components are fully defined (i.e. they are physical too).

CONCLUSION

The manufacturing knowledge, contained in the CAPP system described in this paper, is very diverse. It relates the object classes : workpiece-families, workpiece, features, relations, processes, generic Petri nets, manufacturing direction entities, rules, rule-bases, machines, machine capabilities and various knowledge sources into one common object-oriented model. This model is dynamically instantiated, stepwise built up by relating one or more types of objects together to create new objects. This complex generation mechanism increases the potential to represent manufacturing knowledge in the way humans reason. Moreover, the developed CAPP system is based on a blackboard architecture : several expert modules can be triggered in an arbitrary order to perform a specific planning task, ensuring a very flexible way of performing the CAPP activities. Solutions generated by the assisting expert modules result in objects on the blackboard. These solutions can be adjusted or overruled by the operator, since each blackboard object has an interface to the human expert for manually adding or changing the information residing in the objects. Even more flexibility is obtained by introducing opportunistic process planning, where feature process plans are fitted onto an existing NLPP. The CAPP kernel described in this document, offers the user full control over all separate planning activities as the process plan is being finalised.

ACKNOWLEDGEMENT

Some of the work, reported in this paper, was supported by the projects ESPRIT 6805 "COMPLAN", GOA 95-2 on "Holonic Manufacturing Systems" and IUAP-50.

REFERENCES

1. **Hummel, K.E. and Brooks, S.L.**, XPS-E Revisited : A New Architecture and Implementation Approach for an Automated Process Planning System, *CAM-I Document* DR-88-PP-02, 1988.

2. **Kruth, J.P., Van Zeir, G. and Detand, J.**, An Interactive CAPP Kernel based on a Blackboard System Architecture, *Proc. of the ASME 1996 Design Engineering Technical Conf. and Computers in Engineering Conf.*, Irvine (CA), CD-ROM, 1996.

3. **Carver, N. and Lesser, V.**, The Evolution of Blackboard Control Architectures, *CMPSCI Technical Report* 92-71, 1992.

4. **Corkill, D.D.**, Blackboard Systems, *AI Expert*, Vol.6(9), pp 40-47, 1991.

5. **van Houten, F.J.A.M.**, PART, A Computer Aided Process Planning System, *PhD thesis*, Univ of Twente, 1991.

6. **Jonkers, F.J.**, A software architecture for CAPP systems, *PhD thesis*, University of Twente, ISBN 90-9005041-8, 1992.

7. **Jasperse, H.B.**, A Macro Process Planning System for Machining Operations, *PhD thesis*, Univ of Delft, 1995.

8. **Kruth, J.P. and Detand, J.**, A CAPP system for non-linear process plans, *CIRP Annals* Vol 41(1), pp.489-492, 1992.

9. **Kruth, J.P., Van Zeir, G. and Detand, J.**, Extracting process planning information from various wire-frame and feature based CAD systems, *Computers in Industry*, Vol.30(2), pp.145-162, 1996.

10. **Perremans, P.**, Feature based Description of Modular Fixturing elements : The Key to an Expert System for the Automatic Design of the Physical fixture, *Proc. of the 1994 ASME - Int. Comp. in Eng. Conf.*, Vol.1, pp. 221 - 236, 1994.

11. **Demey, S., Van Brussel, H. and Derache, H.**, Determining Set-ups for mechanical workpieces, *Robotics & Computer-Integrated Manufacturing*, Vol.12(2), pp.195-205, 1996.

12. **Kruth, J.P., Detand, J., Van Zeir, G., Kempenaers, J. and Pinte, J.**, Methods to improve the response time of a CAPP system that generated non-linear process plans, *Advances in Engineering Software*, Vol.25(1), pp. 9 - 17, 1996.

13. **Detand, J.**, A Computer Aided Process Planning System Generating Non-Linear Process Plans, *PhD thesis*, K.U.Leuven, Division PMA, ISBN 90-73802-23-7, 1993.

14. **Carlier, J.**, Computergesteunde werkvoorbereiding voor mechanische werkstukken, (in Dutch) *PhD thesis,* KULeuven, Division PMA, 1983.

15. **Lauwers, B.**, Computer Aided Process Planning and Manufacturing for Electrical Discharge Machining, *PhD thesis*, K.U.Leuven, Division PMA, ISBN 90-73802-25-3, 1993.

16. **Kempenaers, J., Pinte, J., Detand, J. and Kruth J.P.**, A collaborative process planning and scheduling system, *Proc. of the 1994 ASME - Int. Comp. in Eng. Conf.*, Vol.1, pp.221-236, 1994.

ROUTING FLEXIBILITY IN A JUST-IN-TIME SYSTEM

Min-Chun Yu, Timothy J. Greene,
Oklahoma State University

ABSTRACT:

The classic Just-In-Time (JIT) manufacturing system is known for its need for a frozen schedule, balanced work load and inability to handle fluctuations in demand. But today research and development are addressing ways for a JIT system to accommodate greater fluctuations. One approach is to design the JIT system with machines that have a greater flexibility in what work they can perform. An initial step is the development of flexibility measures so that different system configurations and control mechanisms can be easily compared. This paper presents a background to, and a rational for, a flexibility measure for a JIT system.

INTRODUCTION

Customer satisfaction has become the most important issue in today's global market. As customers favor more diversified products and services, companies are now trying to manage their production more efficiently in order to meet customers' demands. The high degree of variety in customers' demands outdates mass production that was used in the past to produce the same type of products in large volume. In order to provide low to medium production volume and higher product variety, new manufacturing technologies and philosophies need to be developed. The emergence of new technologies highlights the concept of manufacturing flexibility in the design, operation, and management of manufacturing systems [Sethi and Sethi, 1990]. As a result, manufacturing flexibility has become an important aspect, along with quality, service and cost, in strategic planning [Clark *et al.*, 1988].

Although flexibility has drawn a lot of attention from researchers, defining it is not an easy task. According to a survey [Sethi and Sethi, 1990], there are at least 50 different terms for various types of flexibility found in the manufacturing literature. Definitions for these terms are not always appropriate and are sometimes inconsistent with one another. To avoid confusion, the routing flexibility studied in this research has adopted the definition presented by Sethi and Sethi [1992] and Das and Nagendra [1993]:

The ability of a manufacturing system to produce parts continuously by alternate routes through the system, where a route is a series of machines visited in order to accomplish a part.

Routing flexibility is the product of alternate routings which are composed of using alternate machines and alternate sequences [Carter, 1986] [Bobrowski and Mabert, 1988] [Sethi and Sethi, 1990]. "Alternate machines" is the instance when one or more machines can be used to substitute for the intended machine to perform identical operations. In a flow shop where each stage contains only one machine, no alternate machine is allowed. If a flow shop has n machines capable of performing a specific operation at a specific stage, it is said that there are n-1 alternate machines available in that stage.

"Alternate sequences", however, is defined as the instance when a part can be performed by different process sequences [Bobrowski and Mabert, 1988]. The sequence of manufacturing is the order of operations in which a part is processed through the machines [French, 1982]. An alternate sequence becomes available when this part can be completed by a swapped order of processes. Research on routing flexibility has mainly focused on the alternate routings incurred by "alternate machines" [Bobrowski and Mabert, 1988].

The limited knowledge of flexibility has created a lot of problems for the management of flexibility [Benjaafar and Ramakrishnan, 1993]. Moreover, the lack of appropriate methodologies for quantifying flexibility enhances the difficulty of justifying the implementation of flexible technologies [Benjaafar and Ramakrishnan, 1993].

Routing Flexibility in Manufacturing Systems

Qualitatively, routing flexibility is defined by Sethi and Sethi [1990] as "*the ability to produce a part by alternate routes through the system.*" These alternate routes are created by the alternate use of different machines, different operations, or different sequences of operations. The flexibility of routing can be found in systems implementing general-purpose machines [Sethi and Sethi, 1990], alternative or identical machines [Yao and Pei, 1990], redundant machine tools [Browne *et al.*, 1984], or the versatility of material handling systems [Sethi and Sethi, 1990].

The purposes of routing flexibility can be identified as follows.

1. To schedule parts more efficiently by better balancing machine loads [Sethi and Sethi, 1990].

2. To continue producing a given set of part types when unanticipated events such as machine breakdowns [Browne *et al.*, 1984], late receipt of machine tooling, a preemptive order of parts, or the discovery of defective parts occur [Sethi and Sethi, 1990].

3. To improve the productivity of a machine shop [Nasr and Elsayed, 1990].

Routing flexibility can exist in any kind of manufacturing system that can provide alternate routings, yet it has mainly been related to push-type systems such as FMSs, flexible transfer lines, and job shops [Buzacott and Shanthikumar, 1980] [Sethi and Sethi, 1990] [Basnet and Mize, 1994]. For pull-type production systems such as Just-In-Time systems, research on routing flexibility has remained untouched despite its existence in this type of system.

Just-In-Time Production Systems

The Just-In-Time (JIT) system is aimed at flexible adaptation to fluctuations in demand mix and quantity in the market through the use of pull mechanisms [Monden, 1993]. The "pull" mechanism allows upstream production stations to produce in response to the demand's pulling force at the downstream stations. In that sense, the system will produce the necessary amount of products only at the necessary time. Therefore, unnecessary waste can be eliminated and inventory levels can be lowered.

According to a survey conducted by Crawford *et al.* [1988], inventory reductions and lead time reductions are the two leading benefits of implementing JIT for the surveyed companies. Excess inventory causes extra manpower, equipment, and floor space that all contribute to high manufacturing costs. Reducing inventory levels not only can lower manufacturing costs but also can eliminate problems that create excess inventory. Shorter lead times enhance the company's ability to react quickly to changes in demand in the middle of a planning period. The capability of swift reaction available when you have a flexible system can compensate for the discrepancies between monthly predetermined production plans and daily dispatched quantities. The ability to accommodate these discrepancies can then greatly reduce the level of inventory, work force, and waste.

The part-flow control in a typical JIT system is simple and straightforward. On the other hand, part flows in systems with alternate routings typically require dedicated control methods to make routing decisions. However, both JIT system and the systems with alternate routings are capable of reducing inventory level and increasing throughput.

The increasing use of general purpose machines has enabled the JIT systems to become more flexible in order to respond to a fluctuating market [Monden, 1993]. Unlike the exclusive-purpose machine, a general-purpose or multi-function machine is capable of processing different types of parts in a manufacturing facility. In other words, a specific operation of a part can be carried out by more than one machine. As a result, alternate routings are allowed in this type of JIT system.

In this paper, issues on routing flexibility for fixed-sequence, pull-type, JIT systems are discussed. An operational measure of routing flexibility is also developed and described.

Previous Research on Routing Flexibility in Manufacturing Systems

A study conducted by Gere [1966], found that production systems equipped with machines capable of performing alternate operations could reduce total lateness for a job shop. Scheduling heuristics result in better performance if alternate operations are allowed in the systems.

An economical evaluation of routing flexibility was conducted by Ghosh and Gaimon [1992]. The results justified the benefits of routing flexibility based on a reduction of total manufacturing costs, WIP inventory, number of bottleneck machines, and system utilization. The only drawback is the increase in the number of setups, but that is considered insignificant when compared to the benefits.

Quantitative measurement of routing flexibility have been developed by a number of authors. The various measures of routing flexibility can be grouped into two categories. The first category includes the majority of the existing measures that were developed based on the number of alternate ways to process a part type. The other category measures routing flexibility in terms of *routing entropy*, a measure of routing uncertainty in a manufacturing system.

Chatterjee *et al.* [1984] proposed a routing flexibility measure which is simply the number of available routes. Chung and Chen [1989] developed a similar measure that measures the routing flexibility of a manufacturing system by calculating the average number of available routes for each part type. The value of routing flexibility will range from 0 (i.e., fixed routed) to any positive value if alternate routings are allowed. Bernardo and Mohamed [1992] utilized the inverse of the number of available routes as a term of the routing flexibility measure. As a result, the values of routing flexibility range from 0, if no alternate routings are allowed, to approximately 1, if a very large number of alternate routes are allowed. These measures assume that each available route has identical value of flexibility so that only the number of available routes is considered.

A measure elaborated by Das and Nagendra [1993] takes into account the difference between various routes. The difference between two routes is expressed by a function of the difference in processing time for each machine. The routing flexibility is the sum of the average differences between each route and all the other routes.

The above measures are primarily a function of the number of available routes and/or the machine information of each route. The parameters utilized in the calculation of routing flexibility must be identified and determined in advance. Therefore, measures in this category are recognized as deterministic measures of routing flexibility because they can be determined before system starts operating.

Entropy has been used to measure the uncertainty or disorder of a system. Kumar [1987] and Yao and Pei [1990] applied the entropy theory to measure the routing flexibility of a manufacturing system. However, their studies have only focused on the system of producing a single part type.

For a manufacturing system, the entropy (i.e., routing flexibility) is determined for every operation based on the number and the availability of processing machines. The equation of the entropy measure is a function of the probability that an operation can be completed on a specific machine. The total routing flexibility of a system is the summation over the entropy measures of all operations. The entropy measure of routing flexibility can be reviewed and calculated at any time during the operation. Therefore, it can be viewed as a dynamic measure of routing flexibility.

JIT System Definition

The flexibility measure developed is for a multi-stage JIT system where the alternate machines are located at the same stage capable of performing identical machining processes as the primary machine but with less efficiency. In this research, the order of machines for processing a part can only be changed within the same stage since the order of processes is fixed due to its flow shop configuration. Therefore the impact of alternate sequences is ignored as applied to these flexibility measure.

A multi-stage JIT system that allows multiple machines at each stage is illustrated in Figure 1. Each stage consists of several machines capable of performing similar operations. Every part must be processed by each of the stages in a fixed sequence.

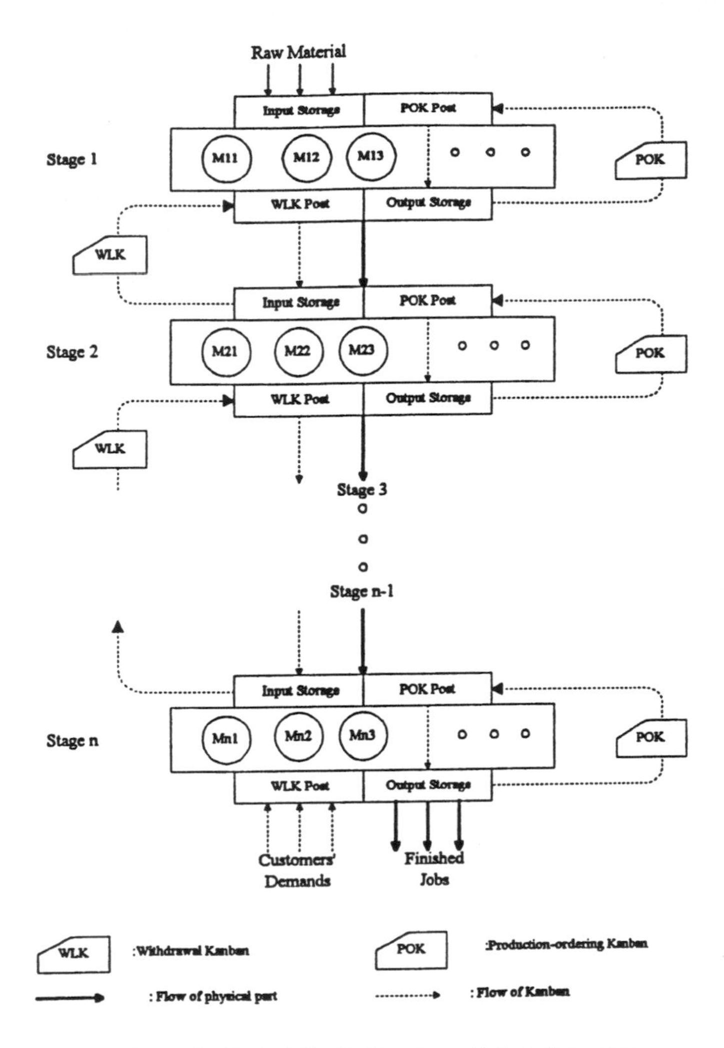

Figure 1. Physical distribution of a multi-stage JIT system

Dual Kanbans are utilized in this JIT system to direct the parts' movement. A Production-Ordering Kanban (POK) specifies the information of the parts that the preceding stage must produce. It is used to authorize the production of parts and the parts' movement within a specific stage. A Withdrawal Kanban (WLK) specifies the part type and the quantity that the subsequent stage needs to withdraw. It is used to control the movement of parts from the output storage at the preceding stage to the input storage at the subsequent stage.

Each stage contains an input buffer that is immediately before the stage and an output buffer that immediately follows. The input buffer includes an input storage area and a POK post while the output buffer is composed of an output storage area and a WLK post.

A given part type can only be processed by a number of selected machines where required tools are available. If an alternate machine is made available for processing a specific part type, required tooling must be prepared and hence longer time is needed. As a result, extra processing time is incurred for an alternate machine to process this part type.

An Operational Measure of Routing Flexibility for JIT Systems

A measure of routing flexibility is developed for fixed-sequence, multi-stage JIT systems that are capable of producing multiple part types. Instead of simply counting the number of available routes, this measure will take into account the loading balance between machines. Therefore, a manufacturing system with overloaded machines will have less routing flexibility than one that is not overloaded when both systems have the same number of available routes.

Notations:

k = Stage number ($k = 1, 2, 3, \ldots.., K$).
i = Part type.
I_k = The number of part types needed to be processed at stage k.
j = Machine number.
J_k = The number of available machines at stage k.
r_k = The number of routes (part type-machine combinations) at a specific stage.
R_{jk} = For a machine j, the number of part types that it can process at stage k.
U_{ijk} = For a machine j, the number of units that it can process that is from part type I at stage k. The value of U_{ijk} can be either 1 or 0.
S_{ik} = For part type i, the number of machines that can process i at stage k.
W_i = The weight of the volume of part type i relative to the other part types.
P_k = The average number of part types a machine can process at stage k.
M_k = The number of machines being used effectively at stage k.
$D_k\langle r \rangle$= The quantitative distance between the current routing combination and the most flexible routings under a total of r routes at stage k.
W_k = The weight of stage k relative to the other stages in the system.
RF_k = The routing flexibility of stage k.
RF = The routing flexibility of the system.

The routing flexibility of a specific stage in a flow shop is a function of the number of routes and the number of part types each machine can process. A route at a stage is a combination of the part type and a machine that can process that part type. Therefore, the routing flexibility of stage k (RF_k) is defined as the percentage of the product of the average number of part types a machine can process (P_k) and the number of machines being effectively used (M_k) to the overall combination of parts and machines ($I_k \times J_k$):

$$RF_k = \frac{P_k \times M_k}{I_k \times J_k} \quad \text{---- (1)}$$

The average number of part types each machine can process at stage k (P_k) can be obtained by dividing the total number of routes by the number of available machines of stage k.

$$P_k = \frac{\sum_{j=1}^{J_k} R_{jk}}{J_k} \quad \text{---- (2)}$$

R_{jk} can be obtained by summing up the number of standard units of the part types that can be processed by machine j. The number of standard units for part type i is determined by $I_k \times W_i$.

$$R_{jk} = \sum_{i=1}^{I_k} I_k \times W_i \times U_{ijk} \quad \text{---- (3)}$$

The number of machines being effectively used at stage k (M_k) is measured by computing the number of machines being used as if each machine can process all part types. It is computed by taking the smallest possible number of machines needed (i.e., $\frac{\sum_{i=1}^{I_k} S_{ik}}{I_k}$)and then adding up the extra number of machines (i.e., $J_k - \frac{\sum_{i=1}^{I_k} S_{ik}}{I_k}$) by multiplying the relative distance from the current to the point of the smallest number of machines (i.e., $\frac{\text{Max}\, D_k\langle r_k\rangle - D_k\langle r_k\rangle}{\text{Max}\, D_k\langle r_k\rangle}$).

$$M_k = \frac{\sum_{i=1}^{I_k} S_{ik}}{I_k} + \left[\frac{Max\, D_k\langle r_k\rangle - D_k\langle r_k\rangle}{Max\, D_k\langle r_k\rangle}\right] \times \left[J_k - \frac{\sum_{i=1}^{I_k} S_{ik}}{I_k}\right] \quad \text{---- (4)}$$

$D_k\langle r_k\rangle$ is used to measure the discrepancy between the current routing and the optimal routing under the same number of routes where a route is a combination of a part type and machine. The optimal routing is the routing when each machine is able to process the same number of part types. Therefore, the difference between each machine's processing capability under the current routing (i.e., R_{jk}) and under the optimal routing (i.e., $\frac{\sum_{j=1}^{J_k} R_{jk}}{J_k}$) is computed and the difference is squared to avoid a negative value. Since the average number of part types a machine can process may not be integer, the optimal routing is simply virtually optimal.

The summation of all the squares of the differences is then computed for representing the overall differences. The square root of the summation is taken to portray the distance between the current routing and the virtually optimal routing under a specific number of routes. The virtually optimal routing under a specific number of routes is the routing where each machine is capable of processing equal number of part types.

$$D_k\langle R_k\rangle = \sqrt{\sum_{j=1}^{J_k} \left[R_{jk} - \left(\frac{\sum_{j=1}^{J_k} R_{jk}}{J_k}\right)\right]^2} \quad \text{---- (5)}$$

After the routing flexibility of every stage is computed, the routing flexibility of the entire system can then be obtained by summing the weighted routing flexibilities of each stage. The weight allocated to each stage (W_k) is the relative importance of that stage. The relative importance may be the degrees of preference and/or relative operational measures among all stages.

$$RF = \sum_{k=1}^{K} W_k \times RF_k \quad \text{---- (6)}$$

In this research, the weight of a stage k (W_k) is the average processing time of that stage relative to the other stages. For stages with higher average processing times, higher routing flexibility can help speed up part flows and hence reduce the effect due to bottleneck stages. Therefore, the systems having higher routing flexibility on stages with higher average processing times are more flexible than those that do not.

W_k can be obtained by dividing the average processing time of stage k by the summation of average processing times of all stages. The average processing time of stage k is the average of mean processing times for all part types processed by the primary machines at stage k.

$$W_k = \frac{AP_k}{\sum_{k=1}^{K} AP_k} \quad \text{---- (7)}$$

A simple example is given in order to illustrate how the above measures are implemented.

Example

Consider a 2-stage flow shop where each stage contains three machines. A total of three part types can be produced by this flow shop. During a typical period, the percentage of demanded volumes (W_{ik}) among the three part types at all stages is 50%, 25%, and 25% respectively. The average processing times that each part type spend at the two stages are listed in Table 1. The routings of each stage are displayed in Figure 2.

Table 1.
The average processing time of each part type

Part type	Stage 1 Average Processing Time	Stage 2 Average Processing Time
1	8	13
2	10	15
3	12	17

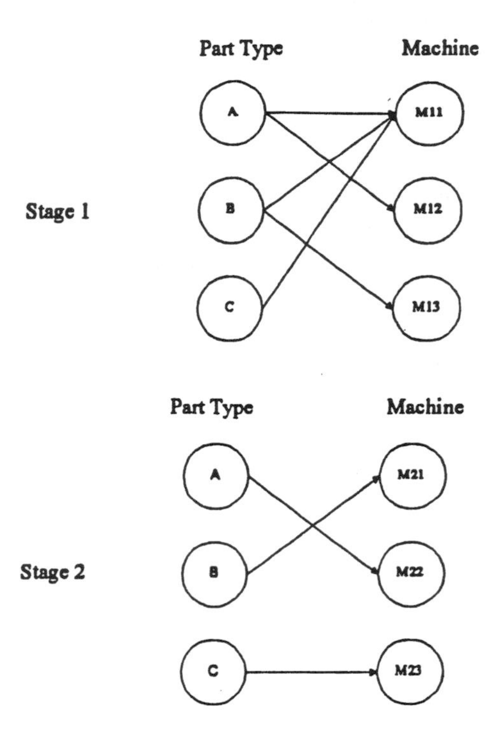

Figure 2. The routings for a two-stage JIT flow shop

The relative quantity of each part type is determined by considering the weight of volume (W_i) for each part type. Therefore, a unit of part type A is equivalent to 1.5 (3×0.5) standard units, while a unit of part type B and a unit of part type C are both equivalent to 0.75 (3×0.25) standard unit.

For stage 1, R_{11}, R_{21}, and R_{31} are calculated by using equation 3. The result shows that machine M11 is capable of processing 3 standard units of part types, while M12 and M13 can process 1.5 and 0.75 standard units respectively. It is said that stage 2 is capable of processing a total of 5.25 standard units of part types.

To process 5.25 standard units of part types at stage 1, the least flexible routing combination is (3, 2.25, 0). One of the three machines will process 3 standard units and another processes 2.25 standard units. The virtually optimal routing is when each machine can process equal number of standard units of part types (i.e., 1.75 standard units).

Similarly, (R_{12}, R_{22}, R_{32}) = (0.75, 1.5, 0.75). The best and worst combinations for stage 2 can be identified as:

(3, 0, 0) --- the least flexible routing
(1, 1, 1) --- the virtually optimal routing

The routing flexibility of each stage as well as the entire system can then be calculated by utilizing the developed formula. The detailed calculations of the measure are carried out as follows.

Stage 1:

$$P_1 = \frac{(3+1.5+0.75)}{3}$$
$$= 1.75$$

(i.e., each machine at stage 1 can process an average of 1.75 part types)

$$D_1\langle r_1 = 5.25\rangle = \sqrt{(3-1.75)^2 + (1.5-1.75)^2 + (0.75-1.75)^2}$$
$$= 1.62$$

(i.e., the distance between the current routing and the virtually optimal routing is 1.62 when the total routes at stage 1 is 5.25)

$$\text{Max}\, D_1\langle r_1 = 5.25\rangle = \sqrt{(3-1.75)^2 + (2.25-1.75)^2 + (0-1.75)^2}$$
$$= 2.21$$

(i.e., the distance between the least flexible routing and the virtually optimal routing is 2.21 when the total routes at stage 1 are 5.25)

$$M_1 = \frac{(3+1.5+0.75)}{3} + \left(\frac{2.21-1.62}{2.21}\right) \times \left[3 - \frac{(3+1.5+0.75)}{3}\right]$$
$$= 2.0828$$

(i.e., a total of 2.0828 machines is being effectively used at stage 1)

$$RF_1 = \frac{1.75 \times 2.0828}{3 \times 3} = 0.405$$

Stage 2:

$$P_2 = \frac{(0.75+1.5+0.75)}{3}$$
$$= 1$$

(i.e., each machine of stage 1 can process an average of 1 part types)

$$D_2\langle r_2 = 3\rangle = \sqrt{(0.75-1)^2 + (1.5-1)^2 + (0.75-1)^2}$$
$$= 0.61$$

(i.e., the distance between the current routing and the virtually optimal routing is 0.61 when the total routes at stage 2 are 3)

$$\text{Max}\, D_2\langle r_2 = 3\rangle = \sqrt{(3-1)^2 + (0-1)^2 + (0-1)^2}$$
$$= 2.449$$

(i.e., the distance between the least flexible routing and the virtually optimal routing is 2.449 when the total routes at stage 2 are 3)

$$M_2 = \frac{(0.75+1.5+0.75)}{3} + \left(\frac{2.449-0.61}{2.449}\right) \times \left[3 - \frac{(0.75+1.5+0.75)}{3}\right]$$
$$= 2.5$$

(i.e., a total of 3 machines is being effectively used at stage 2)

$$RF_2 = \frac{1 \times 2.5}{3 \times 3} = 0.2778$$

To compute the routing flexibility of the system, the weight of the two stages must be determined. The average processing time based weight can be obtained by applying equation 7.

$$W_1 = \frac{10}{(10+15)} = 0.4$$

$$W_2 = \frac{15}{(10+15)} = 0.6$$

Using equation 6, the routing flexibility of the system (RF) can then be obtained by averaging the weighted routing flexibility of the two stages:

$$\begin{aligned} RF &= (0.4 \times 0.405 + 0.6 \times 0.2778) \\ &= 0.3287 \end{aligned}$$

This example shows that a two-stage JIT system has a routing flexibility of 0.3287 when the developed measure is applied. The values of routing flexibility for a production system measured by this method range from approximately 0 to 1.

In sum, there are several advantages that can be obtained by using the developed measure of routing flexibility:

1. It is capable of dealing with multiple part types.

2. Production volumes for each part type are considered.

3. Overloading on specific machines are discouraged since it will result in a smaller value of routing flexibility.

4. The stages with higher average processing times (i.e., bottleneck stages) are given higher weights on their routing flexibility in order to encourage higher flexibility on these stages.

Conclusion

Although routing flexibility is recognized as an important contributor to the system performance of modern manufacturing facility, it has not been properly measured. Most of the existing research on measuring routing flexibility did not provide a measure that can indicate how flexible a system is in directing parts through the system. Others are not capable of measuring the routing flexibility for systems that produce multiple part types.

In this paper, an operational measure of routing flexibility for a fixed-sequence JIT system is presented in order to measure the relativity of routing flexibility between different systems. Multiple part types and balanced machine loads are considered in this measure. Average production volume for each part type is weighted so that the machine loading can be appropriately represented.

Benjaafar, S., and Ramakrishnan, R., The effect of routing and machine flexibility on performance of manufacturing systems. *Second Industrial Engineering Research Conference Proceedings,* 445-450, 1993.

Benjaafar, S., Models for performance evaluation of flexibility in manufacturing systems. *International Journal of Production Research,* 32(6), 1383-1402, 1994.

Bernardo, John J., and Mohamed, Zubair, The measurement and use of operational flexibility in the loading of flexible manufacturing systems. *European Journal of Operational Research,* 60, 144-155, 1992.

Bobrowski, Paul M., and Mabert, Vincent A., Alternate routing strategies in batch manufacturing: an evaluation. *Decision Sciences,* 19(4), 713-733, 1988.

Browne, J., Dubois, D., Rathmill, K., et al, Classification of flexible manufacturing systems. *The FMS Magazine,* 2(2), 114-117, 1984.

Carter, Michael F., Designing flexibility into automated manufacturing systems. *Proceedings of the Second ORSA/TIMS Conference on Flexible Manufacturing Systems: Operations Research Models and Application,* 107-118, 1986.

Chatterjee, A., Cohen, M. A., Maxwell, W. C., et al, Manufacturing flexibility: models and measurement. *Proceedings of the First ORSA/TIMS Special Interest Conference, (Ann Arbor, MI),* 49-64, 1984.

Chu, Chao-Hsien, and Shih, Wei-Ling, Simulation studies in JIT production. *International Journal of Production Research,* 30(11), 2573-2586, 1992.

Chung, C. H., and Chen, I. J, A systematic assessment of the value of flexibility for an FMS. *Proceedings of the Third ORSA/TIMS Conference on Flexible Manufacturing Systems: Operations Research Models and Applications,* 27-32, 1989.

Clark, K., Hayes, R., and Wheelwright, S., *Dynamic Manufacturing: Creating the Learning Organization,* New York, NY: Free Press, 1988.

Crawford, Karlene M., Blackstone, John H. Jr., and Cox, James F., A study of JIT implementation and operating problems. *International Journal of Production Research,* 26(9), 1561-1568, 1988.

Das, S. K., and Nagendra, P., Investigations into the impact of flexibility on manufacturing performance. *International Journal of Production Research,* 31(10), 2337-2354, 1993.

French, Simon, *Sequencing and Scheduling: an Introduction to the Mathematics of the Job-shop,* New York, NY: John Wiley & Sons Inc., 1982.

Ghosh, S., and Gaimon, C., Routing flexibility and production scheduling in a flexible manufacturing system. *European Journal of Operational Research,* 60, 344-364, 1992.

Gunasekaran, A., Goyal, S. K., Martikainen, T., et al, Equipment selection problems in Just-In-Time manufacturing systems. *Journal of Operational Research Society,* 44(4), 345-353, 1993.

Kumar, Vinod, Entropic measures of manufacturing flexibility. *International Journal of Production Research,* 25(7), 957-966, 1987.

Monden, Yasuhiro, *Toyota Production System,* Norcross, GA: Institute of Industrial Engineers, 1993.

Nasr, N., and Elsayed, E. A., Job shop scheduling with alternative machines. *International Journal of Production Research,* 28(9), 1595-1609, 1990.

Sethi, A. K., and Sethi, S. P., Flexibility in manufacturing: a survey. *International Journal of Flexible Manufacturing System,* 2, 289-328, 1990.

Sugimori, Y., Kusunoki, K., Cho, F., et al, Toyota production system and kanban system materialization of Just-In-Time and Respect-For-Human system. *International Journal of Production Research,* 15(6), 553-564, 1977.

Wilhelm, W. E., and Shin, Hyun-Myung, Effectiveness of alternate operations in a flexible manufacturing system. *International Journal of Production Research,* 23(1), 65-79, 1985.

Yao, David D., and Pei, Frances F., Flexible parts routing in manufacturing systems. *IIE Transactions,* 22, 48-55, 1990.

RE-ENGINEERING OF AN AUTOMATED WAREHOUSE SYSTEM USING OBJECT-ORIENTED SIMULATION

Agostino G. Bruzzone, Roberto Mosca
Institute of Technology and Mechanical Plants
University of Genoa
via Opera Pia 15, 16145 Genoa, Italy
email: bruzzone@linux.it

ABSTRACT: This paper proposes real application example of industrial reorganization based on the use of the Simulation. Simulation development and use is presented as a methodological approach to solve the problem of warehouse reorganization. This is the first step to develop new innovative techniques, obtained through a combination of Artificial Intelligence (AI) methodologies; these new techniques linked to the simulator can become the support to the re-engineering process in a highly detailed scenario.

INTRODUCTION

As a general rule, designing automated systems requires large investments to deal with the significant operating complexity. For this reason, the use of simulation techniques during the estimate phase is often fundamental to ensure correct design outcomes. This paper discusses a similar problem, i.e. re-engineering a previously installed warehouse system that does not meet the operating requirements.
The paper proposes an application methodology in which the simulation can be successfully implemented given the general nature of this type of problem. Obviously, for this particular case, attention will be focused on the VV&A phase (Verification, Validation & Accreditation) to ensure the validity of the results obtained (see tab I.) [Ören, 1981]. Another fundamental aspect is the analysis methodology used. In this situation, considering the complexity of the problem, the simulator becomes a tool to test different warehouse management and organizational strategies. To obtain the desired result, optimization techniques are proposed which are based on the integrated use of artificial intelligence tools. In our case, we use a genetic algorithm acting on fuzzy parameters which identifies possible alternatives regarding the layout of warehouse material and the management of the handling operations. The experimental analysis, based on the implemented simulation model, can be used to evaluate this methodology for the particular application case being examined and to highlight the relative characteristics. The research refers to similar cases studied over the last few years by this research team. In particular, these innovative techniques require special efforts for the methodization that guarantees the possible application in an industrial environment in terms of reliability and development time.

THE PLANT PROBLEM

The study refers to a finished parts warehouse of a plant that produces small parts and components. The complexity of the problem derives from the large number of products handled and the various types of customers [Mosca et al. 1993]. In this case, the deliveries can be divided into:

Masterized Material

Personalized Mix based on customer requirements.

Over the last few years, to improve production control and to guarantee optimization of stocks and production, the warehouse in question has developed from a very simple format (forklift trucks with operators guided by written orders) to a new highly computerized system. Turret trucks are used in the warehouse to handle pallets in the horizontal and vertical directions.

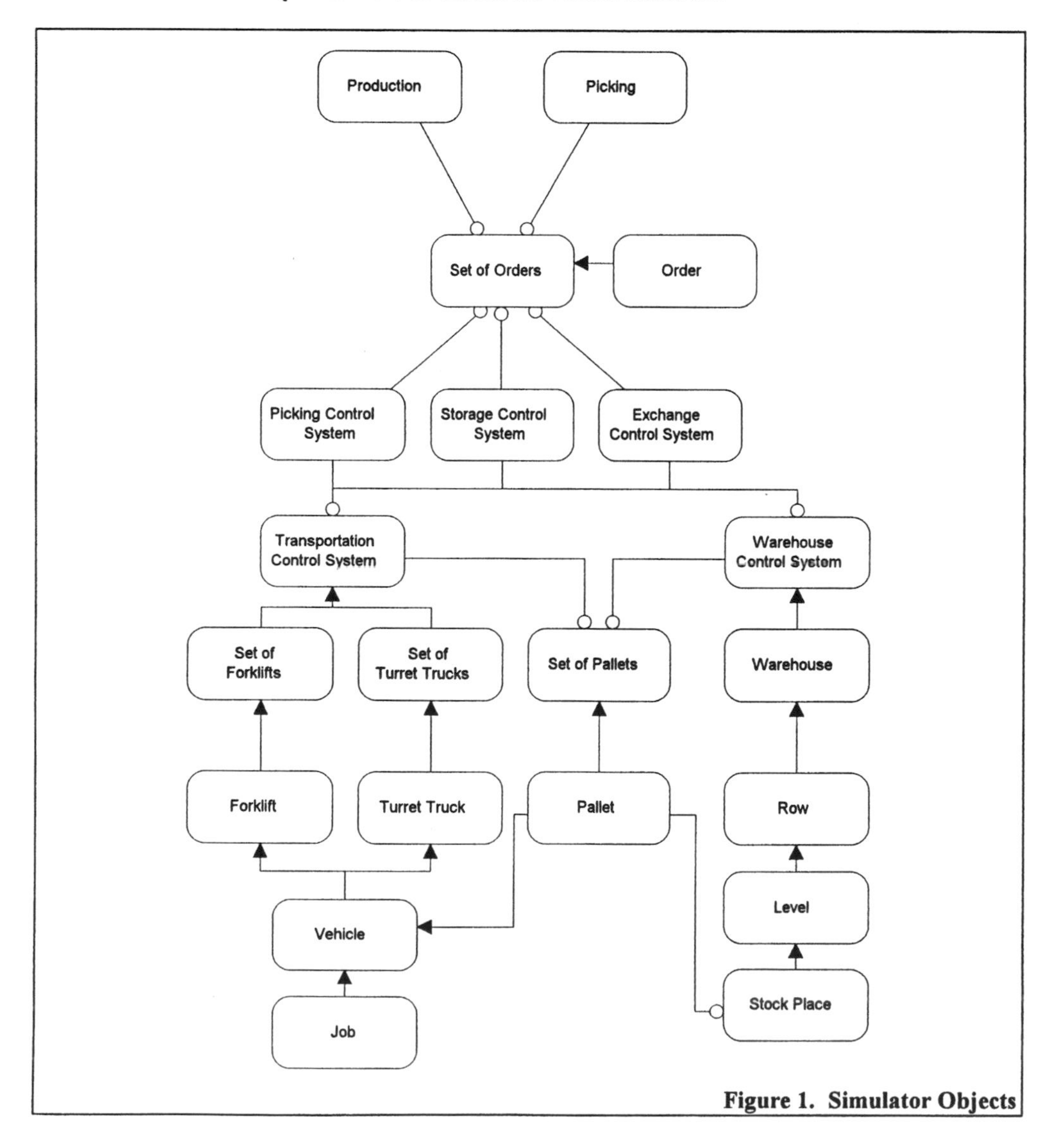

Figure 1. Simulator Objects

These systems, which are particularly cumbersome and relatively slow in carrying out horizontal movements, are flanked by traditional forklift trucks for handling operations from the loading or picking area to the warehouse. Sometimes, when operating on the lower levels of the warehouse aisles, the forklift trucks can store/pick-up the pallets directly, thus avoiding the use of the turret trucks.
The new system introduces wireless, hand-held terminals for communication between the control system and the handling systems. Turret trucks, which can operate on 8 vertical levels, are used together with the forklift trucks.
The new warehouse is managed with a completely computerized system and the missions are all defined based on this approach. In reality, the project installed has led to a major reduction in effective production, with an average handling time that is relatively low with respect to the manual system. An initial analysis led to the following possible causes:

Non-optimized operating procedures
Variation of the picking characteristics due to a market change
Inefficient warehouse management strategy.
Inefficiency by personnel in using the new system.

Since the intervention was assigned to an automation company, the correction strategy focused on speeding up the system by eliminating the bottlenecks related to the single operations, as often occurs. Therefore, bar codes were eliminated, and the central system on databases was implemented to identify the pieces and the locations used, except for different instructions introduced manually. Obviously, this choice led to a ridiculously small improvement in the operating time of the single operation, with a lower average efficiency due to errors and corrections, and with a great reduction in the reliability of such data.
In addition, system functionality was so reduced that it was not possible to maintain such a capacity in line, for which the available data about its effective operations are very limited.
At this point, the re-engineering project was introduced and focused on the following objectives:

Achieving the original efficiency and quality objectives
Identification of the causes of the initial deficiency
Maximum Recovery of the possible hardware and software material
Limiting re-engineering costs.

The simulation is evidently the keystone for finding the solution to this problem since it is the only tool that provides an overall view [Giribone & Bruzzone, 1995].
In fact, in this model we must evaluate:

The management policies of the Forklift and Turret Truck mission
The layout logic of the warehouse components
Dynamic warehouse planning based on the evolving market

THE METHODOLOGICAL DEVELOPMENT PHASES

The study was designed to ensure the success of the approach. Development was structured into phases which may be summarized as follows:

Problem review.
Modeling the Hardware Components.
Modeling the External Forces (Market/Production).
Modeling the Management system.
Identification of the Causes for Insufficiency.
Identification of the Intervention Solutions.

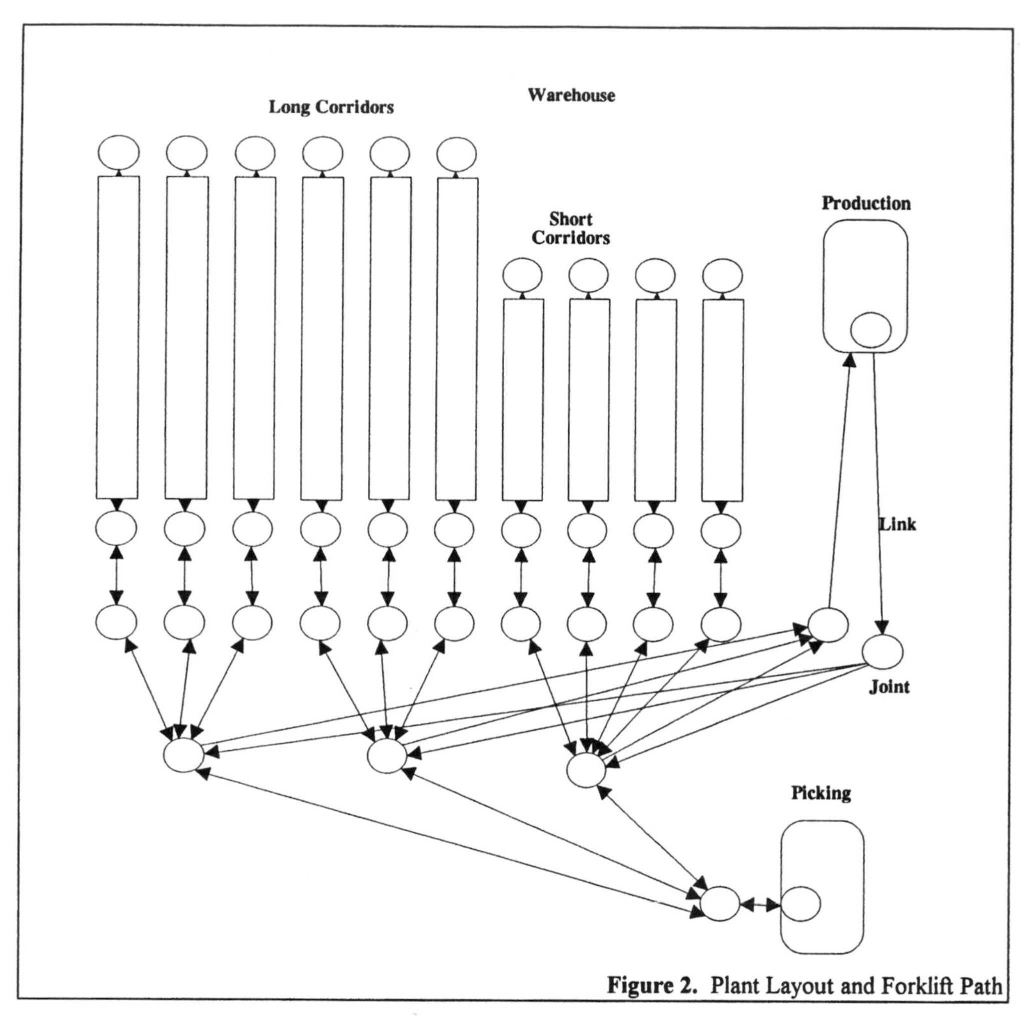

Figure 2. Plant Layout and Forklift Path

Each phase is structured to be carried out, by reviewing the documentation and historical data, a modeling phase, an implementation phase and statistical, graphic and logical validation techniques phase [Bolci, 1994] [Mosca et al. 1994]. Each of these steps is carried out on the basis of operative briefings with the industrial counterpart and on summary meetings [Giribone & Bruzzone, 1996].

THE SIMULATION MODEL

The simulation model was developed in C++, according to the Object-Oriented Design and Analysis approach, and is a discrete-event stochastic simulator [Mosca et al. 1995].
The simulation model is divided into three separate modules:

Physical Components (Forklift and Turret Trucks, Warehouse, Operators, etc.)
External Components (Picking Order Generation, Production)
Management Systems (Reservation, Exchange, Storage Procedures, etc.)

Each of these components is defined with a meta-object based on a hierarchical structure that respects the encapsulation. This is particularly important since numerous operating modes must be simulated:

Market Situation when Stipulating the automation Contract
Current Market Situation
Situation with Upgrade during Evaluation for Production

and similarly, for what concerns management:

Original Manual Operation
Integrated Operation according to the Automation system
New Proposed Modality.

On the other hand, the interventions on the warehouse system are related not only to organizational restructuring hypotheses, but are also architectural and based on physical components (buying an additional turret truck, re-defining the warehouse layout, etc.).
The structure of the objects is reported in figure 1. Encapsulation of the routines guarantees maximum system expandability.
Modeling of the handling systems (forklift trucks and turret trucks) was a particularly complex process because pallet exchanges and mutual interference also had to be managed [Law et al. 1991].

Table I
Operations for the Validation, Verification & Testing procedure

Operations
Data collection during the start-up tests
Review of Warehouse System Specifications
Specification Review of the Simulator developed as a Design support
Construction of the New Simulation Model: Handling & Storage
Definition of the Object operating Methods
Implementation of these Methods
Simulation of Pre-defined Missions handled by the Dbase
Verification, Validation and Accreditation of the components
Construction of the Management and Control Objects
Definition of their Operating Methods
Implementation
Simulation of an operating day handled by the Dbase
Verification, Validation & Accreditation of the strategies

Modeling made reference to a network of possible paths in which each branch is characterized as being with or without interference.
The positions of the mobile systems are defined in terms of arc and curvilinear coordinate along the arc, therefore the network links have a preferential orientation that in any case does not invalidate the movement in the reverse direction [Walsh et al. 1992].
The structure of the network in question is reported in figure 2. This figure shows that there are preferential areas for exchanging the pallet between forklift trucks and turret trucks [Giribone et al. 1994].
The simulator vehicles are driven by a movement_event. The time of each vehicle to reach a critical point, based on its effective speed, becomes a moment_event. The critical points of the system are the joints and the position of other vehicles on the link with interference.
When reaching the joint, if the vehicle interferes with another one, the movement system calculates the interference point and puts the two vehicles in those positions at the time of the interaction.
To avoid interference the control system operates based on rules that provide a relationship between the present density on the link and the characteristic of that link.

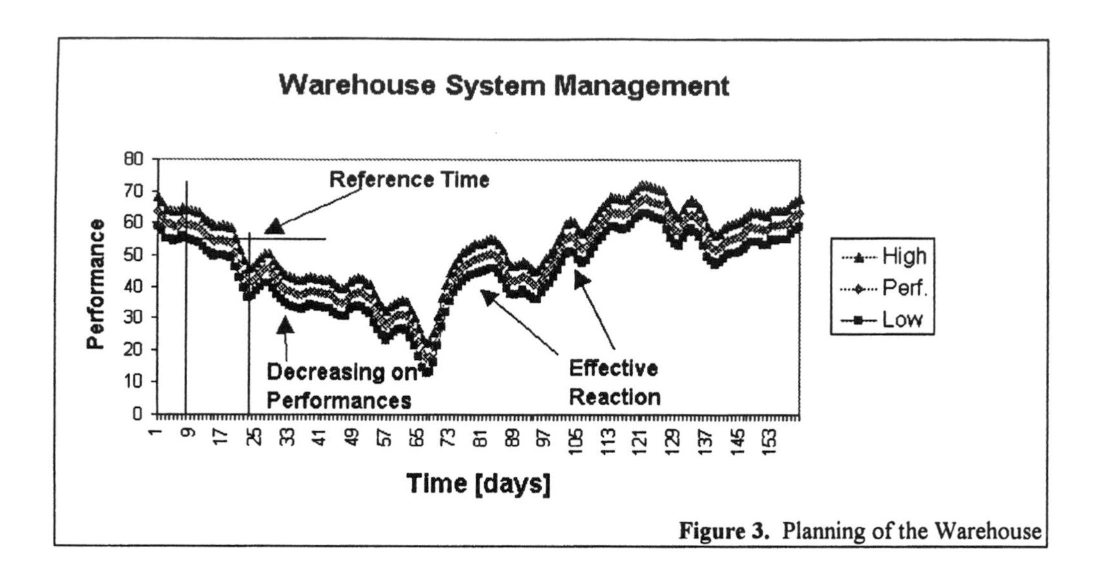

Figure 3. Planning of the Warehouse

THE ANALYSIS PERFORMED

A special warehouse management system, based on the dynamic definition of the reservation areas for the various products, was introduced to evaluate the different management strategies.
This system, unlike the one proposed in the original automation, will solve the market evolution problem by redistributing the quantities in the warehouse step by step based on the needs encountered. This avoids the need for a major redistribution involving significant operating losses.
An algorithm was defined to manage this system. Such an algorithm considers the three main types of handling :

Warehouse Loading (from Production to the Warehouse)
Integrations (handling operations inside the warehouse to integrate the masterization of specific products)
Picking (from Warehouse to packaging)

Though static by nature, the system originally proposed to map the warehouse areas was based on an engineering analysis of the problem and the definition of priorities related to distances. In the present case it is preferred to consider the performances obtained during the simulation as the reference to re-organize the warehouse based on the objectives obtained rather than on the improvement hypothesis.
The performance is measured as follows:

$$P_1(ta) = \sum_{j=1}^{n} Ns_j \frac{\int_{ta-\Delta t}^{ta}\left[(t+\Delta t - ta)(Ts_j(t))\right]dt}{\Delta t^2} + Nv_j \frac{\int_{ta-\Delta t}^{ta}\left[(t+\Delta t - ta)(Tv_j(t))\right]dt}{\Delta t^2} + Np_j \frac{\int_{ta-\Delta t}^{ta}\left[(t+\Delta t - ta)(Tp_j(t))\right]dt}{\Delta t^2} \quad (1)$$

$$P0 \,/\; P_0 \leq P_1(t)\ \forall t,\ \exists\, t_1 \in \{\text{All Simulation Runs}\} \Rightarrow P_0 - P_1(t_1) = 0 \quad (2)$$

$$P(t) = P_0 / P_1(t) \quad (3)$$

P	Performance
n	Number of Products
Δt	size of the integration window
Ns_j	Number of storing operations of the j-th product
Nv_j	Number of integration handling operations of the j-th product
Np_j	Number of picking operations of the j-th product
Ts_j	Unit time for storing the j-th product
Tv_j	Unit time for the integration handling operations of the j-th product
Tp_j	Unit time for picking the j-th product

The key function of the orders is the total handling time. Once ordered, the optimal position is defined by using centers of reference around which to define the areas for each item. This choice is attributed according to fuzzy rules based on the following factors:

Total laying handling time
Total pick-up handling time
Total integration time
Number of laying operations
Number of pick-up operations
Number of integration operations

Starting from these terms, and through FAM (Fuzzy Associative Memory), we obtain the following characterizations:

Proximity to the Picking area
Proximity to the Packing and Shipping area
Proximity to the aisle entrance
Proximity to the storage areas of products with which to create shipments
Proximity to lower levels

A system defined according to a knowledge base systems was used to manage FAM evolution to ensure continuous adaptation of the mapping logic to system development and to create automatic calibration. In this case, rules were defined that, by following the time development of the objective function, change the correlation between the controlled variables and the fuzzy variables [Bruzzone, 1995].
Figure 3 reports the development of the objective function used to determine the system settling time that guarantees to introduce changes to the planning system only when the new layout began to generate results [Bruzzone et al. 1995]; the performance in this figure represents the average ratio between the effective and the theoretic time for each mission; the changes used to stress the model are based on the shift between most and less required products.
The final results of the system demostrate that the new planning system provides a much more interesting performance level than the traditional one.

THE EXPERIMENTAL PROCEDURES

To verify the operation of the system being studied under operating conditions, it was necessary to develop an experimental methodology. The main problems considered were the following:

A) Evolution of the Global Market Situation
B) Evolution of Customer Requests
C) Evolution of Sales of Each Product

Problem A) refers to the ability to consider the capacity of the system to re-organise the warehouse in relation to the global market variations. Fundamentally, these variations refer to two aspects: quality and quantity. The first one corresponds to customer service requirements that can be easily represented by the time that elapses between receiving an order and delivering the product, while the second one obviously influences the general evolution of the demand in terms of quantities requested.
Instead, the second type of problem is related to the natural evolution of customers in terms of volumes and extent of range and especially an evolution in the type of order that each one requires (large volumes of small quantities rather than increasingly complex mixes).
The evolution of the sales of each product involves a continuous and dynamic re-assignment of the products to priority categories and thus what is a mandatory re-arrangement in the warehouse.

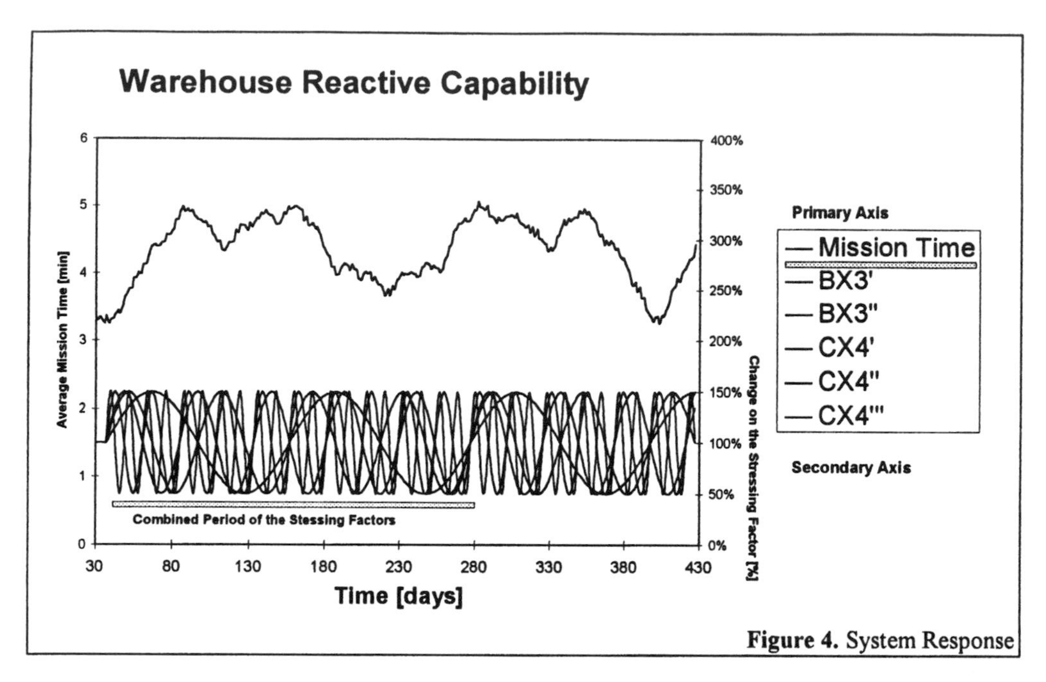

Figure 4. System Response

The basic idea of the modelling proposed refers to the idea of developing a system that can adjust itself based on the data measured which represent the efficiency, time spent by the goods in the warehouse and the speed with which the various orders/customer are prepared. In this way, the warehouse re-organises itself during operation by means of migrations of the most convenient products/customer to the priority areas.

However, this approach involves a series of new problems which must be verified. The main problem, from an operational point of view, is the response time. Obviously, re-planning based on the acquired data involves a certain number of measurements in order to identify a new trend. These observations increase if it is assumed that the monitored parameters also have a highly influential aleatory component (as inevitably occurs in an industrial warehouse). On the other hand, re-organisation based on the proposed system requires a certain amount of time to provide significant results. The sum of these two delays that we indicate with Dm (time to identify the new trends) and Dr (time to become effective) must be satisfactory for what is the specific industrial situation being examined.

Another key aspect is the stability of the management system and its robustness. To verify these aspects, utilising the management techniques proposed here, the experiment must be performed on validated simulation models since it is not possible to obtain an analytical solution to the problem.

In fact, the previously mentioned parameters, relative to the evolution of the market/products/customers, represent some of the variables which make this dynamic mapping a key characteristic to guarantee the efficiency of the management system and its effective reactivity. In order to evaluate the performances of our warehouse, it was decided to design the experiment by using some complementary methods such as the FDM (Frequency Domain Methodology) and the MSpE time development analysis (Mean Square pure Error) [Giribone, et. al. 1996].

We will consider the following factors as independent variables:

AX_1 Percentage increase/decrease of the Volumes requested by the market

AX_2 Percentage increase/decrease of the Average Time between when the Order is received and when it is Dispatched as requested by the market

BX_3 Percentage increase/decrease of the breakdown between Masterised orders and Individual Mixes

CX_4 Percentage increase/decrease of the request of an order

The variables being analysed are associated to the orders and to the customers and can be represented as follows:

$$AX_1(c_i, p_j) = 1.0 + K^a_{ij}\,\mathcal{U}(t - td_{ij}) + H^a_{ij}\,\mathcal{U}(t - tp_{ij})\sin(\omega^a_{ij}(t - tp_{ij})) \quad (4)$$

$$AX_2(c_i, p_j) = 1.0 + K^b_{ij}\,\mathcal{U}(t - td_{ij}) + H^b_{ij}\,\mathcal{U}(t - tp_{ij})\sin(\omega^b_{ij}(t - tp_{ij})) \quad (5)$$

$$BX_3(c_i) = 1.0 + K^c_i\,\mathcal{U}(t - td_i) + H^c_i\,\mathcal{U}(t - tp_i)\sin(\omega^c_i(t - tp_i)) \quad (6)$$

$$CX_4(p_j) = 1.0 + K^d_j\,\mathcal{U}(t - td_j) + H^d_j\,\mathcal{U}(t - tp_j)\sin(\omega^d_j(t - tp_j)) \quad (7)$$

c_i i-th customer
p_j j-th product
td_{ij} Pulse stress starting instant for the j-th product of the i-th customer
tp_{ij} Periodic stress starting instant for the j-th product of the i-th customer
ω^x_y Frequency of the periodic stress

By stressing with the appropriate frequencies and setting the relative oscillations according to what is dictated by the corresponding theory, a sensitivity analysis can be performed that identifies the most influential factors on the system while it is also possible to evaluate the response development [Morrice & Schruben, 1993]. Moreover, this approach can be used to take due account of the behaviour of the system during the transient being examined and thus to evaluate its operating capacity in dynamic terms [Morrice & Gupa, 1994].
Transient development of the monitored variables is used to identify the significance, response times, re-settling, as well as the capacity of the system to keep itself within a predetermined range with respect to the controlled variable. Obviously, since the system (and thus the model) has a strong aleatory nature, replications must be carried out in order to estimate the standard deviation over the measured values. The authors developed a very effective method for this analysis which evaluates the development of the experimental error (MSpE) measured on a carefully selected number of controlled variables [Box et al. 1978].
In this case, the following factors will be considered as controlled variables:

Y_1 Time needed to carry out the average Picking mission
Y_2 Time needed to carry out the average Loading mission
Y_3 Average time the goods remain in the warehouse (divided by classes in relation to flows)
Y_4 Average Daily Handling Operations
Y_5 Average Use of Resources (Fork-lifts and Turret Trucks)

Considering each of the functions examined, it is possible, following the instant *tp*, to identify the fundamental frequencies of the response. In addition, with special filtering (for example, utilising a fast fourier transform and filtering over the spectral density) it is possible to re-construct the development in a way that is not affected by the stochastic components [Jacobson, 1994].
Figure 4 graphically represents the response of the system and, respectively, the development to settling of the system followed by the start of the sinusoidal response in relation to a scenario in which only two classes of customers and three types of products are considered. Obviously, a greater number of different frequencies must be used to obtain a more detailed examination and, to guarantee a relative observation duration, the simulation run duration must be extended and thus the duration of the experiment.

CONCLUSIONS

It is interesting to note that the approach proposed in this system focuses on solving a complex re-engineering problem where the managerial procedures were restructured to correct an initial error and to guarantee the maximum recovery of the investment already made.
By respecting these terms it was also possible to obtain important results. Research has demonstrated the efficiency of planning systems that are capable of proposing dynamic management schemes that continuously adapt. These schemes, by utilizing the simulation, can provide important results in operating terms which, on the other hand, are guaranteed by thorough predictive analyses on the model.

At this point, future research should focus on optimization of the re-planning procedures and on the optimal integration between the KBS (Knowledge-based System) and the redefinition of the matrix that defines the associations; the matrix redefinition must be based on some predictive capabilities. At the present time the authors are evaluating various systems which are based on ANN (Artificial Neural Networks) to model the correlation between the behaviour of the warehouse and the exogenous parameters [Bruzzone et al. 1994].

REFERENCES

1. **Bolci O.** (1994) “Validation, Verification and Testing Techniques throughout the life cycle of simulation Study”, *Proc. of ESS94*, Istanbul
2. **Box, G.E.P., Hunter, W.G and Hunter, J.S.** (1978), "Statistics for experimenters: An introduction to design data analysis and model building", Wiley and Sons, New York;
3. **Bruzzone A.G., P.Giribone, Mosca R.** (1994) "Procedural Approach for a Neural Network Based Analysis of an Industrial Plant: Automatic Bottling Line", *International Journal of Flexible Automation and Integrated Manufacturing*, Vol.2, no 3. pp. 259-273
4. **Bruzzone A.G., Mosca R. & P.Giribone** (1995) "Concurrent Design: Coal Bulk Terminal Design Based On Integrated Simulation Methods", *Proc. of Summer Simulation '95*, Ottawa, July 23-26
5. **Bruzzone A.G.** (1995) "Adaptive Decision Support Techniques and Integrated Simulation as a Tool for Industrial Reorganisation", *Proc. of ESS95*, Erlangen, October 26-28
6. **Giribone P., Mosca R. & A.G.Bruzzone** (1994) "Simulation And Automatic Parking In A Training System For Container Terminal Yard Management", *Proceedings of International Training and Equipement Conference*, The Hague (H)
7. **Giribone P. & A.G.Bruzzone** (1995) "Training System through Simulation and AI Techniques of Process Plant Management", *Proc. of FAIM'95*, Stuttgart, June 28-30
8. **Giribone P. & A.G. Bruzzone** (1996) "Combined Modelling Based on Object Oriented Desing & High Performance Petri Nets for the Analysis of Industrial Processes", *Proc. of MIC'96*, Innsbruck, February
9. **Giribone P, Bruzzone A.G., Macchiavello A.** (1996) “Sensitivity Analysis in Computer Simulation through Dynamic Change of Input Variables: A Case Study” *Proceedings of “Simulation in Industry” ESS'96*, Genoa Italy October
10. **Jacobson, S. H.** (1994), “Harmonic Gradient Estimators: a Survey of Recent Results”, Proc. “Conference on Modelling Simulation - ESM 94”, pp. 853-856
11. **Law, A.M. & Kelton W.D.** (1991) "*Simulation Modeling and Analysis*", Mc Graw-Hill, New York
12. **Morrice, D. J., and Schruben, L. W.** (1993), “Simulation factor screening using harmonic analysis”, Management Science 38:667-687
13. **Morrice, D. J., Gupta, A.** (1994), “Transient sensitivity Analysis in Computer Simulation”, Proc. “Conference on Modelling Simulation - ESM 94”, pp. 858-862
14. **Mosca R., Giribone P. & A.G.Bruzzone** (1993) "Optimum Area Search Techniques Applied to Studies Relative to Plant Problems performed by means of Simulation", *Proceedings Simtec93*, San Francisco, November 8-10
15. **Mosca R., P.Giribone, A.G.Bruzzone** (1994) "Graphic, Analog and Statistical Verification of a Computer Simulation of Port Operations", *Proceedings of HPC94*, San Diego, April 10-15
16. **Mosca R., Giribone P. & A.G.Bruzzone** (1995) "Simulator Object-Oriented Modeling for Educational purposes in the Industrial Plant Sector", *Proceedings of WMC95*, Las Vegas, Nevada (USA), January 15-18
17. **Ören T.I.** (1981) “Concept and Criteria to Assess Acceptability of Simulation Studies: A frame of Reference”, *Communication of the ACM* 24:180-189
18. **Walsh G., D.Tilbury S.Sastry, R.Murray and J.P.Laumond** (1992) "Stabilization of trajectories for systems with non-holonomic contraints", *Proceedings IEEE Conference on Robotics and Automation*, Nice (F)

EFFECTS OF NEGOTIATION MECHANISMS ON PERFORMANCE OF AGENT BASED MANUFACTURING SYSTEMS

Naga K. C. Krothapalli
Department of Industrial Engineering
FAMU-FSU College of Engineering
Tallahassee, FL 32310–6046
E-mail: kroth@eng.fsu.edu

Abhijit V. Deshmukh
Department of Industrial Engineering
FAMU-FSU College of Engineering
Tallahassee, FL 32310–6046
E-mail: deshmukh@eng.fsu.edu

ABSTRACT: This paper proposes new inter–agent negotiation mechanisms for improving the performance of agent based or decentralized manufacturing systems. The focus of this paper is on demonstrating efficiency of different negotiation and collaboration schemes between agents of the same class (parts, machines, etc) and inter–class negotiations using currency metrics. The cooperation and negotiation protocols are modeled using the *Swarm* multi–agent simulation platform. We demonstrate the robustness of the proposed schemes, and compare them with hierarchical scheduling systems.

I. INTRODUCTION

Hierarchical or central control is the most common control architecture used in manufacturing systems. In hierarchical control, a central controller plans the sequence of operations for a group of machines and parts, either *a priori* or on–line. Although hierarchical control has been widely used it has certain limitations when used for large scale flexible systems. [6, 2]

The primary drawback of hierarchically controlled flexible manufacturing systems is their lack of flexibility. Modifications of configuration of hierarchical controlled manufacturing systems are expensive and time consuming as they involve expensive software rewrites. The complexity of hierarchical manufacturing system grows enormously with the increase in number of components. Another major drawback is the potential single point of failure at the central controller. Moreover, these systems require huge databases which can result in data retrieval delays and data consistency maintenance problems.

Recently, decentralized control has been proposed as a promising solution for overcoming the limitations of hierarchical control. [4, 3, 7, 8, 5] In decentralized control the central controller is replaced by several small controllers for individual components of the manufacturing system. Each controller or agent is assigned protocols which will govern the operations of manufacturing entities. Agents communicate or negotiate with the other agents and execute the instructions based on the available local data. The overall performance of a decentralized manufacturing system depends on the negotiations among agents and the data used to make decisions. A major problem in decentralized control is the design of negotiation protocols between agents. These protocols have to be simple, yet be able to carry out all the functions assigned to agent.

The primary focus of this research is on designing decentralized control architectures for discrete part manufacturing systems. We investigate different control protocols for agents in large flexible job shops. A currency like metric is used to help the agents meet their individual objectives, as well as achieve the overall objectives of the system. The control schemes are implemented using an object-oriented simulation platform, *Swarm.* [1] The system behavior is observed under different operating conditions.

II. AGENT BASED MANUFACTURING SYSTEMS

An agent based manufacturing system is a system in which all entities (parts, machines, etc) communicate with other entities and perform their functions without any central controller. The primary motivation for design of these systems is to decentralize the control of manufacturing system, thereby reducing the complexity, increasing flexibility, and enhancing robustness.

In this research, parts and machines are considered as agents with communication capabilities. The material handling devices are assumed to be always available and transportation times are included in machining times. A part agent or machine agent attempts to meet its objective based on the local data. The local data is gathered by communicating with other agents and also from local sensors. The primary objective of a part is to finish all the processing before the due date, while that of a machine is to maximize the utilization rate. Both parts and machines try to reach their objectives by following set rules. The main advantage of this architecture is that agents do not have to rely on a particular entity (central controller) to execute their instructions. Failure of one component will not halt the operation of the entire manufacturing system. All parts can communicate with machines or other parts by broadcasting the messages with appropriate headers as shown in Figure 1. Parts and machines read headers of all the broadcast messages. Machines and parts will read the messages only if the header contains their identification number or type. The following paragraphs outline the construction of machine and part agents.

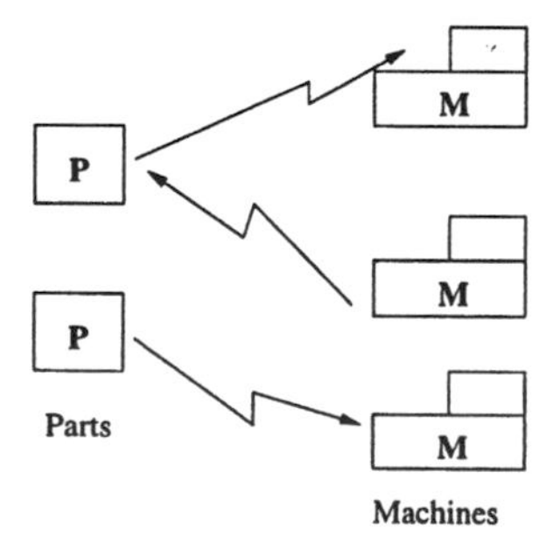

Figure 1: Negotiation between Parts and Machines

2.1 Machine Agent

A machine agent is responsible for all the decisions made by the machine including the selection of parts for processing. The agent is aware of all the attributes of the machine, and its status with the help of sensory data. The following attributes are known to the machine agent:

Machine Identification Number: Each machine has a unique identification number which is used as an identifier in sending and receiving messages. When a part broadcasts a message to a particular machine, that machine's identification number is put as a header in the broadcast message.

Machine Type: Machines are grouped together based on their processing capabilities. A part with a particular process requirement can be processed by any one of the machines in the groups which can perform the given operation.

Speed Factor: Speed factor is defined as ratio of processing speed of a machine with respect to a base machine, which is the slowest machine in the group.

Location: Location of a machine in the manufacturing system is determined by the x and y coordinates on the shop floor.

Buffer Limit: Buffer limit is the maximum number of parts that can wait in queue for a machine. If the buffer limit is reached then the machine can no longer accept part.

Success Rate: The success rate of a machine is defined as the ratio of the number of parts that select the machine out of all the parts that requested bids. The success rate is used in estimating waiting time for parts. If β_W is the bids won in last hour and β_T is the total bids submitted by the machine in that time, then the success rate $\psi = \frac{\beta_W}{\beta_T}$.

Expected Queue Time: Estimated queue time is an estimate of time it would take to process all the parts waiting in the buffer of a machine.

2.2 Part Agent

Part agents communicate with machine agents to get the required processing done before the due date. A part agent communicates by broadcasting messages with the header of agents with which it wants to communicate. Parts are placed on pallets, which have the equipment for broadcasting and receiving messages. A part agent is aware of the following attributes:

Part Identification Number: Each part has a unique identification number which is used in broadcasting message to machines. Every a message sent by a part carries the identification number of the part.

Part Type: The parts are grouped into different groups based the similarity of processing requirements. The routing chart is the same for all parts of a given type.

Part Routing: The part routing chart contains the sequence of operations to be performed on a part type. Operations are represented by numerical controlled (NC) programs. A machine is capable of processing the operations if it can execute the required NC program.

Due Date: Each part enters the manufacturing system with a due date. The objective of each part is to finish all the processing before its due date.

Location: Location of a part is determined by the x and y coordinate on the shop floor. The location is used while requesting transportation.

Nominal Time: Nominal time is defined as an estimate of processing time on base machine.

III. NEGOTIATION MECHANISMS

This paper explores different negotiation protocols for agent based manufacturing systems. The negotiation protocols studied can be categorized into two control schemes.

1. Shortest Processing Time (SPT) Scheme
2. Currency Scheme

3.1 Shortest Processing Time (SPT) Scheme

In the SPT scheme, a part selects the machine with the shortest expected time in system. In this scheme, the parts request for bids and machine tools submit bids. A part makes a decision based on the estimated completion times submitted by machines and selects the machine with least estimated time. The control protocols for part and machine agents are explained in the following paragraphs.

3.1.1 Part Control Protocols When a part enters the manufacturing system it requests bids from machines. Parts wait for a fixed time after which they evaluate all the bids and select the machine with least expected waiting time. In this control scheme, parts are greedy to finish the processing irrespective of their criticality. Once the selection is made, parts send their acceptance to the selected machine and request for transportation as soon as the machine confirms the bid. The part then waits in queue of the selected machine on till it can be machined. During this period part maintains contact with selected machine. If it looses contact with the machine then the part assumes that destination machine has failed and rebroadcasts the request for bids. This procedure makes the control scheme robust to machine failures. The part repeats this procedure until all the processing is completed. Figure 2 shows the control protocols for the part agent.

3.1.2 Machine Control Protocols Machine waits for requests for bids from parts entering the system. When a request for bid has a reference to a NC program which the machine can process then the machine submits a bid with expected completion time, provided the input buffer is not full. The expected completion time ω is given by

$$\omega = Q + \frac{Q}{b}\,\psi\,\beta_K + \frac{m_i}{S} \tag{1}$$

where Q is expected queue time (this takes into account all the parts in the buffer), b is the number of parts in the buffer, β_K is number of bids submitted in last K time units, K is the time a part waits before evaluating bids, ψ is the success rate of machine, m_i is the nominal processing time for the ith operation of the parts, and S is the speed factor.

When a part sends a message requesting confirmation machine bid, the machine once again checks the buffer limit. If the buffer capacity is still not exceeded then the machine sends confirmation to the part. The machine keeps in contact with all the parts that are headed for the input buffer by sending periodic messages. If a machine recognizes any problems or fails, it stops sending messages to the parts which have selected the machine. The machine processes a part when part arrives in front of the queue. The control protocols for machine agents are shown in Figure 3.

3.2 Currency Scheme

In this control scheme, a currency like metric is used to steer the actions of agents in a manufacturing system. Lin and Solberg [5] used currency as a driving factor behind the decisions made by

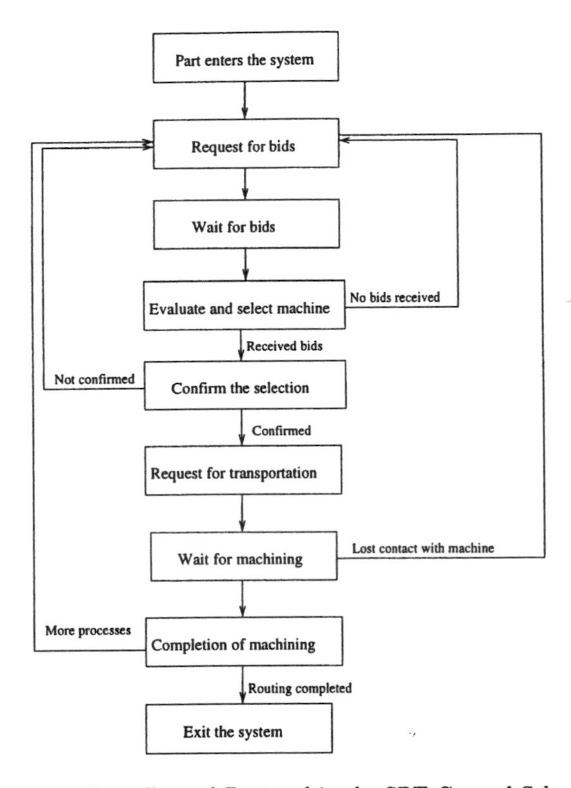

Figure 2: Part Control Protocol in the SPT Control Scheme

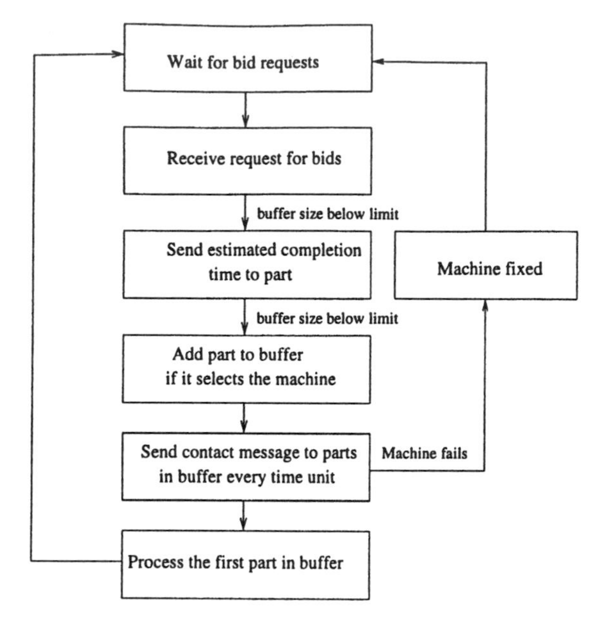

Figure 3: Machine Control Protocol in the SPT Control Scheme

part and machine agents. This control scheme is modeled after the service industry. If a customer wants to get a service offered by an organization, it requires some currency transaction. Consider the machine as the service provider and the part as the customer. The part enters the system with some currency, which depends on the priority of the part. The machine computes the charge based on the expected waiting time and also the recent bid success rate. The part then converts the expected waiting time into equivalent currency, which is based on the due date and adds the machine charge to it. The part only considers the bids which have waiting times less than the due date and the charge under the currency limit. The part selects bids which have the lowest overall charge. The following sections discuss the part and machine agent control protocols in detail.

3.2.1 Part Control Protocols A part enters the manufacturing system with some currency. The amount of currency depends on the due date and nominal processing times. The part with lower ratio of due date over nominal time carries more currency per unit nominal time than the part with higher ratio of due date over nominal time. Part requests bids from machines which can process the NC program in its routing table. After a fixed time, part evaluates the bids table. If there are no bids received then the part rebroadcasts the request for bids. If there are n processes in the routing of a part, then due dates for each process are computed by,

$$d_i = D\left(\frac{m_i}{M}\right) \qquad \forall i = 1, \ldots, n \tag{2}$$

where d_i is the due date for ith process, D is the due date for part, m_i is the nominal time for ith process, M is the total nominal time of all parts.

Similarly, currency spending limits are set for each processing stage. If C is the total currency allocated to a part the currency limit c_i for ith process is given by

$$c_i = C\left(\frac{m_i}{M}\right) \qquad \forall i = 1, \ldots, n \tag{3}$$

The part converts expected completion time into equivalent currency based on completion time and the shape factor. Thus,

$$C_p = 2.0\, tan^{-1}(e^{(A\omega)}) \qquad where\ 0.01 \leq A \leq 0.1 \tag{4}$$

where C_p is the equivalent currency per unit time, A is a shape parameter, ω is the expected completion time submitted by the machine. C_p is multiplied by ω to get the equivalent currency value of the completion time. The currency function, C_p, for different shape factors is shown in Figure 4.

The shape factor A determines the curvature of the currency function. A is a function of the ratio of d_i over m_i and is given below.

$$A = 0.6 - \frac{1 + tanh\left(\frac{d_i}{m_i} - 1\right)}{20} \tag{5}$$

The lower limit for A is set at 0.01 and upper limit is set at 0.1. The shape factor function is shown in Figure 5 and the control protocols for part are shown in Figure 6. The shape factor and the equivalent currency functions are selected to capture the different levels of criticality as the part nears the due date. The specific functions were selected to correspond those in human decision processes.

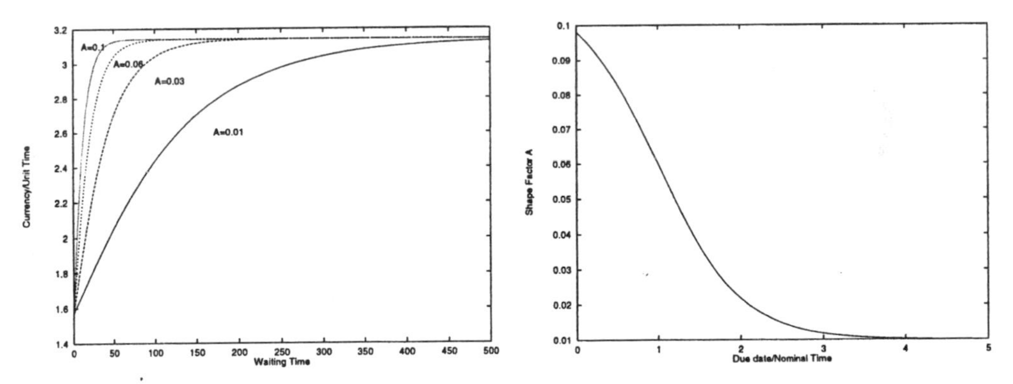

Figure 4: Currency Function for Parts

Figure 5: Shape Factor for Parts

3.2.2 Machine Control Protocols Machine waits for request for bids by parts. If the machine can process the requested NC program and the buffer count is below the limit, then the machine computes estimated completion time and charge for processing. The completion time is computed by Equation 1 similar to the SPT control scheme. The cost of machining is computed using a currency function, which depends on prior success rate and is given by

$$C_m = 4.7 - (2.0 tan^{-1}(e^{(0.11-B)\omega})) \qquad where\ 0.01 \leq B \leq 0.1 \tag{6}$$

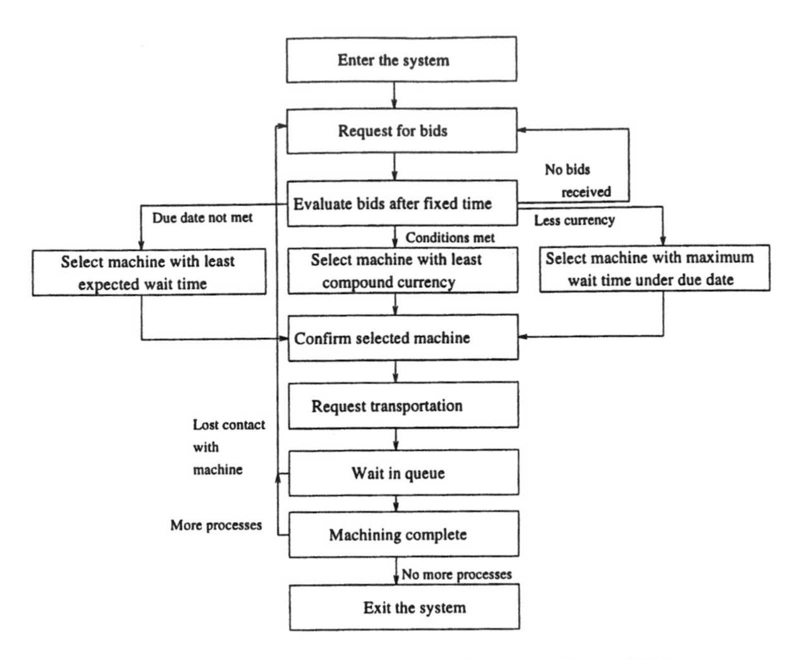

Figure 6: Part Control Protocol in Currency Control Scheme

where C_m is the charge per unit time for processing, B is the shape factor which depends on success rate, and ω is the completion time of part. The currency function of machine is shown in Figure 7.

The shape factor B is given by

$$B = 0.1 - \frac{1 + tanh(1 - 0.1\psi)}{20} \tag{7}$$

where ψ is the success rate of machine. The shape factor function is shown in Figure 8.

If a part selects the machine, then the machine confirms the selection if the buffer count is below the buffer limit. The machine periodically sends contact messages to all parts in queue. If the machine fails then it does not communicate with the parts or detect requests for bids. The machine processes part at the head of the queue. The machine control protocols are shown in Figure 9.

IV. SIMULATION AND RESULTS

A multi–agent manufacturing system is modeled to test the performance of different control schemes. The manufacturing system consists of 40 machines. These machines are grouped into 5 types, each group consists of 8 machines with different speed factors. The system has five part types, each with different routings and are shown in Table 1. The due dates for these parts are set based on the priority of the parts. Parts arrival rates are exponentially distributed. All machines have failure rates and average recovery times, which are used to investigate the robustness of the control schemes to component failures. This paper presents simulation results for the following

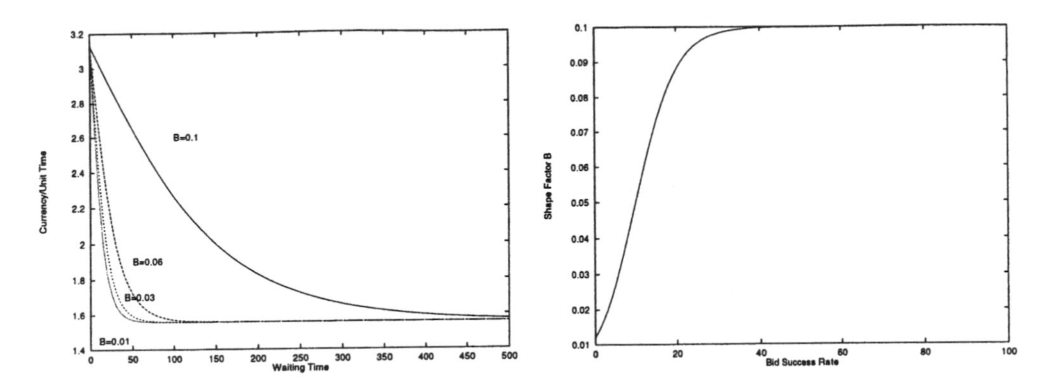

Figure 7: Currency Function for Machines

Figure 8: Currency Shape Factor for Machines

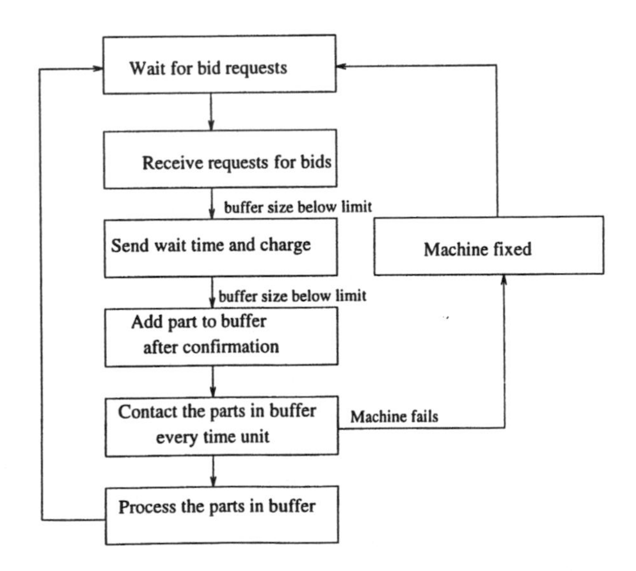

Figure 9: Machine Control Protocol in Currency Control Scheme

three models

Model 1: Hierarchical Control Scheme

Model 2: Agent Based Shortest Processing Time Control Scheme

Model 3: Agent Based Currency Control Scheme

Table 1: Routing Table of Parts

		Part Routing				
Part 1	Process Type	1	2	3	4	5
	Nominal Time	30	50	40	55	35
Part 2	Process Type	2	3	4	5	1
	Nominal Time	30	50	40	55	35
Part 3	Process Type	3	4	5	1	2
	Nominal Time	30	50	40	55	35
Part 4	Process Type	4	5	1	2	3
	Nominal Time	30	50	40	55	35
Part 5	Process Type	2	3	4	5	1
	Nominal Time	30	50	40	55	35

Hierarchical control scheme model is used to compare the performance of multi–agent manufacturing systems models with a reference model. In hierarchical control, parts are assigned to the machines with least number of parts in the buffer. It is assumed that machine failure in a hierarchical manufacturing system causes some delays in reassigning the parts in the buffer of the failed machine. Controller failures are simulated at rates much larger than machine failure rates. Each controller coordinates four machines. When a controller fails, all the machines controlled by it are non-functional.

The following performance measurements are used to compare the different control schemes.

1. Average utilization of machines
2. Throughput of manufacturing system
3. Average tardiness of parts

The manufacturing system models are simulated using *Swarm* [1], an object oriented simulation tool. Parts and machines are designed as instances of classes. In this system, an agent is represented by an object and communication between agents is modeled by messages between objects. Each object has its own state variables, while the generic definition of its behavior is provided by the class. Figure 10 shows a screen capture of the simulation model.

4.1 Design of Experiments and Results

The system performance measures are observed by varying three parameters, system load, failure rates and recovery times. At high load condition the arrival rate is 1/22 parts per time unit, whereas at low load condition the arrival rate is 1/35 parts per time unit. Failure rates are set at

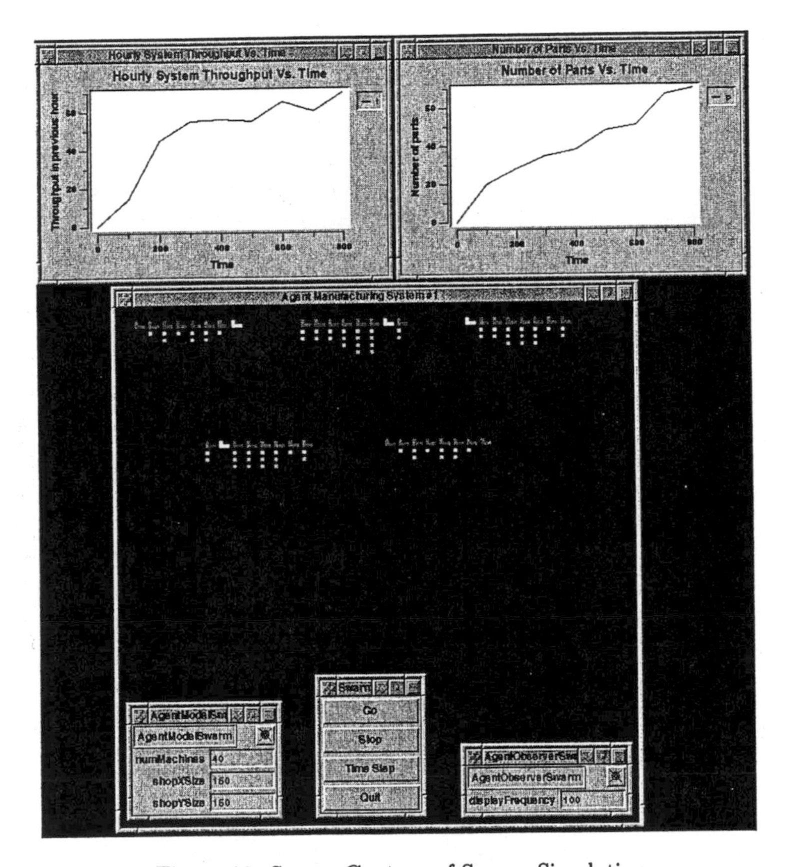

Figure 10: Screen Capture of Swarm Simulation

1000 time units and 2000 time units. And repair times are set at 200 and 100 time units. The 2^k full factorial design of experiments for the three factors is shown in Table 2. Each configuration is replicated 10 times. The simulation run length for each run was 10000 time units.

Table 2: Design of Experiments

	Experiments							
	1	2	3	4	5	6	7	8
Load	−	−	−	−	+	+	+	+
Failure Rate	−	−	+	+	−	−	+	+
Recovery Time	+	−	+	−	+	−	+	−

The results of the simulation experiments are shown in Tables 3, 4, 5. From Table 3, we observe that at high load conditions the agent based systems have better throughput. SPT control scheme has a better throughput than currency control scheme as parts select faster machines. From Table 4, we can observe that average utilization of hierarchical manufacturing system is higher than agent based manufacturing systems as parts are assigned to the machines uniformly based on the buffer size. From Table 5 we observe significant differences in the average tardiness of hierarchical and agent based manufacturing systems. The average tardiness in agent based manufacturing systems is significantly less than hierarchical systems. Currency scheme has smallest tardiness of the three models, since it allocates critical parts to faster machines, based on the due dates and available currency.

Table 3: Average Throughput

		Experiments							
		1	2	3	4	5	6	7	8
Model 1	Mean	43.05	42.15	41.66	43.51	65.49	64.65	66.95	63.54
	SD	0.79	1.71	1.24	0.93	2.31	3.6	2.03	3.1
Model 2	Mean	42.85	42.42	43.24	43.03	68.55	67.91	67.85	69.11
	SD	1.14	1.32	1.00	1.44	1.39	1.3	1.07	1.08
Model 3	Mean	42.55	42.30	42.61	43.06	66.2	66.95	66.06	66.21
	SD	1.00	1.44	1.21	1.05	1.18	1.23	1.05	1.33

Table 4: Average Utilization of Manufacturing System

		Experiments							
		1	2	3	4	5	6	7	8
Model 1	Mean	0.572	0.562	0.555	0.580	0.857	0.846	0.874	0.833
	SD	0.011	0.022	0.015	0.011	0.03	0.046	0.026	0.039
Model 2	Mean	0.459	0.448	0.468	0.458	0.865	0.847	0.859	0.872
	SD	0.015	0.017	0.014	0.019	0.024	0.022	0.02	0.019
Model 3	Mean	0.452	0.443	0.455	0.456	0.828	0.833	0.833	0.825
	SD	0.014	0.019	0.017	0.013	0.021	0.021	0.019	0.023

Table 5: Average Tardiness of Parts

		Experiments							
		1	2	3	4	5	6	7	8
Model 1	Mean	10.12	13.26	18.63	13.88	138.31	201.14	252.60	310.86
	SD	10.94	22.63	20.06	11.72	91.30	179.5	147.23	263.40
Model 2	Mean	0.00	0.00	0.02	0.01	61.17	28.11	89.25	34.73
	SD	0.0.01	0.00	0.02	0.03	20.08	15.65	33.26	15.53
Model 3	Mean	0.07	0.00	0.02	0.01	55.67	40.67	84.69	29.01
	SD	0.13	0.02	0.02	0.03	14.54	13.97	38.65	9.54

V. CONCLUSIONS

Agent–based control schemes offer several advantages over hierarchical control schemes, especially for failure prone manufacturing systems. This paper studied bid construction and evaluation methods based on currency schemes. The currency based protocol performed significantly better for average tardiness criterion. Future research needs to conducted in determining the optimal currency and shape factor relationships.

References

[1] R. Burkhart, M. Askenazi, and N. Minar. *Swarm release documentation.* Santa Fe Institute, 1996.

[2] A.V. Deshmukh, S. Benjaafar, J.J. Talavage, and M.M. Barash. Comparison of centralized and distributed control policies for manufacturing systems. In *2nd Industrial Engineering Research Conference Proceedings*, pages 744–748, 1993.

[3] N.A. Duffie and R.S. Piper. Nonhierarchical control of manufacturing systems. *Journal of Manufacturing Systems*, 5(2):137–139, 1986.

[4] M.M. Lewis, W.and Barash and J.J. Solberg. Computer integrated manufacturing system control: A data flow approach. *Journal of Manufacturing Systems*, 6(3):177–191, 1982.

[5] G. Lin and J.J. Solberg. Performance analysis of an adaptive manufacturing control system. In *Industrial Engineering Research Conference*, pages 749–753. IIE, 1993.

[6] H.V.D. Parunak. Autonomous agent architectures. Technical report, Industrial Technology Institute, Ann Arbor, MI 48106, 1993.

[7] D. Upton, M.M. Barash, and M.M. Matheson. Architectures and auctions in manufacturing. *International Journal of Computer Integrated Manufacturing*, 4(1):23–33, 1991.

[8] D. Veeramani, B. Bhargava, and M.M. Barash. Information system architecture for heterarchical control of large fmss. *International Journal of Computer Integrated Manufacturing*, 6(2):76–92, 1993.

Brief Bio

Naga Krothapalli is a graduate student in the department of Industrial Engineering at the FAMU–FSU College of Engineering. He received his B.Tech in Mechanical Engineering from Nagarjuna University, India in 1994. He is currently enrolled in the Ph.D. program.

His research interests are in decentralization of manufacturing systems and supply chains.

Abhijit Deshmukh is an Assistant Professor in the Department of Industrial Engineering at the FAMU–FSU College of Engineering. He was Assistant Professor in the W. Averell Harriman School for Management and Policy from 1993-95. He received a B.E. in Production Engineering from the University of Bombay in 1987, an M.S. in Industrial Engineering from SUNY at Buffalo in 1989, and a Ph.D. in Industrial Engineering from Purdue University in 1993.

His research interests are in design and control of manufacturing systems. His recent work has been on characterizing complexity in manufacturing systems and modeling decentralized controllers for shop–floor activities. His research has been supported by the National Science Foundation, the Society of Manufacturing Engineers, and several manufacturing companies.

VALIDATION AND SIMULATION OF MANUFACTURING ENGINEERING DATA

Anan Mungwattana[1], John Shewchuk[1], and Albert Jones[2]
[1] Department of Industrial and Systems Engineering, Virginia Polytechnic Institute and State University, Blacksburg, VA 24061-0118, USA;
[2] National Institute of Standards and Technology, Gaithersburg, MD 20899-0001

ABSTRACT: A major problem with the wide variety of manufacturing engineering software applications which are available is that they are not designed to work together. Research is being performed at National Institute of Standards and Technology (NIST) to identify the integration standards and validation issues that must be addressed to implement plug-compatible software environments. One of these efforts is the Manufacturing Engineering Tool Kit (METK), an integrated toolkit providing applications for performing product design, product data and workflow management, operations planning, and engineering data validation using simulation. In this paper, we present an interface for integrating the operations planning application and numerical control programming simulation application. Two categories of manufacturing data validation strategies - object validation and operation validation - are also presented. Finally, the benefits of a generic interface and manufacturing engineering data validation are discussed.

INTRODUCTION

In recent years, many different types of manufacturing engineering software applications have become available. These applications focus on specific engineering functions in the overall product life-cycle. However, they are not designed to work together, which leads to integration problems. Various researchers have tackled these problems, including Skevington and Hsu [1], Nilsson *et al.* [2], McLean and Leong [3], and Koonce *et al.* [4]. A generic solution to such problems would result in significant time and cost savings in the product life-cycle. In order to investigate the problem of integrating manufacturing engineering software applications, the National Institute of Standards and Technology (NIST), along with various academic and industrial collaborators, is developing a Manufacturing Engineering Tool Kit (METK). The METK is a collection of commercial-off-the-shelf (COTS) manufacturing software applications housed together on a high-speed computer workstation [5]. The METK is aimed at identifying the integration standards and validation issues that must be addressed to implement plug-compatible software environments in the future [6]. It is a part of the Computer Aided Manufacturing Engineering (CAME) project, jointly sponsored by the U.S. Navy and NIST [7]. The purpose of the CAME project is to provide an integrated framework, operating environment, common data models, and interface standards for manufacturing engineering software applications.

The METK focus is to develop an integrated toolkit that provides applications for performing the following manufacturing functions: product design, product data and work flow management, operations planning, and engineering data validation using simulation. The system currently includes a product data management application, a CAD application, a generative operations planning application, and a suite of manufacturing simulation software applications. These applications are being used to develop and validate the engineering data required for NC machining operations. The objective is to validate not just the NC program, but the entire operation itself, i.e. obtaining the required machine, tools and fixtures, loading the tools into the tool magazine, fixturing the part at the machine, loading the NC program, and performing the machining process. Thus, the engineering data to be validated includes specifications of the tools, machines, and fixtures used for the operation, as well as the NC program itself. The engineering data is validated by using virtual simulation [8, 9].

This paper is concerned with two aspects relating to implementation of the METK. The first of these is how to integrate the operations planning application with the NC simulation application. The second is how to perform the engineering data validation via the NC simulation application. Section 2 provides details on the METK system structure and operation. Section 3 discusses the integration of the operations planning application with the NC simulation application. The various engineering data validation strategies employed are presented in Section 4. A discussion of the benefits of the integration and data validation strategies, as well as concluding remarks, are presented in Section 5.

MANUFACTURING ENGINEERING TOOL KIT PROJECT

Figure 1 presents a system diagram of the METK. The system currently consists of the following software applications: 1) Matrix (Adra Systems, Inc.), a product data management (PDM) application encapsulating both source file configuration management and workflow management; 2) Pro-Engineer (Parametric Technologies Corporation), a CAD application; 3) ICEM Part (Technomatix Technologies Ltd.), a generative operations planning application, and 4) VNC – Virtual NC (Deneb Robotics, Inc.), an NC simulation application used for verifying NC machining operations. Currently, all of the METK software applications reside and execute on a single Onyx workstation (Silicon Graphics, Inc.) running the IRIX 5.3 operating system.

The process flow for the METK system is shown in Figure 2 and described as follows. First, a product design is created using Pro-Engineer (step 1). Once the product is designed, a solid model geometry of the product is output to a data file and stored by the PDM application. Operations planning is then performed using ICEM Part (2). The data file is retrieved, and this application then uses its internal knowledge-base to generate operations planning data including an operation sheet, NC program, fixture list, tool list, and tool geometry data. These items are then used to drive the NC simulation application, VNC (3), for the purpose of verifying the data. The application can check for simple types of errors, such as missing tools and collisions between the tool and the part, fixture, or table.

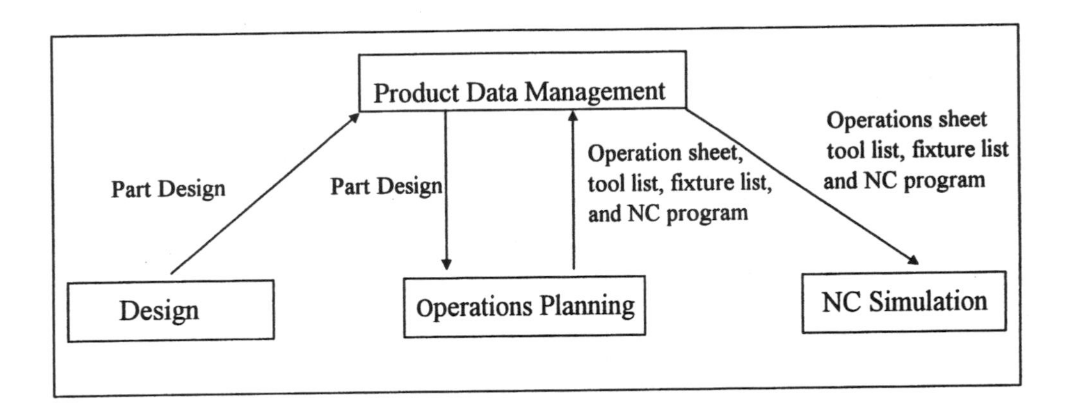

Figure 1. The system diagram of METK.

VNC utilizes three-dimensional models of the machine tool and its controller, fixtures, cutting tools, and part blanks to generate and visualize the NC machining operation. An example virtual manufacturing environment consisting of an EMCO 100 vertical milling machine with its controller is shown in Figure 3.

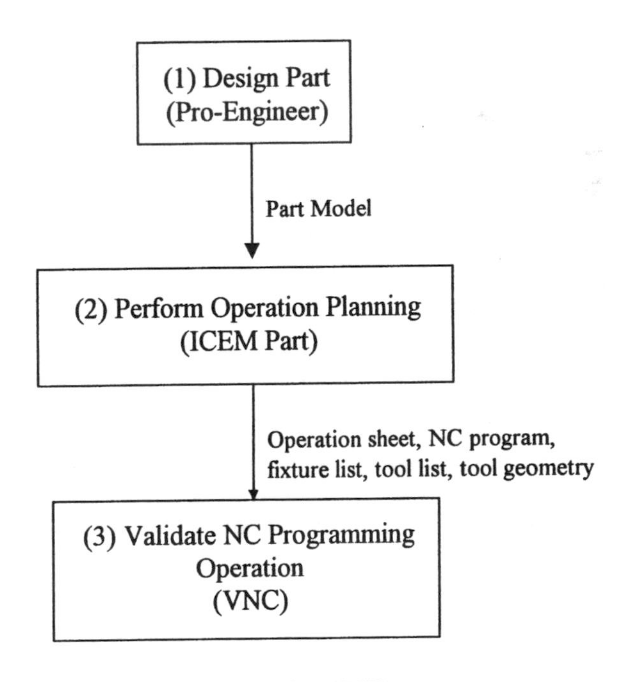

Figure 2. Process flow of the METK system.

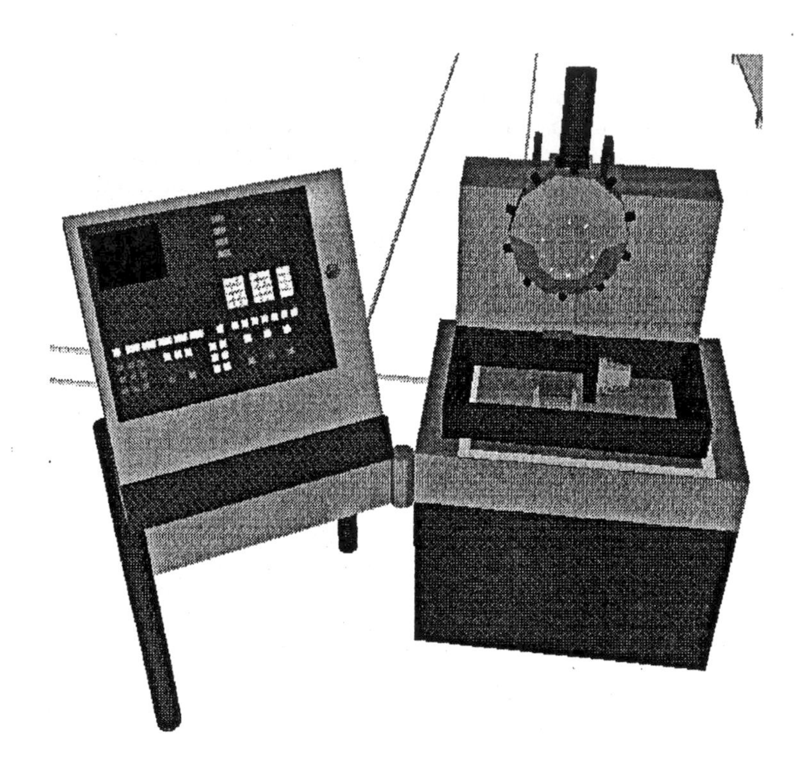

Figure 3. EMCO 100 vertical milling machine.

INTEGRATION OF OPERATIONS PLANNING AND NC SIMULATION APPLICATIONS

Once the operations planning data (operation sheet, NC program, etc.) has been generated, it must be validated using the NC simulation application. However, as anticipated, these applications cannot be directly interfaced with one another. This is due to two reasons. The first is that operations planning output data and simulation input data are fundamentally different in nature: the former is a collection of static data defining what is to be done (machining operation) and how, whereas the latter is a sequence of informational commands (simulation instructions) representing the actual flow of activities. The second is of course that the data formats utilized by the vendors are different.

The approach used to integrate these two applications consists of two steps. The first is to create from the operations planning data an interface using the concept of a process plan [10, 11]. The second is to then develop a process plan interpreter to parse this file into a sequence of commands which can be understood by the NC simulation application. Each of these steps is discussed below.

Process Plan Generation

Generation of the process plan is done using a process plan editor. The process plan is a file containing machining data and instructions. It consists of three sections: Header, Resources, and Procedures. The Header section contains all the administrative information about the plan. This includes the plan identification (ID) number, version number, latest revision date and number, part name, part ID, planner's name, etc. The Resources section lists all of the resources needed to implement the plan, including tools, fixtures, materials, machines, software programs, and (possibly) other process plans. The Procedure section identifies all of the steps necessary to carry out the actual execution plan. Each step has a step number, work description with associated parameters and requirements, and any precedence constraints. An example of a simple process plan, which has been used in our initial experiments, is shown in Figure 4.

HEADER Section
Plan_id=P12345
part_name=Air_frame_test_part
Creation_date=10/24/96
Planner=Mike Iuliano

RESOURCE Section
machine_id=CINC_MILA_T30
tool_name=1/4" TWIST_DRILL
tool_name=1/2" CENTER_DRILL
tool_name=1/8" BALL_NOSE_END_MILL
tool_name=SHANK_END_MILL
fixture_name=vise
workpiece_name=Air_frame_blank
nc_program=Airframe.cnc

PROCDURE Section
Step 1 LOAD_MACHINE
  machine_id=CINC_MILA_T30
  Machine_controller=GE2000
  end_step
Step 2 LOAD_TOOL
  tool_name= TWIST_DRILL
  tool_id=T266
  magazine_slot=1
  end_step
Step 3 LOAD_TOOL
  tool_name= CENTER_DRILL
  tool_id=T271
  magazine_slot=2
  end_step
Step 4 LOAD_TOOL
  tool_name=BALL_NOSE_END_MILL
  tool_id=T268
  magazine_slot=3
  end_step
Step 5 LOAD_TOOL
  tool_name=SHANK_END_MILL
  tool_id=T234
  magazine_slot=4
  end_step
Step 6 LOAD_FIXTURE
  fixture_name=vise
  fixture_id=V178
  ref_frame=x_axis
  x,y,z_offset=152.4, 101.6, 44.45
  units=inches
Step 7 LOAD_WORKPIECE
  workpiece_name=Air_frame_blank
  workpiece_id=W123
  ref_frame=fixture_name
  x,y,z_offset=0,0,0
Step 8 LOAD_NC_PROGRAM
  nc_program=Air_frame.cnc
  end_step
Step 9 RUN_NC_PROGRAM
  nc_program=Air_frame.cnc
  end_step

Figure 4. A simple process plan.

Parsing of Process Plan

Once the process plan has been prepared, it is parsed to obtain the sequence of commands required to execute the NC simulation using VNC. The application which performs this activity is referred to as the process plan interpreter. Six types of commands can be sent to the NC simulation application from the interpreter: various parameters are also sent along with each command (see Figure 4). The commands are as follows:

1. Load machine: causes the NC simulation application to retrieve a machine or workcell model from a specific directory and place it into the working environment. The format for this command is '*Load_machine,machine_name,machine_id*'.
2. Load tool: causes the NC simulation application to create the tools needed to machine the workpiece and load the tools into the machine's tools magazine. The format for this command is '*Load_tool,tool_name,tool_id,slot_no*'.
3. Load fixture: causes the NC simulation application to retrieve the necessary fixture for supporting the workpiece at the machine. The format for this command is '*Load_fixture,fixture_name,fixture_id,axis,offset*'.
4. Load workpiece: causes the NC simulation application to retrieve the workpiece and load it into the fixture. The format for this command is '*Load_workpiece,workpiece_name,workpiece_id, attached_coordinate,offset*'.
5. Load NC program: causes the NC simulation application to retrieve the NC program and load into the machine's controller. The format for this command is '*Load_NC_program, NC_program_name,machine_id*'.
6. Run the simulation: causes the NC simulation application to run the simulation for performing engineering data validation. The format of this command is '*Run_NC_Program,option*'.

For example, the command '*Load_tool,TWIST_DRILLS_225,T7,slot_7*' will cause the NC simulation application to retrieve a model of a twist drill, label it as tool no. 7, and insert it into slot no. 7 of the tool magazine.

MANUFACTURING ENGINEERING DATA VALIDATION STRATEGIES

For validating NC machining operations via simulation, two categories of manufacturing engineering data validation strategies can be identified. The first are those concerned with validating the dimensions, geometry, and possibly other properties of the physical objects utilized in such operations (machine tools and controllers, fixtures, tools, parts, etc.). This is of course of paramount importance in order to ensure that parts fit into their fixtures, tools fit into their toolholders, tools don't collide with parts, etc. For performing virtual simulation of NC machining operations, data used to specify physical objects is required in order to develop the solid models of these objects utilized in the simulation. Consequently, this validation activity is best performed when these models are being created.

The second category of engineering data validation strategies are those concerned with ensuring that the proper sequence of events occurs during the machining operation, i.e., the correct tools are loaded (in the right sequence) and placed (in the right tool slots), the proper fixture is employed, the NC program executes correctly, etc. This validation activity must be performed during execution of the simulation model, as errors resulting from incorrect data of this nature can occur at any time during the machining operation.

Data validation strategies concerned with each of these categories are discussed below.

Object Validation

One of the most critical elements in NC machining operations are the cutting tools. The use of the wrong cutting tools can easily result in broken tools, damaged parts, and damaged machines, not to mention operator injuries. Substantial downtime and cost is almost always associated with rectifying such problems. Thus, it is of paramount importance to validate the cutting tools used. This consists of ensuring that the tools have the correct dimensions and geometry, and are of the correct material. Example tool data includes cutting tool type, tool ID, assembly length, cutter length, cutter diameter, and shank length.

To validate tool dimensional and geometrical data, a set of rules and constraints are specified. Adherence to these rules and constraints ensures that the solid models of the cutting tools created for the simulation will be correct. Some of these rules and constraints are:

- assembly length is greater than or equal to a specified parameter.
- shank length is greater than or equal to a specified parameter and less than assembly length.
- cutter length is greater than or equal to a specified parameter and less than assembly length.
- the sum of shank length and cutter length is less than or equal to assembly length.
- the maximum depth-of-cut is greater than or equal to a specified parameter and less than or equal to cutter length.

Note that the use of such rules and constraints cannot guarantee that the tools created will be correct, but will help to eliminate many sources of error.

Another important object to be validated is the part after machining. This must be done in order to assure that the dimensions and geometry of the part are correct. The benefit of validating the machined part is that all the data created by the operations planning application is then valid, and consequently could be used with confidence in a real manufacturing shop. However, it is not possible to validate the machined part using VNC. Another software application is being investigated for this purpose.

Operation Validation

As specified above, validation of the NC machining operation must be performed during execution of the simulation model. First, commands and their parameters must be validated. This is done by having the interpreter send commands to the simulation application one at a time. After receiving a command, the NC simulation application executes it and sends feedback to the process plan interpreter. Two types of feedback can be sent. The first is a success signal indicating that the assigned task was executed successfully (i.e., error-free): upon receipt of this signal the interpreter sends the next command. The second is an error signal which indicates that the assigned task could not be carried out due to incorrect data or because the command itself was invalid. Along with the error signal, the NC simulation application provides data indicating the cause of the error. Upon receipt of an error signal, the process plan interpreter halts simulation execution and displays the error. The format of an error signal is '*error_code, error_definition*'. Table 1 presents some types of errors encountered and their corresponding codes for each type of command.

Table 1.
Possible types of VNC errors generated during execution of NC simulation program.

Work Elements	VNC Error Definition	VNC Error Code
Load_machine	1. File protcction	110
	2. No such file name	270
Load_tool	1. File protection	110
	2. No such file name	270
	3. No such named device in workcell	260
	4. No such named part in workcell	290
	5. Device not qualified for the operation	360
	6. Attaching a tool into an occupied slot	N/A
Load_fixture	1. File protection	110
	2. No such file name	270
	3. No such named device in workcell	290
	4. No such named part in workcell	290
	5. Device not qualified for the operation	360
Load_workpiece	1. File protection	110
	2. No such file name	270
	3. No such named device in workcell	290
	4. No such named part in workcell	290
	5. Device not qualified for the operation	360
Load_NC_program	1. Compiler error	50
	2. File protection	110
	3. Device is non programmable	210
	4. No such named device in workcell	260
	5. No such file name	270
	6. Device not qualified for the operation	360

After the sequence of commands with their parameters is validated, the machining environment is ready to run the simulation. This step is to verify the NC program to check whether it performs as desired. During the simulation, the sequence of machining operations, as well as collisions between tools and parts, tools and fixtures, tool holders and parts, etc., can also be verified visually. If a collision occurs, it may be due to an invalid NC program or incorrect tool, fixture or part. Detailed investigation must be performed to identify the cause of the problem. Once the simulation is complete, the machined part is visualized to check its geometry as shown in Figure 5.

Figure 5. An example of machined part.

CONCLUSIONS

This paper has described an integration and validation strategy between the operations planning application and the NC simulation application used in the Manufacturing Engineering Tool Kit. Integration was achieved by the intermediate steps of creating a neutral-format process plan, then utilizing a process plan interpreter to parse this plan into a sequence of commands for driving the NC simulation application. With this approach, the process plan serves a generic interface between these applications. The benefit of this approach is that the process plan interpreter can be reused with many types of operations planning applications and NC verification applications.

NC machining operation data validation was divided into object validation and operation validation phases. Object validation is concerned with verifying the dimensions, geometry and possibly other properties of physical objects used in the simulation. In this paper, a strategy for verifying cutting tool geometry and dimension was presented. Operation validation is performed to ensure that proper commands and parameters are sent to the simulation application. Additionally, the sequence of machining operations and possible collisions between objects are checked during simulation execution by visualization. After the simulation, the geometry of the machined part is also checked by visualization and analysis. These validation strategies will help to produce actual parts with the correct geometry and dimensions the first time, resulting in substantial cost and time savings.

Work described in this paper was sponsored by the U.S. Navy Manufacturing Technology program and the NIST System Integration of Manufacturing Applications program. No approval or endorsement of any commercial product by NIST is intended or implied.

REFERENCES

1. **Skevington, C., and Hsu, C.,** "Manufacturing Architecture for Integrated Systems," *Robotics and Computer-Integrated Manufacturing,* Volume 4, Number ¾, 619-623, 1988.
2. **Nilsson, E., Nordhagen, E., and Oftedal, G.,** "Aspects of Systems Integration," *Proceedings of The First International Conference on System Integration,* Morristown, NJ, 435-443, 1990.
3. **McLean, C., and Leong, S.,** "A Process Model For Production System Engineering," *IFIP WG 5.7: Working Conference of Managing Concurrent Manufacturing to Improve Industrial Performance,* Seattle, 33-43, 1995.
4. **Koonce, D., Judd, R., and Parks, C.,** "Manufacturing Systems Engineering and Design: An Intelligent, Multi-Model, Integration Architecture," *International Journal Computer Integrated Manufacturing,* Volume 9, Number 6, 443-453, 1996.
5. **Iuliano, M.,** "Overview of the Manufacturing Engineering Tool Kit Prototype," *NIST Technical Report,* 1995.
6. **McLean, C., Leong, S., and Chen, C.,** "Issues of Manufacturing Engineering Data Validation in Concurrent Engineering Environment," *IFIP WG 5.7: Working Conference of Managing Concurrent Manufacturing to Improve Industrial Performance,* Seattle, 289-299, 1995.
7. **McLean, C.,** "Computer-Aided Manufacturing System Engineering," *IFIP Transaction: Advanced in Production Management Systems,* North Holland, Amsterdam, Netherlands, 341-348, 1993.
8. **Wittenberg, G.,** "A Virtual Reality and Simulation Initiative," *The Industrial Robot,* Volume 20, Number 4, 14-16, 1993.
9. **Cremer, J., Keanery, J., and Ko, H.,** "Simulation and Scenario Support for Virtual Environments," *Computers and Graphics,* Volume 20, Number 2, 199-206, 1996.
10. **McLean, C.,** "Interface Concepts for Plug-Compatible Production Management System," *IFIP WG 5.7: Information Flow in Automated Manufacturing Systems,* North Holland, Amsterdam, Netherlands, 307-318, 1987.
11. **Iuliano, M., Jones, A., and Feng, S.,** "Analysis of AP213 for Usage as a Process Plan Exchange Format," *NIST Technical Report,* 1996.

SELECTION AND COMPARATIVE EVALUATION OF CONVENTIONAL AND RAPID PROTOTYPING PROCESS CHAINS OF THE TOOL AND DIE-MAKING INDUSTRY BASED ON AN INTEGRATED INFORMATION MODEL

Holger Dürr and Uwe Kaschka
Department of Manufacturing Technology
Technische Universität Chemnitz-Zwickau
09107 Chemnitz, GERMANY

ABSTRACT. This paper describes a methodology to select and evaluate process chains in the manufacturing of models, prototypes, moulds and tools, considering the following criteria: quality, time and costs and their integration in the company's information flow.
One of the main goals is to confront conventional manufacturing technologies (e.g. NC-milling) to the new, innovative techniques such as rapid prototyping/rapid tooling, high speed cutting or 5-axis milling and to evaluate the different methods on the basis of a given production task.
The presented two-phase approach not only permits the selection and evaluation of the process chains on the basis of a cost-benefit-analysis but also a rough planning of the technology strategy.
An integrated information model to integrate this methodology into the company's data structure was developed, its components are presented in this paper.

Introduction

The present situation in the manufacturing of tools , dies and moulds is characterized by increased variety, reducing innovation cycles and increasing time pressure. To ensure a company's competitiveness, also on international markets, a fast product introduction to the market and qualities such as flexibility and keeping delivery times are becoming more important.
Based on these demands there can be observed a clear trend in using new manufacturing methods. Technologies such as rapid prototyping (RP), high speed cutting (HSC) and the 5-axis NC milling are prominent. Especially rapid prototyping (also called "generative manufacturing processes" or "layer technologies") offers realistic possibilities to shorten the entire product development time thanks to a faster construction of models and prototypes but it can also reduce the costs of product's modification by using physical models in early stages of design. Additionally, the so called "rapid tooling"-technologies are the main topic of many research works. These technologies should be enabled the direct production of tools and dies in the series material by means of layer manufacturing methods in the future.

A disadvantage however is the user's insufficient knowledge about the possibilities and limitations of the innovative technologies. Permanently new developments and a different degree of industrial applicability of the offered technologies complicate the decision-making. Suitable practical tools for a direct comparison with conventional manufacturing processes are missed. Machine-builders and service providers can not give an objective overview because they are always tend to sell "their own" solution.

This topic has already been dealt with a number of research activities [1-6]. They focus mainly on the evaluation of the several manufacturing technologies without considering the complete process chain of the CAD data preparation leading up to a possible post processing. But in the field of RP technologies

consideration, otherwise we risk losing our time and money savings based on the building process.
In other research approaches, the conventional methods (e.g. the NC-milling) being still relevant for the production of prototypes, tools and dies are neglected.
In general, there is a demand for decision support systems for assistance in complex decision problems in design and process planning focussing on tool and die making industry, which can be integrated into the existing electronic-data-processing environment of a company without any problems.

Process chains in the tool and die making industry

Recent development in the field of RP technologies and HSC extends the opportunities of producing models, tools and dies. Figure 1 provides an overview of the most important process chains.

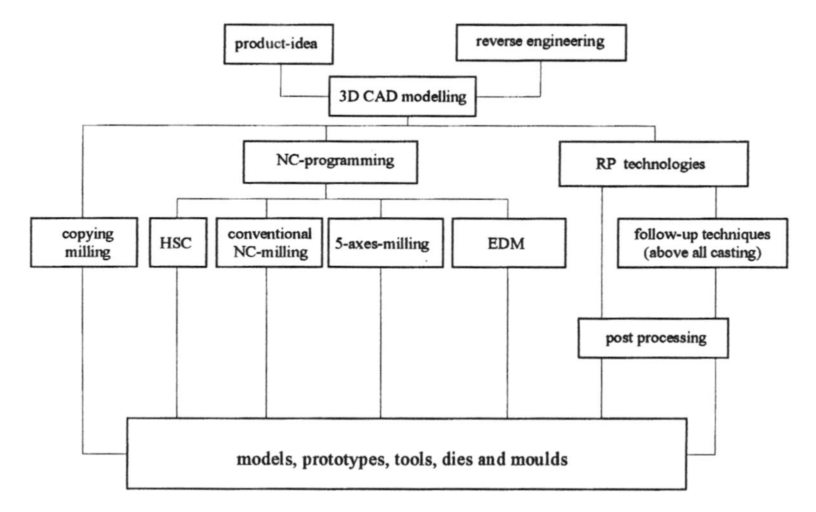

Figure 1. Process chains in the tool and die making industry

Especially the technologies that are known under "rapid prototyping" have become a tremendous success. Starting from the stereolithography (SLA) new methods were developed that are based on the "layer-by-layer" technique. At the moment technologies such as fused deposition modelling (FDM), solid ground curing (SGC), laminated object manufacturing (LOM), selective laser sintering (SLS), direct shell production casting (DSPC), ballistic particle manufacturing (BPM) and model maker 3D-plotting are being frequently used in industry.
We expect even more possibilities for rapid prototyping in the near future because of developments in materials, machines and processes. It is also possible to improve the part's properties by follow-up techniques (casting) or post processing (coating, infiltration). As a consequence not only prototype manufacturing but also the fast and inexpensive fabrication of concept models (rapid modelling), tools and dies (rapid tooling) and of components, small-lot components and replacement parts (rapid manufacturing) can become a field of application for generative technologies [7].

Selection and evaluation of process chains

In selecting the right process chain for a certain component it is important to consider the company's different priorities with regard to time, costs and quality. Additionally there must be considered aspects of business management (operational organisation), manufacturing technology (available machines and devices) as well as information processing (available CAD- and other software systems). In order to achieve an economic result, advantages and disadvantages of techniques and process chains have to be evaluated carefully. Growing information diversity, increasing complexity of demands and partly contradictory objectives require systematic instruments for decision support.

These demands can be fulfilled by applying a two-phase methodology consisting of a selection of the technological strategy and the selection and evaluation of the process chains on the basis of a cost-benefit-analysis (figure 2).

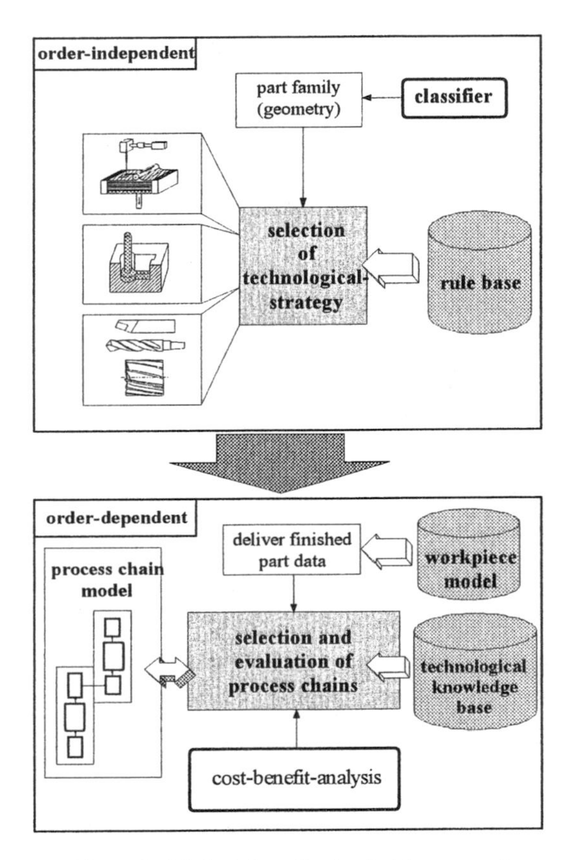

Figure 2. Methodology for the decision support

In the **first phase** the selection of the technological strategy offers the possibility to make a first rough order-independent planning of the available technologies and process chains.
The typical manufacturing processes are classified to distinguish between traditional and new technologies respectively combinations of methods (see table 1).

Technological strategy	Examples of manufacturing methods
conventional	NC-milling, Electrical discharge machining (EDM), copying milling
unconventional	high-speed cutting (HSC)
generative	rapid prototyping technologies such as SLS, SLA, FDM, LOM
combined methods	NC-milling and laser generating, NC-milling and eroding

Table 1
Technological strategies

Based on the results of analyses and an intensive evaluation of process chains there has been developed a set of rules which allows to allocate technological strategies to concrete production tasks.
The development of such a rule base is depend on quantification of "technology determining" criteria. For this decision-making process the criteria are

- part geometry (dimensions, complexity),
- material,
- surface quality,
- dimensional and geometrical accuracy,
- quality of the description of the workpiece,
- number of pieces.

The selection of the technologies is considerably influenced by the component requirements regarding these factors.
In the tool and die making industry especially the criterion of "geometric complexity" is of importance. The more complex the part geometry , the higher is suitability of generative manufacturing processes to produce models and prototypes. This rather vague aspect can be specified in a further stage so the complexity classes can be classified depending on the part's geometry. Now a rule base can allocate the appropriate technological strategy. The most important factors are existence, number, type and position of the complex features.

Condition for an order-depending decision-making for concrete manufacturing sequences is to select and evaluated process chains, the **second phase** in the presented methodology.

The evaluation task is characterized by the following properties:

- to achieve a number of main objectives such as low costs, high quality and time shortening which can be partly contradictory,
- to process criteria that can be quantified (e.g. costs) as well as criteria not to be quantified (e.g. flexibility) and
- to consider criteria whose benefit is difficult to be quantified (benefit caused by reduced time-to-market).

Having analyzed available and applied evaluation methods, there has been chosen a method, which is based on the cost-benefit-analysis [8].
The consequent division of cost- and benefit analysis guarantees a high transparency of the method and the evaluation result. A lot of companies are required to list cost indices separately and in addition existing cost calculation systems can be integrated and financial resources for single expenditures (investments in machine tools and devices) can also be taken into consideration.

All other, mainly qualitative criteria are evaluated by the benefit analysis. The final result is a comparison of the arising costs and the benefit values per process chain (figure 3). That means, that user may select out of a sufficient range of decision variants. There is not any best solution. It is possible to validate results by benchmarking of representative parts.

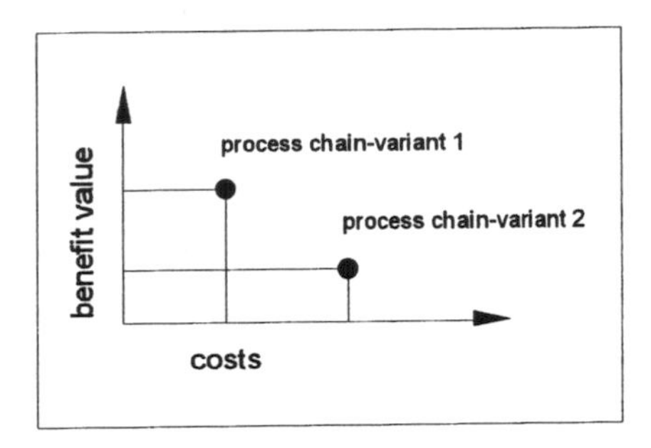

Figure 3. graph of the evaluation results

Cost-benefit-analysis can be studied easily step by step. Considering the rules of probability, making mistakes can almost be excluded. The procedure was applied successfully in various applications in practical use. For evaluation, an expert team is recommended. As a result, parameters (e.g. the level of fulfillment) can be determined in a more objective manner.

Integrated information model

The efficiency of the applied methodology for the selection and evaluation of process chains can be improved by an integration into the complete company's information structure. Data exchange inside the process chain (CAD-CAM) and between company-specific systems (time determination, cost calculation) is to be paid most attention. Losses of time as a result from extensive data transfer methods can be avoided as far as possible (figure 4).

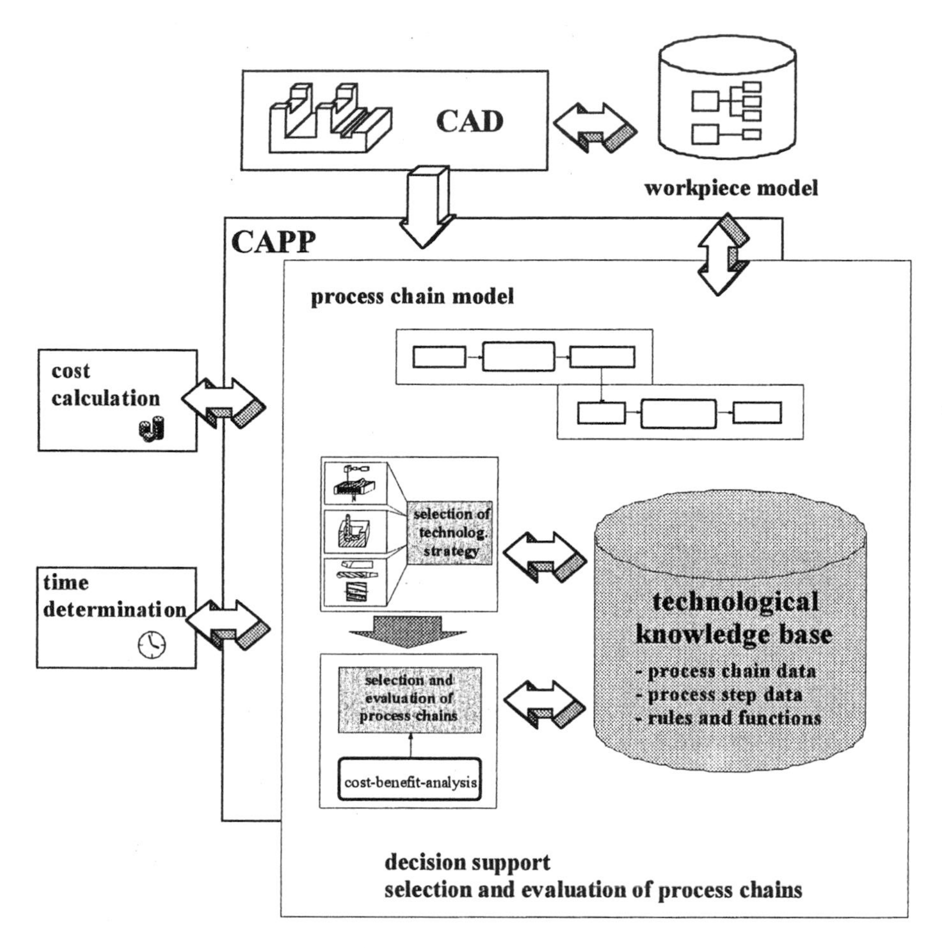

Figure 4. system concept

Starting point for all activities is the description of the manufacturing part. The workpiece is represented in a **workpiece model** based on technical elements (features).
The model contains not only geometrical (e.g. measurements, geometric elements) but also technological data (e.g. tolerance, surface quality).
The workpiece model will implemented in the description language EXPRESS defined in the interface standard STEP.
The tool and die making industry is dominated by prismatic parts with a high number of freeform surfaces. From a technological point of view it is possible to integrate the freeform elements into feature-based workpiece models [10] and this will be the main focus of further work.

Figure 5 shows the **process chain model**, which results from an analysis of relevant manufacturing sequences.

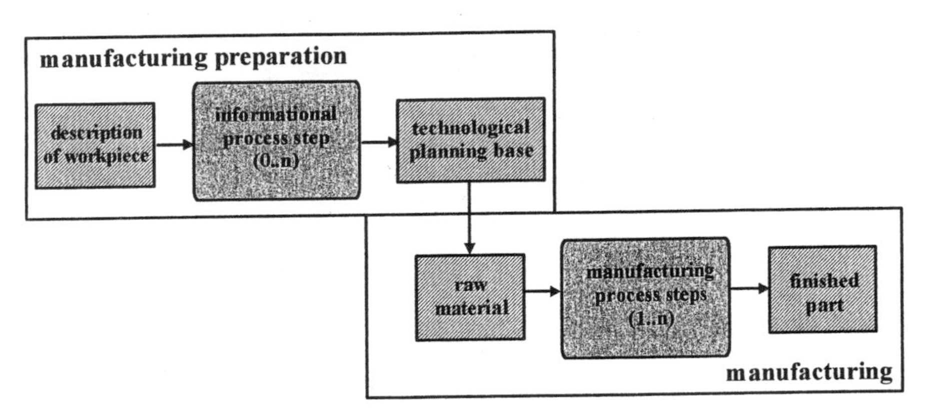

Figure 5. process chain model

A **process chain** in general is a systematic sequence of processes respectively process steps that will lead to a certain goal.
In manufacturing industry the complete product life cycle can be viewed as a process chain starting with the conceptional phase, leading up to design, manufacturing, the actual use, and ending with the possible recycling or disposal of the part.
In the project the main focus is on the fields of manufacturing preparation and manufacturing. For that reason we defined "process chain" as a systematic sequence of that process steps that describe the production of a workpiece starting from the part description (CAD model, technical drawing, sketch,..) leading up to the production planning and manufacturing (figure 5).
In the process chain model there is a separation between informational and manufacturing process steps:
An **informational process step** is that part of the process chain where changes in the information processing unit (e.g. data preparation in the CAD model) can be made or conditions for the manufacturing can be created (e.g. generation of a NC program).
A **manufacturing process step** is that part of the process chain where geometrical and/or technological changes of state of a workpiece can be performed. Exact one technical planning element is being converted. The **technical planning element** describes the relationship between manufacturing method(s) and technical element(s) that are created or changed by it.

Table 2 lists the individual elements of the described process chain using selected examples.

elements of process chain	**examples**
description of workpiece	CAD-model (2D, 3D), drawing, schematic sketch, physical model
information technical process step	CAD-data-preparation, NC-programming, reverse engineering, generation and selection of process chains/manufacturing sequences
technological planning base	production plans, NC-sources, STL-files
raw material	rod, drawn material, powdered material, photocurable liquid polymer
manufacturing process step	NC-milling, high speed cutting (HSC), stereolithography, selective laser sintering, infiltration
finished part	models, prototyps, dies, moulds, tools

table 2
process chain elements

An essential part of the system concept is the **technological knowledge base**.
The knowledge base contains relevant information on technologies and methods of tool and die making industry.
The appearing process chain information are summarised and classified in table 3.

Data types	Explanation	Example
Global process chain data	Information/data, describing the entire process chain	-manufacturing time -manufacturing costs -reference companies -service enterprises
Global process step data	Information/data, describing a process step and which are generally valid (independent of the used process chain) for the process step	-data of technologies and manufacturing methods such as layer manufacturing processes (applicable material, maximum laser power of the device, etc.)
Specific process step data	Information/data, describing a process step and are valid for the corresponding process chain only	-specific parameters of a machine or device inside of a process chain
Product data (of finished part)	Information/data of the final product of each process chain	-achievable product properties (accuracy , surface quality,....)

Table 3
Classification of process chain information

To select and evaluate it is important to have access to that kind of information that refers to complete process chains respectively manufacturing sequences. The definition of "Global process chain data" relates to this requirement.

Formula and functions for the determination of manufacturing time, manufacturing costs and product's quality are assembled in the technological knowledge base. These can be allocated to the individual manufacturing technologies and supply indices for company time determination and cost calculation.

Perspective

The methodology should be tested on the base of selected representative products from the tool and die making industry and converted into a software prototype for consulting and decision support.

The results of the selection and evaluation processes should be examined on their applicability to systems for offer planning and calculation.

References

1. **Tönshoff, H. K., Hennig, K. R.,** Fertigungstechnologien bewerten und auswählen, *VDI-Zeitschrift,* Nr. 6, 30-33, 1995.
2. **Klocke, F., Weck, M., Schell, H., Rüenauver, E.,** Bewertung alternativer Fertigungsfolgen *Zeitschrift für wirtschaftliche Fertigung ZWF,* Nr. 7-8, 359-362, 1996.
3. **Eversheim, W., Klocke, F., Albrecht, T., Nöken, S., Wirtz, H.,** Rapid Prototyping-Unternehmensspezifische Technologiekonzepte, *VDI-Zeitschrift Special Werkzeug- und Formenbau,* Ausgabe November, 20-23, 1995.
4. **Hirsch, B. E., Bauer, J.,** Computerbased Rapid Prototyping System Selection and Support, Proceedings of the International Conference on Rapid Product Development, Stuttgart, Germany, June 10-11, 1996.
5. **Steger, W., Conrad, T.,** Rapid Prototyping: Operative und strategische Bewertung von generativen und konventionellen Fertigungsverfahren, *VDI-Zeitschrift Special Werkzeug- und Formenbau,* Ausgabe November, 12-18, 1995.
6. **Dürr, H., Kunzmann, U.,** Bewertung der Umweltverträglichkeit von Prozeßketten, *Die Maschine dima,* Nr. 10, 47-48, 1996.
7. **Klocke, F., Nöken, S.,** Rapid Prototyping und Rapid Tooling, *VDI-Zeitschrift Special Werkzeug- und Formenbau,* Ausgabe November, 50-54, 1996.
8. **Rinza, P., Schmitz, H.,** Nutzwert-Kosten-Analyse - Eine Entscheidungshilfe, VDI-Verlag GmbH Düsseldorf, Germany, 1992.
9. **Dürr, H., Löbig, S.,** Featuremodellierung unter Berücksichtigung des STEP-Standards, Scientific Reports, 12. Internationale Fachtagung Innovative Technologien, November 1996, Mittweida, Germany
10. **Krause, F.-L., Stiel, C.,** Featurebasierte Konstruktion mit Freiformflächen, *Zeitschrift für wirtschaftliche Fertigung ZWF,* Nr. 3, 101-104, 1996

GETTING STARTED IN RAPID PROTOTYPING

M. Sarwar, S. Hogarth, P.M. Hackney
Center for Rapid Product Development
School of Engineering
University of Northumbria at Newcastle
Tel: 0191 2273644 / 2274589
Fax: 0191 2273854
http://www.unn.ac.uk/~mfx2

ABSTRACT: Currently Industry is confused about the facilities and benefits of rapid prototyping/rapid tooling. Furthermore there appears to be even more difficulty for users to select the appropriate technology and secondary processes.

The paper reviews the current state of the art associated with Time Compression Techniques (TCT's) and the national and international activities associated with the above.
The information should prove extremely useful to managers, designers and manufacturers seeking to achieve a 'Competitive Advantage'.

INTRODUCTION

Rapid prototyping (RP) is a term used to describe a process which produces complex 3D parts by material addition, rather than the traditional material removal methods. There are now several processes available, which all work on much the same principle. That is, a 3D computer aided design (CAD) model of the desired part is split into very thin, 2D, slices forming an industry standard .stl file. The .stl file is then downloaded to the RP machine which builds the part by layering these slices on top of each other to produce a 3D object. The various technologies use different methods of building these layers which will be discussed later. Currently, sheet (LOM), powder (SLS), filament (FDM) and liquid materials (SLA, IJM) are used in RP.

Rapid prototyping is becoming more widespread as 3D CAD is increasing, this is because the RP process requires a 3D model, although most bureaus offer a 2D to 3D conversion service. The advantages over traditional manufacture methods are that complex parts can be produced in hours, often in a design office environment. This means that a designer can hold a part designed yesterday, in his/her hand today.

This greatly enhances the design process, as fit and function problems can be sorted out at he design stage rather than at the production stage, saving time and money. Furthermore lead times are drastically reduced e.g. from 27 weeks to 6 weeks[1]. The RP part can also act as an excellent communication tool, to show others in the manufacturing process the design intent, and allow customers to be shown a part before production has started.

At the moment RP is mainly used for one off production of parts, however it is possible, by various secondary processes, to produce replicas of the RP part more cost effectively. This is useful where the number of parts is of the order of 100, but if more parts are required alternatives need to be found. This is the area of rapid tooling, which can be direct; producing a tool in the RP process for use in moulding, casting; or indirect, using the RP part to produce a tool, via sand casting etc.

RAPID PROTOTYPING PROCESS

The five main types of process available in this country are described in Table 1 below. Only the processes available in the UK will be described here, although there are others world wide and more being developed.

Table 1

A Summary of RP Processes

Process	Stereo-lithography (SLA)	Laminated Object Manufacture (LOM)	Fused Deposition Modelling (FDM)	Selective Laser Sintering (SLS)	Ink jet modelling (IJM)
Manufacturer	3D Systems Inc.	Helisys Inc.	Stratasys Inc.	DTM Corp.	Sanders Prototype Inc.
Part envelope	10"x10"x10" > 20"x20"x23"	10"x15"x14" > 20"x30"x20"	10"x10"x10"	12"dia.x15"	12"x6"x9"
Materials	• Epoxy resins • Acrylate resins	• Paper	• ABS • MABS • Wax • Elastomers	• Nylon • Composite nylon • Polycarbonate • Rapidsteel-metal • Trueform-polymer for RTV	• Build material, thermoplastic • Support material Wax
Merits	• Excellent accuracy and repeatability • Fine detail • Quickcast	• Low initial cost and maintenance costs • low cost part production • Very fast build time • Excellent for large parts	• Small machine • No venting required • No lasers or liquids • No post processing	• A range of materials available • Flexible parts possible • No supports needed	• True desktop system 27"x15"x27" • Non toxic materials • Quick support removal • Good surface finish
Limitations	• Expensive initial and maintenance costs • Expensive for large parts • Support structures • Post processing	• Relatively poor surface finish • Fine detail can be lost • Requires finishing off	• Can not produce fine detail • Thin walls a problem • Thick walls a problem	• Expensive process	• Only small parts can be made • Limited material choice at present
Applications	• Design verification • Low volume prototypes • Tooling • Medical modelling	• Design verification • Low volume prototypes • Tooling • Pattern making	• Design verification • Low volume prototypes • Tooling • Medical modelling	• Design verification • Low volume prototypes • Tooling • Medical modelling	• Design verification • Low volume prototypes

PROCESS DESCRIPTIONS

Stereolithography

Stereolithography (SLA) was introduced to the market in 1988 by 3D Systems, SLA was the first rapid prototyping technology to be developed and is still the most widely used, with approximately 60% of today's RP market.

Stereolithography builds parts onto a platen within a tank of photosensitive liquid polymer resin. A laser is used to trace the outline of the first slice, solidifying the resin on the platform, shown in Figure 1. This process repeats itself with the platen sinking the thickness of one slice each time into the vat. When completed and dried the component can be dressed to suit customer requirements[2].

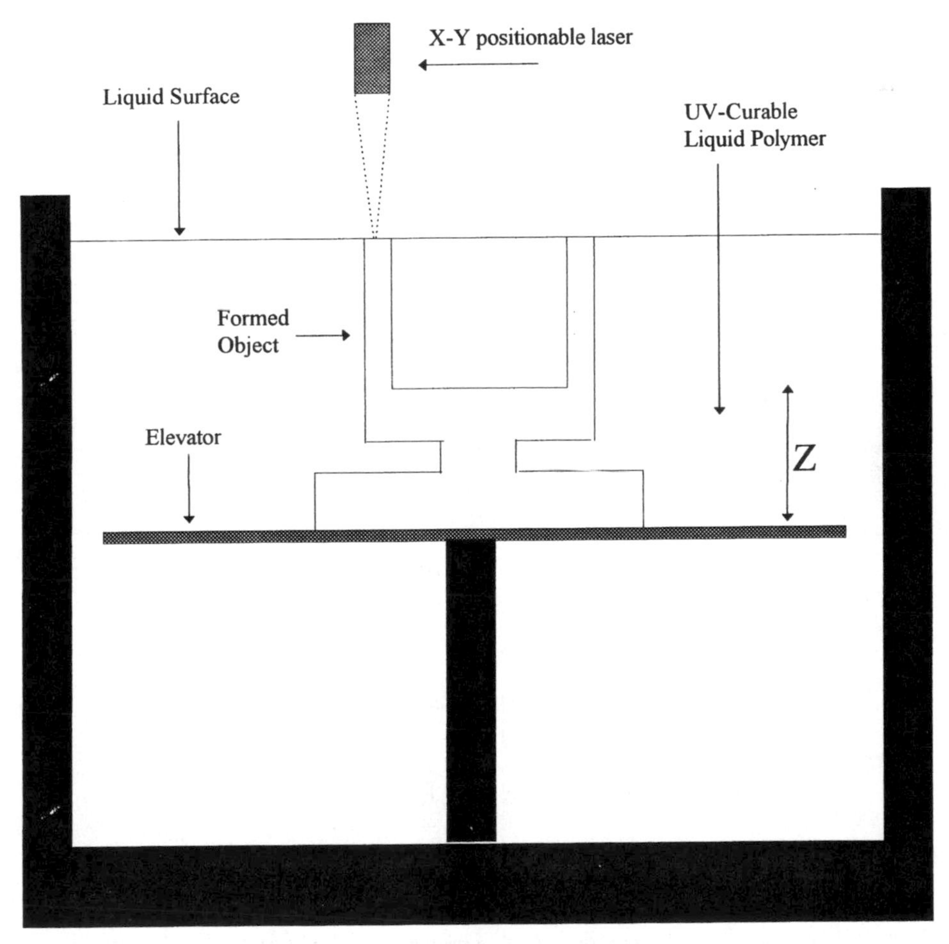

Figure 1. The SLA process

Laminated Object Manufacture

Laminated Object Manufacture (LOM) was developed in 1985 by Helisys' Laminated Object Manufacturing™ Inc. With the first commercial systems being introduced in 1991.

Attached to two rollers, the adhesive material, usually paper pre coated with adhesive, a heat sensitive polymer, is passed over a metal support platform[3]. A single layer of the paper is laminated by temperature controlled heated rollers to the platform. The focused laser beam then precisely cuts the cross-section on the first layer of paper, shown in Figure 2. The laser is mounted on a 2D x-y plotter, only the periphery of the component section needs to be 'drawn' with the laser.

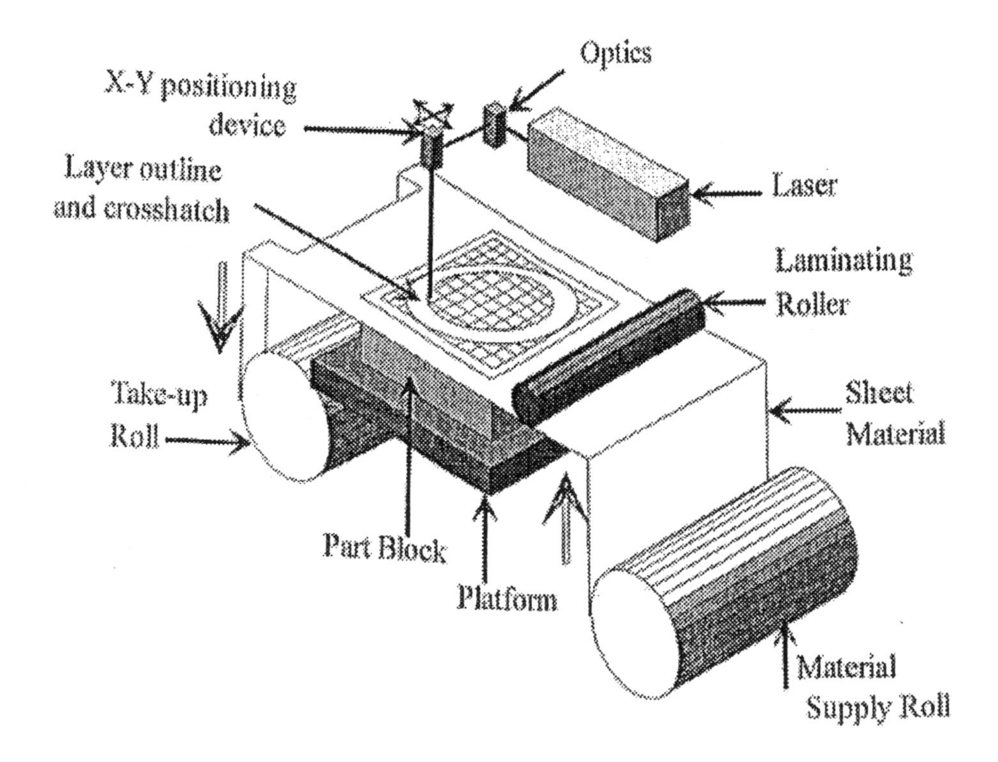

Figure 2. The LOM process

Unlike stereolithography, selective laser sintering and fused deposition modelling where the sections must be completely filled, and because a solid is being cut rather than formed, it takes no longer to create a thick-walled part than a thin-walled one. The laser cuts and cross hatches all the excess material and a boundary for removal later. The platform then moves down the thickness of one slice, this is the thickness of the paper. A new layer of paper is advanced from the feed roll and is bonded to the previously cut layer. The laser cuts hundreds of layers in this way until the part is complete, the laminated stack is then removed from the machine's elevator plate. The cubes are separated from the surface of the object, revealing the finished part.

Selective Laser Sintering

Selective Laser Sintering (SLS) was developed at the University of Texas, Austin, in 1986. Commercial systems are produced by the DTM Corporation, Austin, Texas, who licensed the technology from the university.

SLS uses a laser to create 3D parts from heat fusible powder. A warm up phase is used to heat the fusible powder to approximately two degrees under its sintering point[4]. A thin layer of the powder is then rolled evenly across the work cylinder. SLS fuses the powder into a solid mass by the means of a laser adding just sufficient energy to the powder to cause fusion, shown in Figure 3. The powder within the part boundaries is sintered and formed. After each layer is sintered the work cylinder is lowered one slice thickness and the machine deposits another layer of powder on top of the sintered area. This is then rolled flat, after which the process repeats itself. This continues until the part is complete. Producing the part in this manner means that the developing part does not require supporting structures, as it is supported by unused powder in the container.

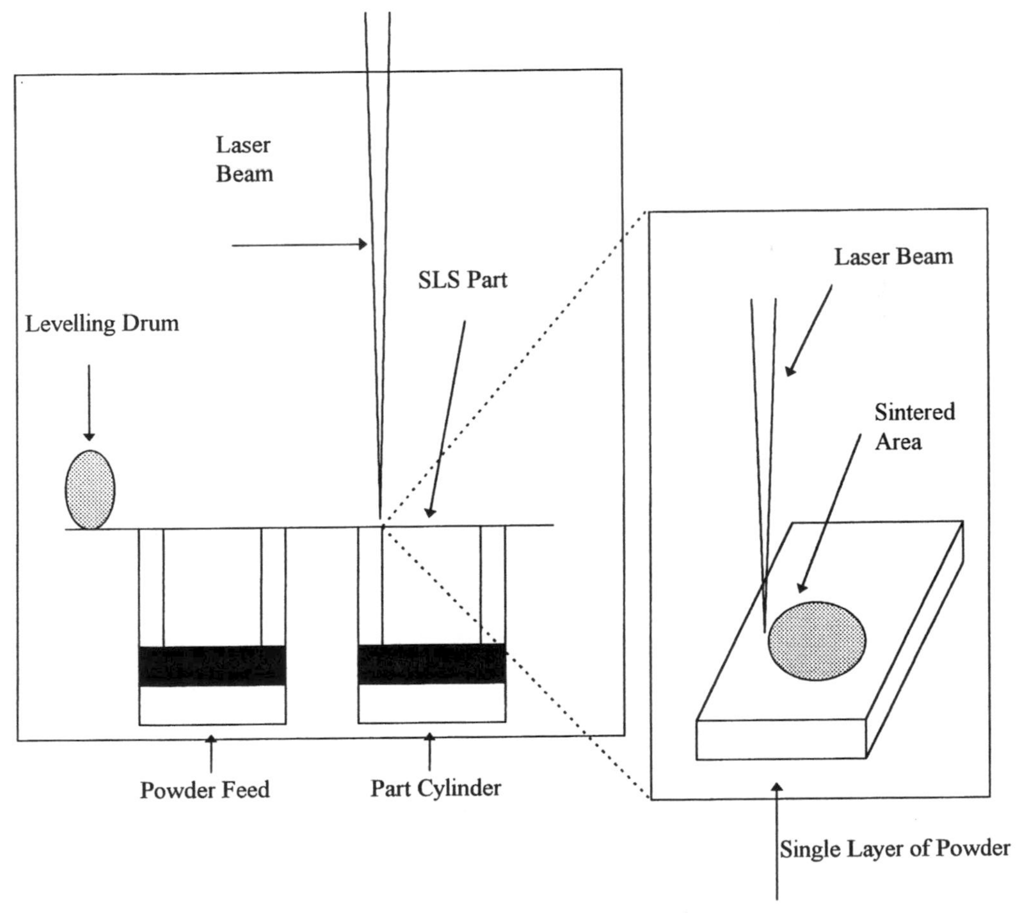

Figure 3. The SLS Process

Fused Deposition Modelling

Fused Deposition Modelling (FDM) was developed in the early 1990's and brought to the market commercially by the Stratasys Corporation.

FDM is one of the most recently patented processes and the most simple of the five processes available. FDM involves the extrusion of thermoplastics just above the melting point using a computer controlled deposition head. The head is fed by a spool of non-toxic thermoplastic material which is heated to just above its solidification temperature. As the x-y controlled head moves across the part, in much the same way as a x-y plotter would do[5]. The material is deposited on the previous layer, solidifying and bonding to it, shown in Figure 4. Thus producing the required part.

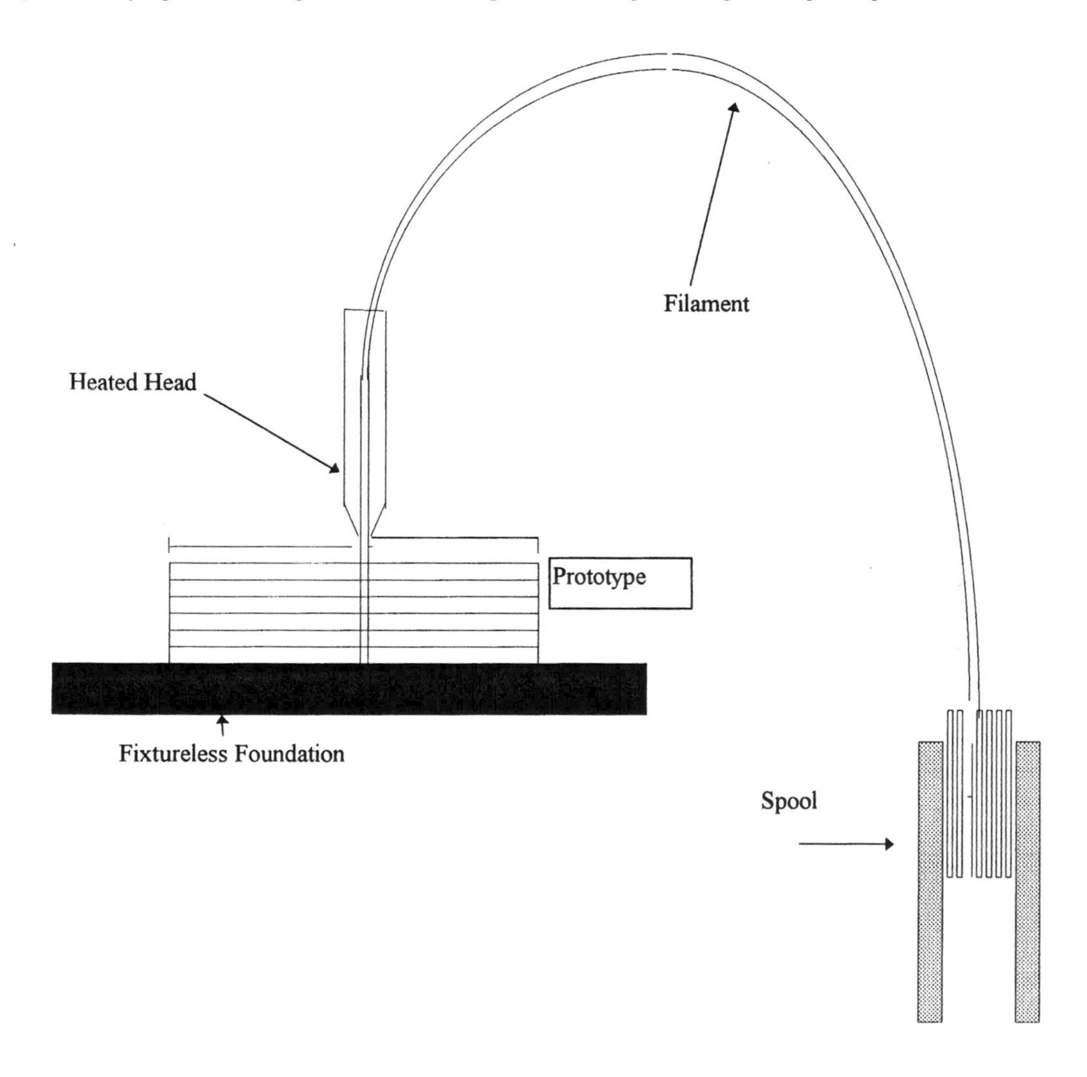

Figure 4. The FDM process

Ink Jet Modelling

Ink jet modelling (IJM) is used by Sanders Prototyping Inc. The 3D plotting system is the closest to a desktop machine currently available, however the physical size of other systems are reducing in size.

The Sanders system is a liquid to solid inkjet plotter with a separate z axis input[6]. There is a dual inkjet subsystem on an x-y drive carriage and deposits both thermoplastic and wax materials on the build substrate. The x-y drive carriage also controls a flatbed milling subsystem for maintaining precise z axis dimensioning of the part by milling off the excess vertical height of the current build layer. This means subsequent build layers will have a known surface reference layer to build upon.

The 3D plotting process consists of the deposition of the build and support materials on a substrate mounted on the z axis build table. The elevation of the table is determined by the control program. The support material is used to support overhangs and cavities in the part during the build sequence. The build and support material are digitally deposited onto the build substrate as a series of uniformly spaced "micro droplets" using a drop on demand process. The droplets adhere to each other during the liquid to solid phase transition to form a uniform mass. The drying process is fast enough to allow milling of the layer directly after the deposition cycle. This process is illustrated in Figure 5.

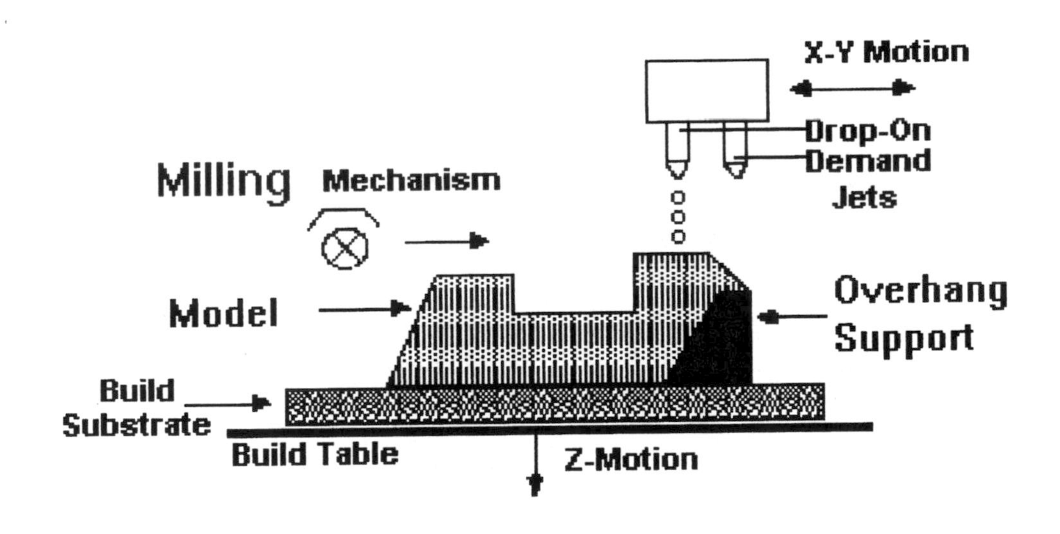

Figure 5. The IJM process

The dual inkjet subsystem delivers droplets of green thermoplastic as the build material and red droplets of wax as the support material. The soluble supports can then be removed by dissolving in a solvent bath which takes minutes.

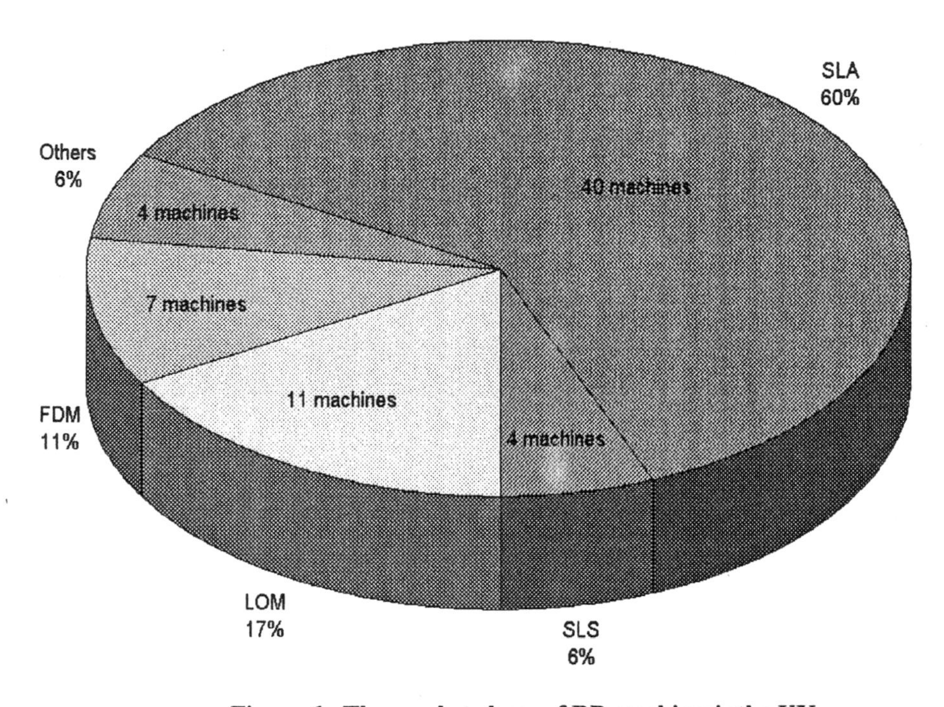

Figure 6. The market share of RP machinesinthe UK

Table 2

RP machines in use in the North East of England

SLA	LOM	SLS	FDM
Styles & JJ Engineering	UNN, Newcastle	AMSYS, Sunderland	Teesside University

Table 3
RP Establishments

	Establishment	Machines				Type	No. of machines
1	3D SYSTEMS	SLA 250/40	SLA 350/10	Actua 2100		Agent / Bureau	3
2	AMSYS	DTM 2000	DTM 2000			Bureau	2
3	ARRK	**SLA 250 (3 off)**	**SLA 350 (2 off)**	SLA 500/30		Bureau	6
4	Automotive Design Centre, Belfast	SLA 250	FDM 1600			Bureau	2
5	British Aerospace, MAD	SLA 250	SLA 500/40			Internal	2
6	Buckinghamshire college, CRDM	SLA 190/20				Academic / Bureau	1
7	CA Models	SLA 350/10				Bureau	1
8	Cardiff University	SLA 250/40	Actua 2100			Academic / Bureau	2
9	Coventry University	FDM 1500	FDM 1600			Academic / Bureau	2
10	DERA	SLA 500/40				Internal / Bureau	1
11	DERC Cardiff	SLA 250/40	LOM 1015			Academic / Bureau	2
12	DRA Malvern	SLA 500				Internal	1
13	Evesham College	FDM 1600				Academic / Bureau	1
14	Ford Motor Co.	LOM 2030E				Internal	1
15	Formation	**SLA 250 (2 off)**	SLA 500/30h			Bureau	3
16	IMI	SLA 250				Bureau	1
17	JJ Engineering	SLA 350/10				Bureau	1
18	Laser Lines	FDM 1650				Agent / Bureau	1
19	Laser Prototypes	STEREOS Desktop				Bureau	1
20	Leeds University	DTM 2000				Academic / Bureau	1
21	Leonardo Computer systems	**JP System 5 (2 off)**				Agent / Academic / bureau	2
22	Lever brothers	SLA 250				Internal	1
23	Laser Integrated Products	LOM 1015	Sanders MM-6B	SLA 250	LOM 2030E*	Agent / Bureau	3
24	Liverpool University	SLA 250/40				Academic / Bureau	1
25	Malcolm Nichols	SLA 500				Bureau	1
26	University of Northumbria	LOM 1015				Academic / Bureau	1
27	Nottingham University	SLA 250	LOM 1015			Academic / Bureau	2
28	Oxon Prototypes	SLA 250				Bureau	1
29	Pace (Rep. Ireland)	SLA 250				Bureau	1
30	PERA	SLA 350/10				Bureau	1

Table 3

RP Establishments

31	Queens University	SLA 250	FDM 1600			Academic / Bureau	2
32	RP solutions	FDM 1600				Bureau	1
33	Rapitypes (Renfrew)	SLA 250/40				Bureau	1
34	Rolls Royce, Bristol	SLA 500/30				Internal / Bureau	1
35	Rover / Warwick University	**SLA 500/30H (2 off)**	SLA 350/10	DTM 2000	LOM 2030E	Academic / Internal / Bureau	5
36	Sandwell college	LOM 1015				Academic / Bureau	1
37	Styles	SLA 250/40	SLA 500/40			Bureau	2
38	Teesside University	FDM ?				Academic / Bureau	1
39	Time RP	Sanders MM-6B	**SLA 250 (2 off)**			Bureau	3
40	UMAK	LOM 1015				Agent / Bureau	1
41	Websters	LOM 1015				Bureau	1
						Total	**68**

WHY UNN CHOSE LAMINATED OBJECT MANUFACTURE (LOM)

When considering which RP technology to introduce to UNN, the situation in the region had to be assessed. Three technologies were available in the North East of England. These being fused deposition modelling, stereolithography and selective laser sintering. From the main technologies, LOM or a desktop type process e.g. IJM, was missing. It was decided that LOM would be chosen as it could produce parts which were unsuitable for other processes. Such as large bulky parts which would be difficult and expensive to model on currently available RP machines.

The LOM technology was found to be most suitable for heavy sections and large parts, this meant the process lent itself to foundry work. That is making patterns for casting. This is an area where RP could make a real difference. Sand moulds have been traditionally, hand crafted wooden patterns, produced by highly skilled workers. The detail and skill involved requires several weeks work. Using LOM, which can quickly produce inexpensive wood like patterns, the labour intensive pattern making process can be dramatically refined. It must be realised, however, that rather than eliminating the traditional methods, LOM streamlines the process of pattern making, since pattern making skills are still needed to finish the model ready for casting.

If a few castings are needed, then the LOM part can be used as a pattern to make direct impressions in sand. The number of castings produced by this method is dependent upon the geometry of the part and the type of sand used by the foundry. Typical figures are a LOM part producing 50 sand moulds.

EXAMPLE OF PART PRODUCED BY LOM

Brook Hansen used LOM to streamline the development of one of their motor housings. The comparison between conventional and RP methods is shown in Table 4.

Table 4

The Brook Hansen experience

Rapid Prototyping Method			Conventional Method		
Item	Cost	Time	Item	Cost	Time
CAD model generation	£10,000	3 weeks	CAD model generation	£10,000	3 Weeks
STL files	£1,200	1 Day	Pattern Frame of body	£60,000	24 Weeks
LOM model	£3,240	4 Days	Pattern end shields	£45,000	19 Weeks
Casting	£2,100	2 Weeks			
Total	£16,000	6 Weeks		£115,000	27 Weeks

This shows an overall time saving of £98,000 (85%) and a product development time saving of 21 (78%) weeks.

PROBLEMS ENCOUNTERED WITH THE LOM PROCESS

The main problems UNN encountered with the process was the production of thin (<3mm) section components. When wall thickness reduces to approximately 3mm over large areas, problems are encountered during removal of the excess material. Namely, the thin section can be mistakenly removed with the cross hatch, if great care is not taken. Also finished products with thin wall sections tend to be delicate, fragile and lacking stiffness, especially if they are part of an assembled LOM part.

Another problem associated with breaking out the part, is the removal of material from enclosed sections such as holes and cavities. This problem is more severe when the hole is orientated along the 'grain' of the paper. Delamination of parts has also been an area of difficulty. To avoid this many coats of sealer may be required, again this is mainly a problem with thinner sections.

CONCLUSIONS

(1) Selection of the LOM process was made primarily to satisfy the regional needs of the North East.
(2) The LOM process lends itself to making patterns for the foundry industry, this has shown to be beneficial when making patterns larger than 6"x 6"x 6".
(3) The LOM process is currently being developed to introduce the concept of bridge tooling and secondary processes.
(4) The major limitations of the LOM process is the wall thickness of 3mm over large areas (wall thickness can be lower, but only for small areas) and an accuracy of 0.5mm with a surface roughness of $2x10^{-6}$ m.
(5) The LOM process does not allow the reproduction of fine detail.
(6) The LOM process is best suited to larger, heavy section parts.

REFERENCES

1. Coole, T.J. et al. Case study of the application of LOM by Brook Hansen for the development of a housing design for small to medium electric motors. In: First National Conference on Rapid Prototyping and Tooling Research, Buckinghamshire College, UK 6th-7th November 1995, edited by G. Bennett. Mechanical Engineering Publications Ltd, 1995, p. 19-28.
2. 3D Systems, Inc. Homepage URL: http://www.3dsystems.com/
3. Helisys, Inc. Homepage URL: http://helisys.com/
4. DTM Technology. Homepage URL: http://www.azizia.dtm-corp.com/
5. Stratasys,Inc. Homepage URL: http://www.stratasys.com/
6. Sanders Prototyping Inc. Homepage URL: http://www.sanders-prototype.com/
7. Bibb, R. Private communication with S.Hogarth. October 1996

ACKNOWLEDGEMENTS

The authors would like to thank Government office North East for providing the funding to support this project.

A GENETIC ALGORITHM FOR OPTIMAL OBJECT PACKING IN A SELECTIVE LASER SINTERING RAPID PROTOTYPING MACHINE

Ilkka Ikonen, William E. Biles
Department of Industrial Engineering
University of Louisville, Louisville, KY 40292

ABSTRACT: In this paper we describe a unique three-dimensional bin packing problem with non-convex objects having cavities and holes. Part packing takes place in an environment where parts float as they would in a weightless environment. The application domain is the selective laser sintering rapid prototyping technique. A genetic algorithm is used as a search method to find a good part-packing solution. The fitness evaluation for the genetic algorithm is based on the actual part geometry as described in the Stereolithography (STL) file. Part-intersection detection utilizes several methods common in computational geometry. Initial results are promising, showing that the genetic algorithm is able to find a good solution for such a difficult packing problem.

KEYWORDS: genetic algorithms, three-dimensional bin packing, non-convex objects, rapid prototyping, selective laser sintering

1. INTRODUCTION

Critical to success of a rapid prototyping (RP) service bureau is its capability to serve several customers from different industries in a timely manner. Typically, build time per manufactured part is reduced if several parts can be fabricated simultaneously in one batch. In RP machines utilizing a laser beam, the time for a laser to manufacture one layer is much less than that required for the build platform to descent and to apply material for a new layer. To make a batch, the RP machine operator must simulate the packing of parts by placing their Stereolithography (STL) or solid model files in a simulated build envelope using CAD or some other appropriate software. Typically a customer of a RP service bureau uses their own CAD system to create the part design and the STL file. When the customer places the order for the RP service bureau, they transfer only the needed STL files via modems or Internet. To improve machine time utilization, as many parts as possible should be packed in each batch. This packing task is a form of the classical bin packing problem, which has been proven to be NP-hard (meaning that solution space grows exponentially with the size of the problem). [Garey, Johnson, 79] There will very likely never be a method to solve these problems optimally in polynomial time. Some optimization methods are, however, able to find a good, if not the optimum, solution for a problem in reasonable time. One of these methods is genetic algorithm (GA), which mimics natural evolution and survival of the fittest to find a good solution.

1.1 Selective Laser Sintering Rapid Prototyping Technology

Like all current rapid prototyping (RP) techniques, the selective laser sintering (SLS) machine produces parts directly from CAD files through an additive manufacturing process. The part is modeled as a solid three-dimensional (3D) CAD model, which is "sliced" into thin (0.003" - 0.02") horizontal slices which are manufactured one at the time. Before slicing, the solid model is converted to an STL file format, which is an industry-standard interface between a 3D model and a rapid prototyping machine. Selective laser sintering uses fine, heat-fusible powder to build a part. The powder is melted by a CO_2 laser beam which follows the shape of the cross section of the part being built. The intensity of the laser beam is controlled to melt the powder only in areas defined in the cross section. Unmelted powder outside part boundaries remains in the build cylinder (15 inches high, 12 inches diameter), providing support for the layers above. Due to this supporting powder, parts can be arranged in any orientation and location inside the cylinder. No support structures are needed to hold them in place. Parts "float" in the powder as they would in a weightless environment.

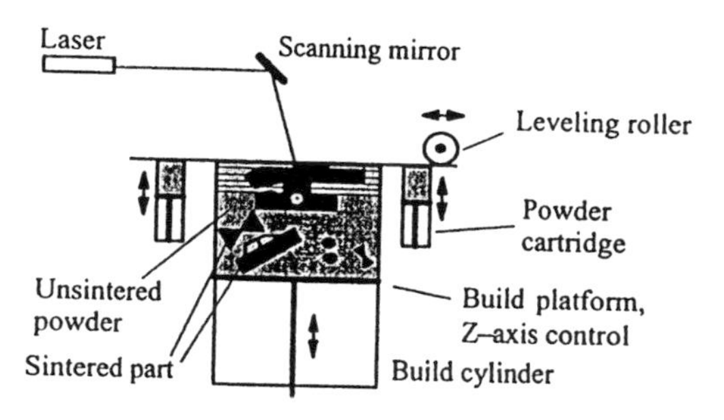

Figure 1: Selective laser sintering process

After the laser has sintered the current layer, the build platform is lowered one layer thickness inside the cylinder and a new layer of powder is spread on the top of the previous one by a leveling roller. The fresh powder is in turn sintered as a new layer is added to the part. Layer-by-layer sintering continues until all parts in the batch are finished. The building platform is elevated from the cylinder, causing loose powder to fall away and reveal the finished prototypes. The selective laser sintering machine has very few restrictions in the part shapes which it can produce. Most prototyped parts vary in size and are complex in shape, having irregular surfaces, cavities and holes.

1.2 STL File Format

Intersection detection in the GA search engine is based on the part geometry as described in the STL file. An STL file format is a tessalated approximation of the actual 3D solid model of a part where each surface is replaced by a set of triangles. A CAD software uses complicated mathematical formulas to describe designed free-form surfaces in 3D. Most commonly used representations are parametric splines, Bezier curves and B-splines. All these formats enable the designer to represent a free-form curve using a parametric representation in each coordinate (x,y,z). These formats give an accurate description of each surface on the part and give the designer freedom to meet product requirements. However, to manufacture a part with current rapid prototyping machines, these formats are not applicable and have to be replaced with an STL file.[Jacobs, 92] The tessalation process is the translation of a free-form surface representation to a representation consisting of a mesh of triangles. Depending on the desired accuracy, surface contours and part size, the number of triangles can grow large. For parts with many cylindrical contours, there can be up to 70 thousand or more triangles. When viewed as an text file, an STL file consists of x,y,z-coordinates indicating the vertexes of triangles.

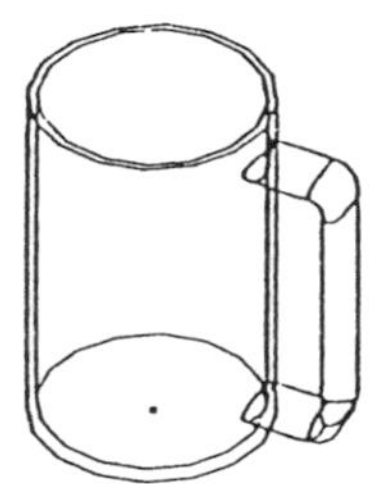

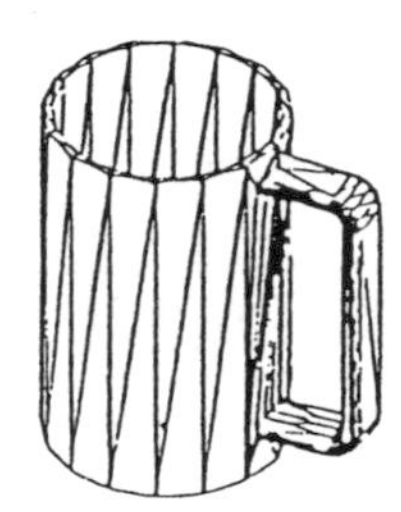

Figure 2: STL file for a coffee mug

For the RP process an STL file is "sliced" into horizontal slices which the machine manufactures one at the time. When a triangle is cut into horizontal slices, points on a horizontal plane are achieved. These points are used to toggle the laser beam on and off, as required to create the cross sections of parts at each layer. The height of each slice defines the step size for the build platform and the layer thickness for the fresh powder.

2. GENETIC ALGORITHMS

Over the last 15 years, genetic algorithms have gained wide popularity among theoretical and application oriented optimization researchers as a search method to solve 'difficult' problems. Many times these problems are hard or impossible to solve using traditional function optimization methods. The basic idea behind a GA search lies in the mechanics of evolution based on natural selection and genetic representation of living beings.[Goldberg, 1989] When used for function optimization, a GA manipulates a population of strings using operators that simulate natural reproduction and mutation. Manipulated strings, or chromosomes, represent solution candidates for the problem to be solved. Even though genetic operators are probabilistic in nature, a GA search is able to avoid unpromising search areas and is able to find good, if not optimum, solutions for many difficult problems. To guide the search, a genetic algorithm requires only a function evaluation which represents how good a solution each candidate is. The basic steps of a GA search are as follows [Davis, 91]:

1) initialize a population of chromosomes;
2) evaluate each chromosome in the population;

3) create new chromosomes by mating current chromosomes, and apply mutation and recombination as the parent chromosomes mate;
4) delete members of the population to make room for the new chromosomes;
5) evaluate new chromosomes and insert them into the population;
6) if time is up, stop and return the best chromosome, else go to step 3.

The method for creating new chromosomes mimics natural sexual reproduction, where chromosomes of two individuals (parents) are combined in some ordered way to produce one or two new individuals (children). Mating is performed by a so called "crossover operator," which is the primary operator for searching promising regions. The mutator is used in random fashion to introduce new gene values into the gene pool to avoid too early convergence to a local optimum.

2.1 Chromosomal Representation

To make the problem solvable for the GA, a solution for it must be coded as a chromosome which the genetic operators is able to manipulate and which can be evaluated by the fitness function. The most natural coding method for a packing problem is an ordered list of integers where each part is represented by a corresponding part number. In most previously studied bin packing and related problems the permutation of packed (scheduled) parts has been sufficient representation. [Cleveland, Smith, 89], [Corcoran, Wainwright, 92], [Jakobs, 96], [Bhuyan et al., 91], [Sysverda, 91] Some researchers have studied coding methods that use more than one chromosome to represent a solution. [Falkenauer, 95], [Falkenauer, Delchambre, 92], [Juliff, 93] In the SLS packing problem there is more information than just the order of parts that must be coded in the chromosome. Included information is the orientation of each part and the location in 3D space where each part is placed. Each part has five points on them where the following part is allowed to attach. These points are called "attachment points". Part orientation and its attachment point are indicated by their own sublists, thus we are using a three-dimensional chromosome representation. An example of a chromosome can be:

List1: 4 3 0 2 1
List2: 6 9 22 1 17
List3: 2 3 0 4 1

This represents packing where part number 4 is packed first in orientation 6 (rotated 270 degrees around x-axis). The first part is attached to the global origin, which is in the center and at the bottom of the cylinder. Then part number 3 is packed in orientation 9 (rotated 45 degrees around y-axis) and is attached to attachment point number 2 on previously packed part. Part number 0 is in orientation number 22 (rotated 270 degrees around z-axis) and attached to attachment point number 3 in previous part. Part number 2 is in orientation 1 (rotated 45 degrees around x-axis) and attached to attachment point number 0 in previous part. The last part, number 1, is packed in orientation 17 (rotated 45 degrees around z-axis) and is attached to attachment point number 4 in the previous part. The chromosomal representation applied here is described more in detail in [Ikonen et al , 97].

2.2 Fitness Function

The genetic algorithm requires a goodness value for each candidate packing to guide the search in a promising direction. The packing evaluator simulates packing of three-dimensional STL files inside the build cylinder of the RP machine. After all parts are packed, the simulator calls an evaluator function which calculates three different values to indicate how good the packing is. Calculated values include: (1) the average distance of parts from the global origin, (2) the amount of intersection between parts, and (3) the amount of intersection between parts and the build cylinder. These values are multiplied by appropriate weighting factors and summed to calculate the total evaluation of the packing plan. The values for the weighting factors depend which one of the three values is the most significant for each packing problem. In most cases, the largest weight is given to the part intersection.

3. INTERSECTION DETECTION

To evaluate the extent to which two or more parts intersect, we must calculate the intersection of the triangles by which they are represented. Two parts intersect only if their triangles intersect. An object-oriented programming approach to the needed routines is given in [Laszlo, 96]. To reduce the required calculations, the bounding boxes of parts and triangles are first checked for intersection. The bounding box of a geometric object is the smallest box that contains the object, where the edges of the box are parallel to major axes. Two objects intersect only if their bounding boxes intersect. If part bounding boxes intersect, pair-wise comparison for all triangle bounding boxes is done. For each intersecting triangle bounding box pair, the following triangle-triangle intersection calculations are performed. Triangle - triangle intersection calculation in 3D requires several steps and combination of various computational geometry methods. All calculations are reduced to simple vector arithmetic tasks, for which vector dot and cross product are the only operators needed. Two triangles intersect if one or two of the edges of one triangle intersect the other triangle. Triangles can be in any orientation and relative location in 3D space, so there is no information to indicate which two edges have to be checked for intersection. Thus, edge - triangle intersection is done for one edge at a time. The vertices of a triangle define a plane in space on which the triangle lies. For an edge to intersect the triangle, it must intersect the plane and the intersection point must lie inside the triangle.

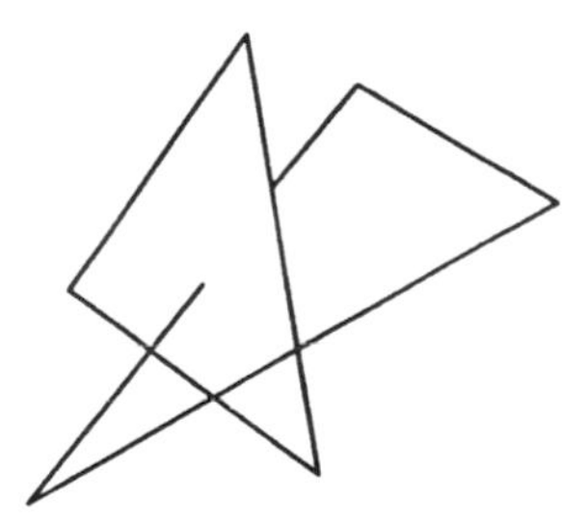

Figure 3: Triangle - triangle intersection in 3D

The edge can be stretched to an infinite line. In most cases, the plane and the line intersect at a point. It is possible for the line to lie on the plane, in which case the intersection is the line itself. If the line and the plane intersect at a point, it can be calculated whether or not this intersection point lies within the line segment specified by the vertices. If so, the edge and the plane intersect and more calculations are required to see if this intersection point lies inside the triangle. To make this calculation easier, both the triangle and the point are projected to a 2D coordinate plane. Containment on this plane is equivalent to the original 3D problem if projected objects are non-degenerate. [Laszlo, 96] To which of the three major coordinate planes (xz, xy, or yz) to project depends on the orientation of the triangle. If the normal vector of the triangle is perpendicular to the z-axis, projection of the triangle is a degenerate line in the xy-plane and the two problems are not the same. In this case, the triangle is projected to the yz-plane. In the same way two other planes are checked. Since the normal vector cannot be perpendicular to all three coordinate axis, at least one of the three projections is non-degenerate.

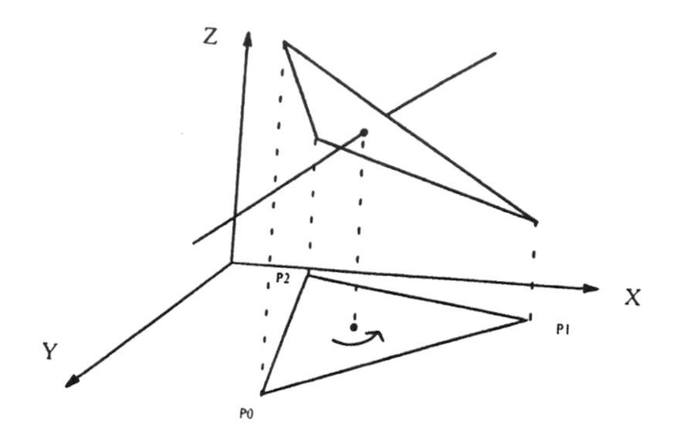

Figure 4: Projection from 3D to 2D

If the projected intersection point lies inside the projected triangle, the edge of one triangle intersects the other triangle. A point lies inside a triangle, or any convex polygon, if it is on the same side of all the edges when an imaginary observer "walks" along the edges of polygon [O'Rourke, 95]. Vertices in the STL file are numbered in a counterclockwise order when looking a triangle from above (outside the part); thus, we use this same order to walk along the edges. Starting from one vertex (point p0) an imaginary observer travels to the next vertex (point p1). Along this edge, the intersection point is either on the left or right-hand side of the observer. If the point is on the right-hand side, we can stop knowing that the edge and the triangle do not intersect. If the point is on the left-hand side, we must travel on to the next vertex (point p2). Again, if the point is on the right we can stop, and if on the left we have to walk the last edge (from point p2 to p0). If the intersection point is also on the left of this edge, the original edge and triangle in 3D space intersect. A parametric value indicating how much of the length of the edge projects through the triangle is returned. These values are summed over all the edges and all the triangles in all parts to give an indication of the total part intersection in a packing. [Ikonen et al , 97]

4. RESULTS

Initial testing was done to find good parameter values for the genetic algorithm and the weighting factors for the fitness function. Packing of rectangular solid blocks was used as a simple test problem. A solid block is a simplest kind of part that can be fabricated in a RP machine and it can also be seen as a bounding box for more complex part shapes. Five blocks size of 2" x 2" x 1" were packed in a cylinder of size 5" (high) and 3" (radius). A small cylinder size was used to create enough selection pressure towards a better solution. If the packing is too easy, the GA is not able to find a good solution. Typical development for the search is shown in Figure 5, which graphs the population best, average and worst for each generation.

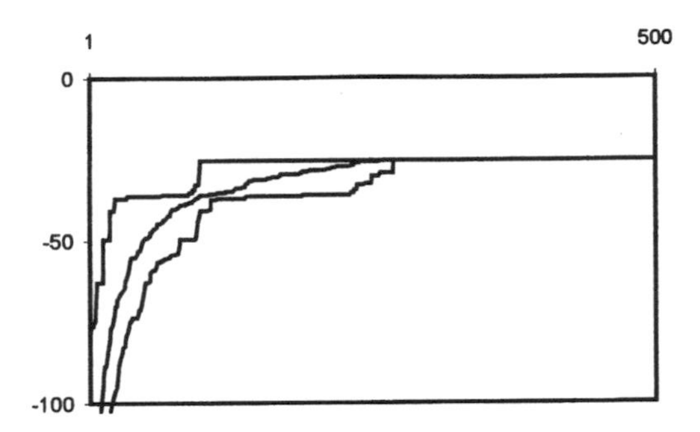

Figure 5: Packing of five solid blocks

The search improves rapidly until it is unable to further improve the population best. At that point the search has converged to a solution. After this possible improvement can still occur if the mutator is able to introduce a good gene values back to population. The quality of the found packing depends heavily on the used weighting factors in the fitness function. For most program runs the GA is able to find a feasible solution in 500 generations when the population size was 50.

After these initial tests the GA was used to solve a more difficult test problem where some parts have holes inside which other, smaller parts, can, and should be packed. Based on the results from several runs with various parameter values it was observed that this is a much harder problem for the GA to solve. In most cases the GA required more generations (up to 1000) with larger population sizes (300). With appropriate selection of weighting factors the GA is able to find a good solution. It was observed that even small changes in the weighting factors in the fitness function affect the quality of the found solution and the weighting factors have to be changed based on the number of triangles in parts. Figure 6 shows a typical search for packing problem with parts having holes.

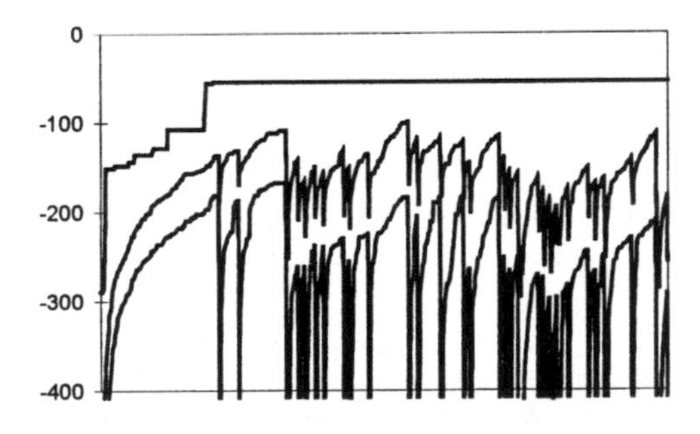

Figure 6: Packing of parts with holes

The search improves steadily until it settles for a solution. The population average and the worst values indicate that the GA keeps searching for a better solution. Figure 7 shows an example for packing with parts with holes and cavities.

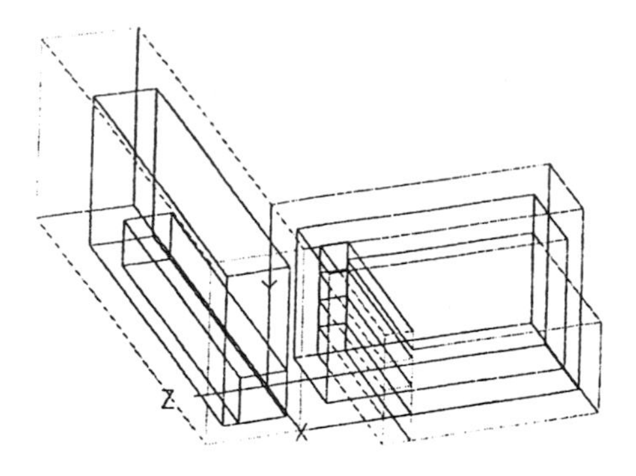

Figure 7: Packing solution for parts with holes

5. CONCLUSIONS AND FUTURE RESEARCH

In this paper we have described a form of the classical bin packing problem as it appears in a selective laser sintering rapid prototyping machine. Supporting powder, the free form of the parts, and a cylindrical bin make this problem unique. We use a genetic algorithm to find a good solution. To evaluate the fitness value for a solution, a packing simulator is programmed to pack the STL files of the parts. Intersection detection inside the build cylinder is based on the actual part geometry. Several computational geometry methods are used to calculate the part intersection. Initial results with solid blocks are promising; the GA is able to find a good solution relatively fast. Packing of parts with holes and cavities is a harder problem for which the GA requires more generations and larger population size. A good solution is found for most test runs.

We are currently working on the improvements to automate several features of the program, such as the scaling of weighting factors for the fitness function and resizing the packing cylinder based on part sizes so that the GA always has enough selection pressure to distinguish solutions. We are also running experiments with more complex parts to study how well the GA is able to solve larger packing problems.

6. REFERENCES

[Bhuyan et al., 91] Bhuan, Jay N.; Raghavan, Vijay V.; Elayavalli, Venkatesh K, Genetic Algorithm for Clustering with an Ordered Representation, in Belew, Richard; Booker, Lashon (Ed.), Proc. Of the 4th International Conference on Genetic Algorithms, 1991, Kaufman Publishers.

[Cleveland, Smith, 89] Cleveland, Gary A, Smith, Stephen F, Using Genetic Algorithms to Schedule Flow Shop Releases, in David Schaffer (Ed.) Proc. of the 3rd Int. Conference on Genetic Algorithms, 1989, Kaufman Publishers.

[Corcoran, Wainwright, 92] Corcoran, Arthur L. III, Wainwright, Roger L., A Genetic Algorithm for Packing in Three Dimensions, Symposium on Applied Computing (SAC 92), Kansas City, March 1-3, 1992, pp. 1021-1030.

[Davis, 91] Davis, Lawrence, Handbook of genetic algorithms, Van Nostrand Reinhold, 1991.

[Falkenauer, 95] Falkenauer, Emanual, Solving Equal Piles with the Grouping Genetic Algorithm, in Proceedings of the 6th Int. Conference on Genetic Algorithms, 1995, Kaufman Publishers.

[Falkenauer, Delchambre, 92] Falkenauer, E.; Delchambre, A., A Genetic Algorithm for Bin Packing and Line Balancing, in Proceedings of the 1992 IEEE International Conference on Robotics and Automation, Nice, France, 1992.

[Garey, Johnson, 79] Garey, Michael R; Johnson,David S, Computers and Intractability: A Guide to the Theory of NP-Completeness, W. H. Freeman and Co., San Fransisco, 1979.

[Ikonen et al., 97] Ikonen, Ilkka, Biles, William E., Kumar, Anup, Wissel, John C., Ragade, Rammohan K., A Genetic Algorithm for Packing Three-Dimensional Non-Convex Objects Having Cavities and Holes, in Proceedings of the 7th International Conference on Genetic Algorithms, East Lansing, MI, July 19-23, 1997

[Jacobs, 92] Jacobs, Paul F., Rapid Prototyping & Manufacturing- Fundamentals of Stereolithography, Society of Manufacturing Engineers, Dearborn, MI, 1992

[Jakobs, 96] Jakobs, Stefan, On Genetic Algorithms for the packing of polygons, European Journal of Operation Research 88, 1996, pp.165-181.

[Juliff, 93] Juliff, Kate, A Multi-Chromosome Genetic Algorithm for Pallet Loading, in Proc. of the 5th International Conference on Genetic Algorithms, Urbana, Ill.,1993, Kaufman Publishers, pp. 467-473.

[Kroger, 95] Kroger, Berthold, Guillotinable bin packing: A genetic approach, European Journal of Operation Research 84, 1995, pp.645-661.

[Laszlo, 96] Laszlo, Michael J., Computational Geometry and Computer Graphics in C++, Prentice-Hall, 1996

[O'Rourke, 95] O'Rourke, Joseph, Computational Geometry in C, Cambridge University Press, 1995.

[Sysverda, 91] Syswerda, Gilbert, Schedule Optimization Using Genetic Algorithms, in [Davis,91].

ADVANCES IN RAPID TOOLING IN PRODUCT INTRODUCTION PROCESSES

Dr Anthony Venus and Dr Amerdeep Riat

CAMM, The Industry Centre, University of Sunderland,
Hylton Riverside West, Wessington Way, Sunderland, SR5 3XB

Tel. (0191) 515-3326 Fax. (0191) 515-3377
E-mail:anthony.venus@sunderland.ac.uk

ABSTRACT: Time compression techniques such as 3D CAD, computer analysis (product functionality and process simulation), and Rapid Prototyping and Tooling (RPT) provide significant opportunities for improving product development and manufacturing strategy. The current research investigates wider issues of manufacturing strategy in addition to the application and use of time compression technologies. This paper describes the infrastructural requirements of RPT with emphasis on the North East of England and investigates the manufacturing business drivers that are influencing further development of RPT time compression technologies.

1. INTRODUCTION

From cars to microchips, heavy engineering to consumer goods, the North East of England relies substantially on volume manufacturing industries to generate wealth and provide employment. The appropriate selection and implementation of RPT technologies provide opportunities for significant reductions in product development time, cost, and a real competitive edge in terms of speed to market. The major RP technology types represented in Europe include: Stereolithography (SLA), Selective Laser Sintering (SLS), Layer Object Manufacturing (LOM), and Fused Deposition Modelling (FDM). The authors have direct experience with the SLS process. Within the University of Sunderland AMSYS Rapid Prototyping has been running a DTM Sinterstation 2000 machine since July 1994 and has built up a successful business as a commercial RPT bureau service. DTM currently has 150 machines worldwide with 35 machines in Europe, four of which are in the UK. Of those in Europe there are 10 RapidTool™ licenses, of which three are in the UK.

1.1 Centre for Achievement in Manufacturing and Management (CAMM)

CAMM is a regional centre established as an industry/university partnership through the formation of a management team comprising a number of major North East of England manufacturers including:

- Nissan NMMUK Ltd.
- Black & Decker Ltd.
- Thorn Lighting Group
- Flymo Ltd.
- Tallent Engineering Ltd.
- Walker Filtration

The interaction of CAMM with the regional industrial community is ensured via this collective membership that is open to all organisations, large and small, across the Northern Region. CAMM is presided over by Professor Peter Wickens OBE, formerly Personnel Director of Nissan UK and now Chairman of Organisation Development International.

"CAMM provides real solutions to address the needs of today's manufacturers by providing tailored programmes to meet a company's exact requirements. This direct source of help will be augmented by a regular programme of activities that will keep CAMM subscribers up to date on the latest thinking in manufacturing best practice, new product development, human resource issues and help them gain a competitive edge in today's world market economy"

Professor Peter Wickens, President of CAMM

CAMM has worked with over 70 companies, many of these forming part of the supply chain to Original Equipment Manufacturers (OEM's). The relationship between CAMM and its members enables lasting business strategies and solutions to be developed. Membership of CAMM brings with it a support package which allows the member company to develop itself and its suppliers as part of a long term strategy to strengthen the manufacturing capability of the North East. RPT methodologies and strategies form a key element in supporting the CAMM community in product and process development in N E England.

2. COMPONENT MANUFACTURING CRITERIA

In the North East there is a strong manufacturing base OEM's. A large number of suppliers to these OEM's are involved in metal casting and plastic moulding or are tooling providers. Part of the CAMM research activity has analysed North East of England manufacturers examining the operational challenges they face (i.e. the implementation of lean production practises), new product development, and supplier development strategies. The emphasis was on SME's in the OEM supply chain.

Riat (1996) investigated the specific operational challenges SME's face, appropriate strategies that enhance their manufacturing operations, and relevant rating systems that systematically assess areas of their manufacturing activities. This research work has developed a generic, 'best practice' model of effective production in SME's that subsequently lead to the development of a robust rating system capable of benchmarking effective manufacturing practice in SME's.

Barber (1997) identified that product development strategies require improvement, especially in the SME sector. This is based upon a detailed analysis of 15 consumer goods companies. The analysis revealed that (i) design tools were poorly understood and ill used, (ii) the design process was poorly managed and given a low priority in many of the companies. Training and education was required in determining strategy and the relevant tools of the product design process (e.g. DFMA, FMEA and QFD) New product development strategies were also found to be inadequate, especially in the case of SME's. The needs of large companies and SME's can be very different with product development criteria (lead time, cost, functionality etc.) and the resources available varied greatly. To improve the performance of a company's product development strategy it is vital to determine what are the key measures, their 'effectiveness' and identify areas for improvement.

Critical to the competitiveness of CAMM member companies is the requirement to reduce product development times continually. For example, in Flymo, these times have halved in recent years from 3 or 4 years to 18 months. The expectation is to reduce them again by half to 9 months in the next few years. This challenge requires a systematic approach to product design and applying pressure to the tooling evolution of RPT. Cost remains a key factor in tooling and foreign companies can offer significant economic advantage by having much lower labour rates. An injection moulding tool cost typically comprises 10% design, 15% material and the remaining 75% is labour (machining). Traditional offshore suppliers (e.g. the Far East and Portugal) are now being challenged by countries like Mexico, Poland and Czechoslovakia. With tooling costing up to several hundred thousand pounds low labour costs become significant. Foreign labour costs are as low as 20 % of the UK. E.g. Poland £5/hour, UK £25-£30/hour. Set against this background the UK tooling industry needs to offer a flexible, responsive and economic service. Effort is required in supporting developments which will bring about these goals. CAMM has worked in helping to strengthen links between some of its OEM members and subcontract tooling companies. The NE has the RP technologies to support effective application, making it well equipped to make a significant impact with manufacturers in the region and beyond. The difficulty of process selection and application is considered below. This will be driven by the real criteria for any specific application and will further develop the infrastructural base in RPT.

3. SELECTING RAPID PROTOTYPING AND TOOLING TECHNOLOGIES

Companies are investing heavily in RPT technologies and therefore defend these investments. Application of newer methods involves higher risk than more traditional methods. The longer established RP approaches are becoming well known and understood with little risk. Newer tooling developments carry with them higher risk as there is less experience and confidence. When product and process are digitally modelled, simulated and optimisation; these should be complemented with effective manufacturing systems. The three generic areas of application of RPT are Concept modelling, Secondary Tooling and Direct Tooling. These are considered in the following section.

3.1 Concept Modelling

RP models offer rapid, cost effective models which provide visualisation and tactile feedback early in the design process allowing engineering design changes to be identified and undertaken. RP models improve communication at all levels by realising the physical form of the CAD data. Much depends on the manufacturing scenario when identifying which process to use. Many RP systems are used for modelling conceptual designs. Producing individual RP models is required in the early design process verification. They are used for form, fit, design review, packaging studies, marketing models and some functional applications. Choosing an appropriate RP technology to use for concept modelling depends on specific criteria. The RP systems have special attributes that are relevant to certain applications. In general these are as follows:

SLA - Accuracy relative to general application across all areas
SLS - Wide range of materials, durability and function. Specific Nylon production
LOM - Size, large dense parts produced economically
FDM - Unique in producing ABS. True desktop process

For one or two functional models a nylon SLS (Venus, 1995) or ABS - FDM prototype will provide reasonably accurate durable parts. If part accuracy and surface finish is important over functionality then Stereolithography would be most appropriate. For large patterns - LOM makes economic sense.

3.1.1 The importance of functionality. Conceptual prototype criteria call for "appearance" or "functionality" dependence. RP models will often suffice for controlled design review environments. However for use in uncontrolled environments, functionality through materials becomes dominant. These may be limited run production, field evaluation or legislation conformity trials. Figure 1 shows the development of a golf trolley prototype incorporating functional SLS nylon components. These were direct from the RP machine and unfinished apart from dye colouring. Figure 2 shows close up detail of one of the components.

For the requirements of production quality parts then higher investment in terms of time and money are required. Tooling systems are developed to enable higher part functionality to be achieved. It is important to determine at an early stage what level of functionality is really required.

Direct RP models are generally of low functionality with SLS offering the best levels in nylon for equivalent thermoplastics. Secondary tooling offers some improvement in functionality with reduced piece part costs for multiple products. Figure 2 shows a lorry cab heating control assembly used for functional trials. On the left is the SLS nylon assembly and on the right the final plastic injection moulded product.

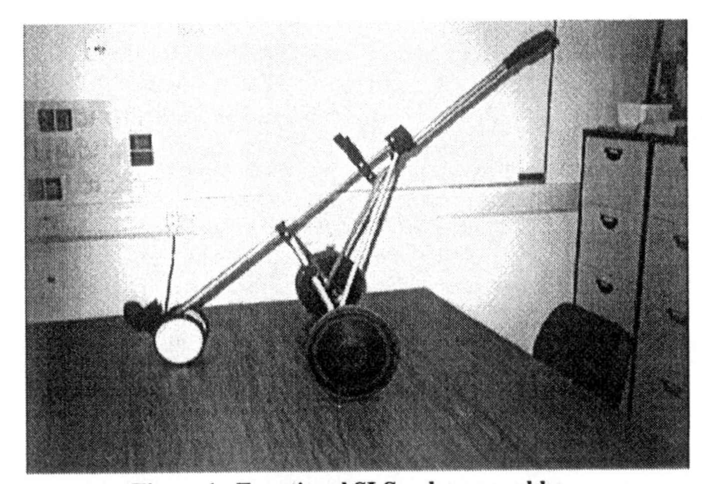

Figure 1. Functional SLS nylon assembly

Figure 2. Close up of SLS nylon components

3.2 Secondary Tooling

Secondary Tooling enables production of single and multiple products in plastic or metal materials. The number of prototypes required, the accuracy needed and the material desired will all influence the choice. For visualisation, testing, or a 'one-off' a single RP 'master' may suffice. For scenarios where larger numbers are required then this will influence the choice. It is not economical to build tens of identical RP parts. Coupling RP processes with the most appropriate tooling option is affected by availability, experience and economics. It is not always obvious which RP process is most appropriate for the combination of criteria required. Although many secondary tooling approaches have been used for some time prior to RP, in many cases they have been given a new lease of life where a master was formerly crafted by hand.

There is now enough experience and confidence with many of the approaches that the risk is low. Multiple parts can be economically produced by using an RP master pattern. Secondary tooling from a master can be split into two subsections; for plastic, and for metal parts. Types of secondary tooling approaches for plastic parts include: Silicone rubber, Epoxy, Spray metal, Cast metal, Ceramic, Kirksite and Keltool. Historically, the two most popular RP master systems used are SLA for accuracy and surface finish, and SLS Trueform^{TM} PM which offers high accuracy and easy finishing.

Silicone rubber tooling (vacuum casting) has become established as the most widespread secondary tooling process but is limited because the part production materials are urethanes. One approach used by the authors is to prepare a modified silicone mould set and use this in an injection mould machine. SLS nylon and Trueform masters have been used to create the part cavity geometry. Using such moulds in injection moulding machines allows true engineering plastics to be used. This approach has been successful in producing polyethylene, polypropylene and 30% glass filled polypropylene, and polyamide parts. Up to around 100 products have been produced from such moulds. Figure 3 shows a simple mould arrangement set into a steel bolster.

Figure 3. Silicone rubber injection mould tool inserts

Material dependent prototypes have in the past been made from thermoplastic or metal machined from solid stock. RP secondary tooling is cheaper, faster and approaches replication of the manufacturing process. Material dependent prototypes will generally be more expensive and take longer than appearance dependent prototypes with smaller quantities. They have higher quality and improved functionality and may challenge production tooling. Keltool (Jacobs, 1996) provides a secondary tooling process to produce robust metal composite tools for small inserts (75 x 75mm). Such tools are showing high volume production capabilities (500,000-1,000,000).

Metal production processes include investment casting, sand casting and die casting. Many of the processes for plastics can be adapted for low melting point (LMP) metals. RPT approaches for metals include silicone rubber (LMP metals), sand directly, cast (for sand, die and inv. wax patterns) and sacrificial master for investment casting. Investment casting has had the largest interest with use of SLA Quickcast, SLS polycarbonate and Trueform, LOM with limitations and FDM wax.

SLA Quickcast (Neuman, 1995) has become established as the most widespread investment casting master pattern system. This is mainly due to the number of SLA machines and therefore the foundries' experience in handling the material as a master pattern. SLS has improved significantly from polycarbonate to Trueform with improved accuracy and surface detail. Each RP process and tooling option has its own advantages, requirements, and trade-offs. Selecting a process combination is simplified by identifying the tooling objectives in advance and giving some prioritisation to these. Key selection criteria include lead-time, quantity, cost, accuracy and functionality. Secondary tooling options can be quicker and less expensive than conventional methods but may not be able to attain 'production quality' parts if these are the priority.

3.3 Direct Tooling

Direct RP tooling is processed on RP machines. Direct tooling may be applied for plastic or metal component production (depending on material capabilities). This includes RapidTool™ LR (Long run) and SLA Tooling - direct Aces Injection Moulding (AIM™).

SLA AIM™ tooling (Decelles and Barritt, 1996) uses SLA parts as core and cavity inserts for injection moulding. It is applicable for injection moulding polyethylene and is capable of producing around 100 in total. Tool failure occurs at the glass transition temperature of the SLA mould (158°F/70°C) where the SLA material sticks to the plastic part and pulls away at ejection.

DTM's RapidTool™ LR (long run - see figure 4) (Arnold-Feret, 1997, Lee, 1996, Gornet, 1996) is the only RP system capable of producing a fully dense durable steel based mould. Parts produced by the RapidTool™ process are fully dense metal composites of approximately 60% steel and 40 % copper. The material exhibits good tensile strength and excellent thermal conductivity. It is currently the most durable Rapid Tooling material available and produces reasonable tool life in excess of 50,000 components. The properties of RapidTool™ LR are shown in table 1. Under development is RapidTool™ SR (short run - see figure 5). This will be an epoxy infiltrated steel structure replacing the copper furnace infiltration cycle and aimed at up to 100 product life (Forderhouse, 1997).

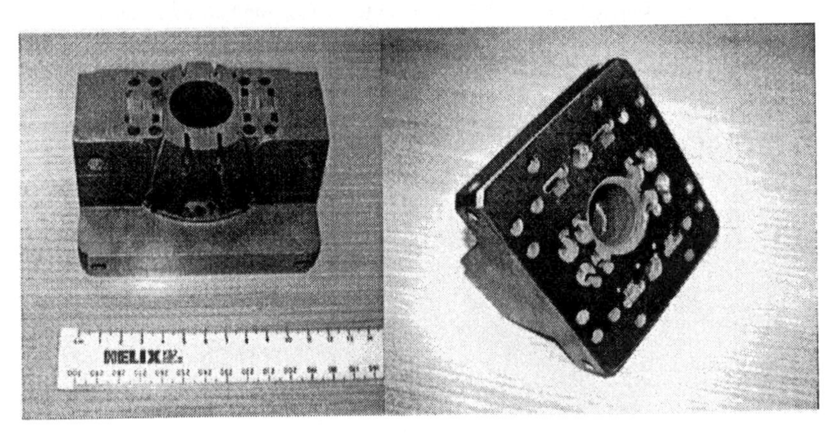

Figure 4. SLS RapidTool™ long run inserts

Table 1
Properties comparison RapidSteel™ with aluminium and P20 steel

Property	Test	RapidSteel™	Al 7075-T6	P20 steel
Average Particle Size (μm)	Laser Diff.	55	-	-
Specific Density (g/cm^3)	ASTM D792	8.23	2.7	7.85
Yield Strength 0.2% (MPa)	ASTM E8	255	502	751
Tensile Strength (MPa)	ASTM E8	475	570	950
Elongation (%) Tensile at break	ASTME8	15	11	20
Young's Modulus (10^3 MPa)	ASTM E8	210	65	210
Hardness (Rockwell B scale)	ASTM E18	75	82	180-210
Thermal conductivity (W/mK)	ASTM E457	185	131	29
Thermal expansion Coeff. (10^6 /K)	ASTM E831	202.3	333.7	23.8

Figure 5. SLS RapidTool™ short run insert

For direct metal production the tooling systems have to endure harsh high temperature environments. DSPC (Direct Shell Production Casting from Soligen, Inc.) produces ceramic shells for casting. DTM's RapidTool™ moulds have been applied as die casting tooling. SLS SandForm™ casting moulds (figure 6) are produced on the SLS machine for direct sand casting, negating the requirement for separate core box production. Complete sand moulds can be produced in days without tooling. However, full consideration of running/gating and feeding is required in the CAD model.

Figure 6. SLS SandForm™ casting moulds

4. CONCLUSIONS

With increasing choices of RP and complementary tooling methodologies the selection criteria must be rigorously applied to determine process routes and identify future needs for development. The criteria applied to tooling design and development are lead time, tool cost, piece part cost, quality, tolerance, and functionality.

All secondary tooling operations exhibit limitations due to the transfer process and finishing of models. Direct tool production has been the aim of many as industry has applied pressure to advance RP systems into this area. This is to enable the creation of prototype and production tooling directly from RP equipment. Rapid Tooling techniques are now being developed for producing production components faster, more efficiently and more economically than traditional approaches.

The marketplace in North East England has demonstrated a significant uptake in RP processes. OEM's require service providers to become more than access points to the technologies. Integration is key. The complementary technology and methodology that best fits a particular organisation needs to be identified, assembled and developed.

Figure 7 shows the strategic contribution of RPT in a manufacturing supply chain environment. In this environment there is a requirement for:

- Assessment of existing technology and skill level
- Identification and implementation of the complementary methodologies
- Increased supply chain integration and knowledge of what practises are effective in SMEs

The problem faced is that of achieving quality that is equivalent to final production parts. This requires focused development to realise full opportunities. Little is known about the manufacturing practises that are relevant and effective in an SME environment. This raises a number of questions.

- Is it true that modern manufacturing practices successfully operated in large companies are equally relevant and effective in the SME environment?
- Is there a different form of effective product development for SME's?

A three year CAMM research program is now commencing addressing the innovation and product introduction strategies of SME's in particular.

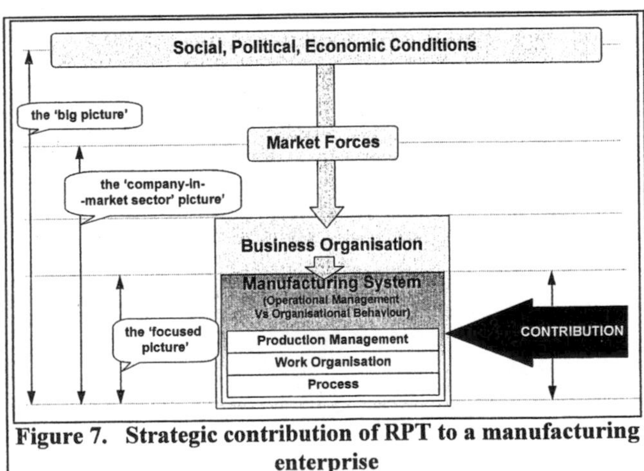

Figure 7. Strategic contribution of RPT to a manufacturing enterprise

The main RP processes have established a history in the North East of England for conceptual model production and are successfully operated across the range of materials which can be selected from the systems. Secondary tooling approaches have been evolved largely from traditional approaches which have been rejuvenated through the development of RP systems. New approaches are continually being identified and developed to provide a diverse range of approaches with different niches, i.e. for metal and plastic tooling. These include metal based systems (Keltool) which approach production lifespans for injection moulding. Direct tooling such as DTM's RapidTool™ offers a bright future, although there has been little experience in identifying the best ways of utilising such technology in established practises. These approaches are relatively unknown in mainstream manufacturing. There is a great deal of scope for research and development effort to take these approaches further. There are currently higher risks in using direct tooling which require prior assessment to be made and careful application.

References:

Arnold-Feret, B.,'**Rapid Prototyping and Rapid Tooling - A Comparison'**, Proc. Rapid Tooling From Rapid Prototyping and Manufacturing, Dallas, Texas, February 6-7, 1997.

Barber, P.,'**An Investigation into the Relationship Between Current Design and Manufacture Practices'**, PhD Thesis under preparation, CAMM, 1996.

Decelles, P. & Barritt, M.,'**Direct AIM PrototypeTooling'**, 3D systems, P/N70275/11-25-96, 1996.

Forderhouse, P. DTM personal communique, 6th February 1997.

Gornet, Graf and Ryder **'Experiences with DTM RapidTool',** Proc. Rapid Prototyping and Manufacturing, April 23-25, 1996, Dearborn, Michigan.

Jacobs, P.,'**Recent Advances in Rapid Tooling from Stereolithography'**, 3D systems publication P/N 70270/10-15-96, 1996.

Lee, Tom, DTM corporation, **'RapidTool'**Proc. Rapid Prototyping and Manufacturing, April 23-25, 1996, Dearborn, Michigan.

Neumann, R,'**3D Systems Stereolithography-Solutions for Investment Casting'**, Proc. 28th Int. Symp. on Automotive Tech. and Automation, 18-22 Sept. 1995, Stuttgart, Germany, pp343-350.

Riat, A.,' **'Best Practice'Lean Production in Small to Medium Sized Manufacturing Enterprises, and its Assessment'**, PhD Thesis, CAMM, November 1996.

Venus, A.D., and Van de Crommert, S.J.,'**The Feasibility of Silicone Rubber as Injection Mould Tooling',** Proc. 2nd National UK Conference on Rapid Prototyping and Tooling Research, Buckinghamshire, 18-19th Nov. 1996

TWO DIMENSIONAL STRESS ANALYSIS OF WORKING CONDITIONS ON THE PATELLA.

F. Nabhani, A. Hart, C.J. Connor, M. Wake
School of Science & Technology, University of Teesside, Middlesbrough. TS1 3BA

ABSTRACT: The stresses imposed on the human patella were examined using a two-dimensional finite element (FE) analysis. Two working constraints, level walking and knee bend, were applied to the models. The analysis was initiated by following previous work, which indicated that the stresses acting on the patella were approximately 180 MegaPascals (Mpa.) at knee bend, and 30 MPa. at level walking. This analysis produced similar results to those previous models, however several discrepancies were noticed to occur in those models.

A technique involving Magnetic Resonance Imaging (MRI) was employed for gathering the geometric data of the patella. Discrepancies that arose from the various models were scrutinised and corrected until a more reliable model was produced. The results of the final set of models for the two conditions showed a reduction in the previously recorded stress levels. Results indicated that the value for a knee bend was reduced from 180 MPa. to 75 MPa., respectively and 30 MPa. to 25 MPa. for level walking.

INTRODUCTION

Before starting any finite element analysis, it is necessary to perform several initial functions. Two fundamental questions arise at this stage,

a) Is the model to be produced in 2D or 3D ?

b) Will the computer platform produce the required analysis, e.g. is a Personnel Computer (PC) adequate or is a workstation system necessary ?

To fully answer the questions it is necessary to first know certain conditions of the FE model. Firstly, how many degrees of freedom are to be examined, and secondly, the boundary conditions for the model. For the stress analysis undergone on the patella the solution was the use a workstation platform, thus a Silicon Graphics Indy™ 4600SC system, together with the SDRC I-DEAS™ modelling software, was used. The computational power of the system and software gave the necessary requirement for this type of analysis.

MATERIAL COMPOSITION

The human patella is a composite of differing bone materials, which act, in unison, as a multi-shock absorber system. In all, three basic shock absorption systems were noted to occur. For the purpose of the analysis it was assumed that the properties of the materials were isotropic. This was indicated by information from the MRI data. Thus, only the Modulus of Elasticity (E) and Poisson's ratio (ν) are required for the analysis. The values used, shown in Table 1, were obtained from previous studies [Minns and Braiden, 1989]. What should be noted is that articular cartilage is a visco-elastic material i.e. stressed over a period of time. As this analysis is of a static nature, the cartilage was also assumed to be isotropic.

Material	Young's Modulus (E) MPa.	Poisson's Ratio (ν)
Hyaline cartilage	7.87	0.473
Subchondral bone	700	0.2
Cancellous bone	410	0.2
Cortical bone	24,100	0.28

Table 1
Material Properties

FORCES ACTING ON THE PATELLA

There are a number of different forces acting directly and indirectly on the patella. For the purpose of the analysis, only the ground to foot reaction force (P_J), ligament and tendon forces (P_T and P_Q respectively), were applied to the 2D models. The forces P_J, P_T and P_Q are shown in Figure 1 as vectors. All forces applied to the patella obliquely.

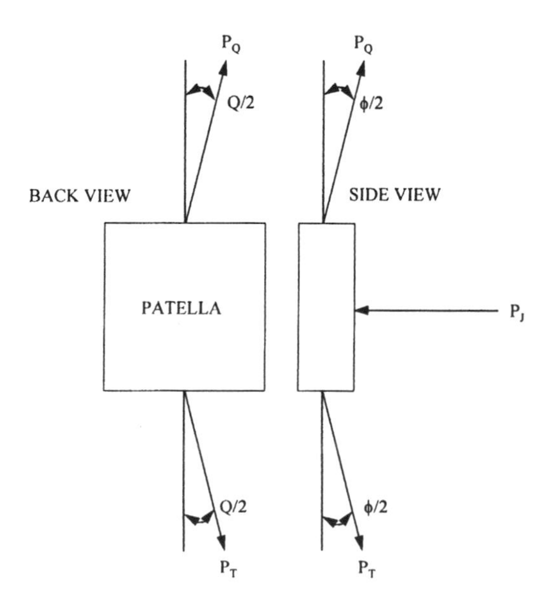

Condition	Flexion Angle	P_Q,P_T (N)	Q Angle	ϕ	P_J (N)
Knees Bend	45°	2433	0°	60°	2433
Level Walking	15°	647	10°	38°	422

Figure 1. Forces acting on the patella

FINITE ELEMENT THEORY

The finite element method was developed as a result of work within the aerospace industry and, with the advent of digital computer systems, the equations developed could be easily solved. Thus Finite Element analysis was created. Although this method was originally employed for solving structural analysis, the general nature of the theory/equations mean that it is possible to extend it to more general engineering problems such as pressure vessels and fluid flow.

In essence, the finite element method is the representation of a body or structure by an assemblage of sub-divisions called finite elements. These elements are interconnected at joints known as node or nodal points. Simple functions are chosen which approximate the distribution of the displacement over each finite element. These functions are known as displacement functions or displacement models. Once the magnitude of the displacement functions are known i.e. the displacement at each nodal point, the solution can then be expressed in terms of a displacement model. There are several forms of displacement models, such as polynomial and trigonometric functions. Since polynomials offer the easiest form of mathematical manipulation they are often employed in the finite element applications.

The theory of the finite element method can be split into two distinct phases. The first phase consists of the study of the individual element. The second is the study of the assemblage of elements which represents the entire body.

MODEL PROCEDURE

When producing any model for a finite element (FE) analysis there are two pre-set initialisations to be carefully considered [Hayes *et al*, 1982; Keyak *et al*, 1993]. These are deciding the overall shape of the model and deciding where the boundary conditions are to occur. In the first procedure the items geometry is the required factor. The geometric data for the models was obtained using Magnetic Resonance Imaging (MRI), in which a human patella was scanned, using an MRI machine, in-vivo, i.e. within the human body.

Once the geometric data had been obtained it was then converted into a numerical data form. Although only a 2D analysis was produced, the data obtained from this method also gave data relating to the entire object in 3D therefore a decision as to which slice was to be used had to be made. It was found by a matter of logic that the slice chosen had to adhere to 3 factors;

1) The slice should adhere to dimensional aspects as small dimensions can cause conflicts or errors in the processing of the model.

2) The ligaments should be positioned over the slice chosen to reduce force re-calibration.

3) The slice should adhere to uniformity of material distribution.

From the analysis of the data slices, it was decided that the centre slice was the most suitable one. The actual stress analysis was performed in four stages. Firstly, dimensional points for the creation of the slice were input. This was achieved by taking the data points obtained from the MRI scan and converting them into a standard file format such as DXF or IGES. Secondly, a series of splines were extended between the points, creating a boundary. Thirdly, a mesh was introduced to the model containing the type of elements and the material properties for the patella. It was decided that plate type elements were to be employed having a thickness of 3 mm. This thickness was a factor relating to the scanning of the patella. Finally the boundary conditions were applied to the model as in accordance with the anatomical data.

RESULTS

To fully appreciate the problems that are known to occur with the previous models, it was decided to simulate, under similar conditions, the previous work. This revealed the inherent faults within the original set of models. The first of these was identified as a geometric factor. This was eliminated in our initial model, yet problems still arose with the associated level of stresses within this model. It was identified that these stresses were too high to be physically possible. Secondly, the boundary conditions between the differing materials for the various bone materials of the patella act in unison, and examination of a patella showed that the material is homogenous throughout and thus acts as a single entity. This was eliminated by establishing a set of constraint equations between the material boundaries such that they materials acted in unison. The result of the analysis of both sets of models indicated no discernable differences between those of the original set.

Finally an examination of the boundary conditions being applied to the model was considered. From previous work [Goodfellow *et al*, 1976; Matthews *et al*, 1977; Townsend *et al*, 1977] it was firmly established that the restraints applied to the models were in their correct position. However, it was shown that the loading of the patella is somewhat different from that of the known anatomy. Therefore, the loadings applied were re-positioned such that it simulated the actual positioning of the tendons.

The stress plots obtained from the analysis indicated several factors of interest. From examination of the level walking model, it was found that the maximum principal stresses were slightly lower than that of the previous models. A very small reduction of 5 MPa taking the stresses from 30 MPa. to approximately 25 Mpa resulted. It was also shown that these stress levels mainly occur in one area which is located on the anterior surface of the patella. This stress zone is extremely localised, indicating that the property of the bony material is the critical factor. In addition it was also seen that there are two stress areas whose position was detached from the main stress area. The value of these stresses were approximately 15 MPa. and 7 MPa. respectively. At first it was assumed that these stresses could be due to stress concentration factors, however the element model proved this assumption to be unfounded.

The Von Mises stress plot, shown in Figure 2, yielded similar results, with the exception of a localised stress area detachment. This has a value of 25 MPa. and was attributed to two factors. Firstly, it is possible that this zone occurred due to stress concentration factors, whilst secondly, and more probable, is the idea that the stresses are a direct result of the summation of the forces of the tendons.

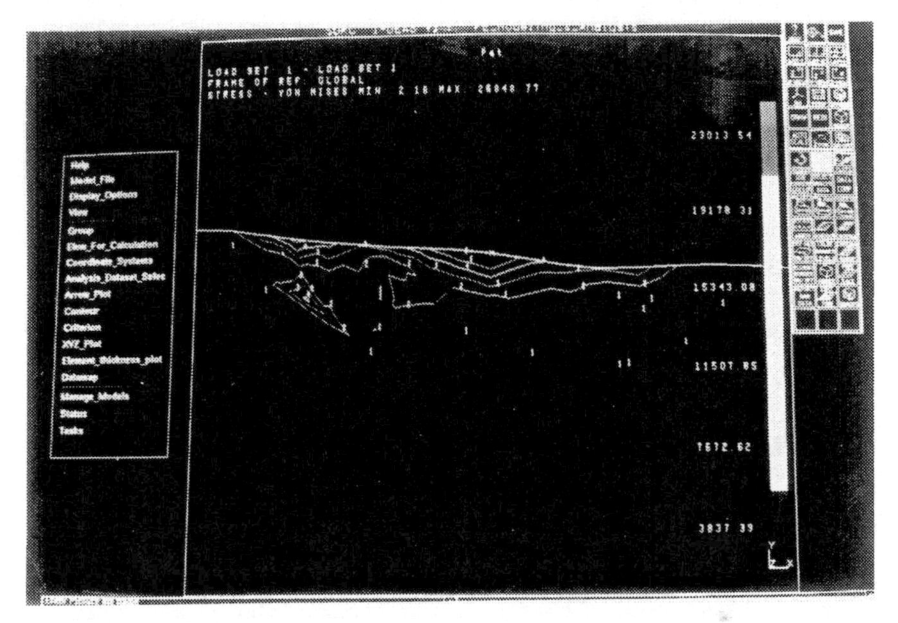

Figure 2. A typical stress plot of the levelwalking condition.

A more in-depth examination of the model, particularly on the hyaline cartilage, revealed two interesting results. Firstly, the maximum principal stresses of the Subchondral bone/Articular Cartilage show that the reaction force of the femur seems to be having no direct effect on the value of the stresses. This indicates that the stress values are only a product of the loading of the patella via the tendons. In addition, it was also indicated that the stress values of the bone/cartilage, although low in comparison with the overall values are at a maximum value of 1 MPa. These stresses are predominately located in the bony material, indicating a material phenomenon. The stress plot of the cartilage alone indicates areas of extremely low stress. This was expected as cartilage cannot sustain high stresses.

The knee bend condition yielded differing results from the level walking condition in several respects. Firstly, the values produced have been dramatically reduced from 180 MPa. in previous studies to 80 MPa., a reduction of 42%. The stresses produced also indicate that the stress zone has become more predominant in one of the previous models detached zones. The value of the stresses in this zone have increased from 15 MPa. to 80 MPa. thus indicating the effects of loading. Also indicated by the plots is that there is now a radial zone of stresses with an approximately value of 20 MPa. This was attributed to two factors. Firstly, the properties of the material. Secondly, the effect of the position of the forces. It was also revealed that there are now signs of the effect of the differing material properties taking effect, especially on the boundaries with that of the articular cartilage.

The Von Mises stresses as illustrated in Figure 3, yielded two new factors. Firstly, the radial zone of maximum stresses has now totally dispersed and is now replaced by an area which is attached to the main area of stresses, an indication of the loading conditions. This stress zone has a value of 40 MPa. Secondly, another stress area, near to the position of the tendons was formed. This was attributed to two factors: a stress concentration factor, and a force factor. The maximum principal stresses of the Subchondral bone/Articular Cartilage and Cartilage yielded two differing results from the previous models conditions. Firstly, the overall stress value has increased from 1 MPa. to 7.5 MPa., an indication of the loading. Secondly, the stresses indicate there is a slight effect from the reaction force of the femur, albeit small. This was indicated in the stresses as they have now been re-positioned in the direction of this force. What was also revealed was that a material difference has once again taken predominance over the forces.

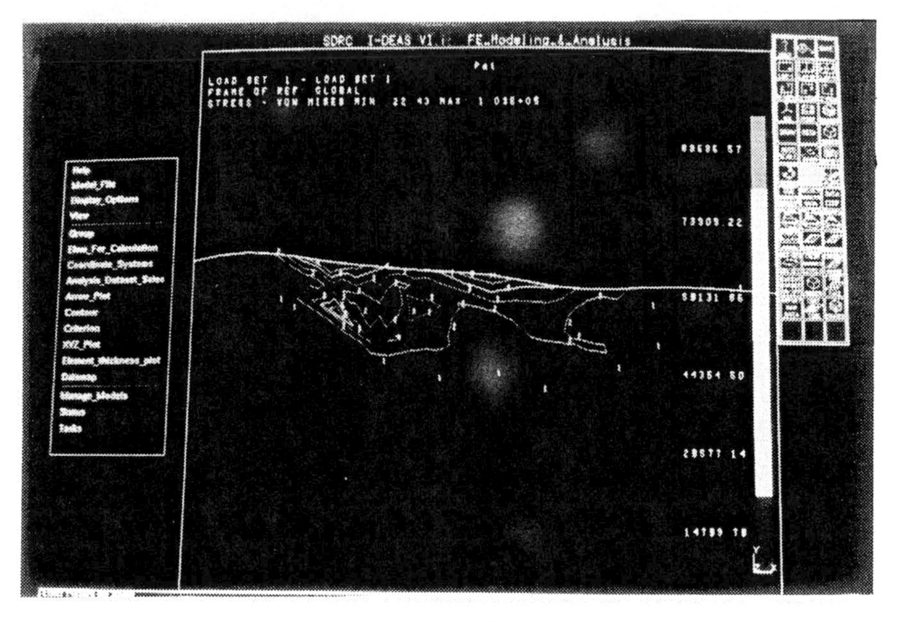

Figure 3. A typical stress plot of the kneesbend condition.

CONCLUSION

It can be concluded from the results of the analysis that although the preliminary work [Minns and Braiden, 1989] does produce useful results, there is an indication that the results are inaccurate. This is primarily due to factors relating to the tendon position and material distribution. It has been shown that this can be overcome using new techniques and models and as a result, stress levels have reduced by 42% .

The results produced are still predominately high, although the contour positions do yield similar results observed, namely, most patella fail at the high stress zone, as indicated on the stress plots. These high stress values can be attributed to the problem when dealing with small 2D models.

What is also revealed is that the differing material properties that exist within the patella do have an overall effect on the stresses, irrespective of modelling the bony materials as one material.

ACKNOWLEDGEMENT

The authors would like to thank Mr. D.S. Muckle, South Cleveland Hospital, and Dr. R.J. Minns, Dryburn Hospital, for their help with this paper and the staff at Darlington Magnetic Resonance Imaging whose help was invaluable to this investigation. The authors would also like to thank the Wishbone Trust for their help in this project.

REFERENCES

Goodfellow J., Hungerford D.S. and Zindel M., Patello-femoral joint mechanics and pathology. I: Functional anatomy of the patello-femoral joint. *J. Bone Jt. Surg.*, 61A, 159. 1982.

Hayes W.C, Snyder B., Levine B.M., Ramaswamy S., In: Finite elements in Biomechanics. Ed: R.H. Gallagher, Wiley, New York. 1982.

Keyak J.H, Fourkas M.G., Meagher J.M. and Skinner H.B., Validation of an automated method of three-dimensional finite element modelling of bone.", *J. Biomed. Eng.*, 15, 505-509. 1993.

Lengsfeld M., Weiβ H., Kienapfel H.,. 3- and 2D FEM-Stress calculations on the patella with and without implant. *Fourth World Biomaterials Conference*, Berlin, Federal Republic of Germany. April 24-28, 1992.

Matthews L.S., Sonstegard D.A. and Henke J.A., Load bearing characteristics of the patello-femoral joint. *Acta. Orthop.*, *Scand.*,48, 511. 1977.

Minns J. and Braiden P.M., A loading and stress analysis of the patella. *Material properties and stress analysis in Biomechanics*, 44-51. 1989.

Townsend P.R., Rose R.M., Radin E.L. and Raux P., The biomechanics of the human patella and it implications for chondomalacia. *J. Biomech.*,10, 403. 1977.

A CASE STUDY OF INTERACTIVE DESIGN OF A PRODUCT AND ITS ASSEMBLY LINE.

Th. Vast[1], B. Raucent[1], F. Petit[2]
[1] Department of Mechanical Engineering
Université catholique de Louvain, Belgium;
[2] Department of Mechanical Engineering
Université catholique de Louvain, Belgium

ABSTRACT: It is well known that a robotic or automatic assembly cell can hardly be economical if the design of the product has not been improved. In a previous paper [1] we presented a methodology for the interactive design of a product and its assembly line.
This paper presents a case study of the design of a household electric fan. The greatest saving arises from the redesign of the base which is mainly composed of molded plastic parts and screws. These parts with low functionality can easily be combined. Saving on the drive system is more difficult. This subassembly is composed of parts with high functionality which can not easily be combined.
Based on the analysis, the product is redesigned and a first draft of the assembly line is proposed.

INTRODUCTION

In today's rapidly moving market, manufacturing firms are compelled to greatly reduce the time to market and in particular the development time in order to develop new products at an increasingly rapid pace. It is now widely admitted that many decisions taken at the design stage are decisive for the entire product life.[2] Studies have shown that large portions of manufacturing costs are determined at early stages of design and that any change at a later stage dramatically increases the production costs. Assuming that product functional requirements are met, the problem is to consider ease of manufacturing and assembly. The best results are achieved when the product design is performed interactively with the production line design. Some years ago Professor Boothroyd and his collaborators proposed a new Design For Assembly method [3] which is certainly one of the most popular and interesting approaches available on the marketplace. In this method, the a priori choice of an homogenous assembly line simplifies the product analysis and requires only relative comparisons of the design efficiency without any line considerations, leading to a design optimized for only one particular assembly method. However, real assembly lines are heterogeneous (e.g. containing manual and robotized stations) rather than homogeneous. The main originalities of the interactive methodology presented in [1] is to investigate the design of heterogeneous lines, comprising manual, automated and robotized workstations. The approach is composed of three steps (see Figure 1): in the first one, the initial design of the product is performed based on product functionality and Direct DFA (general DFA rules). This initial design will be optimized during the second stage. The comparison between different product designs is performed by calculating an economic indicator CIC (Comparable Indicative Cost) for each operation and for different assembly methods (manual, robotized and automated). The CICs are independent of the production volume and can be used to select the design details that could be revised. The result of this second step is an optimized product design and an operating chart with the corresponding assembly method. The third step concerns the assembly line balancing problem and, in general, minor product design adaptations are still allowed at this stage.

This paper describes a case study of the design of a household electric fan. It focuses on the first two steps of the method.

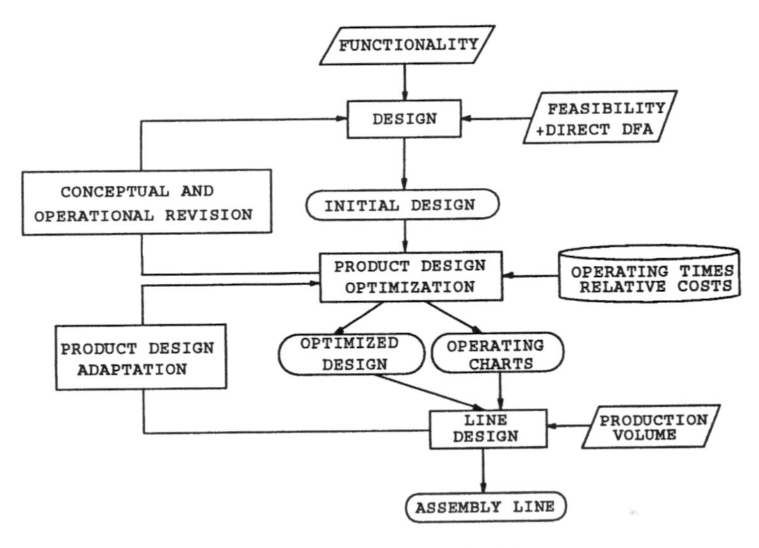

Figure 1. Interactive methodology

METHOD DESCRIPTION

The first stage of the procedure is to find an assembly sequence. Usually it is recommended to take the reverse disassembly order, see for instance [3]. However we have shown [1] that it can be interesting to modify the mounting order and/or the receptor choice. If the initial assembly sequence does not prove satisfactory, it is possible to select another base part or another mounting strategy.

The product is usually made of subassemblies which are assembled on separated assembly line. It is clear that if a product is composed of many complex subassemblies the analysis should start with the more complex subassemblies before analysing the final product assembly. A simple "complexity factor" ν is defined in order to decide whether the analysis should start with the final assembly or a subassembly :

$$\nu = \frac{N_k}{N_t} \tag{1}$$

with N_k the number of components in subassembly k
N_t the number of components in the product

The following different cases can be identified:

$v>0.6$: the product is strongly structured and we recommend to begin the product analysis by the first assembly stage, i. e. by subassembly analysis.

$0.4<v<0.6$: the product is poorly yet sufficiently structured and we recommend to analyse the product as a whole.

$v<0.4$: the product is poorly structured and a complete redesign is recommended to create subassemblies.

If the product is made of multiple-level subassemblies one should use the criteria recursively. The starting point is to calculate the complexity factor for the final assembly (1st level). Depending on this value it is recommended to analyse the final assembly or to calculate the complexity factor for the second level and so on until the last level is reached.

When a first assembly sequence has been determined and sub-assemblies identified, the CIC computation can begin. This evaluation criterion has the dimension of a cost but is not a real assembly cost and is mainly used to summarize the operation characteristics of a given assembly method. At this stage only the equipments and operators necessary during production time are considered.

The Comparative Indicative Cost (CIC) is defined as the cost of preparing and inserting a part (or of performing an assembly operation). For each assembly operation, the following cases are considered :

a) Full manual operation (M+M) : Handling and insertion are performed manually by an operator using, if needed, a hand tool.

b) Assisted manual operation (A+M*) : An assisted manual operation means that thc part preparation is performed automatically using e.g. avibratory bowl feeder.

c) Automatic preparation followed by an automatic insertion (A+D) or (A+R) : The reference case consists of an automatic preparation using a vibratory bowl feeder followed by an automatic insertion performed either by a dedicated machine (A+D) or a robot (A+R).

d) Automatic insertion from manually loaded magazines (K+D) or (K+R) : The reference case is a part tray in which parts are inserted manually without effort in order to prepare an automatic insertion.

After computation, all the information related to a product is summarized in a table (seeTable 1).

This chart contains the operation order number (N), operation name, 0/1 index which indicates if the operation is a candidate for suppression (0) or not (1) depending on the three answers proposed by Boothroyd [3], the number of repetitions of this operation (NR) and the CIC for each assembly method. By looking at this table the designer can immediately identify the parts that cause trouble and see where to concentrate his efforts (high CIC). A conceptual redesign of the product can be performed in order to suppress superfluous parts or to reduce the CIC for each operation and assembly method.

The best design is obtained when the superfluous, or too complicated, operations are eliminated and for an operation the CICs of each assembly method is equal to a minimal value.

				Preparation(1)			Insertion (2)				Combinations OF (1) AND (2)						Other operations		
N	OPERATIONS	0/1	NR	K	A	M	M	M*	D	R	M+M	A+M*	K+D	K+R	A+D	A+R	M	AR	AP
0	SA chassis	1	1	1.49		0.98	0.76	1.26	0.13	3.59	1.74		1.62	5.08					
s1	Thread cable	0	1			0.76	8.57	9.07			9.33								
1	Stator	1	1	1.43		0.93	2.52	3.02			3.45								
2	SA Rotor	1	1	1.41		0.91	0.76	1.26	0.13	3.59	1.66		1.54	5.00					
3	Washer	0	2	2.71	0.20	1.70	1.51	2.52	0.26	7.18	3.22	2.72	2.98	9.90	0.47	7.39			
4	Flexible washer	1	1	2.70	0.61	2.19	0.76	1.26	0.13	3.59	2.95	1.87	2.83	6.29	0.74	4.20			
5	SA Front case	1	1	1.49		0.98	0.76	1.26	0.13	3.59	1.74		1.62	5.08					
6	Screw	0	2	2.52	0.73	1.51	6.07	7.09	0.79	16.59	7.59	7.82	3.31	19.11	1.52	17.32			
7	Front cover	0	1	1.49	3.45	0.98	0.76	1.26	0.13	3.59	1.74	4.71	1.62	5.08	3.58	7.04			
8	Front cover screw	0	3	3.78	0.22	2.27	9.11	10.63	1.19	21.10	11.38	10.85	4.97	24.88	1.41	21.32			
9	Rear grid	1	1	1.41		0.91	0.76	1.27	0.20	3.84	1.67		1.61	5.26					
10	Fixing plate	0	1	1.41		0.91	0.76	1.27	0.20	3.84	1.67		1.61	5.26					
11	Screw	0	4	5.04	0.28	3.02	12.15	14.17	1.98	25.61	15.17	14.45	7.02	30.65	2.26	25.89			
12	Blade	1	1	1.49		0.98	1.77	2.28	0.30	3.84	2.75		1.79	5.33		3.84			
-	Reorientation	1	1														4.54	0.87	5.09
13	Blade screw	0	1	1.26	0.06	0.76	4.05	4.56	0.40	12.26	4.81	4.61	1.66	13.52	0.45	12.32			
-	Reorientation	0	1														4.54	0.87	5.09
14	Front grid	1	1	1.26	3.68	0.76	0.76	1.27	0.13	3.59	1.52	4.95	1.39	4.85	3.81	7.27			
15	Half hoop	0	2	2.82	6.22	1.81	2.53	3.54	0.56	10.92	4.35	9.77	3.38	13.74	6.78	17.14			
16	Screw	0	2	2.52	0.45	1.51	8.10	9.11	1.29	24.00	9.61	9.57	3.81	26.52	1.74	24.46			
-	Reorientation	0	1														4.54	0.87	5.09
17	Rear cover	1	1	1.49	3.50	0.98	0.76	1.27	0.13	3.59	1.74	4.77	1.62	5.08	3.64	7.10			
18	Rear cover screw	0	1	1.26	0.05	0.76	4.05	4.56	0.89	12.26	4.81	4.60	2.15	13.52	0.94	12.31			
-	Reorientation	0	1														4.54	0.87	5.09
19	Engaging lever	1	1	1.26	0.21	0.76	0.76	1.27	0.20	3.59	1.52	1.47	1.46	4.85	0.40	3.80			
20	Lever screw	0	1	1.26	0.05	0.76	4.05	4.56	0.89	12.26	4.81	4.60	2.15	13.52	0.94	12.31			

Table 1

CIC (in BEF, 1 US $ = 30 BEF) for the assembly of the head of the fan

If all the conceptual revisions have been exploited without a significant improvement in assembly performances, the designer can reconsider the mounting chronology chosen at the previous step of the analysis. If any improvement can be made, the last possibility is to make a complete product conceptual revision. The results of this analysis are: a redesigned subassembly, an operating chart containing the succession of the necessary assembly operations with their possible assembly methods, a chart with assembly time operations and a first draft of the assembly line.

CASE STUDY

The device we will study is a household electric fan presented in Figure 2. The initial design, denoted A is made of 137 parts combined in 11 subassemblies. At first sight it is a good candidate for redesign because it contain many screws (about 20% of the number of parts), and 7 different types of screw. It may therefore be easy to reduce the number and the type of parts in order to achieve some savings.

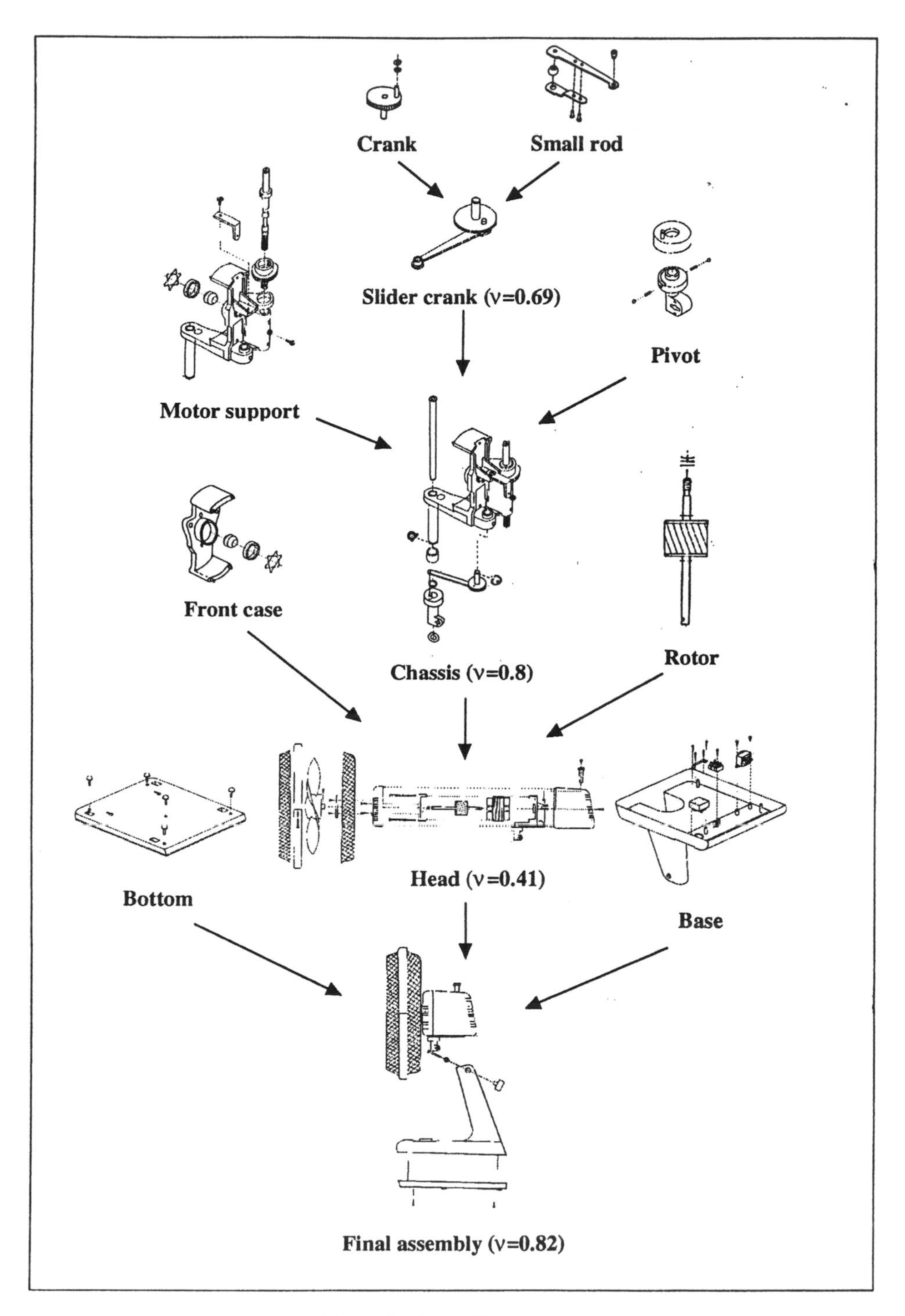

Figure 2. Assembly structure

The redesign sequence can be determined with the evaluation of the complexity factor. The values of the complexity factor for the Final Assembly being 0.82, it is recommended to start the process with its subassemblies i.e. Bottom, Base and Head. As the Head contains subassemblies, we need to evaluate its complexity factor. This being 0.41 the assembly of the Head should be analyzed. When this is done the analysis is resumed with the Head subassemblies i.e. Front case, Chassis, Rotor. As the Chassis contains subassemblies its complexity factor has to be evaluated (0.8). This time the redesign continues straight on with the subassemblies i.e. Motor support, Pivot and Slider crank. The complexity factor of the subassembly Slider crank has to be evaluated (0.69) and thus its components have to be optimized. To end the process it is recommended to return to the final assemblies by analyzing the remaining assembly operations.

For each subassembly the CICs are evaluated, see for example in Table 1 the CIC for the head subassembly. Table 2 gives the number of operations and CIC of each subassembly.

	Number of operation			CIC		
	Design A	Redesign B	Redesign C	Design A	Redesign B	Redesign C
Final assembly	12	6	6	33,24	29,93	29,93
Bottom	6	6	6	6,58	6,58	6,58
Base	15	5	4	15,28	7,27	6,86
Head	35	18	18	44,96	30,41	30,41
Frond case	7	6	6	4,83	4,45	4,45
Rotor	5	4	4	3,74	3,51	3,51
Chassis	10	10	9	14,60	14,60	12,22
Pivot	7	7	0	4,67	4,67	0
Motor support	22	19	19	15,50	13,63	13,63
Slider crank	7	4	0	6,98	3,78	0
Small rod	8	8	0	5,37	5,37	0
Crank	3	2	0	2,38	2,18	0
	137	**95**	**72**	**158.11**	**126,38**	**107,59**

Table 2
Number of operations and CIC for each subassembly

REDESIGN

By analysing the CIC for each subassembly, one can easily detect the parts and operations that may be eliminated. Thus, based on direct DFA principles [3] it is possible to redesign the product. For example, Figure 3 present the initial design A and the redesign B of the blade of the head subassembly.

The first thing to do is to eliminate parts that are candidates for suppression (with a zero in the third column of Table 1). For example the fixing plate (number 10) and its screws (11) can be replaced by plastic snapfit fasteners on the front case (see * in Figure 3). Design rules for snapfit fastener are extensively discussed in [4].

Another source of improvement is to try to reduce the number of reorientations of the product during the assembly process. These operations add no value to the product and only increase the assembly cost. In the initial design the blade assembly requires two reorientations (see Table 1). By replacing the lateral blade screw (13) by an axial nut (17), these expensivereorientations are eliminated.

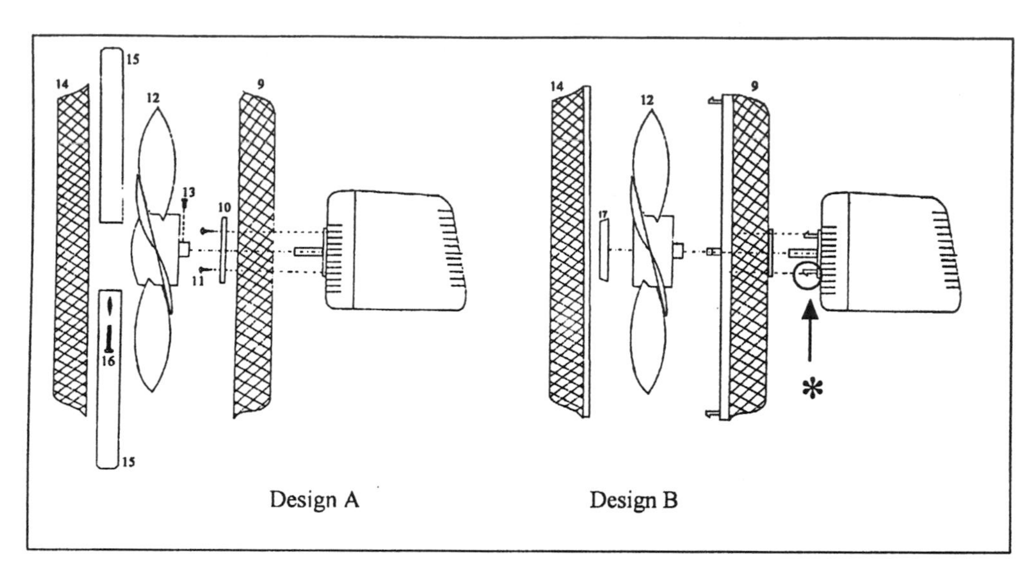

Figure 3 . Assembly of the blade of the subassembly head

The remaining parts can be modified to reduce their CIC and to obtain equivalent costs for the different assembly methods leading to a product that can be assembled either manually, automatically or using a robot with the same cost. It is clear that the parts with the highest CIC must be studied first. The geometry can be modified to facilitate the orientation of the part. The base of the fan comprises a self induction coil. The orientation of this part is difficult and can only be done manually. The only characteristic which proves the correct orientation is a small filament which is inside the part. Table 3 shows that the part has a high preparation cost. To improve the manipulation of the part and to make possible an automatic orientation, the geometry of the part is modified by making the connector asymmetric (see Figure 4). A great benefit was also obtained by standardizing the wiring connectors e.g. using IDT (insulation displacement technique).

				Preparation(1)			Insertion (2)				Combinations OF (1) AND (2)					
N	OPERATIONS	0/1	NR	K	A	M	M	M*	D	R	M+M	A+M*	K+D	K+R	A+D	A+R
2	Initial design	1	1	1.89	9.59	0.98	2.77	3.28	0.26	3.85	3.76	12.87	1.75	5.34	9.85	13.44
2	Redesign	1	1	1.49	0.8	0.98	2.77	3.28	0.26	3.85	3.76	4.08	1.75	5.34	1.06	4.65

Table 3
CIC of the self induction coil

The same techniques have been applied to other parts of the fan. In redesign B all the functionalities present in the initial design are retained but the assembly process is improved. The screws are replaced by plastic snapfit fasteners whenever possible.. In the original design the small rod used for the oscillating motion is equipped with a ball joint and a bearing. These high quality bearings can be suppressed by increasing the tolerance on the passing holes. This modification will decrease the quality but will reduce the assembly time.

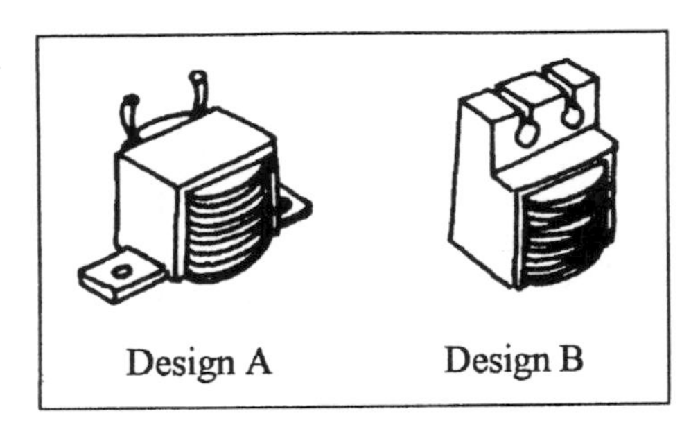

Figure 4 a & b. Redesign of the self-induction coil

Redesign C is more radical. In this case we suppress one functionality. In the initial design, the user can set the starting orientation of the head. Suppression of this function allows to simplify the assembly because the system is made up of small springs and small balls which are difficult to manipulate and to assemble. The reduction of the number of subassemblies can also be a source of savings. In design C, 4 subassemblies are eliminated. This modification will have important consequences on the layout of the assembly line. Small branches of the assembly line will be suppressed.

REDESIGN DISCUSSION

One can observe in Table 2 that in Redesign B one third of the parts have been suppressed but at the same time the CIC reduced only by one fifth. In the base subassembly the number of parts has been divided by 3 but the CIC only by 2. This means that the savings due to the reduction of the number of parts is lower than expected. This is due to the fact that the parts suppressed are mainly "low cost" parts i.e. the assembly cost of such parts is low. For example Table 4 presents the savings according to the type of parts expressed in percent relatively to the initial value.

	Number of operations					*CIC*				
	Design A	Redesign B		Redesign C		Design A	Redesign B		Redesign C	
Screws and washers	39	19%	13	23%	8	21,81	9%	6,97	11%	5,08
Functional parts	36	3%	32	10%	22	32,34	2%	29,54	6%	22,86
Miscellaneous operations	24	4%	19	7%	15	17,81	3%	13,44	4%	11,09
Plastic parts	19	4%	14	4%	14	28,47	5%	21,08	5%	21,08
Subassemblies	11	0%	11	3%	7	24,83	0%	24,72	5%	16,85
Other operations	4	1%	2	1%	2	5,81	1%	3,59	1%	3,59
Total	**137**	31%	**95**	47%	**72**	**158,11**	20%	**126,38**	32%	**107,59**

Table 4
Savings per part family

Design A contains 39 fasteners such as screws, design B containing only 13. This means a reduction of 19% of the total number of parts. The reduction of CIC is only 9%. Clearly screws can easily be suppressed but are "low cost" parts. On the other hand suppression of "functional" parts and plastic parts such as half hoops for the fixing of the two safety grids (see Figure 3 number 15) seems to be more interesting because of their layer dimensions and of their geometric complexity.

Figure 5 presents the CIC per operation before and after redesign. Once again one can easily see that the first parts suppressed are the "cheapest" one. The assembly cost of the remaining parts is more expensive.

In design C the main savings follow from the suppression of subassemblies through that of a functionality. It allows to suppress 47% of parts and save 32% of cost relatively to design A. Comparing designs B and C, it appears that 24% of the parts of design B can be suppressed. This leads to a relative saving of 15%. Based on all this information the full team can decide whether or not to retain the functionality.

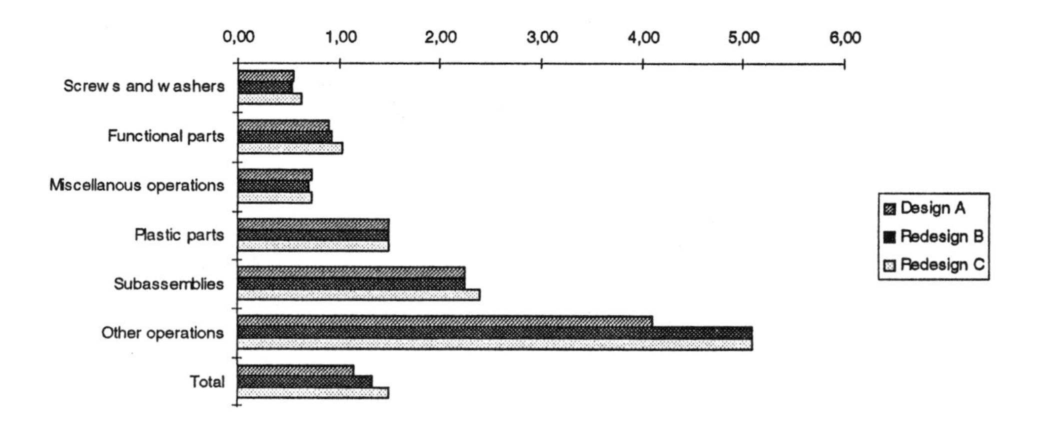

Figure 5. Evolution of CIC per operation

LINE DESIGN

Based on the product redesign and the CIC tables one can easily propose a first draft for the assembly line, see for instance Table 5.

This table shows that some operations cannot be realized by the three assembly methods. For example the placement of the stator must be executed manually due to the difficulty to insert it on the chassis. The assembly line will therefore have a manual station for this operation. An other example is the preparation of the front safety grid which is too expensive with automatic means.

An interesting line would be one combining all the minimal CICs, see the shaded cells in Table 5. However this line is purely theoretical and cannot be implemented in practice. Its only interest is in being used later as a reference line to compare several solutions.

				Preparation(1)			Insertion (2)				Combinations OF (1) AND (2)						Other operations		
N	OPERATIONS	0/1	NR	K	A	M	M	M*	D	R	M+M	A+M*	K+D	K+R	A+D	A+R	M	AR	AP
0	SA chassis	1	1	1.49		0.98	0.76	1.26	0.13	3.59	1.74		1.62	5.08					
s1	Thread cable	0	1			0.76	8.57	9.07			9.33								
1	Stator	1	1	1.43		0.93	2.52	3.02			3.45								
2	SA Rotor	1	1	1.41		0.91	0.76	1.26	0.13	3.59	1.66		1.54	5.00					
3	Washer	0	1	1.36	0.10	0.85	0.76	1.26	0.13	3.59	1.61	1.36	1.49	4.95	0.23	3.69			
4	Flexible washer	1	1	2.70	0.61	2.19	0.76	1.26	0.13	3.59	2.95	1.87	2.83	6.29	0.74	4.20			
5	SA Front case	1	1	1.49		0.98	0.76	1.26	0.13	3.59	1.74		1.62	5.10					
6	Screw	0	2	2.52	0.73	1.51	6.07	7.09	0.79	16.59	7.59	7.82	3.31	19.11	1.52	17.32			
7	Rear grid	1	1	1.26		0.76	2.53	3.04	0.32	3.84	3.29		1.58	5.26					
8	Blade	1	1	1.49		0.98	1.77	2.28	0.30	3.84	2.75		1.79	5.33		3.84			
9	Blade nut	0	1	1.26	0.28	0.76	4.05	4.56	0.40	12.08	4.81	4.84	1.66	13.34	0.68	12.36			
10	Front grid	1	1	1.26	147.2	0.76	2.53	3.04	0.32	3.84	3.29	150.23	1.58	5.10	147.5	151			
-	Reorientation	0	1														4.54	0.87	5.09
11	Rear cover	1	1	1.49	3.50	0.98	0.76	1.27	0.16	3.59	1.74	4.77	1.65	5.08	3.66	7.10			
-	Reorientation	0	1														4.54	0.87	5.09
12	Engaging lever	1	1	1.26	0.21	0.76	0.76	1.27	0.20	3.59	1.52	1.47	1.46	4.85	0.40	3.80			
13	Lever screw	0	1	1.26	0.05	0.76	4.05	4.56	0.89	12.26	4.81	4.60	2.15	13.52	0.94	12.31			

Table 5

CIC (in BEF, 1 US $ = 30 BEF) for the assembly of the head of the fan for Design C

The next step in this analysis will be to perform the line balancing taking into account the required production volume and the under-used equipment cost [5]. This analysis will use the operating costs and times determined in this first analysis to allocate assembly operations to homogeneous stations. Ideal line balancing is reached when the cycle time of all the stations are equal to the throughput time of the line which is a productive constraint. This leads to an ideal continuous flow of products.

As mentioned above, the assembly line can be heterogeneous : all types of assembly methods (manual, automated, robotized) have to be considered for each station. As a consequence, line balancing is not trivial due to the number of possible solutions. It should be keep in mind that the station should be homogeneous. In other words, operations can be clustered into a station only if they are of the same method type, although not always optimal. This type may be determined at a previous step by a particular operation (e.g. an operation that can be performed economically by only one method).

CONCLUSIONS

The case study clearly shows that redesign of the product leads to lower assembly costs. However the saving due to the reduction of the number of parts is lower than that could have been expected because the easiest parts to eliminate are usually the "less expensive" parts. In addition it is shown that more savings can be expected if the functionalities of the product are modified. This modification should of course be discussed by the complete team but CICs can be used to evaluate the cost of a particular functionality.

Selection of assembly methods (manual, automated and robotized) and line balancing will introduce other demands of redesign in order to approach the desired station occupation rate. This analysis will be the subject of a future paper.

REFERENCES

1. **Petit, F., Leroy, A., Raucent, B., et al.,** Interactive design of a product and its assembly line, Proc. of the 6th FAIM Conf., Atlanta, Georgia, USA, May 13-15, 1996,pp 499-508.

2. **Kusiak, A.,** Concurrent Engineering - automation, tools and techniques, J. Wiley, 1992.

3. **Boothroyd, G. and Dewhurst, P.,** Product Design For Assembly, BDI, 1989

4. **Raucent, B., Nederlandt, Ch., and Johnson, D.,** Plastic snapfit fastener design, Journal of Design and Manufacturing, 1995, accepted for publication.

5. **Aguirre, E., Petit, F., Raucent, B. et al.,** Methodology for interactive design of a product and its assembly line. Part 2 : Line balancing, First International Conference IDMME'96, April 1996, Nantes, France.

6. **Aguirre, E., and Raucent, B.,** Economic Comparison of Wire Harness Assembly Systems, Journal of Manufacturing Systems,vol 13, 1994, n° 4, pp. 276-288.

7. **Raucent, B., and Aguirre, E.,** Performances of Wire Harness Assembly Systems, ISIE'94, IEEE Int. Symp. on Industrial Electronics, Santiago, 94, pp. 292-297.

AN INTRODUCTION TO DESIGN FOR ENVIRONMENT AND ASSESSMENT OF DESIGN AIDS

by
Philip M. Wolfe, Tom Caporello, and David Bedworth
Arizona State University

ABSTRACT: There is a growing awareness by individuals, businesses, and governments of the need to protect our environment. Since most pollution is generated through manufacturing activities, this is the logical starting point to implement pollution prevention methodologies. With this realization, in 1992 a few electronics firms originated the concept of design for environment (DFE).

There is a close analogy between DFE and design for quality. People realize that quality must be designed into the product; therefore, it is not difficult to understand that inspecting the end of the pipeline is not a cost effective way to eliminate pollution. DFE is a methodology for developing environmentally compatible products and processes, while maintaining desirable product price/performance and quality characteristics.

Because DFE is a relatively new concept, many engineers are not aware of the design aids that are available even though they are being asked to design green products. To address this problem, this paper discusses some of the tools that are available today: guidelines, checklists, matrices, and life cycle assessment models.

INTRODUCTION

In the 1970's a large amount of political attention in the United States was devoted to the environment which resulted in significant new legislation, such as the National Environmental Policy Act, the Clean Water Act, and the Endangered Species Act. Although environmental statutes existed before, this period was the dawn of emission-based regulation that might be characterized as industrial environmental regulation. Since then extensive environmental regulation has been enacted around the world.

Most of the early U.S. environmental regulation was focused on emissions. However, this type of regulation is largely ineffective because it seeks to control contaminants that have already been generated. Consequently, the focus of environment regulation shifted to process regulation, which led industry to consider how process were performed. The initial regulation was depicted as performing "inspection at the end of the pipeline" while more recent regulation is focused on "inspection in the pipeline". Although this approach is an improvement, it does not place the emphasis on product design where material selection and process specification are made. This realization in the 1990's motivated some of the world's most competitive firms to originate the concept design for environment (DFE).

DFE is the systematic consideration of design performance with respect to environmental, health, and safety objectives over the full product and process life cycle [Fiksel, 1996]. Applying this concept involves integrating environment considerations with other design considerations, such as assembly, quality, and serviceability. Also, the product development, use, and disposal life cycle must be broken down into its component parts.

As firms recognize the importance of environmental responsibility to their long-term success, they realize that the practice of DFE is becoming essential. DFE can provide competitive advantage by reducing the costs of production and waste management, encouraging innovation in product simplification, and attracting new customers. Consumers are very supportive of environmental issues and are increasingly concerned with the environmental friendliness of products that they purchase. In addition, there are several global initiatives that make it imperative that firms address environmental issues if they are to be competitive. The International Organization for Standards (ISO) is developing the ISO 14000 standards for environmental management systems, which are similar to the ISO 9000 quality management systems standards. Also, some countries are initiating regulations to assure that manufactures are responsible for recovery of products and materials at the end of their useful lives, sometimes called "take back".

The Earth Summit of 1992 (the United Nations Conference on Environment and Development) held in Rio de Janeiro produced 27 principles. Four of them are closely related to the topic of this paper [Fiksel, 1996]:

1. Development today must not undermine the development and environment needs of present and future generations.
2. Nations shall use the precautionary approach to protect the environment, meaning that scientific uncertainty shall not be used to postpone cost-effective measures to prevent environmental degradation.
3. In order to achieve sustainable development, environmental protection shall constitute an integral part of the development process.
4. The polluter should, in principle, bear the cost of pollution.

The first principle addresses a fundamental goal of sustainable development: to meet the needs of the present without compromising the ability of future generations to meet their own needs. Today this goal seems idealistic, but the environment has only recently become a major factor in management decision making. As the world becomes more populus, this goal must become more realistic.

Some other factors that are influencing management's adoption of DFE are: product differentiation, customer awareness, regulatory pressures, international standards, and profitability improvement. Although, there is increased awareness of the need for environmental responsibility, few organizations have developed the principles and metrics required to achieve the desired goals. Until this is done, engineers and manufacturing staff performance will not be measured with regard to environmental factors, which will be reflected in minimal consideration of environmental issues as decisions are made. However, with the emergence of ISO 14000, it appears that we are entering a phase of establishing standards and codes of practice, which should lead to realization of environmental responsibility.

COMMON DFE PRACTICES

The system boundaries associated with DFE are much broader than are usually considered in product design. The environment should be considered in the entire life cycle of the product, which includes the production of raw materials, energy used in fabrication, distribution, use, and disposal. Some of the more common DFE practices in industry are: material substitution, substance use reduction, energy use reduction, design for recyclability, design for disassembly, design for material conservation, and waste source reduction. Often these practices are implemented in a variety of forms, such as verbal rules, design manuals, and on-line design aides.

As DFE practices become more formal within a firm, DFE check lists may be developed to aid the design engineer and manufacturing staff. Graedel and Allenby [Graedel, 1996] provide some example check lists under the following categories:

1. Design Considerations in Energy Generation and Use
2. Minimization and Design of Solid Residues in Industry
3. Minimization and Design of Liquid Emissions in Industry
4. Minimization and Design of Gaseous Emissions in Industry
5. Design Considerations in Materials Selection
6. Product Delivery
7. Solid-Residue Generation During Product Use
8. Liquid-Residue Generation During Product Use
9. Gaseous-Residue Generation During Product Use
10. Energy Consumption During Product Use
11. Intentional Dissipative Emissions During Product Use
12. Unintentional Dissipative Emissions During Product Use
13. Design for Maintainability
14. Design for Recycling

Associated with each category is a list of questions.

DFE ASSESSMENT TOOLS

The most comprehensive methodology used to evaluate the environmental responsibility of a product is life-cycle assessment (LCA). It consists of three phases: (1) inventory analysis, (2) impact analysis, and (3) improvement analysis. In the first phase, the flows of energy and materials to and from the product during its life are determined, which may require extensive chemical analysis. In the second phase, the environmental impacts of the flows identified in phase one are evaluated. This process can be very difficult and time consuming because of complexities and lack of scientific knowledge. Environmental science is not at a mature stage. During the third phase, the results of the previous phases are translated into specific actions that will reduce the negative impacts on the environment.

It is difficult to perform a LCA because determining life-cycle inventories is expensive and time-consuming. Impact analysis is difficult and often requires some qualitative judgements. In addition, LCA methodologies currently available are not robust enough to be applied to a wide variety of products. Also, it is difficult to translate the findings into meaningful values. Consequently, LCA is not often used. Even though DFE product and process assessments are difficult, management cannot ignore the environment in making decisions. The remainder of this paper will discuss some of the assessment tools that are available.

Product and Process Assessment Matrices

Graedel and Allenby [Graedel, 1996] are proponents of the "environmentally responsible product assessment matrix "(see Figure 1). One dimension (columns) of the matrix is life-cycle stage and the other dimension (rows) is environmental concern. They state that a suitable assessment methodology should have the following characteristics:

1. It should lend itself to direct comparisons among rated products.
2. It should be usable and consistent across different assessment teams.
3. It should encompass all stages of product, process, or facility life cycles and all relevant environmental concerns.
4. It should be simple enough to permit relatively quick and inexpensive assessments to be made.

When applying this assessment methodology, the DFE assessor studies the product design, manufacturing plan, packaging, usage, and likely disposal and assigns to each element of the matrix an integer rating in the range of 0 to 4 (0 denoting a very negative environmental impact, and 4 denoting lowest impact). These assessments are made using experience, checklists, guidelines, surveys, and other information.

Once values have been assigned to each cell in the matrix, the overall environmentally responsible product rating (R_{ERPT}) is computed as the sum of the matrix cell values:

$$R_{ERPT} = \Sigma\Sigma M_{i,j}$$

Because there are 25 matrix elements, a maximum rating is 100.

The environmentally responsible product-assignment matrix					
	Environmental concern				
Life stage	Materials choice	Energy use	Solid residues	Liquid residues	Gaseous residues
Premanufacture					
Product manufacture					
Product delivery					
Product use					
Refurbishment, recycling, disposal					

Figure 1. The environmentally responsible product-assessment matrix

Graedel and Allenby have also applied this methodology to assessing processes. The columns of the matrix are the same; however, the rows represent the following life stages: resource extraction; process implementation; process operation; complementary process implications; and refurbishment, recycling, and disposal. Otherwise, the environmentally responsible process-assessment rating is computed in the same manner as the product rating.

Although matrix assessment approaches are not as thorough and quantified as a formal LCA, they are more practical. Graedel and Allenby assert that an assessment of modest depth, performed by an objective professional, will succeed in identifying 70 or 80% of the useful DFE actions that could be taken while consuming small levels of time and money. The key success factor is the person that performs the assessment. That person must be experienced and knowledgeable about the types of products, processes, and facilities being reviewed.

EcoSys

In 1994, the Sandia National Laboratories completed the EcoSys prototype product and process assessment system. Considering the complexity of the problem, the system design objectives were ambitious: "The EcoSys life cycle information and expert system was designed to provide an analysis tool for environmental, design, and manufacturing personnel to analyze the design of a product and select optimal materials and processes to reduce environmental impacts throughout the product life cycle." [Watkins, 1994].

EcoSys consists of an environmental decision model, an information management system and an expert system. The environmental decision model applies the principles of LCA, although the prototype explicitly considered only the manufacturing phase of the product life cycle. The information system contains actual inventory data on input materials and parts, and wastes generated, for each process used to manufacture the end item; and the expert system provides the rules used in life cycle analysis impact analysis.

Impact criteria and environmental attributes considered in EcoSys are:

1. Ozone depletion potential	2. Global warming
3. Species extinction	4. Air toxics
5. Water toxics	6. Total waste produced
7. Toxic release inventory (TRI)	8. Municipal solid waste
9. Incineration	10. Treatment
10. Fugitive releases	12 Total resource consumed
11. Non-renewable use	14 Non-reuse
12. Reserves	16. Energy

The prototype data base contained several materials and impact values for these criteria and attributes. A relational database management system is used to manage the data. Since data are maintained by a database management system, queries can be performed by material type to review the associated environmental data.

EcoSys has the capability to consider the bill of materials of a product and the processes used to fabricate and assemble the product. Consequently, environmental impact analysis can be performed for subassemblies of a product and accumulated for the product as a whole. The system provides the ability to accumulate analysis results at any level in the product structure. Also, impact analysis can be performed on alternative processes.

The expert system component of EcoSys will traverse a product structure tree and select the best process were alternative processes are available during fabrication and assembly. If the knowledge base is insufficient or the user wants to override the expert system recommendations, a specific process can be designated.

LCA is very complex; therefore, designing a useful decision support system is difficult. EcoSys is a good, pragmatic design. It contains the environment related elements that a decision maker will need. The component that adds the least value is the expert system. As a result, most people that use the system may not implement this portion. The major deficiency of the system is the lack of ability to obtain the impact data. There are no models provide to calculate the data, so this information must be obtained elsewhere and manually input to the database.

SimaPro

About the time EcoSys was being developed another computer system, SimaPro, was being developed in the Netherlands to perform LCA. This latter system is now a commercial product owned by PRe Consultants [Cleij, 1995]. Using a LCA approach, this system will let you analysis and compare products. The database provided has the capability to store data describing materials, processes used to create the respective materials, and processes used to fabricate a specified product. Also, stored in the database is environmental impact data associated with the respective processes. You can add to the database as the need and data becomes available.

To analyze a product that is being designed, you first must describe the product in terms of quantities of materials used to fabricate the product and any materials and energy consumed during the life of the product. Also, production processes are denoted. If the materials and processes are contained in the database, you only need to identify them. Otherwise you will have to enter the required detail information.

Once a product is specified, a LCA inventory analysis is performed to determine the amount of raw materials and emissions involved in producing the product. These substances are organized by the following categories: raw, air, water, and solid. Next you may evaluate the environmental impact in categories such as greenhouse effect, ozone depletion potential, carcinogen, pesticides, energy, and heavy metal. The impacts can be displayed in several ways, such as a colored bar chart showing the average load of a citizen during a year. The system provides alternative ways to perform the impact evaluation or you can provide additional methods. Also, a similar analysis can be performed for other stages in the life cycle of a product, such as disposal, disassembly, and reuse. In addition, a capability is provided to compare different products. The system has several other features that are not described here.

Considerable effort has be made to make SimaPro a useful DFE tool. One deficiency is the ability to specify processes in sufficient to detail to permit complex emission and environmental impact evaluation. Even so, this system can provide valuable assistance to the design engineer.

SUMMARY

There is an increasing awareness that DFE must be included in the integrated product development process of manufacturing enterprises. However, few firms have implemented DFE concepts. One reason is lack of metrics that can be used to evaluate environmental performance in objective, measurable terms. Another reason is the lack of accounting systems that explicitly denote environmental costs and benefits. A third reason is the lack of robust evaluation tools that are integrated with other tools used by the design engineer. However some stand alone design aids have been developed and related research is increasing. This paper presented examples of the approaches that are being used or proposed. Most firms are using guidelines and checklists. However, these do not support quantitative analysis. Matrix approaches are a compromise between qualitative methods and detailed LCA. Although LCA is difficult to perform, efforts are being made to develop systems that utilize this approach. Of the two systems described, SimaPro provides the most practical value.

REFERENCES

Cleij, Vincent and Goedkoop, SimaPro 3.1 demo version instruction manual, Amersfoort, The Netherlands: PRe Consultants, 1995.

Fiksel, Joseph, *Design for Environment,* New York, N.Y.: McGraw-Hill, 1996.

Graedel, T. E. and Allenby, B. R., *Design for Environment,* Upper Saddle River, NJ: Prentice Hall, 1996.

Watkins, Randall D., Kleban, Stephen D. and Luger, George F., Expert system support for environmentally conscious manufacturing, Albuquerque, N.M.: Sandia National Laboratories,1994.

SHORT-CUT COSTING OF PROCESS VESSELS USING AI AND CONVENTIONAL TECHNIQUES

A M Gerrard[1], J Brass[2]
1 Division of Chemical Engineering, University of Teesside, Middlesbrough TS1 3BA;
2 Department of Computing and Mathematics, Hartlepool College, Hartlepool, TS24 7NT

ABSTRACT: The paper will describe an analysis of a set of real industrial data on the relationship between the cost and size of chemical process pressure vessels. The measures of size include: height, diameter, wall thickness, type, orientation and material of construction, together with the number and size of nozzles. The purpose of the investigation is to provide the company with methods of rapidly estimating the approximate final cost of the item when no detailed design work has been attempted. The approaches to be compared include multi linear regression (MLR), neural networks (NN), fuzzy matching, rational functions and other non-linear models.
MLR was used to give a base point from which to judge the other methods. We have shown that neural networks, having a particularly simple structure, can provide good estimates from a minimum amount of information and the technique of fuzzy matching can yield reasonable accuracies. However, because we had a limited amount of data we were keen to reduce the number of fitted parameters to a minimum. In this regard, the rational and new non-linear forms had much to recommend them.With a small original data set, the non-linear models seemed to be the most attractive method, but if a larger resource was available for analysis then the NN and fuzzy matching ideas had much merit.

INTRODUCTION

Cost estimation is a vitally important activity in manufacturing industry. By accurately forecasting costs, a company is able to set the prices it charges to its customers at an optimum, competitive level in order to maximise sales and to use of production facilities efficiently. This, in turn, leads to the maximum return on investment. If quoted estimates are set too high then customers will take their business elsewhere; if they are set too low, insufficient profits will be made or losses may be incurred. In short, the ability to produce accurate estimates may not only affect a company's profitability, but its very survival.

However, whilst it could be argued that monies spent on cost estimation are an investment towards future profits, it is possible that in some cases no order will ensue no matter how good the tender price. This is often the case with the practice of 'abortive tendering', whereby an intermediary between an end user and a vessel producer will approach a number of companies with the same specification, or in some cases, a number of intermediaries will approach one vessel producer, with similar but not exactly the same requirements. This latter case can result in several similar designs being fully costed where, at best, there is only the possibility of one firm order. Because of this practice, and the high cost of estimating in general, there is a need for an initial costing method that is quick and easy to use, whilst at the same time produces estimates within acceptable limits of accuracy.

The purpose of this study is to use a number of techniques which, instead of costing on a part by part basis, relate costs of new items to the costs of previously estimated or constructed items. Thus, the cost of a new process vessel is found by examining a database of previously costed vessels, in a similar way that a skilled estimator may be able to make a preliminary estimate based upon his past experience. All of the methods investigated are able to mathematically link some input variable(s) to an output variable - for example vessel data (height, diameter, thickness, number of nozzles, etc.) to cost. The techniques used to achieve this task are: regression analysis, neural networks, rational polynomials, some novel non-linear equations and fuzzy matching.

In order to compare the accuracy of the predicted costs produced by each of the methods, the same measure of accuracy is used; the sum of the squares (SS) of the errors; i.e. the total of the squares of the differences between the actual and predicted costs, or residuals

$$SS = \Sigma(\text{predicted cost - actual cost})^2 \qquad (1)$$

The mean percentage errors (%err) of the predicted costs compared to the actual costs and the mean absolute percentage errors (a%err) are also calculated for comparison purposes, where

$$\%err = \frac{\Sigma((\text{predicted cost - actual cost}) / \text{actual cost}) * 100}{\text{number of data points}} \qquad (2)$$

$$a\%err = \frac{\Sigma(\text{ABS}(\text{predicted cost - actual cost}) / \text{actual cost}) * 100}{\text{number of data points}} \qquad (3)$$

In the above equations, ‘actual cost’ is the figure produced by traditional estimation methods, and ‘predicted cost’ is the figure produced by the forecasting method in question, based upon information in the vessel database.

DATA COLLECTION

At the start of this project, the first task to be undertaken was that of data collection. This involved manually examining the company's records on existing estimates, and gathering together as much information as possible on each vessel, including their estimated cost. This was followed by some tidying of the data, for example, the elimination of transport costs which were included in some estimates but not in others. After the removal of incomplete, duplicate and uncorroborated data items, there remained information on 52 fully costed vessels available for analysis. The type of information supplied is listed in the table.

TABLE 1
Raw Data

Reference number	
Date of enquiry	
H	Tan to tan height
D	Internal diameter
T	Shell thickness
M	Material of construction
V	Vertical or horizontal alignment
C	Type (column, drum, etc.)
N	Number and size of nozzles
Description of vessel function	
Cost	

Along with the vessels' basic dimensions, a number of dummy or binary variables were also used to represent factors such as vertical or horizontal orientation, drum or column, etc. The use of these binary variables gives a simple way of including extra information, and at the same time enables different types of vessels to be included in the same database, e.g. columns and drums .

TABLE 2
Dummy (or Binary) Variables

M - Material of construction (1 = mild steel, 2 = others)

V - Orientation - Vertical or horizontal (1 = horizontal, 2= vertical)

C - Type - Column or drum (1 = drum, 2 = column)

Note that the dummy variables were chosen as 1 to represent what would be expected to be the cheaper option and 2 the more expensive option. Logically this should lead, for example with regression, to the respective parameters being greater than one (for linear models) and positive index (for log models). The nozzle information was originally supplied in the form of number of two inch nozzles, number of three inch, number of four inch, etc, up to the number of twenty-four inch manways, although not every measurement between two and twenty-four inches was included, there being 12 categories in all. In order to simplify the nozzle information, this was converted from its original form into three different representations;

$N1 = \Sigma N_i$ the total number of nozzles, irrespective of size

$N2 = \Sigma N_i d_i$ the sum of the nozzles multiplied by their diameters .i.e. a measure of the total perimeter of all the nozzles and

$N3 = \Sigma N_i d_i^2$ the sum of the nozzles multiplied by their diameters squared. i.e. a representation of the aperture size (or the amount of metal to be removed)

THEORY

Conventional cost estimating involves using precise vessel specifications and dimensions to accurately calculate; the amount of metal; the number, length and quality of welding operations required; the cost of any bought-in components (nozzles, pre-formed vessel heads and so on) and the labour required etc. However, in this study we shall use a range of short-cut techniques.

Regression analysis

Initially, a simple linear equation was used

$$\text{Cost} = A + B\,(\text{size}) \qquad (4)$$

Where, A and B are constants (found by regression analysis) and 'size' is some measure of a vessel, height or volume for example.The following power-law equation has also been much used

$$\text{Cost} = A\,(\text{size})^B \qquad (5)$$

The value of B is often approximately 2/3, hence the 'two-thirds power rule'. These equations were then extended to their multi-linear forms to take several independent variables into account

$$\text{Cost} = A + B1\ x1 + B2\ x2 + B3\ x3 + \ldots. \qquad (6)$$

or

$$\text{Cost} = A\ x1^{B1}\ x2^{B2}\ x3^{B3}\ .. \qquad (7)$$

Where x1, x2, etc are the size variables such as height, diameter, etc. and A, B1, B2, etc. are constants.The last equation mentioned can be transformed into its linear equivalent by logarithms. Using logs to effectively code the data, it is possible to use the linear technique of regression to analyse this non-linear relationships, since:

$$\ln(\text{Cost}) = \ln(A) + B1\ \ln(x1) + B2\ \ln(x2) + B3\ \ln(x3) + . \qquad (8)$$

Whilst the above equations are able to handle numeric variables such as height, diameter and thickness, there is an obvious problem when we need to use non-numeric data such as material of construction, vertical or horizontal, and column or drum. This was overcome, as we have noted, by the use of 'dummy' or 'binary' variables.The regression parameters can be found using an appropriate computer software package, e.g. Minitab.

Neural Networks

An alternative method of linking input variables to output quantities is the neural network. These are assembled from a number of individual elements called 'neurons', which are in turn combined to form a number of 'layers'; an input layer, an output layer, and one or more intermediate or hidden layers. Technically the elements in the input layer are not neurons, but rather 'input devices', as they have no effect on the data other than making it available for the next layer . The relationship between the input and output of each neuron in the hidden and output layers is specified by a transfer function, which can add, weight and, in this application, sigmoidally squash its own input signal to produce an output.
For example, using inputs of height, diameter and thickness to produce an output figure for cost would involve three inputs and one output neuron. The number of neurons in the hidden layer(s) can vary, and is generally subject to 'trial-and-improvement' experimentation in order to achieve the best results.

Rational polynomials

A third method of linking input variables to the output in order to model a particular situation, is the use of rational polynomials. By using a ratio of one polynomial divided by another it is possible to model highly non-linear situations, such as the one under investigation.The initial stimulus to investigate rational polynomials, which later led us on to the production of some innovative new forms, was provided by J. W. Ponton (1993a). He argued that, whilst neural networks were a useful tool, the number and complexity of the parameters (or weights) involved made rational polynomials a much simpler alternative. His suggested areas of use including inferential measurement of process variables, fault diagnosis and valve network problems. In brief, Ponton et al (1993b) suggest ratios of low order multivariable polynomials can be used. Three forms are proposed, the simplest being

$$\text{cost} = \frac{a_0 + \Sigma a_i x_i}{1 + \Sigma b_i x_i} \qquad (9)$$

Secondly, they reported that sometimes the addition of a quadratic term helped

$$\text{cost} = \frac{a_0 + \Sigma a_i xi + \Sigma c_i x_i^2}{1 + \Sigma b_i x_i} \qquad (10)$$

Finally, they described a variant which was useful in certain cases when the denominator tended to zero

$$\text{cost} = \frac{a_0 + \Sigma a_i x_i \; (+ \Sigma c_i x_i^2)}{e + (1 + \Sigma b_i x_i)^2} \qquad (11)$$

Where e is a small number, say 0.001. In the above equations, a_i, b_i, c_i (i= 1..n) are all constants where values are to be found and x_i (i= 1..n) are the known input variables.The main advantage to be gained from using rational polynomials rather than neural networks is in the reduction of the number of parameters used, with the associated benefit of a reduction in the computation time to find the best values.

New rational and nonlinear equations

After using equations 9 to 11 a number of new rational forms were tried. Because of the success of log regression, the following four log transformations of Ponton's original equation were used in an attempt to add extra non-linearity

$$\ln(\text{cost}) = \frac{a_0 + \Sigma a_i \ln(x_i)}{1 + \Sigma b_i \ln(x_i)} \tag{12}$$

$$\ln(\text{cost}) = \frac{a_0 + \Sigma a_i \ln(x_i)}{1 + \Sigma b_i x_i} \tag{13}$$

$$\text{cost} = \frac{a_0 + \Sigma a_i \ln(x_i)}{1 + \Sigma b_i \ln(x_i)} \tag{14}$$

$$\text{cost} = \frac{a_0 + \Sigma a_i \ln(x_i)}{1 + \Sigma b_i x_i} \tag{15}$$

These were followed by a further variant, here leaving the denominator as it was in the original, but instead of using addititive terms in the numerator these were replaced by the use of the parameters as powers (as in equation (7)).

$$\text{cost} = \frac{a_0 (\Pi x_i^{a_i})}{1 + \Sigma b_i x_i} \tag{16}$$

Following the success of this form , a number of further variants were used. Firstly a non-linear equation, which uses the nozzle information to increase the accuracy of the results obtained, whilst relying on the minimum number of parameters

$$\text{cost} = a_0 H^{a1} D^{a2} T^{a3} \ldots (1 + a_n N) \tag{17}$$

where N is a measure of nozzles as defined earlier
e.g. $N1 = \Sigma N_i$, or
$N2 = \Sigma N_i d_i$, or
$N3 = \Sigma N_i d_i^2$

We also tried

$$\text{cost} = \frac{a_0 H^{a1} D^{a2} T^{a3} (1 + a4\, N)}{1 + b1M + b2V + b3C} \tag{18}$$

and

$$\text{cost} = \frac{a_0 H^{a1} D^{a2} T^{a3}}{1 + b1M + b2V + b3C + b4\, N} \tag{19}$$

Fuzzy Matching

This is the final method investigated to link size to cost. Fuzzy logic is a development of conventional Boolean logic, whereby not only do we have the two states of true or false, but all of the intermediate stages from entirely true to entirely false (which includes the states of 'partly true', 'partly false', 'almost completely true', 'more or less true', etc.), in a similar way to our having black and white and all the intermediate shades of grey. Fuzzy logic was suggested by Zadeh as a method of representing uncertainty in natural languages. Following his suggestion that "fuzzification" can be applied to any area, we now have fuzzy calculus, fuzzy systems, fuzzy matching, and so on, whereby normally discreet (or crisp) amounts can be handled in a continuous (or fuzzy) manner. In this way, fuzzy matching can be used, for example, to convert handwritten text to ASCII input by finding near matches to each character written, rather than by trying to find an exact match.

Similarly the technique can be used to relate a vessel specification to the database of costed vessels by making an approximate match to one or more of the costed vessels, the hope being that the nearest match in specifications will give an adequate match in terms of cost. Of the papers studied, the most useful were by D. W. Edwards and G. J. Petley (1994). They used fuzzy matching for the estimation of total plant cost and reported that a number of characteristic functions (flat, ramped or curved) may be used to 'offer accurate preliminary capital cost estimates with minimal estimating effort', leading to our use of the technique to estimate the cost of individual vessels.

Various characteristic (or membership) functions (CFs) have been suggested to facilitate the matching process. These include flat, ramped, curved and asymmetric versions. When using the flat matching function, for example, each dimension (or descriptive variable) is compared with the corresponding variables in the database and given a score of one for a match or zero for no match. A match is said to have occurred if the database value falls within the range $t-\beta$ to $t+\beta$ of the new vessel (or target) value t. After completing the matching process with each vessel in the database, the cost of the vessel which gives the highest score would be used as the estimated price of the new vessel. The values of β are systematically varied to get the best fit. If more than one vessel gives the highest score, the price would be the mean of all the highest score vessels' prices.

RESULTS

As mentioned earlier, our task was to use a set of real data to provide us with a rapid means of estimating the costs of process vessels. The data set included 52 vessels and this (relatively) limited resource has influenced how we analysed the data and which methods we chose to favour.

Our first step was to use multi-linear regression (equation 6) to see how well this familiar technique fared. On examination of the H,D,T results(i.e. using only three independent variables), we found that SS=5.13E11 and a%err=92.8% which was rather disappointing. However, when we used the power law form (equation 7) the results showed a marked improvement, with the absolute error being reduced by over a half. This is not a surprise as the economies of scale are well known in vessel manufacture. (The constants B1.....etc. are all fractions).

When we turned to the neural network representation, we found that the SS figure was reduced by more than 45% to 2.72E11 and the a%err was more than halved to 43.3%. The shape of the network had 3 inputs as noted above, with 2 neurons in the hidden layer and just one output to predict the cost. It should be noted that because of our limited data, we had to use all of it for the training and then test the best model on a further eight vessels which became available later.

When all the input data was used (i.e. including size, dummy variables and nozzle information), the SS from regression was 3.84E11 (using H,D,T,M,V,C,N3). With a (7 2 1) neural network using the same inputs gave 0.83E11, and this was reduced to 0.77E11 when N3 was replaced by N2.

Most of the rational polynomial and new nonlinear function results improved on the SS values produced by both linear and log regression. The best results produced in this particular category being 1.3E11 and 32%, from equation 16 using H,D,T,M,V,C,N2. This result is only bettered by the more complex neural networks. We found that the simplest (i.e. equation 9) of the three Ponton forms was the best. Equations 17,18,19 all gave similar prediction accuracies, (better than logged regression but inferior to the neural networks). The simplicity and parsimony of equation 17 is noted. However, the equations 12 to 15 did not give good results and were quickly abandoned.

At first glance, the fuzzy matching results appeared rather disappointing, however, on closer inspection there are a number of pleasing results, particularly in the case of those using less than the maximum number of inputs; e.g. an SS of 2.9E11 from H,D,T (using a curved characteristic function), and 1.5E11 from H,D,T,V,N2 (with a flat characteristic function), this latter example once again only being improved upon by neural networks. We also tried to combine the fuzzy matching approach with the power law model (equation (7), but this intuitively appealing idea failed to improve the predictions. (Edwards also found this when he analysed a total plant cost data base).

CONCLUSIONS

1. The power law transformation of the linear regression equation was superior to the truly linear form.

2. Particularly simple neural networks with only two neurons in the single hidden layer gave good results, which meant the number of constants (weights and biases) to be fitted was not excessive. This was just as well with our limited data!

3. Rational polynomials of low order (equations 9 and 16) were better than the regression models but needed more constants to be fitted.

4. The non-linear equation (17) was attractive because it gave reasonable accuracy with a very small number of fitted parameters. It is also easily comprehensible to the user.

5. The representation of all the nozzle information in the combined forms, N2 and N3 was useful as a cost predictor.

6 It is noted that the neural network and fuzzy matching approaches do not allow extrapolation, whereas the regression, rational polynomial and non-linear forms do allow it.

REFERENCES

Edwards, D. W. and Petley, G., Chemical plant capital cost estimation using fuzzy matching, *Proceedings of ESCAPE 4*, Dublin, Eire,28-30 March 1994.

Ponton, J. W., The use of multivariable rational functions for non-linear data representation and classification, *Computers in chemical engineering*, 17(10) 1993, 1047-1052.

Ponton, J. W. and Klemes, J., Alternatives to neural networks for inferential measurement, *Computers in chemical engineering*, 17(10), 1993, 991-1000,

ACKNOWLEDGEMENT

The authors appreciate the help of David Peel in the work reported here.

NOTATION

A, a_i	Constants in cost equations
B_i, b_i	Constants in cost equations
C	Column or drum dummy variable
CF	Characteristic function
c_i	Constants in cost equations
D	Vessel diameter
e	Small positive constant
H	Vessel height
M	Dummy variable for material
N, N1, N2, N3	Measures of nozzle complexity
SS	sum of squares
T	Vessel thickness
V	Vertical or horizontal dummy variable
x	Independent variable in cost equation

SMALL FIRMS, BIG STOCKS?

Michael Bedwell, Paul Higginbottom and Sharon Sandhu
School of Engineering, Coventry University, CV1 5FB

ABSTRACT: A simple economic model is proposed to compare the relative values of stocks which, under notionally JIT conditions, firms in a given supply chain would be expected to carry. From the model it is hypothesised that, other things being equal, the Stocks/Turnover ratio will be greatest at the retail end of the chain, but elsewhere negatively correlated with turnover. Statistics from the automotive industry partially support this hypothesis; for the major car makers (Stocks/Turnover^2) is indeed greater than that of their suppliers, for whom the predicted correlation is indeed generally negative, though only to the 75% confidence level. However, for the major five first-tier UK manufacturers earning more than £1000m/year, (Stocks/Turnover^2) exceeds that of smaller firms, while the majority of firms earning less than £2.8m/year are self-excluded because they are not legally obliged to disclose their turnover. The paper concludes with a caveat that popular generalisations about lean supply rarely acknowledge this methodological problem.

INTRODUCTION

JIT has been cynically interpreted as "Japanese Induced Terror" [Oliver & Thomas (1991), quoted by MIRA 1994]. This paper will examine how far lean supply can be blamed for passing responsibility for stockholding down to those worst placed to accept it, namely the smaller firms early in the supply hierarchy, without ultimate benefit to the consumer [Eberts, 1995, p83]. A model will be developed showing that, other things being equal, smaller firms have to carry relatively bigger stocks, but that factors other than size may predominate. The argument will be based on some basic notions of economics and of queueing theory, and will be supported by some limited statistical evidence from the vehicle industry; however, it will make little call on the more populist writings promoting lean supply.

SUPPLY CHAIN MODEL

A fundamental tenet of queueing theory is that, other things being equal, both the waiting time and the queue length increase in proportion to the number of operations there are in series [Makower & Williamson, 1975]. In the language of supply management, this means that the more links there are in the chain, the longer the throughput time and the greater the capital tied up in stocks. Both these factors work against the consumers' interest, for it is they who have to do the waiting and they who pick up the bill. And it is a matter of indifference to them where in the supply hierarchy that stock resides or how it is described, the finished goods of one tier being the raw material of the next; ultimately all consumer goods are work-in-progress, from the moment they are extracted from Nature to the moment they are retailed.

A second tenet of queueing theory is that congestion is caused as much by the variance in arrival and process rates as by their mean values. In the manufacturing world, process times are relatively constant, which means that the variance in arrival pattern to each successive process reduces with its distance down the supply chain [Bedwell, August 1995]. The greatest source of variance, and therefore of long lead times and large stocks, is thus the collectivity of the consumers themselves; because they place orders at mutually independent times, the most general model of their aggregated arrival pattern is the Poisson distribution. This has a variance as large as its mean, a characteristic exhibited by no other practical realisation of other statistical distributions. *A priori*, it is thus the retailer who has to carry both the greatest amount of stock and also a capacity much bigger than its mean demand. Responses to this are to integrate horizontally so as to restrict consumers' choice, and to integrate downwards to include activities, in addition to retailing, on which a flexible workforce and other resources can be re-deployed outside peak sales periods.

Since in the above analysis transport and warehousing constitute processes just as much as manufacturing operations, the following model of supply chains in a market economy should logically follow: an oligopoly of large first-tier firms which not only retail but also operate some of the final labour-intensive processes. This would leave the more capital-intensive processes to the second- tier firms, who would in turn purchase from lower-tier companies adding relatively little value. Since at this level there would be few barriers to market entry, there will be many such suppliers competing for the business of the few. In Economics this is formally defined as monopsony, which has been identified as an essential feature of Japancsc supply management[Bratton, 1992].

THE MODEL AND THE REALITY

While the above model makes a complete representation of no identified industry, it does make an instructive framework for comparison. The UK garment industry, for example, is famous for its one dominant retailer, but notorious in the second tier for the continuing existence of sweatshops that transform cloth into clothing. Because of the inherent economies of scale of fibre production, however, it is the third tier of firms who are few and large. In Western economies groceries are again dominated by a few major retailers whose supermarkets have all but extinguished the butcher, baker and greengrocer of earlier times; the reasons are again readily explained by queueing theory, and hinge on the flexibility of labour. While there is little integration with food processors, 'own brand' being an attribute of marketing rather than production, there is ample evidence, even in the literature based on the capitalist ethic, of the monopsonistic exploitation of third-world commodity farmers. [OECD, 1992]

It is from the supermarket that Toyota are reputed to have drawn inspiration for what was to become the exemplar of lean supply, namely the automotive industry [Bratton, 1992 p25]. Here, as the model predicts, horizontal integration is universally apparent from the ever-diminishing number of marques. There are also signs of upwards vertical integration into retailing, a probable precedent having been created by Daewoo's recent direct-selling innovation; in any case the notionally independent showroom concessionaire is already a *de facto* agent [ICC The New Car Industry, 1996 p118]. Yet in automotive manufacturing the trend in the last three decades has been against downwards integration; this is symbolised by the disappearance of the foundry which formerly dominated Ford's Dagenham works, and by that company's declared intent to emulate the Japanese example of delegating to their suppliers all but 20% of the value added to their cars. The reasons for this are less explicit, but among them is the fixed cost of developing sub-assemblies like brake systems; many writers have also identified the power of organised labour, which in the UK of the 60s manifested itself most prominently in demarcation disputes [Jones, 1988].

Thus it seems that the original model, for all its simplicity, can be developed to explain the broad nature of the chains supplying our major consumer products. Such explanation needs to include the supply and demand of labour, monopoly and monopsony power, and the economy of scale in the acquisition of intellectual property. It then emerges that lean supply, as opposed to lean production, arises as a logical outcome of these traditional economic parameters, and not as some exogenous 'Philosophy'. While there are no apparent constraints on any company minimising its work-in-progress by adopting lean production, to become a lean supplier in terms of raw materials and/or finished goods pre-supposes certain conditions: a firm needs to avoid retailing, but must offer one or both of two other attributes: it must own exclusive, value-adding intellectual property, and/or it must be big. For two reasons, only the parameter of size is further examined here: first, it lends itself more readily to measurement, and second, it figures more heavily in the relevant literature, [eg Oliver & Wilkinson,1992]

SMALL FIRMS, BIG STOCKS?

The stock/revenue ratio, or its reciprocal stock turn, has been widely identified as "a critical factor for success" in lean supply [MIRA, 1994, p14]. Accordingly, the hypothesis adopted in the author's previous papers has been that, for independent firms in a given supply chain, the stocks/revenue ratio is negatively correlated with revenue [Bedwell, 1994 & April 1995]. The outcome has been a cautious 'yes' in answer to the above question.

The caution is ascribed to two sources of doubt: (1) about the *de facto* independence of *de jure* separate firms, many auto-assemblers having a large if not controlling share in manufacturers of related components, and (2) whether the sample of firms for which financial statistics are publicly available is demonstrably representative of the whole population. Here, as in the 1995 paper, (1) is controlled by rejecting from the sample any firms shown in the financial directories to be the subsidiaries of a holding company.

However (2) remains problematic, the most fundamental reason for which is that UK company law protects small companies - broadly those with less than £2.8m turnover - from making that turnover public [Companies House, 1995]. It is argued in the Appendix that more than half of automotive component suppliers are thus excluded, a conclusion which is compatible with MIRA's estimate that in European automotive production as a whole, 70% of the turnover is by the biggest 10%, but the number of employees in the smallest 80% is fewer than 100 each - i.e. where the revenue rarely exceeds £5m [MIRA, 1994, pp19, 23] Additionally, the compilers of commercially available sources of statistics impose their own criteria for selection; ICC, for example, consider only about 140 'leaders', one criterion for which is to turn over more than £10 m/year. Yet ICC has estimated the total number of such companies as about 4000 (ICC, 1996, p442). Accordingly, for the analysis which follows, recourse was made to the original source material publicly available from Companies House and conveniently recorded in the FAME computerised information retrieval system; the sample firms were selected by a Boolean search using the following key words: *Automotive AND Components AND Type of Company = Private* [Bureau Van Dijk, 1996].

Stock/Revenue
Years %

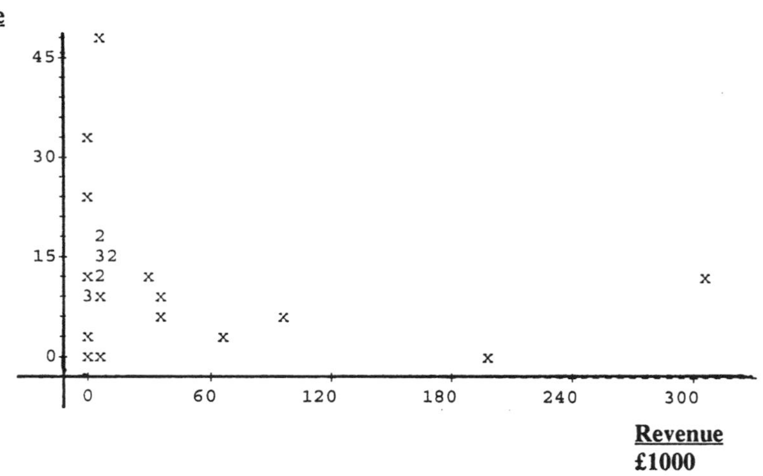

The result of plotting the stock/turnover ratios of the resultant firms is shown above, and does support the hypothesised negative correlation. Further, the mean value of stock/revenue for these firms is significantly greater than the figure of 8.6% calculated by ICC for their sample of 140 much larger organizations. Finally, ICC give 18.5% as the corresponding mean for the auto-assemblers themselves; as predicted by the discussion on retailing liabilities, this is significantly greater than the mean for their direct and indirect suppliers as a whole.

However, a number of reservations remain:-
i) The correlation is only to the 75% confidence level (25df, r = -0.15, t= 0.76, single-tail test)
ii) As shown in the Appendix, the sample is of much smaller size than would be expected from ICC commentary, and contains fewer smaller companies than would be needed to represent the target population correctly.
iii) Contrary to the hypothesis, ICC show that the five dominant British first-tier automotive suppliers, each with turnovers greater than £1000m, have stock/revenue ratios greater than the mean of the other 135 firms covered in their survey. This is perhaps explained by the finding of the Society of Motor Manufacturers and Traders, quoted by ICC: while 84% of the first tier suppliers deliver JIT, only two-thirds of them enjoy JIT supply of their own bought-out components [ICC, 1996, p449].

CONCLUSIONS

1. There is some evidence to support the hypothesis that, apart from the major first-tier suppliers and the car makers themselves, the stock/ revenue ratios of automotive component manufacturers is negatively correlated with their size. But that evidence is not admissible under the established rules of statistical sufficiency.

2. However, neither do the statistics provide evidence to depart from the widely held view that, in the words of one authority "...the larger motor manufacturers are indeed forcing their suppliers to 'eat inventory'" [Oliver & Wilkinson, 1992, p199]. This liability seems to be shouldered not only by small firms but also, surprisingly, by the biggest of the first-tier manufacturers. This runs contrary to the writer's previous speculation that "a relatively few, large and sophisticated systems manufacturers are exerting increasingly oligopolistic power over their car-building customers, and increasingly monopsonistic power over their more numerous but smaller component-making suppliers" [Bedwell, 1994]

3. There is difficulty in establishing reliable data about small and medium enterprises, especially those protected by company disclosure legislation; this is probably why most analyses of lean supply "have focused on..the major OEMs and...first-tier suppliers" [Oliver & Wilkinson, 1992 p 201]. In its references to small firms, the more evangelical literature on lean supply is rife with generalisations and assumptions; before accepting these, we should cross-examine their authors to see how they have addressed this fundamental problem.

REFERENCES

Bedwell, M. Predicting the Bottlenecks, *Proceedings of the 13th International Conference on Production Research*, Jerusalem, August 1995,pp103-5

Bedwell, M. Reduced stocks - whose? *5th International EAEC Congress; Proceedings of the Conference on Lean Product Development and Manufacture.* Strasbourg, 21-23 June 1995

Bedwell, M. JIT and Monopsony. *Proceedings of the Conference on Lean/Agile Manufacturing, 27th ISATA,* Aachen, 1994.

Bratton, J. *Japanization at Work*, Macmillan 1992, p32

Companies House. *Disclosure Requirements.* Leaflet CHN19 of 8/95,Companies House, Central Library, Chamberlain Square, Birmingham B3, August 1995

Eberts R & C, *The Myths of Japanese Quality*, Prentice Hall, ISBN 0-13-180803-6, 1995

FAME - *Financial Analysis Made Easy.* Information Retrieval system by Bureau Van Dijk Ltd., 1 Great Scotland Yard, London SW1A 2HN, 1996

ICC Information Group. Business Ratio Report *Motor Components & Accessories Manufacturers*, 24th edition, ISBN 1-85037-973-4, 1996.

ICC Information Group. Business Ratio Report *The New Car Industry*, 24th edition, ISSN 1358-2127, 1996.

Jones, B. Work and flexible automation in Britain: a review of developments and possibilities, *Work, Employment and Society, 2, 451-86*, 1988

Makower & Williamson, *Operational Research*, Hodder & Stoughton, 3rd Edition, 1975, p94

MIRA (Motor Industry Research Association). *Europe's Automotive Components Suppliers*, Vols I and II (1994).

OECD. *Declaration & Decisions on International Investment and Multinational Enterprises*, OECD, 1992.

Oliver, N & Wilkinson, B. *The Japanization of British Industry*.ISBN 0-631-18676-X Blackwell, 2nd Ed 1992

APPENDIX

We assume the maximum turnover x of the smallest fraction y of firms in a given supply chain to be related by a Pareto distribution of the geometric form

$$y = 1\text{-}A^{-x/X},$$

where A is a constant and X the mean turnover. Substituting the estimates from the ICC publication *Motor Component & Accessory Manufacturers, 1996, p442* that the total population of such firms is about 4000, of which 140 have turnovers greater than £10m (p442) yields

$$y = .61 \text{ where } x = £2.8m,$$

i.e., about 60% of automotive component firms are protected by the rules of disclosure from revealing their profit & loss accounts, and so from having their stock:revenue ratios calculated. We would, however, expect such data to be available for the

$$\{4000\ (1\text{-}0.61) - 140\} = 1820$$

firms whose turnovers are greater than £2.8m but smaller than £10m, in addition to the 140 'leaders' analyzed by ICC.

SIMULATION AND COST ACCOUNTING

A. Redlein, R. Rohrhofer, G. Schildt
Department of Automation
Vienna University of Technology
Treitlgasse 1, A-1040 Vienna, Austria

ABSTRACT: The proof of efficiency of capital-intensive technologies is rather difficult. Classical efficiency calculations depend on clear information about the stream of inflows and outflows. For determining the economic and organizational impact of flexible automation systems we use computer simulation and process-related cost accounting.

First we draw up a computer model of the planned production system. As a next step we create a forecast for the future production based on recent sales figures and on a market analysis. The results of this forecast provide the data to run the computer simulation. The results of the simulation supply the input figures for precise cost accounting that keeps overheads small. These results provide figures on the expected returns.

As the considered alternatives typically are quite similar to each other, the evaluation can usually be reduced to a parameter variation of one model. Parameter variation can also be used to optimize a selected alternative.

PROBLEM

The proof of efficiency of capital-intensive automation systems is very difficult as a static cost comparison is usually not sufficiently reliable. It is also a problem to create an objective prognosis of returns and savings which is a precondition for the application of dynamic investment analyses like the net present value method. In addition the quantification of qualitative advantages and disadvantages is problematic.

Usually a variation in the use of resources leads to varying production results (including different costs and benefits) which correspond to decision alternatives. A change in the system configuration may lead to different personnel requirements or may influence the rate of production throughput. This in turn may have a significant impact on the structure of costs.

SOLUTION

Computer simulation is a powerful tool to show the economic and organizational impact of these complex systems and it offers the opportunity to investigate different alternatives. By using a computer model analysts can identify the possible return related to a specific amount of resources used. The quantity figures (e.g. number of pieces, throughput time) that are supplied by the simulation can then be converted into cost figures by applying process-related cost accounting.

Simulation

This method should only be used when dealing with complex systems, as it requires a lot of time. Computer simulation employs a simplified image of reality to gain additional information about the underlying real system.

Depending on the degree of description its results show:

- the change of capacity utilization
- the reduction of the total processing time
- resource utilization, employment costs, maintenance costs, costs of plant and machinery
- possible errors and problems
- changes in the process timing of the automated system compared to the conventional production process
- organizational changes.

By varying the process parameters of the computer model, the system configuration can be optimized. Effects and problems of a planned production line can thus be explored before the automation system is even built. It is also possible to vary the parameters of existing production systems without the risk of actually reducing production performance. Another advantage of using computer simulation is the fact that the period under review can easily be extended without accumulating additional costs.

A thorough computer simulation requires a detailed collection of data prior to the creation of the model that covers the following elements:

- *Operational sequence*
  The model of the process reflects the sequence and timing of the various phases of work flow.

  The first step is the analysis of the process structure by dividing the process into distinct segments. This can be done by drawing process charts (e.g. figure 1).

main process

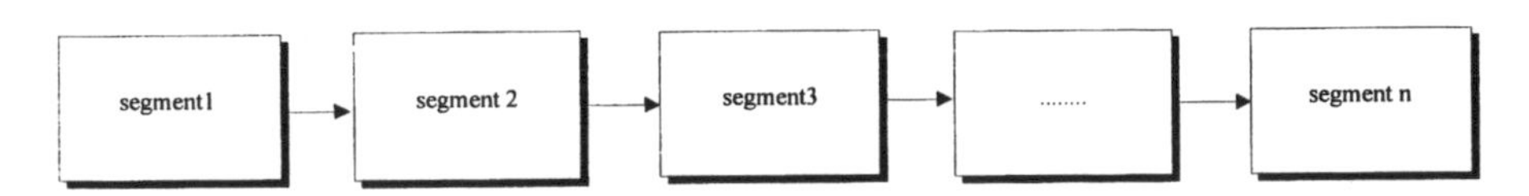

Figure 1: main process structure

The next step is to gather timing information and interaction rules for the production phases. The required information includes processing and transfer times as well as arrival intervals of input goods (external data). These parameters can be constant or variable, in the second case the values can be approximated by applying a probability distribution model.

- *Input times and distribution of final products*
  A market analysis supplies the possible sales figures corresponding with the pieces of the final product to be produced. Using this information, the input times of the raw materials and semi-manufactured goods can be evaluated.

- *Required manipulation time per production unit for each step of the process*
  For this purpose a probability distribution consistent with the measured times has to be found. It is determined by the average and the deviation of the manipulation time.

- *Required resources*
  The number of machines, their processing capacity, required licenses and personnel requirements must be determined.

- *Auxiliary material, expendable supplies and energy requirements*

- *Data on life span*
  These data provide information on the frequency of malfunctions and service intervals which in turn show the downtimes of the plant units. This again leads to the required number of service and maintenance personnel.
  Examples of life span data are:

  - MTBF (Mean Time Between Failure)
  - MTTR (Mean Time To Repair)
  - availability information

- *Potential for improvements*
  The potential of the system that has not yet been fully exhausted and its impact on the process have to be listed here.

Process-Related Cost Accounting

This form of cost accounting uses the results of the simulation or a mathematical calculation relating them to cost information. It provides the relation of costs and benefits for a specific process configuration and compares the cost development over time of the automation system (planned system) with the expenses required for the conventional production system (present system). For this purpose it is necessary to cut down the percentage of indirect costs in order to correlate the costs with the correct process units.

To implement this form of cost accounting, the following data have to be collected:

- *Employment costs*
  The personnel requirements and the salaries have to be established. It is important to consider that a variation of the simulation parameters may lead to a change of these

factors. Therefore these factors have to be determined for each new set of simulation parameters.

- *Agreements regarding employment*
  Information on salaries and wages as well as business and working hours is determined. Arrangements concerning overtime (with or without payment) and/or time compensation schemes must be taken into account because these data might influence the extension of daily working hours and the structure of costs.

- *Costs of resources in relation to time unit or produced unit*
  The representation of these costs employing a probability distribution requires the calculation of the average and the deviation. The expenses for factory supplies and expendable supplies must be considered as well as power supplies (connection load) and power consumption.

- *Acquisition costs for machinery*
  The acquisition costs and the incidental expenses (e.g. shipment costs) have to be listed for each individual machine.

- *Finance charges for the provision of required funds*
  The type of financing has an influence on the result, as the cost burden varies with respect to:

  - cash sale
  - sale on credit
  - leasing
  - payment by instalment

  The terms of payment (the sequence and the amount of individual payments) can have considerable influence on the profitability of the project.

- *Life span data*
  Costs of repair and maintenance equal the payments for service and maintenance agreements or can be derived from the data on frequency and duration of downtimes.

- *Possibilities to sell used machines*
  If plant units can be sold after the estimated working life, the resulting residual has to be considered with the investment costs.

PRACTICAL USE

In order to be able to evaluate a flexible automation system by means of computer simulation and process-related cost accounting, several steps have to be carried out:

- an analysis of the system's structure and a breakdown of the process into units,
- acquisition of the required parameters per unit (e.g. manipulation times, probability of interruptions of the production process due to malfunctions and maintenance),
- personnel requirements have to be determined,
- energy requirements must be listed,

- creation of a simulation model based on a modular concept representing the distinct plant units which can be adapted by varying a set of parameters,
- a survey of related costs,
- these costs have to be broken down in fixed and variable parts so that they can be used in process-related cost accounting,
- simulation of the operation of the flexible production system,
- evaluation of the various approaches by feeding simulation results into process-related cost accounting calculations and thereby converting quantitative figures into cost figures.

PRACTICAL EXAMPLE

Underlying Production Process

As a first step, the quality of the goods is checked. The goods are classified and marked as belonging to one of three quality classes A, B and C with A being the highest level. This is being done simultaneously in several parallel inspection points.

The goods from each of these stations are put on conveyor belts, one for each station. These belts lead to a main conveyor belt at the end of which the different quality classes are separated for further processing.

Technical Parameters

At this plant, the geometric configuration and interaction of the units cannot be changed. Only few parameters, especially the timing scheme, can be adapted. Some changes can be done with respect to short-term or long-term planning.

- *Input rate*
  The input rate and the distribution of the incoming goods is rather easy to adapt for short-term projects. This parameter has great influence on the rate of capacity utilization and therefore on the occurrence of bottlenecks.

- *Manual treatment*
  In our example, process goods of quality C have to be processed manually. The number of workers required for this part of the process has to be determined by the simulation.

- *Design changes*
  The replacement of existing processing units with machinery which has a different function and yields an overall performance improvement (e.g. reducing manual manipulation, increasing input rate and throughput) is a long-term option. The related investment costs are likely to increase whereas running costs can decrease. Several alternative designs have to be evaluated.

Process Breakdown

The process-oriented first stage of the classification is shown in figure 2.

process structure

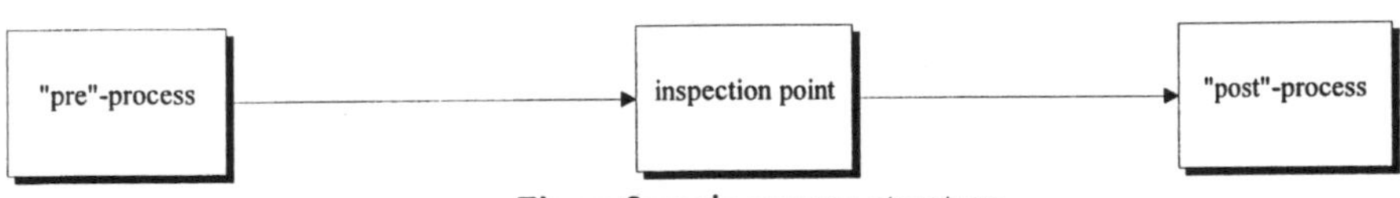

Figure 2: main process structure

Breaking down the post-process results in the process chart shown in figure 3.

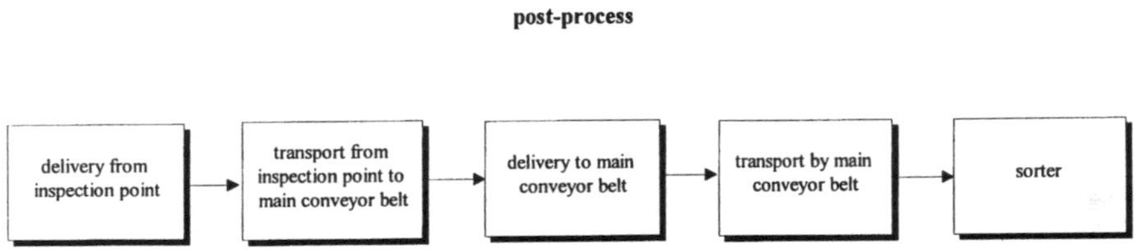

post-process

delivery from inspection point → transport from inspection point to main conveyor belt → delivery to main conveyor belt → transport by main conveyor belt → sorter

Figure 3: post process structure

Based on the process units shown in figure 3, the modular-design elements can be determined:

1. inspection points
2. subordinate conveyor belt
3. main conveyor belt
4. sorter

In addition to the parameters of these elements their links have to be taken into account. Based on these results, the simulation model can be set up.

Setting Up the Simulation Model

In this example, we use ARENA, a simulation software with a graphical user interface, to set up the computer model and to perform the simulation. ARENA translates rules that are described in process charts into the simulation language SIMAN and then executes the derived code. The progress of the simulated processes can be viewed on screen in a graphical representation. In our example, we can create schematic drawings of the plant units where images of the goods move according to the rules set up in the process charts.

Creating Process-Related Cost Accounting

Cost accounting has to provide information on the costs of the underlying process that are related to a specific period of time. Therefore, the following costs have to be identified.

Identification of Costs

Employment costs The employment costs can be divided into two ranges: The first range covers activities that are related to a specific station or person. In our example, the work at the inspection points and the supervision of the whole plant belong to this range. The second range covers work that is not permanently related to a specific person. An example of this range is the manual treatment of goods of quality C, as it is done occasionally, and no specific worker is assigned to it permanently.

To establish employment costs for all these activities, certain questions have to be answered, e.g.

- What is the time an employee needs in order to accomplish a certain task?
- How is this factor distributed over the time under consideration?

These questions have to be answered keeping in mind that personnel requirements involve long-term decisions. The expenditure of labour that has been calculated or is a result of the simulation has to be translated into personnel requirements, and - considering the required qualifications - this in turn leads to personnel costs.

In this case we need one person for each inspection point. The expenditure of time required for manual treatment can be derived from the simulation.

Plant acquisition costs The acquisition costs of the plant serve as a basis to calculate the annual depreciation based on an assumption of the effective life span of the plant. Applying a distribution of expenses that corresponds with the benefits and assuming a constant plant utilization rate, a linear depreciation model can be used. In this case the annual depreciation expense is the acquisition costs, reduced by the expected residual, divided by the effective life span.

In the case of fluctuating output the acquisition costs should be related to the number of pieces produced. In our case the acquisition costs are fixed, we use a linear depreciation model assuming an effective life span of five years.

The types of financing have to be considered as they lead to different modes of payment. In order to compare them, they are discounted with a local interest rate, depending on the purpose of cost accounting.

Expendable supplies

- Electricity
  The power consumption can be calculated from the connected load of the individual machines.
- Compressed air
  Due to the relation of the costs of compressed air to other costs, a rough approximation of the consumed air is sufficient in this system.

Charge material In our simple example, each of the goods gets a label fixed to it to identify its quality class. The costs of these labels have to be listed here.

Maintenance costs The maintenance costs have to be estimated based on the service intervals that have been calculated from the life span data or are a result of the simulation. The risk of fluctuating maintenance costs can be avoided by long-term service and maintenance agreements.

Cost Breakdown

The costs that have been identified in the previous section now have to be calculated for a specific period. It is important to find out which costs depend on the plant's actual output, thus identifying constant and variable costs. In the case of variable costs the reference unit (the parameter that the costs are proportional to) has to be listed.

Employment costs Personnel costs are fixed costs because salaried employees have to be constantly paid, even in a zero output situation. If there is the possibility to engage casual workers, their costs are variable.

Plant acquisition costs The depreciation of the machinery are fixed costs as the utilisation is usually not related to the plant's output (e.g. conveyor belts with light load). Other financing costs of the plant are also fixed costs.

Expendable supplies Energy costs are variable costs, the reference unit is the working time. The air consumption depends on the output.

Charge material It is evident that the costs for the charge material depend on the quantity of goods processed and are therefore variable costs.

Maintenance costs In general service and maintenance costs are variable costs. However, in case of service and maintenance agreements we have fixed costs, as the fee is independent of the actual maintenance intervals.

Implementation of Process-Related Cost Accounting

We use the spreadsheet program MS Excel to implement process-related cost accounting. We have created an interface to the simulation system. The interface consists of an Excel macro that reads the file with the result of the simulation and converts the data into a form that can be processed by Excel.

Analysis Using Process-Related Costs Accounting

The simulation run provides the data shown in figure 4.

These data show:

- the number of goods processed (Gesamteingang)
- the number of goods of each quality class (Gesamtausgang, Anzahl_B, Anzahl_C)
- the downtimes of the previous units (Sorter*_Aus; no errors occurred in this run)
- the downtimes of the simulated plant units (AUSGEBUCHT_**; no errors occurred in this run)
- the downtimes of the following units (Nachanlagen; no errors occurred in this run)
- the throughput time of the goods

SIMAN V - License #9510030
Systems Modeling Corporation
Summary for Replication 1 of 1

Project: Example   Run execution date : 27/ 2/1997
Analyst: Redlein/Rohrhofer   Model revision date: 27/ 2/1997
Replication ended at time: 1573.48

TALLY VARIABLES

Identifier	Average	Variation	Minimum	Maximum	Observations
Ankunftsrate	189.06	1.9031	50.406	1079.3	8
Ausstoss_Sorter_1	38.787	.05230	34.588	42.094	40
Ausstoss_Sorter_2	17.928	.04293	16.808	19.270	87
Ausstoss_Sorter_3	41.369	.10496	34.853	49.678	37
Ausstoss_Sorter_4	21.086	.06450	18.712	23.144	74
Ausstoss_Sorter_5	18.625	.05798	16.829	20.953	84
Sorter1_Aus	--	--	--	--	0
Sorter2_Aus	--	--	--	--	0
Sorter3_Aus	--	--	--	--	0
Sorter4_Aus	--	--	--	--	0
Sorter5_Aus	--	--	--	--	0
Sorter6_Aus	--	--	--	--	0
Sorter7_Aus	--	--	--	--	0
Sorter8_Aus	--	--	--	--	0
Sorter9_Aus	--	--	--	--	0
Sorter10_Aus	--	--	--	--	0

COUNTERS

Identifier	Count	Limit
Gesamteingang	322	Infinite
Anzahl_B	64	Infinite
Anzahl_C	7	Infinite
Nachanlagen	0	Infinite
Gesamtausgang	80	Infinite
Pufferband_	0	Infinite

OUTPUTS

Identifier	Value
AUSGEBUCHT_FEEDER	.00000
AUSGEBUCHT_HTB	.00000
AUSGEBUCHT_MESS	.00000
AUSGEBUCHT_EINLAUF	.00000
AUSGEBUCHT_STAPLER	.00000
AUSGEBUCHT_BF	.00000

Execution time: 1.25 minutes.
Simulation run complete.

Figure 4: Output of a simulation run

The results of the simulation provide the following expenditures:

- The number of goods of quality C is the basis to determine the personnel requirements for manual treatment.
- The number of goods processed and the throughput time determine employment costs.
- The throughput time and the connection load amount to the total power consumption and therefore the energy costs.
- The machine downtimes and the maintenance intervals determine the size of the service and maintenance personnel, from which the personnel costs can be derived.
- The demand on spare parts and expendable parts can also be derived from these data.

Adding up these costs we get the overall costs for the examined period of time. These costs can now be compared to the costs of other alternative plant configurations providing a profound basis for decisions.

Advantages of the Combination of Simulation and Process-Related Cost Accounting

This method makes it possible to evaluate a non-existing plant through simulation over a period of time with respect to:

- throughput times
- costs
- problems
- possible shortcomings
- improvements

As a result it is possible to compare the costs of an existing system to the costs of alternative models and to opt for the most effective economic solution.

For the construction of plants this simulation model provides more information about the specific characteristics of a production process and about the effects of parameter changes on the cost structure. In addition it is possible to learn more about detrimental effects of system configurations and disproportionate cost drivers.

With regard to customer information this simulation model makes it possible for the prospective buyer to evaluate the advantages of his investment. This also helps to raise consultation standards on the part of the seller. Through simulation a realistic impression of a prospective investment can be provided and thus future disappointments can be avoided.

PRODUCT COSTING FOR AUTOMATED FACTORIES

Richard Giglio
Department of Mechanical and Industrial Engineering
University of Massachusetts, Amherst, MA 01002

ABSTRACT: New demands are being made on product costing methodologies by the automation of both design procedures and manufacturing processes. The former requires costing methods which are integrated with CAD systems, while manufacturing automation invalidates traditional costing methods. Recent developments such as Activity Based Costing (ABC) improve costing accuracy but have limitations of their own. We develop a procedure for correcting ABC's primary deficiency, its neglect of capacity costs. A framework is developed for identifying manufacturing environments where ABC's limitations could lead to distortions, and a model is developed to properly account for capacity costs. Implementing this model requires the involvement of nearly all functions within a company, and these changes in a firm's organizational structure are described.

1. BACKGROUND

Computer technology has changed both the methods for designing products and the way in which they are manufactured. With regard to manufacturing methods, computer technology fosters a high degrees of automation which, in turn, decreases the amount of direct labor required while increasing investments in facilities, design and planning. Where direct labor and materials once constituted up to 80% of the cost of many manufactured items, it has now shrunk to as low as 10%-20% for many products such as electronics and consumer products which rely on stamping and injection molding for most components.

Computer technology is also changing the way products are designed. The development of integrated CAD/CAM systems permit both performance analysis and manufacturing consequences to be studied while the designer is determining the features and materials of the product.[1] One natural aspect of these design tools is cost estimators which enable the designer to determine the approximate cost of a particular design early in the creative process, and to examine the economic implications of a variety of designs.[2]

Firms estimate the cost of producing an item to guide decisions: whether or not to make the item, pricing, make-vs.-buy, investments in production facilities. Economic decisions affect the future with today's choices determining tomorrow's income streams. Traditional costing techniques, however, rely on past or current data. When new investment is involved, cost estimation can be circular and iterative. Sales are estimated based on selling price which, in turn, is constrained by production costs. Yet, because of fixed capacity charges, production costs can vary significantly with sales volume which is a function of price. Although costing techniques have improved dramatically in the past decade and they are being automated in many product designer systems, important issues are still not properly addressed.

If per-unit costs are unreliable and fraught with the possibility of misinterpretation, why are they so widely and often unquestioningly used? One reason is that lifelong "training" has conditioned individuals to ask "what will it cost to make this product?", as if that were a question which could reasonably be answered. From childhood on we are faced with economic decisions which involve determining the costs of items, be they lollipops or automobiles. It is therefore not surprising that decision makers feel comfortable asking what it will cost to manufacture an item, even though that may not be a meaningful question unless sales volumes, the product's future capacity utilization and a host of other issues are examined, considerations which most current costing techniques do not address.

A theoretically correct way to choose between economic alternatives is to compare the cash flows which would arise from each alternative or combination of alternatives. These cash flows include investments in plant and equipment, operating costs, increased income from those investments and lost sales from erosion of existing product lines. Although this "with and without" (WW) analyses may be theoretically correct, conducting such analyses is often cumbersome because of difficulties in estimating data and the vast number of combinations of decision alternatives. In particular, the accepted way of conducting WW analyses are not compatible with automated cost modelers. At best, WW analyses are unwieldy and are generally used only for major investment decisions. In many instances estimated per-unit costs which are reasonably accurate under specified conditions often provide a much simpler decision-making framework than WW analyses.

In the past, product costs were determined using cost accounting systems which by their very nature are retrospective and often not suited to guiding decisions about the future. Cost accounting has always struggled with the allocation of indirect costs and this deficiency has become glaring as the ratio of direct to indirect costs has shrunk due to automation. The inadequacies of traditional cost accounting is well documented by Cooper and Kaplan.[3-6] Many authors propose alternative approaches, the most popular being Activity Based Costing (ABC).[7-9] Where appropriate, ABC cost modelers can be imbedded in CAD-based cost modelers systems by employing the expressions developed in Section 2. in place of cost accounting information.

Activity based costing offers a significant improvement to cost accounting by identifying those activities which are the main sources of costs, and employing those activities in cost estimation, rather than allocating overhead expenses by the usual cost-accounting vehicles. ABC assumes that the costs are separable and divisible into transactions. ABC assumes that the cost is linear in activities and that nearly all costs can be accounted for by a transaction-based allocation methodology. Neither opportunity costs due to idle capacities nor the costs of possible additional capacity are considered.

Activity based costing requires the collection of accurate data on direct labor and materials costs followed by an examination of the demands made by particular products on indirect resources. The examination of indirect resources is guided by principles which encourage a focus on expensive resources, on resources where consumption varies significantly by product type, and on resources which are not correlated with the traditional allocation measures, direct labor processing time and material costs. The ABC process guided by these principles is designed to trace costs from resources to activities and then from activities to specific products.

Because ABC drops the fiction that indirect costs are related only to direct labor and direct material costs, incorporating ABC concepts in a cost modeler requires more information and a more global view of the product life cycle than traditional methods. Yet, it is feasible and practical to estimate the required information which can then imbedded in a costing procedures using the expressions of Section 2. ABC can dramatically improve decision-making under certain conditions and there have, in fact, been many successful implementations as reported by Jeans & Morrow, Foster & Gupta, O'Guin and Bailey.[10-13] However, the initial euphoria over the promise of activity based costing has diminished. In many situations the level of improvement from using ABC does not seem to make its implementation justifiable. Such situations are discussed by Roth & Borthick and Piper & Walley.[14,15] Protagonists of ABC generally attribute the causes of such failures to poor system implementation.[16] We believe that in many instances the problems with activity based costing are more attributable to limitations of its conceptual framework than to ineffective implementation. Most important of the limitations is the disregard of capacity and investment considerations. Cost modelers, therefore, must be able to handle capacity considerations when appropriate.

We propose that the term "cost" should never be used unqualified, but that it should be modified according to conditions existing when the decision is to be implemented. We call this paradigm "dynamic costing" because the procedure for developing cost estimates will change in a well defined way as conditions in the firm change. Incorporating this knowledge in a cost modeler requires additional effort, but that effort forces the forward integration of planning to consider pricing and production runs, considerations which should not be ignored when substantial automation exists.

This paper proceeds with a formalized statement of ABC because when capacity is not an issue, ABC is sufficient for product costing. . It then presents a framework for classifying costing situations. Next, procedures are developed for augmenting ABC in situations where the cost of capacity should be considered. Finally implementation issues are discussed.

2. A FORMALIZED STATEMENT OF ACTIVITY BASED COSTING

Following the conceptual framework described by Cooper & Kaplan[9] we develop an analytical representation of the existing form of activity based costing, employing the following notation:

Consider a plant producing n products, denoted by i = 1 to n.

N_i : the planned batch size of the ith product.

At the factory or corporate level, there are k categories of separable expenses denoted by j=1 to k. These expenses represent such "overhead" items as utilities, corporate administration and the cost of the purchasing department.

p: ABC activities indexed by r=1 to p.

C_j: the total of the jth category of factory-or-corporate-wide expenses. These expenditures are needed to accomplish the p activities.

C_{jr}: the cost per unit of the rth activity of the jth factory and corporate expense such that:

$$\sum_{r=1}^{p} C_{jr} = C_j$$

W : the total direct labor hours of the plant per year.

W_i : the amount of direct labor hours required for the production of N_i units of the ith product, such that :

$$\sum_{i=1}^{n} W_i = W$$

C_i^D : the cost of direct labor per hour for producing item i.

D : the total annual direct labor costs for the plant, so:

$$\sum_{i=1}^{n} C_i^D W_i = D$$

M : the total material purchasing costs of the plant annually.

M_i : the cost of the materials required for the production of N_i units of the ith product. Here we assume that M_i includes the material wastage costs due to over-ordering, scrap, etc, and that:

$$\sum_{i=1}^{n} M_i = M$$

X_{ijr} : the units of activity r required for the jth factory or corporate expense in the production of N_i units of item i.

The ABC cost, P_i, for producing N_i units of the ith product consists of three components, the material costs, the costs of direct labor and the costs of manufacturing overheads allocated according to activities: Thus, the ABC cost is:

$$P_i = \frac{M_i + W_i C_i^D + \sum_{j=1}^{k} \sum_{r=1}^{p} X_{ijr} C_{jr}}{N_i} \quad \forall\ i = 1 \text{ to } n \tag{2.1}$$

3. DYNAMIC COSTING

We propose that the procedure for developing cost estimates should change in a well defined way as conditions in the firm change. In general, four conditions provide important information regarding what type of costing procedure should be used:

1. Whether or not there is contention for capacity,
2. Whether the decision is for the short term so that plant capacity can't be changed, or for the long term so new capacity can be installed,
3. The nature of the industry--discrete vs. continuous,
4. The stage in the life cycle of the product of interest.

This paper develops costing models as a function of the first two categories. Subsequent papers will present models for continuous processes and models which employ the concepts of learning curves to further clarify the distinction between long-term and short-term decisions.

Including Capacity Costs In Activity Based Costing

There has been extensive academic literature on the computation and the uses of opportunity costs (Heymann & Bloom provide a comprehensive summary).[15] Activity based costing, however, makes no attempt to include these costs within its framework. Cooper & Kaplan [9] assert that opportunity costs should be treated as a separate line item, as a cost of the period, and not related to individual products. However, such a treatment could contribute to product cost distortions by implicitly charging non responsible products with this additional line item cost and more perniciously by not estimating the actual production costs of the products responsible for such opportunity costs.

All other things being equal, an ABC calculation would give the same figure for a product's cost whether or not the firm was about to run out of capacity or expected to have spare capacity for years to come. Perhaps, constructing new capacity would place a burden on the firm; perhaps the new capacity would enable the product to be made much more cheaply. In either case, a retrospective allocation of existing capacity costs is misleading. The manner in which capacity-related costs should be included depends not only on the type and amount of capacity used for the product being costed, but on the degree to which the capacity is either unique or sharable with other active products.

For non-sharable capacity, there is by definition no contention for its use, and capacity investment costs need not be considered because in the short run the decision to produce or not produce the item can not influence those sunk investment costs. Consequently, those sunk costs should not be included in any costing procedure designed to guide short term decisions.

For sharable capacity we need the concept of contention for capacity. Let:
P_i = total activity related (ABC) cost per unit of product i
T_i = total product cost including capacity costs
u_i = capacity usage per unit of product i
v = value of a unit of capacity.
v is not known beforehand but must be calculated.

(Note: one could add an index for type of capacity, but for simplicity that index is omitted)

$T_i = P_i + u_i v$ = total cost of a unit of product I (3.1)

The procedure for calculating a value for v will depends on whether decisions will be binding for short-term or long-term.

Decision Making for the Short Term.

The reason for developing cost estimates is to guide decisions. It is therefore necessary that cost estimates rank products correctly. For the short term, investment in capacity is a sunk cost. If that capacity is not fully utilized, there need be no opportunity cost associated with that capacity because ignoring capacity costs will not lead to the wrong decision by incorrectly ranking of products (To estimate *actual* rather than *relative* cost, one should use the long range procedure for costing capacity described in the next section).

If there is contention for capacity in the short run, one needs to calculate an opportunity cost associated with making product i as:

$$o_i = s_i - (P_i + u_i v), \quad (3.2)$$

where s_i is the selling price of product i. That is, the opportunity cost equals the foregone profit should the product not be made.

If we choose product i as the base line for comparison, then its opportunity cost is, by definition, zero so:

$o_i = 0 = s_i - (P_i + u_i v)$,
or
$v = (s_i - P_i) / u_i$ (3.3)

The maximum value of capacity is:

$$v_{max} = \max_i \ (s_i - P_i) / u_i . \quad (3.4)$$

Assume this maximum is attained with product i^1. One should produce all of product i^1 that the capacity will permit. If, after satisfying demand, there is more capacity, recalculate v_{max} finding I^2, etc. until for some product i^* there is no additional capacity, and the value of capacity is:

$$v = \frac{S_{I^*} - P_{I^*}}{u_{I^*}}$$

so

$$Ti = P_I + u_I \frac{S_{I^*} - P_{I^*}}{u_{I^*}} \quad (3.5)$$

In other words, *Compared* to product i^* the remaining products will lose money, even though those other products might still be profitable if there were additional capacity.

Decision-Making for the Longer Term

For decision-making for the longer term, relative, short-run costs will not suffice. Absolute costs are necessary because capacity can be expanded or, possibly, contracted. If capacity will not be fully used now and in the future, then there are no related opportunity costs. If, on the other hand, capacity will be fully utilized, then per-unit capacity costs need to be added to activity costs. Notation:

τ_j: the number of years in the future when it is projected that capacity of type j will need to be constructed.

I_j : the per unit cost of capacity of type j.

a : (1-discount factor for the company)

u_{Ij} : the utilization of capacity type j by product I

Then, v, the capacity costs to be added to the ABC costs for product i equal:

$$v = \sum_{j=1}^{J} e^{-a\tau_j} I_j u_{Ij} \quad (3.6)$$

4. IMPLEMENTING DYNAMIC COSTING SYSTEMS

Dynamic costing is an extension of activity-based costing. Consequently, any cost modeler must employ an algorithm which relates costs to a variety of cost drivers and not only direct labor and materials. The procedures for doing this have been well documented [7] although to date, few automated cost modelers employ them. The transition to ABC-based costing procedures, requires a change in philosophy and the involvement of additional people in the development of cost-activity relationships. But since ABC costs are associated with operations and not investments, involvement of strategic planners is not normally involved, and the change in philosophy is evolutionary not revolutionary. It is simply a matter of recognizing that there can be multiple cost drivers, and periodically reviewing processes to identify important drivers and to estimate the quantitative relationships between the drivers and expenditures. Implementing dynamic costing, however, is more difficult requiring the participation of additional people and levels of the organization.

Just as CAD/CAM systems expanded designers horizons by forcing them to consider manufacturing when developing a product design, dynamic costing requires the explicit consideration of market forces and investment policies when designing a product. Furthermore, the panning process must be iterative requiring the interactions between designers, marketers and investment planners early in the product design cycle.

Dynamic costing requires the determination of short-term and longer term demand for various production capacities. This, in turn, requires estimates of sales volumes as a function of selling prices and company policies regarding the relationship between selling prices and production costs. Furthermore, these estimates must be made for all products which share production facilities.

It will take a re-design of a organizational structure to implement procedures which permit the networking of information between different functions and data bases regarding sales projections, investment planning and new product introduction. These changes can not take place overnight, and further papers will propose strategies for accomplishing this transformation. However, even incremental changes to current procedures can be of significant benefit because current procedures often ignore the investment costs of production facilities and research and development expenditures, costs which are substantial and are becoming increasingly important.

REFERENCES

1. **Dong, Zoumin**, "Design for Automated Manufacturing", *Concurrent Engineering: Automation, tools, and Techniques*, Edited by Andrew Kusiak, Wiley, 1993.
2. **Fagade, A**. "Models of Cost Estimators for a Feature Based Sheet Metal Product Modeler", MS Thesis, Department of Mechanical and Industrial Engineering, University of Mass., Amherst.
3. **Kaplan, Robert S.**" One cost system isn't enough " *Harvard Business Review*, January-February 1988, pp. 61-66.
4. **Kaplan, Robert S**. " Limitations of Cost Accounting in Advanced Manufacturing Environments ", *Measures for Manufacturing Excellence*, Harvard Business School Press, 1990.
5. **Kaplan, Robert S**. " New Systems for Measurement and Control ", *The Engineering Economist*, Vol. 36, No 3, Spring 1991, pp. 201-218.

6. **Cooper, Robin and Kaplan, Robert S.**, " How Cost Accounting Distorts Product Costs", *Management Accounting*, April 1988, pp. 20-27.
7. **Cooper, R.** " The Rise of Activity Based Costing - Part Two : When Do I Need an Activity Based Cost System ? " *Journal of Cost Management for the Manufacturing Industry*, Winter 1989, pp. 34-36.
8. **Cooper, R.** " The Rise of Activity Based Costing - Part Four : What do Activity Based Cost Systems do ? " *Journal of Cost Management for the Manufacturing Industry*, Spring 1989, pp. 38-49.
9. **Cooper, Robin and Kaplan, Robert S.** " Measure costs right : make the right decisions ", *Harvard Business Review*, September -October 1988, pp. 96-103.
10. **Jeans, Mike and Morrow, Michael.** " The practicalities of using activity based costing", *Management Accounting*, November 1989, pp. 42-44.
11. **Foster, George and Gupta, Mahendra.** " *Activity Accounting : An Electronics Industry Implementation", Measures for Manufacturing Excellence*, Harvard Business School Press, 1990.
12. **O'Guinn Michael.** " Focus the factory with activity based costing ", *Management Accounting*, February 1990, pp. 36-41.
13. **Bailey, Jim.** " Implementation of ABC systems by UK companies ", *Management Accounting*, February 1991, pp. 30-32.
14. **Roth, Harold P. and Borthick, Faye A.**," Are you distorting costs by violating ABC assumptions ", *Management Accounting*, November 1991, pp. 39-42.
15. **Piper, J. and Walley, P.** " Testing ABC Logic ", *Management Accounting* September 1990, pp. 37-38.
16. **Cooper, R.** " Camelback Communications, Inc." Harvard Business School, Case 9-185-179, 1985.
17. **Heymann, H.G and Bloom, Robert.** " *Opportunity Costs in Finance and Accounting*", Quorum Books, New York 1990.

PERFORMANCE MEASUREMENT OF WORK TEAMS FOR COMPETITIVE MANUFACTURE

A. J. R. Smith[1] and M. Burwood[2]
[1]Department of Mechanical and Manufacturing Engineering,
The University of Melbourne;
[2]Formerly, graduate student, Melbourne Business School,
The University of Melbourne

ABSTRACT: Some companies implementing self-directed work teams and flatter, leaner company structures have demonstrated improved competitiveness; however, many have failed to achieve the expected levels of improvement, or have struggled to instil ongoing commitment to continuous improvement. Failure to maximize improvement from the newer programmes can be due to the continuing use of traditional accounting and reporting systems. These systems produce performance indicators that encourage managers to persist in their previous ways.

From case studies, it has become clear that the success of a measurement system is dependent on its relevance to company strategy and the needs of customers. However, the strategy must first exist in a form that can be easily communicated and understood by all employees. A check list, which combines desirable attributes listed in the literature and findings from the case studies, may be a useful diagnostic and developmental tool for performance measurement systems. Implementation may be assisted by a model based on PDCA cycles.

INTRODUCTION

There has been a realization in the last decade or so that Western manufacturing firms need to change their management practices to meet new competitive demands. During the 1980s manufacturing firms were shocked to lose market share to Asian products, in particular, which were produced at both high quality and low cost. These two dimensions were previously believed to be mutually exclusive. Manufacturing managers began to question many of the traditional beliefs and assumptions applied in their profession.

The above questioning has led to the introduction of a number of operations management programmes such as Just-In-Time, Statistical Process Control, Total Quality Management as well as the use of self-directed work teams and flatter, leaner company structures. All of the newer programmes question the traditional methods of organizing work in manufacturing organizations. However, performance often continues to be measured using measurement systems which were designed to value the traditional methods of scientific management with very clear contributions from a relatively large production and small support workforce.

A new management programme which benefits the company as a whole may actually cause existing measures of performance of individual departments to suffer. When the primary focus of a performance measurement system is on direct labour, say, adopting new management techniques often requires a `leap of faith'. Without this leap of faith there will be a lack of support for the programme from middle management. The likelihood of success will then be significantly reduced, therefore the development of an accurate and relevant performance measuring system is fundamental to any company wishing to change its culture[1].

Traditional systems were set up to deal with the critical issues of the early 20th century. In times of long production runs and labour intensive processes it was found that the most appropriate measures were those which focused on maximizing the effectiveness of labour. In those times the major issues for management were output and cost control. "The early management accounting measures were simple, but seemed to serve well the needs of owners and managers. They focused on conversion costs and produced summary measures such as cost per hour or cost per pound produced for each process and for each worker."[2]

The stock market crash of 1929 was the catalyst for tighter financial regulation in the form of national accounting standards, generally accepted accounting principles (GAAP) and organizations for controlling securities exchange. Companies have, since then, been forced to report their financial performance in very specific ways including balance sheets, profit and loss statements, etc. In addition to this, there has been the increasing complexity of taxation requirements.[3]

At the time that the external requirements were imposed internal requirements were still relatively simple. Information available from the external reporting system combined with managerial experience and engineering records was ample for internal decision making. Unfortunately, the external reporting function is extremely rigid.[3] A fear of auditing has led to a rigid adherence to GAAP. While products and processes have advanced through the decades becoming increasingly diverse and complex, financial information systems have remained set in the early twentieth century. Their major advances have been due to increased external regulation rather than internal requirements.

The two most glaring and oft reported failures of classical costing systems are in the treatment of overheads/burden/indirect costs and inventory. Indirect costs are traditionally apportioned to products (cost objects) on the basis of direct labour (or machine hours) required to produce the products. This completely distorts the situation in companies with varying levels of technology.[4] The matching principle applied within an accounting cycle results in work-in-progress and finished goods inventories being treated as current assets. The high carrying costs are simply made part of overheads and thus increase the reported worth of the inventory. This then encourages overproduction, and does not value inventory reduction programmes such as JIT.

Of even more fundamental concern is the tendency for traditional performance measurement systems to distract managers from considering the total company performance: "The emphasis on direct labour expense and direct labour efficiency causes companies to focus on a relatively unimportant factor of production".[5]

The weakness of relying on traditional financial measures for supporting internal decision making is clearly illustrated in the following case study.

CASE STUDY A

Company A is a manufacturer of chemical products for industrial, construction, automotive and general household applications. There are ten major product groups, and more than one thousand different products which collectively generate revenue of A$35 Million. Approximately 25% of products are exported to South-East Asia. The company is a small player in the total Australasian market, being significantly smaller than the two largest competitors. However, the company is either a market leader or major player in most of the market niches in which it competes.

The company has 220 employees divided into manufacturing staff (140), clerical and administrative staff (25), sales and marketing (35), research and development (10) and distribution (10). There is one manufacturing site in Australia with a distribution centre in each state capital.

The Australian company operates as a semi-autonomous business unit, reporting monthly financial results to its overseas parent. Most operational and expenditure decisions can be made by the Australian management.

Process Improvement

Company A is typical of many manufacturing firms. The company is trying to implement new strategies and tactics in order to generate long term growth and prosperity. A great deal of change has already taken place in the last few years and more is planned for the future.

Within the factory, the structure is now flatter. In the new structure the operators and Manufacturing Director have remained but the positions of Production Managers, Supervisors and Leading Hands have been blended into two layers: Section Managers (four off) and Team Leaders (fourteen off). All four layers in the new structure are required to be more flexible in terms of taking on more duties (both vertically and horizontally) than under the previous system. There is now, also, much less distinction between direct and indirect work responsibilities. The process has enabled significant reductions in factory overhead costs with twelve supervisory and administrative positions being removed, however, direct labour performance has dropped due to the extra duties being performed by team leaders.

Another part of the improvement initiatives has been to give process operators responsibility for doing most of their own quality control testing. There has been a minor dip in productivity in terms of total output per direct labour hour as a result but the savings from reductions in Quality Control staff are seen to more than offset this.

Multi-skilling was introduced as part of the first enterprise bargaining agreement in July 1994. The diverse nature of production processes and union demarcation made it difficult to move operators from one department to another. Training has been conducted to the extent that more than 70% of the work force is competent in two or more different jobs. It is generally considered that multi-skilling has been beneficial to the company. It is now much easier to transfer operators between departments to balance short term peaks and troughs in workload. The production supervisors interviewed felt that as many as seven extra operators would have been required if the current workforce were not multi-skilled.

A number of unsuccessful attempts have been made to introduce programmes for continuous improvement teams. In each case the programme started with teams being selected and projects identified. Some of the projects were completed, most were not. Some claims were made that supervisors hindered their subordinates from participating because of the effect on short term production efficiency.

Two unsuccessful attempts have been made to gain certification to ISO 9000 quality standards. Each attempt failed when the particular Quality Manager resigned. There have been four different Quality Managers in the past two years. Most managers interviewed could see the relevance of quality systems but were sceptical of their success at Company A.

Performance Measurement

Company A's factory has traditionally used a standard cost system with most management decisions being supported by the weekly and monthly variance reports. Each product has standards set for raw material and labour usage. Overheads are allocated per direct labour hour. These figures are used for determining product cost and therefore calculating gross margin. The primary instrument for measuring manufacturing performance is the weekly efficiency report. This is generated by the Costing Accountant and compares actual labour hours to standard labour hours (producing variances).

There are a number of monthly reports generated by the accounting system including a detailed 25 page variance analysis containing information regarding purchase price, material usage, labour rate, labour efficiency variances and volume variances. Monthly controllable costs are reported for each cost centre. Controllable costs include all direct costs which can be attributed to the cost centre other than raw materials and direct labour. A monthly productivity summary is produced by the costing accountant for each department. This is a one page report which contains a graph of volume output per direct labour hour along with monthly and year to date figures for total output and labour efficiency compared with the same period for the previous year.

There are a number of other reports which are not produced by the accounting department. For instance, delivery performance is reported by the Customer Service Department and daily productivity is reported by each of the manufacturing departments.

Managers are finding increasing conflict between what the variance reports suggest they should do compared with what the apparent company strategy requires them to do. The consequence is that variances are being ignored but there are no objective performance measures to replace them. This is causing stress for managers as there are not clearly documented strategies which may be referred to for resolving the dilemmas.

Another area of conflict is the measurement of sales performance. Currently sales people are paid commissions based on gross margins. This puts sales and manufacturing in opposition when it comes to setting product standards. Higher standards will assist manufacturing to meet variance targets, while lower standards will increase gross margins.

Sales managers have deliberately distorted labour standards so that they were able to implement strategic marketing initiatives. This suggests that certain strategies and policies set within the company are in conflict with each other. There are two policies which encourage sales people to distort labour standards, the first is the pricing policy which does not allow goods to be sold below standard cost, the second is the policy of paying sales commission on gross margin. These policies are in direct conflict with strategic initiatives which require that certain products be sold cheaply for various marketing reasons.

IMPROVING PERFORMANCE MEASUREMENT

As already discussed, traditional systems put all indirect costs into one category and allocate it across all products according to a single allocation base such as direct labour hours. This leads to a distortion in the measurement of product cost and subsequently leads to poor decision making, as highlighted in the case study above.

A possible solution is the use of back flush costing.[6] This is a process of delaying the allocation of costs until a point late in the value chain. At this point, actual costs are flushed back through the system, so there are no variances. This method is designed for, and can only be used in, JIT environments because of the reduced need for valuing inventory.

One of the major developments for improving the accuracy and validity of cost data has been Activity-Based Costing (ABC).[7,8] Activity-based accounting accumulates indirect costs into identifiable activities. The cost of each activity is allocated to the product that consumed it. Costing this way is very involved and has only been feasible through the advent of modern computer systems. The benefits of such a system come from demonstrating to management the true cost/profitability of the various products as well as highlighting the major cost areas thus promoting continuous improvement.

McNair discussed how changes to accounting other than ABC may help.[3] For example, many of the 'World-Class' manufacturing techniques have a major effect on the amount of working capital required to run a business. This is recognized in the CAM-I Cost Management System, which seeks to, *inter alia*, assign technology costs directly to products, and recognize holding costs as a non-value-added activity traceable directly to a product.[4]

Senior managers at Company A have been discussing the use of activity based costing for some time. As a result there has been a trend towards more detailed cost centres. It is intended that the current standard costing system be changed eventually to use actual costs for product pricing rather than standard costs, when more resources are available.

Company A appears to suffer from a 'meet standard mentality'[3] where managers are focused on meeting labour standards in order to achieve acceptable efficiency measures. However, in order to do this they often distort the data by failing to correct known inadequacies in labour standards. This means they also fail to support company strategies and objectives, such as new programmes. This is especially evident with the failure of the continuous improvement programmes which appeared to fail due to the negative effect on labour efficiency. Conversely, improvement programmes, such as multi-skilling, which have shown a positive effect on labour efficiency have been quite successful.

As can be seen from the above case study, the setting of labour standards can lead to conflict. So not only are performance figures distorted, energy is expended arguing over the placement of internal goal posts which adds no value to the company. Thus, the real question is: How can firms measure performance in ways that foster competitive improvement? This is not the same question as: What is wrong with cost accounting?[9]

New Methods of Performance Measurement

A number of new methods for performance measurement have been proposed in the literature. There are some consistent themes especially that performance measurement systems should be designed to link operations to business strategy.[9] It seems that the primary goal of performance measurement is to let employees know how their efforts contribute to the achievement of the company's strategic mission. This will help motivate them to achieve that mission.

There is also strong agreement that a performance measurement system should integrate financial and non-financial indicators. It is not always possible to measure plant-level performance in term of dollars given the constraints of economic feasibility and timeliness.

A widely mentioned attribute is that performance measurement systems should focus all business activities on customer requirement.[1,3,9] There is an implied connection to the strategy requirement already discussed. It is assumed by all who mention this factor that customer focus is, or should be, part of every business strategy and therefore of every performance measuring system.

Many researchers agree that simplicity is a key to success.[9-12] The performance report should make obvious to the reader what has to be done. There should not be too many measures and "...they should preferably fit onto one page".[10] This enables the different measures to be viewed jointly, so a balanced view can be taken.

Gaining acceptance by the individuals involved also appears to be important.[13] A comprehensive performance measurement implementation framework suggests that employee involvement is paramount to success.[14]

Some researchers have discussed whether rewards and measurements should be compatible. This factor is not consistent through the research. Some writers are ardent supporters of compatibility[1] while empirical research has shown that companies are divided on this issue[13].

The literature does not contain many tools or techniques of design and implementation of performance measurement systems. One potentially useful diagnostic tool is the Performance Measurement Questionnaire.[9,15] This questionnaire is designed to firstly determine the efficiency and effectiveness of existing internal performance measures, secondly it can measure the alignment of strategy, actions and measures, and thirdly it provides a starting point for improving the performance measuring system.

Another specific tool developed for performance measurement is the Balanced Scorecard concept.[10] It provides answers to four basic questions: How do customers see us? (customer perspective); What must we excel at? (internal perspective); Can we continue to improve and create value? (innovation and learning perspective); and, How do we look to shareholders? (financial perspective). The Balanced Scorecard achieves many of the performance measurement attributes mentioned above.

A more comprehensive framework is the 'Key Performance Indicators Manual'.[14] The manual lists a framework of four general principles for the development and use of Key Performance Indicators (KPIs): 1) Partnership between all the stakeholders including management, employees, unions, customers and suppliers; 2) Empowerment - all employees must be involved in the process of selecting KPIs; 3) Integrated performance improvement strategies should be developed as part of the overall business function; 4) Teamwork - since self-managing teams are an ideal environment for performance indicators to work.

Design and implementation models have been developed from the Balanced Scorecard[16] and KPI Manual's frameworks. These models are summarized in Table 1. It can be seen that there are similarities and differences between the six-step Scorecard and eight-step KPI framework models. Both models recognize the need to align measurement with strategy; to build commitment (in the case of the eight-step model, this is detailed as explaining the purpose and agreeing to process); and, for review and never-ending modification. They differ in particular in their emphasis and timing. The KPI Manual places development of commitment at steps two and three, while the Balanced Scorecard implementation model suggests putting the measures in place first.

TABLE 1
Implementation Models for Performance Measurement

BALANCED SCORECARD	KEY PERFORMANCE INDICATORS MANUAL
Specify goals	Align KPI development to other change strategies
Match measures to strategy	Explain purpose of KPI development to all employees
Identify measures	Establish agreed process for KPI development & use
Predict results	Identify organization-wide critical success factors
Build management commitment	Select KPIs at the team level
Plan the next step	Develop display, reporting & review frameworks
	Facilitate KPI use to assist performance improvement
	Refine & modify KPIs to maintain their relevance

By blending the work of several authors and using their common themes as expressed above, a type of *contingency theory* can be derived. It is suggested that the success of a performance measurement system is contingent on its relevance to company strategy and the needs of customers. The Balanced Scorecard implementation strategy reflects this contingency theory.

Only a few case studies of successful implementations have been cited. Therefore, there is little factual evidence to provide insight into whether the contingency theory actually works.

CASE STUDIES B AND C

In this section, two successful companies are investigated with respect to their performance measurement systems. Information for these case studies was gathered through interviews with senior plant management, together with some documentation.

Company B "is a leading Australian public company with extensive interests in metal, plastic and paper packaging in Australia and internationally. [It] also has a substantial involvement in manufacturing [some packaging materials]". The studied plant belongs to a division with thirty-three factories world-wide.

The plant operates as an autonomous business unit producing and marketing customized packaging for food and beverages, household products and a range of industrial applications. There are literally hundreds of possible size, shape and printing combinations which can be produced.

The plant has 240 employees, approximately 170 are manufacturing and related staff, the rest are sales and administrative staff. Manufacturing operations come under the control of the Manufacturing Manager who reports to the General Manager. There are three layers of personnel below the Manufacturing Manager: Production Coordinators, Leading Hands and Operators.

Company C is a large Australian company which focuses on the development and manufacture of high technology, high value added chemical product systems for technical markets in Australia, South-East Asia and the Pacific. It is one of three major Australian manufacturers which between them control approximately 70% of the local market.

The company is a semi-autonomous subsidiary of a larger national and, ultimately, overseas parent. Whilst most operational and marketing decisions can be made at a local level, there is a shared responsibility with the parent for strategy development.

The studied plant is one of Company C's three manufacturing sites. The factory has 210 employees including 180 manufacturing staff, 15 clerical and administrative staff, 20 research and development staff and 15 warehousing and distribution staff. The forty-strong national technical sales force interact closely with factory manufacturing staff. Factory operations are overseen by the Manufacturing Manager who reports to the General Manager-Technical Markets. The factory is divided into ten production departments, each is led by a Group Leader. Process workers are organized into teams, each team has a team leader.

Performance Measures in Place

Performance measurement at **Company B** has changed considerably in recent times. Previously, the primary focus of performance measurement was to report profitability to senior management. The emphasis now is on delivering relevant, reliable information to the lowest levels of the organization.

Formerly, a standard costing system was in place with direct labour, direct material and overhead allocation costs determined. Many of the old performance indicators have survived, however many new indicators have been developed. External financial reporting is still carried out as previously.

There is an intention to move toward an Activity Based Management system. However, this is considered a low priority as the company feels it can get a better result by committing resources to measuring performance in other ways.

Company B has a strong record for successfully implementing strategic initiatives. Three tiers of performance measures have been developed to support each level of strategy as it is deployed through the organization. The mission statement is the starting point for the development of corporate strategy. The first tier of performance measurement supports the mission statement by setting and measuring specific profit and return of investment goals. Broad corporate and manufacturing strategies are developed from the mission statement and financial goals. The second tier monthly management report deals with the factors considered to be important to supporting these strategies. Finally, third tier performance measures are developed by operational staff to support the operating objectives, under five major headings: Financials, Asset, Performance, Safety, Quality and Delivery.

The General Manager is measured against specific objectives for people issues, market share and profit levels. His subordinates, with the exception of sales staff, are not yet measured against any performance measures (other than annual profit) but may be in the future. They are currently reviewed against more subjective measures for communication, leadership, etc. Sales staff are paid a commission based on revenue.

Variances from standards or budgets are not reported directly. Instead, indicators are reported for the current month and compared to the same period for the previous year, however it was stressed by the General Manager that meeting this standard is not considered to be good performance, improvement is. Indicators are reported together not in isolation so that managers do not place any undue emphasis on any one indicator.

Junior and middle level managers are still developing their own performance measures for the third tier. In the manufacturing departments a number of measures have already been developed. Some measures have been adopted directly from the second tier only reduced in scope to measure a particular section or department rather than the entire company. These include lost time injuries, absenteeism, conversion cost per 1000 square metres and volume produced. There are additional performance measures based on the specific requirements of particular departments. For instance, one of the production processes which is extremely wet is monitored daily in line with Company B's waste reduction programme.

Line managers reportedly prefer daily performance reporting wherever possible. Some performance measures are only reported weekly and monthly as it is not possible to gather and process the data in a daily time frame or they are too erratic from day to day.

The first and second tiers of the performance measurement system are produced completely by the finance department as part of their regular reporting responsibilities. Reporting for the third tier of performance measurement is shared with the finance department and the operating department using the particular measures. The division of responsibility depends on convenience (such as whether data is extracted from machine log books) and required timing. The weekly report is produced by the finance department.

As another initiative to improve performance measurement Company B's plan to run an employee satisfaction survey to find gaps between employee expectations and company objectives.

Prior to 1987, **Company C** relied on a standard costing system to measure performance based on labour efficiency. Costing is still necessary for the purposes of valuing inventory and product pricing, however Company C now uses an actual costing system in some ways similar to the back flush costing system. Alternative measures were developed for internal decision making and to support the various improvement initiatives. These were eventually consolidated into the current system making the standard costing system redundant.

Company C is similar to Company B in that a three tiered approach to performance measurement is used. However, the structure of the second and third tiers is somewhat different. The first tier is purely financial with measures for profit, return on investment total sales and total assets. These come from the external financial reporting system. They are reported monthly to Company C and its Australian parent's management.

The second tier of performance measures are reviewed annually and developed directly from the key strategies. For instance, when effective leadership was a key strategy it was measured through staff appraisal of managers, performance appraisal, achievement of objectives and an employee attitude survey. Knowledge of customer expectations is obtained through market research and market share analysis. The most important criteria to Company C's customers are on-time delivery and stock availability. As a consequence, customer satisfaction is measured through on-time delivery (%), number of stockouts (per month) and a customer complaints system.

Management feels that the keys to manufacturing productivity are optimization of the use of machinery and optimization of the use of labour. As a result, the strategy for increased profitability through productivity is monitored through a range of measures which are generally based on one or two denominators, either batches produced or litres produced. The monthly report contains the measures total litres produced, labour hours per litre, total number of batches produced, average cycle time per batch, and average labour hours per batch. Some of the other measures are reported weekly.

Second tier performance measures are relevant to a range of employees. They are used for monitoring operational performance by senior management, the various work groups and teams and by a consultative committee.

The third tier are more specific operational performance measures which are reported on a weekly basis. Many of the measures are similar to those given in the second tier but contain only the data relevant for that particular team. A typical performance report for a work team will measure the critical dimensions of productivity for that team. For example, the team must produce batches as quickly as possible (Cycle Time/Batch) with as little labour input as possible (Labour Hours/Batch), however, the batches must be of good quality (Rework). Also the team must maintain its commitment to deliver the product to schedule (Batches Finished on Time) and do so in a safe manner (Lost time Injuries).

Performance measures are reported against specific targets. The targets may be known best practice benchmarks or desired performance levels based on capability studies, etc. More importantly the targets are not viewed in isolation, but are balanced against other measures (and their targets) on the same report.

The information system used to generate performance data automatically comes under the control of the finance department, however, some data collection and entry is done by manufacturing personnel. External financial reporting is still carried out as previously.

A significant factor in the success of Company C's performance measures in promoting strategy appears to be the way in which they are used. Each employee is a member of a work group which meets once per week to review performance and discuss changing company goals and objectives. Performance measures are reviewed when strategy is reviewed. Redundant measures are removed and new measures added as required.

Implementation Difficulties and Solutions

The field work uncovered some possible obstacles to the implementation of a new performance measurement system. Research in the field of organizational psychology has long noted that information is a source of power in organizations[17]. New performance measurement systems can result in a transfer of information control, this may result in a power struggle of sorts as was the case in the finance department of Company B.

The control of traditional performance measurement systems in most companies generally lies in the hands of the finance department. It is natural that finance employees will not want to lose control. When this occurred at Company B the General manager was forced to take a direct role in the implementation. The finance department maintained their controlling role but had to be forced to implement the change.

Resistance to change is also likely to occur whenever the new performance measurement system changes (or is perceived to change) rewards. In the case of Company B it was perceived that a change of intrinsic rewards (power) was going to occur. In Company A's case a change in performance measurement is likely to alter the extrinsic rewards (commission) of sales people so resistance can be expected from this area.

Company C learnt some important lessons when it began developing its performance measuring system. The company had felt the need for change and began a number of improvement programmes one of which was the development of key performance indicators and benchmarks. The company acknowledges that, initially, performance measures were developed without a clear framework. The performance measurement system became very complex and difficult to understand, it was not possible to determine which were the really important measures. The mistake was that, initially, no clear link was made to strategy, in fact the company's strategies were not clearly defined, as is the case at Company A.

DISCUSSION

Both Companies B and C have been able to develop successful performance measure systems by deriving them from strategy. Company C provides evidence that before a good performance measuring system could be developed, strategies had to be clearly defined.

The development of strategy at the two best practice companies has some similarities, the most obvious being the three tiers. The reason for three tiers is probably due to the similarity in size of the two companies and not necessarily a standard for strategy development. Both began with a mission statement that was used to develop the second tier of corporate strategies. The corporate strategies were used within the various functions to develop operating strategies and objectives (only manufacturing strategies have been reviewed here).

It seems significant that both companies support each layer of strategy with a layer of performance measures. The clearly defined links from one layer to the next mean that each layer of strategy can only be successful if the previous layer is successful. Therefore, each layer of strategy must be supported equally. If one layer of strategy and/or performance measures is missing, the system is in danger of encouraging the wrong performance.

In order for clearly defined links to be made between each layer of strategy, each layer needs to be expressed in such a way that a subordinate layer can be developed from it. A mission statement must offer something more than just a 'motherhood' statement. The mission statement corresponds to the first tier of performance measurement which is ultimate financial performance. It must then offer something which can be measured.

Both companies exhibit first tier performance measures which are purely traditional financial (ROI, profit, turnover, etc.) but blend in many non-financial measures in the second and third tiers. Both companies include measures for safety, absenteeism, on time delivery, etc. There are also a number of non-traditional 'financial' measures which have been developed. For instance, Company B devotes a section of its monthly second tier report to asset performance with such measures as Stock Turnover Days and Overdue Collection Ratio.

The two best practice companies studied considered that performance measurement systems should be relevant to the needs of customers. In both cases a variety of measures appeared in the second and third tiers to support these needs.

It is clear that the performance measurement system on its own is not enough, the strategy should exist in a form that can be easily communicated and understood by all employees. Both of the best practice companies placed emphasis on employee involvement in the performance measurement and review process. However, the best practice companies studied appeared to still emphasize strategy development over employee involvement. Company C had sought to implement employee involvement programmes before strategy was developed. This was found to be unsuccessful and led the company to believe that strategy must come first.

Both best practice companies considered that employee satisfaction was related to the willingness of employees to implement various improvement programmes. As a consequence, measures are used for employee satisfaction. The significance of this finding is that best practice companies have gone beyond simply involving employees, but use satisfaction as a driver for change and a measure of the effectiveness of the communication of strategy.

The performance measurement systems of best practice companies were kept simple. For instance, Company C found by trial and error that a complex performance measurement system was difficult to understand and use. Company B's management specifically referred to the need to keep management reports to one page, as is occasionally suggested in the literature.

The 'Balanced Scorecard' approach was based on the idea that measures should be presented in a balanced view so that trade-offs between different measures could be gauged[10]. Evidence was found that both best practice companies are taking this approach (although not specifically using the Balanced Scorecard framework). Company B particularly referred to the importance of balancing measures in its second tier monthly report. Clearly, managers need to be able to gauge the effect of their decisions. It can be seen at Company A that, when a narrow view of performance measures is taken (cost and labour efficiency only), important strategic initiatives can be ignored.

The literature is divided on the issue of linking performance measures to rewards for managers. Both best practice companies studied demonstrate such links. Company C integrates performance measures into the annual salary review process for managers. At Company B, currently only the General Manager has a direct link between actual performance measures and his salary review. It is the company's intention to link performance to the salary reviews of all salaried staff, and eventually to link performance measurement to the rewards of process operators through gainsharing. Company A also has provided some evidence of the strong effect from linking rewards to measures, however, in this case the effect was negative. It seems that the linking of rewards to measures will lead to improved performance provided the correct measures are chosen.

Both Companies B and C stressed the need for *targets*, however, it was also stressed that these were used in a much different way to the *standards* of traditional systems. The use of targets has long been a vexed issue, with Deming in his fourteen points for managers warning against their inappropriate use[18]. In the case of Company B the target was based on the previous years performance. Company C found it best to use targets in the benchmarking sense so that a challenging but achievable result could be aimed for.

The best practice companies studied used `academic frameworks' for strategy development. However, neither used any type of academic framework to assist design and implementation of their performance measuring systems. Both companies found, independently, that changes to performance measurement were needed to promote successful strategy implementation.

When designing performance measures, especially productivity measures, both best practice companies chose denominators which were relevant to their particular business. Thus measures in terms of thousands of square metres produced (e.g. Direct Labour/000sqm, External Sales/000sqm) or batches produced or litres produced are widely used. This further highlights the inappropriateness of standard costing for performance measurement.

A Model for Performance Measurement

From the above case studies of best practice companies, it seems that the primary requirement of a performance measurement system is to support the implementation of strategy. The support is required at several levels in the organization and each level of strategy requires a level of performance measures.

Strategy and performance measures combine to form an organization-wide continuous improvement cycle. This can be likened to the four step PDCA (Plan, Do, Check, Act) continuous improvement model[18]. This model is normally applied to process improvement (the lowest level of strategy), however it is also a useful framework for appreciating the relationship between strategy and performance measurement. The cycle begins by planning (strategy development) followed by doing (strategy implementation), then checking (performance measurement), finally action is taken based on lessons learnt (the strategy may be adjusted or replaced as necessary). The cycle keeps repeating.

By applying this model it can be seen that strategy and performance measurement are interdependent. If either is missing or not functioning properly, the continuous improvement cycle stops.

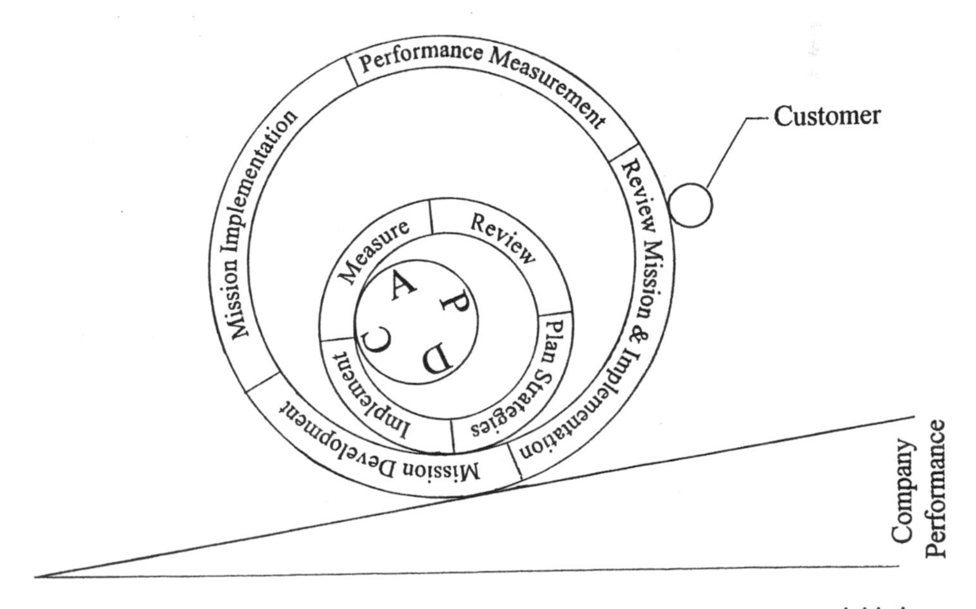

Figure 1. A model for a performance measurement system to support company initiatives

Each level of strategy runs through a separate PDCA cycle. The major difference for each layer is likely to be the cycle time. For instance, the mission statement may run through a very long cycle (in most companies the mission statement is reviewed at most every few years). Corporate strategies are likely to cycle more often while shop floor level improvement initiatives may require review as often as every day.

The layers of improvement cycles should be linked by the progression of strategy from one layer to the next. This can be schematically shown as a set of nested PDCA cycles leading a company up the slope of continual development. As shown in figure 1, the development of a company's mission and supporting performance measures should be driven by customer needs. The inner cycles represent the lower tiers of strategy and measurement which are driven by the level above to ensure congruency.

The findings of the literature review and the case studies of best practice companies have also been condensed by the authors into the following checklist of desirable attributes for performance measurement systems. Use of this checklist along with the model described above should help in the development of a successful performance measurement system.

TABLE 2
A Performance Measurement Checklist

1.	The business must have a sound, clearly developed strategy.
2.	The performance measurement system should be developed for the strategy with a level of performance measures to support each level of strategy.
3.	Involve all employees in the development of measures as well as the eventual operation of the system.
4.	Ensure that strategy supporting measures focus attention on the needs of customers.
5.	Use a range of financial and non-financial measures.
6.	Keep reports simple, preferably to one page.
7.	Do not look at performance measures in isolation.
8.	Consider the use of targets.
9.	Use links to reward systems to help motivate strategy implementation.
10.	Develop denominators based on business at hand.
11.	Be aware of possible obstacles to implementation.

If the performance measurement checklist is applied, a company trying to improve its performance measurement system can see the first requirement is that strategy must be developed in a clearly defined form. When compared with the two best practice companies, certain gaps were noted in Company A's strategy development. With the exception of the mission statement and various engineering objectives, documented strategies did not exist. Once the company strategies are developed they need to be supported with relevant performance measures.

CONCLUSIONS

Successful companies are modifying their performance measurement systems to support their efforts to implement modern manufacturing strategies.

The contingency theory suggested by the literature has been supported by the case studies of best practice companies. Further insights were also revealed to the extent that a checklist of desirable attributes has been developed as a result of this research.

Performance improvement can be achieved by companies that follow a clear structured approach to strategy development and support this development with relevant performance measures. Applying the model for performance measurement described above, a company wishing to improve its performance measuring system needs to develop a total performance improvement system. The system should have strategy development, implementation, performance measurement and review as its key elements. Each layer of strategy should have a corresponding level of performance measures. To succeed the system will require the support and involvement of employees at all levels, as well as a clear demonstration of commitment by senior management.

REFERENCES

1. **Lynch, M. and Cross, K.**, *Measure Up! Performance Measures of Continuous Improvement*, Basil Blackwell Inc., Massachusetts, 1991
2. **Johnson, H.T. and Kaplan, R.S.**, *Relevance Lost: the Rise and Fall of Management Accounting*, Harvard Business School Press, Boston, 1991
3. **McNair, C.J.**, *World-Class Accounting and Finance*, Business One Irwin, Illinois, 1993
4. **Berliner, C. and Brimson, J.A.**, eds, *Cost Management for Today's Advanced Manufacturing*, Harvard Business School Press, Boston, 1988
5. **Kaplan, R.S.**, "Limitations of Cost Accounting Systems", *Measures for Manufacturing Excellence*, (ed. Kaplan, R.S.), p15-38, Harvard Business School Press, Boston, 1989
6. **Hilton, I.H.**, "Against the Flow or Back Flush Costing", *Management Accounting*, p60-66, September, 1989
7. **Cooper, R. and Kaplan, R.S.**, "Measure Costs Right: Make the Right Decisions", *Harvard Business Review*, p96-103, September-October, 1988
8. **Johnson, H.T.**, "Activity-Based Information: A Blueprint for World-C;ass Management Accounting", *Management Accounting*, p23-30, June, 1988
9. **Dixon, J., Nanni, A. and Vollmann, T.**, *The New Performance Challenge: Measuring Operations for World-Class Competition*, Dow Jones-Irwin, Illinois, 1990
10. **Kaplan, R.S. and Norton, D.**, "The Balanced Scorecard - Measures that Drive Performance", *Harvard Business Review*, pbb-79, January-February, 1992
11. **Jones, S.D., Buerkle, M., Hall, A., Rupp, L. and Matt, G.**, "Work Group Performance Measurement and Feedback - An Integrated Comprehensive System for a Manufacturing Department", *Group & Organization Management*, p269-291, v18/3, 1993
12. **Cooper, R. and Turney, P.B.**, "Internally Focused Activity-Based Cost Systems", *Measures for Manufacturing Excellence*, (ed. Kaplan, R.S.), p291-305, Harvard Business School Press, Boston, 1990
13. **Armitage, H.M. and Atkinson, A.A.**, "The Choice of Productivity Measures in Organizations", *Measures for Manufacturing Excellence*, p91-125, Harvard Business School Press, Boston, 1990
14. **Commonwealth of Australia**, *Key Performance Indicators Manual*, Pitman Publishing, South Melbourne, 1995
15. **McMann, P. and Nanni, A.**, "Is your Company Really Measuring Performance?", *Management Accounting*, p55-59, November, 1994
16. **Vitale, M., Mavrinac, S. and Hauser, M.**, "New Process/Final Scorecard: A Strategic Performance Measurement System", *Planning Review*, p12-16, July/August, 1994
17. **Pfeffer, J.**, *Power in Organizations*, Pitman, 1981
18. **Deming, W.E.**, *Quality, Productivity and Competitive Positions*, Centre for Advanced Engineering Study, MIT, 1982

OBJECT ORIENTED MODELING FOR A PERFORMANCE MEASUREMENT SYSTEM FOR SMALL ENTERPRISES

Dr. Orlando Durán
duran@upf.tche.br
FEAR Universidade de Passo Fundo
C.P. 566 or 567 CEP 99001-970
FAX 55-054-3111307
Passo Fundo (RS) Brazil

Dr. Antonio Batocchio
batocchi@fem.unicamp.br
DEF / FEM UNICAMP
FAX 55-019-2393722 C.P. 6122
CEP 13083970
Campinas SP Brazil

ABSTRACT: This paper reports the definition of a modeling technique for performance measurement systems (PMS) based on the object oriented paradigm for analysis and development. The objective of this current research project is to provide small companies with a structured approach for analysis and development of an integrated hierarchy of performance indexes. The integration and the manipulation of contradict business objectives is the main perspective of this project. The proposed methodology is based in a series of iterative steps that synthesizes several object-oriented techniques and notations into a consistent approach. In addition, some quality management techniques are used for complement the global system definition. We aimed with this project to provide a framework for using structured and consistent methodology for define, test and implement the PMS. Only through that, the company will attain sustainable strategic advantages and world-class levels. An exercise to prototype the definition of a PMS in a small make-to-order company using the proposed technique is presented.

INTRODUCTION

Global markets have caused high levels of competition and fundamental changes in the industrial environment. Manufacturers can no longer compete with their traditional and regional competitors but they must be oriented to global customers and must compete with several companies, some of them located at the other side of the world.

Firms must develop strategic objectives which lead to competitive advantages beside the development of organizational structures that allow communicate, monitor and maintain a structure of goals and objectives through the organization , that motivate the strategic objectives fulfillment.

Recent research has identified the need for effective deployment of business objectives down through the organization and the subsequent measurement of performance in critical areas as key elements of sustainable competitive advantage. The integration and manipulation of apparently contradict business objectives is an other challenge that researchers face today.

Any company that aims at attaining agile manufacturing needs to implement radical changes into its structure and behavior. Technology investments are not enough, and any company does need to implement new procedures that allow the organization to manage the activities aiming at higher strategic performance levels.

Thus it is necessary to create a set of performance indicators that reflect the behavior of the activities compared with the planned performance and global company objectives.

Considering the dynamic behavior of performance measurement systems, Berliner and Brimson (1992) suggested that each time the company objectives change, indicators must be reviewed. This review must be continuously and in some cases new indicators have to be added to the system. In other cases any indicator that is no more useful to the system have to be eliminated. Thus there is the need to create structures and methods for continuos

evaluation of the performance measurement system according to the changes in the organizational scenario and its environment.
This paper reports the definition of a modeling technique for performance measurement system (PMS) based on the object oriented paradigm for analysis and development. In the next chapter we discuss some issues about performance measurement. The chapter 3 analyze main features of the object oriented design and analysis method. And chapter 4 shows the application of the proposed methodology for PMS development. Finally, in chapter 5 some discussion is made about the advantages and limitations that the proposed approach has.

PERFORMANCE MEASUREMENT

Frequently, performance measurement are watched as financial evaluations, but these indicators are just one class of performance information, and provide a partial view of company status. Moreover, an apparent financial stability may hide instability or an inefficient global performance. For instance, if one consider the customer service factor, the status of the relation between the company and the customers satisfaction may delay to appear or reflect in financial indicators (Ron, 1995).

An ancient axiom of management tells: "we cannot control what we don't control", hence, evaluation is a predominant task in today's management.

The question is "What we must to measure?", and the generic answer is: a set of measures that encourage the continuos improvement of the activities according to the goals and the mission of the organization. However, the specific answer will depend on particular characteristics of the company. In other words, the nature of the measurement will be closely related to the nature of the system (Fawcett, 1992).

According Hronec (1993), performance indicators are "vital signals" that shows the "health" of the organization. They quantify the degree of efficiency of one process in attaining a desired level. Berliner and Brimson (1992) defined a set of prerequisites for any set of performance indicators:

- coherency with the company goals.
- adaptability with the business needs.
- significant activities evaluation.
- easiness to be applied.
- global approval.
- cost and time efficiency.

Performance perception will be different between one company and other, hence the great number of papers in the literature that report performance measurement systems definition and their implementations(Bullinger et alli.,1995; Hyunbo et alli.1996). However, a common base and a certain degree of similarity may be found among them. Thus one can classify the performance indicators into four categories:

- Business management support.
- The probability or not of attaining the desired goals.
- The necessary feed back into the system aiming maximizing probability of attaining the company goals
- A contribution with the continuos process of goals redefinition.

As it was commented previously, consistency between performance indicators and strategic objectives must exist (lockamy and Cox, 1995). Thus, a methodology that allows the modeling and analysis of performance measurement systems is needed within the organization. This methodology must guarantee that the linkage among the organizational levels exists (strategic, tactical and operational) Only through this linkage integration among the organizational levels will be consolidated (figure 1).

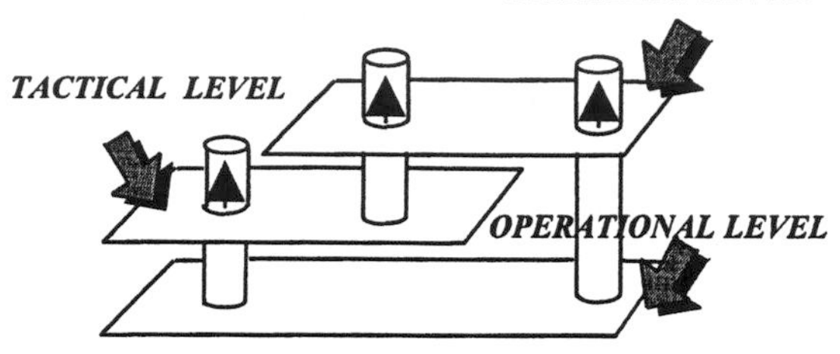

Figure 1. Cause Effect relationship among performance measurement indicators from different levels of the organization.

OBJECT ORIENTED DEVELOPMENT

The early 1990s have seen the emergence of a number of reference models or architectures which propose conceptual frameworks and associated methodology which support re-engineering of manufacturing systems. Aguiar et all. (1996) analyzed a number of initiatives to develop reference models aiming at specifying and analyzing manufacturing systems and their components.

Table 1 contains a summary of several factors on which the traditional modeling paradigm is compared to the object oriented paradigm.

Table 1.- Comparison between traditional and object-oriented paradigms (Mize, Bhuskute, Pratt and Kamath, 1992).

FACTORS	TRADITIONAL PARADIGM	OBJECT ORIENTED PARADIGM
Model Construction:		
Software	Simulation language based on procedural programming style	Simulation environments based on object-oriented programming style
Translation into code	Process is abstract	Process is natural and intuitive
Interface	Usually Textual	Usually graphical with icons and dialog boxes
Level of detail	Usually not much detail due to programming complexity	At user's discretion, but requires detailed object library
Treatment of distinct system elements	Different element types are not distinctly modeled; Aggregation to reduce program complexity	Physical, information, and decision/control elements are modeled distinctly and independently
Effort/time/cost	Moderate costs of model development, but a "throw-away" type	Initial cost of establishing detailed model is very high, but costs of subsequent reuse is relatively low
Model Attributes		
Purpose	Usually a unique model is created for a specific purpose	More general models possible for multiple purpose
Usage	Single usage, throw-away models	Repeated usage and continuos refinement
Flexibility	Highly inflexible; changes almost always result in a complete rewrite of program	Highly flexible, due to the ability to modify fundamental building blocks; quick reconfiguration is possible
Accuracy	Useful for measuring relative differences in alternative configurations	With greater degree of detail and realism, can also estimate absolute performance with greater accuracy
System/Model relationship	Not connected via data links	Detailed model can be imbedded in control structure of the firm, with linkages to databases; continuos model calibration and parameter updating

As Ham et all. (1996) outline there is little consensus among these initiatives, mainly because of the little standardization that exists. In this research we propose the application of a method developed in six steps, that is an adaptation of an approach reported by Ham et all. (1996) and a proposed methodology reported by Hodge and Mock (1992).

Following we resume the series of steps that have to performed to model a performance measurement system:

1. Define objects and classes
2. Define relationship between objects and classes
3. Establish class structure
4. Establish message flow diagram
5. Define relative attributes and methods of classes and objects
6. Establish overall system design

In the next chapter we analyze the proposed methodology with an application example.

APPLICATION OF THE PROPOSED METHODOLOGY

Booch (1986) recommended that in using either functional or object oriented approaches the analysis procedure might start by creating a data flow diagram of the system to capture a model of the problem space.

Other authors called this phase information analysis which aims at modeling entities that exist in the problem domain and the relationship among these entities. These model entities may be considered as data elements that describe what is in the problem domain are represented graphically in an entity relationship diagram (ERD).

An other approach points to represent the structure of a system through the utilization of cause effect diagrams (lebas, 1995; Bititci, 1995). These diagrams provide a platform to visualize a given performance measurement system structure. In addition, and this is quite important, they lead the identification of a structure based on the causal relationship between the different hierarchical levels of the organization. This analysis process shows how each process (shown as a major cause in the diagram) may have an impact on the performance measure.

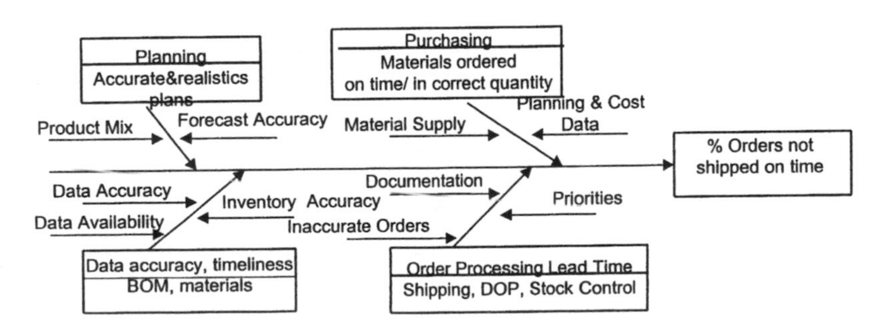

Figure 2. Cause Effect diagram for indicators showing the causality relation between customer satisfaction and other activities along the organization .

Our approach begins with the selection of a strategic performance measure, and after this, to use the cause effect diagram effect for representing causality relations that exist between the selected indicator and the other levels of the organizational structure. The underlying structure behind the diagram can be represented through an entity-relationship diagram. In this diagram data elements are identified and the relationships that exist within the problem domain. Figure 3 shows the entity relationship diagram (ERD) for the situation under study. In this diagram some details were eliminated, and more importance was given to the attributes related to information analysis on performance indicators and the relation among them.

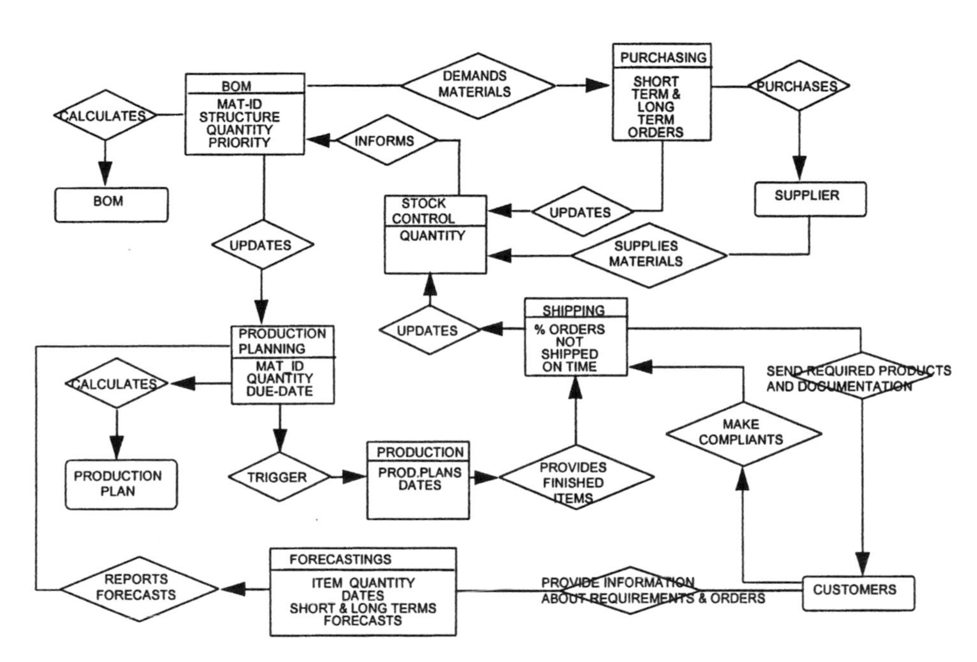

Figure 3. ERD showing the relations among entities related to customer satisfaction indicator

The ERD is a working model that allows identify entities that are present within the domain under analysis. The next step is to translate them into objects. This step attempts to create or define objects that best describe the data elements described in the ERD diagram. Attention must be given to the fact that there are some activities in operating the system under modeling that are intangible operations. These operations are managed by humans or softwares and are called logical objects. The identification of objects is made by answering the question: What are the objects that best describe the entities defined by the ERD? Each data element becomes an object or an attribute of an object. The next list shows the identified objects (and the attributes that are relevant for the analysis) from the ERD:

• BOM_calculator	mat_id structure_of_the_item quantity priority calculates_a_new_BOM
• Production_Planner	mat_id quantity due_date Calculates_a_prod_plan Trigger_a_new_prod_plan
• Stock_Control_Manager	mat_id quantity

	detects_low_inv_levels informs_inv_levels
• Shipping_Module	%orders_not_shipped_on_time send_required_products send_documentation_to_customers
• Production_Manager	prod_plan dates Provides_finished_items
• Purchasing_System	Short_term_orders Long_term_orders Trigger_orders
• Forecasting_Module	mat_id Item_quantity dates short_term forecasts long_term forecasts

Some authors represent this step of the analysis methodology with an object-relationship diagram (ORD), but here, for space economy the ORD is not showed. Once objects are identified in a static structure, the next step is the representation of the dynamic behavior of each one of the objects. Here we used the message flow diagram defining a series of messages among the identified objects. These messages flow within a system as responses of requirements coming from other objects. The requirements are made for compound or calculate a given performance indicator. Figure 4 shows an example of a message flow diagram representing the dynamic behavior of the PMS. In the message flow diagram, an object has several sending messages and several receiving messages. From the PMS design point of view, messages are requirements for information about a given activity or process, which has to be processed by the "client" object composing a performance indicator.

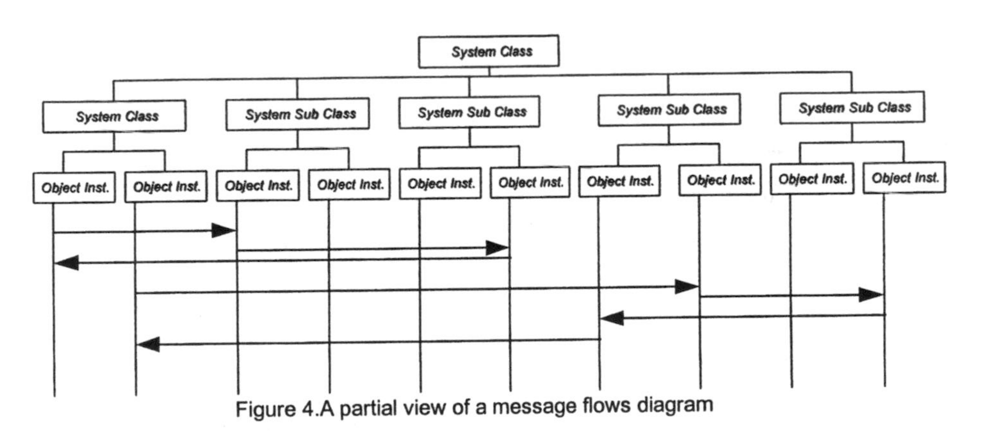

Figure 4.A partial view of a message flows diagram

The implementation phase regard programme writing using a given programming language. Recently, a great number of object oriented programming languages was defined. Is out of the scope of this paper to discuss about PMS implementation. But whatever programming language that were selected the approach must be pointed to create a client-server architecture where any object may serve or be served by another object, aiming the computing of a given performance indicator.

CONCLUSIONS

The proposed methodology is intend to apply a set of techniques or reference models to define a performance measuring system, using the object oriented paradigm. This methodology is a series of steps that provide traceability and consistency between a given domain and the system that intend to measure its performance.

The methodology also provides the flexibility to implement an iterative development process. This allows the user (analyst/designer) perform any kind of modifications pointing to represent the actual state of the system. Through the use of this type of methodologies PMS may be developed in a rapidly and feasible manner.

REFERENCES

AGUIAR, M.W.C., MURGATROYD, I.S. and J.M. ERDWARDS, 1996. Object Oriented resource models: their role in specifying components of integrated manufacturing components of integrated manufacturing systems. Computer Integrated Manufacturing Systems, Vol.9, No. 1 pp.32-48.

BERLINER, C. e J.A. BRIMSON, 1992, Gerenciamento de Custos e indústrias avançadas (Base conceitual CAM-I), T.A.Queiroz, Ed. Ltda., São Paulo, 256 p.

BITITCI, U.S., 1995, Modelling of performance measurement systems in manufacturing enterprises. International Journal of Production Economics, Vol.42, p. 137-147.

BOOCH, G.,1986, Object Oriented Development, IEEE Transactions on Software Engineering, Vol. SE-12, No. 2, 211-221.

BULLINGER, H.J., FREMEREY, F. e FUHRBERG-BAUMANN, J., 1995, Innovative production structures - precondition for a customer-oriented production managment. International Journal of Production Economics, Vol.41, p. 15-22.

FAWCETT, S.E., 1992, Strategic logistics in coordinated global manufacturing success. International Journal of Production Research, Vol. 30, n.4, p. 1081-1099.

HAM, H., JEONG, S. and KIM, Y., 1996, Real Time Shop Floor Control System for PCB auto insertion line based on object-oriented approach. Computers and Industrial Engineering, Vol. 30, No.3, p. 543-555.

HODGE, L.R. and M.T.MOCK, 1992, A proposed object-oriented development methodology, Software Engineering Journal, March, 119-129.

HRONEC, S., 1994, Sinais Vitais. Arthur Andersen-Makron Books, São Paulo, 240 p.

HYUNBO, C., MOOYOUNG, J. e MOONHO KIM, 1996, Enabling technologies of agile manufacturing and its related activities in Korea. Computers and Industrial Engineering, Vol. 30, No.3, p. 323-334.

LEBAS, M.J., 1995, Performance measurement and performance management. International Journal of Production Economics, Vol.41, p. 23-35.

LOCKAMY, A. e COX, J.F., 1995, An empirical study of division and plant performance measurement systems in selected world class manufacturing firms: linkages for competitive advantage. International Journal of Production Research, Vol. 33, n. 1, p. 221-236.

MIZE, J.H., H.C.BHUSKUTE, D.B. PRATT and M.KAMATH, 1992, Modeling of Integrated Manufacturing Systems using an Object-Oriented Approach, IIE Transactions, Vol. 24, No. 3, 14-26.

RON, Ad.J. de, 1995, Measure of Manufacturing performance in advanced manufacturing systems, International Journal of Production Economics, Vol.41, p. 147-160.

STRATEGIC TARGET COSTING

Pui-Mun Lee[1] & William G. Sullivan[2]

[1]Nanyang Technological University
[2]Virginia Tech

STRATEGIC TARGET COSTING

ABSTRACT: Target costing is a cost management concept that originates from the Japanese industries in the 1970s. The ultimate objective for the use of target costing is to help businesses to sustain competitive positions. Current usage of target costing focuses on establishing cost targets for the products on the basis of meeting competitive market prices. The firm then proceeds to achieve target costs for each of the products using design-to-cost, value engineering and other cost reduction techniques. This paper proposes a model for implementing target costing that takes into account the overall portfolio of products/services that an organization has to offer and how it can strategically assign target costs to each of the products/services in the portfolio so as to create the desired economic value for long-term competitiveness. The model calls for the use of activity-based costing to provide better cost information on the products and to maximize the effectiveness of cost reduction plans. The outcome for using the proposed model is to allow firms to apply target costing that will fit into its business strategy.

INTRODUCTION

Effective costing of products and services is considered as an essential and crucial organization's strategy in today's competitive global market. Having the best products and services without competitive prices will not allow any global businesses to survive in the long term. In today's global market, competing on the basis of satisfying the customers and also providing value is the best approach to gain market share and increase revenue. Setting competitive prices is easy as all one needs to do is to look at market prices for the same products and services. However, maintaining competitive prices and at the same time ensuring profitability for the organization is no easy feat. The key to achieve these dual objectives is to reduce costs and increase productivity.

Target costing is a cost management technique that is used to facilitate cost reduction in products, services, and production and support processes. It is an often used cost management technique in Japan and is fast gaining popularity in American corporations and around the world. However, in most organizations, the bulk of target costing is focused on new product development. Less emphasis is placed on existing products and services. Also, target costing usually relied on inadequate cost information obtained from traditional cost accounting methods. Traditional cost accounting provides scant detail about costs of support processes and indirect activities as related to specific products, and relies on apportioning these costs to all products and services through very few generic cost drivers such as direct labor hours or direct material cost. Inadequate cost information will hamper target costing as it may lead to the setting of inappropriate target costs and targeting cost reduction at the wrong areas.

To implement effective target costing, organizations need to have accurate cost information and a holistic approach in which all products and/or services are equally considered for cost reduction and all major activities related to these products/services are identified. Target costing should be applied at the design of new products or redesign of existing products and also at existing processes and support activities of the organizations. Target costs should be realistic and achievable. Reducing material costs and eliminating some features of the products are common approaches to achieve target costs but reducing the cost of activities that support the production of the products is less common because cost information for specific activities is usually not available. However, as many products share similar activities, reducing the cost of activities has a wide impact on many products. Because of these implications, organizations should approach target costing in a holistic manner.

In this paper a strategic target costing model is proposed. The model provides a framework for implementing target costing that will take into account the overall portfolio of products/services that an organization has to offer, and how target costs can be strategically assign to each of the products/services in the portfolio so as to generate the desired economic value for long-term competitiveness. The model calls for the use of activity-based costing to provide better cost information on the products and activities that support the products so as to maximize the effectiveness of cost reduction plans. The outcome for using the proposed model is to allow firms to apply target costing that will fit into its business strategy.

TARGET COSTING AD ACTIVITY BASED COSTING

Target costing has its roots in Japanese industries. It has evolved from the concept of value analysis [1] in the 1960s and was practiced by Toyota and Nissan. The present form of target costing gain widespread acceptance and usage after the 1973 oil crisis. Target costing is a cost management concept that establishes cost goals for new product and process designs, and also cost reduction objectives for existing products and organization activities (Figure 1) [2]. **Target cost** is the ***difference*** between the target price and the target profit, both the latter been established by the organization based on its strategy as well as competitor, market and internal capability analyses [3]. Target costing is an attempt to manage costs and this process derives from Japanese managers' mentality which expects to use cost information to control costs. This mentality is very much different from European or American managers whom generally expects to use cost information to make decisions about pricing or investments [4].

There are several approaches to determine target cost. In the first approach, a target profit is first decided by the organization and then the target cost is defined by subtracting the target profit from the projected sales price of the product. In the second approach, the current estimated cost of the product is derived from cost information and a target cost is set based on current capabilities and projected improvements. In the third approach, a combination of the first and second approaches is employed. The third approach identifies a **drifting cost** for the product. The **drifting cost** is the current estimated cost of the product/service based on existing capabilities. The term "drifting" implies that the estimated current cost is repeatedly revised downwards as cost reduction is applied to bring the drifting cost to the level of the target cost derived from a target profit [5].

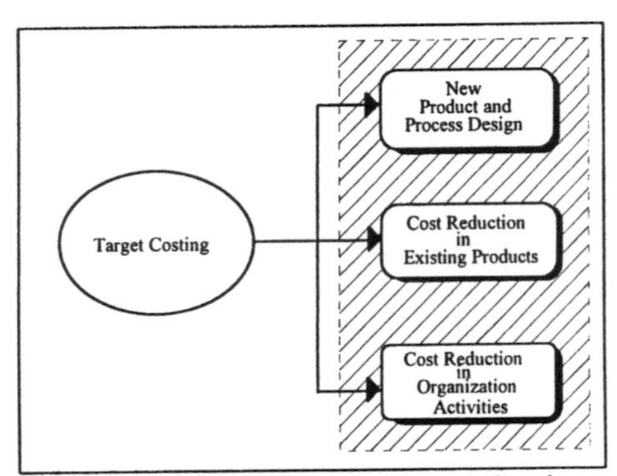

Figure 1: Applications for Target Costing

Activity-based costing (ABC) system has gained popularity within the last 10 years. This system recognizes that many indirect costs are caused not by volume of production, but by the transactions associated with scheduling, ordering, and designing the product. ABC is therefore able to provide more accurate cost information. Starting within the last decade, activity-based accounting developed slowly but its advantages over traditional cost accounting practices were gaining recognition. ABC has now been recognized to provide strategic insights and, thus, should be considered as a tool for decision making rather than as a replacement for an existing cost system [6].

Most of the mechanics of ABC have been described by Kaplan and Cooper [7,8]. The ABC systems are designed by first identifying the activities performed by each support and operating department, and then computing the cost of resources required to support these activities. From this stage, the unit cost of each type of activity can be easily defined through assigned cost drivers. Once the unit costs of activities are determined, support and indirect costs are assigned to products based on the amount of activities (determined by the cost driver(s) used) performed for each individual product.

Traditional cost accounting only attempts to determine a product's cost based on its direct material/labor cost plus a portion of the overhead costs through an apportioning basis that relies only on one or two volume-related bases such as labor hours or material dollars. An ABC system makes use of many cost drivers to ensure that most of the indirect/support costs can be directly linked to the product/service. An ABC system therefore provides more detail and accurate cost information to support target costing.

THE TARGET COSTING MODEL

This proposed model is developed to support the efficient deployment of target costs as well as the target costing process. Current setting of target costs and planning for target costing is well established in many industries but have its deficiencies. This model attempts to address two primary deficiencies in this area of work.

First, in most existing target costing projects, organizations tend to rely on cost information supplied from traditional cost accounting information. These cost information are inadequate and inaccurate because most of the indirect and support costs are arbitrarily assigned to the product. Establishing target cost and conducting target costing using inaccurate cost information will at best caused sub-optimization towards the goal of sustaining competitiveness and in the worst case scenario, will result in erroneous setting of target cost and accompany with time and effort wasted in trying to reduce costs in the wrong areas and/or wrong products. This model incorporates the use of activity-based cost information to establish drifting costs and guide the setting of target costs.

Second, in most existing cases, target cost is derived from a target profit and a target price. The target profit is commonly based on return-on-equity (ROE) or return-on-sales (ROS) objectives while the target price is usually based on a combination of competitors' pricing and market price elasticity. Using ROE or ROS or any other accounting measures to set target profits will distract the organization's primary business objective; which is to generate economic value for the firm. Accounting measures are based on accounting profits, and they do not measured the economic value generated by the firm's overall business activities. A firm may be achieving the desired accounting profits but the amount of money earned may not be enough to cover the cost of renting or utilizing the capital employed to make the products that created those profits. As a result, the firm is not generating any economic value (ie.; it is not growing) and will eventually become noncompetitive. This model suggests that target profits should be computed in relation to projected economic value set by the firm. Projecting economic value for a firm can be done in many ways, such as using engineering economic analysis techniques or the "economic value added" (EVA) concept. EVA, a term trademarked by the originators' firm, Stern Stewart & Co., is developed by Joel M. Stern and G. Bennett Stewart III. EVA is a performance measure that computes the difference between a firm's adjusted net operating income after taxes and its total cost of capital, with some accounting adjustments been incorporated [9].

The proposed target costing model is illustrated in two parts, I and II. Part I focuses on setting target costs and is shown in Figure 2. Part 2 focuses on the process of target costing and is shown in Figure 3.

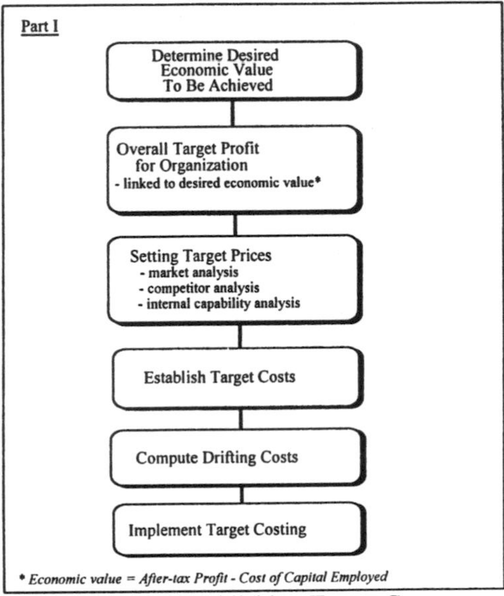

Figure 2: Establishing Target Costs

To establish target costs for the entire portfolio of new products and existing products in the firm, the model proposes that the desired amount of economic value that the firm wishes to attain must first be decided through a strategy session or top management consensus meeting. Following that, the total cost of overall capital being employed by the firm must be calculated. With the desired economic value and the total cost of capital known, the overall target profit amount for the organization can be established. On the other hand, target prices for all products in the portfolio will also be determined through market and competitor analyses. With target prices known, projected product volumes can also be determined. Next, the target profit for each product is computed based on multiplying a profitability percentage (say **P%**) with the target price.

$$\text{Target Profit} = P\% \times \textit{Target Price}$$

The P% is the ratio of overall target profit amount over the total revenue generated based on the target prices and its related product volumes.

$$P\% = \frac{\text{Overall Target Profit Amount}}{\{\text{Target Price for each product X Projected Volume for each product}\}}$$

Once target profit and target price are determined, target cost for each product can then be established. It will just be the difference between target price and target profit. With target cost established, the firm will then estimate the drifting cost for each product based on activity-based cost information. To determine drifting costs using activity-based costing concept, the materials and the activities needed to make the product together with its associated cost drivers must first be identified. The costs of resources used by the activities will be added up together with material cost to arrive at a total drifting cost for the product. It is assumed that the firm has implemented activity-based cost accounting, thus all major activities and cost drivers are readily known. At this stage, the firm is ready to proceed to the second part of the model, which is to implement target costing to align drifting cost to target cost.

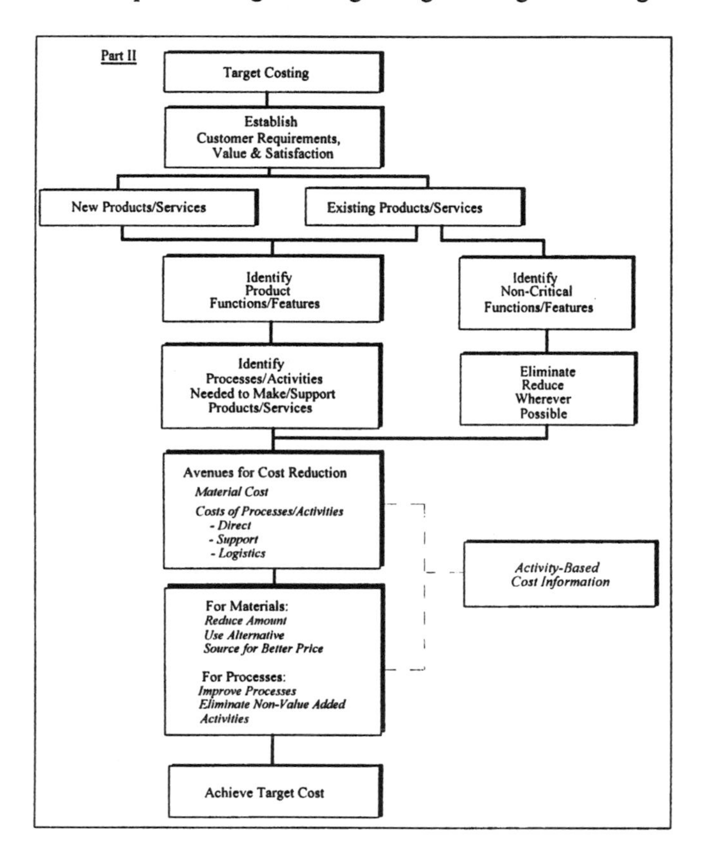

In target costing, products having a drifting cost higher than the target cost will be candidates for cost reduction. For each product (new or existing) requiring realignment to target cost, careful analysis of customer requirements, value and satisfaction measures must first be made. The documentation of these customer's attributes is to ensure that cost reduction will not be done on the expense of diminishing customer's attributes and also to assist the team to focus on appropriate areas for cost reduction.

For new products, the key or critical functions as well as the related processes needed to make them are identified. For existing products, besides identifying critical functions and processes, non-critical functions and processes are also noted. Cost reduction for existing products should first concentrate on reducing or eliminating these non-critical functions and processes. Critical functions are those that have the greatest impact on customer's attributes.

With the aid of activity-based cost information, the cost reduction team will have better insight on what activities and processes are incurring the most costs. Also, instead of just trying to reduce material costs, one also has a whole range of activities (whose costs are known) in which to improve, eliminate or combine to reduce costs. These kinds of added transparency allows the team to direct their effort to reducing cost in areas that will generate a faster rate of cost reduction and to conduct tradeoffs and sensitivity analysis.

CONCLUSION

This paper has presented a model for establishing target cost and implementing target costing. Instead of relying on accounting measures to derive target cost, the model suggested that it should be derived from desired economic value for the firm. In this model, all new and existing products are considered for target costing and are allocated a target cost each. The model also introduced the use of activity-based costing to assist in setting drifting costs and target costing. With activity-based costing, target costing can be carried out with better accuracy and more flexibility. It also helps to inform the cost reduction team of all the drifting costs that will be affected when the costs of some activities (shared by few products) are reduced. By implementing this model, the firm will be able to have a global picture of all cost reduction plans to be undertaken to achieve target costs for all products simultaneously. This capability provides the firm with strategic advantage and in essence, the model is therefore a blueprint for strategic target costing.

REFERENCES

[1] Miles, Lawrence D., *Techniques of Value Analysis and Engineering*, 2nd ed., McGraw-Hill Book Co., New York, 1972

[2] Sakurai, M., "Target Costing and How to Use It," Cost Management, Summer, 1989, pp. 39-50.

[3] Monden, Y. *Cost Reduction Systems: Target Costing and Kaizen Costing*, Productivity Press, Portland, Oregon, USA, 1995.

[4] Tanaka, T. "Target Costing at Toyota," Cost Management, Spring 1993, pp. 4-11

[5] Sakurai, M. *Integrated Cost Management*, Productivity Press, Portland, Oregon, 1996.

[6] Troxel, Richard B. and Weber Jr., Milan G., "The Evolution of Activity-Based Costing," Cost Management, Spring 1990, pp 14-22.

[7] Cooper, Robin, "The rise of activity-based costing - Part One: What is an activity-based cost system?" Cost Management, Summer1988, pp.45-54.

[8] Cooper, Robin & Kaplan, Robert S., "How Cost Accounting Systematically Distorts Product Costs," Accounting and Management Field Study Perspectives, Chp.8, 1987, Harvard Business School Press.

[9] James L. Dodd and Shimin Chen, “EVA: A New Panacea,” B&E Review, July-Sept. 1996, pp. 26-28.

MULTI-CRITERIA EVALUATION OF AUTOMATED MANUFACTURING TECHNOLOGIES

Budi Saleh, Sabah Randhawa and Thomas West
Department of Industrial and Manufacturing Engineering, Oregon State University

ABSTRACT: Evaluating investments in automated technologies is a complex task due to modeling of a continuously changing economic and manufacturing environment, and due to the evaluation of a large set of economic, technical, and strategic factors. This paper describes a process by which complex integrated manufacturing systems can be systematically evaluated and presents a set of attributes for use in such comparisons.

INTRODUCTION

Evaluating investments for automated manufacturing technologies is a complex but critical task faced by management due to the high capital investment required in such investments and due to the complexity and uncertainty associated with estimating and incorporating economic, technical and marketing factors in the decision process. Due to the potential of increasing productivity, efficiency, and quality, many manufacturing organizations have either automated their existing operations or replaced existing operations with automated technologies, such as flexible manufacturing systems. Yet, compared to expectations, many such transitions have ended up in systems being underutilized, dissatisfying both management and production personnel.

Certain characteristics of advanced manufacturing technologies make their justification process more complex than that required for production equipment in the past. Newer technologies require complex control structures with full integration of all manufacturing functions and component interfaces. Furthermore, many of the advantages of newer technologies lie not in the area of cost reduction but in areas such as improved quality and increased flexibility, factors that are much more difficult to quantify. Such factors have often been ignored in the past due to the difficulty in measuring and quantifying some of these factors and the difficulty in establishing dependency relationships among these factors. A real need exists to include such factors in the evaluation and selection of manufacturing technologies.

A number of techniques have been proposed, and some of them used, in the evaluation and justification of manufacturing systems. These include purely economic-based measures such as the net present value and the internal rate of return, weighted scoring models, analytic hierarchy process, utility theory, optimization and simulation models, and expert-based strategic models. The readers are referred to Meredith and Hill [1987], Meredith and Suresh [1986], Proctor and Canada [1992], Randhawa and West [1992a], Son [1992], and Suresh and Meredith [1985] for a discussion of justification concepts and techniques, and further references in this area.

This paper provides a framework to integrate the economic, technical, and strategic factors in the investment decision. The framework, discussed in the following section, integrates concepts from a number of modeling techniques within the weighted scoring model structure.

EVALUATION METHODOLOGY

The overall methodology is shown in Figure 1. The basic components of the system include:

Attribute Selection Process

Developing a list of relevant attributes is physically limited by the time and resources available. It is always desirable to thoroughly consider all the attributes in the decision environment. However, the time and effort required for such complex analysis is always prohibitive, and it is often possible to reduce the dimensionality of the problem without a significant loss of information and accuracy.

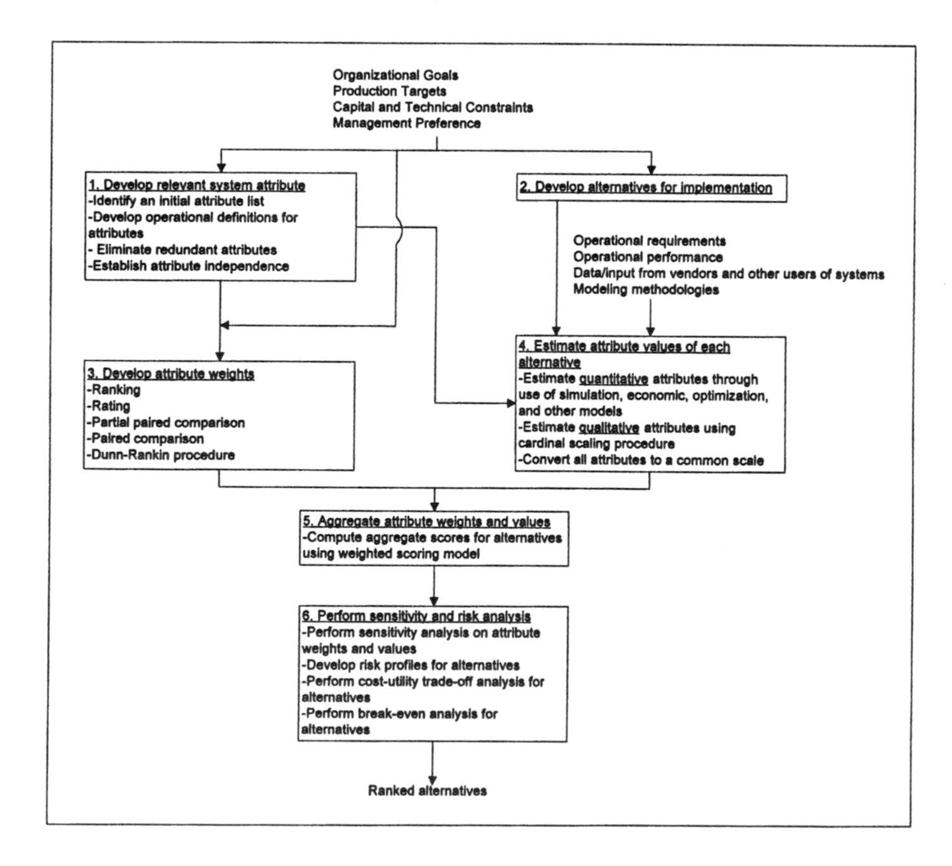

Figure 1. Evaluation methodology

The procedure starts by identifying a set of attributes through literature search and questionnaires, and developing operational definitions for the attributes identified. The attributes chosen as a basis of comparison depend on the specific manufacturing environment. Experience of management and production personnel is used to augment the literature review process, and minimize any omissions and duplications in the attribute set. Independence among attributes, required of most justification techniques including the weighted scoring model, can be enhanced using appropriate statistical procedures such as the factor or principal component analysis.

Alternative Identification

The selection of appropriate alternatives depends on existing technology, budgeting constraints, production goals, and management strategies. Identifying implementation alternatives must consider long term growth as well as realistic projections about the marketing and economic environment. Duplicating the status quo is always a viable alternative. Investing into new technologies, to replace or to augment existing facilities, requires interaction with vendors and other users of the system. Realistic cost and operational estimates for such systems have to be projected from similar applications. These estimates are difficult but critical for a thorough evaluation. Subcontracting is being increasingly used by many manufacturers, particularly as manufacturing assumes an international nature, and should not be overlooked.

Attribute Weighting

The attributes in the justification process are seldom considered equally important. The function of attribute weights is to express the importance of each attribute relative to the other attributes. Several techniques are available for developing weights using judgmental or subjective opinions. These include ranking procedures, ratio weighting, partial paired comparison, paired comparison, Churchman-Ackoff-Arnoff method, and the Dunn-Rankin method.

Attribute Scoring

Manufacturing systems are complex configurations of physical components interacting through extensive information networks and complex control protocols. Operational factors such as equipment utilizations, inventories and throughput rates affect the cash flows used to estimate economic attributes. These factors in turn are affected by factors such as equipment characteristics, scheduling priorities and equipment layout. These complex inter-dependencies and the resulting consequences are difficult to evaluate analytically.

A simulation model of a manufacturing alternative can be used to estimate many of the quantitative attributes. Properly developed, a simulation model can be used to model a wide range of complex manufacturing scenarios using basic modeling constructs (see Randhawa and West [1992b] for such a simulation modeling framework).

Not all quantitative attributes can be estimated using simulation. This is particularly true of systems that are not available in-house but are being considered as viable alternatives. Such evaluations are based on estimates from expert personnel, vendor data, and data obtained from other companies using the same technology in a similar manufacturing environment.

Estimates of qualitative attributes are usually based on a cardinal scaling procedure with the low and high values on the scale representing the minimum and maximum attribute values physically or practically attainable. These estimates generally represent subjective opinions of experts.

Alternative Scoring

The aggregate weighted value for each alternative is most easily obtained using the additive linear model that reduces the n-attribute problem into a single dimension problem using the following expression :

$$A_j = \sum_{i=1}^{n} w_i f_{ij}$$

where A_j is the aggregate score for alternative j, w_i is the weight for attribute I, f_{ij} is the score of attribute I for alternative j, and n is the number of attributes.

Sensitivity and Risk Analysis

This is an important step in the evaluation framework of Figure 1. Attribute weights and some of the attribute values are subjective assessments reflecting judgement of experts and decision makers. A sensitivity analysis involving attribute weights and values consists of investigating the sensitivity of the rankings of the alternatives to small changes in attribute weights or values. If the rankings remain unaffected with changes in weights and/or values, this can be established as evidence that errors in estimation are not important. If one or more weights or values prove to be very sensitive, the analysis will indicate that additional time and effort is required in estimating such values.

Sensitivity analysis is usually performed for each alternative. Once the robustness of the model and the estimated values used in the model is established, cost-utility analysis and break-even analysis can be used to compare multiple alternatives. Cost-utility analysis [Edwards and Newman, 1982] can be used to eliminate "dominated" alternatives (in term of higher cost and/or lower utility) and to establish trade-offs between costs and utility (representing non-quantifiable attributes). Break-even analysis can then provide a comparison of alternatives over time or different output levels.

ATTRIBUTES FOR COMPARING MANUFACTURING ALTERNATIVES

One of the more difficult tasks in the evaluation and justification of manufacturing systems is identifying an appropriate set of attributes for a specific application environment. Table 1 gives a list of typical attributes for a manufacturing application; a brief discussion follows.

System Cost

Total System Cost includes all future costs, acquisition costs, utilization costs, and disposal costs of the system, product or equipment. Figure 2 provides a typical cost structure for a manufacturing application.

TABLE 1
Attributes for Manufacturing Applications

Attribute	Components
System Costs	Research and development, acquisition and installation, operation and maintenance, retirement
Quality	Prevention, appraisal, internal failure, external failure
Flexibility	Machine, material handling, operation, routing, process, volume, product
Safety	
Training Requirements	
Technical and Management Support	Planning, organizing and staffing, directing, control, communication, space and facilities
Vendor Support	Availability, responsiveness, timeliness

Quality

Quality is the extent to which products meets the requirements of users. Similar to a number of other attributes in Table 1, quality has a quantitative as well as a qualitative component. The quantitative components includes costs associated with designing and implementing a quality control program, costs of rework and external failure costs. The qualitative component refers to customer acceptance or customer satisfaction with the product and the quality management system, and its implications for increased market share.

Flexibility

Flexibility is the ability of the manufacturing system to handle planned and unplanned changes economically and quickly. Changes may be a result of variations in either the internal operations and culture of an organization, or variations in the external environment including competition, new products and changes in demand patterns, or both.

The notion of flexibility has been discussed extensively in literature (see for example, Azzone and Bertele [1989], Browne et al. [1984], Gupta and Somers [1992], Son and Park [1987], and Suresh [1990]). Components of flexibility include:

- Machine flexibility: The ability of a machine to perform multiple operations without major changeover time and costs.
- Operation flexibility: The ability of a part to be produced using different processes, or sequence of operations.
- Material handling flexibility: The ability to move different part types efficiently for proper positioning and processing through the manufacturing facility.
- Process flexibility: The ability of the system to produce a set of parts without major setups.
- Product flexibility: The ease with which new parts can be added or substituted for existing parts.
- Volume flexibility: The ability of the system to operate profitably at different output levels.

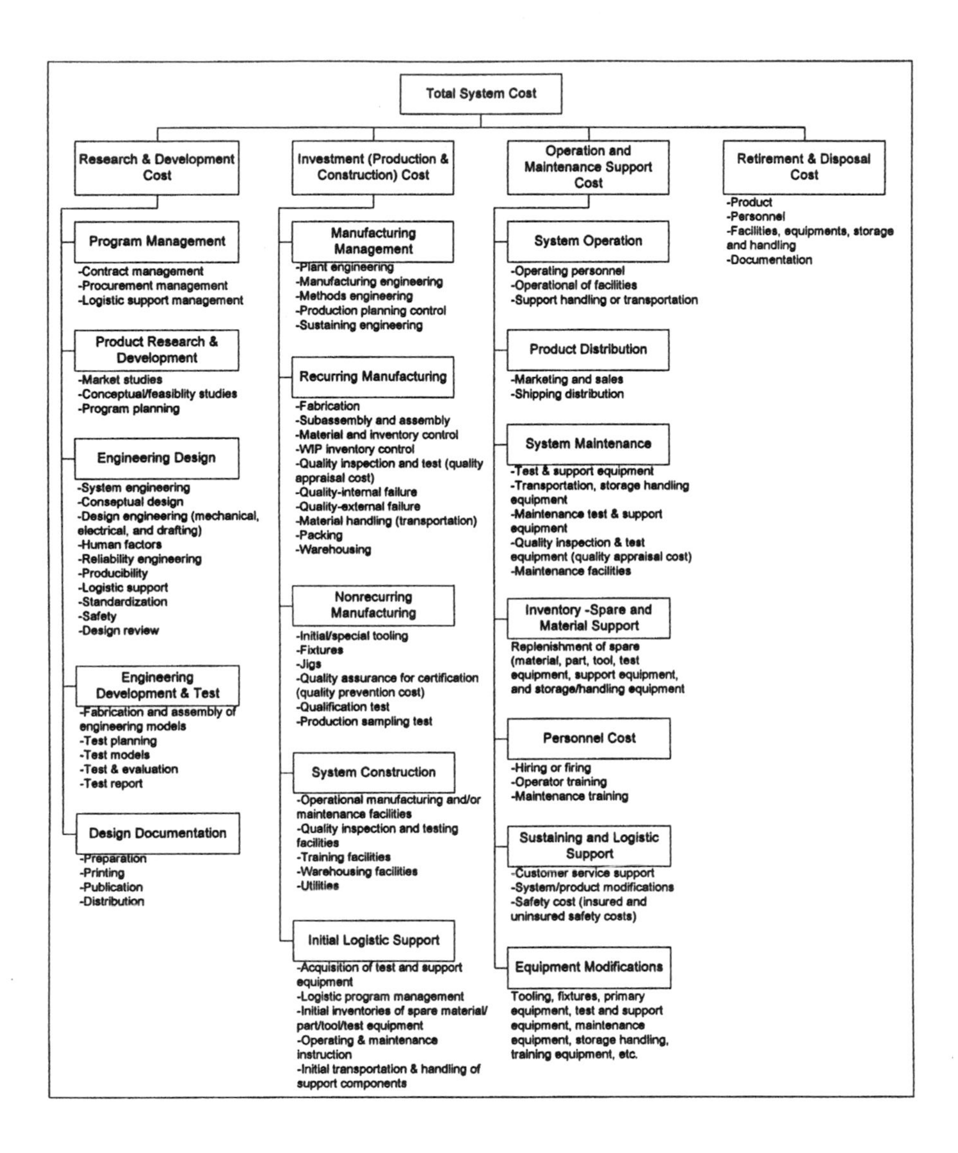

Figure 2. System cost

Other elements of flexibility defined in literature include routing, expansion, market, and program, though there is a considerable overlap between the different flexibility components.

Safety

Compared to conventional manufacturing, advanced manufacturing systems generally lead to a reduced exposure of workers to hazardous situations due to high levels of automation. The reduced exposure, in turn, reduces the workers' safety consciousness with the potential of hazardous situations less recognizable to the workers. The resulting costs to the system include insurance costs and uninsured costs (for example, lost workdays, and costs of repair and replacement).

Training Requirements

Usually technical training for traditional technologies was carried out in a relatively short period of time and for a relatively narrow domain of operations. Due to the complexity and integration inherent in modern manufacturing systems, their technical training requires complete conceptualization and cognitive understanding of the entire system. Issues that need to be considered in developing appropriate training programs include purpose, format, extensiveness, trainers, presentation techniques, timing, and curriculum.[Majchrzak, 1988]

Technical and Management Support

Advanced manufacturing systems result in a wide array of changes in human resource management, from planning and staffing to directing, control and communication. It is generally recognized that advantages, particularly some of the more qualitative benefits, associated with newer technologies cannot be realized without a technical and management support system focused on improving the working environment and enhancing employees morale and performance. For example, effective planning is a pre-requisite for best possible use of resources, timely solution to problems, and ensuring that the employees are aware of organizational goals and policies. Similarly, effective control and communication policies are required to increase creativity, innovation, responsibility, and accountability.

Vendor Support

Vendor support is critical to the success of implementing and operating newer technologies. Factors to be considered include: availability of service, responsiveness to need, professionalism of service, and the quality of service provided.

CONCLUSIONS

Evaluating automated manufacturing technologies requires consideration and integration of numerous factors related to the manufacturing technology, economic and market environment, and organization's goals and objectives. This paper presented a multi-criteria decision framework for evaluating such technologies. Some of more important decision criteria applicable in many manufacturing environments were briefly discussed.

REFERENCES

Azzone, G. and Bertele, U., Measuring the economic effectiveness of flexible automation: A new approach, *International Journal of Production Research*, 27, 735-746, 1989.

Browne, J. et al., Classification of flexible manufacturing systems, *The FMS Magazine*, 2, 114-117, 1984.

Edwards, W. and Newman, J. R., *Multiattribute Evaluation*, Sage Publications, Beverly Hills, CA, 1982.

Gupta, Y. P. and Somers, T. M., The measurement of manufacturing flexibility, *European Journal of Operational Research*, 60, 166-182, 1992.

Majchrzak, A., *The Human Side of Factory Automation: Managerial Human Resource Strategies for Making Automation Succeed*, Jossey-Bass Publishers, San Francisco, CA, 1988.

Meredith, J. R. and Hill, M. M., Justifying new manufacturing systems: An integrated strategic approach, *Sloan Management Review*, 28, 49-61, 1987.

Meredith, J. R. and Suresh, N. C., Justification techniques for advanced manufacturing technologies, *International Journal of Production Research*, 24, 1043-1057, 1986.

Proctor, M. D. and Canada, J. R., Past and present methods of manufacturing investment evaluation: A review of empirical and theoretical literature, *The Engineering Economist*, 38, 45-58, 1992.

Randhawa, S. U. and West, T. M., Evaluating automated manufacturing technologies: Part I -- Concepts and literature review, *Computer-Integrated Manufacturing Systems*, 5, 208-218, 1992.

Randhawa, S. U. and West, T. M., Evaluating automated manufacturing technologies: Part II -- Methodology for evaluation, *Computer-Integrated Manufacturing Systems*, 5, 276-282, 1992.

Son, Y. K., A comprehensive bibliography on justification of advanced manufacturing technologies, *The Engineering Economist*, 38, 59-71, 1992.

Son, Y. K. and Park, C. S., Economic measure of productivity, quality, and flexibility in advanced manufacturing systems, *Journal of Manufacturing Systems*, 6, 193-207, 1987.

Suresh, N. C., Towards an integrated evaluation of flexible automation investments, *International Journal of Production Research*, 28, 1657-1672, 1990.

Suresh, N. C. and Meredith, J. R., Justifying multimachine systems: An integrated strategic approach, *Journal of Manufacturing Systems*, 4, 117-134, 1985.

Performance Measures for a Centralized Material Handling in Flexible Assembly Line Systems

J. Ashayeri
Department of Econometrics
Tilburg University
The Netherlands

Abstract

In a Flexible Assembly Line (FAL) System, parts among work stations are usually transported by an automated material handling system. The automated material handling design decisions have an important impact on the overall performance of a manufacturing system. In this paper we focus on the Automated Guided Vehicle Systems (AGVS). Then by considering a FAL system as a set of tandem flexible workstations, a set of automated inspection stations, linked by an automated material handling system with a central material handling depot, the performance of the flexible assembly line system is modeled by Markov decision theory and queuing theory to quantify the performance of the material handling in the proposed flexible assembly line system.

1. INTRODUCTION

Automated Material Handling Systems (AMHS) and in particular Automated Guided Vehicle Systems (AGVS) are widely used in flexible manufacturing and flexible assembly line systems (see Pourbabai & Ashayeri, 1993). These systems offer many advantages over conventional forms of material handling. However, the design of these systems is complex due to the interrelated decisions that must be taken. A recent overview of the literature on AGVS decision problems can be found in Peters et al. (1996). This paper tries to make a contribution to the research on the AGVS with a central depot in a flexible assembly line environment.

The main belief in manufacturing industries today is that the AGVS play an important role in the "factory of the future". Therefore, we use the AGVS as the transportation medium. This however does not place a restriction on the models that are developed. The AGVS can be replaced by any other kind of transportation medium without affecting the models.

In theory it is often not feasible to deal with the whole of reality at once. Therefore, we study a simplified manufacturing environment. Certain assumptions are made based on which performance models describing the simplified manufacturing environment are developed.

First we formulate models of a discrete time system, i.e. different handling movements are performed in discrete time intervals. Here we start with a flexible assembly line consisting of only one AGV. Then performance models for 2 and K AGV's are developed, where K is the number of work-centers. Second we use continuous time modeling. Again we formulate performance models for 1, 2 and K AGV's respectively. In either case derivations of busy and idle time are given.

2. THE DISCRETE MODEL

In this section, we discuss a discrete time system. Discrete in the sense that after each time interval T a check is made whether a work-center has finished with processing of a job, if so then an AGV is dispatched to the work-center. T can be smallest traveling time denominator for visiting all work-centers.

2.1 One Automated Guided Vehicle

In this case we start with K work-centers, one central depot, 1 AGV and one product, see Figure 1.

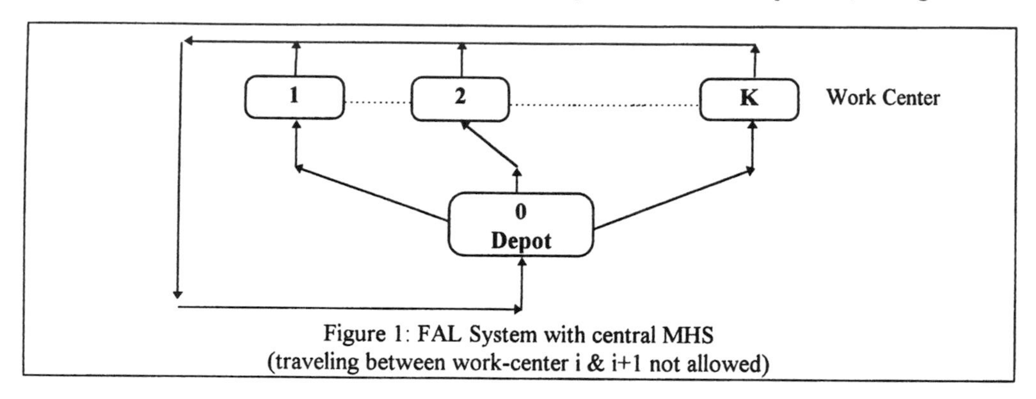

Figure 1: FAL System with central MHS
(traveling between work-center i & i+1 not allowed)

<u>Notations</u>

q_j: The probability that work-center j finishes processing during time interval T.
q'_j: The probability that work-center j finishes processing during time interval T/2.
State i: i=1 if the AGV is traveling to work-center i and back to the depot. i:=0 when the AGV is idle at the depot.
p_{ij}: The transition rate from state i to state j.
T_j: The time required for the AGV to travel from the central depot to work-center j and back to the depot.

The following assumptions are made for the development of performance models:

<u>Assumptions</u>

1. The T_j's are deterministic and equal for each work-center. $T:=T_j$
2. The time to travel to a work-center equals the time to travel back to the depot from the work-center, thus equals T/2. Every T/2 a check is made to see whether a work-center is ready with processing of a job.
3. The time needed for depositing an unprocessed product and for picking a processed product at the work-centers can be neglected in comparison with T.
4. The product has to flow first through work-center 1, then to work-center 2 etc. until work-center K before being unloaded from the system.
5. The AGV travels to a work-center and from there on it goes immediately back to the depot.
6. Raw materials are always available.
7. The probability that the central buffer in the MHS depot is empty or near it's maximum capacity approximates zero.
8. The processing time in work-center j is exponentially distributed with mean processing time $1/\mu_j$.
9. The probability that two work-centers complete in the same time interval T is negligible ($T << 1/\mu_j$).

Given these assumptions, we can study this particular situation. The central depot concept used allows us to model situation by a discrete Markov Chain with time interval T, with the states of the AGV as the Markov states. For this purpose we need to deduce the transition probabilities between the states.

To do this, we notice the following: The probability that work-center j finished processing in a time interval T equals

$$q_j = 1 - e^{-\mu_j * T} \approx \mu_j T$$

and the probability q_j' equals

$$q_j^{'} = 1 - e^{-\mu_j * \frac{T}{2}}$$

This is an immediate consequence of assumption 8, about the Exponential distribution.

The transition probability from state 0 to state 0 is equal to

$$p_{00} = \prod_{j=1}^{j=K} (1 - q_j) := \prod_{j=1}^{j=K} e^{-\mu_j * T}$$

The transition probability from state 0 to state j is proportional to q_j. Also the sum of the transitions probabilities from state 0 to states j equals 1, so

$$\prod_{j=1}^{j=K} (1 - q_j) + x_0 * \sum_{j, j \neq 0} q_j = 1$$

$$x_0 = \frac{(1 - \prod_{j=1}^{j=K} (1 - q_j))}{\sum_{j, j \neq 0} q_j}$$

Thus

$$p_{0j} = x_0 * q_j$$

The transition probability from state i to state 0 equals

$$p_{io} = \prod_{j(\neq i)} (1 - q_j) * (1 - q_i^{'}) := \prod_{j(\neq i)} e^{-\mu_j * T} * e^{-\mu_i * \frac{T}{2}}$$

The same kind of reasoning leads to

$$p_{i0} + x_i * [\sum_{j(\neq 0, i)} q_j + q_i^{'}] = 1$$

$$x_i = \frac{(1 - \prod_{j(\neq i)} (1 - q_j) * (1 - q_i^{'}))}{\sum_{j, j \neq 0, j \neq i} q_j}$$

$$p_{ij} = x_i * q_j \quad (j \neq i)$$

$$p_{ii} = x_i * q_i^{'}$$

With these deductions it is possible to pose a Markov transition matrix for any size of problem. This transition matrix will lead to the steady state solutions ✍$_0$ to ✍$_K$.

Let ☺$_j$ be the expected time that work center j is busy during T given the fact that the work-center was busy at the beginning of T. Let ☺'$_j$ be the time that work-center j is busy during T given the fact that the work-center was idle at the beginning of T. (So during the first T/2 the AGV is on its way to the work center, and the work-center is idle)

Then the expected busy time of work-center i is:

$$E(Busytime) = \sum_{j=0}^{j=i-1} (\pi_j * \tau_i) + \pi_i * \tau_i^{'} + \sum_{j=i+1}^{j=K} (\pi_j * \tau_i)$$

and

$$E(Idletime) = 1 - E(Busytime)$$

Also ☺$_j$ and ☺'$_j$ can be derived as simple conditional expectations.

$$\tau_j = \int_0^T t * \mu * e^{-\mu * t} \, dt + T * \int_T^{\infty} \mu * e^{-\mu * t} \, dt$$

$$\tau_j' = \int_0^{\frac{1}{2}T} t * \mu * e^{-\mu * t}\, dt + \frac{1}{2}T * \int_{\frac{1}{2}T}^{\infty} \mu * e^{-\mu * t}\, dt$$

2.2 A second version of One Automated Guided Vehicle

Assumption 5 is a very restrictive one, so we will relax it. Here we propose a second model for the discrete version with only one AGV in which it is also possible to go from work-center i to work-center i+1. However it is still impossible to go from work-center 1 to for example work-center 3, as a result of assumption 4.

extra assumption

10. The traveling time between work-center i and work-center i+1 can be neglected in comparison to T/2.

This assumption is not a unrealistic one. See Figure 2 for an example. Figure 2 shows a very important advantage of centralized MHS over MHS in a line. When using a MHS in a line for the manufacturing environment such the one illustrated in Figure 2 one would need K+1 AGV's, namely one AGV between K-1 pair of neighbor work-centers, and one AGV between the raw materials and work-center 1, and one AGV between work-center K and the finished products.

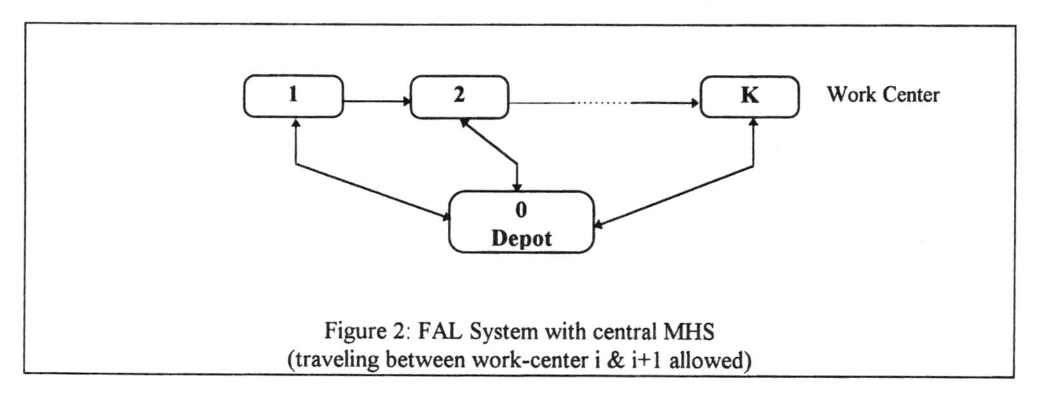

Figure 2: FAL System with central MHS
(traveling between work-center i & i+1 allowed)

In our case under assumption 10, when the AGV arrives at work-center j it is checked if work-center j+1 has finished processing. If so the AGV travels back to the central depot via work-center j+1. When the AGV is idle at the depot, the system checks every T time units whether there is any work-center which has finished processing.

Now we need to give a new definition concerning the states.

State (0,j): The AGV is traveling from the depot to work-center j.
State (j,0): The AGV is traveling from work-center j to the depot, (after traveling from the depot to work-center j).
State (j',0): The AGV is traveling from work-center j-1 to the depot, via work-center j. (State (1',0) does not exist!)
State (0,0): The AGV is idle at the depot within time interval (0,T/2)
State (0,0): The AGV is idle at the depot within time interval (T/2, T)

To provide the transition probabilities we use q_j' as described here above.

The probability to go from state (0,0) to state (0,0) equals

$$P((0,0)\ \ to\ \ (0,0)) = \prod_{i=1}^{i=K}(1-q_i)$$

and from state (0,0) to (0,i)

$$P((0,0)\ \ to\ \ (0,i)) = x_0 * q_i \quad (i \neq 0)$$

(with x_0 a proportional factor)
From state (0,0) the AGV comes in state (0,0) or in any of the states (0,i):

$$\prod_{i=1}^{i=K}(1-q_i) + x_0 * \sum_{i=1}^{i=K} q_i = 1$$

Therefore, the proportional factor x_0 becomes:

$$x_0 = \frac{1-\prod_{i=1}^{i=K}(1-q_i)}{\sum_{i=1}^{i=K} q_i}$$

For the transition from state (j,0) the same kind of deductions can be made:

$$P((j,0)\ \ to\ \ (0,0)) = \prod_{i=0}^{i=j-1}(1-q_i) * (1-q_j^{'}) * \prod_{i=j+1}^{i=K}(1-q_i)$$

$$P((j,0)\ to\ \ (0,i)) = x_j * q_i \quad (i \neq 0, j)$$

$$P((j,0)\ \ to\ \ (0,j)) = x_j * q_j^{'}$$

$$\prod_{i=0}^{i=j-1}(1-q_i) * (1-q_j^{'}) * \prod_{i=j+1}^{i=K}(1-q_i) + x_j[\sum_{i=1}^{i=j-1} q_i + q_j^{'} + \sum_{i=j+1}^{i=K} q_i] = 1$$

$$x_j = \frac{1-\prod_{i=0}^{i=j-1}(1-q_i) * (1-q_j^{'}) * \prod_{i=j+1}^{i=K}(1-q_i)}{\sum_{i=1}^{i=j-1} q_i + q_j^{'} + \sum_{i=j+1}^{i=K} q_i}$$

From state (0,j) there are two possible transitions:

$$P((0,j)\ \ to\ \ (j,0)) = 1-q_{j+1}^{'} \quad (j \neq 0, K)$$

(when work-center j+1 does not finish during the time T/2 the AGV is traveling to work-center j)

$$P((0,j)\ \ to\ \ (j+1^{'},0)) = q_{j+1}^{'} \quad (j \neq 0, K)$$

(otherwise)
There is no following work-center for work-center K, so the AGV always goes straight back to the depot after traveling to work-center K:

$$P((0,K)\ \ to\ \ (K,0)) = 1$$

The last possible transition is:

$$P((j',0)\ \ to\ \ (0,0)) = 1 \quad (j = 2,......,K)$$

This is a result of assumption 9 and the fact that when the AGV is in state (j',0), then work-center j has finished processing in the first T/2 part of the time interval T.

All the other transition probabilities are zero.

Now it is possible to pose the (3K-1)*(3K-1) Markov transition matrix and deduce the steady state solutions $✎_{i,j}$. This is the fraction of time the AGV is in state (i,j).

Using the definitions of $☺_j$ and $☺'_j$ provided in section 2, we derive the following statements for the busy and idle time of work-center i.

$$E(Busytime) = \sum_{j=0}^{j=i-1} (\pi_{j,0} * \tau_i) + \pi_{i,0} * \tau_{i'} + \sum_{j=i+1}^{j=K} (\pi_{j,0} * \tau_i)$$

$$+ \sum_{j'=2}^{j'=i-1} (\pi_{j',0} * \tau_i) + [\pi_{i',0} + \pi_{i+1',0}] * \frac{1}{2} T + \sum_{j'=i+2}^{j=K} (\pi_{j',0} * \tau_i)$$

and

$$E(Idletime) = 1 - E(Busytime)$$

2.3 Two Automated Guided Vehicles

Now, we discuss the same situation as in section 2.2, only we make use of 2 AGV's. So, as in section 2.2 an AGV can travel from work-center i to i+1. Because, we assumed that it is not possible that two work-centers complete in the same time interval, if one AGV is traveling to a work-center (so, that work-center just completed processing) no other AGV is traveling to the same work-center.

A new state definition is given below:

State ((0,0),(0,0))	AGV1 and AGV2 are idle at the depot.
State ((0,j),(0,0)):	AGV1 is traveling from the depot to work-center j and AGV2 is idle at the depot (j⑤0).
State ((0,j),(i,0)):	AGV1 is traveling from the depot to work-center j and AGV2 is traveling from work-center i to the depot (i⑤j).
State ((i,0),(0,j)):	AGV2 is traveling from the depot to work-center j and AGV1 is traveling from work-center i to the depot (i⑤j).
State ((j',0),(0,0)):	AGV1 is traveling from work-center j-1 to the depot via work-center j and AGV2 is idle at the depot (j⑤1).
State ((0,0),(j',0)):	AGV2 is traveling from work-center j-1 to the depot via work-center j and AGV1 is idle at the depot (j⑤1).
State ((j,0),(0,0)):	AGV1 is traveling from work-center j to the depot and AGV2 is idle at the depot.
State ((0,0),(j,0)):	AGV2 is traveling from work-center j to the depot and AGV1 is idle at the depot.

By making use of the derivation in section 2.2 we come to the following transition probabilities. Not that the assumption 10 is still valid.

An important change is the fact that we check every T/2 whether a work-center is ready. To simplify notation, we replace T/2 by T, q'_j by q_j, ☺'$_j$ by ☺$_j$, etc.

The transition probability that the two AGV's both stay at the depot is:

$$P(((0,0),(0,0)) \text{ to } ((0,0),(0,0))) = \prod_{i=1}^{i=K} (1 - q_i)$$

The probability that AGV1 has to go to work-center i while both AGV's were idle at the depot is:

$$P(((0,0),(0,0)) \text{ to } ((0,i),(0,0))) = x_0 * q_i \quad (i \neq 0)$$

From state ((0,0),(0,0)) AGV1 either stays at the depot or travels to work-center i, while AGV2 always remains at the depot:

$$\prod_{i=1}^{i=K} (1 - q_i) + x_0 * \sum_{i=1}^{i=K} q_i = 1$$

$$x_0 = \frac{1 - \prod_{i=1}^{i=K} (1 - q_i)}{\sum_{i=1}^{i=K} q_i}$$

The transition probabilities from the following states:

- *state ((j,0),(0,0))*
- *state ((0,j),(0,0))*
- *state ((j',0),(0,0))*
- *state ((0,0),(j',0))*

are all equal the transition rates from state ((0,0),(0,0)). For instance:

$$P(((j,0),(0,0)) \text{ to } ((0,0),(0,0))) = P(((0,0),(0,0)) \text{ to } ((0,0),(0,0)))$$

$$P(((0,0),(j,0)) \text{ to } ((0,0),(0,0))) = P(((0,0),(0,0)) \text{ to } ((0,0),(0,0)))$$

$$P(((j',0),(0,0)) \text{ to } ((0,0),(0,0))) = P(((0,0),(0,0)) \text{ to } ((0,0),(0,0))) \quad (j' \neq 1)$$

$$P(((0,0),(j',0)) \text{ to } ((0,0),(0,0))) = P(((0,0),(0,0)) \text{ to } ((0,0),(0,0))) \quad (j' \neq 1)$$

This is a result of the fact that in the case that just one AGV is traveling back to the depot, every work-center was processing at the beginning of the time interval T. When it is checked whether a work-center completed processing, the traveling AGV has arrived at the depot. So it is just as if the traveling AGV is idle at the depot, waiting for a new assignment.

Other transition probabilities are:

(j⑤K)

$$P(((0,j),(0,0)) \text{ to } ((j,0),(0,0))) = \prod_{i=1}^{i=j-1}(1-q_i) * \prod_{i=j+1}^{i=K}(1-q_i)$$

(None of the other work-centers finished processing)

$$P(((0,j),(0,0)) \text{ to } ((j,0),(0,i))) = x_j * q_i \quad (i \neq j, j+1)$$

(One of the other work-centers finished processing, but not work-center j+1)

$$P(((0,j),(0,0)) \text{ to } ((j',0),(0,0))) = x_j * q_{j+1}$$

(work-center j+1 finished processing)

These are all possible transitions from state ((0,j),(0,0)), thus:

$$\prod_{i=1}^{i=j-1}(1-q_i) * \prod_{i=j+1}^{i=K}(1-q_i) + x_j[\sum_{i=1}^{i=j-1} q_i + q_{j+1} + \sum_{i=j+2}^{i=K} q_i] = 1$$

$$x_j = \frac{1 - \prod_{i=1}^{i=j-1}(1-q_i) * \prod_{i=j+1}^{i=K}(1-q_i)}{\sum_{i=1}^{i=j-1} q_i + q_{j+1} + \sum_{i=j+2}^{i=K} q_i}$$

When j=K the formulas become as follows:

$$P(((0,K),(0,0)) \text{ to } ((K,0),(0,0))) = \prod_{i=1}^{i=K-1}(1-q_i)$$

(None of the work-centers 1 to K-1 finished processing)

$$P(((0,K),(0,0)) \text{ to } ((K,0),(0,i))) = x_j * q_i \quad (i \neq 0, K)$$

(One of the other work-centers 1 to K-1 finished processing)

These are all possible transitions from state ((0,K),(0,0)), thus:

$$\prod_{i=1}^{i=K-1}(1-q_i) + x_K \sum_{i=1}^{i=K-1} q_i = 1$$

$$x_K = \frac{1 - \prod_{i=1}^{i=K-1}(1-q_i)}{\sum_{i=1}^{i=K-1} q_i}$$

The transition probabilities from the states where both AGV's are in action (states ((0,j),(i,0)) or ((i,0),(0,j))) are equal to the transition probabilities from state ((0,j),(0,0)). This follows from an equivalent reasoning as above.

Now all transition probabilities can be calculated, and again this leads to the steady state solutions ✍$_{(i1,j1),(i2,j2)}$, via a Markov transition matrix. These steady state solutions can again be used the deduce statements for the expectations of the busy and idle time for each work-center. If a AGV is traveling from work-center i then the expected busy time is ☺$_i$. If no AGV is traveling from or to work-center i the expected busy time is also ☺$_i$. However when a AGV is traveling to work-center i then the expected busy time is 0. So it is much easier to calculate the expectation of the idle time first:

$$E(Idletime) = [\,\pi_{(0,i),(0,0))} + \sum_{j=1}^{j=K}(\pi_{((0,i),(j,0))} + \sum_{j=1}^{j=K}(\pi_{((j,0),(0,i))}] * T$$

and

$$E(Busytime) = 1 - E(Idletime)$$

2.4 K Automated Guided Vehicles

The mathematical model for the case of K AGV's is a generalized model from the one with two AGV's. Exactly the same assumptions can be made. But it is far too big a problem to write down here for a reasonable number K. Besides, it is not a very interesting problem, because when K AGV's are available it is better to create a material handling system in line.

3. THE CONTINUOUS MODEL

In this case the MHS makes use of a continuous monitoring system. This means that at any time a work-center completes processing, the AGV can be dispatched immediately. We start with K work-center, one central MHS, one AGV and one product.

3.1 One Automated Guided Vehicle

With only one AGV the traveling times stay deterministic. Therefore, this situation is the same as the discrete version for one AGV, discussed in section 2.1.

3.2 Two Automated Guided Vehicles

Now things are going to change because we use two AGV's. Because of the continuous monitoring there are no longer transition probabilities, but transition rates. One assumption however, has to be changed, in order to keep the Markovian property. Instead of deterministic traveling times, we assume exponentially distributed traveling times T_j. Thus assumption 1 from case 1 is replaced by a new assumption.

<u>*Assumption*</u>

1' The traveling times T_j are exponentially distributed with mean traveling times 1/🖰$_j$. Again these 🖰$_j$ are equal, i.e. the T_j have the same exponential distribution.

All the other assumptions remain unaltered.

Because the traveling time is assumed to be the time for an AGV to travel to a work-center and back to the depot, we restore assumption 5.

Extra assumptions

11. The processing times t_j all have the **same** exponential distribution with mean processing time 1/🖳.
12. The probability of completion by the work-center just served by an AGV before the arrival of the AGV at the MHS is negligible.

The state of the system is changed in the continuous variant inspired by queuing theory. The state is now defined by the number of work-centers waiting to be served by an AGV. In this definition the work-center served at the moment is included in the state. So, for example 3 work-centers waiting (idle) and one work-center served, corresponds to state 4. The AGV's will serve the work-centers in the queue according to first come, first served. All this together can be presented by a M/M/2, FCFS/K/K system. M/M/2 denotes Markovian arrivals, Markovian service times, and two AGV's (server). FCFS/K/K denotes the priority rule is first come, first served, the maximum number permitted into the system is K and also the size of the customer population is K.

We now have to derive the infinitesimal transition rates q_{ij} (see Tijms (1994), chapter 4).

$$q_{j,j+1} = (K-j)*\mu \quad (j \neq K)$$

$$q_{j,j-1} = \lambda \quad (j \neq 0)$$

$$q_{i,j} = 0 \quad (i,j,otherwise)$$

This leads to the following transition diagram.

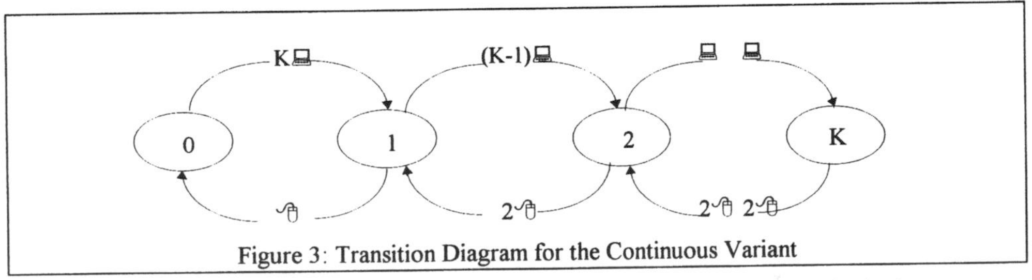

Figure 3: Transition Diagram for the Continuous Variant

Considering the fact that "Transition rate out of set A = Transition rate into set A" in the long run case, and choosing state i, $0 \leq i \leq K-1$ as set A, the following relationships can be derived:

(p_i := Fraction of time the system is in state i.)

$$\lambda * p_1 = K * \mu * p_0$$

$$2*\lambda * p_2 = (K-1)*\mu * p_1$$

$$2*\lambda * p_3 = (K-2)*\mu * p_2$$

$$\ldots\ldots = \ldots\ldots$$

$$2*\lambda * p_i = (K-i+1)*\mu * p_{i-1}$$

$$\ldots\ldots = \ldots\ldots$$

$$2*\lambda * p_K = \mu * p_{K-1}$$

$$\sum_{i=0}^{i=K} p_i = 1$$

These relationships can be used to solve for the probabilities p_i. So now we can calculate the time, in a certain relevant interval of time U, a certain number of work-centers are idle.

The total time of all work-center being idle equals

$$U * \sum_{i=0}^{i=K} i * p_i$$

The fraction of the total time a work-center is idle (not busy) equals

$$\frac{\sum_{i=0}^{i=K} i * p_i}{K}$$

3.3 K Automated Guided Vehicles

It is useless to have more AGV's than the amount K. In this case every work-center is being served. Here again we can consider the same assumptions as in section 3.2. (also with assumption 5). The relationships to calculate p_i are now changed into:

$$\begin{aligned}
\lambda * p_1 &= K * \mu * p_0 \\
2 * \lambda * p_2 &= (K-1) * \mu * p_1 \\
3 * \lambda * p_3 &= (K-2) * \mu * p_2 \\
4 * \lambda * p_4 &= (K-3) * \mu * p_3 \\
\ldots\ldots &= \ldots\ldots \\
i * \lambda * p_i &= (K-i+1) * \mu * p_{i-1} \\
\ldots\ldots &= \ldots\ldots \\
K * \lambda * p_K &= \mu * p_{K-1} \\
\sum_{i=0}^{i=K} p_i &= 1
\end{aligned}$$

Also here we can apply the same derivations for the long-run fractions of idle time we found in section 3.2.

4. CONCLUSIONS

In the manufacturing industry nowadays competition is based on flexibility, service, low cost etc. and a good material handling system can contribute to these competitive priorities. Automated Guided Vehicles fit perfectly into this perception. Therefore, it is very important to have good models to measure the performance of such systems. In this paper we attempted to model a centralized material handling system for a flexible assembly line manufacturing environment. Here, the AGV's are guided by a central host computer which will be aware of all movements of an AGV at any time. These movements are coordinated via a central depot.

To use the proposed models in practice it is most useful to investigate their performance through simulation. Then it will become obvious which of the options, discrete or continuous, will perform better in practice. Continuous models have performed better in other types of models developed for material handling system and we believe it will the case here too, however, only for a model that includes many AGV's, because then the assumption of having exponentially distributed traveling times is more realistic.

For the MHS controller less data are needed if a discrete check is performed. Thus improvement of discrete models is essential. Therefore, further study is needed to look into discrete models with different assumptions. For example, all distances between every work-center and the depot were assumed to be equal. This might not be realistic enough. Here it will be better to look for a more detailed model.

REFERENCE

1. Peters B.A., J.S. Smith, and S. Venkatesh, (1996), A Control of Automated Guided Vehicle Systems, Working Paper to appear in *International Journal of Industrial Engineering*, Dept. of Industrial Engineering, Texas A&M University, USA.
2. Pourbabai B. and J. Ashayeri, (1993), The effect of Material Handling on the Performance of Flexible Assembly Line Systems, in *Flexible Automation and Integrated Manufacturing*, edited by M. M. Ahmad and W. G. Sullivan, Florida: CRC Press Inc., pp. 627-635
3. Tijms H.C., (1994), *Stochastic Models*: An Algorithmic Approach, West Sussex: John Wiley & Sons Ltd.

PROBLEMS IN THE PROCESSING OF PARTICULATE SOLIDS - THE SYSTEMATIC ANALYSIS OF PARTICULATE PROCESSES

A.J.Matchett, Chemical Engineering, University of Teesside, Middlesbrough, England

Abstract
Industrial particulate processes are extremely varied and conventional methods of flowsheeting and process analysis tend to oversimplify and mask real processing problem areas.
The paper presents a method for analysing such systems entitled the Systematic Analysis of Particulate Processes. In this technique, a flowsheet is prepared based upon the experience of the actual particles. Heuristics commonly available to the particulate technology community may then be applied to identify problem areas. Mass and energy balances and economic analysis can then be used to find optimum processing strategies to meet process and product requirements.
The paper presents 2 examples of the application of this technique. A milling process is analysed to produce an ideal milling circuit for a specific situation. A mixing/extrusion process is studied and areas of processing difficulty are identified. Unusual situations are also identified and areas where external expertise must be sought.
It is hoped to develop the system into an automated 'expert' system in the future.

PARTICULATE PROCESSES

The process industries developed largely on the basis of fluid feedstocks and products, typified by the petroleum and petrochemical industries. A sophisticated system of flowsheeting, mass and energy balances has evolved to analyse such systems[1]. However, the processing of systems involving solids as feedstock, intermediate and/or product are seen to be increasingly important[2] : Ennis et al estimate that in 1992, for the US chemicals giant Du Pont, 62% of its 3000 products were powders, crystalline solids, flakes, dispersions or pastes. A further 18% of products incorporated particulates in their processing in some way.
Essentially the same techniques of process analysis are used for solids as for fluids, but there are significant differences between particulate solids and fluids, as pointed out by Jenike[3]. Table 1 compares fluid and solids systems on the basis of their fundamental constituent particles. Table 2 compares and contrasts frictional properties.
Several key differences can be highlighted between the two types of system which have far-reaching effects in terms of process design and development:

1. Particulate systems will not flow in all circumstances, due to limiting friction.
2. There is no particulate equivalent of molecular diffusion - this process cannot be relied upon for mixing in particulate systems.
3. There is a scale of scrutiny at the 'micro' scale, at which particulates are poorly mixed, regardless of their apparent mixedness at the 'macro' level.

Thus, flow cannot be taken for granted in solids systems, whereas in fluid systems the transport and storage of materials is a matter of routine. Various aspects of poor flow are frequent problems in many particulate processing plants, especially those related to flow from hoppers and chutes and conveying.

Table 1 Fundamental Particles in Fluid and Solids Systems

	FLUID	PARTICULATE
Fundamental particle	molecule	particle
Size	extremely small	variable microns to millimetres
Number of particles per sample	very large - Avagadro's number per gram-mole	finite-dependent upon particle and sample size
Inter-particle variation	all molecules of a species are identical in size and shape	variations in size and shape in most systems
Phase	single phase continuous gas or liquid systems	fluid between solids particles - discontinuities at the particulate level
State of motion	all molecules of a fluid are in a state of constant, random motion	particles do not move unless acted upon by an external force
Diffusion	molecular diffusion is present in all fluids	there is no particulate equivalent to molecular diffusion
Effects of gravity	negligible	significant for all but the smallest particles

Table 2 Frictional Properties of Newtonian Fluids and Powders

FLUID	POWDER
Obey Newton's laws of motion	Obey Newton's laws of motion
Friction always opposes motion	Friction always opposes motion
$F = -\mu_f A \frac{dv}{dy}$	$F \leq \mu_s N$
Shear at very low applied stresses	Limiting friction : no shear unless $F \geq \mu_s N$
F proportional to area	F independent of area of contact
F proportional to rate of strain	F independent of rate of strain and velocity for dense beds
F independent of normal stress	F proportional to normal stress
No slip at fluid-solid interfaces	Slip at solid-solid interfaces
Velocity gradients established	Velocity can be localised into shear zones

Furthermore, the finite size of particles with respect to processing create a number of problems not encountered in the processing of continuous fluid phases, related to mixing, segregation and the packing of beds of powders.

Conventional methods of process representation do not take account of these problems. Conventionally, each item of process plant is represented by a box as shown in Figure 1 and Figure 2. In many cases, each box could be the proverbial 'black box' because little is known of what actually happens inside. This situation is further compounded by the wide range of different types of equipment available. For example, there exists a vast array of different types of powder mixer[4][5], each with different geometrys, modes of operation and peculiarities of their own. Furthermore, the equipment tends to have been developed by mechanical engineers rather than process engineers - they have very good bearings, but no-one is quite sure what they actually do to the materials. In such circumstances, process specification is often a matter of 'catalogue engineering'. For new products, the process developer may be reduced to a costly series of pilot trials to find the most suitable equipment.

In contrast, academic research has concentrated upon studies at the individual particle level. This is perhaps best represented by the very complex discrete particle models that have been developed to study the motion of particles in a wide number of situations[6]. In spite of these developments, design procedures often still depend upon earlier work such as hopper design developed by Jenike in the 1960's[3].

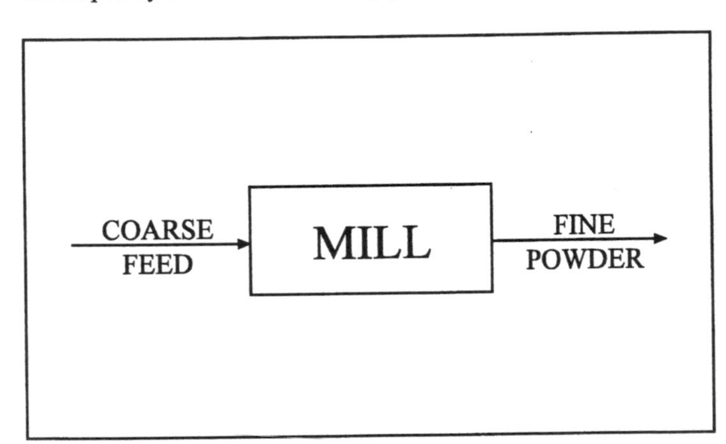

Figure 1 Flowsheet Representation of a Milling Process

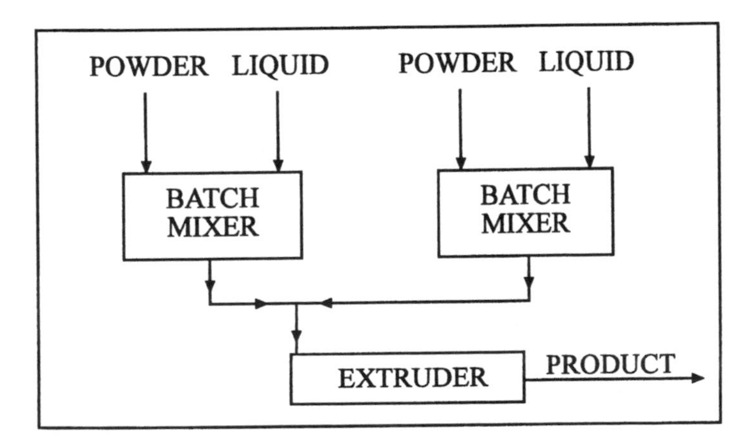

Figure 2 Flowsheet Representation of a Mixing/Extrusion Process

The preceding discussion suggests that a system of analysis of particulate processes is required which is able to take account of the inherent problems in these systems, yet provide a reliable and flexible design tool. Such a tool could be used for:

i) The analysis and optimisation of existing plant
ii) The study of individual items of plants or groups of items in an existing plant
iii) Design of new plant for an existing process
iv) Development of new processes

The following sections of this paper suggest how such an approach might be implemented and illustrates this with some examples.

SYSTEMATIC ANALYSIS OF PARTICULATE PROCESSES

The main feature of the proposed method of analysis is to prepare a flowsheet in terms of what a typical particle experiences upon passage through the process. This will be illustrated with 2 examples:

A comminution process - Figure 1
A mixing/extrusion process - Figure 2

Conceptually, we follow the experience of an 'average' particle upon passage through a piece of equipment or a whole process. Figure 1 shows a milling operation. This is nominally considered as a single piece of equipment, although subsequent analysis will show that it is more complex than that. Figure 2 shows a whole process involving mixing and extrusion.

In preparation of the particulate flowsheet, we must take account of all that happens to a particle. This include stages of flow and of rest or settlement. It has been shown above that conditions of flow cannot be taken for granted in particulate system and these MUST form an essential part of the analysis. Details such as mixing, backmixing and recycling should also be observed.

A Comminution Process

Figure 1 shows a conventional block representation of a milling process. It tells us very little - it does not even tell us what type of mill is used. Additional information such as throughput, power requirements etc, provide indications of the size, shape and other mechanical factors. However, this information does not answer some fundamental questions:

1. What are the process/product requirements?
2. Can the equipment meet these requirements?
3. What is the equipment actually doing to the material?
4. Is this the optimum approach to achieve the process/product requirements?
5. Are there likely to be any problems or problem areas and can these be identified BEFORE the plant is built rather than after

The situation in Figure 1 is for the comminution of a dry, free-flowing powder. Consider the use of a hammer mill for this duty. Figure 3 shows a redrawing of the flowsheet in Figure 1, expressed in terms of what the equipment actually does to the particles. The small boxes represent process steps that the particles go through, which are labelled in italicised script.

The standard script identifies the location within the equipment. Thus, within the mill chamber material undergoes flow, mixing and comminution before meeting the screen at which separation occurs based upon particle size.
It is clear that Figure 3 is far more complex than Figure 1. Even in very simple terms, a hammer mill consists of a feeder, the mill itself and then a cyclone to separate the milled product from the carrier gas used to take the material through the mill. Therefore, even at a very superficial level, Figure 1 is misleading and oversimplifying the complexity of the equipment. Such simplifications can lead the engineer to assume that a process step is either routine or trivial. In too many cases, this turns out to be an erroneous assumption.
A relatively small part of Figure 3 is actually concerned with the breakdown of particles. Furthermore, there are 2 separation stages involved: the screen is present within the mill to recycle oversize material and allow material below a specific size to leave the mill chamber; the cyclone separates the fine powder from the entrained carrier gas. Could one system perform both duties?

Mass and energy balances around the mill are required to provide a sound quantitative basis for further analysis. Several scenarios are then possible dependent upon the flowrates, feed conditions, specifications and value of the product. In all cases, overmilling emerges as a factor in both energy consumption and product quality. Such analyses must then be linked with an economic evaluation to decide upon future courses of action and possible process options.
We will consider a scenario where the prevention of overmilling is a crucial factor. In such a situation, the efficiency of the size separation stage is critical. The backmixing within the mill is also important - backmixing must be reduced to prevent milled material from re-entering the comminution and a 'once-through' or 'plug flow' milling system is required
One approach is to take the size separation stage out of the mill and operate it independently of the mill. The size separation can then be combined with the gas separation stage currently performed by the cyclone. This leads to the specification of a conventional milling circuit[7], BUT we know that such a milling circuit requires a type of mill with direct passage of material through the mill and a minimum of mixing and internal recycling within the mill. Furthermore, if the feed contains a significant proportion of fine material that already meets the product specification, then the feed should be fed into the separator and not the mill. Our ideal milling circuit is then shown in Figure 4.

This approach to analysis can be combined with heuristic rules from a wide range of sources. For example, if the fine powder in Figure 4 has a significant adhesion to surfaces, then this can be combined into any analysis to warn of the dangers of build-up of materials on retaining surfaces. If the separator in Figure 4 were a screen, then this would set up an alarm about the possibilities of screen blinding. Such possibilities would need to be thoroughly investigated before the process was finalised.
There are a great many heuristics within the particulate technology community. Many of these are not formally recorded but exist as experience with individual engineers. They are passed on, often by word of mouth, or even worse, discovered empirically several years later when a new generation of engineers make the same mistakes as the last one.
This systematic approach provides a framework within which such heuristics can be stored and used.

Figure 3 Hammer Mill Comminution as seen from the Particle's Viewpoint

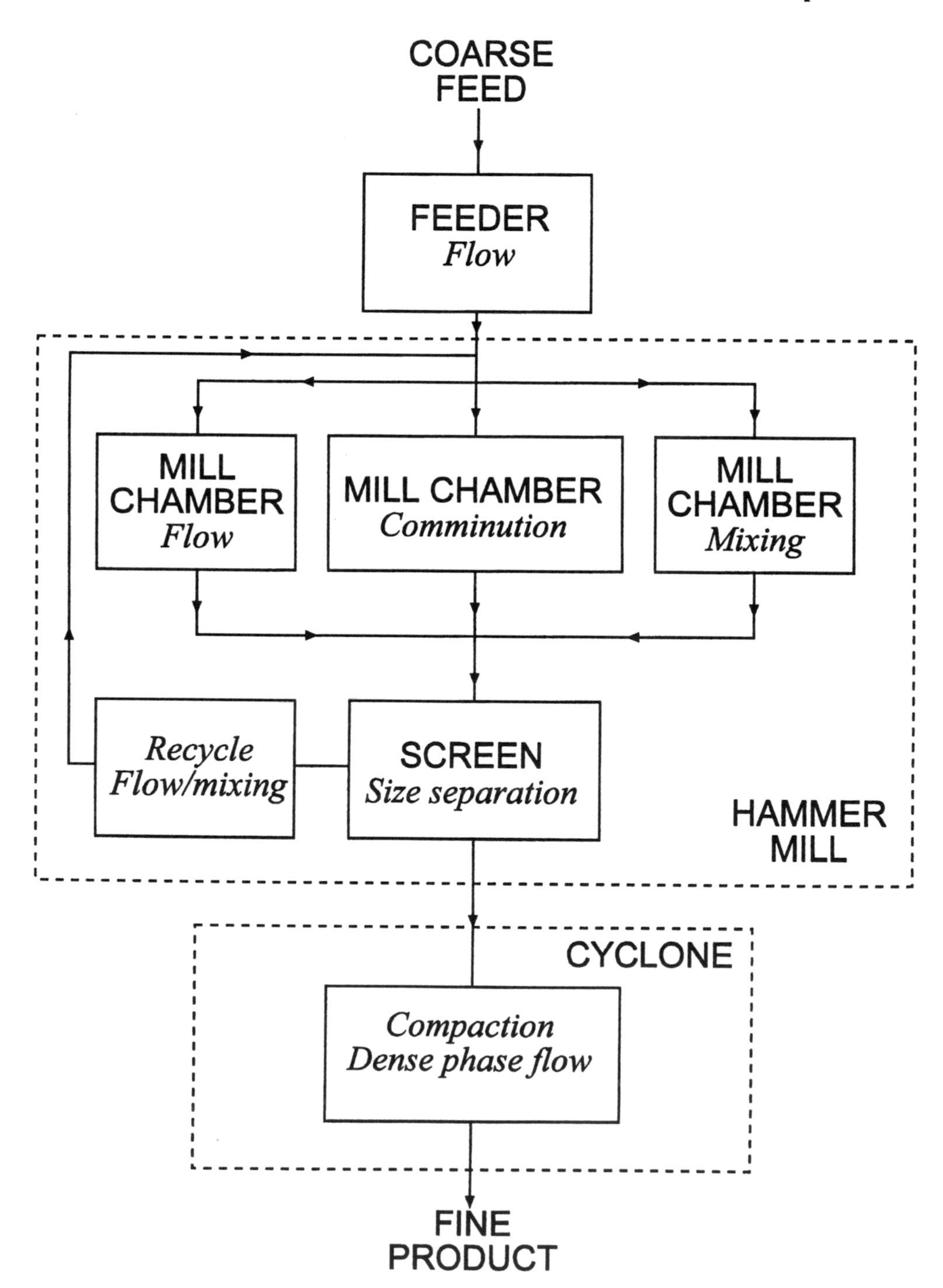

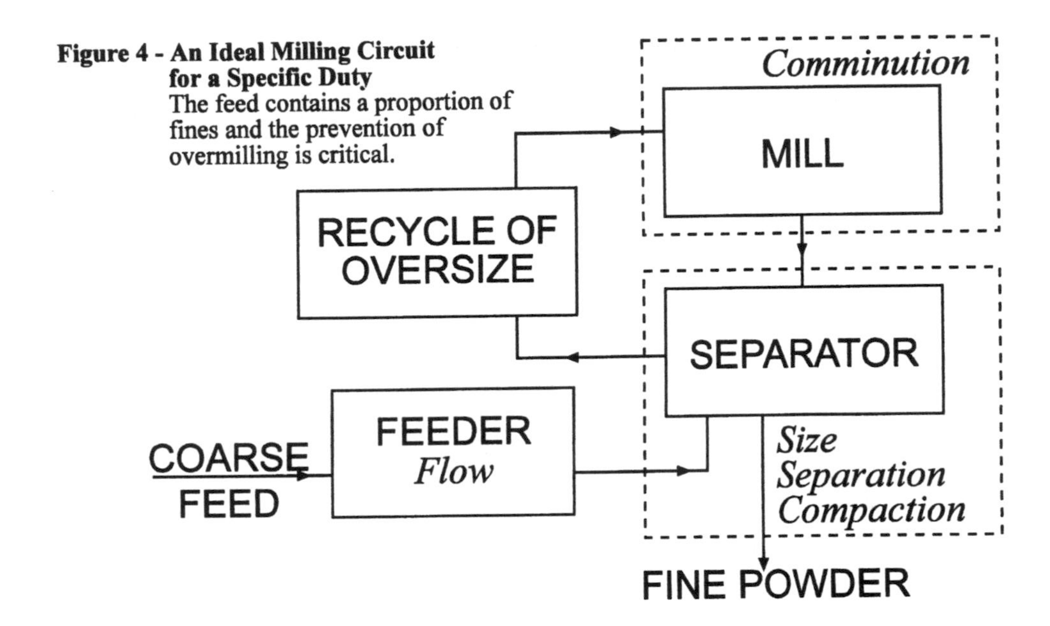

Figure 4 - An Ideal Milling Circuit for a Specific Duty
The feed contains a proportion of fines and the prevention of overmilling is critical.

A Mixing/Extrusion Process

Figure 2 shows the process, consisting of 2 batch mixers which work in parallel to premix a particulate solid with a liquid. This is then extruded to give product. This is based upon a real plant. A flowsheet in terms of the experience of the particles is shown in Figure 5. Again, Figure 5 is more complex that Figure 2.

We note a potential difficulty around the mixer exit and extruder feed. The extruder was originally fed by a small hopper. Thus, material was mixed in the mixer, allowed to settle after the mixing cycle, then fed into the extruder feed hopper, where again it was allowed to settle before finally being fed into the extruder itself. In other words, the material underwent a series of storage and flow cycles between the mixer and the extruder. This suggests, at least, redundancy of operation.

We may now apply another heuristic rule - wet materials are very difficult to handle and move about. They are cohesive and also stick to surfaces. Thus, we would expect difficulties between the mixer and extruder. This was observed to be the case in the actual plant - feed sat in the extruder feed hopper and was only fed into the extruder with difficulties. Furthermore, the opening to the extruder was far too small to consider a mass flow design based upon Jenike[3].

We note another heuristic at this point - flow is usually accompanied by dilation, or is a result of dilation. We see in the extruder a situation of flow accompanied by compaction. This indicates that something unusual is happening in the extruder. It is well known that extrusion is a very specialised field requiring great expertise - thus, the methodology has also indicated an area where expert knowledge should be sought.

The problems of this plant were solved by adopting a philosophy of keeping the material moving and 'live' between the mixer and the extruder such that the zones of material settlement were eliminated.

Figure 5 Mixing/Extrusion Process from the Particle's Viewpoint

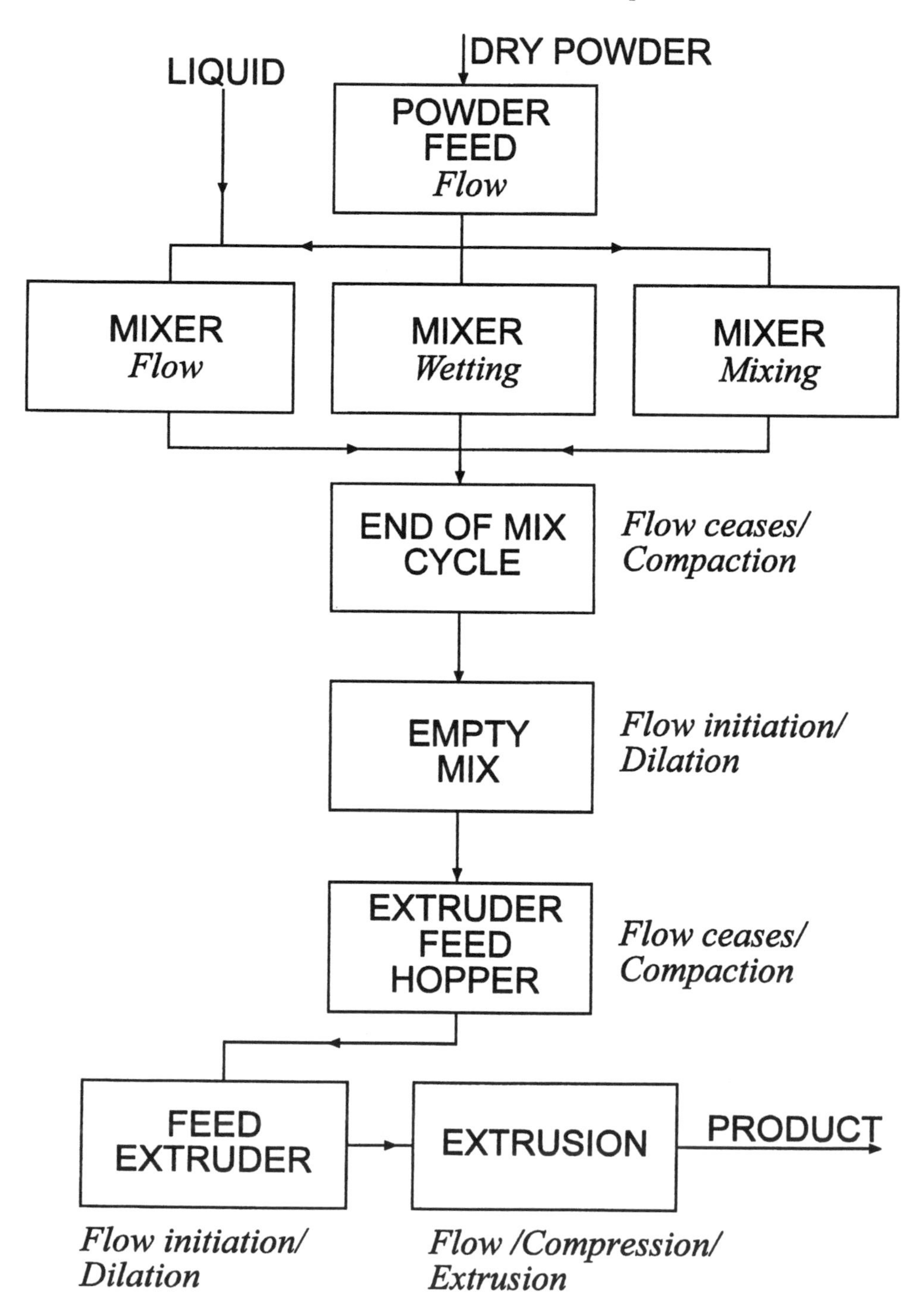

DISCUSSION AND FUTURE DEVELOPMENTS

The 2 examples have shown that the Systematic Analysis of Particulate Processes is able to do the following:

1. Provide a rational and simple method of analysis of particulate processes at a number of levels including:
   - Individual items of equipment
   - Sections of processes and unit operations
   - Whole processing plant
2. The methodology can be used with existing plant and/or to develop completely new processes and products
3. The methodology is intended to be used with mass and energy balances, and economic analyses in order to optimise processes
4. The methodology can identify redundant stages in a process
5. Problem areas can be identified
6. Unusual events are evident
7. Areas where additional expertise is required become apparent

In some processes, there may be so little known that it is difficult to construct a particulate flowchart. However, this in itself is a warning that there is very known about the process. The processes involved are very unusual and therefore great care should be taken. Furthermore, perhaps additional expertise should be sought or some fundamental research needs to be done to obtain the necessary information. The methodology therefore also forms a basis for the planning of process development research.

This methodology and expertise are currently available through the author and is currently being used at Teesside on a number of process studies.
Possible developments include a rigorous system of flowsheeitng with development of recognised standard methods of representation. Ideally, the flowsheet should include particle history, equipment details, physical state of the material(wet, dry, coarse, fine etc.) and identification of interactions between these by use of the heuristic rules. An extension to multi-component solids systems may also be required - different materials in a multi-component system will probably have different processing histories.
It is intended to develop these concepts into an expert system which incorporates the various heuristic rules that are found within the particulate technology community. A system will be developed into which a flowsheet can inputted and the system will then apply the heuristic rules to identify possible problem areas and indicate possible redundancy of process stages. It may even suggest possible process improvements. Mass balances, energy balances and economic analysis will be included at a later date.

References

1. Himmelblau D.M.,'Basic Principles and Calculations in Chemical Engineering', Prentice-Hall International Ed., 1982, USA
2. Ennis B.J.,Green J., Davies R., Chem.Eng.Prog., 1994, **90**, No.4,32
3. Jenike A.W., Bulletin 123, Utah Experimental Station, University of Utah,1963
4. Perry J.H. Chemical Engineers' Handbook, 6th Edition, Chapter 21, McGraw-Hill, 1984
5. Harnby N., Edwards M.F., Neinow A,'Mixing in the Process Industries', Butterworth, 1990

6. Koenders M.A., Stefanovska E., Powder Technology, 1993, **77**, 115
7. Perry J.H. Chemical Engineers' Handbook, 6th Edition, Chapter 8, McGraw-Hill, 1984

Notation

A	Area of contact	m^2
F_f	Fluid frictional force	N
F_s	Solids frictional force	N
N	Normal contact force	N
v	Velocity	ms^{-1}
y	Co-ordinate normal to direction of flow	
μ_f	Fluid viscosity	Pa-s
μ_s	Solids coefficient of friction	-

SEVENTH INTERNATIONAL CONFERENCE
FLEXIBLE AUTOMATION AND INTEGRATED MANUFACTURE

A NEW APPROACH TO SMALL PARTS VIBRATORY CONVEYING

A H Redford, University of Salford, Salford, England
F Dhailami, University of the West of England, Bristol, England

ABSTRACT

In the course of research into flexible assembly systems, a linear vibratory small parts feeder has been developed which, because of its small intrusion into the working envelope of the manipulator, is suitable for this application. As an extension to this work, and to help designers develop orienting tracks for particular applications, a CAD/CAM system has been developed for the design of linear feed tracks with all the necessary orienting features for a range of applications. The intention is to take out of orienting system design the high cost of both designing and manufacturing these features, to enhance their scope of application and to improve their reliability.

This paper describes the feeder design and the CAD/CAM system which has been developed.

INTRODUCTION

In 1982, the Department of Aeronautical and Mechanical Engineering, University of Salford won a contact from the then SERC to look at the economics of small parts feeding for applications in Flexible (Robot) Assembly. The main conclusions of the study were that:

- Invariably presenting parts in pallets or magazines is more expensive than presenting parts using small parts feeders.
- There is no apparent commercial alternative to the vibratory (bowl) feeder.
- Experimental, more generic feeders are either too slow or too expensive and limited in application. i.e. They are not generic
- The vibratory bowl feeder, whilst technically more than capable of meeting the requirements, has an expensive part-dependent feature, the feeder bowl with it's orienting devices.

After looking at the characteristics of the vibratory bowl feeder and comparing these with the specification for applications in flexible assembly is was reasoned that, for flexible assembly, the vibratory bowl feeder:

- Can supply parts at a higher rate than is necessary.
- Has a storage capability greater than that needed
- Is the wrong shape when the requirement might be for many feeders in a limited space
- Does not lend itself to more modern and less expensive means of manufacturing the part-dependent element, the feeder track with it's orienting devices.

With this in mind, a specification was drawn up for a linear vibratory feeder which better meets the requirements and a prototype feeder was designed and manufactured. The feeder was a twin track unit (to reduce feeder costs) and the method of recirculating parts was achieved by using the feeder track vibration to assist the movement of parts down a sloping hopper base and then elevating the parts to the feeder track level using pneumatic elevators.

In a subsequent contract with SERC, a CAD/CAM system was developed to manufacture the features necessary on a linear track to create orienting devices for particular applications. This was developed using a solid modeller and the user could design, develop, modify, store and retrieve information relating to a library of track features. The system was tested and was capable of producing a feeder track with the necessary devices in approximately two hours. The benefits of this were :

- Track manufacture was low cost.
- Features could be manufactured accurately and hence repeatably.
- Old successful designs are 'remembered' and can be retrieved and reused.
- The disadvantages of the new feeder and software were:
- On the twin track feeder it was difficult to control the tracks independently.
- The method of recirculating parts was not reliable.
- The CAD/CAM system, whist very effective was not user-friendly and needed a very skilled operator to use it.

When the DTI/FAMOS project INFACT started in 1989, a materials handling system for a Generic Flexible Assembly Machine was developed. The main element of this work was developing the fixture and pallet transfer system but, additionally, some development work on the linear vibratory feeder was carried out. This was a low-key activity and with the premise that the feeder worked and the feeder tracks could be designed, the only work which was done was to design a single track feeder, redesign the re-circulation system to use a belt conveyor and to try to take some cost out of the unit. This was done and prototype units were developed for INFACT which worked but which still had some obvious disadvantages. These were:

- The feeder drive was purchased from a commercial manufacturer and this was not particularly satisfactory in that it's performance was poor and it's cost high.
- The CAD/CAM system problems still needed to be addressed.
- The basic unit still needed evaluating and designing for economic manufacture.

In 1991 another DTI/FAMOS project ALASCA was started which was to design and develop a Flexible Assembly Machine for the assembly of large products and subassemblies. This also has the requirement for an effective and low cost small parts feeder and work was continued on this. In particular, it was decided that a feeder drive would be developed to be technically superior to that used previously and, more importantly would be of lower cost. It was also decided that the CAD/CAM system would be modified to make it both more user-friendly and more effective.

DRIVE SYSTEM

The vibratory drive system developed has two distinctive technical features:

- The feeder is driven by standard electromagnetic means but by a low voltage power amplifier. This has the significant advantage of low cost and can operate at any reasonable frequency. For it to function a standard electromagnet - iron polepiece combination is not appropriate and the polepiece needs to be either a permanent magnet or a DC electro-magnet.
- Rather than use an oscillator as the input to the power amplifier, a low cost vibration sensor is attached to the feeder track and with a suitable phase shift between the sensor and the amplifier, the feeder track can be made to resonate. As well as reducing power consumption. the big advantage of this is that the frequency of operation is the natural frequency of the drive mass-spring system which results in a very stable and low cost single frequency oscillator.
- The drive system has a vibration damper which significantly reduces vibrations imparted to the surroundings. This should allow the feeder to be fastened directly on to the assembly equipment without any adverse effects such as vibration of manipulators and, in particular, end effectors.

The vibration system is shown schematically in Fig. 1

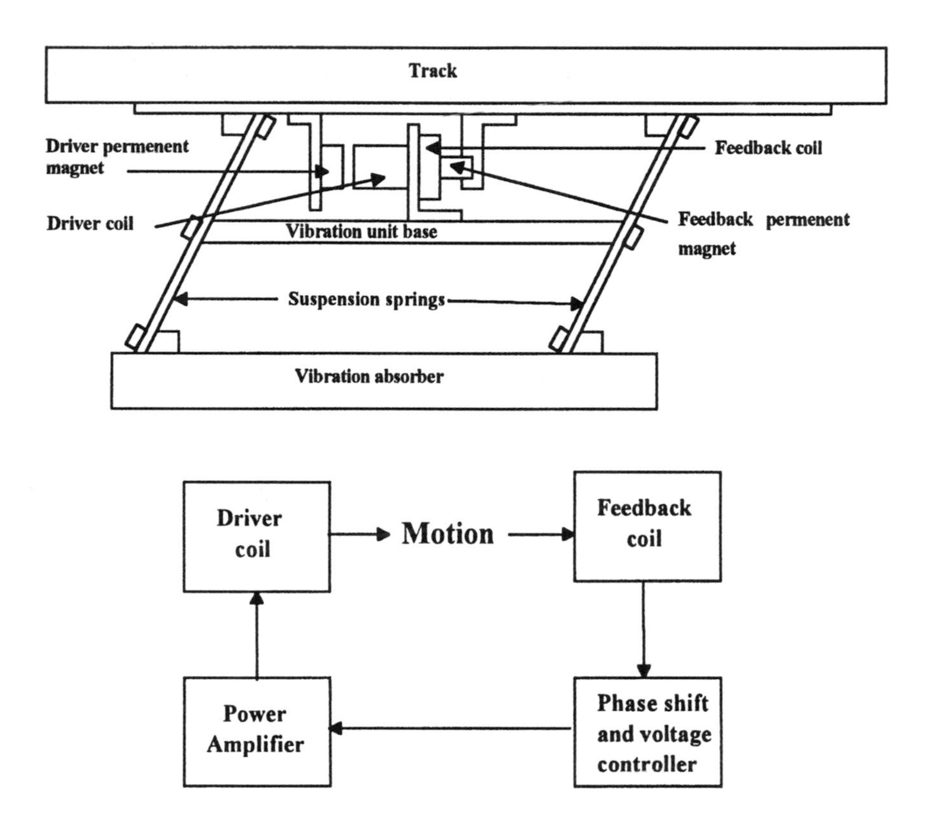

Fig. 1 Vibration System and control circuit

FEEDER STRUCTURE

The vibratory bowl feeder, by definition has a natural re-circulation system for parts which are rejected by orienting devices because they are in the wrong orientation. Clearly, a linear feeder does not have this and the separate parts re-circulation system represents an increase in cost. However, the separate system has two distinct advantages over using the bowl of a bowl feeder as a hopper:

- The supply of parts to the feeder track can be controlled by slowing down or speeding up the belt. This means that parts can be put onto the track such that there is a tendency for them to be spaced out; this is particularly useful when parts tend to tangle or nest.

- Because the hopper size is not constrained by the bowl size, it can be manufactured to any realistic size with only marginal increases in cost and without drastically affecting the size of the vibration unit.

The feeder structure which has to support the drive system, the belt conveyor, the hopper and the feeder controller has been redesigned to allow for more cost effective manufacture. Additionally, the specification has been broadened to allow the feeder to be used in more general high rate applications where a large hopper would be needed. The design allows for hoppers of various sizes to be fitted with only marginal changes in cost.

CAD/CAM SYSTEM

The existing CAD/CAM system was evaluated and it was concluded that as well as not being user-friendly, another feature which needed to be considered. was the method of manufacturing the feeder track. For the existing system, the track is manufactured in one piece and this causes problems.

- Machining has to be carried out a sequence which does not cause important features to be removed by subsequent machining.
- Machining small features near the track wall results in access problems which can only be addressed by reducing the height of the track wall.
- At the intersection of two devices, machining contours leave curved edges and, sometimes, this causes parts to move unsatisfactorily.
- If the track does not function satisfactorily, the whole new track has to be manufactured.
- Individual sections of track containing devices cannot be tested independently.
- There is no mechanism for controlling the mass of the track and this is important for a system using a vibration damper.

As a result of this evaluation it was decided that the feeder track would be fabricated with an L-shaped backing plate to form the track wall and a base for attaching track sections. Individually manufacture track sections would be manufactured to attach to the backing plate as appropriate. This gives important control in track manufacture because:

- The CAD/CAM system is much simpler because all features on track sections can run completely through the section and there are no 'interference' problems.
- There is no track wall to cause access problems.
- The interface between sections is 'square'.
- If a particular section does not function satisfactorily, it can be quickly and inexpensively re-manufactured.

- Tracks can be built in stages with the downstream stages not being manufactured until all upstrean stages function satisfactorily.

- The mass per unit length of each section can be controlled and kept constant by taking the volume of the blank track and subtracting from this the volume of metal removed for the orienting section profile plus the volume of the designated minimum track. The volume remaining is then the excess of material in the track section which can be removed by machining slots in the back of the track section.

A typical track section is shown below in Fig. 2

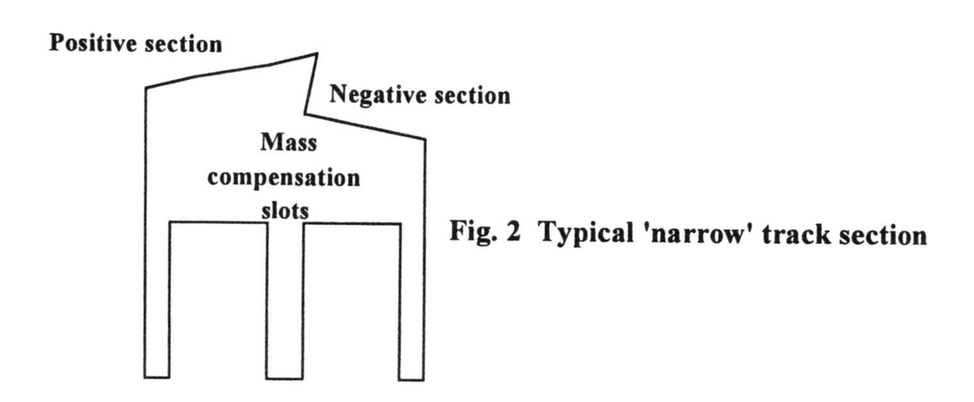

Fig. 2 Typical 'narrow' track section

TRACK DESIGN

Orienting devices in vibratory conveying invariably work on the principle of selection by rejection. i.e. Parts resting on unwanted surfaces and in unwanted orientations pass through mechanical filters (orienting devices) which reject them. During the track design process, the very first and most important step is to decide what combinations of device features should be used for a particular part and what are the dimensions of these device features. To feed and orient a different part, because of its different shape and dimensions, a different track must be designed. Similarly, to feed and orient the same part with a different final orientation will need different track models. Thus, choosing the correct devices and device features is fundamental and, to date, this relies completely on the designer's personal experience.

The creation of a track design software packages allows tracks to be manufactured on a CNC machining centre. Track fabrication is replaced by modern technology and the precision of track making has been improved but disconnecting the track device feature design from the designer's personal experience still remains a problem. The software package can create a library to store information and this will give some help to the designer but it is still insufficient and unsystematic and the question remains.

'Is it possible to set up a database for the device and feature design ? ' This would not only help the designer to design devices, but it would also speed up the whole design process. To answer the question, the key point is that is there any way to classify the parts and is there anything common for part family members during the device feature design.

To set up a database for track design, it is impossible to have a one to one correspondence for every part and its own track. Parts are combinations of different basic shapes, such as, rectangular prisms, cylinders and spheres. Because the combination of these basic shapes are countless, first of all, parts need to be categorised to build up the framework of the database.

How to classify parts into different meaningful categories is not easy. Thankfully, Boothroyd has already done some pioneering work in the field of part classifications. According to part's basic geometry, he codes parts using a three digit code, the first digit of the code is for basic shape, rotational or prismatic, the second digit is for basic symmetry and the third digit for asymmetry. This can be used as the foundations for the setting up of a track design database. Basic shapes are further categorised as being :

- discs (can only rest on ends), short cylinders (can rest on both side and ends and rods (can only rest on side).
- flat (can only rest on two faces), cubic (can rest on all faces) and long (can only rest on four faces).

In Boothroyd's part coding system, parts are classified into different categories based on their shape and the effect of differences in dimensions are ignored. For device and device feature design, this is insufficient. For example, if a part consists of two concentric discs with different diameters and the diameter of one disc is fixed, a diameter change of the other disc will result in different rest positions for the part. When diameter of one disc is very small then when part travels along the track, it will touch the track on one end or on an unstable side/end surface. When diameters of two discs are similar to each other there are still two rest positions but now the part will sit on the track stably on its two end surfaces. It is obvious that when diameters of two discs changed gradually from different to similar, the part can have three rest positions; it can sit on the track on both its end surfaces and on its end/side surface. Clearly, different combinations of devices and device features are needed for these different situations and the Boothroyd's part coding system is insufficient for this activity.

It is a daunting task to build a database which can cover every category, even when the basic part coding system has already been established. In Boothroyd's part coding system, he lists more than 400 different kinds of parts but in reality, only a small percentage of these are in frequent use and it is thought that, for these, a database can be set up and that it will be useful.

A start has been made and Fig. 3 shows device feature selection for end orientations of discs and short cylinders. Because the method of manufacture allows features to be generated very accurately, it was felt that this could be used to advantage by using the feature geometry of the part rather than the position of its centre of mass as the selection/rejection criterion because very small changes in feature size can be distinguished

			TRACK INCLINATION		
			POSITIVE	FLAT	NEGATIVE
ANY WAY ROUND	ANY WAY UP		.	W/STEP	.
			.	.	W/STEP
			.	SLOT	.
	ONE WAY UP	SLOT ON SIDE OR STEP ON SIDE AT BOTTOM	W/RAIL W/STEP	.	.
			.	W/RAIL W/STEP	.
			.	.	W/STEP
			.	SLOT	.
		STEP ON SIDE AT TOP	.	W/STEP	.
			.	.	W/STEP
			.	.	W/STEP SLOT
			.	.	W/STEP SLOT
			.	.	SLOT
			.	SLOT	.
		RECESS ON END AT BOTTOM	SCALLOP E/STEP	.	.
			.	W/STEP	.
			.	.	W/STEP
			.	SLOT	.
ONE WAY ROUND	ANY WAY UP	FLAT ON SIDE	W/STEP	.	.
			.	W/STEP	.
			.	.	W/STEP
			.	SLOT	.
		PROJECTION ON SIDE	.	W/STEP	.
			.	.	W/STEP
			.	.	SLOT
			.	.	E/STEP E/STEP
			.	.	E/STEP
			.	SLOT	.
		SLOT OR STEP ON SIDE	W/RAIL W/STEP	.	.
			.	W/RAIL W/STEP	.
			.	.	W/RAIL W/STEP
			.	SLOT	.

ONE WAY ROUND	ONE WAY UP THE FEATURE DETERMINES BOTH TH WAY ROUND AND THE WAY UP	SLOT OR STEP ON SIDE OR STEP ON FLAT	W/RAIL W/STEP	-	-
			-	W/RAIL W/STEP	-
			-	-	W/RAIL W/STEP
			-	SLOT	-
		SLOT OR STEP ON PROJECTION OUTSIDE DIAMETER	SLOT	-	-
			-	SLOT	-
			-	-	E/RAIL SLOT
				SLOT SLOT E/RAIL	-
			SLOT SLOT	-	-
			-	SLOT SLOT	-
			-	-	SLOT
			-	SLOT	-
		SLOT OR STEP ON PROJECTION INSIDE DIAMETER	W/RAIL W/STEP	-	-
			-	W/RAIL W/STEP	-
			-	-	W/RAIL W/STEP
			-	SLOT	-
	ONE WAY UP FLAT; SLOT ON SIDE; STEP ON SIDE; STEP ON FLAT ONLY DETERMINE WHICH WAY ROUND A FURTHER FEATURE IS NEEDED AND BOTH FEATURES MUST BE ABLE TO BE SEEN IN SILHOUETTE AT THE SAME TIME	SECOND FEATURE SLOT OR STEP ON SIDE	W/RAIL W/STEP	-	-
			-	W/RAIL W/STEP	-
			-	-	W/RAIL W/STEP
			-	-	E/RAIL W/STEP
			-	-	W/STEP SLOT
			-	-	SLOT
			-	SLOT	-
		SECOND FEATURE SLOT ON BASE	W/RAIL W/STEP	-	-
			-	W/RAIL W/STEP	-
			-	-	W/RAIL W/STEP
			-	-	W/STEP LOWRAIL
			-	-	LOWRAIL
			-	SLOT	-
	ONE WAY UP PROJECTION ON SIDE DETERMINES ONLY WHICH WAY ROUND A FURTHER FEATURE IS NEEDED AND BOTH FEATURES MUST BE ABLE TO BE SEEN IN SILHOUETTE AT THE SAME TIME	SECOND FEATURE SLOT OR STEP ON SIDE	SLOT	-	-
			-	SLOT	-
			-	-	SLOT E/RAIL
				SLOT SLOT E/RAIL	-
			SLOT SLOT	-	-
			-	SLOT SLOT	-
			-	-	SLOT SLOT
			-	SLOT	-
		SECOND FEATURE SLOT ON BASE	-	W/STEP	-
			-	-	W/STEP
			-	-	W/STEP LOWRAIL
			-	-	LOWRAIL
			-	SLOT	-

Fig. 3 Device features for stable end orientations of rotational parts with no tangling or nesting

Interestingly, because of the mechanism chosen for part rejection, virtual all the device features shown in Fig. 3 are simple through features which are easily programmable.

PREDICTION OF PARTS FEEDRATE

Before designing and manufacturing an orienting system, it is important to know with some confidence that the required feedrate can be met. To predict the feedrate, the part's shape, size, resting positions and orientations needs to be known. Additionally, any feeding 'difficulties' need to be established and their likely affect determined. These difficulties include very small and very large parts, tangling and nesting parts, sticky and slippy parts, fragile parts, abrasive parts, etc. After entering a part into the CAD/CAM system, the user enters a piece of software which poses relevant questions and which result in a prediction of expected feedrate.

RELATING PART FEATURES TO ORIENTING DEVICES AND DEVICE FEATURES

After establishing that the required rate can be met, it is important to relate part features to orienting devices and device features for the CAD/CAM system to be effective; this needs further questions about the part to identify the features which cause a part to have different orientations and to establish if these features are suitable for separating different orientations in the feeder. i.e. The part's features need relating to the information shown in Fig. 3 to establish if a unique part orientation can be obtained. Work is proceeding on the CAD/CAM system to create the necessary software.

CAD/CAM PROGRAM STRUCTURE

Table 1 shows a block diagram of the program structure. The software is now able to produce modular section of orienting tracks with wall steps edge steps and slots for any chosen track inclination. Where necessary, the software can also accommodate fabricated elements such as rails on the track and wiper blades on the wall in any position and to put in the necessary drilled and tapped holes for fastening. Of course, at this stage, the software is only able to make tracks which have been designed 'from experience' but the simple tracks which have been manufactured to date indicate that the methodology is appropriate and that, as expected, very accurate features can be generated.

CONCLUSIONS

A linear vibratory feeder has been designed and manufactured which facilitates low cost manufacture of feeder orienting tracks using CAD/CAM. The vibratory drive using a vibration absorber gives a smooth conveying speed for parts of 150mm per second and the dissociation of the rate of supplying parts from the rate of motion down the feeder track is particularly useful for parts which tend to tangle and nest.

A CAD/CAM system has been developed for the design and manufacture of linear vibratory feeder tracks. Experience indicates that, given a design, a typical set of track modules can be processed in thirty minutes with a manufacturing time for the modules of one hour.

A start has been made on relating part feature information to orienting device features and size. It appears that relatively few types of feature are needed and that most of these can be through features on modular sections of track.

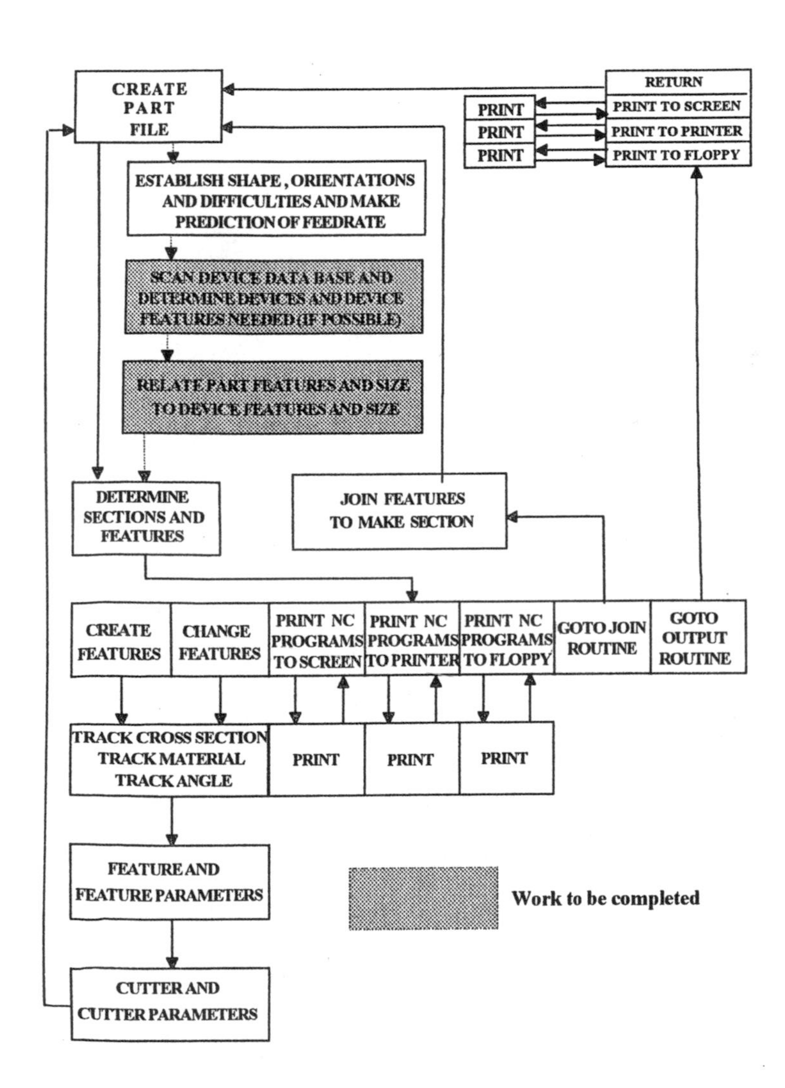

Table 1 CAD/CAM Program Structure

THE APPLICATION OF QFD TO THE MANUFACTURE OF ENGINE BEARINGS

LCol John D. Foster[1] **and Dr. William G. Sullivan**[2]
1. Department of National Defence, Canada; 2. Virginia Polytechnic Institute and State University

ABSTRACT: A Quality Function Deployment project was conducted at the Federal Mogul bearing plant, in Blacksburg Virginia, during the first half of 1996. The project examined the application of Mr. Bob King's matrix of matrices QFD approach, to the design and production of engine main bearings. In particular, a proposed bearing design change and a production process change were evaluated, using this methodology. The multiple matrix approach produced remarkably consistent results and proved to be a very thorough, yet relatively simple and easy to use, QFD technique.

INTRODUCTION

The modern marketplace is an increasingly competitive environment. It has become extremely important to **correctly** identify consumer needs and desires; to **quickly** design, develop, produce and maintain products and services to meet these needs; and to **get it right the first time**. Quality Function Deployment (QFD) is one method that can be used to gain a competitive advantage by: improving product design and production **effectiveness**; improving the **efficiency** of product development; and increasing overall **customer satisfaction**.

This paper summarizes a project which examined the application of QFD in the design and manufacture of engine main bearings, at the Federal Mogul plant in Blacksburg Virginia, using Mr. Bob King's adaptation of Dr. Yogi Akao's QFD methodology. Mr. King has transformed Dr. Akao's approach into a matrix of matrices, with a 'cookbook' style. In this system, a series of matrices are selected to meet the user's objective. Dr. Akao has endorsed Mr. King's method as a simple, yet extremely effective application of QFD.[1]

The matrix of matrices is shown at Figure 1. It consists of seven groups (A to G) of related matrices, focusing on different aspects of product design, development and production. Although Mr. King suggests matrix sequences for specific applications, the user can select individual matrices to satisfy a particular requirement. Several matrices were used to address specific issues concerning the manufacture of engine main bearings at Federal Mogul and these are developed later in this paper.

ENGINE BEARINGS

An engine's main bearings support the crankshaft and allow it to rotate and transmit power to the drive-train. A typical application is shown in Figure 2. The bearings have three primary functions:

1. Reduce friction between the crankshaft and its mounts within the engine block.
2. Support the weight of the shaft and the loads generated during operation.
3. Provide a wear surface between the shaft and the engine block.

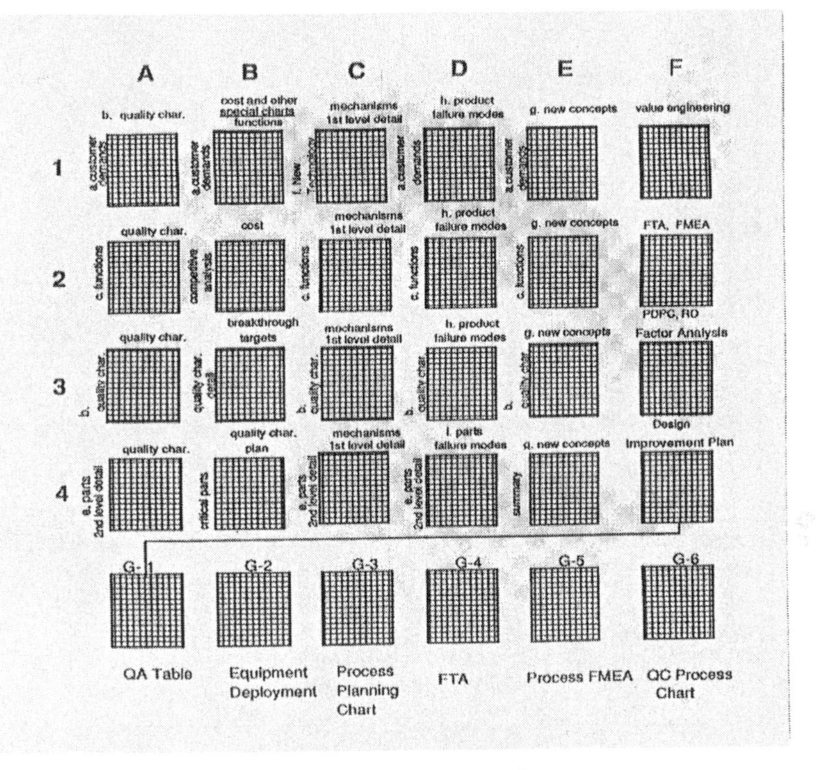

Figure 1. Matrix of matrices[2]

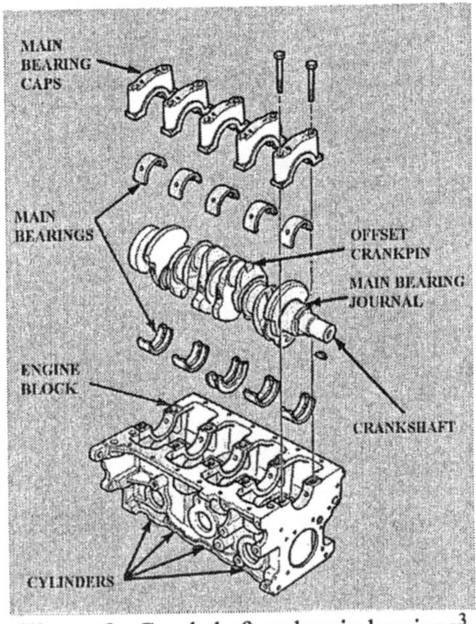

Figure 2. Crankshaft and main bearings[3]

As the shaft rotational speed increases, hydrodynamic pressure increases and an oil wedge lifts the shaft completely clear of the bearing surface. The shaft then rotates on a continuous film of oil. The point of minimum film thickness, h_o, is also the point of maximum pressure. A high quality finish on the bearing surface permits h_o to be minimized, increases load capacity and reduces wear.

QFD APPLICATION

The Voice of the Customer

The first step in applying QFD is to gather information from the customer. Information was collected in the following manner:

1. A survey was completed by two of Federal Mogul's largest customers.
2. An on-site interview with a third Federal Mogul customer, was conducted.
3. A previous Federal Mogul customer survey was incorporated into the QFD process.
4. Federal Mogul employee perceptions of customer needs were also incorporated.

Quality Characteristics

Survey and interview results were analyzed to determine customers' **primary** needs. The results were remarkably consistent and the following bearing quality characteristics were identified:

1. Bearings must be of **High Quality.**
2. Bearings must be easy to use: **Ease of Use.**
3. Wide selection of bearings: **Product Range.**
4. Order responsiveness: **Product Availability.**
5. Customer should receive **Good Service.**

The primary requirements were then broken down into secondary and tertiary quality characteristics as shown in Table 1.

Technical Attributes

Federal Mogul design and production staff were consulted to review the quality characteristics, in order to determine bearing technical attributes. These attributes, listed in Table 2, are correlated with the quality characteristics in the first matrix, often referred to as The House of Quality.

House of Quality Correlations

Quality characteristics at Table 1 and technical attributes at Table 2 are correlated in King's A-1 Chart matrix, shown at Figure 3. In order to streamline the process, customer needs not related to bearing design were not carried through into the House of Quality or subsequent matrices. Customer importance ratings and competitive assessments were determined from survey and interview data. Importance ratings use a 1 to 5 scale, with 1 unimportant and 5 very important. Similarly, the competitive assessment scale ranges from 1 (poor) to 5 (very good). The principle objective of the A-1 Chart matrix is to identify correlations between design quality characteristics (customer wants) and the design technical attributes (the engineering hows). Secondary functions are:

1. To identify the relative importance of each quality characteristic and the relative position of one's own product, in comparison to the competition.
2. To identify the relative importance of the design technical attributes and from this, to select three to six key attributes for deployment to subsequent matrices.
3. To ensure that each quality characteristic is represented by one or more technical attributes, and vice versa. This is evident if there are blank rows or columns in the table.

Table 1
Quality Characteristics

PRIMARY	SECONDARY	TERTIARY
High Quality	Appearance	smooth bearing surface
	Meet Performance Specifications	load capacity
		operating temperature
		duty cycle
		low operating noise
	Dimensional Precision	dimensional consistency
		dimensional accuracy
	Long Lasting	good oil film
		good heat dissipation
		good lubrication
		tolerates minor dirt
		accommodates shaft imperfections
		score resistant
Easy to Use	Easy to Identify	Packaging clearly marked
		Bearings individually marked
	Safe to Handle	Burr free
	Easy to Install	Quick location
		Precise location
Product Range	Size Selection	Shaft size
		Housing bore
		Oil Clearance
		Max wall thickness
		Overall length
	Application Selection	Load range
		Dimensional precision
Product Availability		Orders Filled Quickly
Good Service	Technical Support	Good technical knowledge
		Good customer knowledge
		Timely response
	Customer Service	Courteous service
		Accurate service
		Timely service

The A-1 Chart at Figure 3 indicates that the top six customer needs (quality characteristics) are:

1. The smoothness or uniformity of the bearings wear surface.
2. Bearing dimensional precision.
3. The ability of the bearing to produce a good oil film during hydrodynamic operation.
4. The bearings heat dissipation capability.
5. The ability to locate the bearing in its housing quickly.
6. The ability to locate the bearing in its housing precisely.

Table 2
Technical Attributes

PRIMARY	SECONDARY
Proportional Attributes	length over diameter ratio
	bearing eccentricity
	bearing eccentricity ratio
Dimensional Attributes	wall thickness
	wall thickness tolerance
	lining thickness
	lining surface variation
	back surface variation
	oil hole distance from *ho*
	oil hole diameter
	locating lug position
	locating lug tolerances
	annular groove depth
	annular groove width
	oil clearance
	crush relief
	bearing height
	spreader groove depth
	spreader groove width
	free spread diameter
Feature Attributes	locating lug position
	locating lug tolerances
	annular groove width
	annular groove depth
	spreader groove depth
	spreader groove width
Material Attributes	lining hardness
	lining composition
	backing composition

Similarly the five most important technical attributes are:

1. The oil clearance dimension.
2. The bearing lining composition.
3. The design tolerance of the bearing wall thickness.
4. The extent of surface variations on the bearing lining.
5. The bearing eccentricity.

FUNCTION DEPLOYMENT

The primary functions performed by the bearing are:

1. To reduce friction during boundary lubrication i.e. at startup or low rpm.
2. To reduce friction during full film i.e. hydrodynamic operation.
3. To support the weight of the crankshaft and associated pressures at rest.

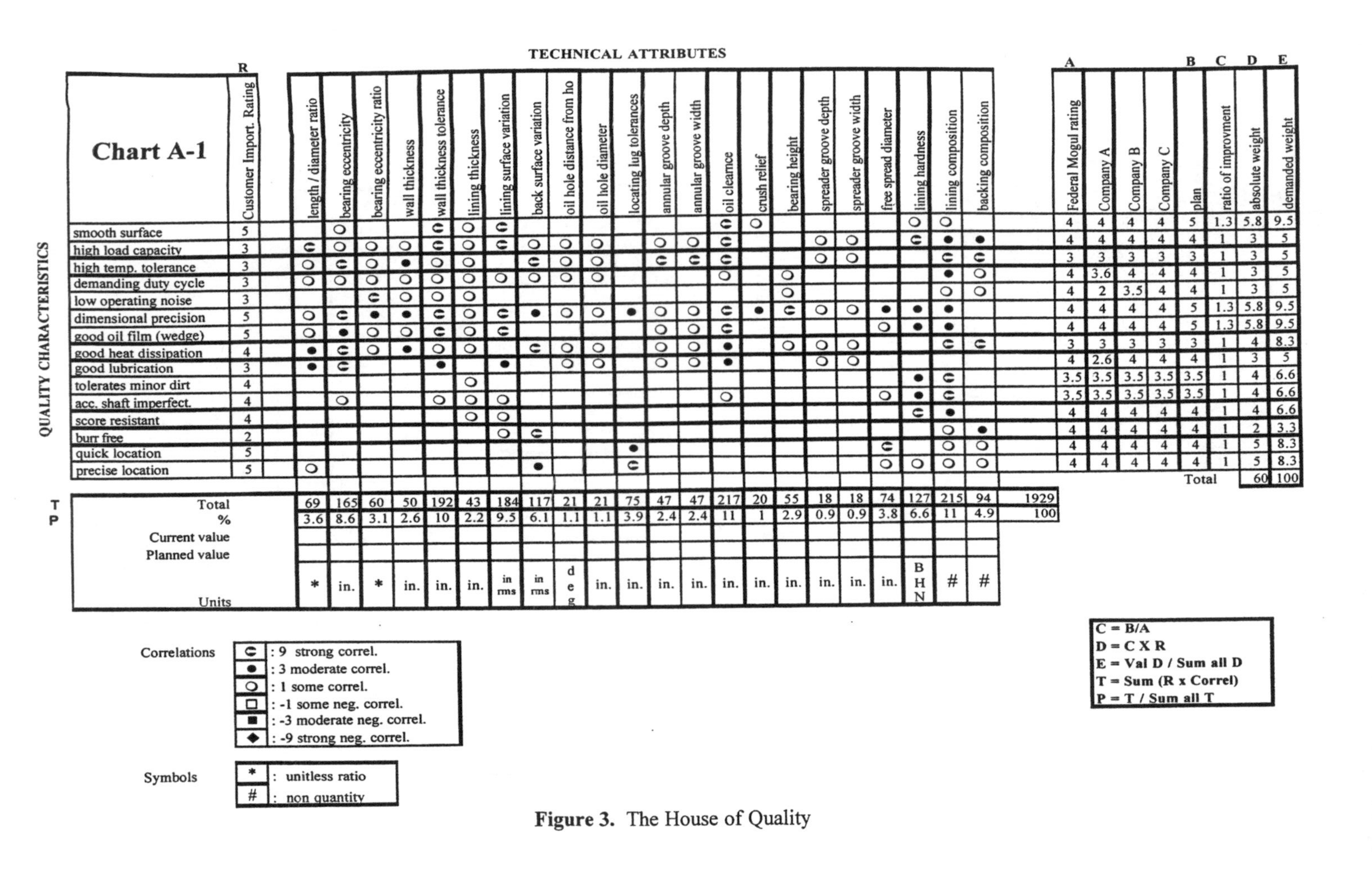

TECHNICAL ATTRIBUTES

Chart A-1 — QUALITY CHARACTERISTICS	R: Customer Import. Rating	length / diameter ratio	bearing eccentricity	bearing eccentricity ratio	wall thickness	wall thickness tolerance	lining thickness	lining surface variation	back surface variation	oil hole distance from ho	oil hole diameter	locating lug tolerances	annular groove depth	annular groove width	oil clearnce	crush relief	bearing height	spreader groove depth	spreader groove width	free spread diameter	lining hardness	lining composition	backing composition	
smooth surface	5		○			◐	○	◐							◐	○					○	○		
high load capacity	3	◐	○	○	○	◐	○	◐	○	○	○		○	○	◐			○	○		◐	●	●	
high temp. tolerance	3	○	◐	○	●	○	○		◐	○	○		◐	◐	◐			○	○			◐	◐	
demanding duty cycle	3	○	○	○	○	○	○	○	○	○	○				○		○					●	○	
low operating noise	3			◐	○	○	○										○					○	○	
dimensional precision	5	○	◐	●	●	◐	○	◐	●	○	○	●	○	○	◐	●	◐	○	○	●	●	●		
good oil film (wedge)	5	○	●	○	○	◐	○	◐					○	○	◐					○	●	●		
good heat dissipation	4	●	◐	○	●	○	○		◐	○	○		○	○	●		○	○	○			◐	◐	
good lubrication	3	●	◐			●		●		○	○		○	○	●			○	○					
tolerates minor dirt	4						○														●	◐		
acc. shaft imperfect.	4		○			○	○	○							○					○	●	◐		
score resistant	4						○	○													◐	●		
burr free	2							○	◐													○	●	
quick location	5											●								◐		○	○	
precise location	5	○							●			◐								○	○	○	○	
T: Total		69	165	60	50	192	43	184	117	21	21	75	47	47	217	20	55	18	18	74	127	215	94	1929
P: %		3.6	8.6	3.1	2.6	10	2.2	9.5	6.1	1.1	1.1	3.9	2.4	2.4	11	1	2.9	0.9	0.9	3.8	6.6	11	4.9	100
Current value																								
Planned value																								
Units		*	in.	*	in.	in.	in.	in rms	in rms	deg	in.	in.	in.	in.	in.	in.	in.	in.	in.	in.	BHN	#	#	

Quality characteristics	A: Federal Mogul rating	Company A	Company B	Company C	B: plan	C: ratio of improvment	D: absolute weight	E: demanded weight
smooth surface	4	4	4	4	5	1.3	5.8	9.5
high load capacity	4	4	4	4	4	1	3	5
high temp. tolerance	3	3	3	3	3	1	3	5
demanding duty cycle	4	3.6	4	4	4	1	3	5
low operating noise	4	2	3.5	4	4	1	3	5
dimensional precision	4	4	4	4	5	1.3	5.8	9.5
good oil film (wedge)	4	4	4	4	5	1.3	5.8	9.5
good heat dissipation	3	3	3	3	3	1	4	8.3
good lubrication	4	2.6	4	4	4	1	3	5
tolerates minor dirt	3.5	3.5	3.5	3.5	3.5	1	4	6.6
acc. shaft imperfect.	3.5	3.5	3.5	3.5	3.5	1	4	6.6
score resistant	4	4	4	4	4	1	4	6.6
burr free	4	4	4	4	4	1	2	3.3
quick location	4	4	4	4	4	1	5	8.3
precise location	4	4	4	4	4	1	5	8.3
Total							60	100

Correlations
- ◐ : 9 strong correl.
- ● : 3 moderate correl.
- ○ : 1 some correl.
- □ : -1 some neg. correl.
- ■ : -3 moderate neg. correl.
- ◆ : -9 strong neg. correl.

Symbols
- * : unitless ratio
- # : non quantity

C = B/A
D = C X R
E = Val D / Sum all D
T = Sum (R x Correl)
P = T / Sum all T

Figure 3. The House of Quality

4. To support the additional engine loads and pressures during operation.
5. Cooling the bearing/shaft interface by direct heat conductance to the engine housing.
6. To assist crankshaft and bearing assembly cooling, through oil flow.
7. To act as a wear surface and thereby minimize wear on the crankshaft.

In the A-2 Chart, at Figure 4, the bearing's functions are deployed to ensure that each of the operating functions is adequately represented by the technical attributes identified in Chart A-1. It also identifies relationships between the technical attributes and the specific product functions which may be most strongly affected by design or redesign of the product. There are two technical attributes with no functional correlation. These are "locating lug tolerances" and "free spread diameter." These attributes relate to the **assembly** of the bearing in the engine and have only minor impact on operational functions. This does not mean they are unimportant, merely that the designers must consider aspects in the bearing's design, other than purely functional performance.

Other relationships in the matrix stand out immediately. For example, even though there is only one correlation between the technical attribute 'back surface variation" and the function "heat transfer" the high negative rating indicates that this is an area of concern. Similarly there are a large number of strong correlations for "lining surface variation", "lining composition", "oil clearance" and "bearing eccentricity", indicating that these are important technical attributes.

TECHNICAL ATTRIBUTE INTERACTION

Chart A-3, the technical attribute interaction matrix for the engine main bearing is shown at Figure 5. From this matrix it is evident that there are a large number of negative interactions. In other words, changes in one attribute may have an adverse affect on one or more other attribute. The impact of this chart on the bearing design process is that the designers must be prepared to do a trade-off analysis, to resolve each conflicting interaction in a systemicly optimal manner.

TECHNICAL ATTRIBUTES

Chart A-2 (FUNCTIONS)	length / diameter ratio	bearing eccentricity	bearing eccentricity ratio	wall thickness	wall thickness tolerance	lining thickness	lining surface variation	back surface variation	oil hole distance from ho	oil hole diameter	locating lug tolerances	annular groove depth	annular groove width	oil clearance	crush relief	bearing height	spreader groove depth	spreader groove width	free spread diameter	lining hardness	lining composition	backing composition
Reduce Friction - Boundary	■	○	□		◆		◆		□	○		●	●	○	●		◐	◐		◆	◐	
Reduce Friction - Full Film	●	◐	●		◆		◆					●	●	◆	●		●	●		○		
Support Load - Static	●	□	○	◐	□	◐	□					□	□	□						◐	◐	◐
Support Load - Hydrodynamic	●	◐	●	○	■	○	◆		○	○		□	□	◆		○	●	●		◐	◐	◐
Cooling - Heat Transfer	●			○		○		◆								○					◐	◐
Cooling - Oil Flow	□	◐			■		◆		○	○		◐	◐	◐			◐	◐				
Act as Wear Surface	●	□	○	●	□	◐	□						□					□		◆	◐	

Correlations

- ◐ : 9 strong correl.
- ● : 3 moderate correl.
- ○ : 1 some correl.
- □ : -1 some neg. correl.
- ■ : -3 moderate neg. correl.
- ◆ : -9 strong neg. correl.

Figure 4. Function deployment

TECHNICAL ATTRIBUTES

Chart A-3	length / diameter ratio	bearing eccentricity	bearing eccentricity ratio	wall thickness	wall thickness tolerance	lining thickness	lining surface variation	back surface variation	oil hole distance from ho	oil hole diameter	locating lug tolerances	annular groove depth	annular groove width	oil clearance	crush relief	bearing height	spreader groove depth	spreader groove width	free spread diameter	lining hardness	lining composition	backing composition
length / diameter ratio		□	□	○	□	○	□				□	□	□	□				□	○	○	○	
bearing eccentricity			□		◆	□	◆	◆						⊙	□	□			□			
bearing eccentricity ratio					□		□							◆					□			
wall thickness						⊙		○				⊙			□		⊙			○	○	
wall thickness tolerance						□	◆	◆				□		◆	□		□			■		
lining thickness							●					●			□		●		□	⊙	⊙	
lining surface variation														●					□			
back surface variation											■			□	□				□			■
oil hole distance from ho																						
oil hole diameter																						
locating lug tolerances																□						
annular groove depth													●				●					
annular groove width																						
oil clearance																□				⊙	⊙	
crush relief																⊙			□	○	○	□
bearing height																			□	□	□	□
spreader groove depth																		⊙		●		
spreader groove width																						
free spread diameter																				□	□	□
lining hardness																					⊙	○
lining composition																						⊙
backing composition																						

(Rows: TECHNICAL ATTRIBUTES)

⊙ : 9 strong correl.
● : 3 moderate correl.
○ : 1 some correl.
□ : -1 some neg. correl.
■ : -3 moderate neg. correl.
◆ : -9 strong neg. correl.

Figure 5. Technical attribute interaction

PARTS DEPLOYMENT

Parts deployment is carried out to identify which product components to focus on. It is important to identify how each part interrelates to the **key** design technical attributes. In this way, critical design considerations and manufacturing processes can be identified.

Although the main bearing is a single component, it has recesses and surfaces that serve specific purposes and require individual design consideration and manufacturing processes. These are identified in Figure 6, (Chart A-4) as the bearing's parts. From the chart it is evident that the bearing lining is strongly interrelated to four of the five key technical attributes. Similarly, the extent of each part's technical attribute correlation, and their relative strengths, can be read directly from the chart.

TECHNICAL ATTRIBUTES

Chart A-4 (PARTS)	bearing eccentricity	lining surface variation	back surface variation	oil clearance	lining composition
Bearing Lining	◎	◎		◎	◎
Bearing Backing	◎		◎	●	●
Locating Lug	○	○	○		○
Oil Hole		○	○	○	
Annular Groove		○			●
Spreader Groove		○		○	○
Crush Relief				◎	●
Inner Chamfer		○			○
Outside Chamfer			○		
Groove Chamfer		○			

Correlation	
	◎ : 9 strong correl.
	● : 3 moderate correl.
	○ : 1 some correl.

Figure 6. Parts deployment

NEW CONCEPT DEPLOYMENT

King's E series consists of four matrices, which are used to evaluate new concepts. The first three, E-1, E-2 and E-3, correlate the new concepts with the product's quality characteristics, functions and technical attributes, respectively. Each new proposal is evaluated from all three perspectives and an assessment is made of the possible impact to each criteria. The final E-4 matrix summarizes the evaluations of the first three. As the matrices are similar in appearance and function, only the first (E-1) and final (E-4) are illustrated in this section. The application of all four are discussed below.

The Federal Mogul project evaluated two new concepts. The first, a **design change**, proposed removing the locating lug, shown in Figure 7. This lug fits into recesses in the engine block, to position the bearing. Although this is a simple function, the lug requires several manufacturing processes. A possible alternative is to produce the bearings symmetrically, with a second oil hole, so that regardless of orientation in the engine block the oil passage will remain unobstructed.

The second new concept, concerned a bearing **production process change**. Federal Mogul's bearings are machined primarily by broaching. This method passes a cutter across the width of the bearing, removing metal to achieve the desired thickness. While this is simple and inexpensive, its precision is limited. Federal Mogul has recently undertaken development of bearings requiring very fine tolerances. To meet these specifications, a rotary machining process is being implemented.

New Concepts and Customer Demands

Chart E-1, starts with the Voice of the Customer. Quality characteristics are evaluated to identify new concepts that will enhance the satisfaction of consumer demands, with a minimum of negative impacts in other areas. The results of this process are shown in Figure 8.

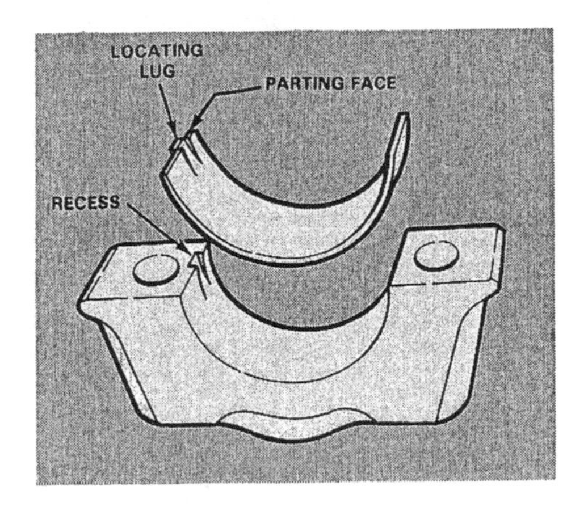

Figure 7. Bearing locating lug[4]

Chart E-1	PRESENT CONCEPT	NEW CONCEPTS	
	Present Concept: 1. Broaching of bearing surface. 2. Bearing positioned using locating lug.	Bore/ream/mill bearings i.e. rotary machining	Eliminate lug: use alternative method or make bearings symmetrical to eliminate requirement
		Anticipated Result	
smooth surface		+	
high load capacity		+	
high temp. tolerance			
demanding duty cycle			
low operating noise		+	+
dimensional precision		+	+
good oil film (wedge)		+	
good heat dissipation			
good lubrication		+	
tolerates minor dirt		-	
acc. shaft imperfect.		-	
score resistant			
burr free		+	+
quick location			-
precise location			-
	+'s	7 +	3 +
	-'s	2 -	2 -

Figure 8. New concept - Customer demands deployment

Production Process Change. It is believed that introduction of the rotary machining process will significantly improve seven of the fifteen quality characteristics. It is also possible that two characteristics will be adversely affected. The extremely precise tolerances achieved may result in bearings that are less 'forgiving' of minor oil contamination and/or slight shaft misalignment.

Design Change. The case for the locating lug design change is less convincing. Here there are three positive affects and two negative. Further, the two negative implications were **emphatically** stressed during interviews and in one of the two questionnaires. In addition to orientation, the locating lug positions the bearing and is used during assembly to greatly speed up the bearing insertion process.

New Concepts and Product Functions

Chart E-2 correlates the effect of the new concepts on the product's functions.

Production Process Change. This analysis suggested that the proposed new machining process would have a positive impact on five of the bearing's seven functions. Further, it was assessed that there are no 'down sides' to this change, as the remaining two functions were unaffected.

Design Change. The proposed removal of the locating lug had no impact, positive or negative, on any of the bearing functions. This is due to the fact that bearing installation is the 'function' of the assembly worker and not the bearing. The concerns with respect to bearing installation were, however, evident in the previous E-1 Chart. This is another example of the way in which the use of multiple matrices can minimize the potential to overlook a design consideration.

New Concepts and Technical Characteristics

Chart E-3 compares 'The Voice of The Engineer' (technical attributes), to the new concepts.

Production Process Change. Analysis indicated that this change would have a positive impact on eleven of the twenty-two technical attributes and would adversely affect only two. This is consistent with the results obtained in considering the bearing functions and the quality characteristics. This would suggest that the translation of customer needs into technical attributes, and the subsequent deployment of The Voice of the Customer through the series of matrices, has been successful.

Design Change. The results obtained for the proposal to remove the locating lug again suggest several problems (three negative impacts) and little benefit (one positive impact), to the change.

New Concept Summary

This is the final matrix in the E-series. It summarizes the data obtained from the three preceding charts and provides a good overview of the net impact of each proposed idea, in terms of the customer's expressed needs, the products actual functions and the technical attributes needed to accommodate the customer and achieve the desired functions. Figure 9 summarizes this process for the Federal Mogul main bearing. As indicated in the chart, the rotary machining process is an extremely positive measure, strongly supported from all three perspectives. The lug elimination proposal, on the other hand, is not supported by the evaluation. If the concept is to be pursued, the design team might consider developing a simpler locating system.

CONCLUSIONS

The application of multiple QFD matrices to the bearing design and production process at Federal Mogul proved to be extremely effective. The consistency of results at all levels suggests that the Voice of the Customer was 'deployed' through successive matrices. In addition, the usefulness of multiple matrices in ensuring that key design factors and interrelationships are not overlooked, was repeatedly observed. Specific conclusions are:

Chart E-4	PRESENT CONCEPT	NEW CONCEPTS	
	Present Concept: 1. Broaching of bearing surface. 2. Bearing positioned using locating lug.	Bore/ream/mill bearings i.e. rotary machining	Eliminate lug: use alternative method or make bearings symmetrical to eliminate requirement
		Anticipated Result	
SUMMARY: Quality Characteristics		7 + 2 -	3 + 2 -
Functions		5 + 0 -	0 + 0 -
Technical Attributes		11 + 0 -	1 + 3 -
	+'s	23 +	4 +
	-'s	2 -	5 -

Figure 9. New concept summary

1. Federal Mogul's new manufacturing process concept (rotary machining vice broaching) has been validated by the QFD process as an extremely effective decision.
2. The proposal to remove the locating lug was demonstrated to require more investigation.
3. QFD is an effective tool that has proven to be adaptable to Federal Mogul's needs.
4. QFD's primary function is to deploy customer needs and demands through the entire product development process. It is, essentially, a design tool and not a quality tool. **Quality** is achieved by ensuring that the customer receives the **qualities** that are desired.
5. The King approach provides an extremely simple, flexible, versatile and comprehensive system for conducting QFD.

REFERENCES

1. **B. King**, *Better Designs in Half The Time*, (GOAL/QPC, Metheun MA, 1989).
2. **B. King**, *Better Designs in Half The Time*, (GOAL/QPC, Metheun MA, 1989).
3. **Federal Mogul**, *Engine Bearing Service Manual*, (Federal Mogul Corporation, Detroit, MI, 1981).
4. **Federal Mogul**, *Engine Bearing Service Manual*, (Federal Mogul Corporation, Detroit, MI, 1981).

QUALITY FUNCTION DEPLOYMENT
- IMPROVING INFORMATION TECHNOLOGY SERVICE WITHIN MANUFACTURING

Brendan Lynch, Ita Richardson,
Department of Computer Science and Information Systems
and
National Centre for Quality Management,
University of Limerick,
National Technological Park,
Castletroy,
Limerick,
Ireland.

ABSTRACT: The objective of this case study is to provide proof that QFD can furnish significantly different results from more traditional improvement techniques. Using techniques such as collecting the voice of the customer, affinity diagrams, Kano questionnaires, importance questionnaires and the house of quality, the authors have undertaken a project with employees within a manufacturing plant.

In this paper, the case study is described in detail, discussing data collection methods. The authors then discuss a correlation of the data, concluding that there is a significant difference between customer ranking and establishing priorities using Quality Function Deployment

1. INTRODUCTION

Quality Function Deployment (QFD), a method of including quality into the design and production of products, was developed by Yoji Akao and first used in Japanese shipyards during the 1970's. Its use spread to the automobile industry in Japan, and in the mid-1980's was introduced to the car industry in the United States. More recently, QFD has been used to improve services provided by organisations. In both manufacturing and the service industry, QFD is being used increasingly within Irish industry, both multi-national and indigenous.

2. WHY QFD?

The primary function of all business enterprises is to survive and grow. Central to survival and growth is the provision of high quality products and services, reliably and at low cost. In order that they have an impact on market share, each organisation must improve the quality, cost and/or time-to-market of their product or service. If the organisation can match a competitor's product or service in any two of quality, cost and time-to-market, and better them in the third, then it can be argued that the rational customer will buy from that organisation [1]. QFD can be used to improve each of these three factors.

2.1 Improvements Gained

QFD is used to translate the affective language of the customer into the scientific and technical language of the engineer and scientist, and in doing so, aims to get the design right first time, and detects possible trade-offs early in the design process. The technique endeavours to supply products that customers will buy by ensuring that the product is what the customer sought. Therefore, the emphasis is put on the design meeting requirements rather than the design meeting specification. "Customers do not want to be involved in legalistic nitpicking about whether specifications have been met. They expect their needs to be satisfied even when those are imperfectly defined."[2]

Based on the influx of Japanese products into the western market, it seems that the Japanese have got it right, and the western world is lagging. It is claimed that, in general, Japanese companies have fewer design changes when any product is in process, and that 90% of these changes are made more than a year before production start-up. In contrast, U.S. companies make a significant number of design changes after the product is in production [3]. Also, Japanese companies use more resources at product development stage with much fewer resources being required for problem solving. For American companies it is exactly the opposite[4]. It is recognised that QFD is at least partially responsible for these differences[5].

It has been argued that QFD can reduce time-to-market in the region of 30%[6,7]. This occurs because customer requirements are collected, in detail, directly from the customer at the start of the process. These requirements are maintained throughout the product life-cycle and any design changes are made early in the process. All planning of the production process and operations to be carried out is done during initial phases, and fewer engineering changes are required after the product is launched.

It is also argued[8] that another factor which contributes to shorter design time is that documentation and communication are improved through the use of QFD.

3. USE OF QFD IN SERVICES

Initially, when QFD concepts spread from the automobile industry in Japan to the automobile industry in the U.S., it was used mainly by other production plants. In more recent years, it has been used to improve services. For example, one case discusses the use of QFD to improve river rescue in Los Angeles[9]. In another case, a group of academics have used QFD to develop a research strategy within their department[10]. Within the University of Limerick, QFD has been used to determine a development strategy for an academic department.

In Ireland, the number of people and organisations using QFD is increasing. The objective of this research is to identify the features and characteristics of QFD that are most suited to Irish Industry, especially the service sector. The research is in two stages, phase one is to use multi-national industries based in Ireland to develop and test suitable instruments, phase two involves transferring the techniques developed in the multinational base into indigenous enterprises.

4. CURRENT RESEARCH

Assuming that Irish companies wish to improve their productivity by whatever means are available, research to date has attempted to establish the following:

- Difficulties to be overcome before implementing QFD within organisations;
- How companies have adapted QFD matrices to suit their particular purpose;
- Whether companies are talking directly to the customer to get requirements;
- What methods companies use to collect the customer requirements;
- What QFD. matrices are used by companies;
- What advantages and disadvantages have accrued due to the use of QFD within the organisations.

4.1 Barriers to implementing QFD

If Quality Function Deployment is to be implemented successfully within Irish industry, perceived barriers to QFD, which currently exist, must be overcome. As part of the workshops held with both groups of industrialists, and to give an indication of the difficulties which people expect to experience when implementing QFD in small and medium sized enterprises in Ireland, the LP method[11] has been used. The groups were asked: WHAT ARE THE BARRIERS TO QFD? The following problems have been identified.

(1) Lack of Time / Money / People:
Those questioned decided that this was the most important barrier to QFD implementation. They believe that because the process is time-consuming and people dependent, that there would be difficulty convincing the accountant that it would be cost-effective.

(2) Cultural and Perception issues:
The group identified that there would be a difficulty with implementing such a 'different' strategy into their organisations, particularly where other quality procedures have failed in the past.

(3) No track record:
People believe that there is an "absence of proof that QFD can effectively replace traditional methods". The requirement from Irish industrialists is to have a number of case studies proving its success within Ireland, not Japan, America, or even Europe.

(4) Complexity:
The argument made about this issue is that QFD can become very complex, particularly if there are many customer requirements. People also have a difficulty with reducing customer requirements to numbers which are subsequently used in calculations.

The research group have identified that these difficulties must be overcome before QFD can be implemented successfully within Irish industry.

4.2. Purpose of Research

One of the aims of the case-study presented here is to deal with concerns (3) and (4) listed in the previous section. It attempts to establish whether there is a significant difference between using QFD and other methods, and that the complexity involved can be reduced through the use of common methods such as interviews and questionnaires.

5. CURRENT SERVICE PROVISION

For the purposes of this case study the authors decided to examine the processes in the Information Technology department within a manufacturing company. At present the department is comprised of a core staff of four supplemented by additional contract programmers. The department supports two different systems, one on a mainframe through which all the companies business is conducted and also a client-server based system. The department also provides a service to both internal (customers within the company who use the service) and external customers (divisions within the company, suppliers and customers). Currently the department has 108 internal users. Of the 108 users, 77 use the mainframe system and 93 use the PC network. Most users use both systems. Initial investigation demonstrated that the current feeling within the company is that the IT department is not providing the level of service that is expected of them. The IT department feel that they are overloaded and can only afford to deal with problems as they arise in a piecemeal fashion. The authors investigated this service gap in more detail with the ultimate aim of developing an action plan for the IT department.

6. METHOD OF RESEARCH

6.1 Interviews:

The initial stage of the research involved collecting the 'voice of the customer'. The purpose of the voice of the customer within quality function deployment (QFD) is to know the customer's expectations, voiced desires, and as yet unperceived turnons. Thus, obtaining the voice of the customer is the focal point of the QFD process.[12] To achieve this the authors conducted interviews with customers of the IT department.

The first stage was to decide the customer group on which to focus. Internal customers of the IT department were selected as the group as they provided the authors with a convenient data source. Because of time and cost constraints it was not possible to conduct a full-scale interview programme. Furthermore it has been found that from interviewing 14-15 customers that 70% of customer requirements could be identified.[13] Customers were selected at random from a list of users provided by the IT department. The interviewee's comments were recorded by taking notes during the duration of the meeting.

Each customer was interviewed at length in an endeavour to attain customer requirements for the chosen group. The interviews were conducted as open-ended interviews. This lack of formality made for a relaxed atmosphere which facilitated the flow of information. Customers were encouraged to talk about their experiences with the IT department and as to how they thought it might be improved. Each interview concluded with a summary by the interviewer of the points covered. This allowed both parties quickly to ensure that the desired points were covered, and that there was no ambiguity in what was recorded. At the conclusion of the interview stage, the authors were left with a list of verbatims which needed to be translated into customer requirements

6.2 Voice of the Customer Tables:

Information gleaned from interviews with the customer group were then arranged into Voice of Customer (VOC) tables. The purpose of the VOC tables was to organise the information obtained from the interviews. They were also used to sort data obtained from customers into their different data types. Most interviews come up with solutions to needs as well as needs themselves so it is important at this stage to separate the solutions from the problems. Qualitative analysis involves judgement rather than the fact involved in quantitative analysis. Its subjective nature means that it is necessarily less "hard", or right and wrong than numerical analysis. When constructing the VOC tables the authors relied heavily on their own opinion of the current situation in the department.

A sample of the final VOC table is shown in Table 1. In this table the customer data is broken up by: source, customer need, and related information. This table was then used to provide the information necessary for constructing an affinity diagram.

I.D.	Demographic	Customer Need	Use									
			What		When		Where		Why		How	
			I/E	Data	I/E	Data	I/E	Data	I/E	Data	I/E	Data
1	Business process Re-engineer	Image of IT	I	aligned with strategy of company					I	seen as a means to apply strategy of company and as support to activities within the company		
1	Business process Re-engineer	Image of IT	I	Approachability of staff	I	At all times			I	Staff in the IT department are unapproachable with problems		
1	Business process Re-engineer	Image of IT	I	Visibility of work being done					I	Outside departments cannot see any constructive work being done within the IT department		

Table 1
Sample VOC Table for IT department.

6.3 Affinity Diagrams:

It has been stated that the affinity diagram method (LP method) "clarifies important but unresolved problems by collecting verbal data from disordered and confused situations and analysing that data by mutual affinity."[14] When the VOC table had been analysed and customer needs had been drawn from it the authors used affinity diagrams to group related data. Groups were set up in which the authors placed data items which had a close relationship with each other. When this process had finished each of the groups which had been arrived at were given a title. These titles and groups of related data were later used in the creation of questionnaires from which quantitative data was obtained.

6.4 Construction of House of Quality

The final phase was the construction of the House of Quality. Data was collected from 15 customers using different questionnaires. The data collected related to the importance of each requirement, the current performance of the IT department and the planned performance from the IT department. Data was also collected on how important each of the customer requirements were in the eyes of the IT department. Some questions remained unanswered on some of the responses. These were taken into account when compiling the overall results.

6.4.1 ***Questionnaires.*** In carrying out this study the authors used different forms of questionnaire. Each questionnaire was used to extract the required information from the users. The objectives of the questionnaires were to obtain quantitative data relating to the qualitative data from the previous stage.

Each questionnaire was designed using the data gathered in the previous stage. The authors were keen to ensure the secrecy of the respondents. Therefore none of the questionnaires requested data of a personal nature. A total of six questionnaires were designed for different purposes. The questionnaires used in the study were:

1. Self Stated Customer Importance
2. Current Performance
3. Planned Performance
4. Kano
5. Rank-Order
6. Self Stated IT Importance

Self stated importance questionnaire. Importance questionnaires were also constructed using the data obtained in the VOC tables. The purpose of these questionnaires was to allow the users to rate on a scale of 1-10 how important they perceived each customer requirement to be. Questionnaires were distributed to the initial sample. Each customer requirement contained in the Kano questionnaire was included in the importance questionnaire.

Current performance questionnaire. A current performance questionnaire was used to obtain data from customers regarding the performance of the IT department in each of the requirements. Each requirement was scored on a scale of 1-10 by the initial sample . The values were normalised to give the departments perceived performance in each requirement. This figure was used later in the construction of the House of Quality.

Planned performance questionnaire. To enable the authors to develop an improvement factor (the value attained by dividing planned performance by current performance), it was necessary to find out the IT departments planned performance in each of the requirements. This questionnaire allowed the IT department to rank on a scale of 1-10 their planned performance for each requirement. This questionnaire was distributed among all the members of the IT department and the results were normalised.

Kano Questionnaires.[15] Many methods are available for investigating the characteristics of customer requirements. As can be seen from figure 1. certain customer requirements display different levels of satisfaction among users. Using the Kano model requirements are broken up into:

- Attractive
- Must-be
- One-Dimensional
- Indifferent
- Reversal
- Questionable

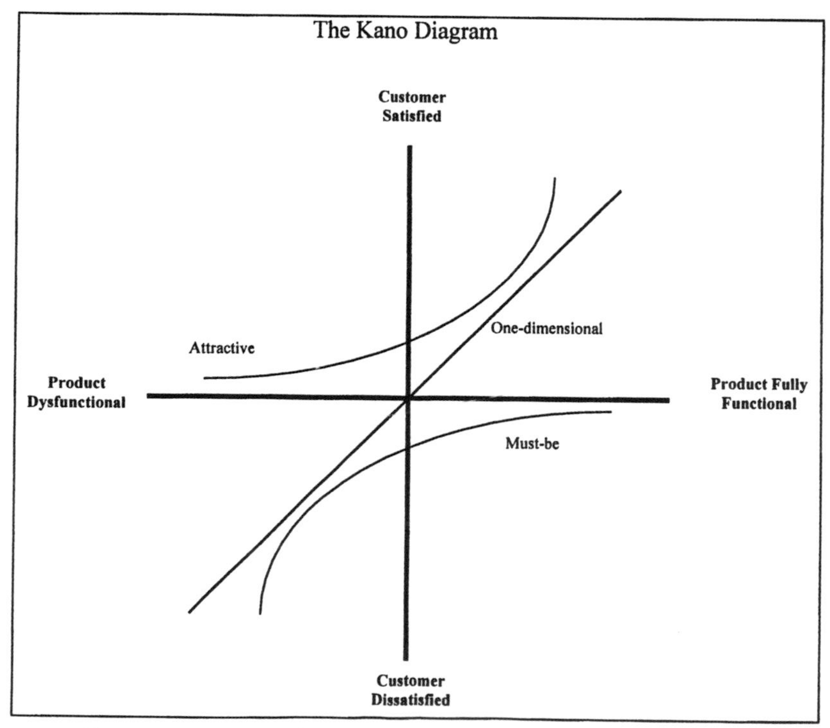

Figure 1 Kano Diagram showing satisfaction levels against product functionality.

Kano believes that customer requirements can be classified along these lines through the use of a customer questionnaire. In a Kano questionnaire the user is asked pairs of questions. The first question (*functional question*) in each pair of questions for a customer requirement refers to a situation where the requirement is met. The second question (*dysfunctional question*) refers to the case where the requirement is not met. When writing the pairs of functional and dysfunctional question for each of the customer requirements it was important to use the data from the previous stage. It was also necessary to ensure that the questions were in customer terms. A sample question is given in Figure 2.

Before using the questionnaire it was tested and questions were revised based on the test. The questionnaire was distributed among people who had been interviewed. This group were chosen as they formed the initial sample group. From the results of this questionnaire it was possible to develop sales point values for the House of Quality.

If the IT department identified new directions/opportunities for the advancement of the company, how would you feel?	1. I would like it ☐ 2. I would expect it ☐ 3. I am indifferent ☐ 4. I could tolerate it. ☐ 5. I would dislike it. ☐
If the IT department did not identify new directions/opportunities for the advancement of the company, how would you feel?	1. I would like it ☐ 2. I would expect it ☐ 3. I am indifferent ☐ 4. I could tolerate it. ☐ 5. I would dislike it. ☐

Figure 2. Sample Kano Question

Sales points are used to arrive at the overall importance value for ranking requirements within the house of quality. The fact that the Kano questionnaires assign values to requirements in the form of an attractive, a must-be etc. makes it hard to assign numerical values to each requirements. To assign a value to each requirement two values were used, a positive and a negative. The values arrived at were representative of the value of meeting the requirement and the cost of not meeting the requirement. The formulas used ignored the responses which were either Reverse or Questionable. The following is the formula used to calculate the positive and negative values:

$$\text{Positive} = \frac{A_i + O_i}{A_i + O_i + M_i + I_i} \qquad \text{Negative} = \frac{O_i + M_i}{A_i + O_i + M_i + I_i}$$

Equation 1

A_i = Number of Attractive responses i

O_i = Number of Optional responses for requirement i

M_i = Number of Must - be responses for requirement i

I_i = Number of Indifferent responses for requirement i

To arrive at a multiplier for the sales point it was decided that the value should represent the both the value of having a requirement present and the cost of not meeting the requirement. The sales point value was obtained by adding the two values arrived at in equation 1.

House of Quality. The final phase was the production of a 'house of quality'. The house is designed on a matrix structure which allows the prioritisation of requirements. According to Mizuno, the matrix analysis method quantifies and arranges matrix diagram data so that the information is easy to visualise and comprehend[16]. The data gathered in the earlier stages is used in the construction of the house of quality. The house of quality allows one to prioritise firstly by WHATs and then prioritising HOW the improvements will be made.

In arriving at the ranking for the 'House of Quality', the authors used the data obtained using the methods described in the previous section. Here the ranking was based on a an overall importance value. To arrive at the overall importance value we used the current importance, performance (both planned and perceived) and the sales points associated with each requirement using the equation:

$$OI_i = \left(PP_i / CP_i\right).CI_i.SP_i$$

Equation 2

OI_i = Overall Importance of requirement *i*

PP_i = Planned Performance of requirement *i*

CP_i = Current Performance in requirement *i*

CI_i = Current Importance of requirement *i*

SP_i = Sales point value of requirement *i*

The following example shows how the ranking for the customer requirement "Inform of software updates or new releases" was ranked and how the overall importance was arrived at.

$$OI_i = 3.64 \quad PP_i = 7.8 \quad CP_i = 4.1 \quad CI_i = 9.1 \quad SP_i = .21 \quad Ranking = 16$$

6.5 Data for comparison purposes.

In addition to using questionnaires to collect data for the construction of the House of Quality, the authors also used questionnaires to obtain data for comparison purposes. The aim was to see the differences that would occur when different approaches are taken.

Rank-Order Questionnaire. This questionnaire was designed to allow users to rank customer requirements in the order that they considered important. It was decided that a separate group of people should be targeted with the rank-order questionnaire than those who completes the Kano and importance questionnaires. The reason for this was to try to reduce the bias caused by completing previous related questionnaires. In designing the questionnaire the customer requirements were jumbled so that they did not appear in the groupings which had been assigned to them at the affinity diagram stage. Respondents were requested to rank the requirements in order of importance as they appeared to them.

Ranking and priorities: When compiling the results for the more traditional ranking method each requirement was assigned a certain value depending on it's ranking. The authors assigned values to requirements that got a ranking between 1 and 15. The authors then assigned weights to each of the requirements based on a descending scale. (A ranking of 1 was weighted as being 15, 2 as 14, etc. ... until 15 which was weighted as 1).

Another reason behind this assignment of weights was due to the fact that some of the respondents had ranked their top 10 requirements and neglected to allocate rankings to the rest of the requirements. Since this was the case, the weightings were averaged for each of the requirements in an effort to normalise the data. A ranking was then developed on the average weightings.

Importance questionnaire for IT department. In conjunction with the planned performance questionnaire the members of the IT department were given a second questionnaire. The design and layout of the questionnaire was the same as the self stated importance questionnaire given to the customers. The IT department were asked to rate on a scale of 1-10 how important they felt each of the requirements were. The results were aggregated to give an importance value for each of the requirements.

7. RESULTS

A comparison between the importances can be seen in Table 2. From this table it can be seen that the means and the standard deviation of the customer importances and the IT importances are very close. However the figure of .61 for the correlation between the two sets of numbers shows that while overall they may be quite close, there exists big differences in what both parties consider to be important.

	Customer Importance Values	IT Importance Values
Min	6.67	4.75
Max	9.73	9.75
Mean	8.60	8.42
Std Deviation	0.70	1.17
Correlation	0.61	

Table 2
Comparison of importance values

This shows that the IT department while overall recognising the importance of the requirements from the customers, they do not recognise the requirements which are most important to their customer. This indicates that the IT department place greater importance on requirements that are not as important to the customer.

The rankings for the top seven customer requirements, using the more traditional rank-order method can be seen from Table 3. Also included in the table is the ranking that each of these requirements achieved under the different ranking systems.

Question Number	Customer Requirements	Rank Order	QFD Rank *(CI values)*	QFD Rank *(ITI values)*
24	Respond quickly to each problem	1	17	19
21	Improve response time of system	2	7	9
20	Improve reliability of system	3	12	13
9	Provide systems that meet requirements	4	21	18
4	Training to use new software	5	20	27
19	Maintain/increase staff level	6	6	7
30	Develop communications structure	7	22	15

Table 3
Rank-order customer requirements rankings

The following table shows the rank when using the QFD methodology with customer importance values. Also included in the table are the corresponding ranks for the rank-order method and the QFD method using the IT importance values.

Question Number	Customer Requirement	QFD Rank *(CI values)*	Rank Order	QFD Rank *(ITI values)*
12	Publish project priorities	1	23	1
11	Publish list of projects	2	22	2
14	Work with other depts. To prioritise projects	3	24	3
13	Current status of projects	4	28	4
17	Provide adequate cover for key personnel	5	17	5
19	Maintain/increase staff level	6	6	7
21	Improve response time of system	7	2	9

Table 4
QFD ranking using customer importance values

As can be seen from the two sets of tables above the results obtained using the two different methods are considerably different. Looking at the top five requirements using the simple ranking method it can be seen that the QFD method places these top requirements considerably further down the ranking list. The difference is more evident when one consider the top five QFD requirements. By comparing these two sets of figures one can see that using the simple ranking system that the highest that any one of the top five ranks is at 17. In the top seven requirements using both methods only one requirement appears on both lists. It is interesting to note that the two contrasting techniques produce vastly different priorities.

Question Number	Customer Requirement	QFD Rank *(ITI values)*
12	Publish project priorities	1
11	Publish list of projects	2
14	Work with other depts. to prioritise projects	3
13	Current status of projects	4
17	Provide adequate cover for key personnel	5
31	Develop forward looking strategy	6
19	Maintain/increase staff level	7

Table 5
QFD ranking using IT department importance values

	Rank-Order	QFD *(CI values)*	QFD *(ITI values)*
Rank Order	-	0.27	0.21
QFD *(CI values)*	0.27	-	0.86
QFD *(ITI values)*	0.21	0.86	-

Table 6
Correlation of requirement ranking.

When the IT importance figures are used instead of the customer importance figures (Table 5) the top five requirements are exactly the same. This suggests that the difference in the importance ratings of the IT department and the customers appears lower down in the ranking of the requirements. It also indicates that even though the IT department recognises the importance of the requirements to the customer that the customer's perception of how well IT performs in each requirement goes unnoticed. Table 6 shows the correlation between the three ranking systems. It can clearly be seen that ranking the customer requirements by using the QFD method, using either the importance as seen by the IT department or the customer, produces a significantly different result than when the rank-order questionnaire is used. It can also be seen from this table that while the correlation between the IT ranking and the customer rankings are very close there is a big difference in the correlation with the rank-order rankings.

8. CONCLUSION

The research is useful in two ways. Firstly, we have indicated to the company that although customers and Information Technology personnel have similar high priorities, this is not maintained for those of lesser importance. It is therefore important that the service department discuss priorities with the customers. Secondly, we have established that there is a significant difference between the results obtained while using Quality Function Deployment and those obtained by asking the customer to rank preferences. To establish which is better, it would be desirable to implement an experimental project where one group worked on those priorities as established by QFD and a second control group working on those priorities established by ranking. The difficulty in setting this up would be to identify two comparable groups to carry out this work. However, the authors believe that the number of reported QFD cases is evidence that QFD has become an accepted method for prioritising in many industries.

---

[1] Calloway, D. and Chadwell, B., "Manufacturing Strategic Plan - QFD and the Winchester Gear Transfer", *Transactions from the 2nd Symposium on QFD*, pp. 370-380, June, 1990, Michigan, U.S.A.

[2] Kenny, A. A., "A New Paradigm for Quality Assurance", *Quality Progress*, pp. 30-32, June, 1988.

[3] American Supplier Institute, Inc., Training Manual, 1989.

[4] American Supplier Institute, Inc., Training Manual, 1989.

[5] Betts, M., "QFD Integrated with Software Engineering", *Transactions from the 2nd Symposium on QFD*, pp. 442-458, June, 1990, Michigan, U.S.A.

[6]Denton, D. K., "Enhance Competitiveness and Customer Satisfaction . . . Here's One Approach", *Industrial Engineering*, pp. 24-30, May, 1990.

[7] Griffin, A., "Metrics for Measuring Product Development Cycle Time", *Journal of Production and Innovation Management*, pp. 112-125, Vol. 19, 1993.

[8] King, R., "Better designs in half the time - Implementing QFD in the USA", *presented at the Growth Opportunity Alliance of Lawrence*, M.A., 1987.

[9] Butler, K., Litwin, R., Marzec, J. *et al* "The Application of Quality Function Deployment to the Los Angeles River Rescue Task Force", pp. 271-280, June, 1993, Michigan, U.S.A.

[10] Chen C. L. and Bullington, S. F., "Development of a Strategic Research plan for an Academic Department Through the use of Quality Function Deployment", *Computers and Industrial Engineering*, pp. 49-52, Vol. 25, Nos 1-4, 1993.

[11] Walden, D., "Using the L-P Method", Internal Workshop, University of Limerick, 15 November, 1995.

[12] http://mijuno.larc.nasa.gov/dfc/qfd/voc.html

[13] Mazur, G. H. and Zultner, R. E., "Voice of Customer Analysis", *Tutorial at 8th Symposium on Quality Function Deployment*, June 9th, 1996, Novi, Michigan.

[14] Mizuno, S., "Management for Quality Improvement: The 7 New QC Tools", 1988, Productivity Press, Inc., Cambridge MA.

[15] "Kano's Methods for Understanding Customer-defined Quality", *The Centre for Quality Management Journal*, Fall 1993, Volume 2, Number 4.

[16] Mizuno, S., "Management for Quality Improvement: The 7 New QC Tools", 1988, Productivity Press, Inc., Cambridge MA.

INTELLIGENT MANUFACTURING AND LATERALISATION

W. F. Gaughran
Department of Manufacturing and Operations Engineering
University *of* Limerick

ABSTRACT: In improving efficiency and intelligence in manufacturing a variety of strategies may be employed. Seldom is the most importamt 'cog' in the wheel of manufacturing given sufficient attention, i.e. the design and manufacturing *engineer*. In training as well as in-career development the recognition of cognitive *lateralisation* is of vital importance. Much of the engineers creative activity is right-cerebral based and involves design and *visualisation*. The stage of development of the engineer's *spatial-ability* will have a significant effect in all engineering and design activities. The classification of such cognitive abilities is discussed, along with strategies and considerations in their development. Research shows that improving an individuals visualisation/spatial-ability skills, will enhanse performance in all engineering, design, math and CAD activities. *Cognitive modelling* is considered in the context of the *'space-factor'*.

Designing and Visualisation

Improving manufacturing and production efficiency has for obvious reasons been a focal point of engineers for some time. Engineers are now very much part of the design team working together to improve efficiency. Most design teams work to a system which impacts on all facets, from concept to consumer and in the light of 'green' issues, the areas which affect design for dismantling and re-use are also becoming a very important a part of the strategy. This approach has given rise to manufacturing strategies such as 'total design' and 'concurrent engineering'. In 'total design' all the production team co-operate to produce the best and most cost-effective solution to design problems. All are concerned with improving, modifying and quality controlling at each phase of production and distribution. All team members are familiar with the source and manufacture of the basic components as well as where and how they will be used in the next and subsequent phases of the product's development.

A similar 'design strategy' is very much part of concurrent engineering (CE). All members of the team are involved and fully cognizant of all relevant issues relating to the effective manufacture of a product. A comparison of the strategic approaches in concurrent engineering and design approaches reveals that these are very similar. To illustrate this the diagrammatic representation of Decter's CE model and Lindbeck's design process model are shown in figure 1. Particular attitudes and ways of thinking are characteristic of these

manufacture will be engineers, however all the engineers will be involved in some way in designing.

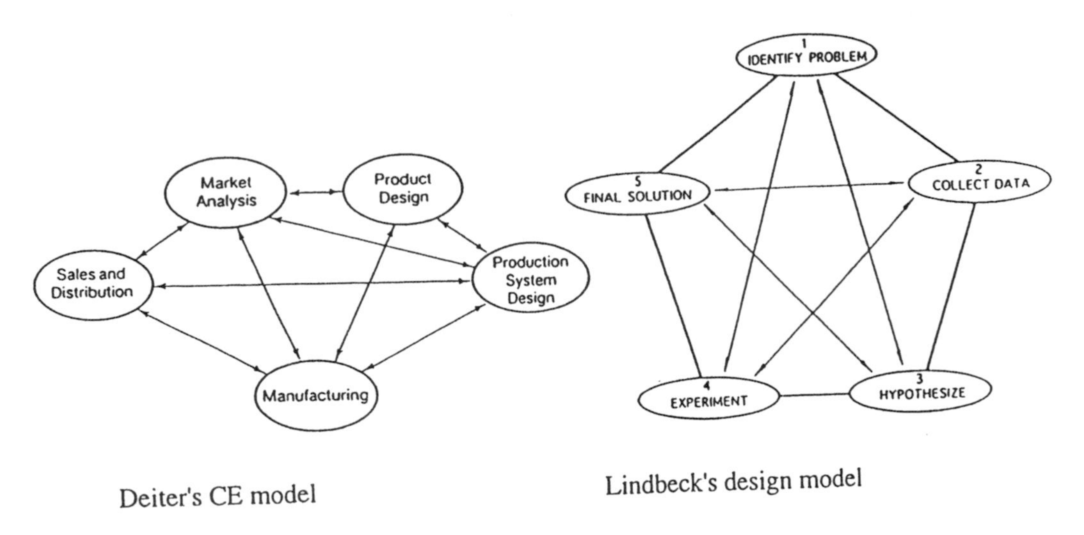

Deiter's CE model

Lindbeck's design model

Fig. 1

It may be said that design is at the core of all facets of engineering. There are many definitions of design but Blumrick's (1979 in Deiter 1991) description would appear to fit well the design activities of most manufacturing engineers; *'Design establishes and defines solutions to and pertinent structures for problems not solved before, or new solutions to problems which have previously been solved in a different way'.* This gives quite a broad brief to the engineer. Regardless of how one defines engineering design it is a creative activity and *the essence of creativity is visualisation.* This however is a cognitive capacity which many engineers lack and may even see as unimportant. Visualising, being creative and designing, in the context of engineering, are normally cognitive functions of the *right cerebral hemisphere*.

The Split Brain and Lateralisation

The human brain is divided into two halves sitting side-by-side and interconnected by the *corpus callosum.* The left cerebral hemisphere is normally identified with language and mathematical logic while the right is generally seen as the part which creates images, which visualises, designs, is creative. These specialisms are basically what *lateralisation* is about. Many do not recognise this as having any great significance to the manufacturing engineer. However if one analyses the attitudes we have and the language which we use we will quickly see that in an off-hand way lateralisation or cerebral specialisms has been recognised for centuries. It is well recognised that the left hemisphere of the brain controls the right hand and visa-versa for the left hand (Figure 2).

'Left' and 'right' have their own connotations in every day language. In the not too distant past many young people who were naturally left-handed were forced to change to right-hand use, for example in writing. In not too many centuries ago being left-handed was seen as sufficient reason to be burnt at the stake. Do such attitudes still exist, even in a subtle way? Being 'right' has to do with 'correctness', 'good', 'proper' and so on, but being left or left-of-centre is completely the opposite.

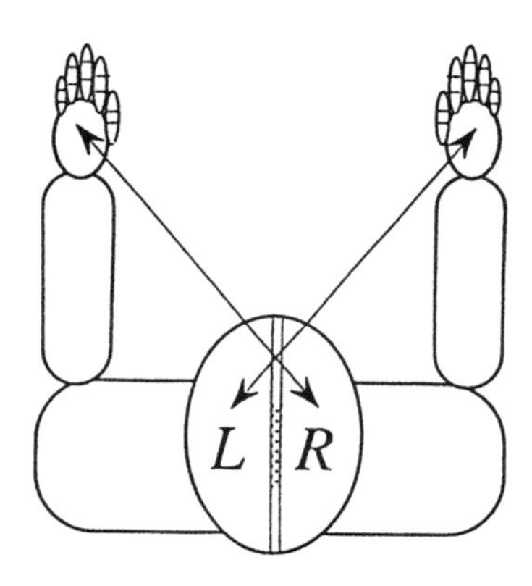

Figure 2

Left - right crossover

A left-handed (right-brain dominant) person is termed a *sinistral.* This word derives from the Latin for left which is 'sinister'. On examination of the Oxford Thesaurus (1991) the synonyms for sinister include; evil, bad, corrupt, sneaky, underhanded, harmful and many others. A right-handed person is termed a *dextral*; this word derives from the Latin 'dexter'. Synonyms for dexterity are; skill, proficiency, deftness, clever, keen and so on. This bias appears in all common languages. The French word for left is *gauche*, which means 'awkward' (from gawk or gallock-hand - left-hand) The French for right is *droit*, from 'good', 'just' or 'proper'. In Irish a left-handed person is called a *ciotóg*; which implies helpless, awkward, contrary. The Irish for a right-handed person is *deasach* from 'deas' - meaning 'appropriate', 'attractive', 'clever', 'witty'. In English, left comes from the Anglo Saxon *lyft*, meaning 'weak' or 'worthless'. Whereas Anglo-Saxon *reht* meant 'straight' or 'just', (Edwards, 1989), The Latin word associated with reht is *rectus* from which we derive 'rectitude' or 'correct'.

All the foregoing illustrate the negativity associated with left-handedness, which indirectly indicates (although sometimes unwittingly) the concentration on left-brain development, at the expense of the right hemisphere. The prejudice is still found in psychological research where some refuse to recognise left-handers and prefer instead to refer to this group as *'non - right-handers'*, Graham (1990).

Why though is it important to focus greater energy on developing the right cerebral hemisphere, particularly in engineers?

Visualisation and Lateralisation

It is widely accepted that the right cerebral hemisphere is responsible for processing design and spatial details and for the cognitive manipulation of such data.

An important ability for any engineer is that of *cognitive modelling*. This involves building and manipulating images so as to define, refine and communicate design problems. Visualisation while used as a general term for building cognitive images or models is also seen as a subset of a more general human intelligence, namely, '*spatial ability*'. Lohman's 1979 definition of spatial ability is; *'the ability to generate, retain and transform abstract visual images'* and Gaughran' (1996); *'the ability to visualise, manipulate and interrelate real or imaginary configurations in space'*. While researchers in the past have identified two or three sub-factors to spatial ability, Gaughran (1990, 1996) recommends five sub-factors. Before these are discussed we will first contextualise spatial ability in relation to the overall human intelligence factor *'g'*, Vernon (1961) and Smith (1964) identify two major subsets to the general intelligence factor *'g'*; these are the *'V:ed'* group and the *'k:m'* group. Where *V:ed* represents the language/logical-mathematical intelligence and *k:m* represents the visuo-spatial/psychomotor intelligence factors. The former is normally a left-brain function and the latter a right-brain activity. See Figure 3.

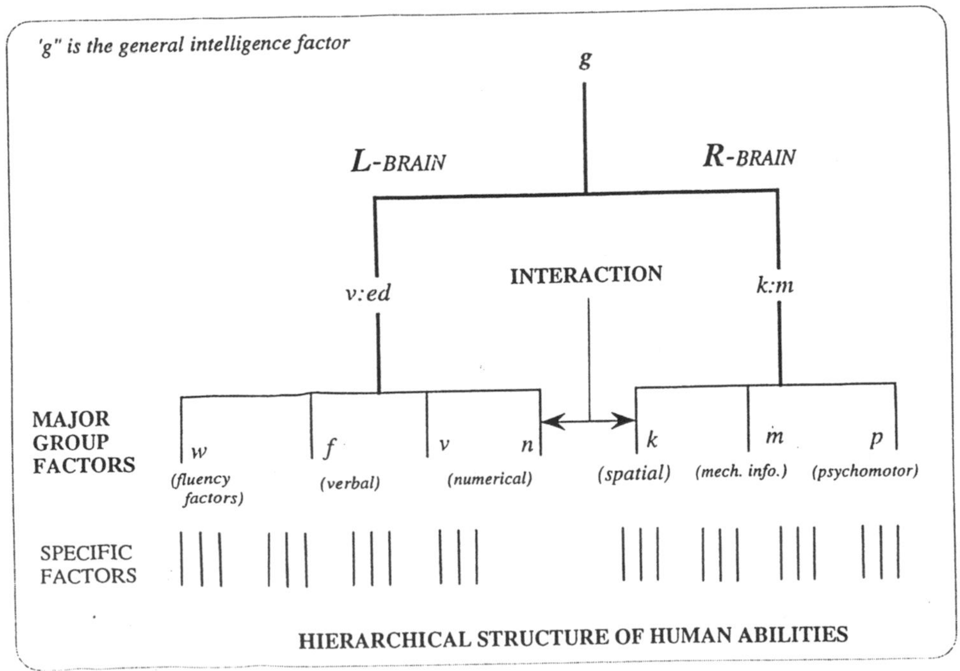

Figure 3

Each of the sub-groups have their own sub-factors, and to have satisfactory methods of developing these, they must be defined and categorised. Some of the sub-factors in the past have been confusing to say the least; for example what Zimmerman (1954) describes as *'spatialising'*, French (1951) appears to call *'orientation'*. Several others including French uses the term *'Visualisation'* to apply to a single sub-factor of spatial ability. The of spatial ability. Gaughran (1992/90) suggested expressing the sub space-factors (SF) in ascending order according to factor-of-difficulty, viz.

(SF-1) *Image Holding and Comparing*: An example would be to build an image of an electric plug-top and compare its topology with a similar one.
(SF-2) *Planar Rotation*: An example would be to rotate the plug-top on a single plane (the vertical or horizontal).
(SF-3) *Spatical Orientation*: Instead of mentally rotating the plug-top as in SF-2 to acquire a different view the 'imager' imagines himself/herself viewing it from a different or various orientation/s.
(SF-4) *Kinetic Imagery*: The ability to manipulate the image of the plug-top in any or all axes in space.
(SF-5) *Dynamic Imagery:* This would include the ability to to 'explode' and re-assemble all the components of the plug-top. See Figure 4.

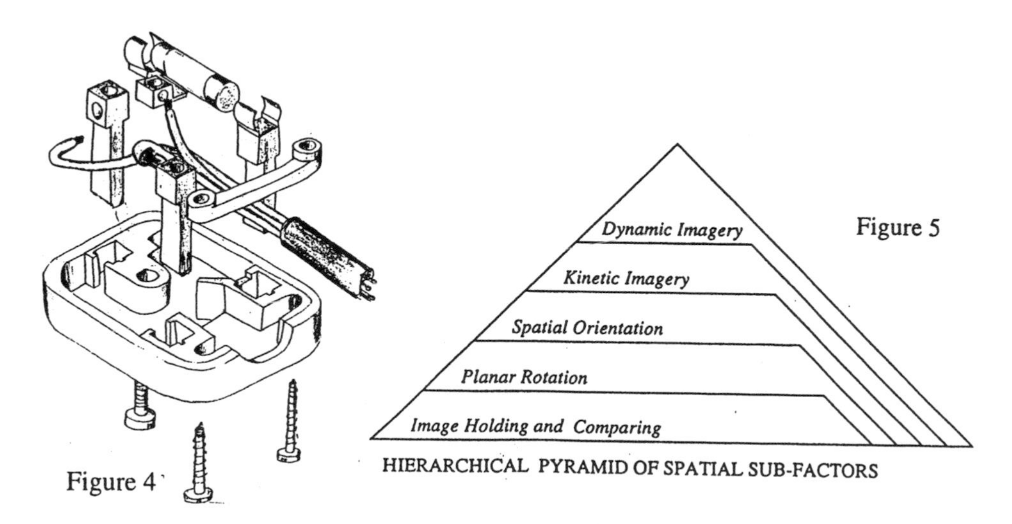

Figure 4

Figure 5

HIERARCHICAL PYRAMID OF SPATIAL SUB-FACTORS

While *SF-1* is a pre-requisite of all other sub-factors, the other space factor may use one, more or as in the case of *SF-5* all the others. Each builds on the other downwards (Fig. 5). *Planar rotation* and *orientation* may interchange or combine in the cognitive manipulation of an image. The logical left-brain may also cooperate, using mathematical algorithms to check size, fit, etc. This is the ideal dual-processing cognitive activity.

Strategies that Work

Some psychologists say that spatial ability is innate and that lateralisation of cerebral functions has been decided at birth. However where *appropriate tutorial intervention* is employed significant advances can be achieved regardless of age (Lord 1987, 1984, Moses 1982, Gaughran 1996). In order for tutorial intervention to be effective cognizance must be taken of individuals with quite limited spatial experience. A sequenced approach is advisable, for example Mamaw and Pilligrino (1984) Figure 6.

Considering the speed of perception in relation to the angle of rotation, Shepard and Metzler (1971) found when using cubelar arrays in perspective that the average rate of rotation is approximately 60° per second, see Figure 7.

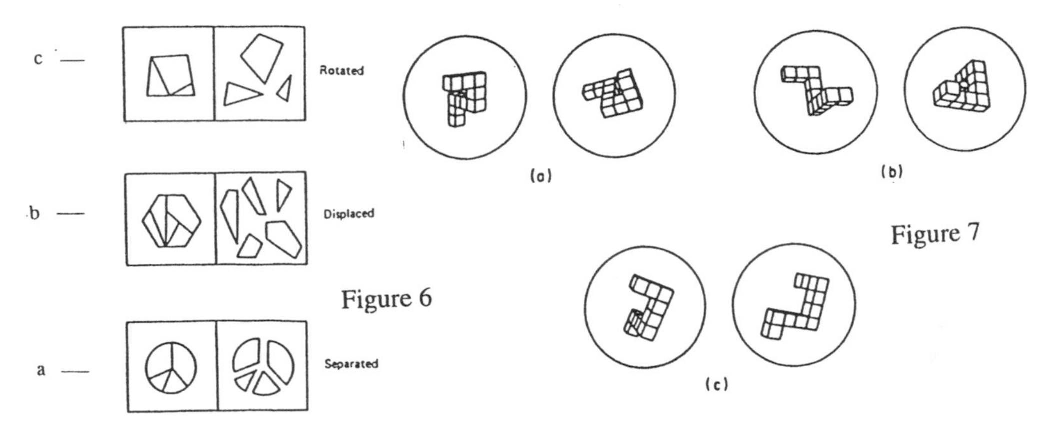

Figure 6

Figure 7

They (Shepard and Metzler) presented perspective view in three different ways: (a) pair that differs by an 80° rotation within the picture plane; (b) differs by rotation in depth and (c) cannot be brought into congruence by any rotation, they are enantiomorphic (mirror images). The mirror image causes difficulty for most learners due to cognitive confusion with left-right orientation. These examples illustrate how a progressive strategy may be employed to increase the factor-of-difficulty and how in three-dimensions the process of encoding, comparison, search, rotation and decision can be followed. The individual with a more advanced spatial ability will quickly reduce the set of alternatives in processing correct solutions.

The above may be supplemented with models and tactile stimuli. Understanding the cognitive process, or at least its sequence, assists in designing meaningful tutorial material. The gradual rotation of sectioned elements assists the beginner and accommodates the more advanced in arriving at an early conclusion. This method provides the learner with a very useful representational 'Bridge', see Figure 8.

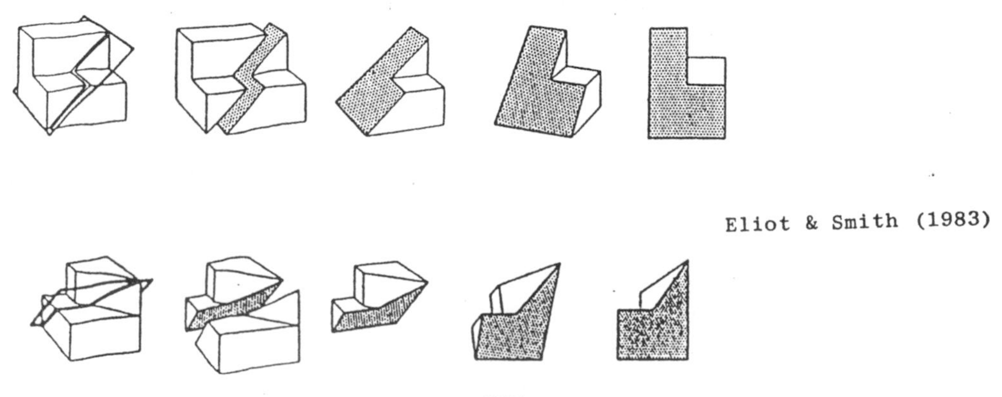

Eliot & Smith (1983)

The first graphic shows the cutting plane penetrating the solid and gradually the cut section is rotated towards the viewer. It appears that 10° is a good rotational interval (Stepsize). They (Seddon, et al) observed that students performance decreased significantly as the stepsize increased. In improving visualisation skills the ability to identify and respond to cues is an essential pre-requisite. An example is seen in Figure 9.

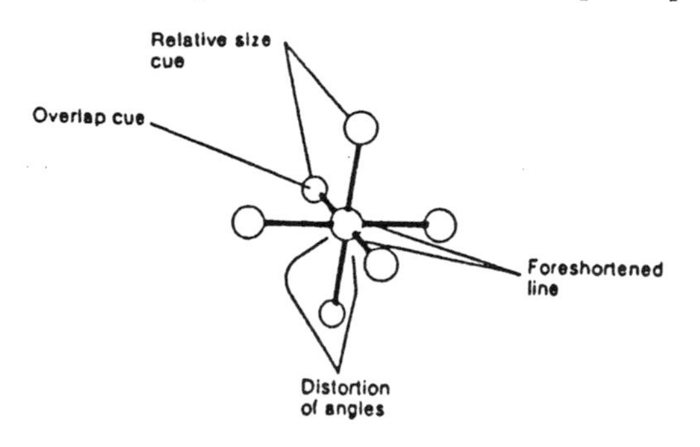

Examples of four different pictorial depth cues.

Figure 9

The designing of tutorial media is complex in that it requires taking cognizance of the variety of intellectual levels and the type of cognitive processes involved. In 1990 the author designed and tested a series of computer based tutorials to determine whether this mode was effective. Elements of the tutorials are shown in Figs 10a and 10b. The graphics elements were mainly 3D and fully interactive.

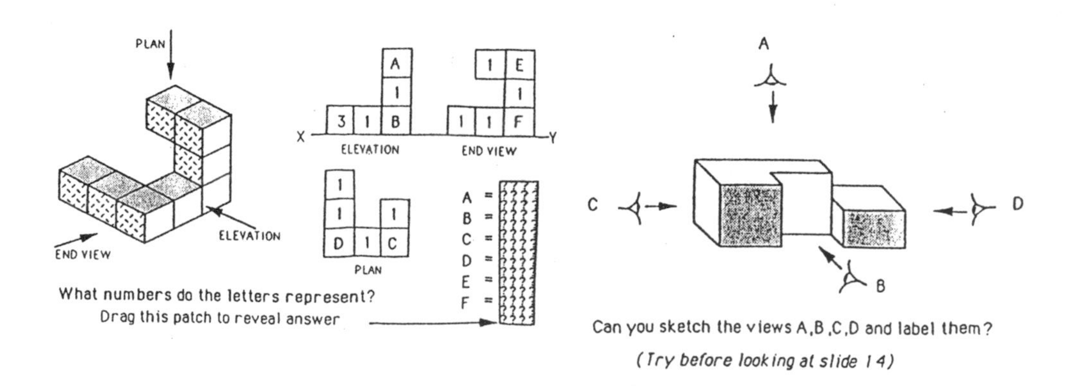

When the experimental group were compared with the control group there were significant increases in spatial ability in all elements SF1 to SF4. SF-5 was not tested because of its complexity and would require much more extended tutorial intervention. The tests and tutorials were carried out over a three week period. The post-testing of the experimental and control groups was conducted three to four days after the tutorials. While the positive results confirm that there is cerebral specialisation in connection with visuo-spatial processing, the significant evidence of cerebral lateralisation is quite interesting.

A cohort of about 120 students were involved of which 12.5% were left-handed. The complete cohort undertook the same pre-test and were later divided for tutorial sessions. They were given exactly the same series of tests on the post-test and the improved score of the control half of the cohort was used as the 'familiarity-factor' and subtracted from the raw scores of the experimental group. This produced an average improvement in score of just under 7% for the experimental group and this, when subjected to statistical analyses proved to be significant.

Points of note:

- In the pre-test the sinistrals scored 12% better than the dextrals
- In the post-test the sinistral's score improvement was nearly double that of the dextrals.
- The top seven scores in the experimental cohort scored above 75% - three of these were sinistrals.

In 1988 a group of children were tested so as to determine whether visually gifted children retain visual images without the intention of deliberately committing them to memory. The gifted children were better able to identify the altered picture and also recalled more of the specific details. The researchers were surprised to find that; *24% of the gifted children were left -handed and that none of the non gifted children were left-handed';* (Resenblatt and Winner, 1988).

Recent studies at the University of Limerick (1992) on whether previous drawing experience affects the computer aided design scores of undergraduates are also interesting. There was a 17% difference in scores by those students with previous advanced drawing experience. The results at the top of the group were particularly interesting. Seven of the top eight performers had additional drawing experience. The eight member of this cohort had an 'A' in higher level mathematics and he was left-handed. Two of the top five were left-handed (sinistrals constituted 12% of the cohort of 75 students).

The foregoing is not intending to say that we need more left-handed engineers but it does illustrate that where handiness is considered it is an indicator of cerebral specialism and lateralisation. It is also worth noting that sinistrals are less specialised cerebrally. It is estimated that while 99% of dextrals have language processing in the left-hemisphere and that only 66% of sinestrals have left-hemispherical language processing. This shows that there is less cerebral specialism among sinistrals. This is bourne out by research on left-handers by Witelson (1985).

Witelson found that on average the corpus callosum in the brain of sinstrals to be an average of 11% larger than in dextrals. See Figure 11.

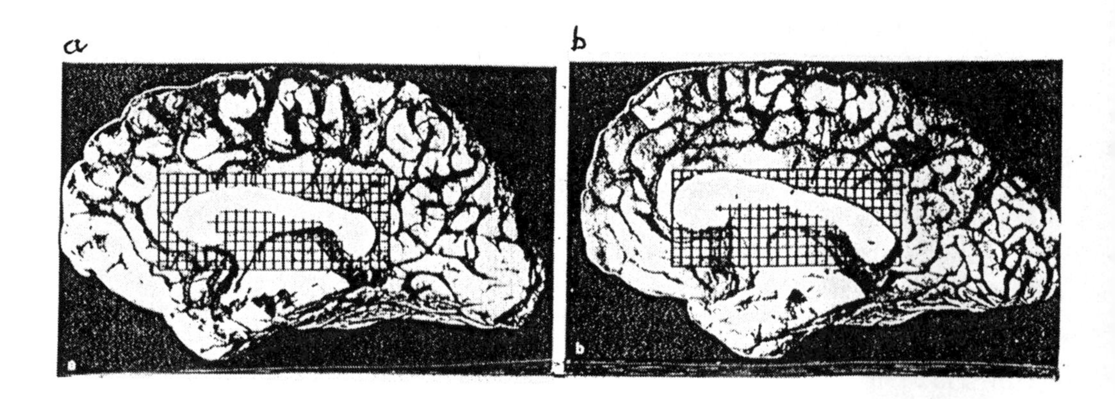

Figure 11 (a) dextral (b) sinistral (note difference in corpus callosum areas)

As the corpus callosum is the nerve mass which communicates between the cerebral hemispheres the interaction between left and right appears greater in sinistrals. Research ha also shown that where lateral specialisation is concentrated on one side that hemispherical asymmetry occurs (the frequently used hemisphere will be larger).

Discussion

Lateralisation involves the two hemispheres in different aspects of cognitive processing. In order to improve engineering performance, cognizance should be taken of the necessity to stimulate right-brain activities such as holistic reasoning and image building or *cognitive modelling* and the important human intelligence of *spatial ability*. While many contend that individuals either possess this ability or they don't (that it is innate) research shows that with *appropriate tutorial intervention,* this ability can be developed and enhanced.

Engineers who are constantly involved in design activity need to be good *visualisers*. They need to be able to build *the cheapest of all prototypes - the cognitive model.* This can only be achieved by recognition of the need to *'spatialise'* design/engineering information. The recognition of cerebral lateralisation should influence the design of undergraduate courses for engineers as well as their in-career development.

"The left hemisphere analyses over time whereas the right hemisphere synthesises over space" (Levy, 1974).

References

1. **Deiter, G.E., 1991,** Engineering Design, McGraw Hill, Boynes 1978.
2. **Edwards, B., 1979** and **1993,** Drawing on the Right Side of the Brain, Collins and Sons, Glasgow.
3. **Eliot and Smith J.M., 1983,** An International Dictionary of Spatial Tests NFEF Nelson, Berks.
4. **French, M.S., 1951,** Conceptual Design for Engineers, Design Cornal London.
5. **Gaughran, W.F., 1992,** C.A.D. in Educational Context, Paper to Computer Education Soceity of Ireland, A.G.M., October, 1992
6. **Gaughran. W.F. 1996,** Design Intelligence on Engineering, Paper to the IMC-13 Conference, Limerick Sept. 1996.
7. **Gaughran, W.F. 1996,** The Graphics Code Visualisation and CAD FAIM Conference, Atlanta May '96.
8. **Gaughran, W.F. 1990,** Developing Spatial Abilities Through Computer Assisted Learning. M.Tech. Thesis, University of Limerick 1990.
9. **Graham, R,B., 1990,** Physiological Phychology, Wadsworth
10. **Levy, J., 1976,** Cerebral Lateralization and Spatial Ability. Behaviour Genetics, T, No. 2, 171-188
11. **Levy, J. 1974,** Phychobiological Implications of Bilateral Asymmetry, in Hemisphere Function in the Human Brain, Wiley, New York.
12. **Lohman, P.F., 1979,** Spatial Ability: A review and reanalysis of the Correlation Literature, Technical Report No. 8, Standord University
13. **Lord, T.R., 1987,** Spatial Teaching. The Science Teacher, February, 1987
14. **Lord, T.R., 1984,** A Pleas for Right Brain Usage. Journal of Research in Science Teaching, 22, V, 395-405.
15. **Mamaw R.J. and Pellegrino J.W., 1984,** Individual Differences in Complex Spatial Processing, Journal of Ed. Psychology, V76, 5, 920-939
16. **Moses, B., 1982,** Visualisation - A Different approach to Problem Solving. School Science and Mathematics, 82-2, 141-147.
17. **Rosenblatt, E. and Winner, E.,** 1988 The Art of Children's Drawings.
18. **Journal of Aesthetic Education,** Vol.22, No. 1, Spring, 1988.
19. **Seddon G.M., Eniaiyeju P.A. and Josoh J., 1984,** The Visualization of Rotation in Diagrams of Three Dimensional Structure American Ed. Research Journal, Spring '89, 21, 1, 25-38
20. **Shepard R.N. And Metzler J., 1971,** Mental Rotation of Three-Dimensional Objects Science, Vol. 171, Feb. '71
21. **Smith, I.M., 1964,** Spatial Ability - Its Educational and Social Significance. N.F.E. R., Nelson, Berks.
22. **Vernon P.E., 1961,** The Structure of Human Abilities Methuen, London.
23. **Zimmerman W.S., 1954,** Hypotheses Concerning the Nature of Spatial Factors Education and Physical Measurement Vol. 14, 306-400
24. **Witelson, 1985,** in Graham R.B.1990, Physiological Phychology, Wadsworth

CREATION OF SOLID MODELS USING MRI SCANNING.

F. Nabhani, A.N. Hart,, C.J. Connor, M. Wake
School of Science and Technology, University of Teesside, Middlesbrough, TS1 3BA.

ABSTRACT: Creating solid models by the implementation of Magnetic Resonance Imaging (MRI) in medicine has proven to be invaluable in diagnosis, but the full potential of MRI in 3D modelling research has yet to be fully realised. In our studies it is shown that it is now possible to obtain accurate data relating to the creation of models, in particular hard tissues such as bone, such that complex models can be constructed. In the studies, the human patella was used as an example and was constructed from in-vivo images.

It has been well documented that the patella is constructed not of a single material but a composite of three forms and the material distribution is necessary when dealing in either 2D or 3D mode such as when the model is to be later used for simulation procedures.

INTRODUCTION

For some time Nuclear Magnetic Resonance has been widely used in the analysis of chemical compounds, however it was not until 1980 that the most well known application i.e. medical imaging was developed. This imaging aspect is more commonly known as Magnetic Resonance Imaging or MRI. Its main and only real practical use has been in the realms of specialised technology areas such as medicine, as an imaging system for producing detailed views of in-vivo tissue discrepancies within the body non-invasively. Its use for the creation of computerised geometrical models, especially in three dimensions has not yet been fully utilised.

The MRI system has the ability to "slice" an object radiologically without having to actually cut tissue, in this particular case, bone [Wehrill *et al,* 1993] , specifically that of the patella [Minns and Braiden, 1989; Gehelman and Hodge, 1992; Koskinen *et al*, 1993; Nabhani *et al*, 1994], revealing the inner material distribution. A similar method has already been developed with the use of digital x-rays of the brain and long bone etc., however this has an inherent problem in identification of bone density distribution and also harmful exposure to radiation. Thus, if an object such as the patella requires examination in which the density of the inner bone or boundaries are extremely close to one another, then it becomes difficult to distinguish these boundaries. As the majority of MRI machines works on the basic principle of free flowing hydrogen atoms, it is possible to view these density differences more clearly.

BASIC THEORY BEHIND MRI

The basic principle of Nuclear Magnetic Resonance is that an atomic nuclei possess some angular momentum or spin. This spin will give rise to a nuclear magnetic dipole moment. This occurs in most materials in a random direction and thus a material can be said to have no net magnetism. However, if a magnetic field is applied to the object, the nuclei will take a specific orientation which is relative to the direction of the field. For the nuclei that have a quantum spin number of 1/2, e.g. hydrogen, the only allowed orientations are parallel (spin up) and anti-parallel (spin down) to the external field. This spin-up state has a slightly lower energy value than that of the spin-down state and although the states are almost equally populated there is a slight excess of nuclei in the lower energy state. This will result in a small bulk magnetisation, M, in the material which is directly parallel to the applied field. If the direction of the magnetisation vector M is tipped away from the applied field direction (z), then M rotates about the z-axis, describing the wall of a cone. This motion is called precession. The tilting of M is achieved by applying a second but smaller magnetic field in the x-y plane, perpendicular to the first applied field and also rotating at the same rate as M. In practice this is usually done by means of surrounding the object within a coil carrying a radio-frequency current. It should be stated however that the frequency of the radio-frequency must match that of the precession frequency so creating this effect. This can then be termed magnetic resonance.

DATA GATHERING

There are only a limited number of ways in which it is possible to gather or process the data produced from a MRI machine. Before scanning can be started it is necessary to first decide upon the type of object and the way it is to be scanned. The object must be examined to verify if any metal, especially ferrous based products, are contained within the material. This is necessary as ferrous material influences the imaging system due to the effect it has upon the magnets used within it. If the object has satisfactorily passed the 'metal test', the next task is to decide upon the way it is to be scanned. If the object has insufficeint hydrogen content it may be necessary to submerge the article within a hydro-carbon medium such as an oil based product, or it must contain a hydrogen enriched substance, which in the case of bone is the bone marrow. Once it has been established that the object to be scanned has sufficient free flowing hydrogen atoms present, it is a simple matter of placing the object into the MRI machine. For data acquisition it is not necessary for the object to be placed in any specific orientation as the modelling process will take this into account. Once the object has been scanned, the data can then be transferred to a suitable reading format such as a disk, film, etc.

DATA PROCESSING

There are two basic methods of processing the MRI data. Firstly, using the MRI 'raw data' from the scanned image, the data is processed through an image processing package with Fast Fourier Transformation analysis such that a vector image i.e. the 'x' and 'y' co-ordinates of each pixel are obtained. An origin is required at this time which is used to reference the image data on a slice by slice basis. The origin position can be set globally e.g. in the MRI machine, and not locally set at the images.

The result of this image processing is a set of data columns which represent the x-y co-ordinates of each slice image of the patella. However, it should be also noted that before the use of the data in creating a 3D image it must also have depth e.g. 'z'. The depth of slice varies from one MRI machine to another and it is possible to cut the patella into sections as little as 1.4 mm thick. Once the data has been given depth it is then possible, using various manipulation techniques, to subtract the unrequired data due to tendons, ligaments, fat, etc. Once removed, the only image data left is that of the patella slice. As it is required that the material boundaries of the slice are known, it is necessary to produce the outline of the boundaries. This leaves an outer image consisting of the overall shape and material boundaries. Using a CAD package and suitable software it is possible to translate the boundaries into an x, y, z format which can be used to generate a 3D model.

The second method involves the use of the x-ray film produced from the MRI scan, (Figure 1). A series of high resolution computer images are produced by digitising the x-ray image. Using suitable software packages, the outer boundaries of the bone can then be digitised on a point to point basis, thus producing a set of data points from which it is possible to construct the borders, (Figure 2). However, due to scaling errors that arise with this method, the results require a magnification factor to be added to them. In addition to this problem, a local origin has to be produced on each scanned image such that each image has a single point which is relative to the other images. This enables the correct orientation and relative position to each other of the planar slices to be achieved and thus a true representation of the object, in 3D, is produced.

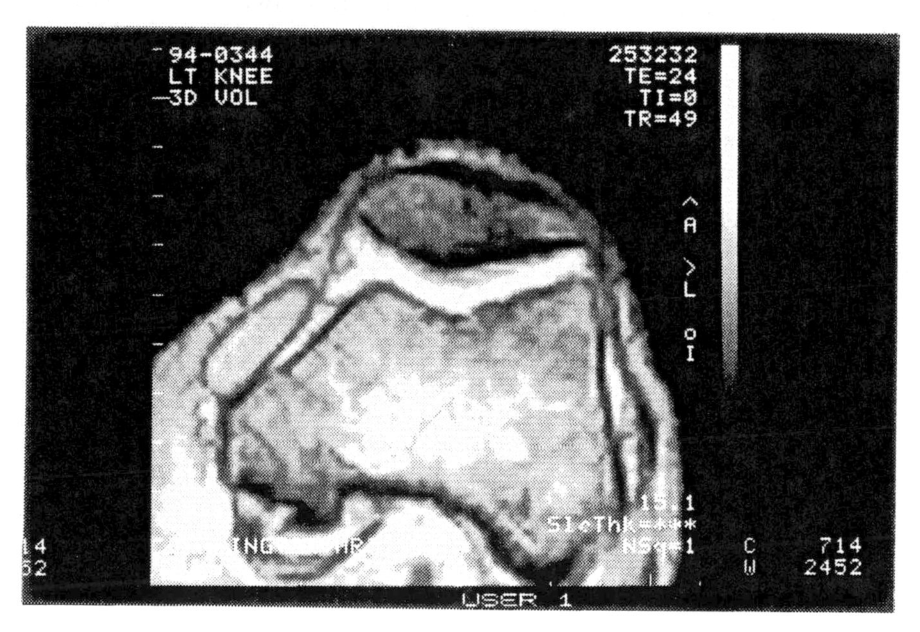

Figure 1. A typical scan image of the patella

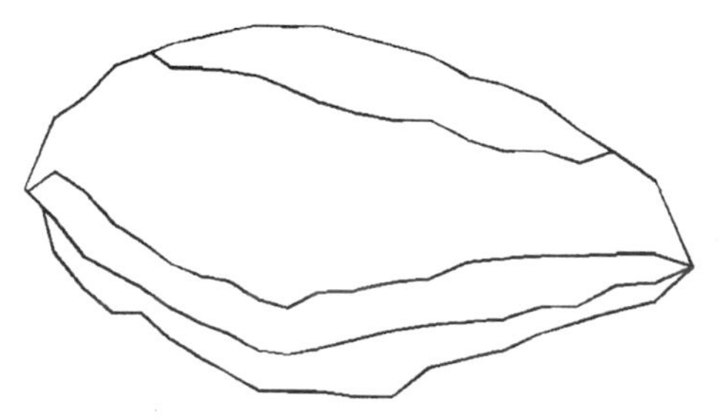

Figure 2. A diagrammatic picture of the patella and its material distribution.

MODELLING USING MRI

Until recently modelling of objects, e.g. bones, has been restricted primarily to two dimensions. With the advent of MRI to gather the geometric data, the possibility of creating complex shapes is now obtainable for it offers a clear photographic slice of the interior of the object, revealing the inner material within the bone. The creation of a 3D model by this method [Higgins and Ojard, 1993; Schubert *et al*, 1993] involves arranging the slices in a consecutive parallel pattern thus replicating the model of the said object in 3D space.

The patella, see figures 3a, 3b, 3c, 3d, comprises not of a single material, but a combination of many differing forms of bone [Minns and Braiden, 1989] which have their individual characteristics. From the modelling aspect this represents a problem, especially if the densities of the bone within the patella are close together. The use of digital x-ray methods result in blurring or joining of the material boundaries on the image. MRI shows promise in more clearly defining the different zones of density within bone for it relies upon free flowing hydrogen atoms and as such, material boundaries are much more clearly defined. This represents a step forward in the production of 3D models which can be used in the determination of model making of the human anatomy.

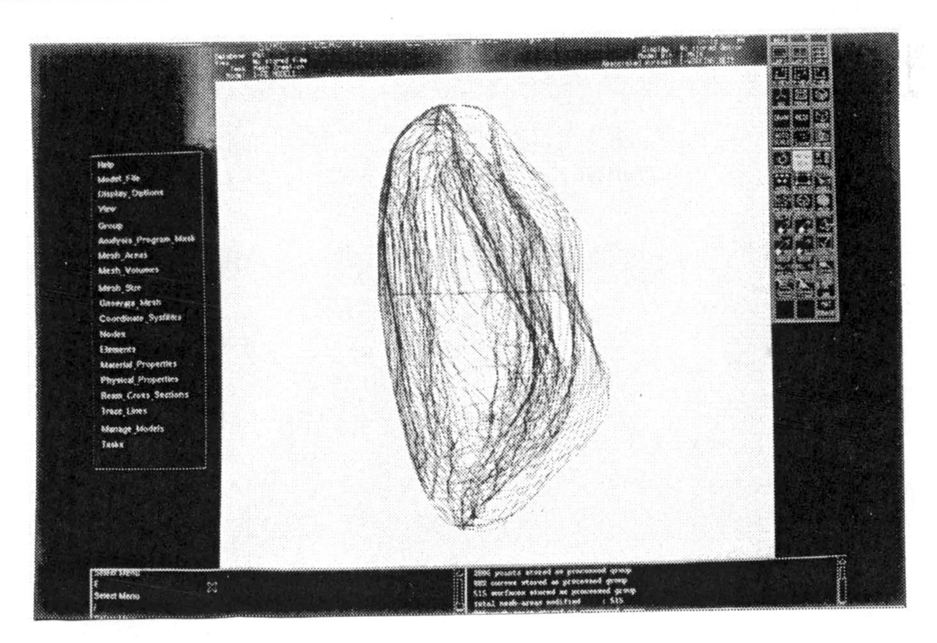

Figure 3a. A 3D wire frame model of the patella, showing material distribution.

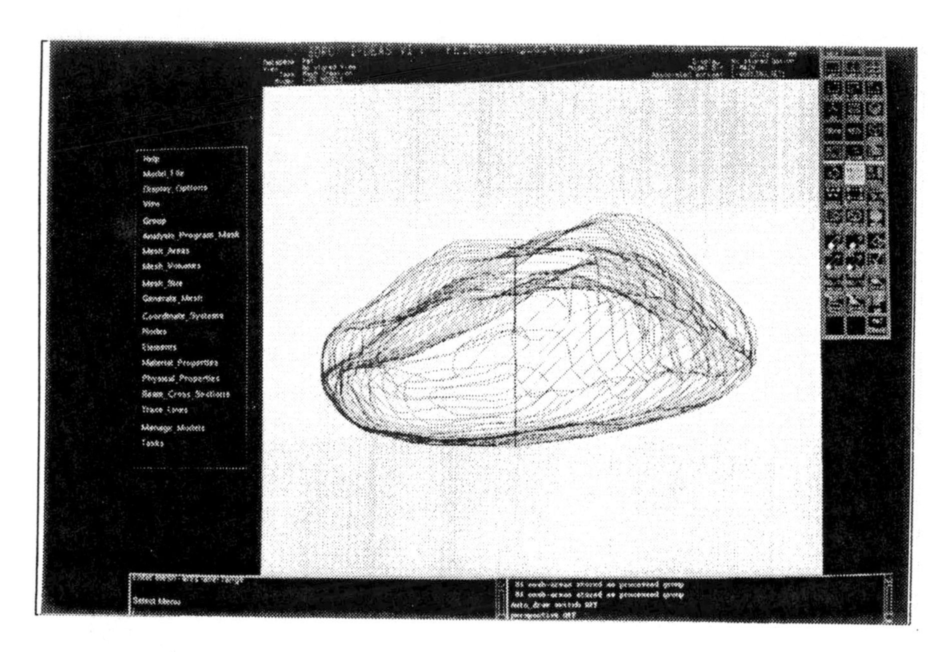

Figure 3b. A 3D wire frame model indicating the outer shell and cartilage distribution of the patella.

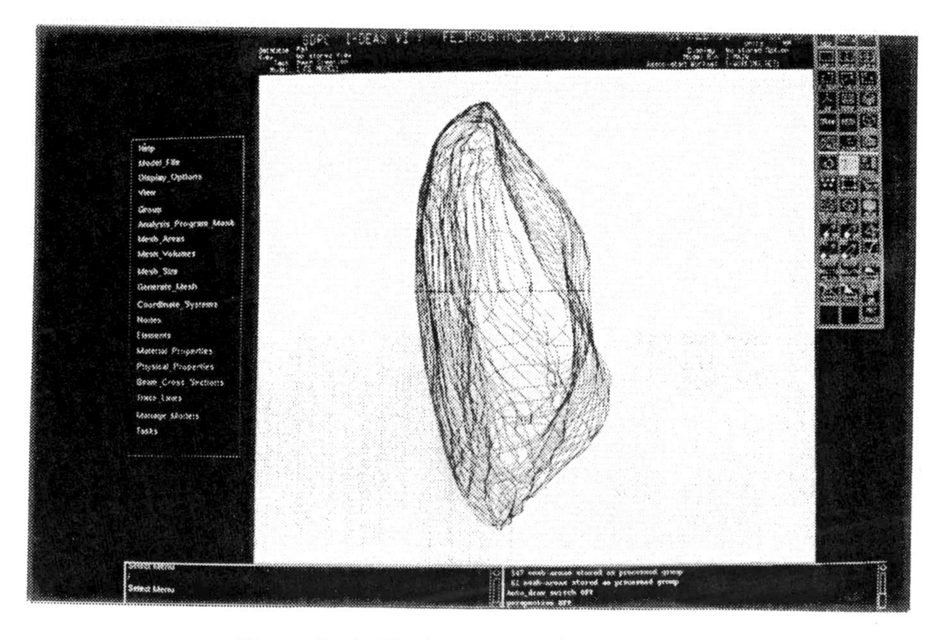

Figure 3c. A 3D wire frame model of the patella showing Cortical bone distribution.

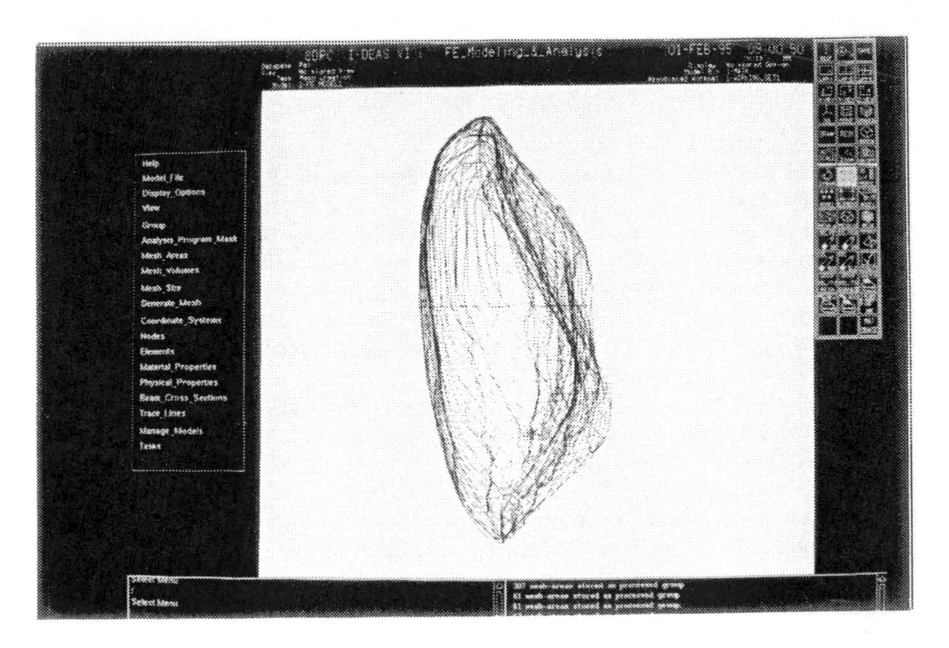

Figure 3d. A wire frame model showing
the Subchondral bone distribution.

CONCLUSION

The use of MRI for modelling purposes is a major step forward as it results in not only 2D models but also more complex 3D models being produced accurately. This will result in a better understanding of the fundamentals of the human body, as it will be possible to create structures in a true orientation from in-vitro and in-vivo studies. MRI does have some inherent problems which cannot be readily rectified, mostly related to the type of material of the object to be modelled. Metal objects do cause imaging problems due to their lack of free hydrogen ions and Ferrous objects do cause some major problems due to their magnetic attraction properties.

DISCUSSION

The MRI system has yet to be fully utilised as a research tool in the modelling of anatomical structures. The application of the MRI system needs to be exploited and developed further if a better understanding of the fundamentals of the human body are to be gained.

ACKNOWLEDGEMENT

The authors of the document would like to thank Mr. D.S. Muckle, South Cleveland Hospital and the staff at Darlington Magnetic Resonance Imaging whose help was invaluable in the investigation. The authors would also like to thank the Wishbone Trust for their help in this project.

References

Gehelman B and Hodge J.C., 'Imaging of the patellafemoral joint',Orthop. Clin. North Am., 23(4):523-43, Oct 1992.

Higgins W.E. and Ojard E.J., 'Interactive morphological watershed analysis for 3D medical images', Comput. Med. Imaging Graph., 17(4-5):387-95, Jul-Oct 1993.

Koskinen S.K., Taimela S., Nelimarkka O. ***et al,***'Magnetic resonance imaging of patellafemoral relationships.', Skeletal Radiol., 22(6):403-410, Aug 1993.

Minns J. and Braiden P.M., 'A loading and stress analysis of the patella', Material properties and stress analysis in Biomechanics, 1989, p. 44-51.

Nabhani F, Sotudeh R., Hart A.N. ***et al,*** 'An Investigation of Stresses acting on the Patella.", International Conference in Mathematical Science, Edinburgh, Scotland., 7-9th Dec. 1994.

Schubert R., Bomans M., Hohne K.H. ***et al,*** 'A new method for representing thehuman anatomy.', Comput. Med. Imaging Graph., 17(4-5):243-9, Jul-Oct 1993.

Wehrill F.W., Ford J.C., Chung H.W. ***et al,*** 'Potential role of nuclear magnetic resonance for the evaluation of trabecular quality', Calcif. Tissue Int. 52 Suppl 1:S162-9, 1993.

PC BASED ROLL STRESS ANALYSIS

J.Gui*, I.G.French*, P.Blackburn, M.M.Ahmad***
*EPICC, University of Teesside, UK
**British Steel PLC, UK

ABSTRACT: One of the limiting factors of rolling mill capacity is that of roll strength. Therefore, in the fast moving world of today, it is essential for mill staff to have a good understanding of what the mill can achieve in terms of product range and capacity.

There is, therefore, a need for mill staff to have appropriate tools available so that well founded decisions can be made quickly when considering new schedules and products.

British Steel saw the opportunity to provide a computer-based tool for use by mill managers and engineers which would permit them to calculate the bending and torsional stresses in mill rolls under a variety of working conditions. Such a tool is designed to form one element of an overall PC software package aimed at optimising rolling schedules and providing a troubleshooting tool for the analysis of the occasional roll failure.

The development of such tools is at the heart of British Steels policy to transfer technology to the front line and forms and integral part of a continuous development policy.

THE SOFTWARE

The software provides users with the means to obtain bending, torsional and combined stresses at a point of interest on a loaded roll. Both generic rolls (plain barrel) and various roll features (e.g. oval or half round) are dealt with. Stress Concentration Factors are applied as necessary in order to give final factored stresses for the various features.

Software Functionality

The software performs the following tasks:

- To allow for the input of the data required to define the roll operation via a customised Graphical User Interface (GUI) capable of accepting either Imperial or Metric units and also capable of converting between them.

- To use inbuilt functions, based on a combination of analytic and empirical results, to obtain the stresses, the stress concentration factors and the factored stresses, generated during a typical roll pass.

- To provide the user with a means of reverse functionality, such that worst case roll operation data can be evaluated based on the input of maximum allowed stresses.

As the system is written for Windows using Microsoft Visual Basic ver. 4.0 the functionality is split into two parts. Firstly an event driven interface part and secondly a bespoke calculational part which produces the results.

- The GUI functionality, using windows components and the API, which is event driven allows the user to enter/select input fields and outputs the calculated results to the VDU.
- The bespoke functionality (triggered by the calculate command button or by the dynamic regulator) manipulates the input data into a form which can be passed to the equation functions which return the stresses, stress concentration factors and factored stresses.

The Software Functionality is shown in Figure1.

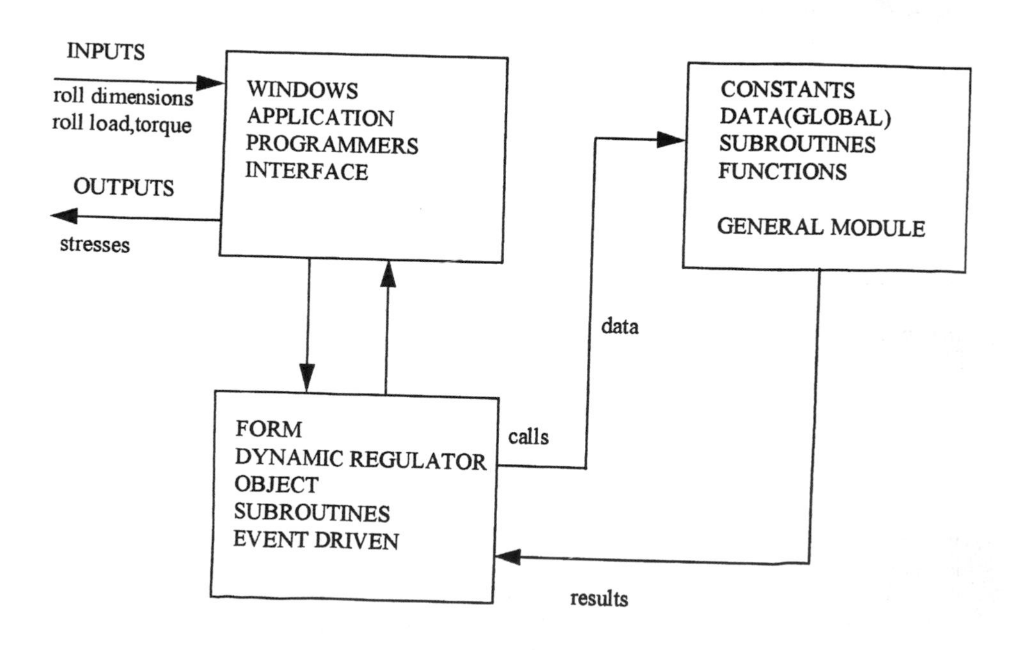

Figure 1 The Software Functionality

The software accepts input of the roll dimensions, rolling loads and torques. This data is then used to calculate stresses and stress concentration factors for a plain roll or a particular feature (pass shape or roll neck) chosen by the user. The software supports reverse functionality and uses a dynamic regulator to get the required outputs and related rolling parameters.

Graphical User Interface (GUI)

A graphical user interface is provided to allow the user to specify various roll inputs, such as the roll dimensions, the rolling load and torque, together with selected features of the roll profile.

The interface itself comprises three basic input-output screens as shown in Figure 2 to Figure 4.

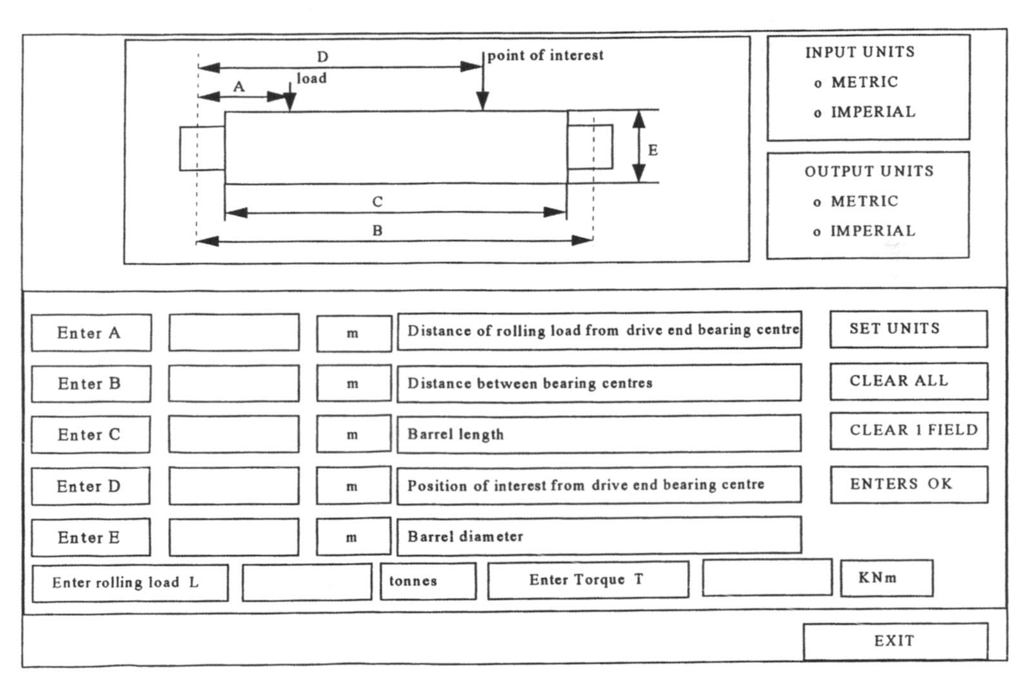

Figure 2 Basic Data Entry Screen

Figure 2, the basic data entry screen, allows entry of the 5 basic roll dimensions (A,B,C,D,E) together with the roll load (L) and torque (T). To maintain simplicity, for the user, data entry is by numeric inputs via a series of textboxes. Further, as in general the tool will be used to provide a diagnostic function in the hands of skilled personnel, there was no functional requirement to provide more than basic error handling and recovery. Consequently, error handling was restricted to simple errors such as incomplete data, divide by zero and so on. However since the sources of information, available to the user, from which input data may be derived, would most likely be of considerably differing ages, there was a requirement to deal with data in either Imperial or Metric units and to be able to convert between these.

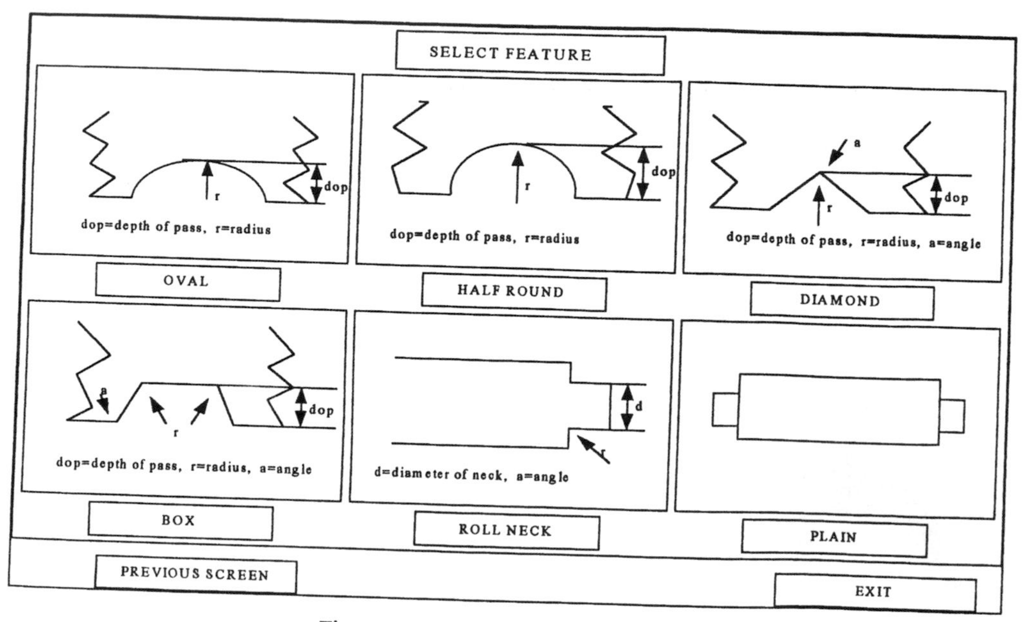

Figure 3 Roll Feature Choose Screen

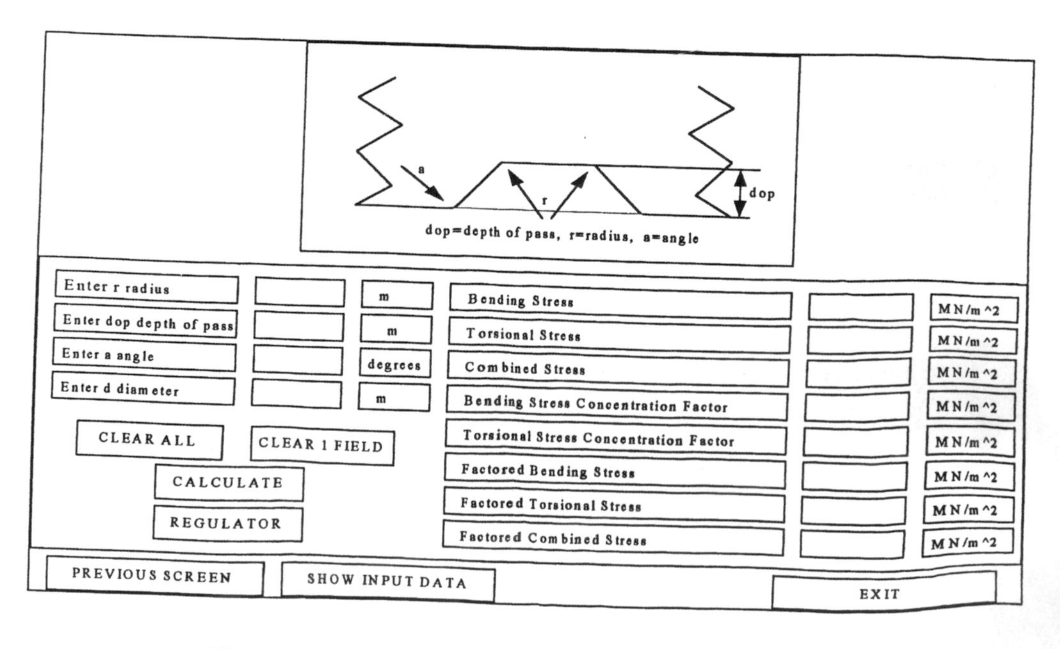

Figure 4 Feature Data Input and Stresses Calculation Screen

Figure 3 shows the roll feature selection screen. In this screen the user is simply required to select, from the given menu of images, the generic roll structure which most closely approximates to the roll pass of interest. Having selected the most appropriate feature the user is presented with Figure 4, the feature data input and stresses calculation screen. The user is then required to input data specific to the chosen roll feature and is provided with output of the stresses contained within the roll. The roll features available are:

- Oval feature: radius and depth of pass needed.
- Half round feature: radius and depth of pass needed.
- Diamond feature: angle, radius and depth of pass needed.
- Roll neck feature: neck radius and neck diameter needed.
- Plain roll: no further inputs needed.

This range of features is felt by British Steel to be sufficient to cater for the majority of roll profiles in common use.

Stress Calculations

The purpose of the GUI is to provide the data to be used in the calculation of the unknown roll stresses. For the software presented here, the calculation of these stresses is based on the equations defined in 'Roll Pass Design', United Steels Co. Ltd. 1960. Stress concentration factors are sourced from 'Formulas for Stress and Strain', Roark, 1989 and 'Stress Concentration Factors', R.E.Peterson,1974.

In using the equations the following assumptions are tacit:

- A roll behaves as a beam simply supported at the bearing centres.
- The barrel of the roll is centred between the bearing centres.
- The rolling load acts as a point load.
- All stresses are calculated elastically.
- The distance to the point of interest will be from the bearing centre to:
    (1) the centre of an oval, half round or diamond,
    (2) either edge radius of a box,
    (3) the neck radius for a roll neck, and
    (4) any point on a plain barrel

The bending moment is thus defined in terms of the user entered dimensions A,B,C,D,E and the rolling load L (Figure 2), as

$$\text{IF } A>=D \quad M = L*D*(B-A)/B \qquad \text{---------------(1)}$$

$$\text{IF } A<D \quad M = L*D*(B-A)/B-L*(D-A) \qquad \text{---------------(2)}$$

and the stresses are evaluated in Imperial units as Bending Stress, δ_b

$$\delta_b = 32*M/\pi D^3 \qquad \text{----------------(3)}$$

Torsional Stress, Ts

$$Ts = 16*T/\pi D^3 \qquad \text{----------------(4)}$$

where T is the input torque

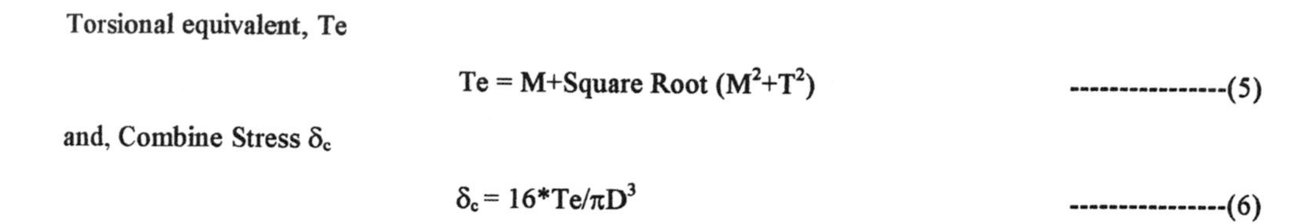

Torsional equivalent, Te

$$Te = M + \text{Square Root}\,(M^2 + T^2) \qquad \text{----------------(5)}$$

and, Combine Stress δ_c

$$\delta_c = 16 * Te/\pi D^3 \qquad \text{----------------(6)}$$

Figure 5, below, shows a schematic representation of the calculation process.

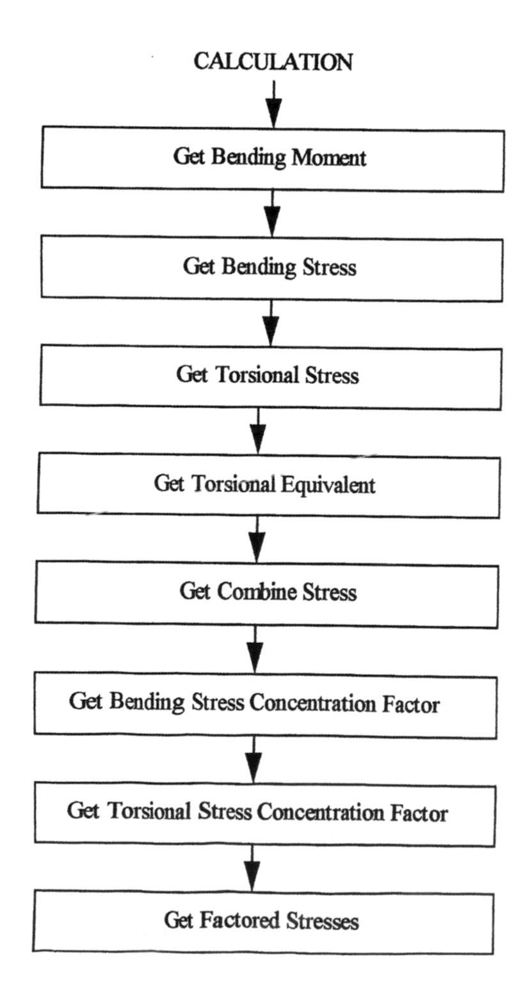

Figure 5 Calculation Block Schematic

Reverse Functionality

If , having entered the data, the outputs are not suitable, the user can return to screen 1, change the input data and then recalculate the outputs. This procedure can then be repeated in a trial and error fashion until all of the outputs fall within limits.

Trial and error procedures of this nature can, however, be very time consuming and often lead to a conservative respecification of the input parameters, thus guaranteeing that outputs fall within tolerances and consequently avoiding the need for a repeat of the process.

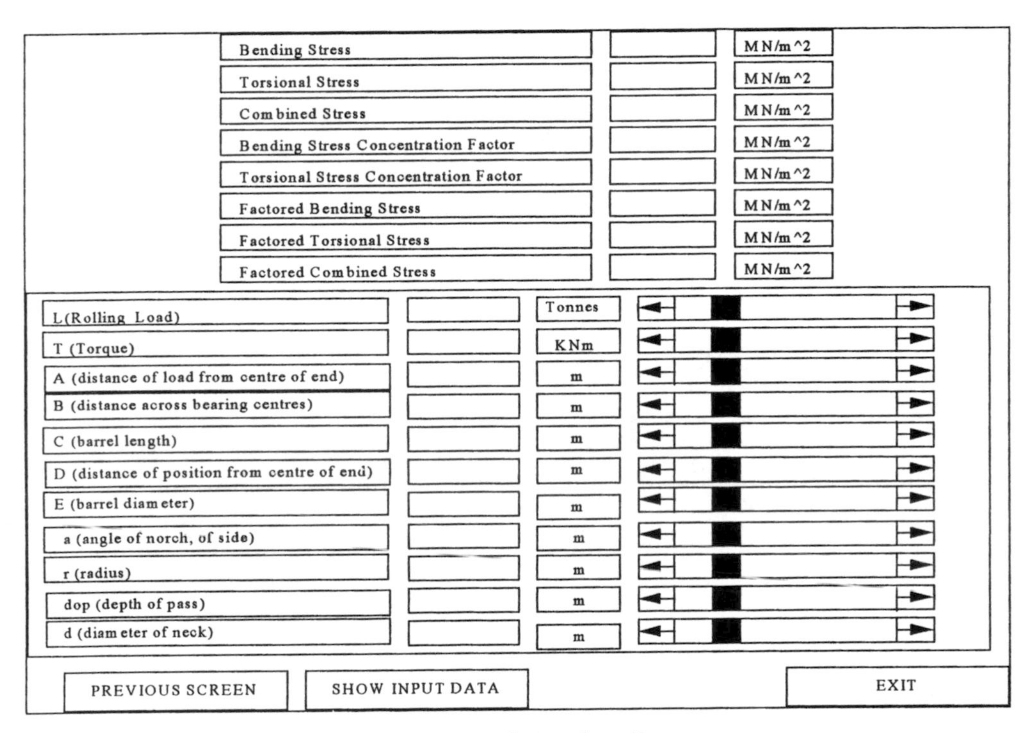

Figure 6 Dynamic Regulator Screen

The 'regulator' screen, shown in Figure 6, provides the user with the means to avoid this trial and error procedure. If the outputs are not suitable, the user can modify the out of range variables by entering the desired values directly from the key board via the regulator screen. The program will then recalculate and update all other variables on the screen. Thus when the outputs are as required the related inputs (roll dimensions, rolling load, torque) can be obtained directly. Using this approach roll passes can be designed to be nonconsenvative since roll stresses can be placed as close as desired to the constraints.

In addition to direct reverse calculation the 'regulator' screen also supports a series of sliders on the input variables. Again, varying any of the input parameters via these sliders will cause all output parameters to be recalculated and updated on the screen. The use of the sliders, therefore, gives the user a rapid means of establishing the sensitivity of roll stresses to changes in input parameters.

CONCLUSION

This paper outlines the functionality of a simple software tool used to support rolling mill managers and engineers within British Steel. The software is seen by British Steel as a time saving device, since the previous system was unwieldy and time consuming. In addition, a greater confidence factor will develop due to the consistent use of units and standard calculation techniques.

REFERENCES

1. United Steels Co. Ltd.(1960), Roll Pass Design.
2. Roark(1989), Formulas for Stress and Strain, 6th Edition, McGraw Hill.
3. R.E.Peterson Wiley(1974), Stress Concentration Factors.
4. J.Gui(1996), 'Mill Roll Strength Requirement Specification Report', British Steel.
5. J.Gui(1996), 'Mill Roll Strength Design Specification Report', British Steel.
6. Microsoft Corporation(1995), Microsoft Visual Basic Version 4.0 Programmer's Guide.
7. Microsoft Corporation(1995), Microsoft Visual Basic Version 4.0 Language Reference.

A REVIEW AND COMPARISON OF EMERGING VIRTUAL 3-D TOOLS FOR PRODUCT DESIGN

Grace M. Bochenek[1], Dr. James M. Ragusa[2]
[1]U.S. Army Tank-automotive & Armaments Command, Vetronics Technology Center, Warren Michigan, USA; [2]University of Central Florida, College of Engineering, Orlando Florida, USA

ABSTRACT: Domestic and international competition has increased pressures for market expansion and product/process reevaluation, and the need for more effective and efficient design tools, capabilities, and methods. This paper presents a review and comparison of five emerging, virtual technologies and immersion environments that are under evaluation at a leading U. S. Government military vehicle research, development, and engineering center. The newer system technologies being tested include: Helmet Mounted Displays (HMDs), Binocular Omni Orientation Monitors (BOOMs), Stereoscopic shutter glasses, Holographic Imaging Systems (HISs), and the CAVE Automatic Virtual Environments (CAVEs). A tentative conclusion reached is that the use of one or more of these technologies has the potential to markedly improve product design environments. However, several shortcomings need to be addressed before each can become a viable alternative to present methods.

INTRODUCTION

The traditional product development process can be described as a sequential, linear process. Shu and Flowers [1994] characterize this activity as a sequence of "throw it over the wall" processes where functional groups make contributions to a project sometimes independent of other elements. As a result, critical decisions which can significantly impact overall product design and development, are frequently made without regard to the effect on and consequences to other functional elements.

But some process improvements are evolving. The need to shorten the development cycle [Cooper & Kleinshmidt, 1994; Cordero, 1991; Iansiti, 1993; Vesey, 1991], to achieve higher quality [Ali, Krapfel, & LaBahn, 1995; Cordero, 1991], and to solicit direct customer feedback [Larson-Mogal, 1994] have stimulated the need for changes to traditional product development methods. As a result, numerous organizations, including the U. S. Army, are investigating the use of concurrent multi-functional teams coupled with emerging computer technologies to create more robust, collaborative virtual design environments. However, this evolution, will require the integration of new strategies for design group interaction with new collaborative design tools if organizations are to become more effective, efficient, and competitive. Throughout this paper, effectiveness is defined as doing the right things and efficiency means doing things right. [Hales, 1993]

The purpose of this paper is to briefly describe how virtual technologies and immersion environment, soon to be tested and evaluated at a U. S. Army combat vehicle development laboratory, can be used to improve interactive and concurrent product design team activities.

Discussed are the product design evolution, virtual environment (VE) three-dimensional (3-D) display technologies, collaborative virtual design systems, VE based design tool maturity, product design success issues, and some conclusions.

PRODUCT DESIGN--EVOLUTION IN ACTION

In the past, products were developed on drawing boards using pencil and paper drawings which only represented two-dimensional (2-D) views of the product design. Product designers or their assistants would spend countless hours manually drawing and revising designs. This process involved much trial and error, serial reviews, and drawing revisions -- a very time and cost intensive process. Once designs were complete, product manufacturers often created physical or functional prototypes so design teams (and maybe their customers) could see, touch, and generally experience a product. [Larson-Mogal, 1994] These prototypes evolved from paper-based descriptions of product concepts, to partial prototypes, to full-scale prototypes, and eventually to a complete functional product.

The onset of computer technology has had a positive impact on the product design process. Today, the process is initiated by developing 2-D and three-dimensional (3-D) solid models of alternative designs on Computer Aided Design (CAD) stations -- sometimes saving up to 70% of total manual design time. [Barfield, Chang, Majchrzak, Eberts & Salvendy, 1987] CAD has provided expanded capabilities for the development and timely revision of new product designs. However, while this technology may enhance the productivity of a single designer, CAD is not very effective or efficient in a collaborative design or group review environment. To overcome this limitation, new tools and approaches are needed.

Emerging technologies which have the potential to significantly impact the product design process are virtual environment (VE) systems. By their nature, these systems are capable of stimulating the human senses of sight, sound, and touch. They allow a person to experience life-like domains and objects that appear to be real but only exist in a computer-based environment. In the eyes of a designer, immersion in such a near-realistic environment can provide visualization of the final product from various perspectives. For instance, an automotive designer could sit in an automobile and visualize the interior layout. According to Kalawsky [1993], the designer in the virtual world could also navigate around the car and look at exterior qualities of the vehicle).

Using VE systems, a designer or a product design team could visualize a final product, assemblies, sub-assemblies, or components, in 3-D -- a concept often called virtual prototyping (VP). Several different but related definitions exist. Garcia, Goecke, and Johnson [1994] define VP as a computer-based simulation of systems and subsystems with a degree of functional realism comparable to a physical prototype that facilitates immersion and navigation. Lee [1994] defined VP as a mechanism for visualization and testing of computer-aided design models on a computer before they are physically created. Dai and Gobel [1994] simply defined VP as an electronic prototype.

While much progress have been made, VE and VP technologies still have a long way to go before they are fully operational and viable for wide-spread commercial design use. However, Machover [1996] predicts that the use of virtual environments for CAD applications will nearly double by 1998. Figure 1 depicts the described evolution of capabilities and the trend in product design capabilities. As is shown, the next logical evolution from 2-D drawing is to 3-D CAD and VE-supported, collaborative design environments.

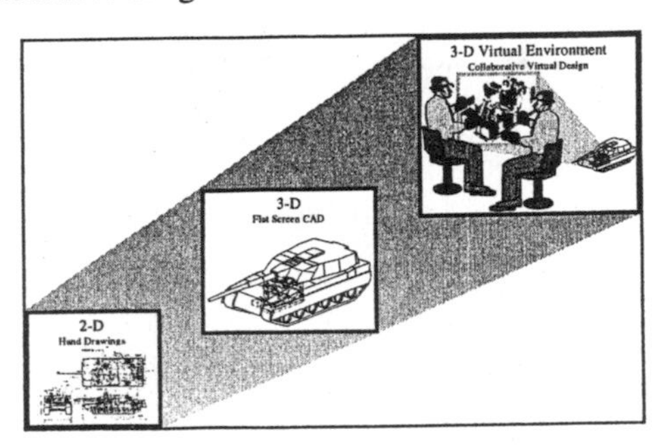

Figure 1. The Trend in Product Design

VE 3-D DISPLAY TECHNOLOGIES: DESCRIPTION AND COMPARISON

Human brain allows us to perceive the exterior, physical world through five primary senses: sight, sound, touch, smell, and taste. [Sekuler and Blake, 1994] A VE system presents a simulated world to the human senses through VE output devices. These devices are the primary feedback to the user. Output technologies for virtual environments present information to users to allow them to perceive appropriate information for interacting in the VE and to perform tasks. Currently, VE technology output devices are limited to the senses of sight, sound and to some degree touch. However, human vision is the primary focus because it is considered the most powerful sensorial channel. The most common visual displays systems are helmet mounted displays (HMDs), binocular omni orientation monitors (BOOMs), and stereoscopic shutter glasses with monitors. Visual display systems use large-screen projectors as in the CAVE Automatic Virtual Environment (CAVE), and holographic optics to create virtual, visual experiences. Figure 2 provides a pictorial image of these five visual display technologies.

Technical characteristics of visual display systems for virtual environments include spatial resolution, depth resolution, responsiveness, field of view, storage and refresh rate, and color [Stuart, 1996]. Each type of 3-D display system offers different technical capabilities and thresholds. For a more in-depth understanding of these technical aspects, the reader is referred to McKenna & Zeltzer [1992]. These researchers provide a detailed description of underlying technology requirements and limitations of several 3-D display systems. This paper will expand on their work providing an applications focus for understanding the usability of these visualization devices for product design activities. Issues such as cost, weight, usability, freedom of movement, level of immersion, and multi-person viewing will be discussed.

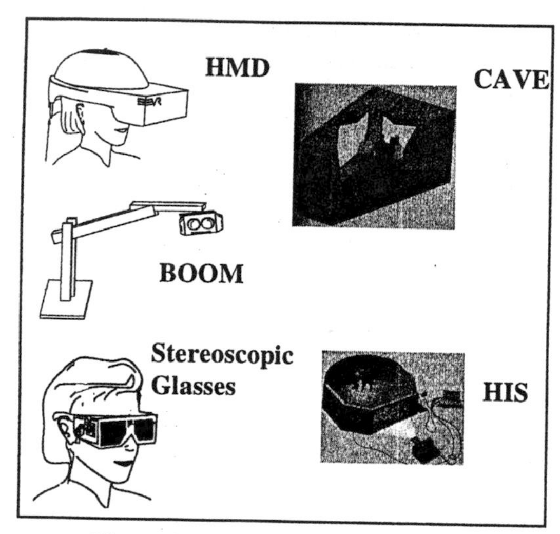

Figure 2. Visual Display Technologies

HMDs are defined by Larijani [1993] as headpieces or head-held braces with viewing or optical devices located or suspended in front of the user's eyes. When wearing this type of visual display device the user experiences full immersion in a 3-D virtual world where they lose their sense of presence in the real world. HMD equipment provides the user with a true virtual experience, but the technology can be limiting. Today's HMDs are characterized by poor quality and resolution of images and high cost. These systems are sensitive and require calibration or physical adjustment for each user's head geometry. Weight is also a crucial factor because extreme weights accompanied with inaccurate center of gravity calculations can cause user fatigue. Other limitations of this technology include restrictive user movement, where movement is dependent on the technical characteristics (distances) allowed by head tracking devices.

BOOMs eliminate some of the inherent properties of HMDs: high weight and low resolution. These devices are based on a mechanical arm that supports an imaging system at one end. The arm is counterbalanced so that the display has zero weight. By using a counterweighted design, problems with optics and display weight, and display size are eliminated [Pimentel & Teixeira, 1994]. Users bring the BOOM's imaging device to their face and use it to view the virtual environment. High costs are associated with this technology, and similar to HMDs, user movement is restricted to the physical dimensions of the mechanical arm.

The cheapest type of visual display device is stereoscopic shutter glasses, a system that adds 3-D depth to the images displayed on a computer monitor. Stereoscopic glasses are synchronized to the alternating display of two separate images on a monitor. When the right side image is shown, the shutter over the left eye is closed, and the one over the right eye is opened. When the second image is shown, the sequence is reversed. The shutter rate of at least 60 times per second per eye is required for the brain to interpret the resulting image as a single stereoscopic view. [Eddings,

1994] An infrared transmitter controls the display frequency. While the primary advantages of this type display device is its low cost and ease of use, the main disadvantage is the lack of user immersion. While the user experiences a 3-D perspective, he/she remains somewhat cognizant of the real world and loses a total sense of true immersion.

The CAVE system combines the concept of stereoscopic shutter glasses with projector technology. The CAVE system was developed at the University of Illinois-Chicago's Electronic Visualization Laboratory in 1992. It is a projection-based virtual environment system that surrounds the user with four screens. The screens are arranged in a cube whose walls are rear projection screens upon which graphics projectors display stereoscopic imagery. The viewer wears simple stereoscopic shutter glasses, and a tracking device provides the drawing system with information about the wearer's viewing position and orientation. In the CAVE, a user sees his/her own body. The CAVE system has a relatively high price tag, but has the advantage of providing improved user mobility and a greater sense of full immersion.

Holographic imaging is an alternative to conventional VE output devices. The U. S. Army is in the early stages of developing a Holographic Imaging System (HIS) for the product design environment. Their current system utilizes a high performance graphics computer, projectors, a holographic optical element, and custom software. Holographic imaging removes some of the disadvantages of current HMD, BOOM, and stereoscopic glasses technologies by eliminating the requirement to wear a physical optical device. The viewer obtains the full effect of the virtual world by sitting or standing in front of the holographic optical imaging device, the key element in the HIS. However, HIS technology only allows single user viewing, and the user's head must be in a static, fixed position to experience a true 3-D effect. Since this technology is in the early stages of research, extensive testing is very limited

Stuart [1996] discusses common issues and comparisons relative to the development of VE systems. Figure 3 provides an extraction and modification of his work in the context of system usability and feasibility.

	HMD	BOOM	Glasses	CAVE	HIS
Cost	Varies with Resolution	Medium	Inexpensive	Expensive	Expensive
Commercial Availability	Yes, Several Vendors	Yes, Limited Vendors	Yes, Several Vendors	Yes, Single Vendor	No, Experimental
Ease of Use					
Level of User Learning	More difficult, user must learn to navigatein the VE	More difficult, user must learn to navigatein the VE	Easy	Easy	Easy
Calibration	Must be calibrated for each individual's head size	None	Adjustment of graphics displays to synchronize right and left eye images	Adjustment of graphics displays to synchronize right and left eye images	Adjustment of projectors
Set-up Time			None, once system is calibrated	None, once system is calibrated	None, once system is calibrated
Physical Characteristics					
Weight	Varies with Resolution	No weight to User	Light	Light	No weight to User
Freedom of Movement	Restricted to area of tracking system	Restrictive	Restrictive	Dependent on CAVE size	Restrictive
Level of Immersion	Fully Immersive	Fully Immersive	Less Immersive, display is virtual but surrounding environment is not	Fully Immersive	Less Immersive, display is virtual but surrounding environment is not
Multi-Person Viewing	Yes, but requires multiple systems	Yes, but requires multiple systems	Yes, but only one person gets the correct viewing perspective	Yes, but only one person gets the correct viewing perspective	No, Single User Viewing

Figure 3. Visual Display Device Comparisons

COLLABORATIVE VIRTUAL DESIGN SYSTEMS

Because of the potential of VE technology, U.S. Army Vetronics Technology Center and University of Central Florida researchers are in the early stages of testing and evaluating the use of virtual environment visualization technologies and immersion environments for collaborative product design. The research plan emulates the traditional design review process except that multi-functional design teams will solve design problems with the assistance of each of the five VE visualization technologies. They will also use a more traditional method --3-D CAD for comparison. These design teams will be formed from functional area representative from several U. S. Army combat vehicle development organizations. Each team will consist of one representative from project management, design engineering, manufacturing engineering, and customer organizations. Each team will then be randomly assigned to an experimental test sequence, each of the visualization technologies. Quantitative and qualitative measures will be taken. Quantitative data will capture task completion times, error detection, and accuracy. Qualitative data will be collected on the visualization technologies capability to provide better quality solutions, improved design review processes, and overall reactions to the systems.

It is hypothesized that one or more of the test VE visualization technologies will greatly assist in collaborative design review activities by creating an improved mechanism for product visualization when compared to more conventional, 3-D CAD systems. This view is supported by Barkan and Iansiti [1993] and Horton and Radcliff [1995] who believe that a VE environment would offer a common focus for multidisciplinary (managerial, technical, marketing, and manufacturing) groups to resolve design issues by sharpening their viewpoints in support of collaborative design evaluation.

These visualization technologies, to one degree or another, offer the potential for true team interaction. Design teammates could simultaneously enter a virtual product design world, and jointly evaluate design issues, ideas, and parameters -- each from their own experience, perspective, viewpoint, and functional responsibility. As a result, it is hypothesized that, team members from various functional elements would thus become more knowledgeable about other activities and, therefore, would become a more integral part of the total decision making process. An objective of the planned testing will be to determine if, in an unique multi-functional virtual design environment, all members share in a concurrent design experience, and if they can relate to the perspectives of others.

In addition to participation and perception, the use of VE for product design could provide design flexibility by allowing the exploration of various options and the opportunity to generate and iterate "what if" exercises early in the design process where mistakes are less expensive to correct. If so, significant cost savings could be achieved in system design and development because many of the problems would be identified and corrected prior to the actual physical product construction. Hopefully, VE technology will enable developers to refine designs before commitments are made, by bringing users into the design process much earlier, and allowing engineers to solve problems in a more collaborative, group setting.

VE-BASED DESIGN TOOLS MATURITY

Given that VE technology has the potential to positively impact the product design process, one must address its feasibility to replace traditional CAD systems. Due to the immaturity of VE technology, the incorporation of a 3-D CAD data files into a virtual world system allows for only simple visual evaluations, accompanied with sound, and to a minor extent touch. However, there exists an abundance of technical problems which must be resolved if VEs are ever going to be considered the next generation of CAD tools. Some of the basic problems involve the ability to easily transfer CAD data into virtual world, manipulate the data in the virtual world, and to maintain files with the same level of detail and integrity as traditional CAD tools.

More specifically, there are problems associated with file translation from standard CAD formats to formats compatible in virtual environments. CAD data files are much more data intensive then what would be required for virtual environment application. A complex CAD data file may contain several million polygons which are a burden in a virtual environment system because each polygon must be processed and rendered in real-time. Data reduction algorithms help increase rendering performance by eliminating duplicate triangles, by removing object back faces that are never seen, and by removing features that are unnecessary for visualization [Division, 1995]. The problem with these data reduction schemes is that a complex data file may be over-reduced yielding an inadequate or ill-defined representation of the product in the virtual environment. In essence, data reduction is not a simple task. It must be completed on a step by step basis, where the virtual environment designer must spend a significant amount of time determining which features of the original CAD data file can be eliminated or reduced.

A major limitation of this technology is that it does not allow design participants to make design changes and directly modify the CAD data file in real-time. For example, if during a VE design evaluation of the interior of an automobile, a design team detects that the angle of steering wheel tilt does not allow visibility of the dashboard for the 95th percentile human, current systems do not allow the interactive movement of the dashboard gauges or changes to the steering wheel tilt. Resolution of these problems require the use of traditional CAD software and re-translation back into the 3-D VE for visualization.

Problems escalate when the design team is geographically dispersed, and the collaborative design is dependent on networking technologies and their associated limitations and reliabilities. Extensive resources are needed to address configuration management, data storage and retrieval, network security, and network performance problems.

Due to the lack of maturity of the technology, virtual environments are not yet being considered as the next generation CAD tool. Instead VEs are simply being used to aid in fleshing out initial concepts, evaluating aerodynamics, checking ergonomics, designing for maintainability, and taking designs to virtual test grounds [Gottschalk & Machlis, 1994]. Currently, VE is not something from which detailed engineering information can be derived from just yet. However, the technology does provide the capability to rotate and look at objects in a 3-D perspective.

PRODUCT DESIGN SUCCESS ISSUES

A single question which prevails in the minds of any product manager is "How does one know when a new product design has reached a level where it has the most potential to succeed in the market place?" Clark and Fujimoto [1991] have concluded that three outcomes of a product development process affect the ability to attract and satisfy customers. These are: product lead time, total product quality, and performance. These factors can be associated with product design effectiveness and efficiency, and eventual product design success.

This planned research attempts to address how product design reviews can best be conducted using a variety of collaborative virtual design environments. The purpose of this research is to provide insight into several product design issues which can be categorized into two categories: technical and managerial.

Technical Issues

Beyond the pure technical capability and technology challenges required to create collaborative virtual design environments there exists a layer of issues revolving around the capabilities of the technologies to assist in a collaborative design process. Some of these technical issues include: Which technologies are best for collaborative virtual product design review evaluation? How do learning curves for these technologies effect individual and team performance? Do these tools actually assist the design team in accomplishing improved design reviews, thereby, yielding better quality solutions and products?

Management Issues

Changes in management style and team communication methods will eventually evolve as these new added capabilities of collaborative virtual design emerge and become common practice. Some of these management issues include: Does this environment improve communication and reduce the number of design iterations and changes? Is the overall productivity of the group enhanced? How do people function in virtual environments? How do organizations develop trust?

CONCLUSIONS

Preliminary results of this study have led the authors to conclude that the application of advanced virtual technologies to the product design process has the potential to significantly improve the efficiency and effectiveness of product design reviews--a major driver of product cost. Our belief is that, the testing and evaluation of VE visualization technologies in collaborative product design environments will provide important insights into the utilization of VEs for group product design tasks. To date, a research strategy and methodology to measure the impact of visualization technologies on design cycle time, design quality, and design performance has been developed. However, several significant technical and managerial issues need to be addressed and solved before the use of one or more visualization technologies can be declared a success.

REFERENCES

Ali, A., Krapfel, R. J. & LaBahn, D. Product innovativeness and entry strategy: impact on cycle time and break-even time, *Journal of Product Innovation Management,* 12, 54-69, 1995.

Barkan, P. & Iansiti, M.. Prototyping: A tool for rapid learning in product development. *Concurrent Engineering: Research and Applications,* 1, 125-134, 1993.

Barfield, W., Chang, T., Majchrzak, A., Eberts, R., & Salvendy, G., Technical and human aspects of computer-aided design (CAD), *Handbook of Human Factors,* New York: John Wiley & Sons, 1987, 1617-1656.

Clark, K. B., & Fujimoto, T., Product development performance, *Boston: Harvard Business School Press,* 1991.

Cooper, R. G., & Kleinschmidt, E. J., Determinants of timeliness in product development, *Journal of Product Innovation Management,* 11(5), 381-396, *1994.*

Cordero, R., Managing for speed to avoid product obsolescence: a survey of techniques,. *Journal of Product Innovation Management,* 8(4), 283-294,1991.

Dai F., and Goebel, M., Virtual prototyping - an approach using VR techniques, *Computers in Engineering,* ASME, 1994.

Division, dVS for unix workstations User guide,*Order N Ltd, 1995.*

Eddings, J., How Virtual Reality Works, *California: Ziff-Davis Press,*1994.

Garcia, L. A. B., Gocke Jr, C. R. P., & Johnson Jr, C. N. P., Virtual Prototyping Concept to Production, *Fort Belvoir, VA: Defense System Management College Press,* March 1994.

Gottschalk, M. A., & Machlis, S., Engineering enters the virtual world, *Design News,* 53-63, November 7, 1994.

Hales, C., Managing engineering design, *UK: Longman Scientific & Technical, 1993.*

Horton, G. I., & Radcliff, D. F., Nature of rapid proof-of-concept prototyping, *Journal of Engineering Design, 6(1), 3-16, 1995.*

Iansiti, M., Real-World R&D: Jumping the product generation gap, *Harvard Business Review,* May-June, 138-147, 1993.

Kalawsky, R. S., The Science of Virtual Reality and Virtual Environments, *New York: Addison-Wesley Publishing Company,1993.*

Larijani, L. C., The Virtual Reality Primer, *New York: McGraw-Hill, Inc.,1994.*

Larson-Mogal, J. S. An immersive paradigm for product design and ergonomic analysis, *Virtual Reality World,* 28-32,1994.

Lee, G., Virtual prototyping on personal computers, *Mechanical Engineering, 117(7), 70-73, July, 1995.*

Machover, C., What virtual reality needs, *Information Display,* 32-34, June, 1996.

McKenna, M., & Zeltser, D., Three Dimensional Visual Display Systems for Virtual Environments, *Presence,* 1(4), 421-458, Fall, 1992.

Pimentel, K., & Teixeira, K., Virtual Reality Through the New Looking Glass, *New York: McGraw-Hill,* 1995.

Sekuler, R., & Blake, R., Perception, *New York: McGraw-Hill,* 1994.

Shu, L., & Flowers, W., Teledesign: groupware user experiments in three-dimensional computer aided design, *Collaborative Computing,* 1, 1-1, 1994.

Stuart, R., The design of virtual environments, *New York: McGraw Hill,* 1996.

Vesey, J. T., Speed to market distinguishes the new competitors, *Research and Technology Management,* 1991.

FACTORY AND LOGISTICS PLANNING WITH VIRTUAL REALITY

Prof. Dr.-Ing. R. D. Schraft, Dipl.-Ing. J. Neugebauer, Dipl.-Ing. K. Grefen
Fraunhofer Institut für Produktionstechnik und Automatisierung (IPA)

ABSTRACT: Layout-planning and material-flow simulation are two significant tasks in the process of todays factory-planning:
Layout-planning and its visualization is done with 2D CAD-Drawings or 3-D-CAD-models. Material-flow will be designed and estimated with the aid of simulation-tools (like TAYLOR or Simple ++). Those tasks affect each other since the results of the capacity planning with material-flow simulation will directly influence layout-planning of the factory and vice versa. The integration of those processes would speed up the whole planning process and increase the reliability of the planning results.

Based on its Virtual Reality simulation system, IPA has now developed an integrated simulation system, combining Virtual Reality simulation with material-flow simulation. The system consists of the three dimensional visualization component and the real-time object simulation component. The object simulation component is directly driven from the outputfiles of a material-flow simulation system. In this paper the structure and the development of the necessary software-components will be discussed and first projects carried out with this simulation-tool will be shown.

INTRODUCTION

Is Virtual Reality just a new interface for visualization or is it a useful communication approach for engineering tasks? What we can be sure of is that human computer interfaces will play a major role in the investigation of product development cycles and manufacturing processes. Concurrent Engineering and integrated product development are generic terms for various communication approaches between designers and manufacturing engineers in the design phase of products. Worldwide communication and sharing of information during collaborative engineering will become an important topic to cope with the challenges of internationalization and reduced innovation cycles and time-to-market. Video conferencing and distributed Virtual Reality systems providing multimodal interfaces will help to make the best use of the resulting increase in available information.

Virtual Reality takes 3D data of a scene and presents this information as stereoscopic computer generated images to the user. The images are updated accordingly in real-time. Advanced input devices for navigation and interaction tasks form an intuitive human computer interface with depth perception in the virtual environment.

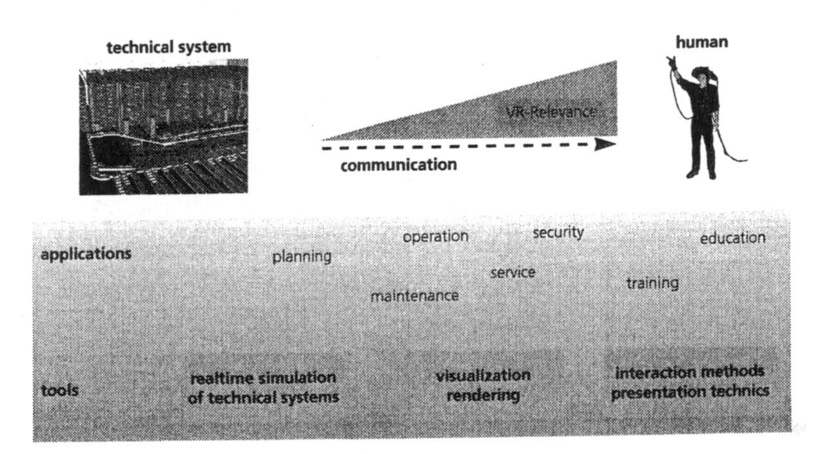

Figure 1. Virtual Reality relevance

FACTORY PLANNING

For the general level of factory planning, material-flow and layout optimization are two significant tasks. Both of them have many different underlying technical and environmental requirements and constrains. Outermost goal for the planning is finding the best overall compromise. Feasibility studies and evaluation of scenarios are common strategies for reaching this demanding solution.

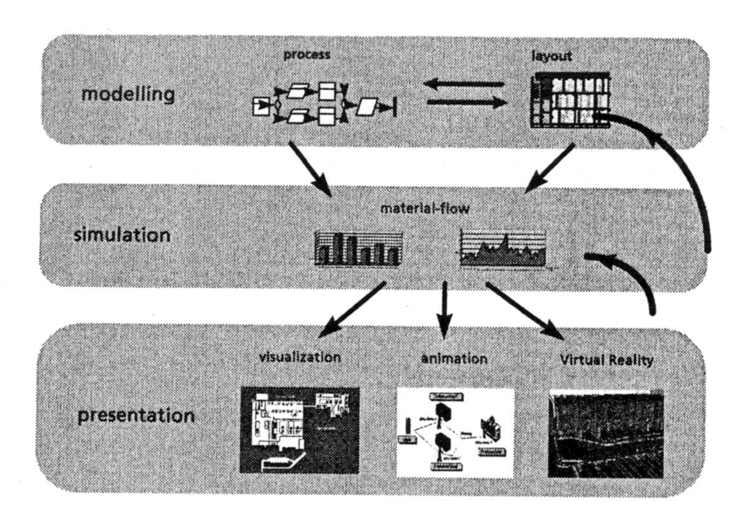

Figure 2. Planning process

For the planning processes on a general level, as well on a detailed level, simulation tools give important input for evaluating possible solutions. Whereas close linkage between material-flow and layout exists in reality most of the available simulation systems address just one side. There also is a lack of data integration between those systems. This leads to a variety of independent solutions with a redundancy and overhead of data-acquisition and data-management. The integration of more sophisticated simulation systems like Virtual Reality systems are hardly possible. Electronic data management systems (EDM) providing an integration platform for both layout and material-flow based data-models give an important contribution for the whole planning process. Based on a project or scenario class description, stored in such an EDM system, a library of technical components could be build up from which new simulations could easily be modeled.

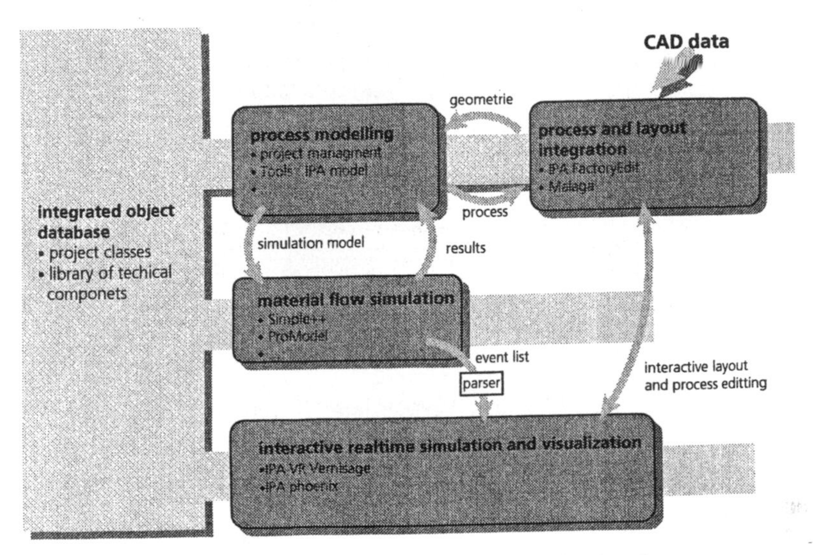

Figure 3. Integration platform for logistic planning

REAL-TIME SIMULATION ENVIRONMENT PHOENIX

Understanding Virtual Reality as a virtual place of work where the user can carry out all steps of development, interactive planning seems to be possible within this environment. Prerequisite to this scenario a real-time simulation environment for the simulation of technical systems has to be developed.

A discrete time based simulation has been build up at Fraunhofer IPA. One part of the simulation model is based on vertical flow of information inside an automated factory, another part of the model is based on material flow.

Starting at the lowermost level we get cycle times of about 20 ms or less, very short data words and a very high event frequency. Functionality of sensors and actors at this level varies in a wide range, depending on their physical behavior.

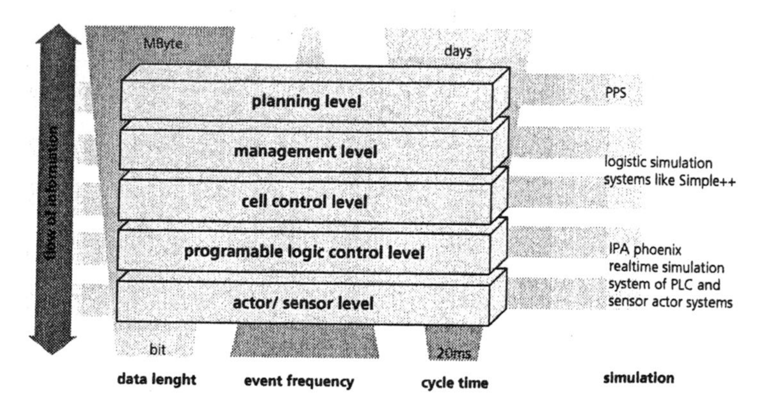

Figure 4. Vertical flow of information

Main tasks in the planning level are global resource management, order management and material management. For all those tasks PPS systems are used.
Main task of the planning level is the administration of orders, stock-management and business management controlling.
With the help of work plans gained from production orders of the planning level the management level produces structured work instructions and occupancy-plans, which are then put at the cell level's disposal. On the basis of this information a more detailed plan for the occupation of machines is done in the cell level, taking the information on the present state into account. When choosing a production order the corresponding PLC control program is downloaded on the machine control.

Only the two lowermost levels, the PLC control level and the sensor and actor level were realized in the simulation software phoenix. Interfaces to programs reflecting upper levels are implemented.

Programmable Control Level

The PLC (programmable logic control) level consists of different PLC components. PLC controllers are bit-oriented computers for the control of technical components. They receive signals from sensors to get a display of the technical process and transmit signals to actors for interacting with the process. The program running a PLC is assembler oriented and programmed in statement lists, ladder diagrams or function block diagrams. The logical behavior of those systems depends strongly on the programming style and the three-dimensional arrangements of sensors. Therefore a simulation of the real behavior of the PLC is necessary for decent results. The realization in the simulation system phoenix is based on the structure of the PLC system S5 from SIEMENS. This system has a characteristic register and memory structure which is widely used and the basis for the DIN 1131-3 for PLC programming languages. The very close link to the real environment allows the use of existing PLC programs in the simulation or the use of programs from the simulation in a real environment. Therefore all described programming languages are available in the simulation system [12].
The PLC program module in the simulation is an executable C-code converted from the PLC language by a preprocessor. The communication with the simulation, sensors and actors is done by shared memory areas corresponding to real environments [13].

Sensor and Actor Level

Sensors and actors are the link between technical process and logical structure of industrial environments. Both are available in various types with different functionality.

An actor is characterized by it's degree of freedom (DOF). For example a conveyer belt's DOF is one, a robot's DOF can be up to six or more. All actors in the simulation have a dynamic behavior. These are velocity, acceleration, maximum and minimum range. Also inverse kinematics are available if needed.
The modeling and simulation of sensors depend on a process chain of ongoing signal proceeding inside the sensor. Starting with measurement functions all other components of the chain like signal generation and electric or pneumatic signal behavior can be modeled. Also output characteristics of sensors and actors can be bit or word oriented. This corresponds with the real environment.

IMPLEMENTATION

General Environment

Logistic environments are very complex with different design aspects like architecture, mechanic or logic. According to these aspects different software tools have been developed. VR4Kin is developed for modeling kinematics and dynamic behavior. The software tool PhoenixEdit is for building up component relationships, sensor descriptions and PLC code editing. Vernisage and phoenix are the integrating tool for final collection, simulation and real-time presentation in an virtual environment.

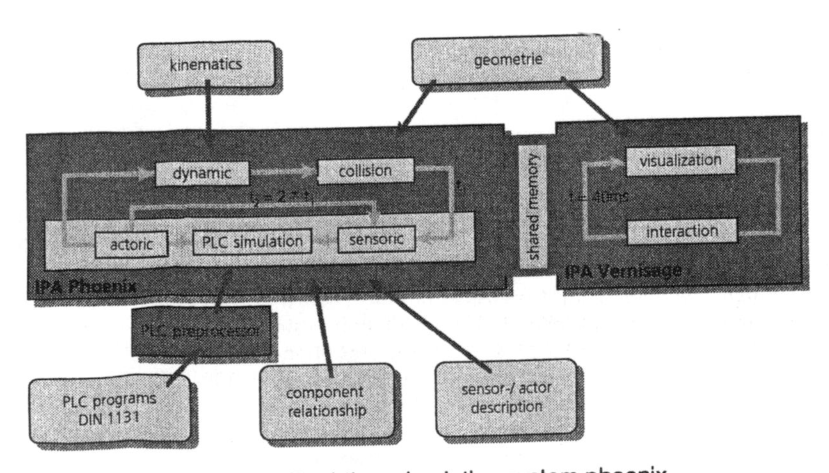

Figure 5. Real-time simulation system phoenix

Modeling Software VR4Kin

VR4Kin is a tool for the modeling and evaluation of kinematic systems. It has been developed as a tool to prepare CAD data for further use in a Virtual Reality environment. The main functions involve the interactive definition of geometrical assemblies and kinematic relationships [14]. With the help of icons and interactive dialogs the user can assemble parts and components to a kinematic system. VR4Kin provides functions for the description of the dynamic behavior of components. Portability and using established standards are goals from software design process. The software is based on: ANSI-C, OSF-Motif and GL/OpenGL.

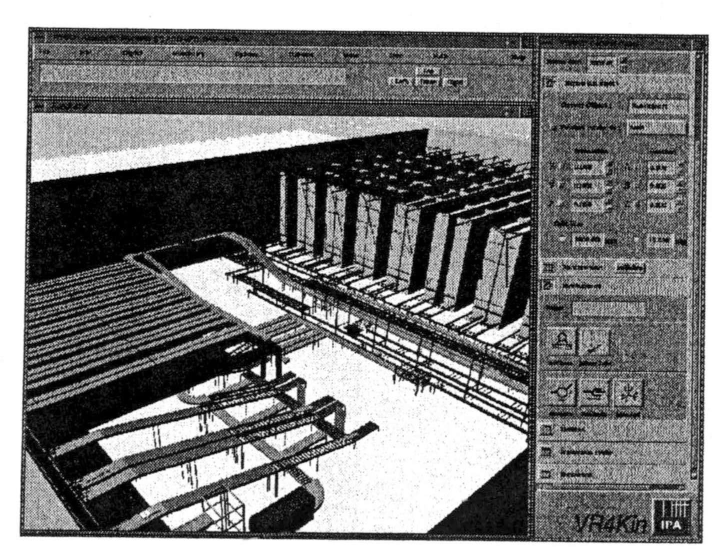

Figure 6. VR4Kin GUI

Simulation Software PhoenixEdit

Phoenix consists of two parts, the phoenix PLC compiler and the phoenix link environment. With the compiler the edited PLC code will be transformed to an executable C-code. Communication with the simulation tool will be done by bit or word oriented shared memory areas. The other part of phoenix is the link environment. Complex technical systems consist of several physical components, controllers, sensors and actors, which define their behavior. Information is running between this parts via links. This links can be defined with the link environment. Each technical component has it's own link bar shown in a resource file generated by the PLC compiler. Based on this resource file the user links up input with output areas, sensors with input and actors with output areas. For minimizing mistakes phoenix checks if the information link is bit or word oriented. Furthermore phoenix allows the user to define and set sensors in a graphical environment based on the tool VR4Kin. Different types of sensors are available corresponding to real sensors like light bares, distance sensors or proximity switches. Positioning can be done graphically or textually.

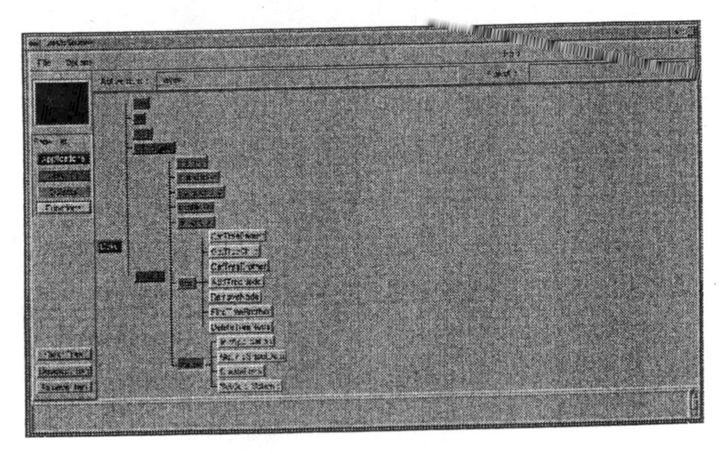

Figure 7. Phoenix link environment

Rendering Software Vernisage

Vernisage has been developed as a basic tool for virtual environments with a great amount of movable objects. The main goal is a real-time oriented display of static and dynamic objects. It is characterized by a familiar window based JAVA-interface, which allows the interactive change of various system configurations. The use of other modeling tools or CAD systems is possible. Data of various of CAD systems can be displayed via data exchange protocols. Vernisage is based on the performer library developed by SGI.

Project Example

In an industrial project modeling and simulation tools have been tested. Figure 5: Project Example shows a part of a logistic center of a shoe manufacturer. Shoe boxes come from the production line are put on carriers, called tablar, by workers. The filled tablars run on conveyer belts to the storage area. The storage warehouse consists of ten automatic storage devices for loading and unloading tablars. If a customer order comes into the system tablars are unloaded from the warehouse and brought to special commission areas, where shoe boxes are be collected. On twenty commission belts the customer orders are collected for further manual handling. The industrial project was done for the product presentation of spacious logistic environments on an exhibition. The graphic model consists of nearly 700.000 textured polygons. The simulation environment consists of about 150 technical components with 450 sensors and 350 information links between the components. Up to 1000 objects are manipulated by full operating rate of the simulation. The simulation cycle time is about 40 ms and accuracy is about 80 ms, which is useable for logistic environment simulation. The Frame rate of the system is 20 frames per second rendered on a SGI INFINITY.

Figure 8. Project example

RESULTS

The simulation in an virtual environment shows great advantages in industrial design process. With the use of a graphical representation the customer of the logistic environment is able to understand the complex circumstances and can therefore discuss changes to reduce the failure rate in the planning process. The intuitive way of presentation allows the off-line programming of complex logistic environments without space or hardware requirements. All tasks in the industrial design process can be done on the graphic computers. A great advantage is the possibility to use a PLC code and information link list in the further realization process. The following advantages derive from the use of Virtual Reality:

- Increasing transparency of logistic processes
- assessment of multiple variables and dynamic dependencies
- increasing safety in planing
- reduction of planing periods
- development of rationalization potentials
- analysis of economic viability

The realization of the simulation shows some very interesting results. The designer of the simulation is faced with the same problems as he would be in a physical realization. A lot of time was needed to synchronize the movements of products, with generation of events and positioning of sensors. This shows that the model behavior is very similar to the real physical behavior.

CONCLUSION

Virtual Reality technology can make significant contributions to the process of planning and design of spacious logistic environments. The tools designed for simulation and modeling tasks the virtual environment has been tested in an industrial project and showed great advantages compared to common planning and design tasks. The easy to use graphical environment has the potential to reduce the failure rate and to accelerate the realization of complex and spacious environments. Further results are the use of sensor and actor simulation in combination with the PLC controller to accelerate the physical realization. PLC code and connection lists of physical components are available in the virtual environment and can be used directly in the real physical environment.

References

1. **Flaig, Th.**, *„Echtzeitorientierte interaktive Simulation mit VR4RobotS am Beispiel eines Industrieprojektes"*, Tagungsunterlagen des 2. VR-Forum ´94, Anwendungen und Trends, Februar 1994, Stuttgart, Deutschland.

2. **Flaig, T; Thrainsson, M.**, *„Virtual Prototyping - New Information Techniques in Product Design"*. In: Rolf D. Schraft (Hrsg.); M. Munir Ahmad (Hrsg.);William G. Sullivan (Hrsg.); Fraunhofer-Institut für Produktionstechnik und Automatisierung (IPA); CIM System Research Center; Department of Industrial and System Engineering: 5th International Conference on Flexible Automation and Intelligent Manufacturing, FAIM ´95. Begell House Inc., 1995. S. 338-348.

3. **Gangl, P.**, *„Simulation - Schlüsseltechnologie der 90er Jahre: Marktsituation, einsatzgebiete und Systemauswahl"*, In. IPA-Technologie-Forum, Simulationsgestützte methoden zur Planung und Optimierung komplexer Produktionsabläufe; 27. Oktober 1993, IZS Stuttgart, Deutschland.

4. **N., N.**, *„TAYLOR II SIMULATION - Simualtion software for personal computers"*, Technical Product Description, F&H Simualtions BV, 1996, Utrecht, Netherlands.

5 **N., N.**, *„IGRIP", „QUEST", „Virtual NC"*, Technical Product Description, Deneb Robotics Inc., 1994, Michigan, USA.

6 **N., N.**, *„Information über ROBCAD"*, Technische Produktinformation, Tecnomatix Automatisierungssysteme GmbH, Offenbach, Deutschland.

7 **N., N.**, *„Prosolvia opens new worlds"*, Technical Product Description, Prosolvia: Clarus AB, Calstedt Research & Technology AB, Pronima AB, 1996, Göteborg, Sweden.

8 **N., N.**, *„AnySIM, 3D-Simulation von der Konstruktion bis zum Vertrieb"*, Technische Produktinformation, Institut für produktionstechnik GmbH, Dornach, Deutschland.

9 **Hirzinger, G., Landzettel, K., Heindl, L., Brunner, B.**, *„ROTEX - Die Telerobotik-Konzepte des ersten Roboters im Weltraums"*, Tagungsunterlagen des 2. VR-Forum ´94, Anwendungen und Trends, Februar 1994, Stuttgart, Deutschland.

10. **Kühnapfel, U.**, *„Grafische Realzeitunterstützung für Fernhandhabungsvorgänge in komplexen Arbeitsumgebungen im Rahmen eines Systems zur Steuerung, Simulation und Off-Line-Programmierung"*, Doktorarbeit am Kernforschungszentrum Karlsruhe; Mai 1992, Karlsruhe, Deutschland.

11. **Henzler, R.**, *„Eine ereignisorientiert Kopplung der Simulatoren SIMPLE++ und VR4RobotS in einer konzeptionellen Simulationsumgebung"*, Diplomarbeit am Fraunhofer IPA, Oktober 1995, Stuttgart, Deutschland.

12. **N.N.**, *„SIMATIC S5"*, Technical Product Description, Siemens AG, Industrial Automation Systems, Nuremberg, Germany.

13. **Neuber, D.**, *„Spezifikation und Implementierung von Datenstrukturen zur echtzeitorientierten Simulation von Speicherprogrammierbaren Steuerungen"*, Diplomarbeit am Fraunhofer IPA, Februar 1995, Stuttgart, Deutschland.

14. **N.N.**, *„VR4Kin, Tool for Modelling and Evaluation of Kinematic Systems"*, Technical Product Description, Fraunhofer IPA, Demonstration Centre Virtual Reality, February 1996, Stuttgart, Germany.

15. **N.N.**, *„VR4, Software Tool for Dynamic and Real-time Oriented Virtual Environments"*, Technical Product Description, Fraunhofer IPA, Demonstration Centre Virtual Reality, February 1996, Stuttgart, Germany.

16. **Grefen, K., Flaig, Th., Neuber, D.**, *„Virtual Prototyping im Anlagenbau"*, IPA-Technologie-Forum, 04. Dezember 1996, Stuttgart

17. **Flaig, Th., Grefen, K., Neuber, D.**, *„Interactive Graphical Planning and Design of Spacious Logistic Environment"*,

18. **Flaig, Th., Wapler, M.**, *„The Hyper Enterprise"*,

WHAT IS REAL-TIME AND HOW SHOULD WE USE IT ?

Antony Eddison MA
Unit Leader - Virtual Environment Research Unit (VEDRU)
Course Leader - MA Advanced Digital Design (MA ADD)
University of Teesside

ABSTRACT: This paper is an overview of the authors current research into the concept of designer style, the conclusions of the research will be discussed in relation to appropriate interaction design criteria between the individual designer and Computer Aided Design (CAD) systems. The applied use of immersive and non-immersive, Real-Time 3 Dimensional (RT3D) techniques during the design process will be discussed with an emphasis on the use of such techniques during the early stages of the design process. We will consider the 'journey' of 'live' 3D concepts created at the sketch stage through concept analysis to production, a recent Industrial design project will act as a vehicle for this examination. Conclusions, implications and areas of further study will then be presented.

Introduction

Most Computer Aided Design (CAD) software typically provide a more or less rigid interface to the user, it is system rather than user driven. The consequences are seen in terms of restricting choice to narrow channels so limiting creativity, it is considered that such systems lead to computer restricted design as opposed to computer aided design. A more desirable system would be characterised by the provision of a more individually tailored set of CAD drawing aids which would consider individual design strategies and make more practical use of the information and sketches produced during the early conceptual stages of the design process.

This is the starting point for this investigation into the concept of designer styles allied to the use of the emerging technologies of Virtual Reality (VR) throughout the design process and particularly, the use of Real-Time 3 Dimensional (RT3D) techniques and procedures during the sketch design and concept generation.

Purpose of the Research

The number and of CAD design tools and type user interface design provided with a particular CAD system is to a large degree determined by the software developers understanding of the task undertaken, the form and context of the information to be displayed and the design strategies adopted by the end user. While a broad consensus exists as to the fundamental structure of the design process, it does not necessarily match the actual style of a given individual and the implications of such a mismatch have not yet been thoroughly examined. It would appear that the structure or architecture of the model of the design process would depend on the size, complexity and the technology employed in the design task and while being a more or less accurate description of the overall processes involved, it would require much clarification and development to be consistent with the design strategies adopted by individuals.

It is commonly accepted that style is neither an objectively determinable property of the designer nor is it a variable that could easily be set independently of other variables, it is a manifestation of often conflicting requirements in an activity not solely determined by the design process and also one outcome of the circumstances of a particular occasion.

The very diverse character of designing general user interfaces used with CAD applications precludes the adoption of a uni-dimensional approach. Whereas the activity of designing user interfaces for specific limited purposes for a well defined user activity, for example a banking or insurance system, concentrates the a limited and well defined number of functional requirements, the activity of designing user interfaces for more general purpose design applications for a much less well understood user is principally dealing with a range of unforeseeable as well as foreseeable possible eventualities and a range of conditions of uncertainty. Such eventualities are not susceptible to description in a deterministic manner, they may even be inconceivable. Designing such systems requires an imaginative, methodical, systematic and probabilistic approach.

Imaginative because the designer needs to be able to pre-conceive scenarios of possible future use and misuse, methodical in order to not overlook possible problems or to neglect options for problem control, systematic to allow for the adaptation of the design in the wider context of its use in the design professions and probabilistic because of the need to assess the overall problems and benefits of alternative solutions and to restrict uncertainty to tolerable areas.

It is evident that designers of GUI's for general purpose design applications require a thorough understanding of the design activity and the end user, an appreciation of the individual style of the user is an important factor for consideration when designing such systems.

So, underlying the notion of *style* is the basic premise that all designers are not the same and that the manner in which any designer works through a design problem towards a proposed solution may be qualitatively different from other designers. If this is shown to be the case and the concept of designer style can be meaningfully discussed then any model of the design process and any system or product, including GUI's, relating to this model must allow for such variations at the level of the group or individual.

A Development of User Interfaces

Research is currently being undertaken by the author to gain a fuller understanding of the concept of an individual designers style and how this may be enhanced by the provision of appropriate interaction tools at the user interface level. The integration of computing, visualisation, client/server, and new interaction techniques is in a state of rapid change and it is expanding the paradigms of current work with computers. It is well understood that one of the most reliable ways in which to make computers easier to learn and use, and to make users more productive, is to improve the communication, visual and otherwise, that takes place in all the elements of the user interface.

The literature stresses that it is important to understand that a user interface is not only a screen design, but also a method of interacting with the application and its data. This combination is often referred to as the look and feel. Look refers to refers to the appearance of the application to a user, whereas feel refers to the way the user interacts with the user and hence, the underlying application. Because of the attention paid to the improvement of interface design recently an attractive look and feel in a products user interface is as important to its success as the products functionality.

Contemporary GUI's enrich communication by using de facto standard tools such as Windows and Motif. The visual quality of a GUI is related to the power of a computer and as UNIX workstations, Pentium PC's and Power PC's continue to fall in price, more and more users from more and more diverse fields are gaining the ability to produce sophisticated visualisation and interaction.

To deal with this need to spread the use of graphics throughout this widening community, there is a need to combine the powerful graphics capability with intuitive user-interface design. However, both applications with visualisation algorithms and the design of appropriate user-interfaces have become extremely sophisticated, and these very different programming tasks can no longer be carried out by the same people.

User-interfaces have become so important for developers that, for many applications, the total amount of code devoted to the user interface exceeds that for the application. It is now not unusual for the user interface code to exceed 80% of the total code for a given application. GUI's are relatively new in information technology and there are a number in common use. The most popular GUI's and the operating systems they access are as follows:

Microsoft Windows (Windows95 & NT)
OSF/Motif (UNIX/SunOS)
Presentation Manager (OS/2)
Sytem 7.5 (MacOS)

These popular GUI's have a similar look and feel and are relatively intuitive so that moving from one to another is not normally too difficult. The main differences between the above GUI's are not at the visual level but in the lower layers, in the underlying windowing technology and the application programming interfaces (API's). Currently some GUI's are based on standards, some are proprietary and some are a combination of the two

Because of the accepted standard for an environment in which to design GUI's, users who use more than one application program benefit from consistency in the behaviour of the user interface as it minimises the mental adaptation when switching between programs. Although a certain level of consistency is guaranteed by using the software development tools, there are certain interface guidelines that cannot be enforced by the software, although style guides do exist and are mostly adhered to.

At this point it is appropriate to develop the discussion to the problem of attempting to tailor such interfaces to a particular type of user.

Motif-based GUI Tools
User-interface software is typically organised in layers. there are four layers comprising the X-based graphics framework for a visual builder: Xlib, Xt and Motif widgets, custom Widgets and GUI builder with each layer building upon the previous layer. The application program may use any combination of these layers to access the desired flexibility, functionality and productivity.

A common approach is to use a GUI builder to interactively build and test the Motif user interface. A GUI builder provides a palette of standard Motif objects and a layout area for arranging these interface objects. Attributes that affect the behaviour and appearance of these objects may be edited and the builder stores this information as resources. Some more advanced GUI builders also provide a built in C interpreter that allows application code to be written, compiled, linked and tested with the user interface design.

Also available as GUI design aids to the programmer are basic sets of interactive techniques from which complete interaction sequences may be built. these libraries, usually in the form of C functions or C++ classes, in conjunction with another graphics display library for 2D or 3D graphics, implement interaction control elements that are commonly called custom widgets. These custom widgets include dialogue boxes, forms, menus and other GUI components.
Most UNIX based GUI builders were designed to operate on top of X window. They make use of routines in the Xlib and Xt libraries and they produce interfaces that conform to the standard Motif style guide.

One of the major concerns today in the area of GUI design is the development of cross-platform GUI tools, those that will design application s and interfaces to run on a variety of platforms. As this is an area currently being developed, there is a great deal of confusion amongst the GUI and application software developers as to which approach is most suitable. There are five primary approaches to cross-platform development:

Layered toolkits
Emulated toolkits
Ported toolkits
Application frameworks
Application level GUI objects

Each approach varies in its ability to provide style guide compliance, environment integration and 'polish' or a smooth user interface. The area of cross-platform development is at a similar position to that of interface development tools several years ago but most developers are struggling with this concept of open cross-platform compatibility as a long term solution.

The evolution from textual command line user interfaces to modern GUI's was made possible by significant hardware improvements, bitmap graphics and the mouse. The are currently a number of developments in interface technology which are exploiting new communication methods including gesture recognition, speech recognition, sound and real-time animation.

As multi media information in computer systems becomes widely used in many application areas it is important that it does not remain technology driven. Their design must become more user centred, and user interfaces must be designed to cover a wider range of target user requirements. More diverse products will emerge for user populations with different needs, interests and abilities.

Achieving a proper balance to meet these diverse needs is complex and ease of use is not always the primary criterion. Designers must make the interfaces natural for the novice user but also devise special conventions for the expert user. Currently this area is solved by two interface approaches, point and click for the novice user and command language for the expert user.

New graphics models for visualisation of complex objects or data sets enable the user to interact with the data displayed on the screen in real-time. For this, the user interfaces and the visualisation must be far more integrated than it is at present to give semantic feedback during interaction. Such an integrated visualisation environment goes beyond the capabilities of traditional visualisation paradigms and requires a completely new visualisation software architecture, this is usually referred to as steering.

With steering, the focus is on direct manipulation of the 3D objects, using a virtual reality (VR) paradigm to promote real-time feedback. This capability also demands that the user has good intuitive control over where and how the exploration is occurring. This the displays, either immersive via a head mounted display or non immersive via a traditional monitor, needs to have the capability of being controlled by a multidimensional input device such as a 6D spaceball. The result is that the user is in the 'driving seat' and navigates through the data model in a real time interactive exploration process.

Such a system must also support the philosophy of not trying to display all of the data at once. Instead the user must be permitted to interactively select the portion of interest and explore the selected data.

Issues currently under development and yet to be solved by the next generations of user interfaces include the matching of the visualisation to the users needs and requirements, integrating visualisation and computation, usability design of application software, managing large data sets, multi-dimensional data ,real-time interaction and the tailoring of GUI's to better suite the expertise and character or style of the end user.

An outline of the concept of style

The concept of style is regularly employed in literature on design, problem solving and creativity and refers to an individuals characteristic manner of performing in a task environment. However, the concept of style is rarely dealt with in detail in isolation which makes a context-independent usage of the term difficult. As such, the concept of style, particularly as it applies to the design profession, has the status of something that is intuitively known but which is very difficult to define formally.

Two distinct trends may be identified when discussing the term style. The first considers style within the psychological framework of cognitive style, the second approach draws a parallel with strategy .

Cognitive Styles

An individuals cognitive style refers to the characteristic manner in which they process and respond to information. In this way, identical stimulus material will not necessarily produce the same cognitive reactions in any two individuals. Theorists in this area of research have attempted to identify dimensions of processing preference that may be used to distinguish individuals. Some nineteen different cognitive styles have been proposed and investigated. Of these, several have emerged as being useful tools to distinguish individuals at a cognitive level.

Field dependence versus field dependence.

This dichotomy is often used as a tool for the evaluation of mental processing. Field independence is defined as the tendency to actively structure situational information without being affected by the irrelevant aspects of the environment. Field dependent processors are less analytical and more global perceivers.

Serialism versus holism.

Serialist thinkers tend to proceed through small logical iterations, making certain that all points are clear and all information is considered. This thought pattern contrasts with holistic thought which proceeds in a more broad fashion, out of context, making logical and non logical connections between available information and possible solutions. This dichotomy distinguishes those who are able to solve a problem through sudden inspiration or intuition (holists) and those who must consciously and rationally work through the problem(rationalists).

Convergence versus divergence

Convergent thinking is classified as the taking in of information and using it to narrow down (converge) the problem state until one correct answer is obtained. Divergent thinking emphasises the generation of a wide range of possible answers until one is obtained that best satisfies the objectives.

Reflective versus impulsive

This dimension is the dichotomy between those who either make considered or rapid decisions when considering complex problems.

Left brain versus right brain

This style is based on the distinction between the left and right brain processing capabilities. The two hemispheres of the brain specialise in the processing of different types of information. The right is heuristic with an emphasis on intuitive, global or pictorial processing. The left is analytic with the emphasis on sequential, linear and symbolic processing. It is argued that right brain thinkers employ few complex decision making rules and the behaviour is anchored to external processes whereas left brain thinkers employ complex decision making rules and their behaviour is based on internal processes.
This dichotomy is commonly discussed in the literature on design methodology, and there is a consensus that designers need to employ both styles of thinking to successfully solve problems.

During the interpretation of cognitive styles it is important to understand that these styles are not mutually exclusive. In this way, any individual may be categorised along any or all of these dichotomies any a profile of styles could be produced. This may be seen as desirable but as there exists only a limited understanding on the casual relationship between cognitive style and problem solving behaviour any dimension obtained in this way would be of dubious merit.

At this processing level it is expected that designers would form a loose group clustered around similar points on style dichotomies. Designers would be capable of clearly seeing patterns in complex relationships (Field independent), be creative (Holistic), be capable of creative composition (Divergent), tend to make more subjective, impulsive judgements than is the norm (Impulsive) and have an emphasis on right brain processing.

But it must be understood that, on any one dimension, few designers will process information completely in one fashion and each stylistic dichotomy has a whole spectrum of processing capabilities between the extremes so an individual who may tend towards divergent thinking would also be capable of convergent thinking.

Given the relationship between cognitive style and problem solving behaviour is not fully understood and the necessary and dubious grouping at this level it is clear that designer style cannot be directly related with cognitive style. Doing so, would impose great individual variation into an artificially cohesive grouping and provide little, if any, information on how an individuals style affects their way of designing.

Strategies
It's possible to examine performance in cognitive problem solving tasks by referring to the strategies an individual employs during the design process. From observation it is possible to distinguish an individuals preferred problem solving strategies. There are six prescriptive strategies that have come to be commonly accepted as useful tools to differentiate individuals at this level, these are:

1. Design Morphology chart - a methodical step by step design sequence.
2. Input-output matrix - a generalised model of design.
3. Convergent thinkers design.
4. Divergent thinkers design.
5. Linear - critical path approach.
6. Cascade - pattern language.

The methodologies for design discussed in the literature may all be seen as the endorsement of specific design strategies which designers would do well to follow in order to successfully solve design problems. It is considered that many of these prescriptive strategies are only compatible with those with certain cognitive style preferences and that many designers cognitive styles are not adequately matched with the design strategies proposed as formal methodologies, it is this mismatch that leads to the distrust or dismissal of such methodologies by designers.

There exists evidence that suggests that designers possess a unique problem solving strategy that may be used to distinguish them from scientists. That is, designers tend to have a solution focused viewpoint, compared to the scientists perspective of deductive problem focused reasoning.

Current belief emphasises the generation - conjecture - analysis model of the design process which states that designers structure problems by interactively exploring aspects of possible solutions. This model seems to be limited in its ability to make distinctions between designers by virtue of its generality and the fact that designers in general modify or deviate from the strategy. It is argued argues that two learning styles are identifiable at this level, comprehension and operation learners, who my be distinguished in terms of the number and types of hypotheses generated during the design process.
It is also argued that designers tend to structure problems in terms of dynamically defined sub problems, not immediately specifiable at the outset of the design process. These sub problems are then tackled in various ways throughout the design cycle. These sub problems may be identified as discrete functional units within the overall design problem by the principle of functioning.

At a lower level than the above strategies there exist a number of strategies and sub strategies that any individual can adopt to solve a design problem, altough strategies at this level are totally situational dependent, with the nature of the specific design problem dictating which strategies are employed.

The six commonly accepted standard strategies at this can be more easily identified when studied in the context of the problem situation rather than a particular designer style.

1. Imagination
2. Analogy
3. Heuristics
4. Random search
5. Systematic search
6. Transformation

On consideration of the above, as with cognitive style, it is unlikely that a satisfactory conceptualisation of designer style can be arrived at simply through an analysis of situational low level and generic higher level strategies. If an individual employs one strategy on one occasion and a second strategy on another and if the more formal prescriptive design strategies are modified, deviated from or simply mistrusted and not used then strategies appear to be less characteristic of a designers cognitive processing than is implied by the notion of designer style.

This initial study on the concept of designer style is currently in progress and further details are beyond the scope of this paper. Although, it is relevant to discuss and use some of the findings from this work in terms of common design styles employed at specific points in the design process and thus the designers needs, which translate into criteria for the design of tools and appropriate user interfaces.

What is Virtual Reality

Studies have been undertaken over the past year within the Institute of Design by the author to integrate VR technologies into the design process. Before we consider one of these studies it would be appropriate to consider just what we mean by VR.

Websters Dictionary gives the definition as, 'Virtual, being in essence or effect but not in fact', and 'reality, a real event, entity or state of affairs'. From this we learn nothing, the term is an oxymoron, it is however, one of the much stated 'buzzwords' or the information age.

The range of commonly accepted alternatives to this term is also diverse, partly due to the many different methods of achieving the same goal. Amongst these descriptions are to be found, 'Artificial Reality', 'Augmented reality', 'Virtual Environment', 'Projected Reality', Synthetic Reality', the list goes on. The term 'Virtual Reality' was first coined by Jaron Lanier in California and he claimed that it meant, "....a computer generated, visual, audible and tactile multimedia experience". The laypersons idea of VR tends to be associated with head mounted displays, gloves, and in some ill informed cases, entire body suits connected directly to the users brain. Big budget movies such as 'The Lawnmower Man' are in part responsible for the mis-information.

Virtual Reality, like many things, is not exactly the way it is represented in the movies and, although the stereotypical features such as head mounted displays and gloves are now common, practical VR or Real-Time 3D is commonly found on a variety of computer systems using large projection screens or simply the computer monitor. The synonymous glove is often replaced by a trackball, joystick or mouse for navigation and manipulation of objects.

According to R. Burlea, "Virtual Reality is a triangle of three I's, Immersion, Interaction and Imagination". As it stands, Virtual Reality is a simulation in which, three dimensional computer graphics and 3D sound are used to give a false impression of a world in which the user may navigate their own way around. The graphics and sound interact with the user in Real-Time to create the illusion, especially in the case of immersive VR, that the user actually inhabits this simulated world. Unfortunately, even with the computer systems available today it is reasonably simple not to confuse a real world with its computer simulated competitor.

Myron Krueger, who coined the term, 'Artificial Reality', condoned this shortcoming in his statement, "The promise of artificial realities is not to reproduce conventional reality or to act in the real world. It is precisely the opportunity to create synthetic realities, for which there are no antecedents that is exciting conceptually and ultimately important economically". It may be assumed then, that VR is a means of providing society with experiences beyond our present comprehension.

Whatever VR is used for in the future, a key consideration for realism must be the effective interaction design. If VR systems are to be considered as tools, as the user works within this environment on a personal level, then their requirements must be considered on a personal level and tailored to fit whatever natural working methods the user may wish to employ.

It was with the following three considerations that the following study was undertaken. How do designers work on a personal level and is there such a thing as designer style, if designer style is a meaningful concept, then can VR technology provide a more personal intuitive interface to CAD than traditional GUI's and, how may RT3D be used effectively for the generation, communication and evaluation of concepts throughout the design process.

Case Study
The author considers it important that the next generation of user interfaces for computer aided design systems and navigation and interaction tools for Internet and WWW applications should be developed around RT3D technology. A project is being carried out with Industrial Design students at the University of Teesside on the generation of intuitive RT3D sketch models for vehicle design which could then be practically used during the remainder of the design process.

As concept creation and concept evaluation are iterative and often a team activity, it was found by earlier projects that any method for the generation of concepts should allow the designer to view a design proposal in 3 dimensions as early on in the design process as possible. Recent developments in 3D systems have reduced the traditional costs and difficulties in doing this at an early stage.

VR and other RT3D techniques are methods by which early design concepts should be able to be rapidly developed to 3D. Both immersive and non immersive techniques may facilitate appropriate types of visualisation of the results, immersive VR in particular, SmartScene in this project, was particularly relevant for the full scale visualisation of concept models for ergonomic and interaction evaluation.

The generation of sketch or conceptual proposals for a solution to a design problem is a major and continuing part of the design process. It's role is to aid enable the designer to arrive quickly and easily at a stage where they can specify a satisfactory design for development for manufacture. A major aim of this project was to use the designers sketches directly as a source of 3D models capable of being evaluated and developed further. These models, generated almost as rapidly as traditional 2D sketches would allow for 3 dimensional evaluation of a concept much earlier in the design process.

The sketch stage was considered both in terms of the processes involved and the designers and management requirements. The process was considered to involve both Concept Design and Design Development, the products of which must be sufficient to suggest a design to an informed observer.

From earlier studies into designer style, the requirements of designers using CAD systems were found to be the following:

Rapidity — The speed at which the concept can be represented
Versatility — The range of tools and techniques available
Visibility — The degree to which the design can be assessed
Modifiability — The degree to which the design may be modified once created

Modelling during this stage must have the following characteristics:

The inputting of geometric information must be quick
The geometry must be simple to specify
Representations must not be over precise
Geometry must be simple to modify
Reviewing alternatives must be simple to achieve
Visual display quality must be adequate for evaluation
The overall interface must be user friendly

In order the assess and evaluate the products of this stage the management needs are defined as:

That the design environment must lend itself to be both single user and multi-user (design team, client, etc.)

That it is able to effectively communicate the completed design proposal to all those necessary

Work in Progress
The project team is now involved with the design and implementation of an intuitive virtual space in which sketch models may be generated and evaluated in an appropriate environment at full size in three dimensions. This work is in collaboration with the Ars Electronica Centre in Linz and is based on CAVE VR technology.

COST EFFECTIVENESS OF RELIABILITY-CENTERED MAINTENANCE

Azim Houshyar, Ph.D.

Department of Industrial & Manufacturing Eng.
Western Michigan University
Kalamazoo, Michigan 49008-5061
USA

ABSTRACT: Modern manufacturing systems and process plants are complex, highly optimized systems, for which seemingly small changes in the operating conditions of a component can have significant impact on the overall performance of the system. In these environments, even though any component breakdown could interrupt the whole process, the cost associated with unnecessary component replacement would lessen the plant's profitability, and should be prevented as well. Therefore, the past maintenance strategies of corrective and/or preventive maintenance, may prove to be costly, presenting the opportunity to predictive maintenance techniques to be utilized. A clear understanding of predictive maintenance can help design and manufacture components with preventive maintenance features in mind. This article reviews the state-of-the-art on predictive maintenance and introduces the new concept of reliability-centered maintenance.

INTRODUCTION

Two of the most important factors that demonstrate the capability of any manufacturing system to deliver its un-interrupted and on-time service are the reliability and maintainability of its machinery and equipment. The former measures the frequency of failures, interruptions, or needed adjustments and corrections, whereas the later measures the duration of those events. A piece of equipment that fails less often, and when it fails, takes less time to be brought back up to its operable conditions is considered to be more reliable and maintainable than a machine that fails more frequently and requires a longer repair time. Reliability and maintainability of any machinery or equipment are functions of the reliability and maintainability of its components, subsystems, and systems. Therefore, to avoid any delays in service delivery, special attention should be given to design more reliable component and their optimum maintenance.

To better maintain the components and to reduce the overall operational/maintenance cost of complex systems, much effort has gone into development of maintenance guidelines and offering recommendations to minimize component degradation. Several universities and industries have been developing and implementing emerging methods in information processing and diagnostics to enhance preventive maintenance technology. Several methodologies including signal validation and instrument surveillance, automated diagnostics of valve systems, residual life estimation of critical components have been developed that help detect any incipient changes in the system performance.

Stevens [5] state that: "Look inside a plant today and you might get the idea that it is the manufacturing function that makes the money, and the maintenance operation that spends the money. But not all maintenance strategies are created equal. How well a maintenance department can capture the cost difference between best and worst practices is a measure of its profitability ... To gain maximum return, many manufacturing companies are now operating in a predictive-maintenance mode".

Texas instrument has had significant savings of their total maintenance and operations budget, by going to predictive maintenance. Their goal with predictive maintenance is to eliminate interruptions, which cause loss of product, and to drive up capacity. General Motors Corp. also drives multiple benefits from predictive maintenance. They have reported good correlation between a predictive maintenance implementation and reduction of maintenance costs, up-time increases, and sometimes have observed a quality impact. BP Chemicals Inc. has used vibration analysis to guide them in cost-effective maintenance on a large granulator, which has saved them a $90,000 contribution to fixed costs and profit. Chevron Inc. has measured run-to-failure mode cost at $18 per horsepower per year, frequency-based maintenance at $13 per horsepower per year, and predictive maintenance at $8 per horsepower per year.

The above-mentioned examples are only a few of many success stories that has been reported by companies who have adopted a predictive maintenance policy. A clear understanding of the notion of *predictive maintenance* and its impact on the overall maintenance cost can help design and manufacture components with preventive maintenance features in mind. This article reviews the concept of predictive maintenance and introduces the philosophy of reliability-centered maintenance.

DISCUSSION

Predictive Maintenance means doing the *right maintenance* at the *right time*. By far the most important required for predictive maintenance is knowing the condition of the machinery and equipment, and its major components at all times, so it would not fail unexpectedly. This concept is in contrast to run-to-failure (also known as corrective, un-scheduled, or reactive maintenance) where breakdowns are handled on an emergency basis as they occur, or preventive maintenance (also known as scheduled maintenance) where service is scheduled at a particular frequency.

Corrective maintenance can result in lengthy downtimes, extensive collateral equipment damage, and safety and environmental hazard. Imagine the cost associated with the failure of a component used in an assembly station that congregates different parts of a car in its final stages of production. It is estimated that each hour of production downtime can cost up to $100,000 an hour in opportunity losses to the company. Therefore, failure of machinery and equipment in those intense stages of production is, simply, not acceptable, and should be avoided.

To prevent such occurrences of costly downtimes, corporation have shown interest in embracing preventive maintenance. The problem associated with the use of preventive maintenance is that it is usually conservative and may be wasteful because some maintenance will be done when not required, and failures may still occur. A typical example of that is the policy of car oil change once every 3000 miles. Realizing that all cars are not driven by the same driver, under the same road and environmental conditions, it makes sense to link the frequency of oil change to the actual conditions of the oil, rather than a magical number of miles or a certain amount of time.

This discussion bring up the concept of predictive maintenance, as a means of optimizing the frequency of repairs based on the physical conditions of the system. It seems natural that in order to save money, any repair should be postponed until the component shows signs of necessity of repair or replacement. For instance, by measuring the viscosity of the oil, the driver can sense the need for an oil change. The question is, then, why predictive maintenance has not been widely adopted by manufacturing systems. One answer that comes to mind is that implementation of any such policy requires access to the technology to be able to determine the status of a system at any given time.

Michal states: "Maintenance is too often thought of as an activity that is done with wrenches and screwdrivers, which fails to recognize the importance of knowing what to do and when to do it and the consequence of not doing it. The information-handling aspect of preventive maintenance is where big improvements can be made, where new technology can have a big impact, and where universities can contribute effectively [4]." Mitcahl lists several information processing methods that are often integrated to develop solutions to a specific problem. These include digital signal processing, applied artificial intelligence methods, data driven models, and physical models of degradation mechanisms.

University of Tennessee has been developing and implementing emerging methods in information processing and diagnostics to enhance preventive maintenance technology . Projects worked on include the development of instrument calibration verification, residual life estimation of electric motors and steam generators, vibration monitoring of plant components, and steam generator tube defect classification by using eddy current test data analysis [4]. They have reported the development of an automated diagnostic system for motor operated valves, stating that valve related problems in U.S. nuclear plants result in about 20% of unscheduled shutdowns, and up to 30% of the industry's annual maintenance budget. Their goal is to develop non-intrusive techniques based on both motor power signature analysis and motor current signature analysis for detecting and diagnosing of faults in motor-operated valves.

Of all the techniques used in predictive maintenance for rotating machinery, *vibration analysis* has been and will continue to be an important key to predictive maintenance predictions [2]. Vibration in itself does not mean that a problem exists. To determine whether vibration is acceptably low, plant personnel must make use of vibration testing and data analysis. Vibration testing to determine a machine's mechanical health can be done by an engineer or knowledge-based software or electronics that can evaluate the amplitude and frequency resulting from a machine's response to imposed impact, as well as from the naturally occurring vibrations caused by operational loads.

Measurement of other parameters that might be more sensitive to deterioration of specific system components, when combined with vibrational analysis, can be more informative of the status of a component or the whole system. Examples are bearing temperature, rate of lubricant oxidation or contamination as demonstrated by wear-debris analysis, thermodynamic-efficiency degradation, and leakage flow through high-pressure-drop sealing locations [2].

In deciding whether the vibration levels are acceptable, base an evaluation on two questions [2]:

- Does the vibration level at a given frequency suggest that something is wrong?
- What is the likelihood of the overall vibration itself causing damage?

An adequate vibration-based predictive maintenance system should provide quantitative answers to both these questions and predict the time-to-failure of specific components as well as that of the entire machine.

RELIABILITY CENTERED MAINTENANCE

Reliability Centered Maintenance (RCM) is a systematic methodology to establish maintenance tasks for critical components in plant with a focus on their reliability characteristics [3]. RCM pursues the identification of applicable and efficient tasks to prevent these components from developing their dominant failure causes, and, in turn, towards achieving proper levels of component availability with low cost.

First step in the RCM analysis encompasses the identification of critical components for the plant from a risk and economical point of view. Furthermore, component failure modes can be ranked using appropriate risk measures which, in turn, allow the determination of the component criticality.

Once critical component selection is finished, the next step encompasses the identification of applicable and efficient tasks to prevent each component from developing its dominant failure causes. Often, the analyst has available many tasks, but only a set of them is required to cope with all the dominant failure causes. For an approach intended to select the most suitable set of tasks to achieve this goal, see Martorell et al. [3].

Reliability centered maintenance requires development of a methodology that provides a logical and self-consistent framework for, preferably a computer model, that uses component reliability figures, inspection data, availability reports, experimental test results, and physical principles, to optimize the repair time of components and systems [1].

SUMMARY

Many corporations are moving toward the use of predictive maintenance as a means of reducing the overall maintenance cost and increasing the machinery uptime. Vibration analysis is found to be one of the main tools of implementing predictive maintenance for the rotating machines. Engaging in a *proactive* mode to eliminate problems, not just predict them, is crucial in the significance of predictive maintenance. Of the total available return from a maintenance cost-reduction program, 20% is in predictive maintenance and 80% is in proactive efforts [5]. Use of reliability centered maintenance methodology fosters successful execution of transformation from other maintenance policies to the predictive maintenance, and is highly recommended. A close relationship between the machine tool builder and the user is momentous in facilitating design of machinery and equipment that are equipped with predictive maintenance technology.

REFERENCES

1. **Berte, F., Klisiewicz W., and Schwarz, F.,** Computer models, methodology help predict plant behavior, *1987 Generation Planbook*, McGraw-Hill, Inc. 1987.
2. **Marccher, W.D.,** Make vibration monitoring live up to its potential, *Power*, July 1994.
3. **Martorell, S., Munoz, A., and Serradell, V.,** An approach to integrating surveillance and maintenance tasks to prevent the dominant failure causes of critical components, *Reliability Eng. & Systems Safety*, 50, 179-187, 1995.
4. **Michal, R.A.,** University makes maintenance a learning experience, *Nuclear News*, January 1995.
5. **Stevens, T.,** The power of prediction, *Industry Week*, July 4, 1994.

Flexible Automation and Intelligent Manufacturing 1997

Due to an editorial error, the page numbers in the table of contents and author index for all papers from pages 571-980 do not reflect the actual page numbers of the papers. Each paper in this page range begins on a page which is one greater than indicated in the table of contents, and each of their authors are referenced incorrectly in the author index. We apologize for any inconvenience caused by this error.

Begell House, Inc.
New York